# Unit Conversion Factors

## Length

1 m = 100 cm =1000 mm = $10^6$ μm = $10^9$ nm
1 km = 1000 m = 0.6214 mi
1 m = 3.281 ft = 39.37 in
1 cm = 0.3937 in
1 in = 2.54 cm (exactly)
1 ft = 30.48 cm (exactly)
1 yd = 91.44 cm (exactly)
1 mi = 5280 ft = 1.609344 km (exactly)
1 Angstrom = $10^{-10}$ m =$10^{-8}$ cm = 0.1 nm
1 nautical mile = 6080 ft = 1.152 mi
1 light-year = $9.461 \cdot 10^{15}$ m

## Area

1 $m^2$ =$10^4$ $cm^2$ =10.76 $ft^2$
1 $cm^2$ = 0.155 $in^2$
1 $in^2$ = 6.452 $cm^2$
1 $ft^2$ =144 $in^2$ = 0.0929 $m^2$
1 hectare = 2.471 acre = 10000 $m^2$
1 acre = 0.4047 hectare = 43560 $ft^2$
1 $mi^2$ = 640 acre
1 $yd^2$ = 0.8361 $m^2$

## Volume

1 liter =1000 $cm^3$ =$10^{-3}$$m^3$ = 0.03531 $ft^3$ = 61.02 $in^3$ = 33.81 fluid ounce
1 $ft^3$ = 0.02832 $m^3$ = 28.32 liter = 7.477 gallon
1 gallon = 3.788 liters
1 quart = 0.9463 liter

## Time

1 min = 60 s
1 h = 3,600 s
1 day = 86,400 s
1 week = 604,800 s
1 year = $3.156 \cdot 10^7$ s

## Angle

1 rad = 57.30° =180°/$\pi$
1° = 0.01745 rad = ($\pi$/180) rad
1 rev = 360° = 2$\pi$ rad
1 rev/min (rpm) = 0.1047 rad/s = 6°/s

## Speed

1 mile per hour (mph) = 0.4470 m/s = 1.466 ft/s = 1.609 km/h
1 m/s = 2.237 mph = 3.281 ft/s
1 km/h = 0.2778 m/s = 0.6214 mph
1 ft/s = 0.3048 m/s
1 knot = 1.151 mph = 0.5144 m/s

## Acceleration

1 m/$s^2$ =100 cm/$s^2$ = 3.281 ft/$s^2$
1 cm/$s^2$ = 0.01 m/$s^2$ = 0.03281 ft/$s^2$
1 ft/$s^2$ = 0.3048 m/$s^2$ = 30.48 cm/$s^2$

## Mass

1 kg =1000 g = 0.0685 slug
1 slug =14.95 kg
1 kg has a weight of 2.205 lb when $g$ = 9.807 m/$s^2$
1 lb has a mass of 0.4546 kg when $g$ = 9.807 m/$s^2$

## Force

1 N = 0.2248 lb
1 lb = 4.448 N
1 stone = 14 lb = 62.27 N

## Pressure

1 Pa =1 N/$m^2$ =$1.450 \cdot 10^{-4}$ lb/$in^2$ = 0.209 lb/$ft^2$
1 atm =$1.013 \cdot 10^5$ Pa =101.3 kPa =14.7 lb/$in^2$ = 2117 lb/$ft^2$ = 760 mm Hg = 29.92 in Hg
1 lb/$in^2$ = 6895 Pa
1 lb/$ft^2$ = 47.88 Pa
1 mm Hg = 1 torr = 133.3 Pa
1 bar =$10^5$ Pa =100 kPa

## Energy

1 J = 0.239 cal
1 cal = 4.186 J
1 Btu = 1055 J = 252 cal
1 kW·h = $3.600 \cdot 10^6$ J
1 ft·lb =1.356 J
1 eV =$1.602 \cdot 10^{-19}$ J

## Power

1 W = 1 J s
1 hp = 746 W = 0.746 kW = 550 ft·lb/s
1 Btu/h = 0.293 W
1 GW =1000 MW =$1.0 \cdot 10^9$ W
1 kW = 1.34 hp

## Temperature

Fahrenheit to Celsius: $T_C = \frac{5}{9}(T_F - 32\ °F)$
Celsius to Fahrenheit: $T_F = \frac{9}{5}T_C + 32\ °C$
Celsius to Kelvin: $T_K = T_C + 273.15\ °C$
Kelvin to Celsius: $T_C = T_K - 273.15\ K$

# University Physics
## with Modern Physics
## Volume One

# University Physics

## with Modern Physics

## Volume One

**Wolfgang Bauer**
*Michigan State University*

**Gary D. Westfall**
*Michigan State University*

The McGraw·Hill Companies

UNIVERSITY PHYSICS, VOLUME 1

Published by McGraw-Hill, a business unit of The McGraw-Hill Companies, Inc., 1221 Avenue of the Americas, New York, NY 10020. 

This book is printed on acid-free paper.

1 2 3 4 5 6 7 8 9 0 WDQ/WDQ 1 0 9 8 7 6 5 4 3 2 1 0

ISBN 978-0-07-336795-8
MHID 0-07-336795-8

Vice President & Editor-in-Chief: *Marty Lange*
Vice President, EDP: *Kimberly Meriwether-David*
Vice-President New Product Launches: *Michael Lange*
Publisher: *Ryan Blankenship*
Sponsoring Editor: *Debra B. Hash*
Senior Developmental Editor: *Mary E. Hurley*
Senior Marketing Manager: *Lisa Nicks*
Senior Project Manager: *Jayne L. Klein*
Lead Production Supervisor: *Sandy Ludovissy*
Lead Media Project Manager: *Judi David*
Senior Designer: *David W. Hash*
(USE) Cover Image: *Swirls of light around sphere, ©Ryichi Okano/Amana Images/Getty Images, Inc.; Solar panels, ©Rob Atkins/Photographer's Choice/Getty Images, Inc.; Bose-Einstein condensate, image courtesy NIST/JILA/CU-Boulder; Scaled chrysophyte, Mallomanas lychenensis, SEM, © Dr. Peter Siver/Visuals Unlimited/Getty Images, Inc.*
Lead Photo Research Coordinator: *Carrie K. Burger*
Photo Research: *Danny Meldung/Photo Affairs, Inc.*
Compositor: *Precision Graphics*
Typeface: *10/12 Minion Pro*
Printer: *World Color Press Inc.*



**Library of Congress Cataloging-in-Publication Data**

Bauer, W. (Wolfgang), 1959-
University physics with modern physics / Wolfgang Bauer, Gary D. Westfall.—1st ed.
p. cm.
Includes index.
ISBN 978-0-07-285736-8 — ISBN 0-07-285736-6 (hard copy : alk. paper) 1. Physics—Textbooks. I. Westfall, Gary D. II. Title.
QC23.2.B38 2011
530—dc22

2009037841

www.mhhe.com

# Brief Contents

# About the Authors

**Wolfgang Bauer** was born in Germany and obtained his Ph.D. in theoretical nuclear physics from the University of Giessen in 1987. After a post-doctoral fellowship at the California Institute of Technology, he joined the faculty at Michigan State University in 1988. He has worked on a large variety of topics in computational physics, from high-temperature superconductivity to supernova explosions, but has been especially interested in relativistic nuclear collisions. He is probably best known for his work on phase transitions of nuclear matter in heavy ion collisions. In recent years, Dr. Bauer has focused much of his research and teaching on issues concerning energy, including fossil fuel resources, ways to use energy more efficiently, and, in particular, alternative and carbon-neutral energy resources. He presently serves as chairperson of the Department of Physics and Astronomy, as well as the Director of the Insitute for Cyber-Enabled Research.

**Gary D. Westfall** started his career at the Center for Nuclear Studies at the University of Texas at Austin, where he completed his Ph.D. in experimental nuclear physics in 1975. From there he went to Lawrence Berkeley National Laboratory (LBNL) in Berkeley, California, to conduct his post-doctoral work in high-energy nuclear physics and then stayed on as a staff scientist. While he was at LBNL, Dr. Westfall became internationally known for his work on the nuclear fireball model and the use of fragmentation to produce nuclei far from stability. In 1981, Dr. Westfall joined the National Superconducting Cyclotron Laboratory (NSCL) at Michigan State University (MSU) as a research professor; there he conceived, constructed, and ran the MSU $4\pi$ Detector. His research using the $4\pi$ Detector produced information concerning the response of nuclear matter as it is compressed in a supernova collapse. In 1987, Dr. Westfall joined the Department of Physics and Astronomy at MSU as an associate professor, while continuing to carry out his research at NSCL. In 1994, Dr. Westfall joined the STAR Collaboration, which is carrying out experiments at the Relativistic Heavy Ion Collider (RHIC) at Brookhaven National Laboratory on Long Island, New York.

**The Westfall/Bauer Partnership** Drs. Bauer and Westfall have collaborated on nuclear physics research and on physics education research for more than two decades. The partnership started in 1988, when both authors were speaking at the same conference and decided to go downhill skiing together after the session. On this occasion, Westfall recruited Bauer to join the faculty at Michigan State University (in part by threatening to push him off the ski lift if he declined). They obtained NSF funding to develop novel teaching and laboratory techniques, authored multimedia physics CDs for their students at the Lyman Briggs School, and co-authored a textbook on CD-ROM, called *cliXX Physik*. In 1992, they became early adopters of the Internet for teaching and learning by developing the first version of their online homework system. In subsequent years, they were instrumental in creating the Learning*Online* Network with CAPA, which is now used at more than 70 universities and colleges in the United States and around the world. Since 2008, Bauer and Westfall have been part of a team of instructors, engineers, and physicists, who investigate the use of peer-assisted learning in the introductory physics curriculum. This project has received funding from the NSF STEM Talent Expansion Program, and its best practices have been incorporated into this textbook.

**Dedication** This book is dedicated to our families. Without their patience, encouragement, and support, we could never have completed it.

# A Note from the Authors

**Physics** is a thriving science, alive with intellectual challenge and presenting innumerable research problems on topics ranging from the largest galaxies to the smallest subatomic particles. Physicists have managed to bring understanding, order, consistency, and predictability to our universe and will continue that endeavor into the exciting future.

However, when we open most current introductory physics textbooks, we find that a different story is being told. Physics is painted as a completed science in which the major advances happened at the time of Newton, or perhaps early in the 20th century. Only toward the end of the standard textbooks is "modern" physics covered, and even that coverage often includes only discoveries made through the 1960s.

**Our main motivation to write this book is to change this perception by appropriately weaving exciting, contemporary physics throughout the text.** Physics is an exciting, dynamic discipline—continuously on the verge of new discoveries and life-changing applications. In order to help students see this, we need to tell the full, exciting story of our science by appropriately integrating contemporary physics into the first-year calculus-based course. Even the very first semester offers many opportunities to do this by weaving recent results from non-linear dynamics, chaos, complexity, and high-energy physics research into the introductory curriculum. Because we are actively carrying out research in these fields, we know that many of the cutting-edge results are accessible in their essence to the first-year student.

Authors in many other fields, such as biology and chemistry, already weave contemporary research into their textbooks, recognizing the substantial changes that are affecting the foundations of their disciplines. This integration of contemporary research gives students the impression that biology and chemistry are the "hottest" research enterprises around. The foundations of physics, on the other hand, are on much firmer ground, but the new advances are just as intriguing and exciting, if not more so. We need to find a way to share the advances in physics with our students.

We believe that talking about the broad topic of energy provides a great opening gambit to capture students' interest. Concepts of energy sources (fossil, renewable, nuclear, and so forth), energy efficiency, alternative energy sources, and environmental effects of energy supply choices (global warming) are very much accessible on the introductory physics level. We find that discussions of energy spark our students' interest like no other current topic, and we have addressed different aspects of energy throughout our book.

In addition to being exposed to the exciting world of physics, students benefit greatly from gaining the ability to **problem solve and think logically about a situation.** Physics is based on a core set of ideas that is fundamental to all of science. We acknowledge this and provide a useful problem-solving method (outlined in Chapter 1) which is used throughout the entire book. This problem-solving method involves a multi-step format that both of us have developed with students in our classes.

With all of this in mind along with the desire to write a captivating textbook, we have created what we hope will be a tool to engage students' imaginations and to better prepare them for future courses in their chosen fields (admittedly, hoping that we would convert at least a few students to physics majors along the way). Having feedback from more than 300 people, including a board of advisors, several contributors, manuscript reviewers, and focus group participants, assisted greatly in this enormous undertaking, as did field testing of our ideas with approximately 4000 students in our introductory physics classes at Michigan State University. We thank you all!

*—Wolfgang Bauer and Gary D. Westfall*

# Contents

# Preface

*University Physics* is intended for use in the calculus-based introductory physics sequence at universities and colleges. It can be used in either a two-semester introductory sequence or a three-semester sequence. The course is intended for students majoring in the biological sciences, the physical sciences, mathematics, and engineering.

## Problem-Solving Skills: Learning to Think Like a Scientist

Perhaps one of the greatest skills students can take from their physics course is the ability to **problem solve and think critically about a situation.** Physics is based on a core set of fundamental ideas that can be applied to various situations and problems. *University Physics* by Bauer and Westfall acknowledges this and provides a problem-solving method class tested by the authors, and used throughout the entire text. The text's problem-solving method involves a multi-step format.

> "The Problem-Solving Guidelines help students improve their problem-solving skills, by teaching them how to break a word problem down to its key components. The key steps in writing correct equations are nicely described and are very helpful for students."
>
> —*Nina Abramzon, California Polytechnic University–Pomona*

> "I often get the discouraging complaint by students, 'I don't know where to start in solving problems.' I think your systematic approach, a clearly laid-out strategy, can only help."
>
> —*Stephane Coutu, The Pennsylvania State University*

### Problem-Solving Method

#### Solved Problem

The book's numbered **Solved Problems** are fully worked problems, each consistently following the seven-step method described in Chapter 1. Each Solved Problem begins with the Problem statement and then provides a complete Solution:

1. **THINK:** Read the problem carefully. Ask what quantities are known, what quantities might be useful but are unknown, and what quantities are asked for in the solution. Write down these quantities, representing them with commonly used symbols. Convert into SI units, if necessary.
2. **SKETCH:** Make a sketch of the physical situation to help visualize the problem. For many learning styles, a visual or graphical representation is essential, and it is often necessary for defining variables.
3. **RESEARCH:** Write down the physical principles or laws that apply to the problem. Use equations that represent these principles and connect the known and unknown quantities to each other. At times, equations may have to be derived, by combining two or more known equations, to solve for the unknown.

SOLVED PROBLEM **6.6** **Power Produced by Niagara Falls**

PROBLEM

Niagara Falls pours an average of 5520 m$^3$ of water over a drop of 49.0 m every second. If all the potential energy of that water could be converted to electrical energy, how much electrical power could Niagara Falls generate?

SOLUTION

THINK

The mass of one cubic meter of water is 1000 kg. The work done by the falling water is equal to the change in its gravitational potential energy. The average power is the work per unit time.

SKETCH

A sketch of a vertical coordinate axis is superimposed on a photo of Niagara Falls in Figure 6.22.

RESEARCH

The average power is given by the work per unit time:

$$\bar{P} = \frac{W}{t}.$$

The work that is done by the water going over Niagara Falls is equal to the change in gravitational potential energy,

$$\Delta U = W.$$

The change in gravitational potential energy of a given mass $m$ of water falling a distance $h$ is given by

$$\Delta U = mgh.$$

SIMPLIFY

We can combine the preceding three equations to obtain

$$\bar{P} = \frac{W}{t} = \frac{mgh}{t} = \left(\frac{m}{t}\right)gh.$$

*Continued—*

**CALCULATE**

We first calculate the mass of water moving over the falls per unit time from the given volume of water per unit time, using the density of water:

$$\frac{m}{t}=\left(5520\,\frac{\text{m}^3}{\text{s}}\right)\left(\frac{1000\text{ kg}}{\text{m}^3}\right)=5.52\cdot 10^6\text{ kg/s}.$$

The average power is then

$$\overline{P}=\left(5.52\cdot 10^6\text{ kg/s}\right)\left(9.81\text{ m/s}^2\right)\left(49.0\text{ m}\right)=2653.4088\text{ MW}.$$

**ROUND**

We round to three significant figures:

$$\overline{P}=2.65\text{ GW}.$$

**DOUBLE-CHECK**

Our result is comparable to the output of large electrical power plants, on the order of 1000 MW (1GW). The combined power generation capability of all of the hydroelectric power stations at Niagara Falls has a peak of 4.4 GW during the high water season in the spring, which is close to our answer. However, you may ask how the water produces power by simply falling over Niagara Falls. The answer is that it doesn't. Instead, a large fraction of the water of the Niagara River is diverted upstream from the falls and sent through tunnels, where it drives power generators. The water that makes it across the falls during the daytime and in the summer tourist season is only about 50% of the flow of the Niagara River. This flow is reduced even further, down to 10%, and more water is diverted for power generation during the nighttime and in the winter.

4. **SIMPLIFY:** Simplify the result algebraically as much as possible. This step is particularly helpful when more than one quantity has to be found.
5. **CALCULATE:** Substitute numbers with units into the simplified equation and calculate. Typically, a number and a physical unit are obtained as the answer.
6. **ROUND:** Consider the number of significant figures that the result should contain. A result obtained by multiplying or dividing should be rounded to the same number of significant figures as the input quantity that had the least number of significant figures. Do not round in intermediate steps, as rounding too early might give a wrong solution. Include the proper units in the answer.
7. **DOUBLE-CHECK:** Consider the result. Does the answer (both the number and the units) seem realistic? Examine the orders of magnitude. Test your solution in limiting cases.

## Examples

Briefer, terser **Examples** (Problem statement and Solution only) focus on a specific point or concept. The briefer Examples also serve as a bridge between fully worked-out Solved Problems (with all seven steps) and the homework problems.

**EXAMPLE 17.4 Rise in Sea Level Due to Thermal Expansion of Water**

The rise in the level of the Earth's oceans is of current concern. Oceans cover $3.6\cdot 10^8$ km$^2$, slightly more than 70% of Earth's surface area. The average ocean depth is 3700 m. The surface ocean temperature varies widely, between 35 °C in the summer in the Persian Gulf and −2 °C in the Arctic and Antarctic regions. However, even if the ocean surface temperature exceeds 20 °C, the water temperature rapidly falls off as a function of depth and reaches 4 °C at a depth of 1000 m (Figure 17.22). The global average temperature of all seawater is approximately 3 °C. Table 17.3 lists a volume expansion coefficient of zero for water at a temperature of 4 °C. Thus, it is safe to assume that the volume of ocean water changes very little at a depth greater than 1000 m. For the top 1000 m of ocean water, let's assume a global average temperature of 10.0 °C and calculate the effect of thermal expansion.

**FIGURE 17.22** Average ocean water temperature as a function of depth below the surface.

**PROBLEM**

By how much would sea level change, solely as a result of the thermal expansion of water, if the water temperature of all the oceans increased by $\Delta T=1.0$ °C?

**SOLUTION**

The volume expansion coefficient of water at 10.0 °C is $\beta=87.5\cdot 10^{-6}$ °C$^{-1}$ (from Table 17.3), and the volume change of the oceans is given by equation 17.9, $\Delta V=\beta V\Delta T$, or

$$\frac{\Delta V}{V}=\beta\Delta T. \quad \text{(i)}$$

We can express the total surface area of the oceans as $A=(0.7)4\pi R^2$, where $R$ is the radius of Earth and the factor 0.7 reflects the fact that about 70% of the surface of the sphere is

## Problem-Solving Practice

**Problem-Solving Practice** provides **Additional Solved Problems,** again following the full seven-step format. This section is found immediately before the end-of-chapter problems to provide a review and to emphasize the fundamental concepts of the chapter. Additional **Problem-Solving Strategies and Guidelines** are also presented here.

> "They provide a useful tool for students to improve their problem-solving skills. The authors did a good job in addressing, for each chapter, the most important steps to approach the solution of the end-of-chapter problems. Students that never had physics before will find this guideline quite beneficial. I liked in particular the connection between the guideline and the solved problem. The detailed description on how to solve these problems will certainly help the students to understand the concepts better."
>
> —*Luca Bertello, University of California–Los Angeles*

**PROBLEM-SOLVING PRACTICE**

**Problem-Solving Guidelines: Kinetic Energy, Work, and Power**

**1.** In all problems involving energy, the first step is to clearly identify the system and the changes in its conditions. If an object undergoes a displacement, check that the displacement is always measured from the same point on the object, such as the front edge or the center of the object. If the speed of the object changes, identify the initial and final speeds at specific points. A diagram is often helpful to show the position and the speed of the object at two different times of interest.

**2.** Be careful to identify the force that is doing work. Also note whether forces doing work are constant forces or variable forces, because they need to be treated differently.

**3.** You can calculate the sum of the work done by individual forces acting on an object or the work done by the net force acting on an object; the result should be the same. (You can use this as a way to check your calculations.)

**4.** Remember that the direction of the restoring force exerted by a spring is always opposite the direction of the displacement of the spring from its equilibrium point.

**5.** The formula for power, $P=\vec{F}\bullet\vec{v}$, is very useful, but applies only for a constant force. When using the more general definition of power, be sure to distinguish between the average power, $\overline{P}=\frac{W}{\Delta t}$, and the instantaneous value of the power, $P=\frac{dW}{dt}$.

**SOLVED PROBLEM 5.2 Lifting Bricks**

**PROBLEM**

A load of bricks at a construction site has a mass of 85.0 kg. A crane raises this load from the ground to a height of 50.0 m in 60.0 s at a low constant speed. What is the average power of the crane?

**SOLUTION**

**THINK**

Raising the bricks at a low constant speed means that the kinetic energy is negligible, so the work in this situation is done against gravity only. There is no acceleration, and friction is negligible. The average power then is just the work done against gravity divided by the time it takes to raise the load of bricks to the stated height.

**SKETCH**

A free-body diagram of the load of bricks is shown in Figure 5.20. Here we have defined a coordinate system in which the $y$-axis is vertical and positive is upward. The tension, $T$, exerted by the cable of the crane is a force in the upward direction, and the weight, $mg$, of the load of bricks is a force downward. Because the load is moving at a constant speed, the sum of the tension and the weight is zero. The load is moved vertically a distance $h$, as shown in Figure 5.21.

**FIGURE 5.20** Free-body diagram of the load of bricks of mass $m$ being lifted by a crane.

**RESEARCH**

The work, $W$, done by the crane is given by

$$W=mgh.$$

The average power, $\overline{P}$, required to lift the load in the given time $\Delta t$ is

$$\overline{P}=\frac{W}{\Delta t}.$$

**SIMPLIFY**

Combining the above two equations gives

$$\overline{P}=\frac{mgh}{\Delta t}.$$

**CALCULATE**

Now we put in the numbers and get

$$\overline{P}=\frac{(85.0\text{ kg})(9.81\text{ m/s}^2)(50.0\text{ m})}{60.0\text{ s}}=694.875\text{ W}.$$

*Continued—*

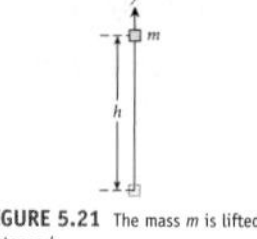

**FIGURE 5.21** The mass $m$ is lifted a distance $h$.

## End-of-Chapter Questions and Problem Sets

Along with providing problem-solving guidelines, examples, and strategies, *University Physics* also offers a **wide variety of end-of-chapter Questions and Problems.** Often professors say, "I don't need a lot of problems, just a handful of really good problems." *University Physics* has both. The end-of-chapter Questions and Problems were developed with the idea of making them interesting to the reader. The authors, along with a panel of excellent writers (who, perhaps more importantly, are also experienced physics instructors) wrote Questions and Problems for each chapter, being sure to provide wide variety in level, content, and style. Included in each chapter are a set of Multiple-Choice Questions, Questions, Problems (by section), and Additional Problems (no section "clue"). One bullet identifies slightly more challenging Problems, and two bullets identify the most challenging Problems. The problem-solving theme from the text is also carried through to the Test Bank: The same group that wrote the end-of-chapter questions and problems also wrote the Test Bank questions, providing consistency in style and coverage.

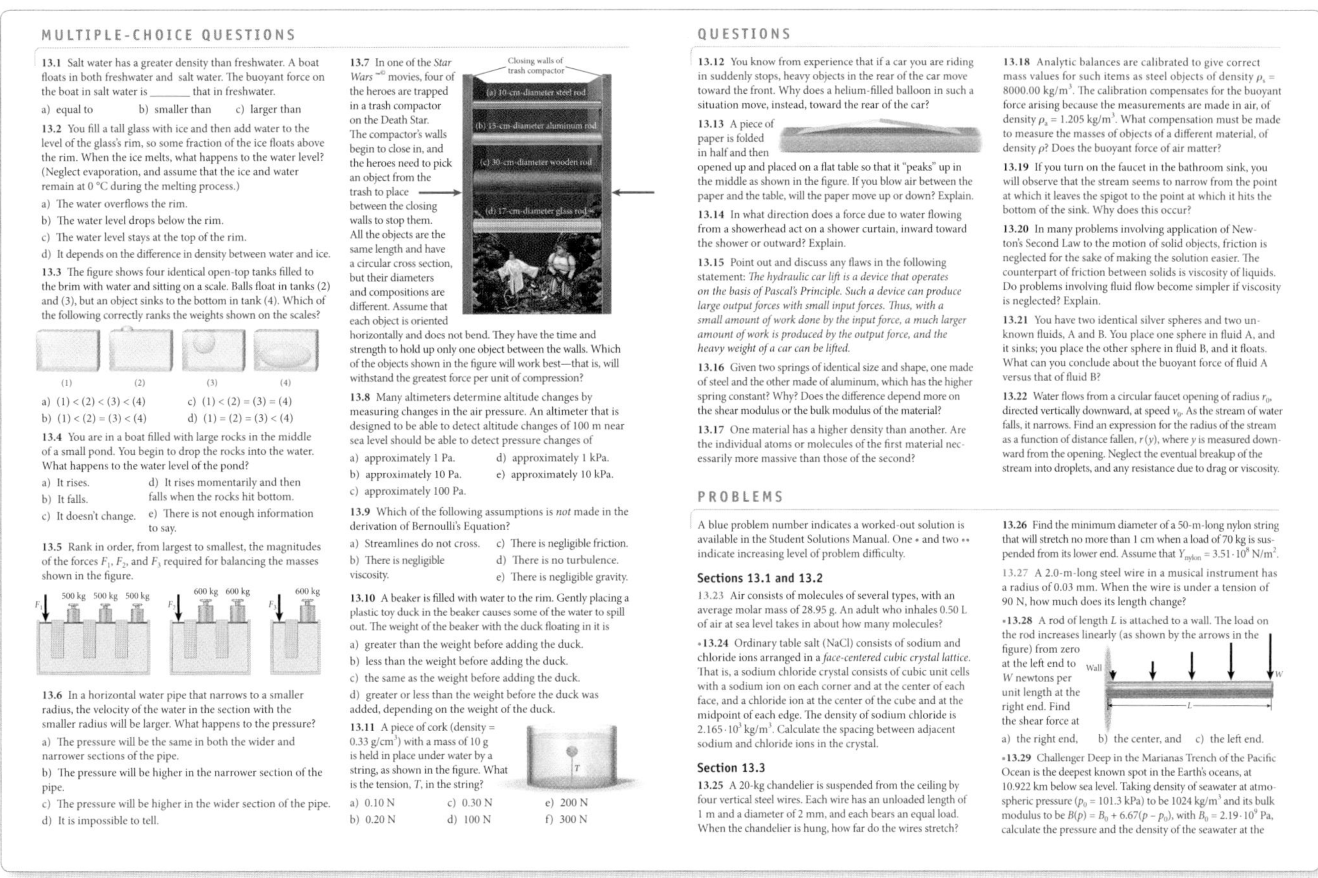

MULTIPLE-CHOICE QUESTIONS

**13.1** Salt water has a greater density than freshwater. A boat floats in both freshwater and salt water. The buoyant force on the boat in salt water is _______ that in freshwater.

a) equal to b) smaller than c) larger than

**13.2** You fill a tall glass with ice and then add water to the level of the glass's rim, so some fraction of the ice floats above the rim. When the ice melts, what happens to the water level? (Neglect evaporation, and assume that the ice and water remain at 0 °C during the melting process.)

a) The water overflows the rim.
b) The water level drops below the rim.
c) The water level stays at the top of the rim.
d) It depends on the difference in density between water and ice.

**13.3** The figure shows four identical open-top tanks filled to the brim with water and sitting on a scale. Balls float in tanks (2) and (3), but an object sinks to the bottom in tank (4). Which of the following correctly ranks the weights shown on the scales?

(1) (2) (3) (4)

a) (1) < (2) < (3) < (4)
b) (1) < (2) = (3) < (4)
c) (1) < (2) = (3) = (4)
d) (1) = (2) = (3) < (4)

**13.4** You are in a boat filled with large rocks in the middle of a small pond. You begin to drop the rocks into the water. What happens to the water level of the pond?

a) It rises.
b) It falls.
c) It doesn't change.
d) It rises momentarily and then falls when the rocks hit bottom.
e) There is not enough information to say.

**13.5** Rank in order, from largest to smallest, the magnitudes of the forces $F_1$, $F_2$, and $F_3$ required for balancing the masses shown in the figure.

$F_1$ 500 kg 500 kg 500 kg $F_2$ 600 kg 600 kg $F_3$ 600 kg

**13.6** In a horizontal water pipe that narrows to a smaller radius, the velocity of the water in the section with the smaller radius will be larger. What happens to the pressure?

a) The pressure will be the same in both the wider and narrower sections of the pipe.
b) The pressure will be higher in the narrower section of the pipe.
c) The pressure will be higher in the wider section of the pipe.
d) It is impossible to tell.

**13.7** In one of the *Star Wars*™© movies, four of the heroes are trapped in a trash compactor on the Death Star. The compactor's walls begin to close in, and the heroes need to pick an object from the trash to place between the closing walls to stop them. All the objects are the same length and have a circular cross section, but their diameters and compositions are different. Assume that each object is oriented horizontally and does not bend. They have the time and strength to hold up only one object between the walls. Which of the objects shown in the figure will work best—that is, will withstand the greatest force per unit of compression?

Closing walls of trash compactor
(a) 10-cm-diameter steel rod
(b) 15-cm-diameter aluminum rod
(c) 30-cm-diameter wooden rod
(d) 17-cm-diameter glass rod

**13.8** Many altimeters determine altitude changes by measuring changes in the air pressure. An altimeter that is designed to be able to detect altitude changes of 100 m near sea level should be able to detect pressure changes of

a) approximately 1 Pa.
b) approximately 10 Pa.
c) approximately 100 Pa.
d) approximately 1 kPa.
e) approximately 10 kPa.

**13.9** Which of the following assumptions is *not* made in the derivation of Bernoulli's Equation?

a) Streamlines do not cross.
b) There is negligible viscosity.
c) There is negligible friction.
d) There is no turbulence.
e) There is negligible gravity.

**13.10** A beaker is filled with water to the rim. Gently placing a plastic toy duck in the beaker causes some of the water to spill out. The weight of the beaker with the duck floating in it is

a) greater than the weight before adding the duck.
b) less than the weight before adding the duck.
c) the same as the weight before adding the duck.
d) greater or less than the weight before the duck was added, depending on the weight of the duck.

**13.11** A piece of cork (density = 0.33 $g/cm^3$) with a mass of 10 g is held in place under water by a string, as shown in the figure. What is the tension, $T$, in the string?

$T$

a) 0.10 N
b) 0.20 N
c) 0.30 N
d) 100 N
e) 200 N
f) 300 N

QUESTIONS

**13.12** You know from experience that if a car you are riding in suddenly stops, heavy objects in the rear of the car move toward the front. Why does a helium-filled balloon in such a situation move, instead, toward the rear of the car?

**13.13** A piece of paper is folded in half and then opened up and placed on a flat table so that it "peaks" up in the middle as shown in the figure. If you blow air between the paper and the table, will the paper move up or down? Explain.

**13.14** In what direction does a force due to water flowing from a showerhead act on a shower curtain, inward toward the shower or outward? Explain.

**13.15** Point out and discuss any flaws in the following statement: *The hydraulic car lift is a device that operates on the basis of Pascal's Principle. Such a device can produce large output forces with small input forces. Thus, with a small amount of work done by the input force, a much larger amount of work is produced by the output force, and the heavy weight of a car can be lifted.*

**13.16** Given two springs of identical size and shape, one made of steel and the other made of aluminum, which has the higher spring constant? Why? Does the difference depend more on the shear modulus or the bulk modulus of the material?

**13.17** One material has a higher density than another. Are the individual atoms or molecules of the first material necessarily more massive than those of the second?

**13.18** Analytic balances are calibrated to give correct mass values for such items as steel objects of density $\rho_s$ = 8000.00 $kg/m^3$. The calibration compensates for the buoyant force arising because the measurements are made in air, of density $\rho_a$ = 1.205 $kg/m^3$. What compensation must be made to measure the masses of objects of a different material, of density $\rho$? Does the buoyant force of air matter?

**13.19** If you turn on the faucet in the bathroom sink, you will observe that the stream seems to narrow from the point at which it leaves the spigot to the point at which it hits the bottom of the sink. Why does this occur?

**13.20** In many problems involving application of Newton's Second Law to the motion of solid objects, friction is neglected for the sake of making the solution easier. The counterpart of friction between solids is viscosity of liquids. Do problems involving fluid flow become simpler if viscosity is neglected? Explain.

**13.21** You have two identical silver spheres and two unknown fluids, A and B. You place one sphere in fluid A, and it sinks; you place the other sphere in fluid B, and it floats. What can you conclude about the buoyant force of fluid A versus that of fluid B?

**13.22** Water flows from a circular faucet opening of radius $r_0$, directed vertically downward, at speed $v_0$. As the stream of water falls, it narrows. Find an expression for the radius of the stream as a function of distance fallen, $r(y)$, where $y$ is measured downward from the opening. Neglect the eventual breakup of the stream into droplets, and any resistance due to drag or viscosity.

PROBLEMS

A blue problem number indicates a worked-out solution is available in the Student Solutions Manual. One • and two •• indicate increasing level of problem difficulty.

**Sections 13.1 and 13.2**

13.23 Air consists of molecules of several types, with an average molar mass of 28.95 g. An adult who inhales 0.50 L of air at sea level takes in about how many molecules?

•**13.24** Ordinary table salt (NaCl) consists of sodium and chloride ions arranged in a *face-centered cubic crystal lattice*. That is, a sodium chloride crystal consists of cubic unit cells with a sodium ion on each corner and at the center of each face, and a chloride ion at the center of the cube and at the midpoint of each edge. The density of sodium chloride is $2.165 \cdot 10^3$ $kg/m^3$. Calculate the spacing between adjacent sodium and chloride ions in the crystal.

**Section 13.3**

**13.25** A 20-kg chandelier is suspended from the ceiling by four vertical steel wires. Each wire has an unloaded length of 1 m and a diameter of 2 mm, and each bears an equal load. When the chandelier is hung, how far do the wires stretch?

**13.26** Find the minimum diameter of a 50-m-long nylon string that will stretch no more than 1 cm when a load of 70 kg is suspended from its lower end. Assume that $Y_{nylon} = 3.51 \cdot 10^8$ $N/m^2$.

13.27 A 2.0-m-long steel wire in a musical instrument has a radius of 0.03 mm. When the wire is under a tension of 90 N, how much does its length change?

•**13.28** A rod of length $L$ is attached to a wall. The load on the rod increases linearly (as shown by the arrows in the figure) from zero at the left end to $W$ newtons per unit length at the right end. Find the shear force at

Wall $W$ $L$

a) the right end, b) the center, and c) the left end.

•**13.29** Challenger Deep in the Marianas Trench of the Pacific Ocean is the deepest known spot in the Earth's oceans, at 10.922 km below sea level. Taking density of seawater at atmospheric pressure ($p_0$ = 101.3 kPa) to be 1024 $kg/m^3$ and its bulk modulus to be $B(p) = B_0 + 6.67(p - p_0)$, with $B_0 = 2.19 \cdot 10^9$ Pa, calculate the pressure and the density of the seawater at the

"The problem-solving technique, to borrow a phrase from my students, 'doesn't suck.' I'm a skeptic when it comes to anybody else's one-size-fits-all approach to problem solving—I've seen too many that just don't work, pedagogically. The approach used by the authors, however, is one in which students are really forced to tap their intuition before they start, to reflect on the relevant first principles. . . .

Wow! There are some really nice problems at the end of the chapter. My compliments to the authors. There was a nice diversity of problems, and most of them required a lot more than simple plug-and-chug. I found many problems I'd be inclined to assign."

—*Brent Corbin, University of California–Los Angeles*

"The text strikes a very good balance of providing mathematical details and rigor together with a clear, intuitive presentation of physics concepts. The balance and variety of problems provided, both as worked-out examples and as end-of-chapter problems, are outstanding. Many features are found in this book that are difficult to find in other standard texts, including proper use of vector notation, explicit evaluation of multiple integrals, e.g., in moment of inertia calculations, and intriguing connections to modern physics."

—*Lisa Everett, University of Wisconsin–Madison*

## Contemporary Topics: Capturing Students' Imaginations

*University Physics* incorporates a wide variety of contemporary topics as well as research-based discussions designed to help students appreciate the beauty of physics and see how physics concepts are related to the development of new technologies in the fields of engineering, medicine, astronomy and more. The "Big Picture" section at the beginning of the text is designed to introduce students to some of the amazing new frontiers of research that are being explored in various fields of physics and the results that have been obtained during the last few years. The authors return to these topics at various points within the book for more in-depth exploration.

The authors of *University Physics* also repeatedly discuss different aspects of the broad topic of energy, by addressing concepts of energy sources (fossil, nuclear, renewable, alternative, and so forth), energy efficiency, and environmental effects of energy supply choices. Alternative energy sources and renewable resources are discussed within the framework of possible solutions to the energy crisis. These discussions provide a great opportunity to capture students' interest and are accessible on the introductory physics level.

The following contemporary physics research topics and topical energy discussions (in green) are found in the text:

**Chapter 1**
Section 1.3 has a subsection called "Metrology" that mentions the new definition of the kilogram and the optical clock at NIST
Section 1.4 mentions the authors' research on heavy-ion collisions

**Chapter 4**
Section 4.2 has a subsection on the Higgs particle
Section 4.7 has a subsection on tribology

**Chapter 5**
Section 5.1 Energy in Our Daily Lives
Section 5.7 Power and Fuel-Efficiency of U.S. Cars

**Chapter 6**
Section 6.8 has a subsection titled "Preview: Atomic Physics" that discusses tunneling of particles
Solved Problem 6.6 Power Produced by Niagara Falls

**Chapter 7**
Example 7.5 Particle Physics
Section 7.8 discusses Sinai billiards and chaotic motion

**Chapter 8**
Example 8.3 mentions electromagnetic propulsion and radiation shielding in context of sending astronauts to Mars

**Chapter 10**
Example 10.7 Death of a Star
Example 10.8 Flybrid

**Chapter 12**
Section 12.1 has a subsection titled "Solar System" that mentions research on objects in Kuiper Belt
Section 12.7 Dark Matter

**Chapter 13**
Section 13.1 briefly discusses nanotubes and nanotechnology
Section 13.2 briefly discusses plasmas and Bose-Einstein condensates
Section 13.6 has a subsection titled "Applications of Bernoulli's Equation" that discusses lift and design of aircraft wings
Section 13.8 Turbulence and Research Frontiers in Fluid Flow

**Chapter 14**
Section 14.7 Chaos

**Chapter 15**
Section 15.5 has a subsection titled "Seismic Waves" that mentions reflection seismology
Figure 15.11b shows nanoscale guitar
Section 15.8 includes Self-Test Opportunity 15.4 on nanoscale guitar string
Section 15.9 Research on Waves

**Chapter 16**
Section 16.4 has a subsection titled "Mach Cone" that mentions creation of shock waves by collisions of nuclei in particle accelerators and has a paragraph on Cherenkov radiation

**Chapter 17**
Section 17.2 has subsections titled "Research at the Low-Temperature Frontier" and "Research at the High-Temperature Frontier"
Section 17.5 Surface Temperature of Earth
Section 17.6 Temperature of the Universe
Example 17.4 Rise in Sea Level Due to Thermal Expansion of Water

**Chapter 18**
Example 18.7 Roof Insulation
Solved Problem 18.2 Warming Costs for Winter
Solved Problem 18.3 Power Carried by the Gulf Stream
Example 18.8 Earth as a Blackbody
Section 18.8 Modes of Thermal Energy Transfer/Global Warming
Section 18.8 has a subsection titled "Heat in Computers"

**Chapter 19**
Example 19.5 The Quark-Gluon Plasma

**Chapter 20**
Example 20.2 Warming a House with a Heat Pump
Example 20.4 Maximum Efficiency of an Electrical Power Plant
Solved Problem 20.1 Efficiency of an Automobile Engine
Section 20.4 Real Engines and Efficiency/Hybrid Cars
Section 20.4 Real Engines and Efficiency/Efficiency and the Energy Crisis
Solved Problem 20.2 Cost to Operate an Electrical Plant
Section 20.7 Entropy Death

**Chapter 21**
Section 21.3 Superconductors
Section 21.5 Electrostatic Force—Coulomb's Law/Electrostatic Precipitator

**Chapter 22**
Example 22.4 Time Projection Chamber (STAR TPC)
Section 22.6 discusses noninvasive imaging of brain electric fields and the brain-computer interface

**Chapter 23**
Section 23.2 describes new batteries
Example 23.2 Battery-Powered Cars

**Chapter 24**
Example 24.4 The National Ignition Facility
Section 24.9 Supercapacitors

**Chapter 25**
Example 25.1 Iontophoresis
Section 25.3 Temperature Dependence, and Superconductivity
Solved Problem 25.2 Brain Probe (ECoG)
Section 25.7 Energy and Power in Electric Circuits discusses high-voltage electric power transmission
Section 25.8 Diodes: One-Way Streets in Circuits
Solved Problem 25.4 Size of Wire for Power Line

**Chapter 27**
Section 27.3 and Example 27.1 STAR TPC
Section 27.3 has a subsection on magnetic levitation
Example 27.3 Cyclotron

**Chapter 28**
Solved Problem 28.1 Electromagnetic Rail Accelerator
Section 28.7 discusses superconducting magnets and the Meissner effect

**Chapter 29**
Section 29.4 Generators and Motors discusses regenerative braking
Section 29.10 discusses computer hard drives and giant magnetoresistance

**Chapter 30**
Section 30.7 Transformers

**Chapter 31**
Example 31.1 Using Solar Panels to Charge an Electric Car

**Chapter 32**
Section 32.2 discusses perfect mirrors and laser surgery

**Chapter 33**
Section 33.8 Laser Tweezers

**Chapter 35**
Example 35.2 discusses the possibility of a muon collider
Section 35.6 discusses particle accelerators
Section 35.8 discusses the Hubble Space Telescope and gravitational lensing, Black holes, LIGO, and LISA
Section 35.9 Relativity in Our Daily Life: GPS

**Chapter 36**
Section 36.2 Blackbody Radiation describes cosmic background radiation
Section 36.3 discusses night-vision devices
Section 36.8 has a subsection on "Bose-Einstein Condensate"

**Chapter 37**
Introduction on quantum computing, nanoscience, and nanotechnology
Section 37.9 discusses two-fermion wave function and computer modeling of physical systems, as well as quantum computing

**Chapter 38**
Section 38.4 has a subsection titled "X-Ray Production" that discusses X-rays in medicine
Section 38.5 discusses the use of lasers in surgery, weapons, measurement, chemical bonds, free-electron lasers, and the NIF laser system

**Chapter 39**
Section 39.1 discusses CERN, LHC, complex atomic-level systems, protein folding, DNA, and the origin of life
Section 39.3 discusses elementary particles and Higgs bosons
Section 39.4 has a subsection titled "Supersymmetry and String Theory"

**Chapter 40**
Section 40.2 discusses neutrino-less double-beta decays
Section 40.3 describes the nuclear shell model and models of nuclear collisions
Section 40.4 Nuclear Energy: Fission and Fusion discusses the use of controlled fusion for energy production, as well as ITER and NIF
Section 40.5 describes core-collapse supernova explosions
Section 40.6 discusses gamma knives (in brain surgery), heavy-ion beam treatments for cancer, and technetium diagnostic scans

"I think this idea is great! It would help the instructor show the students that physics is a live and exciting subject. . . . because it shows that physics is a happening subject, relevant for discovering how the universe works, that it is necessary for developing new technologies, and how it can benefit humanity . . . . The [chapters] contain a lot of interesting modern topics and explain them very clearly."

*—Joseph Kapusta, University of Minnesota*

"Section 17.5 on the surface temperature of the Earth is excellent and is an example of what is *missing* from so many introductory textbooks: examples that are relevant and compelling for the students."

*—John William Gary, University of California–Riverside*

"I think the approach to include modern or contemporary physics throughout the text is great. Students often approach physics as a science of concepts which were discovered long ago. They view engineering as the science which has given them the advances in technology which they see today. It would be great to show students just where these advances do start, with physics."

*—Donna W. Stokes, University of Houston*

## Enhanced Content: Flexibility for Your Student and Course Needs

To instructors who are looking for additional coverage of certain topics and mathematical support for those topics, *University Physics* also offers flexibility. This book includes some topics and some calculus that do not appear in many other texts. However, these topics have been presented in such a way that their exclusion will not affect the overall course. The text as a whole is written at a level appropriate for the typical introductory student. Below is a list of flex-coverage content as well as additional mathematical support:

### Chapter 2

Section 2.3 The concept of the derivative is developed using an approach that is both conceptual and graphical. Examples using the derivative are provided, and students are referred to an appendix for other "refreshers." This is a more extensive approach than is taken in some other texts.

Section 2.4 Acceleration as the time derivative of velocity is introduced by analogy, and the discussion includes an example.

Section 2.6 Integration as the inverse to differentiation is introduced for finding the area under a curve. This more extensive presentation than in many texts is spread over two sections with multiple examples.

Section 2.7 Examples using differentiation are included.

Section 2.8 A derivation on minimum time arguments is shown to lead to a solution that is equivalent to Snell's Law.

End-of-chapter exercises related to this coverage include Questions 20, 22, and 23 and multiple Problems using calculus.

### Chapter 3

Section 3.1 The component-wise derivative of a three-dimensional position vector into three-dimensional velocity and then into three-dimensional acceleration is presented.

Section 3.3 The tangetiality of the velocity vector to the trajectory is covered.

Section 3.4 The maximum height and range of a projectile are found by setting the derivative equal to zero.

Section 3.5 Relative motion is covered (equation 3.27).

End-of-chapter Problem 3.38 covers the derivative.

### Chapter 4

Section 4.8 Example 4.10 on the best angle to pull a sled is a maximum-minimum problem.

### Chapter 5

Section 5.5 Work done by a variable force is covered using definite integrals and the derivation of equation 5.20. The chain rule is also covered.

Section 5.6 Work done by spring force is discussed (equation 5.24).

Section 5.7 Power as the time derivative of work is covered (equation 5.26).

A number of end-of-chapter Problems supplement this coverage, such as Problems 5.34 through 5.37.

### Chapter 6

Section 6.3 Finding the work done by a force includes use of integrals.

Section 6.4 Obtaining force from the potential includes the use of derivatives; partial derivatives and the gradient are also introduced (for example, the Lennard-Jones potential).

A number of end-of-chapter Questions and Problems supplement this coverage, such as Questions 6.24 and 6.25 and Problems 6.34, 6.35, and 6.36.

### Chapter 8

The text introduces volume integrals so that the volume of a sphere and the center of mass of a half sphere can be determined in a worked example.

### Chapter 9

Explicit derivatives of the radical and tangential unit vectors are provided. The text derives the equations of motion for constant angular acceleration, repeating the customary derivation of equations of motion for constant linear acceleration presented in Chapter 2.

### Chapter 10

The volume integral introduced in Chapter 8 is utilized in finding the moment of inertia for different objects. The text derives the expression for angular momentum to determine the relationship between the angular momentum of a system of particles and the torque.

### Chapter 11

Section 11.3 The stability condition is utilized, and the second derivative of potential energy is examined to determine the type of equilibrium through graphical interpretation of the functions.

### Chapter 12

Section 12.1 Unique coverage is provided of the derivation of the gravitational force from a sphere and inside a sphere.

### Chapter 14

Section 14.4 The subsection titled "Small Damping" applies the student's knowledge of simple harmonic oscillators to derive the small damping equation through differentiation. For the case of large damping, the student is again referred to the solution to the differential equation. Example 14.6 walks the student through an example of a damped harmonic oscillator. The solution to this equation is stated explicitly, but the text utilizes calculus to reach the answer. The subsection "Energy Loss in Damped Oscillations" includes a calculation of the rate of energy loss that uses the differential definition of power.

Section 14.5 A thorough discussion of forced harmonic motion takes advantage of the student's understanding of differential equations, graphically analyzes the solution, and then analyzes the outcome.

A number of end-of-chapter Problems supplement this coverage, such as Problems 14.55 and 14.73.

### Chapter 15

Section 15.4 This entire section is unique among introductory physics texts as it utilizes partial differential equations to derive the wave equation.

Several of the end-of-chapter Questions and Problems related to the content of Section 15.4 require an understanding of the calculus used in this section, particularly 15.30 and 15.31.

### Chapter 16

Section 16.4 Covers the Doppler effect as a function of the perpendicular distance.

### Chapter 22

Section 22.9 Solved Problem 22.3 covers the electric field for a nonuniform spherical charge distribution

### Chapter 25

Section 25.5 Solved Problem 25.2, "Brain Probe" covers a case of nonconstant cross-sectional area.

### Chapter 32

Section 32.3 This textbook goes deeper into calculus than many others by demonstrating how calculus can be used in deducing from Newton's Laws of Motion the necessarily parabolic shape of liquid surfaces that have a net circular motion.

Some end-of-chapter Problems, such as Problem 32.43, also use calculus to solve a minimization problem.

### Chapter 34

Section 34.10 This section discusses the quality of diffraction gratings using the concept of dispersion. In many similar physics textbooks, the formula for dispersion is simply given. This textbook, however, uses calculus to derive dispersion in a straightforward manner.

### Chapter 35

Section 35.2 This section has a very instructive and intuitive derivation of light cone variables that goes beyond what is found in most standard textbooks.

Section 35.6 The text features calculus-based derivations for velocity transformation and energy (based on integrating the distance dependence of work). While the energy derivation follows standard integration techniques and is used in most books, the velocity transformation derivation is unique and very instructive.

### Chapter 36

Section 36.2 The level of detail that characterizes the derivation of several radiation laws (Wien, Planck, Boltzmann, Raleigh-Jeans) is not found in many other textbooks.

Section 36.8 The introduction to Bose-Einstein and Fermi-Dirac statistics is important and unique to this textbook. The connection to radiation laws is especially relevant. The end-of-chapter Problems related to Section 36.8, in particular Problems 36.53 through 36.55, are challenging and utilize the relevant mathematics.

### Chapter 37

Most standard textbooks teach quantum mechanics with minimal usage of calculus, using mostly a conceptual approach. This book takes a more formal approach, from Section 35.1 on wave functions. The students are exposed to derivations that are unique to modern physics, starting from the normalization condition of the wave function over operators for momentum and kinetic energy and continuing to solutions for infinite and finite potentials. Hamiltonians are introduced and applied to Schrödinger and Dirac equations. The many-particle wave function is then covered in Section 37.9. The end-of-chapter Problems that utilize calculus include Problems 37.28 through 37.39.

### Chapter 38

This textbook derives the full solution of the hydrogen electron wave function and breaks it down into radial and angular parts. This complete solution allows the student to derive the degeneracy of the quantum levels rather than simply learning a simple formula for calculating the levels without understanding their physical origin.

Section 38.3 The solution of the Schrödinger equation in Section 38.3. is based on the derivations in Chapter 37, and the text further explores the full solution of the hydrogen electron wave function considered earlier in the chapter. The end-of-chapter Problems 38.35, 38.36, and 38.37 use the calculus from the chapter.

### Chapter 39

The definition of the differential scattering cross section is defined in equations 39.3 and 39.4, based on classical (Rutherford) physics. (Many other texts simply show and describe graphs.) The differential cross section from quantum considerations is given in equation 39.6, and the form factor (deviation from Rutherford and deviation from point particle) is given in equations 39.7 and 39.8. Form factors are not often described in other texts. While this discussion adds to the mathematical detail, it could be easily omitted to fit the needs of the course. End-of-chapter Problem 39.32 makes use of calculus to calculate the fraction of particles scattered into a range of angles.

### Chapter 40

The text presents a slightly more detailed discussion than is usually seen in deriving the Fermi energy while covering Fermi's model of the nucleus in Section 40.3. Some end-of-chapter problems could involve simple integration of the exponential function: 40.31, 40.33, 40.52, 40.53, and 40.61.

> "Strongest feature . . . The use of real mathematics, especially calculus, to derive kinematic relations, the relations between quantities in circular motion, the direction of the gravitational force, the magnitude of the tidal force, the maximum extension of a set of piled blocks. Solved problems are always addressed symbolically first. Too often textbooks don't let the math do the work for them."
>
> —*Kieran Mullen, University of Oklahoma*

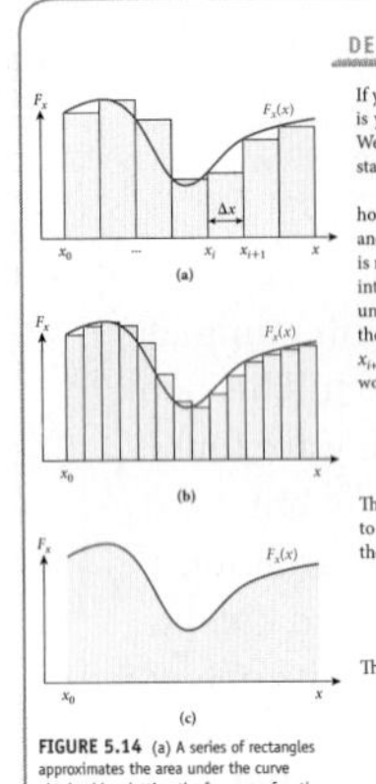

FIGURE 5.14 (a) A series of rectangles approximates the area under the curve obtained by plotting the force as a function of the displacement; (b) a better approximation using rectangles of smaller width; (c) the exact area under the curve.

DERIVATION 5.1

If you have already taken integral calculus, you can skip this section. If equation 5.20 is your first exposure to integrals, the following derivation is a useful introduction. We'll derive the one-dimensional case and use our result for the constant force as a starting point.

In the case of a constant force, we can think of the work as the area under the horizontal line that plots the value of the constant force in the interval between $x_0$ and $x$. For a variable force, the work is the area under the curve $F_x(x)$, but that area is no longer a simple rectangle. In the case of a variable force, we need to divide the interval from $x_0$ to $x$ into many small equal intervals. Then we approximate the area under the curve $F_x(x)$ by a series of rectangles and add their areas to approximate the work. As you can see from Figure 5.14a, the area of the rectangle between $x_i$ and $x_{i+1}$ is given by $F_x(x_i)\cdot(x_{i+1}-x_i)=F_x(x_i)\cdot\Delta x$. We obtain an approximation for the work by summing over all rectangles:

$$W \approx \sum_i W_i = \sum_i F_x(x_i)\cdot\Delta x.$$

Now we space the points $x_i$ closer and closer by using more and more of them. This method makes $\Delta x$ smaller and causes the total area of the series of rectangles to be a better approximation of the area under the curve $F_x(x)$ as in Figure 5.14b. In the limit as $\Delta x \to 0$, the sum approaches the exact expression for the work:

$$W = \lim_{\Delta x\to 0}\left(\sum_i F_x(x_i)\cdot\Delta x\right).$$

This limit of the sum of the areas is exactly how the integral is defined:

$$W = \int_{x_0}^{x} F_x(x')dx'.$$

We have derived this result for the case of on tion of the three-dimensional case proceeds alon in terms of algebra.

## Derivations

Detailed derivations are provided in the text as examples for students, who will eventually need to develop their own derivations as they review Solved Problems, work through Examples, and solve end-of-chapter Problems. The Derivations are identified in the text with numbered headings so that instructors can include these detailed features as necessary to fit the needs of their courses.

"Again the derivation resulting in equation 6.15 is outstanding. Few books that I have seen will show students even once all the math steps in derivations. This is a strength of this book. Also, in the next section, I like very much the generalization of the relation between force and potential energy to three dimensions. It is something that I always do in lecture although most books do not get close."

—*James Stone, Boston University*

# Appendix A

*Mathematics Primer*



## Calculus Primer

A Calculus Primer can be found in the appendices. Since this course sequence is typically given in the first year of study at universities, it assumes knowledge of high school physics and mathematics. It is preferable that students have had a course in calculus before they start this course sequence, but calculus can also be taken in parallel. To facilitate this, the text contains a short calculus primer in an appendix, giving the main results of calculus without the rigorous derivations.

# Building Knowledge: The Text's Learning System

## Chapter Opening Outline

At the beginning of each chapter is an outline presenting the section heads within the chapter. The outline also includes the titles of the Examples and Solved Problems found in the chapter. At a quick glance, students or instructors know if a desired topic, example, or problem is in the chapter.

## What We Will Learn / What We Have Learned

Each chapter of *University Physics* is organized like a good research seminar. It was once said, "Tell them what you will tell them, then tell them, and then tell them what you told them!" Each chapter starts with **What We Will Learn**—a quick summary of the main points, without any equations. And at the end of each chapter, **What We Have Learned/Exam Study Guide** contains key concepts, including major equations, symbols, and key terms. All symbols used in the chapter's formulas are also listed.

WHAT WE WILL LEARN

- A force is a vector quantity that is a measure of how an object interacts with other objects.
- Fundamental forces include gravitational attraction and electromagnetic attraction and repulsion. In daily experience, important forces include tension and normal, friction, and spring forces.
- Multiple forces acting on an object sum to a net force.
- Free-body diagrams are valuable aids in working problems.
- Newton's three laws of motion govern the motion of objects under the influence of forces.
  a) The first law deals with objects for which external forces are balanced.
  b) The second law describes those cases for which external forces are not balanced.
  c) The third law addresses equal (in magnitude) and opposite (in direction) forces that two bodies exert on each other.
- The gravitational mass and the inertial mass of an object are equivalent.
- Kinetic friction opposes the motion of moving objects; static friction opposes the impending motion of objects at rest.
- Friction is important to the unde world motion, but its causes and are still under investigation.
- Applications of Newton's laws of multiple objects, multiple forces, applying the laws to analyze a situ the most important problem-solv physics.

WHAT WE HAVE LEARNED | EXAM STUDY GUIDE

- The net force on an object is the vector sum of the forces acting on the object: $\vec{F}_{net} = \sum_{i=1}^{n} \vec{F}_i$.
- Mass is an intrinsic quality of an object that quantifies both the object's ability to resist acceleration and the gravitational force on the object.
- A free-body diagram is an abstraction showing all forces acting on an isolated object.
- Newton's three laws are as follows:

  Newton's First Law. In the absence of a net force on an object, the object will remain at rest, if it was at rest. If it was moving, it will remain in motion in a straight line with the same velocity.

  Newton's Second Law. If a net external force, $\vec{F}_{net}$, acts on an object with mass $m$, the force will cause an acceleration, $\vec{a}$, in the same direction as the force: $\vec{F}_{net} = m\vec{a}$.

  Newton's Third Law. The forces that two interacting objects exert on each other are always exactly equal in magnitude and opposite in direction: $\vec{F}_{1\to 2} = -\vec{F}_{2\to 1}$.

- Two types of friction occur: static and kinetic friction. Both types of friction are proportional to the normal force, $N$.

  Static friction describes the force of friction between an object at rest on a surface in terms of the coefficient of static friction, $\mu_s$. The static friction force, $f_s$, opposes a force trying to move an object and has a maximum value, $f_{s,max}$, such that $f_s \le \mu_s N = f_{s,max}$.

  Kinetic friction describes the force of friction between a moving object and a surface in terms of the coefficient of kinetic friction, $\mu_k$. Kinetic friction is given by $f_k = \mu_k N$.

  In general, $\mu_s > \mu_k$.

## Conceptual Introductions

Conceptual explanations are provided in the text prior to any mathematical explanations, formulas, or derivations in order to establish for students why the quantity is needed, why it is useful, and why it must be defined accurately. The authors then move from the conceptual explanation and definition to a formula and exact terms.

> "The section on thermal expansion is outstanding and the supporting example problems are very well done. This section can be put up against any text on the market and come out ahead. The authors do very well on basic concepts."
>
> —*Marllin Simon, Auburn University*

## Self-Test Opportunities

Sets of questions follow the coverage of major concepts within the text to encourage students to develop an internal dialogue. These questions will help students think critically about what they have just read, decide whether they have a grasp of the concept, and develop a list of follow-up questions to ask in lecture. The answers to the Self-Tests are found at the end of each chapter.

> "The Self-Test Opportunities are effective for encouraging students to place what they have learned in this chapter in the context of the broader conceptual understanding they have been developing throughout the earlier chapters."
>
> — *Nina Abramzon, California Polytechnic University–Pomona*

**6.3 Self-Test Opportunity**

Why does the lighter-colored ball arrive at the bottom in Figure 6.10 before the other ball?

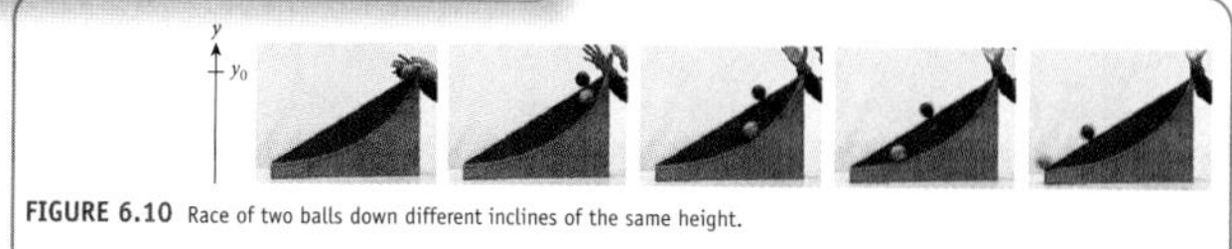


**FIGURE 6.10** Race of two balls down different inclines of the same height.

**6.3** The lighter-colored ball descends to a lower elevation earlier in its motion and thus converts more of its potential energy to kinetic energy early on. Greater kinetic energy means higher speed. Thus, the lighter-colored ball reaches higher speeds earlier and is able to move to the bottom of the track faster, even though its path length is greater.

## In-Class Exercises

In-Class Exercises are designed to be used with personal response system technology. They will appear in the text so that students may begin contemplating the concepts. Answers will only be available to instructors. (Questions and answers are formatted in PowerPoint for universal use with personal response systems.)

**2.2 In-Class Exercise**

Throwing a ball straight up into the air provides an example of free-fall motion. At the instant the ball reaches its maximum height, which of the following statements is true?

a) The ball's acceleration points down, and its velocity points up.

b) The ball's acceleration is zero, and its velocity points up.

c) The ball's acceleration points up, and its velocity points up.

d) The ball's acceleration points down, and its velocity is zero.

e) The ball's acceleration points up, and its velocity is zero.

f) The ball's acceleration is zero, and its velocity points down.

## Visual Program

Familiarity with graphic artwork on the Internet and in video games has raised the bar for the graphical presentations within textbooks, which must now be more sophisticated to excite both students and faculty. Here are some examples of techniques and ideas implemented in the *University Physics:*

- Overlays of line drawings over photographs connect sometimes very abstract physics concepts to students' realities and everyday experiences.
- A three-dimensional look for line drawings adds plasticity to the presentations. Mathematically accurate graphs and plots were created by the authors in software programs such as Mathematica and then used by the graphic artists to ensure complete accuracy along with a visually appealing style.

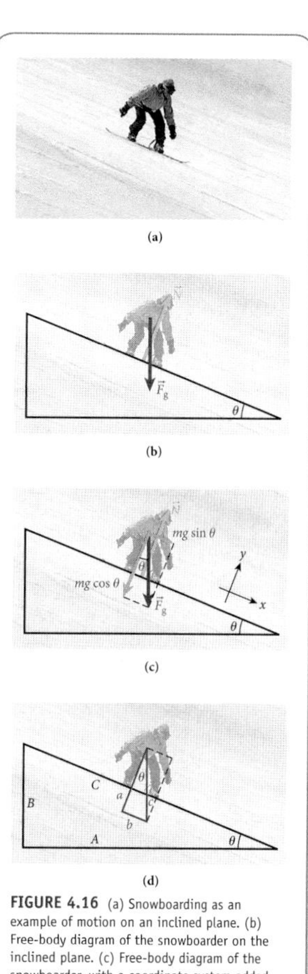


**FIGURE 4.16** (a) Snowboarding as an example of motion on an inclined plane. (b) Free-body diagram of the snowboarder on the inclined plane. (c) Free-body diagram of the snowboarder, with a coordinate system added. (d) Similar triangles in the inclined-plane problem.

McGraw-Hill Higher Education.

# Connect. Learn. Succeed.

McGraw-Hill Higher Education's mission is to help prepare students for the world that awaits. McGraw-Hill provides textbooks, eBooks and other digital instructional content, as well as experiential learning and assignment/assessment platforms, that connect instructors and students to valuable course content—and connect instructors and students to each other.

With the highest quality tools and content, students can engage with their coursework when, where, and however they learn best, enabling greater learning and deeper comprehension.

In turn, students can learn to their full potential and, thus, succeed academically now and in the real world.

Connect

Learn

Succeed

McGraw Hill

*Connect:*
**Instructor Resources**

- McGraw-Hill Connect
- Simulations
- McGraw-Hill Create
- Tegrity Campus 2.0
- Learning Solutions
- Instructor Solutions Manual
- PowerPoint Lecture Outlines
- Clicker Questions
- Electronic Images from the Text
- EzTest Test Bank

*Learn:*
**Course Content**

- Textbooks/Readers
- eBooks
- PowerPoint Presentations
- Enhanced Cartridges
- In-class Simulations
- Lecture Aids
- Custom Publishing

*Succeed:*
**Student Resources**

- Online Homework
- Simulations
- Questions
- eBook

## McGraw-Hill Connect™ Physics

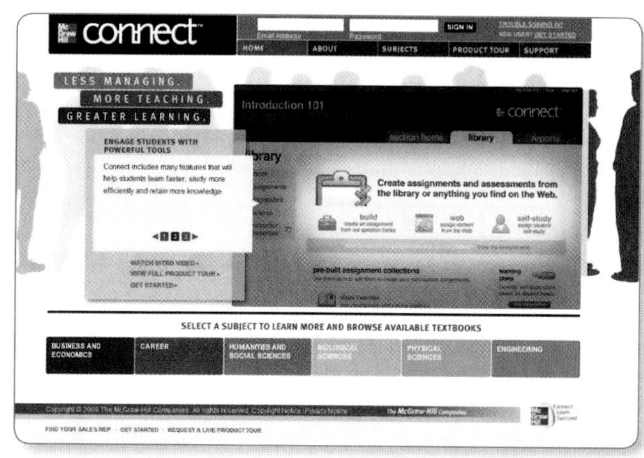

With Connect Physics, instructors can deliver assignments, quizzes, and tests online. The problems directly from the end-of-chapter material in Bauer/Westfall's *University Physics* are presented in an auto-gradable format. The online homework system incorporates new and exciting interactive tools and problem types: a graphing tool; a free-body diagram drawing tool; symbolic entry; a math palette; and multi-part problems. The seven-step problem-solving method from the text is also reflected in Connect Physics: Students will find the familiar seven steps outlined in the guided solutions, with additional detailed help with the math where they need it most. Over 2300 multiple-choice Test Bank questions (written by the authors and the same contributors who provided the text's end-of-chapter problems) are also included.

Instructors can edit existing questions and write entirely new problems; track individual student performance—by question, assignment, concept, or in relation to the class overall—with automatic grading; provide instant feedback to students; and secure storage of detailed grade reports online. Grade reports can be easily integrated with Learning Management Systems (LMS), such as WebCT and Blackboard.

By choosing Connect Physics, instructors are providing their students with a powerful tool for improving academic performance and truly mastering course material. Connect Physics helps reduce the time instructors spend grading homework and closes the loop on student homework and feedback.

Connect Physics allows students to practice important skills at their own pace and on their own schedule. Importantly, students' assessment results and instructors' feedback are all saved online—so students can continually review their progress and plot their course to success.

Instructors also have access to PowerPoint lecture outlines, an Instructor's Solutions Manual, electronic images from the text, clicker questions, quizzes, tutorials, simulations, video clips, and many other resources that are directly tied to specific material in *University Physics*. Students have access to self-quizzes, simulations, tutorials, and more.

Go to **www.mhhe.com/bauerwestfall** to learn more and register.

## ConnectPlus™ Physics

Some instructors may also choose ConnectPlus Physics for their students. ConnectPlus offers an innovative and inexpensive electronic textbook integrated within the homework platform. Like Connect Physics, ConnectPlus Physics provides students with online assignments and assessments AND 24/7 online access to an eBook—an online edition of the *University Physics* text.

As part of the ehomework process, instructors can assign chapter and section readings from the text. With ConnectPlus, links to relevant text topics are also provided where students need them most—accessed directly from the ehomework problem!

**ConnectPlus Physics:**

- Provides students with a ConnectPlus eBook, allowing for anytime, anywhere access to the *University Physics* textbook to aid them in successfully completing their work, wherever and whenever they choose.
- Includes Community Notes for student-to-student or instructor-to-student note sharing to greatly enhance the user learning experience.
- Allows for insertion of lecture discussions or instructor-created additional examples using Tegrity (see page xxvi) to provide additional clarification or varied coverage on a topic.
- Merges media and assessments with the text's narrative to engage students and improve learning and retention. The eBook includes simulations, interactives, videos, and inline assessment questions.
- Pinpoints and connects key physics concepts in a snap using the powerful eBook search engine.
- Manages notes, highlights and bookmarks in one place for simple, comprehensive review.

## TEGRITY

Tegrity Campus is a service that makes class time available all the time by automatically capturing every lecture in a searchable format for students to review when they study and complete assignments. With a simple one-click start-and-stop process, you capture all computer screens and corresponding audio. Students replay any part of any class with easy-to-use browser-based viewing on a PC or Mac. Educators know that the more students can see, hear, and experience class resources, the better they learn. With Tegrity Campus, students quickly recall key moments by using Tegrity Campus's unique search feature. This search helps students efficiently find what they need, when they need it across an entire semester of class recordings. Help turn all your students' study time into learning moments immediately supported by your lecture.

To learn more about Tegrity watch a 2-minute Flash demo at http://tegritycampus.mhhe.com.

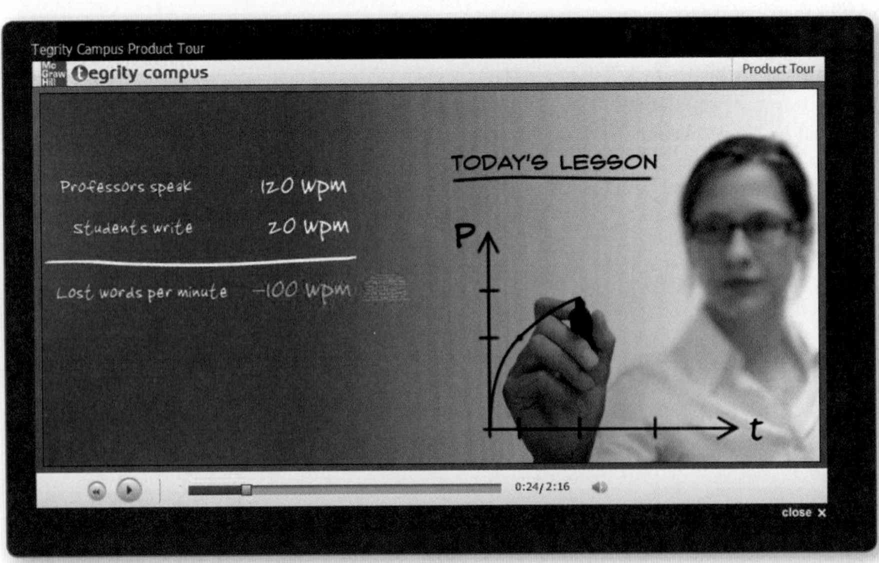

# Additional Resources for Instructors and Students

## Electronic Book Images and Assets for Instructors

*Build instructional materials wherever, whenever, and however you want!*

An online collection of Presentation Tools containing photos, artwork, and other media can be accessed from *University Physics'* Connect Physics companion site to create customized lectures, visually enhanced tests and quizzes, compelling course websites, or attractive printed support materials. Assets are copyrighted by McGraw-Hill Higher Education but can be used by instructors for classroom purposes. The visual resources in this collection are

- **Art** Full-color digital files of all illustrations in the book can be readily incorporated into lecture presentations, exams, or custom-made classroom materials. In addition, all files are pre-inserted into PowerPoint slides for ease of lecture preparation.

**FIGURE 6.16** Total, potential, and kinetic energy for a roller coaster.

- **Photos** The photo collection contains digital files of photographs from the text, which can be reproduced for multiple classroom uses.
- **Solved Problem Library, Example Library, Table Library, and Numbered Equations Library** Access the worked examples, tables, and equations from the text in electronic format for inclusion in your classroom resources.

Also residing on the companion site are

- **PowerPoint Lecture Outlines** Ready-made presentations (created by the text authors) that combine art and lecture notes are provided for each chapter of the text. The outlines include historical information and additional examples.
- **PowerPoint Slides** For instructors who prefer to create their lectures from scratch, all illustrations and available photos are pre-inserted by chapter into PowerPoint slides.

## Computerized Test Bank Online

The same group of instructors who wrote the text's end-of-chapter Questions and Problems also wrote the Test Bank questions to reinforce the important concepts from the text while ensuring consistency in style and coverage. The questions were combined with a set of questions created and tested by the authors of the text. A comprehensive bank of over 2300 test questions in multiple-choice format, organized by section of the text, at a variety of difficulty levels, is provided within a computerized Test Bank powered by McGraw-Hill's flexible electronic testing program: EZ Test Online (www.eztestonline.com). EZ Test Online allows you to easily create paper and online tests or quizzes!

Imagine being able to create and access a test or quiz anywhere, at any time, without installing the testing software. Now, with EZ Test Online, instructors can select questions from multiple McGraw-Hill test banks or create their own, and then either print the test for paper distribution or give it online. Go to www.mhhe.com/bauerwestfall for more information.

## CourseSmart

*CourseSmart* is a new way for faculty to find and review eBooks. It's also a great option for students who are interested in accessing their course materials digitally and saving money. *CourseSmart* offers thousands of the most commonly adopted textbooks across hundreds of courses from a wide variety of higher education publishers. It is the only place for faculty to review and compare the full text of a textbook online, providing immediate access without the environmental impact of requesting a print exam copy. At *CourseSmart,* students can save up to 50% off the cost of a print book, reduce their impact on the environment, and gain access to powerful web tools for learning including full text search, notes and highlighting, and email tools for sharing notes between classmates. For further details contact your sales representative or go to www.coursesmart.com.

## Create

Craft your teaching resources to match the way you teach! With McGraw-Hill Create, www.mcgrawhillcreate.com, you can easily rearrange chapters, combine material from other content sources, and quickly upload content you have written like your course syllabus or teaching notes. Find the content you need in Create by searching through thousands of leading McGraw-Hill textbooks. Arrange your book to fit your teaching style. Create even allows you to personalize your book's appearance by selecting the cover and adding your name, school, and course information. Order a Create book and you'll receive a complimentary print review copy in 3–5 business days or a complimentary electronic review copy (eComp) via email in about one hour. Go to www.mcgrawhillcreate.com today and register. Experience how McGraw-Hill Create empowers you to teach *your* students *your* way.

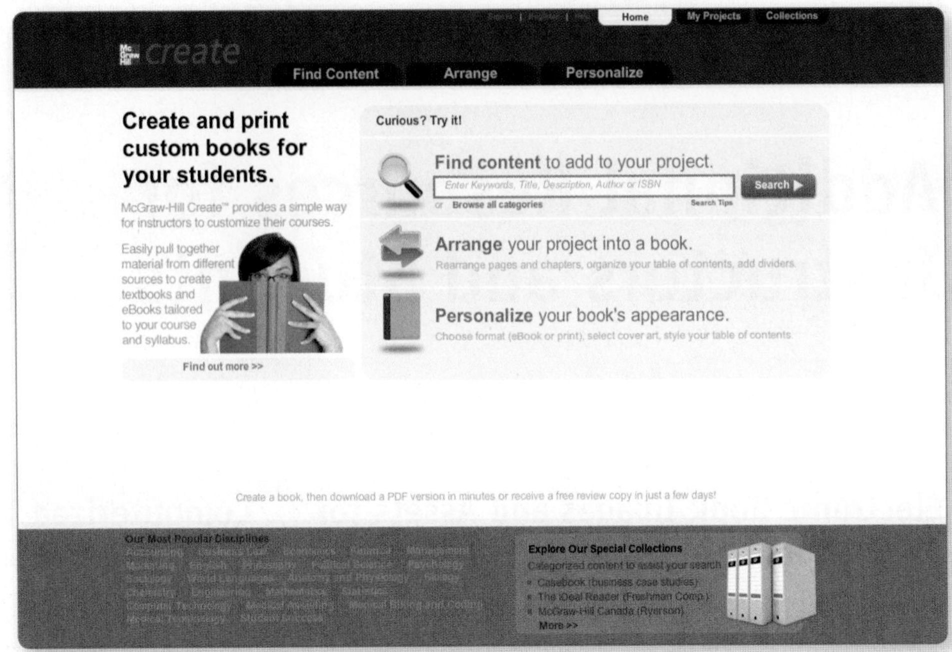


## Personal Response Systems

Personal response systems, or "clickers," bring interactivity into the classroom or lecture hall. Wireless response systems give the instructor and students immediate feedback from the entire class. The wireless response pads are essentially remotes that are easy to use and engage students, allowing instructors to motivate student preparation, interactivity, and active learning. Instructors receive immediate feedback to gauge which concepts students understand. The In-Class Exercises from *University Physics* (formatted in PowerPoint) are available on the companion website.

## Solutions Manuals

The *Instructor's Solutions Manual* includes answers to all of the text's end-of-chapter Questions and complete, worked-out solutions for the end-of-chapter Problems. Chapters 1 through 13 include worked out-solutions that follow the seven-step problem-solving method from the text for all Problems and Additional Problems. Chapters 14 through 40 continue to use the seven-step problem-solving method for all challenging (one bullet) and most challenging (two bullet) Problems and Additional Problems, while switching to a more abbreviated solution for the less challenging (no bullet) Problems and Additional Problems.

The *Student Solutions Manual* contains answers and worked-out solutions to selected end-of-chapter Questions and Problems. Again, the worked-out solutions for Chapters 1 through 13 follow the complete seven-step problem-solving method from the text for Problems and Additional Problems. Chapters 14 through 40 continue to use the seven-step method for challenging (one bullet) and most challenging (two bullet) Problems and Additional Problems, while switching to a more abbreviated solution for the less challenging (no bullet) Problems and Additional Problems.

For more information, contact a McGraw-Hill customer service representative at 800-338-3987 or by e-mail at http://www.mhhe.com/. To locate your sales representative, go to http://www.mhhe.com/ for Find My Sales Rep.

## Disclaimer

As a full-service publisher of quality educational products, McGraw-Hill does much more than just sell textbooks to your students. We create and publish an extensive array of print, video, and digital supplements to support instruction on your campus. Orders of new (versus used) textbooks help us to defray the cost of developing such supplements, which is substantial. Please consult your local McGraw-Hill representative to learn about the availability of the supplements that accompany *University Physics*. If you are not sure who your representative is, you can find him or her by using the tab labeled "My Sales Rep" at www.mhhe.com.

McGraw-Hill's 360° Development process is an ongoing, market-oriented approach to building accurate and innovative print and digital products. It is dedicated to continual large-scale and incremental improvement and is driven by multiple customer feedback loops and checkpoints. This process is initiated during the early planning stages of our new products, intensifies during the development and production stages, and then begins again on publication, in anticipation of the next edition.

A key principle in the development of any physics text is its ability to adapt to teaching specifications in a universal way. The only way to do so is by contacting those universal voices—and learning from their suggestions. We are confident that our book has the most current content the industry has to offer, thus pushing our desire for accuracy to the highest standard possible. In order to accomplish this, we have moved through an arduous road to production. Extensive and open-minded advice is critical in the production of a superior text.

We engaged over 200 instructors and students to provide guidance in the development of this first edition. By investing in this extensive endeavor, McGraw-Hill delivers to you a product suite that has been created, refined, tested, and validated as a successful tool for your course.

## Board of Advisors

A hand-picked group of trusted instructors active in the calculus-based physics course and research groups served as the chief advisors and consultants to the author and editorial team with regard to manuscript development. The Board of Advisors reviewed the manuscript; served as a sounding board for pedagogical, media, and design concerns; helped to respond to issues raised by other reviewers; approved organizational changes; and attended a focus group to confirm the manuscript's readiness for publication.

Nina Abramzon, *California Polytechnic University–Pomona*
Rene Bellweid, *Wayne State University*
David Harrison, *University of Toronto*
John Hopkins, *The Pennsylvania State University*
David C. Ingram, *Ohio University–Athens*
Michael Lisa, *The Ohio State University*
Amy Pope, *Clemson University*
Roberto Ramos, *Drexel University*

## Contributors

A panel of excellent writers created additional Questions and Problems to enhance the variety of exercises found in every chapter:

Carlos Bertulani, *Texas A&M University–Commerce*
Ken Thomas Bolland, *Ohio State University*
John Cerne, *State University of New York–Buffalo**
Ralph Chamberlain, *Arizona State University*
Eugenia Ciocan, *Clemson University**
Fivos Drymiotis, *Clemson University*
Michael Famiano, *Western Michigan University**
Yung Huh, *South Dakota State University*
Pedram Leilabady, *University of North Carolina–Charlotte**
M.A.K. Lodhi, *Texas Tech University*
Charley Myles, *Texas Tech University*
Todd Pedlar, *Luther College**
Corneliu Rablau, *Kettering University*
Roberto Ramos, *Drexel University*
Ian Redmount, *Saint Louis University*
Todd Smith, *University of Dayton**
Donna Stokes, *University of Houston**
Stephen Swingle, *City College of San Francisco*
Marshall Thomsen, *Eastern Michigan University*
Prem Vaishnava, *Kettering University**
John Vasut, *Baylor University**

*These contributors also authored questions for the Test Bank to accompany *University Physics* along with David Bannon of Oregon State University, while Richard Hallstein of Michigan State University organized and reviewed all of the contributions. Additionally, Suzanne Willis of North Illinois University helped us to compile our assets for the *University Physics* ConnectPlus eBook. Jack Cuthbert of Holmes Community College-Ridgeland composed the text's In-Class Exercises into PowerPoint files for use with clickers. Collette Marsh and Deborah Damcott of Harper College edited the PowerPoint Lectures, and finally, but certainly not least, Rob Hagood of Washtenaw Community College and Amy Pope of Clemson University spent innumerable hours reviewing and providing vital input on the quality of our Connect online homework content.

## Class Testing

For five years before the production of *University Physics,* the authors tested and refined the textbook's materials with approximately 4000 of our students at Michigan State University. They collected written feedback and also conducted one-on-one interviews with a representative sample of the students, in addition to classroom testing of the in-class exercises and the PowerPoint slides. Several of the authors' colleagues (Alexandra Gade, Alex Brown, Bernard Pope, Carl Schmidt, Chong-Yu Ruan, C. P. Yuan, Dan Stump, Ed Brown, Hendrik Schatz, Kris Starosta, Lisa Lapidus, Michael Harrison, Michael Moore, Reinhard Schwienhorst, Salemeh Ahmad, S. B. Mahanti, Scott Pratt, Stan Schriber, Tibor Nagy, and Thomas Duguet), who co-taught the introductory physics sequence in parallel sections, also provided invaluable help and insight, and their contributions made this book much stronger.

**1$^{ST}$ ROUND: AUTHORS' MANUSCRIPT**

- √ Multiple rounds of review by college physics instructors
- √ Independent accuracy check of the text, examples, and solved problems by a professional firm employing mathematicians and physicists
- √ Second independent accuracy check of the end-of-chapter problems by a team of solution manual authors
- √ Test bank authors' review
- √ Clicker question author review

**2$^{ND}$ ROUND: TYPESET PAGES**

- √ Authors
- √ First proofreading
- √ Third accuracy check of the text, examples, solved problems, and end-of-chapter problems by the same professional firm employing mathematicians and physicists

**3$^{RD}$ ROUND: REVISED TYPESET PAGES**

- √ Authors
- √ Second proofreading
- √ Accuracy check of any unresolved issues by the professional firm employing mathematicians and physicists
- √ A parallel fourth accuracy check of the end-of-chapter problem and solution content is made after this content is entered into the Connect Online Homework System, which allows for any further issues to be corrected in the printed text and online solutions manuals
- √ Reviews by physics instructors

**4$^{TH}$ ROUND: CONFIRMING TYPESET PAGES**

- √ Final check by authors

**FINAL ROUND: PRINTING**

## Accuracy Assurance

The authors and the publisher acknowledge the fact that inaccuracies can be a source of frustration for both instructors and students. Therefore, throughout the writing and production of this first edition, we have worked diligently to eliminate errors and inaccuracies. Ron Fitzgerald, John Klapstein, and their team at MathResources conducted an independent accuracy check and worked through all end-of-chapter questions and problems in the final draft of the manuscript. They then coordinated the resolution of discrepancies between accuracy checks, ensuring the accuracy of the text, the end-of-book answers, and the solutions manuals. Corrections were then made to the manuscript before it was typeset.

The page proofs of the text were double-proofread against the manuscript to ensure the correction of any errors introduced when the manuscript was typeset. Any issues with the textual examples, solved problems and solutions, end-of-chapter questions and problems, and problem answers were accuracy checked by MathResources again at the page proof stage after the manuscript was typeset. This last round of corrections was then cross-checked against the solutions manuals. The end-of-chapter problems from the text along with their solutions were then double-checked by two independent firms upon entry into the

Connect Physics online homework system, and again any issues were addressed in the text and solutions manuals.

## Developmental Symposia

McGraw-Hill conducted four symposia and reviewer focus groups directly related to the development of *University Physics*. These events were an opportunity for editors, marketing managers, and digital producers from McGraw-Hill to gather information about the needs and challenges of instructors teaching calculus-based physics courses and to confirm the direction of the first edition of *University Physics,* its supplements, and related digital products.

Nina Abramzon, *California State Polytechnic University–Pomona*
Ed Adelson, *The Ohio State University*
Mohan Aggarwal, *Alabama A&M University*
Rene Bellweid, *Wayne State University*
Jason Brown, *Clemson University*
Ronald Brown, *California Polytechnic University San Luis Obispo*
Mike Dubson, *University of Colorado–Boulder*
David Elmore, *Purdue University*
Robert Endorf, *University of Cincinnati*
Gus Evrard, *University of Michigan*
Chris Gould, *University of Southern California*
John B. Gruber, *San Jose State University*
John Hardy, *Texas A&M University*
David Harrison, *University of Toronto*
Richard Heinz, *Indiana University*
Satoshi Hinata, *Auburn University*
John Hopkins, *The Pennsylvania State University*
T. William Houk, *Miami University–Ohio*
David C. Ingram, *Ohio University–Athens*
Elaine Kirkpatrick, *Rose-Hulman Institute of Technology*
David Lamp, *Texas Tech University*
Michael McInerney, *Rose-Hulman Institute of Technology*
Bruce Mellado, *University of Wisconsin–Madison*
C. Fred Moore, *University of Texas–Austin*
Jeffrey Morgan, *University of Northern Iowa*
Kiumars Parvin, *San Jose State University*
Amy Pope, *Clemson University*
Earl Prohofsky, *Purdue University*
Roberto Ramos, *Drexel University*
Dubravka Rupnik, *Louisiana State University*
Homeyra Sadaghiani, *California State Polytechnic University–Pomona*
Sergey Savrasov, *University of California–Davis*
Marllin Simon, *Auburn University*
Leigh Smith, *University of Cincinnati*
Donna Stokes, *University of Houston*
Michael Strauss, *University of Oklahoma*
Gregory Tarlé, *University of Michigan*

## Reviewers of the Text

Numerous instructors participated in over 200 reviews of various drafts of manuscript and the proposed table of contents to provide feedback on the narrative text, content, pedagogical elements, accuracy, organization, problem sets, and general quality. This feedback was summarized by the book team and used to guide the direction of the final-draft manuscript.

Nina Abramzon, *California Polytechnic University–Pomona*
Edward Adelson, *Ohio State University*
Albert Altman, *UMASS Lowell*
Paul Avery, *University of Florida*
David T. Bannon, *Oregon State University*
Marco Battaglia, *UC Berkeley and LBNL*
Douglas R. Bergman, *Rutgers, The State University of New Jersey*
Luca Bertello, *University of California–Los Angeles*
Peter Beyersdorf, *San Jose State University*
Helmut Biritz, *Georgia Institute of Technology*
Ken Thomas Bolland, *Ohio State University*
Richard Bone, *Florida International University*
Dieter Brill, *University of Maryland–College Park*
Branton J. Campbell, *Brigham Young University*
Duncan Carlsmith, *University of Wisconsin–Madison*
Neal Cason, *University of Notre Dame*
K. Kelvin Cheng, *Texas Tech University*
Chris Church, *Miami University of Ohio–Oxford*
Eugenia Ciocan, *Clemson University*
Robert Clare, *University of California–Riverside*
Roy Clarke, *University of Michigan*

J. M. Collins, *Marquette University*
Brent A. Corbin, *University of California–Los Angeles*
Stephane Coutu, *The Pennsylvania State University*
William Dawicke, *Milwaukee School of Engineering*
Mike Dennin, *University of California–Irvine*
John Devlin, *University of Michigan–Dearborn*
John DiNardo, *Drexel University*
Fivos R. Drymiotis, *Clemson University*
Michael DuVernois, *University of Hawaii–Manoa*
David Ellis, *The University of Toledo*
Robert Endorf, *University of Cincinnati*
David Ermer, *Mississippi State University*
Harold Evensen, *University of Wisconsin–Platteville*
Lisa L. Everett, *University of Wisconsin–Madison*
Frank Ferrone, *Drexel University*
Leonard Finegold, *Drexel University*
Ray Frey, *University of Oregon*
J. William Gary, *University of California–Riverside*
Stuart Gazes, *University of Chicago*
Benjamin Grinstein, *University of California–San Diego*
John Gruber, *San Jose State University*
Kathleen A. Harper, *Denison University*
Edwin E. Hach, III, *Rochester Institute of Technology*
John Hardy, *Texas A & M University*
Laurent Hodges, *Iowa State University*
John Hopkins, *The Pennsylvania State University*
George K. Horton, *Rutgers University*
T. William Houk, *Miami University–Ohio*
Eric Hudson, *Massachusetts Institute of Technology*
A. K. Hyder, *University of Notre Dame*
David C. Ingram, *Ohio University–Athens*
Diane Jacobs, *Eastern Michigan University*
Rongying Jin, *The University of Tennessee–Knoxville*
Kate L. Jones, *University of Tennessee*
Steven E. Jones, *Brigham Young University*
Teruki Kamon, *Texas A & M University*
Lev Kaplan, *Tulane University*
Joseph Kapusta, *University of Minnesota*
Kathleen Kash, *Case Western Reserve*
Sanford Kern, *Colorado State University*
Eric Kincanon, *Gonzaga University*
Elaine Kirkpatick, *Rose-Hulman Institute of Technology*
Brian D. Koberlein, *Rochester Institute of Technology*
W. David Kulp, III, *Georgia Institute of Technology*
Fred Kuttner, *University of California–Santa Cruz*
David Lamp, *Texas Tech University*
Andre' LeClair, *Cornell University*
Patrick R. LeClair, *University of Alabama*
Luis Lehner, *Louisiana State University–Baton Rouge*
Michael Lisa, *The Ohio State University*
Samuel E. Lofland, *Rowan University*
Jerome Long, *Virginia Tech*
A. James Mallmann, *Milwaukee School of Engineering*
Pete Markowitz, *Florida International University*
Daniel Marlow, *Princeton University*
Bruce Mason, *Oklahoma University*
Martin McHugh, *Loyola University*
Michael McInerney, *Rose-Hulman Institute of Technology*
David McIntyre, *Oregon State University*
Marina Milner-Bolotin, *Ryerson University–Toronto*
Kieran Mullen, *University of Oklahoma*
Curt Nelson, *Gonzaga University*
Mark Neubauer, *University of Illinois at Urbana–Champaign*
Cindy Neyer, *Tri-State University*
Craig Ogilvie, *Iowa State University*
Bradford G. Orr, *The University of Michigan*
Karur Padmanabhan, *Wayne State University*
Jacqueline Pau, *University of California–Los Angeles*
Leo Piilonen, *Virginia Tech*
Claude Pruneau, *Wayne State University*
Johann Rafelski, *University of Arizona*
Roberto Ramos, *Drexel University*
Lawrence B. Rees, *Brigham Young University*
Andrew J. Rivers, *Northwestern University*
James W. Rohlf, *Boston University*
Philip Roos, *University of Maryland*
Dubravka Rupnik, *Louisiana State University*
Ertan Salik, *California State Polytechnic University–Pomona*
Otto Sankey, *Arizona State University*
Sergey Savrasov, *University of California–Davis*
John Schroeder, *Rensselaer Polytech*
Kunnat Sebastian, *University of Massachusetts–Lowell*
Bjoern Seipel, *Portland State University*
Jerry Shakov, *Tulane University*
Ralph Shiell, *Trent University*
Irfan Siddiqi, *University of California–Berkeley*
Marllin L. Simon, *Auburn University*
Alex Small, *California State Polytechnic University–Pomona*
Leigh Smith, *University of Cincinnati*
Xian-Ning Song, *Richland College*
Jeff Sonier, *Simon Fraser University–Surrey Central*
Chad E. Sosolik, *Clemson University*
Donna W. Stokes, *University of Houston*
James Stone, *Boston University*
Michael G. Strauss, *University of Oklahoma*
Yang Sun, *University of Notre Dame*
Maarij Syed, *Rose-Hulman Institute of Technology*
Douglas C. Tussey, *The Pennsylvania State University*
Somdev Tyagi, *Drexel University*
Erich W. Varnes, *University of Arizona*
Gautam Vemuri, *Indiana University-Purdue University–Indianapolis*
Thad Walker, *University of Wisconsin–Madison*
Fuqiang Wang, *Purdue University*
David J. Webb, *University of California–Davis*
Kurt Wiesenfeld, *Georgia Tech*
Fred Wietfeldt, *Tulane University*
Gary Williams, *University of California–Los Angeles*
Sun Yang, *University of Notre Dame*
L. You, *Georgia Tech*
Billy Younger, *College of the Albemarle*
Andrew Zangwill, *Georgia Institute of Technology*
Jens Zorn, *University of Michigan–Ann Arbor*
Michael Zudov, *University of Minnesota*

## Additional International Reviewers of the Text

El Hassan El Aaoud, *University of Hail, Hail KSA*
Mohamed S. Abdelmonem, *King Fahd University of Petroleum and Minerals, Dhahran, Saudi Arabia*
Sudeb Bhattacharya, *Saha Institute of Nuclear Physics, Kolkata, India*
Shi-Jian Gu, *The Chinese University of Hong Kong, Shatin, N.T., Hong Kong*
Nasser M. Hamdan, *The American University of Sharjah*
Moustafa Hussein, *Arab Academy for Science & Engineering, Egypt*
A.K. Jain, *I.I.T. Roorkee*
Carsten Knudsen, *Technical University of Denmark*
Ajal Kumar, *The University of the South Pacific, Fiji*
Ravindra Kumar Sinha, *Delhi College of Engineering*
Nazir Mustapha, *Al-Imam University*

# Acknowledgments

Reza Nejat, *McMaster University*
K. Porsezian, *Pondicherry University, Puducherry*
Wang Qing-hai, *National University of Singapore*
Kenneth J. Ragan, *McGill University*

A book like the one that you are holding in your hands is impossible to produce without tremendous work by an incredible number of dedicated individuals. First and foremost, we would like to thank the talented marketing and editorial team from McGraw-Hill: Marty Lange, Kent Peterson, Thomas Timp, Ryan Blankenship, Mary Hurley, Liz Recker, Daryl Bruflodt, Lisa Nicks, Dan Wallace, and, in particular, Deb Hash helped us in innumerable ways and managed to reignite our enthusiasm after each revision. Their team spirit, good humor, and unbending optimism kept us on track and always made it fun for us to put in the seemingly endless hours it took to produce the manuscript.

The developmental editors, Richard Heinz and David Chelton, helped us work through the near infinite number of comments and suggestions for improvements from our reviewers. They, as well as the reviewers and our board of advisors, deserve a large share of the credit for improving the quality of the final manuscript. Our colleagues on the faculty of the Department of Physics and Astronomy at Michigan State University—Alexandra Gade, Alex Brown, Bernard Pope, Carl Schmidt, Chong-Yu Ruan, C. P. Yuan, Dan Stump, Ed Brown, Hendrik Schatz, Kris Starosta, Lisa Lapidus, Michael Harrison, Michael Moore, Reinhard Schwienhorst, Salemeh Ahmad, S. B. Mahanti, Scott Pratt, Stan Schriber, Tibor Nagy, and Thomas Duguet—helped us in innumerable ways as well, teaching their classes and sections with the materials developed by us and in the process providing invaluable feedback on what worked and what needed additional refinement. We thank all of them.

We decided to involve a large number of physics instructors from around the country in the authoring of the end-of-chapter problems, in order to ensure that they are of the highest quality, relevance, and didactic value. We thank all of our problem contributors for sharing some of their best work with us, in particular, Richard Hallstein, who took on the task of organizing and processing all contributions.

At the point when we turned in the final manuscript to the publisher, a whole new army of professionals took over and added another layer of refinement, which transformed a manuscript into a book. John Klapstein and the team at MathResources worked through each and every homework problem, each exercise, and each number and equation we wrote down. The photo researchers, in particular, Danny Meldung, improved the quality of the images used in the book immensely, and they made the selection process fun for us. Pamela Crews and the team at Precision Graphics used our original drawings but improved their quality substantially, while at the same time remaining true to our original calculations that went into producing the drawings. Our copyeditor, Jane Hoover, and her team pulled it all together in the end, deciphered our scribbling, and made sure that the final product is as readable as possible. McGraw-Hill's design and production team of Jayne Klein, David Hash, Carrie Burger, Sandy Ludovissy, Judi David, and Mary Jane Lampe expertly guided the book and its ancillary materials through to publication. All of them deserve our tremendous gratitude.

Finally, we could not have made it through the last six years of effort without the support of our families, who had to put up with us working on the book through untold evenings, weekends, and even during many vacations. We hope that all of their patience and encouragement has paid off, and we thank them from the bottom of our hearts for sticking with us during the completion of this book.

*—Wolfgang Bauer*
*—Gary D. Westfall*

# The Big Picture

## *Frontiers of Modern Physics*

This book will attempt to give you insight into some of the astounding recent progress made in physics. Examples from advanced areas of research are accessible with the knowledge available at the introductory level. At many top universities, freshmen and sophomores are already involved in cutting-edge physics research. Often, this participation requires nothing more than the tools developed in this book, a few days or weeks of additional reading, and the curiosity and willingness necessary to learn new facts and skills.

The following pages introduce some of the amazing frontiers of current physics research and describe some results that have been obtained during the last few years. This introduction stays at a qualitative level, skipping all mathematical and other technical details. Chapter numbers in parentheses indicate where more in-depth explorations of the topics can be found.

### Quantum Physics

The year 2005 marked the 100th anniversary of Albert Einstein's landmark papers on Brownian motion (proving that atoms are real; see Chapters 13 and 38), on the theory of relativity (Chapter 35), and on the photoelectric effect (Chapter 36). This last paper introduced one of the ideas forming the basis of quantum mechanics, the physics of matter on the scale of atoms and molecules. Quantum mechanics is a product of the 20th century that led, for example, to the invention of lasers, which are now routinely used in CD, DVD, and Blu-ray players, in price scanners, and even in eye surgery, among many other applications. Quantum mechanics has also provided a more fundamental understanding of chemistry: Physicists are using ultrashort laser pulses less than $10^{-15}$ s in duration to gain an understanding of how chemical bonds develop. The quantum revolution has included exotic discoveries such as antimatter, and there is no end in sight. During the last decade, groups of atoms called *Bose-Einstein condensates* have been formed in electromagnetic traps; this work has opened an entirely new realm of research in atomic and quantum physics (see Chapters 36 through 38).

### Condensed Matter Physics and Electronics

Physics innovations created and continue to drive high-tech industry. Only slightly more than 50 years ago, the first transistor was invented at Bell Labs, ushering in the electronic age. The central processing unit (CPU) of a typical desktop computer or laptop now contains more than 100 million transistor elements. The incredible growth in the power and the scope of applications of computers over the last few decades has been made possible by research in condensed matter physics. Gordon Moore, cofounder of Intel, famously observed that computer processing power doubles every 18 months, a trend that is predicted to continue for at least another decade or more.

Computer storage capacity grows even faster than processing power, with a doubling time of 12 months. The 2007 Nobel Prize in Physics was awarded to Albert Fert and Peter Grünberg for their 1988 discovery of *giant magnetoresistance*. It took only a decade for this discovery to be applied in computer hard disks, enabling storage capacities of hundreds of gigabytes (1 gigabyte = 1 billion pieces of information) and even terabytes (1 terabyte = 1 trillion pieces of information).

Network capacity grows even faster than storage capacity or processing power, doubling every 9 months. You can now go to almost any country on Earth and find wireless access points, from which you can connect your laptop or WiFi-enabled cell phone to the Internet. Yet it is less than a couple of decades since the conception of the World Wide Web by Tim Berners-Lee, who was then working at the particle physics laboratory CERN in Switzerland and who developed this new medium to facilitate collaboration among particle physicists in different parts of the world.

Cell phones and other powerful communication devices have found their way into just about everybody's hands. Modern physics research enables a progressive miniaturization of consumer electronics devices. This process drives a digital convergence, making it possible to equip cell phones with digital cameras, video recorders, e-mail capability, Web browsers, and global positioning system receivers. More functionality is added continuously, while prices continue to fall. Forty years after the first Moon landing, many cell phones now pack more computing power than the Apollo spaceship used for that trip to the Moon.

## Quantum Computing

Physics researchers are still pushing the limits of computing. At present, many groups are investigating ways to build a quantum computer. Theoretically, a quantum computer consisting of $N$ processors would be able to execute $2^N$ instructions simultaneously, whereas a conventional computer consisting of $N$ processors can execute only $N$ instructions at the same time. Thus, a quantum computer consisting of 100 processors would exceed the combined computing power of all currently existing supercomputers. To be sure, many deep problems have to be solved before this vision can become reality, but then again, 50 years ago it seemed utterly impossible to pack 100 million transistors onto a computer chip the size of a thumbnail.

## Computational Physics

The interaction between physics and computers works both ways. Traditionally, physics investigations were either experimental or theoretical in nature. Textbooks appear to favor the theoretical side, because they present the main formulas of physics, but in fact they analyze the conceptual ideas that are encapsulated in the formulas. On the other hand, much research originates on the experimental side, when newly observed phenomena seem to defy theoretical description. However, with the rise of computers, a third branch of physics has become possible: computational physics. Most physicists now rely on computers to process data, visualize data, solve large sets of coupled equations, or study systems for which simple analytical formulations are not known.

The emerging field of chaos and nonlinear dynamics is the prime example of such study. Arguably, MIT atmospheric physicist Edward Lorenz first simulated chaotic behavior with the aid of a computer in 1963, when he solved three coupled equations for a simple model of weather and detected a sensitive dependence on the initial conditions—even the smallest differences in the beginning of the simulation resulted in very large deviations later in time. This behavior is now sometimes called the *butterfly effect*, from the idea that a butterfly flapping its wings in China could change the weather in the United States a few weeks later. This sensitivity implies that long-term deterministic weather prediction is impossible.

## Complexity and Chaos

Systems of many constituents often exhibit very complex behavior, even if the individual constituents follow very simple rules of nonlinear dynamics. Physicists have started to address complexity in many systems, including simple sand piles, traffic jams, the stock market, biological evolution, fractals, and self-assembly of molecules and nanostructures. The science of complexity is another field that was discovered only during the last decade and is experiencing rapid growth.

Chaos and nonlinear dynamics are discussed in Chapter 7 on momentum and in Chapter 14 on oscillations. The models are often quite straightforward, and first-year physics students can make valuable contributions. However, the solution generally requires some computer programming skills. Programming expertise will enable you to contribute to many advanced physics research projects.

## Nanotechnology

Physicists are beginning to acquire the knowledge and skills needed to manipulate matter one atom at a time. During the last two decades, the scanning, tunneling, and atomic force microscopes were invented allowing researchers to see individual atoms (Figure 1). In some cases, individual atoms can be moved around in controlled ways. Nanoscience and nanotechnology are devoted to these types of challenges, whose solution holds promise for great technological advances, from even more miniaturized and thus more powerful electronics to the design of new drugs, or even the manipulation of DNA to cure some diseases.

**FIGURE 1** Individual iron atoms, arranged in the shape of a stadium on a copper surface. The ripples inside the stadium are the result of standing waves formed by electron density distributions. This arrangement was created and then imaged by using a scanning tunneling microscope (Chapter 37).

## Biophysics

Just as physicists moved into the domain of chemists during the 20th century, a rapid interdisciplinary convergence of physics and molecular biology is taking place in the 21st century. Researchers are already able to use laser tweezers (Chapter 34) to move individual biomolecules. X-ray crystallography (Chapter 34) has become sophisticated enough that researchers can obtain pictures of the three-dimensional structures of very complicated proteins. In addition, theoretical biophysics is beginning to be successful in predicting the spatial structure and the associated functionality of these molecules from the sequences of amino acids they contain.

Figure 2 shows a model of the spatial structure of the protein RNA Polymerase II, with the color coding representing the charge distribution (blue is positive, red is negative) at the surface of this biomolecule. (Chapter 23, on electric potentials, will help you understand the determination of charge distributions and their potentials for much simpler arrangements.) Also shown in the figure is a segment of the DNA molecule (yellow spiral structure in the center), indicating how RNA Polymerase II attaches to DNA and functions as a cleaver.

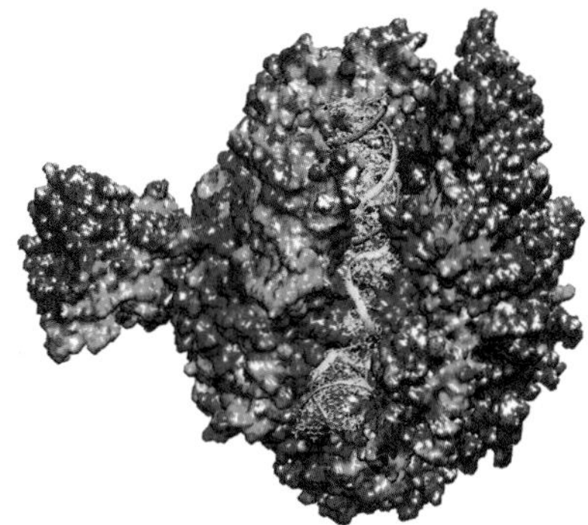

**FIGURE 2** Model of the protein RNA Polymerase II.

## Nuclear Fusion

Seventy years ago, nuclear physicist Hans Bethe and his colleagues figured out how nuclear fusion in the Sun produces the light that makes life on Earth possible. Today,

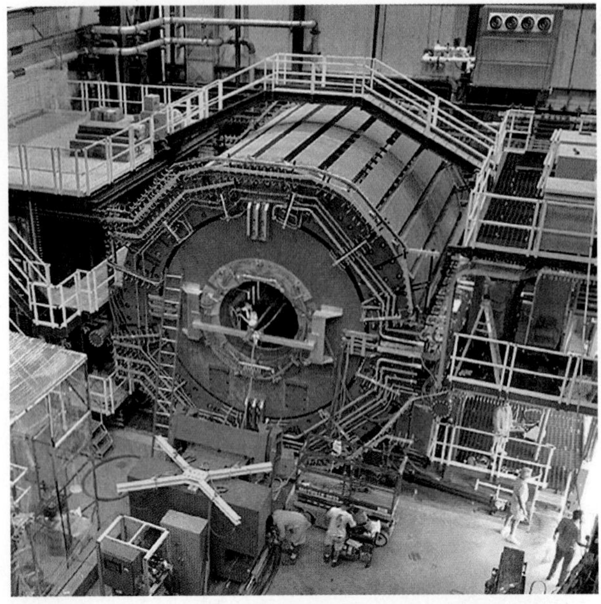

(a)

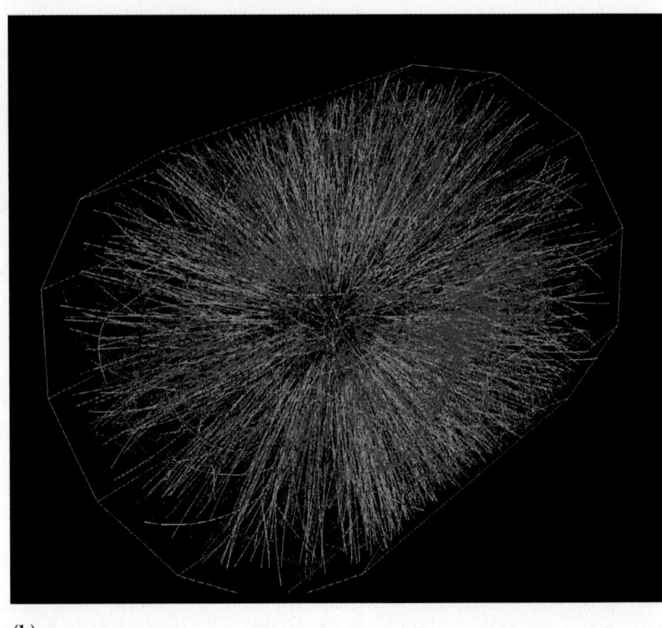

(b)

**FIGURE 3** (a) The STAR detector at RHIC during its construction. (b) Electronically reconstructed tracks of more than 5000 charged subatomic particles produced inside the STAR detector by a high-energy collision of two gold nuclei.

nuclear physicists are working on how to utilize nuclear fusion on Earth for nearly limitless energy production. Chapter 5 introduces the concept of energy, which is discussed further numerous times throughout this book. The international thermonuclear fusion reactor (Chapter 40), presently under construction in southern France by a collaboration involving many industrialized countries, will go a long way toward answering many of the important questions that need to be resolved before the use of fusion is technically feasible and commercially viable.

**FIGURE 4** Aerial view of Geneva, Switzerland, with the location of the underground tunnel of the Large Hadron Collider superimposed in red.

## High-Energy/Particle Physics

Nuclear and particle physicists are probing deeper and deeper into the smallest constituents of matter (Chapters 39 and 40). At Brookhaven National Laboratory on Long Island, for example, the Relativistic Heavy Ion Collider (RHIC) smashes gold nuclei into each other in order to recreate the state of the universe a small fraction of a second after its beginning, the *Big Bang.* Figure 3a is a photo of STAR, a RHIC detector; Figure 3b shows a computer analysis of the tracks left in the STAR detector by the more than 5000 subatomic particles produced in such a collision. Chapters 27 and 28 on magnetism and the magnetic field will explain how these tracks are analyzed in order to find the properties of the particles that produced them.

An even bigger instrument for particle physics research has just been completed at the CERN accelerator laboratory in Geneva, Switzerland. The Large Hadron Collider (LHC) is located in a circular underground tunnel with a circumference of 27 km (16.8 mi), indicated by the red circle in Figure 4. This new instrument is the most expensive research facility ever built, with a cost of more than $8 billion, and it

**FIGURE 5** Some of the 27 individual radio telescopes that are part of the Very Large Array.

was put into operation in September 2008. Particle physicists will use this facility to try to find out what causes different elementary particles to have different masses, to probe what the true elementary constituents of the universe are, and perhaps to look for hidden extra dimensions or other exotic phenomena.

## String Theory

Particle physics has a standard model of all particles and their interactions (Chapter 39), but why this model works so well is not yet understood. String theory is currently thought to be the most likely candidate for a framework that will eventually provide this explanation. Sometimes string theory is hubristically called the *Theory of Everything.* It predicts extra spatial dimensions, which sounds at first like science fiction, but many physicists are trying to find ways in which to test this theory experimentally.

## Astrophysics

Physics and astronomy have extensive interdisciplinary overlap in the areas of investigating the history of the early universe, modeling the evolution of stars, and studying the origin of gravitational waves or cosmic rays of the highest energies. Ever more precise and sophisticated observatories, such as the Very Large Array (VLA) radio telescope in New Mexico (Figure 5), have been built to study these phenomena.

Astrophysicists continue to make astounding discoveries that reshape our understanding of the universe. Only during the last few years has it been discovered that most of the matter in the universe is not contained in stars. The composition of this *dark matter* is still unknown (Chapter 12), but its effects are revealed through *gravitational lensing,* as shown in Figure 6 by the arcs observed in the galaxy cluster Abell 2218, which is 2 billion light-years away from Earth, in the constellation Draco. These arcs are images of even more distant galaxies, distorted by the presence of large quantities of dark matter. This phenomenon is discussed in more detail in Chapter 35 on relativity.

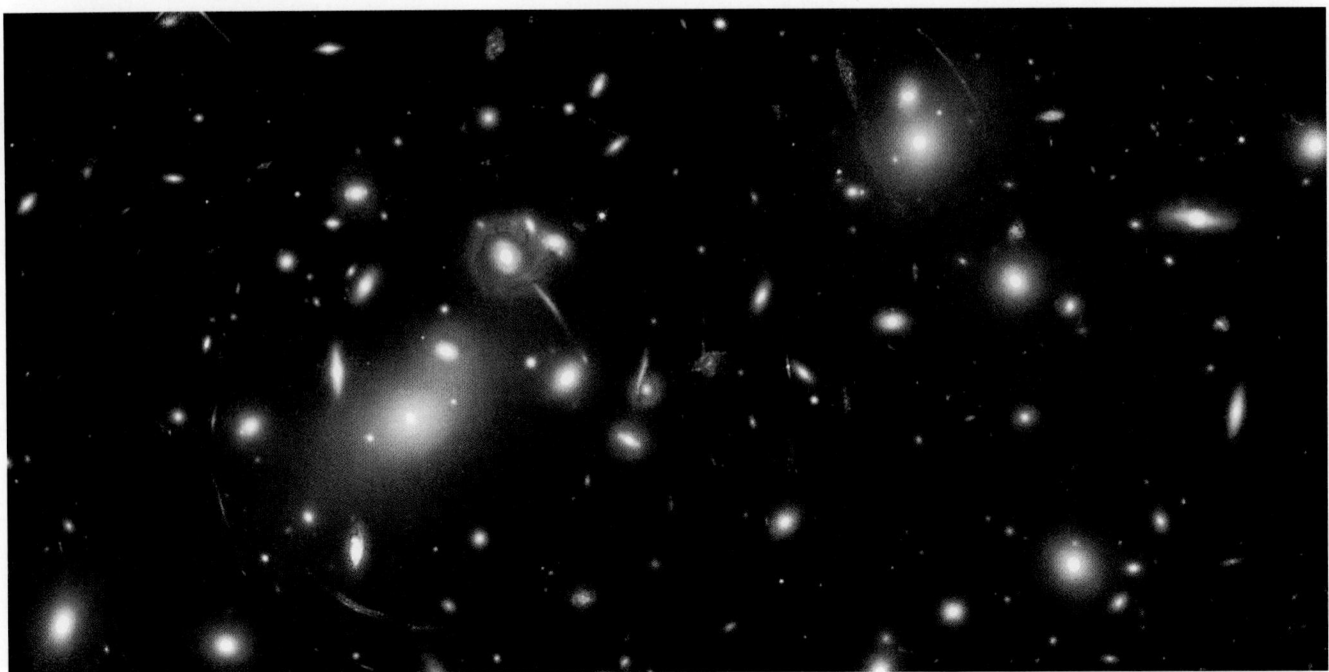

**FIGURE 6** Galaxy cluster Abell 2218, with the arcs that are created by gravitational lensing due to dark matter.

## Symmetry, Simplicity, and Elegance

From the smallest subatomic particles to the universe at large, physical laws govern all structures and dynamics from atomic nuclei to black holes. Physicists have discovered a tremendous amount, but each new discovery opens more exciting unknown territory. Thus, we continue to construct theories to explain any and all physical phenomena. The development of these theories is guided by the need to match experimental facts, as well as by the conviction that symmetry, simplicity, and elegance are key design principles. The fact that laws of nature can be formulated in simple mathematical equations ($F = ma$, $E = mc^2$, and many other, less famous ones) is stunning.

This introduction has tried to convey a bit about the frontiers of modern physics research. This textbook should help you build a foundation for appreciating, understanding, and perhaps even participating in this vibrant research enterprise, which continues to refine and even to reshape our understanding of the world around us.

PART 1 MECHANICS OF POINT PARTICLES

# 1 Overview



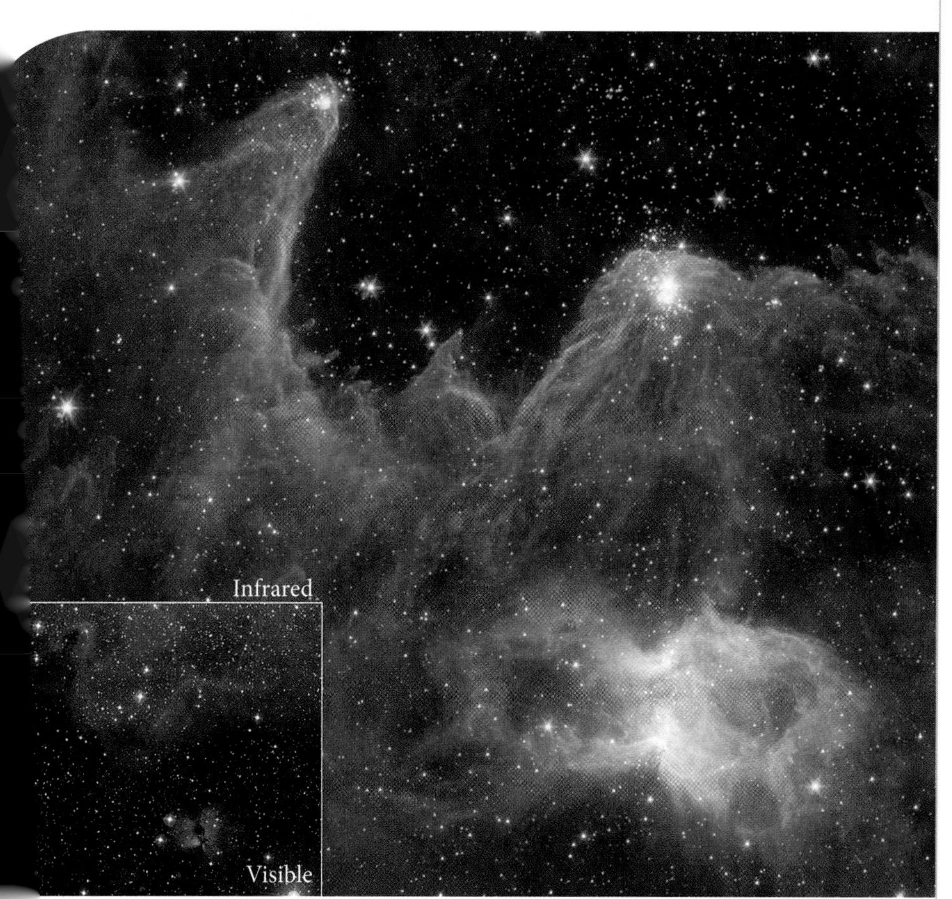


**FIGURE 1.1** An image of the W5 star-forming region taken by the Spitzer Space Telescope using infrared light.

## WHAT WE WILL LEARN

- Studying physics has many benefits.
- The use of scientific notation and the appropriate number of significant figures is important in physics.
- We will become familiar with the international unit system and the definitions of the base units as well as methods of converting among other unit systems.
- We will use available length, mass, and time scales to establish reference points for grasping the vast diversity of systems in physics.
- We present a problem-solving strategy that will be useful in analyzing and understanding problems throughout this course and in science and engineering applications.
- We will work with vectors: vector addition and subtraction, multiplication of vectors by scalars, unit vectors, and length and direction of vectors.

The dramatic image in Figure 1.1 could be showing any of several things: a colored liquid spreading out in a glass of water, or perhaps biological activity in some organism, or maybe even an artist's idea of mountains on some unknown planet. If we said the view was 70 wide, would that help you decide what the picture shows? Probably not—you need to know if we mean for example 70 meters or 70 millionths of an inch or 70 thousand miles.

In fact, this infrared image taken by the Spitzer Space Telescope shows huge clouds of gas and dust about 70 light-years across. (A light-year is the distance traveled by light in 1 year, about 10 quadrillion meters.) These clouds are about 6500 light-years away from Earth and contain newly formed stars embedded in the glowing regions. The technology that enables us to see images such as this one is at the forefront of contemporary astronomy, but it depends in a real way on the basic ideas of numbers, units, and vectors presented in this chapter.

The ideas described in this chapter are not necessarily principles of physics, but they help us to formulate and communicate physical ideas and observations. We will use the concepts of units, scientific notation, significant figures, and vector quantities throughout the course. Once you have understood these concepts, we can go on to discuss physical descriptions of motion and its causes.

## 1.1 Why Study Physics?

Perhaps your reason for studying physics can be quickly summed up as "Because it is required for my major!" While this motivation is certainly compelling, the study of science, and particularly physics, offers a few additional benefits.

Physics is the science on which all other natural and engineering sciences are built. All modern technological advances—from laser surgery to television, from computers to refrigerators, from cars to airplanes—trace back directly to basic physics. A good grasp of essential physics concepts gives you a solid foundation on which to construct advanced knowledge in all sciences. For example, the conservation laws and symmetry principles of physics also hold true for all scientific phenomena and many aspects of everyday life.

The study of physics will help you grasp the scales of distance, mass, and time, from the smallest constituents inside the nuclei of atoms to the galaxies that make up our universe. All natural systems follow the same basic laws of physics, providing a unifying concept for understanding how we fit into the overall scheme of the universe.

Physics is intimately connected with mathematics because it brings to life the abstract concepts used in trigonometry, algebra, and calculus. The analytical thinking and general techniques for problem solving that you learn here will remain useful for the rest of your life.

Science, especially physics, helps remove irrationality from our explanations of the world around us. Prescientific thinking resorted to mythology to explain natural phenomena. For example, the old Germanic tribes believed the god Thor using his hammer caused thunder. You may smile when you read this account, knowing that thunder and lightning come from electric discharges in the atmosphere. However, if you read the daily news, you will find that some misconceptions from prescientific thinking persist even today. You may not find the answer to the meaning of life in this course, but at the very least you will come away with some of the intellectual tools that enable you to weed out inconsistent, logically flawed theories and misconceptions that contradict experimentally verifiable facts. Scientific progress over the last millennium has provided a rational explanation for most of what occurs in the natural world surrounding us.

Through consistent theories and well-designed experiments, physics has helped us obtain a deeper understanding of our surroundings and has given us greater ability to control them. In a time when the consequences of air and water pollution, limited energy resources, and global warming threaten the continued existence of huge portions of life on Earth, the need to understand the results of our interactions with the environment has never been greater. Much of environmental science is based on fundamental physics, and physics drives much of the technology essential to progress in chemistry and the life sciences. You may well be called upon to help decide public policy in these areas, whether as a scientist, an engineer, or simply as a citizen. Having an objective understanding of basic scientific issues is of vital importance in making such decisions. Thus, you need to acquire scientific literacy, an essential tool for every citizen in our technology-driven society.

You cannot become scientifically literate without command of the necessary elementary tools, just as it is impossible to make music without the ability to play an instrument. This is the main purpose of this text: to properly equip you to make sound contributions to the important discussions and decisions of our time. You will emerge from reading and working with this text with a deeper appreciation for the fundamental laws that govern our universe and for the tools that humanity has developed to uncover them, tools that transcend cultures and historic eras.

## 1.2 Working with Numbers

Scientists have established logical rules to govern how they communicate quantitative information to one another. If you want to report the result of a measurement—for example, the distance between two cities, your own weight, or the length of a lecture—you have to specify this result in multiples of a standard unit. Thus, a measurement is the combination of a number and a unit.

At first thought, writing down numbers doesn't seem very difficult. However, in physics, we need to deal with two complications: how to deal with very big or very small numbers, and how to specify precision.

### Scientific Notation

If you want to report a really big number, it becomes tedious to write it out. For example, the human body contains approximately 7,000,000,000,000,000,000,000,000,000 atoms. If you used this number often, you would surely like to have a more compact notation for it. This is exactly what **scientific notation** is. It represents a number as the product of a number greater than 1 and less than 10 (called the *mantissa*) and a power (or exponent) of 10:

$$\text{number} = \text{mantissa} \cdot 10^{\text{exponent}}. \tag{1.1}$$

The number of atoms in the human body can thus be written compactly as $7 \cdot 10^{27}$, where 7 is the mantissa and 27 is the exponent.

Another advantage of scientific notation is that it makes it easy to multiply and divide large numbers. To multiply two numbers in scientific notation, we multiply their mantissas and then add their exponents. If we wanted to estimate, for example, how many atoms are contained in the bodies of all the people on Earth, we could do this calculation rather easily. Earth hosts approximately 7 billion ($=7 \cdot 10^9$) humans. All we have to do to find our answer is to multiply $7 \cdot 10^{27}$ by $7 \cdot 10^9$. We do this by multiplying the two mantissas and adding the exponents:

$$(7 \cdot 10^{27}) \cdot (7 \cdot 10^9) = (7 \cdot 7) \cdot 10^{27+9} = 49 \cdot 10^{36} = 4.9 \cdot 10^{37}. \tag{1.2}$$

### 1.1 In-Class Exercise

The total surface area of Earth is $A = 4\pi R^2 = 4\pi(6370\ \text{km})^2 = 5.099 \cdot 10^{14}\ \text{m}^2$. Assuming there are 7.0 billion humans on the planet, what is the available surface area per person?

a) $7.3 \cdot 10^4\ \text{m}^2$

b) $7.3 \cdot 10^{24}\ \text{m}^2$

c) $3.6 \cdot 10^{24}\ \text{m}^2$

d) $3.6 \cdot 10^4\ \text{m}^2$

In the last step, we follow the common convention of keeping only one digit in front of the decimal point of the mantissa and adjusting the exponent accordingly. (But be advised that we will have to further adjust this answer—read on!)

Division with scientific notation is equally straightforward: If we want to calculate $A / B$, we divide the mantissa of $A$ by the mantissa of $B$ and subtract the exponent of $B$ from the exponent of $A$.

## Significant Figures

When we specified the number of atoms in the average human body as $7 \cdot 10^{27}$, we meant to indicate that we know it is at least $6.5 \cdot 10^{27}$ but smaller than $7.5 \cdot 10^{27}$. However, if we had written $7.0 \cdot 10^{27}$, we would have implied that we know the true number is somewhere between $6.95 \cdot 10^{27}$ and $7.05 \cdot 10^{27}$. This statement is more precise than the previous statement.

As a general rule, the number of digits you write in the mantissa specifies how precisely you claim to know it. The more digits specified, the more precision is implied (see Figure 1.2). We call the number of digits in the mantissa the number of **significant figures.** Here are some rules about using significant figures followed in each case by an example:

- The number of significant figures is the number of reliably known digits. For example, 1.62 has 3 significant figures; 1.6 has 2 significant figures.
- If you give a number as an integer, you specify it with infinite precision. For example, if someone says that he or she has 3 children, this means exactly 3, no less and no more.
- Leading zeros do not count as significant digits. The number 1.62 has the same number of significant digits as 0.00162. There are three significant figures in both numbers. We start counting significant digits from the left at the first nonzero digit.
- Trailing zeros, on the other hand, do count as significant digits. The number 1.620 has four significant figures. Writing a trailing zero implies greater precision!
- Numbers in scientific notation have as many significant figures as their mantissa. For example, the number $9.11 \cdot 10^{-31}$ has three significant figures because that's what the mantissa (9.11) has. The size of the exponent has no influence.
- You can never have more significant figures in a result than you start with in any of the factors of a multiplication or division. For example, 1.23 / 3.4461 is not equal to 0.3569252. Your calculator may give you that answer, but calculators do not automati-

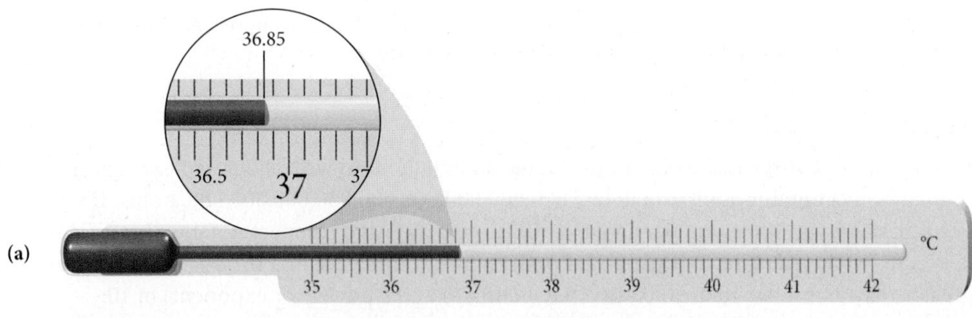


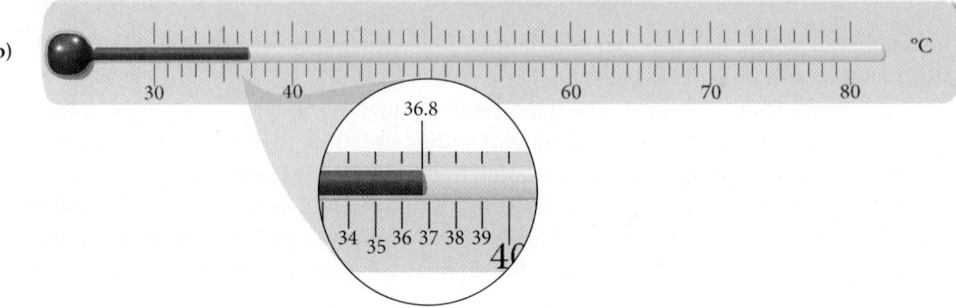


**FIGURE 1.2** Two thermometers measuring the same temperature. (a) The thermometer is marked in tenths of a degree and can be read to four significant figures (36.85 °C); (b) the thermometer is marked in degrees, so it can be read to only three significant figures (36.8 °C).

cally display the correct number of significant figures. Instead, 1.23 / 3.4461 = 0.357. You must round a calculator result to the proper number of significant figures—in this case, three, which is the number of significant figures in the numerator.

- You can only add or subtract when there are significant figures for that place in every number. For example, 1.23 + 3.4461 = 4.68, and not 4.6761 as you may think. This rule, in particular, requires some getting used to.

To finish this discussion of significant figures, let's reconsider the total number of atoms contained in the bodies of all people on Earth. We started with two quantities that were given to only one significant figure. Therefore, the result of our multiplication needs to be properly rounded to one significant digit. The combined number of atoms in all human bodies is thus correctly stated as $5 \cdot 10^{37}$.

### 1.2 In-Class Exercise

How many significant figures are in each of the following numbers?

a) 2.150
b) 0.000215
c) 215.00
d) 0.215000
e) 0.215 + 0.21

## 1.3 SI Unit System

In high school, you may have been introduced to the international system of units and compared it to the British system of units in common use in the United States. You may have driven on a freeway on which the distances are posted both in miles and in kilometers or purchased food where the price was quoted per lb and per kg.

The international system of units is often abbreviated as SI (for Système International). Sometimes the units in this system are called *metric units*. The **SI unit system** is the standard used for scientific work around the world. The base units for the SI system are given in Table 1.1.

The first letters of the first four base units provide another commonly used name for the SI system: the MKSA system. We will use the first three units (meter, kilogram, and second) in the entire first part of this book and in all of mechanics. The current definitions of these base units are as follows:

- 1 meter (m) is the distance that a light beam in vacuum travels in 1/299,792,458 of a second. Originally, the meter was related to the size of the Earth (Figure 1.3).
- 1 kilogram (kg) is defined as the mass of the international prototype of the kilogram. This prototype, shown in its elaborate storage container in Figure 1.4, is kept just outside Paris, France, under carefully controlled environmental conditions.
- 1 second (s) is the time interval during which 9,192,631,770 oscillations of the electromagnetic wave (see Chapter 31) that corresponds to the transition between two specific states of the cesium-133 atom. Until 1967, the standard for the second was 1/86,400 of a mean solar day. However, the atomic definition is more precise and more reliably reproducible.

**FIGURE 1.3** Originally, the meter was defined as 1 ten-millionth of the length of the meridian through Paris from the North Pole to the Equator.

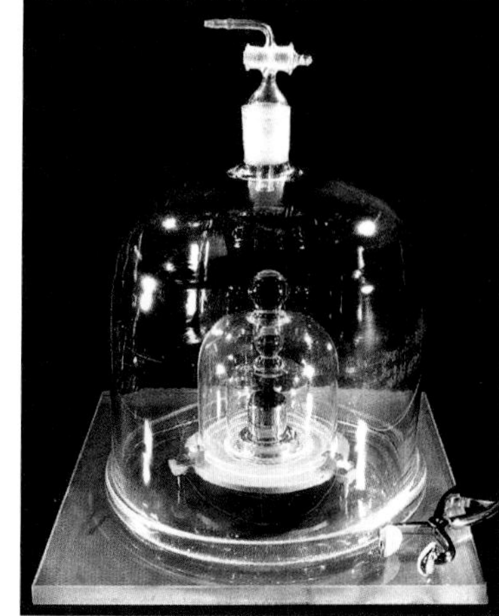

**FIGURE 1.4** Prototype of the kilogram, stored near Paris, France.

***Notation Convention:*** It is common practice to use roman letters for unit abbreviations and italic letters for physical quantities. We follow this convention in this book. For example, m stands for the unit meter, while $m$ is used for the physical quantity mass. Thus, the expression $m = 17.2$ kg specifies that the mass of an object is 17.2 kilograms.

**Table 1.1 Unit Names and Abbreviations for the Base Units of the SI System of Units**

| Unit | Abbreviation | Base Unit for |
|---|---|---|
| meter | m | length |
| kilogram | kg | mass |
| second | s | time |
| ampere | A | current |
| kelvin | K | temperature |
| mole | mol | amount of a substance |
| candela | cd | luminous intensity |

Units for all other physical quantities can be derived from the seven base units of Table 1.1. The unit for area, for example, is $m^2$. The units for volume and mass density are $m^3$ and $kg/m^3$, respectively. The units for velocity and acceleration are m/s and $m/s^2$, respectively. Some derived units were used so often that it became convenient to give them their own names and symbols. Often the name is that of a famous physicist. Table 1.2 lists the 20 derived SI units with special names. In the two rightmost columns of the table, the named unit is listed in terms of other named units and then in terms of SI base units. Also included in this table are the radian and steradian, the dimensionless units of angle and solid angle, respectively.

You can obtain SI-recognized multiples of the base units and derived units by multiplying them by various factors of 10. These factors have universally accepted letter abbreviations that are used as prefixes, shown in Table 1.3. The use of standard prefixes (factors of 10) makes it easy to determine, for example, how many centimeters (cm) are in a kilometer (km):

$$1 \text{ km} = 10^3 \text{ m} = 10^3 \text{ m} \cdot (10^2 \text{ cm/m}) = 10^5 \text{ cm}. \quad (1.3)$$

In comparison, note how tedious it is to figure out how many inches are in a mile:

$$1 \text{ mile} = (5{,}280 \text{ feet/mile}) \cdot (12 \text{ inches/foot}) = 63{,}360 \text{ inches}. \quad (1.4)$$

As you can see, not only do you have to memorize particular conversion factors in the British system, but calculations also become more complicated. For calculations in the SI system, you only have to know the standard prefixes shown in Table 1.3 and how to add or subtract integers in the powers of 10.

The international system of units was adopted in 1799 and is now used daily in almost all countries of the world, the one notable exception being the United States. In the United States, we buy milk and gasoline in gallons, not in liters. Our cars show speed in miles per hour, not meters per second. When we go to the lumberyard, we buy wood

**Table 1.2 Common SI Derived Units**

| Derived or Dimensionless Unit | Name | Symbol | Equivalent | Expressions |
|---|---|---|---|---|
| Absorbed dose | gray | Gy | J/kg | m2 $s^{-2}$ |
| Activity | becquerel | Bq | — | $s^{-1}$ |
| Angle | radian | rad | — | — |
| Capacitance | farad | F | C/V | $m^{-2}\,kg^{-1}\,s^4\,A^2$ |
| Catalytic activity | katal | kat | — | $s^{-1}$ mol |
| Dose equivalent | sievert | Sv | J/kg | $m^2\,s^{-2}$ |
| Electric charge | coulomb | C | — | s A |
| Electric conductance | siemens | S | A/V | $m^{-2}\,kg^{-1}\,s^3\,A^2$ |
| Electric potential | volt | V | W/A | $m^2\,kg\,s^{-3}\,A^{-1}$ |
| Electric resistance | ohm | $\Omega$ | V/A | $m^2\,kg\,s^{-3}\,A^{-2}$ |
| Energy | joule | J | N m | $m^2\,kg\,s^{-2}$ |
| Force | newton | N | — | $m\,kg\,s^{-2}$ |
| Frequency | hertz | Hz | — | $s^{-1}$ |
| Illuminance | lux | lx | $lm/m^2$ | $m^{-2}$ cd |
| Inductance | henry | H | Wb/A | $m^2\,kg\,s^{-2}\,A^{-2}$ |
| Luminous flux | lumen | lm | cd sr | cd |
| Magnetic flux | weber | Wb | V s | $m^2\,kg\,s^{-2}\,A^{-1}$ |
| Magnetic field | tesla | T | $Wb/m^2$ | $kg\,s^{-2}\,A^{-1}$ |
| Power | watt | W | J/s | $m^2\,kg\,s^{-3}$ |
| Pressure | pascal | Pa | $N/m^2$ | $m^{-1}\,kg\,s^{-2}$ |
| Solid angle | steradian | sr | — | — |
| Temperature | degree Celsius | °C | — | K |

**Table 1.3 SI Standard Prefixes**

| Factor | Prefix | Symbol | Factor | Prefix | Symbol |
|---|---|---|---|---|---|
| $10^{24}$ | yotta- | Y | $10^{-24}$ | yocto- | y |
| $10^{21}$ | zetta- | Z | $10^{-21}$ | zepto- | z |
| $10^{18}$ | exa- | E | $10^{-18}$ | atto- | a |
| $10^{15}$ | peta- | P | $10^{-15}$ | femto- | f |
| $10^{12}$ | tera- | T | $10^{-12}$ | pico- | p |
| $10^{9}$ | giga- | G | $10^{-9}$ | nano- | n |
| $10^{6}$ | mega- | M | $10^{-6}$ | micro- | μ |
| $10^{3}$ | kilo- | k | $10^{-3}$ | milli- | m |
| $10^{2}$ | hecto- | h | $10^{-2}$ | centi- | c |
| $10^{1}$ | deka- | da | $10^{-1}$ | deci- | d |

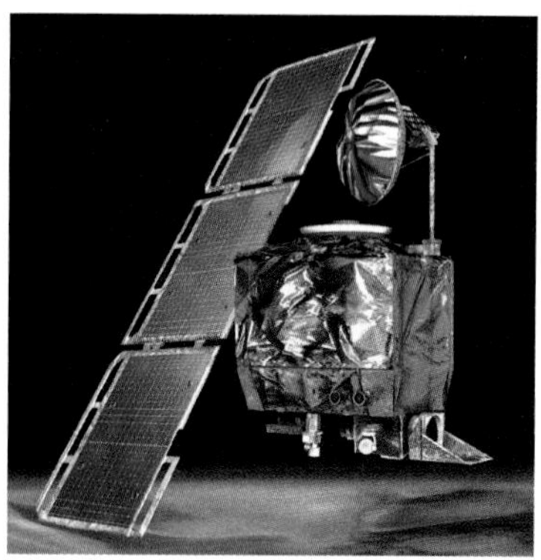

**FIGURE 1.5** The *Mars Climate Orbiter,* a victim of faulty unit conversion.

as two-by-fours (actually 1.5 inches by 3.5 inches, but that's a different story). Since we use British units in our daily lives, this book will indicate British units where appropriate, to establish connections with everyday experiences. But we will use the SI unit system in all calculations to avoid having to work with the British unit conversion factors. However, we will sometimes provide equivalents in British units in parentheses. This is a compromise that will be necessary only until the United States decides to adopt SI units as well.

The use of British units can be costly. The cost can range from a small expense, such as that incurred by car mechanics who need to purchase two sets of wrench socket sets, one metric and one British, to the very expensive loss of the *Mars Climate Orbiter* spacecraft (Figure 1.5) in September 1999. The crash of this spacecraft has been blamed on the fact that one of the engineering teams used British units and the other one SI units. The two teams relied on each other's numbers, without realizing that the units were not the same.

The use of powers of 10 is not completely consistent even within the SI system itself. The notable exception is in the time units, which are not factors of 10 times the base unit (second):

- 365 days form a year,
- a day has 24 hours,
- an hour contains 60 minutes, and
- a minute consists of 60 seconds.

Early metric pioneers tried to establish a completely consistent set of metric time units, but these attempts failed. The not-exactly-metric nature of time units extends to some derived units. For example, a European sedan's speedometer does not show speeds in meters per second, but in kilometers per hour.

## EXAMPLE 1.1 Units of Land Area

The unit of land area used in countries that use the SI system is the hectare, defined as 10,000 $m^2$. In the United States, land area is given in acres; an acre is defined as 43,560 $ft^2$.

### PROBLEM

You just bought a plot of land with dimensions 2.00 km by 4.00 km. What is the area of your new purchase in hectares and acres?

### SOLUTION

The area $A$ is given by

$$A = \text{length} \cdot \text{width} = (2.00\text{ km})(4.00\text{ km}) = \left(2.00 \cdot 10^3\text{ m}\right)\left(4.00 \cdot 10^3\text{ m}\right)$$

$$A = 8.00\text{ km}^2 = 8.00 \cdot 10^6\text{ m}^2.$$

*Continued—*

The area of this plot of land in hectares is then

$$A = 8.00 \cdot 10^6 \text{ m}^2 \frac{1 \text{ hectare}}{10{,}000 \text{ m}^2} = 8.00 \cdot 10^2 \text{ hectare} = 800. \text{ hectare.}$$

To find the area of the land in acres, we need the length and width in British units:

$$\text{length} = 2.00 \text{ km} \frac{1 \text{ mi}}{1.609 \text{ km}} = 1.24 \text{ mi} \frac{5{,}280 \text{ ft}}{1 \text{ mi}} = 6{,}563 \text{ ft}$$

$$\text{width} = 4.00 \text{ km} \frac{1 \text{ mi}}{1.609 \text{ km}} = 2.49 \text{ mi} \frac{5{,}280 \text{ ft}}{1 \text{ mi}} = 13{,}130 \text{ ft.}$$

The area is then

$$A = \text{length} \cdot \text{width} = (1.24 \text{ mi})(2.49 \text{ mi}) = (6{,}563 \text{ ft})(13{,}130 \text{ ft})$$

$$A = 3.09 \text{ mi}^2 = 8.61 \cdot 10^7 \text{ ft}^2.$$

In acres, this is

$$A = 8.61 \cdot 10^7 \text{ ft}^2 \frac{1 \text{ acre}}{43{,}560 \text{ ft}^2} = 1980 \text{ acres.}$$

### Metrology: Research on Measures and Standards

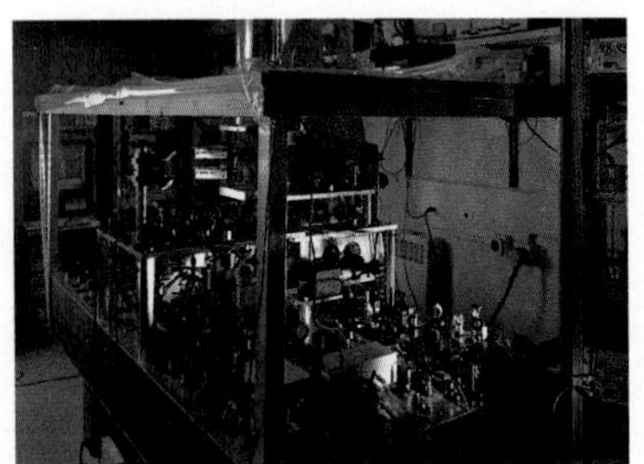

**FIGURE 1.6** Cesium fountain atomic clock at NIST.

The work of defining the standards for the base units of the SI system is by no means complete. A great deal of research is devoted to refining measurement technologies and pushing them to greater precision. This field of research is called **metrology.** In the United States, the laboratory that has the primary responsibility for this work is the National Institute of Standards and Technology (NIST). NIST works in collaboration with similar institutes in other countries to refine accepted standards for the SI base units.

One current research project is to find a definition of the kilogram based on reproducible quantities in nature. This definition would replace the current definition of the kilogram, which is based on the mass of a standard object kept in Sèvres on the outskirts of Paris, as we have seen. The most promising effort in this direction seems to be Project Avogadro, which attempts to define the kilogram using highly purified silicon crystals.

Research on keeping time ever more precisely is one of the major tasks of NIST and similar institutions. Figure 1.6 shows what is currently the most accurate clock at NIST in Boulder, Colorado—the NIST-F1 cesium fountain atomic clock. It is accurate to ± 1 second in 60 million years! However, NIST researchers are working on a new optical clock that promises to be a thousand times more accurate than the NIST-F1.

Greater precision in timekeeping is needed for many applications in our information-based society, where signals can travel around the world in less than 0.2 seconds. The Global Positioning System (GPS) is one example of technology that would be impossible to realize without the precision of atomic clocks and the physics research that enters into their construction. The GPS system also relies on Einstein's theory of relativity, which we will study in Chapter 35.

## 1.4 The Scales of Our World

The most amazing fact about physics is that its laws govern every object, from the smallest to the largest. The scales of the systems for which physics holds predictive power span many orders of magnitude (powers of 10), as we'll see in this section.

***Nomenclature:*** In the following, you will read "on the order of" several times. This phrase means "within a factor of 2 or 3."

### Length Scales

*Length* is defined as the distance measurement between two points in space. Figure 1.7 shows some length scales for common objects and systems that span over 40 orders of magnitude.

Let's start with ourselves. On average, in the United States, a woman is 1.62 m (5 ft 4 in) tall, and a man measures 1.75 m (5 ft 9 in). Thus, human height is on the order of a meter. If you reduce the length scale for a human body by a factor of a million, you arrive at a micrometer. This is the typical diameter of a cell in your body or a bacterium.

If you reduce the length of your measuring stick by another factor of 10,000, you are at a scale of $10^{-10}$ m, the typical diameter of an individual atom. This is the smallest size we can resolve with the aid of the most advanced microscopes.

Inside the atom is its nucleus, with a diameter about 1/10,000 that of the atom, on the order of $10^{-14}$ m. The individual protons and neutrons that make up the atomic nucleus have a diameter of approximately $10^{-15}$ m = 1 fm (a femtometer).

Considering objects larger than ourselves, we can look at the scale of a typical city, on the order of kilometers. The diameter of Earth is just a little bigger than 10,000 km (12,760 km, to be more precise). As discussed earlier, the definition of the meter is now stated in terms of the speed of light. However, the meter was originally defined as 1 ten-millionth of the length of the meridian through Paris from the North Pole to the Equator. If a quarter circle has an arc length of 10 million meters (= 10,000 km), then the circumference of the entire circle would be exactly 40,000 km. Using the modern definition of the meter, the equatorial circumference of the Earth is 40,075 km, and the circumference along the meridian is 40,008 km.

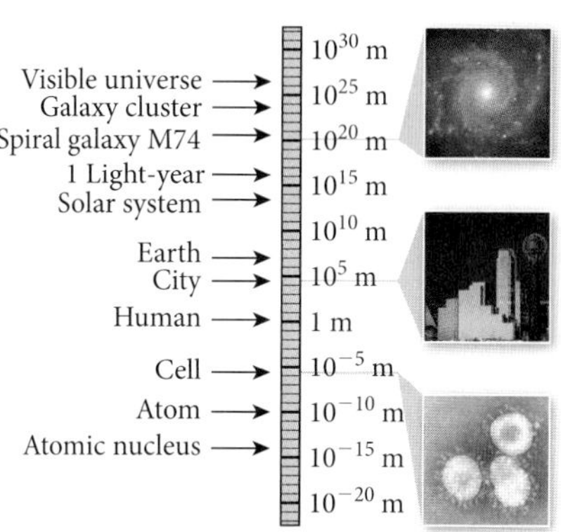


**FIGURE 1.7** Range of length scales for physical systems. The pictures top to bottom are the Spiral Galaxy M 74, the Dallas skyline, and the SARS virus.

The distance from the Earth to the Moon is 384,000 km, and the distance from the Earth to the Sun is greater by a factor of approximately 400, or about 150 million km. This distance is called an *astronomical unit* and has the symbol AU. Astronomers used this unit before the distance from Earth to Sun became known with accuracy, but it is still convenient today. In SI units, an astronomical unit is

$$1 \text{ AU} = 1.495\,98 \cdot 10^{11} \text{ m}. \tag{1.5}$$

The diameter of our solar system is conventionally stated as approximately $10^{13}$ m, or 60 AU.

We have already remarked that light travels in vacuum at a speed of approximately 300,000 km/s. Therefore, the distance between Earth and Moon is covered by light in just over 1 second, and light from the Sun takes approximately 8 minutes to reach Earth. In order to cover distance scales outside our solar system, astronomers have introduced the (non-SI, but handy) unit of the light-year, the distance that light travels in 1 year in vacuum:

$$1 \text{ light-year} = 9.46 \cdot 10^{15} \text{ m}. \tag{1.6}$$

The nearest star to our Sun is just over 4 light-years away. The Andromeda Galaxy, the sister galaxy of our Milky Way, is about 2.5 million light-years = $2 \cdot 10^{22}$ m away.

Finally, the radius of the visible universe is approximately 14 billion light-years = $1.5 \cdot 10^{26}$ m. Thus, about 41 orders of magnitude span between the size of an individual proton and that of the entire visible universe.

## Mass Scales

*Mass* is the amount of matter in an object. When you consider the range of masses of physical objects, you obtain an even more awesome span of orders of magnitude (Figure 1.8) than for lengths.

Atoms and their parts have incredibly small masses. The mass of an electron is only $9.11 \cdot 10^{-31}$ kg. A proton's mass is $1.67 \cdot 10^{-27}$ kg, roughly a factor of 2000 more than the mass of an electron. An individual atom of lead has a mass of $3.46 \cdot 10^{-25}$ kg.

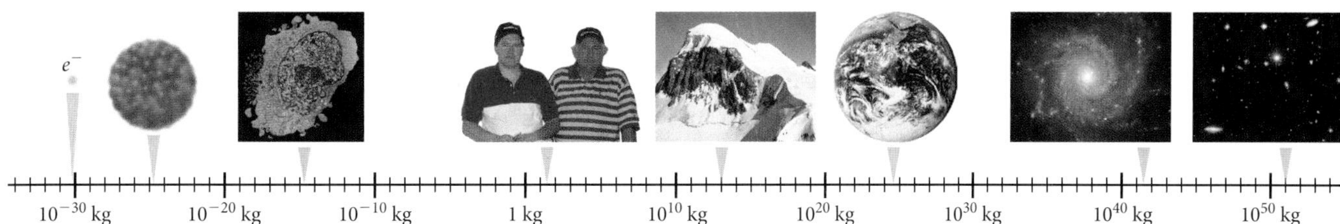


**FIGURE 1.8** Range of mass scales for physical systems.

The mass of a single cell in the human body is on the order of $10^{-15}$ kg to $10^{-14}$ kg. Even a fly has more than 10 billion times the mass of a cell, at approximately $10^{-4}$ kg.

A car's mass is on the order of $10^3$ kg, and that of a passenger plane is on the order of $10^5$ kg.

A mountain typically has a mass of $10^{12}$ kg to $10^{14}$ kg, and the combined mass of all the water in all of the Earth's oceans is estimated to be on the order of $10^{19}$ kg to $10^{20}$ kg.

The mass of the entire Earth can be specified fairly precisely at $6.0 \cdot 10^{24}$ kg. The Sun has a mass of $2.0 \cdot 10^{30}$ kg, or above 300,000 times the mass of the Earth. Our entire galaxy, the Milky Way, is estimated to have 200 billion stars in it and thus has a mass around $3 \cdot 10^{41}$ kg. Finally, the entire universe contains billions of galaxies. Depending on the assumptions about dark matter, a currently active research topic (see Chapter 12), the mass of the universe as a whole is roughly $10^{51}$ kg. However, you should recognize that this number is an estimate and may be off by a factor of up to 100.

Interestingly, some objects have no mass. For example, photons, the "particles" that light is made of, have zero mass.

### Time Scales

*Time* is the duration between two events. Human time scales lie in the range from a second (the typical duration of a human heartbeat) to a century (about the life expectancy of a person born now). Incidentally, human life expectancy is increasing at an ever faster rate. During the Roman Empire, 2000 years ago, a person could expect to live only 25 years. In 1850, actuary tables listed the mean lifetime of a human as 39 years. Now that number is 80 years. Thus, it took almost 2000 years to add 50% to human life expectancy, but in the last 150 years, life expectancy has doubled again. This is perhaps the most direct evidence that science has basic benefits for all of us. Physics contributes to this progress in the development of more sophisticated medical imaging and treatment equipment, and today's fundamental research will enter clinical practice tomorrow. Laser surgery, cancer radiation therapy, magnetic resonance imaging, and positron emission tomography are just a few examples of technological advances that have helped increase life expectancy.

In their research, the authors of this book study ultrarelativistic heavy-ion collisions. These collisions occur during time intervals on the order of $10^{-22}$ s, more than a million times shorter than the time intervals we can measure directly. During this course, you will learn that the time scale for the oscillation of visible light is $10^{-15}$ s, and that of audible sound is $10^{-3}$ s.

The longest time span we can measure *indirectly* or *infer* is the age of the universe. Current research puts this number at 13.7 billion years, but with an uncertainty of up to 0.2 billion years.

We cannot leave this topic without mentioning one interesting fact to ponder during your next class lecture. Lectures typically last 50 minutes at most universities. A century, by comparison, has $100 \cdot 365 \cdot 24 \cdot 60 \approx 50{,}000{,}000$ minutes. So a lecture lasts about 1 millionth of a century, leading to a handy (non-SI) time unit, the microcentury = duration of one lecture.

## 1.5 General Problem-Solving Strategy

Physics involves more than solving problems—but problem solving is a big part of it. At times, while you are laboring over your homework assignments, it may seem that's all you do. However, repetition and practice are important parts of learning.

A basketball player spends hours practicing the fundamentals of free throw shooting. Many repetitions of the same action enable a player to become very reliable at this task. You need to develop the same philosophy toward solving mathematics and physics problems: You have to practice good problem-solving techniques. This work will pay huge dividends, not just during the remainder of this physics course, not just during exams, not even just in your other science classes, but also throughout your entire career.

What constitutes a good problem-solving strategy? Everybody develops his or her own routines, procedures, and shortcuts. However, here is a general blueprint that should help get you started:

1. **THINK** Read the problem carefully. Ask yourself what quantities are known, what quantities might be useful but are unknown, and what quantities are asked for in the solution. Write down these quantities and represent them with their commonly used symbols. Convert into SI units, if necessary.
2. **SKETCH** Make a sketch of the physical situation to help you visualize the problem. For many learning styles, a visual or graphical representation is essential, and it is often essential for defining the variables.
3. **RESEARCH** Write down the physical principles or laws that apply to the problem. Use equations representing these principles to connect the known and unknown quantities to each other. In some cases, you will immediately see an equation that has only the quantities that you know and the one unknown that you are supposed to calculate, and nothing else. More often you may have to do a bit of deriving, combining two or more known equations into the one that you need. This requires some experience, more than any of the other steps listed here. To the beginner, the task of deriving a new equation may look daunting, but you will get better at it the more you practice.
4. **SIMPLIFY** Do not plug numbers into your equation yet! Instead, simplify your result algebraically as much as possible. For example, if your result is expressed as a ratio, cancel out common factors in the numerator and the denominator. This step is particularly helpful if you need to calculate more than one quantity.
5. **CALCULATE** Put the numbers with units into the equation and get to work with a calculator. Typically, you will obtain a number and a physical unit as your answer.
6. **ROUND** Determine the number of significant figures that you want to have in your result. As a rule of thumb, a result obtained by multiplying or dividing should be rounded to the same number of significant figures as in the input quantity that is given with the least number of significant figures. You should not round in intermediate steps, as rounding too early might give you a wrong solution.
7. **DOUBLE-CHECK** Step back and look at the result. Judge for yourself if the answer (both the number and the units) seems realistic. You can often avoid handing in a wrong solution by making this final check. Sometimes the units of your answer are simply wrong, and you know you must have made an error. Or sometimes the order of magnitude comes out totally wrong. For example, if your task is to calculate the mass of the Sun (we will do this later in this book), and your answer comes out near $10^6$ kg (only a few thousand tons), you know you must have made a mistake somewhere.

Let's put this strategy to work in the following example.

## SOLVED PROBLEM 1.1 Volume of a Cylinder

### PROBLEM

Nuclear waste material in a physics laboratory is stored in a cylinder of height $4\frac{13}{16}$ inches and circumference $8\frac{3}{16}$ inches. What is the volume of this cylinder, measured in metric units?

### SOLUTION

In order to practice problem-solving skills, we will go through each of the steps of the strategy outlined above.

### THINK

From the question, we know that the height of the cylinder, converted to cm, is

$$\begin{aligned} h &= 4\tfrac{13}{16}\text{ in} = 4.8125\text{ in} \\ &= (4.8125\text{ in})\cdot(2.54\text{ cm/in}) \\ &= 12.22375\text{ cm}. \end{aligned}$$

*Continued—*

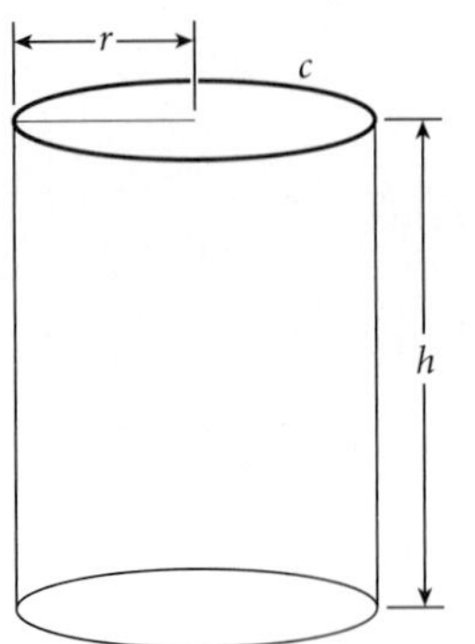


**FIGURE 1.9** Sketch of right cylinder.

Also, the circumference of the cylinder is specified as

$$c = 8\tfrac{3}{16} \text{ in} = 8.1875 \text{ in}$$
$$= (8.1875 \text{ in}) \cdot (2.54 \text{ cm/in})$$
$$= 20.79625 \text{ cm}.$$

Obviously, the dimensions given were rounded to the nearest 16th of an inch. Thus, it makes no sense to specify five digits after the decimal point for the dimensions when converted to centimeters, as the output of our calculator seems to suggest. A more realistic way to specify the given quantities is $h = 12.2$ cm and $c = 20.8$ cm, where the numbers have three significant figures, just like the numbers given originally.

### SKETCH

Next, we produce a sketch, something like Figure 1.9. Note that the given quantities are shown with their symbolic representations, not with their numerical values. The circumference is represented by the thicker circle (oval, actually, in this projection).

### RESEARCH

Now we have to find the volume of the cylinder in terms of its height and its circumference. This relationship is not commonly listed in collections of geometric formulas. Instead, the volume of a cylinder is given as the product of base area and height:

$$V = \pi r^2 h.$$

Once we find a way to connect the radius and the circumference, we'll have the formula we need. The top and bottom areas of a cylinder are circles, and for a circle we know that

$$c = 2\pi r.$$

### SIMPLIFY

Remember: We do not plug in numbers yet! To simplify our numerical task, we can solve the second equation for $r$ and insert this result into the first equation:

$$c = 2\pi r \Rightarrow r = \frac{c}{2\pi}$$

$$V = \pi r^2 h = \pi \left( \frac{c}{2\pi} \right)^2 h = \frac{c^2 h}{4\pi}.$$

### CALCULATE

Now it's time to get out the calculator and put in the numbers:

$$V = \frac{c^2 h}{4\pi}$$
$$= \frac{(20.8 \text{ cm})^2 \cdot (12.2 \text{ cm})}{4\pi}$$
$$= 420.026447 \text{ cm}^3.$$

### ROUND

The output of the calculator has again made our result look much more precise than we can claim realistically. We need to round. Since the input quantities are given to only three significant figures, our result needs to be rounded to three significant figures. Our final answer is $V = 420.$ cm$^3$.

### DOUBLE-CHECK

Our last step is to check that the answer is reasonable. First, we look at the unit we got for our result. Cubic centimeters are a unit for volume, so our result passes its first test. Now let's look at the magnitude of our result. You might recognize that the height and circumference given for the cylinder are close to the corresponding dimensions of a soda can. If you look on a can of your favorite soft drink, it will list the contents as 12 fluid ounces and also give you the information that this is 355 mL. Because 1 mL = 1 cm$^3$, our answer is reasonably close to the volume of the soda can. Note that this does *not* tell us that our calculation is correct, but it shows that we are not way off.

Suppose the researchers decide that a soda can is not large enough for holding the waste in the lab and replace it with a larger cylindrical container with a height of 44.6 cm and a circumference of 62.5 cm. If we want to calculate the volume of this replacement cylinder, we don't need to do all of Solved Problem 1.1 over again. Instead, we can go directly to the algebraic formula we derived in the Simplify step and substitute our new data into that, ending up with a volume of 13,900 $\text{cm}^3$ when rendered to 3 significant figures. This example illustrates the value of waiting to substitute in numbers until algebraic simplification has been completed.

In Solved Problem 1.1, you can see that we followed the seven steps outlined in our general strategy. It is tremendously helpful to train your brain to follow a certain procedure in attacking all kinds of problems. This is not unlike following the same routine whenever you are shooting free throws in basketball, where frequent repetition helps you build the muscle memory essential for consistent success, even when the game is on the line.

Perhaps more than anything, an introductory physics class should enable you to develop methods to come up with your own solutions to a variety of problems, eliminating the need to accept "authoritative" answers uncritically. The method we used in Solved Problem 1.1 is extremely useful, and we will practice it again and again in this book. However, to make a simple point, one that does not require the full set of steps used for a solved problem, we'll sometimes use an illustrative example.

## EXAMPLE 1.2 Volume of a Barrel of Oil

### PROBLEM

The volume of a barrel of oil is 159 L. We need to design a cylindrical container that will hold this volume of oil. The container needs to have a height of 1.00 m to fit in a transportation container. What is the required circumference of the cylindrical container?

### SOLUTION

Starting with the equation we derived in the Simplify step in Solved Problem 1.1, we can relate the circumference, $c$, and the height, $h$, of the container to the volume, $V$, of the container:

$$V = \frac{c^2 h}{4\pi}.$$

Solving for the circumference, we get

$$c = \sqrt{\frac{4\pi V}{h}}.$$

The volume in SI units is

$$V = 159\ \text{L} \frac{1000\ \text{mL}}{\text{L}} \frac{1\ \text{cm}^3}{1\ \text{mL}} \frac{\text{m}^3}{10^6\ \text{cm}^3} = 0.159\ \text{m}^3.$$

The required circumference is then

$$c = \sqrt{\frac{4\pi V}{h}} = \sqrt{\frac{4\pi\left(0.159\ \text{m}^3\right)}{1.00\ \text{m}}} = 1.41\ \text{m}.$$

As you may have already realized from the preceding problem and example, a good command of algebra is essential for success in an introductory physics class. For engineers and scientists, most universities and colleges also require calculus, but at many schools an introductory physics class and a calculus class can be taken concurrently. This first chapter does not contain any calculus, and subsequent chapters will review the relevant calculus concepts as you need them. However, there is another field of mathematics that is used extensively in introductory physics: trigonometry. Virtually every chapter of this book uses right triangles in some way. Therefore, it is a good idea to review the formulas for sine, cosine, and the like, as well as the indispensable Pythagorean theorem. Let's look at another solved problem, which makes use of trigonometric concepts.

## SOLVED PROBLEM 1.2 View from the Willis Tower

### PROBLEM

It goes without saying that one can see farther from a tower than from ground level; the higher the tower, the farther one can see. The Willis Tower in Chicago has an observation deck, which is 412 m above ground. How far can one see out over Lake Michigan from this observation deck under perfect weather conditions? (Assume eye level is at 413 m above the level of the lake.)

### SOLUTION

#### THINK

As we have stressed before, this is the most important step in the problem-solving process. A little preparation at this stage can save a lot of work at a later stage. Perfect weather conditions are specified, so fog or haze is not a limiting factor. What else could determine how far one can see? If the air is clear, one can see mountains that are quite far away. Why mountains? Because they are very tall. But the landscape around Chicago is flat. What then could limit the viewing range? Nothing, really; one can see all the way to the horizon. And what is the deciding factor for where the horizon is? It is the curvature of the Earth. Let's make a sketch to make this a little clearer.

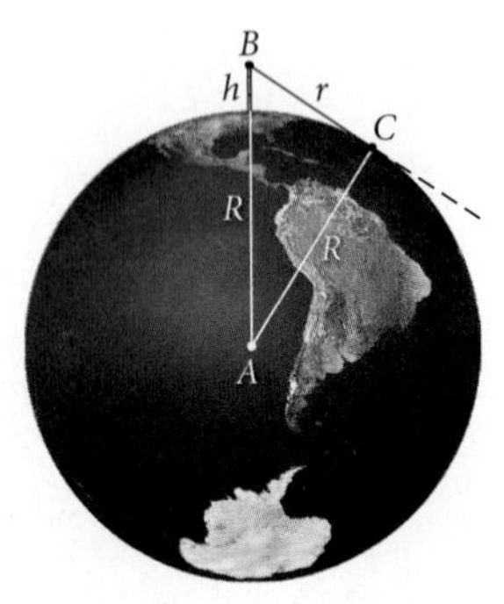


**FIGURE 1.10** Distance from the top of the Willis Tower (*B*) to the horizon (*C*).

#### SKETCH

Our sketch does not have to be elaborate, but it needs to show a simple version of the Willis Tower on the surface of the Earth. It is not important that the sketch be to scale, and we elect to greatly exaggerate the height of the tower relative to the size of the Earth. See Figure 1.10.

It seems obvious from this sketch that the farthest point (point *C*) that one can see from the top of the Willis Tower (point *B*) is where the line of sight just touches the surface of the Earth tangentially. Any point on Earth's surface farther away from the Willis Tower is hidden from view (below the dashed line segment). The viewing range is then given by the distance $r$ between that surface point *C* and the observation deck (point *B*) on top of the tower, at height $h$. Included in the sketch is also a line from the center of Earth (point *A*) to the foot of the Willis Tower. It has length $R$, which is the radius of Earth. Another line of the same length, $R$, is drawn to the point where the line of sight touches the Earth's surface tangentially.

#### RESEARCH

As you can see from the sketch, a line drawn from the center of the Earth to the point where the line of sight touches the surface (*A* to *C*) will form a right angle with that line of sight (*B* to *C*); that is, the three points *A*, *B*, and *C* form the corners of a right triangle. This is the key insight, which enables us to use trigonometry and the Pythagorean theorem to attack the solution of this problem. Examining the sketch in Figure 1.10, we find

$$r^2 + R^2 = (R+h)^2.$$

#### SIMPLIFY

Remember, we want to find the distance to the horizon, for which we used the symbol $r$ in the previous equation. Isolating that variable on one side of our equation gives

$$r^2 = (R+h)^2 - R^2.$$

Now we can simplify the square and obtain

$$r^2 = R^2 + 2hR + h^2 - R^2 = 2hR + h^2.$$

Finally, we take the square root and obtain our final algebraic answer:

$$r = \sqrt{2hR + h^2}.$$

#### CALCULATE

Now we are ready to insert numbers. The accepted value for the radius of Earth is $R = 6370$ km $= 6.37 \cdot 10^6$ m, and $h = 413$ m $= 4.13 \cdot 10^2$ m was given in the problem. This leads to

$$r = \sqrt{2(4.13 \cdot 10^2 \text{ m})(6.37 \cdot 10^6 \text{ m}) + (4.13 \cdot 10^2 \text{ m})^2} = 7.25382 \cdot 10^4 \text{ m}.$$

**ROUND**

The Earth's radius was given to three-digit precision, as was the elevation of the eye level of the observer. So we round to three digits and give our final result as

$$r = 7.25 \cdot 10^4 \text{ m} = 72.5 \text{ km}.$$

**DOUBLE-CHECK**

Always check the units first. Since the problem asked "how far," the answer needs to be a distance, which has the dimension of length and thus the base unit meter. Our answer passes this first check. What about the magnitude of our answer? Since the Willis Tower is almost 0.5 kilometer high, we expect the viewing range to be at least several kilometers; so a multi-kilometer range for the answer seems reasonable. Lake Michigan is slightly more than 80 km wide, if you look toward the east from Chicago. Our answer then implies that you cannot see the Michigan shore of Lake Michigan if you stand on top of the Willis Tower. Experience shows that this is correct, which gives us additional confidence in our answer.

### 1.3 In-Class Exercise

What is the maximum distance from which a sailor on top of a mast of ship 1, who is 34 m above the ocean's surface, can see another sailor on top of a mast of ship 2, who is 26 m above the ocean's surface?

a) 0.17 km

b) 0.89 km

c) 4.5 km

d) 21 km

e) 39 km

## Problem-Solving Guidelines: Limits

In Solved Problem 1.2, we found a handy formula, $r = \sqrt{2hR + h^2}$, for how far one can see on the surface of Earth from an elevation *h*, where *R* is the radius of Earth. There is another test that we can perform to check the validity of this formula. We did not include it in the Double-Check step, because it deserves seperate consideration. This general problem-solving technique is examining the limits of an equation.

What does "examining the limits" mean? In terms of Solved Problem 1.2, it means that instead of just inserting the given number for *h* into our formula and computing the solution, we can also step back and think about what should happen to the distance *r* one can see if *h* becomes very large or very small. Obviously, the smallest that *h* can become is zero. In this case, *r* will also approach zero. This is expected, of course; if your eye level is at ground level, you cannot see very far. On the other hand, we can ponder what happens if *h* becomes large compared to the radius of Earth (see Figure 1.11). (Yes, it is impossible to build a tower that tall but *h* could also stand for the altitude of a satellite above ground.) In that case, we expect that the viewing range would eventually simply be the height *h*. Our formula also bears out this expectation, because as *h* becomes large compared to *R*, the first term in the square root can be neglected, and we find $\lim_{h \to \infty} \sqrt{2hR + h^2} = h$.

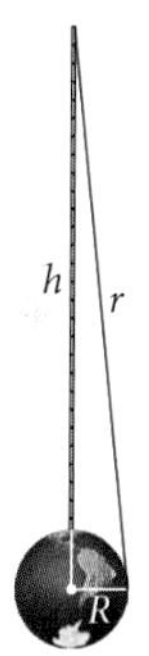

**FIGURE 1.11** Viewing range in the limit of very large *h*.

What we have illustrated by this example is a general guideline: If you derive a formula, you can check its validity by substituting extreme values of the variables in the formula and checking if these limits agree with common sense. Often the formula simplifies drastically at a limit. If your formula has a limiting behavior that is correct, it does not necessarily mean that the formula itself is correct, but it gives you additional confidence in its validity.

## Problem-Solving Guidelines: Ratios

Another very common class of physics problems asks what happens to a quantity that depends on a certain parameter if that parameter changes by a given factor. These problems provide excellent insight into physical concepts and take almost no time to do. This is true, in general, *if* two conditions are met: First, you have to know what formula to use; and second, you have to know how to solve this general class of problems. But that is a big *if*. Studying will equip your memory with the correct formulas, but you need to acquire the skill of solving problems of this general type.

Here is the trick: Write down the formula that connects the dependent quantity to the parameter that changes. Write it twice, once with the dependent quantity and the parameters indexed (or labeled) with 1 and once with them indexed with 2. Then form ratios of the indexed quantities by dividing the right-hand sides and the left-hand sides of the two equations. Next, insert the factor of change for the parameter (expressed as a ratio) and do the calculation to find the factor of change for the dependent quantity (also expressed as a ratio).

Here is an example that demonstrates this method.

### EXAMPLE 1.3 Change in Volume

**PROBLEM**

If the radius of a cylinder increases by a factor of 2.73, by what factor does the volume change? Assume that the height of the cylinder stays the same.

**SOLUTION**

The formula that connects the volume of a cylinder, $V$, and its radius, $r$, is

$$V = \pi r^2 h.$$

The way the problem is phrased, $V$ is the dependent quantity and $r$ is the parameter it depends on. The height of the cylinder, $h$, also appears in the equation but remains constant, according to the problem statement.

Following the general problem-solving guideline, we write the equation twice, once with 1 as indexes and once with 2:

$$V_1 = \pi r_1^2 h$$
$$V_2 = \pi r_2^2 h.$$

Now we divide the second equation by the first, obtaining

$$\frac{V_2}{V_1} = \frac{\pi r_2^2 h}{\pi r_1^2 h} = \left(\frac{r_2}{r_1}\right)^2.$$

As you can see, $h$ did not receive an index because it stayed constant in this problem; it canceled out in the division.

The problem states that the change in radius is given by:

$$r_2 = 2.73 r_1.$$

We substitute for $r_2$ in our ratio:

$$\frac{V_2}{V_1} = \left(\frac{r_2}{r_1}\right)^2 = \left(\frac{2.73 r_1}{r_1}\right)^2 = 2.73^2 = 7.4529,$$

or

$$V_2 = 7.45 V_1,$$

where we have rounded the solution to the three significant digits that the quantity given in the problem had. Thus, the answer is that the volume of the cylinder increases by a factor of 7.45 when you increase its radius by a factor of 2.73.

## Problem-Solving Guidelines: Estimation

Sometimes you don't need to solve a physics problem exactly. When an estimate is all that is asked for, knowing the order of magnitude of some quantity is enough. For example, an answer of $1.24 \cdot 10^{20}$ km is for most purposes not much different from $1 \cdot 10^{20}$ km. In such cases, you can round off all the numbers in a problem to the nearest power of 10 and carry out the necessary arithmetic. For example, the calculation in Solved Problem 1.1 reduces to

$$\frac{(20.8 \text{ cm})^2 \cdot (12.2 \text{ cm})}{4\pi} \approx \frac{(2 \cdot 10^1 \text{ cm})^2 \cdot (10 \text{ cm})}{10} = \frac{4 \cdot 10^3 \text{ cm}^3}{10} = 400 \text{ cm}^3,$$

which is pretty close to our answer of 420. $\text{cm}^3$. Even an answer of 100 $\text{cm}^3$ (rounding 20.8 cm to 10 cm) has the correct order of magnitude for the volume. Notice that you can often round the number $\pi$ to 3 or round $\pi^2$ to 10. With practice, you can find more tricks of approximation like these that can make estimations simpler and faster.

The technique of gaining useful results through careful estimation was made famous by the 20th-century physicist Enrico Fermi (1901–1954), who estimated the energy released by the Trinity nuclear explosion on July 16, 1945, near Socorro, New Mexico, by observing how far a piece of paper was blown by the wind from the blast. There is a class of estimation problems called *Fermi problems* that can yield interesting results when reasonable assumptions are made about quantities that are not known exactly.

Estimates are useful to gain insight into a problem before turning to more complicated methods of calculating a precise answer. For example, one could estimate how many tacos people eat and how many taco stands there are in town before investing in a complete business plan to construct a taco stand. In order to practice estimation skills, let's estimate the number of dentists working in the United States.

### EXAMPLE 1.4 Number of Dentists

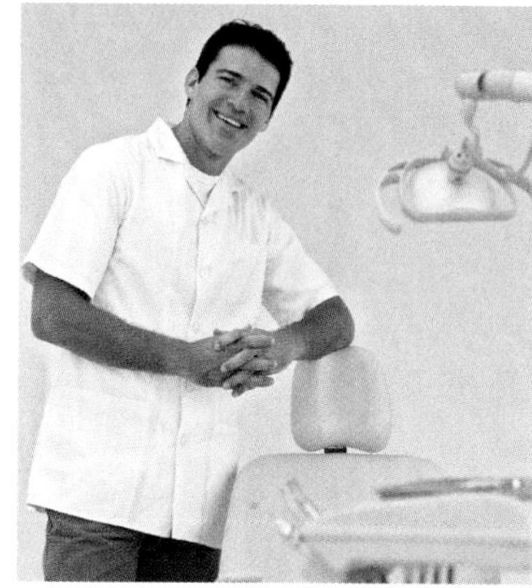

**PROBLEM**

How many dentists are practicing in the United States?

**SOLUTION**

To make this estimate, we start with the following facts and assumptions: There are about 300 million people in the United States. One half of them have regular dentist visits, mainly for routine teeth cleaning. Each person who sees a dentist regularly visits the dentist twice a year. Each office visit lasts $\frac{1}{2}$ hour on average, with the occasional fix of a cavity included. Thus, the total number of hours of dental care being provided in the United States in 1 year is

$$\left(300 \cdot 10^6 \text{ people}\right)\left(0.5\right)\left(2 \text{ visits/person}\right)\left(0.5 \text{ hours/visit}\right) = 1.5 \cdot 10^8 \text{ hours.}$$

We assume that a dentist works 40 hours per week and 50 weeks a year. Thus, a dentist works

$$\left(40 \text{ hours/week}\right)\left(50 \text{ weeks/year}\right) = 2000 \text{ hours/year.}$$

Thus, the required number of dentists is

$$\text{Number of dentists} = \frac{1.5 \cdot 10^8 \text{ hours}}{2000 \text{ hours}} = 75{,}000 \text{ dentists.}$$

The Department of Labor's Bureau of Labor Statistics reports that there were 161,000 licensed dentists in the United States in 2006. Thus, our estimate of the number of dentists practicing in the United States was within a factor of 3 of the actual number.

## 1.6 Vectors

**Vectors** are mathematical descriptions of quantities that have both magnitude and direction. The magnitude of a vector is a nonnegative number, often combined with a physical unit. Many vector quantities are important in physics and, indeed, in all of science. Therefore, before we start this study of physics, you need to be familiar with vectors and some basic vector operations.

Vectors have a starting point and an ending point. For example, consider a flight from Seattle to New York. To represent the change of the plane's position, we can draw an arrow from the plane's departure point to its destination (Figure 1.12). (Real flight paths are not exactly straight lines due to the fact that the Earth is a sphere and due to airspace restrictions and air traffic regulations, but a straight line is a reasonable approximation for our purpose.) This arrow represents a *displacement vector,* which always goes from somewhere to somewhere else. Any vector quantity has a magnitude and a direction. If the vector represents a physical quantity, such as displacement, it will also have a physical unit. A quantity that can be represented without giving a direction is called a **scalar.** A scalar quantity has just a magnitude and possibly a physical unit. Examples of scalar quantities are time and temperature.

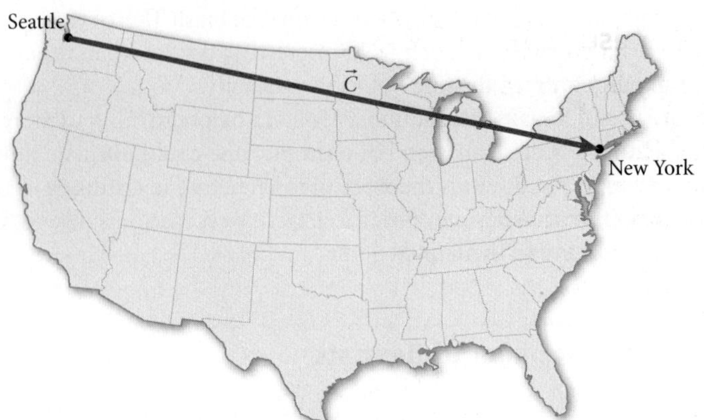


**FIGURE 1.12** Flight path from Seattle to New York as an example of a vector.

This book denotes a vector quantity by a letter with a small horizontal arrow pointing to the right above it. For example, in the drawing of the trip from Seattle to New York (Figure 1.12), the displacement vector has the symbol $\vec{C}$. In the rest of this section, you will learn how to work with vectors: how to add and subtract them and how to multiply them with scalars. In order to perform these operations, it is very useful to introduce a coordinate system in which to represent vectors.

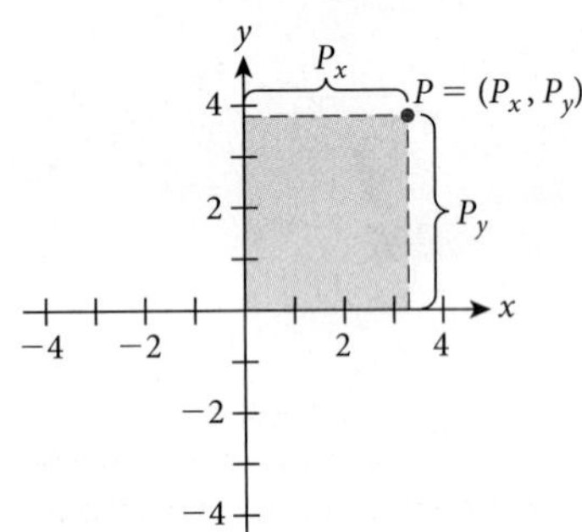


**FIGURE 1.13** Representation of a point $P$ in two-dimensional space in terms of its Cartesian coordinates.

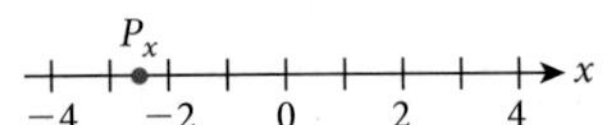


**FIGURE 1.14** Representation of a point $P$ in a one-dimensional Cartesian coordinate system.

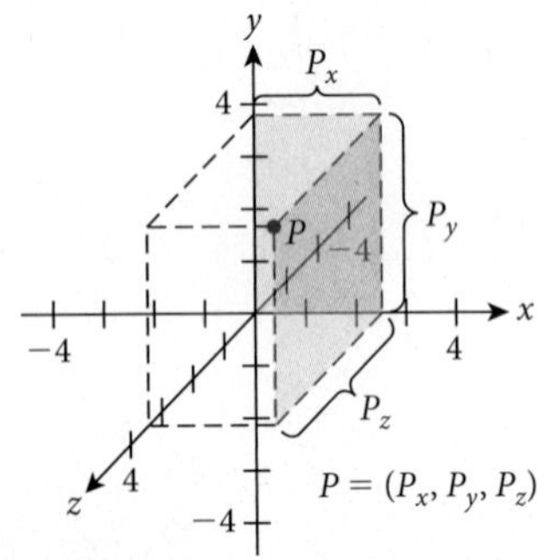


**FIGURE 1.15** Representation of a point $P$ in a three-dimensional space in terms of its Cartesian coordinates.

## Cartesian Coordinate System

A **Cartesian coordinate system** is defined as a set of two or more axes with angles of 90° between each pair. These axes are said to be orthogonal to each other. In a two-dimensional space, the coordinate axes are typically labeled $x$ and $y$. We can then uniquely specify any point $P$ in the two-dimensional space by giving its coordinates $P_x$ and $P_y$ along the two coordinate axes, as shown in Figure 1.13. We will use the notation $(P_x, P_y)$ to specify a point in terms of its coordinates. In Figure 1.13, for example, the point $P$ has the position (3.3, 3.8), because its $x$-coordinate has a value of 3.3 and its $y$-coordinate has a value of 3.8. Note that each coordinate is a number and can have a positive or negative value or be zero.

We can also define a one-dimensional coordinate system, for which any point is located on a single straight line, conventionally called the $x$-axis. Any point in this one-dimensional space is then uniquely defined by specifying one number, the value of the $x$-coordinate, which again can be negative, zero, or positive (Figure 1.14). The point $P$ in Figure 1.14 has the $x$-coordinate $P_x = -2.5$.

Clearly, one- and two-dimensional coordinate systems are easy to draw, because the surface of paper has two dimensions. In a true three-dimensional coordinate system, the third coordinate axis would have to be perpendicular to the other two and would thus have to stick straight out of the plane of the page. In order to draw a three-dimensional coordinate system, we have to rely on conventions that make use of the techniques for perspective drawings. We represent the third axis by a line that is at a 45° angle with the other two (Figure 1.15).

In a three-dimensional space, we have to specify three numbers to uniquely determine the coordinates of a point. We use the notation $P = (P_x, P_y, P_z)$to accomplish this. It is possible to construct Cartesian coordinate systems with more than three orthogonal axes, although they are almost impossible to visualize. Modern string theories (described in Chapter 39), for example, are usually constructed in 10-dimensional spaces. However, for the purposes of this book and for almost all of physics, three dimensions are sufficient. As a matter of fact, for most applications, the essential mathematical and physical understanding can be obtained from two-dimensional representations.

## Cartesian Representation of Vectors

The example of the flight from Seattle to New York established that vectors are characterized by two points: start and finish, represented by the tail and head of an arrow, respectively. Using the Cartesian representation of points, we can define the Cartesian representation of a displacement vector as the difference in the coordinates of the end point and the starting point. Since the difference between the two points for a vector is all that matters, we can shift the vector around in space as much as we like. As long as we do not change the length or direction of the arrow, the vector remains mathematically the same. Consider the two vectors in Figure 1.16.

Figure 1.16a shows the displacement vector $\vec{A}$ that points from point $P=(-2,-3)$ to point $Q=(3,1)$. With the notation just introduced, the **components** of $\vec{A}$ are the coordinates of point Q minus those of point $P$, $\vec{A}=(3-(-2),1-(-3))=(5,4)$. Figure 1.16b shows another vector from point $R=(-3,-1)$ to point $S=(2,3)$. The difference between these coordinates is $(2-(-3),3-(-1)=(5,4)$, which is the same as the vector $\vec{A}$ pointing from $P$ to $Q$.

For simplicity, we can shift the beginning of a vector to the origin of the coordinate system, and the components of the vector will be the same as the coordinates of its end point (Figure 1.17). As a result, we see that we can represent a vector in Cartesian coordinates as

$$\vec{A}=(A_x,A_y) \text{ in two-dimensional space} \tag{1.7}$$

$$\vec{A}=(A_x,A_y,A_z) \text{ in three-dimensional space} \tag{1.8}$$

where $A_x$, $A_y$, and $A_z$ are numbers. Note that the notation for a point in Cartesian coordinates is similar to the notation for a vector in Cartesian coordinates. Whether the notation specifies a point or a vector will be clear from the context of the reference.

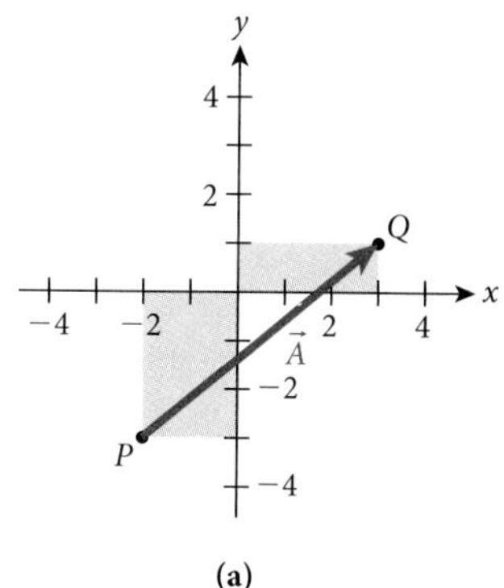


(a)

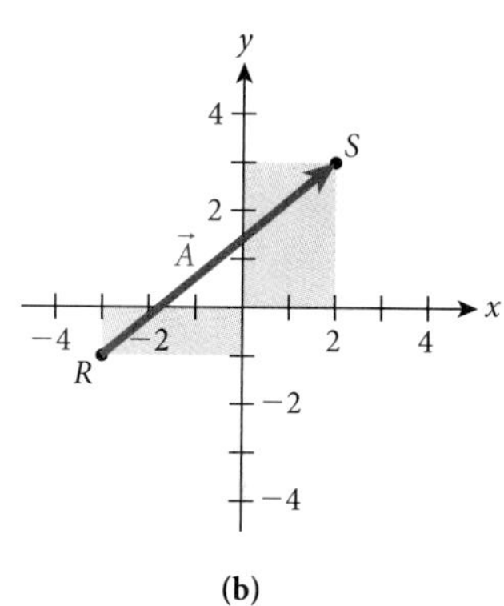


(b)

**FIGURE 1.16** Cartesian representations of a vector $\vec{A}$. (a) Displacement vector from $P$ to $Q$; (b) displacement vector from $R$ to $S$.

## Graphical Vector Addition and Subtraction

Suppose that the direct flight from Seattle to New York shown in Figure 1.12 was not available, and you had to make a connection through Dallas (Figure 1.18). Then the displacement vector $\vec{C}$ for the flight from Seattle to New York is the sum of a displacement vector $\vec{A}$ from Seattle to Dallas and a displacement vector $\vec{B}$ from Dallas to New York:

$$\vec{C}=\vec{A}+\vec{B}. \tag{1.9}$$

This example shows the general procedure for vector addition in a graphical way: Move the tail of vector $\vec{B}$ to the head of vector $\vec{A}$; then the vector from the tail of vector $\vec{A}$ to the head of vector $\vec{B}$ is the sum vector, or **resultant,** of the two.

If you add two real numbers, the order does not matter: 3+5=5+3. This property is called the *commutative property of addition.* Vector addition is also commutative:

$$\vec{A}+\vec{B}=\vec{B}+\vec{A}. \tag{1.10}$$

Figure 1.19 demonstrates this commutative property of vector addition graphically. It shows the same vectors as in Figure 1.18, but also shows the beginning of vector $\vec{A}$ moved to the tip of vector $\vec{B}$ (dashed arrows)—note that the resultant vector is the same as before.

Next, the inverse (or reverse or negative) vector, $-\vec{C}$, of the vector $\vec{C}$ is a vector with the same length as $\vec{C}$ but pointing in the opposite direction (Figure 1.20). For the vector representing the flight from Seattle to New York, for example, the inverse vector is the return trip. Clearly, if you add $\vec{C}$ and its inverse vector, $-\vec{C}$, you end up at the point you started from. Thus, we find

$$\vec{C}+(-\vec{C})=\vec{C}-\vec{C}=(0,0,0), \tag{1.11}$$

and the magnitude is zero, $|\vec{C}-\vec{C}|=0$. This seemingly simple identity shows that we can treat vector subtraction as vector addition, by simply adding the inverse vector. For example, the vector $\vec{B}$ in Figure 1.19 can be obtained as $\vec{B}=\vec{C}-\vec{A}$. Therefore, vector addition and subtraction follow exactly the same rules as the addition and subtraction of real numbers.

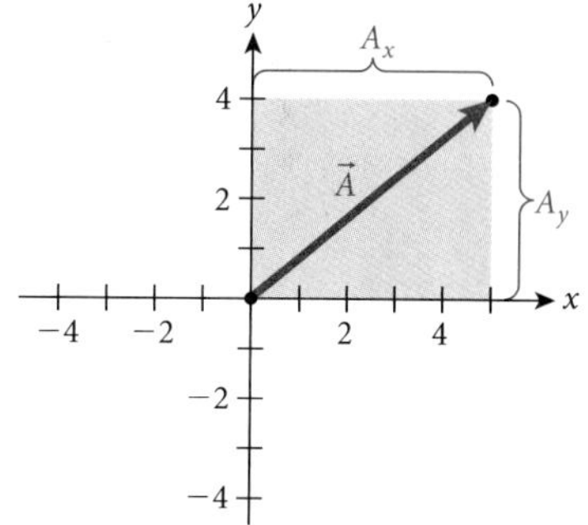


**FIGURE 1.17** Cartesian components of vector $\vec{A}$ in two dimensions.

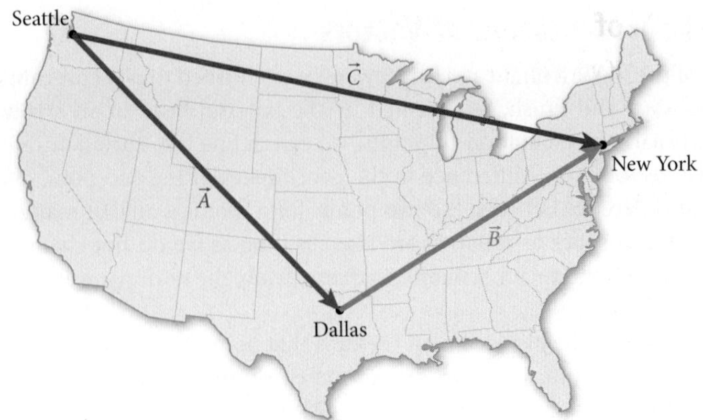


**FIGURE 1.18** Direct flight versus one-stop flight as an example of vector addition.

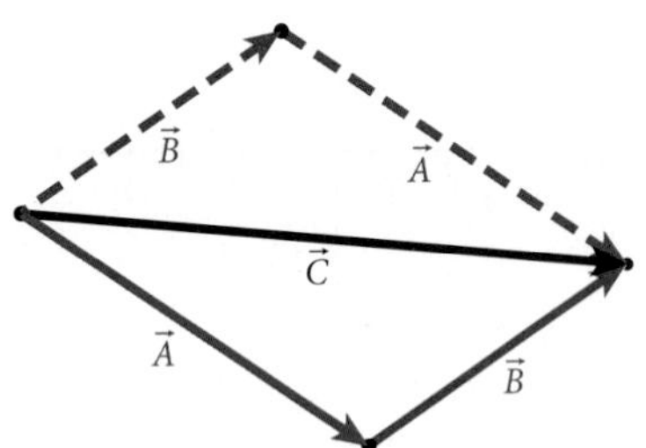


**FIGURE 1.19** Commutative property of vector addition.

## Vector Addition Using Components

Graphical vector addition illustrates the concepts very well, but for practical purposes the component method of vector addition is much more useful. (This is because calculators are easier to use and much more precise than rulers and graph paper.) Let's consider the component method for addition of three-dimensional vectors. The equations for two-dimensional vectors are special cases that arise by neglecting the $z$-components. Similarly, the one-dimensional equation can be obtained by neglecting all $y$- and $z$-components.

If you add two three-dimensional vectors, $\vec{A}=(A_x, A_y, A_z)$ and $\vec{B}=(B_x, B_y, B_z)$, the resulting vector is

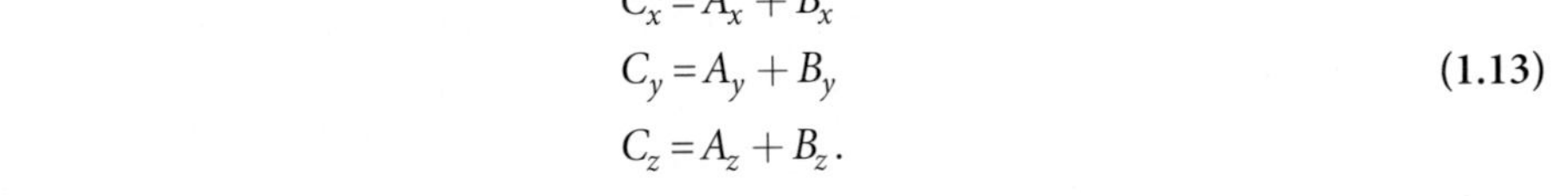

$$\vec{C}=\vec{A}+\vec{B}=(A_x, A_y, A_z)+(B_x, B_y, B_z)=(A_x+B_x, A_y+B_y, A_z+B_z). \quad (1.12)$$

In other words, the components of the sum vector are the sum of the components of the individual vectors:

$$\begin{aligned} C_x &= A_x + B_x \\ C_y &= A_y + B_y \\ C_z &= A_z + B_z. \end{aligned} \quad (1.13)$$

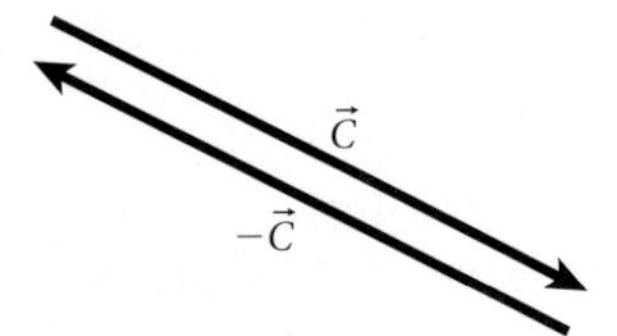


**FIGURE 1.20** Inverse vector $-\vec{C}$ of a vector $\vec{C}$.

The relationship between graphical and component methods is illustrated in Figure 1.21. Figure 1.21a shows two vectors $\vec{A}=(4, 2)$ and $\vec{B}=(3, 4)$ in two-dimensional space, and Figure 1.21b displays their sum vector $\vec{C}=(4+3,2+4)=(7,6)$. Figure 1.21b clearly shows that $C_x = A_x + B_x$, because the whole is equal to the sum of its parts.

In the same way, we can take the difference $\vec{D}=\vec{A}-\vec{B}$, and the Cartesian components of the difference vector are given by

$$\begin{aligned} D_x &= A_x - B_x \\ D_y &= A_y - B_y \\ D_z &= A_z - B_z. \end{aligned} \quad (1.14)$$

**FIGURE 1.21** Vector addition by components. (a) Components of vectors $\vec{A}$ and $\vec{B}$; (b) the components of the resultant vector are the sums of the components of the individual vectors.

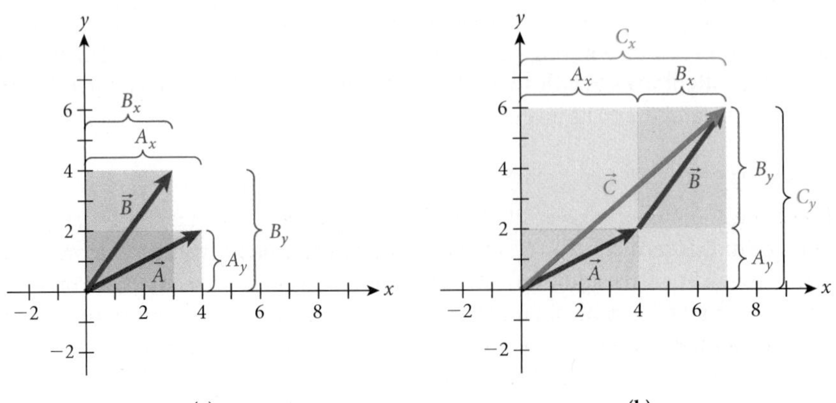

## Multiplication of a Vector with a Scalar

What is $\vec{A}+\vec{A}+\vec{A}$? If your answer to this question is $3\vec{A}$, you already understand multiplying a vector with a scalar. The vector that results from multiplying the vector $\vec{A}$ with the scalar 3 is a vector that points in the same direction as the original vector $\vec{A}$ but is 3 times as long.

Multiplication of a vector with an arbitrary positive scalar—that is, a positive number—results in another vector that points in the same direction but has a magnitude that is the product of the magnitude of the original vector and the value of the scalar. Multiplication of a vector by a negative scalar results in a vector pointing in the opposite direction to the original with a magnitude that is the product of the magnitude of the original vector and the magnitude of the scalar.

Again, the component notation is useful. For the multiplication of a vector $\vec{A}$ with a scalar $s$, we obtain:

$$\vec{E}=s\vec{A}=s(A_x,A_y,A_z)=(sA_x,sA_y,sA_z). \qquad (1.15)$$

In other words, each component of the vector $\vec{A}$ is multiplied by the scalar in order to arrive at the components of the product vector:

$$\begin{aligned} E_x &= sA_x \\ E_y &= sA_y \\ E_z &= sA_z. \end{aligned} \qquad (1.16)$$

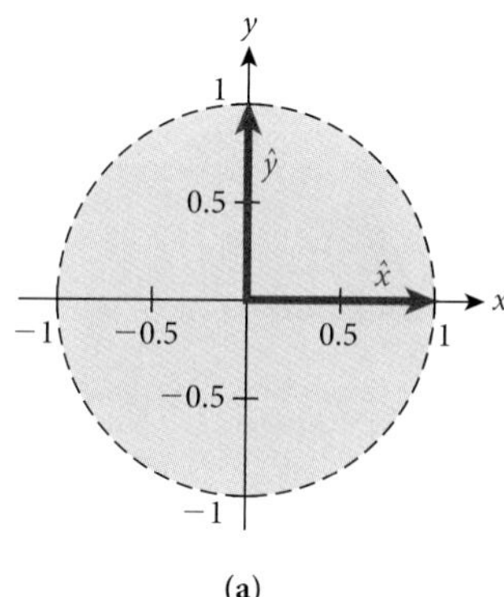


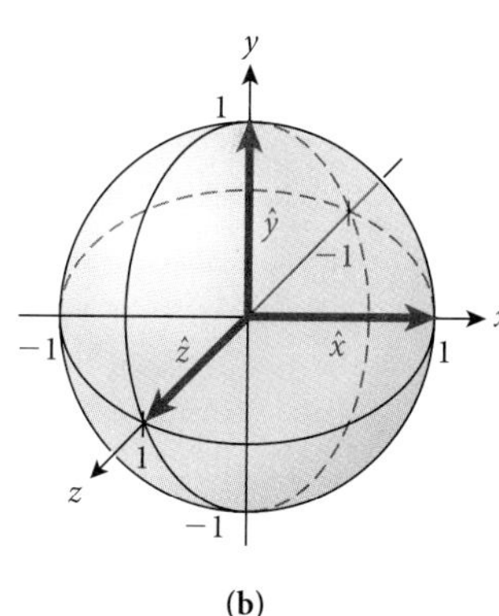


**FIGURE 1.22** Cartesian unit vectors in (a) two and (b) three dimensions.

## Unit Vectors

There is a set of special vectors that make much of the math associated with vectors easier. Called **unit vectors,** they are vectors of magnitude 1 directed along the main coordinate axes of the coordinate system. In two dimensions, these vectors point in the positive $x$-direction and the positive $y$-direction. In three dimensions, a third unit vector points in the positive $z$-direction. In order to distinguish these as unit vectors, we give them the symbols $\hat{x}$, $\hat{y}$, and $\hat{z}$. Their component representation is

$$\begin{aligned} \hat{x} &= (1,0,0) \\ \hat{y} &= (0,1,0) \\ \hat{z} &= (0,0,1). \end{aligned} \qquad (1.17)$$

Figure 1.22a shows the unit vectors in two dimensions, and Figure 1.22b shows the unit vectors in three dimensions.

What is the advantage of unit vectors? We can write any vector as a sum of these unit vectors instead of using the component notation; each unit vector is multiplied by the corresponding Cartesian component of the vector:

$$\begin{aligned} \vec{A} &= (A_x,A_y,A_z) \\ &= (A_x,0,0)+(0,A_y,0)+(0,0,A_z) \\ &= A_x(1,0,0)+A_y(0,1,0)+A_z(0,0,1) \\ &= A_x\hat{x}+A_y\hat{y}+A_z\hat{z}. \end{aligned} \qquad (1.18)$$

In two dimensions, we have

$$\vec{A}=A_x\hat{x}+A_y\hat{y}. \qquad (1.19)$$

This unit vector representation of a general vector will be particularly useful later in the book for multiplying two vectors.

## Vector Length and Direction

If we know the component representation of a vector, how can we find its length (magnitude) and the direction it is pointing in? Let's look at the most important case: a vector in two dimensions. In two dimensions, a vector $\vec{A}$ can be specified uniquely by giving the two Cartesian components, $A_x$ and $A_y$. We can also specify the same vector by giving two other numbers: its length $A$ and its angle $\theta$ with respect to the positive $x$-axis.

Let's take a look at Figure 1.23 to see how we can determine $A$ and $\theta$ from $A_x$ and $A_y$. Figure 1.23a shows the result of equation 1.19 in graphical representation. The vector $\vec{A}$ is

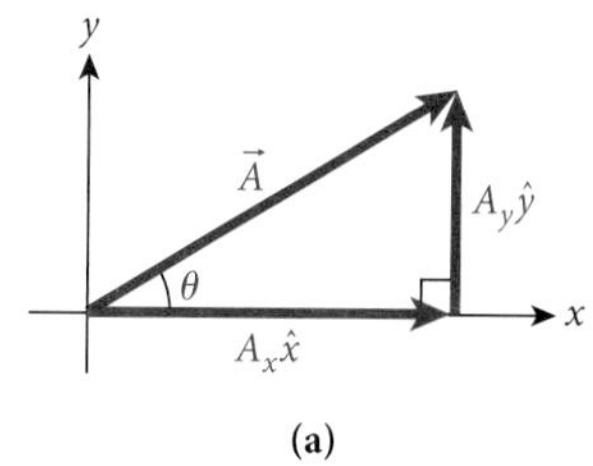


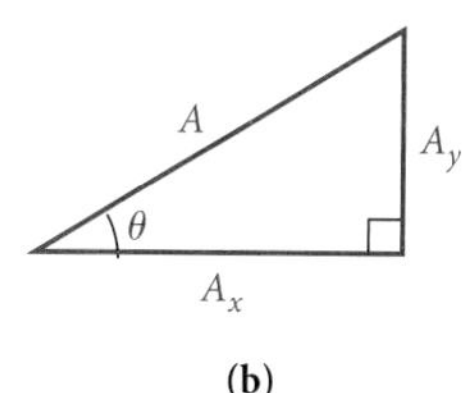


**FIGURE 1.23** Length and direction of a vector. (a) Cartesian components $A_x$ and $A_y$; (b) length $A$ and angle $\theta$.

### 1.4 In-Class Exercise

Into which quadrant do each of the following vectors point?

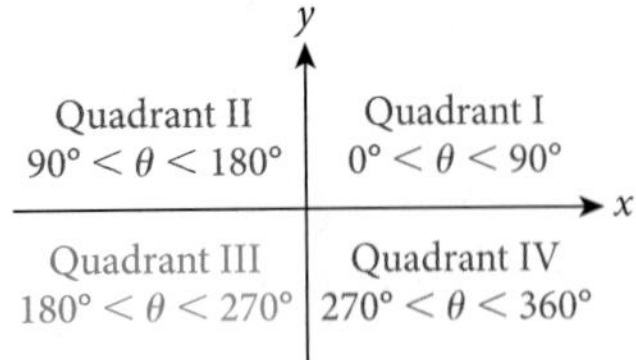


a) $A = (A_x, A_y)$ with $A_x = 1.5$ cm, $A_y = -1.0$ cm

b) a vector of length 2.3 cm and angle of 131°

c) the inverse vector of $B = (0.5$ cm, $1.0$ cm$)$

d) the sum of the unit vectors in the $x$ and $y$ directions

the sum of the vectors $A_x\hat{x}$ and $A_y\hat{y}$. Since the unit vectors $\hat{x}$ and $\hat{y}$ are by definition orthogonal to each other, these vectors form a 90° angle. Thus, the three vectors $\vec{A}$, $A_x\hat{x}$, and $A_y\hat{y}$ form a right triangle with side lengths $A$, $A_x$, and $A_y$, as shown in Figure 1.23b.

Now we can employ basic trigonometry to find $\theta$ and $A$. Using the Pythagorean theorem results in

$$A = \sqrt{A_x^2 + A_y^2}. \tag{1.20}$$

We can find the angle $\theta$ from the definition of the tangent function

$$\theta = \tan^{-1}\frac{A_y}{A_x}. \tag{1.21}$$

In using equation 1.21 you must be careful that $\theta$ is in the correct quadrant. We can also invert equations 1.20 and 1.21 to obtain the Cartesian components of a vector of given length and direction:

$$A_x = A\cos\theta \tag{1.22}$$

$$A_y = A\sin\theta. \tag{1.23}$$

You will encounter these trigonometric relations again and again throughout introductory physics. If you need to refamiliarize yourself with trigonometry, consult the mathematics primer provided in Appendix A.

## WHAT WE HAVE LEARNED | EXAM STUDY GUIDE

- Large and small numbers can be represented using scientific notation, consisting of a mantissa and a power of ten.
- Physical systems are described by the SI system of units. These units are based on reproducible standards and provide convenient methods of scaling and calculation. The base units of the SI system include meter (m), kilogram (kg), second (s), and ampere (A).
- Physical systems have widely varying sizes, masses, and time scales, but the same physical laws govern all of them.
- A number (with a specific number of significant figures) or a set of numbers (such as components of a vector) must be combined with units to describe physical quantities.
- Vectors in three dimensions can be specified by their three Cartesian components, $\vec{A} = (A_x, A_y, A_z)$. Each of these Cartesian components is a number.
- Vectors can be added or subtracted. In Cartesian components, $\vec{C} = \vec{A} + \vec{B} = (A_x, A_y, A_z) + (B_x, B_y, B_z) = (A_x + B_x, A_y + B_y, A_z + B_z)$.
- Multiplication of a vector with a scalar results in another vector in the same or opposite direction but of different magnitude, $\vec{E} = s\vec{A} = s(A_x, A_y, A_z) = (sA_x, sA_y, sA_z)$.
- Unit vectors are vectors of length 1. The unit vectors in Cartesian coordinate systems are denoted by $\hat{x}$, $\hat{y}$, and $\hat{z}$.
- The length and direction of a two-dimensional vector can be determined from its Cartesian components: $A = \sqrt{A_x^2 + A_y^2}$ and $\theta = \tan^{-1}(A_y / A_x)$.
- The Cartesian components of a two-dimensional vector can be calculated from the vector's length and angle with respect to the $x$-axis: $A_x = A\cos\theta$ and $A_y = A\sin\theta$.

## KEY TERMS

scientific notation, p. 9
significant figures, p. 10
SI unit system, p. 11
metrology, p. 14
vector, p. 23
scalar, p. 23
Cartesian coordinate system, p. 24
components, p. 25
resultant, p. 25
unit vectors, p. 27

## PROBLEM-SOLVING PRACTICE

### SOLVED PROBLEM 1.3 Hiking

You are hiking in the Florida Everglades heading southwest from your base camp, for 1.72 km. You reach a river that is too deep to cross; so you make a 90° right turn and hike another 3.12 km to a bridge. How far away are you from your base camp?

**SOLUTION**

**THINK**

If you are hiking, you are moving in a two-dimensional plane: the surface of Earth (because the Everglades are flat and the distance of the hike is small compared with the distance over which the altitude changes significantly due to the curvature of the Earth). Thus, we can use two-dimensional vectors to characterize the various segments of the hike. Making one straight-line hike, then performing a turn, followed by another straight-line hike amounts to a problem of vector addition that is asking for the length of the resultant vector.

**SKETCH**

Figure 1.24 presents a coordinate system in which the $y$-axis points north and the $x$-axis points east, as is conventional. The first portion of the hike, in the southwestern direction, is indicated by the vector $\vec{A}$, and the second portion by the vector $\vec{B}$. The figure also shows the resultant vector, $\vec{C}=\vec{A}+\vec{B}$, for which we want to determine the length.

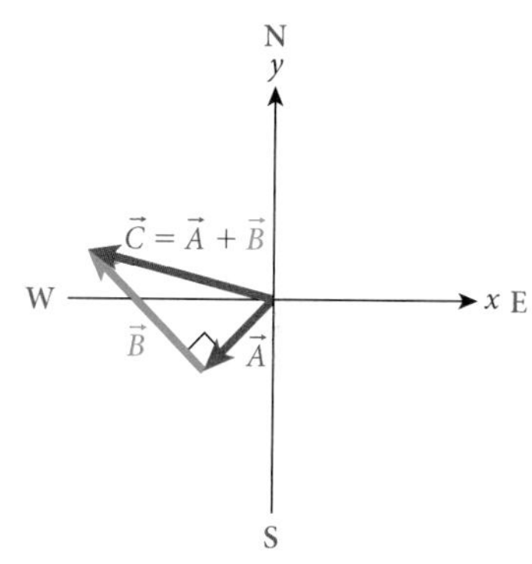


**FIGURE 1.24** Hike with a 90° turn.

**RESEARCH**

If you have drawn the sketch with sufficient accuracy, making the lengths of the vectors in your drawing to be proportional to the lengths of the segments of the hike (as was done in Figure 1.24), then you can measure the length of the vector $\vec{C}$ to determine the distance from your base camp at the end of the second segment of the hike. However, the given distances are specified to three significant digits, so the answer should also have three significant digits. Thus, we cannot rely on the graphical method but must use the component method of vector addition.

In order to calculate the components of the vectors, we need to know their angles relative to the positive $x$-axis. For the vector $\vec{A}$, which points southwest, this angle is $\theta_A$=225°, as shown in Figure 1.25. The vector $\vec{B}$ has an angle of 90° relative to $\vec{A}$, and thus $\theta_B$=135° relative to the positive $x$-axis. To make this point clearer, the starting point of $\vec{B}$ has been moved to the origin of the coordinate system in Figure 1.25. (Remember: We can move vectors around at will. As long as we leave the direction and length of a vector the same, the vector remains unchanged.)

Now we have everything in place to start our calculation. We have the lengths and directions of both vectors, allowing us to calculate their Cartesian components. Then, we will add their components to calculate the components of the vector $\vec{C}$, from which we can calculate the length of this vector.

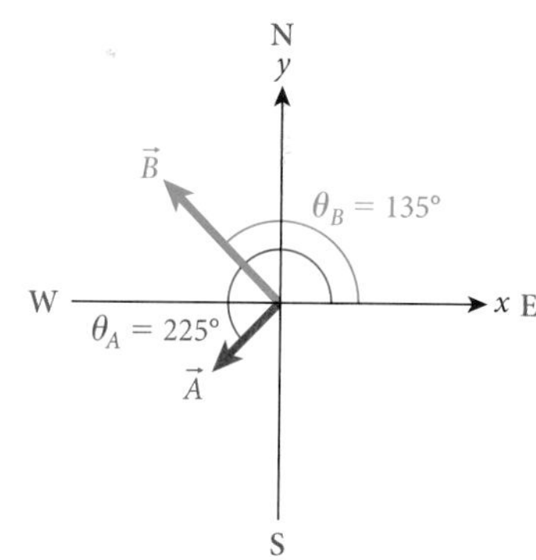


**FIGURE 1.25** Angles of the two hike segments.

**SIMPLIFY**

The components of the vector $\vec{C}$ are:

$$C_x = A_x + B_x = A\cos\theta_A + B\cos\theta_B$$
$$C_y = A_y + B_y = A\sin\theta_A + B\sin\theta_B.$$

Thus, the length of the vector $\vec{C}$ is (compare with equation 1.20)

$$C=\sqrt{C_x^2+C_y^2}=\sqrt{(A_x+B_x)^2+(A_y+B_y)^2}$$
$$=\sqrt{(A\cos\theta_A+B\cos\theta_B)^2+(A\sin\theta_A+B\sin\theta_B)^2}.$$

*Continued—*

**CALCULATE**

Now all that is left is to put in the numbers to obtain the vector length:

$$C=\sqrt{\left((1.72\text{ km})\cos 225°+(3.12\text{ km})\cos 135°\right)^2+\left((1.72\text{ km})\sin 225°+(3.12\text{ km})\sin 135°\right)^2}$$

$$=\sqrt{\left(1.72\cdot(-\sqrt{1/2})+3.12\cdot(-\sqrt{1/2})\right)^2+\left(1.72\cdot(-\sqrt{1/2})+3.12\cdot\sqrt{1/2}\right)^2}\text{ km}.$$

Entering these numbers into a calculator, we obtain:

$$C=3.562695609\text{ km}.$$

**ROUND**

Because the initial distances were given to three significant figures, our final answer should also have (at most) the same precision. Rounding to three significant figures yields our final answer:

$$C=3.56\text{ km}.$$

**DOUBLE-CHECK**

This problem was intended to provide practice with vector concepts. However, if you forget for a moment that the displacements are vectors and note that they form a right triangle, you can immediately calculate the length of side $C$ from the Pythagorean theorem as follows:

$$C=\sqrt{A^2+B^2}=\sqrt{1.72^2+3.12^2}\text{ km}=3.56\text{ km}.$$

Here we also rounded our result to three significant figures, and we see that it agrees with the answer obtained using the longer procedure of vector addition.

## MULTIPLE-CHOICE QUESTIONS

**1.1** Which of the following is the frequency of high C?

a) 376 g b) 483 m/s c) 523 Hz d) 26.5 J

**1.2** If $\vec{A}$ and $\vec{B}$ are vectors and $\vec{B}=-\vec{A}$, which of the following is true?

a) The magnitude of $\vec{B}$ is equal to the negative of the magnitude of $\vec{A}$.

b) $\vec{A}$ and $\vec{B}$ are perpendicular.

c) The direction angle of $\vec{B}$ is equal to the direction angle of $\vec{A}$ plus 180°.

d) $\vec{A}+\vec{B}=2\vec{A}$.

**1.3** Compare three SI units: millimeter, kilogram, and microsecond. Which is the largest?

a) millimeter
b) kilogram
c) microsecond
d) The units are not comparable.

**1.4** What is(are) the difference(s) between 3.0 and 3.0000?

a) 3.0000 could be the result from an intermediate step in a calculation; 3.0 has to result from a final step.

b) 3.0000 represents a quantity that is known more precisely than 3.0.

c) There is no difference.

d) They convey the same information, but 3.0 is preferred for ease of writing.

**1.5** A speed of 7 mm/μs is equal to:

a) 7000 m/s b) 70 m/s c) 7 m/s d) 0.07 m/s

**1.6** A hockey puck, whose diameter is approximately 3 inches, is to be used to determine the value of $\pi$ to three significant figures by carefully measuring its diameter and its circumference. For this calculation to be done properly, the measurements must be made to the nearest ______.

a) hundredth of a mm
b) tenth of a mm
c) mm
d) cm
e) in

**1.7** What is the sum of $5.786\cdot 10^3$ m and $3.19\cdot 10^4$ m?

a) $6.02\cdot 10^{23}$ m
b) $3.77\cdot 10^4$ m
c) $8.976\cdot 10^3$ m
d) $8.98\cdot 10^3$ m

**1.8** What is the number of carbon atoms in 0.5 nanomoles of carbon? One mole contains $6.02\cdot 10^{23}$atoms.

a) $3.2\cdot 10^{14}$ atoms
b) $3.19\cdot 10^{14}$ atoms
c) $3.\cdot 10^{14}$ atoms
d) $3.2\cdot 10^{17}$ atoms
e) $3.19\cdot 10^{17}$ atoms
f) $3.\cdot 10^{17}$ atoms

**1.9** The resultant of the two-dimensional vectors (1.5 m, 0.7 m), (–3.2 m,1.7 m), and (1.2 m,–3.3 m) lies in quadrant ______.

a) I b) II c) III d) IV

**1.10** By how much does the volume of a cylinder change if the radius is halved and the height is doubled?

a) The volume is quartered.
b) The volume is cut in half.
c) There is no change in the volume.
d) The volume doubles.
e) The volume quadruples.

## QUESTIONS

**1.11** In Europe, cars' gas consumption is measured in liters per 100 kilometers. In the United States, the unit used is miles per gallon.

a) How are these units related?

b) How many miles per gallon does your car get if it consumes 12.2 liters per 100 kilometers?

c) What is your car's gas consumption in liters per 100 kilometers if it gets 27.4 miles per gallon?

d) Can you draw a curve plotting miles per gallon versus liters per 100 kilometers ? If yes, draw the curve.

**1.12** If you draw a vector on a sheet of paper, how many components are required to describe it? How many components does a vector in real space have? How many components would a vector have in a four-dimensional world?

**1.13** Since vectors in general have more than one component and thus more than one number is used to describe them, they are obviously more difficult to add and subtract than single numbers. Why then work with vectors at all?

**1.14** If $\vec{A}$ and $\vec{B}$ are vectors specified in magnitude-direction form, and $\vec{C} = \vec{A} + \vec{B}$ is to be found and to be expressed in magnitude-direction form, how is this done? That is, what is the procedure for adding vectors that are given in magnitude-direction form?

**1.15** Suppose you solve a problem and your calculator's display reads 0.0000000036. Why not just write this down? Is there any advantage to using the scientific notation?

**1.16** Since the British system of units is more familiar to most people in the United States, why is the international (SI) system of units used for scientific work in the United States?

**1.17** Is it possible to add three equal-length vectors and obtain a vector sum of zero? If so, sketch the arrangement of the three vectors. If not, explain why not.

**1.18** Is mass a vector quantity? Why or why not?

**1.19** Two flies sit exactly opposite each other on the surface of a spherical balloon. If the balloon's volume doubles, by what factor does the distance between the flies change?

**1.20** What is the ratio of the volume of a cube of side $r$ to that of a sphere of radius $r$? Does your answer depend on the particular value of $r$?

**1.21** Consider a sphere of radius $r$. What is the length of a side of a cube that has the same surface area as the sphere?

**1.22** The mass of the Sun is $2 \cdot 10^{30}$ kg, and the Sun contains more than 99% of all the mass in the solar system. Astronomers estimate there are approximately 100 billion stars in the Milky Way and approximately 100 billion galaxies in the universe. The Sun and other stars are predominantly composed of hydrogen; a hydrogen atom has a mass of approximately $2 \cdot 10^{-27}$ kg.

a) Assuming that the Sun is an average star and the Milky Way is an average galaxy, what is the total mass of the universe?

b) Since the universe consists mainly of hydrogen, can you estimate the total number of atoms in the universe?

**1.23** A futile task is proverbially said to be "like trying to empty the ocean with a teaspoon." Just how futile is such a task? Estimate the number of teaspoonfuls of water in the Earth's oceans.

**1.24** The world's population passed 6.5 billion in 2006. Estimate the amount of land area required if each person were to stand in such a way as to be unable to touch another person. Compare this area to the land area of the United States, 3.5 million square miles, and to the land area of your home state (or country).

**1.25** Advances in the field of nanotechnology have made it possible to construct chains of single metal atoms linked one to the next. Physicists are particularly interested in the ability of such chains to conduct electricity with little resistance. Estimate how many gold atoms would be required to make such a chain long enough to wear as a necklace. How many would be required to make a chain that encircled the Earth? If 1 mole of a substance is equivalent to roughly $6.022 \cdot 10^{23}$ atoms, how many moles of gold are required for each necklace?

**1.26** One of the standard clichés in physics courses is to talk about approximating a cow as a sphere. How large a sphere makes the best approximation to an average dairy cow? That is, estimate the radius of a sphere that has the same mass and density as a dairy cow.

**1.27** Estimate the mass of your head. Assume that its density is that of water, 1000 kg/m$^3$.

**1.28** Estimate the number of hairs on your head.

## PROBLEMS

A blue problem number indicates a worked-out solution is available in the Student Solutions Manual. One • and two •• indicate increasing level of problem difficulty.

### Section 1.2

**1.29** How many significant figures are in each of the following numbers?

a) 4.01 b) 4.010 c) 4 d) 2.00001 e) 0.00001 f) 2.1 – 1.10042 g) $7.01 \cdot 3.1415$

**1.30** Two different forces, acting on the same object, are measured. One force is 2.0031 N and the other force, in the same direction, is 3.12 N. These are the only forces acting on the object. Find the total force on the object *to the correct number of significant figures.*

**1.31** Three quantities, the results of measurements, are to be added. They are 2.0600, 3.163, and 1.12. What is their sum *to the correct number of significant figures*?

**1.32** Given the equation $w = xyz$, and $x = 1.1 \cdot 10^3$, $y = 2.48 \cdot 10^{-2}$, and $z = 6.000$, what is $w$, in scientific notation and with the correct number of significant figures?

**1.33** Write this quantity in scientific notation: one ten-millionth of a centimeter.

**1.34** Write this number in scientific notation: one hundred fifty-three million.

## Section 1.3

**1.35** How many inches are in 30.7484 miles?

**1.36** What metric prefixes correspond to the following powers of 10?

a) $10^3$ b) $10^{-2}$ c) $10^{-3}$

**1.37** How many millimeters in a kilometer?

**1.38** A hectare is a hundred ares and an are is a hundred square meters. How many hectares are there in a square kilometer?

**1.39** The unit of pressure in the SI system is the pascal. What would be the SI name for 1 one-thousandth of a pascal?

**1.40** The masses of four sugar cubes are measured to be 25.3 g, 24.7 g, 26.0 g, and 25.8 g. Express the answers to the following questions in scientific notation, with standard SI units and an appropriate number of significant figures.

a) If the four sugar cubes were crushed and all the sugar collected, what would be the total mass, in kilograms, of the sugar?

b) What is the average mass, in kilograms, of these four sugar cubes?

•**1.41** What is the surface area of a right cylinder of height 20.5 cm and radius 11.9 cm?

## Section 1.4

**1.42** You step on your brand-new digital bathroom scale, and it reads 125.4 pounds. What is your mass in kilograms?

**1.43** The distance from the center of the Moon to the center of the Earth ranges from approximately 356,000 km to 407,000 km. What are these distances in miles? Be certain to round your answers to the appropriate number of significant figures.

**1.44** In Major League baseball, the pitcher delivers his pitches from a distance of 60 feet, 6 inches from home plate. What is the distance in meters?

**1.45** A flea hops in a straight path along a meter stick, starting at 0.7 cm and making successive jumps, which are measured to be 3.2 cm, 6.5 cm, 8.3 cm, 10.0 cm, 11.5 cm, and 15.5 cm. Express the answers to the following questions in scientific notation, with units of meters and an appropriate number of significant figures. What is the total distance covered by the flea in these six hops? What is the average distance covered by the flea in a single hop?

•**1.46** One cubic centimeter of water has a mass of 1 gram. A milliliter is equal to a cubic centimeter. What is the mass, in kilograms, of a liter of water? A metric ton is a thousand kilograms. How many cubic centimeters of water are in a metric ton of water? If a metric ton of water were held in a thin-walled cubical tank, how long (in meters) would each side of the tank be?

•**1.47** The speed limit on a particular stretch of road is 45 miles per hour. Express this speed limit in millifurlongs per microfortnight. A furlong is $\frac{1}{8}$ mile, and a fortnight is a period of 2 weeks. (A microfortnight is in fact used as a unit in a particular type of computing system called the VMS system.)

•**1.48** According to one mnemonic rhyme, "A pint's a pound, the world around." Investigate this statement of equivalence by calculating the weight of a pint of water, assuming that the density of water is 1000 kg/m$^3$ and that the weight of 1 kg of a substance is 2.2 pounds. The volume of 1 fluid ounce is 29.6 mL.

## Section 1.5

**1.49** If the radius of a planet is larger than that of Earth by a factor of 8.7, how much bigger is the surface area of the planet than Earth's?

**1.50** If the radius of a planet is larger than that of Earth by a factor of 5.8, how much bigger is the volume of the planet than Earth's?

**1.51** How many cubic inches are in 1.56 barrels?

**1.52** A car's gasoline tank has the shape of a right rectangular box with a square base whose sides measure 62 cm. Its capacity is 52 L. If the tank has only 1.5 L remaining, how deep is the gasoline in the tank, assuming the car is parked on level ground?

•**1.53** The volume of a sphere is given by the formula $\frac{4}{3}\pi r^3$, where $r$ is the radius of the sphere. The average density of an object is simply the ratio of its mass to its volume. Using the numerical data found in Table 12.1, express the answers to the following questions in scientific notation, with SI units and an appropriate number of significant figures.

a) What is the volume of the Sun?

b) What is the volume of the Earth?

c) What is the average density of the Sun?

d) What is the average density of the Earth?

•**1.54** A tank is in the shape of an inverted cone, having height $h = 2.5$ m and base radius $r = 0.75$ m. If water is poured into the tank at a rate of 15 L/s, how long will it take to fill the tank?

•**1.55** Water flows into a cubical tank at a rate of 15 L/s. If the top surface of the water in the tank is rising by 1.5 cm every second, what is the length of each side of the tank?

••**1.56** The atmosphere has a weight that is, effectively, about 15 pounds for every square inch of Earth's surface. The average density of air at the Earth's surface is about 1.275 kg/m$^3$. If the atmosphere were uniformly dense (it is not—the density varies quite significantly with altitude), how thick would it be?

## Section 1.6

1.57 A position vector has a length of 40 m and is at an angle of 57° above the $x$-axis. Find the vector's components.

**1.58** In the triangle shown in the figure, the side lengths are $a = 6.6$ cm, $b = 13.7$ cm, and $c = 9.2$ cm. What is the value of the angle $\gamma$? (*Hint:* See Appendix A for the law of cosines.)

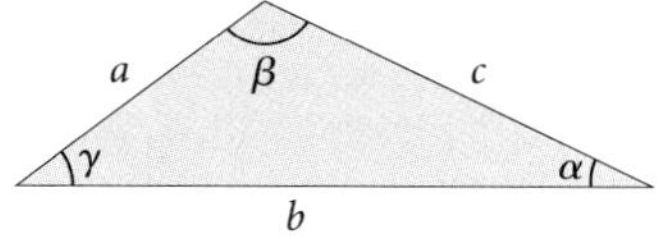


**1.59** Write the vectors $\vec{A}$, $\vec{B}$, and $\vec{C}$ in Cartesian coordinates.

**1.60** Calculate the length and direction of the vectors $\vec{A}$, $\vec{B}$, and $\vec{C}$.

1.61 Add the three vectors $\vec{A}$, $\vec{B}$, and $\vec{C}$ graphically.

**1.62** Determine the difference vector $\vec{E} = \vec{B} - \vec{A}$ graphically.

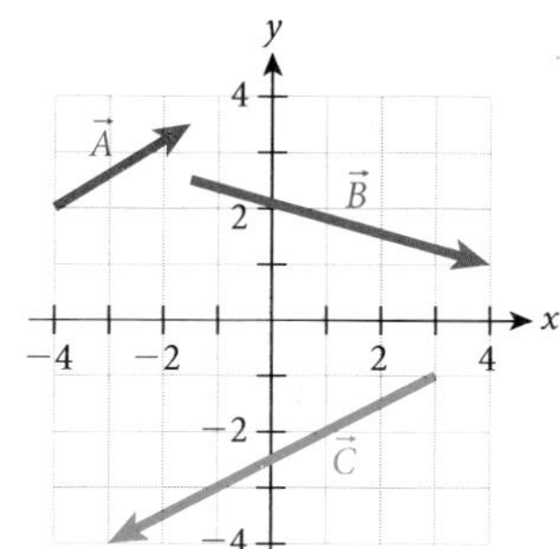


Figure for problems 1.59 through 1.64

**1.63** Add the three vectors $\vec{A}$, $\vec{B}$, and $\vec{C}$ using the component method, and find their sum vector $\vec{D}$.

**1.64** Use the component method to determine the length of the vector $\vec{F} = \vec{C} - \vec{A} - \vec{B}$.

1.65 Find the components of the vectors $\vec{A}$, $\vec{B}$, $\vec{C}$, and $\vec{D}$, where their lengths are given by $A = 75$, $B = 60$, $C = 25$, $D = 90$ and their angles are as shown in the figure. Write the vectors in terms of unit vectors.

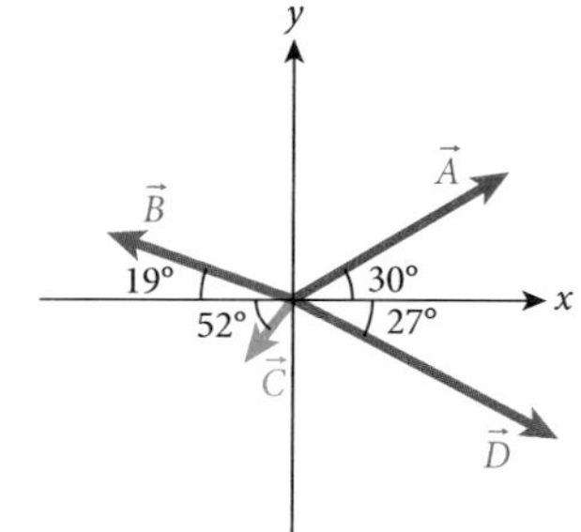


•**1.66** Use the components of the vectors from Problem 1.65 to find

a) the sum $\vec{A} + \vec{B} + \vec{C} + \vec{D}$ in terms of its components

b) the magnitude and direction of the sum $\vec{A} - \vec{B} + \vec{D}$

•**1.67** The Bonneville Salt Flats, located in Utah near the border with Nevada, not far from interstate I-80, cover an area of over 30,000 acres. A race car driver on the Flats first heads north for 4.47 km, then makes a sharp turn and heads southwest for 2.49 km, then makes another turn and heads east for 3.59 km. How far is she from where she started?

•**1.68** A map in a pirate's log gives directions to the location of a buried treasure. The starting location is an old oak tree. According to the map, the treasure's location is found by proceeding 20 paces north from the oak tree and then 30 paces northwest. At this location, an iron pin is sunk in the ground. From the iron pin, walk 10 paces south and dig. How far (in paces) from the oak tree is the spot at which digging occurs?

••1.69 The next page of the pirate's log contains a set of directions that differ from those on the map in Problem 1.68. These say the treasure's location is found by proceeding 20 paces north from the old oak tree and then 30 paces northwest. After finding the iron pin, one should "walk 12 paces nor'ward and dig downward 3 paces to the treasure box." What is the vector that points from the base of the old oak tree to the treasure box? What is the length of this vector?

••**1.70** The Earth's orbit has a radius of $1.5 \cdot 10^{11}$ m, and that of Venus has a radius of $1.1 \cdot 10^{11}$ m. Consider these two orbits to be perfect circles (though in reality they are ellipses with slight eccentricity). Write the direction and length of a vector from Earth to Venus (take the direction from Earth to Sun to be 0°) when Venus is at the maximum angular separation in the sky relative to the Sun.

••**1.71** A friend walks away from you a distance of 550 m, and then turns (as if on a dime) an unknown angle, and walks an additional 178 m in the new direction. You use a laser range-finder to find out that his final distance from you is 432 m. What is the angle between his initial departure direction and the direction to his final location? Through what angle did he turn? (There are two possibilities.)

## Additional Problems

**1.72** The radius of Earth is 6378. km. What is its circumference to three significant figures?

1.73 Estimate the product of 4,308,229 and 44 to one significant figure (show your work and do not use a calculator), and express the result in standard scientific notation.

**1.74** Find the vector $\vec{C}$ that satisfies the equation $3\hat{x} + 6\hat{y} - 10\hat{z} + \vec{C} = -7\hat{x} + 14\hat{y}$.

**1.75** Sketch the vectors with the components $\vec{A} = (A_x, A_y) = (30.0 \text{ m}, -50.0 \text{ m})$ and $\vec{B} = (B_x, B_y) = (-30.0 \text{ m}, 50.0 \text{ m})$, and find the magnitudes of these vectors.

**1.76** What angle does $\vec{A} = (A_x, A_y) = (30.0 \text{ m}, -50.0 \text{ m})$ make with the positive $x$-axis? What angle does it make with the negative $y$-axis?

1.77 Sketch the vectors with the components $\vec{A} = (A_x, A_y) = (-30.0 \text{ m}, -50.0 \text{ m})$ and $\vec{B} = (B_x, B_y) = (30.0 \text{ m}, 50.0 \text{ m})$, and find the magnitudes of these vectors.

**1.78** What angle does $\vec{B} = (B_x, B_y) = (30.0 \text{ m}, 50.0 \text{ m})$ make with the positive $x$-axis? What angle does it make with the positive $y$-axis?

**1.79** A position vector has components $x = 34.6$ m and $y = -53.5$ m. Find the vector's length and angle with the $x$-axis.

**1.80** For the planet Mars, calculate the distance around the Equator, the surface area, and the volume. The radius of Mars is $3.39 \cdot 10^6$ m.

1.81 Find the magnitude and direction of each of the following vectors, which are given in terms of their $x$- and $y$-components: $\vec{A} = (23.0, 59.0)$, and $\vec{B} = (90.0, -150.0)$.

**1.82** Find the magnitude and direction of $-\vec{A}+\vec{B}$, where $\vec{A}=(23.0,59.0)$, $\vec{B}=(90.0,-150.0)$.

**1.83** Find the magnitude and direction of $-5\vec{A}+\vec{B}$, where $\vec{A}=(23.0,59.0)$, $\vec{B}=(90.0,-150.0)$.

**1.84** Find the magnitude and direction of $-7\vec{B}+3\vec{A}$, where $\vec{A}=(23.0,59.0)$, $\vec{B}=(90.0,-150.0)$.

•**1.85** Find the magnitude and direction of (a) $9\vec{B}-3\vec{A}$ and (b) $-5\vec{A}+8\vec{B}$, where $\vec{A}=(23.0,59.0)$, $\vec{B}=(90.0,-150.0)$.

•**1.86** Express the vectors $\vec{A}=(A_x,A_y)=(-30.0\text{ m},-50.0\text{ m})$ and $\vec{B}=(B_x,B_y)=(30.0\text{ m},50.0\text{ m})$ by giving their magnitude and direction as measured from the positive $x$-axis.

•**1.87** The force $F$ a spring exerts on you is directly proportional to the distance $x$ you stretch it beyond its resting length. Suppose that when you stretch a spring 8 cm, it exerts a 200 N force on you. How much force will it exert on you if you stretch it 40 cm?

•**1.88** The distance a freely falling object drops, starting from rest, is proportional to the square of the time it has been falling. By what factor will the distance fallen change if the time of falling is three times as long?

•**1.89** A pilot decides to take his small plane for a Sunday afternoon excursion. He first flies north for 155.3 miles, then makes a 90° turn to his right and flies on a straight line for 62.5 miles, then makes another 90° turn to his right and flies 47.5 miles on a straight line.

a) How far away from his home airport is he at this point?

b) In which direction does he need to fly from this point on to make it home in a straight line?

c) What was the farthest distance he was away from the home airport during the trip?

•**1.90** As the photo shows, during a total eclipse, the Sun and the Moon appear to the observer to be almost exactly the same size. The radii of the Sun and Moon are $r_S=6.96\cdot10^8$ m and $r_M=1.74\cdot10^6$ m, respectively. The distance between the Earth and the Moon is $d_{EM}=3.84\cdot10^8$ m.

Total solar eclipse.

a) Determine the distance from the Earth to the Sun at the moment of the eclipse.

b) In part (a), the implicit assumption is that the distance from the observer to the Moon's center is equal to the distance between the centers of the Earth and the Moon. By how much is this assumption incorrect, if the observer of the eclipse is on the Equator at noon? *Hint:* Express this quantitatively, by calculating the relative error as a ratio: (assumed observer-to-Moon distance—actual observer-to-Moon distance)/(actual observer-to-Moon distance).

c) Use the corrected observer-to-Moon distance to determine a corrected distance from Earth to the Sun.

•**1.91** A hiker travels 1.5 km north and turns to a heading of 20° north of west, traveling another 1.5 km along that heading. Subsequently, he then turns north again and travels another 1.5 km. How far is he from his original point of departure, and what is the heading relative to that initial point?

•**1.92** Assuming that 1 mole ($6.02\cdot10^{23}$ molecules) of an ideal gas has a volume of 22.4 L at standard temperature and pressure (STP) and that nitrogen, which makes up roughly 80% of the air we breathe, is an ideal gas, how many nitrogen molecules are there in an average 0.5 L breath at STP?

•**1.93** On August 27, 2003, Mars approached as close to Earth as it will for over 50,000 years. If its angular size (the planet's radius, measured by the angle the radius subtends) on that day was measured by an astronomer to be 24.9 seconds of arc, and its radius is known to be 6784 km, how close was the approach distance? Be sure to use an appropriate number of significant figures in your answer.

•**1.94** A football field's length is exactly 100 yards, and its width is $53\frac{1}{3}$ yards. A quarterback stands at the exact center of the field and throws a pass to a receiver standing at one corner of the field. Let the origin of coordinates be at the center of the football field and the $x$-axis point along the longer side of the field, with the $y$-direction parallel to the shorter side of the field.

a) Write the direction and length of a vector pointing from the quarterback to the receiver.

b) Consider the other three possibilities for the location of the receiver at corners of the field. Repeat part (a) for each.

•**1.95** The circumference of the Cornell Electron Storage Ring is 768.4 m. Express the diameter in inches, to the proper number of significant figures.

••**1.96** Roughly 4% of what you exhale is carbon dioxide. Assume that 22.4 L is the volume of 1 mole ($6.02\cdot10^{23}$ molecules) of carbon dioxide and that you exhale 0.5 L per breath.

a) Estimate how many carbon dioxide molecules you breathe out each day.

b) If each mole of carbon dioxide has a mass of 44 g, how many kilograms of carbon dioxide do you exhale in a year?

••**1.97** The Earth's orbit has a radius of $1.5\cdot10^{11}$ m, and that of Mercury has a radius of $4.6\cdot10^{10}$ m. Consider these orbits to be perfect circles (though in reality they are ellipses with slight eccentricity). Write down the direction and length of a vector from Earth to Mercury (take the direction from Earth to Sun to be 0°) when Mercury is at the maximum angular separation in the sky relative to the Sun.

# 2 Motion in a Straight Line

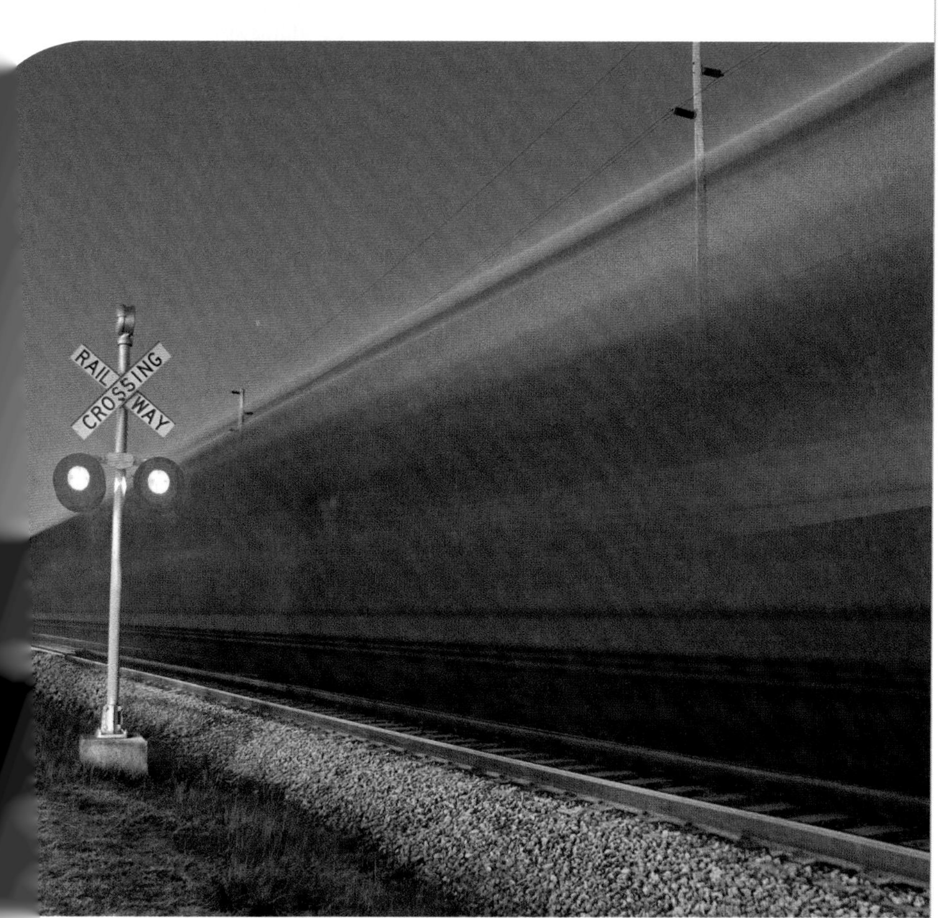


**FIGURE 2.1** A fast-moving train passes a railroad crossing.

## WHAT WE WILL LEARN

- We will learn to describe the motion of an object traveling in a straight line or in one dimension.
- We will learn to define position, displacement, and distance.
- We will learn to describe motion of an object in a straight line with constant acceleration.
- An object can be in free fall in one dimension where it undergoes constant acceleration due to gravity.
- We will define the concepts of instantaneous velocity and average velocity.
- We will define the concepts of instantaneous acceleration and average acceleration.
- We will learn to calculate the position, velocity, and acceleration of an object moving in a straight line.

You can see that the train in Figure 2.1 is moving very fast by noticing its image is blurred compared with the stationary crossing signal and telephone pole. But can you tell if the train is speeding up, slowing down, or zipping by at constant speed? A photograph can convey an object's speed because the object moves during the exposure time, but a photograph cannot show a change in speed, or acceleration. Yet acceleration is extremely important in physics, at least as important as speed itself.

In this chapter, we look at the terms used in physics to describe an object's motion: displacement, velocity, and acceleration. We examine motion along a straight line (one-dimensional motion) in this chapter and motion on a curved path (motion in a plane, or two-dimensional motion) in the next chapter. One of the greatest advantages of physics is that its laws are universal, so the same general terms and ideas apply to a wide range of situations. Thus, we can use the same equations to describe the flight of a baseball and the lift-off of a rocket into space from Earth to Mars. In this chapter, we will use some of the problem-solving techniques discussed in Chapter 1 along with some new ones.

As you continue in this course, you will see that almost everything moves relative to other objects on some scale or other, whether it is a comet plunging through space at several kilometers per second or the atoms in a seemingly stationary object vibrating millions of times per second. The terms we introduce in this chapter will be part of your study for the rest of the course and afterward.

## 2.1 Introduction to Kinematics

The study of physics is divided into several large parts, one of which is mechanics. **Mechanics,** or the study of motion and its causes, is usually subdivided. In this chapter and the next, we examine the kinematics aspect of mechanics. **Kinematics** is the study of the motion of objects. These objects may be, for example, cars, baseballs, people, planets, or atoms. For now, we will set aside the question of what causes this motion. We will return to that question when we study forces.

We will also not consider rotation in this chapter, but concentrate on only translational motion (motion without rotation). Furthermore, we will neglect all internal structure of a moving object and consider it to be a point particle, or pointlike object. That is, to determine the equations of motion for an object, we imagine it to be located at a single point in space at each instant of time. What point of an object should we choose to represent its location? Initially, we will simply use the geometric center, the middle. (Chapter 8, on systems of particles and extended objects, will give a more precise definition for the **point location of an object** called the *center of mass*.)

## 2.2 Position Vector, Displacement Vector, and Distance

The simplest motion we can investigate is that of an object moving in a straight line. Examples of this motion include a person running the 100-m dash, a car driving on a straight segment of road, and a stone falling straight down off a cliff. In later chapters, we will consider motion in two or more dimensions and will see that the same concepts that we derive for one-dimensional motion still apply.

If an object is located on a particular point on a line, we can denote this point with its **position vector,** as described in Section 1.6. Throughout this book, we use the symbol $\vec{r}$ to denote the position vector. Since we are working with motion in only one dimension in this chapter, the position vector has only one component. If the motion is in the horizontal direction, this one component is the $x$-component. (For motion in the vertical direction, we will use the $y$-component; see Section 2.7.) One number, the $x$-coordinate or $x$-component of the position vector (with a corresponding unit), uniquely specifies the position vector in one-dimensional motion. Some valid ways of writing a position are $x = 4.3$ m, $x = 7\frac{3}{8}$ inches, and $x = -2.04$ km; it is understood that these specifications refer to the $x$-component of the position vector. Note that a position vector's $x$-component can have a positive or a negative value, depending on the location of the point and the axis direction that we choose to be positive. The value of the $x$-component also depends on where we define the origin of the coordinate system—the zero of the straight line.

The position of an object can change as a function of time, $t$; that is, the object can move. We can therefore formally write the position vector in function notation: $\vec{r} = \vec{r}(t)$. In one dimension, this means that the $x$-component of the vector is a function of time, $x = x(t)$. If we want to specify the position at some specific time $t_1$, we use the notation $x_1 \equiv x(t_1)$.

## Position Graphs

Before we go any further, let's graph an object's position as a function of time. Figure 2.2a illustrates the principle involved by showing several frames of a video of a car driving down a road. The video frames were taken at time intervals of $\frac{1}{3}$ second.

We are free to choose the origins of our time measurements and of our coordinate system. In this case, we choose the time of the second frame to be $t = \frac{1}{3}$ s and the position of the center of the car in the second frame as $x = 0$. We can now draw our coordinate axes and graph over the frames (Figure 2.2b). The position of the car as a function of time lies on a straight line. Again, keep in mind that we are representing the car by a single point.

When drawing graphs, it is customary to plot the independent variable—in this case, the time $t$—on the horizontal axis and to plot $x$, which is called the dependent variable because its value depends on the value of $t$, on the vertical axis. Figure 2.3 is a graph of the car's position as a function of time drawn in this customary way. (Note that if Figure 2.2b were rotated 90° counterclockwise and the pictures of the car removed, the two graphs would be the same.)

(a)

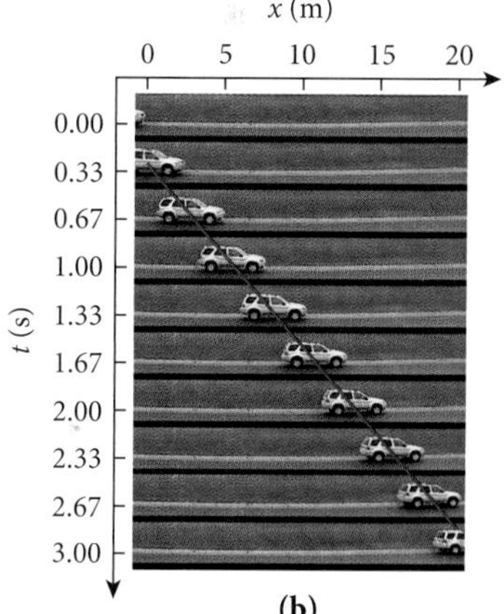

(b)

**FIGURE 2.2** (a) Series of video frames of a moving car, taken every $\frac{1}{3}$ second; (b) same series, but with a coordinate system and a red line connecting the centers of the car in each frame.

## Displacement

Now that we have specified the position vector, let's go one step further and define displacement. **Displacement** is simply the difference between the final position vector, $\vec{r}_2 \equiv \vec{r}(t_2)$, at the end of a motion and the initial position vector, $\vec{r}_1 \equiv \vec{r}(t_1)$. We write the displacement vector as

$$\Delta\vec{r} = \vec{r}_2 - \vec{r}_1. \tag{2.1}$$

We use the notation $\Delta\vec{r}$ for the displacement vector to indicate that it is a difference between two position vectors. Note that the displacement vector is independent of the location of the origin of the coordinate system. Why? Any shift of the coordinate system will add to the position vector $\vec{r}_2$ the same amount that it adds to the position vector $\vec{r}_1$; thus the difference between the position vectors, or $\Delta\vec{r}$, will not change.

Just like the position vector, the displacement vector in one dimension has only an $x$-component, which is the difference between the $x$-components of the final and initial position vectors:

$$\Delta x = x_2 - x_1. \tag{2.2}$$

Also just like position vectors, displacement vectors can be positive or negative. In particular, the displacement vector $\Delta\vec{r}_{ba}$ for going from point $a$ to point $b$ is exactly the negative of $\Delta\vec{r}_{ab}$ going from point $b$ to point $a$:

$$\Delta\vec{r}_{ba} = \vec{r}_b - \vec{r}_a = -(\vec{r}_a - \vec{r}_b) = -\Delta\vec{r}_{ab}. \tag{2.3}$$

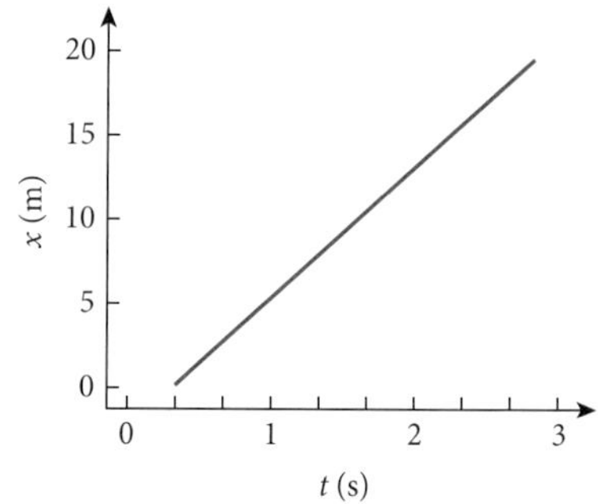


**FIGURE 2.3** Same graph as Figure 2.2b, but rotated so that the time axis is horizontal, and without the pictures of the car.

And it is probably obvious to you at this point that this relationship also holds for the $x$-component of the displacement vector, $\Delta x_{ba} = x_b - x_a = -(x_a - x_b) = -\Delta x_{ab}$.

## Distance

The **distance,** $\ell$, that a moving object travels is the absolute value of the displacement vector:

$$\ell = \left|\Delta \vec{r}\right|. \tag{2.4}$$

For one-dimensional motion, this distance is also the absolute value of the $x$-component of the displacement vector, $\ell = \left|\Delta x\right|$. (For multidimensional motion, we calculate the length of the displacement vector as shown in Chapter 1). The distance is always greater than or equal to zero and is measured in the same units as position and displacement. However, distance is a scalar quantity, not a vector. If the displacement is not in a straight line or if it is not all in the same direction, the displacement must be broken up into segments that are approximately straight and unidirectional and then the distances for the various segments are added to get the total distance. The following solved problem illustrates the difference between distance and displacement.

### SOLVED PROBLEM 2.1 Trip Segments

The distance between Des Moines and Iowa City is 170.5 km (106.0 miles) along Interstate 80, and as you can see from the map (Figure 2.4), the route is a straight line to a good approximation. Approximately halfway between the two cities, where I80 crosses highway US63, is the city of Malcom, 89.9 km (54.0 miles) from Des Moines.

**FIGURE 2.4** Route I80 between Des Moines and Iowa City.

#### PROBLEM

If we drive from Malcom to Des Moines and then go to Iowa City, what are the total distance and total displacement for this trip?

#### SOLUTION

**THINK**

Distance and displacement are not identical. If the trip consisted of one segment in one direction, the distance would just be the absolute value of the displacement, according to equation 2.4. However, this trip is composed of segments with a direction change, so we need to be careful. We'll treat each segment individually and then add up the segments in the end.

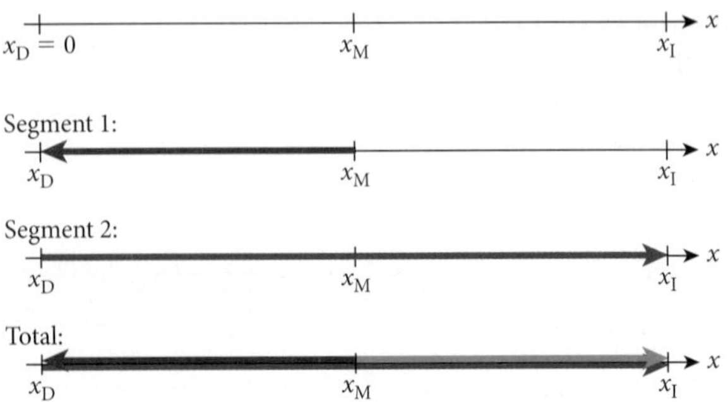


**FIGURE 2.5** Coordinate system and trip segments for the Malcom to Des Moines to Iowa City trip.

**SKETCH**

Because I80 is almost a straight line, it is sufficient to draw a straight horizontal line and make this our coordinate axis. We enter the positions of the three cities as $x_I$ (Iowa City), $x_M$ (Malcom), and $x_D$ (Des Moines). We always have the freedom to define the origin of our coordinate system, so we elect to put it at Des Moines, thus setting $x_D = 0$. As is conventional, we define the positive direction to the right, in the eastward direction. See Figure 2.5.

We also draw arrows for the displacements of the two segments of the trip. We represent segment 1 from Malcom to Des Moines by a red arrow, and segment 2 from Des Moines to Iowa City by a blue arrow. Finally, we draw a diagram for the total trip as the sum of the two trips.

## RESEARCH

With our assignment of $x_D = 0$, Des Moines is the origin of the coordinate system. According to the information given us, Malcom is then at $x_M = +89.9$ km and Iowa City is at $x_I = +170.5$ km. Note that we write a plus sign in front of the numbers for $x_M$ and $x_I$ to remind us that these are components of position vectors and can have positive or negative values.

For the first segment, the *displacement* is given by

$$\Delta x_1 = x_D - x_M.$$

Thus the distance driven for this segment is

$$\ell_1 = |\Delta x_1| = |x_D - x_M|.$$

In the same way, the displacement and distance for the second segment are

$$\Delta x_2 = x_I - x_D$$
$$\ell_2 = |\Delta x_2| = |x_I - x_D|.$$

For the sum of the two segments, the total trip, we use simple addition to find the displacement,

$$\Delta x_{\text{total}} = \Delta x_1 + \Delta x_2,$$

and the total distance,

$$\ell_{\text{total}} = \ell_1 + \ell_2.$$

## SIMPLIFY

We can simplify the equation for the total displacement a little bit by inserting the expressions for the displacements for the two segments:

$$\begin{aligned}\Delta x_{\text{total}} &= \Delta x_1 + \Delta x_2 \\ &= (x_D - x_M) + (x_I - x_D) \\ &= x_I - x_M.\end{aligned}$$

This is an interesting result—for the total displacement of the entire trip, it does not matter at all that we went to Des Moines. All that matters is where the trip started and where it ended. The total displacement is a result of a one-dimensional vector addition, as indicated in the bottom part of Figure 2.5 by the green arrow.

## CALCULATE

Now we can insert the numbers for the positions of the three cities in our coordinate system. We then obtain for the net displacement in our trip

$$\Delta x_{\text{total}} = x_I - x_M = (+170.5 \text{ km}) - (+89.9 \text{ km}) = +80.6 \text{ km}.$$

For the total distance driven, we get

$$\ell_{\text{total}} = |89.9 \text{ km}| + |170.5 \text{ km}| = 260.4 \text{ km}.$$

(Remember, the distance between Des Moines and Malcom, or $\Delta x_1$, and that between Des Moines and Iowa City, or $\Delta x_2$, were given in the problem; so we do not have to calculate them again from the differences in the position vectors of the cities.)

## ROUND

The numbers for the distances were initially given to a tenth of a kilometer. Since our entire calculation only amounted to adding or subtracting these numbers, it is not surprising that we end up with numbers that are also accurate to a tenth of a kilometer. No further rounding is needed.

*Continued—*

### 2.1 Self-Test Opportunity

Suppose we had chosen to put the origin of the coordinate system in Solved Problem 2.1 at Malcom instead of Des Moines. Would the final result of our calculation change? If yes, how? If no, why not?

**DOUBLE-CHECK**

As is customary, we first make sure that the units of our answer came out properly. Since we are looking for quantities with the dimension of length, it is comforting that our answers have the units of kilometers. At first sight, it may be surprising that the net displacement for the trip is only 80.6 km, much smaller than the total distance traveled. This is a good time to recall that the relationship between the absolute value of the displacement and the distance (equation 2.4) is valid only if the moving object does *not* change direction (but it did in this example).

This discrepancy is even more apparent for a round trip. In that case, the total distance driven is twice the distance between the two cities, but the total displacement is zero, because the starting point and the end point of the trip are identical.

This result is a general one: If the initial and final positions are the same, the total displacement is 0. As straightforward as this seems for the trip example, it is a potential pitfall in many exam questions. You need to remember that displacement is a vector, whereas distance is a positive scalar.

## 2.3 Velocity Vector, Average Velocity, and Speed

Just as distance (a scalar) and displacement (a vector) mean different things in physics, their rates of change with time are also different. Although the words "speed" and "velocity" are often used interchangeably in everyday speech, in physics "speed" refers to a scalar and "velocity" to a vector.

We define $v_x$, the $x$-component of the velocity vector, as the change in position (i.e., the displacement component) in a given time interval divided by that time interval, $\Delta x/\Delta t$. Velocity can change from moment to moment. The velocity calculated by taking the ratio of displacement per time interval is the average of the velocity over this time interval, or the $x$-component of the **average velocity,** $\bar{v}_x$:

$$\bar{v}_x = \frac{\Delta x}{\Delta t}. \qquad (2.5)$$

***Notation:*** A bar above a symbol is the notation for averaging over a finite time interval.

In calculus, a time derivative is obtained by taking a limit as the time interval approaches zero. We use the same concept here to define the **instantaneous velocity,** usually referred to simply as the **velocity,** as the time derivative of the displacement. For the $x$-component of the velocity vector, this implies

$$v_x = \lim_{\Delta t \to 0} \bar{v}_x = \lim_{\Delta t \to 0} \frac{\Delta x}{\Delta t} \equiv \frac{dx}{dt}. \qquad (2.6)$$

We can now introduce the velocity vector, $\vec{v}$, as the vector for which each component is the time derivative of the corresponding component of the position vector,

$$\vec{v} = \frac{d\vec{r}}{dt}, \qquad (2.7)$$

with the understanding that the derivative operation applies to each of the components of the vector. In the one-dimensional case, this velocity vector $\vec{v}$ has only an $x$-component, $v_x$, and the velocity is equivalent to a single velocity component in the spatial $x$-direction.

Figure 2.6 presents three graphs of the position of an object with respect to time. Figure 2.6a shows that we can calculate the average velocity of the object by finding the change in position of the object between two points and dividing by the time it takes to go from $x_1$ to $x_2$. That is, the average velocity is given by the displacement, $\Delta x_1$, divided by the time interval, $\Delta t_1$, or $\bar{v}_1 = \Delta x_1/\Delta t_1$. In Figure 2.6b, the average velocity, $\bar{v}_2 = \Delta x_2/\Delta t_2$, is determined over a smaller time interval, $\Delta t_2$. In Figure 2.6c, the instantaneous velocity, $v(t_3) = dx/dt|_{t=t_3}$, is represented by the slope of the blue line tangent to the red curve at $t = t_3$.

Velocity is a vector, pointing in the same direction as the vector of the infinitesimal displacement, $dx$. Because the position $x(t)$ and the displacement $\Delta x(t)$ are functions of

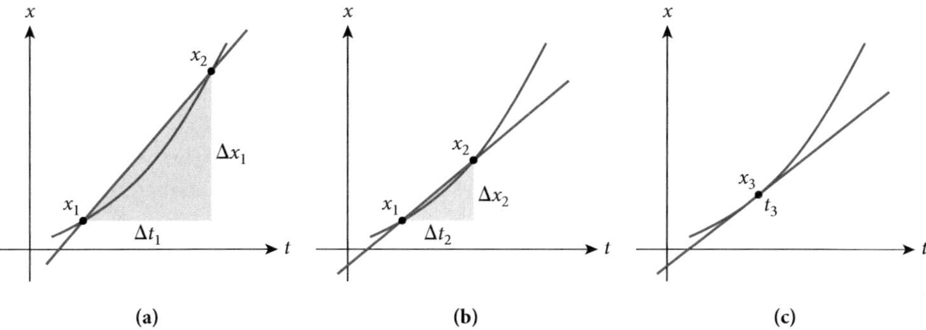


**FIGURE 2.6** Instantaneous velocity as the limit of the ratio of displacement to time interval: (a) an average velocity over a large time interval; (b) an average velocity over a smaller time interval; and (c) the instantaneous velocity at a specific time, $t_3$.

time, so is the velocity. Because the velocity vector is defined as the time derivative of the displacement vector, all the rules of differentiation introduced in calculus hold. If you need a refresher, consult Appendix A.

## EXAMPLE 2.1 Time Dependence of Velocity

### PROBLEM

During the time interval from 0.0 to 10.0 s, the position vector of a car on a road is given by $x(t) = a + bt + ct^2$, with $a = 17.2$ m, $b = -10.1$ m/s, and $c = 1.10$ m/s$^2$. What is the car's velocity as a function of time? What is the car's average velocity during this interval?

### SOLUTION

According to the definition of velocity in equation 2.6, we simply take the time derivative of the position vector function to arrive at our solution:

$$v_x = \frac{dx}{dt} = \frac{d}{dt}(a + bt + ct^2) = b + 2ct = -10.1 \text{ m/s} + 2 \cdot (1.10 \text{ m/s}^2)t.$$

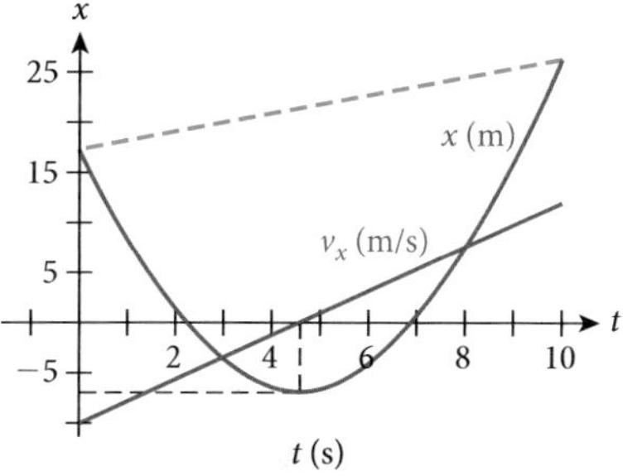


**FIGURE 2.7** Graph of the position $x$ and velocity $v_x$ as a function of the time $t$. The slope of the dashed line represents the average velocity for the time interval from 0 to 10 s.

It is instructive to graph this solution. In Figure 2.7, the position as a function of time is shown in blue, and the velocity as a function of time is shown in red. Initially, the velocity has a value of −10.1 m/s, and at $t = 10$ s, the velocity has a value of +11.9 m/s.

Note that the velocity is initially negative, is zero at 4.59 s (indicated by the vertical dashed line in Figure 2.7), and then is positive after 4.59 s. At $t = 4.59$ s, the position graph $x(t)$ shows an extremum (a minimum in this case), just as expected from calculus, since

$$\frac{dx}{dt} = b + 2ct_0 = 0 \Rightarrow t_0 = -\frac{b}{2c} = -\frac{-10.1 \text{ m/s}}{2.20 \text{ m/s}^2} = 4.59 \text{ s}.$$

From the definition of average velocity, we know that to determine the average velocity during a time interval, we need to subtract the position at the beginning of the interval from the position at the end of the interval. By inserting $t = 0$ and $t = 10$ s into the equation for the position vector as a function of time, we obtain $x(t = 0) = 17.2$ m and $x(t = 10 \text{ s}) = 26.2$ m. Therefore,

$$\Delta x = x(t = 10) - x(t = 0) = 26.2 \text{ m} - 17.2 \text{ m} = 9.0 \text{ m}.$$

We then obtain for the average velocity over this time interval:

$$\bar{v}_x = \frac{\Delta x}{\Delta t} = \frac{9.0 \text{ m}}{10 \text{ s}} = 0.90 \text{ m/s}.$$

The slope of the green dashed line in Figure 2.7 is the average velocity over this time interval.

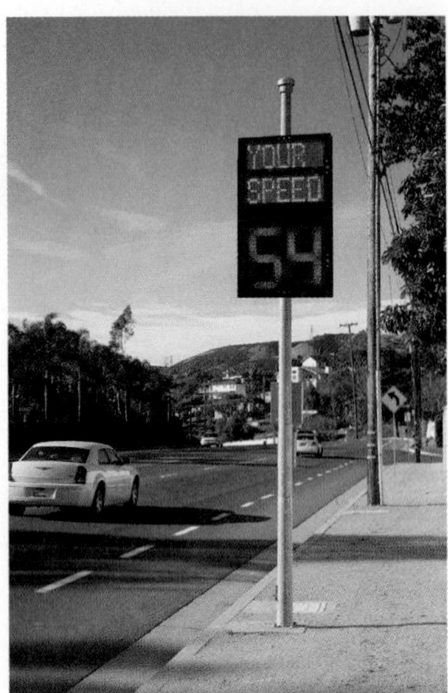

FIGURE 2.8 Measuring the speeds of passing cars.

## Speed

**Speed** is the absolute value of the velocity vector. For a moving object, speed is always positive. "Speed" and "velocity" are used interchangeably in everyday contexts, but in physical terms they are very different. Velocity is a vector, which has a direction. For one-dimensional motion, the velocity vector can point either in the positive or negative direction; in other words, its component can have either sign. Speed is the absolute magnitude of the velocity vector and is thus a scalar quantity:

$$\text{speed} \equiv v = |\vec{v}| = |v_x|. \tag{2.8}$$

The last part of this equation makes use of the fact that the velocity vector has only an $x$-component for one-dimensional motion.

In everyday experience, we recognize that speed can never be negative: Speed limits are always posted as positive numbers, and the radar of passing cars also always displays positive numbers (Figure 2.8).

Earlier, distance was defined as the absolute value of displacement for each straight-line segment *in which the movement does not reverse direction* (see the discussion following equation 2.4). The average speed when a distance $\ell$ is traveled during a time interval $\Delta t$ is

$$\text{average speed} \equiv \bar{v} = \frac{\ell}{\Delta t}. \tag{2.9}$$

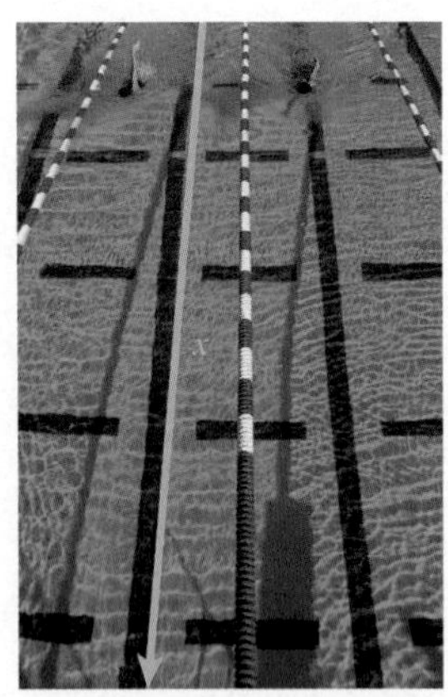
FIGURE 2.9 Choosing an $x$-axis in a swimming pool.

### EXAMPLE 2.2 Speed and Velocity

Suppose a swimmer completes the first 50 m of the 100-m freestyle in 38.2 s. Once she reaches the far side of the 50-m-long pool, she turns around and swims back to the start in 42.5 s.

**PROBLEM**

What are the swimmer's average velocity and average speed for (a) the leg from the start to the far side of the pool, (b) the return leg, and (c) the total lap?

**SOLUTION**

We start by defining our coordinate system, as shown in Figure 2.9. The positive $x$-axis points toward the bottom of the page.

*(a) First leg of the swim:*

The swimmer starts at $x_1 = 0$ and swims to $x_2 = 50$ m. It takes her $\Delta t = 38.2$ s to accomplish this leg. Her average velocity for leg 1 then, according to our definition, is

$$\bar{v}_{x1} = \frac{x_2 - x_1}{\Delta t} = \frac{50\text{ m} - 0\text{ m}}{38.2\text{ s}} = \frac{50}{38.2}\text{ m/s} = 1.31\text{ m/s}.$$

Her average speed is the distance divided by time interval, which, in this case, is the same as the absolute value of her average velocity, or $|\bar{v}_{x1}| = 1.31$ m/s.

*(b) Second leg of the swim:*

We use the same coordinate system for leg 2 as for leg 1. This choice means that the swimmer starts at $x_1 = 50$ m and finishes at $x_2 = 0$, and it takes $\Delta t = 42.5$ s to do so. Her average velocity for this leg is

$$\bar{v}_{x2} = \frac{x_2 - x_1}{\Delta t} = \frac{0\text{ m} - 50\text{ m}}{42.5\text{ s}} = \frac{-50}{42.5}\text{ m/s} = -1.18\text{ m/s}.$$

Note the negative sign for the average velocity for this leg. The average speed is again the absolute magnitude of the average velocity, or $|\bar{v}_{x2}| = |-1.18\text{ m/s}| = 1.18$ m/s.

*(c) The entire lap:*

We can find the average velocity in two ways, demonstrating that they result in the same answer. First, because the swimmer started at $x_1 = 0$ and finished at $x_2 = 0$, the difference is 0. Thus, the net displacement is 0, and consequently the average velocity is also 0.

We can also find the average velocity for the whole lap by taking the time-weighted sum of the components of the average velocities of the individual legs:

$$\bar{v}_x = \frac{\bar{v}_{x1} \cdot \Delta t_1 + \bar{v}_{x2} \cdot \Delta t_2}{\Delta t_1 + \Delta t_2} = \frac{(1.31 \text{ m/s})(38.2 \text{ s}) + (-1.18 \text{ m/s})(42.5 \text{ s})}{(38.2 \text{ s}) + (42.5 \text{ s})} = 0.$$

What do we find for the average speed? The average speed, according to our definition, is the total distance divided by the total time. The total distance is 100 m and the total time is 38.2 s plus 42.5 s, or 80.7 s. Thus,

$$\bar{v} = \frac{\ell}{\Delta t} = \frac{100 \text{ m}}{80.7 \text{ s}} = 1.24 \text{ m/s}.$$

We can also use the time-weighted sum of the average speeds, leading to the same result. Note that the average speed for the entire lap is between that for leg 1 and that for leg 2. It is not exactly halfway between these two values, but is closer to the lower value because the swimmer spent more time completing leg 2.

## 2.4. Acceleration Vector

Just as the average velocity is defined as the displacement per time interval, the $x$-component of **average acceleration** is defined as the velocity change per time interval:

$$\bar{a}_x = \frac{\Delta v_x}{\Delta t}. \tag{2.10}$$

Similarly, the $x$-component of the **instantaneous acceleration** is defined as the limit of the average acceleration as the time interval approaches 0:

$$a_x = \lim_{\Delta t \to 0} \bar{a}_x = \lim_{\Delta t \to 0} \frac{\Delta v_x}{\Delta t} \equiv \frac{dv_x}{dt}. \tag{2.11}$$

We can now define the acceleration vector as

$$\vec{a} = \frac{d\vec{v}}{dt}, \tag{2.12}$$

where again the derivative operation is understood to act component-wise, just as in the definition of the velocity vector.

Figure 2.10 illustrates this relationship among velocity, time interval, average acceleration, and instantaneous acceleration as the limit of the average acceleration (for a decreasing time interval). In Figure 2.10a, the average acceleration is given by the velocity change, $\Delta v_1$, divided by the time interval $\Delta t_1$: $\bar{a}_1 = \Delta v_1/\Delta t_1$. In Figure 2.10b, the average acceleration is determined over a smaller time interval, $\Delta t_2$. In Figure 2.10c, the instantaneous acceleration, $a(t_3) = dv/dt|_{t=t_3}$, is represented by the slope of the blue line tangent to the red curve at $t = t_3$. Figure 2.10 looks very similar to Figure 2.6, and this is not a coincidence. The similarity emphasizes that the mathematical operations and physical relationships that connect the velocity and acceleration vectors are the same as those that connect the position and velocity vectors.

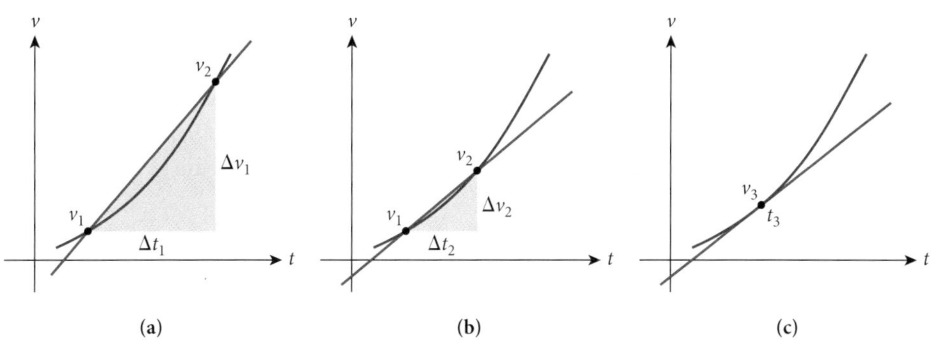


**FIGURE 2.10** Instantaneous acceleration as the limit of the ratio of velocity change to time interval: (a) average acceleration over a large time interval; (b) average acceleration over a smaller time interval; and (c) instantaneous acceleration in the limit as the time interval goes to zero.

### 2.1 In-Class Exercise

When you're driving a car along a straight road, you may be traveling in the positive or negative direction and you may have a positive acceleration or a negative acceleration. Match the following combinations of velocity and acceleration with the list of outcomes.

a) positive velocity, positive acceleration

b) positive velocity, negative acceleration

c) negative velocity, positive acceleration

d) negative velocity, negative acceleration

1) slowing down in positive direction

2) speeding up in negative direction

3) speeding up in positive direction

4) slowing down in negative direction

The acceleration is the time derivative of the velocity, and the velocity is the time derivative of the displacement. The acceleration is therefore the second derivative of the displacement:

$$a_x = \frac{d}{dt}v_x = \frac{d}{dt}\left(\frac{d}{dt}x\right) = \frac{d^2}{dt^2}x. \tag{2.13}$$

There is no word in everyday language for the absolute value of the acceleration.

Note that we often refer to the *deceleration* of an object as a decrease in the speed the object over time, which corresponds to acceleration in the opposite direction of the motion of the object.

In one-dimensional motion, an acceleration, which is a change in velocity, necessarily entails a change in the magnitude of the velocity—that is, the speed. However, in the next chapter, we will consider motion in more than one spatial dimension, where the velocity vector can also change its direction, not just its magnitude. In Chapter 9, we will examine motion in a circle with constant speed; in that case, there is a constant acceleration that keeps the object on a circular path but leaves the speed constant.

As the in-class exercise shows, even in one dimension, a positive acceleration does not necessarily mean speeding up and a negative acceleration does not mean that it must be slowing down. Rather, the combination of velocity and acceleration determines the motion. *If the velocity and acceleration are in the same direction, the object moves faster; if they are in opposite directions, it slows down.* We will examine this relationship further in the next chapter.

## 2.5 Computer Solutions and Difference Formulas

In some situations, the acceleration changes as a function of time, but the exact functional form is not known beforehand. However, we can still calculate velocity and acceleration even if the position is known only at certain points in time. The following example illustrates this procedure.

### EXAMPLE 2.3 World Record for the 100-m Dash

In the 1991 Track and Field World Championships in Tokyo, Japan, Carl Lewis of the United States set a new world record in the 100-m dash. Figure 2.11 lists the times at which he arrived at the 10-m mark, the 20-m mark, and so on, as well as values for his average velocity and average acceleration, calculated from the formulas in equations 2.5 and 2.10. From Figure 2.11, it is clear that after about 3 s, Lewis reached an approximately constant average velocity between 11 and 12 m/s.

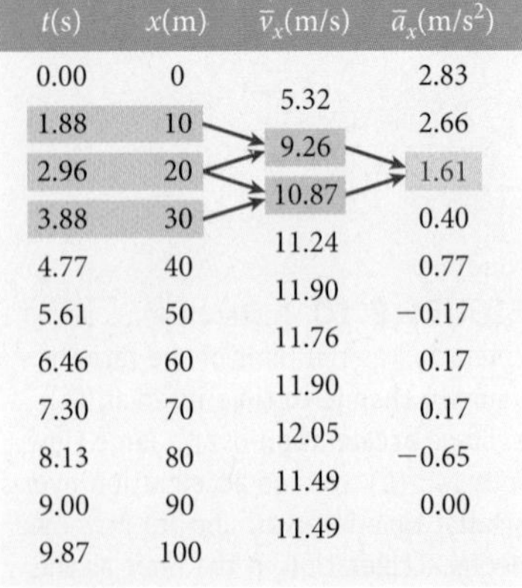

| $t$(s) | $x$(m) | $\bar{v}_x$(m/s) | $\bar{a}_x$(m/s²) |
|---|---|---|---|
| 0.00 | 0 | | 2.83 |
| | | 5.32 | |
| 1.88 | 10 | | 2.66 |
| | | 9.26 | |
| 2.96 | 20 | | 1.61 |
| | | 10.87 | |
| 3.88 | 30 | | 0.40 |
| | | 11.24 | |
| 4.77 | 40 | | 0.77 |
| | | 11.90 | |
| 5.61 | 50 | | −0.17 |
| | | 11.76 | |
| 6.46 | 60 | | 0.17 |
| | | 11.90 | |
| 7.30 | 70 | | 0.17 |
| | | 12.05 | |
| 8.13 | 80 | | −0.65 |
| | | 11.49 | |
| 9.00 | 90 | | 0.00 |
| | | 11.49 | |
| 9.87 | 100 | | |

**FIGURE 2.11** Time, position, average velocity, and average acceleration during Carl Lewis's world record 100-m dash.

Figure 2.11 also indicates how the values for average velocity and average acceleration were obtained. Take, for example, the upper two green boxes, which contain the times and positions for two measurements. From these we get $\Delta t = 2.96\text{ s} - 1.88\text{ s} = 1.08\text{ s}$ and $\Delta x = 20\text{ m} - 10\text{ m} = 10\text{ m}$. The average velocity in this time interval is then $\bar{v} = \Delta x/\Delta t = 10\text{ m}/1.08\text{ s} = 9.26\text{ m/s}$. We have rounded this result to three significant digits, because the times were given to that accuracy. The accuracy for the distances can be assumed to be even better, because these data were extracted from video analysis that showed the times at which Lewis crossed marks on the ground.

In Figure 2.11 the calculated average velocity is positioned halfway between the lines for time and distance, indicating that it is a good approximation for the instantaneous velocity in the middle of the time interval.

The average velocities for other time intervals were obtained in the same way. Using the numbers in the second and third green boxes in Figure 2.11, we obtained an average velocity of 10.87 m/s for the time interval from 2.96 s to 3.88 s. With two velocity values, we can use the difference formula for the acceleration to calculate the average acceleration. Here we assume that the instantaneous velocity at a time corresponding to halfway between the first two green boxes (2.42 s) is equal to the average velocity during the interval between the first two green boxes, or 9.26 m/s. Similarly, we take the instantaneous velocity at 3.42 s (midway between the second and third boxes) to be 10.87 m/s. Then, the average acceleration between 2.42 s and 3.42 s is

$$\bar{a}_x = \Delta v_x / \Delta t = (10.87 \text{ m/s} - 9.26 \text{ m/s})/(3.42 \text{ s} - 2.42 \text{ s}) = 1.61 \text{ m/s}^2.$$

From the entries in Figure 2.11 obtained in this way, we can see that Lewis did most of his accelerating between the start of the race and the 30-m mark, where he attained his maximum velocity between 11 and 12 m/s. He then ran with about that velocity until he reached the finish line. This result is clearer in a graphical display of his position versus time during the race (Figure 2.12a). The red dots represent the data points from Figure 2.11, and the green straight line represents a constant velocity of 11.58 m/s. In Figure 2.12b, Lewis's velocity is plotted as a function of time. The green line again represents a constant velocity of 11.58 m/s, fitted to the last six points, where Lewis is no longer accelerating but running at a constant velocity.

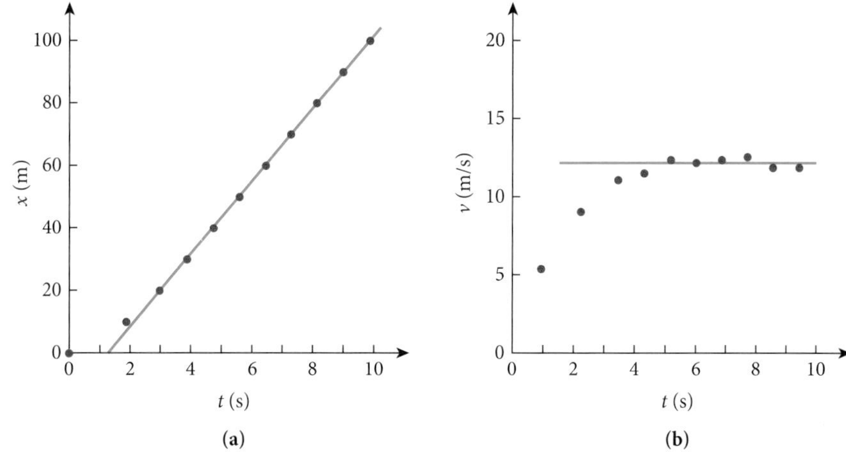


**FIGURE 2.12** Analysis of Carl Lewis's 100-m dash in 1991: (a) his position as a function of time; (b) his velocity as a function of time.

The type of numerical analysis that treats average velocities and accelerations as approximations for the instantaneous values of these quantities is very common in all types of scientific and engineering applications. It is indispensable in situations where the precise functional dependencies on time are not known and researchers must rely on numerical approximations for derivatives obtained via the difference formulas. Most practical solutions of scientific and engineering problems found with the aid of computers make use of difference formulas such as those introduced here.

The entire field of numerical analysis is devoted to finding better numerical approximations that will enable more precise and faster computer calculations and simulations of natural processes. Difference formulas similar to those introduced here are as important for the everyday work of scientists and engineers as the calculus-based analytic expressions. This importance is a consequence of the computer revolution in science and technology, which, however, does not make the contents of the textbook any less important. In order to devise a valid solution to an engineering or science problem, you have to understand the basic underlying physical principles, no matter what calculation techniques you use. This fact is well recognized by cutting-edge movie animators and creators of special digital effects, who have to take basic physics classes to ensure that the products of their computer simulations look realistic to the audience.

## 2.6 Finding Displacement and Velocity from Acceleration

The fact that integration is the inverse operation to differentiation is known as the *Fundamental Theorem of Calculus.* It allows us to reverse the differentiation process leading from displacement to velocity to acceleration and instead integrate the equation for velocity (2.6) to obtain displacement and the equation for acceleration (2.13) to obtain velocity. Let's start with the equation for the $x$-component of the velocity:

$$v_x(t)=\frac{dx(t)}{dt}\Rightarrow$$

$$\int_{t_0}^{t} v_x(t')dt'=\int_{t_0}^{t}\frac{dx(t')}{dt'}dt'=x(t)-x(t_0)\Rightarrow$$

$$x(t)=x_0+\int_{t_0}^{t} v_x(t')dt'. \tag{2.14}$$

***Notation:*** Here again we have used the convention that $x(t_0) = x_0$, the initial position. Further, we have used the notation $t'$ in the definite integrals in equation 2.14. This prime notation reminds us that the integration variable is a dummy variable, which serves to identify the physical quantity we want to integrate. Throughout this book, we will reserve the prime notation for dummy integration variables in definite integrals. (Note that some books use a prime to denote a spatial derivative, but to avoid possible confusion, this book will not do so.)

In the same way, we integrate equation 2.13 for the $x$-component of the acceleration to obtain an expression for the $x$-component of the velocity:

$$a_x(t)=\frac{dv_x(t)}{dt}\Rightarrow$$

$$\int_{t_0}^{t} a_x(t')dt'=\int_{t_0}^{t}\frac{dv_x(t')}{dt'}dt'=v_x(t)-v_x(t_0)\Rightarrow$$

$$v_x(t)=v_{x0}+\int_{t_0}^{t} a_x(t')dt'. \tag{2.15}$$

Here $v_x(t_0) = v_{x0}$ is the initial velocity component in the $x$-direction. Just as the derivative operation is understood to act component-wise, integration follows the same convention, so we can write the integral relationships for the vectors from those for the components in equations 2.14 and 2.15. Formally, we then have

$$\vec{r}(t)=\vec{r}_0+\int_{t_0}^{t}\vec{v}(t')dt' \tag{2.16}$$

and

$$\vec{v}(t)=\vec{v}_0+\int_{t_0}^{t}\vec{a}(t')dt'. \tag{2.17}$$

This result means that for any given time dependence of the acceleration vector, we can calculate the velocity vector, provided we are given the initial value of the velocity vector. We can also calculate the displacement vector, if we know its initial value and the time dependence of the velocity vector.

In calculus, you probably learned that the geometrical interpretation of the definite integral is an area under a curve. This is true for equations 2.14 and 2.15. We can interpret the area under the curve of $v_x(t)$ between $t_0$ and $t$ as the difference in the position between these two times, as shown in Figure 2.13a. Figure 2.13b shows that the area under the curve of $a_x(t)$ in the time interval between $t_0$ and $t$ is the velocity difference between these two times.

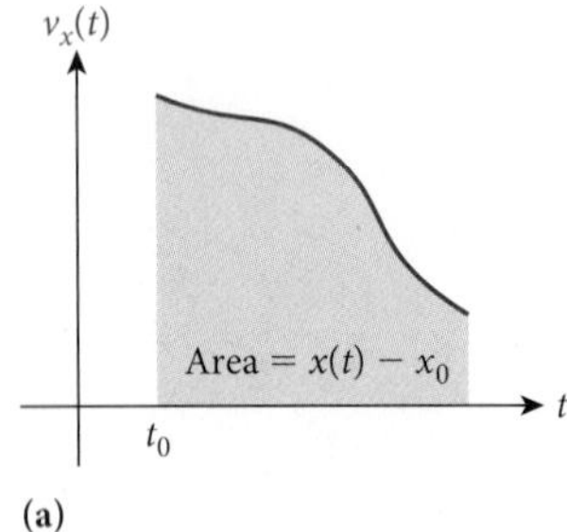

(a)

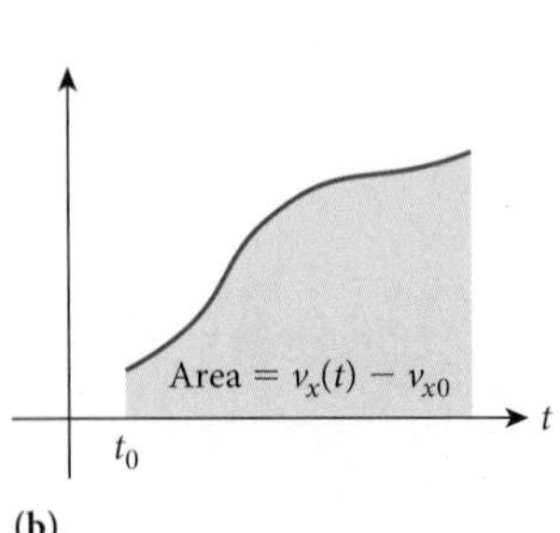

(b)

**FIGURE 2.13** Geometrical interpretation of the integrals of (a) velocity and (b) acceleration with respect to time.

## 2.7 Motion with Constant Acceleration

In many physical situations, the acceleration experienced by an object is approximately, or perhaps even exactly, constant. We can derive useful equations for these special cases of motion with constant acceleration. If the acceleration, $a_x$, is a constant, then the time integral used to obtain the velocity in equation 2.15 results in

$$v_x(t) = v_{x0} + \int_0^t a_x dt' = v_{x0} + a_x \int_0^t dt' \Rightarrow$$

$$v_x(t) = v_{x0} + a_x t, \tag{2.18}$$

where we have taken the lower limit of the integral to be $t_0 = 0$ for simplicity. This means that the velocity is a linear function of time.

$$x = x_0 + \int_0^t v_x(t')dt' = x_0 + \int_0^t (v_{x0} + a_x t')dt'$$

$$= x_0 + v_{x0}\int_0^t dt' + a_x \int_0^t t'dt' \Rightarrow$$

$$x(t) = x_0 + v_{x0}t + \tfrac{1}{2}a_x t^2. \tag{2.19}$$

Thus, with a constant acceleration, the velocity is always a linear function of time, and the position is a quadratic function of time. Three other useful equations can be derived using equations 2.18 and 2.19 as a starting point. After listing these three equations, we'll work through their derivations.

The average velocity in the time interval from 0 to $t$ is the average of the velocities at the beginning and end of the time interval:

$$\bar{v}_x = \tfrac{1}{2}(v_{x0} + v_x). \tag{2.20}$$

The average velocity from equation 2.20 leads to an alternative way to express the position:

$$x = x_0 + \bar{v}_x t. \tag{2.21}$$

Finally, we can write an equation for the square of the velocity that does not contain the time explicitly:

$$v_x^2 = v_{x0}^2 + 2a_x(x - x_0). \tag{2.22}$$

### DERIVATION 2.1

Mathematically, to obtain the time average of a quantity over a certain interval $\Delta t$, we have to integrate this quantity over the time interval and then divide by the time interval:

$$\bar{v}_x = \frac{1}{t}\int_0^t v_x(t')dt' = \frac{1}{t}\int_0^t (v_{x0} + at')dt'$$

$$= \frac{v_{x0}}{t}\int_0^t dt' + \frac{a}{t}\int_0^t t'dt' = v_{x0} + \tfrac{1}{2}at$$

$$= \tfrac{1}{2}v_{x0} + \tfrac{1}{2}(v_{x0} + at)$$

$$= \tfrac{1}{2}(v_{x0} + v_x).$$

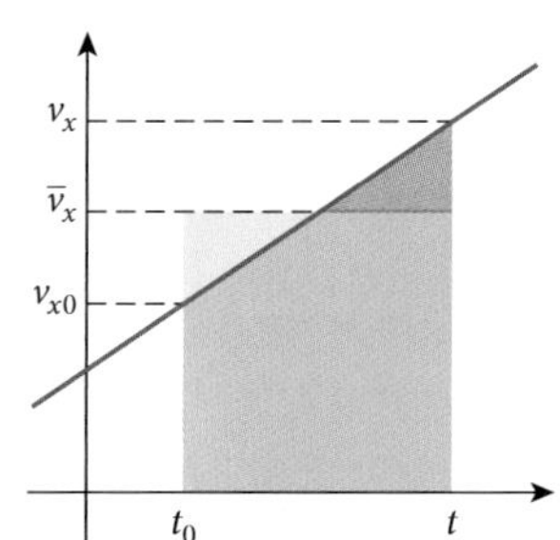


**FIGURE 2.14** Graph of velocity versus time for motion with constant acceleration.

This averaging procedure for the time interval $t_0$ to $t$ is illustrated in Figure 2.14. You can see that the area of the trapezoid formed by the blue line representing $v(t)$ and the two vertical lines at $t_0$ and $t$ is equal to the area of the square formed by the horizontal line to $\bar{v}_x$ and the two vertical lines. The base line for both areas is the horizontal $t$-axis. It is more apparent that these two areas are equal if you note that the yellow triangle (part of the square) and the orange triangle (part of the trapezoid) are equal in size. [Algebraically, the

*Continued—*

area of the square is $\bar{v}_x(t - t_0)$, and the area of the trapezoid is $\frac{1}{2}(v_{x0} + v_x)(t - t_0)$. Setting these two areas equal to each other gives us equation 2.20 again.]

To derive the equation for the position, we take $t_0 = 0$ and use the expression $\bar{v}_x = v_{x0} + \frac{1}{2}a_x t$ and multiply both sides by the time:

$$\bar{v}_x = v_{x0} + \tfrac{1}{2}a_x t$$
$$\Rightarrow \bar{v}_x t = v_{x0}t + \tfrac{1}{2}a_x t^2.$$

Now we compare this result to the expression we already obtained for $x$ (equation 2.19) and find:

$$x = x_0 + v_{x0}t + \tfrac{1}{2}a_x t^2 = x_0 + \bar{v}_x t.$$

For the derivation of equation 2.22 for the square of the velocity, we solve $v_x = v_{x0} + a_x t$ for the time, getting $t = (v_x - v_{x0})/a_x$. We then substitute into the expression for the position, which is equation 2.19:

$$\begin{aligned}x &= x_0 + v_{x0}t + \tfrac{1}{2}a_x t^2\\ &= x_0 + v_{x0}\left(\frac{v_x - v_{x0}}{a_x}\right) + \tfrac{1}{2}a_x\left(\frac{v_x - v_{x0}}{a_x}\right)^2\\ &= x_0 + \frac{v_x v_{x0} - v_{x0}^2}{a_x} + \frac{1}{2}\frac{v_x^2 + v_{x0}^2 - 2v_x v_{x0}}{a_x}.\end{aligned}$$

Next, we subtract $x_0$ from both sides of the equation and then multiply by $a_x$:

$$a_x(x - x_0) = v_x v_{x0} - v_{x0}^2 + \tfrac{1}{2}(v_x^2 + v_{x0}^2 - 2v_x v_{x0})$$
$$\Rightarrow a_x(x - x_0) = \tfrac{1}{2}v_x^2 - \tfrac{1}{2}v_{x0}^2$$
$$\Rightarrow v_x^2 = v_{x0}^2 + 2a_x(x - x_0).$$

Here are the five kinematical equations we have obtained for the *special case of motion with constant acceleration* (where initial time when $x = x_0$, $v = v_0$ has been chosen to be 0):

$$\begin{aligned}&\text{(i)} && x = x_0 + v_{x0}t + \tfrac{1}{2}a_x t^2\\ &\text{(ii)} && x = x_0 + \bar{v}_x t\\ &\text{(iii)} && v_x = v_{x0} + a_x t\\ &\text{(iv)} && \bar{v}_x = \tfrac{1}{2}(v_x + v_{x0})\\ &\text{(v)} && v_x^2 = v_{x0}^2 + 2a_x(x - x_0)\end{aligned} \qquad (2.23)$$

These five equations allow us to solve many kinds of problems for motion in one dimension with constant acceleration. However, remember that if the acceleration is not constant, these equations will not give the correct solutions.

Many real-life problems involve motion along a straight line with constant acceleration. In these situations, equations 2.23 provide the template for answering any question about the motion. The following solved problem and example will illustrate how useful these kinematical equations are. However, keep in mind that physics is not simply about finding an appropriate equation and plugging in numbers, but is instead about understanding concepts. Only if you understand the underlying ideas will you be able to extrapolate from specific examples to becoming skilled at solving more general problems.

## SOLVED PROBLEM 2.2 Airplane Takeoff

As an airplane rolls down a runway to reach takeoff speed, it is accelerated by its jet engines. On one particular flight, one of the authors of this book measured the acceleration produced by the plane's jet engines. Figure 2.15 shows the measurements.

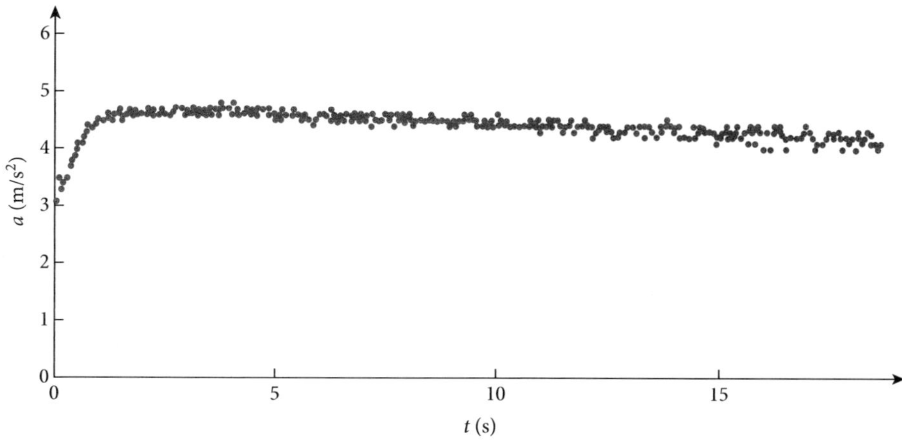

**FIGURE 2.15** Data for the acceleration of a jet plane before takeoff.

You can see that the assumption of constant acceleration is not quite correct in this case. However, an average acceleration of $a_x = 4.3\ \text{m/s}^2$ over the 18.4 s (measured with a stopwatch) it took the airplane to take off is a good approximation.

**PROBLEM**

Assuming a constant acceleration of $a_x = 4.3\ \text{m/s}^2$ starting from rest, what is the airplane's takeoff velocity after 18.4 s? How far down the runway has the plane moved by the time it takes off?

**SOLUTION**

**THINK**

An airplane moving along a runway prior to takeoff is a nearly perfect example of one-dimensional accelerated motion. Because we are assuming constant acceleration, we know that the velocity increases linearly with time, and the displacement increases as the second power of time. Since the plane starts from rest, the initial value of the velocity is 0. As usual, we can define the origin of our coordinate system at any location; it is convenient to locate it at the point of the plane's standing start.

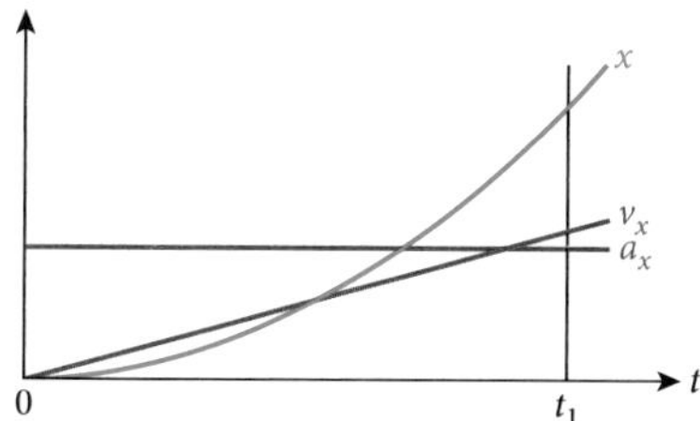

**FIGURE 2.16** The acceleration, velocity, and displacement of the plane before takeoff.

**SKETCH**

The sketch in Figure 2.16 shows how we expect the velocity and displacement to increase for this case of constant acceleration, where the initial conditions are set at $v_{x0} = 0$ and $x_0 = 0$. Note that no scales have been placed on the axes, because displacement, velocity, and acceleration are measured in different units. Thus, the points at which the three curves intersect are completely arbitrary.

**RESEARCH**

Finding the takeoff velocity is actually a straightforward application of equation 2.23(iii):

$$v_x = v_{x0} + a_x t.$$

Similarly, the distance down the runway that the airplane moves before taking off can be obtained from equation 2.23(i):

$$x = x_0 + v_{x0}t + \tfrac{1}{2}a_x t^2.$$

**SIMPLIFY**

The airplane accelerates from a standing start, so the initial velocity is $v_{x0} = 0$, and by our choice of coordinate system origin, we have set $x_0 = 0$. Therefore, the equations for takeoff velocity and distance simplify to

$$v_x = a_x t$$
$$x = \tfrac{1}{2}a_x t^2.$$

*Continued—*

**CALCULATE**

The only thing left to do is to put in the numbers:

$$v_x = (4.3\ \text{m/s}^2)(18.4\ \text{s}) = 79.12\ \text{m/s}$$

$$x = \tfrac{1}{2}(4.3\ \text{m/s}^2)(18.4\ \text{s})^2 = 727.904\ \text{m}.$$

**ROUND**

The acceleration was specified to two significant digits, and the time to three. Multiplying these two numbers must result in an answer that has two significant digits. Our final answers are therefore

$$v_x = 79\ \text{m/s}$$

$$x = 7.3 \cdot 10^2\ \text{m}.$$

Note that the measured time to takeoff of 18.4 s was probably not actually that precise. If you have ever tried to determine the moment at which a plane starts to accelerate down the runway, you will have noticed that it is almost impossible to determine that point in time with an accuracy to 0.1 s.

**DOUBLE-CHECK**

As this book has repeatedly stressed, the most straightforward check of any answer to a physics problem is to make sure the units fit the situation. This is the case here, because we obtained displacement in units of meters and velocity in units of meters per second. Solved problems in the rest of this book may sometimes skip this simple test; however, if you want to do a quick check for algebraic errors in your calculations, it can be valuable to first look at the units of the answer.

Now let's see if our answers have the appropriate orders of magnitude. A takeoff displacement of 730 m (~0.5 mi) is reasonable, because it is on the order of the length of an airport runway. A takeoff velocity of $v_x = 79$ m/s translates into

$$(79\ \text{m/s})(1\ \text{mi}/1609\ \text{m})(3600\ \text{s}/1\ \text{h}) \approx 180\ \text{mph}.$$

This answer also appears to be in the right ballpark.

**ALTERNATIVE SOLUTION**

Many problems in physics can be solved in several ways, because it is often possible to use more than one relationship between the known and unknown quantities. In this case, once we obtained the final velocity, we could use this information and solve kinematical equation 2.23(v) for $x$. This results in

$$v_x^2 = v_{x0}^2 + 2a_x(x - x_0) \Rightarrow$$

$$x = x_0 + \frac{v_x^2 - v_{x0}^2}{2a_x} = 0 + \frac{(79\ \text{m/s})^2}{2(4.3\ \text{m/s}^2)} = 7.3 \cdot 10^2\ \text{m}.$$

Thus, we arrive at the same answer for the distance in a different way, giving us additional confidence that our solution makes sense.

Solved Problem 2.2 was a fairly easy one; solving it amounted to little more than plugging in numbers. Nevertheless, it shows that the kinematical equations we derived can be applied to real-world situations and lead to answers that have physical meaning. The following short example, this time from motor sports, addresses the same concepts of velocity and acceleration, but in a slightly different light.

**FIGURE 2.17** Top fuel NHRA race car.

## EXAMPLE 2.4 Top Fuel Racing

Accelerating from rest, a top fuel race car (Figure 2.17) can reach 333.2 mph (= 148.9 m/s), a record established in 2003, at the end of a quarter mile (= 402.3 m). For this example, we will assume constant acceleration.

**PROBLEM 1**
What is the value of the race car's constant acceleration?

**SOLUTION 1**
Since the initial and final values of the velocity are given and the distance is known, we are looking for a relationship between these three quantities and the acceleration, the unknown. In this case, it is most convenient to use kinematics equation 2.23(v) and solve for the acceleration, $a_x$:

$$v_x^2 = v_{x0}^2 + 2a_x(x - x_0) \Rightarrow a_x = \frac{v_x^2 - v_{x0}^2}{2(x - x_0)} = \frac{(148.9 \text{ m/s})^2}{2(402.3 \text{ m})} = 27.6 \text{ m/s}^2.$$

**PROBLEM 2**
How long does it take the race car to complete a quarter-mile run from a standing start?

**SOLUTION 2**
Because the final velocity is 148.9 m/s, the average velocity is [using equation 2.23(iv)]: $\bar{v}_x = \frac{1}{2}(148.9 \text{ m/s} + 0) = 74.45 \text{ m/s}$. Relating this average velocity to the displacement and time using equation 2.23(ii), we obtain:

$$x = x_0 + \bar{v}_x t \Rightarrow t = \frac{x - x_0}{\bar{v}_x} = \frac{402.3 \text{ m}}{74.45 \text{ m/s}} = 5.40 \text{ s}.$$

Note that we could have obtained the same result by using kinematics equation 2.23(iii), because we already calculated the acceleration in Solution 1.

If you are a fan of top fuel racing, however, you know that the real record time for the quarter mile is slightly lower than 4.5 s. The reason our calculated answer is somewhat higher is that our assumption of constant acceleration is not quite correct. The acceleration of the car at the beginning of the race is actually higher than the value we calculated above, and the actual acceleration is lower than our value toward the end of the race.

## Free Fall

The acceleration due to the gravitational force is constant, to a good approximation, near the surface of Earth. If this statement is true, it must have observable consequences. Let's assume it is true and work out the consequences for motion of objects under the influence of a gravitational attraction to Earth. Then we'll compare our results with experimental observations and see if constant acceleration due to gravity makes sense.

The acceleration due to gravity near the surface of the Earth has the value $g = 9.81 \text{ m/s}^2$. We call the vertical axis the $y$-axis and define the positive direction as up. Then the acceleration vector $\vec{a}$ has only a nonzero $y$-component, which is given by

$$a_y = -g. \tag{2.24}$$

This situation is a specific application of motion with constant acceleration, which we discussed earlier in this section. We modify equations 2.23 by substituting for acceleration from equation 2.24. We also use $y$ instead of $x$ to indicate that the displacement takes place in the $y$-direction. We obtain:

$$\begin{aligned}
&\text{(i)} && y = y_0 + v_{y0}t - \tfrac{1}{2}gt^2 \\
&\text{(ii)} && y = y_0 + \bar{v}_y t \\
&\text{(iii)} && v_y = v_{y0} - gt \\
&\text{(iv)} && \bar{v}_y = \tfrac{1}{2}(v_y + v_{y0}) \\
&\text{(v)} && v_y^2 = v_{y0}^2 - 2g(y - y_0)
\end{aligned} \tag{2.25}$$

Motion under the sole influence of a gravitational acceleration is called **free fall,** and equations 2.25 allow us to solve problems for objects in free fall.

## 2.2 In-Class Exercise

Throwing a ball straight up into the air provides an example of free-fall motion. At the instant the ball reaches its maximum height, which of the following statements is true?

a) The ball's acceleration points down, and its velocity points up.

b) The ball's acceleration is zero, and its velocity points up.

c) The ball's acceleration points up, and its velocity points up.

d) The ball's acceleration points down, and its velocity is zero.

e) The ball's acceleration points up, and its velocity is zero.

f) The ball's acceleration is zero, and its velocity points down.

## 2.3 In-Class Exercise

A ball is thrown upward with a speed $v_1$, as shown in Figure 2.19. The ball reaches a maximum height of $y = h$. What is the ratio of the speed of the ball, $v_2$, at $y = h/2$ in Figure 2.19b, to the initial upward speed of the ball, $v_1$, at $y = 0$ in Figure 2.19a?

a) $v_2/v_1 = 0$

b) $v_2/v_1 = 0.50$

c) $v_2/v_1 = 0.71$

d) $v_2/v_1 = 0.75$

e) $v_2/v_1 = 0.90$

Now let's consider an experiment that tested the assumption of constant gravitational acceleration. The authors went to the top of a building of height 12.7 m and dropped a computer from rest ($v_{y0} = 0$) under controlled conditions. The computer's fall was recorded by a digital video camera. Because the camera records at 30 frames per second, we know the time information. Figure 2.18 displays 14 frames, equally spaced in time, from this experiment, with the time after release marked on the horizontal axis for each frame. The yellow curve superimposed on the frames has the form

$$y = 12.7 \text{ m} - \tfrac{1}{2}(9.81 \text{ m/s}^2)t^2,$$

which is what we expect for initial conditions $y_0 = 12.7$ m, $v_{y0} = 0$ and the assumption of a constant acceleration, $a_y = -9.81$ m/s$^2$. As you can see, the computer's fall follows this curve almost perfectly. This agreement is, of course, not a conclusive proof, but it is a strong indication that the gravitational acceleration is constant near the surface of Earth, and that it has the stated value.

In addition, the value of the gravitational acceleration is the same for all objects. This is by no means a trivial statement. Objects of different sizes and masses, if released from the same height, should hit the ground at the same time. Is this consistent with our everyday experience? Well, not quite! In a common lecture demonstration, a feather and a coin are dropped from the same height. It is easy to observe that the coin reaches the floor first, while the feather slowly floats down. This difference is due to air resistance. If this experiment is done in an evacuated glass tube, the coin and the feather fall at the same rate. We will return to air resistance in Chapter 4, but for now we can conclude that the gravitational acceleration near the surface of Earth is constant, has the absolute value of $g = 9.81$ m/s$^2$, and is the same for all objects provided we can neglect air resistance. In Chapter 4, we will examine the conditions under which the assumption of zero air resistance is justified.

To help you understand the answer to the in-class exercise, consider throwing a ball straight up, as illustrated in Figure 2.19. In Figure 2.19a, the ball is thrown upward with a velocity $\vec{v}$. When the ball is released, it experiences only the force of gravity and thus accelerates downward with the acceleration due to gravity, which is given by $\vec{a} = -g\hat{y}$. As the ball travels upward, the acceleration due to gravity acts to slow the ball down. In Figure 2.19b, the ball is moving upward and has reached half of its maximum height, $h$. The ball has slowed down, but its acceleration is still the same. The ball reaches its maximum height in Figure 2.19c. Here the velocity of the ball is zero, but the acceleration is still $\vec{a} = -g\hat{y}$. The ball now begins to move downward, and the acceleration due to gravity acts to speed it up. In Figure 2.19d, the velocity of the ball is downward and the acceleration is still the same. Finally, the ball returns to $y = 0$ in Figure 2.19e. The velocity of the ball now has the same magnitude as when the ball was initially thrown upward, but its direction is now downward. The acceleration is still the same. Note that the acceleration remains constant and downward even though the velocity changes from upward to zero to downward.

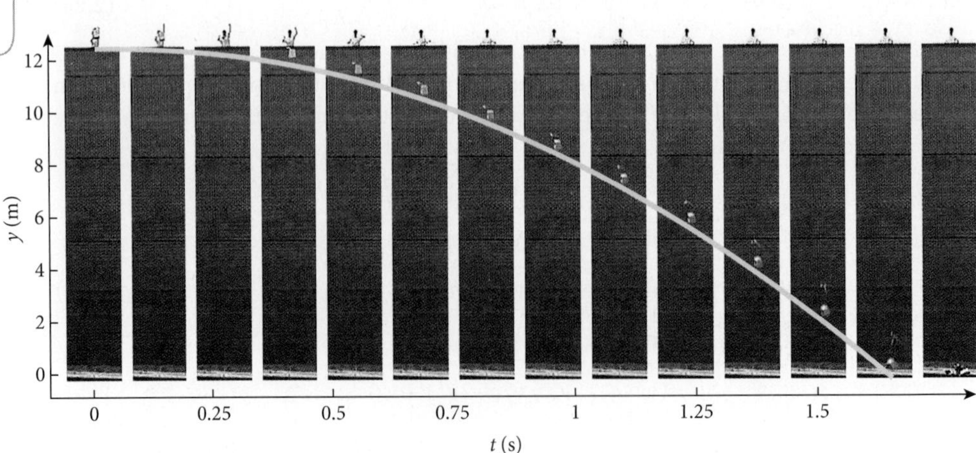


**FIGURE 2.18** Free-fall experiment: dropping a computer off the top of a building.

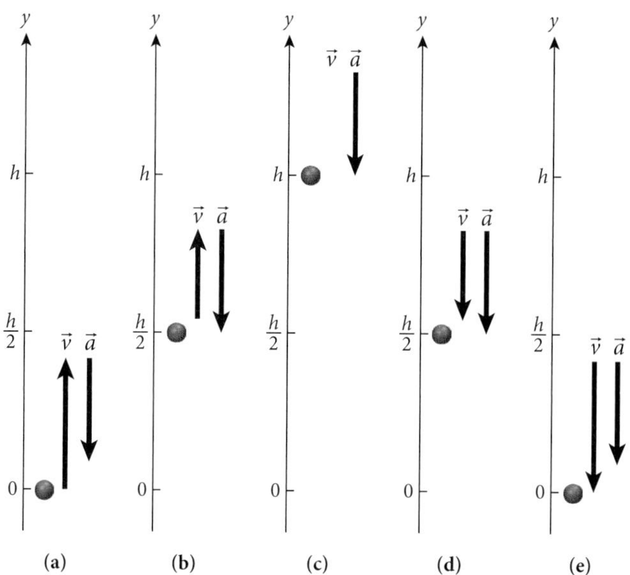


**FIGURE 2.19** The velocity vector and acceleration vector of a ball thrown straight up in the air. (a) The ball is initially thrown upward at $y = 0$. (b) The ball going upward at a height of $y = h/2$. (c) The ball at its maximum height of $y = h$. (d) The ball coming down at $y = h/2$. (e) The ball back at $y = 0$ going downward.

## EXAMPLE 2.5 Reaction Time

It takes time for a person to react to any external stimulus. For example, at the beginning of a 100-m dash in a track-and-field meet, a gun is fired by the starter. A slight time delay occurs before the runners come out of the starting blocks, due to their finite reaction time. In fact, it counts as a false start if a runner leaves the blocks less than 0.1 s after the gun is fired. Any shorter time indicates that the runner has "jumped the gun."

There is a simple test, shown in Figure 2.20, that you can perform to determine your reaction time. Your partner holds a meter stick, and you get ready to catch it when your partner releases it, as shown in the left frame of the figure. From the distance $h$ that the meter stick falls after it is released until you grab it (shown in the right frame), you can determine your reaction time.

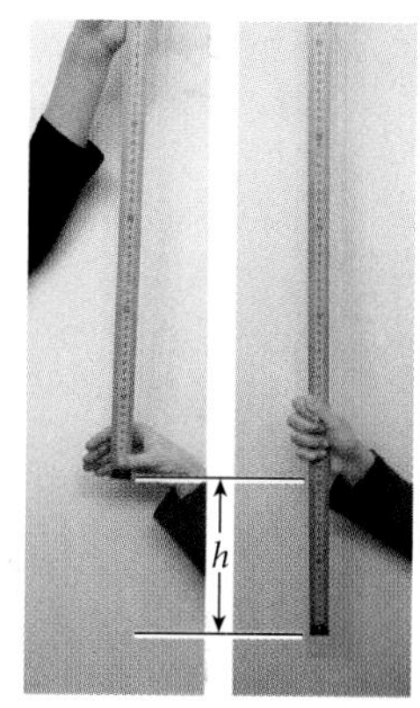


**FIGURE 2.20** Simple experiment to measure reaction time.

### PROBLEM

If the meter stick falls 0.20 m before you catch it, what is your reaction time?

### SOLUTION

This situation is a free-fall scenario. For these problems, the solution invariably comes from one of equations 2.25. The problem we want to solve here involves the time as an unknown.

We are given the displacement, $h = y_0 - y$. We also know that the initial velocity of the meter stick is zero because it is released from rest. We can use kinematical equation 2.25(i): $y = y_0 + v_{y0}t - \frac{1}{2}gt^2$. With $h = y_0 - y$ and $v_{0=0}$, this equation becomes

$$y = y_0 - \tfrac{1}{2}gt^2$$

$$\Rightarrow h = \tfrac{1}{2}gt^2$$

$$\Rightarrow t = \sqrt{\frac{2h}{g}} = \sqrt{\frac{2 \cdot 0.20 \text{ m}}{9.81 \text{ m/s}^2}} = 0.20 \text{ s.}$$

Your reaction time was 0.20 s. This reaction time is typical. For comparison, when Usain Bolt established a world record of 9.69 s for the 100-m dash in August 2008, his reaction time was measured to be 0.165 s.

### 2.2 Self-Test Opportunity

Sketch a graph of reaction time as a function of the distance the meter stick drops. Discuss whether this method is more precise for reaction times around 0.1 s or those around 0.3 s.

Let's consider one more free-fall scenario, this time with two moving objects.

### 2.4 In-Class Exercise

If the reaction time of person B determined with the meter stick method is twice as long as that of person A, then the displacement $h_B$ measured for person B relative to the displacement $h_A$ for person A is

a) $h_B = 2h_A$

b) $h_B = \frac{1}{2}h_A$

c) $h_B = \sqrt{2}h_A$

d) $h_B = 4h_A$

e) $h_B = \sqrt{\frac{1}{2}}h_A$

## SOLVED PROBLEM 2.3 Melon Drop

Suppose you decide to drop a melon from rest from the first observation platform of the Eiffel Tower. The initial height $h$ from which the melon is released is 58.3 m above the head of your French friend Pierre, who is standing on the ground right below you. At the same instant you release the melon, Pierre shoots an arrow straight up with an initial velocity of 25.1 m/s. (Of course, Pierre makes sure the area around him is cleared and gets out of the way quickly after he shoots his arrow.)

### PROBLEM

(a) How long after you drop the melon will the arrow hit it? (b) At what height above Pierre's head does this collision occur?

### SOLUTION

#### THINK

At first sight, this problem looks complicated. We will solve it using the full set of steps and then examine a shortcut we could have taken. Obviously, the dropped melon is in a free fall. However, because the arrow is shot straight up, the arrow is also in free fall, only with an upward initial velocity.

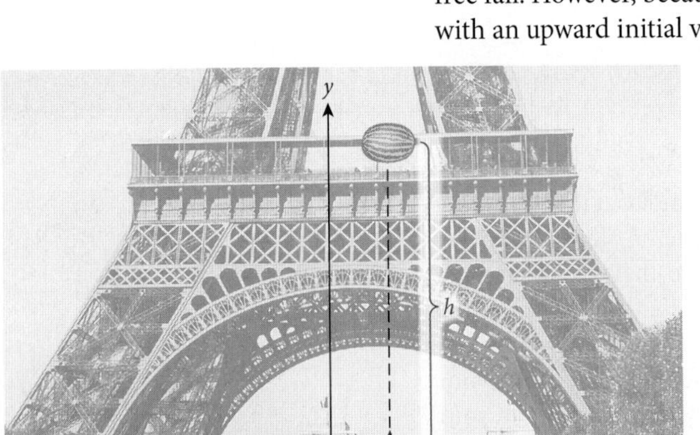


**FIGURE 2.21** The melon drop (melon and person are not drawn to scale!).

#### SKETCH

We set up our coordinate system with the $y$-axis pointing vertically up, as is conventional, and we locate the origin of the coordinate system at Pierre's head (Figure 2.21). Thus, the arrow is released from an initial position $y = 0$, and the melon from $y = h$.

#### RESEARCH

We use the subscript m for the melon and a for the arrow. We start from the general free-fall equation, $y = y_0 + v_{y0}t - \frac{1}{2}gt^2$, *and use the initial conditions given for the melon* ($v_{y0} = 0$, $y_0 = h = 58.3$ m) and for the arrow ($v_{y0} \equiv v_{a0} = 25.1$ m/s, $y_{0=0}$) to set up the two equations of free-fall motion:

$$y_m(t) = h - \tfrac{1}{2}gt^2$$

$$y_a(t) = v_{a0}t - \tfrac{1}{2}gt^2.$$

The key insight is that at $t_c$, the moment when the melon and arrow collide, their coordinates are identical:

$$y_a(t_c) = y_m(t_c).$$

#### SIMPLIFY

Inserting $t_c$ into the two equations of motion and setting them equal results in

$$h - \tfrac{1}{2}gt_c^2 = v_{a0}t_c - \tfrac{1}{2}gt_c^2 \Rightarrow$$

$$h = v_{a0}t_c \Rightarrow$$

$$t_c = \frac{h}{v_{a0}}.$$

We can now insert this value for the time of collision in either of the two free-fall equations and obtain the height above Pierre's head at which the collision occurs. We select the equation for the melon:

$$y_m(t_c) = h - \tfrac{1}{2}gt_c^2.$$

#### CALCULATE

(a) All that is left to do is to insert the numbers given for the height of release of the melon and the initial velocity of the arrow, which results in

$$t_c = \frac{58.3 \text{ m}}{25.1 \text{ m/s}} = 2.32271 \text{ s}$$

for the time of impact.

(b) Using the number we obtained for the time, we find the position at which the collision occurs:

$$y_m(t_c) = 58.3\text{ m} - \tfrac{1}{2}(9.81\text{ m/s}^2)(2.32271\text{ s})^2 = 31.8376\text{ m}.$$

## ROUND

Since the initial values of the release height and the arrow velocity were given to three significant figures, we have to limit our final answers to three digits. Thus, the arrow will hit the melon after 2.32 s, and this will occur at a position 31.8 m above Pierre's head.

## DOUBLE-CHECK

Could we have obtained the answers in an easier way? Yes, if we had realized that both melon and arrow fall under the influence of the same gravitational acceleration, and thus their free-fall motion does not influence the distance between them. This means that the time it takes them to meet is simply the initial distance between them divided by their initial velocity difference. With this realization, we could have written $t = h/v_{a0}$ right away and been done. However, thinking in terms of relative motion in this way takes some practice, and we will return to it in more detail in the next chapter.

Figure 2.22 shows the complete graph of the positions of arrow and melon as functions of time. The dashed portions of both graphs indicate where the arrow and melon would have gone, had they not collided.

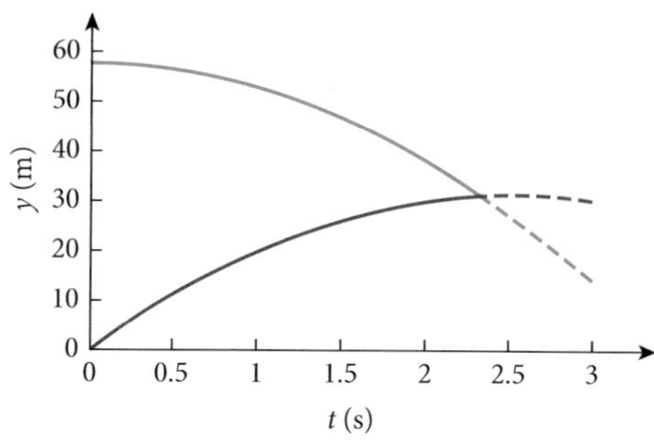


**FIGURE 2.22** Position as a function of time for the arrow (red curve) and the melon (green curve).

## ADDITIONAL QUESTION

What are the velocities of melon and arrow at the moment of the collision?

## SOLUTION

We obtain the velocity by taking the time derivative of the position. For arrow and melon, we get

$$y_m(t) = h - \tfrac{1}{2}gt^2 \Rightarrow v_m(t) = \frac{dy_m(t)}{dt} = -gt$$

$$y_a(t) = v_{a0}t - \tfrac{1}{2}gt^2 \Rightarrow v_a(t) = \frac{dy_a(t)}{dt} = v_{a0} - gt.$$

Now, inserting the time of the collision, 2.32 s, will produce the answers. Note that, unlike the positions of the arrow and the melon, the velocities of the two objects are not the same right before contact!

$$v_m(t_c) = -(9.81\text{ m/s}^2)(2.32\text{ s}) = -22.8\text{ m/s}$$

$$v_a(t_c) = (25.1\text{ m/s}) - (9.81\text{ m/s}^2)(2.32\text{ s}) = 2.34\text{ m/s}.$$

Furthermore, you should note that the difference between the two velocities is still 25.1 m/s, just as it was at the beginning of the trajectories.

We finish this problem by plotting (Figure 2.23) the velocities as a function of time. You can see that the arrow starts out with a velocity that is 25.1 m/s greater than that of the melon. As time progresses, arrow and melon experience the same change in velocity under the influence of gravity, meaning that their velocities maintain the initial difference.

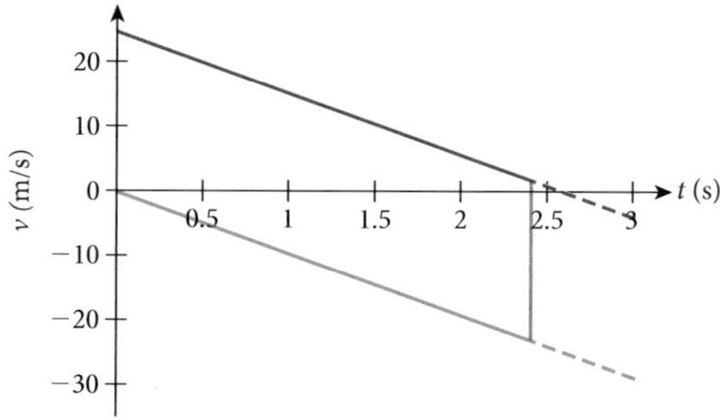


**FIGURE 2.23** Velocities of the arrow (red curve) and melon (green curve) as a function of time.

### 2.5 In-Class Exercise

If the melon in Solved Problem 2.3 is thrown straight up with an initial velocity of 5 m/s at the same time that the arrow is shot upward, how long does it take before the collision occurs?

a) 2.32 s

b) 2.90 s

c) 1.94 s

d) They do not collide before the melon hits the ground.

### 2.3 Self-Test Opportunity

As you can see from the answer to Solved Problem 2.3, the velocity of the arrow is only 2.34 m/s when it hits the melon. This means that by the time the arrow hits the melon, its initial velocity has dropped substantially because of the effect of gravity. Suppose that the initial velocity of the arrow was smaller by 5.0 m/s. What would change? Would the arrow still hit the melon?

## 2.8 Reducing Motion in More Than One Dimension to One Dimension

Motion in one spatial dimension is not all there is to kinematics. We can also investigate more general cases, in which objects move in two or three spatial dimensions. We will do this in the next few chapters. However, in some cases, motion in more than one dimension can be reduced to one-dimensional motion. Let's consider a very interesting case of motion in two dimensions for which each segment can be described by motion in a straight line.

### EXAMPLE 2.6 Aquathlon

The triathlon is a sporting competition that was invented by the San Diego Track Club in the 1970s and became an event in the Sydney Olympic Games in 2000. Typically, it consists of a 1.5-km swim, followed by a 40-km bike race, and finishing with a 10-km foot race. To be competitive, athletes must be able to swim the 1.5-km distance in less than 20 minutes, do the 40-km bike race in less than 70 minutes, and run the 10-km race in less than 35 minutes.

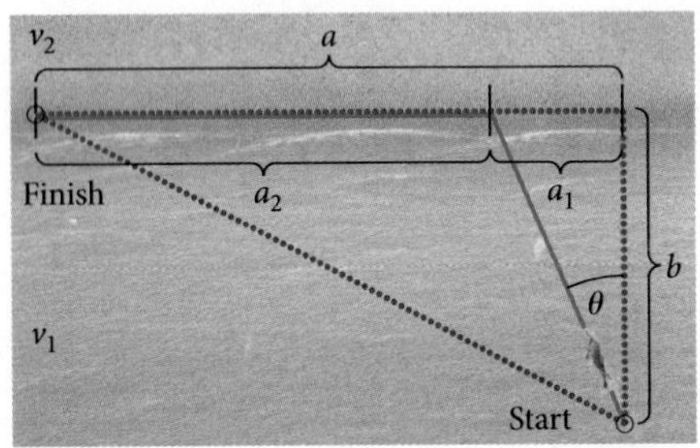


**FIGURE 2.24** Geometry of the aquathlon course.

However, for this example, we'll consider a competition where thinking is rewarded in addition to athletic prowess. The competition consists of only two legs: a swim followed by a run. (This competition is sometimes called an *aquathlon.*) Athletes start a distance $b = 1.5$ km from shore, and the finish line is a distance $a = 3$ km to the left along the shoreline (Figure 2.24). Suppose you can swim with a speed of $v_1 = 3.5$ km/h and can run across the sand with a speed of $v_2 = 14$ km/h.

**PROBLEM**

What angle $\theta$ will result in the shortest finish time under these conditions?

**SOLUTION**

Clearly, the dotted red line marks the shortest distance between start and finish. This distance is $\sqrt{a^2+b^2} = \sqrt{1.5^2+3^2}$ km $= 3.354$ km. Because this entire path is in water, the time it takes to complete the race in this way is

$$t_{\text{red}} = \frac{\sqrt{a^2+b^2}}{v_1} = \frac{3.354 \text{ km}}{3.5 \text{ km/h}} = 0.958 \text{ h}.$$

Because you can run faster than you can swim, we can also try the approach indicated by the dotted blue line: swim straight to shore and then run. This takes

$$t_{\text{blue}} = \frac{b}{v_1} + \frac{a}{v_2} = \frac{1.5 \text{ km}}{3.5 \text{ km/h}} + \frac{3 \text{ km}}{14 \text{ km/h}} = 0.643 \text{ h}.$$

Thus, the blue path is better than the red one. But is it the best? To answer this question, we need to search the angle interval from 0 (blue path) to $\tan^{-1}(3/1.5) = 63.43°$ (red path). Let's consider the green path, with an arbitrary angle $\theta$ with respect to the straight line to shore (that is, the normal to the shoreline). On the green path, you have to swim a distance of $\sqrt{a_1^2+b^2}$ and then run a distance of $a_2$, as indicated in Figure 2.24. The total time for this path is

$$t = \frac{\sqrt{a_1^2+b^2}}{v_1} + \frac{a_2}{v_2}.$$

To find the minimum time, we can express the time in terms of the distance $a_1$ only, take the derivative of that time with respect to that distance, set that derivative equal to zero, and solve for the distance. Using $a_1$, we can then calculate the angle $\theta$ that the athlete must swim. We can express the distance $a_2$ in terms of the given distance $a$ and $a_1$:

$$a_2 = a - a_1.$$

We can then express the time to complete the race in terms of $a_1$:

$$t(a_1) = \frac{\sqrt{a_1^2 + b^2}}{v_1} + \frac{a - a_1}{v_2}. \quad (2.26)$$

Taking the derivative with respect to $a_1$ and setting that result equal to zero gives us

$$\frac{dt(a_1)}{da_1} = \frac{a_1}{v_1\sqrt{a_1^2 + b^2}} - \frac{1}{v_2} = 0.$$

Rearranging yields

$$\frac{a_1}{v_1\sqrt{a_1^2 + b^2}} = \frac{1}{v_2},$$

which we can rewrite as

$$\frac{v_1\sqrt{a_1^2 + b^2}}{a_1} = v_2.$$

Squaring both sides and rearranging terms produces

$$v_1^2\left(a_1^2 + b^2\right) = a_1^2 v_2^2 \Rightarrow a_1^2 v_1^2 + b^2 v_1^2 = a_1^2 v_2^2 \Rightarrow b^2 v_1^2 = a_1^2\left(v_2^2 - v_1^2\right).$$

Solving for $a_1$ gives us

$$a_1 = \frac{bv_1}{\sqrt{v_2^2 - v_1^2}}.$$

In Figure 2.24, we can see that $\tan\theta = a_1/b$, so we can write

$$\tan\theta = \frac{a_1}{b} = \frac{\dfrac{bv_1}{\sqrt{v_2^2 - v_1^2}}}{b} = \frac{v_1}{\sqrt{v_2^2 - v_1^2}}.$$

We can simplify this result by looking at the triangle for the angle $\theta$ in Figure 2.25. The Pythagorean theorem tells us that the hypotenuse of the triangle is

$$\sqrt{v_1^2 + v_2^2 - v_1^2} = v_2.$$

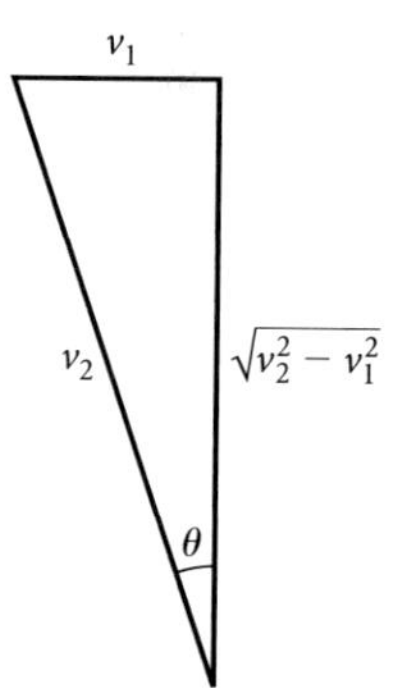


**FIGURE 2.25** Relation between $v_1$, $v_2$, and $\theta$.

We can then write

$$\sin\theta = \frac{v_1}{v_2}.$$

This result is very interesting because the distances $a$ and $b$ do not appear in it at all! Instead the sine of the optimum angle is simply the ratio of the speeds in water and on land. For the given values of the two speeds, this angle is

$$\theta_m = \sin^{-1}\frac{3.5}{14} = 14.48°.$$

Inserting $a_1 = b\tan\theta_m$ into equation 2.26, we find $t(\theta_m) = 0.629$ h. This time is approximately 49 seconds faster than swimming straight to shore and then running (blue path).

Strictly speaking, we have not quite shown that this angle results in the minimum time. To accomplish this, we also need to show that the second derivative of the time with respect to the angle is larger than zero. However, since we did find one extremum, and since its value is smaller than those of the boundaries, we know that this extremum is a true minimum.

*Continued—*

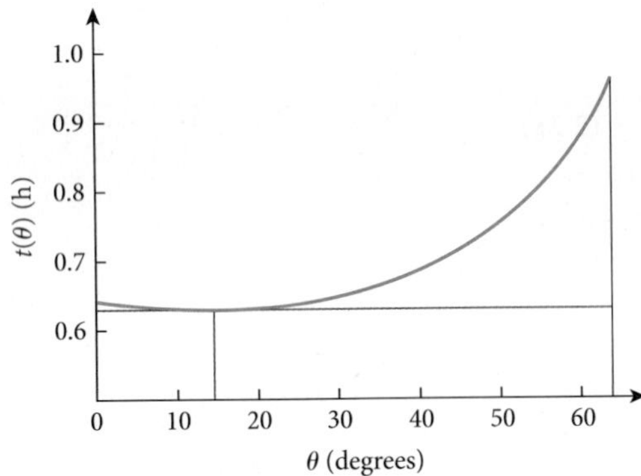


**FIGURE 2.26** Duration of the race as a function of initial angle.

Finally, Figure 2.26 plots the time, in hours, needed to complete the race for all angles between 0° and 63.43°, indicated by the green curve. This plot is obtained by substituting $a_1 = b\tan\theta$ into equation 2.26, which gives us

$$t(\theta)=\frac{\sqrt{(b\tan\theta)^2+b^2}}{v_1}+\frac{a-b\tan\theta}{v_2}=\frac{b\sec\theta}{v_1}+\frac{a-b\tan\theta}{v_2},$$

using the identity $\tan^2\theta + 1 = \sec^2\theta$. A vertical red line marks the maximum angle, corresponding to a straight-line swim from start to finish. The vertical blue line marks the optimum angle that we calculated, and the horizontal blue line marks the duration of the race for this angle.

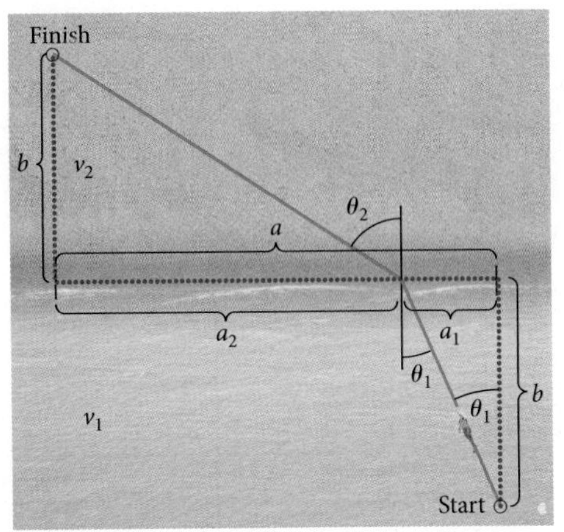


**FIGURE 2.27** Modified aquathlon, with finish away from the shoreline.

Having completed Example 2.6, we can address a more complicated question: If the finish line is not at the shoreline, but at a perpendicular distance $b$ away from the shoreline, as shown in Figure 2.27, what are the angles $\theta_1$ and $\theta_2$ that a competitor needs to select to achieve the minimum time?

We proceed in a fashion very similar to our approach in Example 2.6. However, now we have to realize that the time depends on two angles, $\theta_1$ and $\theta_2$. These two angles are not independent of each other. We can better see the relationship between $\theta_1$ and $\theta_2$ by changing the orientation of the lower triangle in Figure 2.27, as shown in Figure 2.28.

Now we see that the two right triangles $a_1bc_1$ and $a_2bc_2$ have a common side $b$, which helps us relate the two angles to each other. We can express the time to complete the race as

$$t=\frac{\sqrt{a_1^2+b^2}}{v_1}+\frac{\sqrt{a_2^2+b^2}}{v_2}.$$

Again realizing that $a_2 = a - a_1$, we can write

$$t(a_1)=\frac{\sqrt{a_1^2+b^2}}{v_1}+\frac{\sqrt{(a-a_1)^2+b^2}}{v_2}.$$

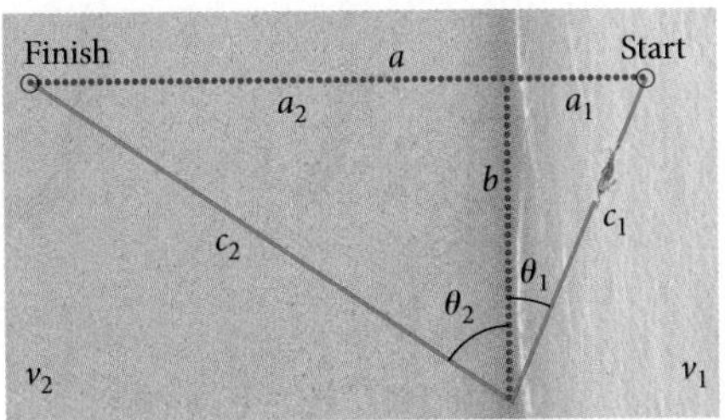


**FIGURE 2.28** Same as previous figure, but with lower triangle reflected along the horizontal axis.

Taking the derivative of the time to complete the race with respect to $a_1$ and setting that result equal to zero, we obtain

$$\frac{dt(a_1)}{da_1}=\frac{a_1}{v_1\sqrt{a_1^2+b^2}}-\frac{a-a_1}{v_2\sqrt{(a-a_1)^2+b^2}}=0.$$

We can rearrange this equation to get

$$\frac{a_1}{v_1\sqrt{a_1^2+b^2}}=\frac{a-a_1}{v_2\sqrt{(a-a_1)^2+b^2}}=\frac{a_2}{v_2\sqrt{a_2^2+b^2}}.$$

Looking at Figure 2.28 and referring to our previous result from Figure 2.25, we can see that

$$\sin\theta_1=\frac{a_1}{\sqrt{a_1^2+b^2}}$$

and

$$\sin\theta_2=\frac{a_2}{\sqrt{a_2^2+b^2}}.$$

We can insert these two results into the previous equation and finally find the shortest time to finish the race we require

$$\frac{\sin\theta_1}{v_1}=\frac{\sin\theta_2}{v_2}. \tag{2.27}$$

We can now see that our previous result, where we forced the running to take place along the beach, is a special case of this more general result equation 2.27, with $\theta_2 = 90°$.

Just as for that special case, we find that the relationship between the angles does not depend on the values of the displacements $a$ and $b$, but depends only on the speeds with which the competitor can move in the water and on land. The angles are still related to $a$ and $b$ by the overall constraint that the competitor has to get from the start to the finish. However, for the path of minimum time, the change in direction at the boundary between water and land, as expressed by the two angles $\theta_1$ and $\theta_2$, is determined exclusively by the ratio of the speeds $v_1$ and $v_2$.

The initial condition specified that the perpendicular distance $b$ from the starting point to the shoreline be the same as the perpendicular distance between the shoreline and the finish. We did this to keep the algebra relatively brief. However, in the final formula you see that there are no more references to $b$; it has canceled out. Thus, equation 2.27 is even valid in the case that the two perpendicular distances have different values. The ratio of angles for the path of minimum time is exclusively determined by the two speeds in the different media.

Interestingly, we will encounter the same relationship between two angles and two speeds when we study light changing direction at the interface between two media through which light moves with different speeds. In Chapter 32, we will see that light also moves along the path of minimum time and that the result obtained in equation 2.27 is known as Snell's Law.

Finally, we make an observation that may appear trivial, but is not: If a competitor started at the point marked "Finish" in Figure 2.27 and ended up at the point marked "Start," he or she would have to take exactly the same path as the one we just calculated for the reverse direction. Snell's Law holds in both directions.

## WHAT WE HAVE LEARNED | EXAM STUDY GUIDE

- $x$ is the $x$-component of the position vector. Displacement is the change in position: $\Delta x = x_2 - x_1$.
- Distance is the absolute value of displacement, $\ell = |\Delta x|$, and is a positive scalar for motion in one direction.
- The average velocity of an object in a given time interval is given by $\bar{v}_x = \dfrac{\Delta x}{\Delta t}$.
- The $x$-component of (instantaneous) velocity *vector* is the derivative of the $x$-component of position *vector* as a function of time, $v_x = \dfrac{dx}{dt}$.
- Speed is the absolute value of the velocity: $v = |v_x|$.
- The $x$-component of (instantaneous) acceleration *vector* is the derivative of the $x$-component of the velocity *vector* as a function of time, $a_x = \dfrac{dv_x}{dt}$.
- For constant accelerations, five kinematical equations describe motion in one dimension:
  (i) $x = x_0 + v_{x0}t + \frac{1}{2}a_x t^2$
  (ii) $x = x_0 + \bar{v}_x t$
  (iii) $v_x = v_{x0} + a_x t$
  (iv) $\bar{v}_x = \frac{1}{2}(v_x + v_{x0})$
  (v) $v_x^2 = v_{x0}^2 + 2a_x(x - x_0)$
  where $x_0$ is the initial position, $v_{x0}$ is the initial velocity, and the initial time $t_0$ is set to zero.
- For situations involving free fall (constant acceleration), we replace the acceleration $a$ with $-g$ and $x$ with $y$ to obtain
  (i) $y = y_0 + v_{y0}t - \frac{1}{2}gt^2$
  (ii) $y = y_0 + \bar{v}_y t$
  (iii) $v_y = v_{y0} - gt$
  (iv) $\bar{v}_y = \frac{1}{2}(v_y + v_{y0})$
  (v) $v_y^2 = v_{y0}^2 - 2g(y - y_0)$
  where $y_0$ is the initial position, $v_{y0}$ is the initial velocity, and the $y$-axis points up.

## KEY TERMS

mechanics, p. 36
kinematics, p. 36
position vector, p. 37
displacement, p. 37
distance, p. 38
average velocity, p. 40
instantaneous velocity, p. 40
speed, p. 42
average acceleration, p. 43
instantaneous acceleration, p. 43
free fall, p. 51

## NEW SYMBOLS

$\Delta x = x_2 - x_1$, displacement in one dimension

$\bar{v}_x = \frac{\Delta x}{\Delta t}$, average velocity in one dimension in a time interval $\Delta t$

$v_x = \frac{dx}{dt}$, instantaneous velocity in one dimension

$\bar{a}_x = \frac{\Delta v}{\Delta t}$, average acceleration in one dimension in a time interval $\Delta t$

$a_x = \frac{dv_x}{dt}$, instantaneous acceleration in one dimension

## ANSWERS TO SELF-TEST OPPORTUNITIES

**2.1** The result would not change, because a shift in the origin of the coordinate system has no influence on the net displacements or distances.

**2.2**

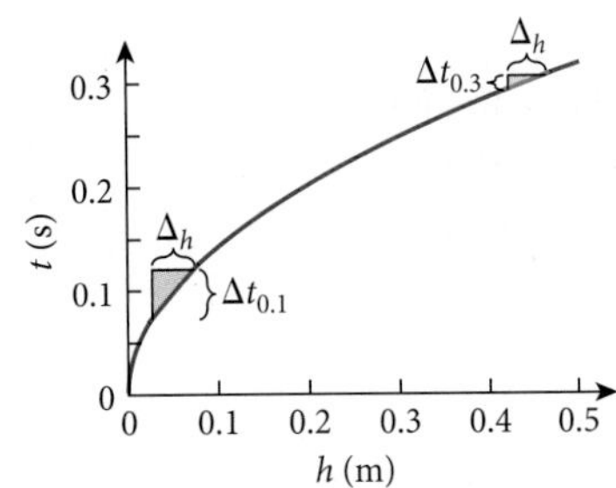


This method is more precise for longer reaction times, because the slope of the curve decreases as a function of the height, $h$. (In the graph, the slope at 0.1 s is indicated by the blue line, and that at 0.3 s by the green line.) So, a given uncertainty, $\Delta h$, in measuring the height results in a smaller uncertainty, $\Delta t$, in the value for the reaction time for longer reaction times.

**2.3** The arrow would still hit the melon, but at the time of collision the arrow would already have negative velocity and thus be moving downward again. Thus, the melon would catch the arrow as it was falling. The collision would occur a bit later, after $t = 58.3 \text{ m}/(20.1 \text{ m/s}) = 2.90$ s. The altitude of the collision would be quite a bit lower, at $y_m(t_c) = 58.3 \text{ m} - \frac{1}{2}(9.81 \text{ m/s}^2)(2.90 \text{ s})^2 = 1.70$ m above Pierre's head.

## PROBLEM-SOLVING PRACTICE

### SOLVED PROBLEM 2.4 Racing with a Head Start

Cheri has a new Dodge Charger with a Hemi engine and has challenged Vince, who owns a tuned VW GTI, to a race at a local track. Vince knows that Cheri's Charger is rated to go from 0 to 60 mph in 5.3 s, whereas his VW needs 7.0 s. Vince asks for a head start and Cheri agrees to give him exactly 1.0 s.

**PROBLEM**

How far down the track is Vince before Cheri gets to start the race? At what time does Cheri catch Vince? How far away from the start are they when this happens? (Assume constant acceleration for each car during the race.)

**SOLUTION**

**THINK**

This race is a good example of one-dimensional motion with constant acceleration. The temptation is to look over the kinematical equations 2.23 and see which one we can apply. However, we have to be a bit more careful here, because the time delay between Vince's start and Cheri's start adds a complication. In fact, if you try to solve this problem using the kinematical equations directly, you will not get the right answer. Instead, this problem requires some careful definition of the time coordinates for each car.

**SKETCH**

For our sketch, we plot time on the horizontal axis and position on the vertical axis. Both cars move with constant acceleration from a standing start, so we expect simple parabolas for their paths in this diagram.

Since Cheri's car has the greater acceleration, her parabola (blue curve in Figure 2.29) has the larger curvature and thus the steeper rise. Therefore, it is clear that Cheri will catch Vince at some point, but it is not yet clear where this point is.

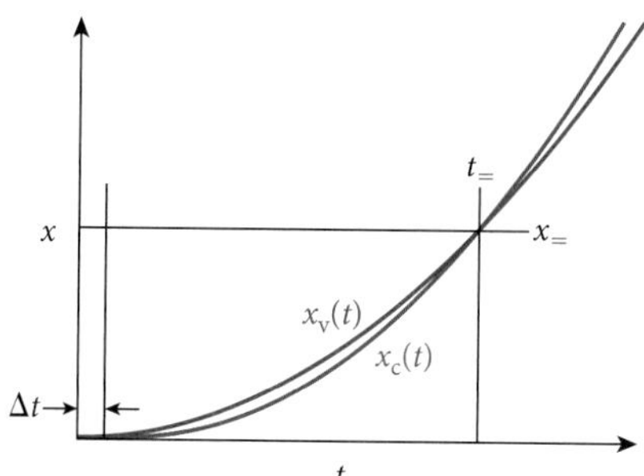


**FIGURE 2.29** Position versus time for the race between Cheri and Vince.

## RESEARCH

We set up the problem quantitatively. We call the time delay before Cheri can start $\Delta t$, and we use the indices (subscripts) C for Cheri's Charger and V for Vince's VW. We put the coordinate system's origin at the starting line. Thus, both cars have initial position $x_C(t=0) = x_V(t=0) = 0$. Since both cars are at rest at the start, their initial velocities are zero. The equation of motion for Vince's VW is

$$x_V(t) = \tfrac{1}{2}a_V t^2.$$

Here we use the symbol $a_V$ for the acceleration of the VW. We can calculate its value from the 0 to 60 mph time given by the problem statement, but we will postpone this step until it is time to put in the numbers.

To obtain the equation of motion for Cheri's Charger we have to be careful, because she is forced to wait $\Delta t$ after Vince has taken off. We can account for this delay with a shifted time: $t' = t - \Delta t$. Once $t$ reaches the value of $\Delta t$, the time $t'$ has the value 0 and then Cheri can take off. So her Charger's equation of motion is

$$x_C(t) = \tfrac{1}{2}a_C t'^2 = \tfrac{1}{2}a_C(t-\Delta t)^2 \qquad \text{for } t \geq \Delta t.$$

Just like $a_V$, the constant acceleration $a_C$ for Cheri's Charger will be evaluated below.

## SIMPLIFY

When Cheri catches Vince, their coordinates will have the same value. We will call the time at which this happens $t_=$, and the coordinate where it happens $x_= \equiv x(t_=)$. Because the two coordinates are the same, we have

$$x_= = \tfrac{1}{2}a_V t_=^2 = \tfrac{1}{2}a_C(t_= - \Delta t)^2.$$

We can solve this equation for $t_=$ by dividing out the common factor of $\frac{1}{2}$ and then taking the square root of both sides of the equation:

$$\sqrt{a_V}\,t_= = \sqrt{a_C}\,(t_= - \Delta t) \Rightarrow$$
$$t_=\left(\sqrt{a_C} - \sqrt{a_V}\right) = \Delta t\sqrt{a_C} \Rightarrow$$
$$t_= = \frac{\Delta t\sqrt{a_C}}{\sqrt{a_C} - \sqrt{a_V}}.$$

Why do we use the positive root and discard the negative root here? The negative root would lead to a physically impossible solution: We are interested in the time that the two cars meet *after* they have left the start, not in a negative value that would imply a time before they left.

## CALCULATE

Now we can get a numerical answer for each of the questions that were asked. First, let's figure out the values of the cars' constant accelerations from the 0 to 60 mph specifications given. We use $a = (v_x - v_{x0})/t$ and get

$$a_V = \frac{60 \text{ mph}}{7.0 \text{ s}} = \frac{26.8167 \text{ m/s}}{7.0 \text{ s}} = 3.83095 \text{ m/s}^2$$
$$a_C = \frac{60 \text{ mph}}{5.3 \text{ s}} = \frac{26.8167 \text{ m/s}}{5.3 \text{ s}} = 5.05975 \text{ m/s}^2.$$

Again, we postpone rounding our results until we have completed all steps in our calculations. However, with the values for the accelerations, we can immediately calculate how far Vince travels during the time $\Delta t = 1.0$ s:

$$x_V(1.0 \text{ s}) = \tfrac{1}{2}(3.83095 \text{ m/s}^2)(1.0 \text{ s})^2 = 1.91548 \text{ m}.$$

*Continued—*

Now we can calculate the time when Cheri catches up to Vince:

$$t_{=} = \frac{\Delta t\sqrt{a_C}}{\sqrt{a_C} - \sqrt{a_V}} = \frac{(1.0 \text{ s})\sqrt{5.05975 \text{ m/s}^2}}{\sqrt{5.05975 \text{ m/s}^2} - \sqrt{3.83095 \text{ m/s}^2}} = 7.70055 \text{ s}.$$

At this time, both cars have traveled to the same point. Thus, we can insert this value into either equation of motion to find this position:

$$x_{=} = \tfrac{1}{2}a_V t_{=}^2 = \tfrac{1}{2}(3.83095 \text{ m/s}^2)(7.70055 \text{ s})^2 = 113.585 \text{ m}.$$

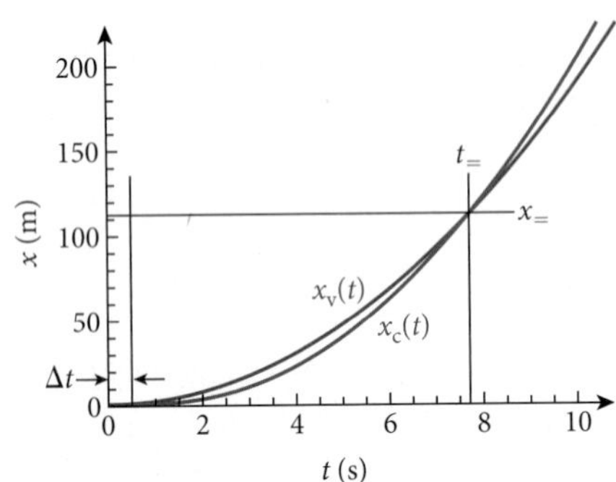


**FIGURE 2.30** Plot of the parameters and equations of motion for the race between Cheri and Vince.

### ROUND

The initial data were specified only to a precision of two significant digits. Rounding our results to the same precision, we finally arrive at our answers: Vince receives a head start of 1.9 m, and Cheri will catch him after 7.7 s. At that time, they will be $1.1 \cdot 10^2$ m into their race.

### DOUBLE-CHECK

It may seem strange to you that Vince's car was able to move only 1.9 m, or approximately half the length of the car, during the first second. Have we done anything wrong in our calculation? The answer is no; from a standing start, cars move only a comparatively short distance during the first second of acceleration. The following solved problem contains visible proof for this statement.

Figure 2.30 shows a plot of the equations of motion for both cars, this time with the proper units.

## SOLVED PROBLEM 2.5 Accelerating Car

**FIGURE 2.31** Video sequence of a car accelerating from a standing start.

### PROBLEM

You are given the image sequence shown in Figure 2.31, and you are told that there is a time interval of 0.333 s between successive frames. Can you determine how fast this car (2007 Ford Escape Hybrid, length 174.9 in) was accelerating from rest? Also, can you give an estimate for the time it takes this car to go from 0 to 60 mph?

### SOLUTION

### THINK

Acceleration is measured in dimensions of length per time per time. To find a number for the value of the acceleration, we need to know the time and length scales in Figure 2.31. The time scale is straightforward because we were given the information that 0.333 s passes between successive frames. We can get the length scale from the specified vehicle dimensions. For example, if we focus on the length of the car and compare it to the overall width of the frame, we can find the distance the car covered between the first and the last frame (which are 3.000 s apart).

### SKETCH

We draw vertical yellow lines over Figure 2.31, as shown in Figure 2.32. We put the center of the car at the line between the front and rear windows (the exact location is irrelevant, as long as we are consistent). Now we can use a ruler and measure the perpendicular distance between the two yellow lines, indicated by the yellow double-headed arrow in the figure. We can also measure the length of the car, as indicated by the red double-headed arrow.

### RESEARCH

Dividing the measured length of the yellow double-headed arrow by that of the red one gives us a ratio of 3.474. Since the two vertical yellow lines mark the position of the center of the car at 0.000 s and 3.000 s, we know that the car covered a distance of 3.474 car lengths in this time interval. The car length was given as 174.9 in = 4.442 m. Thus, the total distance covered is $d = 3.474 \cdot \ell_{car} = 3.474 \cdot 4.442\ \text{m} = 15.4315\ \text{m}$ (remember that we round to the proper number of significant digits at the end).

**SIMPLIFY**

We have two choices on how to proceed. The first one is more complicated: We could measure the position of the car in each frame and then use difference formulas like those in Example 2.3. The other and much faster way to proceed is to assume constant acceleration and then use the measurements of the car's positions in only the first and last frames. We'll use the second way, but in the end we'll need to double-check that our assumption of constant acceleration is justified.

For a constant acceleration from a standing start, we simply have

$$x = x_0 + \tfrac{1}{2}at^2 \Rightarrow$$
$$d = x - x_0 = \tfrac{1}{2}at^2 \Rightarrow$$
$$a = \frac{2d}{t^2}.$$

This is the acceleration we want to find. Once we have the acceleration, we can give an estimate for the time from 0 to 60 mph by using $v_x = v_{x0} + at \Rightarrow t = (v_x - v_{x0})/a$. A standing start means $v_{x0} = 0$, and so we have

$$t(0-60\ \text{mph}) = \frac{60\ \text{mph}}{a}.$$

**CALCULATE**

We insert the numbers for the acceleration:

$$a = \frac{2d}{t^2} = \frac{2 \cdot (15.4315\ \text{m})}{(3.000\ \text{s})^2} = 3.42922\ \text{m/s}^2.$$

Then, for the time from 0 to 60 mph, we obtain

$$t(0-60\ \text{mph}) = \frac{60\ \text{mph}}{a} = \frac{(60\ \text{mph})(1609\ \text{m/mi})(1\ \text{h}/3600\ \text{s})}{3.42922\ \text{m/s}^2} = 7.82004\ \text{s}.$$

**ROUND**

The length of the car sets our length scale, and it was given to four significant figures. The time was given to three significant figures. Are we entitled to quote our results to three significant digits? The answer is no, because we also performed measurements on Figure 2.32, which are probably accurate to only two digits at best. In addition, you can see field-of-vision and lens distortions in the image sequence: In the first few frames, you see a little bit of the front of the car, and in the last few frames, a little bit of the back. Taking all this into account, our results should be quoted to two significant figures. So our final answer for the acceleration is

$$a = 3.4\ \text{m/s}^2.$$

For the time from 0 to 60 mph, we give

$$t(0-60\ \text{mph}) = 7.8\ \text{s}.$$

**DOUBLE-CHECK**

The numbers we have found for the acceleration and the time from 0 to 60 mph are fairly typical for cars or small SUVs; see also Solved Problem 2.4. Thus, we have confidence that we are not off by orders of magnitude.

What we must also double-check, however, is the assumption of constant acceleration. For constant acceleration from a standing start, the points $x(t)$ should fall on a parabola $x(t) = \tfrac{1}{2}at^2$. Therefore, if we plot $x$ on the horizontal axis and $t$ on the vertical axis as in Figure 2.33, the points $t(x)$ should follow a square root dependence: $t(x) = \sqrt{2x/a}$. This functional dependence is shown by the red curve in Figure 2.33. We can see that the same point of the car is hit by the curve in every frame, giving us confidence that the assumption of constant acceleration is reasonable.

**FIGURE 2.32** Determination of the length scale for Figure 2.31.

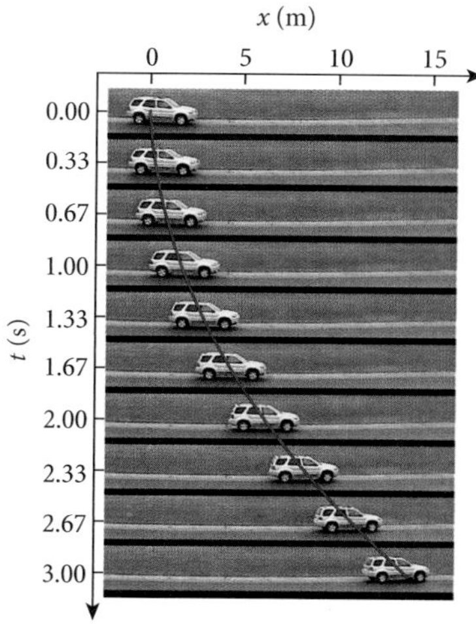


**FIGURE 2.33** Graphical analysis of the accelerating car problem.

## MULTIPLE-CHOICE QUESTIONS

**2.1** Two athletes jump straight up. Upon leaving the ground, Adam has half the initial speed of Bob. Compared to Adam, Bob jumps

a) 0.50 times as high.
b) 1.41 times as high.
c) twice as high.
d) three times as high.
e) four times as high.

**2.2** Two athletes jump straight up. Upon leaving the ground, Adam has half the initial speed of Bob. Compared to Adam, Bob is in the air

a) 0.50 times as long.
b) 1.41 times as long.
c) twice as long.
d) three times as long.
e) four times as long.

**2.3** A car is traveling due west at 20.0 m/s. Find the velocity of the car after 3.00 s if its acceleration is 1.0 $m/s^2$ due west. Assume the acceleration remains constant.

a) 17.0 m/s west
b) 17.0 m/s east
c) 23.0 m/s west
d) 23.0 m/s east
e) 11.0 m/s south

**2.4** A car is traveling due west at 20.0 m/s. Find the velocity of the car after 37.00 s if its constant acceleration is 1.0 $m/s^2$ due east. Assume the acceleration remains constant.

a) 17.0 m/s west
b) 17.0 m/s east
c) 23.0 m/s west
d) 23.0 m/s east
e) 11.0 m/s south

**2.5** An electron, starting from rest and moving with a constant acceleration, travels 1.0 cm in 2.0 ms. What is the magnitude of this acceleration?

a) 25 $km/s^2$
b) 20 $km/s^2$
c) 15 $km/s^2$
d) 10 $km/s^2$
e) 5.0 $km/s^2$

**2.6** A car travels 22.0 m/s north for 30.0 min and then reverses direction and travels 28.0 m/s for 15.0 min. What is the car's total displacement?

a) $1.44 \cdot 10^4$ m b) $6.48 \cdot 10^4$ m c) $3.96 \cdot 10^4$ m d) $9.98 \cdot 10^4$ m

**2.7** Which of these statement(s) is (are) true?

1. An object can have zero acceleration and be at rest.
2. An object can have nonzero acceleration and be at rest.
3. An object can have zero acceleration and be in motion.

a) 1 only b) 1 and 3 c) 1 and 2 d) 1, 2, and 3

**2.8** A car moving at 60 km/h comes to a stop in 4.0 s. What was its average deceleration?

a) 2.4 $m/s^2$ b) 15 $m/s^2$ c) 4.2 $m/s^2$ d) 41 $m/s^2$

**2.9** You drop a rock from a cliff. If air resistance is neglected, which of the following statements is (are) true?

1. The speed of the rock will increase.
2. The speed of the rock will decrease.
3. The acceleration of the rock will increase.
4. The acceleration of the rock will decrease.

a) 1 b) 1 and 4 c) 2 d) 2 and 3

**2.10** A car travels at 22.0 mph for 15.0 min and 35.0 mph for 30.0 min. How far does it travel overall?

a) 23.0 m b) $3.70 \cdot 10^4$ m c) $1.38 \cdot 10^3$ m d) $3.30 \cdot 10^2$ m

## QUESTIONS

**2.11** Consider three ice skaters: Anna moves in the positive $x$-direction without reversing. Bertha moves in the negative $x$-direction without reversing. Christine moves in the positive $x$-direction and then reverses the direction of her motion. For which of these skaters is the magnitude of the average velocity smaller than the average speed over some time interval?

**2.12** You toss a small ball vertically up in the air. How are the velocity and acceleration vectors of the ball oriented with respect to one another during the ball's flight up and down?

**2.13** After you apply the brakes, the acceleration of your car is in the opposite direction to its velocity. If the acceleration of your car remains constant, describe the motion of your car.

**2.14** Two cars are traveling at the same speed, and the drivers hit the brakes at the same time. The deceleration of one car is double that of the other. By what factor does the time required for that car to come to a stop compare with that for the other car?

**2.15** If the acceleration of an object is zero and its velocity is nonzero, what can you say about the motion of the object? Sketch velocity versus time and acceleration versus time graphs for your explanation.

**2.16** Can an object's acceleration be in the opposite direction to its motion? Explain.

**2.17** You and a friend are standing at the edge of a snow-covered cliff. At the same time, you both drop a snowball over the edge of the cliff. Your snowball is twice as heavy as your friend's. Neglect air resistance. (a) Which snowball will hit the ground first? (b) Which snowball will have the greater speed?

**2.18** You and a friend are standing at the edge of a snow-covered cliff. At the same time, you throw a snowball straight upward with a speed of 8.0 m/s over the edge of the cliff and your friend throws a snowball straight downward over the edge of the cliff with the same speed. Your snowball is twice as heavy as your friend's. Neglecting air resistance, which snowball will hit the ground first, and which will have the greater speed?

**2.19** A car is slowing down and comes to a complete stop. The figure shows an image sequence of this process. The time between successive frames is 0.333 s, and the car is the same as the one in Solved Problem 2.5. Assuming constant acceleration, what is its value? Can you give some estimate of the error in your answer? How well justified is the assumption of constant acceleration?

**2.20** A car moves along a road with a constant velocity. Starting at time $t = 2.5$ s, the driver accelerates with constant acceleration. The resulting position of the car as a function of time is shown by the blue curve in the figure.

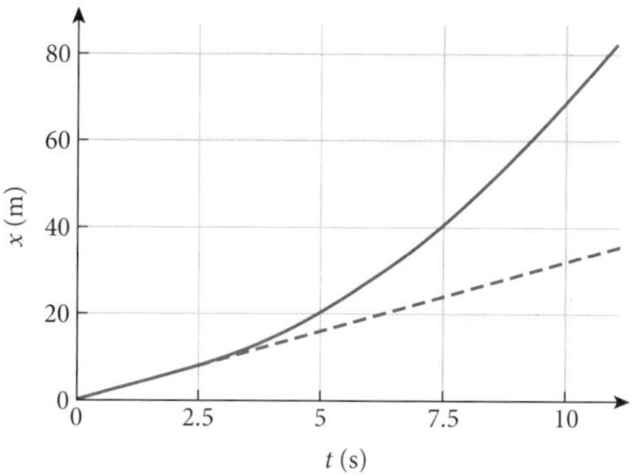


a) What is the value of the constant velocity of the car before 2.5 s? (*Hint:* The dashed blue line is the path the car would take in the absence of the acceleration.)

b) What is the velocity of the car at $t = 7.5$ s? Use a graphical technique (i.e., draw a slope).

c) What is the value of the constant acceleration?

**2.21** You drop a rock over the edge of a cliff from a height $h$. Your friend throws a rock over the edge from the same height with a speed $v_0$ vertically downward, at some time $t$ after you drop your rock. Both rocks hit the ground at the same time. How long after you dropped your rock did your friend throw hers? Express your answer in terms of $v_0$, $g$, and $h$.

**2.22** The position of a particle as a function of time is given as $x(t) = \frac{1}{4}x_0e^{3\alpha t}$, where $\alpha$ is a positive constant.

a) At what time is the particle at $2x_0$?

b) What is the speed of the particle as a function of time?

c) What is the acceleration of the particle as a function of time?

d) What are the SI units for $\alpha$?

**2.23** The position versus time for an object is given as $x = At^4 - Bt^3 + C$.

a) What is the instantaneous velocity as a function of time?

b) What is the instantaneous acceleration as a function of time?

**2.24** A wrench is thrown vertically upward with speed $v_0$. How long after its release is it halfway to its maximum height?

## PROBLEMS

A blue problem number indicates a worked-out solution is available in the Student Solutions Manual. One • and two •• indicate increasing level of problem difficulty.

### Section 2.2

**2.25** A car travels north at 30 m/s for 10 min. It then travels south at 40 m/s for 20 min. What are the total distance the car travels and its displacement?

**2.26** You ride your bike along a straight line from your house to a store 1000 m away. On your way back, you stop at a friend's house which is halfway between your house and the store.

a) What is your displacement?

b) What is the total distance traveled? After talking to your friend, you continue to your house. When you arrive back at your house,

c) What is your displacement?

d) What is the distance you have traveled?

### Section 2.3

**2.27** Running along a rectangular track 50 m × 40 m you complete one lap in 100 s. What is your average velocity for the lap?

**2.28** An electron moves in the positive $x$-direction a distance of 2.42 m in $2.91 \cdot 10^{-8}$ s, bounces off a moving proton, and then moves in the opposite direction a distance of 1.69 m in $3.43 \cdot 10^{-8}$ s.

a) What is the average velocity of the electron over the entire time interval?

b) What is the average speed of the electron over the entire time interval?

**2.29** The graph describes the position of a particle in one dimension as a function of time. Answer the following questions.

a) In which time interval does the particle have its maximum speed? What is that speed?

b) What is the average velocity in the time interval between −5 s and +5 s?

c) What is the average speed in the time interval between −5 s and +5 s?

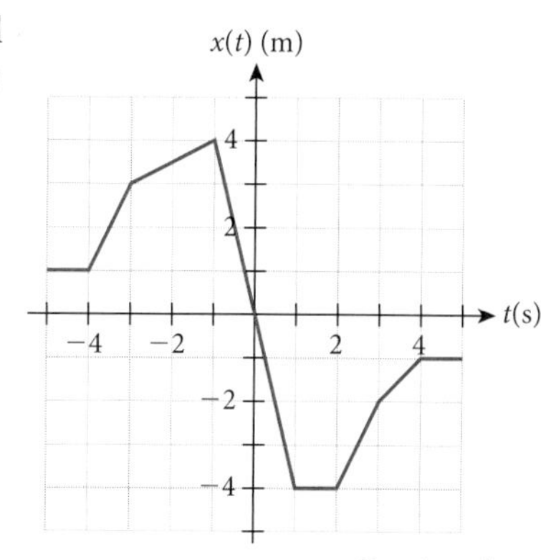


*Continued—*

d) What is the ratio of the velocity in the interval between 2 s and 3 s to that in the interval between 3 s and 4 s?

e) At what time(s) is the particle's velocity zero?

**2.30** The position of a particle moving along the $x$-axis is given by $x = (11 + 14t - 2.0t^2)$, where $t$ is in seconds and $x$ is in meters. What is the average velocity during the time interval from $t = 1.0$ s to $t = 4.0$ s?

•**2.31** The position of a particle moving along the $x$-axis is given by $x = 3.0t^2 - 2.0t^3$, where $x$ is in meters and $t$ is in seconds. What is the position of the particle when it achieves its maximum speed in the positive $x$-direction?

**2.32** The rate of continental drift is on the order of 10 mm/yr. Approximately how long did it take North America and Europe to reach their current separation of about 3000 mi?

**2.33** You and a friend are driving to the beach during spring break. You travel 16.0 km east and then 80.0 km south in a total time of 40 minutes. (a) What is the average speed of the trip? (b) What is the average velocity?

•**2.34** The trajectory of an object is given by the equation

$$x(t) = (4.35 \text{ m}) + (25.9 \text{ m/s})t - (11.79 \text{ m/s}^2)t^2$$

a) For which time $t$ is the displacement $x(t)$ at its maximum?

b) What is this maximum value?

## Section 2.4

**2.35** A bank robber in a getaway car approaches an intersection at a speed of 45 mph. Just as he passes the intersection, he realizes that he needed to turn. So he steps on the brakes, comes to a complete stop, and then accelerates driving straight backward. He reaches a speed of 22.5 mph moving backward. Altogether his deceleration and reacceleration in the opposite direction take 12.4 s. What is the average acceleration during this time?

**2.36** A car is traveling west at 22.0 m/s. After 10.0 s, its velocity is 17.0 m/s in the same direction. Find the magnitude and direction of the car's average acceleration.

**2.37** Your friend's car starts from rest and travels 0.500 km in 10.0 s. What is the magnitude of the constant acceleration required to do this?

**2.38** A fellow student found in the performance data for his new car the velocity-versus-time graph shown in the figure.

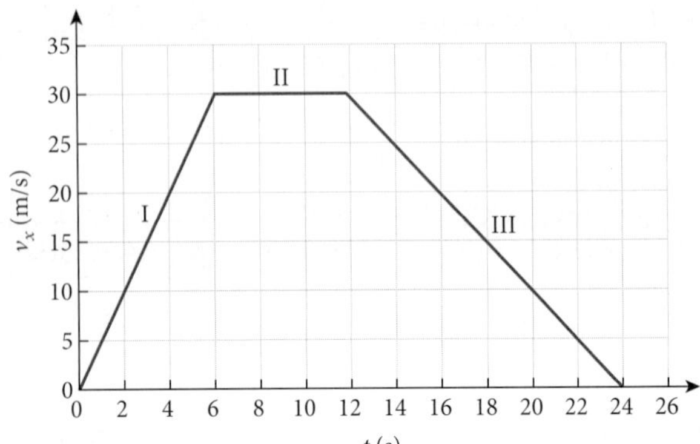


a) Find the average acceleration of the car during each of the segments I, II, and III.

b) What is the total distance traveled by the car from $t = 0$ s to $t = 24$ s?

•**2.39** The velocity of a particle moving along the $x$-axis is given, for $t > 0$, by $v_x = (50.0t - 2.0t^3)$ m/s, where $t$ is in seconds. What is the acceleration of the particle when (after $t = 0$) it achieves its maximum displacement in the positive $x$-direction?

•**2.40.** The 2007 world record for the men's 100-m dash was 9.77 s. The third-place runner crossed the finish line in 10.07 s. When the winner crossed the finish line, how far was the third-place runner behind him?

a) Compute an answer that assumes that each runner ran at his average speed for the entire race.

b) Compute another answer that uses the result of Example 2.3, that a world-class sprinter runs at a speed of 12 m/s after an initial acceleration phase. If both runners in this race reach this speed, how far behind is the third-place runner when the winner finishes?

•**2.41** The position of an object as a function of time is given as $x = At^3 + Bt^2 + Ct + D$. The constants are $A = 2.1$ m/s$^3$, $B = 1.0$ m/s$^2$, $C = -4.1$ m/s, and $D = 3$ m.

a) What is the velocity of the object at $t = 10.0$ s?

b) At what time(s) is the object at rest?

c) What is the acceleration of the object at $t = 0.50$ s?

d) Plot the acceleration as a function of time for the time interval from $t = -10.0$ s to $t = 10.0$ s.

## Section 2.5

••**2.42** An F-14 Tomcat fighter jet is taking off from the deck of the USS *Nimitz* aircraft carrier with the assistance of a steam-powered catapult. The jet's location along the flight deck is measured at intervals of 0.20 s. These measurements are tabulated as follows:

| $t$ (s) | 0.00 | 0.20 | 0.40 | 0.60 | 0.80 | 1.00 | 1.20 | 1.40 | 1.60 | 1.80 | 2.00 |
|---|---|---|---|---|---|---|---|---|---|---|---|
| $x$ (m) | 0.0 | 0.7 | 3.0 | 6.6 | 11.8 | 18.5 | 26.6 | 36.2 | 47.3 | 59.9 | 73.9 |

Use difference formulas to calculate the jet's average velocity and average acceleration for each time interval. After completing this analysis, can you say if the F-14 Tomcat accelerated with approximately constant acceleration?

## Section 2.6

**2.43** A particle starts from rest at $x = 0$ and moves for 20 s with an acceleration of +2.0 cm/s$^2$. For the next 40 s, the acceleration of the particle is −4.0 cm/s$^2$. What is the position of the particle at the end of this motion?

**2.44** A car moving in the $x$-direction has an acceleration $a_x$ that varies with time as shown in the figure. At the moment $t = 0$ s, the car is located at $x = 12$ m and has a velocity of 6 m/s in the positive $x$-direction. What is the velocity of the car at $t = 5.0$ s?

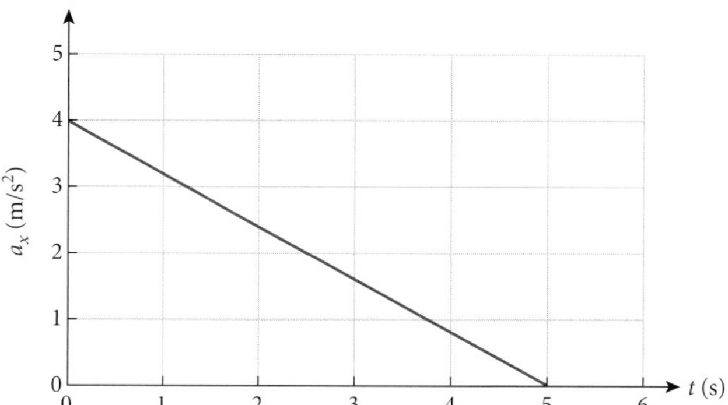


2.45 The velocity as a function of time for a car on an amusement park ride is given as $v = At^2 + Bt$ with constants $A = 2.0$ m/s$^3$ and $B = 1.0$ m/s$^2$. If the car starts at the origin, what is its position at $t = 3.0$ s?

**2.46** An object starts from rest and has an acceleration given by $a = Bt^2 - \frac{1}{2}Ct$, where $B = 2.0$ m/s$^4$ and $C = -4.0$ m/s$^3$.

a) What is the object's velocity after 5.0 s?

b) How far has the object moved after $t = 5.0$ s?

•**2.47** A car is moving along the $x$-axis and its velocity, $v_x$, varies with time as shown in the figure. If $x_0 = 2.0$ m at $t_0 = 2.0$ s, what is the position of the car at $t = 10.0$ s?

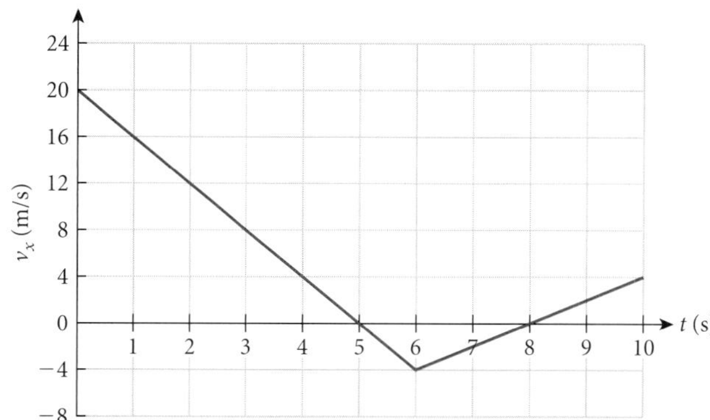


•**2.48** A car is moving along the $x$-axis and its velocity, $v_x$, varies with time as shown in the figure. What is the displacement, $\Delta x$, of the car from $t = 4$ s to $t = 9$ s ?

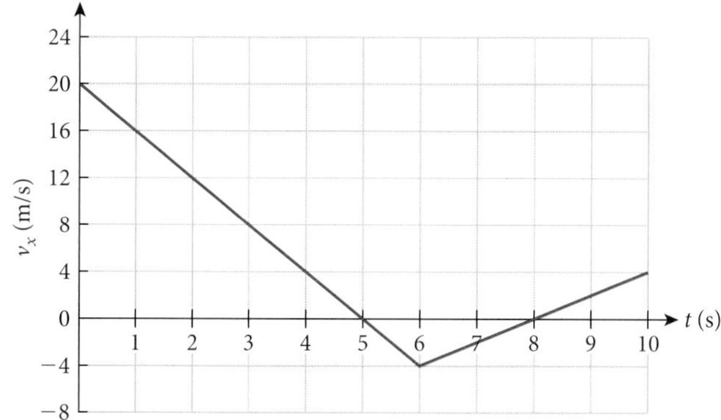


•2.49 A motorcycle starts from rest and accelerates as shown in the figure. Determine (a) the motorcycle's speed at $t = 4.00$ s and at $t = 14.0$ s, and (b) the distance traveled in the first 14.0 s.

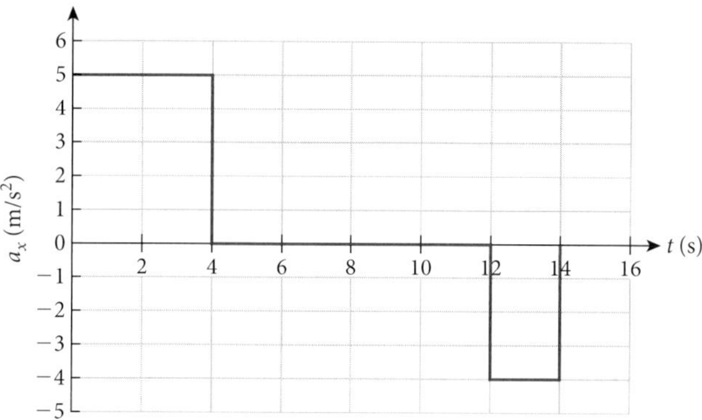


## Section 2.7

**2.50** How much time does it take for a car to accelerate from a standing start to 22.2 m/s if the acceleration is constant and the car covers 243 m during the acceleration?

**2.51** A car slows down from a speed of 31.0 m/s to a speed of 12.0 m/s over a distance of 380. m.

a) How long does this take, assuming constant acceleration?

b) What is the value of this acceleration?

**2.52** A runner of mass 57.5 kg starts from rest and accelerates with a constant acceleration of 1.25 m/s$^2$ until she reaches a velocity of 6.3 m/s. She then continues running with this constant velocity.

a) How far has she run after 59.7 s?

b) What is the velocity of the runner at this point?

2.53 A fighter jet lands on the deck of an aircraft carrier. It touches down with a speed of 70.4 m/s and comes to a complete stop over a distance of 197.4 m. If this process happens with constant deceleration, what is the speed of the jet 44.2 m before its final stopping location?

**2.54** A bullet is fired through a board 10.0 cm thick, with a line of motion perpendicular to the face of the board. If the bullet enters with a speed of 400 m/s and emerges with a speed of 200 m/s, what is its acceleration as it passes through the board?

**2.55** A car starts from rest and accelerates at 10.0 m/s$^2$. How far does it travel in 2.00 s?

**2.56** An airplane starts from rest and accelerates at 12.1 m/s$^2$. What is its speed at the end of a 500-m runway?

2.57 Starting from rest, a boat increases its speed to 5.00 m/s with constant acceleration.

a) What is the boat's average speed?

b) If it takes the boat 4.00 s to reach this speed, how far has it traveled?

**2.58** A ball is tossed vertically upward with an initial speed of 26.4 m/s. How long does it take before the ball is back on the ground?

**2.59** A stone is thrown upward, from ground level, with an initial velocity of 10.0 m/s.

a) What is the velocity of the stone after 0.50 s?

b) How high above ground level is the stone after 0.50 s?

**2.60** A stone is thrown downward with an initial velocity of 10.0 m/s. The acceleration of the stone is constant and has the value of the free-fall acceleration, 9.81 m/s$^2$. What is the velocity of the stone after 0.500 s?

**2.61** A ball is thrown directly downward, with an initial speed of 10.0 m/s, from a height of 50.0 m. After what time interval does the ball strike the ground?

**2.62** An object is thrown vertically upward and has a speed of 20 m/s when it reaches two thirds of its maximum height above the launch point. Determine its maximum height.

**2.63** What is the velocity at the midway point of a ball able to reach a height $y$ when thrown with an initial velocity $v_0$?

•**2.64** Runner 1 is standing still on a straight running track. Runner 2 passes him, running with a constant speed of 5.1 m/s. Just as runner 2 passes, runner 1 accelerates with a constant acceleration of 0.89 m/s$^2$. How far down the track does runner 1 catch up with runner 2?

•**2.65** A girl is riding her bicycle. When she gets to a corner, she stops to get a drink from her water bottle. At that time, a friend passes by her, traveling at a constant speed of 8.0 m/s.

a) After 20 s, the girl gets back on her bike and travels with a constant acceleration of 2.2 m/s$^2$. How long does it take for her to catch up with her friend?

b) If the girl had been on her bike and rolling along at a speed of 1.2 m/s when her friend passed, what constant acceleration would she need to catch up with her friend in the same amount of time?

•**2.66** A speeding motorcyclist is traveling at a constant speed of 36.0 m/s when he passes a police car parked on the side of the road. The radar, positioned in the police car's rear window, measures the speed of the motorcycle. At the instant the motorcycle passes the police car, the police officer starts to chase the motorcyclist with a constant acceleration of 4.0 m/s$^2$.

a) How long will it take the police officer to catch the motorcyclist?

b) What is the speed of the police car when it catches up to the motorcycle?

c) How far will the police car be from its original position?

•**2.67** Two train cars are on a straight, horizontal track. One car starts at rest and is put in motion with a constant acceleration of 2.0 m/s$^2$. This car moves toward a second car that is 30 m away and moving at a constant speed of 4.0 m/s.

a) Where will the cars collide?

b) How long will it take for the cars to collide?

•**2.68** The planet Mercury has a mass that is 5% of that of Earth, and its gravitational acceleration is $g_{\text{mercury}} = 3.7$ m/s$^2$.

a) How long does it take for a rock that is dropped from a height of 1.75 m to hit the ground on Mercury?

b) How does this time compare to the time it takes the same rock to reach the ground on Earth, if dropped from the same height?

c) From what height would you have to drop the rock on Earth so that the fall-time on both planets is the same?

•**2.69** Bill Jones has a bad night in his bowling league. When he gets home, he drops his bowling ball in disgust out the window of his apartment, from a height of 63.17 m above the ground. John Smith sees the bowling ball pass by his window when it is 40.95 m above the ground. How much time passes from the time when John Smith sees the bowling ball pass his window to when it hits the ground?

•**2.70** Picture yourself in the castle of Helm's Deep from the *Lord of the Rings*. You are on top of the castle wall and are dropping rocks on assorted monsters that are 18.35 m below you. Just when you release a rock, an archer located exactly below you shoots an arrow straight up toward you with an initial velocity of 47.4 m/s. The arrow hits the rock in midair. How long after you release the rock does this happen?

•**2.71** An object is thrown vertically and has an upward velocity of 25 m/s when it reaches one fourth of its maximum height above its launch point. What is the initial (launch) speed of the object?

•**2.72** In a fancy hotel, the back of the elevator is made of glass so that you can enjoy a lovely view on your ride. The elevator travels at an average speed of 1.75 m/s. A boy on the 15th floor, 80.0 m above the ground level, drops a rock at the same instant the elevator starts its ascent from the 1st to the 5th floor. Assume the elevator travels at its average speed for the entire trip and neglect the dimensions of the elevator.

a) How long after it was dropped do you see the rock?

b) How long does it take for the rock to reach ground level?

••**2.73** You drop a water balloon straight down from your dormitory window 80.0 m above your friend's head. At 2.00 s after you drop the balloon, not realizing it has water in it your friend fires a dart from a gun, which is at the same height as his head, directly upward toward the balloon with an initial velocity of 20.0 m/s.

a) How long after you drop the balloon will the dart burst the balloon?

b) How long after the dart hits the balloon will your friend have to move out of the way of the falling water? Assume the balloon breaks instantaneously at the touch of the dart.

### Additional Problems

**2.74** A runner of mass 56.1 kg starts from rest and accelerates with a constant acceleration of 1.23 m/s$^2$ until she reaches a velocity of 5.10 m/s. She then continues running at this constant velocity. How long does the runner take to travel 173 m?

**2.75** A jet touches down on a runway with a speed of 142.4 mph. After 12.4 s, the jet comes to a complete stop. Assuming constant acceleration of the jet, how far down the runway from where it touched down does the jet stand?

**2.76** On the graph of position as a function of time, mark the points where the velocity is zero, and the points where the acceleration is zero.

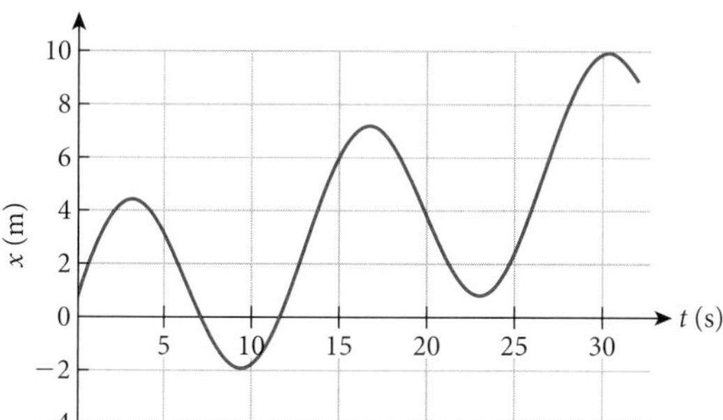


2.77 An object is thrown upward with a speed of 28.0 m/s. How long does it take it to reach its maximum height?

**2.78** An object is thrown upward with a speed of 28.0 m/s. How high above the projection point is it after 1.00 s?

**2.79** An object is thrown upward with a speed of 28 m/s. What maximum height above the projection point does it reach?

**2.80** The minimum distance necessary for a car to brake to a stop from a speed of 100.0 km/h is 40 m on a dry pavement. What is the minimum distance necessary for this car to brake to a stop from a speed of 130.0 km/h on dry pavement?

2.81 A car moving at 60 km/h comes to a stop in $t = 4.0$ s. Assume uniform deceleration.

a) How far does the car travel while stopping?

b) What is its deceleration?

**2.82** You are driving at 29.1 m/s when the truck ahead of you comes to a halt 200.0 m away from your bumper. Your brakes are in poor condition and you decelerate at a constant rate of 2.4 $m/s^2$.

a) How close do you come to the bumper of the truck?

b) How long does it take you to come to a stop?

**2.83** A train traveling at 40.0 m/s is headed straight toward another train, which is at rest on the same track. The moving train decelerates at 6.0 $m/s^2$, and the stationary train is 100.0 m away. How far from the stationary train will the moving train be when it comes to a stop?

**2.84** A car traveling at 25.0 m/s applies the brakes and decelerates uniformly at a rate of 1.2 $m/s^2$.

a) How far does it travel in 3.0 s?

b) What is its velocity at the end of this time interval?

c) How long does it take for the car to come to a stop?

d) What distance does the car travel before coming to a stop?

2.85 The fastest speed in NASCAR racing history was 212.809 mph (reached by Bill Elliott in 1987 at Talladega). If the race car decelerated from that speed at a rate of 8.0 $m/s^2$, how far would it travel before coming to a stop?

**2.86** You are flying on a commercial airline on your way from Houston, Texas, to Oklahoma City, Oklahoma. Your pilot announces that the plane is directly over Austin, Texas, traveling at a constant speed of 245 mph, and will be flying directly over Dallas, Texas, 362 km away. How long will it be before you are directly over Dallas, Texas?

**2.87** The position of a race car on a straight track is given as $x = at^3 + bt^2 + c$, where $a = 2.0\ m/s^3$, $b = 2.0\ m/s^2$, and $c = 3.0$ m.

a) What is the car's position between $t = 4.0$ s and $t = 9.0$ s?

b) What is the average speed between $t = 4.0$ s and $t = 9.0$ s?

**2.88** A girl is standing at the edge of a cliff 100 m above the ground. She reaches out over the edge of the cliff and throws a rock straight upward with a speed 8.0 m/s.

a) How long does it take the rock to hit the ground?

b) What is the speed of the rock the instant before it hits the ground?

•2.89 A double speed trap is set up on a freeway. One police cruiser is hidden behind a billboard, and another is some distance away under a bridge. As a sedan passes by the first cruiser, its speed is measured to be 105.9 mph. Since the driver has a radar detector, he is alerted to the fact that his speed has been measured, and he tries to slow his car down gradually without stepping on the brakes and alerting the police that he knew he was going too fast. Just taking the foot off the gas leads to a constant deceleration. Exactly 7.05 s later the sedan passes the second police cruiser. Now its speed is measured to be only 67.1 mph, just below the local freeway speed limit.

a) What is the value of the deceleration?

b) How far apart are the two cruisers?

•**2.90** During a test run on an airport runway, a new race car reaches a speed of 258.4 mph from a standing start. The car accelerates with constant acceleration and reaches this speed mark at a distance of 612.5 m from where it started. What was its speed after one-fourth, one-half, and three-fourths of this distance?

•**2.91** The vertical position of a ball suspended by a rubber band is given by the equation

$$y(t) = (3.8\ \text{m})\sin(0.46\, t/\text{s} - 0.31) - (0.2\ \text{m/s})t + 5.0\ \text{m}$$

a) What are the equations for velocity and acceleration for this ball?

b) For what times between 0 and 30 s is the acceleration zero?

•**2.92** The position of a particle moving along the $x$-axis varies with time according to the expression $x = 4t^2$, where $x$ is in meters and $t$ is in seconds. Evaluate the particle's position

a) at $t = 2.00$ s.

b) at 2.00 s + $\Delta t$.

c) Evaluate the limit of $\Delta x/\Delta t$ as $\Delta t$ approaches zero, to find the velocity at $t = 2.00$ s.

•**2.93** In 2005, Hurricane Rita hit several states in the southern United States. In the panic to escape her wrath, thousands of people tried to flee Houston, Texas by car. One car full of college students traveling to Tyler, Texas, 199 miles north of Houston, moved at an average speed of 3.0 m/s for one-fourth of the time, then at 4.5 m/s for another one-fourth of the time, and at 6.0 m/s for the remainder of the trip.

a) How long did it take the students to reach their destination?

b) Sketch a graph of position versus time for the trip.

•**2.94** A ball is thrown straight upward in the air at a speed of 15.0 m/s. Ignore air resistance.

a) What is the maximum height the ball will reach?

b) What is the speed of the ball when it reaches 5.00 m?

c) How long will it take to reach 5.00 m above its initial position on the way up?

d) How long will it take to reach 5.00 m above its initial position on its way down?

•**2.95** The Bellagio Hotel in Las Vegas, Nevada, is well known for its Musical Fountains, which use 192 HyperShooters to fire water hundreds of feet into the air to the rhythm of music. One of the HyperShooters fires water straight upward to a height of 240 ft.

a) What is the initial speed of the water?

b) What is the speed of the water when it is at half this height on its way down?

c) How long will it take for the water to fall back to its original height from half its maximum height?

•**2.96** You are trying to improve your shooting skills by shooting at a can on top of a fence post. You miss the can, and the bullet, moving at 200 m/s, is embedded 1.5 cm into the post when it comes to a stop. If constant acceleration is assumed, how long does it take for the bullet to stop?

•**2.97** You drive with a constant speed of 13.5 m/s for 30.0 s. You then accelerate for 10.0 s to a speed of 22.0 m/s. You then slow to a stop in 10.0 s. How far have you traveled?

•**2.98** A ball is dropped from the roof of a building. It hits the ground and it is caught at its original height 5.0 s later.

a) What was the speed of the ball just before it hits the ground?

b) How tall was the building? You are watching from a window 2.5 m above the ground. The window opening is 1.2 m from the top to the bottom.

c) At what time after the ball was dropped did you first see the ball in the window?

# Motion in Two and Three Dimensions

# 3

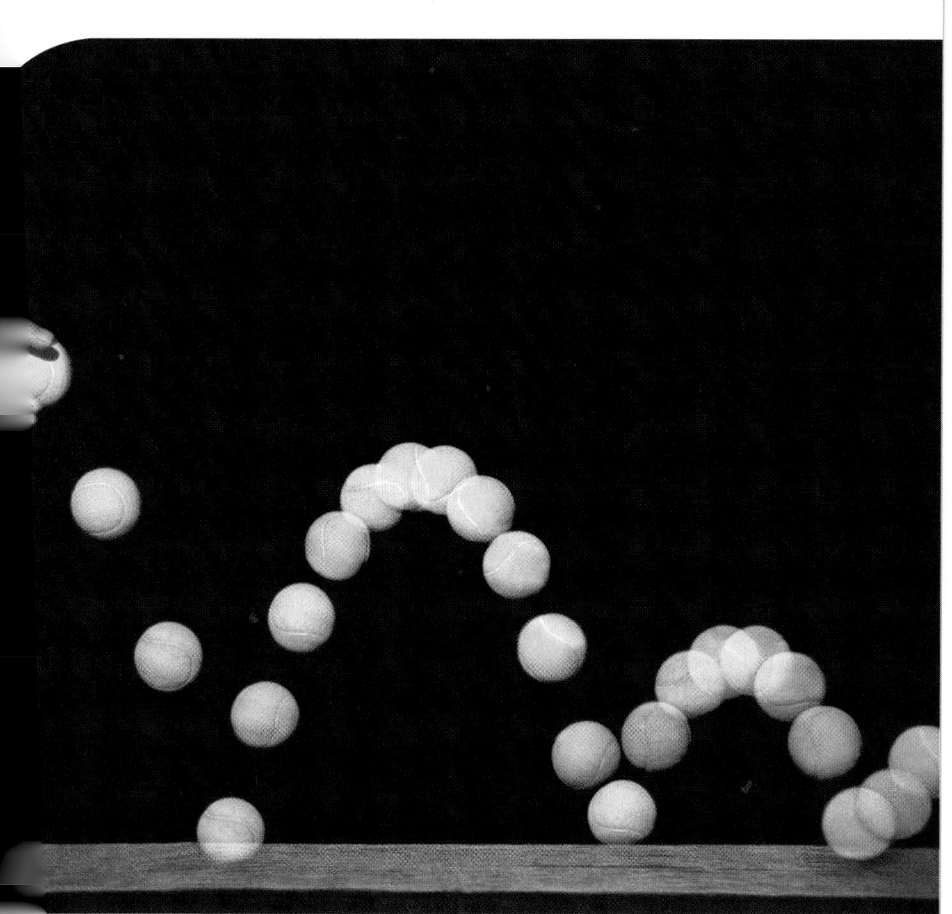

**FIGURE 3.1** Multiple exposure sequence of a bouncing ball.

## WHAT WE WILL LEARN

- You will learn to handle motion in two and three dimensions using methods developed for one-dimensional motion.
- You will determine the parabolic path of ideal projectile motion.
- You will be able to calculate the maximum height and maximum range of an ideal projectile trajectory in terms of the initial velocity vector and the initial position.
- You will learn to describe the velocity vector of a projectile at any time during its flight.
- You will appreciate that the realistic trajectories of objects like baseballs are affected by air friction and are not exactly parabolic.
- You will learn to transform velocity vectors from one reference frame to another.

Everyone has seen a bouncing ball, but have you ever looked closely at the path it takes? If you could slow the ball down, as in the photo in Figure 3.1, you would see the symmetrical arch of each bounce, which gets smaller until the ball stops. This path is characteristic of a kind of two-dimensional motion known as *projectile motion*. You can see the same parabolic shape in water fountains, fireworks, basketball shots—any kind of isolated motion where the force of gravity is relatively constant and the moving object is dense enough that air resistance (a force that tends to slow down objects moving though air) can be ignored.

This chapter extends Chapter 2's discussion of displacement, velocity, and acceleration to two-dimensional motion. The definitions of these vectors in two dimensions are very similar to the one-dimensional definitions, but we can apply them to a greater variety of real-life situations. Two-dimensional motion is still more restricted than general motion in three dimensions, but it applies to a wide range of common and important motions that we will consider throughout this course.

## 3.1 Three-Dimensional Coordinate Systems

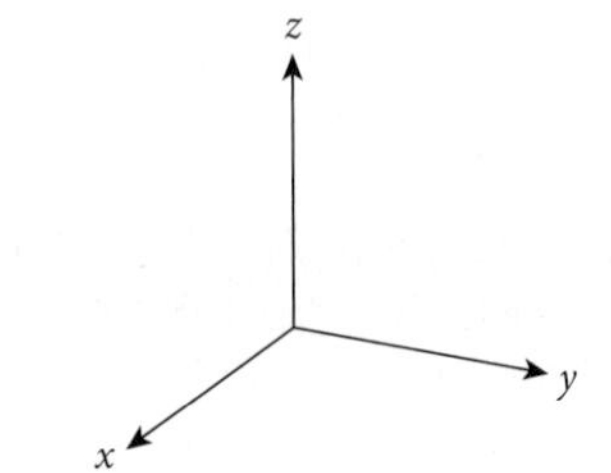


**FIGURE 3.2** A right-handed *xyz* Cartesian coordinate system.

Having studied motion in one dimension, we next tackle more complicated problems in two and three spatial dimensions. To describe this motion, we will work in Cartesian coordinates. In a three-dimensional Cartesian coordinate system, we choose the $x$- and $y$-axes to lie in the horizontal plane and the $z$-axis to point vertically upward (Figure 3.2). The three coordinate axes are at 90° (orthogonal) to one another, as required for a Cartesian coordinate system.

The convention that is followed without exception in this book is that the Cartesian coordinate system is **right-handed.** This convention means that you can obtain the relative orientation of the three coordinate axes using your right hand. To determine the positive directions of the three axes, hold your right hand with the thumb sticking straight up and the index finger pointing straight out; they will naturally have a 90° angle relative to each other. Then stick out your middle finger so that it is at a right angle with both the index finger and the thumb (Figure 3.3). The three axes are assigned to the fingers as shown in Figure 3.3a: thumb is $x$, index finger is $y$, and middle finger is $z$. You can rotate your right

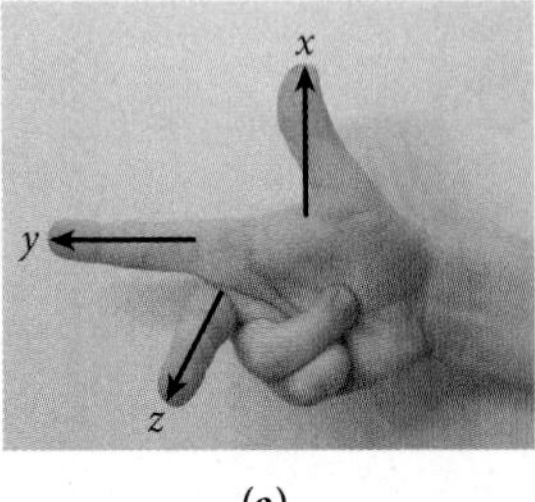


(a)

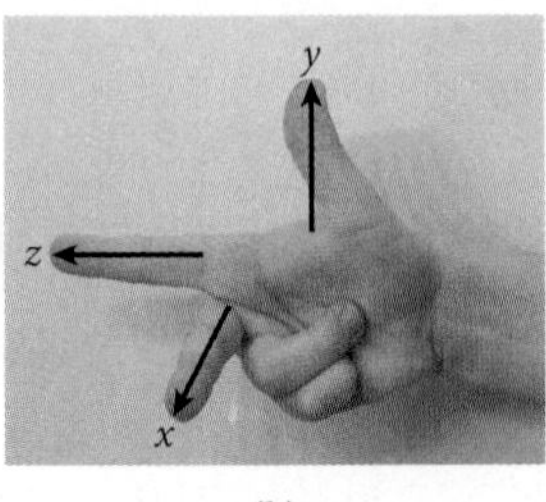


(b)

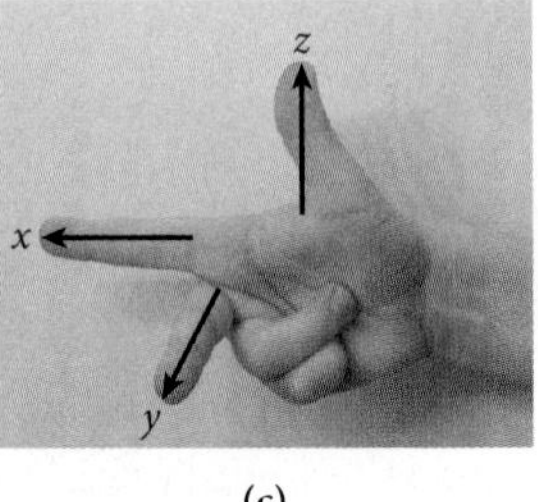


(c)

**FIGURE 3.3** All three possible realizations of a right-handed Cartesian coordinate system.

hand in any direction, but the relative orientation of thumb and fingers stays the same. If you want, you can exchange the letters on the fingers, as shown in Figure 3.3b and Figure 3.3c. However, $z$ always has to follow $y$, which always has to follow $x$. Figure 3.3 shows all possible combinations of the right-handed assignment of the axes to the fingers. You really only have to remember one of them, because your hand can always be orientated in three-dimensional space in such a way that the axes assignments on your fingers can be brought into alignment with the schematic coordinate axes shown in Figure 3.2.

With this set of Cartesian coordinates, a position vector can be written in component form as

$$\vec{r} = (x, y, z) = x\hat{x} + y\hat{y} + z\hat{z}. \tag{3.1}$$

A velocity vector is

$$\vec{v} = (v_x, v_y, v_y) = v_x\hat{x} + v_y\hat{y} + v_z\hat{z}. \tag{3.2}$$

For one-dimensional vectors, the time derivative of the position vector defines the velocity vector. This is also the case for more than one dimension:

$$\vec{v} = \frac{d\vec{r}}{dt} = \frac{d}{dt}(x\hat{x} + y\hat{y} + z\hat{z}) = \frac{dx}{dt}\hat{x} + \frac{dy}{dt}\hat{y} + \frac{dz}{dt}\hat{z}. \tag{3.3}$$

In the last step of this equation, we used the sum and product rules of differentiation, as well as the fact that the unit vectors are constant vectors (fixed directions along the coordinate axes and constant magnitude of 1). Comparing equations 3.2 and 3.3, we see that

$$v_x = \frac{dx}{dt}, \quad v_y = \frac{dy}{dt}, \quad v_z = \frac{dz}{dt}. \tag{3.4}$$

The same procedure leads us from the velocity vector to the acceleration vector by taking another time derivative:

$$\vec{a} = \frac{d\vec{v}}{dt} = \frac{dv_x}{dt}\hat{x} + \frac{dv_y}{dt}\hat{y} + \frac{dv_z}{dt}\hat{z}. \tag{3.5}$$

We can therefore write the Cartesian components of the acceleration vector:

$$a_x = \frac{dv_x}{dt}, \quad a_y = \frac{dv_y}{dt}, \quad a_z = \frac{dv_z}{dt}. \tag{3.6}$$

## 3.2 Velocity and Acceleration in a Plane

The most striking difference between velocity along a line and velocity in two or more dimensions is that the latter can change direction as well as magnitude. Because acceleration is defined as a change in velocity—any change in velocity—divided by a time interval, there can be acceleration even when the magnitude of the velocity does not change.

Consider, for example, a particle moving in two dimensions (that is, in a plane). At time $t_1$, the particle has velocity $\vec{v}_1$, and at a later time $t_2$, the particle has velocity $\vec{v}_2$. The change in velocity of the particle is $\Delta\vec{v} = \vec{v}_2 - \vec{v}_1$. The average acceleration, $\vec{a}_{\text{ave}}$, for the time interval $\Delta t = t_2 - t_1$ is given by

$$\vec{a}_{\text{ave}} = \frac{\Delta\vec{v}}{\Delta t} = \frac{\vec{v}_2 - \vec{v}_1}{t_2 - t_1}. \tag{3.7}$$

Figure 3.4 shows three different cases for the change in velocity of a particle moving in two dimensions over a given time interval. Figure 3.4a shows the initial and final velocities of the particle having the same direction, but the magnitude of the final velocity is greater than the magnitude of the initial velocity. The resulting change in velocity and the average acceleration are in the same direction as the velocities. Figure 3.4b again shows the initial and final velocities pointing in the same direction, but the magnitude of the final velocity is less than the magnitude of the initial velocity. The resulting change in velocity and the average acceleration are in the opposite direction from the velocities. Figure 3.4c illustrates the

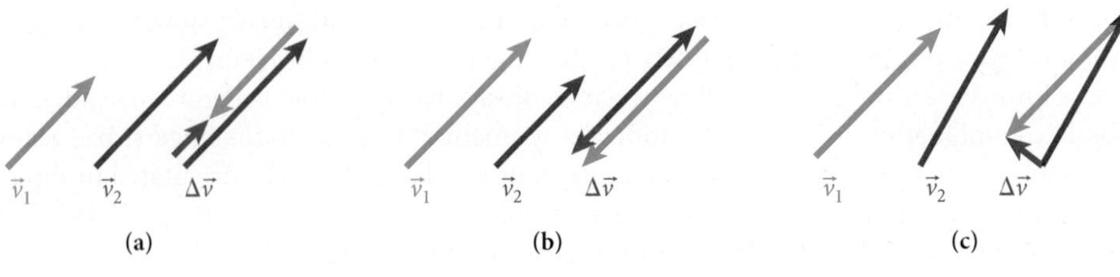


**FIGURE 3.4** At time $t_1$, a particle has a velocity $\vec{v}_1$. At a later time $t_2$, the particle has a velocity $\vec{v}_2$. The average acceleration is given by $\vec{a}_{\text{ave}} = \Delta\vec{v} / \Delta t = (\vec{v}_2 - \vec{v}_1)/(t_2 - t_1)$. (a) A time interval corresponding to $|\vec{v}_2| > |\vec{v}_1|$, with $\vec{v}_2$ and $\vec{v}_1$ in the same direction. (b) A time interval corresponding to $|\vec{v}_2| < |\vec{v}_1|$, with $\vec{v}_2$ and $\vec{v}_1$ in the same direction. (c) A time interval with $|\vec{v}_2| = |\vec{v}_1|$, but with $\vec{v}_2$ in a different direction from $\vec{v}_1$.

case when the initial and final velocities have the same magnitude but the direction of the final velocity vector is different from the direction of the initial velocity vector. Even though the magnitudes of the initial and final velocity vectors are the same, the change in velocity and the average acceleration are not zero and can be in a direction not obviously related to the initial or final velocity directions.

Thus, in two dimensions, an acceleration vector arises if an object's velocity vector changes in magnitude or direction. Any time an object travels along a curved path, in two or three dimensions, it must have acceleration. We will examine the components of acceleration in more detail in Chapter 9, when we discuss circular motion.

## 3.3 Ideal Projectile Motion

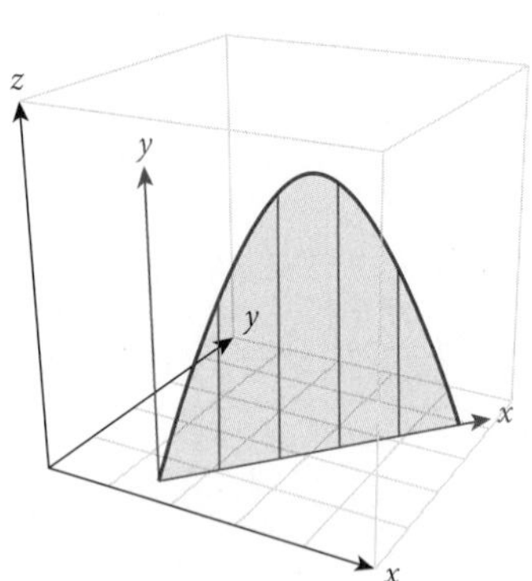


**FIGURE 3.5** Trajectory in three dimensions reduced to a trajectory in two dimensions.

**FIGURE 3.6** Photograph of a free throw with the parabolic trajectory of the basketball superimposed.

In some special cases of three-dimensional motion, the horizontal projection of the trajectory, or flight path, is a straight line. This situation occurs whenever the accelerations in the horizontal $xy$-plane are zero, so the object has constant velocity components, $v_x$ and $v_y$, in the horizontal plane. Such a case is shown in Figure 3.5 for a baseball tossed in the air. In this case, we can assign new coordinate axes such that the $x$-axis points along the horizontal projection of the trajectory and the $y$-axis is the vertical axis. In this special case, the motion in three dimensions can in effect be described as a motion in two spatial dimensions. A large class of real-life problems falls into this category, especially problems that involve ideal projectile motion.

An **ideal projectile** is any object that is released with some initial velocity and then moves only under the influence of gravitational acceleration, which is assumed to be constant and in the vertical downward direction. A basketball free throw (Figure 3.6) is a good example of ideal projectile motion, as is the flight of a bullet or the trajectory of a car that becomes airborne. **Ideal projectile motion** neglects air resistance and wind speed, spin of the projectile, and other effects influencing the flight of real-life projectiles. For realistic situations in which a golf ball, tennis ball, or baseball moves in air, the actual trajectory is not well described by ideal projectile motion and requires a more sophisticated analysis. We will discuss these effects in Section 3.5, but will not go into quantitative detail.

Let's begin with ideal projectile motion, with no effects due to air resistance or any other forces besides gravity. We work with two Cartesian components: $x$ in the horizontal direction and $y$ in the vertical (upward) direction. Therefore, the position vector for projectile motion is

$$\vec{r} = (x, y) = x\hat{x} + y\hat{y}, \tag{3.8}$$

and the velocity vector is

$$\vec{v} = (v_x, v_y) = v_x\hat{x} + v_x\hat{y} = \left(\frac{dx}{dt}, \frac{dy}{dt}\right) = \frac{dx}{dt}\hat{x} + \frac{dy}{dt}\hat{y}. \tag{3.9}$$

Given our choice of coordinate system, with a vertical $y$-axis, the acceleration due to gravity acts downward, in the negative $y$-direction; there is no acceleration in the horizontal direction:

$$\vec{a} = (0, -g) = -g\hat{y}. \tag{3.10}$$

For this special case of a constant acceleration only in the $y$-direction and with zero acceleration in the $x$-direction, we have a free-fall problem in the vertical direction and motion with constant velocity in the horizontal direction. The kinematical equations for the $x$-direction are those for an object moving with constant velocity:

$$x = x_0 + v_{x0}t \tag{3.11}$$

$$v_x = v_{x0}. \tag{3.12}$$

Just as in Chapter 2, we use the notation $v_{x0} \equiv v_x(t = 0)$ for the initial value of the $x$-component of the velocity. The kinematical equations for the $y$-direction are those for free-fall motion in one dimension:

$$y = y_0 + v_{y0}t - \tfrac{1}{2}gt^2 \tag{3.13}$$

$$y = y_0 + \bar{v}_y t \tag{3.14}$$

$$v_y = v_{y0} - gt \tag{3.15}$$

$$\bar{v}_y = \tfrac{1}{2}(v_y + v_{y0}) \tag{3.16}$$

$$v_y^2 = v_{y0}^2 - 2g(y - y_0). \tag{3.17}$$

For consistency, we write $v_{y0} \equiv v_y(t=0)$. With these seven equations for the $x$- and $y$-components, we can solve any problem involving an ideal projectile. Notice that since two-dimensional motion can be split into separate one-dimensional motions, these equations are written in component form, without the use of unit vectors.

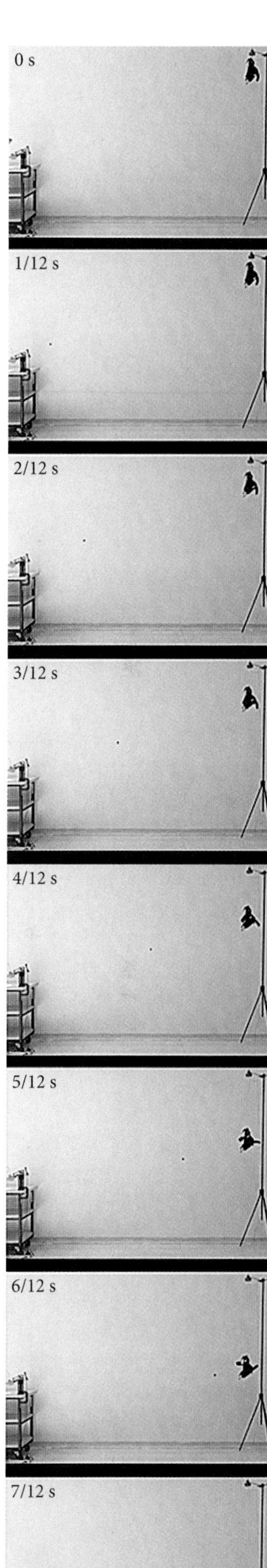


### EXAMPLE 3.1 Shoot the Monkey

Many lecture demonstrations illustrate that motion in the $x$-direction and motion in the $y$-direction are indeed independent of each other, as assumed in the derivation of the equations for projectile motion. One popular demonstration, called "shoot the monkey," is shown in Figure 3.7. The demonstration is motivated by a story. A monkey has escaped from the zoo and has climbed a tree. The zookeeper wants to shoot the monkey with a tranquilizer dart in order to recapture it, but she knows that the monkey will let go of the branch it is holding onto at the sound of the gun firing. Her challenge is therefore to hit the monkey in the air as it is falling.

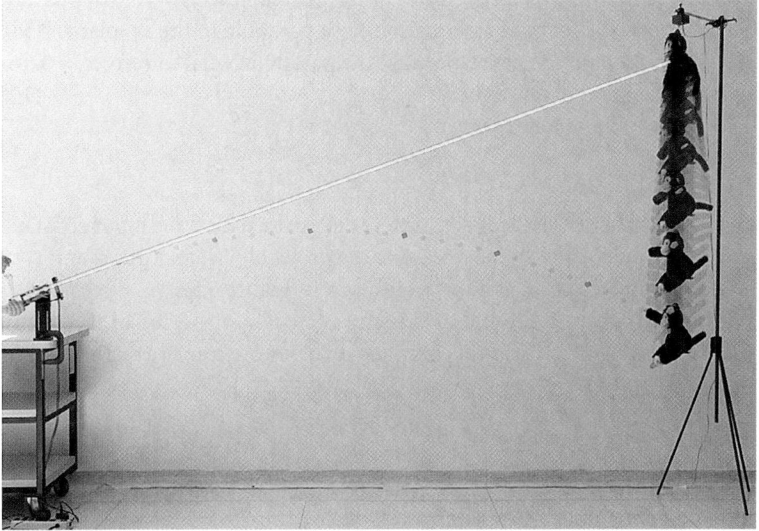

**FIGURE 3.7** The shoot-the-monkey lecture demonstration. On the right are some of the individual frames of the video, with information on their timing in the upper-left corners. On the left, these frames have been combined into a single image with a superimposed yellow line indicating the initial aim of the projectile launcher.

*Continued—*

**PROBLEM**

Where does the zookeeper need to aim to hit the falling monkey?

**SOLUTION**

The zookeeper must aim directly at the monkey, as shown in Figure 3.7, assuming that the time for the sound of the gun firing to reach the monkey is negligible and the speed of the dart is fast enough to cover the horizontal distance to the tree. As soon as the dart leaves the gun, it is in free fall, just like the monkey. Because both the monkey and the dart are in free fall, they fall with the same acceleration, independent of the dart's motion in the $x$-direction and of the dart's initial velocity. The dart and the monkey will meet at a point directly below the point from which the monkey dropped.

**DISCUSSION**

Any sharpshooter can tell you that, for a fixed target, you need to correct your gun sight for the free-fall motion of the projectile on the way to the target. As you can infer from Figure 3.7, even a bullet fired from a high-powered rifle will not fly in a straight line but will drop under the influence of gravitational acceleration. Only in a situation like the shoot-the-monkey demonstration, where the target is in free fall as soon as the projectile leaves the muzzle, can one aim directly at the target without making corrections for the free-fall motion of the projectile.

## Shape of a Projectile's Trajectory

Let's now examine the **trajectory** of a projectile in two dimensions. To find $y$ as a function of $x$, we solve the equation $x = x_0 + v_{x0}t$ for the time, $t = (x - x_0)/v_{x0}$, and then substitute for $t$ in the equation $y = y_0 + v_{y0}t - \frac{1}{2}gt^2$:

$$y = y_0 + v_{y0}t - \tfrac{1}{2}gt^2 \Rightarrow$$

$$y = y_0 + v_{y0}\frac{x - x_0}{v_{x0}} - \tfrac{1}{2}g\left(\frac{x - x_0}{v_{x0}}\right)^2 \Rightarrow$$

$$y = \left(y_0 - \frac{v_{y0}x_0}{v_{x0}} - \frac{gx_0^2}{2v_{x0}^2}\right) + \left(\frac{v_{y0}}{v_{x0}} + \frac{gx_0}{2v_{x0}^2}\right)x - \frac{g}{2v_{x0}^2}x^2. \tag{3.18}$$

Thus, the trajectory follows an equation of the general form $y = c + bx + ax^2$, with constants $a$, $b$, and $c$. This is the form of an equation for a parabola in the $xy$-plane. It is customary to set the $x$-component of the initial point of the parabola equal to zero: $x_0 = 0$. In this case, the equation for the parabola becomes

$$y = y_0 + \frac{v_{y0}}{v_{x0}}x - \frac{g}{2v_{x0}^2}x^2. \tag{3.19}$$

The trajectory of the projectile is completely determined by three input constants. These constants are the initial height of the release of the projectile, $y_0$, and the $x$- and $y$-components of the initial velocity vector, $v_{x0}$ and $v_{y0}$, as shown in Figure 3.8.

We can also express the initial velocity vector $\vec{v}_0$, in terms of its magnitude, $v_0$, and direction, $\theta_0$. Expressing $\vec{v}_0$ in this manner involves the transformation

$$v_0 = \sqrt{v_{x0}^2 + v_{y0}^2}$$

$$\theta_0 = \tan^{-1}\frac{v_{y0}}{v_{x0}}. \tag{3.20}$$

In Chapter 1, we discussed this transformation from Cartesian coordinates to length and angle of the vector, as well as the inverse transformation:

$$v_{x0} = v_0 \cos\theta_0$$

$$v_{y0} = v_0 \sin\theta_0. \tag{3.21}$$

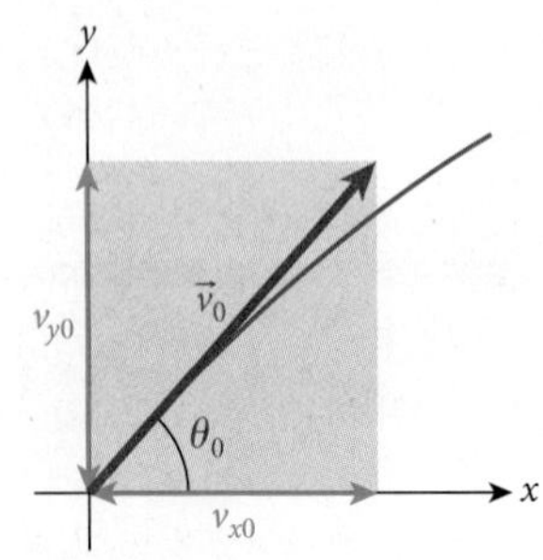


**FIGURE 3.8** Initial velocity vector $\vec{v}_0$ and its components, $v_{x0}$ and $v_{y0}$.

Expressed in terms of the magnitude and direction of the initial velocity vector, the equation for the path of the projectile becomes

$$y = y_0 + (\tan\theta_0)x - \frac{g}{2v_0^2\cos^2\theta_0}x^2. \quad (3.22)$$

The fountain shown in Figure 3.9 is in the Detroit Metropolitan Wayne County (DTW) airport. You can clearly see that the water shot out of many pipes traces almost perfect parabolic trajectories.

Note that because a parabola is symmetric, a projectile takes the same amount of time and travels the same distance from its launch point to the top of its trajectory as from the top of its trajectory back to launch level. Also, the speed of a projectile at a given height on its way up to the top of its trajectory is the same as its speed at that same height going back down.

**FIGURE 3.9** A fountain with water following parabolic trajectories.

## Time Dependence of the Velocity Vector

From equation 3.12, we know that the $x$-component of the velocity is constant in time: $v_x = v_{x0}$. This result means that a projectile will cover the same horizontal distance in each time interval of the same duration. Thus, in a video of projectile motion, such as a basketball player shooting a free throw as in Figure 3.6, or the path of the dart in the shoot-the-monkey demonstration in Figure 3.7, the horizontal displacement of the projectile from one frame of the video to the next will be constant.

The $y$-component of the velocity vector changes according to equation 3.15, $v_y = v_{y0} - gt$; that is, the projectile falls with constant acceleration. Typically, projectile motion starts with a positive value, $v_{y0}$. The apex (highest point) of the trajectory is reached at the point where $v_y = 0$ and the projectile moves only in the horizontal direction. At the apex, the $y$-component of the velocity is zero, and it changes sign from positive to negative.

We can indicate the instantaneous values of the $x$- and $y$-components of the velocity vector on a plot of $y$ versus $x$ for the flight path of a projectile (Figure 3.10). The $x$-components, $v_x$, of the velocity vector are shown by green arrows, and the $y$-components, $v_y$, by red arrows. Note the identical lengths of the green arrows, demonstrating the fact that $v_x$ remains constant. Each blue arrow is the vector sum of the $x$- and $y$-velocity components and depicts the instantaneous velocity vector along the path. Note that the direction of the velocity vector is always tangential to the trajectory. This is because the slope of the velocity vector is

$$\frac{v_y}{v_x} = \frac{dy/dt}{dx/dt} = \frac{dy}{dx},$$

which is also the local slope of the flight path. At the top of the trajectory, the green and blue arrows are identical because the velocity vector has only an $x$-component—that is, it points in the horizontal direction.

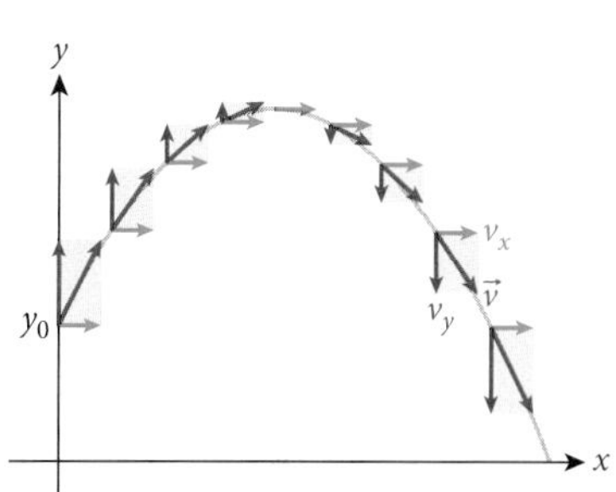


**FIGURE 3.10** Graph of a parabolic trajectory with the velocity vector and its Cartesian components shown at constant time intervals.

Although the vertical component of the velocity vector is equal to zero at the top of the trajectory, the gravitational acceleration has the same constant value as on any other part of the trajectory. Beware of the common misconception that the gravitational acceleration is equal to zero at the top of the trajectory. The gravitational acceleration has the same constant value everywhere along the trajectory.

Finally, let's explore the functional dependence of the absolute value of the velocity vector on time and/or the $y$-coordinate. We start with the dependence of $|\vec{v}|$ on $y$. We use the fact that the absolute value of a vector is given as the square root of the sum of the squares of the components. Then we use kinematical equation 3.12 for the $x$-component and kinematical equation 3.17 for the $y$-component. We obtain

$$|\vec{v}| = \sqrt{v_x^2 + v_y^2} = \sqrt{v_{x0}^2 + v_{y0}^2 - 2g(y - y_0)} = \sqrt{v_0^2 - 2g(y - y_0)}. \quad (3.23)$$

Note that the initial launch angle does not appear in this equation. The absolute value of the velocity—the speed—depends only on the initial value of the speed and the difference between the $y$-coordinate and the initial launch height. Thus, if we release a projectile from

### 3.1 In-Class Exercise

At the top of the trajectory of any projectile, which of the following statement(s), if any, is (are) true?

a) The acceleration is zero.

b) The $x$-component of the acceleration is zero.

c) The $y$-component of the acceleration is zero.

d) The speed is zero.

e) The $x$-component of the velocity is zero.

f) The $y$-component of the velocity is zero.

### 3.1 Self-Test Opportunity

What is the dependence of $|\vec{v}|$ on the $x$-coordinate?

a certain height above ground and want to know the speed with which it hits the ground, it does not matter if the projectile is shot straight up, or horizontally, or straight down. Chapter 5 will discuss the concept of kinetic energy, and then the reason for this seemingly strange fact will become more apparent.

## 3.4 Maximum Height and Range of a Projectile

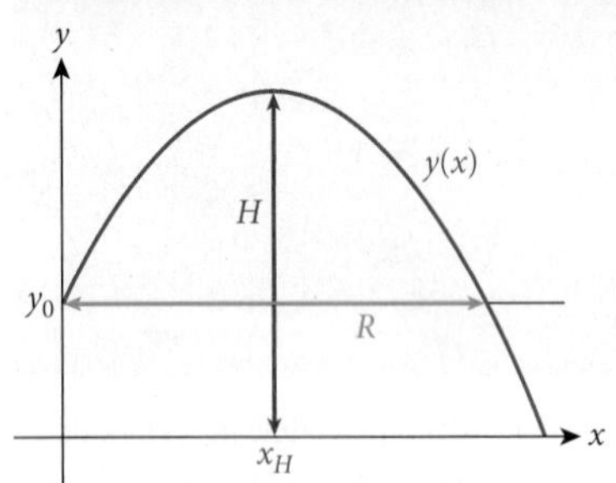


**FIGURE 3.11** The maximum height (red) and range (green) of a projectile.

When launching a projectile, for example, throwing a ball, we are often interested in the **range** ($R$), or how far the projectile will travel horizontally before returning to its original vertical position, and the **maximum height** ($H$) it will reach. These quantities $R$ and $H$ are illustrated in Figure 3.11. We find that the maximum height reached by the projectile is

$$H = y_0 + \frac{v_{y0}^2}{2g}. \tag{3.24}$$

We'll derive this equation below. We'll also derive this equation for the range:

$$R = \frac{v_0^2}{g}\sin 2\theta_0, \tag{3.25}$$

where $v_0$ is the absolute value of the initial velocity vector and $\theta_0$ is the launch angle. The maximum range, for a given fixed value of $v_0$, is reached when $\theta_0 = 45°$.

### DERIVATION 3.1

Let's investigate the maximum height first. To determine its value, we obtain an expression for the height, differentiate it, set the result equal to zero, and solve for the maximum height. Suppose $v_0$ is the initial speed and $\theta_0$ is the launch angle. We take the derivative of the path function $y(x)$, equation 3.22, with respect to $x$:

$$\frac{dy}{dx} = \frac{d}{dx}\left( y_0 + (\tan\theta_0)x - \frac{g}{2v_0^2\cos^2\theta_0}x^2 \right) = \tan\theta_0 - \frac{g}{v_0^2\cos^2\theta_0}x.$$

Now we look for the point $x_H$ where the derivative is zero:

$$0 = \tan\theta_0 - \frac{g}{v_0^2\cos^2\theta_0}x_H$$

$$\Rightarrow x_H = \frac{v_0^2\cos^2\theta_0\tan\theta_0}{g} = \frac{v_0^2}{g}\sin\theta_0\cos\theta_0 = \frac{v_0^2}{2g}\sin 2\theta_0.$$

In the second line above, we used the trigonometric identities $\tan\theta = \sin\theta/\cos\theta$ and $2\sin\theta\cos\theta = \sin 2\theta$. Now we insert this value for $x$ into equation 3.22 and obtain the maximum height, $H$:

$$H \equiv y(x_H) = y_0 + x_H\tan\theta_0 - \frac{g}{2v_0^2\cos^2\theta_0}x_H^2$$

$$= y_0 + \frac{v_0^2}{2g}\sin 2\theta_0\tan\theta_0 - \frac{g}{2v_0^2\cos^2\theta_0}\left(\frac{v_0^2}{2g}\sin 2\theta_0\right)^2$$

$$= y_0 + \frac{v_0^2}{g}\sin^2\theta_0 - \frac{v_0^2}{2g}\sin^2\theta_0$$

$$= y_0 + \frac{v_0^2}{2g}\sin^2\theta_0.$$

Because $v_{y0} = v_0\sin\theta_0$, we can also write

$$H = y_0 + \frac{v_{y0}^2}{2g},$$

which is equation 3.24.

The range, $R$, of a projectile is defined as the horizontal distance between the launching point and the point where the projectile reaches the same height from which it started, $y(R) = y_0$. Inserting $x = R$ into equation 3.22:

$$y_0 = y_0 + R\tan\theta_0 - \frac{g}{2v_0^2\cos^2\theta_0}R^2$$

$$\Rightarrow \tan\theta_0 = \frac{g}{2v_0^2\cos^2\theta_0}R$$

$$\Rightarrow R = \frac{2v_0^2}{g}\sin\theta_0\cos\theta_0 = \frac{v_0^2}{g}\sin 2\theta_0,$$

which is equation 3.25.

Note that the range, $R$, is twice the value of the $x$-coordinate, $x_H$, at which the trajectory reached its maximum height: $R = 2x_H$.

Finally, we consider how to maximize the range of the projectile. One way to maximize the range is to maximize the initial velocity, because the range increases with the absolute value of the initial velocity, $v_0$. The question then is, given a specific initial speed, what is the dependence of the range on the launch angle $\theta_0$? To answer this question, we take the derivative of the range (equation 3.25) with respect to the launch angle:

$$\frac{dR}{d\theta_0} = \frac{d}{d\theta_0}\left(\frac{v_0^2}{g}\sin 2\theta_0\right) = 2\frac{v_0^2}{g}\cos 2\theta_0.$$

Then we set this derivative equal to zero and find the angle for which the maximum value is achieved. The angle between 0° and 90° for which $\cos 2\theta_0 = 0$ is 45°. So the maximum range of an ideal projectile is given by

$$R_{\max} = \frac{v_0^2}{g}. \tag{3.26}$$

We could have obtained this result directly from the formula for the range because, according to that formula (equation 3.25), the range is at a maximum when $\sin 2\theta_0$ has its maximum value of 1, and it has this maximum when $2\theta_0 = 90°$, or $\theta_0 = 45°$.

Most sports involving balls provide numerous examples of projectile motion. We next consider a few examples where the effects of air resistance and spin do not dominate the motion, and so the findings are reasonably close to what happens in reality. In the next section, we'll look at what effects air resistance and spin can have on a projectile.

### 3.2 Self-Test Opportunity

Another way to arrive at a formula for the range uses the fact that it takes just as much time for the projectile to reach the top of the trajectory as to come down, because of the symmetry of a parabola. We can calculate the time to reach the top of the trajectory, where $v_{y0} = 0$, and then multiply this time by two and then by the horizontal velocity component to arrive at the range. Can you derive the formula for calculating the range in this way?

## SOLVED PROBLEM 3.1 Throwing a Baseball

When listening to a radio broadcast of a baseball game, you often hear the phrase "line drive" or "frozen rope" for a ball hit really hard and at a low angle with respect to the ground. Some announcers even use "frozen rope" to describe a particularly strong throw from second or third base to first base. This figure of speech implies movement on a straight line—but we know that the ball's actual trajectory is a parabola.

### PROBLEM

What is the maximum height that a baseball reaches if it is thrown from second base to first base and from third base to first base, released from a height of 6.0 ft, with a speed of 90 mph, and caught at the same height?

### SOLUTION

#### THINK

The dimensions of a baseball infield are shown in Figure 3.12. (In this problem, we'll need to perform lots of unit conversions. Generally, this book uses SI units, but baseball is

*Continued—*

full of British units.) The baseball infield is a square with sides 90 ft long. This is the distance between second and first base, and we get $d_{12} = 90\text{ ft} = 90 \cdot 0.3048\text{ m} = 27.4\text{ m}$. The distance from third to first base is the length of the diagonal of the infield square: $d_{13} = d_{12}\sqrt{2} = 38.8\text{ m}$.

A speed of 90 mph (the speed of a good Major League fastball) translates into

$$v_0 = 90\text{ mph} = 90 \cdot 0.4469\text{ m/s} = 40.2\text{ m/s}.$$

As with most trajectory problems, there are many ways to solve this problem. The most straightforward way follows from our considerations of range and maximum height. We can equate the base-to-base distance with the range of the projectile because the ball is released and caught at the same height, $y_0 = 6\text{ ft} = 6 \cdot 0.3048\text{ m} = 1.83\text{ m}$.

## SKETCH

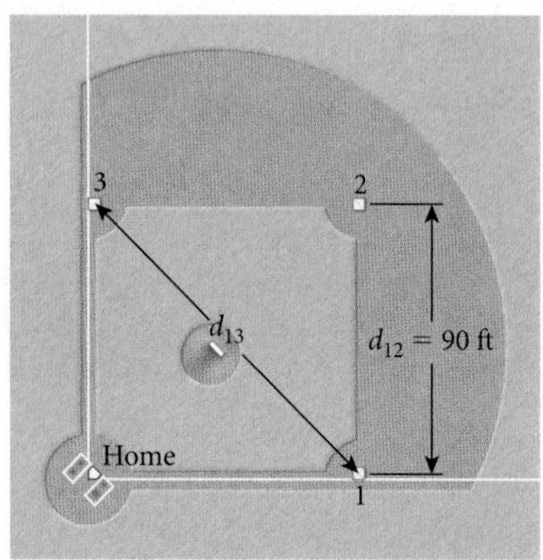


**FIGURE 3.12** Dimensions of a baseball infield.

## RESEARCH

In order to obtain the initial launch angle of the ball, we use equation 3.25, setting the range equal to the distance between first and second base:

$$d_{12} = \frac{v_0^2}{g}\sin 2\theta_0 \Rightarrow \theta_0 = \tfrac{1}{2}\sin^{-1}\left(\frac{d_{12}g}{v_0^2}\right).$$

However, we already have an equation for the maximum height:

$$H = y_0 + \frac{v_0^2 \sin^2\theta_0}{2g}.$$

## SIMPLIFY

Substituting our expression for the launch angle into the equation for the maximum height results in

$$H = y_0 + \frac{v_0^2 \sin^2\left(\tfrac{1}{2}\sin^{-1}\left(\dfrac{d_{12}g}{v_0^2}\right)\right)}{2g}.$$

## CALCULATE

We are ready to insert numbers:

$$H = 1.83\text{ m} + \frac{(40.2\text{ m/s})^2 \sin^2\left(\tfrac{1}{2}\sin^{-1}\left(\dfrac{(27.4\text{ m})(9.81\text{ m/s}^2)}{(40.2\text{ m/s})^2}\right)\right)}{2(9.81\text{ m/s}^2)} = 2.40367\text{ m}.$$

## ROUND

The initial precision specified was two significant digits. So we round our final result to

$$H = 2.4\text{ m}.$$

Thus, a 90-mph throw from second to first base is 2.39 m − 1.83 m = 0.56 m—that is, almost 2 ft—above a straight line at the middle of its trajectory. This number is even bigger for the throw from third to first base, for which we find an initial angle of 6.8° and a maximum height of 3.0 m, or 1.2 m (almost 4 ft) above the straight line connecting the points of release and catch.

**DOUBLE-CHECK**

Common sense says that the longer throw from third to first needs to have a greater maximum height than the throw from second to first, and our answers agree with that. If you watch a baseball game from the stands or on television, these calculated heights may seem too large. However, if you watch a game from ground level, you'll see that the other infielders really do have to get some height on the ball to make a good throw to first base.

Let's consider one more example from baseball and calculate the trajectory of a batted ball (see Figure 3.13).

**FIGURE 3.13** The motion of a batted baseball can be treated as projectile motion.

## EXAMPLE 3.2 Batting a Baseball

During the flight of a batted baseball, in particular, a home run, air resistance has a quite noticeable impact. For now though, we want to neglect it. Section 3.5 will discuss the effect of air resistance.

**PROBLEM**

If the ball comes off the bat with a launch angle of 35° and an initial speed of 110 mph, how far will the ball fly? How long will it be in the air? What will its speed be at the top of its trajectory? What will its speed be when it lands?

**SOLUTION**

Again, we need to convert to SI units first: $v_0 = 110$ mph = 49.2 m/s. We first find the range:

$$R = \frac{v_0^2}{g}\sin 2\theta_0 = \frac{(49.2\text{ m/s})^2}{9.81\text{ m/s}^2}\sin 70° = 231.5\text{ m}.$$

This distance is about 760 feet, which would be a home run in even the biggest ballpark. However, this calculation does not take air resistance into account. If we took friction due to air resistance into account, the distance would be reduced to approximately 400 feet. (See Section 3.5 on realistic projectile motion.)

In order to find the baseball's time in the air, we can divide the range by the horizontal component of the velocity assuming the ball is hit at about ground level.

$$t = \frac{R}{v_0 \cos\theta_0} = \frac{231.5\text{ m}}{(49.2\text{ m/s})\,(\cos 35°)} = 5.74\text{ s}.$$

Now we will calculate the speeds at the top of the trajectory and at landing. At the top of the trajectory, the velocity has only a horizontal component, which is $v_0 \cos\theta_0 = 40.3$ m/s. When the ball lands, we can calculate its speed using equation 3.23: $|\vec{v}| = \sqrt{v_0^2 - 2g(y - y_0)}$. Because we assume that the altitude at which it lands is the same as the one from which it was launched, we see that the speed is the same at the landing point as at the launching point, 49.2 m/s.

A real baseball would not quite follow the trajectory calculated here. If instead we launched a small steel ball bearing with the same angle and speed, neglecting air resistance would have led to a very good approximation, and the trajectory parameters just found would be verified in such an experiment. The reason we can comfortably neglect air resistance for the steel ball bearing is that it has a much higher mass density and smaller surface area than a baseball, so drag effects (which depend on cross-sectional area) are small compared to gravitational effects.

Baseball is not the only sport that provides examples of projectile motion. Let's consider an example from football.

## SOLVED PROBLEM 3.2 Hang Time

When a football team is forced to punt the ball away to the opponent, it is very important to kick the ball as far as possible but also to attain a sufficiently long hang time—that is, the ball should remain in the air long enough that the punt-coverage team has time to run down field and tackle the receiver right after the catch.

### PROBLEM

What are the initial angle and speed with which a football has to be punted so that its hang time is 4.41 s and it travels a distance of 49.8 m (= 54.5 yd)?

### SOLUTION

**THINK**

A punt is a special case of projectile motion for which the initial and final values of the vertical coordinate are both zero. If we know the range of the projectile, we can figure out the hang time from the fact that the horizontal component of the velocity vector remains at a constant value; thus, the hang time must simply be the range divided by this horizontal component of the velocity vector. The equations for hang time and range give us two equations in the two unknown quantities, $v_0$ and $\theta_0$, that we are looking for.

**SKETCH**

This is one of the few cases in which a sketch does not seem to provide additional information.

**RESEARCH**

We have already seen (equation 3.25) that the range of a projectile is given by

$$R = \frac{v_0^2}{g}\sin 2\theta_0 .$$

As already mentioned, the hang time can be most easily computed by dividing the range by the horizontal component of the velocity:

$$t = \frac{R}{v_0 \cos\theta_0} .$$

Thus, we have two equations in the two unknowns, $v_0$ and $\theta_0$. (Remember, $R$ and $t$ were given in the problem statement.)

**SIMPLIFY**

We solve both equations for $v_0^2$ and set them equal:

$$R = \frac{v_0^2}{g}\sin 2\theta_0 \Rightarrow v_0^2 = \frac{gR}{\sin 2\theta_0} \searrow$$

$$t = \frac{R}{v_0\cos\theta_0} \Rightarrow v_0^2 = \frac{R^2}{t^2\cos^2\theta_0} \nearrow$$

$$\frac{gR}{\sin 2\theta_0} = \frac{R^2}{t^2\cos^2\theta_0} .$$

Now, we can solve for $\theta_0$. Using $\sin 2\theta_0 = 2\sin\theta_0\cos\theta_0$, we find

$$\frac{g}{2\sin\theta_0\cos\theta_0} = \frac{R}{t^2\cos^2\theta_0}$$

$$\Rightarrow \tan\theta_0 = \frac{gt^2}{2R}$$

$$\Rightarrow \theta_0 = \tan^{-1}\left(\frac{gt^2}{2R}\right).$$

Next, we substitute this expression in either of the two equations we started with. We select the equation for hang time and solve it for $v_0$:

$$t = \frac{R}{v_0 \cos\theta_0} \Rightarrow v_0 = \frac{R}{t \cos\theta_0}.$$

**CALCULATE**

All that remains is to insert numbers into the equations we have obtained:

$$\theta_0 = \tan^{-1}\left(\frac{(9.81 \text{ m/s}^2)(4.41 \text{ s})^2}{2(49.8 \text{ m})}\right) = 62.4331°$$

$$v_0 = \frac{49.8 \text{ m}}{(4.41 \text{ s})(\cos 1.08966)} = 24.4013 \text{ m/s}.$$

**ROUND**

The range and hang time were specified to three significant figures, so we state our final results to this precision:

$$\theta_0 = 62.4°$$

and

$$v_0 = 24.4 \text{ m/s}.$$

**DOUBLE-CHECK**

We know that the maximum range is reached with a launch angle of 45°. The punted ball here is launched at an initial angle that is significantly steeper, at 62.4°. Thus, the ball does not travel as far as it could go with the value of the initial speed that we computed. Instead, it travels higher and thus maximizes the hang time. If you watch good college or pro punters practice their skills during football games, you'll see that they try to kick the ball with an initial angle larger than 45°, in agreement with what we found in our calculations.

### 3.2 In-Class Exercise

The same range as in Solved Problem 3.2 could be achieved with the same initial speed of 24.4 m/s but a launch angle different from 62.4°. What is the value of this angle?

a) 12.4°
b) 27.6°
c) 45.0°
d) 55.2°

### 3.3 In-Class Exercise

What is the hang time for that other launch angle found in In-Class Exercise 3.2?

a) 2.30 s
b) 3.14 s
c) 4.41 s
d) 5.14 s

## 3.5 Realistic Projectile Motion

If you are familiar with tennis or golf or baseball, you know that the parabolic model for the motion of a projectile is only a fairly crude approximation to the actual trajectory of any real ball. However, by ignoring some factors that affect real projectiles, we were able to focus on the physical principles that are most important in projectile motion. This is a common technique in science: Ignore some factors involved in a real situation in order to work with fewer variables and come to an understanding of the basic concept. Then go back and consider how the ignored factors affect the model. Let's briefly consider the most important factors that affect real projectile motion: air resistance, spin, and surface properties of the projectile.

The first modifying effect that we need to take into account is air resistance. Typically, we can parameterize air resistance as a velocity-dependent acceleration. The general analysis exceeds the scope of this book; however, the resulting trajectories are called *ballistic curves.*

Figure 3.14 shows the trajectories of baseballs launched at an initial angle of 35° with respect to the horizontal at initial speeds of 90 and 110 mph. Compare the trajectory shown for the launch speed of 110 mph with the result we calculated in Example 3.2: The real range of this ball is only slightly more than 400 ft, whereas we found 760 ft when we neglected air resistance. Obviously, for a long fly ball, neglecting air resistance is not valid.

Another important effect that the parabolic model neglects is the spin of the projectile as it moves through the air. When a quarterback throws a "spiral" in football, for example, the spin is important for the stability of the flight motion and prevents the ball from rotating end-over-end. In tennis, a ball with topspin drops much faster than a ball without noticeable spin, given the same initial values of speed and launch angle. Conversely, a tennis ball with underspin, or backspin, "floats" deeper into the court. In golf, backspin is sometimes

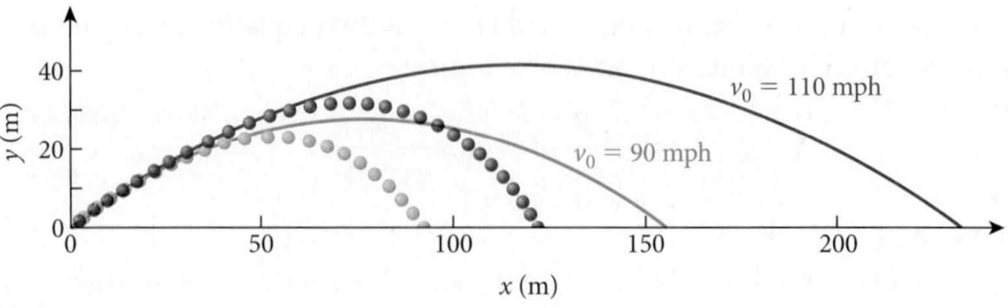


**FIGURE 3.14** Trajectories of baseballs initially launched at an angle of 35° above the horizontal at speeds of 90 mph (green) and 110 mph (red). Solid curves neglect air resistance and backspin; dotted curves reflect air resistance and backspin.

desired, because it causes a steeper landing angle and thus helps the ball come to rest closer to its landing point than a ball hit without backspin. Depending on the magnitude and direction of rotation, sidespin of a golf ball can cause a deviation from a straight-line path along the ground (draws and fades for good players or hooks and slices for the rest of us).

In baseball, sidespin is what enables a pitcher to throw a curveball. By the way, there is no such thing as a "rising fastball" in baseball. However, balls thrown with severe backspin do not drop as fast as the batter expects and are thus sometimes perceived as rising—an optical illusion. In the graph of ballistic baseball trajectories in Figure 3.14, an initial backspin of 2000 rpm was assumed.

Curving and practically all other effects of spin on the trajectory of a moving ball are a result of the air molecules bouncing with higher speeds off the side of the ball (and the boundary layer of air molecules) that is rotating in the direction of the flight motion (and thus has a higher velocity relative to the incoming air molecules) than off the side of the ball rotating against the flight direction. We will return to this topic in Chapter 13 on fluid motion.

The surface properties of projectiles also have significant effects on their trajectories. Golf balls have dimples to make them fly farther. Balls that are otherwise identical to typical golf balls but have a smooth surface can be driven only about half as far. This surface effect is also the reason why sandpaper found in a pitcher's glove leads to ejection of that player from the game, because a baseball that is roughened on parts of its surface moves differently from one that is not.

## 3.6 Relative Motion

To study motion, we have allowed ourselves to shift the origin of the coordinate system by properly choosing values for $x_0$ and $y_0$. In general, $x_0$ and $y_0$ are constants that can be chosen freely. If this choice is made intelligently, it can help make a problem more manageable. For example, when we calculated the path of the projectile, $y(x)$, we set $x_0 = 0$ to simplify our calculations. The freedom to select values for $x_0$ and $y_0$ arises from the fact that our ability to describe any kind of motion does not depend on the location of the origin of the coordinate system.

So far, we have examined physical situations where we have kept the origin of the coordinate system at a fixed location during the motion of the object we wanted to consider. However, in some physical situations, it is impractical to choose a reference system with a fixed origin. Consider, for example, a jet plane landing on an aircraft carrier that is going forward at full throttle at the same time. You want to describe the plane's motion in a coordinate system fixed to the carrier, even though the carrier is moving. The reason why this is important is that the plane needs to come to rest *relative* to the carrier at some fixed location on the deck. The reference frame from which we view motion makes a big difference in how we describe the motion, producing an effect known as **relative velocity.**

Another example of a situation for which we cannot neglect relative motion is a transatlantic flight flying from Detroit, Michigan, to Frankfurt, Germany, which takes 8 h and 10 min. Using the same aircraft and going in the reverse direction, from Frankfurt to Detroit, takes 9 h and 10 min, a full hour longer. The primary reason for this difference is that the prevailing wind at high altitudes, the jet stream, tends to blow from west to east at speeds as high as 67 m/s (150 mph). Even though the airplane's speed relative to the air around it is the same in both directions, that air is moving with its own speed. Thus, the relationship of the coordinate system of the air inside the jet stream to the coordinate system in which the locations of Detroit and Frankfurt remain fixed is important in understanding the difference in flight times.

For a more easily analyzed example of a moving coordinate system, let's consider motion on a moving walkway, as is typically found in airport terminals. This system is an example of one-dimensional relative motion. Suppose that the walkway surface moves with a certain velocity, $v_{wt}$, relative to the terminal. We use the subscripts w for walkway and t for terminal. Then a coordinate system that is fixed to the walkway surface has exactly velocity $v_{wt}$ relative to a coordinate system attached to the terminal. The man shown in Figure 3.15 is walking with a velocity $v_{mw}$ as measured in a coordinate system on the walkway, and he has a velocity $v_{mt} = v_{mw} + v_{wt}$ with respect to the terminal. The two velocities $v_{mw}$ and $v_{wt}$ add as vectors since the corresponding displacements add as vectors. (We will show this explicitly when we generalize to three dimensions.) For example, if the walkway moves with $v_{wt} = 1.5$ m/s and the man moves with $v_{mw} = 2.0$ m/s, then he will progress through the terminal with a velocity of $v_{mt} = v_{mw} + v_{wt} = 2.0 \text{ m/s} + 1.5 \text{ m/s} = 3.5 \text{ m/s}$.

**FIGURE 3.15** Man walking on a moving walkway, demonstrating one-dimensional relative motion.

One can achieve a state of no motion relative to the terminal by walking in the direction opposite of the motion of the walkway with a velocity that is exactly the negative of the walkway velocity. Children often try to do this. If a child were to walk with $v_{mw} = -1.5$ m/s on this walkway, her velocity would be zero relative to the terminal.

It is essential for this discussion of relative motion that the two coordinate systems have a velocity relative to each other that is constant in time. In this case, we can show that the accelerations measured in both coordinate systems are identical: $v_{wt} = \text{const.} \Rightarrow dv_{wt}/dt = 0$. From $v_{mt} = v_{mw} + v_{wt}$, we then obtain:

$$\frac{dv_{mt}}{dt} = \frac{d(v_{mw} + v_{wt})}{dt} = \frac{dv_{mw}}{dt} + \frac{dv_{wt}}{dt} = \frac{dv_{mw}}{dt} + 0$$
$$\Rightarrow a_t = a_w. \tag{3.27}$$

Therefore, the accelerations measured in both coordinate systems are indeed the same. This type of velocity addition is also known as a **Galilean transformation.** Before we go on to the two- and three-dimensional cases, note that this type of transformation is valid only for speeds that are small compared to the speed of light. Once the speed approaches the speed of light, we must use a different transformation, which we discuss in detail in Chapter 35 on the theory of relativity.

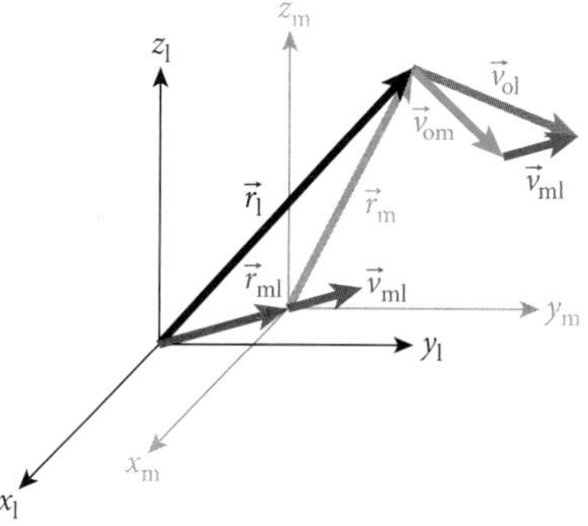


**FIGURE 3.16** Reference frame transformation of a velocity vector and a position vector at some particular time.

Now let's generalize this result to more than one spatial dimension. We assume that we have two coordinate systems: $x_l$, $y_l$, $z_l$ and $x_m$, $y_m$, $z_m$. (Here we use the subscripts l for the coordinate system that is at rest in the laboratory and m for the one that is moving.) At time $t = 0$, suppose the origins of both coordinate systems are located at the same point, with their axes exactly parallel to one another. As indicated in Figure 3.16, the origin of the moving $x_m y_m z_m$ coordinate system moves with a constant translational velocity $\vec{v}_{ml}$ (blue arrow) relative to the origin of the laboratory $x_l y_l z_l$ coordinate system. After a time $t$, the origin of the moving $x_m y_m z_m$ coordinate system is thus located at the point $\vec{r}_{ml} = \vec{v}_{ml} t$.

We can now describe the motion of any object in either coordinate system. If the object is located at coordinate $\vec{r}_l$ in the $x_l y_l z_l$ coordinate system and at coordinate $\vec{r}_m$ in the $x_m y_m z_m$ coordinate system, then the position vectors are related to each other via simple vector addition:

$$\vec{r}_l = \vec{r}_m + \vec{r}_{ml} = \vec{r}_m + \vec{v}_{ml} t. \tag{3.28}$$

A similar relationship holds for the object's velocities, as measured in the two coordinate systems. If the object has velocity $\vec{v}_{ol}$ in the $x_l y_l z_l$ coordinate system and velocity $\vec{v}_{om}$ in the $x_m y_m z_m$ coordinate system, these two velocities are related via:

$$\vec{v}_{ol} = \vec{v}_{om} + \vec{v}_{ml}. \tag{3.29}$$

This equation can be obtained by taking the time derivative of equation 3.28, because $\vec{v}_{ml}$ is constant. Note that the two inner subscripts on the right-hand side of this equation are the same (and will be in any application of this equation). This makes the equation understandable on an intuitive level, because it says that the velocity of the *o*bject in the *l*ab frame (subscript ol) is equal to the sum of the velocity with which the *o*bject moves relative to the *m*oving frame (subscript om) and the velocity with which the *m*oving frame moves relative to the *l*ab frame (subscript ml).

Taking another time derivative produces the accelerations. Again, because $\vec{v}_{\text{ml}}$ is constant and thus has a derivative equal to zero, we obtain, just as in the one-dimensional case,

$$\vec{a}_{\text{l}} = \vec{a}_{\text{m}}. \quad (3.30)$$

The magnitude and direction of the acceleration for an object is the same in both coordinate systems.

## EXAMPLE 3.3 Airplane in a Crosswind

Airplanes move relative to the air that surrounds them. Suppose a pilot points his plane in the northeast direction. The airplane moves with a speed of 160. m/s relative to the wind, and the wind is blowing at 32.0 m/s in a direction from east to west (measured by an instrument at a fixed point on the ground).

### PROBLEM

What is the velocity vector—speed and direction—of the airplane relative to the ground? How far off course does the wind blow this plane in 2.0 h?

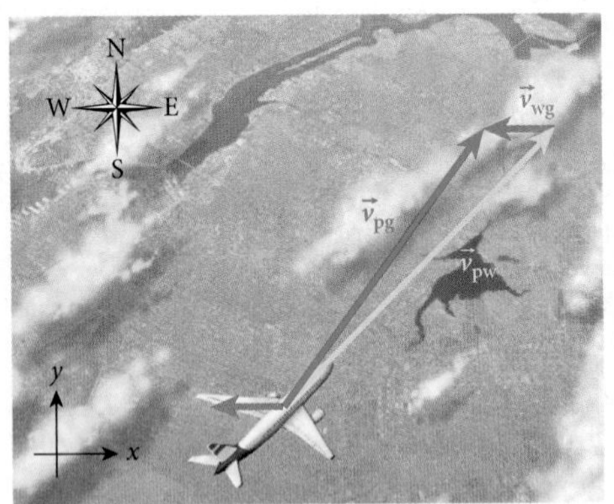


**FIGURE 3.17** Velocity of an airplane with respect to the wind (yellow), the velocity of the wind with respect to the ground (orange), and the resultant velocity of the airplane with respect to the ground (green).

### SOLUTION

Figure 3.17 shows a vector diagram of the velocities. The airplane heads in the northeast direction, and the yellow arrow represents its velocity vector relative to the wind. The velocity vector of the wind is represented in orange and points due west. Graphical vector addition results in the green arrow that represents the velocity of the plane relative to the ground. To solve this problem, we apply the basic transformation of equation 3.29 embodied in the equation

$$\vec{v}_{\text{pg}} = \vec{v}_{\text{pw}} + \vec{v}_{\text{wg}}.$$

Here $\vec{v}_{\text{pw}}$ is the velocity of the plane with respect to the wind and has these components:

$$v_{\text{pw},x} = v_{\text{pw}} \cos\theta = 160 \text{ m/s} \cdot \cos 45° = 113 \text{ m/s}$$
$$v_{\text{pw},y} = v_{\text{pw}} \sin\theta = 160 \text{ m/s} \cdot \sin 45° = 113 \text{ m/s}.$$

The velocity of the wind with respect to the ground, $\vec{v}_{\text{wg}}$, has these components:

$$v_{\text{wg},x} = -32 \text{ m/s}$$
$$v_{\text{wg},y} = 0.$$

We next obtain the components of the airplane's velocity relative to a coordinate system fixed to the ground, $\vec{v}_{\text{pg}}$:

$$v_{\text{pg},x} = v_{\text{pw},x} + v_{\text{wg},x} = 113 \text{ m/s} - 32 \text{ m/s} = 81 \text{ m/s}$$
$$v_{\text{pg},y} = v_{\text{pw},y} + v_{\text{pw},y} = 113 \text{ m/s}.$$

The absolute value of the velocity vector and its direction in the ground-based coordinate system are therefore

$$v_{\text{pg}} = \sqrt{v_{\text{pg},x}^2 + v_{\text{pg},y}^2} = 139 \text{ m/s}$$
$$\theta = \tan^{-1}\left(\frac{v_{\text{pg},y}}{v_{\text{pg},x}}\right) = 54.4°.$$

Now we need to find the course deviation due to the wind. To find this quantity, we can multiply the plane's velocity vectors in each coordinate system by the elapsed time of 2 h = 7200 s, then take the vector difference, and finally obtain the magnitude of the vector difference. The answer can be obtained more easily if we use equation 3.29 multiplied by the elapsed time to reflect that the course deviation, $\vec{r}_T$, due to the wind is the wind velocity, $\vec{v}_{wg}$, times 7200 s:

$$\left|\vec{r}_T\right| = \left|\vec{v}_{wg}\right| t = 32.0 \text{ m/s} \cdot 7200 \text{ s} = 230.4 \text{ km}.$$

**DISCUSSION**

The Earth itself moves a considerable amount in 2 h, as a result of its own rotation and its motion around the Sun, and you might think we have to take these motions into account. That the Earth moves is true, but is irrelevant for the present example: The airplane, the air, and the ground all participate in this rotation and orbital motion, which is superimposed on the relative motion of the objects described in the problem. Thus, we simply perform our calculations in a coordinate system in which the Earth is at rest and not rotating.

Another interesting consequence of relative motion can be seen when observing rain while in a moving car. You may have wondered why the rain seems to come almost straight at you as you are driving. The following example answers this question.

### EXAMPLE 3.4 Driving through Rain

Let's supppose rain is falling straight down on a car, as indicated by the white lines in Figure 3.18. A stationary observer outside the car would be able to measure the velocities of the rain (blue arrow) and of the moving car (red arrow).

However, if you are sitting inside the moving car, the outside world of the stationary observer (including the street, as well as the rain) moves with a relative velocity of $\vec{v} = -\vec{v}_{car}$. The velocity of this relative motion has to be added to all outside events as observed from inside the moving car. This motion results in a velocity vector $\vec{v}\,'_{rain}$ for the rain as observed from inside the moving car (Figure 3.19); mathematically, this vector is a sum, $\vec{v}\,'_{rain} = \vec{v}_{rain} - \vec{v}_{car}$, where $\vec{v}_{rain}$ and $\vec{v}_{car}$ are the velocity vectors of the rain and the car as observed by the stationary observer.

**FIGURE 3.18** The velocity vectors of a moving car and of rain falling straight down on the car, as viewed by a stationary observer.

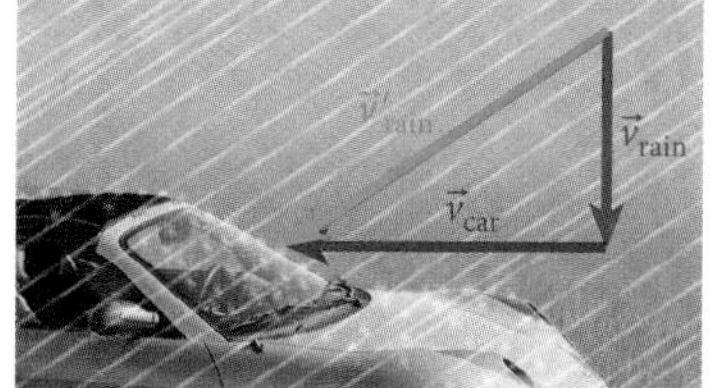


**FIGURE 3.19** The velocity vector $\vec{v}\,'_{rain}$ of rain, as observed from inside the moving car.

## WHAT WE HAVE LEARNED | EXAM STUDY GUIDE

- In two or three dimensions, any change in the magnitude or direction of an object's velocity corresponds to acceleration.
- Projectile motion of an object can be separated into motion in the $x$-direction, described by the equations

(1) $x = x_0 + v_{x0}t$

(2) $v_x = v_{x0}$

and motion in the $y$-direction, described by

(3) $y = y_0 + v_{y0}t - \frac{1}{2}gt^2$

(4) $y = y_0 + \bar{v}_y t$

(5) $v_y = v_{y0} - gt$

(6) $\bar{v}_y = \frac{1}{2}(v_y + v_{y0})$

(7) $v_y^2 = v_{y0}^2 - 2g(y - y_0)$

- The relationship between the $x$- and $y$-coordinates for ideal projectile motion can be described by a parabola given by the formula $y = y_0 + (\tan\theta_0)x - \frac{g}{2v_0^2\cos^2\theta_0}x^2$, where $y_0$ is the initial vertical position, $v_0$ is the initial speed of the projectile, and $\theta_0$ is the initial angle with respect to the horizontal at which the projectile is launched.
- The range $R$ of a projectile is given by
$$R = \frac{v_0^2}{g}\sin 2\theta_0.$$
- The maximum height $H$ reached by an ideal projectile is given by $H = y_0 + \frac{v_{y0}^2}{2g}$, where $v_{y0}$ is the vertical component of the initial velocity.
- Projectile trajectories are not parabolas when air resistance is taken into account. In general, the trajectories of realistic projectiles do not reach the maximum predicted height, and have a significantly shorter range.
- The velocity $\vec{v}_{\text{ol}}$ of an object with respect to a stationary laboratory reference frame can be calculated using a Galilean transformation of the velocity, $\vec{v}_{\text{ol}} = \vec{v}_{\text{om}} + \vec{v}_{\text{ml}}$, where $\vec{v}_{\text{om}}$ is the velocity of the object with respect to a moving reference frame and $\vec{v}_{\text{ml}}$ is the constant velocity of the moving reference frame with respect to the laboratory frame.

## KEY TERMS

right-handed coordinate system, p. 72
ideal projectile, p. 74
ideal projectile motion, p. 74
trajectory, p. 76
range, p. 78
maximum height, p. 78
relative velocity, p. 84
Galilean transformation, p. 85

## NEW SYMBOLS

$\vec{v}_{\text{ol}}$, relative velocity of an object with respect to a laboratory frame of reference

## ANSWERS TO SELF-TEST OPPORTUNITIES

**3.1** Use equation 3.23 and $t = (x - x_0)/v_{x0} = (x - x_0)/(v_0\cos\theta_0)$ to find

$$|\vec{v}| = \sqrt{v_0^2 - 2g(x - x_0)(\tan\theta_0) + g^2(x - x_0)^2/(v_0\cos\theta_0)^2}$$

**3.2** The time to reach the top is $v_y = v_{y0} - gt_{\text{top}} = 0 \Rightarrow t_{\text{top}} = v_{y0}/g = v_0\sin\theta/g$. The total flight time is $t_{\text{total}} = 2t_{\text{top}}$ because of the symmetry of the parabolic projectile trajectory. The range is the product of the total flight time and the horizontal velocity component: $R = t_{\text{total}}v_{x0} = 2t_{\text{top}}v_0\cos\theta = 2(v_0\sin\theta/g)v_0\cos\theta = v_0^2\sin(2\theta)/g$.

## PROBLEM-SOLVING PRACTICE

### Problem-Solving Guidelines

**1.** In all problems involving moving reference frames, it is important to clearly distinguish which object has what motion in which frame and relative to what. It is convenient to use subscripts consisting of two letters, where the first letter stands for a particular object and the second letter for the object it is moving relative to. The moving walkway situation discussed at the opening of Section 3.6 is a good example of this use of subscripts.

**2.** In all problems concerning ideal projectile motion, the motion in the $x$-direction is independent of that in the $y$-direction. To solve these, you can almost always use the seven kinematical equations (3.11 through 3.17), which describe motion with constant velocity in the horizontal direction and free-fall motion with constant acceleration in the vertical direction. In general, you should avoid cookie-cutter–style application of formulas, but in exam situations, these seven kinematic equations can be your first line of defense. Keep in mind, however, that these equations work only in situations in which the horizontal acceleration component is zero and the vertical acceleration component is constant.

## SOLVED PROBLEM 3.3 Time of Flight

You may have participated in Science Olympiad during middle school or high school. In one of the events in Science Olympiad, the goal is to hit a horizontal target at a fixed distance with a golf ball launched by a trebuchet. Competing teams build their own trebuchets. Your team has constructed a trebuchet that is able to launch the golf ball with an initial speed of 17.2 m/s, according to extensive tests performed before the competition.

### PROBLEM

If the target is located at the same height as the elevation from which the golf ball is released and at a horizontal distance of 22.42 m away, how long will the golf ball be in the air before it hits the target?

### SOLUTION

#### THINK

Let's first eliminate what does not work. We cannot simply divide the distance between trebuchet and target by the initial speed, because this would imply that the initial velocity vector is in the horizontal direction. Since the projectile is in free fall in the vertical direction during its flight, it would certainly miss the target. So we have to aim the golf ball with an angle larger than zero relative to the horizontal. But at what angle do we need to aim?

If the golf ball, as stated, is released from the same height as the height of the target, then the horizontal distance between the trebuchet and the target is equal to the range. Because we also know the initial speed, we can calculate the release angle. Knowing the release angle and the initial speed lets us determine the horizontal component of the velocity vector. Since this horizontal component does not change in time, the flight time is simply given by the range divided by the horizontal component of the velocity.

#### SKETCH

We don't need a sketch at this point because it would simply show a parabola, as for all projectile motion. However, we do not know the initial angle yet, so we will need a sketch later.

#### RESEARCH

The range of a projectile is given by equation 3.25:

$$R = \frac{v_0^2}{g}\sin 2\theta_0.$$

If we know the value of this range and the initial speed, we can find the angle:

$$\sin 2\theta_0 = \frac{gR}{v_0^2}.$$

Once we have the value for the angle, we can use it to calculate the horizontal component of the initial velocity:

$$v_{x0} = v_0 \cos\theta_0.$$

Finally, as noted previously, we obtain the flight time as the ratio of the range and the horizontal component of the velocity:

$$t = \frac{R}{v_{x0}}.$$

#### SIMPLIFY

If we solve the equation for the angle, $\sin 2\theta_0 = Rg/v_0^2$, we see that it has two solutions: one for an angle of less than 45° and one for an angle of more than 45°. Figure 3.20 plots the function $\sin 2\theta_0$ (in red) for all possible values of the initial angle $\theta_0$ and shows where that curve crosses the plot of $gR/v_0^2$ (blue horizontal line). We call the two solutions $\theta_a$ and $\theta_b$.

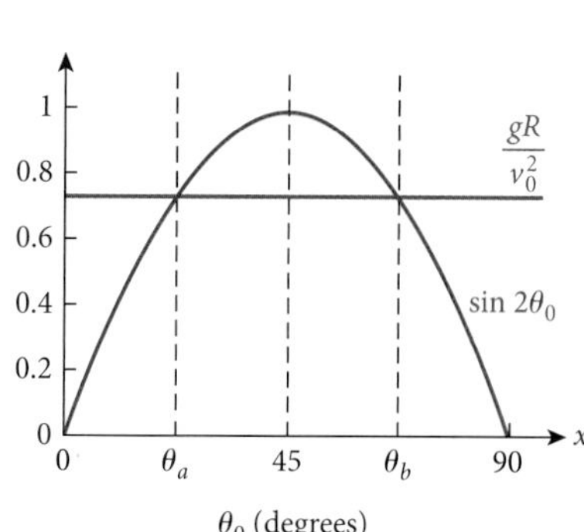


**FIGURE 3.20** Two solutions for the initial angle.

*Continued—*

Algebraically, these solutions are given as

$$\theta_{a,b} = \tfrac{1}{2}\sin^{-1}\left(\frac{Rg}{v_0^2}\right).$$

Substituting this result into the formula for the horizontal component of the velocity results in

$$t = \frac{R}{v_{x0}} = \frac{R}{v_0\cos\theta_0} = \frac{R}{v_0\cos\left(\tfrac{1}{2}\sin^{-1}\left(\dfrac{Rg}{v_0^2}\right)\right)}.$$

**CALCULATE**

Inserting numbers, we find:

$$\theta_{a,b} = \tfrac{1}{2}\sin^{-1}\left(\frac{(22.42\text{ m})(9.81\text{ m/s}^2)}{(17.2\text{ m/s})^2}\right) = 24.0128° \text{ or } 65.9872°$$

$$t_a = \frac{R}{v_0\cos\theta_a} = \frac{22.42\text{ m}}{(17.2\text{ m/s})(\cos 24.0128°)} = 1.42699\text{ s}$$

$$t_b = \frac{R}{v_0\cos\theta_b} = \frac{22.42\text{ m}}{(17.2\text{ m/s})(\cos 65.9872°)} = 3.20314\text{ s}.$$

**ROUND**

The range was specified to four significant figures, and the initial speed to three. Therefore, we also state our final results to three significant figures:

$$t_a = 1.43\text{ s}, \quad t_b = 3.20\text{ s}.$$

Note that both solutions are valid in this case, and the team can select either one.

**DOUBLE-CHECK**

Back to the approach that does not work: simply taking the distance from the trebuchet to the target and dividing it by the speed. This incorrect procedure leads to $t_{\text{min}} = d/v_0 = 1.30$ s. We write $t_{\text{min}}$ to symbolize this value to indicate that it is some lower boundary representing the case in which the initial velocity vector points horizontally and in which we neglect the free-fall motion of the projectile. Thus, $t_{\text{min}}$ serves as an absolute lower boundary, and it is reassuring to note that the shorter time we obtained above is a little larger than this lowest possible, but physically unrealistic, value.

## SOLVED PROBLEM 3.4 Moving Deer

The zookeeper who captured the monkey in Example 3.1 now has to capture a deer. We found that she needed to aim directly at the monkey for that earlier capture. She decides to fire directly at her target again, indicated by the bull's-eye in Figure 3.21.

**FIGURE 3.21** The red arrow indicates the velocity of the deer in the zookeeper's reference frame.

**PROBLEM**

Where will the tranquilizer dart hit if the deer is $d$ = 25 m away from the zookeeper and running from her right to her left with a speed of $v_d$ = 3.0 m/s? The tranquilizer dart leaves her rifle horizontally with a speed of $v_0$ = 90. m/s.

**SOLUTION**

**THINK**

The deer is moving at the same time as the dart is falling, which introduces two complications. It is easiest to think about this problem in the moving reference frame of the deer.

In that frame, the sideways horizontal component of the dart's motion has a constant velocity of $-\vec{v}_d$. The vertical component of the motion is again a free-fall motion. The total displacement of the dart is then the vector sum of the displacements caused by both of these motions.

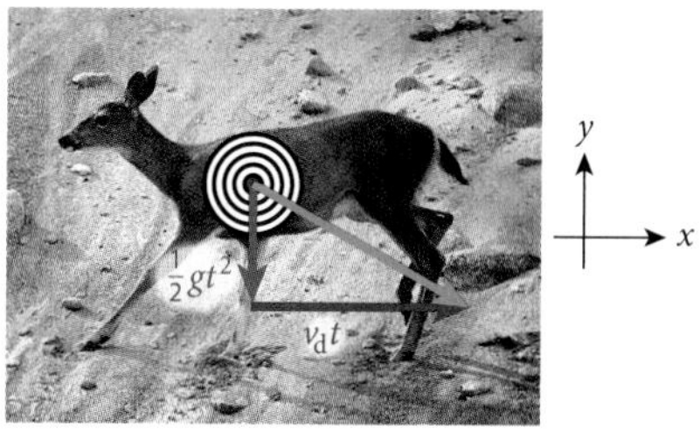

**FIGURE 3.22** Displacement of the tranquilizer dart in the deer's reference frame.

## SKETCH

We draw the two displacements in the reference frame of the deer (Figure 3.22). The blue arrow is the displacement due to the free-fall motion, and the red arrow is the sideways horizontal motion of the dart in the reference frame of the deer. The advantage of drawing the displacements in this moving reference frame is that the bull's-eye is attached to the deer and is moving with it.

## RESEARCH

First, we need to calculate the time it takes the tranquilizer dart to move 25 m in the direct line of sight from the gun to the deer. Because the dart leaves the rifle in the horizontal direction, the initial forward horizontal component of the dart's velocity vector is 90 m/s. For projectile motion, the horizontal velocity component is constant. Therefore, for the time the dart takes to cross the 25-m distance, we have

$$t = \frac{d}{v_0}.$$

During this time, the dart falls under the influence of gravity, and this vertical displacement is

$$\Delta y = -\tfrac{1}{2}gt^2.$$

Also, during this time, the deer has a sideways horizontal displacement in the reference frame of the zookeeper of $x = -v_d t$ (the deer moves to the left, hence the negative value of the horizontal velocity component). Therefore, the displacement of the dart in the reference frame of the deer is (see Figure 3.22)

$$\Delta x = v_d t.$$

## SIMPLIFY

Substituting the expression for the time into the equations for the two displacements results in

$$\Delta x = v_d \frac{d}{v_0} = \frac{v_d}{v_0} d$$

$$\Delta y = -\tfrac{1}{2}gt^2 = -\frac{d^2 g}{2v_0^2}.$$

## CALCULATE

We are now ready to put in the numbers:

$$\Delta x = \frac{(3.0 \text{ m/s})}{(90. \text{ m/s})}(25 \text{ m}) = 0.8333333 \text{ m}$$

$$\Delta y = -\frac{(25 \text{ m})^2 (9.81 \text{ m/s}^2)}{2(90. \text{ m/s})^2} = -0.378472 \text{ m}.$$

## ROUND

Rounding our results to two significant figures gives:

$$\Delta x = 0.83 \text{ m}$$

$$\Delta y = -0.38 \text{ m}.$$

The net effect is the vector sum of the sideways horizontal and vertical displacements, as indicated by the green diagonal arrow in Figure 3.22: The dart will miss the deer and hit the ground behind the deer.

*Continued—*

**DOUBLE-CHECK**

Where should the zookeeper aim? If she wants to hit the running deer, she has to aim approximately 0.38 m above and 0.83 m to the left of her intended target. A dart fired in this direction will hit the deer, but not in the center of the bull's-eye. Why? With this aim, the initial velocity vector does not point in the horizontal direction. This lengthens the flight time, as we just saw in Solved Problem 3.3. A longer flight time translates into a larger displacement in both $x$- and $y$-directions. This correction is small, but calculating it is a bit too involved to show here.

## MULTIPLE-CHOICE QUESTIONS

**3.1** An arrow is shot horizontally with a speed of 20 m/s from the top of a tower 60 m high. The time to reach the ground will be

a) 8.9 s
b) 7.1 s
c) 3.5 s
d) 2.6 s
e) 1.0 s

**3.2** A projectile is launched from the top of a building with an initial velocity of 30 m/s at an angle of 60° above the horizontal. The magnitude of its velocity at $t = 5$ s after the launch is

a) –23.0 m/s
b) 7.3 m/s
c) 15.0 m/s
d) 27.5 m/s
e) 50.4 m/s

**3.3** A ball is thrown at an angle between 0° and 90° with respect to the horizontal. Its velocity and acceleration vectors are parallel to each other at

a) 0°
b) 45°
c) 60°
d) 90°
e) none of the above

**3.4** An outfielder throws the baseball to first base, located 80 m away from the fielder, with a velocity of 45 m/s. At what launch angle above the horizontal should he throw the ball for the first baseman to catch the ball in 2 s at the same height?

a) 50.74°
b) 25.4°
c) 22.7°
d) 18.5°
e) 12.6°

**3.5** A 50-g ball rolls off a countertop and lands 2 m from the base of the counter. A 100-g ball rolls off the same counter top with the same speed. It lands ________ from the base of the counter.

a) less than 1 m
b) 1 m
c) 2 m
d) 4 m
e) more than 4 m

**3.6** For a given initial speed of an ideal projectile, there is (are) ______ launch angle(s) for which the range of the projectile is the same.

a) only one
b) two different
c) more than two but a finite number of
d) only one if the angle is 45° but otherwise two different
e) an infinite number of

**3.7** A cruise ship moves southward in still water at a speed of 20.0 km/h, while a passenger on the deck of the ship walks toward the east at a speed of 5.0 km/h. The passenger's velocity with respect to Earth is

a) 20.6 km/h, at an angle of 14.04° east of south.
b) 20.6 km/h, at an angle of 14.04° south of east.
c) 25.0 km/h, south.
d) 25.0 km/h, east.
e) 20.6 km/h, south.

**3.8** Two cannonballs are shot from different cannons at angles $\theta_{01} = 20°$ and $\theta_{02} = 30°$, respectively. Assuming ideal projectile motion, the ratio of the launching speeds, $v_{01}/v_{02}$, for which the two cannonballs achieve the same range is

a) 0.742 m
b) 0.862 m
c) 1.212 m
d) 1.093 m
e) 2.222 m

**3.9** The acceleration due to gravity on the Moon is 1.62 m/s$^2$, approximately a sixth of the value on Earth. For a given initial velocity $v_0$ and a given launch angle $\theta_0$, the ratio of the range of an ideal projectile on the Moon to the range of the same projectile on Earth, $R_{\text{Moon}}/R_{\text{Earth}}$, will be

a) 6 m
b) 3 m
c) 12 m
d) 5 m
e) 1 m

**3.10** A baseball is launched from the bat at an angle $\theta_0 = 30°$ with respect to the positive $x$-axis and with an initial speed of 40 m/s, and it is caught at the same height from which it was hit. Assuming ideal projectile motion (positive $y$-axis upward), the velocity of the ball when it is caught is

a) $(20.00\,\hat{x} + 34.64\,\hat{y})$ m/s.
b) $(-20.00\,\hat{x} + 34.64\,\hat{y})$ m/s.
c) $(34.64\,\hat{x} - 20.00\,\hat{y})$ m/s.
d) $(34.64\,\hat{x} + 20.00\,\hat{y})$ m/s.

**3.11** In ideal projectile motion, the velocity and acceleration of the projectile at its maximum height are, respectively,

a) horizontal, vertical downward.
b) horizontal, zero.
c) zero, zero.
d) zero, vertical downward.
e) zero, horizontal.

**3.12** In ideal projectile motion, when the positive $y$-axis is chosen to be vertically upward, the $y$-component of the acceleration of the object during the ascending part of the motion and the $y$-component of the acceleration during the descending part of the motion are, respectively,

a) positive, negative.
b) negative, positive.
c) positive, positive.
d) negative, negative.

**3.13** In ideal projectile motion, when the positive $y$-axis is chosen to be vertically upward, the $y$-component of the velocity of the object during the ascending part of the motion and the $y$-component of the velocity during the descending part of the motion are, respectively,

a) positive, negative.
b) negative, positive.
c) positive, positive.
d) negative, negative.

## QUESTIONS

**3.14** A ball is thrown from ground at an angle between 0° and 90°. Which of the following remain constant: $x$, $y$, $v_x$, $v_y$, $a_x$, $a_y$?

**3.15** A ball is thrown straight up by a passenger in a train that is moving with a constant velocity. Where would the ball land—back in his hands, in front of him, or behind him? Does your answer change if the train is accelerating in the forward direction? If yes, how?

**3.16** A rock is thrown at an angle 45° below the horizontal from the top of a building. Immediately after release will its acceleration be greater than, equal to, or less than the acceleration due to gravity?

**3.17** Three balls of different masses are thrown horizontally from the same height with different initial speeds, as shown in the figure. Rank in order, from the shortest to the longest, the times the balls take to hit the ground.

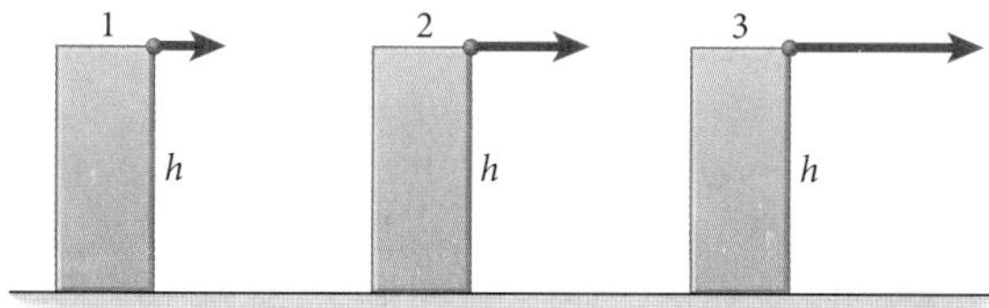


**3.18** To attain maximum height for the trajectory of a projectile, what angle would you choose between 0° and 90°, assuming that you can launch the projectile with the same initial speed independent of the launch angle. Explain your reasoning.

**3.19** An airplane is traveling at a constant horizontal speed $v$, at an altitude $h$ above a lake when a trapdoor at the bottom of the airplane opens and a package is released (falls) from the plane. The airplane continues horizontally at the same altitude and velocity. Neglect air resistance.

a) What is the distance between the package and the plane when the package hits the surface of the lake?

b) What is the horizontal component of the velocity vector of the package when it hits the lake?

c) What is the speed of the package when it hits the lake?

**3.20** Two cannonballs are shot in sequence from a cannon, into the air, with the same muzzle velocity, at the same launch angle. Based on their trajectory and range, how can you tell which one is made of lead and which one is made of wood. If the same cannonballs where launched in vacuum, what would the answer be?

**3.21** One should never jump off a moving vehicle (train, car, bus, etc.). Assuming, however, that one does perform such a jump, from a physics standpoint, what would be the best direction to jump in order to minimize the impact of the landing? Explain.

**3.22** A boat travels at a speed of $v_{BW}$ relative to the water in a river of width $D$. The speed at which the water is flowing is $v_W$.

a) Prove that the time required to cross the river to a point exactly opposite the starting point and then to return is $T_1 = 2D/\sqrt{v_{BW}^2 - v_W^2}$.

b) Also prove that the time for the boat to travel a distance $D$ downstream and then return is $T_1 = 2Dv_B/(v_{BW}^2 - v_W^2)$.

**3.23** A rocket-powered hockey puck is moving on a (frictionless) horizontal air-hockey table. The $x$- and $y$-components of its velocity as a function of time are presented in the graphs below. Assuming that at $t = 0$ the puck is at $(x_0, y_0) = (1, 2)$, draw a detailed graph of the trajectory $y(x)$.

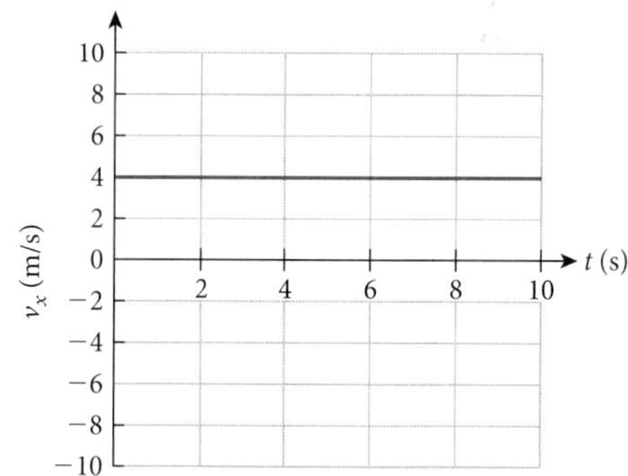


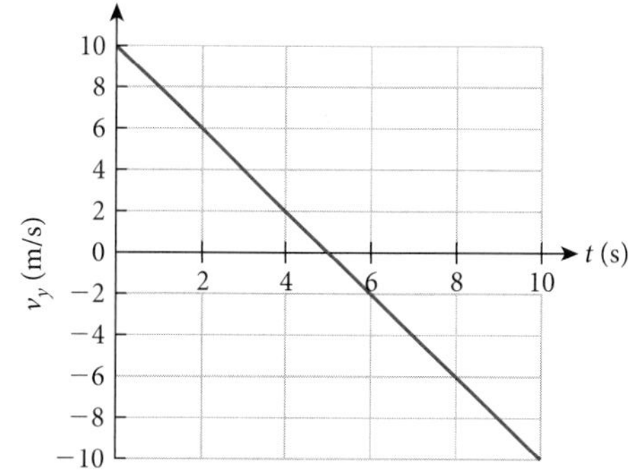

**3.24** In a three-dimensional motion, the $x$-, $y$-, and $z$-coordinates of the object as a function of time are given by

$$x(t)=\frac{\sqrt{2}}{2}t,\ y(t)=\frac{\sqrt{2}}{2}t,\ \text{and}\ z(t)=-4.9t^2+\sqrt{3}t.$$

Describe the motion and the trajectory of the object in an $xyz$ coordinate system.

**3.25** An object moves in the $xy$-plane. The $x$- and $y$-coordinates of the object as a function of time are given by the following equations: $x(t) = 4.9t^2 + 2t + 1$ and $y(t) = 3t + 2$. What is the velocity vector of the object as a function of time? What is its acceleration vector at a time $t = 2$ s?

**3.26** A particle's motion is described by the following two parametric equations:

$$x(t)=5\cos(2\pi t)$$
$$y(t)=5\sin(2\pi t)$$

where the displacements are in meters and $t$ is the time, in seconds.

a) Draw a graph of the particle's trajectory (that is, a graph of $y$ versus $x$).

b) Determine the equations that describe the $x$- and $y$-components of the *velocity*, $v_x$ and $v_y$, as functions of time.

c) Draw a graph of the particle's speed as a function of time.

**3.27** In a proof-of-concept experiment for an antiballistic missile defense system, a missile is fired from the ground of a shooting range toward a stationary target on the ground. The system detects the missile by radar, analyzes in real time its parabolic motion, and determines that it was fired from a distance $x_0 = 5000$ m, with an initial speed of 600 m/s at a launch angle $\theta_0 = 20°$. The defense system then calculates the required time delay measured from the launch of the missile and fires a small rocket situated at $y_0 = 500$ m with an initial velocity of $v_0$ m/s at a launch angle $\alpha_0 = 60°$ in the $yz$-plane, to intercept the missile. Determine the initial speed $v_0$ of the intercept rocket and the required time delay.

**3.28** A projectile is launched at an angle of 45° above the horizontal. What is the ratio of its horizontal range to its maximum height? How does the answer change if the initial speed of the projectile is doubled?

**3.29** In a projectile motion, the horizontal range and the maximum height attained by the projectile are equal.

a) What is the launch angle?

b) If everything else stays the same, how should the launch angle, $\theta_0$, of a projectile be changed for the range of the projectile to be halved?

**3.30** An air-hockey puck has a model rocket rigidly attached to it. The puck is pushed from one corner along the long side of the 2-m long air-hockey table, with the rocket pointing along the short side of the table, and at the same time the rocket is fired. If the rocket thrust imparts an acceleration of 2 m/s$^2$ to the puck, and the table is 1 m wide, with what minimum initial velocity should the puck be pushed to make it to the opposite short side of the table without bouncing off either long side of the table? Draw the trajectory of the puck for three initial velocities: $v < v_{\min}$, $v = v_{\min}$, and $v > v_{\min}$. Neglect friction and air resistance.

**3.31** On a battlefield, a cannon fires a cannonball up a slope, from ground level, with an initial velocity $v_0$ at an angle $\theta_0$ above the horizontal. The ground itself makes an angle $\alpha$ above the horizontal ($\alpha < \theta_0$). What is the range $R$ of the cannonball, measured along the inclined ground? Compare your result with the equation for the range on horizontal ground (equation 3.25).

**3.32** Two swimmers with a soft spot for physics engage in a peculiar race that models a famous optics experiment: the Michelson-Morley experiment. The race takes place in a river 50 m wide that is flowing at a steady rate of 3 m/s. Both swimmers start at the same point on one bank and swim at the same speed of 5 m/s *with respect to the stream.* One of the swimmers swims directly across the river to the closest point on the opposite bank and then turns around and swims back to the starting point. The other swimmer swims *along* the river bank, first upstream a distance exactly equal to the width of the river and then downstream back to the starting point. Who gets back to the starting point first?

## PROBLEMS

A blue problem number indicates a worked-out solution is available in the Student Solutions Manual. One • and two •• indicate increasing level of problem difficulty.

### Section 3.2

**3.33** What is the magnitude of an object's average velocity if an object moves from a point with coordinates $x = 2.0$ m, $y = -3.0$ m to a point with coordinates $x = 5.0$ m, $y = -9.0$ m in a time interval of 2.4 s?

**3.34** A man in search of his dog drives first 10 mi northeast, then 12 mi straight south, and finally 8 mi in a direction 30° north of west. What are the magnitude and direction of his resultant displacement?

**3.35** During a jaunt on your sailboat, you sail 2 km east, 4 km southeast, and an additional distance in an unknown direction. Your final position is 6 km directly east of the starting point. Find the magnitude and direction of the third leg of your journey.

**3.36** A truck travels 3.02 km north and then makes a 90° left turn and drives another 4.30 km. The whole trip takes 5.00 min.

a) With respect to a two-dimensional coordinate system on the surface of Earth such that the $y$-axis points north, what is the net displacement vector of the truck for this trip?

b) What is the magnitude of the average velocity for this trip?

•3.37 A rabbit runs in a garden such that the $x$- and $y$-components of its displacement as function of times are given by $x(t) = -0.45t^2 - 6.5t + 25$ and $y(t) = 0.35t^2 + 8.3t + 34$. (Both $x$ and $y$ are in meters and $t$ is in seconds.)

a) Calculate the rabbit's position (magnitude and direction) at $t = 10$ s.

b) Calculate the rabbit's velocity at $t = 10$ s.

c) Determine the acceleration vector at $t = 10$ s.

••**3.38** Some rental cars have a GPS unit installed, which allows the rental car company to check where you are at all times and thus also know your speed at any time. One of these rental cars is driven by an employee in the company's lot and, during the time interval from 0 to 10 s, is found to have a position vector as a function of time of

$$\vec{r}(t) = \Big((24.4 \text{ m}) - t(12.3 \text{ m/s}) + t^2(2.43 \text{ m/s}^2), \\ (74.4 \text{ m}) + t^2(1.80 \text{ m/s}^2) - t^3(0.130 \text{ m/s}^3)\Big)$$

a) What is the distance of this car from the origin of the coordinate system at $t = 5.00$ s?

b) What is the velocity vector as a function of time?

c) What is the speed at $t = 5.00$ s?

Extra credit: Can you produce a plot of the trajectory of the car in the $xy$-plane?

## Section 3.3

**3.39** A skier launches off a ski jump with a horizontal velocity of 30.0 m/s (and no vertical velocity component). What are the magnitudes of the horizontal and vertical components of her velocity the instant before she lands 2 s later?

**3.40** An archer shoots an arrow from a height of 1.14 m above ground with an initial velocity of 47.5 m/s and an initial angle of 35.2° above the horizontal. At what time after the release of the arrow from the bow will the arrow be flying exactly horizontally?

3.41 A football is punted with an initial velocity of 27.5 m/s and an initial angle of 56.7°. What is its hang time (the time until it hits the ground again)?

**3.42** You serve a tennis ball from a height of 1.8 m above the ground. The ball leaves your racket with a speed of 18.0 m/s at an angle of 7.00° above the horizontal. The horizontal distance from the court's baseline to the net is 11.83 m, and the net is 1.07 m high. Neglect spin imparted on the ball as well as air resistance effects. Does the ball clear the net? If yes, by how much? If not, by how much did it miss?

**3.43** Stones are thrown horizontally with the same velocity from two buildings. One stone lands twice as far away from its building as the other stone. Determine the ratio of the heights of the two buildings.

**3.44** You are practicing throwing darts in your dorm. You stand 3.0 m from the wall on which the board hangs. The dart leaves your hand with a horizontal velocity at a point 2.0 m above the ground. The dart strikes the board at a point 1.65 m from the ground. Calculate:

a) the time of flight of the dart;

b) the initial speed of the dart;

c) the velocity of the dart when it hits the board.

•3.45 A football player kicks a ball with a speed of 22.4 m/s at an angle of 49° above the horizontal from a distance of 39 m from the goal line.

a) By how much does the ball clear or fall short of clearing the crossbar of the goalpost if that bar is 3.05 m high?

b) What is the vertical velocity of the ball at the time it reaches the goalpost?

•**3.46** An object fired at an angle of 35° above the horizontal takes 1.5 s to travel the last 15 m of its vertical distance and the last 10 m of its horizontal distance. With what velocity was the object launched?

•**3.47** A conveyor belt is used to move sand from one place to another in a factory. The conveyor is tilted at an angle of 14° from the horizontal and the sand is moved without slipping at the rate of 7 m/s. The sand is collected in a big drum 3 m below the end of the conveyor belt. Determine the horizontal distance between the end of the conveyor belt and the middle of the collecting drum.

•**3.48** Your friend's car is parked on a cliff overlooking the ocean on an incline that makes an angle of 17.0° below the horizontal. The brakes fail, and the car rolls from rest down the incline for a distance of 29.0 m to the edge of the cliff, which is 55.0 m above the ocean, and, unfortunately, continues over the edge and lands in the ocean.

a) Find the car's position relative to the base of the cliff when the car lands in the ocean.

b) Find the length of time the car is in the air.

•3.49 An object is launched at a speed of 20.0 m/s from the top of a tall tower. The height $y$ of the object as a function of the time $t$ elapsed from launch is $y(t) = -4.9t^2 + 19.32t + 60$, where $h$ is in meters and $t$ is in seconds. Determine:

a) the height $H$ of the tower;

b) the launch angle;

c) the horizontal distance traveled by the object before it hits the ground.

•**3.50** A projectile is launched at a 60° angle above the horizontal on level ground. The change in its velocity between launch and just before landing is found to be $\Delta\vec{v} \equiv \vec{v}_{\text{landing}} - \vec{v}_{\text{launch}} = -20\hat{y}$ m/s. What is the initial velocity of the projectile? What is its final velocity just before landing?

••3.51 The figure shows the paths of a tennis ball your friend drops from the window of her apartment and of the rock you throw from the ground at the same instant. The rock and the ball collide at $x = 50$ m, $y = 10$ m and $t = 3$ s. If the ball was dropped from a height of 54 m, determine the velocity of the rock initially and at the time of its collision with the ball.

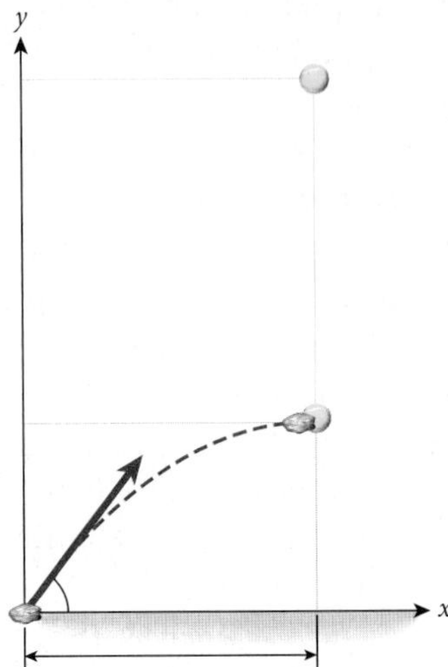


## Section 3.4

3.52 For a science fair competition, a group of high school students build a kicker-machine that can launch a golf ball from the origin with a velocity of 11.2 m/s and initial angle of 31.5° with respect to the horizontal.

a) Where will the golf ball fall back to the ground?

b) How high will it be at the highest point of its trajectory?

c) What is the ball's velocity vector (in Cartesian components) at the highest point of its trajectory?

d) What is the ball's acceleration vector (in Cartesian components) at the highest point of its trajectory?

3.53 If you want to use a catapult to throw rocks and the maximum range you need these projectiles to have is 0.67 km, what initial speed do your projectiles have to have as they leave the catapult?

3.54 What is the maximum height above ground a projectile of mass 0.79 kg, launched from ground level, can achieve if you are able to give it an initial speed of 80.3 m/s?

•3.55 During one of the games, you were asked to punt for your football team. You kicked the ball at an angle of 35° with a velocity of 25 m/s. If your punt goes straight down the field, determine the average velocity at which the running back of the opposing team standing at 70 m from you must run to catch the ball at the same height as you released it. Assume that the running back starts running as the ball leaves your foot and that the air resistance is negligible.

•3.56 By trial and error, a frog learns that it can leap a maximum horizontal distance of 1.3 m. If, in the course of an hour, the frog spends 20% of the time resting and 80% of the time performing identical jumps of that maximum length, in a straight line, what is the distance traveled by the frog?

•3.57 A circus juggler performs an act with balls that he tosses with his right hand and catches with his left hand. Each ball is launched at an angle of 75° and reaches a maximum height of 90 cm above the launching height. If it takes the juggler 0.2 s to catch a ball with his left hand, pass it to his right hand and toss it back into the air, what is the maximum number of balls he can juggle?

••3.58 In an arcade game, a ball is launched from the corner of a smooth inclined plane. The inclined plane makes a 30° angle with the horizontal and has a width of $w = 50$ cm. The spring-loaded launcher makes an angle of 45° with the lower edge of the inclined plane. The goal is to get the ball in a small hole at the opposite corner of the inclined plane. With what initial velocity should you launch the ball to achieve this goal?

••3.59 A copy-cat daredevil tries to reenact Evel Knievel's 1974 attempt to jump the Snake River Canyon in a rocket-powered motorcycle. The canyon is $L = 400$ m wide, with the opposite rims at the same height. The height of the launch ramp at one rim of the canyon is $h = 8$ m above the rim, and the angle of the end of the ramp is 45° with the horizontal.

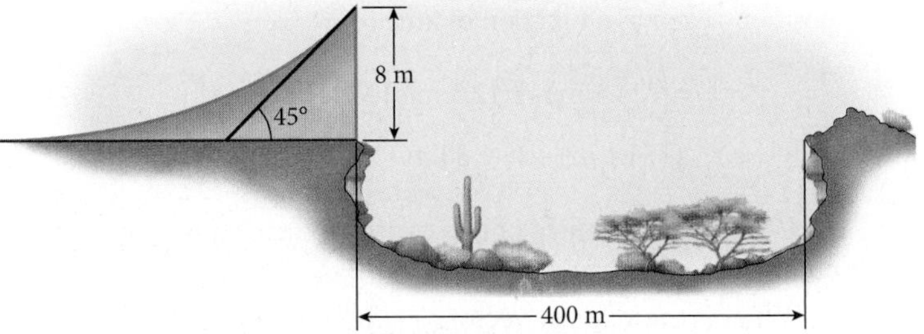


a) What is the minimum launch speed required for the daredevil to make it across the canyon? Neglect the air resistance and wind.

b) Famous after his successful first jump, but still recovering from the injuries sustained in the crash caused by a strong bounce upon landing, the daredevil decides to jump again but to add a landing ramp with a slope that will match the angle of his velocity at landing. If the height of the landing ramp at the opposite rim is 3 m, what is the new required launch speed, and how far from the launch rim and at what height should the edge of the landing ramp be?

## Section 3.5

3.60 A golf ball is hit with an initial angle of 35.5° with respect to the horizontal and an initial velocity of 83.3 mph. It lands a distance of 86.8 m away from where it was hit. By how much did the effects of wind resistance, spin, and so forth reduce the range of the golf ball from the ideal value?

## Section 3.6

3.61 You are walking on a moving walkway in an airport. The length of the walkway is 59.1 m. If your velocity relative to the walkway is 2.35 m/s and the walkway moves with a velocity of 1.77 m/s, how long will it take you to reach the other end of the walkway?

**3.62** The captain of a boat wants to travel directly across a river that flows due east with a speed of 1.00 m/s. He starts from the south bank of the river and heads toward the north bank. The boat has a speed of 6.10 m/s with respect to the water. What direction (in degrees) should the captain steer the boat? Note that 90° is east, 180° is south, 270° is west, and 360° is north.

**3.63** The captain of a boat wants to travel directly across a river that flows due east. He starts from the south bank of the river and heads toward the north bank. The boat has a speed of 5.57 m/s with respect to the water. The captain steers the boat in the direction 315°. How fast is the water flowing? Note that 90° is east, 180° is south, 270° is west, and 360° is north.

•**3.64** The air speed indicator of a plane that took off from Detroit reads 350 km/h and the compass indicates that it is heading due east to Boston. A steady wind is blowing due north at 40 km/h. Calculate the velocity of the plane with reference to the ground. If the pilot wishes to fly directly to Boston (due east) what must the compass read?

•3.65 You want to cross a straight section of a river that has a uniform current of 5.33 m/s and is 127. m wide. Your motorboat has an engine that can generate a speed of 17.5 m/s for your boat. Assume that you reach top speed right away (that is, neglect the time it takes to accelerate the boat to top speed).

a) If you want to go directly across the river with a 90° angle relative to the riverbank, at what angle relative to the riverbank should you point your boat?

b) How long will it take to cross the river in this way?

c) In which direction should you aim your boat to achieve minimum crossing time?

d) What is the minimum time to cross the river?

e) What is the minimum speed of your boat that will still enable you to cross the river with a 90° angle relative to the riverbank?

•**3.66** During a long airport layover, a physicist father and his 8-year-old daughter try a game that involves a moving walkway. They have measured the walkway to be 42.5 m long. The father has a stopwatch and times his daughter. First, the daughter walks with a constant speed in the same direction as the conveyor. It takes 15.2 s to reach the end of the walkway. Then, she turns around and walks with the same speed relative to the conveyor as before in the opposite direction. The return leg takes 70.8 s. What is the speed of the walkway conveyor relative to the terminal, and with what speed was the girl walking?

•**3.67** An airplane has an air speed of 126.2 m/s and is flying due north, but the wind blows from the northeast to the southwest at 55.5 m/s. What is the plane's actual ground speed?

## Additional Problems

**3.68** A cannon is fired from a hill 116.7 m high at an angle of 22.7° with respect to the horizontal. If the muzzle velocity is 36.1 m/s, what is the speed of a 4.35-kg cannonball when it hits the ground 116.7 m below?

3.69 A baseball is thrown with a velocity of 31.1 m/s at an angle of $\theta = 33.4°$ above horizontal. What is the horizontal component of the ball's velocity at the highest point of the ball's trajectory?

**3.70** A rock is thrown horizontally from the top of a building with an initial speed of $v = 10.1$ m/s. If it lands $d = 57.1$ m from the base of the building, how high is the building?

**3.71** A car is moving at a constant 19.3 m/s, and rain is falling at 8.9 m/s straight down. What angle $\theta$ (in degrees) does the rain make with respect to the horizontal as observed by the driver?

**3.72** You passed the salt and pepper shakers to your friend at the other end of a table of height 0.85 m by sliding them across the table. They both missed your friend and slid off the table, with velocities of 5 m/s and 2.5 m/s, respectively.

a) Compare the times it takes the shakers to hit the floor.

b) Compare the distance that each shaker travels from the edge of the table to the point it hits the floor.

3.73 A box containing food supplies for a refugee camp was dropped from a helicopter flying horizontally at a constant elevation of 500 m. If the box hit the ground at a distance of 150 m horizontally from the point of its release, what was the speed of the helicopter? With what speed did the box hit the ground?

**3.74** A car drives straight off the edge of a cliff that is 60 m high. The police at the scene of the accident note that the point of impact is 150 m from the base of the cliff. How fast was the car traveling when it went over the cliff?

**3.75** At the end of the spring term, a high school physics class celebrates by shooting a bundle of exam papers into the town landfill with a homemade catapult. They aim for a point that is 30 m away and at the same height from which the catapult releases the bundle. The initial horizontal velocity component is 3.9 m/s. What is the initial velocity component in the vertical direction? What is the launch angle?

**3.76** Salmon often jump upstream through waterfalls to reach their breeding grounds. One salmon came across a waterfall 1.05 m in height, which she jumped in 2.1 s at an angle of 35° to continue upstream. What was the initial speed of her jump?

3.77 A firefighter, 60 m away from a burning building, directs a stream of water from a ground-level fire hose at an angle of 37° above the horizontal. If the water leaves the hose at 40.3 m/s, which floor of the building will the stream of water strike? Each floor is 4 m high.

**3.78** A projectile leaves ground level at an angle of 68° above the horizontal. As it reaches its maximum height, $H$, it has traveled a horizontal distance, $d$, in the same amount of time. What is the ratio $H/d$?

**3.79** The McNamara Northwest terminal at the Metro Detroit Airport has moving walkways for the convenience of the passengers. Robert walks beside one walkway and takes 30.0 s to cover its length. John simply stands on the walkway and covers the same distance in 13.0 s. Kathy walks on the walkway with the same speed as Robert's. How long does Kathy take to complete her stroll?

**3.80** Rain is falling vertically at a constant speed of 7.0 m/s. At what angle from the vertical do the raindrops appear to be falling to the driver of a car traveling on a straight road with a speed of 60 km/h?

**3.81** To determine the gravitational acceleration at the surface of a newly discovered planet, scientists perform a projectile motion experiment. They launch a small model rocket at an initial speed of 50 m/s and an angle of 30° above the horizontal and measure the (horizontal) range on flat ground to be 2165 m. Determine the value of $g$ for the planet.

**3.82** A diver jumps from a 40 m high cliff into the sea. Rocks stick out of the water for a horizontal distance of 7 m from the foot of the cliff. With what minimum horizontal speed must the diver jump off the cliff in order to clear the rocks and land safely in the sea?

**3.83** An outfielder throws a baseball with an initial speed of 32 m/s at an angle of 23° to the horizontal. The ball leaves his hand from a height of 1.83 m. How long is the ball in the air before it hits the ground?

•**3.84** A rock is tossed off the top of a cliff of height 34.9 m. Its initial speed is 29.3 m/s, and the launch angle is 29.9° with respect to the horizontal. What is the speed with which the rock hits the ground at the bottom of the cliff?

•**3.85** During the 2004 Olympic Games, a shot putter threw a shot put with a speed of 13.0 m/s at an angle of 43° above the horizontal. She released the shot put from a height of 2 m above the ground.

a) How far did the shot put travel in the horizontal direction?

b) How long was it until the shot put hit the ground?

•**3.86** A salesman is standing on the Golden Gate Bridge in a traffic jam. He is at a height of 71.8 m above the water below. He receives a call on his cell phone that makes him so mad that he throws his phone horizontally off the bridge with a speed of 23.7 m/s.

a) How far does the cell phone travel horizontally before hitting the water?

b) What is the speed with which the phone hits the water?

•**3.87** A security guard is chasing a burglar across a rooftop, both running at 4.2 m/s. Before the burglar reaches the edge of the roof, he has to decide whether or not to try jumping to the roof of the next building, which is 5.5 m away and 4.0 m lower. If he decides to jump horizontally to get away from the guard, can he make it? Explain your answer.

•**3.88** A blimp is ascending at the rate of 7.5 m/s at a height of 80 m above the ground when a package is thrown from its cockpit horizontally with a speed of 4.7 m/s.

a) How long does it take for the package to reach the ground?

b) With what velocity (magnitude and direction) does it hit the ground?

•**3.89** Wild geese are known for their lack of manners. One goose is flying northward at a level altitude of $h_g = 30.0$ m above a north-south highway, when it sees a car ahead in the distance moving in the southbound lane and decides to deliver (drop) an "egg." The goose is flying at a speed of $v_g = 15.0$ m/s, and the car is moving at a speed of $v_c =$ 100.0 km/h.

a) Given the details in the figure, where the separation between the goose and the front bumper of the car, $d =$ 104.1 m, is specified at the instant when the goose takes action, will the driver have to wash the windshield after this encounter? (The center of the windshield is $h_c = 1.00$ m off the ground.)

b) If the delivery is completed, what is the relative velocity of the "egg" with respect to the car at the moment of the impact?

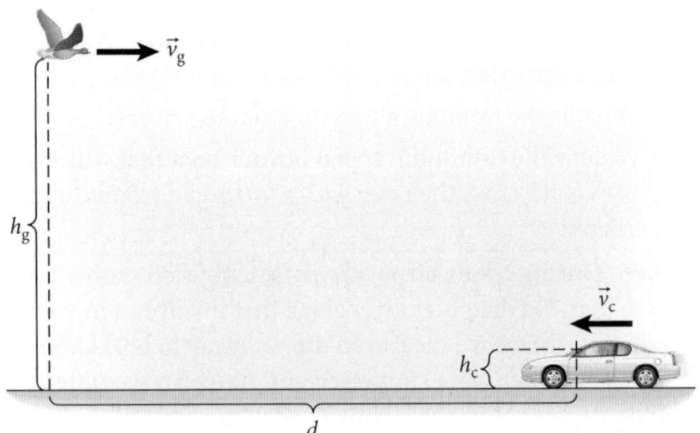


•**3.90** You are at the mall on the top step of a down escalator when you lean over laterally to see your 1.8 m tall physics professor on the bottom step of the adjacent up escalator. Unfortunately, the ice cream you hold in your hand falls out of its cone as you lean. The two escalators have identical angles of 40° with the horizontal, a vertical height of 10 m, and move at the same speed of 0.4 m/s. Will the ice cream land on your professor's head? Explain. If it does land on his head, at what time and at what vertical height does that happen? What is the relative speed of the ice cream with respect to the head at the time of impact?

•**3.91** A basketball player practices shooting three-pointers from a distance of 7.50 m from the hoop, releasing the ball at a height of 2.00 m above ground. A standard basketball hoop's rim top is 3.05 m above the floor. The player shoots the ball at an angle of 48° with the horizontal. At what initial speed must he shoot to make the basket?

•**3.92** Wanting to invite Juliet to his party, Romeo is throwing pebbles at her window with a launch angle of 37° from the horizontal. He is standing at the edge of the rose garden 7.0 m below her window and 10.0 m from the base of the wall. What is the initial speed of the pebbles?

•3.93 An airplane flies horizontally above the flat surface of a desert at an altitude of 5 km and a speed of 1000 km/h. If the airplane is to drop a care package that is supposed to hit a target on the ground, where should the plane be with respect to the target when the package is released? If the target covers a circular area with a diameter of 50 m, what is the "window of opportunity" (or margin of error allowed) for the release time?

•**3.94** A plane diving with constant speed at an angle of 49° with the vertical, releases a package at an altitude of 600 m. The package hits the ground 3.5 s after release. How far horizontally does the package travel?

••**3.95** Ten seconds after being fired, a cannonball strikes a point 500 m horizontally from and 100 m vertically above the point of launch.

a) With what initial velocity was the cannonball launched?

b) What maximum height was attained by the ball?

c) What is the magnitude and direction of the ball's velocity just before it strikes the given point?

••**3.96** Neglect air resistance for the following. A soccer ball is kicked from the ground into the air. When the ball is at a height of 12.5 m, its velocity is $(5.6\hat{x} + 4.1\hat{y})$ m/s.

a) To what maximum height will the ball rise?

b) What horizontal distance will be traveled by the ball?

c) With what velocity (magnitude and direction) will it hit the ground?

# 4 Force

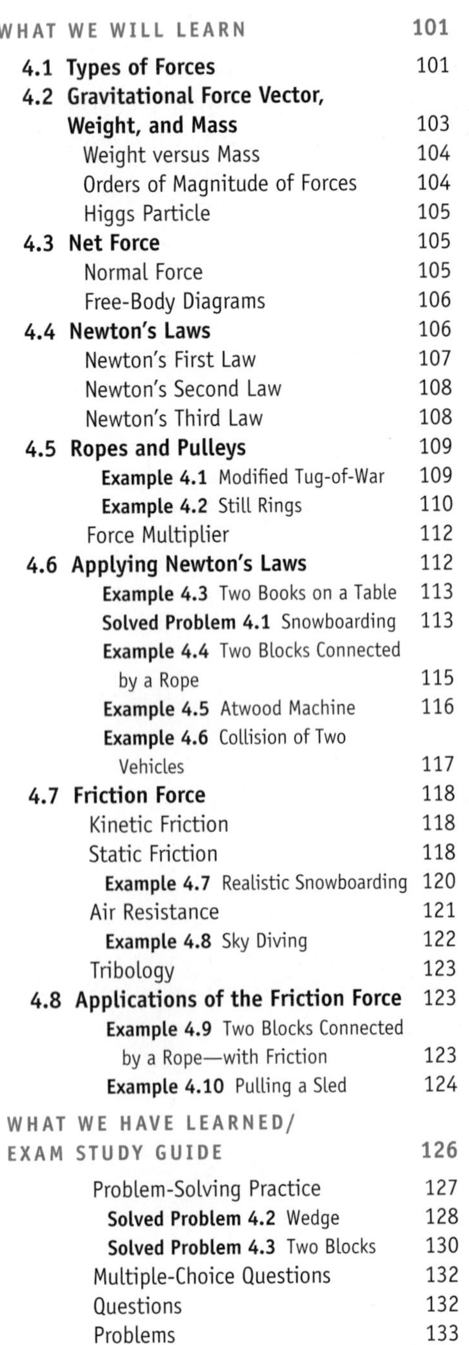

**FIGURE 4.1** The Space Shuttle Columbia lifts off from the Kennedy Space Center.

## WHAT WE WILL LEARN

- A force is a vector quantity that is a measure of how an object interacts with other objects.
- Fundamental forces include gravitational attraction and electromagnetic attraction and repulsion. In daily experience, important forces include tension and normal, friction, and spring forces.
- Multiple forces acting on an object sum to a net force.
- Free-body diagrams are valuable aids in working problems.
- Newton's three laws of motion govern the motion of objects under the influence of forces.
  a) The first law deals with objects for which external forces are balanced.
  b) The second law describes those cases for which external forces are not balanced.
  c) The third law addresses equal (in magnitude) and opposite (in direction) forces that two bodies exert on each other.
- The gravitational mass and the inertial mass of an object are equivalent.
- Kinetic friction opposes the motion of moving objects; static friction opposes the impending motion of objects at rest.
- Friction is important to the understanding of real-world motion, but its causes and exact mechanisms are still under investigation.
- Applications of Newton's laws of motion involve multiple objects, multiple forces, and friction; applying the laws to analyze a situation is among the most important problem-solving techniques in physics.

The launch of a Space Shuttle is an awesome sight. Huge clouds of smoke obscure the shuttle until it rises up high enough to be seen above them, with brilliant exhaust flames exiting the main engines. The boosters provide a force of 30.16 meganewtons (6.781 million pounds), enough to shake the ground for miles around. This tremendous force accelerates the shuttle (over 2 million kilograms, or 4.5 million pounds) sufficiently for lift-off. Several engine systems are used to accelerate the shuttle further to the final speed needed to achieve orbit—approximately 8 km/s.

The Space Shuttle has been called one of the greatest technological achievements of the 20th century, but the basic principles of force, mass, and acceleration that govern its operation have been known for over 300 years. First stated by Isaac Newton in 1687, the laws of motion apply to all interactions between objects. Just as kinematics describes how objects move, Newton's laws of motion are the foundation of **dynamics,** which describes what makes objects move. We will study dynamics for the next several chapters.

In this chapter, we examine Newton's laws of motion and explore the various kinds of forces they describe. The process of identifying the forces that act on an object, determining the motion caused by those forces, and interpreting the overall vector result is one of the most common and important types of analysis in physics, and we will use it numerous times throughout this book. Many of the kinds of forces introduced in this chapter, such as contact forces, friction forces, and weight, will play a role in many of the concepts and principles discussed later.

## 4.1 Types of Forces

You are probably sitting on a chair as you are reading this page. The chair exerts a force on you, which prevents you from falling to the ground. You can feel this force from the chair on the underside of your legs and your backside. Conversely, you exert a force on the chair.

If you pull on a string, you exert a force on the string, and that string, in turn, can exert a force on something tied to it's other end. This force is an example of a **contact force,** in which one object has to be in contact with another to exert a force on it, as is the previous example of you sitting on your chair. If you push or pull on an object, you exert a contact force on it. Pulling on an object, such as a rope or a string, gives rise to the contact force called **tension.** Pushing on an object causes the contact force called **compression.** The force that acts on you when you sit on a chair is called a **normal force,** where the word *normal* means "perpendicular to the surface." We will examine normal forces in more detail a little later in this chapter.

The **friction force** is another important contact force that we will study in more detail in this chapter. If you push a glass across the surface of a table, it comes to rest rather quickly. The force that causes the glass's motion to stop is the friction force, sometimes also simply called *friction.* Interestingly, the exact nature and microscopic origin of the friction force is still under intense investigation, as we'll see.

A force is needed to compress a spring as well as to extend it. The **spring force** has the special property that it depends linearly on the change in length of the spring. Chapter 5 will introduce the spring force and describe some of its properties. Chapter 14 will focus on oscillations, a special kind of motion resulting from the action of spring forces.

Contact forces, friction forces, and spring forces are the results of the **fundamental forces** of nature acting between the constituents of objects. The **gravitational force,** often simply called *gravity,* is one example of a fundamental force. If you hold an object in your hand and let go of it, it falls downward. We know what causes this effect: the gravitational attraction between the Earth and the object. The gravitational acceleration was introduced in Chapter 2, and this chapter describes how it is related to the gravitational force. Gravity is also responsible for holding the Moon in orbit around the Earth and the Earth in orbit around the Sun. In a famous story (which may even be true!), Isaac Newton was reported to have had this insight in the 17th century, after sitting under an apple tree and being hit by an apple falling off the tree: The same type of gravitational force acts between celestial objects as operates between terrestrial objects. However, keep in mind that the gravitational force discussed in this chapter is a limited instance, valid only near the surface of Earth, of the more general gravitational force. Near the surface of Earth, a constant gravitational force acts on all objects, which is sufficient to solve practically all trajectory problems of the kind covered in Chapter 3. The more general form of the gravitational interaction, however, is inversely proportional to the square of the distance between the two objects exerting the gravitational force on each other. Chapter 12 is devoted to this force.

(a)

(b)

(c)

**FIGURE 4.2** Some common types of forces. (a) A grinding wheel works by using the force of friction to remove the outer surface of an object. (b) Springs are often used as shock absorbers in cars to reduce the force transmitted to the wheels by the ground. (c) Some dams are among the largest structures ever built. They are designed to resist the force exerted by the water they hold back.

Another fundamental force that can act at a distance is the **electromagnetic force,** which, like the gravitational force, is inversely proportional to the square of the distance over which it acts. The most apparent manifestation of this force is the attraction or repulsion between two magnets, depending on their relative orientation. The entire Earth also acts as a huge magnet, which makes compass needles orient themselves toward the North Pole. The electromagnetic force was the big physics discovery of the 19th century, and its refinement during the 20th century led to many of the high-tech conveniences (basically everything that plugs into an electrical outlet or uses batteries) we enjoy today. Chapters 21 through 31 will provide an extensive tour of the electromagnetic force and its many manifestations.

In particular, we will see that all of the contact forces listed above (normal force, tension, friction, spring force) are fundamentally consequences of the electromagnetic force. Why then study these contact forces in the first place? The answer is that phrasing a problem in terms of contact forces gives us great insight and allows us to formulate simple solutions to real-world problems whose solutions would otherwise require the use of supercomputers if we tried to analyze them in terms of the electromagnetic interactions between atoms.

The other two fundamental forces—called the **strong nuclear force** and the **weak nuclear force**—act only on the length scales of atomic nuclei and between elementary particles. These forces between elementary particles will be discussed in Chapter 39 on particle physics and Chapter 40 on nuclear physics. In general, forces can be defined as the means for objects to influence each other (Figure 4.2).

Most of the forces mentioned here have been known for hundreds of years. However, the ways in which scientists and engineers use forces continues to evolve as new materials and new designs are created. For example, the idea of a bridge to cross a river or deep ravine has been used for thousands of years, starting with simple forms such as a log dropped across a stream or a series of ropes strung across a gorge. Over time, engineers developed the idea of an arch bridge that can support a heavy roadway and a load of traffic using compressive forces. Many such bridges were built from stone or steel, materials that can support compression well (Figure 4.3a). In the late 19th and 20th centuries, bridges were built with the roadway suspended from steel cables supported by tall piers (Figure 4.3b). The cables

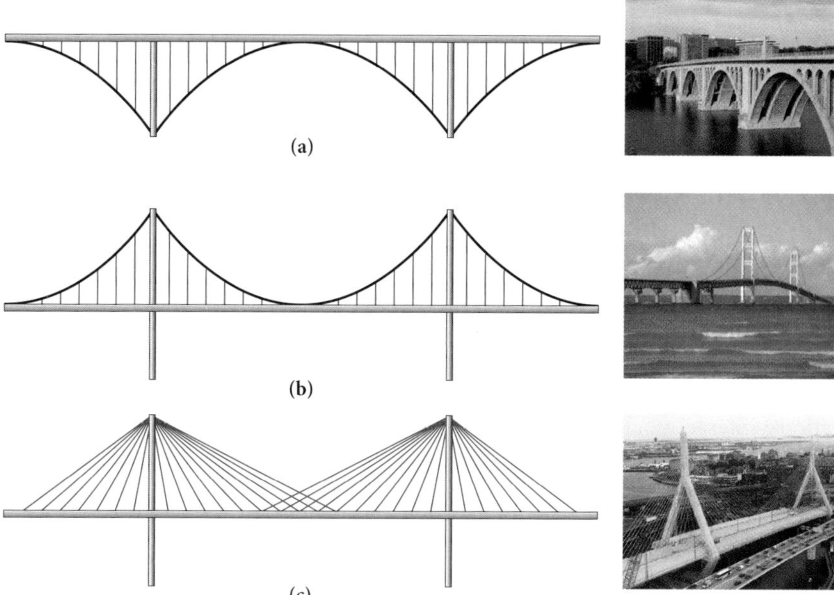


**FIGURE 4.3** Different ways to use forces. (a) Arch bridges (such as Francis Scott Key Bridge in Washington, DC) support a road by compressive forces, with each end of the arch anchored in place. (b) Suspension bridges (such as Mackinac Bridge in Michigan) support the roadway by tension forces in the cables, which are in turn supported by compressive forces in the tall piers sunk into the ground below the water. (c) Cable-stayed bridges (such as Zakim Bridge in Boston) also use tension forces in cables to support the roadway, but the load is distributed over many more cables, which do not need to be as strong and difficult to build as in suspension bridges.

supported tension, and these bridges could be lighter as well as longer than previous bridge designs. In the late 20th century, cable-stayed bridges began to appear, with the roadway supported by cables attached directly to the piers (Figure 4.3c). These bridges are not generally as long as suspension bridges but are less expensive and time-consuming to build.

## 4.2 Gravitational Force Vector, Weight, and Mass

After this general introduction to forces, it is time to get more quantitative. Let's start with an obvious fact: Forces have a direction. For example, if you are holding a laptop computer in your hand, you can easily tell that the gravitational force acting on the computer points downward. This direction is the direction of the **gravitational force vector** (Figure 4.4). Again, to characterize a quantity as a vector quantity throughout this book, a small right-pointing arrow appears above the symbol for the quantity. Thus, the gravitational force vector acting on the laptop is denoted by $\vec{F}_g$ in the figure.

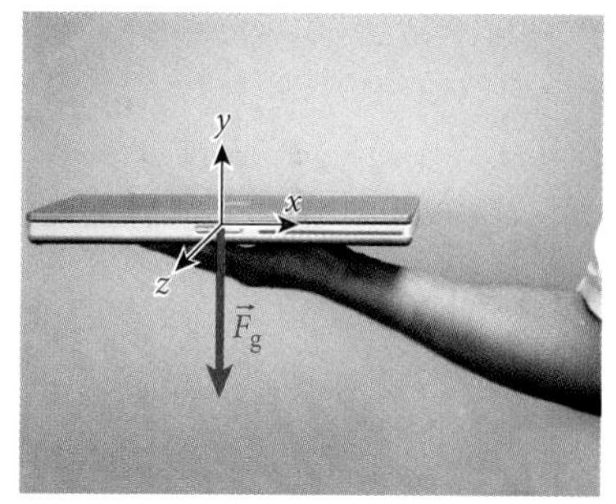


**FIGURE 4.4** Force vector of gravity acting on a laptop computer, in relation to the conventional right-handed Cartesian coordinate system.

Figure 4.4 also shows a convenient Cartesian coordinate system, which follows the convention introduced in Chapter 3 in which up is the positive $y$-direction (and down the negative $y$-direction). The $x$- and $z$-directions then lie in the horizontal plane, as shown. As always, we use a right-handed coordinate system. Also, we restrict ourselves to two-dimensional coordinate systems with $x$- and $y$-axes wherever possible.

In the coordinate system in Figure 4.4, the force vector of the gravitational force acting on the laptop is pointing in the negative $y$-direction:

$$\vec{F}_g = -F_g\hat{y}. \quad (4.1)$$

Here we see that the force vector is the product of its magnitude, $F_g$, and its direction, $-\hat{y}$. The magnitude $F_g$ is called the **weight** of the object.

Near the surface of the Earth (within a few hundred meters above the ground), the magnitude of the gravitational force acting on an object is given by the product of the mass of the object, $m$, and the Earth's gravitational acceleration, $g$:

$$F_g = mg. \quad (4.2)$$

We have used the magnitude of the Earth's gravitational acceleration in previous chapters: It has the value $g = 9.81$ m/s$^2$. Note that this constant value is valid only up to a few hundred meters above the ground, as we will see in Chapter 12.

With equation 4.2, we find that the unit of force is the product of the unit of mass (kg) and the unit of acceleration (m/s$^2$), which makes the unit of force kg m/s$^2$. (Perhaps it is worth repeating that we represent units with roman letters and physical quantities with italic letters. Thus, m is the unit of length; *m* stands for the physical quality of mass.) Since dealing with forces is so common in physics, the unit of force has received its own name, the newton (N), after Sir Isaac Newton, the British physicist who made a key contribution to the analysis of forces.

$$1 \text{ N} \equiv 1 \text{ kg m/s}^2. \tag{4.3}$$

## Weight versus Mass

Before discussing forces in greater detail, we need to clarify the concept of mass. Under the influence of gravity, an object has a weight that is proportional to its **mass,** which is (intuitively) the amount of matter in the object. This weight is the magnitude of a force that acts on an object due to its gravitational interaction with the Earth (or another object). Near the surface of Earth, the magnitude of this force is $F_g = mg$, as stated in equation 4.2. The mass in this equation is also called the **gravitational mass** to indicate that it is responsible for the gravitational interaction. However, mass has a role in dynamics as well.

Newton's laws of motion, which will be introduced later in this chapter, deal with inertial mass. To understand the concept of inertial mass, consider the following examples: It is a lot easier to throw a tennis ball than a shot put. It is also easier to pull open a door made of lightweight materials like foam-core with wood veneer than one made of a heavy material like iron. The more massive objects seem to resist being put into motion more than the less massive ones do. This property of an object is referred to as its **inertial mass.** However, the gravitational mass and the inertial mass are identical, so most of the time we refer simply to the mass of an object.

For a laptop computer with mass $m = 3.00$ kg, for example, the magnitude of the gravitational force is $F_g = mg = (3.00 \text{ kg})(9.81 \text{ m/s}^2) = 29.4 \text{ kg m/s}^2 = 29.4 \text{ N}$. Now we can write an equation for the force vector that contains both the magnitude and the direction of the gravitational force acting on the laptop computer (see Figure 4.4):

$$\vec{F}_g = -mg\hat{y}. \tag{4.4}$$

To summarize, the mass of an object is measured in kilograms and the weight of an object is measured in newtons. The mass and the weight of an object are related to each other by multiplying the mass (in kilograms) by the gravitational acceleration constant, $g = 9.81 \text{ m/s}^2$, to arrive at the weight (in newtons). For example, if your mass is 70.0 kg, then your weight is 687 N. In the United States, the pound (lb) is still a widely used unit. The conversion between pounds and kilograms is 1 lb = 0.4536 kg. Thus, your mass of 70.0 kg is 154 lb if you express it in British units. In everyday language, you might say that your weight is 154 pounds, which is not correct. Unfortunately, engineers in the United States also use the unit of pound-force ($\text{lb}_f$) as a force unit, which is often shortened to just pound. However, 1 pound-force is 1 pound times the gravitational acceleration constant, just as 1 newton equals 1 kilogram times $g$. This means that

$$1 \text{ lb}_f = (1 \text{ lb}) \cdot g = (0.4536 \text{ kg})(9.81 \text{ m/s}^2) = 4.45 \text{ N}.$$

Confusing? Yes! This is one more reason to steer away from using British units. Use kilograms for mass and newtons for weight, which is a force!

## Orders of Magnitude of Forces

The concept of a force is a central theme of this book, and we will return to it again and again. For this reason, it is instructive to look at the orders of magnitude that different forces can have. Figure 4.5 gives an overview of magnitudes of some typical forces with the aid of a logarithmic scale, similar to those used in Chapter 1 for length and mass.

A humans' body weight is in the range between 100 and 1000 N and is represented by the young soccer player in Figure 4.5. The earphone to the player's left in Figure 4.5 symbolizes the force exerted by sound on our eardrums, which can be as large as $10^{-4}$ N, but is still detectable when it is as small as $10^{-13}$ N. (Chapter 16 will focus on sound.) A single electron is kept in orbit around a proton by an electrostatic force of approximately $10^{-9} \text{ N} \equiv 1 \text{ nN}$, which will be introduced in Chapter 21 on electrostatics. Forces as small as $10^{-15} \text{ N} \equiv 1 \text{ fN}$ can be measured in the lab; these forces are typical of those needed to stretch the double-helical DNA molecule.

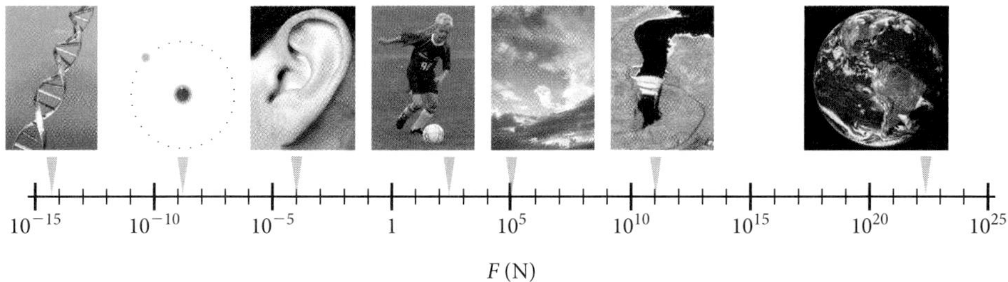


**FIGURE 4.5** Typical magnitudes for different forces.

The Earth's atmosphere exerts quite a sizeable force on our bodies, on the order of $10^5$ N, which is approximately 100 times the average body weight. Chapter 13 on solids and fluids will expand on this topic and will also show how to calculate the force of water on a dam. For example, the Hoover Dam (shown in Figure 4.5) has to withstand a force close to $10^{11}$ N, a huge force, more than 30 times the weight of the Empire State Building. But this force pales, of course, in comparison to the gravitational force that the Sun exerts on Earth, which is $3 \cdot 10^{22}$ N. (Chapter 12 on gravity will describe how to calculate this force.)

## Higgs Particle

As far as our studies are concerned, mass is an intrinsic, given, property of an object. The origin of mass is still under intense study in nuclear and particle physics. Different elementary particles have been observed to vary widely in mass. For example, some of these particles are several thousand times more massive than others. Why? We don't really know. In recent years, particle physicists have theorized that the so-called **Higgs particle** (named after Scottish physicist Peter Higgs, who first proposed it) may be responsible for the creation of mass in all other particles, with the mass of a particular type of particle depending on how it interacts with the Higgs particle. A search is underway at the largest particle accelerators to find the Higgs particle, which is thought to be one of the central missing pieces in the standard model of particle physics. However, a complete discussion of the origin of mass is beyond the scope of this book.

# 4.3 Net Force

Because forces are vectors, we must add them as vectors, using the methods developed in Chapter 1. We define the **net force** as the vector sum of all force vectors that act on an object:

$$\vec{F}_{\text{net}} = \sum_{i=1}^{n} \vec{F}_i = \vec{F}_1 + \vec{F}_2 + \cdots + \vec{F}_n. \tag{4.5}$$

Following the rules for the addition of vectors using components, the Cartesian components of the net force are given by

$$\begin{aligned} F_{\text{net},x} &= \sum_{i=1}^{n} F_{i,x} = F_{1,x} + F_{2,x} + \cdots + F_{n,x} \\ F_{\text{net},y} &= \sum_{i=1}^{n} F_{i,y} = F_{1,y} + F_{2,y} + \cdots + F_{n,y} \\ F_{\text{net},z} &= \sum_{i=1}^{n} F_{i,z} = F_{1,z} + F_{2,z} + \cdots + F_{n,z}. \end{aligned} \tag{4.6}$$

To explore the concept of the net force, let's return again to the example of the laptop held up by a hand.

## Normal Force

So far we have only looked at the gravitational force acting on the laptop computer. However, other forces are also acting on it. What are they?

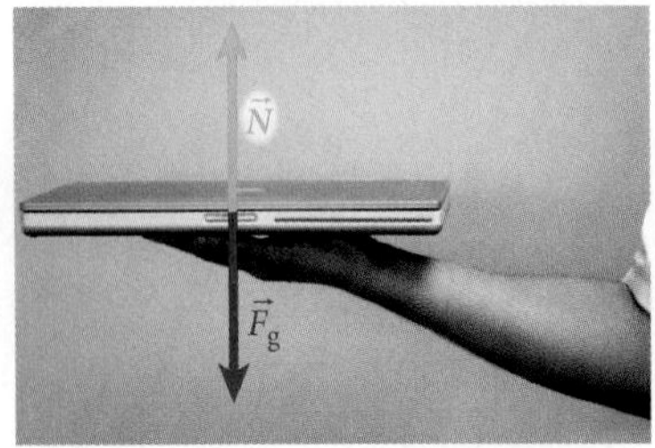


**FIGURE 4.6** Force of gravity acting downward and normal force acting upward exerted by the hand holding the laptop computer.

In Figure 4.6, the force exerted on the laptop computer by the hand is represented by the yellow arrow labeled $\vec{N}$. (Careful—the magnitude of the normal force is represented by the italic letter *N*, whereas the force unit, the newton, is represented by the roman letter N.) Note in the figure that the magnitude of the vector $\vec{N}$ is exactly equal to that of the vector $\vec{F}_g$ and that the two vectors point in opposite directions, or $\vec{N} = -\vec{F}_g$. This situation is not an accident. We will see shortly that there is no net force on an object at rest. If we calculate the net force acting on the laptop computer, we obtain

$$\vec{F}_{net} = \sum_{i=1}^{n} \vec{F}_i = \vec{F}_g + \vec{N} = \vec{F}_g - \vec{F}_g = 0.$$

In general, we can characterize the *normal force,* $\vec{N}$, as a contact force that acts at the surface between two objects. The normal force is always directed perpendicular to the plane of the contact surface. (Hence the name—*normal* means "perpendicular.") The normal force is just large enough to prevent objects from penetrating through each other and is not necessarily equal to the force of gravity in all situations.

For the hand holding the laptop computer, the contact surface between the hand and the computer is the bottom surface of the computer, which is aligned with the horizontal plane. By definition, the normal force has to point perpendicular to this plane, or vertically upward in this case.

Free-body diagrams greatly ease the task of determining net forces on objects.

## Free-Body Diagrams

We have represented the entire effect that the hand has in holding up the laptop computer by the force vector $\vec{N}$. We do not need to consider the influence of the arm, the person to whom the arm belongs, or the entire rest of the world when we want to consider the forces acting on the laptop computer. We can just eliminate them from our consideration, as illustrated in Figure 4.7a, where everything but the laptop computer and the two force vectors has been removed. For that matter, a realistic representation of the laptop computer is not necessary either; it can be shown as a dot, as in Figure 4.7b. This type of drawing of an object, in which all connections to the rest of the world are ignored and only the force vectors that act on it are drawn, is called a **free-body diagram.**

### 4.1 Self-Test Opportunity

Draw the free-body diagrams for a golf ball resting on a tee, your car parked on the street, and you sitting on a chair.

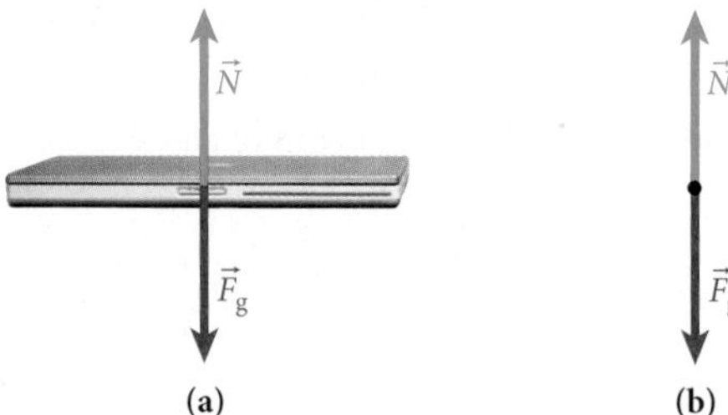


**FIGURE 4.7** (a) Forces acting on a real object, a laptop computer; (b) abstraction of the object as a free body being acted on by two forces.

## 4.4 Newton's Laws

So far this chapter has introduced several types of forces without really explaining how they work and how we can deal with them. The key to working with forces involves understanding Newton's laws. We discuss these laws in this section and then present several examples showing how they apply to practical situations.

Sir Isaac Newton (1642–1727) was perhaps the most influential scientist who ever lived. He is generally credited with being the founder of modern mechanics, as well as calculus (along with the German mathematician Gottfried Leibniz). The first few chapters of this book are basically about Newtonian mechanics. Although he formulated his three famous laws in the 17th century, these laws are still the foundation of our understanding of forces. To begin this discussion, we simply list Newton's three laws, published in 1687.

**Newton's First Law:**

If the net force on an object is equal to zero, the object will remain at rest if it was at rest. If it was moving, it will remain in motion in a straight line with the same constant velocity.

**Newton's Second Law:**

If a net external force, $\vec{F}_{\text{net}}$, acts on an object with mass $m$, the force will cause an acceleration, $\vec{a}$, in the same direction as the force:

$$\vec{F}_{\text{net}} = m\vec{a}.$$

**Newton's Third Law:**

The forces that two interacting objects exert on each other are always exactly equal in magnitude and opposite in direction:

$$\vec{F}_{1\to 2} = -\vec{F}_{2\to 1}.$$

## Newton's First Law

The earlier discussion of net force mentioned that zero net external force is the necessary condition for an object to be at rest. We can use this condition to find the magnitude and direction of any unknown forces in a problem. That is, if we know that an object is at rest and we know its weight force, then we can use the condition $\vec{F}_{\text{net}} = 0$ to solve for other forces acting on the object. This kind of analysis led to the magnitude and direction of the force $\vec{N}$ in the example of the laptop computer being held at rest.

We can use this way of thinking as a general principle: If object 1 rests on object 2, then the normal force $\vec{N}$ equal to the object's weight keeps object 1 at rest, and therefore the net force on object 1 is zero. If $\vec{N}$ were larger than the object's weight, object 1 would lift-off into the air. If $\vec{N}$ were smaller than the object's weight, object 1 would sink into object 2.

Newton's First Law says there are two possible states for an object with no net force on it: An object at rest is said to be in **static equilibrium.** An object moving with constant velocity is said to be in **dynamic equilibrium.**

Before we move on, it is important to state that the equation $\vec{F}_{\text{net}} = 0$ as a condition for static equilibrium really represents one equation for each dimension of the coordinate space that we are considering. Thus, in three-dimensional space, we have three independent equilibrium conditions:

$$F_{\text{net},x} = \sum_{i=1}^{n} F_{i,x} = F_{1,x} + F_{2,x} + \cdots + F_{n,x} = 0$$

$$F_{\text{net},y} = \sum_{i=1}^{n} F_{i,y} = F_{1,y} + F_{2,y} + \cdots + F_{n,y} = 0$$

$$F_{\text{net},z} = \sum_{i=1}^{n} F_{i,z} = F_{1,z} + F_{2,z} + \cdots + F_{n,z} = 0.$$

However, Newton's First Law also addresses the case when an object is already in motion with respect to some particular reference frame. For this case, the law specifies that the acceleration is zero, provided the net external force is zero. Newton's abstraction claims something that seemed at the time in conflict with everyday experience. Today, however, we have the benefit of having seen television pictures of objects floating in a spaceship, moving with unchanged velocities until an astronaut pushes them and thus exerts a force on them. This visual experience is in complete accord with what Newton's First Law claims, but in Newton's time this experience was not the norm.

Consider a car that is out of gas and needs to be pushed to the nearest gas station on a horizontal street. As long as you push the car, you can make it move. However, as soon as you stop pushing, the car slows down and comes to a stop. It seems that as long as you push the car, it moves at constant velocity, but as soon as you stop exerting a force on it, it stops moving. This idea that a constant force is required to move something with a constant speed was the Aristotelian view, which originated from the ancient Greek philosopher Aristotle

(384–322 BC) and his students. Galileo (1564–1642) proposed a law of inertia and theorized that moving objects slowed down because of friction. Newton's First Law builds on this law of inertia.

What about the car that slows down once you stop pushing? This situation is not a case of zero net force. Instead, a force *is* acting on the car to slow it down—the force of friction. Because the friction force acts as a nonzero net force, the example of the car slowing down turns out to be an example not of Newton's First Law, but of Newton's Second Law. We will work more with friction later in this chapter.

Newton's First Law is sometimes also called the *law of inertia.* Inertial mass was defined earlier (Section 4.2), and the definition implied that inertia is an object's resistance to a change in its motion. This is exactly what Newton's First Law says: To change an object's motion, you need to apply an external net force—the motion won't change by itself, neither in magnitude nor in direction.

## Newton's Second Law

The second law relates the concept of acceleration, for which we use the symbol $\vec{a}$, to the force. We have already considered acceleration as the time derivative of the velocity and the second time derivative of the position. Newton's Second Law tells us what causes acceleration.

**Newton's Second Law:**

If a net external force, $\vec{F}_{\text{net}}$, acts on an object with mass $m$, the force will cause an acceleration, $\vec{a}$, in the same direction as the force:

$$\vec{F}_{\text{net}} = m\vec{a}. \tag{4.7}$$

This formula, $F = ma$, is arguably the second most famous equation in all of physics. (We will encounter the most famous, $E = mc^2$, later in the book.) Equation 4.7 tells us that the magnitude of the acceleration of an object is proportional to the magnitude of the net external force acting on it. It also tells us that for a given external force, the magnitude of the acceleration is inversely proportional to the mass of the object. All things being equal, more massive objects are harder to accelerate than less massive ones.

However, equation 4.7 tells us even more, because it is a vector equation. It says that the acceleration vector experienced by the object with mass $m$ is in the same direction as the net external force vector that is acting on the object to cause this acceleration. Because it is a vector equation, we can immediately write the equations for the three spatial components:

$$F_{\text{net},x} = ma_x, \qquad F_{\text{net},y} = ma_y, \qquad F_{\text{net},z} = ma_z.$$

This result means that $F = ma$ holds independently for each Cartesian component of the force and acceleration vectors.

## Newton's Third Law

If you have ever ridden a skateboard, you must have made the following observation: If you are standing at rest on the skateboard, and you step off over the front or back, the skateboard shoots off into the opposite direction. In the process of stepping off, the skateboard exerts a force on your foot, and your foot exerts a force onto the skateboard. This experience seems to suggest that these forces point in opposite directions, and it provides one example of a general truth, quantified in Newton's Third Law.

**Newton's Third Law:**

The forces that two interacting objects exert on each other are always exactly equal in magnitude and opposite in direction:

$$\vec{F}_{1\to 2} = -\vec{F}_{2\to 1}. \tag{4.8}$$

Note that these two forces do not act on the same body but are the forces with which two bodies act on each other.

Newton's Third Law seems to present a paradox. For example, if a horse pulls forward on a wagon with the same force with which the wagon pulls backward on the horse, then how do horse and wagon move anywhere? The answer is that these forces act on different objects in the system. The wagon experiences the pull from the horse and moves forward. The horse feels the pull from the wagon and pushes hard enough against the ground to overcome this force and move forward. A free-body diagram of an object can show only half of such an action-reaction pair of forces.

Newton's Third Law is a consequence of the requirement that internal forces—that is, forces that act between different components of the same system—must add to zero; otherwise, their sum would contribute to a net external force and cause an acceleration, according to Newton's Second Law. No object or group of objects can accelerate themselves without interacting with external objects. The story of Baron Münchhausen, who claimed to have pulled himself out of a swamp by simply pulling very hard on his own hair, is unmasked by Newton's Third Law as complete fiction.

We'll consider some examples of the use of Newton's laws to solve problems, but we'll discuss how ropes and pulleys convey forces. Many problems involving Newton's laws involve forces on a rope (or a string), often one wrapped around a pulley.

## 4.5 Ropes and Pulleys

Problems that involve ropes and pulleys are very common. In this chapter, we consider only massless (idealized) ropes and pulleys. Whenever a rope is involved, the direction of the force due to the pull on the rope acts exactly in the direction along the rope. The force with which we pull on the massless rope is transmitted through the entire rope unchanged. The magnitude of this force is referred to as the *tension* in the rope. Every rope can withstand only a certain maximum force, but for now we will assume that all applied forces are below this limit. Ropes cannot support a *compression* force.

If a rope is guided over a pulley, the direction of the force is changed, but the magnitude of the force is still the same everywhere inside the rope. In Figure 4.8, the right end of the green rope was tied down and someone pulled on the other end with a certain force, 11.5 N, as indicated by the inserted force measurement devices. As you can clearly see, the magnitude of the force on both sides of the pulley is the same. (The weight of the force measurement devices is a small real-world complication, but enough pulling force was used that it is reasonably safe to neglect this effect.)

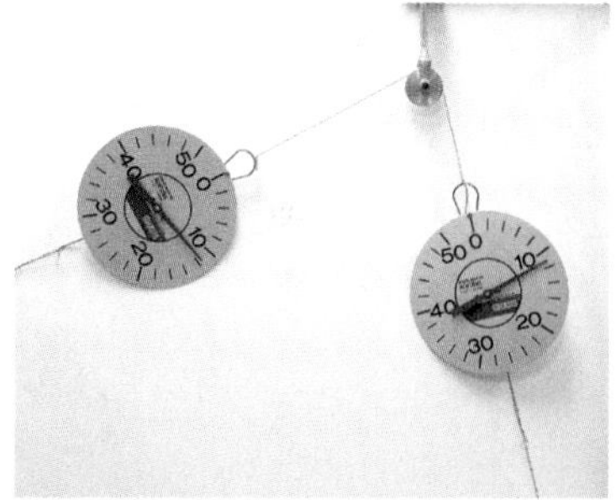

**FIGURE 4.8** A rope passing over a pulley with force measurement devices attached, showing that the magnitude of the force is constant throughout the rope.

### EXAMPLE 4.1 Modified Tug-of-War

In a tug-of-war competition, two teams try to pull each other across a line. If neither team is moving, then the two teams exert equal and opposite forces on a rope. This is an immediate consequence of Newton's Third Law. That is, if the team shown in Figure 4.9 pulls on the rope with a force of magnitude $F$, the other team necessarily has to pull on the rope with a force of the same magnitude but in the opposite direction.

**FIGURE 4.9** Men compete in Tug O'War contest at Braemar Games Highland Gathering, Scotland, UK.

**PROBLEM**

Now let's consider the situation where three ropes are tied together at one point, with a team pulling on each rope. Suppose team 1 is pulling due west with a force of 2750 N, and team 2 is pulling due north with a force of 3630 N. Can a third team pull in such a way that the three-team tug-of-war ends at a standstill, that is, no team is able to move the rope? If yes, what is the magnitude and direction of the force needed to accomplish this?

**SOLUTION**

The answer to the first question is yes, no matter with what force and in what direction teams 1 and 2 pull. This is the case because the two forces will always add up to a combined force, and all that team 3 has to do is pull with a force equal to and opposite in direction to

*Continued—*

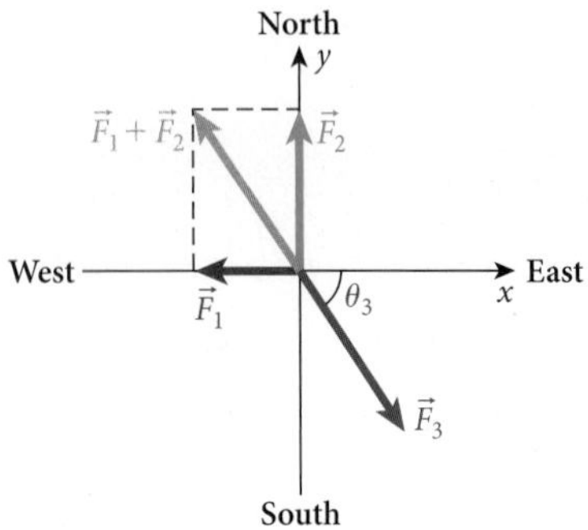


**FIGURE 4.10** Addition of force vectors in the three-team tug-of-war.

that combined force. Then all three forces will add up to zero, and by Newton's First Law, the system has achieved static equilibrium. Nothing will accelerate, so if the ropes start at rest, nothing will move.

Figure 4.10 represents this physical situation. The vector addition of forces exerted by teams 1 and 2 is particularly simple, because the two forces are perpendicular to each other. We choose a conventional coordinate system with its origin at the point where all the ropes meet, and we designate north to be in the positive *y*-direction and west to be in the negative *x*-direction. Thus, the force vector for team 1, $\vec{F}_1$, points in the negative *x*-direction and the force vector for team 2, $\vec{F}_2$, points in the positive *y*-direction. We can then write the two force vectors and their sum as follows:

$$\vec{F}_1 = -(2750 \text{ N})\hat{x}$$
$$\vec{F}_2 = (3630 \text{ N})\hat{y}$$
$$\vec{F}_1 + \vec{F}_2 = -(2750 \text{ N})\hat{x} + (3630 \text{ N})\hat{y}.$$

The addition was made easier by the fact that the two forces pointed along the chosen coordinate axes. However, more general cases of two forces would still be added in terms of their components. Because the sum of all three forces has to be zero for a standstill, we obtain the force that the third team has to exert:

$$0 = \vec{F}_1 + \vec{F}_2 + \vec{F}_3$$
$$\Leftrightarrow \vec{F}_3 = -(\vec{F}_1 + \vec{F}_2)$$
$$= (2750 \text{ N})\hat{x} - (3630 \text{ N})\hat{y}.$$

This force vector is also shown in Figure 4.10. Having the Cartesian components of the force vector we were looking for, we can get the magnitude and direction by using trigonometry:

$$F_3 = \sqrt{F_{3,x}^2 + F_{3,y}^2} = \sqrt{(2750 \text{ N})^2 + (-3630 \text{ N})^2} = 4554 \text{ N}$$

$$\theta_3 = \tan^{-1}\left(\frac{F_{3,y}}{F_{3,x}}\right) = \tan^{-1}\left(\frac{-3630 \text{ N}}{2750 \text{ N}}\right) = -52.9^\circ.$$

These results complete our answer.

Because this type of problem occurs frequently, let's work through another example.

## EXAMPLE 4.2 Still Rings

A gymnast of mass 55 kg hangs vertically from a pair of parallel rings (Figure 4.11a).

### PROBLEM 1

If the ropes supporting the rings are vertical and attached to the ceiling directly above, what is the tension in each rope?

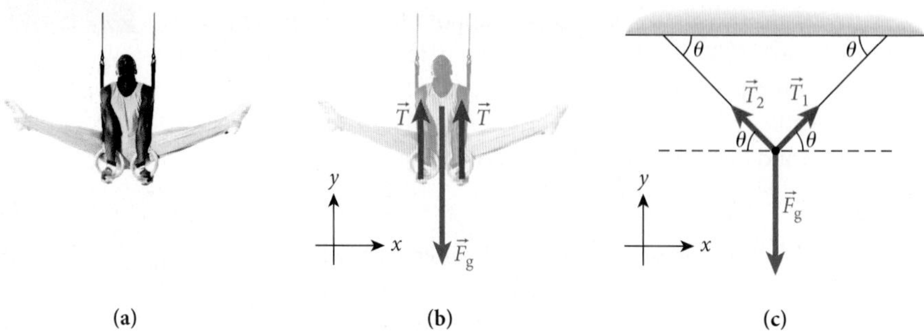


**FIGURE 4.11** (a) Still rings in men's gymnastics. (b) Free-body diagram for problem 1. (c) Free-body diagram for problem 2.

**SOLUTION 1**

In this example, we define the $x$-direction to be horizontal and the $y$-direction to be vertical. The free-body diagram is shown Figure 4.11b. For now, there are no forces in the $x$-direction. In the $y$-direction, we have $\sum_i F_{y,i} = T_1 + T_2 - mg = 0$. Because both ropes support the gymnast equally, the tension has to be the same in both ropes, $T_1 = T_2 \equiv T$, and we get

$$T + T - mg = 0$$
$$\Rightarrow T = \tfrac{1}{2}mg = \tfrac{1}{2}(55 \text{ kg}) \cdot (9.81 \text{ m/s}^2) = 270 \text{ N}.$$

**PROBLEM 2**

If the ropes are attached so that they make an angle $\theta = 45°$ with the ceiling (Figure 4.11c), what is the tension in each rope?

**SOLUTION 2**

In this part, forces do occur in both $x$- and $y$-directions. We will work in terms of a general angle and then plug in the specific angle, $\theta = 45°$, at the end.

In the $x$-direction, we have for our equilibrium condition:

$$\sum_i F_{x,i} = T_1 \cos\theta - T_2 \cos\theta = 0.$$

In the $y$-direction, our equilibrium condition is

$$\sum_i F_{y,i} = T_1 \sin\theta + T_2 \sin\theta - mg = 0.$$

From the equation for the $x$-direction, we again get $T_1 = T_2 \equiv T$, and from the equation for the $y$-direction, we then obtain:

$$2T \sin\theta - mg = 0 \Rightarrow T = \frac{mg}{2\sin\theta}.$$

Putting in the numbers, we obtain the tension in each rope:

$$T = \frac{(55 \text{ kg})(9.81 \text{ m/s}^2)}{2\sin 45°} = 382 \text{ N}.$$

**PROBLEM 3**

How does the tension in the ropes change as the angle $\theta$ between the ceiling and the ropes becomes smaller and smaller?

**SOLUTION 3**

As the angle $\theta$ between the ceiling and the ropes becomes smaller, the tension in the ropes, $T = mg/2\sin\theta$, gets larger. As $\theta$ approaches zero, the tension becomes infinitely large. In reality, of course, the gymnast has only finite strength and cannot hold his position for small angles.

## 4.1 In-Class Exercise

Choose the set of three coplanar vectors that sum to a net force of zero: $\vec{F}_1 + \vec{F}_2 + \vec{F}_3 = 0$.

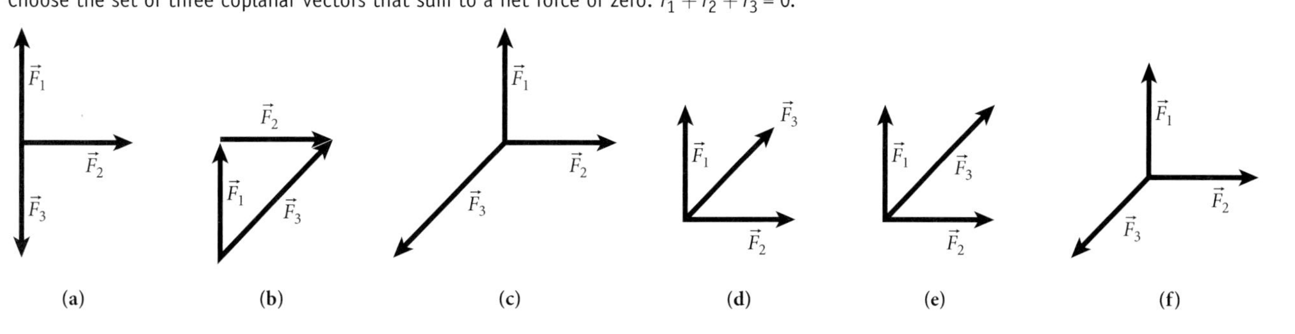

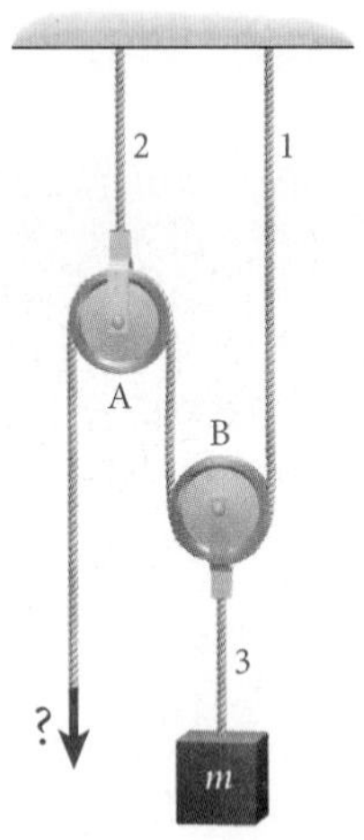


FIGURE 4.12 Rope guided over two pulleys.

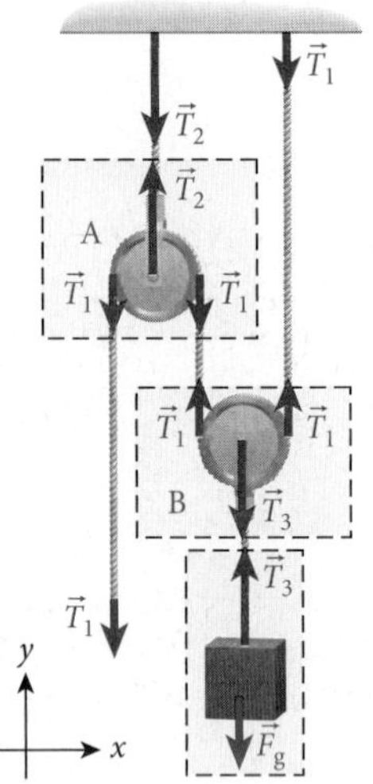

FIGURE 4.13 Free-body diagrams for the two pulleys and the mass to be lifted.

FIGURE 4.14 Pulley with three loops.

## Force Multiplier

Ropes and pulleys can be combined to lift objects that are too heavy to lift otherwise. In order to see how this can be done, consider Figure 4.12. The system shown consists of rope 1, which is tied to the ceiling (upper right) and then guided over pulleys B and A. Pulley A is also tied to the ceiling with rope 2. Pulley B is free to move vertically and is attached to rope 3. The object of mass $m$, which we want to lift, is hanging from the other end of rope 3. We assume that the two pulleys have negligible mass and that rope 1 can glide across the pulleys without friction.

What force do we need to apply to the free end of rope 1 to keep the system in static equilibrium? We will call the tension force in rope 1, $\vec{T}_1$, that in rope 2, $\vec{T}_2$, and that in rope 3, $\vec{T}_3$. Again, the key idea is that the magnitude of this tension force is the same everywhere in a given rope.

Figure 4.13 again shows the system in Figure 4.12, but with dashed lines and shaded areas, indicating the free-body diagrams of the two pulleys and the object of mass $m$. We start with the mass $m$. For the condition of zero net force to be fulfilled, we need

$$\vec{T}_3 + \vec{F}_g = 0$$

or

$$F_g = mg = T_3.$$

From the free-body diagram of pulley B, we see that the tension force applied from rope 1 acts on both sides of pulley B. This tension has to balance the tension from rope 3, giving us

$$2T_1 = T_3.$$

Combining the last two equations, we see that

$$T_1 = \tfrac{1}{2}mg.$$

This result means that the force we need to apply to suspend the object of mass $m$ in this way is only half as large as the force we would have to use to simply hold it up with a rope, without pulleys. This change in force is why a pulley is called a *force multiplier.*

Even greater force multiplication is achieved if rope 1 passes a total of $n$ times over the same two pulleys. In this case, the force needed to suspend the object of mass $m$ is

$$T = \frac{1}{2n}mg. \tag{4.9}$$

Figure 4.14 shows the situation for the lower pulley in Figure 4.13 with $n = 3$. This arrangement results in $2n = 6$ force arrows of magnitude $T$ pointing up, able to balance a downward force of magnitude $6T$, as expressed by equation 4.9.

### 4.2 In-Class Exercise

Using a pulley pair with two loops, we can lift a weight of 440 N. If we add two loops to the pulley, with the same force, we can lift

a) half the weight.

b) twice the weight.

c) one-fourth the weight.

d) four times the weight.

e) the same weight.

## 4.6 Applying Newton's Laws

Now let's look at how Newton's laws allow us to solve various kinds of problems involving force, mass, and acceleration. We will make frequent use of free-body diagrams and will assume massless ropes and pulleys. We will also neglect friction for now but will consider it in Section 4.7.

## EXAMPLE 4.3 Two Books on a Table

We have considered the simple situation of one object (the laptop computer) supported from below and held at rest. Now let's look at two objects at rest: two books on a table (Figure 4.15a).

### PROBLEM

What is the magnitude of the force that the table exerts on the lower book?

### SOLUTION

We start with a free-body diagram of the book on top, book 1 (Figure 4.15b). This situation is the same as that of the laptop computer held steady by the hand. The gravitational force due to the Earth's attraction that acts on the upper book is indicated by $\vec{F}_1$. It has the magnitude $m_1g$, where $m_1$ is the mass of the upper book, and points straight down. The magnitude of the normal force, $\vec{N}_1$, that the lower book exerts on the upper book from below is then $N_1 = F_1 = m_1g$, from the condition of zero net force on the upper book (Newton's First Law). The force $\vec{N}_1$ points straight up, as shown in the free-body diagram, $\vec{N}_1 = -\vec{F}_1$.

Newton's Third Law now allows us to calculate the force that the upper book exerts on the lower book. This force is equal in magnitude and opposite in direction to the force that the lower book exerts on the upper one:

$$\vec{F}_{1\to2} = -\vec{N}_1 = -(-\vec{F}_1) = \vec{F}_1.$$

This relationship says that the force that the upper book exerts on the lower one is exactly equal to the gravitational force acting on the upper book—that is, its weight. You may find this result trivial at this point, but the application of this general principle allows us to analyze and do calculations for complicated situations.

Now consider the free-body diagram of the lower book, book 2 (Figure 4.15c). This free-body diagram allows us to calculate the normal force that the table exerts on the lower book. We sum up all the forces acting on this book:

$$\vec{F}_{1\to2} + \vec{N}_2 + \vec{F}_2 = 0 \Rightarrow \vec{N}_2 = -(\vec{F}_{1\to2} + \vec{F}_2) = -(\vec{F}_1 + \vec{F}_2),$$

where $\vec{N}_2$ is the normal force exerted by the table on the lower book, $\vec{F}_{1\to2}$ is the force exerted by the upper book on the lower book, and $\vec{F}_2$ is the gravitational force on the lower book. In the last step, we used the result we obtained from the free-body diagram of book 1. This result means that the force that the table exerts on the lower book is exactly equal in magnitude and opposite in direction to the sum of the weights of the two books.

(a)

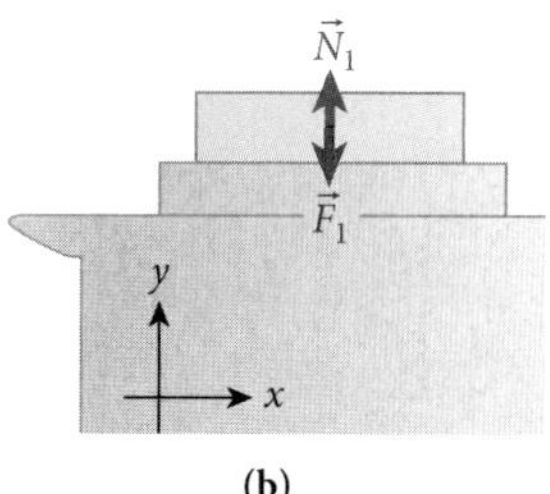


(b)

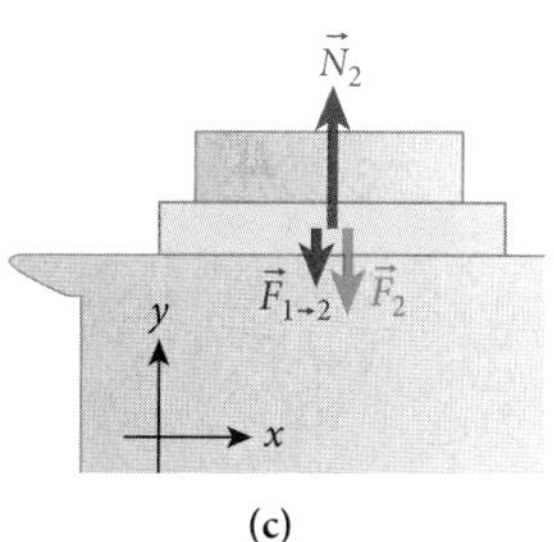


(c)

**FIGURE 4.15** (a) Two books on top of a table. (b) Free-body diagram for book 1. (c) Free-body diagram for book 2.

The use of Newton's Second Law enables us to perform a wide range of calculations involving motion and acceleration. The following problem is a classic example: Consider an object of mass $m$ located on a plane that is inclined by an angle $\theta$ relative to the horizontal. Assume that there is no frictional force between the plane and the object. What can Newton's Second Law tell us about this situation?

## SOLVED PROBLEM 4.1 Snowboarding

### PROBLEM

A snowboarder (mass 72.9 kg, height 1.79 m) glides down a slope with an angle of 22° with respect to the horizontal (Figure 4.16a). If we can neglect friction, what is his acceleration?

### SOLUTION

#### THINK

The motion is restricted to moving along the plane because the snowboarder cannot sink into the snow, and he cannot lift off from the plane. (At least not without jumping!) It is

*Continued—*

(a)

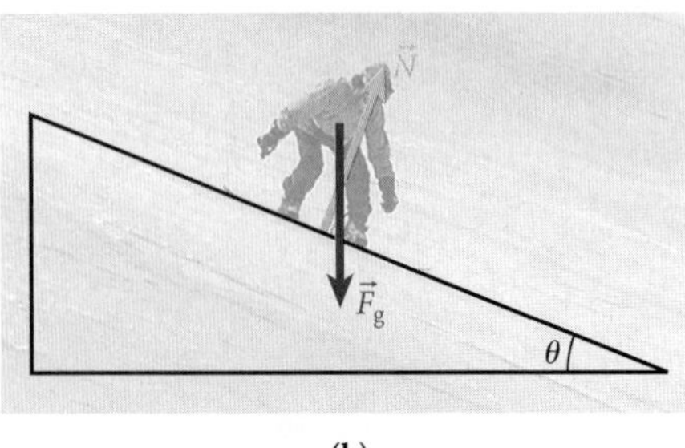


(b)

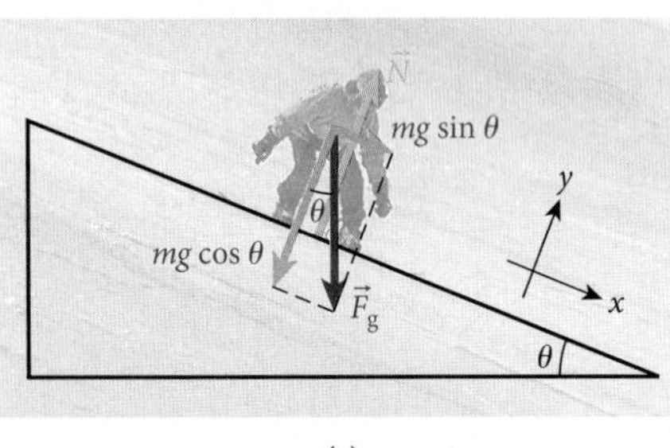


(c)

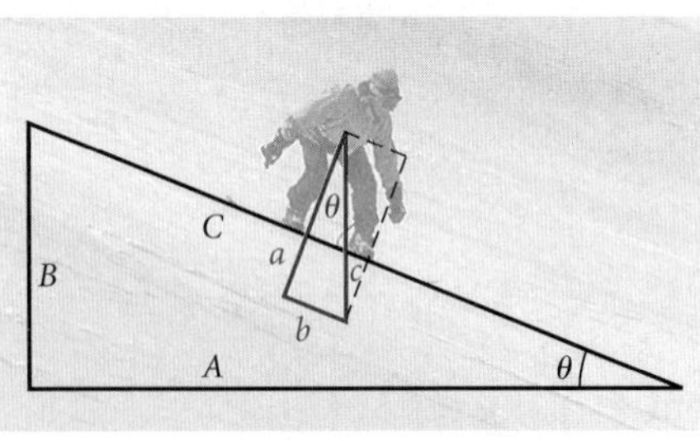


(d)

**FIGURE 4.16** (a) Snowboarding as an example of motion on an inclined plane. (b) Free-body diagram of the snowboarder on the inclined plane. (c) Free-body diagram of the snowboarder, with a coordinate system added. (d) Similar triangles in the inclined-plane problem.

always advisable to start with a free-body diagram. Figure 4.16b shows the force vectors for gravity, $\vec{F}_g$, and the normal force, $\vec{N}$. Note that the normal force vector is directed perpendicular to the contact surface, as required by the definition of the normal force. Also observe that the normal force and the force of gravity do not point in exactly opposite directions and thus do not cancel each other out completely.

## SKETCH

Now we pick a convenient coordinate system. As shown in Figure 4.16c, we choose a coordinate system with the $x$-axis along the direction of the inclined plane. This choice ensures that the acceleration is only in the $x$-direction. Another advantage of this choice of coordinate system is that the normal force is pointing exactly in the $y$-direction. The price we pay for this convenience is that the gravitational force vector does not point along one of the major axes in our coordinate system but has an $x$- and a $y$-component. The red arrows in the figure indicate the two components of the gravitational force vector. Note that the angle of inclination of the plane, $\theta$, also appears in the rectangle constructed from the two components of the gravity force vector, which is the diagonal of that rectangle. You can see this relationship by considering the similar triangles with sides $abc$ and $ABC$ in Figure 4.16d. Because $a$ is perpendicular to $C$ and $c$ is perpendicular to $A$, it follows that the angle between $a$ and $c$ is the same as the angle between $A$ and $C$.

## RESEARCH

The $x$- and $y$-components of the gravitational force vector are found from trigonometry:

$$F_{g,x} = F_g \sin\theta = mg\sin\theta$$

$$F_{g,y} = -F_g \cos\theta = -mg\cos\theta.$$

## SIMPLIFY

Now we do the math in a straightforward way, separating calculations by components.

First, there is no motion in the $y$-direction, which means that, according to Newton's First Law, all the external force components in the $y$-direction have to add up to zero:

$$F_{g,y} + N = 0 \Rightarrow$$

$$-mg\cos\theta + N = 0 \Rightarrow$$

$$N = mg\cos\theta.$$

Our analysis of the motion in the $y$-direction has given us the magnitude of the normal force, which balances the component of the weight of the snowboarder perpendicular to the slope. This is a very typical result. The normal force almost always balances the net force perpendicular to the contact surface that is contributed by all other forces. Thus, objects do not sink into or lift off from surfaces.

The information we are interested in comes from looking at the $x$-direction. In this direction, there is only one force component, the $x$-component of the gravitational force. Therefore, according to Newton's Second Law, we obtain

$$F_{g,x} = mg\sin\theta = ma_x \Rightarrow$$

$$a_x = g\sin\theta.$$

Thus, we now have the acceleration vector in the specified coordinate system:

$$\vec{a} = (g\sin\theta)\hat{x}.$$

Note that the mass, $m$, dropped out of our answer. The acceleration does not depend on the mass of the snowboarder; it depends only on the angle of inclination of the plane. Thus, the mass of the snowboarder given in the problem statement turns out to be just as irrelevant as his height.

**CALCULATE**

Putting in the given value for the angle leads to

$$a_x = (9.81 \text{ m/s}^2)(\sin 22°) = 3.67489 \text{ m/s}^2.$$

**ROUND**

Because the angle of the slope was given to only two-digit accuracy, it makes no sense to give our result to a greater precision. The final answer is

$$a_x = 3.7 \text{ m/s}^2.$$

**DOUBLE-CHECK**

The units of our answer, m/s$^2$, are those of acceleration. The number we obtained is positive, which means a positive acceleration down the slope in the coordinate system we have chosen. Also the number is less than 9.81, which is comforting. It means that our calculated acceleration is less than that for free fall. As a final step, let's check for consistency of our answer, $a_x = g \sin\theta$, in limiting cases. In the case where $\theta \to 0°$, the sine also converges to zero, and the acceleration vanishes. This result is consistent because we expect no acceleration of the snowboarder if he rests on a horizontal surface. As $\theta \to 90°$, the sine approaches 1, and the acceleration is the acceleration due to gravity, as we expect as well. In this limiting case, the snowboarder would be in free fall.

Inclined-plane problems, like the one we have just solved, are very common and provide practice with concepts of component decomposition of forces. Another common type of problem involves redirection of forces via pulleys and ropes. The next example shows how to proceed in a simple case.

## EXAMPLE 4.4 Two Blocks Connected by a Rope

In this classic problem, a hanging mass generates an acceleration for a second mass on a horizontal surface (Figure 4.17a). One block, of mass $m_1$, lies on a horizontal frictionless surface and is connected via a massless rope (for simplicity, oriented in the horizontal direction) running over a massless pulley to another block, with mass $m_2$, hanging from the rope.

**PROBLEM**

What is the acceleration of block $m_1$, and what is the acceleration of block $m_2$?

**SOLUTION**

Again we start with a free-body diagram for each object. For block $m_1$, the free-body diagram is shown in Figure 4.17b. The gravitational force vector points straight down

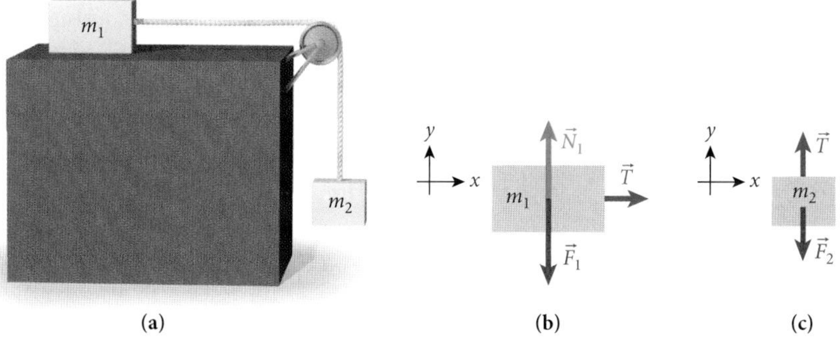


**FIGURE 4.17** (a) Block hanging vertically from a rope that runs over a pulley and is connected to a second block on a horizontal frictionless surface. (b) Free-body diagram for block $m_1$. (c) Free-body diagram for block $m_2$.

*Continued—*

and has magnitude $F_1 = m_1g$. The force due to the rope, $\vec{T}$, acts along the rope and thus is in the horizontal direction, which we have chosen as the $x$-direction. The normal force, $\vec{N}_1$, acting on $m_1$ acts perpendicular to the contact surface. Because the surface is horizontal, $\vec{N}_1$ acts in the vertical direction. From the requirement of zero net force in the $y$-direction, we get $N_1 = F_1 = m_1g$ for the magnitude of the normal force. The magnitude of the tension force in the rope, $T$, remains to be determined. For the component of acceleration in the $x$-direction, Newton's Second Law gives us

$$m_1 a = T.$$

Now we turn to the free-body diagram for mass $m_2$ (Figure 4.17c). The force due to the rope, $\vec{T}$, that acts on $m_1$ also acts on $m_2$, but the redirection due to the pulley causes the force to act in a different direction. However, we are interested in the magnitude of the tension, $T$, and this value is the same for both masses.

For the $y$-component of the net force acting on $m_2$, Newton's Second Law gives us

$$T - F_2 = T - m_2 g = -m_2 a.$$

The magnitude of the acceleration $\vec{a}$ for $m_2$ that appears in this equation is the same as $a$ in the equation of motion for $m_1$, because the two masses are tied to each other by a rope and experience the same magnitude of acceleration. This is a key insight: If two objects are tied to each other in this way, they must experience the same magnitude of acceleration, provided the rope is kept under tension. The negative sign on the right side of this equation indicates that $m_2$ accelerates in the negative $y$-direction.

We can now combine the two equations for the two masses to eliminate the magnitude of the string force, $T$, and obtain the common acceleration of the two masses:

$$m_1 a = T = m_2 g - m_2 a \Rightarrow$$

$$a = g\left(\frac{m_2}{m_1 + m_2}\right).$$

This result makes sense: In the limit where $m_1$ is very large compared to $m_2$, there will be almost no acceleration, whereas if $m_1$ is very small compared to $m_2$, then $m_2$ will accelerate with almost the acceleration due to gravity, as if $m_1$ were not there.

Finally, we can calculate the magnitude of the tension by reinserting our result for the acceleration into one of the two equations we obtained using Newton's Second Law:

$$T = m_1 a = g\left(\frac{m_1 m_2}{m_1 + m_2}\right).$$

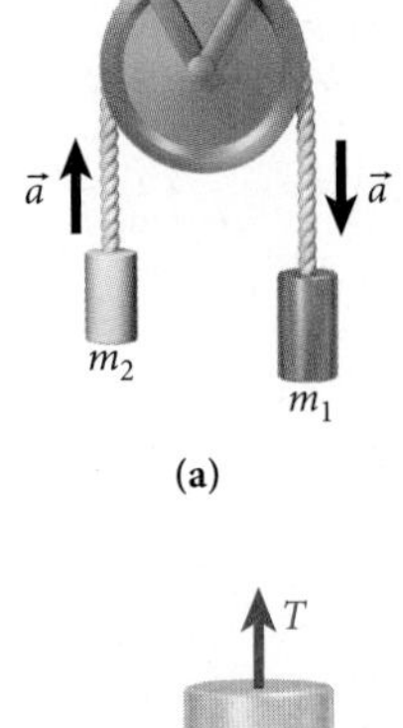


(a)

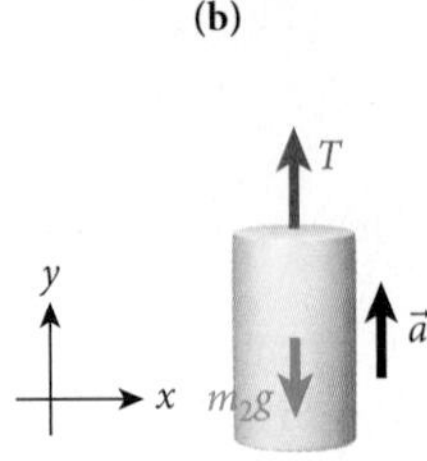


(b)



(c)

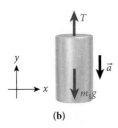

**FIGURE 4.18** (a) Atwood machine with the direction of positive acceleration assumed to be as indicated. (b) Free-body diagram for the weight on the right side of the Atwood machine. (c) Free-body diagram for the weight on the left side of the Atwood machine.

In Example 4.4, it is clear in which direction the acceleration will occur. In more complicated cases, the direction in which objects begin to accelerate may not be clear at the beginning. You just have to define a direction as positive and use this assumption consistently throughout your calculations. If the acceleration value you obtain in the end turns out to be negative, this result means that the objects accelerate in the direction opposite to the one you initially assumed. The calculated value will remain correct. Example 4.5 illustrates such a situation.

## EXAMPLE 4.5 Atwood Machine

The Atwood machine consists of two hanging weights (with masses $m_1$ and $m_2$) connected via a rope running over a pulley. For now, we consider a friction-free case, where the pulley does not move, and the rope glides over it. (In Chapter 10 on rotation, we will return to this problem and solve it with friction present, which causes the pulley to rotate.) We also assume that $m_1 > m_2$. In this case, the acceleration is as shown in Figure 4.18a. (The formula derived in the following is correct for any case. If $m_1 < m_2$, then the value of the acceleration, $a$, will have a negative sign, which will mean that the acceleration direction is opposite to what we assumed in working the problem.)

We start with free-body diagrams for $m_1$ and $m_2$, as shown in Figure 4.18b and Figure 4.18c. For both free-body diagrams, we elect to point the positive $y$-axis upward, and both diagrams show our choice for the direction of the acceleration. The rope exerts a tension $T$, of magnitude still to be determined, upward on both $m_1$ and $m_2$. With our choices of the coordinate system and the direction of the acceleration, the downward acceleration of $m_1$ is acceleration in a negative direction. This leads to an equation that can be solved for $T$:

$$T - m_1 g = -m_1 a \Rightarrow T = m_1 g - m_1 a = m_1(g - a).$$

From the free-body diagram for $m_2$ and the assumption that the upward acceleration of $m_2$ corresponds to acceleration in a positive direction, we get

$$T - m_2 g = m_2 a \Rightarrow T = m_2 g + m_2 a = m_2(g + a).$$

Equating the two expressions for $T$, we obtain

$$m_1(g - a) = m_2(g + a),$$

which leads to an expression for the acceleration:

$$(m_1 - m_2)g = (m_1 + m_2)a \Rightarrow$$

$$a = g\left(\frac{m_1 - m_2}{m_1 + m_2}\right).$$

From this equation, you can see that the magnitude of the acceleration, $a$, is always smaller than $g$ in this situation. If the masses are equal, we obtain the expected result of no acceleration. By selecting the proper combination of masses, we can generate any value of the acceleration between zero and $g$ that we desire.

### 4.2 Self-Test Opportunity

For the Atwood machine, can you write a formula for the magnitude of the tension in the rope?

### 4.3 In-Class Exercise

If you double both masses in an Atwood machine, the resulting acceleration will be

a) twice as large.

b) half as large.

c) the same.

d) one-quarter as large.

e) four times as large.

## EXAMPLE 4.6 Collision of Two Vehicles

Suppose that an SUV with mass $m = 3260$ kg has a head-on collision with a subcompact car of mass $m = 1194$ kg and exerts a force of magnitude $2.9 \cdot 10^5$ N on the subcompact.

### PROBLEM

What is the magnitude of the force that the subcompact exerts on the SUV in the collision?

### SOLUTION

As paradoxical as it may seem at first, the little subcompact exerts just as much force on the SUV as the SUV exerts on it. This equality is a straightforward consequence of Newton's Third Law, equation 4.8. So, the answer is $2.9 \cdot 10^5$ N.

### DISCUSSION

The answer may be straightforward, but it is by no means intuitive. The subcompact will usually sustain much more damage in such a collision, and its passengers will have a much bigger chance of getting hurt. However, this difference is due to Newton's Second Law, which says that the same force applied to a less massive object yields a higher acceleration than when it is applied to a more massive one. Even in a head-on collision between a mosquito and a car on the freeway, the forces exerted on each body are equal; the difference in damage to the car (none) and to the mosquito (obliteration) is due to their different resulting accelerations. We will revisit this idea in Chapter 7 on momentum and collisions.

### 4.4 In-Class Exercise

For the collision in Example 4.6, if we call the absolute value of the acceleration experienced by the SUV $a_{SUV}$ and that of the subcompact car $a_{car}$, we find that approximately

a) $a_{SUV} \approx \frac{1}{9}a_{car}$.

b) $a_{SUV} \approx \frac{1}{3}a_{car}$.

c) $a_{SUV} \approx a_{car}$.

d) $a_{SUV} \approx 3a_{car}$.

e) $a_{SUV} \approx 9a_{car}$.

## 4.7 Friction Force

So far, we have neglected the force of friction and considered only frictionless approximations. However, in general, we have to include friction in most of our calculations when we want to describe physically realistic situations.

We could conduct a series of very simple experiments to learn about the basic characteristics of friction. Here are the findings we would obtain:

- If an object is at rest, it takes an external force with a certain threshold magnitude and acting parallel to the contact surface between the object and the surface to overcome the friction force and make the object move.
- The friction force that has to be overcome to make an object at rest move is larger than the friction force that has to be overcome to keep the object moving at a constant velocity.
- The magnitude of the friction force acting on a moving object is proportional to the magnitude of the normal force.
- The friction force is independent of the size of the contact area between object and surface.
- The friction force depends on the roughness of the surfaces; that is, a smoother interface generally provides less friction force than a rougher one.
- The friction force is independent of the velocity of the object.

These statements about friction are not principles in the same way as Newton's laws. Instead, they are general observations based on experiments. For example, you might think that the contact of two extremely smooth surfaces would yield very low friction. However, in some cases, extremely smooth surfaces actually fuse together as a cold weld. Investigations into the nature and causes of friction continue, as we discuss later in this section.

From these findings, it is clear that we need to distinguish between the case where an object is at rest relative to its supporting surface (static friction) and the case where an object is moving across the surface (kinetic friction). The case in which an object is moving across a surface is easier to treat, and so we consider kinetic friction first.

### Kinetic Friction

The above general observations can be summarized in the following approximate formula for the magnitude of the kinetic friction force, $f_k$:

$$f_k = \mu_k N. \tag{4.10}$$

Here $N$ is the magnitude of the normal force and $\mu_k$ is the **coefficient of kinetic friction.** This coefficient is always equal to or greater than zero. (The case where $\mu_k = 0$ corresponds to a frictionless approximation. In practice, however, it can never be reached perfectly.) In almost all cases, $\mu_k$ is also less than 1. (Some special tire surfaces used for car racing, though, have a coefficient of friction with the road that can significantly exceed 1.) Some representative coefficients of kinetic friction are shown in Table 4.1.

The direction of the kinetic friction force is *always opposite to the direction of motion* of the object relative to the surface it moves on.

If you push an object with an external force parallel to the contact surface, and the force has a magnitude exactly equal to that of the force of kinetic friction on the object, then the total net external force is zero, because the external force and the friction force cancel each other. In that case, according to Newton's First Law, the object will continue to slide across the surface with constant velocity.

### Static Friction

If an object is at rest, it takes a certain threshold amount of external force to set it in motion. For example, if you push lightly against a refrigerator, it will not move. As you push harder and harder, you reach a point where the refrigerator finally slides across the kitchen floor.

**Table 4.1 Typical Coefficients of Both Static and Kinetic Friction Between Material 1 and Material 2***

| Material 1 | Material 2 | $\mu_s$ | $\mu_k$ |
|---|---|---|---|
| rubber | dry concrete | 1 | 0.8 |
| rubber | wet concrete | 0.7 | 0.5 |
| steel | steel | 0.7 | 0.6 |
| wood | wood | 0.5 | 0.3 |
| waxed ski | snow | 0.1 | 0.05 |
| steel | oiled steel | 0.12 | 0.07 |
| Teflon | steel | 0.04 | 0.04 |
| curling stone | ice | | 0.017 |

*Note that these values are approximate and depend strongly on the condition of surface that exists between the two materials.

For any external force acting on an object that remains at rest, the friction force is exactly equal in magnitude and opposite in direction to the component of that external force that acts along the contact surface between the object and its supporting surface. However, the magnitude of the static friction force has a maximum value: $f_s \leq f_{s,max}$. This maximum magnitude of the static friction force is proportional to the normal force, but with a different proportionality constant than the coefficient of kinetic friction: $f_{s,max} = \mu_s N$. We can write for the magnitude of the force of static friction

$$f_s \leq \mu_s N = f_{s,max}, \tag{4.11}$$

where $\mu_s$ is called the **coefficient of static friction.** Some typical coefficients of static friction are shown in Table 4.1. In general, for any object on any supporting surface, the maximum static friction force is greater than the force of kinetic friction. You may have experienced this when trying to slide a heavy object across a surface: As soon as the object starts moving, a lot less force is required to keep the object in constant sliding motion. We can write this finding as a mathematical inequality between the two coefficients:

$$\mu_s > \mu_k. \tag{4.12}$$

Figure 4.19 presents a graph showing how the friction force depends on an external force, $F_{ext}$, applied to an object. If the object is initially at rest, a small external force results in a small force of friction, rising linearly with the external force until it reaches a value of $\mu_s N$. Then it drops rather quickly to a value of $\mu_k N$, when the object is set in motion. At this point, the external force has a value of $F_{ext} = \mu_s N$, resulting in a sudden acceleration of the object. This dependence of the friction force on the external force is shown in Figure 4.19 as a red line.

On the other hand, if we start with a large external force and the object is already in motion, then we can reduce the external force below a value of $\mu_s N$, but still above $\mu_k N$, and the object will keep moving and accelerating. Thus, the friction coefficient retains a value of $\mu_k$ until the external force is reduced to a value of $\mu_k N$. At this point (and only at this point!), the object will move with a constant velocity, because the external force and the friction force are equal in magnitude. If we reduce the external force further, the object decelerates (horizontal segment of the blue line left of the red diagonal in Figure 4.19), because the kinetic friction force is bigger than the external force. Eventually, the object comes to rest due to the kinetic friction, and the external force is not sufficient to move it anymore. Then static friction takes over, and the friction force is reduced proportionally to the external force until both reach zero. The blue line in Figure 4.19 illustrates this dependence of the friction force on the external force. Where the blue line and the red line overlap, this is indicated by alternating blue and red squares. The most interesting part about Figure 4.19 is that the blue and red lines do not coincide between $\mu_k N$ and $\mu_s N$.

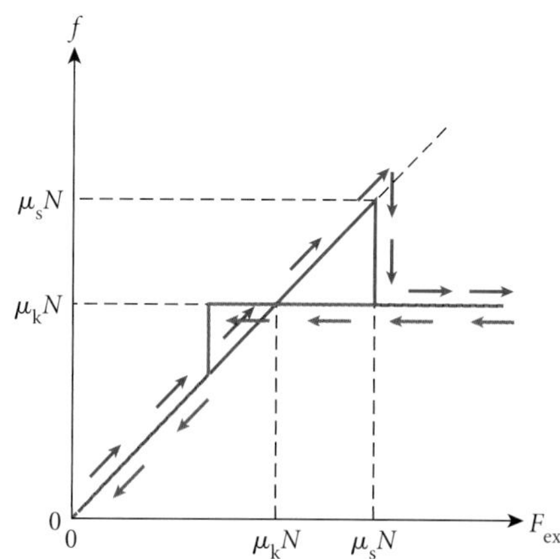


**FIGURE 4.19** Magnitudes of the forces of friction as a function of the magnitude of an external force.

Let's return to the attempt to move a refrigerator across the kitchen floor. Initially, the refrigerator sits on the floor, and the static friction force resists your effort to move it. Once you push hard enough, the refrigerator jars into motion. In this process, the friction force follows the red path in Figure 4.19. Once the refrigerator moves, you can push less hard and still keep it moving. If you push with less force so that it moves with constant velocity, the external force you apply follows the blue path n Figure 4.19 until it is reduced to $F_{\text{ext}} = \mu_k N$. Then the friction force and the force you apply to the fridge add up to zero, and there is no net force acting on the refrigerator, allowing it to move with constant velocity.

## EXAMPLE 4.7 Realistic Snowboarding

Let's reconsider the snowboarding situation from Solved Problem 4.1, but now include friction. A snowboarder moves down a slope for which $\theta = 22°$. Suppose the coefficient of kinetic friction between his board and the snow is 0.21, and his velocity, which is along the direction of the slope, is measured as 8.3 m/s at a given instant.

### PROBLEM 1

Assuming a constant slope, what will be the speed of the snowboarder along the direction of the slope, 100 m farther down the slope?

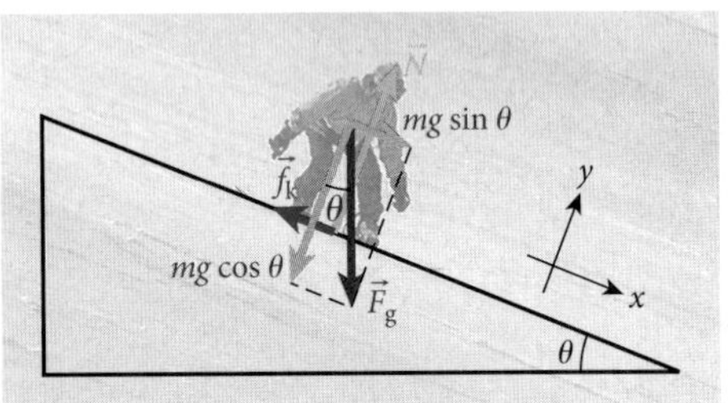


**FIGURE 4.20** Free-body diagram of a snowboarder, including the friction force.

### SOLUTION 1

Figure 4.20 shows a free-body diagram for this problem. The gravitational force points downward and has magnitude $mg$, where $m$ is the mass of the snowboarder and his equipment. We choose convenient $x$- and $y$-axes parallel and perpendicular to the slope, respectively, as indicated in Figure 4.20. The angle $\theta$ that the slope makes with the horizontal (22° in this case) also appears in the decomposition of the components of the gravitational force parallel and perpendicular to the slope. (This analysis is a general feature of any inclined-plane problem.) The force component along the plane is then $mg \sin\theta$, as shown in Figure 4.20. The normal force is given by $N = mg\cos\theta$, and the force of kinetic friction is $f_k = -\mu_k mg\cos\theta$, with the minus sign indicating that the force is acting in the negative $x$-direction, in our chosen coordinate system.

We thus get for the total force component in the $x$-direction:

$$mg\sin\theta - \mu_k mg\cos\theta = ma_x \Rightarrow$$
$$a_x = g(\sin\theta - \mu_k\cos\theta).$$

Here we have used Newton's Second Law, $F_x = ma_x$, in the first line. The mass of the snowboarder drops out, and the acceleration, $a_x$, along the slope is a constant. Inserting the numbers given in the problem statement, we obtain

$$a \equiv a_x = (9.81 \text{ m/s}^2)(\sin 22° - 0.21\cos 22°) = 1.76 \text{ m/s}^2.$$

Thus, we see that this situation is a problem of motion in a straight line in one direction with constant acceleration. We can apply the relationship between the squares of the initial and final velocities and the acceleration that we have derived for one-dimensional motion with constant acceleration:

$$v^2 = v_0^2 + 2a(x - x_0).$$

With $v_0 = 8.3$ m/s and $x - x_0 = 100$ m, we calculate the final speed:

$$\begin{aligned} v &= \sqrt{v_0^2 + 2a(x - x_0)} \\ &= \sqrt{(8.3 \text{ m/s})^2 + 2\cdot(1.76 \text{ m/s}^2)(100 \text{ m})} \\ &= 20.5 \text{ m/s}. \end{aligned}$$

### PROBLEM 2

How long does it take the snowboarder to reach this speed?

**SOLUTION 2**
Since we now know the acceleration and the final speed and were given the initial speed, we use

$$v = v_0 + at \Rightarrow t = \frac{v - v_0}{a} = \frac{(20.5 - 8.3)\text{ m/s}}{1.76\text{ m/s}^2} = 6.95\text{ s}.$$

**PROBLEM 3**
Given the same coefficient of friction, what would the angle of the slope have to be for the snowboarder to glide with constant velocity?

**SOLUTION 3**
Motion with constant velocity implies zero acceleration. We have already derived an equation for the acceleration as a function of the slope angle. We set this expression equal to zero and solve the resulting equation for the angle $\theta$:

$$\begin{aligned} &a_x = g(\sin\theta - \mu_\text{k}\cos\theta) = 0 \\ &\Rightarrow \sin\theta = \mu_\text{k}\cos\theta \\ &\Rightarrow \tan\theta = \mu_\text{k} \\ &\Rightarrow \theta = \tan^{-1}\mu_\text{k} \end{aligned}$$

Because $\mu_\text{k} = 0.21$ was given, the angle is $\theta = \tan^{-1} 0.21 = 12°$. With a steeper slope, the snowboarder will accelerate, and with a shallower slope, the snowboarder will slow down until he comes to a stop.

## Air Resistance

So far we have ignored the friction due to moving through the air. Unlike the force of kinetic friction that you encounter when dragging or pushing one object across the surface of another, air resistance increases as speed increases. Thus, we need to express the friction force as a function of the velocity of the object relative to the medium it moves through. The direction of the force of air resistance is opposite to the direction of the velocity vector.

In general, the magnitude of the friction force due to air resistance, or **drag force,** can be expressed as $F_\text{frict} = K_0 + K_1 v + K_2 v^2 + \ldots$, with the constants $K_0, K_1, K_2, \ldots$ to be determined experimentally. For the drag force on macroscopic objects moving at relatively high speeds, we can neglect the linear term in the velocity. The magnitude of the drag force is then approximately

$$F_\text{drag} = Kv^2. \tag{4.13}$$

This equation means that the force due to air resistance is proportional to the square of the speed.

When an object falls through air, the force from air resistance increases as the object accelerates until it reaches a so-called **terminal speed.** At this point, the upward force of air resistance and the downward force due to gravity equal each other. Thus, the net force is zero, and there is no more acceleration. Because there is no more acceleration, the falling object has constant terminal speed:

$$F_\text{g} = F_\text{drag} \Rightarrow mg = Kv^2.$$

Solving this for the terminal speed, we obtain

$$v = \sqrt{\frac{mg}{K}}. \tag{4.14}$$

Note that the terminal speed depends on the mass of the object, whereas when we neglected air resistance, the mass of the object did not affect the object's motion. In the absence of air resistance, all objects fall at the same rate, but the presence of air resistance explains why heavy objects fall faster than light ones that have the same (drag) constant $K$.

To compute the terminal speed for a falling object, we need to know the value of the constant $K$. This constant depends on many variables, including the size of the cross-sectional area, $A$, exposed to the air stream. In general terms, the bigger the area, the bigger is the constant $K$. $K$ also depends linearly on the air density, $\rho$. All other dependences on the shape of the object, on its inclination relative to the direction of motion, on air viscosity, and compressibility are usually collected in a drag coefficient, $c_d$:

$$K = \tfrac{1}{2} c_d A \rho. \tag{4.15}$$

Equation 4.15 has the factor $\frac{1}{2}$ to simplify calculations involving the energy of objects undergoing free fall with air resistance. We will return to this subject when we discuss kinetic energy in Chapter 5.

Creating a low drag coefficient is an important consideration in automotive design, because it has a strong influence on the maximum speed of a car and its fuel consumption. Numerical computations are useful, but the drag coefficient is usually optimized experimentally by putting car prototypes into wind tunnels and testing the air resistance at different speeds. The same wind tunnel tests are also used to optimize the performance of equipment and athletes in events such as downhill ski racing and bicycle racing.

For motion in very viscous media or at low velocities, the linear velocity term of the friction force cannot be neglected. In this case, the friction force can be approximated by the form $F_{frict} = K_1 v$. This form applies to most biological processes, including large biomolecules or even microorganisms such as bacteria moving through liquids. This approximation of the friction force is also useful when analyzing the sinking of an object in a fluid, for example, a small stone or fossil shell in water.

### 4.5 In-Class Exercise

An unused coffee filter reaches its terminal speed very quickly if you let it fall. Suppose you release a single coffee filter from a height of 1 m. From what height do you have to release a stack of two coffee filters at the same instant so that they will hit the ground at the same time as the single coffee filter? (You can safely neglect the time needed to reach terminal speed.)

a) 0.5 m

b) 0.7 m

c) 1 m

d) 1.4 m

e) 2m

## EXAMPLE 4.8 Sky Diving

An 80-kg skydiver falls through air with a density of 1.15 kg/m$^3$. Assume that his drag coefficient is $c_d = 0.57$. When he falls in the spread-eagle position, as shown in Figure 4.21a, his body presents an area $A_1 = 0.94$ m$^2$ to the wind, whereas when he dives head first, with arms close to the body and legs together, as shown in Figure 4.21b, his area is reduced to $A_2 = 0.21$ m$^2$.

### PROBLEM

What are the terminal speeds in both cases?

### SOLUTION

We use equation 4.14 for the terminal speed and equation 4.15 for the air resistance constant, rearrange the formulas, and insert the given numbers:

$$v = \sqrt{\frac{mg}{K}} = \sqrt{\frac{mg}{\frac{1}{2} c_d A \rho}}$$

$$v_1 = \sqrt{\frac{(80 \text{ kg})(9.81 \text{ m/s}^2)}{\frac{1}{2} 0.57(0.94 \text{ m}^2)(1.15 \text{ kg/m}^3)}} = 50.5 \text{ m/s}$$

$$v_2 = \sqrt{\frac{(80 \text{ kg})(9.81 \text{ m/s}^2)}{\frac{1}{2} 0.57(0.21 \text{ m}^2)(1.15 \text{ kg/m}^3)}} = 107 \text{ m/s}.$$

These results show that, by diving head first, the skydiver can reach higher velocities during free fall than when he uses the spread-eagle position. Therefore, it is possible to catch up to a person who has fallen out of an airplane, assuming that the person is not diving head first, too. However, in general, this technique cannot be used to save such a person because it would be nearly impossible to hold onto him or her during the sudden deceleration shock caused by the rescuer's parachute opening.

(a)

(b)

**FIGURE 4.21** (a) Skydiver in the high-resistance position. (b) Skydiver in low-resistance position.

## Tribology

What causes friction? The answer to this question is not at all easy or obvious. When surfaces rub against each other, different atoms (more on atoms in Chapter 13) from the two surfaces make contact with each other in different ways. Atoms get dislocated in the process of dragging surfaces across each other. Electrostatic interaction (more on this in Chapter 21) between the atoms on the surfaces causes additional static friction. A true microscopic understanding of friction is beyond the scope of this book and is currently the focus of great research activity.

The science of friction has a name: **tribology.** The laws of friction we have discussed were already known 300 years ago. Their discovery is generally credited to Guillaume Amontons and Charles Augustin de Coulomb, but even Leonardo da Vinci may have known them. Yet amazing things are still being discovered about friction, lubrication, and wear.

Perhaps the most interesting advance in tribology that occurred in the last two decades was the development of atomic and friction force microscopes. The basic principle that these microscopes employ is the dragging of a very sharp tip across a surface with analysis by cutting-edge computer and sensor technology. Such friction force microscopes can measure friction forces as small as 10 pN = $10^{-11}$ N. Shown in Figure 4.22 is a cut-away schematic drawing of one of these instruments, constructed by physicists at the University of Leiden, Netherlands. State-of-the-art microscopic simulations of friction are still not able to completely explain it, and so this research area is of great interest in the field of nanotechnology.

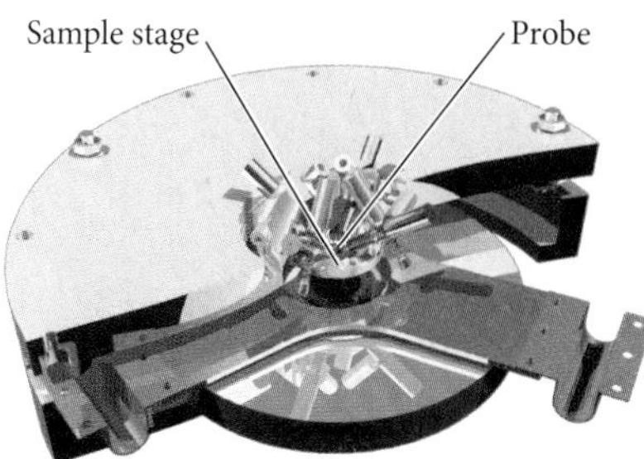


**FIGURE 4.22** Cut-away drawing of a microscope used to study friction forces by dragging a probe in the form of sharp point across the surface to be studied.

Friction is responsible for the breaking off of small particles from surfaces that rub against each other, causing wear. This phenomenon is of particular importance in high-performance car engines, which require specially formulated lubricants. Understanding the influence of small surface impurities on the friction force is of great interest in this context. Research into lubricants continues to try to find ways to reduce the coefficient of kinetic friction, $\mu_k$, to a value as close to zero as possible. For example, modern lubricants include *buckyballs*—molecules consisting of 60 carbon atoms arranged in the shape of a soccer ball, which were discovered in 1985. These molecules act like microscopic ball bearings.

Solving problems involving friction is also important to car racing. In the Formula 1 circuit, using the right tires that provide optimally high friction is essential for winning races. While friction coefficients are normally in the range between 0 and 1, it is not unusual for top fuel race cars to have tires that have friction coefficients with the track surface of 3 or even larger.

## 4.8 Applications of the Friction Force

With Newton's three laws, we can solve a huge class of problems. Knowing about static and kinetic friction allows us to approximate real-world situations and come to meaningful conclusions. Because it is helpful to see various applications of Newton's laws, we will solve several practice problems. These examples are designed to demonstrate a range of techniques that are useful in the solution of many kinds of problems.

### EXAMPLE 4.9 Two Blocks Connected by a Rope—with Friction

We solved this problem in Example 4.4, with the assumptions that $m_1$ slides without friction across the horizontal support surface and that the rope slides without friction across the pulley. Now we will allow for friction between $m_1$ and the surface it slides across. For now, we will still assume that the rope slides without friction across the pulley. (Chapter 10 will present techniques that let us deal with the pulley being set into rotational motion by the rope moving across it.)

**PROBLEM 1**

Let the coefficient of static friction between block 1 (mass $m_1$ = 2.3 kg) and its support surface have a value of 0.73 and the coefficient of kinetic friction have a value of 0.60. (Refer back to Figure 4.17.) If block 2 has mass $m_2$ = 1.9 kg, will block 1 accelerate from rest?

*Continued—*

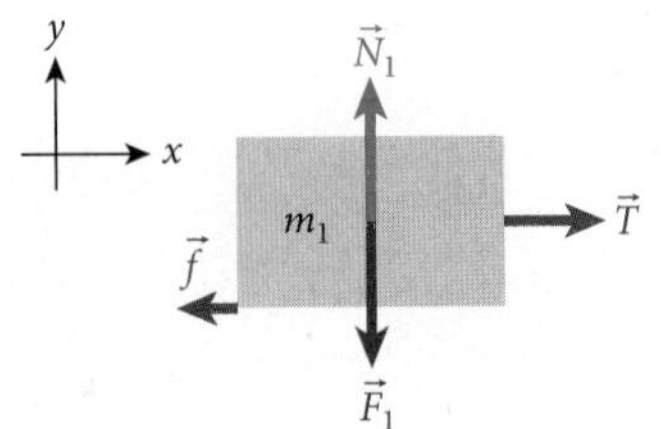


**FIGURE 4.23** Free-body diagram for $m_1$, including the force of friction.

**SOLUTION 1**

All the force considerations from Example 4.4 remain the same, except that the free-body diagram for block $m_1$ (Figure 4.23) now has a force arrow corresponding to the friction force, $f$. Keep in mind that in order to draw the direction of the friction force, you need to know in which direction movement would occur in the absence of friction. Because we have already solved the frictionless case, we know that $m_1$ would move to the right. Because the friction force is directed opposite to the movement, the friction vector thus points to the left.

The equation we derived in Example 4.4 by applying Newton's Second Law to $m_1$ changes from $m_1 a = T$ to

$$m_1 a = T - f.$$

Combining this with the equation we obtained in Example 4.4 via application of Newton's Second Law to $m_2$, $T - m_2 g = -m_2 a$, and again eliminating $T$ gives us

$$m_1 a + f = T = m_2 g - m_2 a \Rightarrow$$

$$a = \frac{m_2 g - f}{m_1 + m_2}.$$

So far, we have avoided specifying any further details about the friction force. We now do so by first calculating the maximum magnitude of the static friction force, $f_{s,max} = \mu_s N_1$. For the magnitude of the normal force, we already found $N_1 = m_1 g$, giving us the formula for the maximum static friction force:

$$f_{s,\max} = \mu_s N_1 = \mu_s m_1 g = (0.73)(2.3 \text{ kg})(9.81 \text{ m/s}^2) = 16.5 \text{ N}.$$

We need to compare this value to that of $m_2 g$ in the numerator of our equation for the acceleration, $a = (m_2 g - f)/(m_1 + m_2)$. If $f_{s,max} \geq m_2 g$, then the static friction force will assume a value exactly equal to $m_2 g$, causing the acceleration to be zero. In other words, there will be no motion, because the pull due to block $m_2$ hanging from the rope is not sufficient to overcome the force of static friction between block $m_1$ and its supporting surface. If $f_{s,max} < m_2 g$, then there will be positive acceleration, and the two blocks will start moving. In the present case, because $m_2 g = (1.9 \text{ kg})(9.81 \text{ m/s}^2) = 18.6 \text{ N}$, the blocks will start moving.

**PROBLEM 2**

What is the value of the acceleration?

**SOLUTION 2**

As soon as the static friction force is overcome, kinetic friction takes over. We can then use our equation for the acceleration, $a = (m_2 g - f)/(m_1 + m_2)$, substitute $f = \mu_k N_1 = \mu_k m_1 g$, and obtain

$$a = \frac{m_2 g - \mu_k m_1 g}{m_1 + m_2} = g\left(\frac{m_2 - \mu_k m_1}{m_1 + m_2}\right).$$

Inserting the numbers, we find

$$a = (9.81 \text{ m/s}^2)\left[\frac{(1.9 \text{ kg}) - 0.6 \cdot (2.3 \text{ kg})}{(2.3 \text{ kg}) + (1.9 \text{ kg})}\right] = 1.21 \text{ m/s}^2.$$

## EXAMPLE 4.10 Pulling a Sled

Suppose you are pulling a sled across a level snow-covered surface by exerting constant force on a rope, at an angle $\theta$ relative to the ground.

**PROBLEM 1**

If the sled, including its load, has a mass of 15.3 kg, the coefficients of friction between the sled and the snow are $\mu_s = 0.076$ and $\mu_k = 0.070$, and you pull with a force of 25.3 N on the rope at an angle of 24.5° relative to the horizontal ground, what is the sled's acceleration?

## SOLUTION 1

Figure 4.24 shows the free-body diagram for the sled, with all the forces acting on it. The directions of the force vectors are correct, but the magnitudes are not necessarily drawn to scale. The acceleration of the sled will, if it occurs at all, be directed along the horizontal, in the $x$-direction. In terms of components, Newton's Second Law gives:

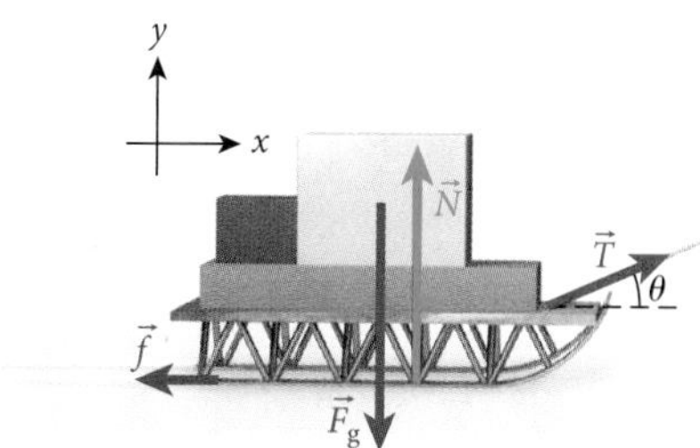


**FIGURE 4.24** Free-body diagram of the sled and its load.

$$x\text{-component:}\quad ma = T\cos\theta - f$$

and

$$y\text{-component:}\quad 0 = T\sin\theta - mg + N.$$

For the friction force, we will use the form $f = \mu N$ for now, without specifying whether it is kinetic or static friction, but in the end we will have to return to this point. The normal force can be calculated from the above equation for the $y$-component and then substituted into the equation for the $x$-component:

$$\begin{aligned} N &= mg - T\sin\theta \\ ma &= T\cos\theta - \mu(mg - T\sin\theta) \Rightarrow \\ a &= \frac{T}{m}(\cos\theta + \mu\sin\theta) - \mu g. \end{aligned}$$

We see that the normal force is less than the weight of the sled, because the force pulling on the rope has an upward $y$-component. The vertical component of the force pulling on the rope also contributes to the acceleration of the sled since it affects the normal force and hence the horizontal frictional force.

When putting in the numbers, we first use the value of the coefficient of static friction to see if enough force is applied by pulling on the rope to generate a positive acceleration. If the resulting value for $a$ turns out to be negative, this means that there is not enough pulling force to overcome the static friction force. With the given value of $\mu_s$ (0.076), we obtain

$$a' = \frac{25.3\text{ N}}{15.3\text{ kg}}(\cos 24.5° + 0.076\sin 24.5°) - 0.076(9.81\text{ m/s}^2) = 0.81\text{ m/s}^2.$$

Because this calculation results in a positive value for $a'$, we know that the force is strong enough to overcome the friction force. We now use the given value for coefficient of kinetic friction to calculate the actual acceleration of the sled:

$$a = \frac{25.3\text{ N}}{15.3\text{ kg}}(\cos 24.5° + 0.070\sin 24.5°) - 0.070(9.81\text{ m/s}^2) = 0.87\text{ m/s}^2.$$

## PROBLEM 2

What angle of the rope with the horizontal will produce the maximum acceleration of the sled for the given value of the magnitude of the pulling force, $T$? What is that maximum value of $a$?

## SOLUTION 2

In calculus, to find the extremum of a function, we have to take the first derivative and find the value of the independent variable for which that derivative is zero:

$$\frac{d}{d\theta}a = \frac{d}{d\theta}\left(\frac{T}{m}(\cos\theta + \mu\sin\theta) - \mu g\right) = \frac{T}{m}(-\sin\theta + \mu\cos\theta).$$

Searching for the root of this equation results in

$$\begin{aligned} \left.\frac{da}{d\theta}\right|_{\theta=\theta_{max}} &= \frac{T}{m}(-\sin\theta_{max} + \mu\cos\theta_{max}) = 0 \\ &\Rightarrow \sin\theta_{max} = \mu\cos\theta_{max} \Rightarrow \\ &\theta_{max} = \tan^{-1}\mu. \end{aligned}$$

*Continued—*

Inserting the given value for the coefficient of kinetic friction, 0.070, into this equation results in $\theta_{max} = 4.0°$. This means that the rope should be oriented almost horizontally. The resulting value of the acceleration can be obtained by inserting the numbers into the equation for $a$ we used in Solution 1:

$$a_{max} \equiv a(\theta_{max}) = 0.97 \text{ m/s}^2.$$

*Note:* A zero first derivative is only a necessary condition for a maximum, not a sufficient one. You can convince yourself that we have indeed found the maximum by first realizing that we obtained only one root of the first derivative, meaning that the function $a(\theta)$ has only one extremum. Also, because the value of the acceleration we calculated at this point is bigger than the value we previously obtained for 24.5°, we are assured that our single extremum is indeed a maximum. Alternatively, we could have taken the second derivative and found that it is negative at the point $\theta_{max} = 4.0°$; then, we could have compared the value of the acceleration obtained at that point with those at $\theta = 0°$ and $\theta = 90°$.

## WHAT WE HAVE LEARNED | EXAM STUDY GUIDE

- The net force on an object is the vector sum of the forces acting on the object: $\vec{F}_{net} = \sum_{i=1}^{n} \vec{F}_i$.
- Mass is an intrinsic quality of an object that quantifies both the object's ability to resist acceleration and the gravitational force on the object.
- A free-body diagram is an abstraction showing all forces acting on an isolated object.
- Newton's three laws are as follows:

  Newton's First Law. In the absence of a net force on an object, the object will remain at rest, if it was at rest. If it was moving, it will remain in motion in a straight line with the same velocity.

  Newton's Second Law. If a net external force, $\vec{F}_{net}$, acts on an object with mass $m$, the force will cause an acceleration, $\vec{a}$, in the same direction as the force: $\vec{F}_{net} = m\vec{a}$.

  Newton's Third Law. The forces that two interacting objects exert on each other are always exactly equal in magnitude and opposite in direction: $\vec{F}_{1\to 2} = -\vec{F}_{2\to 1}$.
- Two types of friction occur: static and kinetic friction. Both types of friction are proportional to the normal force, $N$.

  Static friction describes the force of friction between an object at rest on a surface in terms of the coefficient of static friction, $\mu_s$. The static friction force, $f_s$, opposes a force trying to move an object and has a maximum value, $f_{s,max}$, such that $f_s \le \mu_s N = f_{s,max}$.

  Kinetic friction describes the force of friction between a moving object and a surface in terms of the coefficient of kinetic friction, $\mu_k$. Kinetic friction is given by $f_k = \mu_k N$.

  In general, $\mu_s > \mu_k$.

## KEY TERMS

dynamics, p. 101
contact force, p. 101
tension, p. 101
compression, p. 101
normal force, p. 101
friction force, p. 102
spring force, p. 102
fundamental forces, p. 102
gravitational force, p. 102
electromagnetic force, p. 102
strong nuclear force, p. 102
weak nuclear force, p. 102
gravitational force vector, p. 103
weight, p. 103
mass, p. 104
gravitational mass, p. 104
inertial mass, p. 104
Higgs particle, p. 105
net force, p. 105
free-body diagram, p. 106
Newton's First Law, p. 107
Newton's Second Law, p. 107
Newton's Third Law, p. 107
static equilibrium, p. 107
dynamic equilibrium, p. 107
coefficient of kinetic friction, p. 118
coefficient of static friction, p. 119
drag force, p. 121
terminal speed, p. 121
tribology, p. 123

## NEW SYMBOLS AND EQUATIONS

$\vec{F}_g = -mg\hat{y}$, gravitational force vector

$\vec{F}_{net} = \sum_{i=1}^{n} \vec{F}_i = \vec{F}_1 + \vec{F}_2 + \cdots + \vec{F}_n$, net force

$\vec{N}$, normal force

$\vec{F}_{net} = 0$, Newton's First Law, condition for static equilibrium

$\vec{F}_{net} = m\vec{a}$, Newton's Second Law

$\vec{F}_{1\to 2} = -\vec{F}_{2\to 1}$, Newton's Third Law

$\vec{T}$, string tension

$f_k$, force of kinetic friction

$\mu_k$, coefficient of kinetic friction

$f_s$, force of static friction

$\mu_s$, coefficient of static friction

$F_{drag}$, force due to air resistance, or drag force

$c_d$, drag coefficient

## ANSWERS TO SELF-TEST OPPORTUNITIES

**4.1**

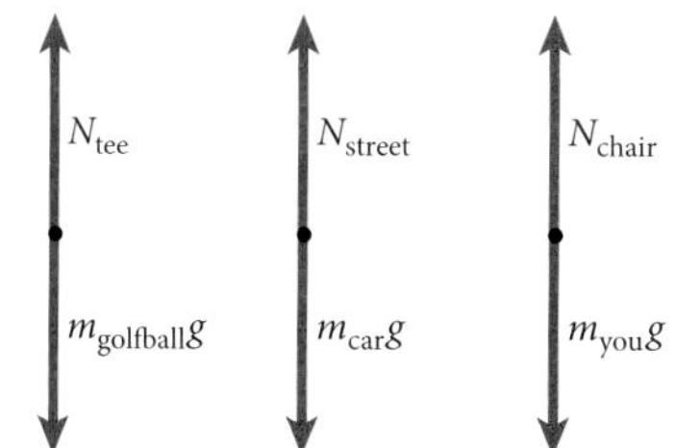


**4.2** Using $T = m_2(g + a)$ and inserting the value for the acceleration, $a = g\left(\frac{m_1 - m_2}{m_1 + m_2}\right)$, we find

$$T = m_2(g+a) = m_2\left(g + g\frac{m_1 - m_2}{m_1 + m_2}\right) = m_2 g\left(\frac{m_1 + m_2}{m_1 + m_2} + \frac{m_1 - m_2}{m_1 + m_2}\right)$$

$$= 2g\frac{m_1 m_2}{m_1 + m_2}.$$

## PROBLEM-SOLVING PRACTICE

### Problem-Solving Guidelines: Newton's Laws

Analyzing a situation in terms of forces and motion is a vital skill in physics. One of the most important techniques is the proper application of Newton's laws. The following guidelines can help you solve mechanics problems in terms of Newton's three laws. These are part of the seven-step strategy for solving all types of physics problems and are most relevant to the Sketch, Think, and Research steps.

**1.** An overall sketch can help you visualize the situation and identify the concepts involved, but you also need a separate free-body diagram for each object to identify which forces act on that particular object and no others. Drawing correct free-body diagrams is the key to solving all problems in mechanics, whether they involve static (nonmoving) situations or kinetic (moving) ones. Remember that the $m\vec{a}$ from Newton's Second Law should not be included as a force in any free-body diagram.

**2.** Choosing the coordinate system is important—often the choice of coordinate system makes the difference between very simple equations and very difficult ones. Placing an axis along the same direction as an object's acceleration, if there is any, is often very helpful. In a statics problem, orienting an axis along a surface, whether horizontal or inclined, is often useful. Choosing the most advantageous coordinate system is an acquired skill gained through experience as you work many problems.

**3.** Once you have chosen your coordinate directions, determine whether the situation involves acceleration in either direction. If no acceleration occurs in the $y$-direction, for example, then Newton's First Law applies in that direction, and the sum of forces (the net force) equals zero. If acceleration does occur in a given direction, for example, the $x$-direction, then Newton's Second Law applies in that direction, and the net force equals the object's mass times its acceleration.

**4.** When you decompose a force vector into components along the coordinate directions, be careful about which direction involves the sine of a given angle and which direction involves the cosine. Do not generalize from past problems and think that all components in the $x$-direction involve the cosine; you will find problems where the $x$-component involves the sine. Rely instead on clear definitions of angles and coordinate directions and the geometry of the given situation. Often the same angle appears at different points and between different lines in a problem. This usually results in similar triangles, often involving right angles. If you create a sketch of a problem with a general angle $\theta$, try to use an angle that is not close to 45°, because it is hard to distinguish between such an angle and its complement in your sketch.

**5.** Always check your final answer. Do the units make sense? Are the magnitudes reasonable? If you change a variable to approach some limiting value, does your answer make a valid prediction about what happens? Sometimes you can estimate the answer to a problem by using order-of-magnitude approximations, as discussed in Chapter 1; such an estimate can often reveal whether you made an arithmetical mistake or wrote down an incorrect formula.

**6.** The friction force always opposes the direction of motion and acts parallel to the contact surface; the static friction force opposes the direction in which the object would move, if the friction force were not present. Note that the kinetic friction force is *equal* to the product of the coefficient of friction and the normal force, whereas the static friction force is *less than or equal* to that product.

## SOLVED PROBLEM 4.2 Wedge

A wedge of mass $m = 37.7$ kg is held in place on a fixed plane that is inclined by an angle $\theta = 20.5°$ with respect to the horizontal. A force $F = 309.3$ N in the horizontal direction pushes on the wedge, as shown in Figure 4.25a. The coefficient of kinetic friction between the wedge and the plane is $\mu_k = 0.171$. Assume that the coefficient of static friction is low enough that the net force will move the wedge.

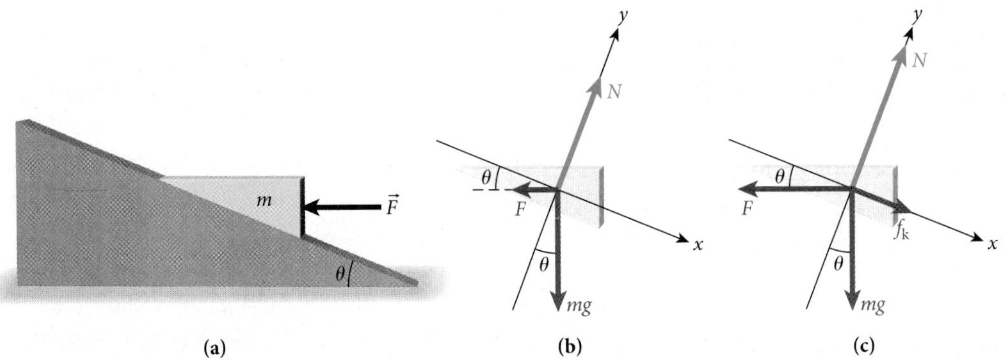


**FIGURE 4.25** (a) A wedge-shaped block being pushed on an inclined plane. (b) Free-body diagram of the wedge, including the external force, the force of gravity, and the normal force. (c) Free-body diagram including the external force, the force of gravity, the normal force, and the friction force.

### PROBLEM

What is the acceleration of the wedge along the plane when it is released and free to move?

### SOLUTION

#### THINK

We want to know the acceleration $a$ of the wedge of mass $m$ along the plane, which requires us to determine the component of the net force that acts on the wedge parallel to the surface of the inclined plane. Also, we need to find the component of the net force that acts on the wedge perpendicular to the plane, to allow us to determine the force of kinetic friction.

The forces acting on the wedge are gravity, the normal force, the force of kinetic friction $f_k$, and the external force $F$. The coefficient of kinetic friction, $\mu_k$, is given, so we can calculate the friction force once we determine the normal force. Before we can continue with our analysis of the forces, we must determine in which direction the wedge will move after it is released as a result of force $F$. Once we know in which direction the wedge will go, we can determine the direction of the friction force and complete our analysis.

To determine the net force before the wedge begins to move, we need a free-body diagram with just the forces $F$, $N$, and $mg$. Once we determine the direction of motion, we can determine the direction of the friction force, using a second free-body diagram with the friction force added.

#### SKETCH

A free-body diagram showing the forces acting on the wedge before it is released is presented in Figure 4.25b. We have defined a coordinate system in which the $x$-axis is parallel to the surface of the inclined plane, with the positive $x$-direction pointing down the plane. The sum of the forces in the $x$-direction is

$$mg\sin\theta - F\cos\theta = ma.$$

We need to determine if the mass will move to the right (positive $x$-direction, or down the plane) or to the left (negative $x$-direction, or up the plane). We can see from the equation that the quantity $mg \sin\theta - F\cos\theta$ will determine the direction of the motion. With the given numerical values, we have

$$\begin{aligned} mg\sin\theta - F\cos\theta &= (37.7\text{ kg})(9.81\text{ m/s}^2)(\sin 20.5°) - (309.3\text{ N})(\cos 20.5°) \\ &= -160.193\text{ N} \end{aligned}$$

Thus, the mass will move up the plane (to the left, or in the negative $x$-direction). Now we can redraw the free-body diagram as shown in Figure 4.25c by inserting the arrow for the force of kinetic friction, $f_k$, pointing down the plane (in the positive $x$-direction), because the friction force always opposes the direction of motion.

## RESEARCH

Now we can write the components of the forces in the $x$- and $y$-directions based on the final free-body diagram. For the $x$-direction, we have

$$mg\sin\theta - F\cos\theta + f_k = ma. \qquad \text{(i)}$$

For the $y$-direction we have

$$N - mg\cos\theta - F\sin\theta = 0.$$

From this equation, we can get the normal force $N$ that we need to calculate the friction force:

$$f_k = \mu_k N = \mu_k\left(mg\cos\theta + F\sin\theta\right). \qquad \text{(ii)}$$

## SIMPLIFY

Having related all the known and unknown quantities to each other, we can get an expression for the acceleration of the mass using equations i and ii:

$$mg\sin\theta - F\cos\theta + \mu_k\left(mg\cos\theta + F\sin\theta\right) = ma.$$

We can rearrange this expression:

$$\begin{aligned} &mg\sin\theta - F\cos\theta + \mu_k mg\cos\theta + \mu_k F\sin\theta = ma \\ &\left(mg + \mu_k F\right)\sin\theta + \left(\mu_k mg - F\right)\cos\theta = ma, \end{aligned}$$

and then solve for the acceleration:

$$a = \frac{\left(mg + \mu_k F\right)\sin\theta + \left(\mu_k mg - F\right)\cos\theta}{m}. \qquad \text{(iii)}$$

## CALCULATE

Now we put in the numbers and get a numerical result. The first term in the numerator of equation iii is

$$\left((37.7\text{ kg})(9.81\text{ m/s}^2) + (0.171)(309.3\text{ N})\right)(\sin 20.5°) = 148.042\text{ N}.$$

Note that we have not rounded this result yet. The second term in the numerator of equation iii is

$$\left((0.171)(37.7\text{ kg})(9.81\text{ m/s}^2) - (309.3\text{ N})\right)(\cos 20.5°) = -230.476\text{ N}.$$

Again we have not yet rounded the result. Now we calculate the acceleration using equation iii:

$$a = \frac{(148.042\text{ N}) + (-230.476\text{ N})}{37.7\text{ kg}} = -2.1866\text{ m/s}^2.$$

*Continued—*

**ROUND**

Because all of the numerical values were initially given with three significant figures, we report our final result as

$$a = -2.19 \text{ m/s}^2.$$

**DOUBLE-CHECK**

Looking at our answer, we see that the acceleration is negative, which means it is in the negative $x$-direction. We had determined that the mass would move to the left (up the plane, or in the negative $x$-direction), which agrees with the sign of the acceleration in our final result. The magnitude of the acceleration is a fraction of the acceleration of gravity (9.81 m/s$^2$), which makes physical sense.

## SOLVED PROBLEM 4.3 Two Blocks

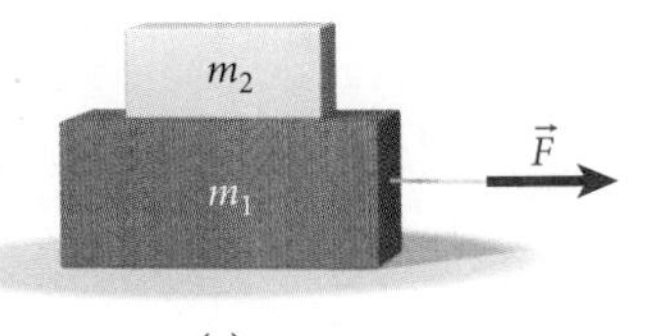

(a)

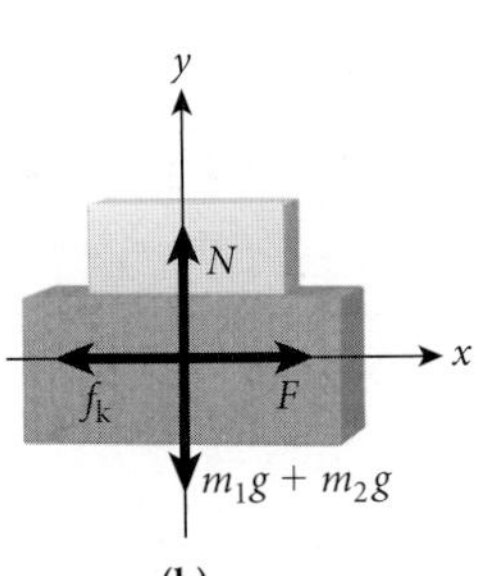

(b)

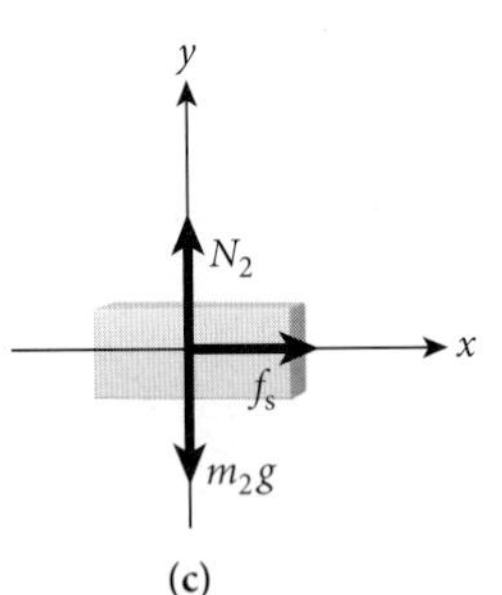

(c)

**FIGURE 4.26** (a) Two stacked blocks being pulled to the right. (b) Free-body diagram for the two blocks moving together. (c) Free-body diagram for the upper block.

Two rectangular blocks are stacked on a table as shown in Figure 4.26a. The upper block has a mass of 3.40 kg, and the lower block has a mass of 38.6 kg. The coefficient of kinetic friction between the lower block and the table is 0.260. The coefficient of static friction between the blocks is 0.551. A string is attached to the lower block, and an external force $\vec{F}$ is applied horizontally, pulling on the string as shown.

**PROBLEM**

What is the maximum force that can be applied to the string without having the upper block slide off?

**SOLUTION**

**THINK**

To begin this problem, we note that as long as the force of static friction between the two blocks is not overcome, the two blocks will travel together. Thus, if we pull gently on the lower block, the upper block will stay in place on top of it, and the two blocks will slide as one. However, if we pull hard on the lower block, the force of static friction between the blocks will not be sufficient to keep the upper block in place and it will begin to slide off the lower block.

The forces acting in this problem are the external force $F$ pulling on the string, the force of kinetic friction $f_k$ between the lower block and the surface on which the blocks are sliding, the weight $m_1g$ of the lower block, the weight $m_2g$ of the upper block, and the force of static friction $f_s$ between the blocks and the normal forces.

**SKETCH**

We start with a free-body diagram of the two blocks moving together (Figure 4.26b), because we will treat the two blocks as one system for the first part of this analysis. We define the $x$-direction to be parallel to the surface on which the blocks are sliding and parallel to the external force pulling on the string, with the positive direction to the right, in the direction of the external force. The sum of the forces in the $x$-direction is

$$F - f_k = (m_1 + m_2)a. \quad \text{(i)}$$

The sum of the forces in the $y$-direction is

$$N - (m_1g + m_2g) = 0. \quad \text{(ii)}$$

Equations i and ii describe the motion of the two blocks together.

Now we need a second free-body diagram to describe the forces acting on the upper block. The forces in the free-body diagram for the upper block (Figure 4.26c) are the normal force $N_2$ exerted by the lower block, the weight $m_2 g$, and the force of static friction $f_s$. The sum of the forces in the $x$-direction is

$$f_s = m_2 a. \qquad \text{(iii)}$$

The sum of the forces in the $y$-direction is

$$N_2 - m_2 g = 0. \qquad \text{(iv)}$$

## RESEARCH

The maximum value of the force of static friction between the upper and lower blocks is given by

$$f_s = \mu_s N_2 = \mu_s (m_2 g).$$

where we have used equations iii and iv. Thus, the maximum acceleration that the upper block can have without sliding is

$$a_{max} = \frac{f_s}{m_2} = \frac{\mu_s m_2 g}{m_2} = \mu_s g. \qquad \text{(v)}$$

This maximum acceleration for the upper block is also the maximum acceleration for both blocks together. From equation ii, we get the normal force between the lower block and the sliding surface:

$$N = m_1 g + m_2 g. \qquad \text{(vi)}$$

The force of kinetic friction between the lower block and the sliding surface is then

$$f_k = \mu_k (m_1 g + m_2 g). \qquad \text{(vii)}$$

## SIMPLIFY

We can now relate the maximum acceleration to the maximum force, $F_{max}$, that can be exerted without the upper block sliding off, using equations v–vii:

$$F_{max} - \mu_k (m_1 g + m_2 g) = (m_1 + m_2) \mu_s g.$$

We solve this for the maximum force to obtain

$$F_{max} = \mu_k (m_1 g + m_2 g) + (m_1 + m_2) \mu_s g = g(m_1 + m_2)(\mu_k + \mu_s).$$

## CALCULATE

Putting in the given numerical values, we get

$$F_{max} = (9.81 \text{ m/s}^2)(38.6 \text{ kg} + 3.40 \text{ kg})(0.260 + 0.551) = 334.148 \text{ N}.$$

## ROUND

All of the numerical values were given to three significant figures, so we report our answer as

$$F_{max} = 334 \text{ N}.$$

## DOUBLE-CHECK

The answer is a positive value, implying a force to the right, which agrees with the free-body diagram in Figure 4.26b.

The maximum acceleration is

$$a_{max} = \mu_s g = (0.551)(9.81 \text{ m/s}^2) = 5.41 \text{ m/s}^2,$$

which is a fraction of the acceleration due to gravity, which seems reasonable. If there were no friction between the lower block and the surface on which it slides, the force required to accelerate both blocks would be

$$F = (m_1 + m_2) a_{max} = (38.6 \text{ kg} + 3.40 \text{ kg})(5.41 \text{ m/s}^2) = 227 \text{ N}.$$

Thus, our answer of 334 N for the maximum force seems reasonable because it is higher than the force calculated when there is no friction.

## MULTIPLE-CHOICE QUESTIONS

**4.1** A car of mass $M$ travels in a straight line at constant speed along a level road with a coefficient of friction between the tires and the road of $\mu$ and a drag force of $D$. The magnitude of the net force on the car is

a) $\mu Mg$.
b) $\mu Mg + D$.
c) $\sqrt{(\mu Mg)^2 + D^2}$.
d) zero.

**4.2** A person stands on the surface of the Earth. The mass of the person is $m$ and the mass of the Earth is $M$. The person jumps upward, reaching a maximum height $h$ above the Earth. When the person is at this height $h$, the magnitude of the force exerted on the Earth by the person is

a) $mg$.
b) $Mg$.
c) $M^2g/m$.
d) $m^2g/M$.
e) zero.

**4.3** Leonardo da Vinci discovered that the magnitude of the friction force is usually simply proportional to the magnitude of the normal force; that is, the friction force does not depend on the width or length of the contact area. Thus, the main reason to use wide tires on a race car is that they

a) look cool.
b) have more apparent contact area.
c) cost more.
d) can be made of softer materials.

**4.4** The Tornado is a carnival ride that consists of a vertical cylinder that rotates rapidly about its vertical axis. As the Tornado rotates, the riders are pressed against the inside wall of the cylinder by the rotation, and the floor of the cylinder drops away. The force that points upward, preventing the riders from falling downward, is

a) friction force.
b) a normal force.
c) gravity.
d) a tension force.

**4.5** When a bus makes a sudden stop, passengers tend to jerk forward. Which of Newton's laws can explain this?

a) Newton's First Law
b) Newton's Second Law
c) Newton's Third Law
d) It cannot be explained by Newton's laws.

**4.6** Only two forces, $\vec{F}_1$ and $\vec{F}_2$, are acting on a block. Which of the following can be the magnitude of the net force, $\vec{F}$, acting on the block (indicate all possibilities)?

a) $F > F_1 + F_2$
b) $F = F_1 + F_2$
c) $F < F_1 + F_2$
d) none of the above

**4.7** Which of the following observations about the friction force is (are) incorrect?

a) The magnitude of the kinetic friction force is always proportional to the normal force.
b) The magnitude of the static friction force is always proportional to the normal force.
c) The magnitude of the static friction force is always proportional to the external applied force.
d) The direction of the kinetic friction force is always opposite the direction of the relative motion of the object with respect to the surface the object moves on.
e) The direction of the static friction force is always opposite that of the impending motion of the object relative to the surface it rests on.
f) All of the above are correct.

**4.8** A horizontal force equal to the object's weight is applied to an object resting on a table. What is the acceleration of the moving object when the coefficient of kinetic friction between the object and floor is 1 (assuming the object is moving in the direction of the applied force).

a) zero
b) 1 m/s$^2$
c) Not enough information is given to find the acceleration.

**4.9** Two blocks of equal mass are connected by a massless horizontal rope and resting on a frictionless table. When one of the blocks is pulled away by a horizontal external force $\vec{F}$, what is the ratio of the net forces acting on the blocks?

a) 1:1
b) 1:1.41
c) 1:2
d) none of the above

**4.10** If a cart sits motionless on level ground, there are no forces acting on the cart.

a) true
b) false
c) maybe

**4.11** An object whose mass is 0.092 kg is initially at rest and then attains a speed of 75.0 m/s in 0.028 s. What average net force acted on the object during this time interval?

a) $1.2 \cdot 10^2$ N
b) $2.5 \cdot 10^2$ N
c) $2.8 \cdot 10^2$ N
d) $4.9 \cdot 10^2$ N

**4.12** You push a large crate across the floor at constant speed, exerting a horizontal force $F$ on the crate. There is friction between the floor and the crate. The force of friction has a magnitude that is

a) zero.
b) $F$.
c) greater than $F$.
d) less than $F$.
e) impossible to quantify without further information.

## QUESTIONS

**4.13** You are at the shoe store to buy a pair of basketball shoes that have the greatest traction on a specific type of hardwood. To determine the coefficient of static friction, $\mu$, you place each shoe on a plank of the wood and tilt the plank to an angle $\theta$, at which the shoe just starts to slide. Obtain an expression for $\mu$ as a function of $\theta$.

**4.14** A heavy wooden ball is hanging from the ceiling by a piece of string that is attached from the ceiling to the top of the ball. A similar piece of string is attached to the bottom of the ball. If the loose end of the lower string is pulled down sharply, which is the string that is most likely to break?

**4.15** A car pulls a trailer down the highway. Let $F_t$ be the magnitude of the force on the trailer due to the car, and let $F_c$ be the magnitude of the force on car due to the trailer. If the car and trailer are moving at a constant velocity across level ground, then $F_t = F_c$. If the car and trailer are accelerating up a hill, what is the relationship between the two forces?

**4.16** A car accelerates down a level highway. What is the force in the direction of motion that accelerates the car?

**4.17** If the forces that two interacting objects exert on each other are always exactly equal in magnitude and opposite in direction, how is it possible for an object to accelerate?

**4.18** True or false: A physics book on a table will not move at all if and only if the net force is zero.

**4.19** A mass slides on a ramp that is at an angle of $\theta$ above the horizontal. The coefficient of friction between the mass and the ramp is $\mu$.

a) Find an expression for the magnitude and direction of the acceleration of the mass as it slides up the ramp.

b) Repeat part (a) to find an expression for the magnitude and direction of the acceleration of the mass as it slides down the ramp.

**4.20** A shipping crate that weighs 340 N is initially stationary on a loading dock. A forklift arrives and lifts the crate with an upward force of 500 N, accelerating the crate upward. What is the magnitude of the force due to gravity acting on the shipping crate while it is accelerating upward?

**4.21** A block is sliding on a (near) frictionless slope with an incline of 30°. Which force is greater in magnitude, the net force acting on the block or the normal force acting on the block?

**4.22** A tow truck of mass $M$ is using a cable to pull a shipping container of mass $m$ across a horizontal surface as shown in the figure. The cable is attached to the container at the front bottom corner and makes an angle $\theta$ with the vertical as shown. The coefficient of kinetic friction between the surface and the crate is $\mu$.

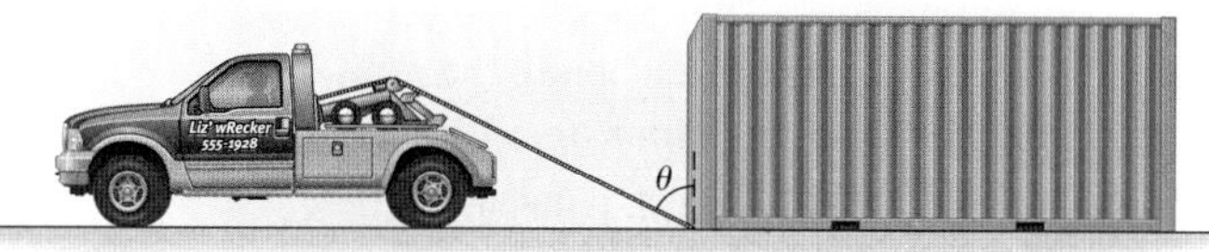


a) Draw a free-body diagram for the container.

b) Assuming that the truck pulls the container at a constant speed, write an equation for the magnitude $T$ of the string tension in the cable.

## PROBLEMS

A blue problem number indicates a worked-out solution is available in the Student Solutions Manual. One • and two •• indicate increasing level of problem difficulty.

### Section 4.2

4.23 The gravitational acceleration on the Moon is a sixth of that on Earth. The weight of an apple is 1.00 N on Earth.

a) What is the weight of the apple on the Moon?

b) What is the mass of the apple?

### Section 4.4

**4.24** A 423.5-N force accelerates a go-cart and its driver from 10.4 m/s to 17.9 m/s in 5.00 s. What is the mass of the go-cart plus driver?

**4.25** You have just joined an exclusive health club, located on the top floor of a skyscraper. You reach the facility by using an express elevator. The elevator has a precision scale installed so that members can weigh themselves before and after their workouts. A member steps into the elevator and gets on the scale before the elevator doors close. The scale shows a weight of 183.7 lb. Then the elevator accelerates upward with an acceleration of 2.43 m/s$^2$, while the member is still standing on the scale. What is the weight shown by the scale's display while the elevator is accelerating?

**4.26** An elevator cabin has a mass of 358.1 kg, and the combined mass of the people inside the cabin is 169.2 kg. The cabin is pulled upward by a cable, with a constant acceleration of 4.11 m/s$^2$. What is the tension in the cable?

4.27 An elevator cabin has a mass of 363.7 kg, and the combined mass of the people inside the cabin is 177 kg. The cabin is pulled upward by a cable, in which there is a tension force of 7638 N. What is the acceleration of the elevator?

**4.28** Two blocks are in contact on a frictionless, horizontal tabletop. An external force, $\vec{F}$, is applied to block 1, and the two blocks are moving with a constant acceleration of 2.45 m/s$^2$.

a) What is the magnitude, $F$, of the applied force?

b) What is the contact force between the blocks?

c) What is the net force acting on block 1? Use $M_1 = 3.20$ kg and $M_2 = 5.70$ kg.

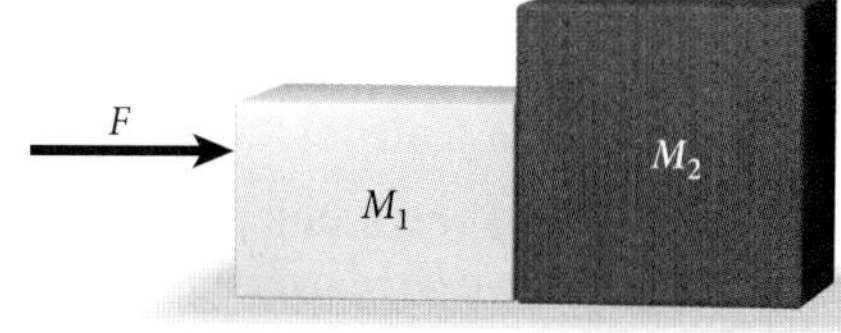

•**4.29** The density (mass per unit volume) of ice is 917 kg/m$^3$, and the density of seawater is 1024 kg/m$^3$. Only 10.4% of the volume of an iceberg is above the water's surface. If the volume of a particular iceberg that is above water is 4205.3 m$^3$, what is the magnitude of the force that the seawater exerts on this iceberg?

•**4.30** In a physics laboratory class, three massless ropes are tied together at a point. A pulling force is applied along each rope: $F_1$ = 150 N at 60°, $F_2$ = 200 N at 100°, $F_3$ = 100 N at 190°. What is the magnitude of a fourth force and the angle at which it acts to keep the point at the center of the system stationary? (All angles are measured from the positive $x$-axis.)

## Section 4.5

**4.31** Four weights, of masses $m_1$ = 6.50 kg, $m_2$ = 3.80 kg, $m_3$ = 10.70 kg, and $m_4$ = 4.20 kg, are hanging from the ceiling as shown in the figure. They are connected with ropes. What is the tension in the rope connecting masses $m_1$ and $m_2$?

**4.32** A hanging mass, $M_1$ = 0.50 kg, is attached by a light string that runs over a frictionless pulley to a mass $M_2$ = 1.50 kg that is initially at rest on a frictionless table. Find the magnitude of the acceleration, $a$, of $M_2$.

•**4.33** A hanging mass, $M_1$ = 0.50 kg, is attached by a light string that runs over a frictionless pulley to the front of a mass $M_2$ = 1.50 kg that is initially at rest on a frictionless table. A third mass $M_3$ = 2.50 kg, which is also initially at rest on a frictionless table, is attached to the back of $M_2$ by a light string.

a) Find the magnitude of the acceleration, $a$, of mass $M_3$.

b) Find the tension in the string between masses $M_1$ and $M_2$.

•**4.34** A hanging mass, $M_1$ = 0.40 kg, is attached by a light string that runs over a frictionless pulley to a mass $M_2$ = 1.20 kg that is initially at rest on a frictionless ramp. The ramp is at an angle of $\theta$ = 30° above the horizontal, and the pulley is at the top of the ramp. Find the magnitude and direction of the acceleration, $a_2$, of $M_2$.

•**4.35** A force table is a circular table with a small ring that is to be balanced in the center of the table. The ring is attached to three hanging masses by strings of negligible mass that pass over frictionless pulleys mounted on the edge of the table. The magnitude and direction of each of the three horizontal forces acting on the ring can be adjusted by changing the amount of each hanging mass and the position of each pulley, respectively. Given a mass $m_1$ = 0.040 kg pulling in the positive $x$-direction, and a mass $m_2$ = 0.030 kg pulling in the positive $y$-direction, find the mass ($m_3$) and the angle ($\theta$, counterclockwise from the positive $x$-axis) that will balance the ring in the center of the table.

•**4.36** A monkey is sitting on a wood plate attached to a rope whose other end is passed over a tree branch, as shown in the figure. The monkey holds the rope and tries to pull it down. The combined mass of the monkey and the wood plate is 100 kg.

a) What is the minimum force the monkey has to apply to lift-off the ground?

b) What applied force is needed to move the monkey with an upward acceleration of 2.45 m/s$^2$?

c) Explain how the answers would change if a second monkey on the ground pulled on the rope instead.

## Section 4.6

**4.37** A bosun's chair is a device used by a boatswain to lift himself to the top of the mainsail of a ship. A simplified device consists of a chair, a rope of negligible mass, and a frictionless pulley attached to the top of the mainsail. The rope goes over the pulley, with one end attached to the chair, and the boatswain pulls on the other end, lifting himself upward. The chair and boatswain have a total mass $M$ = 90 kg.

a) If the boatswain is pulling himself up at a constant speed, with what magnitude of force must he pull on the rope?

b) If, instead, the boatswain moves in a jerky fashion, accelerating upward with a maximum acceleration of magnitude $a$ = 2.0 m/s$^2$, with what maximum magnitude of force must he pull on the rope?

**4.38.** A granite block of mass 3311 kg is suspended from a pulley system as shown in the figure. The rope is wound around the pulleys 6 times. What is the force with which you would have to pull on the rope to hold the granite block in equilibrium?

**4.39** Arriving on a newly discovered planet, the captain of a spaceship performed the following experiment to calculate the gravitational acceleration for the planet: He placed masses of 100 g and 200 g on an Atwood device made of massless string and a frictionless pulley and measured that it took 1.52 s for each mass to travel 1.00 m from rest.

a) What is the gravitational acceleration for the planet?

b) What is the tension in the string?

•**4.40** A store sign of mass 4.25 kg is hung by two wires that each make an angle of $\theta$ = 42.4° with the ceiling. What is the tension in each wire?

•**4.41** A crate of oranges slides down an inclined plane without friction. If it is released from rest and reaches a speed of 5.832 m/s after sliding a distance of 2.29 m, what is the angle of inclination of the plane with respect to the horizontal?

•**4.42** A load of bricks of mass $M = 200$ kg is attached to a crane by a cable of negligible mass and length $L = 3.0$ m. Initially, when the cable hangs vertically downward, the bricks are a horizontal distance $D = 1.5$ m from the wall where the bricks are to be placed. What is the magnitude of the horizontal force that must be applied to the load of bricks (without moving the crane) so that the bricks will rest directly above the wall?

•4.43 A large ice block of mass $M = 80$ kg is held stationary on a frictionless ramp. The ramp is at an angle of $\theta = 36.9°$ above the horizontal.

a) If the ice block is held in place by a tangential force along the surface of the ramp (at angle $\theta$ above the horizontal), find the magnitude of this force.

b) If, instead, the ice block is held in place by a horizontal force, directed horizontally toward the center of the ice block, find the magnitude of this force.

•**4.44** A mass $m_1 = 20$ kg on a frictionless ramp is attached to a light string. The string passes over a frictionless pulley and is attached to a hanging mass $m_2$. The ramp is at an angle of $\theta = 30°$ above the horizontal. $m_1$ moves up the ramp uniformly (at constant speed). Find the value of $m_2$.

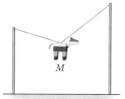

•**4.45** A piñata of mass $M = 8.0$ kg is attached to a rope of negligible mass that is strung between the tops of two vertical poles. The horizontal distance between the poles is $D = 2.0$ m, and the top of the right pole is a vertical distance $h = 0.50$ m higher than the top of the left pole. The piñata is attached to the rope at a horizontal position halfway between the two poles and at a vertical distance $s = 1.0$ m below the top of the left pole. Find the tension in each part of the rope due to the weight of the piñata.

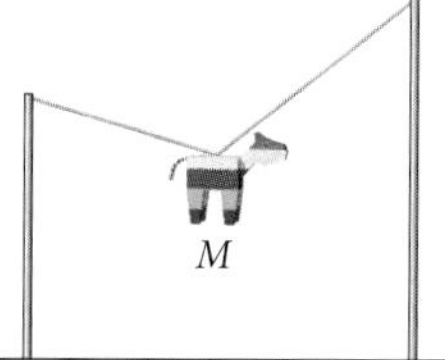


••**4.46** A piñata of mass $M = 12$ kg hangs on a rope of negligible mass that is strung between the tops of two vertical poles. The horizontal distance between the poles is $D = 2.0$ m, the top of the right pole is a vertical distance $h = 0.50$ m higher than the top of the left pole, and the total length of the rope between the poles is $L = 3.0$ m. The piñata is attached to a ring, with the rope passing through the center of the ring. The ring is frictionless, so that it can slide freely along the rope until the piñata comes to a point of static equilibrium.

a) Determine the distance from the top of the left (lower) pole to the ring when the piñata is in static equilibrium.

b) What is the tension in the rope when the piñata is at this point of static equilibrium?

••4.47 Three objects with masses $m_1 = 36.5$ kg, $m_2$ 19.2 kg, and $m_3 = 12.5$ kg are hanging from ropes that run over pulleys. What is the acceleration of $m_1$?

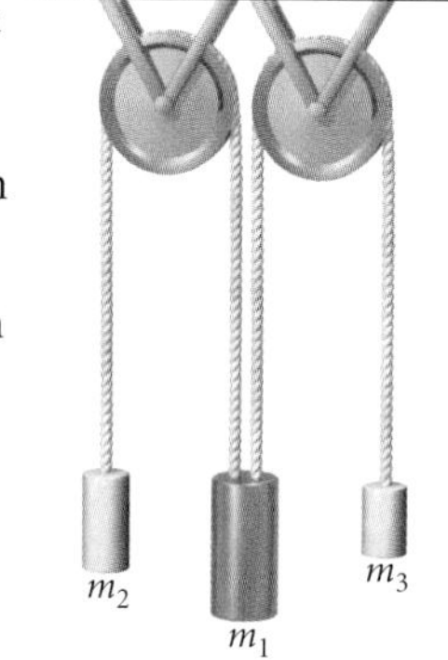


••**4.48** A rectangular block of width $w = 116.5$ cm, depth $d = 164.8$ cm, and height $h = 105.1$ cm is cut diagonally from one upper corner to the opposing lower corners so that a triangular surface is generated, as shown in the figure. A paperweight of mass $m = 16.93$ kg is sliding down the incline without friction. What is the magnitude of the acceleration that the paperweight experiences?

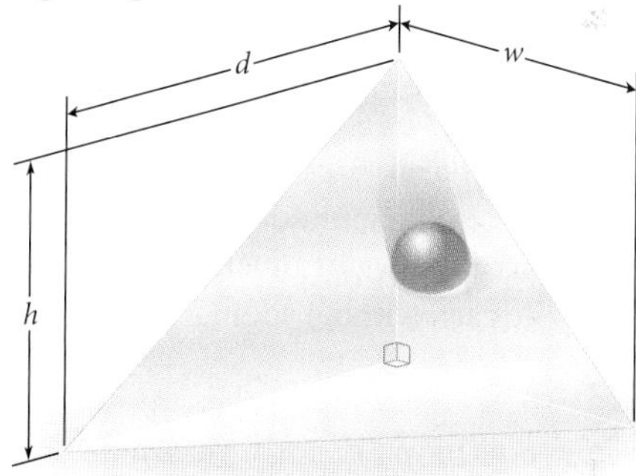


••**4.49** A large cubical block of ice of mass $M = 64$ kg and sides of length $L = 0.40$ m is held stationary on a frictionless ramp. The ramp is at an angle of $\theta = 26°$ above the horizontal. The ice cube is held in place by a rope of negligible mass and length $l = 1.6$ m. The rope is attached to the surface of the ramp and to the upper edge of the ice cube, a distance $L$ above the surface of the ramp. Find the tension in the rope.

••**4.50** A bowling ball of mass $M_1 = 6.0$ kg is initially at rest on the sloped side of a wedge of mass $M_2 = 9.0$ kg that is on a frictionless horizontal floor. The side of the wedge is sloped at an angle of $\theta = 36.9°$ above the horizontal.

a) With what magnitude of horizontal force should the wedge be pushed to keep the bowling ball at a constant height on the slope?

b) What is the magnitude of the acceleration of the wedge, if no external force is applied?

## Section 4.7

4.51 A skydiver of mass 82.3 kg (including outfit and equipment) floats downward suspended from her parachute, having reached terminal speed. The drag coefficient is 0.533, and the area of her parachute is 20.11 m$^2$. The density of air is 1.14 kg/m$^3$. What is the air's drag force on her?

4.52 The elapsed time for a top fuel dragster to start from rest and travel in a straight line a distance of $\frac{1}{4}$ mile (402 m) is 4.41 s. Find the minimum coefficient of friction between the tires and the track needed to achieve this result. (Note that the minimum coefficient of friction is found from the simplifying assumption that the dragster accelerates with constant acceleration.)

4.53 An engine block of mass $M$ is on the flatbed of a pickup truck that is traveling in a straight line down a level road with an initial speed of 30 m/s. The coefficient of static friction between the block and the bed is $\mu_s = 0.540$. Find the minimum distance in which the truck can come to a stop without the engine block sliding toward the cab.

•4.54 A box of books is initially at rest a distance $D = 0.540$ m from the end of a wooden board. The coefficient of static friction between the box and the board is $\mu_s = 0.320$, and the coefficient of kinetic friction is $\mu_k = 0.250$. The angle of the board is increased slowly, until the box just begins to slide; then the board is held at this angle. Find the speed of the box as it reaches the end of the board.

•4.55 A block of mass $M_1 = 0.640$ kg is initially at rest on a cart of mass $M_2 = 0.320$ kg with the cart initially at rest on a level air track. The coefficient of static friction between the block and the cart is $\mu_s = 0.620$, but there is essentially no friction between the air track and the cart. The cart is accelerated by a force of magnitude $F$ parallel to the air track. Find the maximum value of $F$ that allows the block to accelerate with the cart, without sliding on top of the cart.

## Section 4.8

4.56 Coffee filters behave like small parachutes, with a drag force that is proportional to the velocity squared, $F_{drag} = Kv^2$. A single coffee filter, when dropped from a height of 2.0 m, reaches the ground in a time of 3.0 s. When a second coffee filter is nestled within the first, the drag force remains the same, but the weight is doubled. Find the time for the combined filters to reach the ground. (Neglect the brief period when the filters are accelerating up to their terminal speed.)

4.57 Your refrigerator has a mass of 112.2 kg, including the food in it. It is standing in the middle of your kitchen, and you need to move it. The coefficients of static and kinetic friction between the fridge and the tile floor are 0.46 and 0.37, respectively. What is the magnitude of the force of friction acting on the fridge, if you push against it horizontally with a force of each magnitude?

a) 300 N b) 500 N c) 700N

•4.58 On the bunny hill at a ski resort, a towrope pulls the skiers up the hill with constant speed of 1.74 m/s. The slope of the hill is 12.4° with respect to the horizontal. A child is being pulled up the hill. The coefficients of static and kinetic friction between the child's skis and the snow are 0.152 and 0.104, respectively, and the child's mass is 62.4 kg, including the clothing and equipment. What is the force with which the towrope has to pull on the child?

•4.59 A skier starts with a speed of 2.0 m/s and skis straight down a slope with an angle of 15.0° relative to the horizontal. The coefficient of kinetic friction between her skis and the snow is 0.100. What is her speed after 10.0 s?

••4.60 A block of mass $m_1 = 21.9$ kg is at rest on a plane inclined at $\theta = 30.0°$ above the horizontal. The block is connected via a rope and massless pulley system to another block of mass $m_2 = 25.1$ kg, as shown in the figure. The coefficients of static and kinetic friction between block 1 and the inclined plane are $\mu_s = 0.109$ and $\mu_k = 0.086$, respectively. If the blocks are released from rest, what is the displacement of block 2 in the vertical direction after 1.51 s? Use positive numbers for the upward direction and negative numbers for the downward direction.

••4.61 A wedge of mass $m = 36.1$ kg is located on a plane that is inclined by an angle $\theta = 21.3°$ with respect to the horizontal. A force $F = 302.3$ N in the horizontal direction pushes on the wedge, as shown in the figure. The coefficient of kinetic friction between the wedge and the plane is 0.159. What is the acceleration of the wedge along the plane?

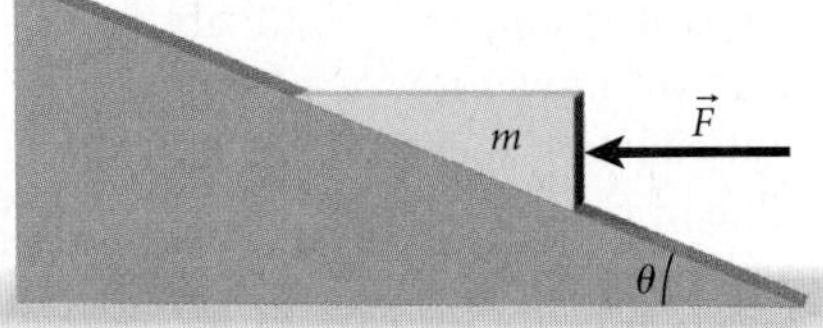


••4.62 A chair of mass $M$ rests on a level floor, with a coefficient of static friction $\mu_s = 0.560$ between the chair and the floor. A person wishes to push the chair across the floor. He pushes on the chair with a force $F$ at an angle $\theta$ below the horizontal. What is the maximum value of $\theta$ for which the chair will not start to move across the floor?

••4.63 As shown in the figure, blocks of masses $m_1 = 250.0$ g and $m_2 = 500.0$ g are attached by a massless string over a frictionless and massless pulley. The coefficients of static and kinetic friction between the block and inclined plane are 0.250 and 0.123, respectively. The angle of the incline is $\theta = 30.0°$, and the blocks are at rest initially.

a) In which direction do the blocks move?

b) What is the acceleration of the blocks?

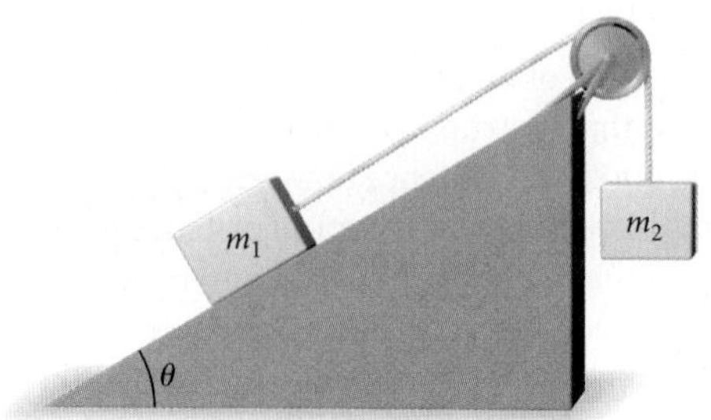

••**4.64** A block of mass $M = 500.0$ g sits on a horizontal tabletop. The coefficients of static and kinetic friction are 0.53 and 0.41, respectively, at the contact surface between table and block. The block is pushed on with a 10.0 N external force at an angle $\theta$ with the horizontal.

a) What angle will lead to the maximum acceleration of the block for a given pushing force?

b) What is the maximum acceleration?

## Additional Problems

**4.65** A car without ABS (antilock brake system) was moving at 15.0 m/s when the driver hit the brake to make a sudden stop. The coefficients of static and kinetic friction between the tires and the road are 0.550 and 0.430, respectively.

a) What was the acceleration of the car during the interval between braking and stopping?

b) How far did the car travel before it stopped?

**4.66** A 2.0-kg block ($M_1$) and a 6.0-kg block ($M_2$) are connected by a massless string. Applied forces, $F_1 = 10$ N and $F_2 = 5.0$ N, act on the blocks, as shown in the figure.

a) What is the acceleration of the blocks?

b) What is the tension in the string?

c) What is the net force acting on $M_1$? (Neglect friction between the blocks and the table.)

**4.67** An elevator contains two masses: $M_1 = 2.0$ kg is attached by a string (string 1) to the ceiling of the elevator, and $M_2 = 4.0$ kg is attached by a similar string (string 2) to the bottom of mass 1.

a) Find the tension in string 1($T_1$) if the elevator is moving upward at a constant velocity of $v = 3.0$ m/s.

b) Find $T_1$ if the elevator is accelerating upward with an acceleration of $a = 3.0$ m/s$^2$.

**4.68** What coefficient of friction is required to stop a hockey puck sliding at 12.5 m/s initially over a distance of 60.5 m?

**4.69** A spring of negligible mass is attached to the ceiling of an elevator. When the elevator is stopped at the first floor, a mass $M$ is attached to the spring, stretching the spring a distance $D$ until the mass is in equilibrium. As the elevator starts upward toward the second floor, the spring stretches an additional distance $D/4$. What is the magnitude of the acceleration of the elevator? Assume the force provided by the spring is linearly proportional to the distance stretched by the spring.

**4.70** A crane of mass $M = 10{,}000$ kg lifts a wrecking ball of mass $m = 1200$ kg directly upward.

a) Find the magnitude of the normal force exerted on the crane by the ground while the wrecking ball is moving upward at a constant speed of $v = 1.0$ m/s.

b) Find the magnitude of the normal force if the wrecking ball's upward motion slows at a constant rate from its initial speed $v = 1.0$ m/s to a stop over a distance $D = 0.25$ m.

**4.71** A block of mass 20.0 kg supported by a vertical massless cable is initially at rest. The block is then pulled upward with a constant acceleration of 2.32 m/s$^2$.

a) What is the tension in the cable?

b) What is the net force acting on the mass?

c) What is the speed of the block after it has traveled 2.00 m?

**4.72** Three identical blocks, A, B, and C, are on a horizontal frictionless table. The blocks are connected by strings of negligible mass, with block B between the other two blocks. If block C is pulled horizontally by a force of magnitude $F = 12$ N, find the tension in the string between blocks B and C.

•**4.73** A block of mass $m_1 = 3.00$ kg and a block of mass $m_2 = 4.00$ kg are suspended by a massless string over a frictionless pulley with negligible mass, as in an Atwood machine. The blocks are held motionless and then released. What is the acceleration of the two blocks?

•**4.74** Two blocks of masses $m_1$ and $m_2$ are suspended by a massless string over a frictionless pulley with negligible mass, as in an Atwood machine. The blocks are held motionless and then released. If $m_1 = 3.5$ kg, what value does $m_2$ have to have in order for the system to experience an acceleration $a = 0.4\,g$? (*Hint:* There are two solutions to this problem.)

•**4.75** A tractor pulls a sled of mass $M = 1000$ kg across level ground. The coefficient of kinetic friction between the sled and the ground is $\mu_k = 0.600$. The tractor pulls the sled by a rope that connects to the sled at an angle of $\theta = 30°$ above the horizontal. What magnitude of tension in the rope is necessary to move the sled horizontally with an acceleration $a = 2.0$ m/s$^2$?

•**4.76** A 2.00 kg block is on a plane inclined at 20.0° with respect to the horizontal. The coefficient of static friction between the block and the plane is 0.60.

a) How many forces are acting on the block?

b) What is the normal force?

c) Is this block moving? Explain.

•**4.77** A block of mass 5.00 kg is sliding at a constant velocity down an inclined plane that makes an angle of 37° with respect to the horizontal.

a) What is the friction force?

b) What is the coefficient of kinetic friction?

•**4.78** A skydiver of mass 83.7 kg (including outfit and equipment) falls in the spread-eagle position, having reached terminal speed. Her drag coefficient is 0.587, and her surface area that is exposed to the air stream is 1.035 $m_2$. How long does it take her to fall a vertical distance of 296.7 m? (The density of air is 1.14 kg/m$^3$.)

•**4.79** A 0.50 kg physics textbook is hanging from two massless wires of equal length attached to a ceiling. The tension on each wire is measured as 15.4 N. What is the angle of the wires with the horizontal?

•4.80 In the figure, an external force $\vec{F}$ is holding a bob of mass 500 g in a stationary position. The angle that the massless rope makes with the vertical is $\theta = 30°$.

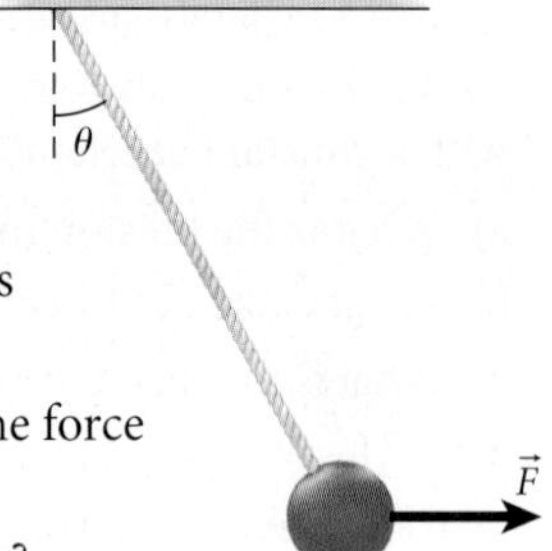


a) What is the magnitude, $F$, of the force needed to maintain equilibrium?

b) What is the tension in the rope?

•4.81 In a physics class, a 2.70-g ping-pong ball was suspended from a massless string. The string makes an angle of $\theta = 15.0°$ with the vertical when air is blown horizontally at the ball at a speed of 20.5 m/s. Assume that the friction force is proportional to the squared speed of the air stream.

a) What is the proportionality constant in this experiment?

b) What is the tension in the string?

•4.82 A nanowire is a (nearly) one-dimensional structure with a diameter on the order of a few nanometers. Suppose a 100.0-nm-long nanowire made of pure silicon (density of Si = 2.33 g/cm$_3$) has a diameter of 5.0 nm. This nanowire is attached at the top and hanging down vertically due to the force of gravity.

a) What is the tension at the top?

b) What is the tension in the middle?

(*Hint:* Treat the nanowire as a cylinder of diameter 5.0 nm and length 100.0 nm, made of silicon.)

•4.83 Two blocks are stacked on a frictionless table, and a horizontal force $F$ is applied to the top block (block 1). Their masses are $m_1 = 2.50$ kg and $m_2 = 3.75$ kg. The coefficients of static and kinetic friction between the blocks are 0.456 and 0.380, respectively.

a) What is the maximum applied force $F$ for which $m_1$ will not slide off $m_2$?

b) What are the accelerations of $m_1$ and $m_2$ when $F = 24.5$ N is applied to $m_1$?

•4.84 Two blocks ($m_1 = 1.23$ kg and $m_2 = 2.46$ kg) are glued together and are moving downward on an inclined plane having an angle of 40.0° with respect to the horizontal. Both blocks are lying flat on the surface of the inclined plane. The coefficients of kinetic friction are 0.23 for $m_1$ and 0.35 for $m_2$. What is the acceleration of the blocks?

•4.85 A marble block of mass $m_1 = 567.1$ kg and a granite block of mass $m_2 = 266.4$ kg are connected to each other by a rope that runs over a pulley, as shown in the figure. Both blocks are located on inclined planes, with angles $\alpha = 39.3°$ and $\beta = 53.2°$. Both blocks move without friction, and the rope glides over the pulley without friction. What is the acceleration of the marble block? Note that the positive $x$-direction is indicated in the figure.

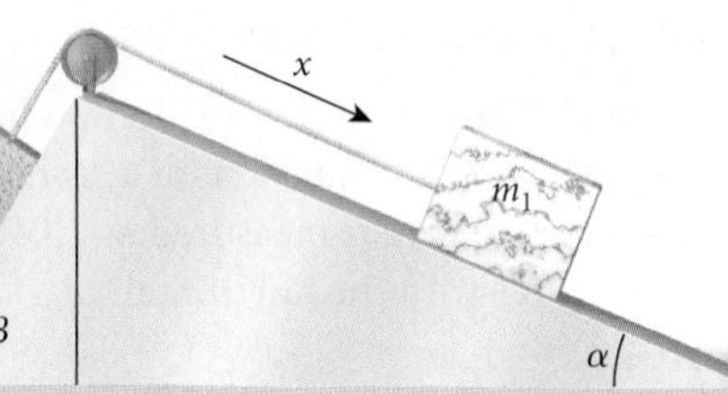


••4.86 A marble block of mass $m_1 = 559.1$ kg and a granite block of mass $m_2 = 128.4$ kg are connected to each other by a rope that runs over a pulley as shown in the figure. Both blocks are located on inclined planes with angles $\alpha = 38.3°$ and $\beta = 57.2°$. The rope glides over the pulley without friction, but the coefficient of friction between block 1 and the inclined plane is $\mu_1 = 0.13$, and that between block 2 and the inclined plane is $\mu_2 = 0.31$. (For simplicity, assume that the coefficients of static and kinetic friction are the same in each case.) What is the acceleration of the marble block? Note that the positive $x$-direction is indicated in the figure.

••4.87 As shown in the figure, two masses, $m_1 = 3.50$ kg and $m_2 = 5.00$ kg, are on a frictionless tabletop and mass $m_3 = 7.60$ kg is hanging from $m_1$. The coefficients of static and kinetic friction between $m_1$ and $m_2$ are 0.60 and 0.50, respectively.

a) What are the accelerations of $m_1$ and $m_2$?

b) What is the tension in the string between $m_1$ and $m_3$?

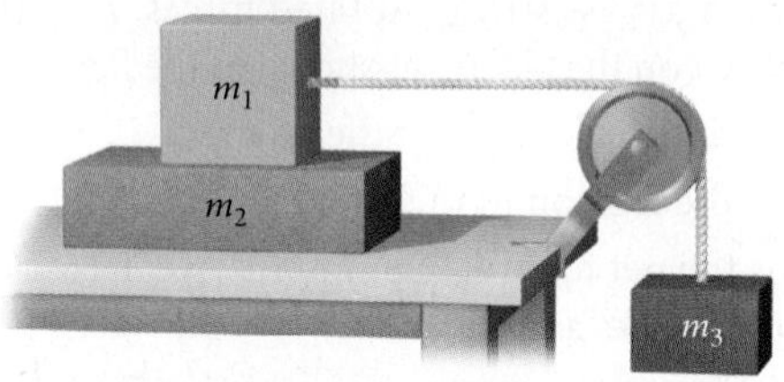


••4.88 A block of mass $m_1 = 2.30$ kg is placed in front of a block of mass $m_2 = 5.20$ kg, as shown in the figure. The coefficient of static friction between $m_1$ and $m_2$ is 0.65, and there is negligible friction between the larger block and the tabletop.

a) What forces are acting on $m_1$?

b) What is the minimum external force $F$ that can be applied to $m_2$ so that $m_1$ does not fall?

c) What is the contact force between $m_1$ and $m_2$?

d) What is the net force acting on $m_2$ when the force found in part (b) is applied?

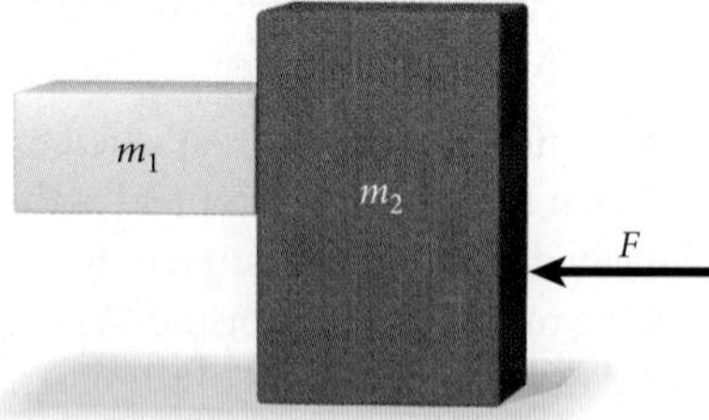


••4.89 A suitcase of weight $Mg = 450$ N is being pulled by a small strap across a level floor. The coefficient of kinetic friction between the suitcase and the floor is $\mu_k = 0.640$.

a) Find the optimal angle of the strap above the horizontal. (The optimal angle minimizes the force necessary to pull the suitcase at constant speed.)

b) Find the minimum tension in the strap needed to pull the suitcase at constant speed.

••4.90 As shown in the figure, a block of mass $M_1 = 0.450$ kg is initially at rest on a slab of mass $M_2 = 0.820$ kg, and the slab is initially at rest on a level table. A string of negligible mass is connected to the slab, runs over a frictionless pulley on the edge of the table, and is attached to a hanging mass $M_3$. The block rests on the slab but is not tied to the string, so friction provides the only horizontal force on the block. The slab has a coefficient of kinetic friction $\mu_k = 0.340$ and a coefficient of static friction $\mu_s = 0.560$ with both the table and the block. When released, $M_3$ pulls on the string and accelerates the slab, which accelerates the block. Find the maximum mass of $M_3$ that allows the block to accelerate with the slab, without sliding on top of the slab.

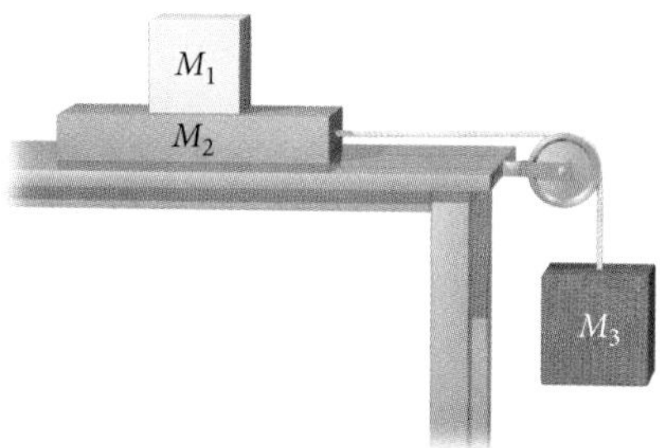


••4.91 As shown in the figure, a block of mass $M_1 = 0.250$ kg is initially at rest on a slab of mass $M_2 = 0.420$ kg, and the slab is initially at rest on a level table. A string of negligible mass is connected to the slab, runs over a frictionless pulley on the edge of the table, and is attached to a hanging mass $M_3 = 1.80$ kg. The block rests on the slab but is not tied to the string, so friction provides the only horizontal force on the block. The slab has a coefficient of kinetic friction $\mu_k = 0.340$ with both the table and the block. When released, $M_3$ pulls on the string, which accelerates the slab so quickly that the block starts to slide on the slab. Before the block slides off the top of the slab:

a) Find the magnitude of the acceleration of the block.

b) Find the magnitude of the acceleration of the slab.

# 5 Kinetic Energy, Work, and Power

**FIGURE 5.1** A composite image of NASA satellite photographs taken at night. Photos were taken from November, 1994 through March, 1995.

## WHAT WE WILL LEARN

- Kinetic energy is the energy associated with the motion of an object.
- Work is energy transferred to an object or transferred from an object due to the action of an external force. Positive work transfers energy to the object, and negative work transfers energy from the object.
- Work is the scalar product of the force vector and the displacement vector.
- The change in kinetic energy due to applied forces is equal to the work done by the forces.
- Power is the rate at which work is done.
- The power provided by a constant force acting on an object is the scalar product of the velocity vector of that object and the force vector.

Figure 5.1 is a composite image of satellite photographs taken at night, showing which parts of the world use the most energy for nighttime illumination. Not surprisingly, the United States, Western Europe, and Japan stand out. The amount of light emitted by a region during the night is a good measure of the amount of energy that region consumes.

In physics, energy has a fundamental significance: Practically no physical activity takes place without the expenditure or transformation of energy. Calculations involving the energy of a system are of primary importance in all of science and engineering. As we'll see in this chapter, problem-solving methods involving energy provide an alternative to working with Newton's laws and are often simpler and easier to use.

This chapter presents the concepts of kinetic energy, work, and power and introduces some techniques that use these ideas, such as the work–kinetic energy theorem, to solve several types of problems. Chapter 6 will introduce additional types of energy and expand the work–kinetic energy theorem to cover these; it will also discuss one of the great ideas in physics and, indeed, in all of science: the law of conservation of energy.

## 5.1 Energy in Our Daily Lives

No physical quantity has a greater importance in our daily lives than energy. Energy consumption, energy efficiency, and energy "production" are all of the utmost economic importance and the focus of heated discussions about national policies and international agreements. (The word production is in quotes because energy is not produced but rather is converted from a less usable form to a more usable form.) Energy also has an important role in each individual's daily routine: energy intake through food calories and energy consumption through cellular processes, activities, work, and exercise. Weight loss or weight gain is ultimately due to an imbalance between energy intake and use.

Energy has many forms and requires several different approaches to cover completely. Thus, energy is a recurring theme throughout this book. We start in this chapter and the next by investigating forms of mechanical energy: kinetic energy and potential energy. Thermal energy, another form of energy, is one of the central pillars of thermodynamics. Chemical energy is stored in chemical compounds, and chemical reactions can either consume energy from the environment (endothermic reactions) or yield usable energy to the surroundings (exothermic reactions). Our petroleum economy makes use of chemical energy and its conversion to mechanical energy and heat, which is another form of energy (or energy transfer).

**FIGURE 5.2** Wind farms harvest renewable energy.

In Chapter 31, we will see that electromagnetic radiation contains energy. This energy is the basis for one renewable form of energy—solar energy. Almost all other renewable energy sources on Earth can be traced back to solar energy. Solar energy is responsible for the wind that drives large wind turbines (Figure 5.2). The Sun's radiation is also responsible for evaporating water from the Earth's surface and moving it into the clouds, from which it falls down as rain and eventually joins rivers that can be dammed (Figure 5.3) to extract energy. Biomass, another form of renewable energy, depends on the ability of plants and animals to store solar energy during their metabolic and growth processes.

(a)

(b)

**FIGURE 5.3** Dams provide renewable electrical energy. (a) The Grand Coulee Dam on the Columbia River in Washington. (b) The Itaipú Dam on the Paraná River in Brazil and Paraguay.

(a)

(b)

**FIGURE 5.4** (a) Solar farm with an adjustable array of mirrors; (b) solar panel.

In fact, the energy radiated onto the surface of Earth by the Sun exceeds the energy needs of the entire human population by a factor of more than 10,000. It is possible to convert solar energy directly into electrical energy by using photovoltaic cells (Figure 5.4b). Current research efforts to increase the efficiency and reliability of these photocells while reducing their cost are intense. Versions of solar cells are already being used for some practical purposes, for example, in patio and garden lights. Experimental solar farms like the one in Figure 5.4a are in operation as well. The chapter on quantum physics (Chapter 36) will discuss in detail how photocells work. Problems with using solar energy are that it is not available at night, has seasonal variations, and is strongly reduced in cloudy conditions or bad weather. Depending on the installation and conversion methods used, present solar devices convert only 10–15% of solar energy into electrical energy; increasing this fraction is a key goal of present research activity. Materials with 30% or higher yield of electrical energy from solar energy have been developed in the laboratory but are still not deployed on an industrial scale. Biomass, in comparison, has much lower efficiencies of solar energy capture, on the order of 1% or less.

In Chapter 35 on relativity, we will see that energy and mass are not totally separate concepts but are related to each other via Einstein's famous formula $E = mc^2$. When we study nuclear physics (Chapter 40), we will find that splitting massive atomic nuclei (such as uranium or plutonium) liberates energy. Conventional nuclear power plants are based on this physical principle, called *nuclear fission.* We can also obtain useful energy by merging atomic nuclei with very small masses (hydrogen, for example) into more massive nuclei, a process called *nuclear fusion.* The Sun and most other stars in the universe use nuclear fusion to generate energy.

The energy from nuclear fusion is thought by many to be the most likely means of satisfying the long-term energy needs of modern industrialized society. Perhaps the most likely approach to achieving progress toward controlled fusion reactions is the proposed international nuclear fusion reactor facility ITER ("the way" in Latin), which will be constructed in France. But there are other promising approaches to solving the problem of how to use nuclear fusion, for example, the National Ignition Facility (NIF), opened in May 2009 at Lawrence Livermore National Laboratory in California. We will discuss these technologies in greater detail in Chapter 40.

Related to energy are work and power. We all use these words informally, but this chapter will explain how these quantities relate to energy in precise physical and mathematical terms.

You can see that energy occupies an essential place in our lives. One of the goals of this book is to give you a solid grounding in the fundamentals of energy science. Then you will be able to participate in some of the most important policy discussions of our time in an informed manner.

A final question remains: What is energy? In many textbooks, energy is defined as the ability to do work. However, this definition only shifts the mystery without giving a deeper explanation. And the truth is that there is no deeper explanation. In his famous *Feynman Lectures on Physics,* the Nobel laureate and physics folk hero Richard Feynman wrote in 1963: "It is important to realize that in physics today, we have no knowledge of what energy *is.* We do not have a picture that energy comes in little blobs of a definite amount. It is not that way. However, there are formulas for calculating some numerical quantity, and when we add it all together it gives '28'—always the same number. It is an abstract thing in that it does not tell us the mechanism or the *reasons* for the various formulas." More than four decades later, this has not changed. The concept of energy and, in particular, the law of energy conservation (see Chapter 6), are extremely useful tools for figuring out the behavior of systems. But no one has yet given an explanation as to the true nature of energy.

## 5.2 Kinetic Energy

The first kind of energy we'll consider is the energy associated with the motion of a moving object: **kinetic energy.** Kinetic energy is defined as one-half the product of a moving object's mass and the square of its speed:

$$K = \tfrac{1}{2}mv^2. \tag{5.1}$$

Note that, by definition, kinetic energy is always positive or equal to zero, and it is only zero for an object at rest. Also note that kinetic energy, like all forms of energy, is a scalar, not a vector quantity. Because it is the product of mass (kg) and speed squared (m/s · m/s), the units of kinetic energy are kg m$^2$/s$^2$. Because energy is such an important quantity, it has its own SI unit, the **joule (J).** The SI force unit, the newton, is 1 N = 1 kg m$^2$/s$^2$, and we can make a useful conversion:

$$\text{Energy unit:}\quad 1\,\text{J} = 1\,\text{N m} = 1\,\text{kg m}^2/\text{s}^2. \tag{5.2}$$

Let's look at a few sample energy values to get a feeling for the size of the joule. A car of mass 1310 kg being driven at the speed limit of 55 mph (24.6 m/s) has a kinetic energy of

$$K_{\text{car}} = \tfrac{1}{2}mv^2 = \tfrac{1}{2}(1310\text{ kg})(24.6\text{ m/s})^2 = 4.0 \cdot 10^5\text{ J}.$$

The mass of the Earth is $6.0 \cdot 10^{24}$ kg, and it orbits the Sun with a speed of $3.0 \cdot 10^4$ m/s. The kinetic energy associated with this motion is $2.7 \cdot 10^{33}$ J. A person of mass 64.8 kg jogging at 3.50 m/s has a kinetic energy of 400 J, and a baseball (mass of "5 ounces avoirdupois" = 0.142 kg) thrown at 80 mph (35.8 m/s) has a kinetic energy of 91 J. On the atomic scale, the average kinetic energy of an air molecule is $6.1 \cdot 10^{-21}$ J, as we will see in Chapter 19. The typical magnitudes of kinetic energies of some moving objects are presented in Figure 5.5. You can see from these examples that the range of energies involved in physical processes is very large.

**FIGURE 5.5** Range of kinetic energies displayed on a logarithmic scale. The kinetic energies (left to right) of an air molecule, a red blood cell traveling through the aorta, a mosquito in flight, a thrown baseball, a moving car, and the Earth orbiting the Sun are compared with the energy released from a 15-Mt nuclear explosion and that of a supernova, which emits particles with a total kinetic energy of approximately $10^{46}$ J.

Some other frequently used energy units are the electron-volt (eV), the food calorie (Cal), and the mega-ton of TNT (Mt):

$$1\text{ eV} = 1.602 \cdot 10^{-19}\text{ J}$$
$$1\text{ Cal} = 4186\text{ J}$$
$$1\text{ Mt} = 4.18 \cdot 10^{15}\text{ J}.$$

On the atomic scale, 1 electron-volt (eV) is the kinetic energy that an electron gains when accelerated by an electric potential of 1 volt. The energy content of the food we eat is usually (and mistakenly) given in terms of calories but should be given in food calories. As we'll see when we study thermodynamics, 1 food calorie is equal to 1 kilocalorie. On a larger scale, 1 Mt is the energy released by exploding 1 million metric tons of the explosive TNT, an energy release achieved only by nuclear weapons or by catastrophic natural events such as the impact of a large asteroid. For comparison, in 2007, the annual energy consumption by all humans on Earth reached $5 \cdot 10^{20}$ J. (All of these concepts will be discussed further in subsequent chapters.)

For motion in more than one dimension, we can write the total kinetic energy as the sum of the kinetic energies associated with the components of velocity in each spatial direction. To show this, we start with the definition of kinetic energy (equation 5.1) and then use $v^2 = v_x^2 + v_y^2 + v_z^2$:

$$K = \tfrac{1}{2}mv^2 = \tfrac{1}{2}m\left(v_x^2 + v_y^2 + v_z^2\right) = \tfrac{1}{2}mv_x^2 + \tfrac{1}{2}mv_y^2 + \tfrac{1}{2}mv_z^2. \tag{5.3}$$

(*Note:* Kinetic energy is a scalar, so these components are not added like vectors but simply by taking their algebraic sum.) Thus, we can think of kinetic energy as the sum of the kinetic energies associated with the motion in the $x$-direction, $y$-direction, and $z$-direction. This concept is particularly useful for ideal projectile problems, where the motion consists of free fall in the vertical direction ($y$-direction) and motion with constant velocity in the horizontal direction ($x$-direction).

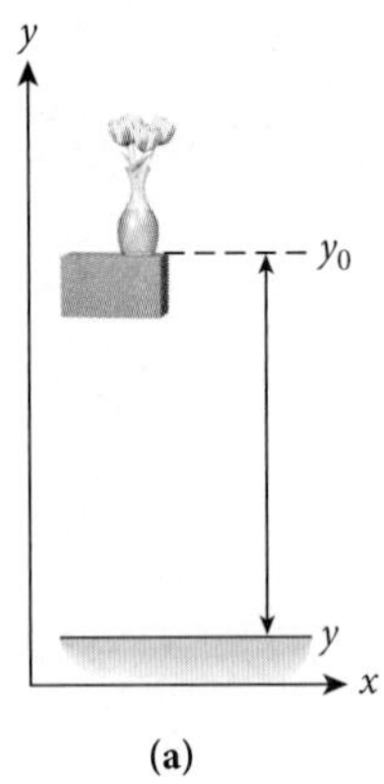

(a)

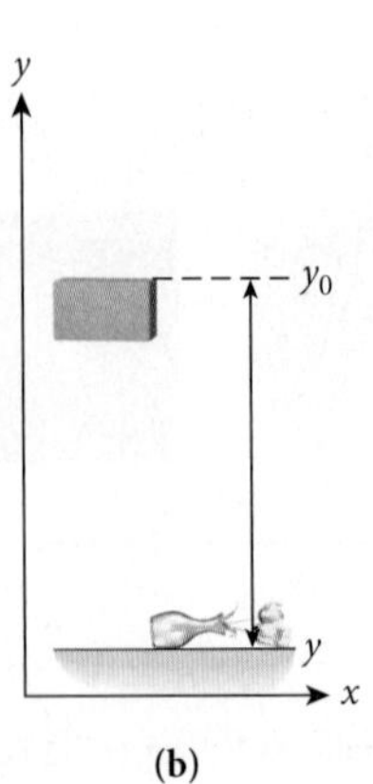

(b)

**FIGURE 5.6** (a) A vase is released from rest at a height of $y_0$. (b) The vase falls to the floor, which has a height of $y$.

## EXAMPLE 5.1 Falling Vase

### PROBLEM

A crystal vase (mass = 2.40 kg) is dropped from a height of 1.30 m and falls to the floor, as shown in Figure 5.6. What is its kinetic energy just before impact? (Neglect air resistance for now.)

### SOLUTION

Once we know the velocity of the vase just before impact, we can put it into the equation defining kinetic energy. To obtain this velocity, we recall the kinematics of free-falling objects. In this case, it is most straightforward to use the relationship between the initial and final velocities and heights that we derived in Chapter 2 for free-fall motion:

$$v_y^2 = v_{y0}^2 - 2g(y - y_0).$$

(Remember that the $y$-axis must be pointing up to use this equation.) Because the vase is released from rest, the initial velocity components are $v_{x0} = v_{y0} = 0$. Because there is no acceleration in the $x$-direction, the $x$-component of velocity remains zero during the fall of the vase: $v_x = 0$. Therefore, we have

$$v^2 = v_x^2 + v_y^2 = 0 + v_y^2 = v_y^2.$$

We then obtain

$$v^2 = v_y^2 = 2g(y_0 - y).$$

We use this result in equation 5.1:

$$K = \tfrac{1}{2}mv^2 = \tfrac{1}{2}m\big(2g(y_0 - y)\big) = mg(y_0 - y).$$

Inserting the numbers given in the problem gives us the answer:

$$K = (2.40\ \text{kg})(9.81\ \text{m/s}^2)(1.30\ \text{m}) = 30.6\ \text{J}.$$

## 5.3 Work

In Example 5.1, the vase started out with zero kinetic energy, just before it was released. After falling a distance of 1.30 m, it had acquired a kinetic energy of 30.6 J. The greater the height from which the vase is released, the greater the speed the vase will attain (ignoring air resistance), and therefore the greater its kinetic energy becomes. In fact, as we found in Example 5.1, the kinetic energy of the vase depends linearly on the height from which it falls: $K = mg(y_0 - y)$.

The gravitational force, $\vec{F}_g = -mg\hat{y}$, accelerates the vase and therefore gives it its kinetic energy. We can see from the equation above that the kinetic energy also depends linearly on the magnitude of the gravitational force. Doubling the mass of the vase would double the gravitational force acting on it and thus double its kinetic energy.

Because the speed of an object can be increased or decreased by accelerating or decelerating it, respectively, its kinetic energy also changes in this process. For the vase, we have just seen that the force of gravity is responsible for this change. We account for a change in the kinetic energy of an object caused by a force with the concept of work, $W$.

### Definition

**Work** is the energy transferred to or from an object due to the action of a force. Positive work is a transfer of energy to the object, and negative work is a transfer of energy from the object.

The vase gained kinetic energy from positive work done by the gravitational force and so $W_g = mg\,(y_0 - y)$.

Note that this definition is not restricted to kinetic energy. The relationship between work and energy described in this definition holds in general for different forms of energy besides kinetic energy. This definition of work is not exactly the same as the meaning attached to the word **work** in everyday language. The work being considered in this chapter is mechanical work in connection with energy transfer. However, the work, physical as well as mental, that we commonly speak of does not necessarily involve the transfer of energy.

## 5.4 Work Done by a Constant Force

Suppose we let the vase of Example 5.1 slide, from rest, along an inclined plane that has an angle $\theta$ with respect to the horizontal (Figure 5.7). For now, we neglect the friction force, but we will come back to it later. As we showed in Chapter 4, in the absence of friction, the acceleration along the plane is given by $a = g\sin\theta = g\cos\alpha$. (Here the angle $\alpha = 90° - \theta$ is the angle between the gravitational force vector and the displacement vector; see Figure 5.7.)

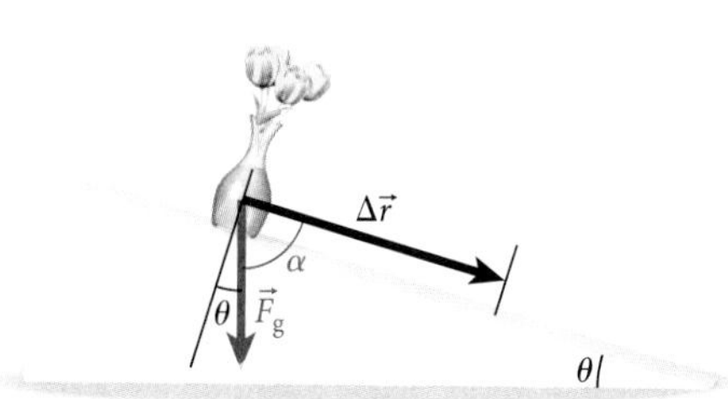


**FIGURE 5.7** Vase sliding without friction on an inclined plane.

We can determine the kinetic energy the vase has in this situation as a function of the displacement, $\Delta\vec{r}$. Most conveniently, we can perform this calculation by using the relationship between the squares of initial and final velocities, the displacement, and the acceleration, which we obtained for one-dimensional motion in Chapter 2:

$$v^2 = v_0^2 + 2a\Delta r.$$

## 5.1 Self-Test Opportunity

Draw the free-body diagram for the vase that is sliding down the inclined plane.

We set $v_0 = 0$ because we are again assuming that the vase is released from rest, that is, with zero kinetic energy. Then we use the expression for the acceleration, $a = g\cos\alpha$, that we just obtained. Now we have

$$v^2 = 2g\cos\alpha\Delta r \Rightarrow K = \tfrac{1}{2}mv^2 = mg\Delta r\cos\alpha.$$

The kinetic energy transferred to the vase was the result of positive work done by the gravitational force and so

$$\Delta K = mg\Delta r\cos\alpha = W_g. \tag{5.4}$$

Let's look at two limiting cases of equation 5.4:

- For $\alpha = 0$, both the gravitational force and the displacement are in the negative $y$-direction. Thus, these vectors are parallel, and we have the result we already derived for the case of the vase falling under the influence of gravity, $W_g = mg\Delta r$.
- For $\alpha = 90°$, the gravitational force is still in the negative $y$-direction, but the vase cannot move in the negative $y$-direction because it is sitting on the horizontal surface of the plane. Hence, there is no change in the kinetic energy of the vase, and there is no work done by the gravitational force on the vase; that is, $W_g = 0$. The work done on the vase by the gravitational force is also zero, if the vase moves at a constant speed along the surface of the plane.

Because $mg = \left|\vec{F}_g\right|$ and $\Delta r = \left|\Delta\vec{r}\right|$, we can write the work done on the vase as $W = \left|\vec{F}\right|\left|\Delta\vec{r}\right|\cos\alpha$. From the two limiting cases we have just discussed, we gain confidence that we can use the equation we have just derived for motion on an inclined plane as the definition of the work done by a constant force:

$$W = \left|\vec{F}\right|\left|\Delta\vec{r}\right|\cos\alpha, \quad \text{where } \alpha \text{ is the angle between } \vec{F} \text{ and } \Delta\vec{r}.$$

This equation for the work done by a constant force acting over some spatial displacement holds for all constant force vectors, arbitrary displacement vectors, and angles between the two. Figure 5.8 shows three cases for the work done by a force $\vec{F}$ acting over a displacement $\vec{r}$. In Figure 5.8a, the maximum work is done because $\alpha = 0$ and $\vec{F}$ and $\vec{r}$ are in the same direction. In Figure 5.8b, $\vec{F}$ is at an arbitrary angle $\alpha$ with respect to $\vec{r}$. In Figure 5.8c, no work is done because $\vec{F}$ is perpendicular to $\vec{r}$.

(a)

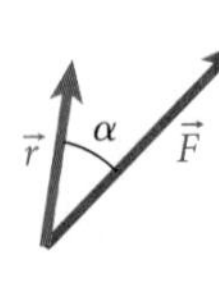


(b)

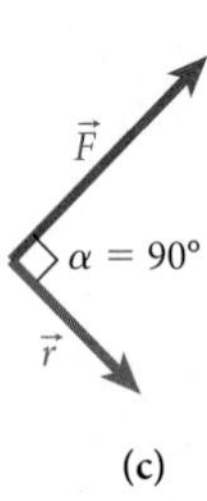


(c)

**FIGURE 5.8** (a) $\vec{F}$ is parallel to $\vec{r}$ and $W = \left|\vec{F}\right|\left|\vec{r}\right|$. (b) The angle between $\vec{F}$ and $\vec{r}$ is $\alpha$ and $W = \left|\vec{F}\right|\left|\vec{r}\right|\cos\alpha$. (c) $\vec{F}$ is perpendicular to $\vec{r}$ and $W = 0$.

## Mathematical Insert: Scalar Product of Vectors

In Section 1.6, we saw how to multiply a vector with a scalar. Now we will define one way of multiplying a vector with a vector and obtain the **scalar product.** The scalar product of two vectors $\vec{A}$ and $\vec{B}$ is defined as

$$\vec{A}\bullet\vec{B} = \left|\vec{A}\right|\left|\vec{B}\right|\cos\alpha, \tag{5.5}$$

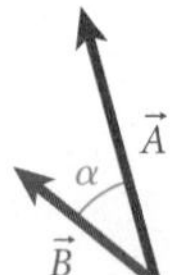


**FIGURE 5.9** Two vectors $\vec{A}$ and $\vec{B}$ and the angle $\alpha$ between them.

where $\alpha$ is the angle between the vectors $\vec{A}$ and $\vec{B}$, as shown in Figure 5.9. Note the use of the larger dot (•) as the multiplication sign for the scalar product between vectors, in contrast to the smaller dot (·) that is used for the multiplication of scalars. Because of the dot, the scalar product is often referred to as the *dot product.*

If two vectors form a 90° angle, then the scalar product has the value zero. In this case, the two vectors are orthogonal to each other. The scalar product of a pair of orthogonal vectors is zero.

If $\vec{A}$ and $\vec{B}$ are given in Cartesian coordinates as $\vec{A} = (A_x, A_y, A_z)$ and $\vec{B} = (B_x, B_y, B_z)$, then their scalar product can be shown to be equal to:

$$\vec{A}\bullet\vec{B} = (A_x, A_y, A_z)\bullet(B_x, B_y, B_z) = A_xB_x + A_yB_y + A_zB_z. \tag{5.6}$$

From equation 5.6, we can see that the scalar product has the commutative property:

$$\vec{A}\bullet\vec{B} = \vec{B}\bullet\vec{A}. \tag{5.7}$$

This result is not surprising, since the commutative property also holds for the multiplication of two scalars.

For the scalar product of any vector with itself, we have, in component notation, $\vec{A} \bullet \vec{A} = A_x^2 + A_y^2 + A_z^2$. Then, from equation 5.5, we find $\vec{A} \bullet \vec{A} = |\vec{A}|\,|\vec{A}|\cos\alpha = |\vec{A}|\,|\vec{A}| = |\vec{A}|^2$ (because the angle between the vector $\vec{A}$ and itself is zero, and the cosine of that angle has the value 1). Combining these two equations, we obtain the expression for the length of a vector that was introduced in Chapter 1:

$$|\vec{A}| = \sqrt{A_x^2 + A_y^2 + A_z^2}. \tag{5.8}$$

We can also use the definition of the scalar product to compute the angle between two arbitrary vectors in three-dimensional space:

$$\vec{A} \bullet \vec{B} = |\vec{A}|\,|\vec{B}|\cos\alpha \Rightarrow \cos\alpha = \frac{\vec{A} \bullet \vec{B}}{|\vec{A}|\,|\vec{B}|} \Rightarrow \alpha = \cos^{-1}\left(\frac{\vec{A} \bullet \vec{B}}{|\vec{A}|\,|\vec{B}|}\right). \tag{5.9}$$

For the scalar product, the same distributive property that is valid for the conventional multiplication of numbers holds:

$$\vec{A} \bullet (\vec{B} + \vec{C}) = \vec{A} \bullet \vec{B} + \vec{A} \bullet \vec{C}. \tag{5.10}$$

The following example puts the scalar product to use.

## EXAMPLE 5.2 Angle Between Two Position Vectors

### PROBLEM

What is the angle $\alpha$ between the two position vectors shown in Figure 5.10, $\vec{A} = (4.00, 2.00, 5.00)$ cm and $\vec{B} = (4.50, 4.00, 3.00)$ cm?

### SOLUTION

To solve this problem, we have to put the numbers for the components of each of the two vectors into equation 5.8 and equation 5.6 then use equation 5.9:

$$|\vec{A}| = \sqrt{4.00^2 + 2.00^2 + 5.00^2}\ \text{cm} = 6.71\ \text{cm}$$

$$|\vec{B}| = \sqrt{4.50^2 + 4.00^2 + 3.00^2}\ \text{cm} = 6.73\ \text{cm}$$

$$\vec{A} \bullet \vec{B} = A_x B_x + A_y B_y + A_z B_z = (4.00 \cdot 4.50 + 2.00 \cdot 4.00 + 5.00 \cdot 3.00)\ \text{cm}^2 = 41.0\ \text{cm}^2$$

$$\Rightarrow \alpha = \cos^{-1}\left(\frac{41.0\ \text{cm}^2}{6.71\ \text{cm} \cdot 6.73\ \text{cm}}\right) = 24.7°.$$

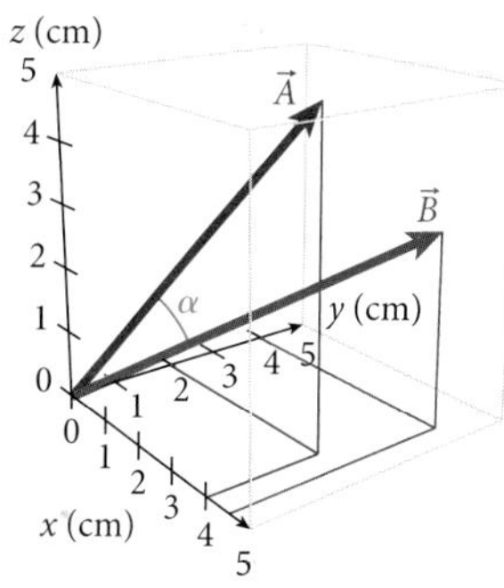


**FIGURE 5.10** Calculating the angle between two position vectors.

### Scalar Product for Unit Vectors

Section 1.6 introduced unit vectors in the three-dimensional Cartesian coordinate system: $\hat{x} = (1,0,0)$, $\hat{y} = (0,1,0)$, and $\hat{z} = (0,0,1)$. With our definition (5.6) of the scalar product, we find

$$\hat{x} \bullet \hat{x} = \hat{y} \bullet \hat{y} = \hat{z} \bullet \hat{z} = 1 \tag{5.11}$$

and

$$\begin{aligned} \hat{x} \bullet \hat{y} = \hat{x} \bullet \hat{z} = \hat{y} \bullet \hat{z} = 0 \\ \hat{y} \bullet \hat{x} = \hat{z} \bullet \hat{x} = \hat{z} \bullet \hat{y} = 0. \end{aligned} \tag{5.12}$$

Now we see why the unit vectors are called that: Their scalar products with themselves have the value 1. Thus, the unit vectors have length 1, or unit length, according to equation 5.8. In addition, any pair of different unit vectors has a scalar product that is zero, meaning that these vectors are orthogonal to each other. Equations 5.11 and 5.12 thus state that the unit vectors $\hat{x}$, $\hat{y}$, and $\hat{z}$ form an orthonormal set of vectors, which makes them extremely useful for the description of physical systems.

### 5.2 Self-Test Opportunity

Show that equations 5.11 and 5.12 are correct by using equation 5.6 and the definitions of the unit vectors.

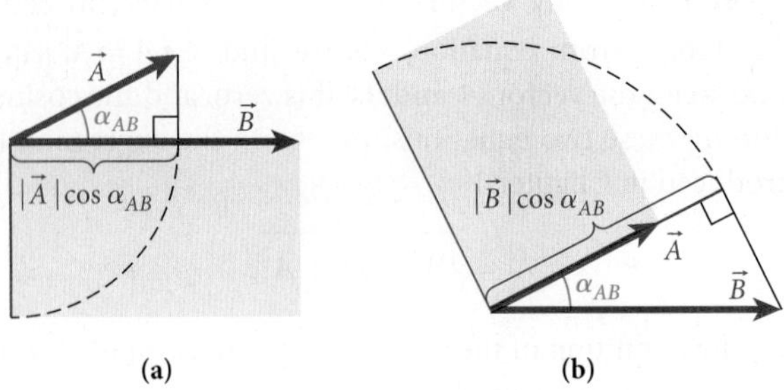


**FIGURE 5.11** Geometrical interpretation of the scalar product as an area. (a) The projection of $\vec{A}$ onto $\vec{B}$. (b) The projection of $\vec{B}$ onto $\vec{A}$.

## Geometrical Interpretation of the Scalar Product

In the definition of the scalar product $\vec{A}\bullet\vec{B}=|\vec{A}|\,|\vec{B}|\cos\alpha$ (equation 5.5), we can interpret $|\vec{A}|\cos\alpha$ as the projection of the vector $\vec{A}$ onto the vector $\vec{B}$ (Figure 5.11a). In this drawing, the line $|\vec{A}|\cos\alpha$ is rotated by 90° to show the geometrical interpretation of the scalar product as the area of a rectangle with sides $|\vec{A}|\cos\alpha$ and $|\vec{B}|$. In the same way, we can interpret $|\vec{B}|\cos\alpha$ as the projection of the vector $\vec{B}$ onto the vector $\vec{A}$ and construct a rectangle with side lengths $|\vec{B}|\cos\alpha$ and $|\vec{A}|$ (Figure 5.11b). The areas of the two yellow rectangles in Figure 5.11 are identical and are equal to the scalar product of the two vectors $\vec{A}$ and $\vec{B}$.

Finally, if we substitute from equation 5.9 for the cosine of the angle between the two vectors, the projection $|\vec{A}|\cos\alpha$ of the vector $\vec{A}$ onto the vector $\vec{B}$ can be written as

$$|\vec{A}|\cos\alpha=|\vec{A}|\frac{\vec{A}\bullet\vec{B}}{|\vec{A}|\,|\vec{B}|}=\frac{\vec{A}\bullet\vec{B}}{|\vec{B}|},$$

and the projection $|\vec{B}|\cos\alpha$ of the vector $\vec{B}$ onto the vector $\vec{A}$ can be expressed as

$$|\vec{B}|\cos\alpha=\frac{\vec{A}\bullet\vec{B}}{|\vec{A}|}.$$

### 5.1 In-Class Exercise

Consider an object undergoing a displacement $\Delta\vec{r}$ and experiencing a force $\vec{F}$. In which of the three cases shown below is the work done by the force on the object zero?

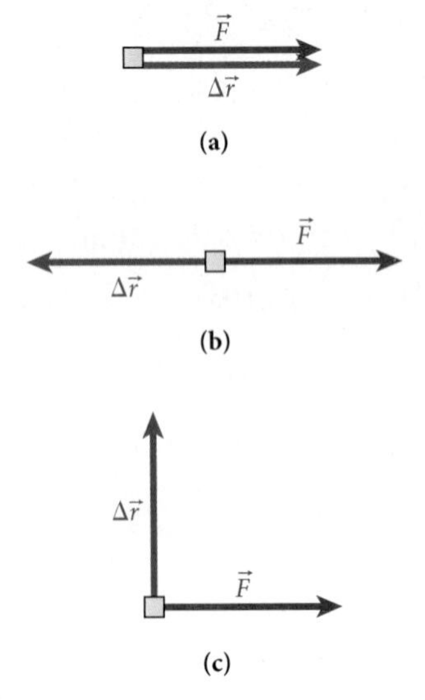


Using a scalar product, we can write the work done by a constant force as

$$W=\vec{F}\bullet\Delta\vec{r}. \tag{5.13}$$

This equation is the main result of this section. It says that the work done by a constant force $\vec{F}$ in displacing an object by $\Delta\vec{r}$ is the scalar product of the two vectors. In particular, if the displacement is perpendicular to the force, the scalar product is zero, and no work is done.

Note that we can use any force vector and any displacement vector in equation 5.13. If there is more than one force acting on an object, the equation holds for any of the individual forces, and it holds for the net force. The mathematical reason for this generalization lies in the distributive property of the scalar product, equation 5.10. To verify this statement, we can look at a constant net force that is the sum of individual constant forces, $\vec{F}_{\text{net}}=\sum_i\vec{F}_i$. According to equation 5.13, the work done by this net force is

$$W_{\text{net}}=\vec{F}_{\text{net}}\bullet\Delta\vec{r}=\left(\sum_i\vec{F}_i\right)\bullet\Delta\vec{r}=\sum_i\left(\vec{F}_i\bullet\Delta\vec{r}\right)=\sum_i W_i.$$

In other words, the net work done by the net force is equal to the sum of the work done by the individual forces. We have demonstrated this additive property of work only for constant forces, but it is also valid for variable forces (or, strictly speaking only for conservative forces, which we'll encounter in Chapter 6). But to repeat the main point: Equation 5.13 is valid for each individual force as well as the net force. We will typically consider the net force when calculating the work done on an object, but we will omit the index "net" to simplify the notation.

## One-Dimensional Case

In all cases of motion in one dimension, the work done to produce the motion is given by

$$\begin{aligned} W &= \vec{F} \bullet \Delta\vec{r} \\ &= \pm F_x \cdot \left|\Delta\vec{r}\right| = F_x \Delta x \\ &= F_x (x - x_0). \end{aligned} \tag{5.14}$$

The force $\vec{F}$ and displacement $\Delta\vec{r}$ can point in the same direction, $\alpha = 0 \Rightarrow \cos\alpha = 1$, resulting in positive work, or they can point in opposite directions, $\alpha = 180° \Rightarrow \cos\alpha = -1$, resulting in negative work.

## Work–Kinetic Energy Theorem

The relationship between kinetic energy of an object and the work done by the forces acting on it, called the **work–kinetic energy theorem,** is expressed formally as

$$\Delta K \equiv K - K_0 = W. \tag{5.15}$$

Here, $K$ is the kinetic energy that an object has after work $W$ has been done on it and $K_0$ is the kinetic energy before the work is done. The definitions of $W$ and $K$ are such that equation 5.15 is equivalent to Newton's Second Law. To see this equivalence, consider a constant force acting in one dimension on an object of mass $m$. Newton's Second Law is then $F_x = ma_x$, and the (also constant!) acceleration, $a_x$, of the object is related to the difference in the squares of its initial and final velocities via $v_x^2 - v_{x0}^2 = 2a_x(x - x_0)$, which is one of the five kinematical equations we derived in Chapter 2. Multiplication of both sides of this equation by $\frac{1}{2}m$ yields

$$\tfrac{1}{2}mv_x^2 - \tfrac{1}{2}mv_{x0}^2 = ma_x(x - x_0) = F_x \Delta x = W. \tag{5.16}$$

Thus, we see that, for this one-dimensional case, the work–kinetic energy theorem is equivalent to Newton's Second Law.

Because of the equivalence we have just established, if more than one force is acting on an object, we can use the net force to calculate the work done. Alternatively, and more commonly in energy problems, if more than one force is acting on an object, we can calculate the work done by each force, and then $W$ in equation 5.15 represents their sum.

The work–kinetic energy theorem specifies that the change in kinetic energy of an object is equal to the work done on the object by the forces acting on it. We can rewrite equation 5.15 to solve for $K$ or $K_0$:

$$K = K_0 + W$$

or

$$K_0 = K - W.$$

By definition, the kinetic energy cannot be less than zero; so, if an object has $K_0 = 0$, the work–kinetic energy theorem implies that $K = K_0 + W = W \geq 0$.

While we have only verified the work–kinetic energy theorem for a constant force, it is also valid for variable forces, as we will see below. Is it valid for all kinds of forces? The short answer is no! Friction forces are one kind of force that violate the work–kinetic energy theorem. We will discuss this point further in Chapter 6.

### 5.3 Self-Test Opportunity

Show the equivalence between Newton's Second Law and the work–kinetic energy theorem for the case of a constant force acting in three-dimensional space.

## Work Done by the Gravitational Force

With the work–kinetic energy theorem at our disposal, we can now take another look at the problem of an object falling under the influence of the gravitational force, as in Example 5.1. On the way down, the work done by the gravitational force on the object is

$$W_g = +mgh, \tag{5.17}$$

where $h = \left|y - y_0\right| = \left|\Delta\vec{r}\right| > 0$. The displacement $\Delta\vec{r}$ and the force of gravity $\vec{F}_g$ point in the same direction, resulting in a positive scalar product and therefore positive work. This situation is

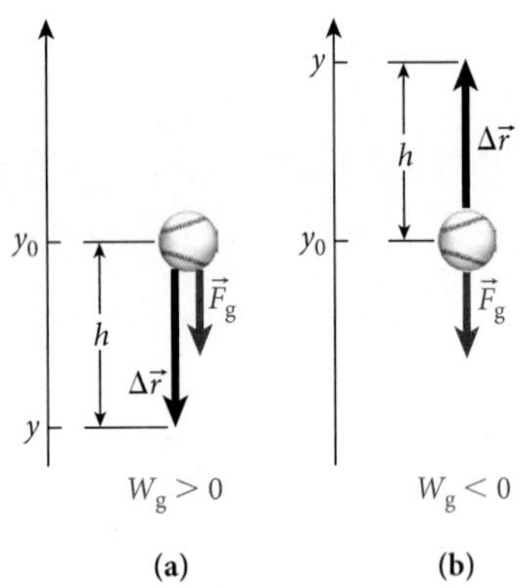


**FIGURE 5.12** Work done by the gravitational force. (a) The object during free fall. (b) Tossing an object upward.

illustrated in Figure 5.12a. Since the work is positive, the gravitational force increases the kinetic energy of the object.

We can reverse this situation and toss the object vertically upward, making it a projectile and giving it an initial kinetic energy. This kinetic energy will decrease until the projectile reaches the top of its trajectory. During this time, the displacement vector $\Delta\vec{r}$ points up, in the opposite direction to the force of gravity (Figure 5.12b). Thus, the work done by the gravitational force during the object's upward motion is

$$W_g = -mgh. \tag{5.18}$$

Therefore, the work done by the gravitational force reduces the kinetic energy of the object during its upward motion. This conclusion is consistent with the general formula for work done by a constant force, $W = \vec{F} \bullet \Delta\vec{r}$, because the displacement (pointing upward) of the object and the gravitational force (pointing downward) are in opposite directions.

## Work Done in Lifting and Lowering an Object

Now let's consider the situation in which a vertical external force is applied to an object—for example, by attaching the object to a rope and lifting it up or lowering it down. The work–kinetic energy theorem now has to include the work done by the gravitational force, $W_g$, and the work done by the external force, $W_F$:

$$K - K_0 = W_g + W_F.$$

For the case where the object is at rest both initially, $K_0 = 0$, and finally, $K = 0$, we have

$$W_F = -W_g.$$

The work done by force in lifting or lowering the object is then

$$W_F = -W_g = mgh \text{ (for lifting) or } W_F = -W_g = -mgh \text{ (for lowering).} \tag{5.19}$$

### EXAMPLE 5.3 Weightlifting

In the sport of weightlifting, the task is to pick up a very large mass, lift it over your head, and hold it there at rest for a moment. This action is an example of doing work by lifting and lowering a mass.

**PROBLEM 1**

The German lifter Ronny Weller won the silver medal at the Olympic Games in Sydney, Australia, in 2000. He lifted 257.5 kg in the "jerk" competition. Assuming he lifted the mass to a height of 1.83 m and held it there, what was the work he did in this process?

**SOLUTION 1**

This problem is an application of equation 5.19 for the work done against the gravitational force. The work Weller did was

$$W = mgh = (257.5 \text{ kg})(9.81 \text{ m/s}^2)(1.83 \text{ m}) = 4.62 \text{ kJ}.$$

**PROBLEM 2**

Once Weller had successfully completed the lift and was holding the mass with outstretched arms above his head, what was the work done by him in lowering the weight slowly (with negligible kinetic energy) back down to the ground?

**SOLUTION 2**

This calculation is the same as that in Solution 1 except the sign of the displacement changes. Thus, the answer is that –4.62 kJ of work is done in bringing the weight back down—exactly the opposite of what we obtained for Problem 1!

Now is a good time to remember that we are dealing with strictly mechanical work. Every lifter knows that you can feel the muscles "burn" just as much when holding the mass overhead or lowering the mass (in a controlled way) as when lifting it. (In Olympic competitions, by the way, the weightlifters just drop the mass after the successful lift.) However, this physiological effect is not mechanical work, which is what we are presently interested in. Instead it is the conversion of chemical energy, stored in different molecules such as sugars, into the energy needed to contract the muscles.

You may think that Olympic weightlifting is not the best example to consider because the force used to lift the mass is not constant. This is true, but as we discussed previously, the work–kinetic energy theorem applies to nonconstant forces. Additionally, even when a crane lifts a mass very slowly and with constant speed, the lifting force is still not exactly constant, because a slight initial acceleration is needed to get the mass from zero speed to a finite value and a deceleration occurs at the end of the lifting process.

## Lifting with Pulleys

When we studied pulleys and ropes in Chapter 4, we learned that pulleys act as force multipliers. For example, with the setup shown in Figure 5.13, the force needed to lift a pallet of bricks of mass $m$ by pulling on the rope is only half the gravitational force, $T = \frac{1}{2}mg$. How does the work done in pulling up the pallet of bricks with ropes and pulleys compare to the work of lifting it without such mechanical aids?

Figure 5.13 shows the initial and final positions of the pallet of bricks and the ropes and pulleys used to lift it. Lifting it without mechanical aids would require the force $\vec{T}_2$, as indicated, whose magnitude is given by $T_2 = mg$. The work done by force $\vec{T}_2$ in this case is $W_2 = \vec{T}_2 \bullet \vec{r}_2 = T_2 r_2 = mgr_2$. Pulling on the rope with force $\vec{T}_1$ of magnitude $T_1 = \frac{1}{2}T_2 = \frac{1}{2}mg$ accomplishes the same thing. However, now the displacement is twice as long, $r_1 = 2r_2$, as you can see by examining Figure 5.13. Thus, the work done in this case is $W_1 = \vec{T}_1 \bullet \vec{r}_1 = (\frac{1}{2}T_2)(2r_2) = mgr_2 = W_2$.

The same amount of work is done in both cases. It is necessary to compensate for the reduced force by pulling the rope through a longer distance. This result is general for the use of pulleys or lever arms or any other mechanical force multiplier: The total work done is the same as it would be if the mechanical aid were not used. Any reduction in the force is always going to be compensated for by a proportional lengthening of the displacement.

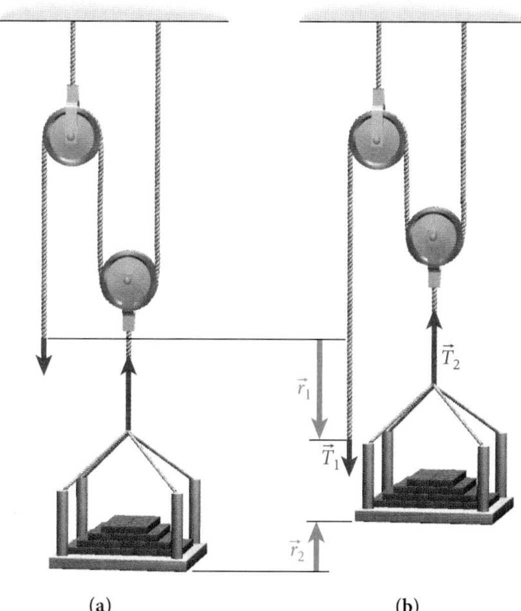


**FIGURE 5.13** Forces and displacements for the process of lifting a pallet of bricks at a work site with the aid of a rope-and-pulley mechanism. (a) Pallet in the initial position. (b) Pallet in the final position.

## 5.5 Work Done by a Variable Force

Suppose the force acting on an object is not constant. What is the work done by such a force? In a case of motion in one dimension with a variable $x$-component of force, $F_x(x)$, the work is

$$W = \int_{x_0}^{x} F_x(x')dx'. \tag{5.20}$$

(The integrand has $x'$ as a dummy variable to distinguish it from the integral limits.) Equation 5.20 shows that the work $W$ is the area under the curve of $F_x(x)$ (see Figure 5.14 in the following derivation).

### DERIVATION 5.1

If you have already taken integral calculus, you can skip this section. If equation 5.20 is your first exposure to integrals, the following derivation is a useful introduction. We'll derive the one-dimensional case and use our result for the constant force as a starting point.

In the case of a constant force, we can think of the work as the area under the horizontal line that plots the value of the constant force in the interval between $x_0$ and $x$. For a variable force, the work is the area under the curve $F_x(x)$, but that area is no longer a simple rectangle. In the case of a variable force, we need to divide the interval from $x_0$ to $x$ into many small equal intervals. Then we approximate the area under the curve $F_x(x)$ by a series of rectangles and add their areas to approximate the work. As you can see from Figure 5.14a, the area of the rectangle between $x_i$ and $x_{i+1}$ is given by $F_x(x_i)\cdot(x_{i+1} - x_i) = F_x(x_i)\cdot\Delta x$. We obtain an approximation for the work by summing over all rectangles:

$$W \approx \sum_i W_i = \sum_i F_x(x_i)\cdot\Delta x.$$

Now we space the points $x_i$ closer and closer by using more and more of them. This method makes $\Delta x$ smaller and causes the total area of the series of rectangles to be a better approximation of the area under the curve $F_x(x)$ as in Figure 5.14b. In the limit as $\Delta x \to 0$, the sum approaches the exact expression for the work:

$$W = \lim_{\Delta x \to 0}\left(\sum_i F_x(x_i)\cdot\Delta x\right).$$

This limit of the sum of the areas is exactly how the integral is defined:

$$W = \int_{x_0}^{x} F_x(x')dx'.$$

We have derived this result for the case of one-dimensional motion. The derivation of the three-dimensional case proceeds along similar lines but is more involved in terms of algebra.

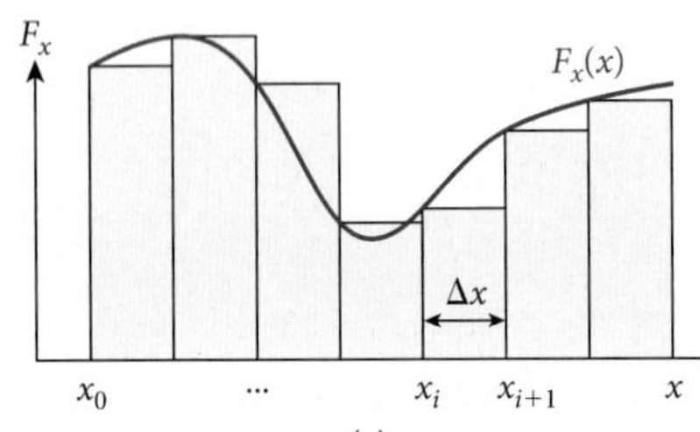


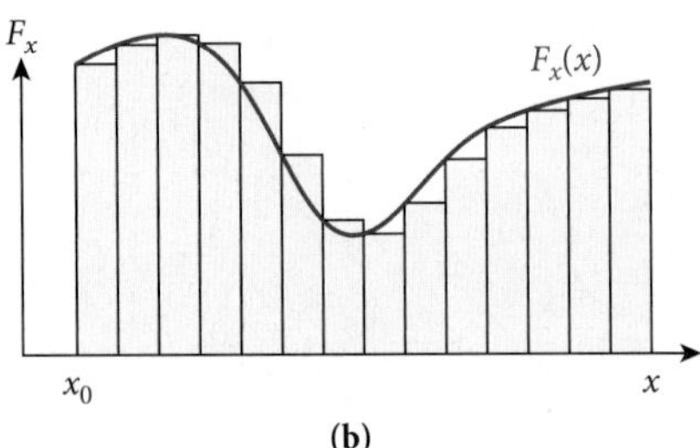


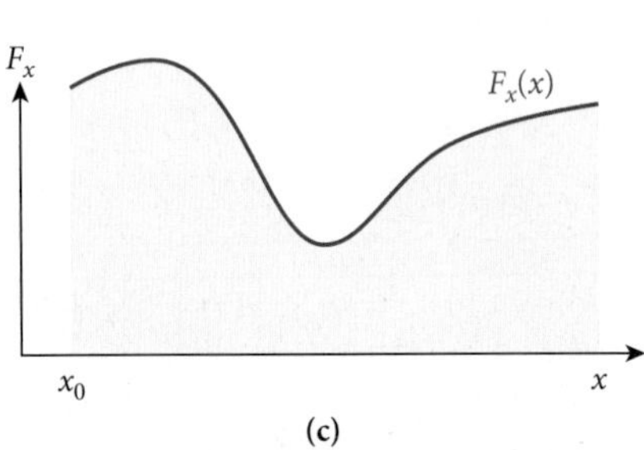


**FIGURE 5.14** (a) A series of rectangles approximates the area under the curve obtained by plotting the force as a function of the displacement; (b) a better approximation using rectangles of smaller width; (c) the exact area under the curve.

As promised earlier, we can verify that the work–kinetic energy theorem (equation 5.15) is valid when the force is variable. We show this result for one-dimensional motion for simplicity, but the work–kinetic energy theorem also holds for variable forces and displacements in more than one dimension. We assume a variable force in the $x$-direction, $F_x(x)$, as in equation 5.20, which we can express as

$$F_x(x) = ma,$$

using Newton's Second Law. We use the chain rule of calculus to obtain

$$a=\frac{dv}{dt}=\frac{dv}{dx}\frac{dx}{dt}.$$

We can then use equation 5.20 and integrate over the displacement to get the work done:

$$W=\int_{x_0}^{x}F_x(x')dx'=\int_{x_0}^{x}ma\,dx'=\int_{x_0}^{x}m\frac{dv}{dx'}\frac{dx'}{dt}dx'.$$

We now change the variable of integration from displacement ($x$) to velocity ($v$):

$$W=\int_{x_0}^{x}m\frac{dx'}{dt}\frac{dv}{dx'}dx'=\int_{v_0}^{v}mv'dv'=m\int_{v_0}^{v}v'dv',$$

where $v'$ is a dummy variable of integration. We carry out the integration and obtain the promised result:

$$W=m\int_{v_0}^{v}v'dv'=m\left[\frac{v'^2}{2}\right]_{v_0}^{v}=\frac{1}{2}mv^2-\frac{1}{2}mv_0^2=K-K_0=\Delta K.$$

### 5.2 In-Class Exercise

An $x$-component of a force has the dependence $F_x(x) = -c \cdot x^3$ on the displacement, $x$, where the constant $c = 19.1 \text{ N/m}^3$. How much work does it take to change the displacement from 0.810 m to 1.39 m?

a) 12.3 J
b) 0.452 J
c) −15.8 J
d) −3.76 J
e) 0.00 J

## 5.6 Spring Force

Let's examine the force that is needed to stretch or compress a spring. We start with a spring that is neither stretched nor compressed from its normal length and take the end of the spring in this condition to be located at the equilibrium position, $x_0$, as shown in Figure 5.15a. If we pull the end of this spring a bit toward the right using an external force, $\vec{F}_{\text{ext}}$, the spring gets longer. In the stretching process, the spring generates a force directed to the left, that is, pointing toward the equilibrium position, and increasing in magnitude with increasing length of the spring. This force is conventionally called the **spring force,** $\vec{F}_{\text{s}}$.

Pulling with an external force of a given magnitude stretches the spring to a certain displacement from equilibrium, at which point the spring force is equal in magnitude to the external force (Figure 5.15b). Doubling this external force doubles the displacement from equilibrium (Figure 5.15c). Conversely, pushing with an external force toward the left compresses the spring from its equilibrium length, and the resulting spring force points to the right, again toward the equilibrium position (Figure 5.15d). Doubling the amount of compression (Figure 5.15e) also doubles the spring force, just as with stretching.

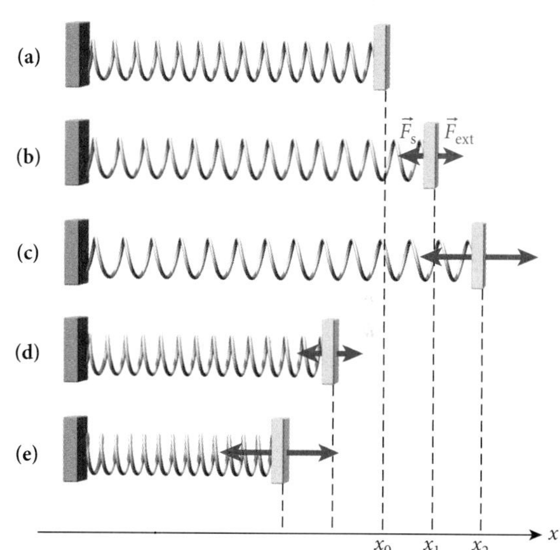


**FIGURE 5.15** Spring force. The spring is in its equilibrium position in (a), is stretched in (b) and (c), and is compressed in (d) and (e). In each nonequilibrium case, the external force acting on the end of the spring is shown as a red arrow, and the spring force as a blue arrow.

We can summarize these observations by noting that the magnitude of the spring force is proportional to the magnitude of the displacement of the end of the spring from its equilibrium position, and that the spring force always points toward the equilibrium position and thus is in the direction opposite to the displacement vector:

$$\vec{F}_{\text{s}}=-k(\vec{x}-\vec{x}_0). \quad (5.21)$$

As usual, this vector equation can be written in terms of components; in particular, for the $x$-component, we can write

$$F_{\text{s}}=-k\left(x-x_0\right). \quad (5.22)$$

The constant $k$ is by definition always positive. The negative sign in front of $k$ indicates that the spring force is always directed opposite to the direction of the displacement from the equilibrium position. We can choose the equilibrium position to be $x_0 = 0$, allowing us to write

$$F_{\text{s}}=-kx. \quad (5.23)$$

This simple force law is called **Hooke's Law,** after the British physicist Robert Hooke (1635–1703), a contemporary of Newton and the Curator of Experiments for the Royal Society. Note that for a displacement $x > 0$, the spring force points in the negative direction, and $F_s < 0$. The converse is also true; if $x < 0$, then $F_s > 0$. Thus, in all cases, the spring force points toward the equilibrium position, $x = 0$. At exactly the equilibrium position, the spring force is zero, $F_s(x = 0) = 0$. As a reminder from Chapter 4, zero force is one of the defining conditions for equilibrium. The proportionality constant, $k$, that appears in Hooke's Law is called the **spring constant** and has units of N/m = kg/s$^2$. The spring force is an important example of a **restoring force:** It always acts to restore the end of the spring to its equilibrium position.

Linear restoring forces that follow Hooke's Law can be found in many systems in nature. Examples are the forces on an atom that has moved slightly out of equilibrium in a crystal lattice, the forces due to shape deformations in atomic nuclei, and any other force that leads to oscillations in a physical system (discussed in further detail in Chapters 14 through 16). In Chapter 6, we will see that we can usually approximate the force in many physical situations by a force that follows Hooke's Law.

Of course, Hooke's Law is not valid for all spring displacements. Everyone who has played with a spring knows that if it is stretched too much, it will deform and then not return to its equilibrium length when released. If stretched even further, it will eventually break into two parts. Every spring has an elastic limit—a maximum deformation—below which Hooke's Law is still valid; however, where exactly this limit lies depends on the material characteristics of the particular spring. For our considerations in this chapter, we assume that springs are always inside the elastic limit.

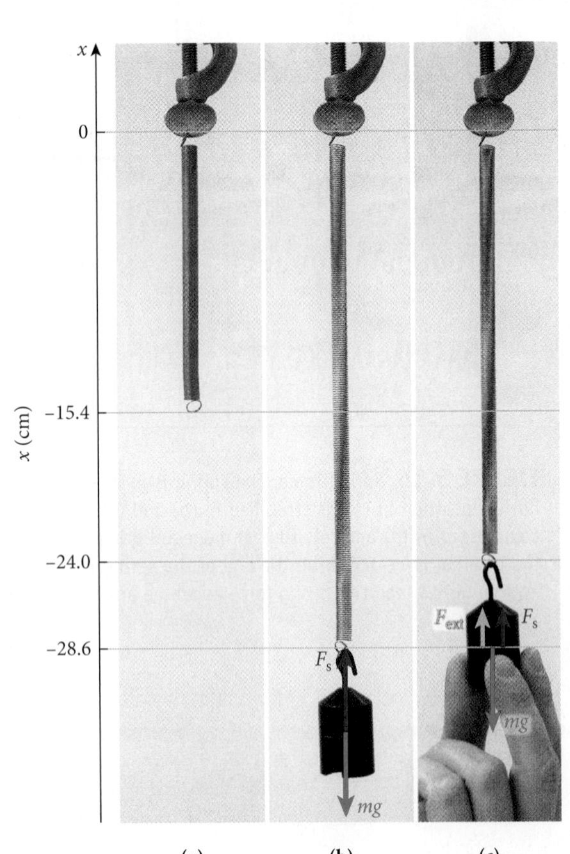


**FIGURE 5.16** Mass on a spring. (a) The spring without any mass attached. (b) The spring with the mass hanging freely. (c) The mass pushed upward by an external force.

## EXAMPLE 5.4 Spring Constant

### PROBLEM 1

A spring has a length of 15.4 cm and is hanging vertically from a support point above it (Figure 5.16a). A weight with a mass of 0.200 kg is attached to the spring, causing it to extend to a length of 28.6 cm (Figure 5.16b). What is the value of the spring constant?

### SOLUTION 1

We place the origin of our coordinate system at the top of the spring, with the positive direction upward, as is customary. Then, $x_0 = -15.4$ cm and $x = -28.6$ cm. According to Hooke's Law, the spring force is

$$F_s = -k(x - x_0).$$

Also, we know the force exerted on the spring was provided by the weight of the 0.200-kg mass: $F = -mg = -(0.200 \text{ kg})(9.81 \text{ m/s}^2) = -1.962$ N. Again, the negative sign indicates the direction. Now we can solve the force equation for the spring constant:

$$k = -\frac{F_s}{x - x_0} = -\frac{-1.962 \text{ N}}{(-0.286 \text{ m}) - (-0.154 \text{ m})} = 14.9 \text{ N/m}.$$

Note that we would have obtained exactly the same result if we had put the origin of the coordinate system at another point or if we had elected to designate the downward direction as positive.

### PROBLEM 2

How much force is needed to hold the weight at a position 4.6 cm above –28.6 cm (Figure 5.16c)?

### SOLUTION 2

At first sight, this problem might appear to require a complicated calculation. However, remember that the mass has stretched the spring to a new equilibrium position. To move the mass from that position takes an external force. If the external force moves the mass up 4.6 cm, then it has to be exactly equal in magnitude and opposite in direction to the spring force resulting from a displacement of 4.6 cm. Thus, all we have to do to find the external force is to use Hooke's Law for the spring force (choosing new equilibrium position to be at $x_0 = 0$):

$$F_{\text{ext}} + F_s = 0 \Rightarrow F_{\text{ext}} = -F_s = kx = (0.046 \text{ m})(14.9 \text{ N/m}) = 0.68 \text{ N}.$$

### 5.4 Self-Test Opportunity

A block is hanging vertically from a spring at the equilibrium displacement. The block is then pulled down a bit and released from rest. Draw the free-body diagram for the block in each of the following cases:

a) The block is at the equilibrium displacement.

b) The block is at its highest vertical point.

c) The block is at its lowest vertical point.

At this point, it is worthwhile to generalize the observations made in Example 5.4: Adding a constant force—for example, by suspending a mass from the spring—only shifts the equilibrium position. (This generalization is true for all forces that depend linearly on displacement.) Moving the mass, up or down, away from the new equilibrium position then results in a force that is linearly proportional to the displacement from the new equilibrium position. Adding another mass will only cause an additional shift to a new equilibrium position. Of course, adding more mass cannot be continued without limit. At some point, the addition of more and more mass will overstretch the spring. Then the spring will not return to its original length once the mass is removed, and Hooke's Law is no longer valid.

## Work Done by the Spring Force

The displacement of a spring is a case of motion in one spatial dimension. Thus, we can apply the one-dimensional integral of equation 5.20 to find the work done by the spring force in moving from $x_0$ to $x$. The result is

$$W_s = \int_{x_0}^{x} F_s(x')dx' = \int_{x_0}^{x} (-kx')dx' = -k\int_{x_0}^{x} x'dx'.$$

The work done by the spring force is then

$$W_s = -k\int_{x_0}^{x} x'dx' = -\tfrac{1}{2}kx^2 + \tfrac{1}{2}kx_0^2. \tag{5.24}$$

If we set $x_0 = 0$ and start at the equilibrium position, as we did in arriving at Hooke's Law (equation 5.23), the second term on the right side in equation 5.24 becomes zero and we obtain

$$W_s = -\tfrac{1}{2}kx^2. \tag{5.25}$$

Note that because the spring constant is always positive, the work done by the spring force is always negative for displacements from equilibrium. Equation 5.24 shows that the work done by the spring force is positive if the starting spring displacement is farther from equilibrium than the ending displacement. External work of magnitude $\frac{1}{2}kx^2$ will stretch or compress it out of its equilibrium position.

## SOLVED PROBLEM 5.1 Compressing a Spring

A massless spring located on a smooth horizontal surface is compressed by a force of 63.5 N, which results in a displacement of 4.35 cm from the initial equilibrium position. As shown in Figure 5.17, a steel ball of mass 0.075 kg is then placed in front of the spring and the spring is released.

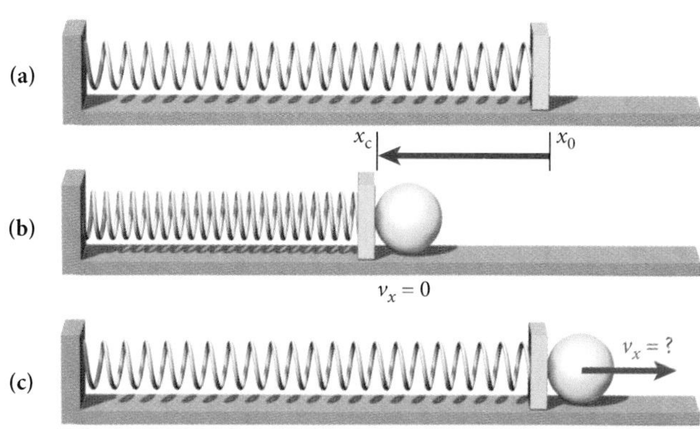


**FIGURE 5.17** (a) Spring in its equilibrium position; (b) compressing the spring; (c) relaxing the compression and accelerating the steel ball.

### PROBLEM

What is the speed of the steel ball when it is shot off by the spring, that is, right after it loses contact with the spring? (Assume there is no friction between the surface and the steel ball; the steel ball will then simply slide across the surface and will not roll.)

*Continued—*

## SOLUTION

### THINK

If we compress a spring with an external force, we do work against the spring force. Releasing the spring by withdrawing the external force enables the spring to do work on the steel ball, which acquires kinetic energy in this process. Calculating the initial work done against the spring force enables us to figure out the kinetic energy that the steel ball will have and thus will lead us to the speed of the ball.

### SKETCH

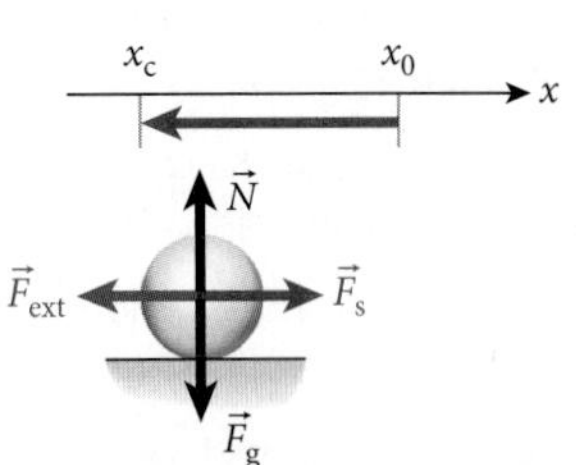


**FIGURE 5.18** Free-body diagram of the steel ball before the external force is removed.

We draw a free-body diagram at the instant before the external force is removed (see Figure 5.18). At this instant, the steel ball is at rest in equilibrium, because the external force and the spring force exactly balance each other. Note that the diagram also includes the support surface and shows two more forces acting on the ball: the force of gravity, $\vec{F}_g$, and the normal force from the support surface, $\vec{N}$. These two forces cancel each other out and thus do not enter into our calculations, but it is worthwhile to note the complete set of forces that act on the ball.

We set the $x$-coordinate of the ball at its left edge, which is where the ball touches the spring. This is the physically relevant location, because it measures the elongation of the spring from its equilibrium position.

### RESEARCH

The motion of the steel ball starts once the external force is removed. Without the blue arrow in Figure 5.18, the spring force is the only unbalanced force in the situation, and it accelerates the ball. (This acceleration is not constant over time as is the case for free-fall motion, for example, but rather changes in time.) However, the beauty of applying energy considerations is that we do not need to know the acceleration to calculate the final speed.

As usual, we are free to choose the origin of the coordinate system, and we put it at $x_0$, the equilibrium position of the spring. This implies that we set $x_0 = 0$. The relation between the $x$-component of the spring force at the moment of release and the initial compression of the spring $x_c$ is

$$F_s(x_c) = -kx_c.$$

Because $F_s(x_c) = -F_{ext}$, we find

$$kx_c = F_{ext}.$$

The magnitude of this external force, as well as the value of the displacement, were given, and so we can calculate the value of the spring constant from this equation. Note that with our choice of the coordinate system, $F_{ext} < 0$, because its vector arrow points in the negative $x$-direction. In addition, $x_c < 0$, because the displacement from equilibrium is in the negative direction.

We can now calculate the work $W$ needed to compress this spring. Since the force that the ball exerts on the spring is always equal and opposite to the force that the spring exerts on the ball, the definition of work allows us to set

$$W = -W_s = \tfrac{1}{2}kx_c^2.$$

According to the work–kinetic energy theorem, this work is related to the change in the kinetic energy of the steel ball via

$$K = K_0 + W = 0 + W = \tfrac{1}{2}kx_c^2.$$

Finally, the ball's kinetic energy is, by definition,

$$K = \tfrac{1}{2}mv_x^2,$$

which allows us to determine the ball's speed.

**SIMPLIFY**

We solve the equation for the kinetic energy for the speed, $v_x$, and then use the $K = \frac{1}{2}kx_c^2$ to obtain

$$v_x = \sqrt{\frac{2K}{m}} = \sqrt{\frac{2(\frac{1}{2}kx_c^2)}{m}} = \sqrt{\frac{kx_c^2}{m}} = \sqrt{\frac{F_{ext}x_c}{m}}.$$

(In the third step, we canceled out the factors 2 and $\frac{1}{2}$, and in the fourth step, we used $kx_c = F_{ext}$.)

**CALCULATE**

Now we are ready to insert the numbers: $x_c = -0.0435$ m, $m = 0.075$ kg, and $F_{ext} = -63.5$ N. Our result is

$$v_x = \sqrt{\frac{(-63.5\text{ N})(-0.0435\text{ m})}{0.075\text{ kg}}} = 6.06877\text{ m/s}.$$

Note that we choose the positive root for the $x$-component of the ball's velocity. By examining Figure 5.17, you can see that this is the appropriate choice, because the ball will move in the positive $x$-direction after the spring is released.

**ROUND**

Rounding to the two-digit accuracy to which the mass was specified, we state our result as

$$v_x = 6.1\text{ m/s}.$$

**DOUBLE-CHECK**

We are limited in the checking we can perform to verify that our answer makes sense until we study motion under the influence of the spring force in more detail in Chapter 14. However, our answer passes the minimum requirements in that it has the proper units and the order of magnitude seems in line with typical velocities for balls propelled from spring-loaded toy guns.

### 5.3 In-Class Exercise

How much work would it take to compress the spring of Solved Problem 5.1 from 4.35 cm to 8.15 cm?

a) 4.85 J

b) 1.38 J

c) −3.47 J

d) −1.38 J

e) 3.47 J

## 5.7 Power

We can now readily calculate the amount of work required to accelerate a 1550-kg (3410-lb) car from a standing start to a speed of 26.8 m/s (60.0 mph). The work done is simply the difference between the final and initial kinetic energies. The initial kinetic energy is zero, and the final kinetic energy is

$$K = \tfrac{1}{2}mv^2 = \tfrac{1}{2}(1550\text{ kg})(26.8\text{ m/s})^2 = 557\text{ kJ},$$

which is also the amount of work required. However, the work requirement is not that interesting to most of us—we'd be more interested in how quickly the car is able to reach 60 mph. That is, we'd like to know the rate at which the car can do this work.

**Power** is the rate at which work is done. Mathematically, this means that the power, $P$, is the time derivative of the work, $W$:

$$P = \frac{dW}{dt}. \quad (5.26)$$

It is also useful to define the average power, $\bar{P}$ as

$$\bar{P} = \frac{W}{\Delta t}. \quad (5.27)$$

The SI unit of power is the **watt** (W). [Beware of confusing the symbol for work, *W (italicized)*, and the abbreviation for the unit of power, W (nonitalicized).]

$$1\text{ W} = 1\text{ J/s} = 1\text{ kg m}^2/\text{s}^3. \tag{5.28}$$

Conversely, one joule is also one watt times one second. This relationship is reflected in a very common unit of energy (not power!), the **kilowatt-hour** (kWh):

$$1\text{ kWh} = (1000\text{ W})(3600\text{ s}) = 3.6 \cdot 10^6\text{ J} = 3.6\text{ MJ}.$$

The unit kWh appears on utility bills and quantifies the amount of electrical energy that has been consumed. Kilowatt-hours can be used to measure any kind of energy. Thus, the kinetic energy of the 1550-kg car moving with a speed of 26.8 m/s, which we calculated as 557 kJ, can be expressed with equal validity as

$$(557{,}000\text{ J})(1\text{ kWh}/3.6 \cdot 10^6\text{ J}) = 0.155\text{ kWh}.$$

The two most common non-SI power units are the horsepower (hp) and the foot-pound per second (ft lb/s): 1 hp = 550 ft lb/s = 746 W.

### 5.4 In-Class Exercise

Is each of the following statements true or false?

a) Work cannot be done in the absence of motion.

b) More power is required to lift a box slowly than to lift a box quickly.

c) A force is required to do work.

## Power for a Constant Force

For a constant force, we found that the work is given by $W = \vec{F} \bullet \Delta\vec{r}$ and the differential work as $dW = \vec{F} \bullet d\vec{r}$. In this case, the time derivative is

$$P = \frac{dW}{dt} = \frac{\vec{F} \bullet d\vec{r}}{dt} = \vec{F} \bullet \vec{v} = Fv\cos\alpha, \tag{5.29}$$

where $\alpha$ is the angle between the force vector and the velocity vector. Therefore, for a constant force, the power is the scalar product of the force vector and the velocity vector.

### EXAMPLE 5.5 Accelerating a Car

**PROBLEM**

Returning to the example of an accelerating car, let's assume that the car, of mass 1550 kg, can reach a speed of 60 mph (26.8 m/s) in 7.1 s. What is the average power needed to accomplish this?

**SOLUTION**

We already found that the car's kinetic energy at 60 mph is

$$K = \tfrac{1}{2}mv^2 = \tfrac{1}{2}(1550\text{ kg})(26.8\text{ m/s})^2 = 557\text{ kJ}.$$

The work to get the car to the speed of 60 mph is then

$$W = \Delta K = K - K_0 = 557\text{ kJ}.$$

The average power needed to get to 60 mph in 7.1 s is therefore

$$\bar{P} = \frac{W}{\Delta t} = \frac{5.57 \cdot 10^5\text{ J}}{7.1\text{ s}} = 78.4\text{ kW} = 105\text{ hp}.$$

If you own a car with a mass of at least 1550 kg that has an engine with 105 hp, you know that it cannot possibly reach 60 mph in 7.1 s. An engine with at least 180 hp is needed to accelerate a car of mass 1550 kg (including the driver, of course) to 60 mph in that time interval.

Our calculation in Example 5.5 is not quite correct for several reasons. First, not all of the power output of the engine is available to do useful work such as accelerating the car. Second, friction and air resistance forces act on a moving car, but were ignored in Example 5.5. Chapter 6 will address work and energy in the presence of friction forces (rolling friction and air resistance in this case). Finally, a car's rated horsepower is a peak specification, truly only at the most beneficial rpm-domain of the engine. As you accelerate the car from rest, this peak output of the engine is not maintainable as you shift through the gears.

The average mass, power, and fuel efficiency (for city driving) of mid-sized cars sold in the United States from 1975 to 2007 are shown in Figure 5.19. The mass of a car is important in city driving because of the many instances of acceleration in stop-and-go conditions. We can combine the work–kinetic energy theorem (equation 5.15) and the definition of average power (equation 5.27) to get

$$\overline{P} = \frac{W}{\Delta t} = \frac{\Delta K}{\Delta t} = \frac{\frac{1}{2}mv^2}{\Delta t} = \frac{mv^2}{2\Delta t}. \quad (5.30)$$

You can see that the average power required to accelerate a car from rest to a speed $v$ in a given time interval, $\Delta t$, is proportional to the mass of the car. The energy consumed by the car is equal to the average power times the time interval. Thus, the larger the mass of a car, the more energy is required to accelerate it in a given amount of time.

Following the 1973 oil embargo, the average mass of mid-sized cars decreased from 2100 kg to 1500 kg between 1975 and 1982. During that same period, the average power decreased from 160 hp to 110 hp, and the fuel efficiency increased from 10 to 18 mpg. From 1982 to 2007, however, the average mass and fuel efficiency of mid-sized cars stayed roughly constant, while the power increased steadily. Apparently, buyers of mid-sized cars in the United States have valued increased power over increased efficiency.

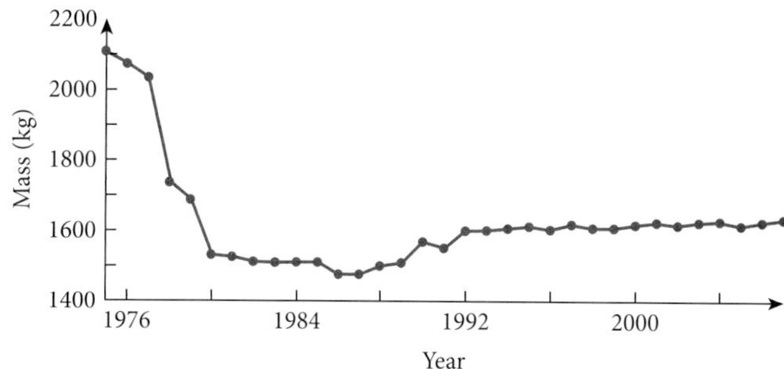


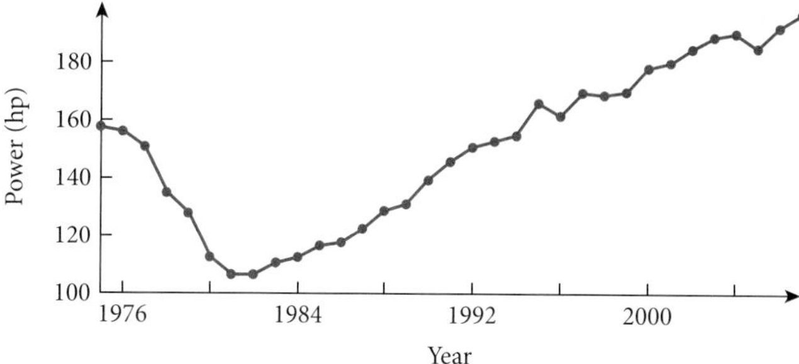


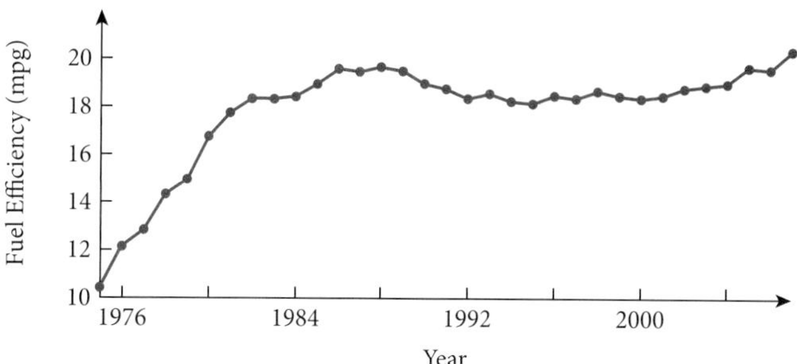


**FIGURE 5.19** The mass, power, and fuel efficiency of mid-sized cars sold in the United States from 1975 to 2007. The fuel efficiency is that for typical city driving.

## WHAT WE HAVE LEARNED | EXAM STUDY GUIDE

- Kinetic energy is the energy associated with the motion of an object, $K = \frac{1}{2}mv^2$.
- The SI unit of work and energy is the joule: $1\text{ J} = 1\text{ kg m}^2/\text{s}^2$.
- Work is the energy transferred to an object or transferred from an object due to the action of a force. Positive work is a transfer of energy to the object, and negative work is a transfer of energy from the object.
- Work done by a constant force is $W = |\vec{F}||\Delta\vec{r}|\cos\alpha$, where $\alpha$ is the angle between $\vec{F}$ and $\Delta\vec{r}$.
- Work done by a variable force in one dimension is $W = \int_{x_0}^{x} F_x(x')dx'$.
- Work done by the gravitational force in the process of lifting an object is $W_g = -mgh < 0$, where $h = |y - y_0|$; the work done by the gravitational force in lowering an object is $W_g = +mgh > 0$.
- The spring force is given by Hooke's Law: $F_s = -kx$.
- Work done by the spring force is $W = -k\int_{x_0}^{x} x'dx' = -\frac{1}{2}kx^2 + \frac{1}{2}kx_0^2$.
- The work–kinetic energy theorem is $\Delta K \equiv K - K_0 = W$.
- Power, $P$, is the time derivative of the $P = \frac{dW}{dt}$.
- The average power, $\overline{P}$, is $\overline{P} = \frac{W}{\Delta t}$.
- The SI unit of power is the watt (W): 1 W = 1 J/s.
- Power for a constant force is $P = \frac{dW}{dt} = \frac{\vec{F}\bullet d\vec{r}}{dt} = \vec{F}\bullet\vec{v} = Fv\cos\alpha_{Fv}$, where $\alpha$ is the angle between the force vector and the velocity vector.

## KEY TERMS

kinetic energy, p. 143
joule, p. 143
work, p. 145
scalar product, p. 146
work–kinetic energy theorem, p. 149
spring force, p. 153
Hooke's Law, p. 154
spring constant, p. 154
restoring force, p. 154
power, p. 157
watt, p. 158
kilowatt-hour, p. 158

## NEW SYMBOLS AND EQUATIONS

$K = \frac{1}{2}mv^2$, kinetic energy

$W = \vec{F} \bullet \Delta\vec{r}$, work done by a constant force

$W = \int_{x_0}^{x} F_x(x')dx'$, work done by a variable force

$\Delta K = W$, work–kinetic energy theorem

$F_s = -kx$, Hooke's Law

$W_s = -\frac{1}{2}kx^2$, work done by a spring

$P = \frac{dW}{dt}$, power

## ANSWERS TO SELF-TEST OPPORTUNITIES

**5.1**

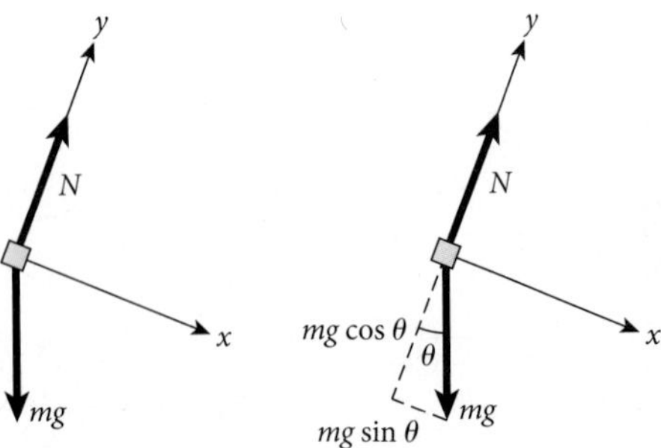


**5.2** Equation 5.11

$$\hat{x}\bullet\hat{x} = (1,0,0)\bullet(1,0,0) = 1\cdot1+0\cdot0+0\cdot0 = 1$$
$$\hat{y}\bullet\hat{y} = (0,1,0)\bullet(0,1,0) = 0\cdot0+1\cdot1+0\cdot0 = 1$$
$$\hat{z}\bullet\hat{z} = (0,0,1)\bullet(0,0,1) = 0\cdot0+0\cdot0+1\cdot1 = 1$$

Equation 5.12

$$\hat{x}\bullet\hat{y} = (1,0,0)\bullet(0,1,0) = 1\cdot0+0\cdot1+0\cdot0 = 0$$
$$\hat{x}\bullet\hat{z} = (1,0,0)\bullet(0,0,1) = 1\cdot0+0\cdot0+0\cdot1 = 0$$
$$\hat{y}\bullet\hat{z} = (0,1,0)\bullet(0,0,1) = 0\cdot0+1\cdot0+0\cdot1 = 0$$
$$\hat{y}\bullet\hat{x} = (0,1,0)\bullet(1,0,0) = 0\cdot1+1\cdot0+0\cdot0 = 0$$
$$\hat{z}\bullet\hat{x} = (0,0,1)\bullet(1,0,0) = 0\cdot1+0\cdot0+1\cdot0 = 0$$
$$\hat{z}\bullet\hat{y} = (0,0,1)\bullet(0,1,0) = 0\cdot0+0\cdot1+1\cdot0 = 0$$

**5.3** $\vec{F} = m\vec{a}$ can be re-written as

$F_x = ma_x$
$F_y = ma_y$
$F_z = ma_z$

for each component

$v_x^2 - v_{x0}^2 = 2a_x(x - x_0)$
$v_y^2 - v_{y0}^2 = 2a_y(y - y_0)$
$v_z^2 - v_{z0}^2 = 2a_z(z - z_0)$

multiply by $\frac{1}{2}m$

$\frac{1}{2}mv_x^2 - \frac{1}{2}mv_{x0}^2 = ma_x(x - x_0)$
$\frac{1}{2}mv_y^2 - \frac{1}{2}mv_{y0}^2 = ma_y(y - y_0)$
$\frac{1}{2}mv_z^2 - \frac{1}{2}mv_{z0}^2 = ma_z(z - z_0)$

add the three equations

$$\tfrac{1}{2}m\left(v_x^2 + v_y^2 + v_z^2\right) - \tfrac{1}{2}m\left(v_{x0}^2 + v_{y0}^2 + v_{z0}^2\right) =$$
$$ma_x(x - x_0) + ma_y(y - y_0) + ma_z(z - z_0)$$

$$K = \frac{1}{2}m\left(v_x^2 + v_y^2 + v_z^2\right) = \frac{1}{2}mv^2$$
$$K_0 = \frac{1}{2}m\left(v_{x0}^2 + v_{y0}^2 + v_{z0}^2\right) = \frac{1}{2}mv_0^2$$
$$\Delta\vec{r} = (x - x_0)\hat{x} + (y - y_0)\hat{y} + (z - z_0)\hat{z}$$
$$\vec{F} = ma_x\hat{x} + ma_y\hat{y} + ma_z\hat{z}$$
$$K - K_0 = \Delta K = \vec{F} \bullet \Delta\vec{r} = W$$

**5.4**

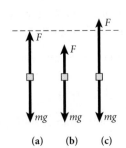

F F F
mg mg mg
(a) (b) (c)

# PROBLEM-SOLVING PRACTICE

## Problem-Solving Guidelines: Kinetic Energy, Work, and Power

**1.** In all problems involving energy, the first step is to clearly identify the system and the changes in its conditions. If an object undergoes a displacement, check that the displacement is always measured from the same point on the object, such as the front edge or the center of the object. If the speed of the object changes, identify the initial and final speeds at specific points. A diagram is often helpful to show the position and the speed of the object at two different times of interest.

**2.** Be careful to identify the force that is doing work. Also note whether forces doing work are constant forces or variable forces, because they need to be treated differently.

**3.** You can calculate the sum of the work done by individual forces acting on an object or the work done by the net force acting on an object; the result should be the same. (You can use this as a way to check your calculations.)

**4.** Remember that the direction of the restoring force exerted by a spring is always opposite the direction of the displacement of the spring from its equilibrium point.

**5.** The formula for power, $P = \vec{F} \bullet \vec{v}$, is very useful, but applies only for a constant force. When using the more general definition of power, be sure to distinguish between the average power, $\bar{P} = \frac{W}{\Delta t}$, and the instantaneous value of the power, $P = \frac{dW}{dt}$.

## SOLVED PROBLEM 5.2 Lifting Bricks

### PROBLEM

A load of bricks at a construction site has a mass of 85.0 kg. A crane raises this load from the ground to a height of 50.0 m in 60.0 s at a low constant speed. What is the average power of the crane?

### SOLUTION

### THINK

Raising the bricks at a low constant speed means that the kinetic energy is negligible, so the work in this situation is done against gravity only. There is no acceleration, and friction is negligible. The average power then is just the work done against gravity divided by the time it takes to raise the load of bricks to the stated height.

### SKETCH

A free-body diagram of the load of bricks is shown in Figure 5.20. Here we have defined a coordinate system in which the $y$-axis is vertical and positive is upward. The tension, $T$, exerted by the cable of the crane is a force in the upward direction, and the weight, $mg$, of the load of bricks is a force downward. Because the load is moving at a constant speed, the sum of the tension and the weight is zero. The load is moved vertically a distance $h$, as shown in Figure 5.21.

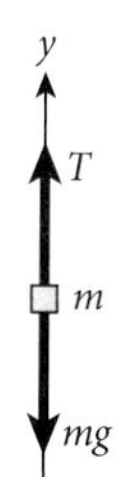


**FIGURE 5.20** Free-body diagram of the load of bricks of mass $m$ being lifted by a crane.

### RESEARCH

The work, $W$, done by the crane is given by

$$W = mgh.$$

The average power, $\bar{P}$, required to lift the load in the given time $\Delta t$ is

$$\bar{P} = \frac{W}{\Delta t}.$$

### SIMPLIFY

Combining the above two equations gives

$$\bar{P} = \frac{mgh}{\Delta t}.$$

### CALCULATE

Now we put in the numbers and get

$$\bar{P} = \frac{(85.0 \text{ kg})(9.81 \text{ m/s}^2)(50.0 \text{ m})}{60.0 \text{ s}} = 694.875 \text{ W}.$$

Continued—

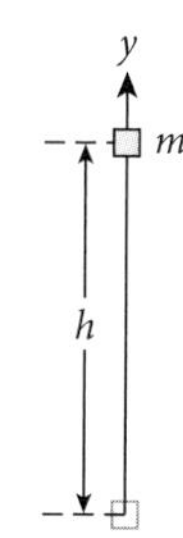


**FIGURE 5.21** The mass $m$ is lifted a distance $h$.

**ROUND**

We report our final result as

$$\bar{P} = 695 \text{ W}$$

because all the numerical values were initially given with three significant figures.

**DOUBLE-CHECK**

To double-check our result for the required average power, we convert the average power in watts to horsepower:

$$\bar{P} = (695 \text{ W})\frac{1 \text{ hp}}{746 \text{ W}} = 0.932 \text{ hp}.$$

Thus, a 1-hp motor is sufficient to lift the 85.0-kg load 50 m in 60 s, which seems not completely unreasonable, although surprisingly small. Because motors are not 100% efficient, in reality, the crane would have to have a motor with a somewhat higher power rating to lift the load.

## SOLVED PROBLEM 5.3 Shot Put

**PROBLEM**

Shot put competitions use metal balls with a mass of 16 lb (= 7.26 kg). A competitor throws the shot at an angle of 43.3° and releases it from a height of 1.82 m above where it lands, and it lands a horizontal distance of 17.7 m from the point of release. What is the kinetic energy of the shot as it leaves the thrower's hand?

**SOLUTION**

**THINK**

We are given the horizontal distance, $x_s = 17.7$ m, the height of release, $y_0 = 1.82$ m, and the angle of the initial velocity, $\theta_0 = 43.3°$, but not the initial speed, $v_0$. If we can figure out the initial speed from the given data, then calculating the initial kinetic energy will be straightforward because we also know the mass of the shot: $m = 7.26$ kg.

Because the shot is very heavy, air resistance can be safely ignored. This situation is an excellent realization of ideal projectile motion. After the shot leaves the thrower's hand, the only force on the shot is the force of gravity, and the shot will follow a parabolic trajectory until it lands on the ground. Thus, we'll solve this problem by application of the rules about ideal projectile motion.

**SKETCH**

The trajectory of the shot is shown in Figure 5.22.

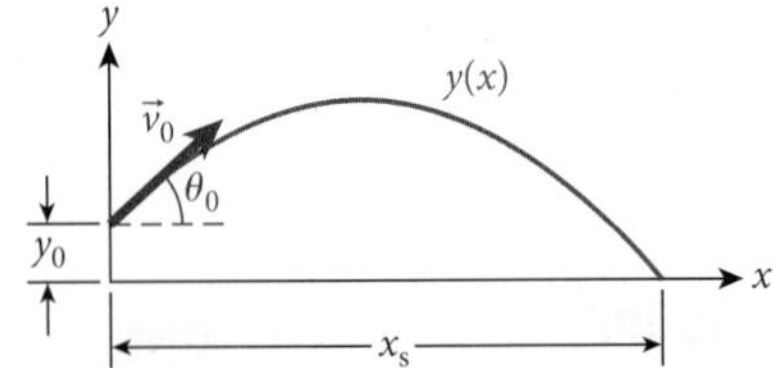


**FIGURE 5.22** Parabolic trajectory of a thrown shot.

**RESEARCH**

The initial kinetic energy $K$ of the shot of mass $m$ is given by

$$K = \tfrac{1}{2}mv_0^2.$$

Now we need to decide how to obtain $v_0$. We are given the distance, $x_s$, to where the shot hits the ground, but this is *not* equal to the range, $R$ (for which we obtained a formula in Chapter 3), because the range formula assumes that the heights of the start and end of the trajectory are equal. Here the initial height of the shot is $y_0$, and the final height is zero. Therefore, we have to use the full expression for the trajectory of an ideal projectile from Chapter 3:

$$y = y_0 + x\tan\theta_0 - \frac{x^2 g}{2v_0^2\cos^2\theta_0}.$$

This equation describes the $y$-component of the trajectory as a function of the $x$-component.

In this problem, we know that $y(x = x_s) = 0$, that is, that the shot touches the ground at $x = x_s$. Substituting for $x$ when $y = 0$ in the equation for the trajectory results in

$$0 = y_0 + x_s \tan\theta_0 - x_s^2 \frac{g}{2v_0^2 \cos^2\theta_0}.$$

**SIMPLIFY**

We solve this equation for $v_0^2$:

$$y_0 + x_s \tan\theta_0 = \frac{x_s^2 g}{2v_0^2 \cos^2\theta_0} \Rightarrow$$

$$2v_0^2 \cos^2\theta_0 = \frac{x_s^2 g}{y_0 + x_s \tan\theta_0} \Rightarrow$$

$$v_0^2 = \frac{x_s^2 g}{2\cos^2\theta_0 (y_0 + x_s \tan\theta_0)}.$$

Now, substituting for $v_0^2$ in the expression for the initial kinetic energy gives us

$$K = \tfrac{1}{2} m v_0^2 = \frac{m x_s^2 g}{4\cos^2\theta_0 (y_0 + x_s \tan\theta_0)}.$$

**CALCULATE**

Putting in the given numerical values, we get

$$K = \frac{(7.26\text{ kg})(17.7\text{m})^2 (9.81\text{ m/s}^2)}{4(\cos^2 43.3°)[1.82\text{ m} + (17.7\text{m})(\tan 43.3°)]} = 569.295\text{ J}.$$

**ROUND**

All of the numerical values given for this problem had three significant figures, so we report our answer as

$$K = 569\text{ J}.$$

**DOUBLE-CHECK**

Since we have an expression for the initial speed, $v_0^2 = x_s^2 g / (2\cos^2\theta_0 (y_0 + x_s \tan\theta_0))$, we can find the horizontal and vertical components of the initial velocity vector:

$$v_{x0} = v_0 \cos\theta_0 = 9.11\text{ m/s}$$

$$v_{y0} = v_0 \sin\theta_0 = 8.59\text{ m/s}.$$

As we discussed in Section 5.2, we can split up the total kinetic energy in ideal projectile motion into contributions from the motion in horizontal and vertical directions (see equation 5.3). The kinetic energy due to the motion in the $x$-direction remains constant. The kinetic energy due to the motion in the $y$-direction is initially

$$\tfrac{1}{2} m v_{y0}^2 \sin^2\theta_0 = 268\text{ J}.$$

At the top of the shot's trajectory, the vertical velocity component is zero, as in all projectile motion. This also means that the kinetic energy associated with the vertical motion is zero at this point. All 268 J of the initial kinetic energy due to the $y$-component of the motion has been used up to do work against the force of gravity (see Section 5.3). This work is (refer to equation 5.18) $-268\text{ J} = -mgh$, where $h = y_{max} - y_0$ is the maximum height of the trajectory. We thus find the value of $h$:

$$h = \frac{268\text{ J}}{mg} = \frac{268\text{ J}}{(7.26\text{ kg})(9.81\text{ m/s}^2)} = 3.76\text{ m}.$$

*Continued—*

Let's use known concepts of projectile motion to find the maximum height for the initial velocity we have determined. In Section 3.4, the maximum height $H$ of an object in projectile motion was shown to be

$$H = y_0 + \frac{v_{y0}^2}{2g}.$$

Putting in the numbers gives $v_{y0}^2 / 2g = 3.76$ m. This value is the same as that obtained by applying energy considerations.

## MULTIPLE-CHOICE QUESTIONS

**5.1** Which of the following is a correct unit of energy?

a) kg m/s$^2$
b) kg m$^2$/s
c) kg m$^2$/s$^2$
d) kg$^2$ m/s$^2$
e) kg$^2$ m$^2$/s$^2$

**5.2** An 800-N box is pushed up an inclined plane that is 4.0 m long. It requires 3200 J of work to get the box to the top of the plane, which is 2.0 m above the base. What is the magnitude of the average friction force on the box? (Assume the box starts at rest and ends at rest.)

a) 0 N
b) not zero but less than 400 N
c) greater than 400 N
d) 400 N
e) 800 N

**5.3** An engine pumps water continuously through a hose. If the speed with which the water passes through the hose nozzle is $v$ and if $k$ is the mass per unit length of the water jet as it leaves the nozzle, what is the kinetic energy being imparted to the water?

a) $\frac{1}{2}kv^3$
b) $\frac{1}{2}kv^2$
c) $\frac{1}{2}kv$
d) $\frac{1}{2}v^2/k$
e) $\frac{1}{2}v^3/k$

**5.4** A 1500-kg car accelerates from 0 to 25 m/s in 7.0 s. What is the average power delivered by the engine (1 hp = 746 W)?

a) 60 hp
b) 70 hp
c) 80 hp
d) 90 hp
e) 180 hp

**5.5** Which of the following is a correct unit of power?

a) kg m/s$^2$
b) N
c) J
d) m/s$^2$
e) W

**5.6** How much work is done when a 75-kg person climbs a flight of stairs 10 m high at constant speed?

a) $7.35 \cdot 10^5$ J
b) 750 J
c) 75 J
d) 7500 J
e) 7350 J

**5.7** How much work do movers do (horizontally) in pushing a 150-kg crate 12.3 m across a floor at constant speed if the coefficient of friction is 0.70?

a) 1300 J
b) 1845 J
c) $1.3 \cdot 10^4$ J
d) $1.8 \cdot 10^4$ J
e) 130 J

**5.8** Eight books, each 4.6 cm thick and of mass 1.8 kg, lie on a flat table. How much work is required to stack them on top of one another?

a) 141 J
b) 23 J
c) 230 J
d) 0.81 J
e) 14 J

**5.9** A particle moves parallel to the $x$-axis. The net force on the particle increases with $x$ according to the formula $F_x$ = (120 N/m)$x$, where the force is in newtons when $x$ is in meters. How much work does this force do on the particle as it moves from $x = 0$ to $x = 0.50$ m?

a) 7.5 J
b) 15 J
c) 30 J
d) 60 J
e) 120 J

**5.10** A skydiver is subject to two forces: gravity and air resistance. Falling vertically, she reaches a constant terminal speed at some time after jumping from a plane. Since she is moving at a constant velocity from that time until her chute opens, we conclude from the work–kinetic energy theorem that, over that time interval,

a) the work done by gravity is zero.
b) the work done by air resistance is zero.
c) the work done by gravity equals the negative of the work done by air resistance.
d) the work done by gravity equals the work done by air resistance.
e) her kinetic energy increases.

## QUESTIONS

5.11 If the net work done on a particle is zero, what can be said about the particle's speed?

5.12. Paul and Kathleen start from rest at the same time at height $h$ at the top of two differently configured water slides. The slides are nearly frictionless. a) Which slider arrives first at the bottom? b) Which slider is traveling faster at the bottom? What physical principle did you use to answer this?

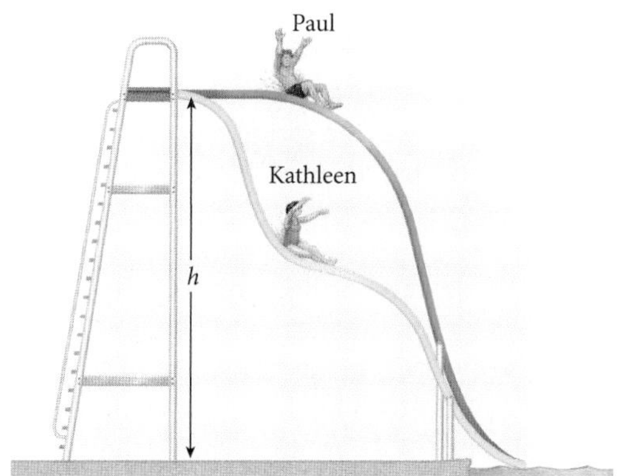


5.13 Does the Earth do any work on the Moon as the Moon moves in its orbit?

5.14 A car, of mass $m$, traveling at a speed $v_1$ can brake to a stop within a distance $d$. If the car speeds up by a factor of 2, $v_2 = 2v_1$, by what factor is its stopping distance increased, assuming that the braking force $F$ is approximately independent of the car's speed?

## PROBLEMS

A blue problem number indicates a worked-out solution is available in the Student Solutions Manual. One • and two •• indicate increasing level of problem difficulty.

### Section 5.2

5.15 The damage done by a projectile on impact is correlated with its kinetic energy. Calculate and compare the kinetic energies of these three projectiles:

a) a 10.0 kg stone at 30.0 m/s

b) a 100.0 g baseball at 60.0 m/s

c) a 20.0 g bullet at 300 m/s

5.16 A limo is moving at a speed of 100 km/h. If the mass of the limo, including passengers, is 1900 kg, what is its kinetic energy?

5.17 Two railroad cars, each of mass 7000 kg and traveling at 90 km/h, collide head on and come to rest. How much mechanical energy is lost in this collision?

5.18 Think about the answers to these questions next time you are driving a car:

a) What is the kinetic energy of a 1500-kg car moving at 15 m/s?

b) If the car changed its speed to 30 m/s, how would the value of its kinetic energy change?

5.19 A 200-kg moving tiger has a kinetic energy of 14,400 J. What is the speed of the tiger?

•5.20 Two cars are moving. The first car has twice the mass of the second car but only half as much kinetic energy. When both cars increase their speed by 5.0 m/s, they then have the same kinetic energy. Calculate the original speeds of the two cars.

•5.21 What is the kinetic energy of an ideal projectile of mass 20.1 kg at the apex (highest point) of its trajectory, if it was launched with an initial speed of 27.3 m/s and at an initial angle of 46.9° with respect to the horizontal?

### Section 5.4

5.22 A force of 5 N acts through a distance of 12 m in the direction of the force. Find the work done.

5.23 Two baseballs are thrown off the top of a building that is 7.25 m high. Both are thrown with initial speed of 63.5 mph. Ball 1 is thrown horizontally, and ball 2 is thrown straight down. What is the difference in the speeds of the two balls when they touch the ground? (Neglect air resistance.)

5.24 A 95-kg refrigerator rests on the floor. How much work is required to move it at constant speed for 4.0 m along the floor against a friction force of 180 N?

5.25 A hammerhead of mass $m = 2.0$ kg is allowed to fall onto a nail from a height $h = 0.4$ m. Calculate the maximum amount of work it could do on the nail.

5.26 You push your couch a distance of 4.0 m across the living room floor with a horizontal force of 200.0 N. The force of friction is 150.0 N. What is the work done by you, by the friction force, by gravity, and by the net force?

•5.27 Supppose you pull a sled with a rope that makes an angle of 30° to the horizontal. How much work do you do if you pull with 25.0 N of force and the sled moves 25 m?

•5.28 A father pulls his son, whose mass is 25.0 kg and who is sitting on a swing with ropes of length 3.00 m, backward until the ropes make an angle of 33.6° with respect to the vertical. He then releases his son from rest. What is the speed of the son at the bottom of the swinging motion?

•5.29 A constant force, $\vec{F} = (4.79, -3.79, 2.09)$ N, acts on an object of mass 18.0 kg, causing a displacement of that object by $\vec{r} = (4.25, 3.69, -2.45)$ m. What is the total work done by this force?

•5.30 A mother pulls her daughter, whose mass is 20.0 kg and who is sitting on a swing with ropes of length 3.50 m, backward until the ropes make an angle of 35.0° with respect to the vertical. She then releases her daughter from rest. What is the speed of the daughter when the ropes make an angle of 15.0° with respect to the vertical?

•5.31 A ski jumper glides down a 30° slope for 80 ft before taking off from a negligibly short horizontal ramp. If the jumper's takeoff speed is 45 ft/s, what is the coefficient of kinetic friction between skis and slope? Would the value of the coefficient of friction be different if expressed in SI units? If yes, by how much would it differ?

•5.32 At sea level, a nitrogen molecule in the air has an average kinetic energy of $6.2 \cdot 10^{-21}$ J. Its mass is $4.7 \cdot 10^{-26}$ kg. If the molecule could shoot straight up without colliding with other molecules, how high would it rise? What percentage of the Earth's radius is this height? What is the molecule's initial speed? (Assume that you can use $g = 9.81$ m/s$^2$; although we'll see in Chapter 12 that this assumption may not be justified for this situation.)

••5.33 A bullet moving at a speed of 153 m/s passes through a plank of wood. After passing through the plank, its speed is 130 m/s. Another bullet, of the same mass and size but moving at 92 m/s, passes through an identical plank. What will this second bullet's speed be after passing through the plank? Assume that the resistance offered by the plank is independent of the speed of the bullet.

## Section 5.5

•5.34 A particle of mass $m$ is subjected to a force acting in the $x$-direction. $F_x = (3.0 + 0.50x)$ N. Find the work done by the force as the particle moves from $x = 0$ to $x = 4.0$ m.

•5.35 A force has the dependence $F_x(x) = -kx^4$ on the displacement $x$, where the constant $k = 20.3$ N/m$^4$. How much work does it take to change the displacement from 0.73 m to 1.35 m?

•5.36 A body of mass $m$ moves along a trajectory $\vec{r}(t)$ in three-dimensional space with constant kinetic energy. What geometric relationship has to exist between the body's velocity vector, $\vec{v}(t)$, and its acceleration vector, $\vec{a}(t)$, in order to accomplish this?

•5.37 A force given by $F(x) = 5x^3\hat{x}$(in N/m$^3$) acts on a 1-kg mass moving on a frictionless surface. The mass moves from $x = 2$ m to $x = 6$ m.

a) How much work is done by the force?

b) If the mass has a speed of 2 m/s at $x = 2$ m, what is its speed at $x = 6$ m?

## Section 5.6

5.38 An ideal spring has the spring constant $k = 440$ N/m. Calculate the distance this spring must be stretched from its equilibrium position for 25 J of work to be done.

5.39 A spring is stretched 5.00 cm from its equilibrium position. If this stretching requires 30.0 J of work, what is the spring constant?

5.40 A spring with spring constant $k$ is initially compressed a distance $x_0$ from its equilibrium length. After returning to its equilibrium position, the spring is then stretched a distance $x_0$ from that position. What is the ratio of the work that needs to be done on the spring in the stretching to the work done in the compressing?

•5.41 A spring with a spring constant of 238.5 N/m is compressed by 0.231 m. Then a steel ball bearing of mass 0.0413 kg is put against the end of the spring, and the spring is released. What is the speed of the ball bearing right after it loses contact with the spring? (The ball bearing will come off the spring exactly as the spring returns to its equilibrium position. Assume that the mass of the spring can be neglected.)

## Section 5.7

5.42 A horse draws a sled horizontally across a snow-covered field. The coefficient of friction between the sled and the snow is 0.195, and the mass of the sled, including the load, is 202.3 kg. If the horse moves the sled at a constant speed of 1.785 m/s, what is the power needed to accomplish this?

5.43 A horse draws a sled horizontally on snow at constant speed. The horse can produce a power of 1.060 hp. The coefficient of friction between the sled and the snow is 0.115, and the mass of the sled, including the load, is 204.7 kg. What is the speed with which the sled moves across the snow?

5.44 While a boat is being towed at a speed of 12 m/s, the tension in the towline is 6.0 kN. What is the power supplied to the boat through the towline?

5.45 A car of mass 1214.5 kg is moving at a speed of 62.5 mph when it misses a curve in the road and hits a bridge piling. If the car comes to rest in 0.236 s, how much average power (in watts) is expended in this interval?

5.46 An engine expends 40 hp in moving a car along a level track at a speed of 15 m/s. How large is the total force acting on the car in the opposite direction of the motion of the car?

•5.47 A bicyclist coasts down a 7.0° slope at a steady speed of 5.0 m/s. Assuming a total mass of 75 kg (bicycle plus rider), what must the cyclist's power output be to pedal up the same slope at the same speed?

•5.48 A car of mass 942.4 kg accelerates from rest with a constant power output of 140.5 hp. Neglecting air resistance, what is the speed of the car after 4.55 s?

•5.49 A small blimp is used for advertising purposes at a football game. It has a mass of 93.5 kg and is attached by a towrope to a truck on the ground. The towrope makes an angle of 53.3° downward from the horizontal, and the blimp hovers at a constant height of 19.5 m above the ground. The truck moves on a straight line for 840.5 m on the level surface of the stadium parking lot at a constant velocity of 8.9 m/s. If the drag coefficient ($K$ in $F = Kv^2$) is 0.5 kg/m, how much work is done by the truck in pulling the blimp (assuming there is no wind)?

••5.50 A car of mass $m$ accelerates from rest along a level straight track, not at constant acceleration but with constant engine power, $P$. Assume that air resistance is negligible.

a) Find the car's velocity as a function of time.

b) A second car starts from rest alongside the first car on the same track, but maintains a constant acceleration. Which car takes the initial lead? Does the other car overtake it? If yes, write a formula for the distance from the starting point at which this happens.

c) You are in a drag race, on a straight level track, with an opponent whose car maintains a constant acceleration of 12.0 m/s$^2$. Both cars have identical masses of 1000 kg. The cars start together from rest. Air resistance is assumed to be negligible. Calculate the minimum power your engine needs for you to win the race, assuming the power output is constant and the distance to the finish line is 0.250 mi.

## Additional Problems

5.51 At the 2004 Olympics Games in Athens, Greece, the Iranian athlete Hossein Reza Zadeh won the super-heavy-weight class gold medal in weightlifting. He lifted 472.5 kg (1041 lb) combined in his two best lifts in the competition. Assuming that he lifted the weights a height of 196.7 cm, what work did he do?

**5.52** How much work is done against gravity in lifting a 6-kg weight through a distance of 20 cm?

**5.53** A certain tractor is capable of pulling with a steady force of 14 kN while moving at a speed of 3.0 m/s. How much power in kilowatts and in horsepower is the tractor delivering under these conditions?

**5.54** A shot-putter accelerates a 7.3-kg shot from rest to 14 m/s. If this motion takes 2.0 s, what average power was supplied?

5.55 An advertisement claims that a certain 1200-kg car can accelerate from rest to a speed of 25 m/s in 8.0 s. What average power must the motor supply in order to cause this acceleration? Ignore losses due to friction.

**5.56** A car of mass $m = 1250$ kg is traveling at a speed of $v_0 = 105$ km/h (29.2 m/s). Calculate the work that must be done by the brakes to completely stop the car.

**5.57** An arrow of mass $m = 88$ g (0.088 kg) is fired from a bow. The bowstring exerts an average force of $F = 110$ N on the arrow over a distance $d = 78$ cm (0.78 m). Calculate the speed of the arrow as it leaves the bow.

**5.58** The mass of a physics textbook is 3.4 kg. You pick the book up off a table and lift it 0.47 m at a constant speed of 0.27 m/s.

a) What is the work done by gravity on the book?

b) What is the power you supplied to accomplish this task?

5.59 A sled, with mass $m$, is given a shove up a frictionless incline, which makes a 28° angle with the horizontal. Eventually, the sled comes to a stop at a height of 1.35 m above where it started. Calculate its initial speed.

**5.60** A man throws a rock of mass $m = 0.325$ kg straight up into the air. In this process, his arm does a total amount of work $W_{net} = 115$ J on the rock. Calculate the maximum distance, $h$, above the man's throwing hand that the rock will travel. Neglect air resistance.

**5.61** A car does the work $W_{car} = 7.0 \cdot 10^4$ J in traveling a distance $x = 2.8$ km at constant speed. Calculate the average force $F$ (from all sources) acting on the car in this process.

**•5.62** A softball, of mass $m = 0.25$ kg, is pitched at a speed $v_0 = 26.4$ m/s. Due to air resistance, by the time it reaches home plate it has slowed by 10%. The distance between the plate and the pitcher is $d = 15$ m. Calculate the average force of air resistance, $F_{air}$, that is exerted on the ball during its movement from the pitcher to the plate.

•5.63 A flatbed truck is loaded with a stack of sacks of cement whose combined mass is 1143.5 kg. The coefficient of static friction between the bed of the truck and the bottom sack in the stack is 0.372, and the sacks are not tied down but held in place by the force of friction between the bed and the bottom sack. The truck accelerates uniformly from rest to 56.6 mph in 22.9 s. The stack of sacks is 1 m from the end of the truck bed. Does the stack slide on the truck bed? The coefficient of kinetic friction between the bottom sack and the truck bed is 0.257. What is the work done on the stack by the force of friction between the stack and the bed of the truck?

**•5.64** A driver notices that her 1000-kg car slows from $v_0 = 90$ km/h (25 m/s) to $v = 70$ km/h (19.4 m/s) in $t = 6.0$ s moving on level ground in neutral gear. Calculate the power needed to keep the car moving at a constant speed, $v_{ave} = 80$ km/h (22.2 m/s).

**•5.65** The 125-kg cart in the figure starts from rest and rolls with negligible friction. It is pulled by three ropes as shown. It moves 100 m horizontally. Find the final velocity of the cart.

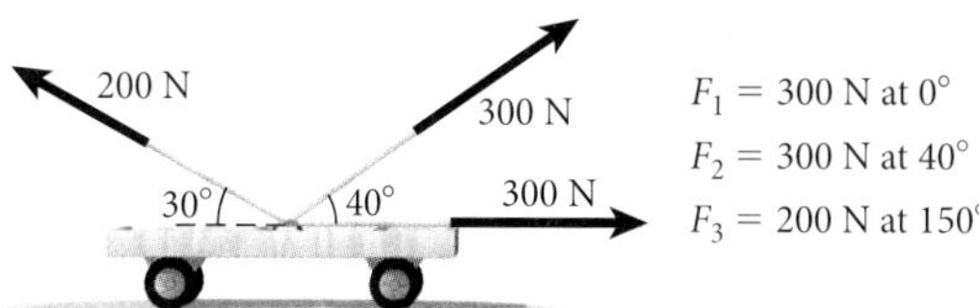


**•5.66** Calculate the power required to propel a 1000.0-kg car at 25.0 m/s up a straight slope inclined 5.0° above the horizontal. Neglect friction and air resistance.

•5.67 A grandfather pulls his granddaughter, whose mass is 21.0 kg and who is sitting on a swing with ropes of length 2.50 m, backward and releases her from rest. The speed of the granddaughter at the bottom of the swinging motion is 3.00 m/s. What is the angle (in degrees, measured relative to the vertical) from which she is released?

**•5.68** A 65-kg hiker climbs to the second base camp on Nanga Parbat in Pakistan, at an altitude of 3900 m, starting from the first base camp at 2200 m. The climb is made in 5.0 h. Calculate (a) the work done against gravity, (b) the average power output, and (c) the rate of energy input required, assuming the energy conversion efficiency of the human body is 15%.

# 6 Potential Energy and Energy Conservation

WHAT WE WILL LEARN 169

**FIGURE 6.1** Niagara Falls.

## WHAT WE WILL LEARN

- Potential energy, $U$, is the energy stored in the configuration of a system of objects that exert forces on one another.
- When a conservative force does work on an object that travels on a path and returns to where it started (a closed path), the total work is zero. A force that does not fulfill this requirement is called a nonconservative force.
- A potential energy can be associated with any conservative force. The change in potential energy due to some spatial rearrangement of a system is equal to the negative of the work done by the conservative force during this spatial rearrangement.
- The mechanical energy, $E$, is the sum of kinetic energy and potential energy.
- The total mechanical energy is conserved (it remains constant over time) for any mechanical process within an isolated system that involves only conservative forces.
- In an isolated system, the total energy—that is, the sum of all forms of energy, mechanical or other—is always conserved. This holds with both conservative and nonconservative forces.
- Small perturbations about a stable equilibrium point result in small oscillations around the equilibrium point; for an unstable equilibrium point, small perturbations result in an accelerated movement away from the equilibrium point.

Niagara Falls is one of the most spectacular sights in the world, with about 5500 cubic meters of water dropping 49 m (160 ft) every *second!* Horseshoe Falls on the Canadian border, shown in Figure 6.1, has a length of 790 m (2592 ft); American Falls on the United States side extends another 305 m (1001 ft) long. Together, they are one of the great tourist attractions of North America. However, Niagara Falls is more than a scenic wonder. It is also one of the largest sources of electric power in the world, producing over 2500 megawatts (see Solved Problem 6.6). Humans have used the energy of falling water since ancient times, using it to turn large paddlewheels for mills and factories. Today, the conversion of energy in falling water to electrical energy by hydroelectric dams is a major source of energy throughout the world.

As we saw in Chapter 5, energy is a fundamental concept in physics that governs many of the interactions involving forces and motions of objects. In this chapter, we continue our study of energy, introducing several new forms of energy and new laws that govern its use. We will return to the laws of energy in the chapters on thermodynamics, building on much of the material presented here. First, though, we will continue our study of mechanics, relying heavily on the ideas discussed here.

## 6.1 Potential Energy

Chapter 5 examined in detail the relationship between kinetic energy and work, and one of the main points was that work and kinetic energy can be converted into one another. Now, this section introduces another kind of energy called *potential energy.*

**Potential energy,** $U$, is the energy stored in the configuration of a system of objects that exert forces on one another. For example, we have seen that work is done by an external force in lifting a load against the force of gravity, and this work is given by $W = mgh$, where $m$ is the mass of the load and $h = y - y_0$ is the height to which the load is lifted above its initial position. (In this chapter, we will assume the $y$-axis points upward unless specified differently.) This lifting can be accomplished without changing the kinetic energy, as in the case of a weightlifter who lifts a mass above his head and holds it there. There is energy stored in holding the mass above the head. If the weightlifter lets go of the mass this energy can be converted back into kinetic energy as the mass accelerates and falls to the ground. We can express the gravitational potential energy as

$$U_g = mgy. \tag{6.1}$$

The change in the gravitational potential energy of the mass is then

$$\Delta U_g \equiv U_g(y) - U_g(y_0) = mg(y - y_0) = mgh. \quad (6.2)$$

(Equation 6.1 is valid only near the surface of the Earth, where $F_g = mg$, and in the limit that Earth is infinitely massive relative to the object. We will encounter a more general expression for $U_g$ in Chapter 12.) In Chapter 5, we also calculated the work done by the gravitational force on an object that is lifted through a height $h$ to be $W_g = -mgh$. From this, we see that the work done by the gravitational force and the gravitational potential energy for an object lifted from rest to a height $h$ are related by

$$\Delta U_g = -W_g. \quad (6.3)$$

Let's consider the gravitational potential energy in a specific situation: A weightlifter who lifts a barbell with mass $m$. The weightlifter starts with the barbell on the floor, as shown in Figure 6.2a. At $y = 0$, the gravitational potential energy can be defined to be $U_g = 0$. The weightlifter then picks up the barbell, lifts it to a height of $y = h/2$, and holds it there, as shown in Figure 6.2b. The gravitational potential energy is now $U_g = mgh/2$, and the work done by gravity on the barbell is $W_g = -mgh/2$. The weightlifter next lifts the barbell over his head to a height of $y = h$, as shown in Figure 6.2c. The gravitational potential energy is now $U_g = mgh$, and the work done by gravity during this part of the lift is $W_g = -mgh/2$. Having completed the lift, the weightlifter lets go of the barbell, and it falls to the floor, as illustrated in Figure 6.2d. The gravitational potential energy of the barbell on the floor is again $U_g = 0$, and the work done by gravity during the fall is $W_g = mgh$.

Equation 6.3 is true even for complicated paths involving horizontal as well as vertical motion of the object, because the gravitational force does no work during horizontal segments of the motion. In horizontal motion, the displacement is perpendicular to the force of gravity (which always points vertically down), and thus the scalar product between the force and displacement vectors is zero; hence, no work is done.

Lifting any mass to a higher elevation involves doing work against the force of gravity and generates an increase in gravitational potential energy of the mass. This energy can be stored for later use. This principle is employed, for example, at many hydroelectric dams. The excess electricity generated by the turbines is used to pump water to a reservoir at a higher elevation. There it constitutes a reserve that can be tapped in times of high energy demand and/or low water supply. Stated in general terms, if $\Delta U_g$ is positive, there exists the potential (hence the name *potential energy*) to allow $\Delta U_g$ to be negative in the future, thereby extracting positive work, since $W_g = -\Delta U_g$.

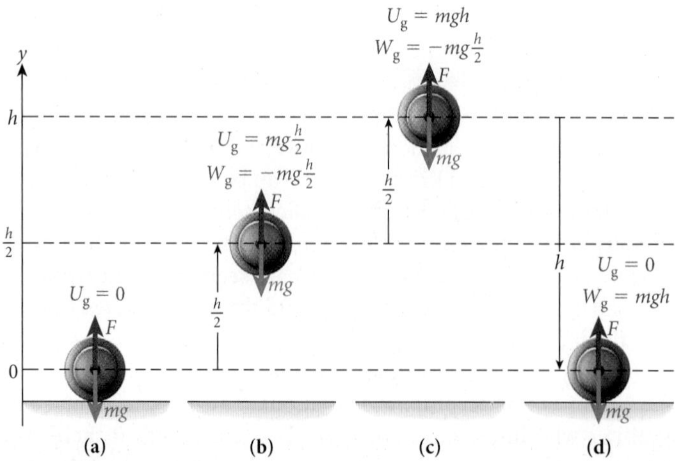


**FIGURE 6.2** Lifting a barbell and potential energy (the diagram shows the barbell in side view and omits the weightifter). The weight of barbell is *mg*, and the normal force exerted by the floor or the weightlifter to hold the weight up is *F*. (a) The barbell is initially on the floor. (b) The weightlifter lifts the barbell of mass *m* to a height of *h*/2 and holds it there. (c) The weightlifter lifts the barbell an additional distance *h*/2, to a height of *h*, and holds it there. (d) The weightlifter lets the barbell drop to the floor.

## 6.2 Conservative and Nonconservative Forces

Before we can calculate the potential energy from a given force, we have to ask: Can all kinds of forces be used to store potential energy for later retrieval? If not, what kinds of forces can we use? To answer this question, we need to consider what happens to the work done by a force when the direction of the path taken by an object is reversed. In the case of the gravitational force, we have already seen what happens. As shown in Figure 6.3, the work done by $F_g$ when an object of mass $m$ is lifted from elevation $y_A$ to $y_B$ has the same magnitude, but the opposite sign, to the work done by $F_g$ when lowering the same object from elevation $y_B$ to $y_A$. This means that the total work done by $F_g$ in lifting the object from some elevation to a different one and then returning it to the same elevation is zero. This fact is the basis for the definition of a conservative force (refer to Figure 6.4a).

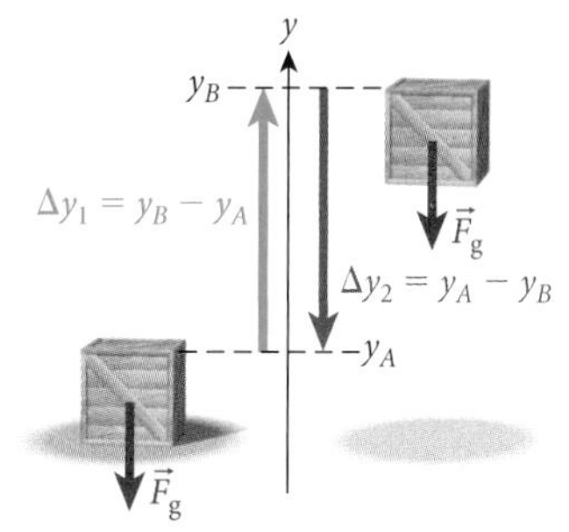


**FIGURE 6.3** Gravitational force vectors and displacement for lifting and lowering a box.

### Definition

A **conservative force** is any force for which the work done over any closed path is zero. A force that does not fulfill this requirement is called a **nonconservative force.**

For conservative forces, we can immediately state two consequences of this definition:

1. If we know the work, $W_{A\to B}$, done by a conservative force on an object as the object moves along a path from point $A$ to point $B$, then we also know the work, $W_{B\to A}$, that the same force does on the object as it moves along the path in the reverse direction, from point $B$ to point $A$ (see Figure 6.4b):

$$W_{B\to A} = -W_{A\to B} \text{ (for conservative forces).} \tag{6.4}$$

The proof of this statement is obtained from the condition of zero work over a closed loop. Because the path from $A$ to $B$ to $A$ forms a closed loop, the sum of the work contributions from the loop has to equal zero. In other words,

$$W_{A\to B} + W_{B\to A} = 0,$$

from which equation 6.4 follows immediately.

2. If we know the work, $W_{A\to B,\text{path 1}}$, done by a conservative force on an object moving along path 1 from point $A$ to point $B$, then we also know the work, $W_{A\to B,\text{path 2}}$, done by the same force on the object when it uses any other path 2 to go from point $A$ to point $B$ (see Figure 6.4c). The work is the same; the work done by a conservative force is independent of the path taken by the object:

$$W_{A\to B,\text{path 2}} = W_{A\to B,\text{path 1}} \tag{6.5}$$
(for arbitrary paths 1 and 2, for conservative forces).

This statement is also easy to prove from the definition of a conservative force as a force for which the work done over any closed path is zero. The path from point $A$ to point $B$ on path 1 and then back from $B$ to $A$ on path 2 is a closed loop; therefore, $W_{A\to B,\text{path 2}} + W_{B\to A,\text{path 1}} = 0$. Now we use equation 6.4 the path direction, $W_{B\to A,\text{path 1}} = -W_{A\to B,\text{path 1}}$. Combining these two results gives us $W_{A\to B,\text{path 2}} - W_{A\to B,\text{path 1}} = 0$, from which equation 6.5 follows.

One physical application of the mathematical results just given involves riding a bicycle from one point, such as your home, to another point, such as the swimming pool. Assuming that your home is located at the foot of a hill and the pool at the top, we can use the Figure 6.4c to illustrate this example, with point $A$ representing your home and point $B$ the pool. What the above statements regarding conservative forces mean is that you do the same amount of work riding your bike from home to the pool, independent of the route you select. You can take a shorter and steeper route or a flatter and longer route; you can even take a route that goes up and down between points $A$ and $B$. The total work will be the same. As with almost all real-world examples, however, there are some complications here: It matters whether

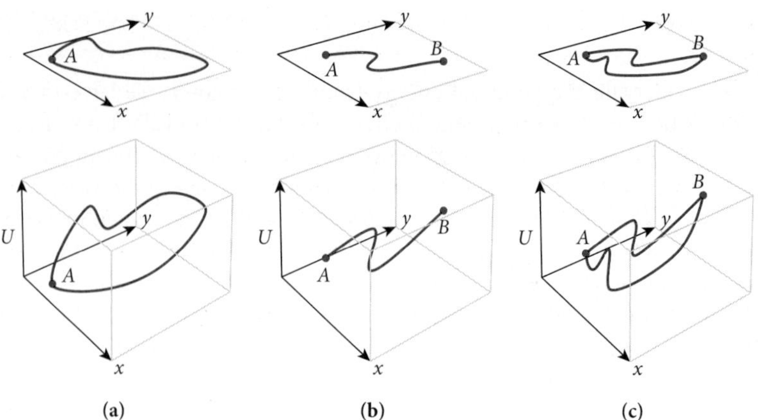


**FIGURE 6.4** Various paths for the potential energy related to a conservative force as a function of positions $x$ and $y$, with $U$ proportional to $y$. The two-dimensional plots are projections of the three-dimensional plots onto the $xy$-plane. (a) Closed loop. (b) A path from point $A$ to point $B$. (c) Two different paths between points $A$ and $B$.

you use the handbrakes; there is air resistance and tire friction to consider; and your body also performs other metabolic functions during the ride, in addition to moving your mass and that of the bicycle from point $A$ to $B$. But, this example can help you develop a mental picture of the concepts of path-independence of work and conservative forces.

The gravitational force, as we have seen, is an example of a conservative force. Another example of a conservative force is the spring force. Not all forces are conservative, however. Which forces are nonconservative?

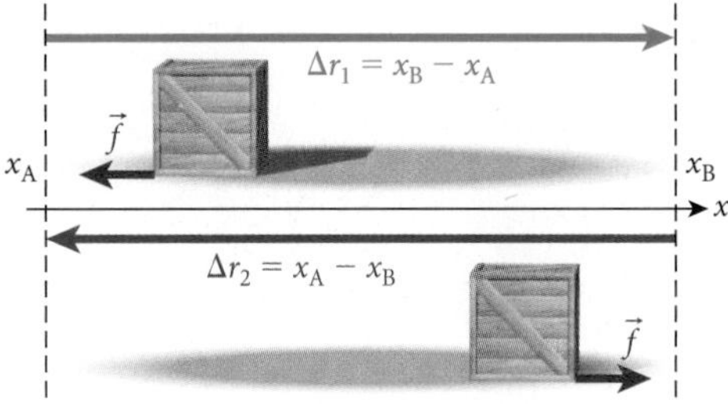


**FIGURE 6.5** Friction force vector and displacement vector for the process of sliding a box back and forth across a surface with friction.

## Friction Forces

Let's consider what happens in sliding a box across a horizontal surface, from point $A$ to point $B$ and then back to point $A$, if the coefficient of kinetic friction between the box and the surface is $\mu_k$ (Figure 6.5). As we have learned, the friction force is given by $f = \mu_k N = \mu_k mg$ and always points in the direction opposite to that of the motion. Let's use results from Chapter 5 to find the work done by this friction force. Since the friction force is constant, the amount of work it does is found by simply taking the scalar product between the friction force and displacement vectors.

For the motion from $A$ to $B$, we use the general scalar product formula for work done by a constant force:

$$W_{f1} = \vec{f} \bullet \Delta \vec{r}_1 = -f \cdot (x_B - x_A) = -\mu_k mg \cdot (x_B - x_A).$$

We have assumed that the positive $x$-axis is pointing to the right, as is conventional, so the friction force points in the negative $x$-direction. For the motion from $B$ back to $A$, then, the friction force points in the positive $x$-direction. Therefore, the work done for this part of the path is

$$W_{f2} = \vec{f} \bullet \Delta \vec{r}_2 = f \cdot (x_A - x_B) = \mu_k mg \cdot (x_A - x_B).$$

This result leads us to conclude that the total work done by the friction force while the box slides across the surface on the closed path from point $A$ to point $B$ and back to point $A$ is not zero, but instead

$$W_f = W_{f1} + W_{f2} = -2\mu_k mg(x_B - x_A) < 0. \tag{6.6}$$

There appears to be a contradiction between this result and the work–kinetic energy theorem. The box starts with zero kinetic energy and at a certain position, and it ends up with zero kinetic energy and at the same position. According to the work–kinetic energy theorem, the total work done should be zero. This leads us to conclude that the friction force does not do work in the way that a conservative force does. Instead, the friction force converts kinetic and/or potential energy into internal excitation energy of the two objects that exert friction on each other (the box and the support surface, in this case). This internal excitation energy can take the form of vibrations or thermal energy or even chemical or electrical energy. The main point is that the conversion of kinetic and/or potential energy to internal excitation energy is not reversible; that is, the internal excitation energy cannot be fully converted back into kinetic and/or potential energy.

Thus, we see that the friction force is an example of a nonconservative force. Because the friction force always acts in a direction opposite to the displacement, the dissipation of energy due to the friction force is always negative, whether or not the path is closed. Work done by a conservative force, $W$, can be positive or negative, but the dissipation from the friction force, $W_f$, is always negative, withdrawing kinetic and/or potential energy and converting it into internal excitation energy. Using the symbol $W_f$ for this dissipated energy is a reminder that we use the same procedures to calculate it as to calculate work for conservative forces.

The decisive fact is that the friction force switches direction as a function of the direction of motion and causes dissipation. The friction force vector is always antiparallel to the velocity vector; any force with this property cannot be conservative. Dissipation converts kinetic energy into internal energy of the object, which is another important characteristic of a nonconservative force. In Section 6.7, we will examine this point in more detail.

Another example of a nonconservative force is the force of air resistance. It is also velocity dependent and always points in the direction opposite to the velocity vector, just like the force of kinetic friction. Yet another example of a nonconservative force is the damping force (discussed in Chapter 14). It, too, is velocity dependent and opposite in direction to the velocity.

### 6.1 In-Class Exercise

A person pushes a box with mass 10.0 kg a distance of 5.00 m across a floor. The coefficient of kinetic friction between the box and the floor is 0.250. The person then picks up the box, raising it to a height of 1.00 m, carries the box back to the starting point, and puts it back down on the floor. How much work has the person done on the box?

a) 0 J
b) 12.5 J
c) 98.1 J
d) 123 J
e) 25.0 J
f) 246 J

## 6.3 Work and Potential Energy

In considering the work done by the gravitational force and its relationship to gravitational potential energy in Section 6.1, we found that the change in potential energy is equal to the negative of the work done by the force, $\Delta U_g = -W_g$. This relationship is true for all conservative forces. In fact, we can use it to define the concept of potential energy.

For any conservative force, the change in potential energy due to some spatial rearrangement of a system is equal to the negative of the work done by the conservative force during this spatial rearrangement:

$$\Delta U = -W. \tag{6.7}$$

We have already seen that work is given by

$$W = \int_{x_0}^{x} F_x(x')dx'. \tag{6.8}$$

Combining equations 6.7 and 6.8 gives us the relationship between the conservative force and the potential energy:

$$\Delta U = U(x) - U(x_0) = -\int_{x_0}^{x} F_x(x')dx'. \tag{6.9}$$

We could use equation 6.9 to calculate the potential energy change due to the action of any given conservative force. Why should we bother with the concept of potential energy when we can deal directly with the conservative force itself? The answer is that the change in potential energy depends only on the beginning and final states of the system and is independent of the path taken to get to the final state. Often, we have a simple expression for the potential energy (and thus its change) prior to working a problem! In contrast, evaluating the integral on the right-hand side of equation 6.9 could be quite complicated. And, in fact, the computational savings is not the only rationale, as the use of energy considerations is based on an underlying physical law (the law of energy conservation, to be introduced in Section 6.4).

In Chapter 5, we evaluated this integral for the force of gravity and for the spring force. The result for the gravitational force is

$$\Delta U_g = U_g(y) - U_g(y_0) = -\int_{y_0}^{y} (-mg)dy' = mg\int_{y_0}^{y} dy' = mgy - mgy_0. \tag{6.10}$$

This is in accord with the result we found in Section 6.1. Consequently, the gravitational potential energy is

$$U_g(y) = mgy + \text{constant}. \tag{6.11}$$

Note that we are able to determine the potential energy at coordinate $y$ only to within an additive constant. The only physically observable quantity, the work done, is related to the *difference* in the potential energy. If we add an arbitrary constant to the value of the potential energy everywhere, the difference in the potential energies remains unchanged.

In the same way, we find for the spring force that

$$\begin{aligned}\Delta U_s &= U_s(x) - U_s(x_0)\\ &= -\int_{x_0}^{x} F_s(x')dx'\\ &= -\int_{x_0}^{x} (-kx')dx'\\ &= k\int_{x_0}^{x} x'dx'\end{aligned}$$

$$\Delta U_s = \tfrac{1}{2}kx^2 - \tfrac{1}{2}kx_0^2. \tag{6.12}$$

Thus, the potential energy associated with elongating a spring from its equilibrium position, at $x = 0$, is

$$U_s(x) = \tfrac{1}{2}kx^2 + \text{constant}. \tag{6.13}$$

Again, the potential energy is determined only to within an additive constant. However, keep in mind that physical situations will often force a choice of this additive constant.

## 6.4 Potential Energy and Force

How can we find the conservative force when we have information on the corresponding potential energy? In calculus, taking the derivative is the inverse operation of integrating, and integration is used in equation 6.9 for the change in potential energy. Therefore, we take the derivative of that expression to obtain the force from the potential energy:

$$F_x(x) = -\frac{dU(x)}{dx}. \tag{6.14}$$

Equation 6.14 is an expression for the force from the potential energy for the case of motion in one dimension. As you can see from this expression, any constant that you add to the potential energy will not have any influence on the result you obtain for the force, because taking the derivative of a constant term results in zero. This is further evidence that the potential energy can be determined only within an additive constant.

We will not consider motion is three-dimensional situations until later in this book. However, for completeness, we can state the expression for the force from the potential energy for the case of three-dimensional motion:

$$\vec{F}(\vec{r}) = -\left(\frac{\partial U(\vec{r})}{\partial x}\hat{x} + \frac{\partial U(\vec{r})}{\partial y}\hat{y} + \frac{\partial U(\vec{r})}{\partial z}\hat{z}\right). \tag{6.15}$$

Here, the force components are given as partial derivatives with respect to the corresponding coordinates. If you major in engineering or science, you will encounter partial derivatives in many situations.

### 6.2 In-Class Exercise

The potential energy, $U(x)$, is shown as a function of position, $x$, in the figure. In which region is the magnitude of the force the highest?

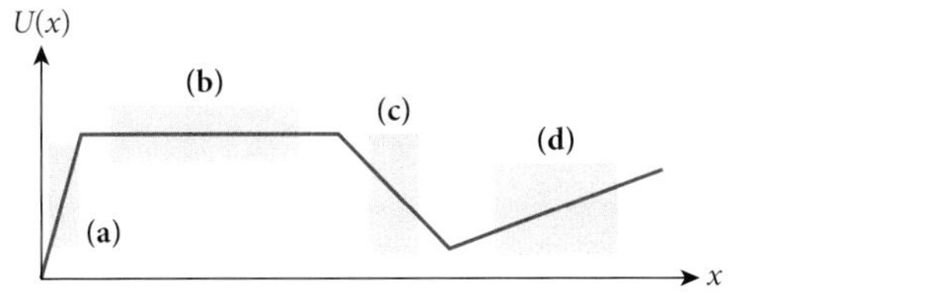


## Lennard-Jones Potential

Empirically, the potential energy associated with the interaction of two atoms in a molecule as a function of the separation of the atoms has a form that is called a *Lennard-Jones potential.* This potential energy as a function of the separation, $x$, is given by

$$U(x)=4U_0\left(\left(\frac{x_0}{x}\right)^{12}-\left(\frac{x_0}{x}\right)^{6}\right). \tag{6.16}$$

Here $U_0$ is a constant energy and $x_0$ is a constant length. The Lennard-Jones potential is one of the most important concepts in atomic physics and is used for most numerical simulations of molecular systems.

### EXAMPLE 6.1 Molecular Force

**PROBLEM**

What is the force resulting from the Lennard-Jones potential?

**SOLUTION**

We simply take the negative of the derivative of the potential energy with respect to $x$:

$$\begin{aligned}F_x(x)&=-\frac{dU(x)}{dx}\\&=-\frac{d}{dx}\left(4U_0\left(\left(\frac{x_0}{x}\right)^{12}-\left(\frac{x_0}{x}\right)^{6}\right)\right)\\&=-4U_0x_0^{12}\frac{d}{dx}\left(\frac{1}{x^{12}}\right)+4U_0x_0^{6}\frac{d}{dx}\left(\frac{1}{x^{6}}\right)\\&=48U_0x_0^{12}\frac{1}{x^{13}}-24U_0x_0^{6}\frac{1}{x^{7}}\\&=\frac{24U_0}{x_0}\left(2\left(\frac{x_0}{x}\right)^{13}-\left(\frac{x_0}{x}\right)^{7}\right).\end{aligned}$$

**PROBLEM**

At what value of $x$ does the Lennard-Jones potential have its minimum?

**SOLUTION**

Since we just found that the force is the derivative of the potential energy function, all we have to do is find the point(s) where $F(x) = 0$. This leads to

$$F_x(x)\Big|_{x=x_{\min}}=\frac{24U_0}{x_0}\left(2\left(\frac{x_0}{x_{\min}}\right)^{13}-\left(\frac{x_0}{x_{\min}}\right)^{7}\right)=0.$$

*Continued—*

This condition can be fulfilled only if the expression in the larger parentheses is zero; thus

$$2\left(\frac{x_0}{x_{\min}}\right)^{13}=\left(\frac{x_0}{x_{\min}}\right)^{7}.$$

Multiplying both sides by $x^{13}x_0^{-7}$ then yields

$$2x_0^6 = x_{\min}^6$$

or

$$x_{\min} = 2^{1/6}x_0 \approx 1.1225x_0.$$

Mathematically, it is not enough to show that the derivative is zero to establish that the potential indeed has a minimum at this coordinate. We should also make sure that the second derivative is positive. You can do this as an exercise.

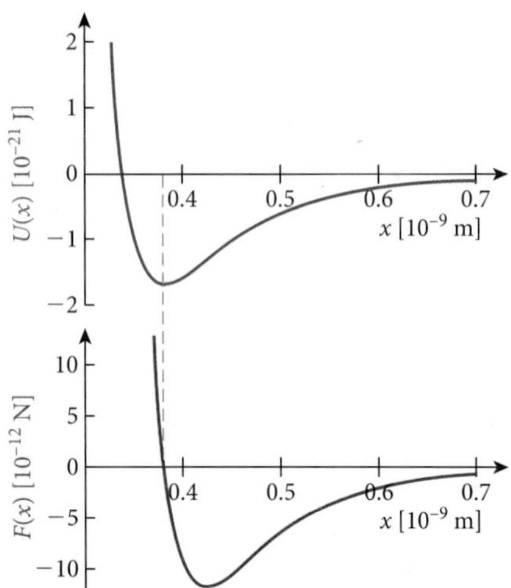


**FIGURE 6.6** (a) Dependence of the potential energy on the $x$-coordinate of the potential energy function in equation 6.16. (b) Dependence of the force on the $x$-coordinate of the potential energy function in equation 6.16.

Figure 6.6a, shows the shape of the Lennard-Jones potential, plotted from equation 6.16, with $x_0 = 0.34$ nm and $U_0 = 1.70 \cdot 10^{-21}$ J, for the interaction of two argon atoms as a function of the separation of the centers of the two atoms. Figure 6.6b plots the corresponding molecular force, using the expression we found in Example 6.1. The vertical gray dashed line marks the coordinate where the potential has a minimum and where consequently the force is zero. Also note that close to the minimum point of the potential (within ±0.1 nm), the force can be closely approximated by a linear function, $F_x(x) \approx -k(x - x_{\min})$. This means that close to the minimum, the molecular force due to the Lennard-Jones potential behaves like a spring force.

Chapter 5 mentioned that forces similar to the spring force appear in many physical systems, and the connection between potential energy and force just described tells us why. Look, for example, at the skateboarder in the half-pipe in Figure 6.7. The curved surface of the half-pipe approximates the shape of the Lennard-Jones potential close to the minimum. If the skateboarder is at $x = x_{\min}$, he can remain there at rest. If he is to the left of the minimum, where $x < x_{\min}$, then the half-pipe exerts a force on him, which points to the right, $F_x > 0$; the further to the left he moves, the bigger the force becomes. On the right side of the half-pipe, for $x > x_{\min}$, the force points to the left, that is, $F_x < 0$. Again, these observations can be summarized with a force expression that approximately follows Hooke's Law: $F_x(x) = -k(x - x_{\min})$.

In addition, we can reach this same conclusion mathematically, by writing a Taylor expansion for $F_x(x)$ around $x_{\min}$:

$$F_x(x)=F_x(x_{\min})+\left(\frac{dF_x}{dx}\right)_{x=x_{\min}}\cdot(x-x_{\min})+\frac{1}{2}\left(\frac{d^2F_x}{dx^2}\right)_{x=x_{\min}}\cdot(x-x_{\min})^2+\cdots.$$

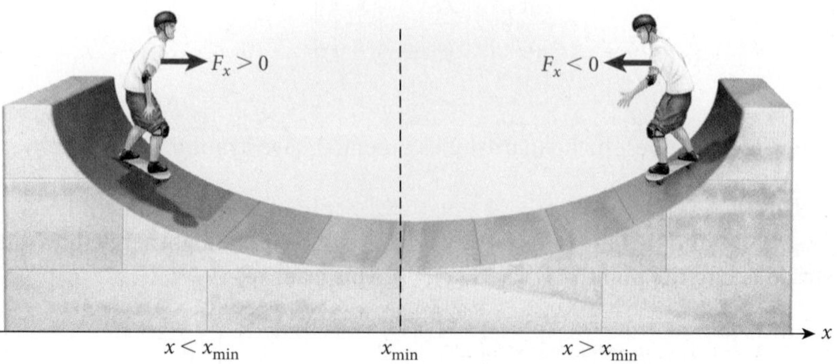


**FIGURE 6.7** Skateboarder in a half-pipe.

Since we are expanding around the potential energy minimum and since we have just shown that the force is zero there, we have $F_x(x_{min}) = 0$. If there is a potential minimum at $x = x_{min}$, then the second derivative of the potential must be positive. Since, according to equation 6.14, the force is $F_x(x) = -dU(x)/dx$, this means that the derivative of the force is $dF_x(x)/dx = -d^2U(x)/dx^2$. At the minimum of the potential, we thus have $(dF_x/dx)_{x=x_{min}} < 0$. Expressing the value of the first derivative of the force at coordinate $x_{min}$ as some constant, $(dF_x/dx)_{x=x_{min}} = -k$ (with $k > 0$ ), we find $F_x(x) = -k(x - x_{min})$, if we are sufficiently close to $x_{min}$ that we can neglect terms proportional to $(x - x_{min})^2$ and higher powers.

These physical and mathematical arguments establish why it is important to study Hooke's Law and the resulting equations of motion in detail. In this chapter, we study the work done by the spring force. In Chapter 14 on oscillations, we will analyze the motion of an object under the influence of the spring force.

### 6.1 Self-Test Opportunity

Some forces in nature depend on the inverse of the distance between objects squared. How does the potential energy associated with such a force depend on the distance between the objects?

## 6.5 Conservation of Mechanical Energy

We have defined potential energy in reference to a system of objects. We will examine different kinds of general systems in later chapters, but here we focus on one particular kind of system: an **isolated system,** which by definition is a system of objects that exert forces on one another but for which no force external to the system causes energy changes within the system. This means that no energy is transferred into or out of the system. This very common situation is highly important in science and engineering and has been extensively studied. One of the fundamental concepts of physics involves energy within an isolated system.

To investigate this concept, we begin with a definition of **mechanical energy,** *E*, as the sum of kinetic energy and potential energy:

$$E = K + U. \tag{6.17}$$

(Later, when we move beyond mechanics, we will add other kinds of energy to this sum and call it the *total energy*.)

For any mechanical process that occurs inside an isolated system and involves only conservative forces, the total mechanical energy is conserved. This means that the total mechanical energy remains constant in time:

$$\Delta E = \Delta K + \Delta U = 0. \tag{6.18}$$

An alternative way of writing this result (which we'll derive below) is

$$K + U = K_0 + U_0, \tag{6.19}$$

where $K_0$ and $U_0$ are the initial kinetic energy and potential energy, respectively. This relationship, which is called the law of **conservation of mechanical energy,** does not imply that the kinetic energy of the system cannot change, or that the potential energy alone remains constant. Rather it states that their changes are exactly compensating and thus offset each other. It is worth repeating that conservation of mechanical energy is valid only for conservative forces and for an isolated system, for which the influence of external forces can be neglected.

### DERIVATION 6.1

As we have already seen in equation 6.7, if a conservative force does work, then the work causes a change in potential energy:

$$\Delta U = -W.$$

(If the force under consideration is not conservative, this relationship does not hold in general, and conservation of mechanical energy is not valid.)

In Chapter 5, we learned that the relationship between the change in kinetic energy and the work done by a force is (equation 5.15):

$$\Delta K = W.$$

*Continued—*

Combining these two results, we obtain

$$\Delta U = -\Delta K \Rightarrow \Delta U + \Delta K = 0.$$

Using $\Delta U = U - U_0$ and $\Delta K = K - K_0$, we find

$$0 = \Delta U + \Delta K = U - U_0 + K - K_0 = U + K - (U_0 + K_0) \Rightarrow$$
$$U + K = U_0 + K_0.$$

Note that Derivation 6.1 did not make any reference to the particular path along which the force did the work that caused the rearrangement. In fact, you do not need to know any detail about the work or the force, other than that the force is conservative. Nor do you need to know how many conservative forces are acting. If more than one conservative force is present, you interpret $\Delta U$ as the sum of all the potential energy changes and $W$ as the total work done by all of the conservative forces, and the derivation is still valid.

The law of energy conservation enables us to easily solve a huge number of problems that involve only conservative forces, problems that would have been very hard to solve without this law. Later in this chapter, the more general work-energy theorem for mechanics, which includes nonconservative forces, will be presented. This law will enable us to solve an even wider range of problems, including those involving friction.

Equation 6.19 introduces our first conservation law, the law of conservation of mechanical energy. Chapters 18 and 20 will extend this law to include thermal energy (heat) as well. Chapter 7 will present a conservation law for linear momentum. When we discuss rotation in Chapter 10, we will encounter a conservation law for angular momentum. In studying electricity and magnetism, we will find a conservation law for net charge (Chapter 21), and in looking at elementary particle physics (Chapter 39), we will find conservation laws for several other quantities. This list is intended to give you a flavor of a central theme of physics—the discovery of conservation laws and their use in determining the dynamics of various systems.

Before we solve a sample problem, one more remark on the concept of an isolated system is in order. In situations that involve the motion of objects under the influence of the Earth's gravitational force, the isolated system to which we apply the law of conservation of energy actually consists of the moving object plus the entire Earth. However, in using the approximation that the gravitational force is a constant, we assume that the Earth is infinitely massive (and that the moving object is close to the surface of Earth). Therefore, no change in the kinetic energy of the Earth can result from the rearrangement of the system. Thus, we calculate all changes in kinetic energy and potential energy only for the "junior partner"—the object moving under the influence of the gravitational force. This force is conservative and internal to the system consisting of Earth plus moving object, so all the conditions for the utilization of the law of energy conservation are fulfilled.

Specific examples of situations that involve objects moving under the influence of the gravitational force are projectile motion and pendulum motion occurring near the Earth's surface.

**FIGURE 6.8** Illustration for a possible projectile path (red parabola) from the courtyard to the camp below and in front of the castle gate. The blue line indicates the horizontal.

## SOLVED PROBLEM 6.1 The Catapult Defense

Your task is to defend Neuschwanstein Castle from attackers (Figure 6.8). You have a catapult with which you can lob a rock with a launch speed of 14.2 m/s from the courtyard over the castle walls onto the attackers' camp in front of the castle at an elevation 7.20 m below that of the courtyard.

### PROBLEM

What is the speed with which a rock will hit the ground at the attackers' camp? (Neglect air resistance.)

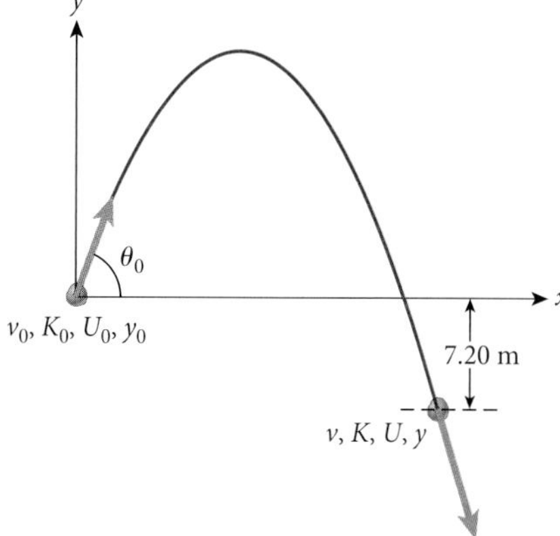


**FIGURE 6.9** Trajectory of the rock launched by the catapult.

## SOLUTION

### THINK

We can solve this problem by applying the conservation of mechanical energy. Once the catapult launches a rock, only the conservative force of gravity is acting on the rock. Thus, the total mechanical energy is conserved, which means the sum of the kinetic and potential energies of the rock always equals the total mechanical energy.

### SKETCH

The trajectory of the rock is shown in Figure 6.9, where the initial speed of the rock is $v_0$, the initial kinetic energy $K_0$, the initial potential energy $U_0$, and the initial height $y_0$. The final speed is $v$, the final kinetic energy $K$, the final potential energy $U$, and the final height $y$.

### RESEARCH

We can use conservation of mechanical energy to write

$$E = K + U = K_0 + U_0,$$

where $E$ is the total mechanical energy. The kinetic energy of the projectile can be expressed as

$$K = \tfrac{1}{2}mv^2,$$

where $m$ is the mass of the projectile and $v$ is its speed when it hits the ground. The potential energy of the projectile can be expressed as

$$U = mgy,$$

where $y$ is the vertical component of the position vector of the projectile when it hits the ground.

### SIMPLIFY

We substitute for $K$ and $U$ in $E = K + U$ to get

$$E = \tfrac{1}{2}mv^2 + mgy = \tfrac{1}{2}mv_0^2 + mgy_0.$$

The mass of the rock, $m$, cancels out, and we are left with

$$\tfrac{1}{2}v^2 + gy = \tfrac{1}{2}v_0^2 + gy_0.$$

We solve this for the speed:

$$v = \sqrt{v_0^2 + 2g(y_0 - y)}. \tag{6.20}$$

### CALCULATE

According to the problem statement, $y_0 - y = 7.20$ m and $v_0 = 14.2$ m/s. Thus, for the final speed, we find

$$v = \sqrt{(14.2 \text{ m/s})^2 + 2(9.81 \text{ m/s}^2)(7.20 \text{ m})} = 18.51766724 \text{ m/s}.$$

### ROUND

The relative height was given to three significant figures, so we report our final answer as

$$v = 18.5 \text{ m/s}.$$

### DOUBLE-CHECK

Our answer for the speed of the rock when it hits the ground in front of the castle is 18.5 m/s, compared with the initial launch speed of 14.2 m/s, which seems reasonable. This speed has to be bigger due to the gain from the difference in gravitational potential energy, and it is comforting that our answer passes this simple test.

Because we were only interested in the speed at impact, we did not even need to know the initial launch angle $\theta_0$ to solve the problem. All launch angles will give the same result

*Continued—*

(for a given launch speed), which is a somewhat surprising finding. (Of course, if you were in this situation, you would obviously want to aim high enough to clear the castle wall and accurately enough to strike the attackers' camp.)

We can also solve this problem using the concepts of projectile motion, which is useful to double-check our answer and to show the power of applying the concept of energy conservation. We start by writing the components of the initial velocity vector $\vec{v}_0$:

$$v_{x0} = v_0 \cos\theta_0$$

and

$$v_{y0} = v_0 \sin\theta_0.$$

The final $x$-component of the velocity, $v_x$, is equal to the initial $x$-component of the initial velocity $v_{x0}$,

$$v_x = v_{x0} = v_0 \cos\theta_0.$$

The final component of the velocity in the $y$-direction can be obtained from a result of the analysis of projectile motion in Chapter 3:

$$v_y^2 = v_{y0}^2 - 2g(y - y_0).$$

Therefore, the final speed of the rock as it hits the ground is

$$\begin{aligned} v &= \sqrt{v_x^2 + v_y^2} \\ &= \sqrt{(v_0 \cos\theta_0)^2 + \left(v_{y0}^2 - 2g(y - y_0)\right)} \\ &= \sqrt{v_0^2 \cos^2\theta_0 + v_0^2 \sin^2\theta_0 - 2g(y - y_0)}. \end{aligned}$$

Remembering that $\sin^2\theta + \cos^2\theta = 1$, we can further simplify and get:

$$v = \sqrt{v_0^2(\cos^2\theta_0 + \sin^2\theta_0) - 2g(y - y_0)} = \sqrt{v_0^2 - 2g(y - y_0)} = \sqrt{v_0^2 + 2g(y_0 - y)}.$$

This is the same as equation 6.20, which we obtained using energy conservation. Even though the final result is the same, the solution process based on energy conservation was by far easier than that based on kinematics.

### 6.2 Self-Test Opportunity

In Solved Problem 6.1, we neglected air resistance. Discuss qualitatively how our final answer would have changed if we had included the effects of air resistance.

As you can see from Solved Problem 6.1, applying the conservation of mechanical energy provides us with a powerful technique for solving problems that seem rather complicated at first sight.

In general, we can determine final speed as a function of the elevation in situations where the gravitational force is at work. For instance, consider the image sequence in Figure 6.10. Two balls are released at the same time from the same height at the top of two ramps with different shapes. At the bottom end of the ramps, both balls reach the same lower elevation. Therefore, in both cases, the height difference between the initial and final points is the same. Both balls also experience normal forces in addition to the gravitational force; however, the normal forces do no work because they are perpendicular to the contact surface, by definition, and the motion is parallel to the surface. Thus, the scalar product of the normal force and displacement vectors is zero. (There is a small friction force, but it is negligible in this case.)

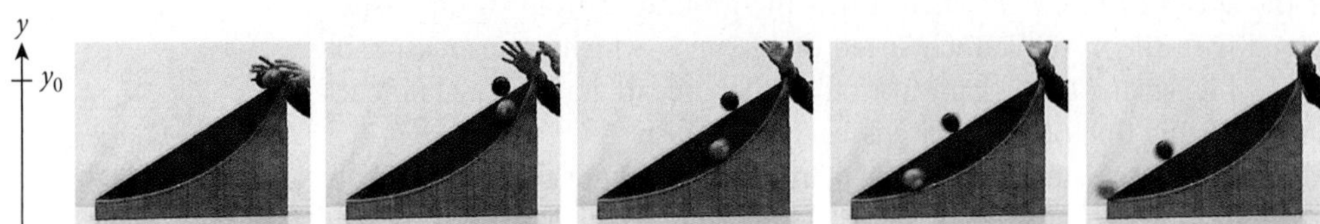


**FIGURE 6.10** Race of two balls down different inclines of the same height.

Energy conservation considerations (see equation 6.20 in Solved Problem 6.1) tell us that the speed of both balls at the bottom end of the ramps has to be the same:

$$v = \sqrt{2g(y_0 - y)}.$$

This equation is a special case of equation 6.20 with $v_0 = 0$. Note that, depending on the curve of the bottom ramp, this result could be rather difficult to obtain using Newton's Second Law. However, even though the velocities at the top and the bottom of the ramps are the same for both balls, you cannot conclude from this result that both balls arrive at the bottom at the same time. The image sequence clearly shows that this is not the case.

### 6.3 Self-Test Opportunity

Why does the lighter-colored ball arrive at the bottom in Figure 6.10 before the other ball?

## 6.6 Work and Energy for the Spring Force

In Section 6.3, we found that the potential energy stored in a spring is: $U_s = \frac{1}{2}kx^2$, where $k$ is the spring constant and $x$ is the displacement from the equilibrium position. Here we choose the additive constant to be zero, corresponding to having $U_s = 0/x$ at $k = 0$. Using the principle of energy conservation, we can find the velocity $v$ as a function of the position. First, we can write, in general, for the total mechanical energy:

$$E = K + U_s = \tfrac{1}{2}mv^2 + \tfrac{1}{2}kx^2. \tag{6.21}$$

Once we know the total mechanical energy, we can solve this equation for the velocity. What is the total mechanical energy? The point of maximum elongation of a spring from the equilibrium position is called the **amplitude,** $A$. When the displacement reaches the amplitude, the velocity is briefly zero. At this point, the total mechanical energy of an object oscillating on a spring is

$$E = \tfrac{1}{2}kA^2.$$

However, conservation of mechanical energy means that this is the value of the energy for any point in the spring's oscillation. Inserting the above expression for $E$ into equation 6.21 yields

$$\tfrac{1}{2}kA^2 = \tfrac{1}{2}mv^2 + \tfrac{1}{2}kx^2. \tag{6.22}$$

From equation 6.22, we can get an expression for the speed as a function of the position:

$$v = \sqrt{(A^2 - x^2)\frac{k}{m}}. \tag{6.23}$$

Note that we did not rely on kinematics to get this result, as that approach is rather challenging—another piece of evidence that using principles of conservation (in this case, conservation of mechanical energy) can yield powerful results. We will return to the equation of motion for a mass on a spring in Chapter 14.

### SOLVED PROBLEM 6.2 Human Cannonball

In a favorite circus act, called the "human cannonball," a person is shot from a long barrel, usually with a lot of smoke and a loud bang added for theatrical effect. Before the Italian Zacchini brothers invented the compressed air cannon for shooting human cannonballs in the 1920s, the Englishman George Farini used a spring-loaded cannon for this purpose in the 1870s.

Suppose someone wants to recreate Farini's spring-loaded human cannonball act with a spring inside a barrel. Assume the barrel is 4.00 m long, with a spring that extends the entire length of the barrel. Further, the barrel is upright, so it points vertically toward the ceiling of the circus tent. The human cannonball is lowered into the barrel and compresses the spring to some degree. An external force is added to compress the spring even further, to a length of only 0.70 m. At a height of 7.50 m above the top of the barrel is a spot on the tent that the human cannonball, of height 1.75 m and mass 68.4 kg, is supposed to touch at the top of his trajectory. Removing the external force releases the spring and fires the human cannonball vertically upward.

*Continued—*

## PROBLEM 1

What is the value of the spring constant needed to accomplish this stunt?

## SOLUTION 1

### THINK

Let's apply energy conservation considerations to solve this problem. Potential energy is stored in the spring initially and then converted to gravitational potential energy at the top of the human cannonball's flight. As a reference point for our calculations, we select the top of the barrel and place the origin of our coordinate system there. To accomplish the stunt, enough energy has to be provided, through compressing the spring, that the top of the head of the human cannonball is elevated to a height of 7.50 m above the zero point we have chosen. Since the person has a height of 1.75 m, his feet need to be elevated by only $h = 7.50$ m$-1.75$ m $= 5.75$ m. We can specify all position values for the human cannonball on the $y$-coordinate as the position of the bottom of his feet.

### SKETCH

To clarify this problem, let's apply energy conservation at different instants of time. Figure 6.11a shows the initial equilibrium position of the spring. In Figure 6.11b, the external force $\vec{F}$ and the weight of the human cannonball compress the spring by 3.30 m to a length of 0.70 m. When the spring is released, the cannonball accelerates and has a velocity $\vec{v}_c$ as he passes the spring's equilibrium position (see Figure 6.11c). From this position, he has to rise 5.75 m and arrive at the spot (Figure 6.11e) with zero velocity.

### RESEARCH

We are free to choose the zero point for the gravitational potential energy arbitrarily. We elect to set the gravitational potential to zero at the equilibrium position of the spring without a load, as shown in Figure 6.11a.

At the instant depicted in Figure 6.11b, the human cannonball has zero kinetic energy and potential energies from the spring force and gravity. Therefore, the total energy at this instant is

$$E = \tfrac{1}{2}ky_b^2 + mgy_b.$$

At the instant shown in Figure 6.11c, the human cannonball has only kinetic energy and zero potential energy:

$$E = \tfrac{1}{2}mv_c^2.$$

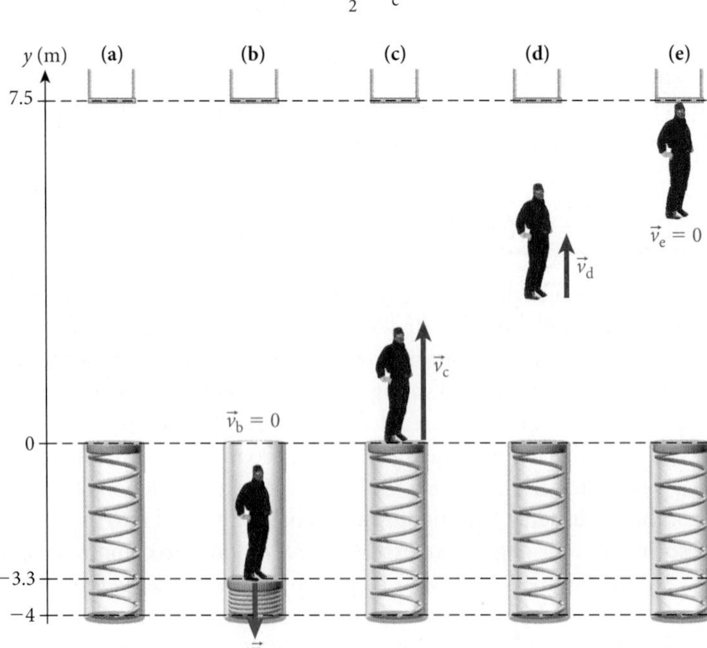


**FIGURE 6.11** The human cannonball stunt at five different instants of time.

Right after this instant, the human cannonball leaves the spring, flies through the air as shown in Figure 6.11d, and finally reaches the top (Figure 6.11e). At the top, he has only gravitational potential energy and no kinetic energy (because the spring is designed to allow him to reach the top with no residual speed):

$$E = mgy_e.$$

**SIMPLIFY**

Energy conservation requires that the total energy remain the same. Setting the first and third expressions written above for $E$ equal, we obtain

$$\tfrac{1}{2}ky_b^2 + mgy_b = mgy_e.$$

We can rearrange this equation to obtain the spring constant:

$$k = 2mg\frac{y_e - y_b}{y_b^2}.$$

**CALCULATE**

According to the given information and the origin of the coordinate system we selected, $y_b = -3.30$ m and $y_e = 5.75$ m. Thus, we find for the spring constant needed:

$$k = 2(68.4 \text{ kg})(9.81 \text{ m/s}^2)\frac{5.75 \text{ m} - (-3.30 \text{ m})}{(3.30 \text{ m})^2} = 1115.26 \text{ N/m}.$$

**ROUND**

All of the numerical values used in the calculation have three significant figures, so our final answer is

$$k = 1.12 \cdot 10^3 \text{ N/m}.$$

**DOUBLE-CHECK**

When the spring is compressed initially, the potential energy stored in it is

$$U = \tfrac{1}{2}ky_b^2 = \tfrac{1}{2}(1.12 \cdot 10^3 \text{ N/m})(3.30 \text{ m})^2 = 6.07 \text{ kJ}.$$

The gravitational potential energy gained by the human cannonball is

$$U = mg\Delta y = (68.4 \text{ kg})(9.81 \text{ m/s}^2)(9.05 \text{ m}) = 6.07 \text{ kJ},$$

which is the same as the energy stored in the spring initially. Our calculated value for the spring constant makes sense.

Note that the mass of the human cannonball enters into the equation for the spring constant. We can turn this around and state that the same cannon with the same spring will shoot people of different masses to different heights.

**PROBLEM 2**

What is the speed that the human cannonball reaches as he passes the equilibrium position of the spring?

**SOLUTION 2**

We have already determined that our choice of origin implies that at this instant the human cannonball has only kinetic energy. Setting this kinetic energy equal to the potential energy reached at the top, we find

$$\tfrac{1}{2}mv_c^2 = mgy_e \Rightarrow$$

$$v_c = \sqrt{2gy_e} = \sqrt{2(9.81 \text{ m/s}^2)(5.75 \text{ m})} = 10.6 \text{ m/s}.$$

This speed corresponds to 23.7 mph.

### 6.3 In-Class Exercise

What is the maximum acceleration that the human cannonball of Solved Problem 6.2 experiences?

a) $1.00g$
b) $2.14g$
c) $3.25g$
d) $4.48g$
e) $7.30g$

### 6.4 In-Class Exercise

A ball of mass $m$ is thrown vertically into the air with an initial speed $v$. Which of the following equations correctly describes the maximum height, $h$, of the ball?

a) $h = \sqrt{\dfrac{v}{2g}}$
b) $h = \dfrac{g}{\frac{1}{2}v^2}$
c) $h = \dfrac{2mv}{g}$
d) $h = \dfrac{mv^2}{g}$
e) $h = \dfrac{v^2}{2g}$

### 6.4 Self-Test Opportunity

Graph the potential and kinetic energies of the human cannonball of Solved Problem 6.2 as a function of the $y$-coordinate. For what value of the displacement is the speed of the human cannonball at a maximum? (*Hint:* This occurs not exactly at $y = 0$ but at a value of $y < 0$.)

## EXAMPLE 6.2 Bungee Jumper

A bungee jumper locates a suitable bridge that is 75.0 m above the river below, as shown in Figure 6.12. The jumper has a mass of $m = 80.0$ kg and a height of $L_{jumper} = 1.85$ m. We can think of a bungee cord as a spring. The spring constant of the bungee cord is $k = 50.0$ N/m. Assume that the mass of the bungee cord is negligible compared with the jumper's mass.

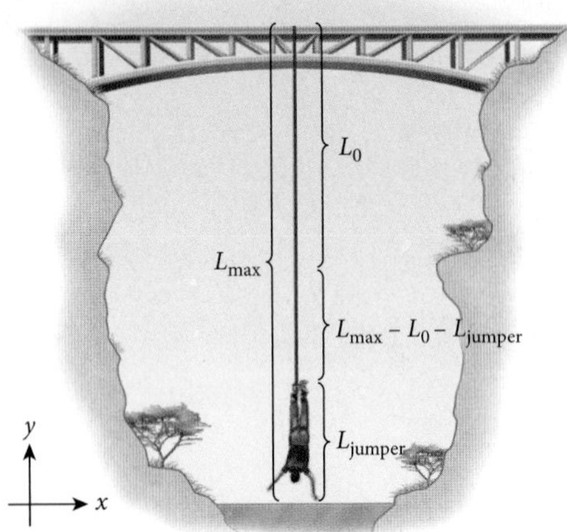


**FIGURE 6.12** A bungee jumper needs to calculate how long a bungee cord he can safely use.

### PROBLEM

The jumper wants to know the maximum length of bungee cord he can safely use for this jump.

### SOLUTION

We are looking for the unstretched length of the bungee cord, $L_0$, as the jumper would measure it standing on the bridge. The distance from the bridge to the water is $L_{max} = 75.0$ m. Energy conservation tells us that the gravitational potential energy that the jumper has, as he dives off the bridge, will be converted to potential energy stored in the bungee cord. The jumper's gravitational potential energy on the bridge is

$$U_g = mgy = mgL_{max},$$

assuming that gravitational potential energy is zero at the level of the water. Before he starts his jump, he has zero kinetic energy, and so his total energy when he is on top of the bridge is

$$E_{top} = mgL_{max}.$$

At the bottom of the jump, where the jumper's head just touches the water, the potential energy stored in the bungee cord is

$$U_s = \tfrac{1}{2}ky^2 = \tfrac{1}{2}k\left(L_{max} - L_{jumper} - L_0\right)^2,$$

where $L_{max} - L_{jumper} - L_0$ is the length the bungee cord stretches beyond its unstretched length. (Here we have to subtract the jumper's height from the height of the bridge to obtain the maximum length, $L_{max} - L_{jumper}$, to which the bungee cord is allowed to stretch, assuming that it is tied around his ankles.) Since the bungee jumper is momentarily at rest at this lowest point of his jump, the kinetic energy is zero at that point, and the total energy is then

$$E_{bottom} = \tfrac{1}{2}k\left(L_{max} - L_{jumper} - L_0\right)^2.$$

From the conservation of mechanical energy, we know that $E_{top} = E_{bottom}$, and so we find

$$mgL_{max} = \tfrac{1}{2}k\left(L_{max} - L_{jumper} - L_0\right)^2.$$

Solving for the required unstretched length of bungee cord gives us

$$L_0 = L_{max} - L_{jumper} - \sqrt{\frac{2mgL_{max}}{k}}.$$

Putting in the given numbers, we get

$$L_0 = (75.0\text{ m}) - (1.85\text{ m}) - \sqrt{\frac{2(80.0\text{ kg})(9.81\text{ m/s}^2)(75.0\text{ m})}{50.0\text{ N/m}}} = 24.6\text{ m}.$$

For safety, the jumper would be wise to use a bungee cord shorter than this and to test it with a dummy mass similar to his.

### 6.5 In-Class Exercise

At the moment of maximum stretching of the bungee cord in Example 6.2, what is the net acceleration that the jumper experiences (in terms of $g = 9.81$ m/s$^2$)?

a) 0$g$

b) 1.0$g$, directed downward

c) 1.0$g$, directed upward

d) 2.1$g$, directed downward

e) 2.1$g$, directed upward

### 6.5 Self-Test Opportunity

Can you derive an expression for the acceleration that the bungee jumper experiences at the maximum stretching of the bungee cord? How does this acceleration depend on the spring constant of the bungee cord?

## Potential Energy of an Object Hanging from a Spring

We saw in Solved Problem 6.2 that the initial potential energy of the human cannonball has contributions from the spring force and the gravitational force. In Example 5.4, we established that hanging an object of mass $m$ from a spring with spring constant $k$ shifts the equilibrium position of the spring from zero to $y_0$, given by the equilibrium condition,

$$ky_0 = mg \Rightarrow y_0 = \frac{mg}{k}. \quad (6.24)$$

Figure 6.13 shows the forces that act on an object suspended from a spring when it is in different positions. This figure shows two different choices for the origin of the vertical coordinate axis: In Figure 6.13a, the vertical coordinate is called $y$ and has zero at the equilibrium position of the end of the spring without the mass hanging from it; in Figure 6.13b the new equilibrium point, $y_0$, with the object suspended from the spring, is calculated according to equation 6.24. This new equilibrium point is the origin of the axis and the vertical coordinate is called $s$. The end of the spring is located at $s = 0$. The system is in equilibrium because the force exerted by the spring on the object balances the gravitational force acting on the object:

$$\vec{F}_s(y_0) + \vec{F}_g = 0.$$

In Figure 6.13c, the object has been displaced downward away from the new equilibrium position, so $y = y_1$ and $s = s_1$. Now there is a net upward force tending to restore the object to the new equilibrium position:

$$\vec{F}_{net}(s_1) = \vec{F}_s(y_1) + \vec{F}_g.$$

If instead, the object is displaced upward, above the new equilibrium position, as shown in Figure 6.13d, there is a net downward force that tends to restore the object to the new equilibrium position:

$$\vec{F}_{net}(s_2) = \vec{F}_s(y_2) + \vec{F}_g.$$

We can calculate the potential energy of the object and the spring for these two choices of the coordinate system and show that they differ by only a constant. We start by defining the

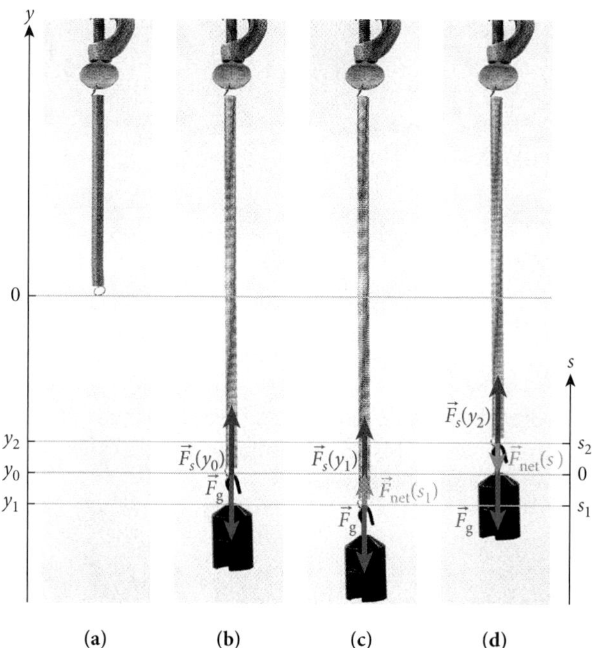


**FIGURE 6.13** (a) A spring is hanging vertically with its end in the equilibrium position at $y = 0$. (b) An object of mass $m$ is hanging at rest from the same spring, with the end of the spring now at $y = y_0$ or $s = 0$. (c) The end of the spring with the object attached is at $y = y_1$ or $s = s_1$. (d) The end of the spring with the object attached is at $y = y_2$ or $s = s_2$.

potential energy of the object connected to the spring, taking $y$ as the variable and assuming that the potential energy is zero at $y = 0$:

$$U(y) = \tfrac{1}{2}ky^2 + mgy.$$

Using the relation $y = s - y_0$, we can express this potential energy in terms of the variable $s$:

$$U(s) = \tfrac{1}{2}k(s - y_0)^2 + mg(s - y_0).$$

Rearranging gives us

$$U(s) = \tfrac{1}{2}ks^2 - ksy_0 + \tfrac{1}{2}ky_0^2 + mgs - mgy_0.$$

Substituting $ky_0 = mg$, from equation 6.24, into this equation, we get

$$U(s) = \tfrac{1}{2}ks^2 - (mg)s + \tfrac{1}{2}(mg)y_0 + mgs - mgy_0.$$

Thus, we find that the potential energy in terms of $s$ is

$$U(s) = \tfrac{1}{2}ks^2 - \tfrac{1}{2}mgy_0. \tag{6.25}$$

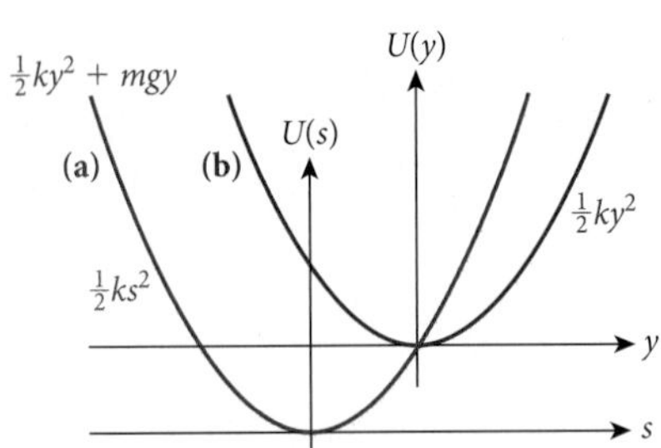


**FIGURE 6.14** Potential energy functions for the two vertical coordinate axes used in Figure 6.13.

Figure 6.14 shows the potential energy functions for these two coordinate axes. The blue curve in Figure 6.14 shows the potential energy as a function of the vertical coordinate $y$, with the choice of zero potential energy at $y = 0$ corresponding to the spring hanging vertically without the object connected to it. The new equilibrium position, $y_0$, is determined by the displacement that occurs when an object of mass $m$ is attached to the spring, as calculated using equation 6.24. The red curve in Figure 6.14 represents the potential energy as a function of the vertical coordinate $s$, with the equilibrium position chosen to be $s = 0$. The potential energy curves $U(y)$ and $U(s)$ are both parabolas, which are offset from each other by a simple constant.

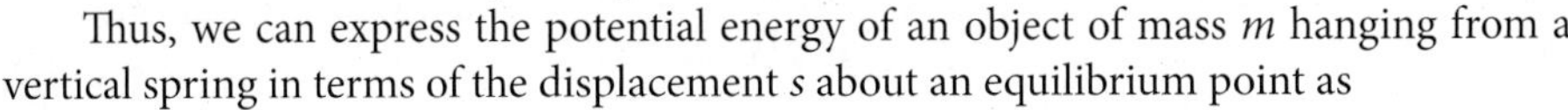

Thus, we can express the potential energy of an object of mass $m$ hanging from a vertical spring in terms of the displacement $s$ about an equilibrium point as

$$U(s) = \tfrac{1}{2}ks^2 + C,$$

where $C$ is a constant. For many problems, we can choose zero as the value of this constant, allowing us to write

$$U(s) = \tfrac{1}{2}ks^2.$$

This result allows us to use the same spring force potential for different masses attached to the end of a spring by simply shifting the origin to the new equilibrium position. (Of course, this only works if we do not attach too much mass to the end of the spring and overstretch it beyond its elastic limit.)

Finally, with the introduction of the potential energy we can extend and augment the work–kinetic energy theorem of Chapter 5. By including the potential energy as well, we find the **work-energy theorem**

$$W = \Delta E = \Delta K + \Delta U \tag{6.26}$$

where $W$ is the work done by an *external* force, $\Delta K$ is the change of kinetic energy, and $\Delta U$ is the change in potential energy. This relationship means that external work done to a system can change the total energy of the system.

## 6.7 Nonconservative Forces and the Work-Energy Theorem

Is energy conservation violated in the presence of nonconservative forces? The word *nonconservative* seems to imply that it is violated, and, indeed, the total *mechanical* energy is not conserved. Where, then, does the energy go? Section 6.2 showed that the friction force does not do work but instead dissipates mechanical energy into internal excitation energy, which can be vibration energy, deformation energy, chemical energy, or electrical energy, depending on the material of which the object is made and on the particular form of the friction force. In Section 6.2, $W_f$ is defined to be the total energy dissipated by nonconservative forces into internal energy and then into other energy forms besides mechanical energy. If we add this type of energy to the total mechanical energy, we obtain the **total energy:**

$$E_{total} = E_{mechanical} + E_{other} = K + U + E_{other}. \quad (6.27)$$

Here $E_{other}$ stands for all other forms of energy that are not kinetic or potential energies. The change in the other energy forms is exactly the negative of the energy dissipated by the friction force in going from the initial to the final state of the system:

$$\Delta E_{other} = -W_f.$$

The total energy is conserved—that is, stays constant in time—even for nonconservative forces. This is the most important point in this chapter:

**The total energy—the sum of all forms of energy, mechanical or other—is *always* conserved in an isolated system.**

We can also write this law of energy conservation in the form that states that the change in the total energy of an isolated system is zero:

$$\Delta E_{total} = 0. \quad (6.28)$$

Since we do not yet know what exactly this internal energy is and how to calculate it, it may seem that we cannot use energy considerations when at least one of the forces acting is nonconservative. However, this is not the case. For the case in which only conservative forces are acting, we found that (see equation 6.18) the total mechanical energy is conserved, or $\Delta E = \Delta K + \Delta U = 0$, where $E$ refers to the total mechanical energy. In the presence of nonconservative forces, combining equations 6.28 and 6.26 gives

$$W_f = \Delta K + \Delta U. \quad (6.29)$$

This relationship is a generalization of the work-energy theorem. In the absence of nonconservative forces, $W_f = 0$, and equation 6.29 reduces to the law of the conservation of mechanical energy, equation 6.19. When applying either of these two equations, you must select two times—a beginning and an end. Usually this choice is obvious, but sometimes care must be taken, as demonstrated in the following solved problem.

## SOLVED PROBLEM 6.3 Block Pushed Off a Table

Consider a block on a table. This block is pushed by a spring attached to the wall, slides across the table, and then falls to the ground. The block has a mass $m = 1.35$ kg. The spring constant is $k = 560$ N/m, and the spring has been compressed by 0.11 m. The block slides a distance $d = 0.65$ m across the table of height $h = 0.75$ m. The coefficient of kinetic friction between the block and the table is $\mu_k = 0.16$.

### PROBLEM

What speed will the block have when it lands on the floor?

### SOLUTION

### THINK

At first sight, this problem does not seem to be one to which we can apply mechanical energy conservation, because the nonconservative force of friction is in play. However, we can utilize the work-energy theorem, equation 6.29. To be certain that the block actually leaves the table, though, we first calculate the total energy imparted to the block by the spring and make sure that the potential energy stored in the compressed spring is sufficient to overcome the friction force.

### SKETCH

Figure 6.15a shows the block of mass $m$ pushed by the spring. The mass slides on the table a distance $d$ and then falls to the floor, which is a distance $h$ below the table.

*Continued—*

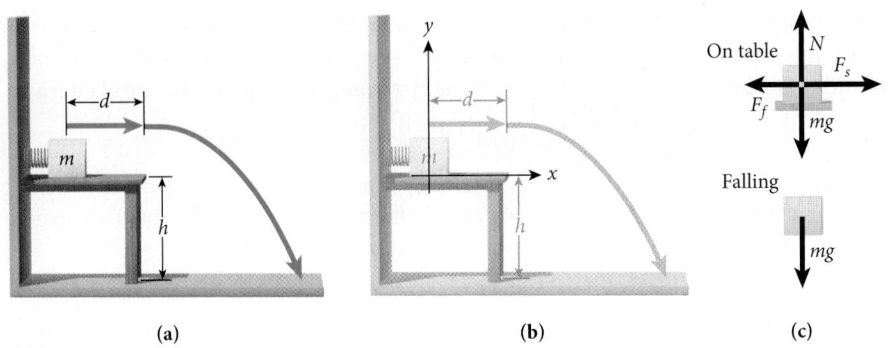


**FIGURE 6.15** (a) Block of mass *m* is pushed off a table by a spring. (b) A coordinate system is superimposed on the block and table. (c) Free-body diagrams of the block while moving on the table and falling.

We choose the origin of our coordinate system such that the block starts at $x = y = 0$, with the $x$-axis running along the bottom surface of the block and the $y$-axis running through its center (Figure 6.15b). The origin of the coordinate system can be placed at any point, but it is important to fix an origin, because all potential energies have to be expressed relative to some reference point.

## RESEARCH

*Step 1:* Let's analyze the problem situation without the friction force. In this case, the block initially has potential energy from the spring and no kinetic energy, since it is at rest. When the block hits the floor, it has kinetic energy and negative gravitational potential energy. Conservation of mechanical energy results in

$$\begin{aligned} K_0 + U_0 &= K + U \Rightarrow \\ 0 + \tfrac{1}{2}kx_0^2 &= \tfrac{1}{2}mv^2 - mgh. \end{aligned} \tag{i}$$

Usually, we would solve this equation for the speed and put in the numbers later. However, because we will need them again, let's evaluate the two expressions for potential energy:

$$\tfrac{1}{2}kx_0^2 = 0.5(560 \text{ N/m})(0.11 \text{ m})^2 = 3.39 \text{ J}$$

$$mgh = (1.35 \text{ kg})(9.81 \text{ m/s}^2)(0.75 \text{ m}) = 9.93 \text{ J}.$$

Now, solving equation i for the speed results in

$$v = \sqrt{\frac{2}{m}\left(\tfrac{1}{2}kx_0^2 + mgh\right)} = \sqrt{\frac{2}{1.35 \text{ kg}}(3.39 \text{ J} + 9.93 \text{ J})} = 4.44 \text{ m/s}.$$

*Step 2:* Now we include friction. Our considerations remain almost unchanged, except that we have to include the energy dissipated by the nonconservative force of friction. We find the force of friction using the upper free-body diagram in Figure 6.15c. We can see that the normal force is equal to the weight of the block and write

$$N = mg.$$

The friction force is given by

$$F_\text{k} = \mu_\text{k} N = \mu_\text{k} mg.$$

We can then write the energy dissipated by the friction force as

$$W_\text{f} = -\mu_\text{k} mgd.$$

In applying the generalization of the work-energy theorem, we choose the initial time to be when the block is about to start moving (see Figure 6.15a) and the final time to be when the block reaches the edge of the table and is about to start the free-fall portion of its path. Let $K_\text{top}$ be the kinetic energy at the final time, chosen to make sure that the block makes it to the end of the table. Using equation 6.29 and the value we calculated above for the block's initial potential energy, we find:

$$W_f = \Delta K + \Delta U = K_{top} - \tfrac{1}{2}kx_0^2 = -\mu_k mgd$$
$$K_{top} = \tfrac{1}{2}kx_0^2 - \mu_k mgd$$
$$= 3.39 \text{ J} - (0.16)(1.35 \text{ kg})(9.81 \text{ m/s}^2)(0.65 \text{ m})$$
$$= 3.39 \text{ J} - 1.38 \text{ J} = 2.01 \text{ J}.$$

Because the kinetic energy $K_{top} > 0$, the block can overcome friction and slide off the table. Now we can calculate the block's speed when it hits the floor.

**SIMPLIFY**

For this part of the problem, we choose the initial time to be when the block is at the table's edge to exploit the calculations we have already done. The final time is when the block hits the floor. (If we chose the beginning to be as shown in Figure 6.15a, our result would be the same.)

$$W_f = \Delta K + \Delta U = 0$$
$$\tfrac{1}{2}mv^2 - K_{top} + 0 - mgh = 0,$$
$$v = \sqrt{\frac{2}{m}(K_{top} + mgh)}.$$

**CALCULATE**

Putting in the numerical values gives us

$$v = \sqrt{\frac{2}{1.35 \text{ kg}}(2.01 \text{ J} + 9.93 \text{ J})} = 4.20581608 \text{ m/s}.$$

**ROUND**

All of the numerical values were given to three significant figures, so we have

$$v = 4.21 \text{ m/s}.$$

**DOUBLE-CHECK**

As you can see, the main contribution to the speed of the block at impact originates from the free-fall portion of its path. Why did we go through the intermediate step of figuring out the value of $K_{top}$, instead of simply using the formula, $v = \sqrt{2(\tfrac{1}{2}kx_0^2 - \mu_k mgd + mgh)/m}$, that we got from a generalization of the work-energy theorem? We needed to calculate $K_{top}$ first to ensure that it is positive, meaning that the energy imparted to the block by the spring is sufficient to exceed the work to be done against the friction force. If $K_{top}$ had turned out to be negative, the block would have stopped on the table. For example, if we had attempted to solve the problem described in Solved Problem 6.3 with a coefficient of kinetic friction between the block and the table of $\mu_k = 0.50$ instead of $\mu_k = 0.16$, we would have found that

$$K_{top} = 3.39 \text{ J} - 4.30 \text{ J} = -0.91 \text{ J},$$

which is impossible.

As Solved Problem 6.3 shows, energy considerations are still a powerful tool for performing otherwise very difficult calculations, even in the presence of nonconservative forces. However, the principle of conservation of mechanical energy cannot be applied quite as straightforwardly when nonconservative forces are present, and you have to account for the energy dissipated by these forces.

### 6.6 In-Class Exercise

A curling stone of mass 19.96 kg is given an initial velocity on the ice of 2.46 m/s. The coefficient of kinetic friction between the stone and the ice is 0.0109. How far does the stone slide before it stops?

a) 18.7 m
b) 28.3 m
c) 34.1 m
d) 39.2 m
e) 44.5 m

## 6.8 Potential Energy and Stability

Let's return to the relationship between force and potential energy. Perhaps it may help you gain physical insight into this relationship if you visualize the potential energy curve as the track of a roller coaster. This analogy is not a perfect one, because a roller coaster moves in a two-dimensional plane or even in three-dimensional space, not in one dimension, and there is some small amount of friction between the cars and the track. Still, it is a good approximation to assume that there is conservation of mechanical energy. The motion of the roller coaster car can then be described by a potential energy function.

**FIGURE 6.16** Total, potential, and kinetic energy for a roller coaster.

Shown in Figure 6.16 are plots of the potential energy (yellow line following the outline of the track), the total energy (horizontal orange line), and the kinetic energy (difference between these two, the red line) as a function of position for a segment of a roller coaster ride. You can see that the kinetic energy has a minimum at the highest point of the track, where the speed of the cars is smallest, and the speed increases as the cars roll down the incline. All of these effects are a consequence of the conservation of total mechanical energy.

Figure 6.17 shows graphs of a potential energy function (part a) and the corresponding force (part b). Because the potential energy can be determined only within an additive constant, the zero value of the potential energy in Figure 6.17a is set at the lowest value. However, for all physical considerations, this is irrelevant. On the other hand, the zero value for the force cannot be chosen arbitrarily.

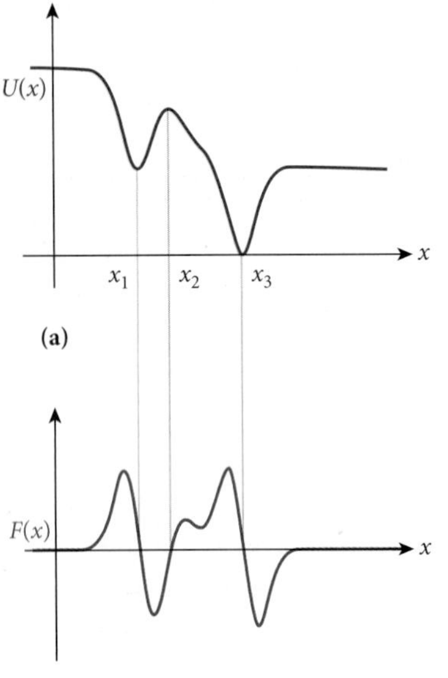


**FIGURE 6.17** (a) Potential energy as a function of position; (b) the force corresponding to this potential energy function, as a function of position.

### Equilibrium Points

Three special points on the $x$-coordinate axis of Figure 6.17b are marked by vertical gray lines. These points indicate where the force has a value of zero. Because the force is the derivative of the potential energy with respect to the $x$-coordinate, the potential energy has an extremum—a maximum or minimum value—at such points. You can clearly see that the potential energies at $x_1$ and $x_3$ represent minima, and the potential energy at $x_2$ is a maximum. At all three points, an object would experience no acceleration, because it is located at an extremum where the force is zero. Because there is no force, Newton's Second Law tells us that there is no acceleration. Thus, these points are equilibrium points.

The equilibrium points in Figure 6.17 represent two different kinds. Points $x_1$ and $x_3$ represent stable equilibrium points, and $x_2$ is an unstable equilibrium point. What distinguishes stable and unstable equilibrium points is the response to perturbations (small changes in position around the equilibrium position).

#### Definition

At **stable equilibrium points,** small perturbations result in small oscillations around the equilibrium point. At **unstable equilibrium points,** small perturbations result in an accelerating movement away from the equilibrium point.

The roller coaster analogy may be helpful here: If you are sitting in a roller coaster car at point $x_1$ or $x_3$ and someone gives the car a push, it will just rock back and forth on the track, because you are sitting at a local point of lowest energy. However, if the car gets the same small push while sitting at $x_2$, then it will result in the car rolling down the slope.

What makes an equilibrium point stable or unstable from a mathematical standpoint is the value of the second derivative of the potential energy function, or the curvature. Negative curvature means a local maximum of the potential energy function, and therefore an unstable equilibrium point; a positive curvature indicates a stable equilibrium point. Of course, there is also the situation between a stable and unstable equilibrium, between a positive and negative curvature. This is a point of metastable equilibrium, with zero local curvature, that is, a value of zero for the second derivative of the potential energy function.

## Turning Points

Figure 6.18a shows the same potential energy function as Figure 6.17, but with the addition of horizontal lines for four different values of the total mechanical energy ($E_1$ through $E_4$). For each value of this total energy and for each point on the potential energy curve, we can calculate the value of the kinetic energy by simple subtraction. Let's first consider the largest value of the total mechanical energy shown in the figure, $E_1$ (blue horizontal line):

$$K_1(x) = E_1 - U(x). \tag{6.30}$$

The kinetic energy, $K_1(x)$, is shown in Figure 6.18b by the blue curve, which is clearly an upside-down version of the potential energy curve in Figure 6.18a. However, its absolute height is not arbitrary but results from equation 6.30. As previously mentioned, we can always add an arbitrary additive constant to the potential energy, but then we are forced to add the same additive constant to the total mechanical energy, so that their difference, the kinetic energy, remains unchanged.

For the other values of the total mechanical energy in Figure 6.18, an additional complication arises: the condition that the kinetic energy has to be larger than or equal to zero. This condition means that the kinetic energy is not defined in a region where $E_i - U(x)$ is negative. For the total mechanical energy of $E_2$, the kinetic energy is greater than zero only for $x \geq a$, as indicated in Figure 6.18b by the green curve. Thus, an object moving with total energy $E_2$ from right to left in Figure 6.18 will reach the point $x = a$ and have zero velocity there. Referring to Figure 6.17, you see that the force at that point is positive, pushing the object to the right, that is, making it turn around. This is why such a point is called a **turning point.** Moving to the right, this object will pick up kinetic energy and follow the same kinetic energy curve from left to right, making its path reversible. This behavior is a consequence of the conservation of total mechanical energy.

### Definition

**Turning points** are points where the kinetic energy is zero and where a net force moves the object away from the point.

An object with total energy equal to $E_4$ in Figure 6.18 has two turning points in its path: $x = e$ and $x = f$. The object can move only between these two points. It is trapped in this interval and cannot escape. Perhaps the roller coaster analogy is again helpful: A car released from point $x = e$ will move through the dip in the potential energy curve to the right until it reaches the point $x = f$, where it will reverse direction and move back to $x = e$, never having enough total mechanical energy to escape the dip. The region in which the object is trapped is often referred to as a *potential well.*

Perhaps the most interesting situation is that for which the total energy is $E_3$. If an object moves in from the right in Figure 6.18 with energy $E_3$, it will be reflected at the turning point where $x = d$, in complete analogy to the situation at $x = a$ for the object with energy $E_2$. However, there is another allowed part of the path farther to the left, in the interval $b \leq x \leq c$. If the object starts out in this interval, it remains trapped in a dip, just like the object with energy $E_4$. Between the allowed intervals $b \leq x \leq c$ and $x \geq d$ is a *forbidden region* that an object with total mechanical energy $E_3$ cannot cross.

## Preview: Atomic Physics

In studying atomic physics, we will again encounter potential energy curves. Particles with energies such as $E_4$ in Figure 6.18, which are trapped between two turning points, are said to be in *bound states.* One of the most interesting phenomena in atomic and nuclear physics, however, occurs in situations like the one shown in Figure 6.18 for a total mechanical

### 6.7 In-Class Exercise

Which of the four drawings represents a stable equilibrium point for the ball on its supporting surface?

(a)

(b)

(c)

(d)

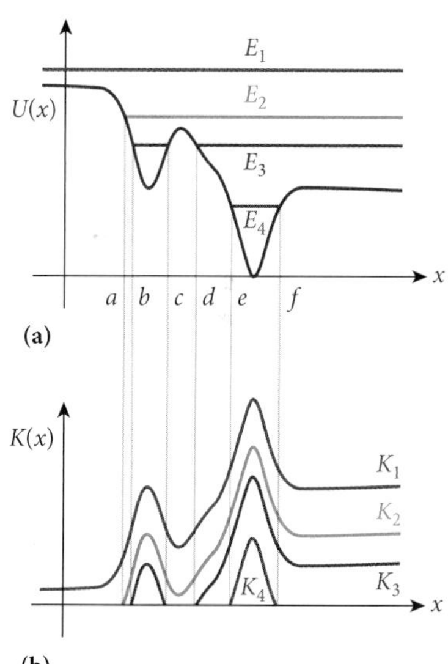


**FIGURE 6.18** (a) The same potential energy function as in Figure 6.17. Shown are lines representing four different values of the total energy, $E_1$ through $E_4$. (b) The corresponding kinetic energy functions for these four total energies and the potential energy function in the upper part. The gray vertical lines mark the turning points.

energy of $E_3$. From our considerations of classical mechanics in this chapter, we expect that an object sitting in a bound state between $b \le x \le c$ cannot escape. However, in atomic and nuclear physics applications, a particle in such a bound state has a small probability of escaping out of this potential well and through the classically forbidden region into the region $x \ge d$. This process is called *tunneling*. Depending on the height and width of the barrier, the tunneling probability can be quite large, leading to a fast escape, or quite small, leading to a very slow escape. For example, the isotope $^{235}$U of the element uranium, used in nuclear fission power plants and naturally occurring on Earth, has a half-life of over 700 million years, which is the average time that elapses until an alpha particle (a tightly bound cluster of two neutrons and two protons in the nucleus) tunnels through its potential barrier, causing the uranium nucleus to decay. In contrast, the isotope $^{238}$U has a half-life of 4500 million years. Thus, much of the original $^{235}$U present on Earth has decayed away. The fact that $^{235}$U comprises only 0.7% of all naturally occurring uranium means that the first sign that a nation is attempting to use nuclear power, for any purpose, is the acquisition of equipment that can separate $^{235}$U from the much more abundant (99.3%) $^{238}$U, which is not suitable for nuclear fission power production.

The previous paragraph is included to whet your appetite for things to come. In order to understand the processes of atomic and nuclear physics, you'll need to be familiar with quite a few more concepts. However, the basic considerations of energy introduced here will remain virtually unchanged.

## WHAT WE HAVE LEARNED | EXAM STUDY GUIDE

- Potential energy, $U$, is the energy stored in the configuration of a system of objects that exert forces on one another.
- Gravitational potential energy is defined as $U_g = mgy$.
- The potential energy associated with elongating a spring from its equilibrium position at $x = 0$ is $U_s(x) = \frac{1}{2}kx^2$.
- A conservative force is a force for which the work done over any closed path is zero. A force that does not fulfill this requirement is called a nonconservative force.
- For any conservative force, the change in potential energy due to some spatial rearrangement of a system is equal to the negative of the work done by the conservative force during this spatial rearrangement.
- The relationship between a potential energy and the corresponding conservative force is
$$\Delta U = U(x) - U(x_0) = -\int_{x_0}^{x} F_x(x')dx'.$$
- In one-dimensional situations, the force component can be obtained from the potential energy using $F_x(x) = -\dfrac{dU(x)}{dx}$.
- The mechanical energy, $E$, is the sum of kinetic energy and potential energy: $E = K + U$.
- The total mechanical energy is conserved for any mechanical process inside an isolated system that involves only conservative forces: $\Delta E = \Delta K + \Delta U = 0$. An alternative way of expressing this conservation of mechanical energy is $K + U = K_0 + U_0$.
- The total energy—the sum of all forms of energy, mechanical or other—is always conserved in an isolated system. This holds for conservative as well as nonconservative forces: $E_{total} = E_{mechanical} + E_{other} = K + U + E_{other} =$ constant.
- Energy problems involving nonconservative forces can be solved using the work-energy theorem: $W_f = \Delta K + \Delta U$.
- At stable equilibrium points, small perturbations result in small oscillations around the equilibrium point; at unstable equilibrium points, small perturbations result in an accelerating movement away from the equilibrium point.
- Turning points are points where the kinetic energy is zero and where a net force moves the object away from the point.

## KEY TERMS

potential energy, p. 169
conservative force, p. 171
nonconservative force, p. 171
isolated system, p. 177
mechanical energy, p. 177
conservation of mechanical energy, p. 177
amplitude, p. 181
work-energy theorem, p. 186
total energy, p. 187
stable equilibrium points, p. 190
unstable equilibrium points, p. 190
turning points, p. 191

## NEW SYMBOLS AND EQUATIONS

$U$, potential energy

$W_f$, energy dissipated by a friction force

$U_g = mgy$, gravitational potential energy

$U_s(x) = \frac{1}{2}kx^2$, potential energy of a spring

$K + U = K_0 + U_0$, conservation of mechanical energy

$A$, amplitude

$W_f = \Delta K + \Delta U$, work-energy theorem

## ANSWERS TO SELF-TEST OPPORTUNITIES

**6.1** The potential energy is proportional to the inverse of the distance between the two objects. Examples of these forces are the force of gravity (see Chapter 12) and the electrostatic force (see Chapter 21).

**6.2** To handle this problem with added air resistance, we would have introduced the work done by air resistance, which can be treated as a friction force. We would have modified our statement of energy conservation to reflect the fact that work, $W_f$, is done by the friction force

$$W_f + K + U = K_0 + U_0.$$

The solution would have be done numerically because the work done by friction in this case would depend on the distance that the rock actually travels through the air.

**6.3** The lighter-colored ball descends to a lower elevation earlier in its motion and thus converts more of its potential energy to kinetic energy early on. Greater kinetic energy means higher speed. Thus, the lighter-colored ball reaches higher speeds earlier and is able to move to the bottom of the track faster, even though its path length is greater.

**6.4** The speed is at a maximum where the kinetic energy is at a maximum:

$$K(y) = U(-3.3\text{ m}) - U(y) = (3856\text{ J}) - (671\text{ J/m})y - (557.5\text{ J/m}^2)y^2$$

$$\frac{d}{dy}K(y) = -(671\text{ J/m}) - (1115\text{ J/m}^2)y = 0 \Rightarrow y = -0.602\text{ m}$$

$$v(-0.602\text{ m}) = \sqrt{2K(-0.602\text{ m})/m} = 10.89\text{ m/s}.$$

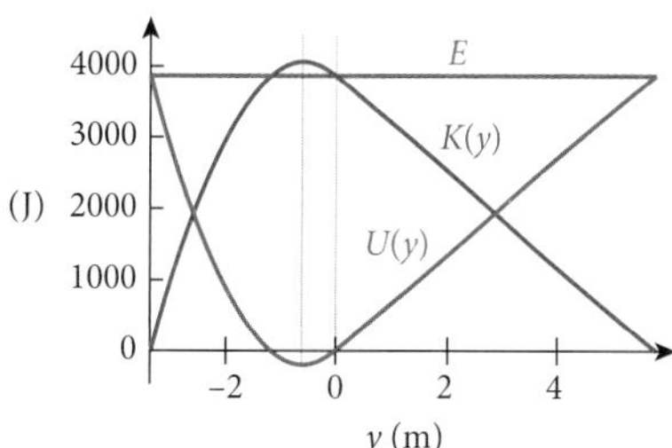


Note that the value at which the speed is maximum is the equilibrium position of the spring once it is loaded with the human cannonball.

**6.5** The net force at the maximum stretching is $F = k(L_{max} - L_{jumper} - L_0) - mg$. Therefore, the acceleration at this point is $a = k(L_{max} - L_{jumper} - L_0)/m - g$. Inserting the expression we found for $L_0$ gives

$$a = \sqrt{\frac{2gL_{max}k}{m}} - g.$$

The maximum acceleration increases with the square root of the spring constant. If one wants to jump from a great height, $L_{max}$, a very soft bungee cord is needed.

## PROBLEM-SOLVING PRACTICE

### Problem-Solving Guidelines: Conservation of Energy

**1.** Many of the problem-solving guidelines given in Chapter 5 apply to problems involving conservation of energy as well. It is important to identify the system and determine the state of the objects in it at different key times, such as the beginning and end of each kind of motion. You should also identify which forces in the situation are conservative or nonconservative, because they affect the system in different ways.

**2.** Try to keep track of each kind of energy throughout the problem situation. When does the object have kinetic energy? Does gravitational potential energy increase or decrease? Where is the equilibrium point for a spring?

**3.** Remember that you can choose where potential energy is zero, so try to determine what choice will simplify the calculations.

**4.** A sketch is almost always helpful, and often a free-body diagram is useful as well. In some cases, drawing graphs of potential energy, kinetic energy, and total mechanical energy is a good idea.

## SOLVED PROBLEM 6.4 Trapeze Artist

### PROBLEM

A circus trapeze artist starts her motion with the trapeze at rest at an angle of 45.0° relative to the vertical. The trapeze ropes have a length of 5.00 m. What is her speed at the lowest point in her trajectory?

*Continued—*

## SOLUTION

### THINK

Initially, the trapeze artist has only gravitational potential energy. We can choose a coordinate system such that $y = 0$ is at her trajectory's lowest point, so the potential energy is zero at that lowest point. When the trapeze artist is at the lowest point, her kinetic energy will be a maximum. We can then equate the initial gravitational potential energy to the final kinetic energy of the trapeze artist.

### SKETCH

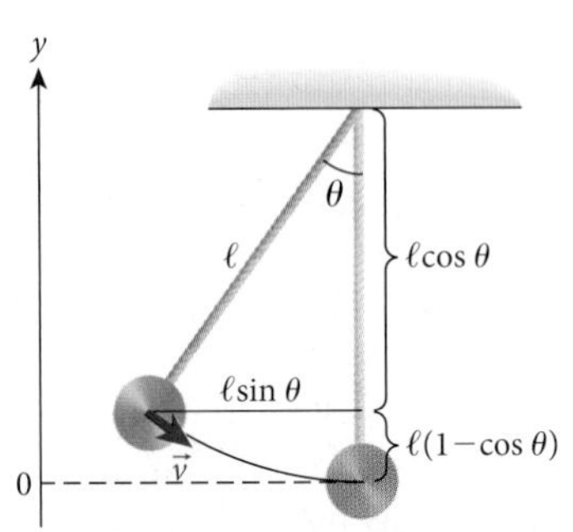


**FIGURE 6.19** Geometry of a trapeze artist's swing or trajectory.

We represent the trapeze artist in Figure 6.19 as an object of mass $m$ suspended by a rope of length $\ell$. We indicate the position of the trapeze artist at a given value of the angle $\theta$ by the blue circle. The lowest point of the trajectory is reached at $\theta = 0$, and we indicate this in Figure 6.19 by a gray circle.

The figure shows that the trapeze artist is at a distance $\ell$ (the length of the rope) below the ceiling at the lowest point and at a distance $\ell \cos\theta$ below the ceiling for all other values of $\theta$. This means that she is at a height $h = \ell - \ell\cos\theta = \ell(1 - \cos\theta)$ above the lowest point in the trajectory when the trapeze forms an angle $\theta$ with the vertical.

### RESEARCH

The trapeze is pulled back to an initial angle $\theta_0$ relative to the vertical and thus at a height $h = \ell(1 - \cos\theta_0)$ above the lowest point in the trajectory, according to our analysis of Figure 6.19. The potential energy at this maximum deflection, $\theta_0$, is therefore

$$E = K + U = 0 + U = mg\ell(1 - \cos\theta_0).$$

This is also the value for the total mechanical energy, because the trapeze artist has zero kinetic energy at the point of maximum deflection. For any other deflection, the energy is the sum of kinetic and potential energies:

$$E = mg\ell(1 - \cos\theta) + \tfrac{1}{2}mv^2.$$

### SIMPLIFY

Solving this equation for the speed, we obtain

$$mg\ell(1 - \cos\theta_0) = mg\ell(1 - \cos\theta) + \tfrac{1}{2}mv^2 \Rightarrow$$

$$mg\ell(\cos\theta - \cos\theta_0) = \tfrac{1}{2}mv^2 \Rightarrow$$

$$|v| = \sqrt{2g\ell(\cos\theta - \cos\theta_0)}.$$

Here, we are interested in the speed for $|v(\theta = 0)|$, which is

$$|v(\theta = 0)| = \sqrt{2g\ell(\cos 0 - \cos\theta_0)} = \sqrt{2g\ell(1 - \cos\theta_0)}.$$

### CALCULATE

The initial condition is $\theta_0 = 45°$. Inserting the numbers, we find

$$|v(0°)| = \sqrt{2(9.81 \text{ m/s}^2)(5.00 \text{ m})(1 - \cos 45°)} = 5.360300809 \text{ m/s}.$$

### ROUND

All of the numerical values were specified to three significant figures, so we report our answer as

$$|v(0°)| = 5.36 \text{ m/s}.$$

### DOUBLE-CHECK

First, the obvious check of units: m/s is the SI unit for velocity and speed. The speed of the trapeze artist at the lowest point is 5.36 m/s (12 mph), which seems in line with what we see in the circus.

We can perform another check on the formula $|v(\theta = 0)| = \sqrt{2g\ell(1 - \cos\theta_0)}$ by considering the limiting cases for the initial angle $\theta_0$ to see if they yield reasonable results. In this case the limiting values for $\theta_0$ are 90°, where the trapeze starts out horizontal, and

0°, where it starts out vertical. If we use $\theta_0 = 0$, the trapeze is just hanging at rest, and we expect zero speed, an expectation borne out by our formula. On the other hand, if we use $\theta_0 = 90°$, or $\cos\theta_0 = \cos 90° = 0$, we obtain the limiting result $\sqrt{2g\ell}$, which is the same result as a free fall from the ceiling to the bottom of the trapeze swing. Again, this limit is as expected, which gives us additional confidence in our solution.

## SOLVED PROBLEM 6.5 Sledding on Mickey Mouse Hill

### PROBLEM

A boy on a sled starts from rest and slides down snow-covered Mickey Mouse Hill, as shown in Figure 6.20. Together the boy and sled have a mass of 23.0 kg. Mickey Mouse Hill makes an angle $\theta = 35.0°$ with the horizontal. The surface of the hill is 25.0 m long. When the boy and the sled reach the bottom of the hill, they continue sliding on a horizontal snow-covered field. The coefficient of kinetic friction between the sled and the snow is 0.100. How far do the boy and sled move on the horizontal field before stopping?

**FIGURE 6.20** A boy sleds down Mickey Mouse Hill.

### SOLUTION

**THINK**

The boy and sled start with zero kinetic energy and finish with zero kinetic energy and have gravitational potential energy at the top of Mickey Mouse Hill. As boy and sled go down the hill, they gain kinetic energy. At the bottom of the hill, their potential energy is zero, and they have kinetic energy. However, the boy and sled are continuously losing energy to friction. Thus, the change in potential energy will equal the energy lost to friction. We must take into account the fact that the friction force will be different when the sled is on Mickey Mouse Hill than when it is on the flat field.

**SKETCH**

A sketch of the boy sledding on Mickey Mouse Hill is shown in Figure 6.21.

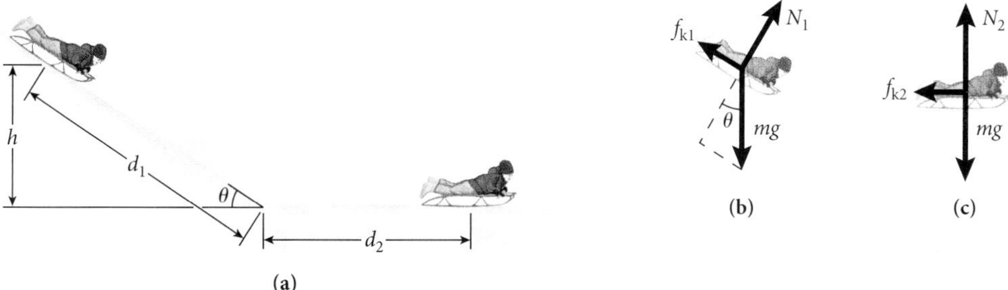


**FIGURE 6.21** (a) Sketch of the sled on Mickey Mouse Hill and on the flat field showing the angle of incline and distances. (b) Free-body diagram for the sled on Mickey Mouse Hill. (c) Free-body diagram for the sled on the flat field.

**RESEARCH**

The boy and sled start with zero kinetic energy and finish with zero kinetic energy. We call the length of Mickey Mouse Hill $d_1$, and the distance that the boy and sled travel on the flat field $d_2$ as shown in Figure 6.21a. Assuming that the gravitational potential energy of the boy and sled is zero at the bottom of the hill, the change in the gravitational potential energy from the top of Mickey Mouse Hill to the flat field is

$$\Delta U = mgh,$$

where $m$ is the mass of the boy and sled together and

$$h = d_1 \sin\theta.$$

*Continued—*

The force of friction is different on the slope and on the flat field because the normal force is different. From Figure 6.21b, the force of friction on Mickey Mouse Hill is

$$f_{k1} = \mu_k N_1 = \mu_k mg \cos\theta.$$

From Figure 6.21c, the force of friction on the flat field is

$$f_{k2} = \mu_k N_2 = \mu_k mg.$$

The energy dissipated by friction, $W_f$, is equal to the energy dissipated by friction while sliding on Mickey Mouse Hill, $W_1$, plus the energy dissipated while sliding on the flat field, $W_2$:

$$W_f = W_1 + W_2.$$

The energy dissipated by friction on Mickey Mouse Hill is

$$W_1 = f_{k1} d_1,$$

and the energy dissipated by friction on the flat field is

$$W_2 = f_{k2} d_2.$$

**SIMPLIFY**

According to the preceding three equations, the total energy dissipated by friction is given by

$$W_f = f_{k1} d_1 + f_{k2} d_2.$$

Substituting the two expressions for the friction forces into this equation gives us

$$W_f = (\mu_k mg \cos\theta) d_1 + (\mu_k mg) d_2.$$

The change in potential energy is obtained by combining the equation $\Delta U = mgh$ with the expression obtained for the height, $h = d_1 \sin\theta$:

$$\Delta U = mgd_1 \sin\theta.$$

Since the sled is at rest at the top of the hill and at the end of the ride as well, we have $\Delta K = 0$, and so according to equation 6.29, in this case, $\Delta U = W_f$. Now we can equate the change in potential energy with the energy dissipated by friction:

$$mgd_1 \sin\theta = (\mu_k mg \cos\theta) d_1 + (\mu_k mg) d_2.$$

Canceling out $mg$ on both sides and solving for the distance the boy and sled travel on the flat field, we get

$$d_2 = \frac{d_1(\sin\theta - \mu_k \cos\theta)}{\mu_k}.$$

**CALCULATE**

Putting in the given numerical values, we get

$$d_2 = \frac{(25.0 \text{ m})(\sin 35.0° - 0.100 \cdot \cos 35.0°)}{0.100} = 122.9153 \text{ m}.$$

**ROUND**

All of the numerical values were specified to three significant figures, so we report our answer as

$$d_2 = 123. \text{ m}.$$

**DOUBLE-CHECK**

The distance that the sled moves on the flat field is a little longer than a football field, which certainly seems possible after coming off a steep hill of length 25 m.

We can double-check our answer by assuming that the friction force between the sled and the snow is the same on Mickey Mouse Hill as it is on the flat field,

$$f_k = \mu_k mg.$$

The change in potential energy would then equal the approximate energy dissipated by friction for the entire distance the sled moves:

$$mgd_1 \sin\theta = \mu_k mg(d_1 + d_2).$$

The approximate distance traveled on the flat field would then be

$$d_2 = \frac{d_1(\sin\theta - \mu_k)}{\mu_k} = \frac{(25.0 \text{ m})(\sin 35.0° - 0.100)}{0.100} = 118. \text{ m}.$$

This result is close to but less than our answer of 123 m, which we expect because the friction force on the flat field is higher than the friction force on Mickey Mouse Hill. Thus, our answer seems reasonable.

The concepts of power introduced in Chapter 5 can be combined with the conservation of mechanical energy to obtain some interesting insight into electrical power generation from the conversion of gravitational potential energy.

## SOLVED PROBLEM 6.6 Power Produced by Niagara Falls

### PROBLEM

Niagara Falls pours an average of 5520 $m^3$ of water over a drop of 49.0 m every second. If all the potential energy of that water could be converted to electrical energy, how much electrical power could Niagara Falls generate?

### SOLUTION

#### THINK

The mass of one cubic meter of water is 1000 kg. The work done by the falling water is equal to the change in its gravitational potential energy. The average power is the work per unit time.

#### SKETCH

A sketch of a vertical coordinate axis is superimposed on a photo of Niagara Falls in Figure 6.22.

**FIGURE 6.22** Niagara Falls, showing an elevation of $h$ for the drop of the water going over the falls.

#### RESEARCH

The average power is given by the work per unit time:

$$\bar{P} = \frac{W}{t}.$$

The work that is done by the water going over Niagara Falls is equal to the change in gravitational potential energy,

$$\Delta U = W.$$

The change in gravitational potential energy of a given mass $m$ of water falling a distance $h$ is given by

$$\Delta U = mgh.$$

#### SIMPLIFY

We can combine the preceding three equations to obtain

$$\bar{P} = \frac{W}{t} = \frac{mgh}{t} = \left(\frac{m}{t}\right)gh.$$

*Continued—*

**CALCULATE**

We first calculate the mass of water moving over the falls per unit time from the given volume of water per unit time, using the density of water:

$$\frac{m}{t}=\left(5520\ \frac{\text{m}^3}{\text{s}}\right)\left(\frac{1000\ \text{kg}}{\text{m}^3}\right)=5.52\cdot 10^6\ \text{kg/s}.$$

The average power is then

$$\overline{P}=\left(5.52\cdot 10^6\ \text{kg/s}\right)\left(9.81\ \text{m/s}^2\right)\left(49.0\ \text{m}\right)=2653.4088\ \text{MW}.$$

**ROUND**

We round to three significant figures:

$$\overline{P}=2.65\ \text{GW}.$$

**DOUBLE-CHECK**

Our result is comparable to the output of large electrical power plants, on the order of 1000 MW (1GW). The combined power generation capability of all of the hydroelectric power stations at Niagara Falls has a peak of 4.4 GW during the high water season in the spring, which is close to our answer. However, you may ask how the water produces power by simply falling over Niagara Falls. The answer is that it doesn't. Instead, a large fraction of the water of the Niagara River is diverted upstream from the falls and sent through tunnels, where it drives power generators. The water that makes it across the falls during the daytime and in the summer tourist season is only about 50% of the flow of the Niagara River. This flow is reduced even further, down to 10%, and more water is diverted for power generation during the nighttime and in the winter.

## MULTIPLE-CHOICE QUESTIONS

**6.1** A block of mass 5.0 kg slides without friction at a speed of 8.0 m/s on a horizontal table surface until it strikes and sticks to a mass of 4.0 kg attached to a horizontal spring (with spring constant of $k = 2000.0$ N/m), which in turn is attached to a wall. How far is the spring compressed before the masses come to rest?

a) 0.40 m
b) 0.54 m
c) 0.30 m
d) 0.020 m
e) 0.67 m

**6.2** A pendulum swings in a vertical plane. At the bottom of the swing, the kinetic energy is 8 J and the gravitational potential energy is 4 J. At the highest position of its swing, the kinetic and gravitational potential energies are

a) kinetic energy = 0 J and gravitational potential energy = 4 J.

b) kinetic energy = 12 J and gravitational potential energy = 0 J.

c) kinetic energy = 0 J and gravitational potential energy = 12 J.

d) kinetic energy = 4 J and gravitational potential energy = 8 J.

e) kinetic energy = 8 J and gravitational potential energy = 4 J.

**6.3** A ball of mass 0.5 kg is released from rest at point *A*, which is 5 m above the bottom of a tank of oil, as shown in the figure. At *B*, which is 2 m above the bottom of the tank, the ball has a speed of 6 m/s. The work done on the ball by the force of fluid friction is

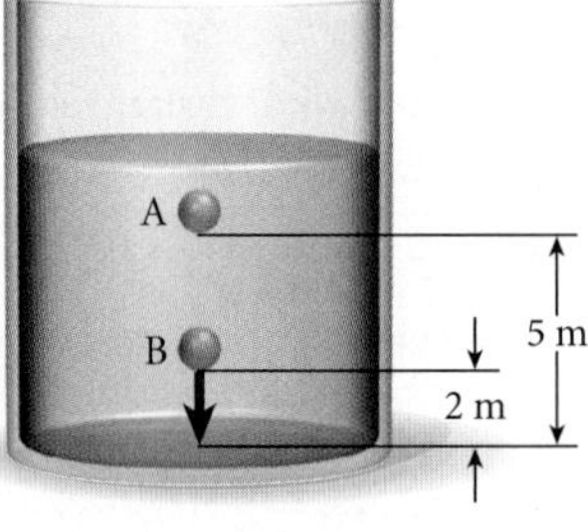


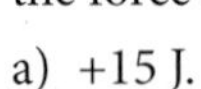

a) +15 J.
b) +9 J.
c) −15 J.
d) −9 J.
e) −5.7 J.

**6.4** A child throws three identical marbles from the same height above the ground so that they land on the *flat* roof of a building. The marbles are launched with the same initial speed. The first marble, marble A, is thrown at an angle of 75° above horizontal, while marbles B and C are thrown with launch angles of 60° and 45°, respectively. Neglecting air resistance, rank the marbles according to the speeds with which they hit the roof.

a) A < B < C
b) C < B < A
c) A and C have the same speed; B has a lower speed.
d) B has the highest speed; A and C have the same speed.
e) A, B, and C all hit the roof with the same speed.

**6.5** Which of the following is *not* a valid potential energy function for the spring force $F = -kx$?

a) $(\frac{1}{2})kx^2$
b) $(\frac{1}{2})kx^2 + 10$ J
c) $(\frac{1}{2})kx^2 - 10$ J
d) $-(\frac{1}{2})kx^2$
e) None of the above is valid.

**6.6** You use your hand to stretch a spring to a displacement $x$ from its equilibrium position and then slowly bring it back to that position. Which is true?

a) The spring's $\Delta U$ is positive.
b) The spring's $\Delta U$ is negative.
c) The hand's $\Delta U$ is positive.
d) The hand's $\Delta U$ is negative.
e) None of the above statements is true.

**6.7** In question 6, what is the work done by the hand?

a) $-(\frac{1}{2})kx^2$
b) $+(\frac{1}{2})kx^2$
c) $(\frac{1}{2})mv^2$, where $v$ is the speed of the hand
d) zero
e) none of the above

**6.8** Which of the following is *not* a unit of energy?

a) newton-meter
b) joule
c) kilowatt-hour
d) kg m$^2$/ s$^2$
e) all of the above

**6.9** A spring has a spring constant of 80 N/m. How much potential energy does it store when stretched by 1.0 cm?

a) $4.0 \cdot 10^{-3}$ J
b) 0.40 J
c) 80 J
d) 800 J
e) 0.8 J

## QUESTIONS

**6.10** Can the kinetic energy of an object be negative? Can the potential energy of an object be negative?

**6.11** a) If you jump off a table onto the floor, is your mechanical energy conserved? If not, where does it go? b) A car moving down the road smashes into a tree. Is the mechanical energy of the car conserved? If not, where does it go?

**6.12** How much work do you do when you hold a bag of groceries while standing still? How much work do you do when carrying the same bag a distance $d$ across the parking lot of the grocery store?

**6.13** An arrow is placed on a bow, the bowstring is pulled back, and the arrow is shot straight up into the air; the arrow then comes back down and sticks into the ground. Describe all of the changes in work and energy that occur.

**6.14** Two identical billiard balls start at the same height and the same time and roll along different tracks, as shown in the figure.

a) Which ball has the highest speed at the end?
b) Which one will get to the end first?

**6.15** A girl of mass 49.0 kg is on a swing, which has a mass of 1.0 kg. Suppose you pull her back until her center of mass is 2.0 m above the ground. Then you let her go, and she swings out and returns to the same point. Are all forces acting on the girl and swing conservative?

**6.16** Can a potential energy function be defined for the force of friction?

**6.17** Can the potential energy of a spring be negative?

**6.18** One end of a rubber band is tied down and you pull on the other end to trace a complicated *closed* trajectory. If you were to measure the elastic force $F$ at every point and took its scalar product with the local displacements, $\vec{F} \bullet \Delta\vec{r}$, and then summed all of these, what would you get?

**6.19** Can a unique potential energy function be identified with a particular conservative force?

**6.20** In skydiving, the vertical velocity component of the skydiver is typically zero at the moment he or she leaves the plane; the vertical component of the velocity then increases until the skydiver reaches terminal speed (see Chapter 4). Let's make a simplified model of this motion. We assume that the horizontal velocity component is zero. The vertical velocity component increases linearly, with acceleration $a_y = -g$, until the skydiver reaches terminal velocity, after which it stays constant. Thus, our simplified model assumes free fall without air resistance followed by falling at constant speed. Sketch the kinetic energy, potential energy, and total energy as a function of time for this model.

**6.21** A projectile of mass $m$ is launched from the ground at $t = 0$ with a speed $v_0$ and at an angle $\theta_0$ above the horizontal. Assuming that air resistance is negligible, write the kinetic, potential, and total energies of the projectile as explicit functions of time.

**6.22** The energy height, $H$, of an aircraft of mass $m$ at altitude $h$ and with speed $v$ is defined as its total energy (with the zero of the potential energy taken at ground level) divided by its weight. Thus, the energy height is a quantity with units of length.

a) Derive an expression for the energy height, $H$, in terms of the quantities $m$, $h$, and $v$.

b) A Boeing 747 jet with mass $3.5 \cdot 10^5$ kg is cruising in level flight at 250.0 m/s at an altitude of 10.0 km. Calculate the value of its energy height.

*Note:* The energy height is the maximum altitude an aircraft can reach by "zooming" (pulling into a vertical climb without changing the engine thrust). This maneuver is not recommended for a 747, however.

**6.23** A body of mass $m$ moves in one dimension under the influence of a force, $F(x)$, which depends only on the body's position.

a) Prove that Newton's Second Law and the law of conservation of energy for this body are exactly equivalent.

b) Explain, then, why the law of conservation of energy is considered to be of greater significance than Newton's Second Law.

**6.24** The molecular bonding in a diatomic molecule such as the nitrogen ($N_2$) molecule can be modeled by the Lennard-Jones potential, which has the form

$$U(x)=4U_0\left(\left(\frac{x_0}{x}\right)^{12}-\left(\frac{x_0}{x}\right)^{6}\right),$$

where $x$ is the separation distance between the two nuclei and $x_0$, and $U_0$ are constants. Determine, in terms of these constants, the following:

a) the corresponding force function;

b) the equilibrium separation $x_0$, which is the value of $x$ for which the two atoms experience zero force from each other; and

c) the nature of the interaction (repulsive or attractive) for separations larger and smaller than $x_0$.

**6.25** A particle of mass $m$ moving in the $xy$-plane is confined by a two-dimensional potential function, $U(x, y) = \frac{1}{2}k(x^2 + y^2)$.

a) Derive an expression for the net force, $\vec{F} = F_x\hat{x} + F_y\hat{y}$.

b) Find the equilibrium point on the $xy$-plane.

c) Describe qualitatively the effect of net force.

d) What is the magnitude of the net force on the particle at the coordinate (3,4) in cm if $k = 10$ N/cm?

e) What are the turning points if the particle has 10 J of total mechanical energy?

**6.26** For a rock dropped from rest from a height $h$, to calculate the speed just before it hits the ground, we use the conservation of mechanical energy and write $mgh = \frac{1}{2}mv^2$. The mass cancels out, and we solve for $v$. A very common error made by some beginning physics students is to assume, based on the appearance of this equation, that they should set the kinetic energy equal to the potential energy at the same point in space. For example, to calculate the speed $v_1$ of the rock at some height $y_1 < h$, they often write $mgy_1 = \frac{1}{2}mv_1^2$ and solve for $v_1$. Explain why this approach is wrong.

## PROBLEMS

A blue problem number indicates a worked-out solution is available in the Student Solutions Manual. One • and two •• indicate increasing level of problem difficulty.

### Section 6.1

**6.27** What is the gravitational potential energy of a 2.0-kg book 1.5 m above the floor?

**6.28** a) If the gravitational potential energy of a 40.0-kg rock is 500 J relative to a value of zero on the ground, how high is the rock above the ground?

b) If the rock were lifted to twice its original height, how would the value of its gravitational potential energy change?

**6.29** A rock of mass 0.773 kg is hanging from a string of length 2.45 m on the Moon, where the gravitational acceleration is a sixth of that on Earth. What is the change in gravitational potential energy of this rock when it is moved so that the angle of the string changes from 3.31° to 14.01°? (Both angles are measured relative to the vertical.)

**6.30** A 20.0-kg child is on a swing attached to ropes that are $L = 1.50$ m long. Take the zero of the gravitational potential energy to be at the position of the child when the ropes are horizontal.

a) Determine the child's gravitational potential energy when the child is at the lowest point of the circular trajectory.

b) Determine the child's gravitational potential energy when the ropes make an angle of 45° relative to the vertical.

c) Based on these results, which position has the higher potential energy?

### Section 6.3

**6.31** A 1500-kg car travels 2.5 km up an incline at constant velocity. The incline has an angle of 3° with respect to the horizontal. What is the change in the car's potential energy? What is the net work done on the car?

**6.32** A constant force of 40.0 N is needed to keep a car traveling at constant speed as it moves 5.0 km along a road. How much work is done? Is the work done on or by the car?

**6.33** A piñata of mass 3.27 kg is attached to a string tied to a hook in the ceiling. The length of the string is 0.81 m, and the piñata is released from rest from an initial position in which the string makes an angle of 56.5° with the vertical. What is the work done by gravity by the time the string is in a vertical position for the first time?

### Section 6.4

**•6.34** A particle is moving along the $x$-axis subject to the potential energy function $U(x) = 1/x + x^2 + x - 1$.

a) Express the force felt by the particle as a function of $x$.

b) Plot this force and the potential energy function.

c) Determine the net force on the particle at the coordinate $x = 2$ m.

**•6.35** Calculate the force $F(y)$ associated with each of the following potential energies:

a) $U = ay^3 - by^2$ b) $U = U_0 \sin(cy)$

**•6.36** The potential energy of a certain particle is given by $U = 10x^2 + 35z^3$. Find the force vector exerted on the particle.

## Section 6.5

**6.37** A ball is thrown up in the air, reaching a height of 5 m. Using energy conservation considerations, determine its initial speed.

**6.38** A cannonball of mass 5.99 kg is shot from a cannon at an angle of 50.21° relative to the horizontal and with an initial speed of 52.61 m/s. As the cannonball reaches the highest point of its trajectory, what is the gain in its potential energy relative to the point from which it was shot?

**6.39** A basketball of mass 0.624 kg is shot from a vertical height of 1.2 m and at a speed of 20.0 m/s. After reaching its maximum height, the ball moves into the hoop on its downward path, at 3.05 m above the ground. Using the principle of energy conservation, determine how fast the ball is moving just before it enters the hoop.

**•6.40** A classmate throws a 1.0-kg book from a height of 1.0 m above the ground straight up into the air. The book reaches a maximum height of 3.0 m above the ground and begins to fall back. Assume that 1.0 m above the ground is the reference level for zero gravitational potential energy. Determine

a) the gravitational potential energy of the book when it hits the ground.

b) the velocity of the book just before hitting the ground.

**•6.41** Suppose you throw a 0.052-kg ball with a speed of 10.0 m/s and at an angle of 30.0° above the horizontal from a building 12.0 m high.

a) What will be its kinetic energy when it hits the ground?

b) What will be its speed when it hits the ground?

**•6.42** A uniform chain of total mass $m$ is laid out straight on a frictionless table and held stationary so that one-third of its length, $L = 1$ m, is hanging vertically over the edge of the table. The chain is then released. Determine the speed of the chain at the instant when only one-third of its length remains on the table.

**•6.43** a) If you are at the top of a toboggan run that is 40 m high, how fast will you be going at the bottom, provided you can ignore friction between the sled and the track?

b) Does the steepness of the run affect how fast you will be going at the bottom?

c) If you do not ignore the small friction force, does the steepness of the track affect the value of the speed at the bottom?

## Section 6.6

**6.44** A block of mass 0.773 kg on a spring with spring constant 239.5 N/m oscillates vertically with amplitude 0.551 m. What is the speed of this block at a distance of 0.331 m from the equilibrium position?

**6.45** A spring with $k = 10$ N/cm is initially stretched 1 cm from its equilibrium length.

a) How much more energy is needed to further stretch the spring to 5 cm beyond its equilibrium length?

b) From this new position, how much energy is needed to compress the spring to 5 cm shorter than its equilibrium position?

**•6.46** A 5.00-kg ball of clay is thrown downward from a height of 3.00 m with a speed of 5.00 m/s onto a spring with $k =$ 1600. N/m. The clay compresses the spring a certain maximum amount before momentarily stopping.

a) Find the maximum compression of the spring.

b) Find the total work done on the clay during the spring's compression.

**•6.47** A horizontal slingshot consists of two light, identical springs (with spring constants of 30 N/m) and a light cup that holds a 1-kg stone. Each spring has an equilibrium length of 50 cm. When the springs are in equilibrium, they line up vertically. Suppose that the cup containing the mass is pulled to $x = 0.7$ m to the left of the vertical and then released. Determine

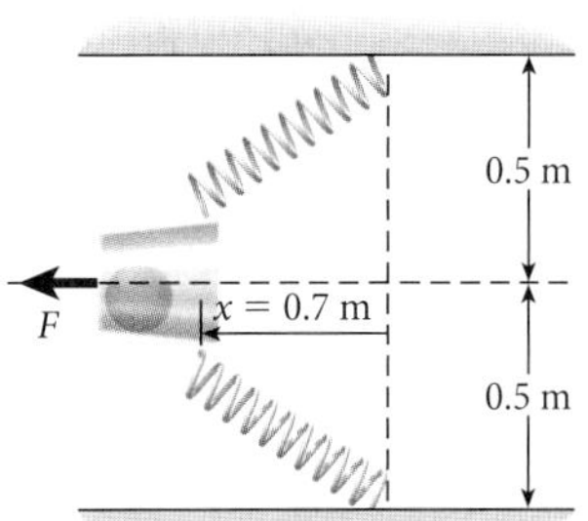


a) the system's total mechanical energy.

b) the speed of the stone at $x = 0$.

**•6.48** Suppose the stone in Problem 6.47 is instead launched vertically and the mass is a lot smaller ($m = 0.1$ kg). Take the zero of the gravitational potential energy to be at the equilibrium point.

a) Determine the total mechanical energy of the system.

b) How fast is the stone moving as it passes the equilibrium point?

## Section 6.7

**6.49** A 80-kg fireman slides down a 3-m pole by applying a frictional force of 400 N against the pole with his hands. If he slides from rest, how fast is he moving once he reaches the ground?

**6.50** A large air-filled 0.1-kg plastic ball is thrown up into the air with an initial speed of 10 m/s. At a height of 3 m, the ball's speed is 3 m/s. What fraction of its original energy has been lost to air friction?

**6.51** How much mechanical energy is lost to friction if a 55.0-kg skier slides down a ski slope at constant speed of 14.4 m/s? The slope is 123.5 m long and makes an angle of 14.7° with respect to the horizontal.

**•6.52** A truck of mass 10,212 kg moving at a speed of 61.2 mph has lost its brakes. Fortunately, the driver finds a runaway lane, a gravel-covered incline that uses friction to stop a truck in such a situation; see the figure.

$\Delta x$

$\theta$

In this case, the incline makes an angle of $\theta = 40.15°$ with the horizontal, and the gravel has a coefficient of friction of 0.634 with the tires of the truck. How far along the incline ($\Delta x$) does the truck travel before it stops?

•**6.53** A snowboarder of mass 70.1 kg (including gear and clothing), starting with a speed of 5.1 m/s, slides down a slope at an angle $\theta = 37.1°$ with the horizontal. The coefficient of kinetic friction is 0.116. What is the net work done on the snowboarder in the first 5.72 s of descent?

•**6.54** The greenskeepers of golf courses use a stimpmeter to determine how "fast" their greens are. A stimpmeter is a straight aluminum bar with a V-shaped groove on which a golf ball can roll. It is designed to release the golf ball once the angle of the bar with the ground reaches a value of $\theta = 20.0°$. The golf ball (mass = 1.62 oz = 0.0459 kg) rolls 30.0 in down the bar and then continues to roll along the green for several feet. This distance is called the "reading." The test is done on a level part of the green, and stimpmeter readings between 7 and 12 ft are considered acceptable. For a stimpmeter reading of 11.1 ft, what is the coefficient of friction between the ball and the green? (The ball is rolling and not sliding, as we usually assume when considering friction, but this does not change the result in this case.)

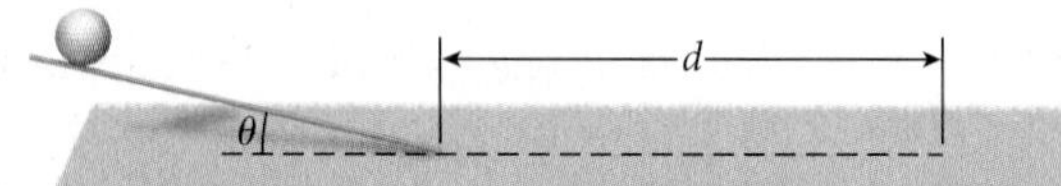


•**6.55** A 1-kg block is pushed up and down a rough plank of length $L = 2$ m, inclined at 30° above the horizontal. From the bottom, it is pushed a distance $L/2$ up the plank, then pushed back down a distance $L/4$, and finally pushed back up the plank until it reaches the top end. If the coefficient of kinetic friction between the block and plank is 0.3, determine the work done by the block against friction.

••**6.56** A 1-kg block initially at rest at the top of a 4-m incline with a slope of 45° begins to slide down the incline. The upper half of the incline is frictionless, while the lower half is rough, with a coefficient of kinetic friction $\mu_k = 0.3$.

a) How fast is the block moving midway along the incline, before entering the rough section?

b) How fast is the block moving at the bottom of the incline?

••**6.57** A spring with a spring constant of 500 N/m is used to propel a 0.50-kg mass up an inclined plane. The spring is compressed 30 cm from its equilibrium position and launches the mass from rest across a horizontal surface and onto the plane. The plane has a length of 4 m and is inclined at 30°. Both the plane and the horizontal surface have a coefficient of kinetic friction with the mass of 0.35. When the spring is compressed, the mass is 1.5 m from the bottom of the plane.

a) What is the speed of the mass as it reaches the bottom of the plane?

b) What is the speed of the mass as it reaches the top of the plane?

c) What is the total work done by friction from the beginning to the end of the mass's motion?

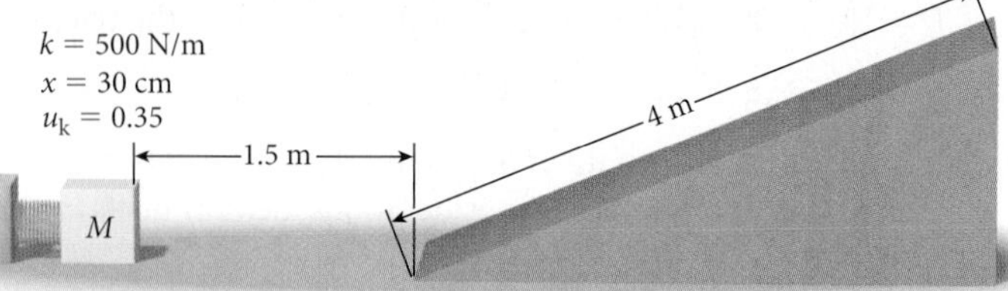


••**6.58** The sled shown in the figure leaves the starting point with a velocity of 20 m/s. Use the work-energy theorem to calculate the sled's speed at the end of the track or the maximum height it reaches if it stops before reaching the end.

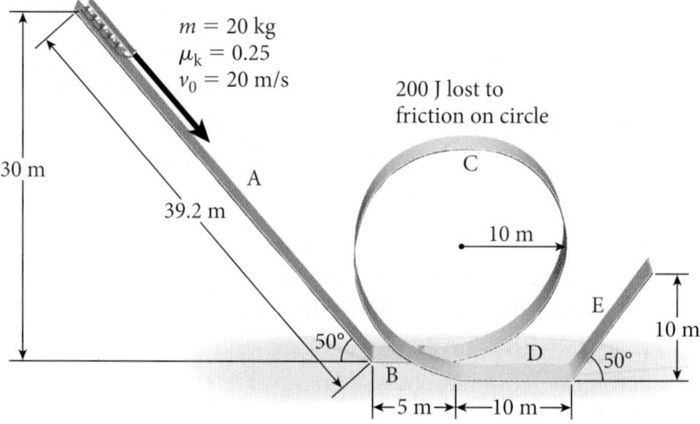


## Section 6.8

•**6.59** On the segment of roller coaster track shown in the figure, a cart of mass 237.5 kg moves from left to right and arrives at $x = 0$ with a speed of 16.5 m/s. Assuming that dissipation of energy due to friction is small enough to be ignored, where is the turning point of this trajectory?

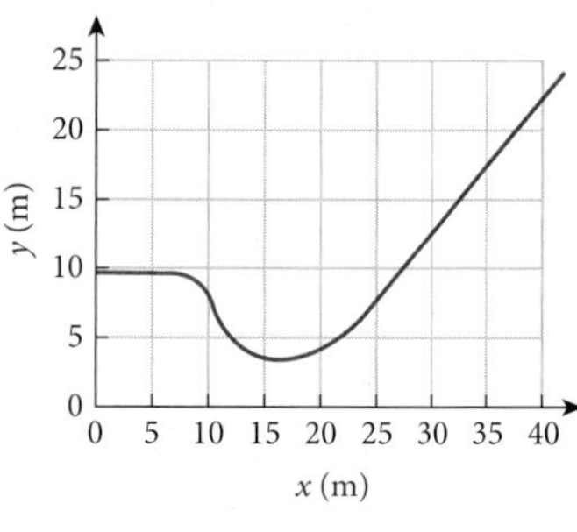


•**6.60** A 70-kg skier moving horizontally at 4.5 m/s encounters a 20° incline.

a) How far up the incline will the skier move before she momentarily stops, ignoring friction?

b) How far up the incline will the skier move if the coefficient of kinetic friction between the skies and snow is 0.1?

•**6.61** A 0.2-kg particle is moving along the $x$-axis, subject to the potential energy function shown in the figure, where $U_A = 50$ J, $U_B = 0$ J, $U_C = 25$ J, $U_D = 10$ J, and $U_E = 60$ J along the path. If the particle was initially at $x = 4$ m and had a total mechanical energy of 40 J, determine:

a) the particle's speed at $x = 3$ m;

b) the particle's speed at $x = 4.5$ m, and

c) the particle's turning points.

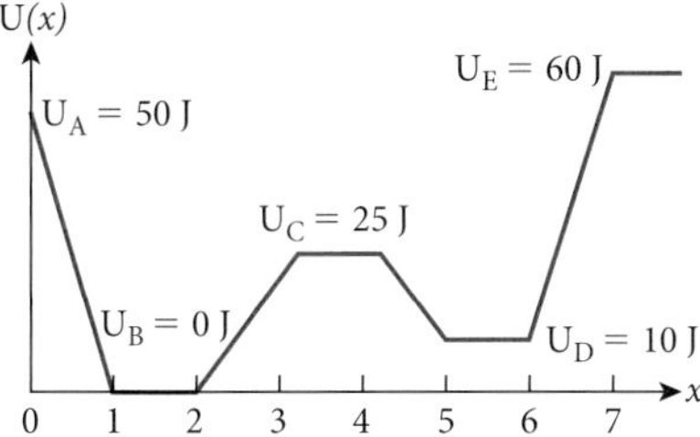


## Additional Problems

**6.62** A ball of mass 1.84 kg is dropped from a height $y_1 =$ 1.49 m and then bounces back up to a height of $y_2 = 0.87$ m. How much mechanical energy is lost in the bounce? The effect of air resistance has been experimentally found to be negligible in this case, and you can ignore it.

6.63 A car of mass 987 kg is traveling on a horizontal segment of a freeway with a speed of 64.5 mph. Suddenly, the driver has to hit the brakes hard to try to avoid an accident up ahead. The car does not have an ABS (antilock braking system), and the wheels lock, causing the car to slide some distance before it is brought to a stop by the friction force between the car's tires and the road surface. The coefficient of kinetic friction is 0.301. How much mechanical energy is lost to heat in this process?

**6.64** Two masses are connected by a light string that goes over a light, frictionless pulley, as shown in the figure. The 10-kg mass is released and falls through a vertical distance of 1 m before hitting the ground. Use conservation of mechanical energy to determine:

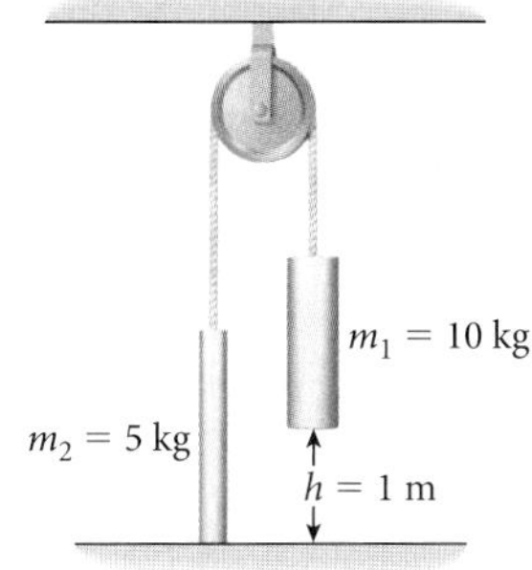


a) how fast the 5-kg mass is moving just before the 10-kg mass hits the ground; and

b) the maximum height attained by the 5-kg mass.

**6.65** In 1896 in Waco, Texas, William George Crush, owner of the K-T (or "Katy") Railroad, parked two locomotives at opposite ends of a 6.4-km-long track, fired them up, tied their throttles open, and then allowed them to crash head-on at full speed in front of 30,000 spectators. Hundreds of people were hurt by flying debris; several were killed. Assuming that each locomotive weighed $1.2 \cdot 10^6$ N and its acceleration along the track was a constant 0.26 m/s$^2$, what was the total kinetic energy of the two locomotives just before the collision?

**6.66** A baseball pitcher can throw a 9.00-oz baseball with a speed measured by a radar gun to be 90.0 mph. Assuming that the force exerted by the pitcher on the ball acts over a distance of two arm lengths, each 28.0 in, what is the average force exerted by the pitcher on the ball?

6.67 A 1.5-kg soccer ball has a speed of 20 m/s when it is 15 m above the ground. What is the total energy of the ball?

**6.68** If it takes an average force of 5.5 N to push a 4.5-g dart 6.0 cm into a dart gun, assuming the barrel is frictionless, how fast will the dart exit the gun?

**6.69** A high jumper approaches the bar at 9.0 m/s. What is the highest altitude the jumper can reach, if he does not use any additional push off the ground and is moving at 7.0 m/s as he goes over the bar?

**6.70** A roller coaster is moving at 2.0 m/s at the top of the first hill ($h = 40$ m). Ignoring friction and air resistance, how fast will the roller coaster be moving at the top of a subsequent hill, which is 15 m high?

6.71 You are on a swing with a chain 4.0 m long. If your maximum displacement from the vertical is 35°, how fast will you be moving at the bottom of the arc?

**6.72** A truck is descending a winding mountain road. When the truck is 680 m above sea level and traveling at 15 m/s, its brakes fail. What is the maximum possible speed of the truck at the foot of the mountain, 550 m above sea level?

**6.73** Tarzan swings on a taut vine from his tree house to a limb on a neighboring tree, which is located a horizontal distance of 10.0 m from and 4.00 m below his starting point. Amazingly the vine neither stretches nor breaks; Tarzan's trajectory is thus a portion of a circle. If Tarzan starts with zero speed, what is his speed when he reaches the limb?

**6.74** The graph shows the component ($F \cos \theta$) of the net force that acts on a 2.0-kg block as it moves along a flat horizontal surface. Find

a) the net work done on the block.

b) the final speed of the block if it starts from rest at $s = 0$.

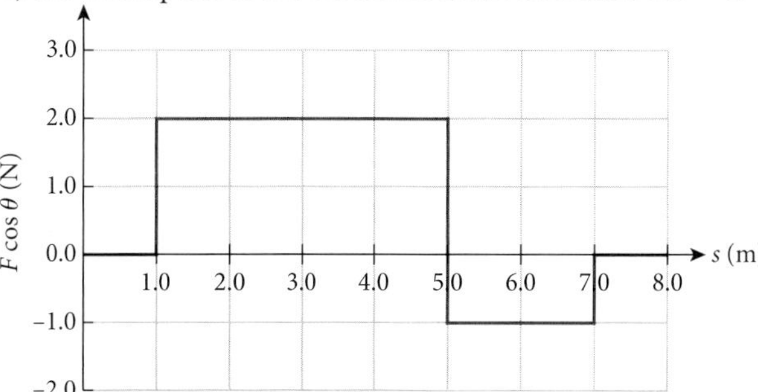


•6.75 A 3.00-kg model rocket is launched vertically upward with sufficient initial speed to reach a height of $1.00 \cdot 10^2$ m, even though air resistance (a nonconservative force) performs $-8.00 \cdot 10^2$ J of work on the rocket. How high would the rocket have gone, if there were no air resistance?

•**6.76** A 0.50-kg mass is attached to a horizontal spring with $k = 100$ N/m. The mass slides across a frictionless surface. The spring is stretched 25 cm from equilibrium, and then the mass is released from rest.

*Continued—*

a) Find the mechanical energy of the system.

b) Find the speed of the mass when it has moved 5 cm.

c) Find the maximum speed of the mass.

•**6.77** You have decided to move a refrigerator (mass = 81.3 kg, including all the contents) to the other side of the room. You slide it across the floor on a straight path of length 6.35 m, and the coefficient of kinetic friction between floor and fridge is 0.437. Happy about your accomplishment, you leave the apartment. Your roommate comes home, wonders why the fridge is on the other side of the room, picks it up (you have a strong roommate!), carries it back to where it was originally, and puts it down. How much net mechanical work have the two of you done together?

•**6.78** A 1-kg block compresses a spring for which $k$ = 100 N/m by 20 cm and is then released to move across a horizontal, frictionless table, where it hits and compresses another spring, for which $k$ = 50 N/m. Determine

a) the total mechanical energy of the system,

b) the speed of the mass while moving freely between springs, and

c) the maximum compression of the second spring.

•**6.79** A 1-kg block is resting against a light, compressed spring at the bottom of a rough plane inclined at an angle of 30°; the coefficient of kinetic friction between block and plane is $\mu_k = 0.1$. Suppose the spring is compressed 10 cm from its equilibrium length. The spring is then released, and the block separates from the spring and slides up the incline a distance of only 2 cm beyond the spring's normal length before it stops. Determine

a) the change in total mechanical energy of the system and

b) the spring constant $k$.

•**6.80** A 0.1-kg ball is dropped from a height of 1 m and lands on a light (approximately massless) cup mounted on top of a light, vertical spring initially at its equilibrium position. The maximum compression of the spring is to be 10 cm.

a) What is the required spring constant of the spring?

b) Suppose you ignore the change in the gravitational energy of the ball during the 10-cm compression. What is the percentage difference between the calculated spring constant for this case and the answer obtained in part (a)?

•**6.81** A mass of 1 kg attached to a spring with a spring constant of 100 N/m oscillates horizontally on a smooth frictionless table with an amplitude of 0.5 m. When the mass is 0.25 m away from equilibrium, determine:

a) its total mechanical energy;

b) the system's potential energy and the mass's kinetic energy;

c) the mass's kinetic energy when it is at the equilibrium point.

d) Suppose there was friction between the mass and the table so that the amplitude was cut in half after some time. By what factor has the mass's maximum kinetic energy changed?

e) By what factor has the maximum potential energy changed?

•**6.82** Bolo, the human cannonball, is projected from a 3.5 m long barrel. If Bolo ($m$ = 80 kg) has a speed of 12 m/s at the top of his trajectory, 15 m above the ground, what was the average force exerted on him while in the barrel?

•**6.83** A 1.0-kg mass is suspended vertically from a spring with $k$ = 100 N/m and oscillates with an amplitude of 0.2 m. At the top of its oscillation, the mass is hit in such a way that it instantaneously moves down with a speed of 1 m/s. Determine

a) its total mechanical energy,

b) how fast it is moving as it crosses the equilibrium point, and

c) its new amplitude.

•**6.84** A runner reaches the top of a hill with a speed of 6.5 m/s. He descends 50 m and then ascends 28 m to the top of the next hill. His speed is now 4.5 m/s. The runner has a mass of 83 kg. The total distance that the runner covers is 400 m, and there is a constant resistance to motion of 9.0 N. Use energy considerations to find the work done by the runner over the total distance.

•**6.85** A package is dropped on a horizontal conveyor belt. The mass of the package is $m$, the speed of the conveyor belt is $v$, and the coefficient of kinetic friction between the package and the belt is $\mu_k$.

a) How long does it take for the package to stop sliding on the belt?

b) What is the package's displacement during this time?

c) What is the energy dissipated by friction?

d) What is the total work supplied by the system?

•**6.86** A father exerts a $2.40 \cdot 10^2$ N force to pull a sled with his daughter on it (combined mass of 85.0 kg) across a horizontal surface. The rope with which he pulls the sled makes an angle of 20.0° with the horizontal. The coefficient of kinetic friction is 0.200, and the sled moves a distance of 8.00 m. Find

a) the work done by the father,

b) the work done by the friction force, and

c) the total work done by all the forces.

•**6.87** A variable force acting on a 0.1-kg particle moving in the $xy$-plane is given by $F(x, y) = (x^2\,\hat{x} + y^2\,\hat{y})$ N, where $x$ and $y$ are in meters. Suppose that due to this force, the particle moves from the origin, $O$, to point $S$, with coordinates (10 m,10 m). The coordinates of points $P$ and $Q$ are (0 m,10 m) and (10 m,0 m), respectively. Determine the work performed by the force as the particle moves along each of the following paths:

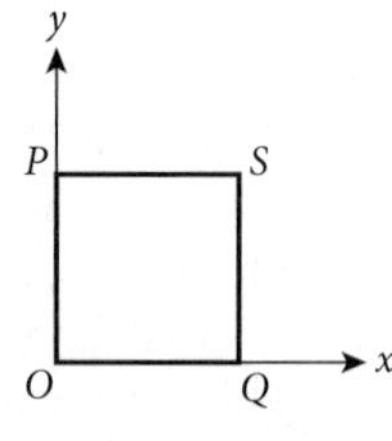


a) $OPS$ b) $OQS$ c) $OS$ d) $OPSQO$ e) $OQSPO$

••**6.88** In problem 6.87, suppose there was friction between the 0.1-kg particle and the $xy$-plane, with $\mu_k = 0.1$. Determine the net work done by all forces on this particle when it takes each of the following paths:

a) $OPS$ b) $OQS$ c) $OS$ d) $OPSQO$ e) $OQSPO$

# Momentum and Collisions

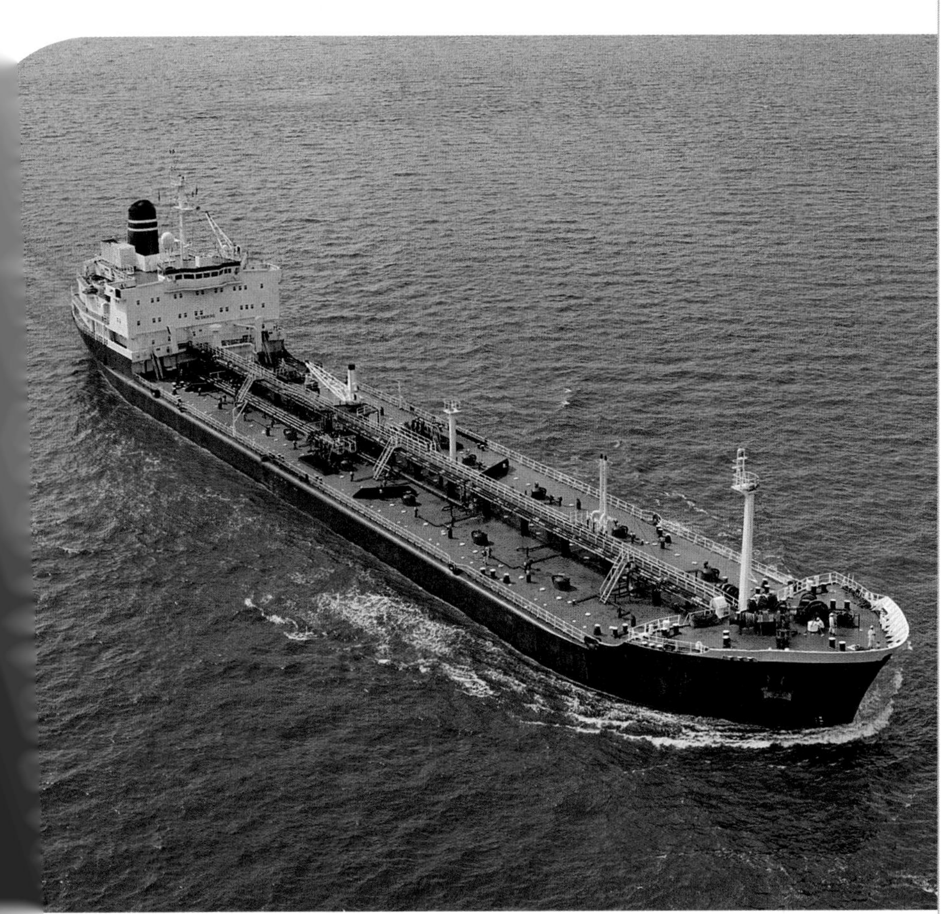

**FIGURE 7.1** A supertanker.

## WHAT WE WILL LEARN

- The momentum of an object is the product of its velocity and mass. Momentum is a vector quantity and points in the same direction as the velocity vector.
- Newton's Second Law can be phrased more generally as follows: The net force on an object equals the time derivative of the object's momentum.
- A change of momentum, called *impulse,* is the time integral of the net force that causes the momentum change.
- In all collisions, the momentum is conserved.
- Besides conservation of momentum, elastic collisions also have the property that the total kinetic energy is conserved.
- In totally inelastic collisions, the maximum amount of kinetic energy is removed, and the colliding objects stick to each other. Total kinetic energy is not conserved, but momentum is.
- Collisions that are neither elastic nor totally inelastic are partially inelastic, and the change in kinetic energy is proportional to the square of the coefficient of restitution.
- The physics of collisions has a direct connection to the research frontier of chaos science.

Supertankers for transporting oil around the world are the largest ships ever built (Figure 7.1). They can have a mass (including cargo) of up to 650,000 tons and carry over 2 million barrels (84 million gallons = 318 million liters) of oil. However, their large size creates practical problems. Supertankers are too big to enter most shipping ports and have to stop at offshore platforms to offload their oil. In addition, piloting a ship of this size is extremely difficult. For example, when the captain gives the order to reverse engines and come to a stop, the ship can continue to move forward for more than 3 miles!

The physical quantity that makes a large moving object difficult to stop is *momentum,* the subject of this chapter. Momentum is a fundamental property associated with an object's motion, similar to kinetic energy. Moreover, both momentum and energy are subjects of important conservation laws in physics. However, momentum is a vector quantity, whereas energy is a scalar. Thus, working with momentum requires taking account of angles and components, as we did for force in Chapter 4.

The importance of momentum becomes most evident when we deal with collisions between two or more objects. In this chapter, we examine various collisions in one and two dimensions. In later chapters, we will make use of conservation of momentum in many different situations on vastly different scales—from bursts of elementary particles to collisions of galaxies.

## 7.1 Linear Momentum

For the terms *force, position, velocity,* and *acceleration,* the precise physical definitions are quite close to the words' usage in everyday language. With the term *momentum,* the situation is more analogous to that of *energy,* for which there is only a vague connection between conversational use and precise physical meaning. You sometimes hear that the campaign of a particular political candidate gains momentum or that legislation gains momentum in Congress. Often, sports teams or individual players are said to gain or lose momentum. What these statements imply is that the objects said to gain momentum have become harder to stop. However, Figure 7.2 shows that even objects with large momentum can be stopped!

### Definition of Momentum

In physics, **momentum** is defined as the product of an object's mass and its velocity:

$$\vec{p} = m\vec{v}. \tag{7.1}$$

As you can see, the lowercase letter $\vec{p}$ is the symbol for linear momentum. The velocity $\vec{v}$ is a vector and is multiplied by a scalar quantity, the mass $m$. The product is thus a vector as well. The momentum vector, $\vec{p}$, and the velocity vector, $\vec{v}$, are parallel to each other; that is,

**FIGURE 7.2** Rocket-sled crash test of a fighter plane. Tests like these can be used to improve the design of critical structures such as nuclear reactors so they can withstand the impact of a plane crash.

they point in the same direction. As a simple consequence of equation 7.1, the magnitude of the momentum is

$$p = mv.$$

The momentum is also referred to as *linear momentum* to distinguish it from angular momentum, a concept we will study in Chapter 10 on rotation. The units of momentum are kg m/s. Unlike the unit for energy, the unit for momentum does not have a special name. The magnitude of momentum spans a large range. Momenta of various objects, from a subatomic particle to a planet orbiting the Sun, are given in Table 7.1.

## Momentum and Force

Let's take the time derivative of equation 7.1. We use the product rule of differentiation to obtain

$$\frac{d}{dt}\vec{p} = \frac{d}{dt}(m\vec{v}) = m\frac{d\vec{v}}{dt} + \frac{dm}{dt}\vec{v}.$$

For now, we assume that the mass of the object does not change, and therefore the second term is zero. Because the time derivative of the velocity is the acceleration, we have

$$\frac{d}{dt}\vec{p} = m\frac{d\vec{v}}{dt} = m\vec{a} = \vec{F},$$

according to Newton's Second Law. The relationship

$$\vec{F} = \frac{d}{dt}\vec{p} \tag{7.2}$$

is an equivalent form of Newton's Second Law. This form is more general than $\vec{F} = m\vec{a}$ because it also holds in cases where the mass is not constant in time. This distinction will become important when we examine rocket motion in Chapter 8. Because equation 7.2 is a vector equation, we can also write it in Cartesian components:

$$F_x = \frac{dp_x}{dt}; \quad F_y = \frac{dp_y}{dt}; \quad F_z = \frac{dp_z}{dt}.$$

**Table 7.1 Momenta of Various Objects**

| Object | Momentum (kg m/s) |
|---|---|
| Alpha ($\alpha$) particle from $^{238}$U decay | $9.53 \cdot 10^{-20}$ |
| 90-mph fastball | 5.75 |
| Charging rhinoceros | $3 \cdot 10^{4}$ |
| Car moving on freeway | $5 \cdot 10^{4}$ |
| Supertanker at cruising speed | $4 \cdot 10^{9}$ |
| Moon orbiting Earth | $7.58 \cdot 10^{25}$ |
| Earth orbiting Sun | $1.78 \cdot 10^{29}$ |

### 7.1 In-Class Exercise

A typical scene from a Saturday afternoon college football game: A linebacker of mass 95 kg runs with a speed of 7.8 m/s, and a wide receiver of mass 74 kg runs with a speed of 9.6 m/s. We denote the magnitude of the momentum and kinetic energy of the linebacker by $p_l$ and $K_l$, respectively, and the magnitude of the momentum and kinetic energy of the wide receiver by $p_w$ and $K_w$. Which set of inequalities is correct?

a) $p_l > p_w$, $K_l > K_w$

b) $p_l < p_w$, $K_l > K_w$

c) $p_l > p_w$, $K_l < K_w$

d) $p_l < p_w$, $K_l < K_w$

## Momentum and Kinetic Energy

In Chapter 5, we established the relationship, $K = \frac{1}{2}mv^2$ (equation 5.1), between the kinetic energy $K$, the speed $v$, and the mass $m$. We can use $p = mv$ to obtain

$$K = \frac{mv^2}{2} = \frac{m^2v^2}{2m} = \frac{p^2}{2m}.$$

This equation gives us an important relationship between kinetic energy, mass, and momentum:

$$K = \frac{p^2}{2m}. \tag{7.3}$$

At this point, you may wonder why we need to reformulate the concepts of force and kinetic energy in terms of momentum. This reformulation is far more than a mathematical game. We will see that momentum is conserved in collisions and disintegrations, and this principle will provide an extremely helpful way to find solutions to complicated problems. These relationships of momentum with force and kinetic energy will be very useful in working such problems. First, though, we need to explore the physics of changing momentum in a little more detail.

## 7.2 Impulse

The change in momentum is defined as the difference between the final (index f) and initial (index i) momenta:

$$\Delta\vec{p} \equiv \vec{p}_\text{f} - \vec{p}_\text{i}.$$

To see why this definition is useful, we have to do a bit of math. Let's start by exploring the relationship between force and momentum just a little further. We can integrate each component of the equation $\vec{F} = d\vec{p}/dt$ over time. For the integral over $F_x$, for example, we obtain:

$$\int_{t_\text{i}}^{t_\text{f}} F_x\,dt = \int_{t_\text{i}}^{t_\text{f}} \frac{dp_x}{dt}\,dt = \int_{p_{x,\text{i}}}^{p_{x,\text{f}}} dp_x = p_{x,\text{f}} - p_{x,\text{i}} \equiv \Delta p_x.$$

This equation requires some explanation. In the second step, we performed a substitution of variables to transform an integration over time into an integration over momentum. Figure 7.3a illustrates this relationship: The area under the $F_x(t)$ curve is the change in momentum, $\Delta p_x$. We can obtain similar equations for the $y$- and $z$-components.

Combining them into one vector equation yields the following result:

$$\int_{t_\text{i}}^{t_\text{f}} \vec{F}\,dt = \int_{t_\text{i}}^{t_\text{f}} \frac{d\vec{p}}{dt}\,dt = \int_{\vec{p}_\text{i}}^{\vec{p}_\text{f}} d\vec{p} = \vec{p}_\text{f} - \vec{p}_\text{i} \equiv \Delta\vec{p}.$$

The time integral of the force is called the **impulse,** $\vec{J}$:

$$\vec{J} \equiv \int_{t_\text{i}}^{t_\text{f}} \vec{F}\,dt. \tag{7.4}$$

This definition immediately gives us the relationship between the impulse and the momentum change:

$$\vec{J} = \Delta\vec{p}. \tag{7.5}$$

From equation 7.5, we can calculate the momentum change over some time interval, if we know the time dependence of the force. If the force is constant or has some form that we can integrate, then we can simply evaluate the integral of equation 7.4. However, we can also define an average force,

$$\vec{F}_\text{ave} = \frac{\int_{t_\text{i}}^{t_\text{f}} \vec{F}\,dt}{\int_{t_\text{i}}^{t_\text{f}} dt} = \frac{1}{t_\text{f} - t_\text{i}} \int_{t_\text{i}}^{t_\text{f}} \vec{F}\,dt = \frac{1}{\Delta t}\int_{t_\text{i}}^{t_\text{f}} \vec{F}\,dt. \tag{7.6}$$

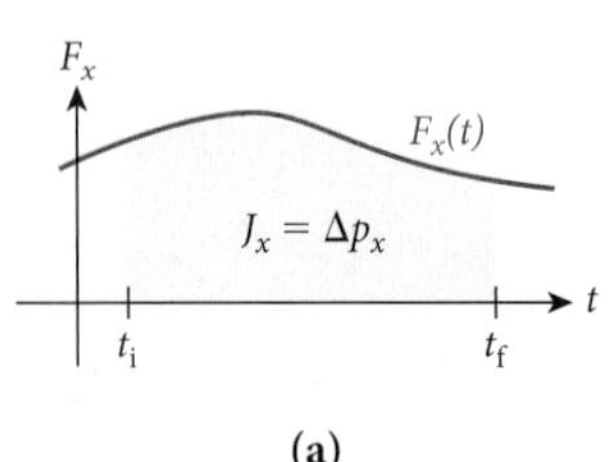


(a)

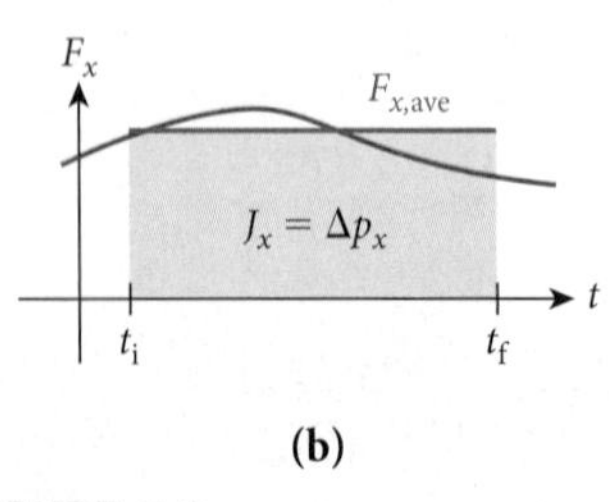


(b)

**FIGURE 7.3** (a) The impulse (yellow area) is the time integral of the force; (b) same impulse resulting from an average force.

This integral gives us

$$\vec{J} = \vec{F}_{\text{ave}}\Delta t. \tag{7.7}$$

You may think this transformation is trivial in that it conveys the same information as equation 7.5. After all, the integration is still there, hidden in the definition of the average force. This is true, but sometimes we are only interested in the average force. Measuring the time interval, $\Delta t$, over which a force acts as well as the resulting impulse an object receives tells us the average force that the object experiences during that time interval. Figure 7.3b illustrates the relationship between the time-averaged force, the momentum change, and the impulse.

## EXAMPLE 7.1 Baseball Home Run

A major league pitcher throws a fastball that crosses home plate with a speed of 90.0 mph (40.23 m/s) and an angle of 5.0° below the horizontal. A batter slugs it for a home run, launching it with a speed of 110.0 mph (49.17 m/s) at an angle of 35.0° above the horizontal (Figure 7.4). The mass of a baseball is required to be between 5 and 5.25 oz; let's say that the mass of the ball hit here is 5.10 oz (0.145 kg).

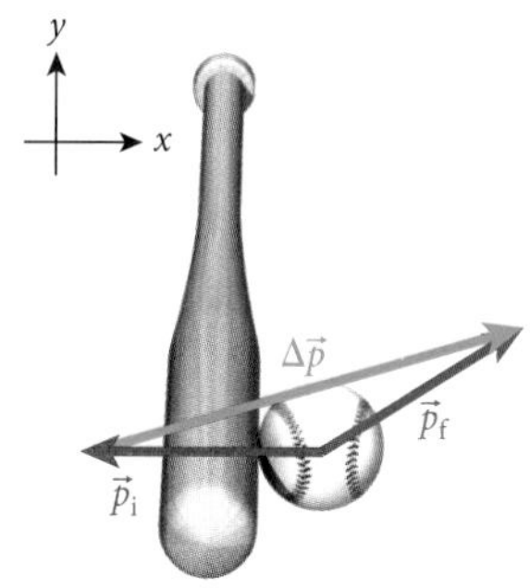


**FIGURE 7.4** Baseball being hit by a bat. Initial (red) and final (blue) momentum vectors, as well as the impulse (green) vector (or change in momentum vector) are shown.

### PROBLEM 1

What is the magnitude of the impulse the baseball receives from the bat?

### SOLUTION 1

The impulse is equal to the momentum change of the baseball. Unfortunately, there is no shortcut; we must calculate $\Delta\vec{v} \equiv \vec{v}_f - \vec{v}_i$ for the $x$- and $y$-components separately, add them as vectors, and finally multiply by the mass of the baseball:

$$\Delta v_x = (49.17 \text{ m/s})(\cos 35.0°) - (40.23 \text{ m/s})(\cos 185.0°) = 80.35 \text{ m/s}$$

$$\Delta v_y = (49.17 \text{ m/s})(\sin 35.0°) - (40.23 \text{ m/s})(\sin 185.0°) = 31.71 \text{ m/s}$$

$$\Delta v = \sqrt{\Delta v_x^2 + \Delta v_y^2} = \sqrt{(80.35)^2 + (31.71)^2} \text{ m/s} = 86.38 \text{ m/s}$$

$$\Delta p = m\Delta v = (0.145 \text{ kg})(86.38 \text{ m/s}) = 12.5 \text{ kg m/s}.$$

**Avoiding a Common Mistake:**

It is tempting to just add the magnitudes of the initial and final momentum vectors, because they point approximately in opposite directions. This method would lead to $\Delta p_{\text{wrong}} = m(v_1 + v_2) = 12.96$ kg m/s. As you can see, this answer is pretty close to the correct one, only about 3% off. It can serve as a first estimate, if you realize that the vectors point in almost opposite directions and that, in such a case, vector subtraction implies an addition of the two magnitudes. However, to get the correct answer, you have to go through the calculations above.

### PROBLEM 2

High-speed video shows that the ball-bat contact lasts only about 1 ms (0.001 s). Suppose for the home run we're considering, the contact lasted 1.20 ms. What was the magnitude of the average force exerted on the ball by the bat during that time?

### SOLUTION 2

The force can be calculated by simply using the formula for the impulse:

$$\Delta\vec{p} = \vec{J} = \vec{F}_{\text{ave}}\Delta t$$

$$\Rightarrow F_{\text{ave}} = \frac{\Delta p}{\Delta t} = \frac{12.5 \text{ kg m/s}}{0.00120 \text{ s}} = 10.4 \text{ kN}.$$

This force is approximately the same as the weight of an entire baseball team! The collision of the bat and the ball results in significant compression of the baseball, as shown in Figure 7.5.

**FIGURE 7.5** A baseball being compressed as it is hit by a baseball bat.

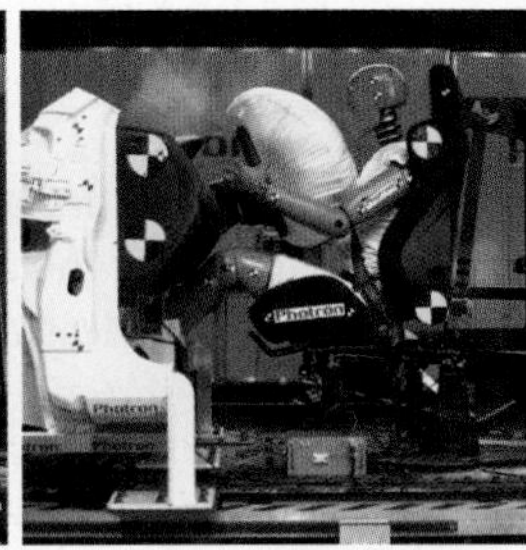

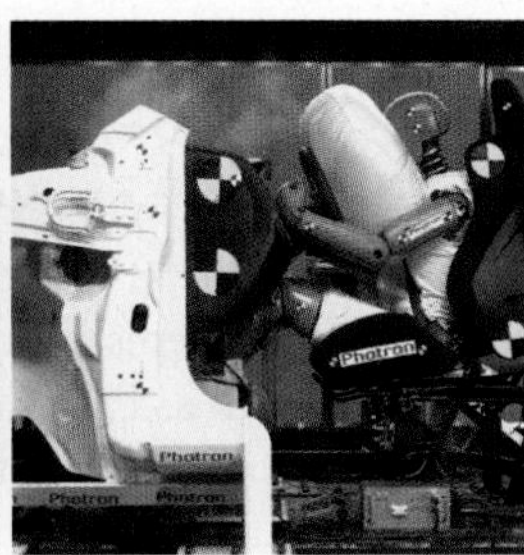

**FIGURE 7.6** Time sequence of a crash test, showing the role of air bags, seat belts, and crumple zones in reducing the forces acting on the driver during a crash. The air bag can be seen deploying in the second photograph of the sequence.

Some important safety devices, such as air bags and seat belts in cars, make use of equation 7.7 relating impulse, average force, and time. If the car you are driving has a collision with another vehicle or a stationary object, the impulse—that is, the momentum change of your car—is rather large, and it can be delivered over a very short time interval. Equation 7.7 then results in a very large average force:

$$\vec{F}_{\text{ave}} = \frac{\vec{J}}{\Delta t}.$$

If no seat belts or air bags were installed in your car, a sudden stop could cause your head to hit the windshield and experience the impulse during a very short time of only a few milliseconds. This could result in a big average force acting on your head, causing injury or even death. Air bags and seat belts are designed to make the time over which the momentum change occurs as long as possible. Maximizing this time and having the driver's body decelerate in contact with the air bag minimize the force acting on the driver, greatly reducing injuries (see Figure 7.6).

### 7.2 In-Class Exercise

If the baseball in Example 7.1 was hit so that it had the same speed of 110 mph after it left the bat but left the bat with an angle of 38° above the horizontal, the impulse the baseball received would have been

a) bigger.

b) smaller.

c) the same.

### 7.1 Self-Test Opportunity

Can you think of other everyday articles that are designed to minimize the average force for a given impulse?

### 7.3 In-Class Exercise

Several cars are designed with active crumple zones in the front that get severely damaged during head-on collisions. The purpose of this design is to

a) reduce the impulse experienced by the driver during the collision.

b) increase the impulse experienced by the driver during the collision.

c) reduce the collision time and thus reduce the force acting on the driver.

d) increase the collision time and thus reduce the force acting on the driver.

e) make the repair as expensive as possible.

## 7.3 Conservation of Linear Momentum

Suppose two objects collide with each other. They might then rebound away from each other, like two billiard balls on a billiard table. This kind of collision is called an **elastic collision** (at least it is approximately elastic, as we will see later). Another example of a collision is that of a subcompact car with an 18-wheeler, where the two vehicles stick to each other. This kind of collision is called a **totally inelastic collision.** Before seeing exactly what is meant by the terms *elastic* and *inelastic collisions,* let's look at the momenta, $\vec{p}_1$ and $\vec{p}_2$, of two objects during a collision.

We find that the sum of the two momenta after the collision is the same as the sum of the two momenta before the collision (index i1 indicates the initial value for object 1, just before the collision, and index f1 indicates the final value for the same object):

$$\vec{p}_{\text{f}1} + \vec{p}_{\text{f}2} = \vec{p}_{\text{i}1} + \vec{p}_{\text{i}2}. \tag{7.8}$$

This equation is the basic expression of the law of **conservation of total momentum,** the most important result of this chapter and the second conservation law we have encountered (the first being the law of conservation of energy in Chapter 6). Let's first go through its derivation and then consider its consequences.

## DERIVATION 7.1

During a collision, object 1 exerts a force on object 2. Let's call this force $\vec{F}_{1\to 2}$. Using the definition of impulse and its relationship to the momentum change, we get for the momentum change of object 2 during the collision:

$$\int_{t_i}^{t_f} \vec{F}_{1\to 2}\,dt = \Delta \vec{p}_2 = \vec{p}_{f2} - \vec{p}_{i2}.$$

Here we neglect external forces; if they exist, they are usually negligible compared to $\vec{F}_{1\to 2}$ during the collision. The initial and final times are selected to bracket the time of the collision process. In addition, the force $\vec{F}_{2\to 1}$, which object 2 exerts on object 1, is also present. The same argument as before leads to

$$\int_{t_i}^{t_f} \vec{F}_{2\to 1}\,dt = \Delta \vec{p}_1 = \vec{p}_{f1} - \vec{p}_{i1}.$$

Newton's Third Law (see Chapter 4) tells us that these forces are equal and opposite to each other, $\vec{F}_{1\to 2} = -\vec{F}_{2\to 1}$, or

$$\vec{F}_{1\to 2} + \vec{F}_{2\to 1} = 0.$$

Integration of this equation results in

$$0 = \int_{t_i}^{t_f} (\vec{F}_{2\to 1} + \vec{F}_{1\to 2})dt = \int_{t_i}^{t_f} \vec{F}_{2\to 1}\,dt + \int_{t_i}^{t_f} \vec{F}_{1\to 2}\,dt = \vec{p}_{f1} - \vec{p}_{i1} + \vec{p}_{f2} - \vec{p}_{i2}.$$

Collecting the initial momentum vectors on one side, and the final momentum vectors on the other gives us equation 7.8:

$$\vec{p}_{f1} + \vec{p}_{f2} = \vec{p}_{i1} + \vec{p}_{i2}.$$

Equation 7.8 expresses the principle of conservation of linear momentum. The sum of the final momentum vectors is exactly equal to the sum of the initial momentum vectors. Note that this equation does not depend on any particular conditions for the collision. It is valid for all two-body collisions, elastic or inelastic.

You may object that other external forces may be present. In a collision of billiard balls, for example, there is a friction force due to each ball rolling or sliding across the table. In a collision of two cars, friction acts between the tires and the road. However, what characterizes a collision is the occurrence of a very large impulse due to a very large contact force during a relatively short time. If you integrate the external forces over the collision time, you obtain only very small or moderate impulses. Thus, these external forces can usually be safely neglected in calculations of collision dynamics, and we can treat two-body collisions as if only internal forces are at work. We will assume we are dealing with an isolated system, which is a system with no external forces.

In addition, the same argument holds if there are more than two objects taking part in the collision or if there is no collision at all. As long as the net external force is zero, the total momentum of the interaction of objects will be conserved:

$$\text{if } \vec{F}_{\text{net}} = 0 \text{ then } \sum_{k=1}^{n} \vec{p}_k = \text{constant}. \tag{7.9}$$

Equation 7.9 is the general formulation of the law of conservation of momentum. We will return to this general formulation in Chapter 8 when we talk about systems of particles. For the remainder of this chapter, we consider only idealized cases in which the net external force is negligibly small, and thus the total momentum is always conserved in all processes.

## 7.4 Elastic Collisions in One Dimension

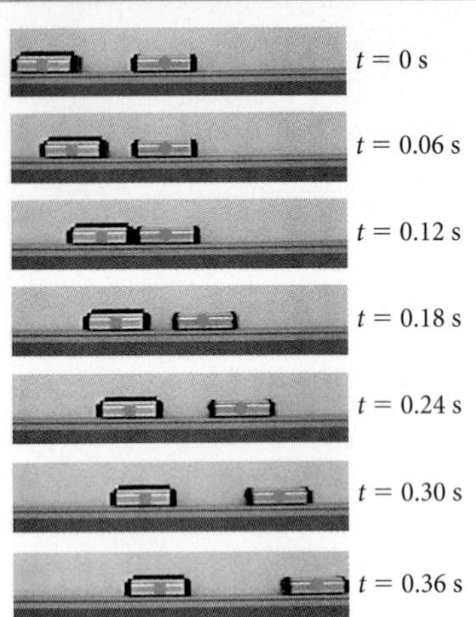


**FIGURE 7.7** Video sequence of a collision between two carts of non-equal masses on an air track. The cart with the orange dot carries a black metal bar to increase its mass.

Figure 7.7 shows the collision of two carts on an almost frictionless track. The collision was videotaped, and the figure includes seven frames of this video, taken at intervals of 0.06 s. The cart marked with the green circle is initially at rest. The cart marked with the orange square has a larger mass and is approaching from the left. The collision happens in the frame marked with the time $t = 0.12$ s. You can see that after the collision, both carts move to the right, but the lighter cart moves with a significantly higher speed. (The speed is proportional to the horizontal distance between the carts' markings in adjacent video frames.) Next, we'll derive equations that can be used to determine the velocities of the carts after the collision.

What exactly is an elastic collision? As with so many concepts in physics, it is an idealization. In practically all collisions, at least some kinetic energy is converted into other forms of energy that are not conserved. The other forms can be heat or sound or the energy to deform an object, for example. However, an elastic collision is defined as one in which the total kinetic energy of the colliding objects is conserved. This definition does not mean that each object involved in the collision retains its kinetic energy. Kinetic energy can be transferred from one object to the other, but *in an elastic collision, the sum of the kinetic energies has to remain constant.*

We'll consider objects moving in one dimension and use the notation $p_{i1,x}$ for the initial momentum, and $p_{f1,x}$ for the final momentum of object 1. (We use the subscript $x$ to remind ourselves that these could equally well be the $x$-components of the two- or three-dimensional momentum vector.) In the same way, we denote the initial and final momenta of object 2 by $p_{i2,x}$ and $p_{f2,x}$. Because we are restricted to collisions in one dimension, the equation for conservation of kinetic energy can be written as

$$\frac{p_{f1,x}^2}{2m_1}+\frac{p_{f2,x}^2}{2m_2}=\frac{p_{i1,x}^2}{2m_1}+\frac{p_{i2,x}^2}{2m_2}. \tag{7.10}$$

(For motion in one dimension, the square of the $x$-component of the vector is also the square of the absolute value of the vector.) The equation for conservation of momentum in the $x$-direction can be written as

$$p_{f1,x}+p_{f2,x}=p_{i1,x}+p_{i2,x}. \tag{7.11}$$

(Remember that momentum is conserved in any collision in which the external forces are negligible.)

Let's look more closely at equations 7.10 and 7.11. What is known, and what is unknown? Typically, we know the two masses and components of the initial momentum vectors, and we want to find the final momentum vectors after the collision. This calculation can be done because equations 7.10 and 7.11 give us two equations for two unknowns, $p_{f1,x}$ and $p_{f2,x}$. This is by far the most common use of these equations, but it is also possible, for example, to calculate the two masses if the initial and final momentum vectors are known.

Let's find the components of the final momentum vectors:

$$\begin{aligned} p_{f1,x}&=\left(\frac{m_1-m_2}{m_1+m_2}\right)p_{i1,x}+\left(\frac{2m_1}{m_1+m_2}\right)p_{i2,x}\\ p_{f2,x}&=\left(\frac{2m_2}{m_1+m_2}\right)p_{i1,x}+\left(\frac{m_2-m_1}{m_1+m_2}\right)p_{i2,x}. \end{aligned} \tag{7.12}$$

Derivation 7.2 shows how this result is obtained. It will help you solve similar problems.

### DERIVATION 7.2

We start with the equations for energy and momentum conservation and collect all quantities connected with object 1 on the left side and all those connected with object 2 on the right. Equation 7.10 for the (conserved) kinetic energy then becomes:

$$\frac{p_{f1,x}^2}{2m_1}-\frac{p_{i1,x}^2}{2m_1}=\frac{p_{i2,x}^2}{2m_2}-\frac{p_{f2,x}^2}{2m_2}$$

or

$$m_2(p_{f1,x}^2 - p_{i1,x}^2) = m_1(p_{i2,x}^2 - p_{f2,x}^2). \quad \text{(i)}$$

By rearranging equation 7.11 for momentum conservation we obtain

$$p_{f1,x} - p_{i1,x} = p_{i2,x} - p_{f2,x}. \quad \text{(ii)}$$

Next, we divide the left and right sides of equation (i) by the corresponding sides of equation (ii). To do this division, we use the algebraic identity $a^2 - b^2 = (a + b)(a - b)$. This process results in

$$m_2(p_{i1,x} + p_{f1,x}) = m_1(p_{i2,x} + p_{f2,x}). \quad \text{(iii)}$$

Now we can solve equation (ii) for $p_{f1,x}$ and substitute the expression $p_{i1,x} + p_{i2,x} - p_{f2,x}$ into equation (iii):

$$m_2(p_{i1,x} + [p_{i1,x} + p_{i2,x} - p_{f2,x}]) = m_1(p_{i2,x} + p_{f2,x})$$

$$2m_2 p_{i1,x} + m_2 p_{i2,x} - m_2 p_{f2,x} = m_1 p_{i2,x} + m_1 p_{f2,x}$$

$$p_{f2,x}(m_1 + m_2) = 2m_2 p_{i1,x} + (m_2 - m_1) p_{i2,x}$$

$$p_{f2,x} = \frac{2m_2 p_{i1,x} + (m_2 - m_1) p_{i2,x}}{m_1 + m_2}.$$

This result is one of the two desired components of equation 7.12. We can obtain the other component by solving equation (ii) for $p_{f2,x}$ and substituting the expression $p_{i1,x} + p_{i2,x} - p_{f1,x}$ into equation (iii). We can also obtain the result for $p_{f1,x}$ from the result for $p_{f2,x}$ that we just derived by exchanging the indices 1 and 2. It is, after all, arbitrary which object is labeled 1 or 2, and so the resulting equations should be symmetric under the exchange of the two labels. Use of this type of symmetry principle is very powerful and very convenient. (But it does take some getting used to at first!)

With the result for the final momenta, we can also obtain expressions for the final velocities by using $p_x = mv_x$:

$$v_{f1,x} = \left(\frac{m_1 - m_2}{m_1 + m_2}\right) v_{i1,x} + \left(\frac{2m_2}{m_1 + m_2}\right) v_{i2,x}$$
$$v_{f2,x} = \left(\frac{2m_1}{m_1 + m_2}\right) v_{i1,x} + \left(\frac{m_2 - m_1}{m_1 + m_2}\right) v_{i2,x}. \quad \text{(7.13)}$$

These equations for the final velocities look, at first sight, very similar to those for the final momenta (equation 7.12). However, there is one important difference: In the second term of the right-hand side of the equation for $v_{f1,x}$ the numerator is $2m_2$ instead of $2m_1$; and conversely, the numerator is now $2m_1$ instead of $2m_2$ in the first term of the equation for $v_{f2,x}$.

As a last point in this general discussion, let's find the relative velocity, $v_{f1,x} - v_{f2,x}$, after the collision:

$$v_{f1,x} - v_{f2,x} = \left(\frac{m_1 - m_2 - 2m_1}{m_1 + m_2}\right) v_{i1,x} + \left(\frac{2m_2 - (m_2 - m_1)}{m_1 + m_2}\right) v_{i2,x} \quad \text{(7.14)}$$
$$= -v_{i1,x} + v_{i2,x} = -(v_{i1,x} - v_{i2,x}).$$

We see that *in elastic collisions, the relative velocity simply changes sign*, $\Delta v_f = -\Delta v_i$. We will return to this result later in this chapter. You should not try to memorize the general expressions for momentum and velocity in equations 7.13 and 7.14, but instead study the method we used to derive them. Next, we examine two special cases of these general results.

## Special Case 1: Equal Masses

If $m_1 = m_2$, the general expressions in equation 7.12 simplify considerably, because the terms proportional to $m_1 - m_2$ are equal to zero and the ratios $2m_1/(m_1 + m_2)$ and $2m_2/(m_1 + m_2)$ become unity. We then obtain the extremely simple result

$$p_{f1,x} = p_{i2,x}$$
$$p_{f2,x} = p_{i1,x}. \quad \text{(for the special case where } m_1 = m_2\text{)} \quad \text{(7.15)}$$

This result means that in any elastic collision of two objects of equal mass moving in one dimension, the two objects simply *exchange* their momenta. The initial momentum of object 1 becomes the final momentum of object 2. The same is true for the velocities:

$$\begin{aligned} v_{f1,x} &= v_{i2,x} \\ v_{f2,x} &= v_{i1,x}. \end{aligned} \quad \text{(for the special case where } m_1 = m_2\text{)} \qquad (7.16)$$

## Special Case 2: One Object Initially at Rest

Now suppose the two objects in a collision are not necessarily the same in mass, but one of the two is initially at rest, that is, has zero momentum. Without loss of generality, we can say that object 1 is the one at rest. (Remember that the equations are invariant under exchange of the indices 1 and 2.) By using the general expressions in equation 7.12 and setting $p_{i1,x} = 0$, we get

$$\begin{aligned} p_{f1,x} &= \left(\frac{2m_1}{m_1+m_2}\right)p_{i2,x} \\ p_{f2,x} &= \left(\frac{m_2-m_1}{m_1+m_2}\right)p_{i2,x}. \end{aligned} \quad \text{(for the special case where } p_{i1,x} = 0\text{)} \qquad (7.17)$$

In the same way, we obtain for the final velocities

$$\begin{aligned} v_{f1,x} &= \left(\frac{2m_2}{m_1+m_2}\right)v_{i2,x} \\ v_{f2,x} &= \left(\frac{m_2-m_1}{m_1+m_2}\right)v_{i2,x}. \end{aligned} \quad \text{(for the special case where } p_{i1,x} = 0\text{)} \qquad (7.18)$$

If $v_{i2,x} > 0$, object 2 moves from left to right, with the conventional assignment of the positive $x$-axis pointing to the right. This situation is shown in Figure 7.7. Depending on which mass is larger, the collision can have one of four outcomes:

1. $m_2 > m_1 \Rightarrow (m_2 - m_1)/(m_2 + m_1) > 0$: The final velocity of object 2 points in the same direction but is reduced in magnitude.
2. $m_2 = m_1 \Rightarrow (m_2 - m_1)/(m_2 + m_1) = 0$: Object 2 is at rest, and object 1 moves with the initial velocity of object 2.
3. $m_2 < m_1 \Rightarrow (m_2 - m_1)/(m_2 + m_1) < 0$: Object 2 bounces back; the direction of its velocity vector changes.
4. $m_2 \ll m_1 \Rightarrow (m_2 - m_1)/(m_2 + m_1) \approx -1$ and $2m_2/(m_1 + m_2) \approx 0$: Object 1 remains at rest, and object 2 approximately reverses its velocity. This situation occurs, for example, in the collision of a ball with the ground. In this collision, object 1 is the entire Earth and object 2 is the ball. If the collision is sufficiently elastic, the ball bounces back with the same speed it had right before the collision, but in the opposite direction—up instead of down.

### 7.4 In-Class Exercise

Suppose an elastic collision occurs in one dimension, like the one shown in Figure 7.7, where the cart marked with the green dot is initially at rest and the cart with the orange dot initially has $v_{\text{orange}} > 0$, that is, is moving from left to right. What can you say about the masses of the two carts?

a) $m_{\text{orange}} < m_{\text{green}}$

b) $m_{\text{orange}} > m_{\text{green}}$

c) $m_{\text{orange}} = m_{\text{green}}$

### 7.5 In-Class Exercise

In the situation shown in Figure 7.7, suppose the mass of the cart with the orange dot is very much larger than that of the cart with the green dot. What outcome do you expect?

a) The outcome is about the same as the one shown in the figure.

b) The cart with the orange dot moves with almost unchanged velocity after the collision, and the cart with the green dot moves with a velocity almost twice as large as the initial velocity of the cart with the orange dot.

c) Both carts move with almost the same speed that the cart with the orange dot had before the collision.

d) The cart with the orange dot stops, and the cart with the green dot moves to the right with the same speed that the cart with the orange dot had originally.

### 7.6 In-Class Exercise

In the situation shown in Figure 7.7, if the mass of the cart with the green dot (originally at rest) is very much larger than that of the cart with the orange dot, what outcome do you expect?

a) The outcome is about the same as shown in the figure.

b) The cart with the orange dot moves with an almost unchanged velocity after the collision, and the cart with the green dot moves with a velocity almost twice as large as the initial velocity of the cart with the orange dot.

c) Both carts move with almost the same speed that the cart with the orange dot had before the collision.

d) The cart with the green dot moves with a very low speed slightly to the right, and the cart with the orange dot bounces back to the left with almost the same speed it had originally.

### 7.2 Self-Test Opportunity

The figure shows a high-speed video sequence of the collision of a golf club with a golf ball. The ball experiences significant deformation, but this deformation is sufficiently restored before the ball leaves the club's face. Thus, this collision can be approximated as a one-dimensional elastic collision. Discuss the speed of the ball relative to that of the club after the collision and how the cases we have discussed apply to this result.

Collision of a golf club with a golf ball.

### EXAMPLE 7.2 Average Force on a Golf Ball

A driver is a golf club used to hit a golf ball a long distance. The head of a driver typically has a mass of 200. g. A skilled golfer can give the club head a speed of around 40.0 m/s. The mass of a golf ball is 45.0 g. The ball stays in contact with the face of the driver for 0.500 ms.

**PROBLEM**

What is the average force exerted on the golf ball by the driver?

**SOLUTION**

The golf ball is initially at rest. Because the driver head and the ball are in contact for only a short time, we can consider the collision between them to be an elastic collision. We can use equation 7.18 to calculate the speed of the golf ball, $v_{f1,x}$, after the collision with the driver head

$$v_{f1,x} = \left(\frac{2m_2}{m_1 + m_2}\right) v_{i2,x},$$

where $m_1$ is the mass of the golf ball, $m_2$ is the mass of the driver head, and $v_{i2,x}$ is the speed of the driver head. The speed of the golf ball leaving the face of the driver head in this case is

$$v_{f1,x} = \frac{2(0.200 \text{ kg})}{0.0450 \text{ kg} + 0.200 \text{ kg}}(40.0 \text{ m/s}) = 65.3 \text{ m/s}.$$

Note that if the driver head were much more massive than the golf ball, the golf ball would attain twice the speed of the driver head. However, in that case, the golfer would have a difficult time giving the club head a substantial speed. The momentum change of the golf ball is

$$\Delta p = m\Delta v = mv_{f1,x}.$$

The impulse is then

$$\Delta p = F_{\text{ave}}\Delta t,$$

where $F_{\text{ave}}$ is the average force exerted by the driver head and $\Delta t$ is the time that the driver head and golf ball are in contact. The average force is then

$$F_{\text{ave}} = \frac{\Delta p}{\Delta t} = \frac{mv_{f1,x}}{\Delta t} = \frac{(0.045 \text{ kg})(65.3 \text{ m/s})}{0.500 \cdot 10^{-3} \text{ s}} = 5880 \text{ N}.$$

Thus, the driver exerts a very large force on the golf ball. This force compresses the golf ball significantly, as shown in the video sequence in Self-Test Opportunity 7.2. Also, note that the driver does not propel the ball in the horizontal direction and imparts spin to the ball. Thus, an accurate description of striking a golf ball with a driver requires a more detailed analysis.

## 7.5 Elastic Collisions in Two or Three Dimensions

### 7.7 In-Class Exercise

Choose the correct statement:

a) In an elastic collision of an object with a wall, energy may or may not be conserved.

b) In an elastic collision of an object with a wall, momentum may or may not be conserved.

c) In an elastic collision of an object with a wall, the incident angle is equal to the final angle.

d) In an elastic collision of an object with a wall, the original momentum vector does not change as a result of the collision.

e) In an elastic collision of an object with a wall, the wall cannot change the momentum of the object because momentum is conserved.

### 7.8 In-Class Exercise

Choose the correct statement:

a) When a moving object strikes a stationary object, the angle between the velocity vectors of the two objects after the collision is always 90°.

b) For a real-life collision between a moving object and a stationary object, the angle between the velocity vectors of the two objects after the collision is never less than 90°.

c) When a moving object has a head-on collision with a stationary object, the angle between the two velocity vectors after the collision is 90°.

d) When a moving object collides head-on and elastically with a stationary object of the same mass, the object that was moving stops and the other object moves with the original velocity of the moving object.

e) When a moving object collides elastically with a stationary object of the same mass, the angle between the two velocity vectors after the collision cannot be 90°.

### Collisions with Walls

To begin our discussion of two- and three-dimensional collisions, we consider the elastic collision of an object with a solid wall. In Chapter 4 on forces, we saw that a solid surface exerts a force on any object that attempts to penetrate the surface. Such forces are normal forces; they are directed perpendicular to the surface (Figure 7.8). If a normal force acts on an object colliding with a wall, the normal force can only transmit an impulse that is perpendicular to the wall; the normal force has no component parallel to the wall. Thus, the momentum component of the object directed along the wall does not change, $p_{f,\parallel} = p_{i,\parallel}$. In addition, for an elastic collision, we have the condition that the kinetic energy of the object colliding with the wall has to remain the same. This makes sense because the wall stays at rest (it is connected to the Earth and has a much bigger mass than the ball). The kinetic energy of the object is $K = p^2/2m$, so we see that $p_f^2 = p_i^2$.

Because $p_f^2 = p_{f,\parallel}^2 + p_{f,\perp}^2$ and $p_i^2 = p_{i,\parallel}^2 + p_{i,\perp}^2$, we get $p_{f,\perp}^2 = p_{i,\perp}^2$. The only two outcomes possible for the collision are then $p_{f,\perp} = p_{i,\perp}$ and $p_{f,\perp} = -p_{i,\perp}$. Only for the second solution does the perpendicular momentum component point away from the wall after the collision, so it is the only physical solution.

To summarize, when an object collides elastically with a wall, the length of the object's momentum vector remains unchanged, as does the momentum component directed along the wall; the momentum component perpendicular to the wall changes sign, but retains the same absolute value. The angle of incidence, $\theta_i$, on the wall (Figure 7.8) is then also the same as the angle of reflection, $\theta_f$:

$$\theta_i = \cos^{-1}\frac{p_{i,\perp}}{p_i} = \cos^{-1}\frac{p_{f,\perp}}{p_f} = \theta_f. \quad (7.19)$$

We will see this same relationship again when we study light and its reflection off a mirror in Chapter 32.

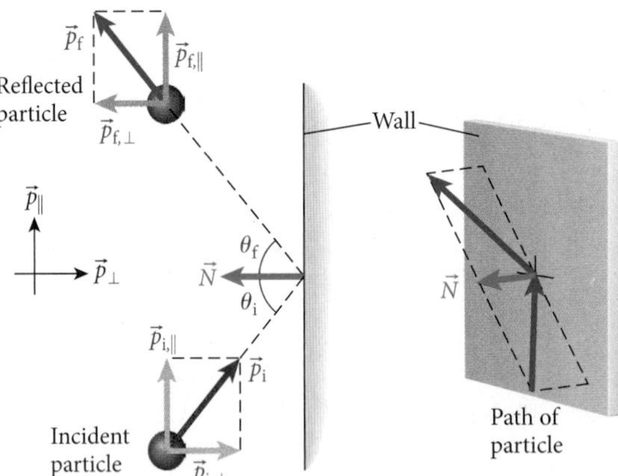


**FIGURE 7.8** Elastic collision of an object with a wall. The symbol ⊥ represents the component of the momentum perpendicular to the wall and the symbol ∥ represents the component of the momentum parallel to the wall.

### Collisions of Two Objects in Two Dimensions

We have just seen that problems involving elastic collisions in one dimension are always solvable if we have the initial velocity or momentum conditions for the two colliding objects, as well as their masses. Again, this is true because we have two equations for the two unknown quantities, $p_{f1,x}$ and $p_{f2,x}$.

For collisions in two dimensions, each of the final momentum vectors has two components. Thus, this situation gives us four unknown quantities to determine. How many equations do we have at our disposal? Conservation of kinetic energy again provides one of them. Conservation of linear momentum provides independent equations for the $x$- and $y$-directions.

Therefore, we have only three equations for the four unknown quantities. Unless an additional condition is specified for the collision, there is no unique solution for the final momenta.

For collisions in three dimensions, the situation is even worse. Here we need to determine two vectors with three components each, for a total of six unknown quantities. However, we have only four equations: one from energy conservation and three from the conservation equations for the $x$-, $y$-, and $z$-components of momentum.

Incidentally, this fact is what makes the game of billiards or pool interesting from a physics perspective. The final momenta of two balls after a collision are determined by where on their spherical surfaces the balls hit each other. Speaking of billiard ball collisions, an interesting observation can be made. Suppose object 2 is initially at rest and both objects have the same mass. Then conservation of momentum results in

$$\vec{p}_{\mathrm{f1}} + \vec{p}_{\mathrm{f2}} = \vec{p}_{\mathrm{i1}}$$

$$(\vec{p}_{\mathrm{f1}} + \vec{p}_{\mathrm{f2}})^2 = (\vec{p}_{\mathrm{i1}})^2$$

$$p_{\mathrm{f1}}^2 + p_{\mathrm{f2}}^2 + 2\vec{p}_{\mathrm{f1}} \bullet \vec{p}_{\mathrm{f2}} = p_{\mathrm{i1}}^2.$$

Here we squared the equation for momentum conservation and then used the properties of the scalar product. On the other hand, conservation of kinetic energy leads to

$$\frac{p_{\mathrm{f1}}^2}{2m} + \frac{p_{\mathrm{f2}}^2}{2m} = \frac{p_{\mathrm{i1}}^2}{2m}$$

$$p_{\mathrm{f1}}^2 + p_{\mathrm{f2}}^2 = p_{\mathrm{i1}}^2,$$

for $m_1 = m_2 \equiv m$. If we subtract this result from the previous result, we obtain

$$2\vec{p}_{\mathrm{f1}} \bullet \vec{p}_{\mathrm{f2}} = 0. \qquad (7.20)$$

However, the scalar product of two vectors can be zero only if the two vectors are perpendicular to each other or if one of them has length zero. The latter condition is in effect in a head-on collision of two billiard balls, after which the cue ball remains at rest ($\vec{p}_{\mathrm{f1}} = 0$) and the other ball moves away with the momentum that the cue ball had initially. After all non–head-on collisions, both balls move, and they move in directions that are perpendicular to each other.

You can do a simple experiment to see if the 90° angle between final velocity vectors works out quantitatively. Put two coins on a piece of paper, as shown in Figure 7.9. Mark the position of one of them (the target coin) on the paper by drawing a circle around it. Then flick the other coin with your finger into the target coin (Figure 7.9a). The coins will bounce off each other and slide briefly, before friction forces bring them to rest (Figure 7.9b). Then draw a line from the final position of the target coin back to the circle that you drew, as shown in Figure 7.9c, and thereby deduce the trajectory of the other coin. The images of parts (a) and (b) are superimposed on that in part (c) of the figure to show the motion of the coins before and after the collision, indicated by the red arrows. Measuring the angle between the two black lines in Figure 7.9c results in $\theta = 80°$, so the theoretically derived result of $\theta = 90°$ is not quite correct for this experiment. Why?

What we neglected in our derivation is the fact that—for collisions between coins or billiard balls—some of each object's kinetic energy is associated with rotation and the transfer of energy due to that motion, as well as the fact that such a collision is not quite elastic. However, the 90° rule just derived is a good first approximation for two colliding coins. You

### 7.3 Self-Test Opportunity

The experiment in Figure 7.9 specifies the angle between the two coins after the collision, but not the individual deflection angles. In order to obtain these individual angles, you also have to know how far off center on each object is the point of impact, called the *impact parameter.* Quantitatively, the impact parameter, $b$, is the distance that the original trajectory would need to be moved parallel to itself for a head-on collision (see the figure). Can you produce a sketch of the dependence of the deflection angles on the impact parameter? (*Hint:* You can do this experimentally, as shown in Figure 7.9, or you can think about limiting cases first and then try to interpolate between them.)

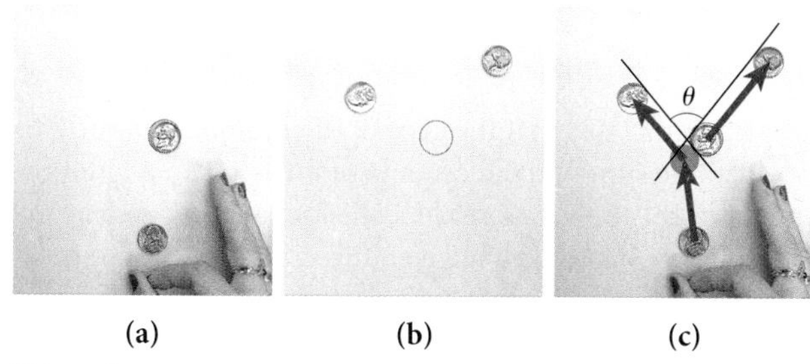


**FIGURE 7.9** Collision of two nickels.

### 7.4 Self-Test Opportunity

Suppose we do exactly the same experiment as shown in Figure 7.9, but replace one of the nickels with a less heavy dime or a heavier quarter. What changes? (*Hint:* Again, you can explore the answer by doing the experiment.)

can perform a similar experiment of this kind on any billiard table; you will find that the angle of motion between the two billiard balls is not quite 90°, but this approximation will give you a good idea where your cue ball will go after you hit the target ball.

## SOLVED PROBLEM 7.1 Curling

The sport of curling is all about collisions. A player slides a 19.0-kg (42.0-lb) granite "stone" about 35–40 m down the ice into a target area (concentric circles with cross hairs). Teams take turns sliding stones, and the stone closest to the bull's-eye in the end wins. Whenever a stone of one team is closest to the bull's-eye, the other team attempts to knock that stone out of the way, as shown in Figure 7.10.

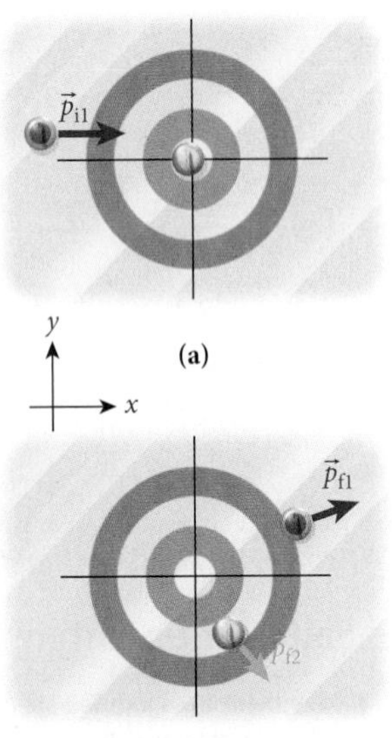


**FIGURE 7.10** Overhead view of a collision of two curling stones: (a) just before the collision; (b) just after the collision.

### PROBLEM

The red curling stone shown in Figure 7.10 has an initial velocity of 1.60 m/s in the $x$-direction and is deflected after colliding with the yellow stone to an angle of 32.0° relative to the $x$-axis. What are the two final momentum vectors right after this elastic collision, and what is the sum of the stones' kinetic energies?

### SOLUTION

#### THINK

Momentum conservation tells us that the sum of the momentum vectors of both stones before the collision is equal to the sum of the momentum vectors of both stones after the collision. Energy conservation tells us that in an elastic collision, the sum of the kinetic energies of both stones before the collision is equal to the sum of the kinetic energies of both stones after the collision. Before the collision, the red stone (stone 1) has momentum and kinetic energy because it is moving, while the yellow stone (stone 2) is at rest and has no momentum or kinetic energy. After the collision, both stones have momentum and kinetic energy. We must calculate the momentum in terms of $x$- and $y$-components.

#### SKETCH

A sketch of the momentum vectors of the two stones before and after the collision is shown in Figure 7.11a. The $x$- and $y$-components of the momentum vectors after the collision of the two stones are shown in Figure 7.11b.

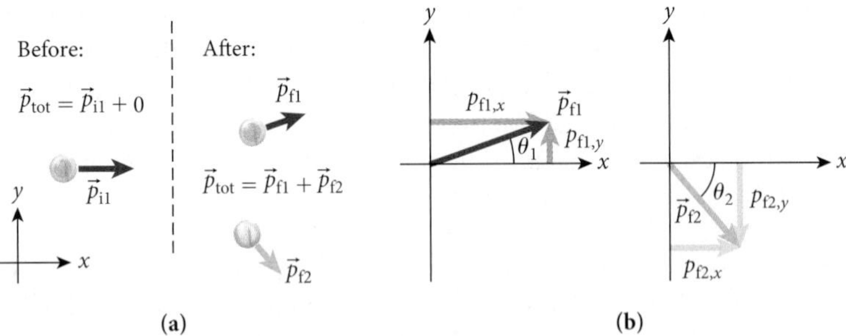


**FIGURE 7.11** (a) Sketch of momentum vectors before and after the two stones collide. (b) The $x$- and $y$-components of the momentum vectors of the two stones after the collision.

#### RESEARCH

Momentum conservation dictates that the sum of the momenta of the two stones before the collision must equal the sum of the momenta of the two stones after the collision. We know the momenta of both stones before the collision, and our task is to calculate their momenta after the collision, based on the given directions of those momenta. For the $x$-components, we can write

$$p_{i1,x} + 0 = p_{f1,x} + p_{f2,x}.$$

For the $y$-components, we can write

$$0+0=p_{\text{f}1,y}+p_{\text{f}2,y}.$$

The problem specifies that stone 1 is deflected at $\theta_1 = 32.0°$. According to the 90°-rule that we derived for perfectly elastic collisions between equal masses, stone 2 has to be deflected at $\theta_2 = -58.0°$. Therefore, in the $x$-direction we obtain,

$$p_{\text{i}1,x}=p_{\text{f}1,x}+p_{\text{f}2,x}=p_{\text{f}1}\cos\theta_1+p_{\text{f}2}\cos\theta_2. \quad \text{(i)}$$

And in the $y$-direction we have:

$$0=p_{\text{f}1,y}+p_{\text{f}2,y}=p_{\text{f}1}\sin\theta_1+p_{\text{f}2}\sin\theta_2. \quad \text{(ii)}$$

Because we know the two angles and the initial momentum of stone 1, we need to solve a system of two equations for two unknown quantities, which are the magnitudes of the final momenta, $p_{\text{f}1}$ and $p_{\text{f}2}$.

## SIMPLIFY

We solve this system of equations by direct substitution. We can solve the $y$-component equation (ii) for $p_{\text{f}1}$

$$p_{\text{f}1}=-p_{\text{f}2}\frac{\sin\theta_2}{\sin\theta_1} \quad \text{(iii)}$$

and substitute into the $x$-component equation (i) to get

$$p_{\text{i}1,x}=\left(-p_{\text{f}2}\frac{\sin\theta_2}{\sin\theta_1}\right)\cos\theta_1+p_{\text{f}2}\cos\theta_2.$$

We can rearrange this equation to get

$$p_{\text{f}2}=\frac{p_{\text{i}1,x}}{\cos\theta_2-\sin\theta_2\cot\theta_1}.$$

## CALCULATE

First, we calculate the magnitude of the initial momentum of stone 1:

$$p_{\text{i}1,x}=mv_{\text{i}1,x}=(19.0\text{ kg})(1.60\text{ m/s})=30.4\text{ kg m/s}.$$

We can then calculate the magnitude of the final momentum of stone 2:

$$p_{\text{f}2}=\frac{30.4\text{ kg m/s}}{(\cos-58.0°)-(\sin-58.0°)(\cot 32.0°)}=16.10954563\text{ kg m/s}.$$

The magnitude of the final momentum of stone 1 is

$$p_{\text{f}1}=-p_{\text{f}2}\frac{\sin(-58.0°)}{\sin(32.0°)}=25.78066212\text{ kg m/s}.$$

Now we can answer the question concerning the sum of the kinetic energies of the two stones after the collision. Because this collision is elastic, we can simply calculate the initial kinetic energy of the red stone (the yellow one was at rest). Thus, our answer is

$$K=\frac{p_{\text{i}1}^2}{2m}=\frac{(30.4\text{ kg m/s})^2}{2(19.0\text{ kg})}=24.32\text{ J}.$$

## ROUND

Because all the numerical values were specified to three significant figures, we report the magnitude of the final momentum of the first stone as

$$p_{\text{f}1}=25.8\text{ kg m/s}.$$

*Continued—*

### 7.5 Self-Test Opportunity

Double-check the results for the final momenta of the two stones in Solved Problem 7.1 by calculating the individual kinetic energies of the two stones after the collision to verify that their sum is indeed equal to the initial kinetic energy.

The direction of the first stone is +32.0° with respect to the horizontal. We report the magnitude of the final momentum of the second stone as

$$p_{f2} = 16.1 \text{ kg m/s}$$

The direction of the second stone is −58.0° with respect to the horizontal.

The total kinetic energy of the two stones after the collision is

$$K = 24.3 \text{ J}.$$

## 7.6 Totally Inelastic Collisions

In all collisions that are not completely elastic, the conservation of kinetic energy no longer holds. These collisions are called *inelastic,* because some of the initial kinetic energy gets converted into internal energy of excitation, deformation, vibration, or (eventually) heat. At first sight, this conversion of energy may make the task of calculating the final momentum or velocity vectors of the colliding objects appear more complicated. However, that is not the case; in particular, the algebra becomes considerably easier for the limiting case of totally inelastic collisions.

A totally inelastic collision is one in which the colliding objects stick to each other after colliding. This result implies that both objects have the same velocity vector after the collision: $\vec{v}_{f1} = \vec{v}_{f2} \equiv \vec{v}_f$. (Thus, the *relative velocity* between the two colliding objects is zero after the collision.) Using $\vec{p} = m\vec{v}$ and conservation of momentum, we get the final velocity vector:

$$\vec{v}_f = \frac{m_1\vec{v}_{i1} + m_2\vec{v}_{i2}}{m_1 + m_2}. \tag{7.21}$$

This useful formula enables you to solve practically all problems involving totally inelastic collisions. Derivation 7.3 shows how it was obtained.

### DERIVATION 7.3

We start with the conservation law for total momentum (equation 7.8):

$$\vec{p}_{f1} + \vec{p}_{f2} = \vec{p}_{i1} + \vec{p}_{i2}.$$

Now we use $\vec{p} = m\vec{v}$ and get

$$m_1\vec{v}_{f1} + m_2\vec{v}_{f2} = m_1\vec{v}_{i1} + m_2\vec{v}_{i2}.$$

The condition that the collision is totally inelastic implies that the final velocities of the two objects are the same. Thus, we have equation 7.21:

$$m_1\vec{v}_f + m_2\vec{v}_f = m_1\vec{v}_{i1} + m_2\vec{v}_{i2}$$
$$(m_1 + m_2)\vec{v}_f = m_1\vec{v}_{i1} + m_2\vec{v}_{i2}$$
$$\vec{v}_f = \frac{m_1\vec{v}_{i1} + m_2\vec{v}_{i2}}{m_1 + m_2}.$$

### 7.9 In-Class Exercise

In a totally inelastic collision between a moving object and a stationary object, the two objects will

a) stick together.

b) bounce off each other, losing energy.

c) bounce off each other, without losing energy.

Note that the condition of a totally inelastic collision implies only that the final velocities are the same for both objects. In general, the final momentum vectors of the objects can have quite different magnitudes.

We know from Newton's Third Law (see Chapter 4) that the forces two objects exert on each other during a collision are equal in magnitude. However, the changes in velocity—that is, the accelerations that the two objects experience in a totally inelastic collision—can be drastically different. The following example illustrates this phenomenon.

## EXAMPLE 7.3 Head-on Collision

Consider a head-on collision of a full-size SUV, with mass $M = 3023$ kg, and a compact car, with mass $m = 1184$ kg. Each vehicle has an initial speed of $v = 22.35$ m/s (50 mph), and they are moving in opposite directions (Figure 7.12). So, as shown in the figure, we can say $v_x$ is the initial velocity of the compact car and $-v_x$ is the initial velocity of the SUV. The two cars crash into each other and become entangled, a case of a totally inelastic collision.

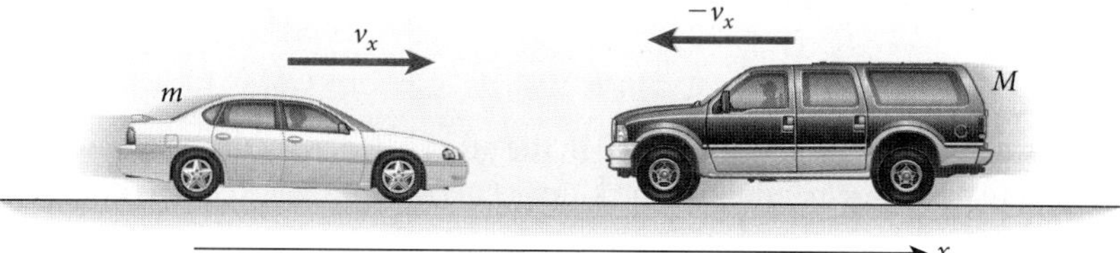


**FIGURE 7.12** Head-on collision of two vehicles with different masses and identical speeds.

### PROBLEM

What are the changes of the two cars' velocities in the collision? (Neglect friction between the tires and the road.)

### SOLUTION

We first calculate the final velocity that the combined mass has immediately after the collision. To do this calculation, we simply use equation 7.21 and get

$$v_{f,x} = \frac{mv_x - Mv_x}{m+M} = \left(\frac{m-M}{m+M}\right)v_x$$
$$= \frac{1184 \text{ kg} - 3023 \text{ kg}}{1184 \text{ kg} + 3023 \text{ kg}}(22.35 \text{ m/s}) = -9.77 \text{ m/s}.$$

Thus, the velocity change for the SUV turns out to be

$$\Delta v_{\text{SUV},x} = -9.77 \text{ m/s} - (-22.35 \text{ m/s}) = 12.58 \text{ m/s}.$$

However, the velocity change for the compact car is

$$\Delta v_{\text{compact},x} = -9.77 \text{ m/s} - (22.35 \text{ m/s}) = -32.12 \text{ m/s}.$$

We obtain the corresponding average accelerations by dividing the velocity changes by the time interval, $\Delta t$, during which the collision takes place. This time interval is obviously the same for both cars, which means that the magnitude of the acceleration experienced by the body of the driver of the compact car is bigger than that experienced by the body of the driver of the SUV by a factor of 32.12/12.58 = 2.55.

From this result alone, it is clear that it is safer to be in the SUV in this head-on collision than in the compact car. Keep in mind that this result is true even though Newton's Third Law says that the forces exerted by the two vehicles on each other are the same (compare Example 4.6).

### 7.6 Self-Test Opportunity

You can start with Newton's Third Law and make use of the fact that the forces the two cars exert on each other are equal. Use the values of the masses given in Example 7.3. What is the ratio of the accelerations of the two cars that you obtain?

## Ballistic Pendulum

A **ballistic pendulum** is a device that can be used to measure muzzle speeds of projectiles shot from firearms. It consists of a block of material into which the bullet is fired. This block is suspended so that it forms a pendulum (Figure 7.13). From the deflection angle of the pendulum and the known masses of the bullet, $m$, and the block, $M$, we can calculate the speed of the bullet right before it hits the block.

To obtain an expression for the speed of the bullet in terms of the deflection angle, we have to calculate the speed of the bullet-plus-block combination right after the bullet is embedded in the block. This collision is a prototypical totally inelastic collision, and thus we

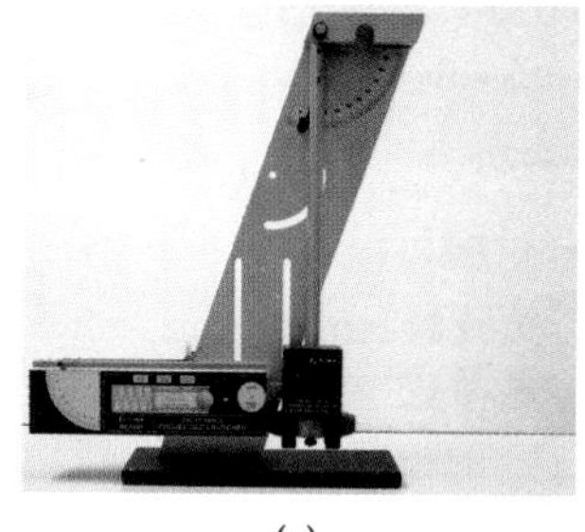

(a)

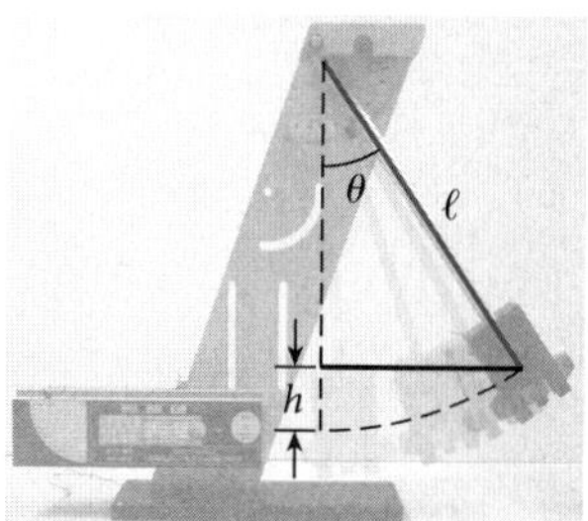


(b)

**FIGURE 7.13** Ballistic pendulum used in an introductory physics laboratory.

can apply equation 7.21. Because the pendulum is at rest before the bullet hits it, the speed of the block-plus-bullet combination is

$$v = \frac{m}{m+M} v_\text{b},$$

where $v_\text{b}$ is the speed of the bullet before it hits the block and $v$ is the speed of the combined masses right after impact. The kinetic energy of the bullet is $K_\text{b} = \frac{1}{2}mv_\text{b}^2$ just before it hits the block, whereas right after the collision, the block-plus-bullet combination has the kinetic energy

$$K = \tfrac{1}{2}(m+M)v^2 = \tfrac{1}{2}(m+M)\left(\frac{m}{m+M}v_\text{b}\right)^2 = \tfrac{1}{2}mv_\text{b}^2 \frac{m}{m+M} = \frac{m}{m+M}K_\text{b}. \qquad (7.22)$$

Clearly, kinetic energy is not conserved in the process whereby the bullet embeds in the block. (With a real ballistic pendulum, the kinetic energy is transferred into deformation of bullet and block. In this demonstration version, kinetic energy is transferred into frictional work between the bullet and the block.) Equation 7.22 shows that the total kinetic energy (and with it the total mechanical energy) is reduced by a factor of $m/(m + M)$. However, after the collision, the block-plus-bullet combination retains its remaining total energy in the ensuing pendulum motion, converting all of the initial kinetic energy of equation 7.22 into potential energy at the highest point:

$$U_\text{max} = (m+M)gh = K = \tfrac{1}{2}\left(\frac{m^2}{m+M}\right)v_\text{b}^2. \qquad (7.23)$$

### 7.7 Self-Test Opportunity

If you use a bullet with half the mass of a .357 Magnum caliber round and the same speed, what is your deflection angle?

As you can see from Figure 7.13b, the height $h$ and angle $\theta$ are related via $h = \ell(1 - \cos\theta)$, where $\ell$ is the length of the pendulum. (We found the same relationship in Solved Problem 6.4.) Substituting this result into equation 7.23 yields

$$(m+M)g\ell(1-\cos\theta) = \tfrac{1}{2}\left(\frac{m^2}{m+M}\right)v_\text{b}^2 \Rightarrow$$

$$v_b = \frac{m+M}{m}\sqrt{2g\ell(1-\cos\theta)}. \qquad (7.24)$$

### 7.10 In-Class Exercise

A ballistic pendulum is used to measure the speed of a bullet shot from a gun. The mass of the bullet is 50.0 g, and the mass of the block is 20.0 kg. When the bullet strikes the block, the combined mass rises a vertical distance of 5.00 cm. What was the speed of the bullet as it struck the block?

a) 397 m/s
b) 426 m/s
c) 457 m/s
d) 479 m/s
e) 503 m/s

It is clear from equation 7.24 that practically any bullet speed can be measured with a ballistic pendulum, provided the mass of the block, $M$, is chosen appropriately. For example, if you shoot a .357 Magnum caliber round ($m$ = 0.125 kg) into a block ($M$ = 40.0 kg) suspended by a 1.00 m long rope, the deflection is 25.4°, and equation 7.24 lets you deduce that the muzzle speed of this bullet fired from the particular gun that you used is 442 m/s (which is a typical value for this type of ammunition).

## Kinetic Energy Loss in Totally Inelastic Collisions

As we have just seen, total kinetic energy is not conserved in totally inelastic collisions. How much kinetic energy is lost in the general case? We can find this loss by taking the difference between the total initial kinetic energy, $K_\text{i}$, and the total final kinetic energy, $K_\text{f}$:

$$K_\text{loss} = K_\text{i} - K_\text{f}.$$

The total initial kinetic energy is the sum of the individual kinetic energies of the two objects before the collision:

$$K_i = \frac{p_\text{i1}^2}{2m_1} + \frac{p_\text{i2}^2}{2m_2}.$$

The total final kinetic energy for the case in which the two objects stick together and move as one, with the total mass of $m_1 + m_2$ and velocity $\vec{v}_\text{f}$, using equation 7.21 is

$$K_\text{f} = \tfrac{1}{2}(m_1+m_2)v_\text{f}^2$$

$$= \tfrac{1}{2}(m_1+m_2)\left(\frac{m_1\vec{v}_\text{i1}+m_2\vec{v}_\text{i2}}{m_1+m_2}\right)^2$$

$$= \frac{(m_1\vec{v}_\text{i1}+m_2\vec{v}_\text{i2})^2}{2(m_1+m_2)}.$$

Now we can take the difference between the final and initial kinetic energies and obtain the kinetic energy loss:

$$K_{\text{loss}} = K_{\text{i}} - K_{\text{f}} = \frac{1}{2}\frac{m_1 m_2}{m_1 + m_2}(\vec{v}_{\text{i}1} - \vec{v}_{\text{i}2})^2. \tag{7.25}$$

The derivation of this result involves a bit of algebra and is omitted here. What matters, though, is that the difference in the initial velocities—that is, the initial relative velocity—enters into the equation for energy loss. We will explore the significance of this fact in the following section and again in Chapter 8, when we consider center-of-mass motion.

### 7.11 In-Class Exercise

Suppose mass 1 is initially at rest and mass 2 moves initially with speed $v_{\text{i},2}$. In a totally inelastic collision between the two objects, the loss of kinetic energy in terms of the initial kinetic energy is larger for

a) $m_1 \ll m_2$

b) $m_1 \gg m_2$

c) $m_1 = m_2$

## SOLVED PROBLEM 7.2 Forensic Science

Figure 7.14a shows a sketch of a traffic accident. The white pickup truck (car 1) with mass $m_1 = 2209$ kg was traveling north and hit the westbound red car (car 2) with mass $m_2 = 1474$ kg. When the two vehicles collided, they became entangled (stuck together). Skid marks on the road reveal the exact location of the collision, and the direction in which the two vehicles were sliding immediately afterward. This direction was measured to be 38° relative to the initial direction of the white pickup truck. The white pickup truck had the right of way, because the red car had a stop sign. The driver of the red car, however, claimed that the driver of the white pickup truck was moving at a speed of at least 50 mph (22 m/s), although the speed limit was 25 mph (11 m/s). Furthermore, the driver of the red car claimed that he had stopped at the stop sign and was then driving through the intersection at a speed of less than 25 mph when the white pickup truck hit him. Since the driver of the white pickup truck was speeding, he would legally forfeit the right of way and be declared responsible for the accident.

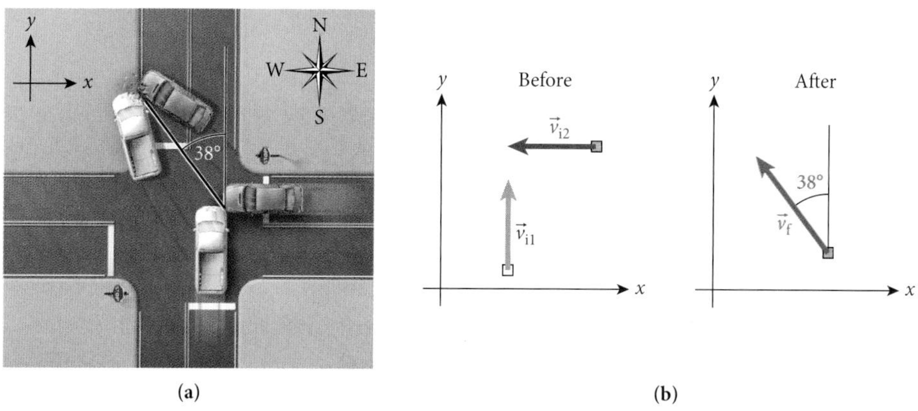


**FIGURE 7.14** (a) Sketch of the accident scene. (b) Velocity vectors of the two cars before and after the collision.

### PROBLEM

Can the version of the accident as described by the driver of the red car be correct?

### SOLUTION

#### THINK

This collision is clearly a totally inelastic collision, and so we know that the velocity of the entangled cars after the collision is given by equation 7.21. We are given the angles of the initial velocities of both cars and the angle of the final velocity vector of the entangled cars. However, we are not given the magnitudes of these three velocities. Thus, we have one equation and three unknowns. However, to solve this problem, we only need to determine the ratio of the magnitude of the initial velocity of car 1 to the magnitude of the initial velocity of car 2. Here the initial velocity is the velocity of the car just before the collision occurred.

*Continued—*

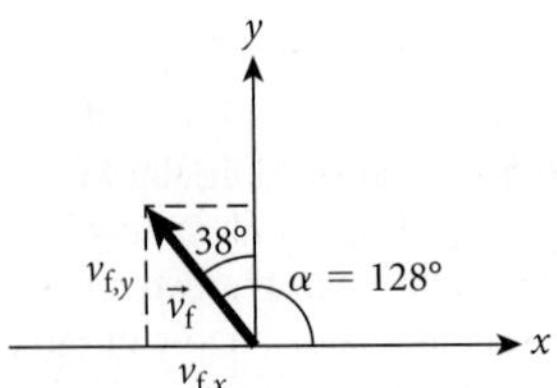


**FIGURE 7.15** Components of the velocity vector of the two entangled cars after the collision.

**SKETCH**

A sketch of the velocity vectors of the two cars before and after the collision is shown in Figure 7.14b. A sketch of the components of the velocity vector of the two stuck together after the collision is shown in Figure 7.15.

**RESEARCH**

Using the coordinate system shown in Figure 7.14a, the white pickup truck (car 1) has only a $y$-component for its velocity vector, $\vec{v}_{i1} = v_{i1}\hat{y}$, where $v_{i1}$ is the initial speed of the white pickup. The red car (car 2) has a velocity component only in the negative $x$-direction, $\vec{v}_{i2} = -v_{i2}\hat{x}$. The final velocity $\vec{v}_f$ of the cars when stuck together after the collision, expressed in terms of the initial velocities, is given by

$$\vec{v}_f = \frac{m_1\vec{v}_{i1} + m_2\vec{v}_{i2}}{m_1 + m_2}.$$

**SIMPLIFY**

Substituting the initial velocities into this equation for the final velocity gives

$$\vec{v}_f = v_{f,x}\hat{x} + v_{f,y}\hat{y} = \frac{-m_2 v_{i2}}{m_1 + m_2}\hat{x} + \frac{m_1 v_{i1}}{m_1 + m_2}\hat{y}.$$

The components of the final velocity vector, $v_{f,x}$ and $v_{f,y}$, are shown in Figure 7.15. From trigonometry, we obtain an expression for the tangent of the angle of the final velocity as the ratio of its $y$- and $x$-components:

$$\tan\alpha = \frac{v_{f,y}}{v_{f,x}} = \frac{\dfrac{m_1 v_{i1}}{m_1 + m_2}}{\dfrac{-m_2 v_{i2}}{m_1 + m_2}} = -\frac{m_1 v_{i1}}{m_2 v_{i2}}.$$

Thus, we can find the initial speed of car 1 in terms of the initial speed of car 2:

$$v_{i1} = -\frac{m_2 \tan\alpha}{m_1} v_{i2}.$$

We have to be careful with the value of the angle $\alpha$. It is *not* 38°, as you might conclude from a casual examination of Figure 7.14b. Instead, it is 38° + 90° =128°, as shown in Figure 7.15, because angles must be measured relative to the positive $x$-axis when using the tangent formula, $\tan\alpha = v_{f,y} / v_{f,x}$.

**CALCULATE**

With this result and the known values of the masses of the two cars, we find

$$v_{i1} = -\frac{(1474 \text{ kg})(\tan 128°)}{2209 \text{ kg}} v_{i2} = 0.854066983 v_{i2}.$$

**ROUND**

The angle at which the two cars moved after the collision was specified to two significant digits, so we report our answer to two significant digits:

$$v_{i1} = 0.85 v_{i2}.$$

The white pickup truck (car 1) was moving at a slower speed than the red car (car 2). The story of the driver of the red car is not consistent with the facts. Apparently, the driver of the white pickup truck was not speeding at the time of the collision, and the cause of the accident was the driver of the red car running the stop sign.

### 7.8 Self-Test Opportunity

To double-check the result for Solved Problem 7.2, estimate what the angle of deflection would have been if the white pickup truck had been traveling with a speed of 50 mph and the red car had been traveling with a speed of 25 mph just before the collision.

## Explosions

In totally inelastic collisions, two or more objects merge into one and move in unison with the same total momentum after the collision as before it. The reverse process is also possible. If an object moves with initial momentum $\vec{p}_i$ and then explodes into fragments, the process of the explosion only generates internal forces among the fragments. Because an explosion takes place over a very short time, the impulse due to external forces can usually be neglected. In such a situation, according to Newton's Third Law, the total momentum is conserved. This result implies that the sum of the momentum vectors of the fragments has to add up to the initial momentum vector of the object:

$$\vec{p}_{\rm i}=\sum_{k=1}^{n}\vec{p}_{{\rm f}k}. \tag{7.26}$$

This equation relating the momentum of the exploding object just before the explosion to the sum of the momentum vectors of the fragments after the explosion is exactly the same as the equation for a totally inelastic collision except that the indices for the initial and final states are exchanged. In particular, if an object breaks up into two fragments, equation 7.26 matches equation 7.21, with the indices i and f exchanged:

$$\vec{v}_{\rm i}=\frac{m_1\vec{v}_{\rm f1}+m_2\vec{v}_{\rm f2}}{m_1+m_2}. \tag{7.27}$$

This relationship allows us, for example, to reconstruct the initial velocity if we know the fragments' velocities and masses. Further, we can calculate the energy released in a breakup that yields two fragments from equation 7.25, again with the indices i and f exchanged:

$$K_{\rm release}=K_{\rm f}-K_{\rm i}=\frac{1}{2}\frac{m_1m_2}{m_1+m_2}\left(\vec{v}_{\rm f1}-\vec{v}_{\rm f2}\right)^2. \tag{7.28}$$

## EXAMPLE 7.4 Decay of a Radon Nucleus

Radon is a gas that is produced by the radioactive decay of naturally occurring heavy nuclei, such as thorium and uranium. Radon gas can be breathed into the lungs, where it can decay further. The nucleus of a radon atom has a mass of 222 u, where u is an atomic mass unit (to be introduced in Chapter 40). Assume that the nucleus is at rest when it decays into a polonium nucleus with mass 218 u and a helium nucleus with mass 4 u (called an alpha particle), releasing 5.59 MeV of kinetic energy.

### PROBLEM

What are the kinetic energies of the polonium nucleus and the alpha particle?

### SOLUTION

The polonium nucleus and the alpha particle are emitted in opposite directions. We will say that the alpha particle is emitted with speed $v_{x1}$ in the positive $x$-direction and the polonium nucleus is emitted with speed $v_{x2}$ in the negative $x$-direction. The mass of the alpha particle is $m_1$ = 4u, and the mass of the polonium nucleus is $m_2$ = 118 u. The initial velocity of the radon nucleus is zero, so we can use equation 7.27 to write

$$\vec{v}_{\rm i}=0=\frac{m_1\vec{v}_1+m_2\vec{v}_2}{m_1+m_2},$$

which gives us

$$m_1v_{x1}=-m_2v_{x2}. \tag{i}$$

Using equation 7.28, we can express the kinetic energy released by the decay of the radon nucleus

$$K_{\rm release}=\frac{1}{2}\frac{m_1m_2}{m_1+m_2}\left(v_{x1}-v_{x2}\right)^2. \tag{ii}$$

We can then use equation (i) to express the velocity of the polonium nucleus in terms of the velocity of the alpha particle:

$$v_{x2}=-\frac{m_1}{m_2}v_{x1}.$$

*Continued—*

Substituting from equation (i) into equation (ii) gives us

$$K_{\text{release}} = \frac{1}{2}\frac{m_1 m_2}{m_1+m_2}\left(v_{x1} + \frac{m_1}{m_2}v_{x1}\right)^2 = \frac{1}{2}\frac{m_1 m_2 v_{x1}^2}{m_1+m_2}\left(\frac{m_1+m_2}{m_2}\right)^2. \qquad \text{(iii)}$$

Rearranging equation (iii) leads to

$$K_{\text{release}} = \frac{1}{2}\frac{m_1(m_1+m_2)v_{x1}^2}{m_2} = \frac{1}{2}m_1 v_{x1}^2 \frac{m_1+m_2}{m_2} = K_1\left(\frac{m_1+m_2}{m_2}\right),$$

where $K_1$ is the kinetic energy of the alpha particle. This kinetic energy is then

$$K_1 = K_{\text{release}}\left(\frac{m_2}{m_1+m_2}\right).$$

Putting in the numerical values, we get the kinetic energy of the alpha particle:

$$K_1 = (5.59 \text{ MeV})\left(\frac{218}{4+218}\right) = 5.49 \text{ MeV}.$$

The kinetic energy, $K_2$, of the polonium nucleus is then

$$K_2 = K_{\text{release}} - K_1 = 5.59 \text{ MeV} - 5.49 \text{ MeV} = 0.10 \text{ MeV}.$$

The alpha particle gets most of the kinetic energy when the radon nucleus decays, and this is sufficient energy to damage surrounding tissue in the lungs.

## EXAMPLE 7.5 Particle Physics

The conservation laws of momentum and energy are essential for analyzing the products of particle collisions at high energies, such as those produced at Fermilab's Tevatron, near Chicago, Illinois, currently the world's highest-energy proton-antiproton accelerator. (An accelerator called LHC, for Large Hadron Collider began operation in 2009 at the CERN Laboratory in Geneva, Switzerland, and it will be more powerful than Tevatron. However, the LHC is a proton-proton accelerator.)

In the Tevatron accelerator, particle physicists collide protons and antiprotons at total energies of 1.96 TeV (hence the name). Remember that 1 eV = $1.602 \cdot 10^{-19}$ J; so 1.96 TeV = $1.96 \cdot 10^{12}$ eV = $3.1 \cdot 10^{-7}$ J. The Tevatron is set up so that the protons and antiprotons move in the collider ring in opposite directions, with, for all practical purposes, exactly opposite momentum vectors. The main particle detectors, D-Zero and CDF, are located at the interaction regions, where protons and antiprotons collide.

Figure 7.16a shows an example of such a collision. In this computer-generated display of one particular collision event at the D-Zero detector, the proton's initial momentum vector points straight into the page and that of the antiproton points straight out of the page. Thus, the total initial momentum of the proton-antiproton system is zero. The explosion produced by this collision produces several fragments, almost all of which are registered by the detector. These measurements are indicated in the display (Figure 7.16a). Superimposed on this event display are the momentum vectors of the corresponding parent particles of these fragments, with their lengths and directions based on computer analysis of the detector response. (The momentum unit GeV/*c*, commonly used in high-energy physics and shown in the figure, is the energy unit GeV divided by the speed of light.) In Figure 7.16b, the momentum vectors have been summed graphically, giving a nonzero vector, indicated by the thicker green arrow.

However, conservation of momentum absolutely requires that the sum of the momentum vectors of all particles produced in this collision must be zero. Thus, conservation of momentum allows us to assign the missing momentum represented by the green arrow to a particle that escaped undetected. With the aid of this missing-momentum analysis, physicists at Fermilab were able to show that the event displayed here was one in which an elusive particle known as a top quark was produced.

### 7.9 Self-Test Opportunity

The length of each vector arrow in Figure 7.16b is proportional to the magnitude of the momentum vector of each individual particle. Can you determine the momentum of the nondetected particle (green arrow)?

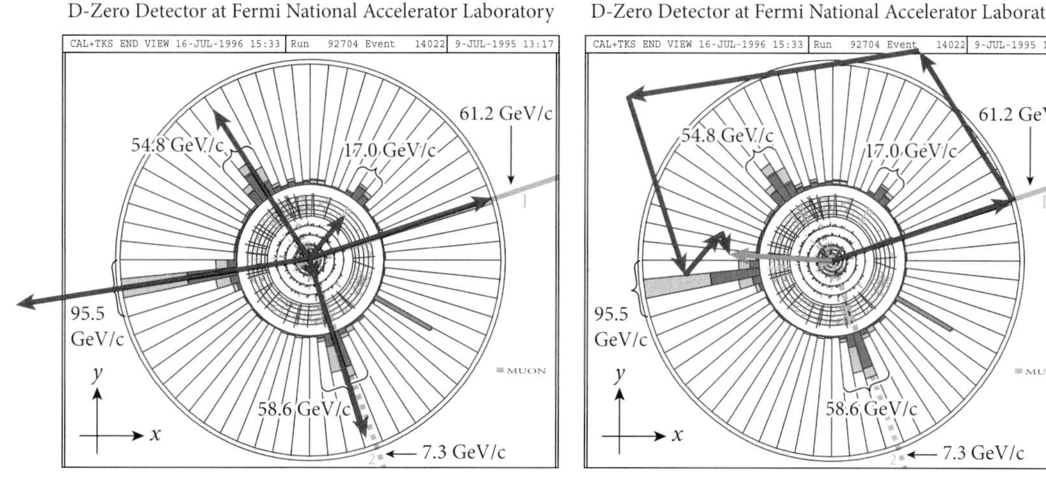


**FIGURE 7.16** Event display generated by the D-Zero collaboration and education office at Fermilab, showing a top-quark event. (a) Momentum vectors of the detected particles produced by the event; (b) graphical addition of the momentum vectors, showing that they add up to a nonzero sum, indicated by the thicker green arrow.

## 7.7 Partially Inelastic Collisions

What happens if a collision is neither elastic nor totally inelastic? Most real collisions are somewhere between these two extremes, as we saw in the coin collision experiment in Figure 7.9. Therefore, it is important to look at partially inelastic collisions in more detail.

We have already seen that the relative velocity of the two objects in a one-dimensional elastic collision simply changes sign. This is also true in two- and three-dimensional elastic collisions, although we will not prove it here. The relative velocity becomes zero in totally inelastic collisions. Thus, it seems logical to define the elasticity of a collision in a way that involves the ratio of initial and final relative velocities.

The **coefficient of restitution,** symbolized by $\epsilon$, is the ratio of the magnitudes of the final and initial relative velocities in a collision:

$$\epsilon = \frac{|\vec{v}_{\text{f}1} - \vec{v}_{\text{f}2}|}{|\vec{v}_{\text{i}1} - \vec{v}_{\text{i}2}|}. \tag{7.29}$$

This definition gives a coefficient of restitution of $\epsilon = 1$ for elastic collisions and $\epsilon = 0$ for totally inelastic collisions.

First, let's examine what happens in the limit where one of the two colliding objects is the ground (for all intents and purposes, infinitely massive) and the other one is a ball. We can see from equation 7.29 that if the ground does not move when the ball bounces, $\vec{v}_{\text{i}1} = \vec{v}_{\text{f}1} = 0$, and we can write for the speed of the ball:

$$v_{\text{f}2} = \epsilon v_{\text{i}2}.$$

If we release the ball from some height, $h_\text{i}$, we know that it reaches a speed of $v_\text{i} = \sqrt{2gh_\text{i}}$ immediately before it collides with the ground. If the collision is elastic, the speed of the ball just after the collision is the same, $v_\text{f} = v_\text{i} = \sqrt{2gh_\text{i}}$, and it bounces back to the same height from which it was released. If the collision is totally inelastic, as in the case of a ball of putty that falls to the ground and then just stays there, the final speed is zero. For all cases in between, we can find the coefficient of restitution from the height $h_\text{f}$ that the ball returns to:

$$h_\text{f} = \frac{v_\text{f}^2}{2g} = \frac{\epsilon^2 v_\text{i}^2}{2g} = \epsilon^2 h_\text{i} \Rightarrow$$

$$\epsilon = \sqrt{h_\text{f} / h_\text{i}}.$$

### 7.10 Self-Test Opportunity

The maximum kinetic energy loss is obtained in the limit of a totally inelastic collision. What fraction of this maximum possible energy loss is obtained in the case where $\epsilon = \frac{1}{2}$?

Using this formula to measure the coefficient of restitution, we find $\epsilon = 0.58$ for baseballs, using typical relative velocities that would occur in ball-bat collisions in Major League games.

In general, we can state (without proof) that the kinetic energy loss in partially inelastic collisions is

$$K_{\text{loss}} = K_{\text{i}} - K_{\text{f}} = \frac{1}{2}\frac{m_1 m_2}{m_1 + m_2}(1-\epsilon^2)(\vec{v}_{\text{i}1} - \vec{v}_{\text{i}2})^2. \quad (7.30)$$

In the limit $\epsilon \to 1$, we obtain $K_{\text{loss}} = 0$—that is, no loss in kinetic energy, as required for elastic collisions. In addition, in the limit $\epsilon \to 0$, equation 7.30 matches the energy release for totally inelastic collisions already shown in equation 7.28.

## Partially Inelastic Collision with a Wall

If you play racquetball or squash, you know that the ball loses energy when you hit it against the wall. While the angle with which a ball bounces off the wall in an elastic collision is the same as the angle with which it hit the wall, the final angle is not so clear for a partially inelastic collision (Figure 7.17).

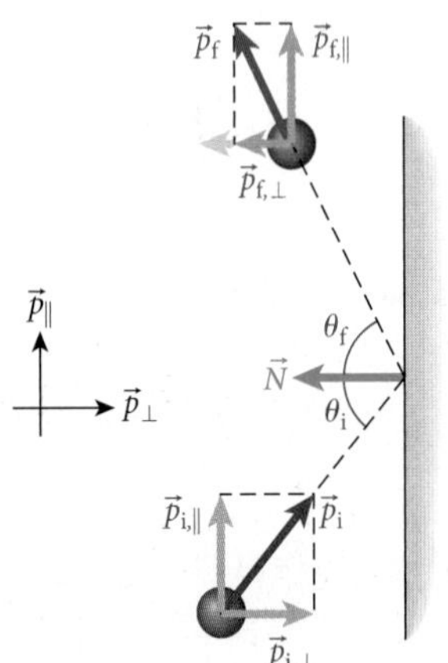


**FIGURE 7.17** Partially inelastic collision of a ball with a wall.

The key to obtaining a first approximation to this angle is to consider only the normal force, which acts in a direction perpendicular to the wall. Then the momentum component directed along the wall remains unchanged, just as in an elastic collision. However, the momentum component perpendicular to the wall is not simply inverted but is also reduced in magnitude by the coefficient of restitution: $p_{\text{f},\perp} = -\epsilon p_{\text{i},\perp}$. This approximation gives us an angle of reflection relative to the normal that is larger than the initial angle:

$$\theta_\text{f} = \cot^{-1}\frac{p_{\text{f},\perp}}{p_{\text{f},\parallel}} = \cot^{-1}\frac{\epsilon p_{\text{i},\perp}}{p_{\text{i},\parallel}} > \theta_\text{i}. \quad (7.31)$$

The magnitude of the final momentum vector is also changed and reduces to

$$p_\text{f} = \sqrt{p_{\text{f},\parallel}^2 + p_{\text{f},\perp}^2} = \sqrt{p_{\text{i},\parallel}^2 + \epsilon^2 p_{\text{i},\perp}^2} < p_\text{i}. \quad (7.32)$$

If we want a quantitative description, we need to include the effect of a friction force between ball and wall, acting for the duration of the collision. (This is why squash balls and racquetballs leave marks on the walls.) Further, the collision with the wall also changes the rotation of the ball, and thus additionally alters the direction and kinetic energy of the ball as it bounces off. However, equations 7.31 and 7.32 still provide a very reasonable first approximation to partially inelastic collisions with walls.

## 7.8 Billiards and Chaos

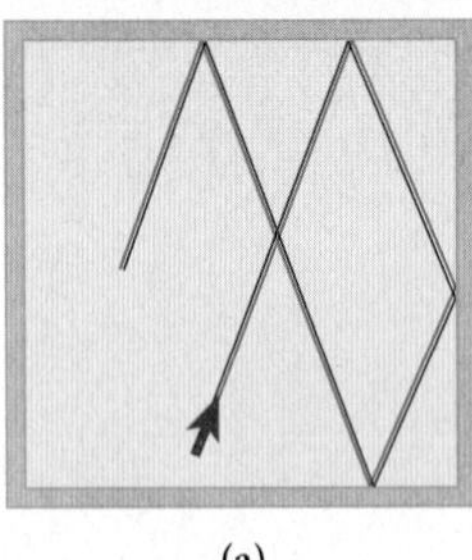

(a)

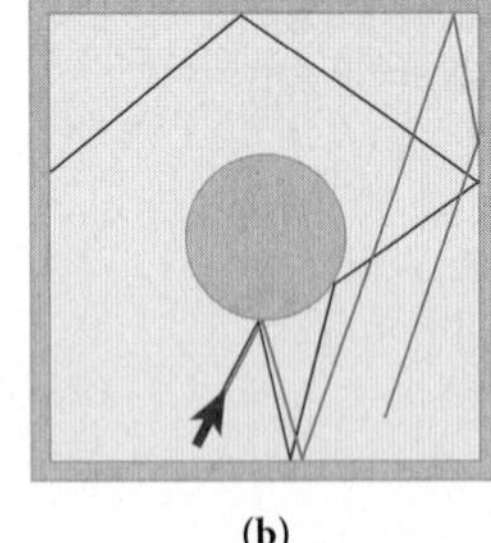

(b)

**FIGURE 7.18** Collisions of particles with walls for two particles starting out very close to each other and with the same momentum: (a) regular billiard table and; (b) Sinai billiard.

Let's look at billiards in an abstract way. The abstract billiard system is a rectangular (or even square) billiard table, on which particles can bounce around and have elastic collisions with the sides. Between collisions, these particles move on straight paths without energy loss. When two particles start off close to each other, as in Figure 7.18a, they stay close to each other. The figure shows the paths (red and green lines) of two particles, which start close to each other with the same initial momentum (indicated by the red arrow) and clearly stay close.

The situation becomes qualitatively different when a circular wall is added in the middle of the billiard table. Now each collision with the circle drives the two paths farther apart. In Figure 7.18b, you can see that one collision with the circle was enough to separate the red and green lines for good. This type of billiard system is called a *Sinai billiard,* named after the Russian academician Yakov Sinai (b. 1935) who first studied it in 1970. The Sinai billiard exhibits **chaotic motion,** or motion that follows the laws of physics—it is not random—but cannot be predicted because it changes significantly with slight changes in conditions, including starting conditions. Surprisingly, Sinai billiard systems are still not fully explored. For example, only in the last decade have the decay properties of these systems been described. Researchers interested in the physics of chaos are continually gaining new knowledge of these systems.

Here is one example from the authors' own research. If you block off the pockets and cut a hole into the side wall of a conventional billiard table and then measure the time it takes a ball to hit this hole and escape, you obtain a power-law decay time distribution: $N(t) = N(t = 0)t^{-1}$, where $N(t = 0)$ is the number of balls used in the experiment and $N(t)$ is the number of balls remaining after time $t$. However, if you did the same for the Sinai billiard, you would obtain an exponential time dependence: $N(t) = N(t = 0)e^{-t/T}$.

These types of investigations are not simply theoretical speculations. Try the following experiment: Place a billiard ball on the surface of a table and hold onto the ball. Then hold a second billiard ball as exactly as you can directly above the first one and release it from a height of a few inches (or centimeters). You will see that the upper ball cannot be made to bounce on the lower one more than three or four times before going off in some uncontrollable direction. Even if you could fix the location of the two balls to atomic precision, the upper ball would bounce off the lower one after only ten to fifteen times. This result means that the ability to predict the outcome of this experiment extends to only a few collisions. This limitation of predictability goes to the heart of chaos science. It is one of the main reasons, for example, that exact long-term weather forecasting is impossible. After all, air molecules bounce off each other, too.

## Laplace's Demon

Marquis Pierre-Simon Laplace (1749–1827) was an eminent French physicist and mathematician of the 18th century. He lived during the time of the French Revolution and other major social upheavals, characterized by the struggle for self-determination and freedom. No painting symbolizes this struggle better than *Liberty Leading the People* (1830) by Eugène Delacroix (Figure 7.19).

**FIGURE 7.19** *Liberty Leading the People,* Eugène Delacroix (Louvre, Paris, France).

Laplace had an interesting idea, now known as *Laplace's Demon*. He reasoned that everything is made of atoms, and all atoms obey differential equations governed by the forces acting on them. If the initial positions and velocities of all atoms, together with all force laws, were fed into a huge computer (he called this an "intellect"), then "for such an intellect nothing could be uncertain and the future just like the past would be present before its eyes." This reasoning implies that everything is predetermined; we are only cogs in a huge clockwork, and nobody has free will. Ironically, Laplace came up with this idea at a time when quite a few people believed that they could achieve free will, if only they could overthrow those in power.

The downfall of Laplace's Demon comes from the combination of two principles of physics. One is from chaos science, which points out that long-term predictability depends sensitively on knowledge of the initial conditions, as seen in the experiment with the bouncing billiard balls. This principle applies to molecules interacting with one another, as, for example, air molecules do. The other physics principle notes the impossibility of specifying both the position and the momentum of any object exactly at the same time. This is the uncertainty relation in quantum physics (discussed in Chapter 36). Thus, free will is still alive and well—the predictability of large or complex systems such as the weather or the human brain over the long term is impossible. The combination of chaos theory and quantum theory ensures that Laplace's Demon or any computer cannot possibly calculate and predict our individual decisions.

## WHAT WE HAVE LEARNED | EXAM STUDY GUIDE

- Momentum is defined as the product of an object's mass and its velocity: $\vec{p} = m\vec{v}$.
- Newton's Second Law can be written as $\vec{F} = d\vec{p}/dt$.
- Impulse is the change in an object's momentum and is equal to the time integral of the applied external force: $\vec{J} = \Delta\vec{p} = \int_{t_i}^{t_f} \vec{F}dt$.
- In a collision of two objects, momentum can be exchanged, but the sum of the momenta of the colliding objects remains constant: $\vec{p}_{f1} + \vec{p}_{f2} = \vec{p}_{i1} + \vec{p}_{i2}$. This is the law of conservation of total momentum.
- Collisions can be elastic, totally inelastic, or partially inelastic.

- In elastic collisions, the total kinetic energy also remains constant:

$$\frac{p_{f1}^2}{2m_1}+\frac{p_{f2}^2}{2m_2}=\frac{p_{i1}^2}{2m_1}+\frac{p_{i2}^2}{2m_2}.$$

- For one-dimensional elastic collisions in general, the final velocities of the two colliding objects can be expressed as a function of the initial velocities:

$$v_{f1,x}=\left(\frac{m_1-m_2}{m_1+m_2}\right)v_{i1,x}+\left(\frac{2m_2}{m_1+m_2}\right)v_{i2,x}$$
$$v_{f2,x}=\left(\frac{2m_1}{m_1+m_2}\right)v_{i1,x}+\left(\frac{m_2-m_1}{m_1+m_2}\right)v_{i2,x}.$$

- In totally inelastic collisions, the colliding objects stick together after the collision and have the same velocity: $\vec{v}_f=(m_1\vec{v}_{i1}+m_2\vec{v}_{i2})/(m_1+m_2)$.
- All partially inelastic collisions are characterized by a coefficient of restitution, defined as the ratio of the magnitudes of the final and the initial relative velocities: $\epsilon=|\vec{v}_{f1}-\vec{v}_{f2}|/|\vec{v}_{i1}-\vec{v}_{i2}|$. The kinetic energy loss in a partially inelastic collision is then given by

$$\Delta K=K_i-K_f=\frac{1}{2}\frac{m_1m_2}{m_1+m_2}(1-\epsilon^2)(\vec{v}_{i1}-\vec{v}_{i2})^2.$$

## KEY TERMS

momentum, p. 206
impulse, p. 208
elastic collision, p. 210
totally inelastic collision, p. 210
conservation of total momentum, p. 210
ballistic pendulum, p. 221
coefficient of restitution, p. 227
chaotic motion, p. 228

## NEW SYMBOLS AND EQUATIONS

$\vec{p}=m\vec{v}$, momentum

$\vec{J}=\Delta\vec{p}=\int_{t_i}^{t_f}\vec{F}dt$, impulse

$\vec{p}_{f1}+\vec{p}_{f2}=\vec{p}_{i1}+\vec{p}_{i2}$, conservation of total momentum

$\epsilon=\frac{|\vec{v}_{f1}-\vec{v}_{f2}|}{|\vec{v}_{i1}-\vec{v}_{i2}|}$, coefficient of restitution

## ANSWERS TO SELF-TEST OPPORTUNITIES

**7.1** rubber door stop
padded baseball glove
padded dashboard in car
water filled barrels in front of bridge abutments on highways
pads on basketball goal support
pads on goalpost supports on football field
portable pad for gym floor
padded shoe inserts

**7.2** The collision is that of a stationary golf ball being struck by a moving golf club head. The head of the driver has more mass than the golf ball. If we take the golf ball to be $m_1$ and the drive head to be $m_2$, then $m_2 > m_1$ and

$$v_{f1,x}=\left(\frac{2m_2}{m_1+m_2}\right)v_{i2,x}.$$

So the speed of the golf ball after impact will be a factor of

$$\left(\frac{2m_2}{m_1+m_2}\right)=\frac{2}{\frac{m_1}{m_2}+1}>1$$

greater than the speed of the golf club head. If the golf club head is much more massive than the golf ball, then

$$\left(\frac{2m_2}{m_1+m_2}\right)\approx 2.$$

**7.3** Take two coins, each with radius $R$.
Collide with impact parameter $b$.
The scattering angle is $\theta$.
$b=0\rightarrow\theta=180°$
$b=2R\rightarrow\theta=0°$

The complete function is

$$\theta=180°-2\sin^{-1}\left(\frac{b}{2R}\right).$$

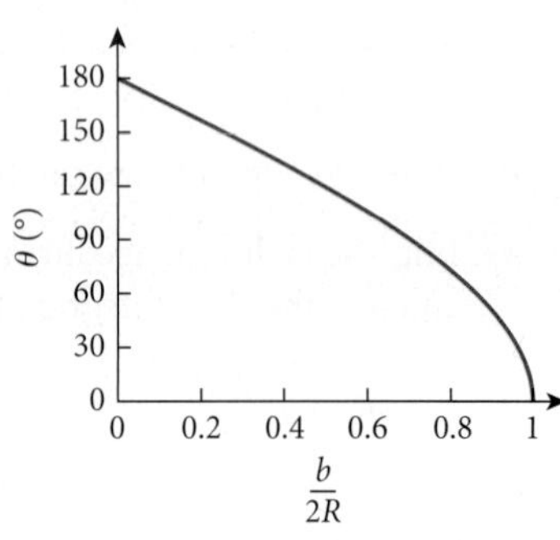

**7.4** Suppose the larger coin is at rest and we shoot a smaller coin at the larger coin. For $b = 0$, we get $\theta = 180°$ and for $b = R_1 + R_2$, we get $\theta = 0°$.

For a very small moving coin incident on a very large stationary coin with radius $R$, we get

$b = R\cos(\theta/2)$

$$\cos^{-1}\left(\frac{b}{R}\right) = \frac{\theta}{2}$$

$$\theta = 2\cos^{-1}\left(\frac{b}{R}\right).$$

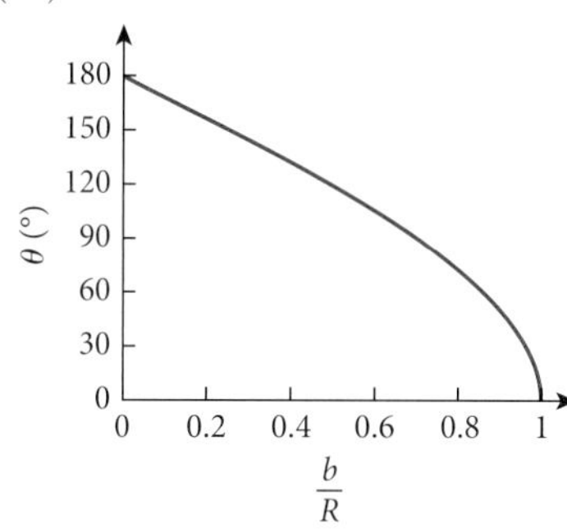


Suppose the smaller coin is at rest and we shoot a larger coin at the smaller coin. For $b = 0$, we get $\theta = 0$ because the larger coin continues forward in a head-on collision. For $b = R_1 + R_2$, we get $\theta = 0$.

**7.5** The initial kinetic energy is

$K_i = \frac{1}{2}mv^2 = \frac{1}{2}(19.0 \text{ kg})(1.60 \text{ m/s})^2 = 24.3 \text{ J}.$

**7.6** $F = ma$, and both experience the same force

$m_{SUV}a_{SUV} = m_{compact}a_{compact}$

$$\frac{a_{compact}}{a_{SUV}} = \frac{m_{SUV}}{m_{compact}} = \frac{3023 \text{ kg}}{1184 \text{ kg}} = 2.553.$$

**7.7** $$v_b = \frac{m+M}{m}\sqrt{2g\ell(1-\cos\theta)} \Rightarrow$$

$$\theta = \cos^{-1}\left(1 - \frac{1}{2g\ell}\left(\frac{mv_b}{m+M}\right)^2\right)$$

$$\theta = \cos^{-1}\left(1 - \frac{1}{2(9.81 \text{ m/s}^2)(1.00 \text{ m})}\left(\frac{(0.0625 \text{ kg})(442 \text{ m/s})}{0.0625 + 40.0 \text{ kg}}\right)^2\right)$$

$\theta = 12.6°$

or roughly half the original angle.

**7.8** Again we have to be careful with the tangent. The $x$-component is negative and the $y$-component is positive. So we must end up in quadrant II with $90° < \alpha < 180°$.

$$\tan\alpha = \frac{-m_1 v_{i1}}{m_2 v_{i2}}$$

$$\alpha = \tan^{-1}\left(\frac{-m_1 v_{i1}}{m_2 v_{i2}}\right) = \tan^{-1}\left(\frac{-2209 \text{ kg}}{1474 \text{ kg}}\frac{50 \text{ mph}}{25 \text{ mph}}\right)$$

$\alpha = 108°$ or 18° from the vertical, compared with 38° from the vertical in the actual collision.

**7.9** The length of the missing momentum vector is 57 pixels. The length of the 95.5 GeV/c vector is 163 pixels. The missing momentum is then (163/57)95.5 GeV/c = 33 GeV/c.

**7.10** Loss for $\epsilon = 0.5$ is

$$K_{loss,0.5} = K_i - K_f = \frac{1}{2}\frac{m_1 m_2}{m_1 + m_2}(1 - (0.5)^2)(\vec{v}_{i1} - \vec{v}_{i2})^2$$

$$= (0.75)\frac{1}{2}\frac{m_1 m_2}{m_1 + m_2}(\vec{v}_{i1} - \vec{v}_{i2})^2 = 0.75\, K_{loss,max}.$$

## PROBLEM-SOLVING PRACTICE

### Problem-Solving Guidelines: Conservation of Momentum

**1.** Conservation of momentum applies to isolated systems with no external forces acting on them—always be sure the problem involves a situation that satisfies or approximately satisfies these conditions. Also, be sure that you take account of every interacting part of the system; conservation of momentum applies to the entire system, not just one object.

**2.** If the situation you're analyzing involves a collision or explosion, identify the momenta immediately before and immediately after the event to use in the conservation of momentum equation. Remember that a change in total momentum equals an impulse, but the impulse can be either an instantaneous force acting over an instant of time or an average force acting over a time interval.

**3.** If the problem involves a collision, you need to recognize what kind of collision it is. If the collision is perfectly elastic, kinetic energy is conserved, but this is not true for other kinds of collisions.

**4.** Remember that momentum is a vector and is conserved in the $x$-, $y$-, and $z$-directions separately. For a collision in more than one dimension, you may need additional information to analyze momentum changes completely.

## SOLVED PROBLEM 7.3 Egg Drop

### PROBLEM

An egg in a special container is dropped from a height of 3.70 m. The container and egg together have a mass of 0.144 kg. A net force of 4.42 N will break the egg. What is the minimum time over which the egg/container can come to a stop without breaking the egg?

### SOLUTION

#### THINK

When the egg/container is released, it accelerates with the acceleration due to gravity. When the egg/container strikes the ground, its velocity goes from the final velocity due to the gravitational acceleration to zero. As the egg/container comes to a stop, the force stopping it times the time interval (the impulse) will equal the mass of the egg/container times the change in speed. The time interval over which the velocity change takes place will determine whether the force exerted on the egg by the collision with the floor will break the egg.

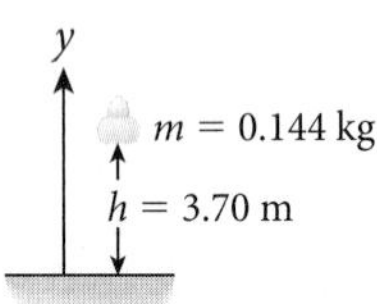


**FIGURE 7.20** An egg in a special container is dropped from a height of 3.70 m.

#### SKETCH

The egg/container is dropped from rest from a height of $h = 3.70$ m (Figure 7.20).

#### RESEARCH

From the discussion of kinematics in Chapter 2, we know the final speed, $v_y$, of the egg/container resulting from free fall from a height of $y_0$ to a final height of $y$, starting with an initial velocity $v_{y0}$, is given by

$$v_y^2 = v_{y0}^2 - 2g(y - y_0). \tag{i}$$

We know that $v_{y0} = 0$ because the egg/container was released from rest. We define the final height to be $y = 0$ and the initial height to be $y_0 = h$, as shown in Figure 7.20. Thus, equation (i) for the final speed in the $y$-direction reduces to

$$v_y = \sqrt{2gh}. \tag{ii}$$

When the egg/container strikes the ground, the impulse, $\vec{J}$, exerted on it is given by

$$\vec{J} = \Delta\vec{p} = \int_{t_1}^{t_2} \vec{F}dt, \tag{iii}$$

where $\Delta\vec{p}$ is the change in momentum of the egg/container and $\vec{F}$ is the force exerted to stop it. We assume the force is constant, so we can rewrite the integral in equation (iii) as

$$\int_{t_1}^{t_2} \vec{F}dt = \vec{F}(t_2 - t_1) = \vec{F}\Delta t.$$

The momentum of the egg/container will change from $p = mv_y$ to $p = 0$ when it strikes the ground, so we can write

$$\Delta p_y = 0 - (-mv_y) = mv_y = F_y\Delta t, \tag{iv}$$

where the term $-mv_y$ is negative because the velocity of the egg/container just before impact is in the negative $y$-direction.

#### SIMPLIFY

We can now solve equation (iv) for the time interval and substitute the expression for the final velocity from equation (ii):

$$\Delta t = \frac{mv_y}{F_y} = \frac{m\sqrt{2gh}}{F_y}. \tag{v}$$

### CALCULATE

Inserting the numerical values, we get

$$\Delta t = \frac{(0.144\text{ kg})\sqrt{2(9.81\text{ m/s}^2)(3.70\text{ m})}}{4.42\text{ N}} = 0.277581543\text{ s}.$$

### ROUND

All of the numerical values in this problem were given with three significant figures, so we report our answer as

$$\Delta t = 0.278\text{ s}.$$

### DOUBLE-CHECK

Slowing the egg/container from its final velocity to zero over a time interval of 0.278 s seems reasonable. Looking at equation (v) we see that the force exerted on the egg as it hits the ground is given by

$$F = \frac{mv_y}{\Delta t}.$$

For a given height, we could reduce the force exerted on the egg in several ways. First, we could make $\Delta t$ larger by making some kind of crumple zone in the container. Second, we could make the egg/container as light as possible. Third, we could construct the container so that it had a large surface area and thus significant air resistance, which would reduce the value of $v_y$ from that for frictionless free fall.

## SOLVED PROBLEM 7.4 Collision with a Parked Car

### PROBLEM

A moving truck strikes a car parked in the breakdown lane of a highway. During the collision, the vehicles stick together and slide to a stop. The moving truck has a total mass of 1982 kg (including the driver), and the parked car has a total mass of 966.0 kg. If the vehicles slide 10.5 m before coming to rest, how fast was the truck going? The coefficient of sliding friction between the tires and the road is 0.350.

### SOLUTION

### THINK

This situation is a totally inelastic collision of a moving truck with a parked car. The kinetic energy of the truck/car combination after the collision is reduced by the energy dissipated by friction while the truck/car combination is sliding. The kinetic energy of the truck/car combination can be related to the initial speed of the truck before the collision.

### SKETCH

Figure 7.21 is a sketch of the moving truck, $m_1$, and the parked car, $m_2$. Before the collision, the truck is moving with an initial speed of $v_{i1,x}$. After the truck collides with the car, the two vehicles slide together with a speed of $v_{f,x}$.

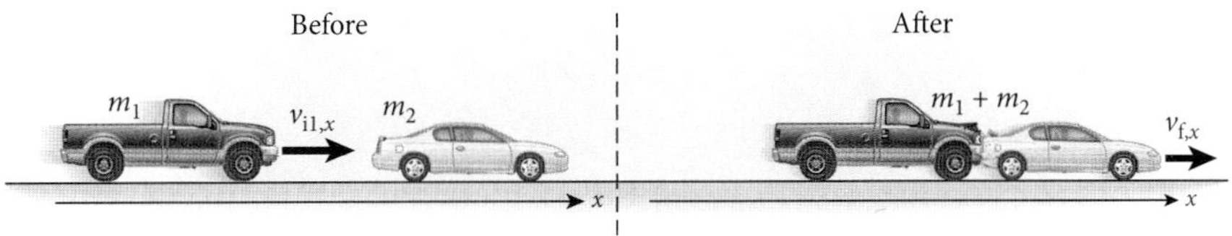


**FIGURE 7.21** The collision between a moving truck and a parked car.

*Continued—*

## RESEARCH

Momentum conservation tells us that the velocity of the two vehicles just after the totally inelastic collision is given by

$$v_{f,x} = \frac{m_1 v_{i1,x}}{m_1 + m_2}.$$

The kinetic energy of the combined truck/car combination just after the collision is

$$K = \tfrac{1}{2}(m_1 + m_2)v_{f,x}^2, \tag{i}$$

where, as usual, $v_{f,x}$ is the final speed.

We finish solving this problem by using the work-energy theorem from Chapter 6, $W_f = \Delta K + \Delta U$. In this situation, the energy dissipated by friction, $W_f$, on the truck/car system is equal to the change in kinetic energy, $\Delta K$, of the truck/car system, since $\Delta U = 0$. We can thus write

$$W_f = \Delta K.$$

The change in kinetic energy is equal to zero (since the truck and car finally stop) minus the kinetic energy of the truck/car system just after the collision. The truck/car system slides a distance $d$. The $x$-component of the frictional force slowing down the truck/car system is given by $f_x = -\mu_k N$, where $\mu_k$ is the coefficient of kinetic friction and $N$ is the magnitude of the normal force. The normal force has a magnitude equal to the weight of the truck/car system, or $N = (m_1 + m_2)g$. The energy dissipated is equal to the $x$-component of the friction force times the distance that the truck/car system slides along the $x$-axis, so we can write

$$W_f = f_x d = -\mu_k (m_1 + m_2) gd. \tag{ii}$$

## SIMPLIFY

We can substitute for the final speed in equation (i) for the kinetic energy and obtain

$$K = \tfrac{1}{2}(m_1 + m_2)v_{f,x}^2 = \tfrac{1}{2}(m_1 + m_2)\left(\frac{m_1 v_{i1,x}}{m_1 + m_2}\right)^2 = \frac{(m_1 v_{i1,x})^2}{2(m_1 + m_2)}.$$

Combining this equation with the work-energy theorem and equation (ii) for the energy dissipated by friction, we get

$$\Delta K = W_f = 0 - \frac{(m_1 v_{i1,x})^2}{2(m_1 + m_2)} = -\mu_k (m_1 + m_2) gd.$$

Solving for $v_{i1,x}$ finally leads us to

$$v_{i1,x} = \frac{(m_1 + m_2)}{m_1}\sqrt{2\mu_k gd}.$$

## CALCULATE

Putting in the numerical values results in

$$v_{i1,x} = \frac{1982 \text{ kg} + 966 \text{ kg}}{1982 \text{ kg}}\sqrt{2(0.350)(9.81 \text{ m/s}^2)(10.5 \text{ m})} = 12.62996079 \text{ m/s}.$$

## ROUND

All of the numerical values given in this problem were specified to three significant figures. Therefore, we report our result as

$$v_{i1,x} = 12.6 \text{ m/s}.$$

## DOUBLE-CHECK

The initial speed of the truck was 12.6 m/s (28.2 mph), which is within the range of normal speeds for vehicles on highways and thus certainly within the expected magnitude for our result.

## MULTIPLE-CHOICE QUESTIONS

**7.1** In many old Western movies, a bandit is knocked back 3 m after being shot by a sheriff. Which statement best describes what happened to the sheriff after he fired his gun?

a) He remained in the same position.
b) He was knocked back a step or two.
c) He was knocked back approximately 3 m.
d) He was knocked forward slightly.
e) He was pushed upward.

**7.2** A fireworks projectile is traveling upward as shown on the right in the figure just before it explodes. Sets of possible momentum vectors for the shell fragments immediately after the explosion are shown below. Which sets could actually occur?

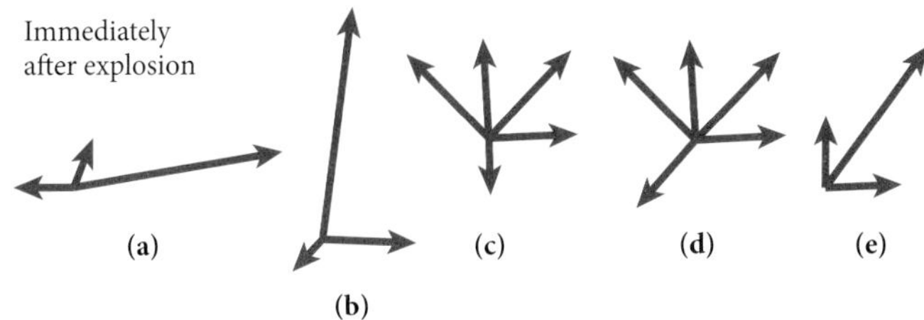


**7.3** The figure shows sets of possible momentum vectors before and after a collision, with no external forces acting. Which sets could actually occur?

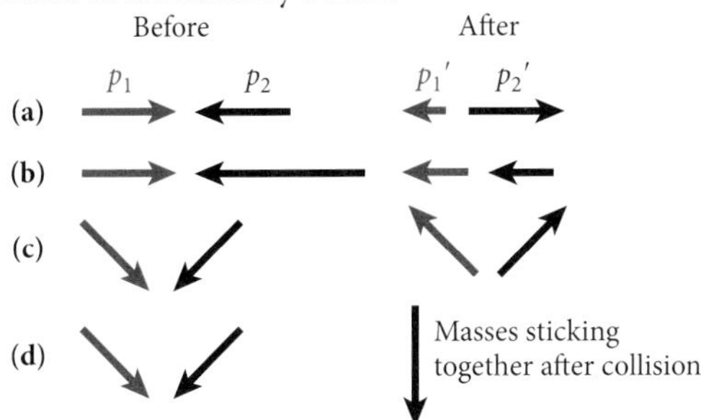


**7.4** The value of the momentum for a system is the same at a later time as at an earlier time if there are no

a) collisions between particles within the system.
b) inelastic collisions between particles within the system.
c) changes of momentum of individual particles within the system.
d) internal forces acting between particles within the system.
e) external forces acting on particles of the system.

**7.5** Consider these three situations:

(i) A ball moving to the right at speed $v$ is brought to rest.
(ii) The same ball at rest is projected at speed $v$ toward the left.
(iii) The same ball moving to the left at speed $v$ speeds up to $2v$.

In which situation(s) does the ball undergo the largest change in momentum?

a) situation (i)
b) situation (ii)
c) situation (iii)
d) situations (i) and (ii)
e) all three situations

**7.6** Consider two carts, of masses $m$ and $2m$, at rest on a frictionless air track. If you push the lower-mass cart for 35 cm and then the other cart for the same distance and with the same force, which cart undergoes the larger change in momentum?

a) The cart with mass $m$ has the larger change.
b) The cart with mass $2m$ has the larger change.
c) The change in momentum is the same for both carts.
d) It is impossible to tell from the information given.

**7.7** Consider two carts, of masses $m$ and $2m$, at rest on a frictionless air track. If you push the lower-mass cart for 3 s and then the other cart for the same length of time and with the same force, which cart undergoes the larger change in momentum?

a) The cart with mass $m$ has the larger change.
b) The cart with mass $2m$ has the larger change.
c) The change in momentum is the same for both carts.
d) It is impossible to tell from the information given.

**7.8** Which of the following statements about car collisions are true and which are false?

a) The essential safety benefit of crumple zones (parts of the front of a car designed to receive maximum deformation during a head-on collision) results from absorbing kinetic energy, converting it into deformation, and lengthening the effective collision time, thus reducing the average force experienced by the driver.

b) If car 1 has mass $m$ and speed $v$, and car 2 has mass $0.5m$ and speed $1.5v$, then both cars have the same momentum.

c) If two identical cars with identical speeds collide head on, the magnitude of the impulse received by each car and each driver is the same as if one car at the same speed had collided head on with a concrete wall.

d) Car 1 has mass $m$, and car 2 has mass $2m$. In a head-on collision of these cars while moving at identical speeds in opposite directions, car 1 experiences a bigger acceleration than car 2.

e) Car 1 has mass $m$, and car 2 has mass $2m$. In a head-on collision of these cars while moving at identical speeds in opposite directions, car 1 receives an impulse of bigger magnitude than that received by car 2.

**7.9** A fireworks projectile is launched upward at an angle above a large flat plane. When the projectile reaches the top of its flight, at a height of $h$ above a point that is a horizontal distance $D$ from where it was launched, the projectile explodes into two equal pieces. One piece reverses its velocity and travels directly back to the launch point. How far from the launch point does the other piece land?

a) $D$
b) $2D$
c) $3D$
d) $4D$

# QUESTIONS

**7.10** An astronaut becomes stranded during a space walk after her jet pack malfunctions. Fortunately, there are two objects close to her that she can push to propel herself back to the International Space Station (ISS). Object A has the same mass as the astronaut, and Object B is 10 times more massive. To achieve a given momentum toward the ISS by pushing one of the objects away from the ISS, which object should she push? That is, which one requires less work to produce the same impulse? Initially, the astronaut and the two objects are at rest with respect to the ISS. (*Hint:* Recall that work is force times distance and think about how the two objects move when they are pushed.)

**7.11** Consider a ballistic pendulum (see Section 7.6) in which a bullet strikes a block of wood. The wooden block is hanging from the ceiling and swings up to a maximum height after the bullet strikes it. Typically, the bullet becomes embedded in the block. Given the same bullet, the same initial bullet speed, and the same block, would the maximum height of the block change if the bullet did not get stopped by the block but passed through to the other side? Would the height change if the bullet and its speed were the same but the block was steel and the bullet bounced off it, directly backward?

**7.12** A bungee jumper is concerned that his elastic cord might break if it is overstretched and is considering replacing the cord with a high-tensile-strength steel cable. Is this a good idea?

**7.13** A ball falls straight down onto a block that is wedge-shaped and is sitting on frictionless ice. The block is initially at rest (see the figure). Assuming the collision is perfectly elastic, is the total momentum of the block/ball system conserved? Is the total kinetic energy of the block/ball system exactly the same before and after the collision? Explain.

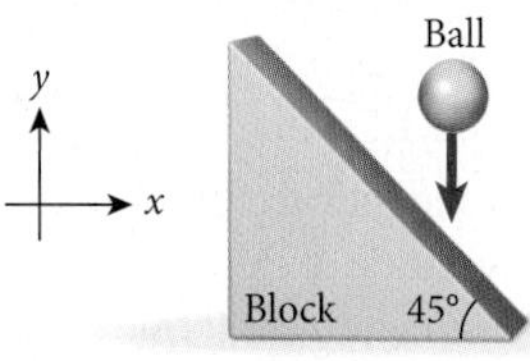


**7.14** To solve problems involving projectiles traveling through the air by applying the law of conservation of momentum requires evaluating the momentum of the system *immediately* before and *immediately* after the collision or explosion. Why?

**7.15** Two carts are riding on an air track as shown in the figure. At time $t = 0$, cart B is at the origin traveling in the positive $x$-direction with speed $v_B$, and cart A is at rest as shown in the diagram below. The carts collide, but do not stick.

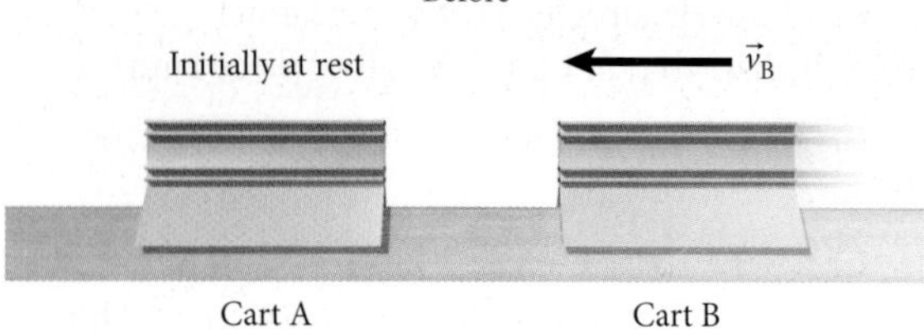


Each of the graphs depicts a possible plot of a physical parameter with respect to time. Each graph has two curves, one for each cart, and each curve is labeled with the cart's letter. For each property (a)–(e) specify the graph that could be a plot of the property; if a graph for a property is not shown, choose alternative 9.

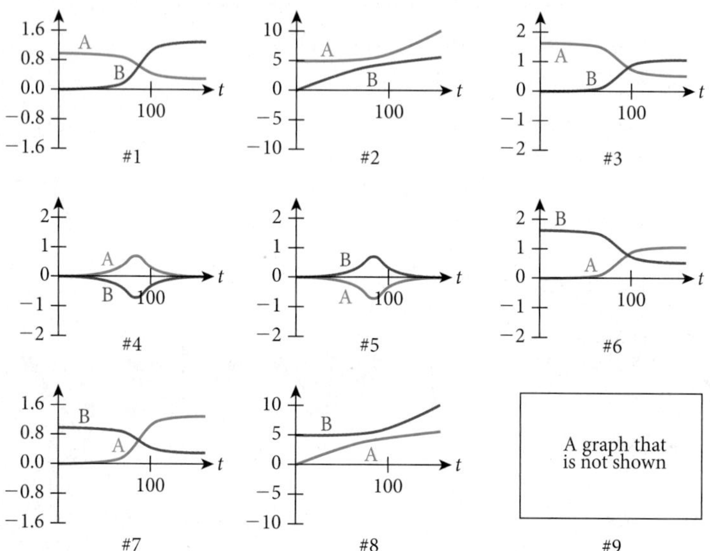


a) the forces *exerted by* the carts
b) the positions of the carts
c) the velocities of the carts
d) the accelerations of the carts
e) the momenta of the carts

**7.16** Using momentum and force principles, explain why an air bag reduces injury in an automobile collision.

**7.17** A rocket works by expelling gas (fuel) from its nozzles at a high velocity. However, if we take the system to be the rocket and fuel, explain qualitatively why a stationary rocket is able to move.

**7.18** When hit in the face, a boxer will "ride the punch"; that is, if he anticipates the punch, he will allow his neck muscles to go slack. His head then moves back easily from the blow. From a momentum-impulse standpoint, explain why this is much better than stiffening his neck muscles and bracing himself against the punch.

**7.19** An open train car moves with speed $v_0$ on a flat frictionless railroad track, with no engine pulling it. It begins to rain. The rain falls straight down and begins to fill the train car. Does the speed of the car decrease, increase, or stay the same? Explain.

# PROBLEMS

A blue problem number indicates a worked-out solution is available in the Student Solutions Manual. One • and two •• indicate increasing level of problem difficulty.

## Section 7.1

**7.20** Rank the following objects from highest to lowest in terms of momentum and from highest to lowest in terms of energy.

a) an asteroid with mass $10^6$ kg and speed 500 m/s

b) a high-speed train with a mass of 180,000 kg and a speed of 300 km/h

c) a 120-kg linebacker with a speed of 10 m/s

d) a 10-kg cannonball with a speed of 120 m/s

e) a proton with a mass of $6 \cdot 10^{-27}$ kg and a speed of $2 \cdot 10^8$ m/s

**7.21** A car of mass 1200 kg, moving with a speed of 72 mph on a highway, passes a small SUV with a mass $1\frac{1}{2}$ times bigger, moving at 2/3 of the speed of the car.

a) What is the ratio of the momentum of the SUV to that of the car?

b) What is the ratio of the kinetic energy of the SUV to that of the car?

**7.22** The electron-volt, eV, is a unit of energy (1 eV = $1.602 \cdot 10^{-19}$ J, 1 MeV = $1.602 \cdot 10^{-13}$ J). Since the unit of momentum is an energy unit divided by a velocity unit, nuclear physicists usually specify momenta of nuclei in units of MeV/$c$, where $c$ is the speed of light ($c = 2.998 \cdot 10^9$ m/s). In the same units, the mass of a proton ($1.673 \cdot 10^{-27}$ kg) is given as 938.3 MeV/$c^2$. If a proton moves with a speed of 17,400 km/s, what is its momentum in units of MeV/$c$?

**7.23** A soccer ball with a mass of 442 g bounces off the crossbar of a goal and is deflected upward at an angle of 58.0° with respect to horizontal. Immediately after the deflection, the kinetic energy of the ball is 49.5 J. What are the vertical and horizontal components of the ball's momentum immediately after striking the crossbar?

**•7.24** A billiard ball of mass $m = 0.250$ kg hits the cushion of a billiard table at an angle of $\theta_1 = 60.0°$ at a speed of $v_1 = 27.0$ m/s. It bounces off at an angle of $\theta_2 = 71.0°$ and a speed of $v_2 = 10.0$ m/s.

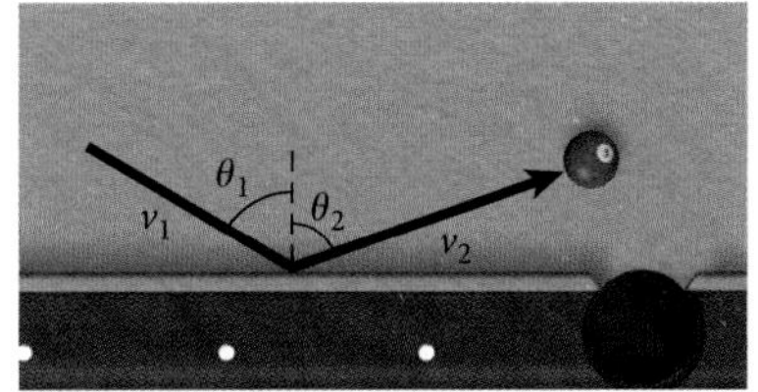


a) What is the magnitude of the change in momentum of the billiard ball?

b) In which direction does the change of momentum vector point?

## Section 7.2

**7.25** In the movie *Superman,* Lois Lane falls from a building and is caught by the diving superhero. Assuming that Lois, with a mass of 50 kg, is falling at a terminal velocity of 60 m/s, how much force does Superman exert on her if it takes 0.1 s to slow her to a stop? If Lois can withstand a maximum acceleration of 7 $g$'s, what minimum time should it take Superman to stop her after he begins to slow her down?

**7.26** One of the events in the Scottish Highland Games is the sheaf toss, in which a 9.09-kg bag of hay is tossed straight up into the air using a pitchfork. During one throw, the sheaf is launched straight up with an initial speed of 2.7 m/s.

a) What is the impulse exerted on the sheaf by gravity during the upward motion of the sheaf (from launch to maximum height)?

b) Neglecting air resistance, what is the impulse exerted by gravity on the sheaf during its downward motion (from maximum height until it hits the ground)?

c) Using the total impulse produced by gravity, determine how long the sheaf is airborne.

**7.27** An 83-kg running back leaps straight ahead toward the end zone with a speed of 6.5 m/s. A 115-kg linebacker, keeping his feet on the ground, catches the running back and applies a force of 900 N in the opposite direction for 0.75 s before the running back's feet touch the ground.

a) What is the impulse that the linebacker imparts to the running back?

b) What change in the running back's momentum does the impulse produce?

c) What is the running back's momentum when his feet touch the ground?

d) If the linebacker keeps applying the same force after the running back's feet have touched the ground, is this still the only force acting to change the running back's momentum?

**7.28** A baseball pitcher delivers a fastball that crosses the plate at an angle of 7.25° relative to the horizontal and a speed of 88.5 mph. The ball (of mass 0.149 kg) is hit back over the head of the pitcher at an angle of 35.53° with respect to the horizontal and a speed of 102.7 mph. What is the magnitude of the impulse received by the ball?

**•7.29** Although they don't have mass, photons—traveling at the speed of light—have momentum. Space travel experts have thought of capitalizing on this fact by constructing *solar sails*—large sheets of material that would work by reflecting photons. Since the momentum of the photon would be reversed, an impulse would be exerted on it by the solar sail, and—by Newton's Third Law—an impulse would also be exerted on the sail, providing a force. In space near the Earth, about $3.84 \cdot 10^{21}$ photons are incident per square meter per second. On average, the momentum of each photon is $1.3 \cdot 10^{-27}$ kg m/s. For a 1000-kg spaceship starting from rest and attached to a square sail 20 m wide, how fast could the ship be moving after 1 hour? One week? One month? How long would it take the ship to attain a speed of 8000 m/s, roughly the speed of the space shuttle in orbit?

**•7.30** In a severe storm, 1 cm of rain falls on a flat horizontal roof in 30 min. If the area of the roof is 100 m$^2$ and the terminal velocity of the rain is 5 m/s, what is the average force exerted on the roof by the rain during the storm?

•**7.31** NASA has taken an increased interest in near Earth asteroids. These objects, popularized in recent blockbuster movies, can pass very close to Earth on a cosmic scale, sometimes as close as 1 million miles. Most are small—less than 500 m across—and while an impact with one of the smaller ones could be dangerous, experts believe that it may not be catastrophic to the human race. One possible defense system against near Earth asteroids involves hitting an incoming asteroid with a rocket to divert its course. Assume a relatively small asteroid with a mass of $2.1 \cdot 10^{10}$ kg is traveling toward the Earth at a modest speed of 12 km/s.

a) How fast would a large rocket with a mass of 80,000 kg have to be moving when it hit the asteroid head on in order to stop the asteroid?

b) An alternative approach would be to divert the asteroid from its path by a small amount to cause it to miss Earth. How fast would the rocket of part (a) have to be traveling at impact to divert the asteroid's path by 1°? In this case, assume that the rocket hits the asteroid while traveling along a line perpendicular to the asteroid's path.

•**7.32** In nanoscale electronics, electrons can be treated like billiard balls. The figure shows a simple device currently under study in which an electron elastically collides with a rigid wall (a ballistic electron transistor). The green bars represent electrodes that can apply a vertical force of $8.0 \cdot 10^{-13}$ N to the electrons. If an electron initially has velocity components $v_x = 1.00 \cdot 10^5$ m/s and $v_y = 0$ and the wall is at 45°, the deflection angle $\theta_D$ is 90°. How long does the vertical force from the electrodes need to be applied to obtain a deflection angle of 120°?

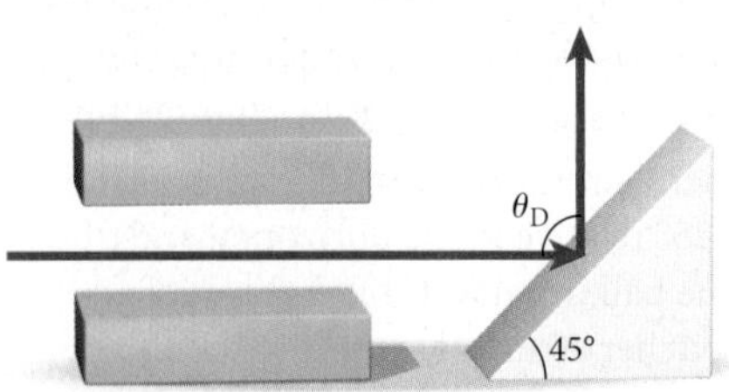


•**7.33** The largest railway gun ever built was called Gustav and was used briefly in World War II. The gun, mount, and train car had a total mass of $1.22 \cdot 10^6$ kg. The gun fired a projectile that was 80.0 cm in diameter and weighed 7502 kg. In the firing illustrated in the figure, the gun has been elevated 20.0° above the horizontal. If the railway gun was at rest before firing and moved to the right at a speed of 4.68 m/s immediately after firing, what was the speed of the projectile as it left the barrel (muzzle velocity)? How far will the projectile travel if air resistance is neglected? Assume that the wheel axles are frictionless.

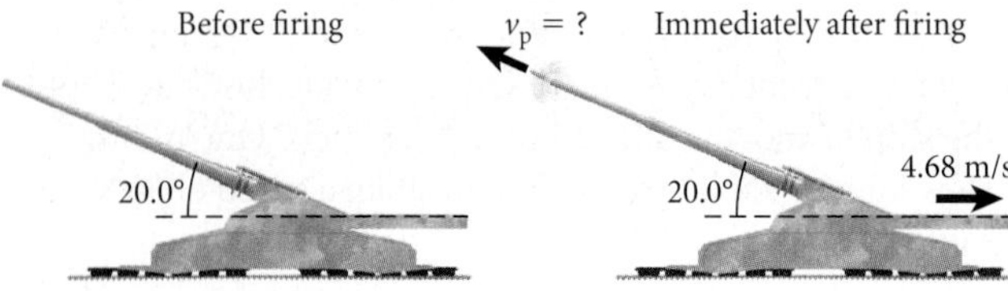


••**7.34** A 6-kg clay ball is thrown directly against a perpendicular brick wall at a velocity of 22 m/s and shatters into three pieces, which all fly backward, as shown in the figure. The wall exerts a force on the ball of 2640 N for 0.1 s. One piece of mass 2 kg travels backward at a velocity of 10 m/s and an angle of 32° above the horizontal. A second piece of mass 1 kg travels at a velocity of 8 m/s and an angle of 28° below the horizontal. What is the velocity of the third piece?

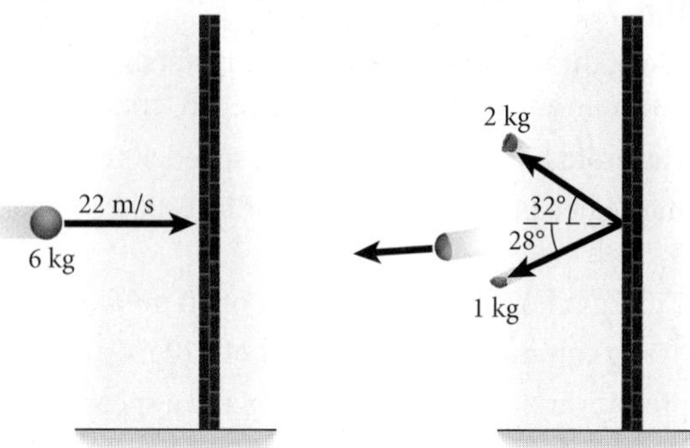


## Section 7.3

**7.35** A sled initially at rest has a mass of 52.0 kg, including all of its contents. A block with a mass of 13.5 kg is ejected to the left at a speed of 13.6 m/s. What is the speed of the sled and the remaining contents?

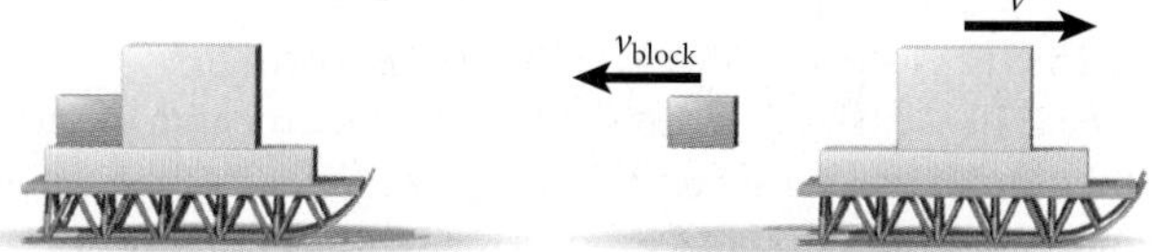


**7.36** Stuck in the middle of a frozen pond with only your physics book, you decide to put physics in action and throw the 5-kg book. If your mass is 62 kg and you throw the book at 13 m/s, how fast do you then slide across the ice? (Assume the absence of friction.)

**7.37** Astronauts are playing baseball on the International Space Station. One astronaut with a mass of 50.0 kg, initially at rest, hits a baseball with a bat. The baseball was initially moving toward the astronaut at 35.0 m/s, and after being hit, travels back in the same direction with a speed of 45.0 m/s. The mass of a baseball is 0.14 kg. What is the recoil velocity of the astronaut?

•**7.38** An automobile with a mass of 1450 kg is parked on a moving flatbed railcar; the flatbed is 1.5 m above the

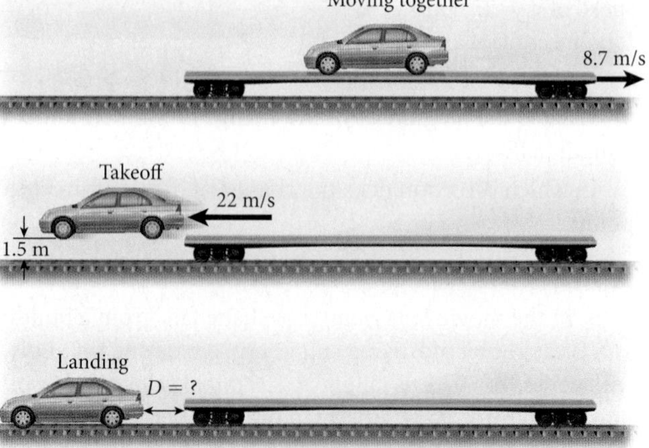

ground. The railcar has a mass of 38,500 kg and is moving to the right at a constant speed of 8.7 m/s on a frictionless rail. The automobile then accelerates to the left, leaving the railcar at a speed of 22 m/s with respect to the ground. When the automobile lands, what is the distance $D$ between it and the left end of the railcar? See the figure.

•**7.39** Three people are floating on a 120-kg raft in the middle of a pond on a warm summer day. They decide to go swimming, and all jump off the raft at the same time and from evenly spaced positions around the perimeter of the raft. One person, with a mass of 62 kg, jumps off the raft at a speed of 12 m/s. The second person, of mass 73 kg, jumps off the raft at a speed of 8 m/s. The third person, of mass 55 kg, jumps off the raft at a speed of 11 m/s. At what speed does the raft drift from its original position?

•**7.40** A missile is shot straight up into the air. At the peak of its trajectory, it breaks up into three pieces of equal mass, all of which move horizontally away from the point of the explosion. One piece travels in a horizontal direction of 28° east of north with a speed of 30 m/s. The second piece travels in a horizontal direction of 12° south of west with a speed of 8 m/s. What is the velocity of the remaining piece? Give both speed and angle.

••**7.41** Once a favorite playground sport, dodgeball is becoming increasingly popular among adults of all ages who want to stay in shape and who even form organized leagues. Gutball is a less popular variant of dodgeball in which players are allowed to bring their own (typically nonregulation) equipment and in which cheap shots to the face are permitted. In a gutball tournament against people half his age, a physics professor threw his 0.4-kg soccer ball at a kid throwing a 0.6-kg basketball. The balls collided in midair (see the figure), and the basketball flew off with an energy of 95 J at an angle of 32° relative to its initial path. Before the collision, the energy of the soccer ball was 100 J and the energy of the basketball was 112 J. At what angle and speed did the soccer ball move away from the collision?

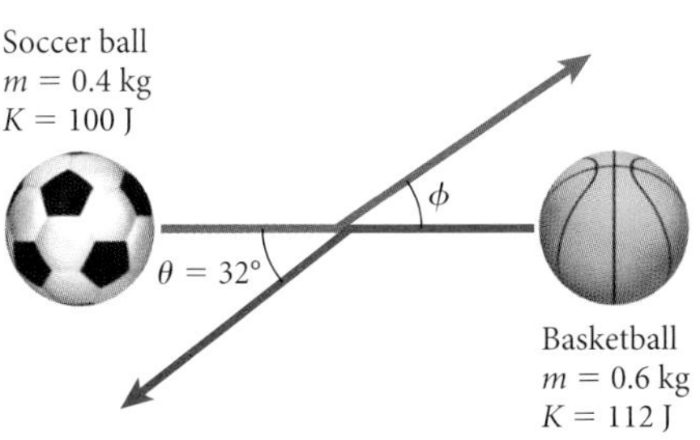


## Section 7.4

**7.42** Two bumper cars moving on a frictionless surface collide elastically. The first bumper car is moving to the right with a speed of 20.4 m/s and rear-ends the second bumper car, which is also moving to the right but with a speed of 9.00 m/s. What is the speed of the first bumper car after the collision? The mass of the first bumper car is 188 kg, and the mass of the second bumper car is 143 kg. Assume that the collision takes place in one dimension.

**7.43** A satellite with a mass of 274 kg approaches a large planet at a speed $v_{i,1}$ =13.5 km/s. The planet is moving at a speed $v_{i,2}$ =10.5 km/s in the opposite direction. The satellite partially orbits the planet and then moves away from the planet in a direction opposite to its original direction (see the figure). If this interaction is assumed to approximate an elastic collision in one dimension, what is the speed of the satellite after the collision? This so-called *slingshot effect* is often used to accelerate space probes for journeys to distance parts of the solar system (see Chapter 12).

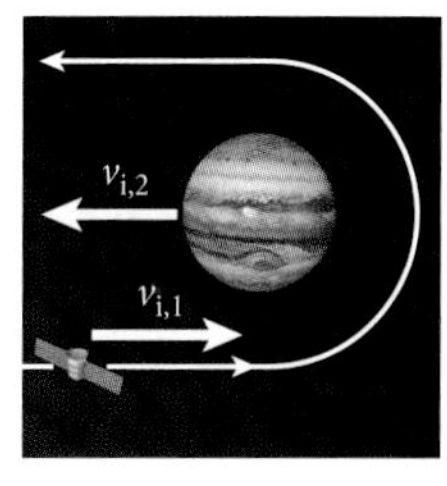


•**7.44** You see a pair of shoes tied together by the laces and hanging over a telephone line. You throw a 0.25-kg stone at one of the shoes (mass = 0.37 kg), and it collides elastically with the shoe with a velocity of 2.3 m/s in the horizontal direction. How far up does the shoe move?

•**7.45** Block A and block B are forced together, compressing a spring (with spring constant $k$ = 2500 N/m) between them 3 cm from its equilibrium length. The spring, which has negligible mass, is not fastened to either block and drops to the surface after it has expanded. What are the speeds of block A and block B at this moment? (Assume that the friction between the blocks and the supporting surface is so low that it can be neglected.)

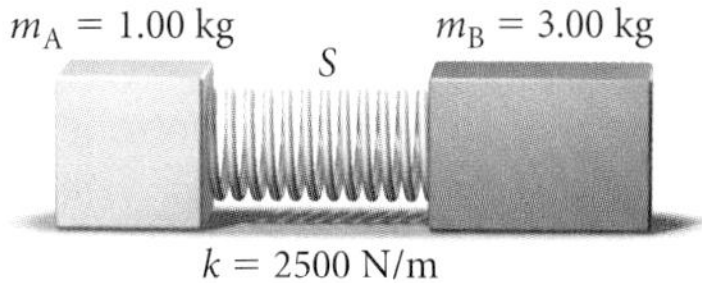


•**7.46** An alpha particle (mass = 4 u) has a head-on, elastic collision with a nucleus (mass = 166 u) that is initially at rest. What percentage of the kinetic energy of the alpha particle is transferred to the nucleus in the collision?

•**7.47** You notice that a shopping cart 20.0 m away is moving with a velocity of 0.70 m/s toward you. You launch a cart with a velocity of 1.1 m/s directly at the other cart in order to intercept it. When the two carts collide elastically, they remain in contact for approximately 0.2 s. Graph the position, velocity, and force for both carts as a function of time.

•**7.48** A 0.280-kg ball has an elastic, head-on collision with a second ball that is initially at rest. The second ball moves off with half the original speed of the first ball.

a) What is the mass of the second ball?

b) What fraction of the original kinetic energy ($\Delta K/K$) is transferred to the second ball?

••**7.49** Cosmic rays from space that strike Earth contain some charged particles with energies billions of times higher than any that can be produced in the biggest accelerator. One model that was proposed to account for these particles is shown schematically in the figure. Two very strong sources of magnetic fields move toward each other and repeatedly reflect the charged particles trapped between them. (These magnetic field sources can be approximated as infinitely heavy walls from which charged particles get

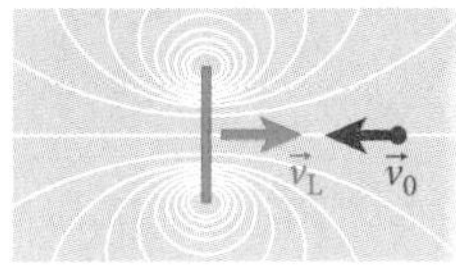


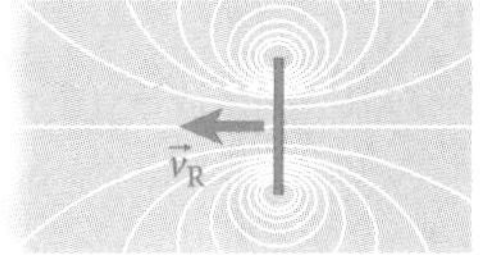

reflected elastically.) The high-energy particles that strike the Earth would have been reflected a large number of times to attain the observed energies. An analogous case with only a few reflections demonstrates this effect. Suppose a particle has an initial velocity of –2.21 km/s (moving in the negative $x$-direction, to the left), the left wall moves with a velocity of 1.01 km/s to the right, and the right wall moves with a velocity of 2.51 km/s to the left. What is the velocity of the particle after six collisions with the left wall and five collisions with the right wall?

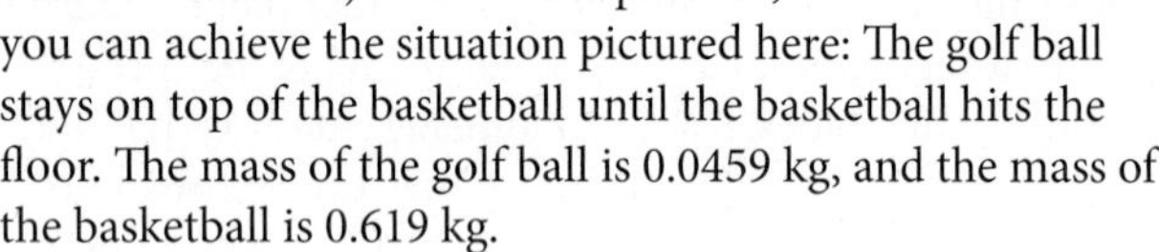

••**7.50** Here is a popular lecture demonstration that you can perform at home. Place a golf ball on top of a basketball, and drop the pair from rest so they fall to the ground. (For reasons that should become clear once you solve this problem, do not attempt to do this experiment inside, but outdoors instead!) With a little practice, you can achieve the situation pictured here: The golf ball stays on top of the basketball until the basketball hits the floor. The mass of the golf ball is 0.0459 kg, and the mass of the basketball is 0.619 kg.

a) If the balls are released from a height where the bottom of the basketball is at 0.701 m above the ground, what is the absolute value of the basketball's momentum just before it hits the ground?

b) What is the absolute value of the momentum of the golf ball at this instant?

c) Treat the collision of the basketball with the floor and the collision of the golf ball with the basketball as totally elastic collisions in one dimension. What is the absolute magnitude of the momentum of the golf ball after these collisions?

d) Now comes the interesting question: How high, measured from the ground, will the golf ball bounce up after its collision with the basketball?

## Section 7.5

**7.51** A hockey puck with mass 0.250 kg traveling along the blue line (a blue-colored straight line on the ice in a hockey rink) at 1.5 m/s strikes a stationary puck with the same mass. The first puck exits the collision in a direction that is 30° away from the blue line at a speed of 0.75 m/s (see the figure). What is the direction and magnitude of the velocity of the second puck after the collision? Is this an elastic collision?

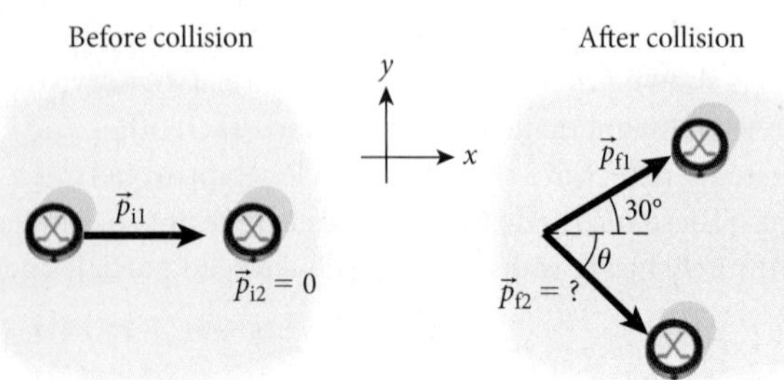


•**7.52** A ball with mass $m = 0.210$ kg and kinetic energy $K_1 = 2.97$ J collides elastically with a second ball of the same mass that is initially at rest. After the collision, the first ball moves away at an angle of $\theta_1 =$ 30.6° with respect to the horizontal, as shown in the figure. What is the kinetic energy of the first ball after the collision?

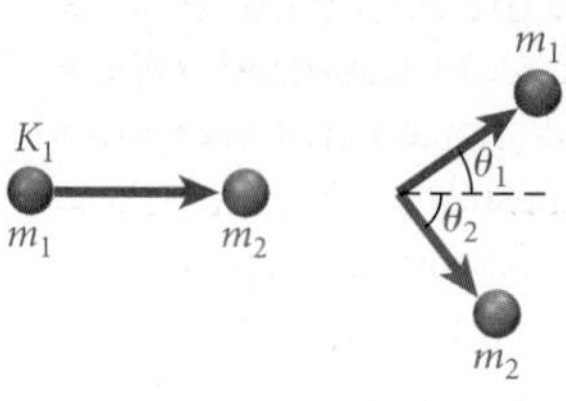


•**7.53** When you open the door to an air-conditioned room, you mix hot gas with cool gas. Saying that a gas is hot or cold actually refers to its average energy; that is, the hot gas molecules have a higher kinetic energy than the cold gas molecules. The difference in kinetic energy in the mixed gases decreases over time as a result of elastic collisions between the gas molecules, which redistribute the energy. Consider a two-dimensional collision between two nitrogen molecules ($N_2$, molecular weight = 28 g/mol). One molecule moves at 30° with respect to the horizontal with a velocity of 672 m/s. This molecule collides with a second molecule moving in the negative horizontal direction at 246 m/s. What are the molecules' final velocities if the one that is initially more energetic moves in the vertical direction after the collision?

•**7.54** A ball falls straight down onto a wedge that is sitting on frictionless ice. The ball has a mass of 3.00 kg, and the wedge has a mass of 5.00 kg. The ball is moving a speed of 4.50 m/s when it strikes the wedge, which is initially at rest (see the figure). Assuming that the collision is instantaneous and perfectly elastic, what are the velocities of the ball and the wedge after the collision?



••**7.55** Betty Bodycheck ($m_B = 55.0$ kg, $v_B = 22.0$ km/h in the positive $x$-direction) and Sally Slasher ($m_S = 45.0$ kg, $v_S = 28.0$ km/h in the positive $y$-direction) are both racing to get to a hockey puck. Immediately after the collision, Betty is heading in a direction that is 76.0° counterclockwise from her original direction, and Sally is heading back and to her right in a direction that is 12.0° from the $x$-axis. What are Betty and Sally's final kinetic energies? Is their collision elastic?

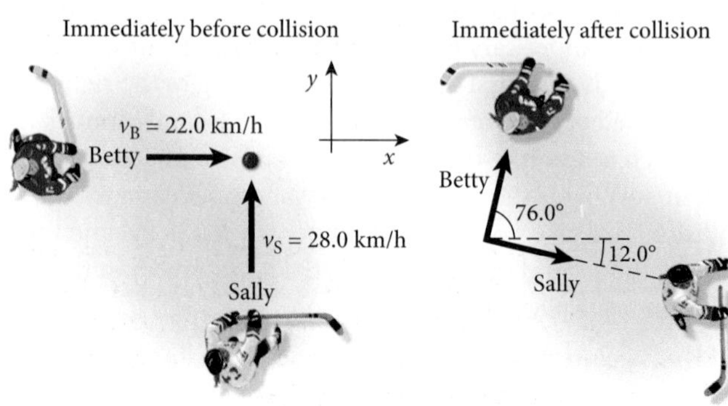

••**7.56** Current measurements and cosmological theories suggest that only about 4% of the total mass of the universe is composed of ordinary matter. About 22% of the mass is composed of *dark matter,* which does not emit or reflect light and can only be observed through its gravitational interaction with its surroundings (see Chapter 12). Suppose a galaxy with mass $M_G$ is moving in a straight line in the $x$-direction. After it interacts with an invisible clump of dark matter with mass $M_{DM}$, the galaxy moves with 50% of its initial speed in a straight line in a direction that is rotated by an angle $\theta$ from its initial velocity. Assume that initial and final velocities are given for positions where the galaxy is very far from the clump of dark matter, that the gravitational attraction can be neglected at those positions, and that the dark matter is initially at rest. Determine $M_{DM}$ in terms of $M_G$, $v_0$, and $\theta$.

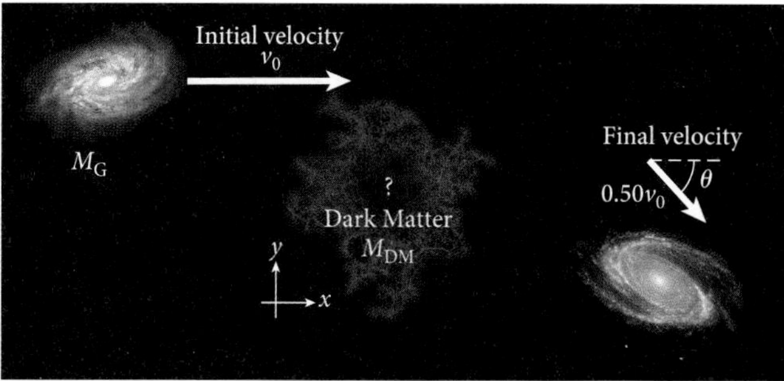


## Section 7.6

**7.57** A 1439-kg railroad car traveling at a speed of 12 m/s strikes an identical car at rest. If the cars lock together as a result of the collision, what is their common speed (in m/s) afterward?

**7.58** Bats are extremely adept at catching insects in midair. If a 50-g bat flying in one direction at 8 m/s catches a 5-g insect flying in the opposite direction at 6 m/s, what is the speed of the bat immediately after catching the insect?

•**7.59** A small car of mass 1000 kg traveling at a speed of 33 m/s collides head on with a large car of mass 3000 kg traveling in the opposite direction at a speed of 30 m/s. The two cars stick together. The duration of the collision is 100 ms. What acceleration (in $g$) do the occupants of the small car experience? What acceleration (in $g$) do the occupants of the large car experience?

•**7.60** To determine the muzzle velocity of a bullet fired from a rifle, you shoot the 2.0-g bullet into a 2.0-kg wooden block. The block is suspended by wires from the ceiling and is initially at rest. After the bullet is embedded in the block, the block swings up to a maximum height of 0.5 cm above its initial position. What is the velocity of the bullet on leaving the gun's barrel?

•**7.61** A 2000-kg Cadillac and a 1000-kg Volkswagen meet at an intersection. The stoplight has just turned green, and the Cadillac, heading north, drives forward into the intersection. The Volkswagen, traveling east, fails to stop. The Volkswagen crashes into the left front fender of the Cadillac; then the cars stick together and slide to a halt. Officer Tom, responding to the accident, sees that the skid marks are directed 55° north of east from the point of impact. The driver of the Cadillac, who keeps a close eye on the speedometer, reports that he was traveling at 30 m/s when the accident occurred. How fast was the Volkswagen going just before the impact?

•**7.62** Two balls of equal mass collide and stick together as shown in the figure. The initial velocity of ball B is twice that of ball A.

a) Calculate the angle above the horizontal of the motion of mass A + B after the collision.

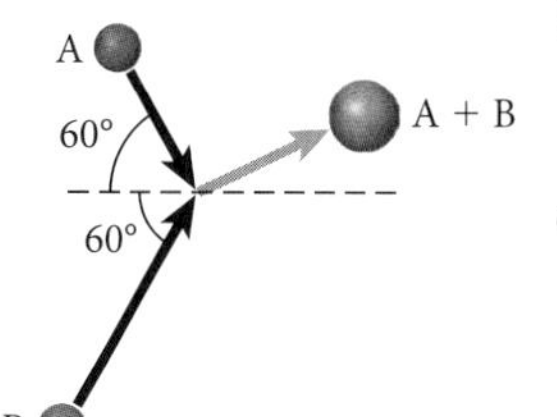


b) What is the ratio of the final velocity of the mass A + B to the initial velocity of ball A, $v_f/v_A$?

c) What is the ratio of the final energy of the system to the initial energy of the system, $E_f/E_i$? Is the collision elastic or inelastic?

••**7.63** Tarzan swings on a vine from a cliff to rescue Jane, who is standing on the ground surrounded by snakes. His plan is to push off the cliff, grab Jane at the lowest point of his swing, and carry them both to the safety of a nearby tree (see the figure). Tarzan's mass is 80 kg, Jane's mass is 40 kg, the height of the lowest limb of the target tree is 10 m, and Tarzan is initially standing on a cliff of height 20 m. The length of the vine is 30 m. With what speed should Tarzan push off the cliff if he and Jane are to make it to the tree limb successfully?

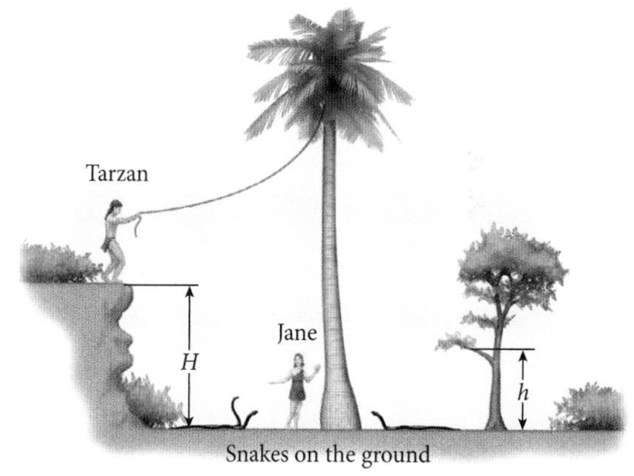


••**7.64** A 3000-kg Cessna airplane flying north at 75 m/s at an altitude of 1600 m over the jungles of Brazil collided with a 7000-kg cargo plane flying 35° north of west with a speed of 100 m/s. As measured from a point on the ground directly below the collision, the Cessna wreckage was found 1000 m away at an angle of 25° south of west, as shown in

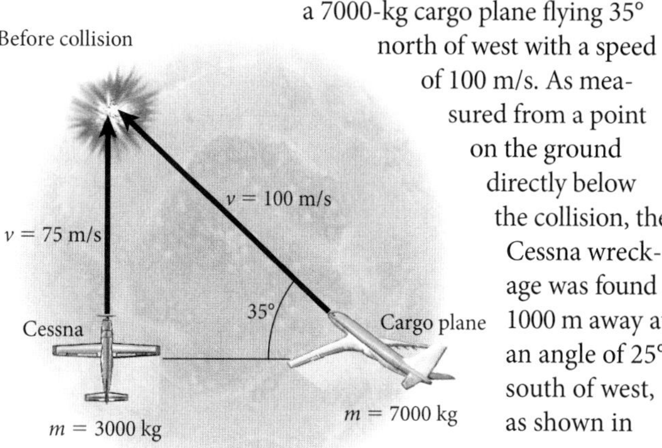

the figure. The cargo plane broke into two pieces. Rescuers located a 4000-kg piece 1800 m away from the same point at an angle of 22° east of north. Where should they look for the other piece of the cargo plane? Give a distance and a direction from the point directly below the collision.

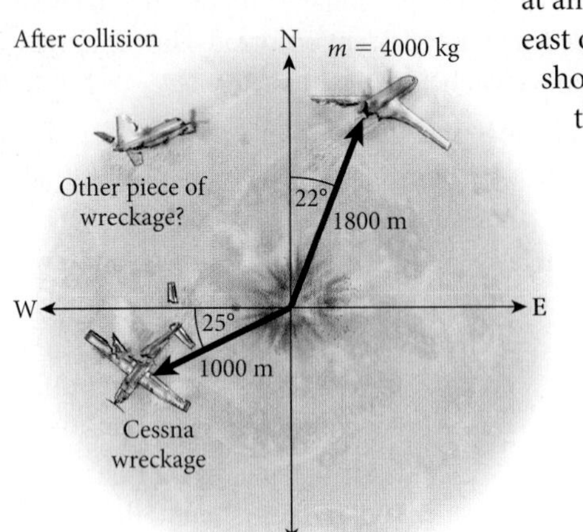


## Section 7.7

7.65 The objects listed in the table are dropped from a height of 85 cm. The height they reach after bouncing has been recorded. Determine the coefficient of restitution for each object.

| Object | $H$ (cm) | $h_1$ (cm) | $\epsilon$ |
|---|---|---|---|
| range golf ball | 85.0 | 62.6 | |
| tennis ball | 85.0 | 43.1 | |
| billiard ball | 85.0 | 54.9 | |
| hand ball | 85.0 | 48.1 | |
| wooden ball | 85.0 | 30.9 | |
| steel ball bearing | 85.0 | 30.3 | |
| glass marble | 85.0 | 36.8 | |
| ball of rubber bands | 85.0 | 58.3 | |
| hollow, hard plastic ball | 85.0 | 40.2 | |

**7.66** A golf ball is released from rest from a height of 0.811 m above the ground and has a collision with the ground, for which the coefficient of restitution is 0.601. What is the maximum height reached by this ball as it bounces back up after this collision?

**•7.67** A billiard ball of mass 0.162 kg has a speed of 1.91 m/s and collides with the side of the billiard table at an angle of 35.9°. For this collision, the coefficient of restitution is 0.841. What is the angle relative to the side (in degrees) at which the ball moves away from the collision?

**•7.68** A soccer ball rolls out of a gym through the center of a doorway into the next room. The adjacent room is 6 m by 6 m with the 2-m wide doorway located at the center of the wall. The ball hits the center of a side wall at 45°. If the coefficient of restitution for the soccer ball is 0.7, does the ball bounce back out of the room? (Note that the ball rolls without slipping, so no energy is lost to the floor.)

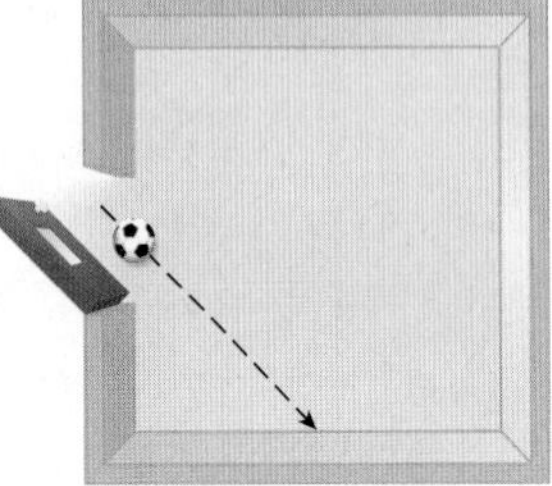

**•7.69** In a *Tom and Jerry*™ cartoon, Tom the cat is chasing Jerry the mouse outside in a yard. The house is in a neighborhood where all the houses are exactly the same, each with the same size yard and the same 2-m-high fence around each yard. In the cartoon, Tom rolls Jerry into a ball and throws him over the fence. Jerry moves like a projectile, bounces at the center of the next yard, and continues to fly toward the next fence, which is 7.5 m away. If Jerry's original height above ground when he was thrown is 5 m, his original range is 15 m, and his coefficient of restitution is 0.8, does he make it over the next fence?

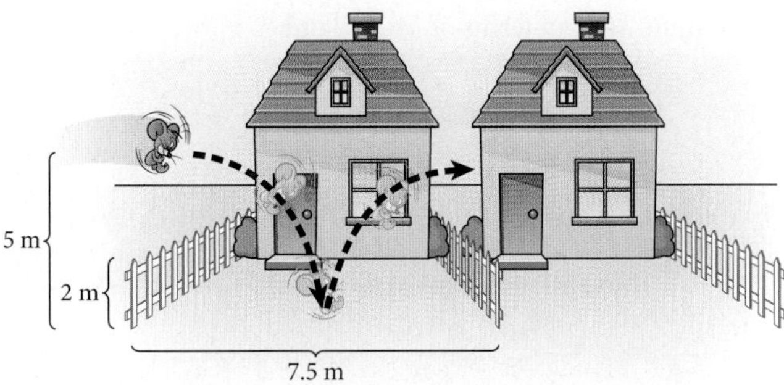


**••7.70** Two Sumo wrestlers are involved in an inelastic collision. The first wrestler, Hakurazan, has a mass of 135 kg and moves forward along the positive $x$-direction at a speed of 3.5 m/s. The second wrestler, Toyohibiki, has a mass of 173 kg and moves straight toward Hakurazan at a speed of 3.0 m/s. Immediately after the collision, Hakurazan is deflected to his right by 35° (see the figure). In the collision, 10% of the wrestlers' initial total kinetic energy is lost. What is the angle at which Toyohibiki is moving immediately after the collision?

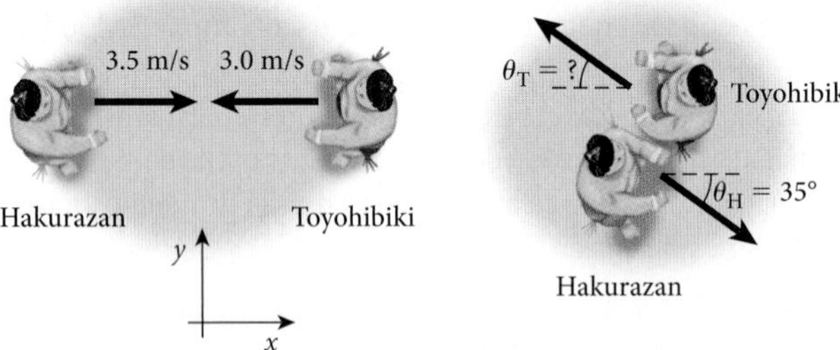


**••7.71** A hockey puck ($m = 170.$ g and $v_0 = 2.0$ m/s) slides without friction on the ice and hits the rink board at 30° with respect to the normal. The puck bounces off the board at a 40° angle with respect to the normal. What is the coefficient of restitution for the puck? What is the ratio of the puck's final kinetic energy to its initial kinetic energy?

## Additional Problems

7.72 How fast would a 5-g fly have to be traveling to slow a 1900-kg car traveling at 55 mph by 5 mph if the fly hit the car in a totally inelastic head-on collision?

7.73 Attempting to score a touchdown, an 85-kg tailback jumps over his blockers, achieving a horizontal speed of 8.9 m/s. He is met in midair just short of the goal line by a 110-kg linebacker traveling in the opposite direction at a speed of 8.0 m/s. The linebacker grabs the tailback.

a) What is the speed of the entangled tailback and linebacker just after the collision?

b) Will the tailback score a touchdown (provided that no other player has a chance to get involved, of course)?

**7.74** The nucleus of radioactive thorium-228, with a mass of about $3.8 \cdot 10^{-25}$ kg, is known to decay by emitting an alpha particle with a mass of about $6.68 \cdot 10^{-27}$ kg. If the alpha particle is emitted with a speed of $1.8 \cdot 10^7$ m/s, what is the recoil speed of the remaining nucleus (which is the nucleus of a radon atom)?

**7.75** A 60-kg astronaut inside a 7-m-long space capsule of mass 500 kg is floating weightlessly on one end of the capsule. He kicks off the wall at a velocity of 3.5 m/s toward the other end of the capsule. How long does it take the astronaut to reach the far wall?

**7.76** Moessbauer spectroscopy is a technique for studying molecules by looking at a particular atom within them. For example, Moessbauer measurements of iron (Fe) inside hemoglobin, the molecule responsible for transporting oxygen in the blood, can be used to determine the hemoglobin's flexibility. The technique starts with X-rays emitted from the nuclei of $^{57}$Co atoms. These X-rays are then used to study the Fe in the hemoglobin. The energy and momentum of each X-ray are 14 keV and 14 keV/$c$ (see Example 7.5 for an explanation of the units). A $^{57}$Co nucleus recoils as an X-ray is emitted. A single $^{57}$Co nucleus has a mass of $9.52 \cdot 10^{-26}$ kg. What are the final momentum and kinetic energy of the $^{57}$Co nucleus? How do these compare to the values for the X-ray?

7.77 Assume the nucleus of a radon atom, $^{222}$Rn, has a mass of $3.68 \cdot 10^{-25}$ kg. This radioactive nucleus decays by emitting an alpha particle with an energy of $8.79 \cdot 10^{-13}$ J. The mass of an alpha particle is $6.65 \cdot 10^{-27}$ kg. Assuming that the radon nucleus was initially at rest, what is the velocity of the nucleus that remains after the decay?

**7.78** A skateboarder of mass 35.0 kg is riding her 3.50-kg skateboard at a speed of 5.00 m/s. She jumps backward off her skateboard, sending the skateboard forward at a speed of 8.50 m/s. At what speed is the skateboarder moving when her feet hit the ground?

**7.79** During an ice-skating extravaganza, *Robin Hood on Ice,* a 50.0-kg archer is standing still on ice skates. Assume that the friction between the ice skates and the ice is negligible. The archer shoots a 0.100-kg arrow horizontally at a speed of 95.0 m/s. At what speed does the archer recoil?

**7.80** Astronauts are playing catch on the International Space Station. One 55.0-kg astronaut, initially at rest, throws a baseball of mass 0.145 kg at a speed of 31.3 m/s. At what speed does the astronaut recoil?

7.81 A bungee jumper with mass 55.0 kg reaches a speed of 13.3 m/s moving straight down when the elastic cord tied to her feet starts pulling her back up. After 0.0250 s, the jumper is heading back up at a speed of 10.5 m/s. What is the average force that the bungee cord exerts on the jumper? What is the average number of $g$'s that the jumper experiences during this direction change?

**7.82** A 3.0-kg ball of clay with a speed of 21 m/s is thrown against a wall and sticks to the wall. What is the magnitude of the impulse exerted on the ball?

**7.83** The figure shows before and after scenes of a cart colliding with a wall and bouncing back. What is the cart's change of momentum? (Assume that right is the positive direction in the coordinate system.)

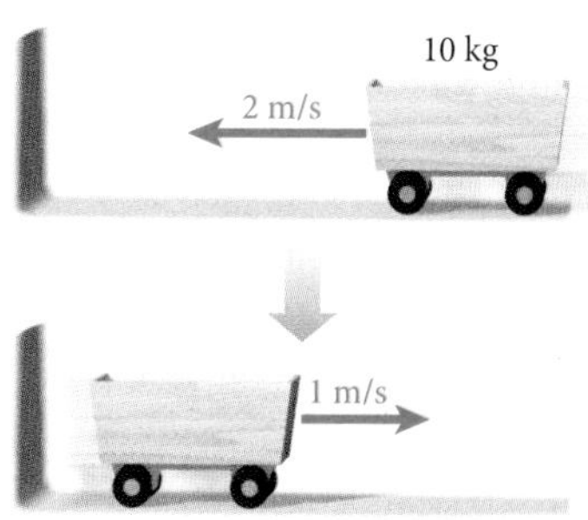


**7.84** Tennis champion Venus Williams is capable of serving a tennis ball at around 127 mph.

a) Assuming that her racquet is in contact with the 57-g ball for 0.25 s, what is the average force of the racket on the ball?

b) What average force would an opponent's racquet have to exert in order to return Williams's serve at a speed of 50 mph, assuming that the opponent's racquet is also in contact with the ball for 0.25 s?

7.85 Three birds are flying in a compact formation. The first bird, with a mass of 100 g is flying 35° east of north at a speed of 8 m/s. The second bird, with a mass of 123 g, is flying 2° east of north at a speed of 11 m/s. The third bird, with a mass of 112 g, is flying 22° west of north at a speed of 10 m/s. What is the momentum vector of the formation? What would be the speed and direction of a 115-g bird with the same momentum?

**7.86** A golf ball of mass 45 g moving at a speed of 120 km/h collides head on with a French TGV high-speed train of mass 380,000 kg that is traveling at 300 km/h. Assuming that the collision is elastic, what is the speed of the golf ball after the collision? (Do not try to conduct this experiment!)

**7.87** In bocce, the object of the game is to get your balls (each with mass $M = 1.00$ kg) as close as possible to the small white ball (the *pallina,* mass $m = 0.045$ kg). Your first throw positioned your ball 2.00 m to the left of the pallina. If your next throw has a speed of $v = 1.00$ m/s and the coefficient of kinetic friction is $\mu_k = 0.20$, what are the final distances of your two balls from the pallina in each of the following cases?

a) You throw your ball from the left, hitting your first ball.

b) You throw your ball from the right, hitting the pallina. (*Hint:* Use the fact that $m \ll M$.)

•**7.88** A bored boy shoots a soft pellet from an air gun at a piece of cheese with mass 0.25 kg that sits, keeping cool for dinner guests, on a block of ice. On one particular shot, his 1.2-g pellet gets stuck in the cheese, causing it to slide 25 cm before coming to a stop. According to the package the gun came in, the muzzle velocity is 65 m/s. What is the coefficient of friction between the cheese and the ice?

•7.89 Some kids are playing a dangerous game with fireworks. They strap several firecrackers to a toy rocket and launch it into the air at an angle of 60° with respect to the ground. At the top of its trajectory, the contraption explodes, and the rocket breaks into two equal pieces. One of the pieces has half the speed that the rocket had before it exploded and travels straight upward with respect to the ground. Determine the speed and direction of the second piece.

•7.90 A soccer ball with mass 0.265 kg is initially at rest and is kicked at an angle of 20.8° with respect to the horizontal. The soccer ball travels a horizontal distance of 52.8 m after it is kicked. What is the impulse received by the soccer ball during the kick? Assume there is no air resistance.

•7.91 Tarzan, King of the Jungle (mass = 70.4 kg), grabs a vine of length 14.5 m hanging from a tree branch. The angle of the vine was 25.9° with respect to the vertical when he grabbed it. At the lowest point of his trajectory, he picks up Jane (mass = 43.4 kg) and continues his swinging motion. What angle relative to the vertical will the vine have when Tarzan and Jane reach the highest point of their trajectory?

•7.92 A bullet with mass 35.5 g is shot horizontally from a gun. The bullet embeds in a 5.90-kg block of wood that is suspended by strings. The combined mass swings upward, gaining a height of 12.85 cm. What was the speed of the bullet as it left the gun? (Air resistance can be ignored here.)

•7.93 A 170-g hockey puck moving in the positive $x$-direction at 30 m/s is struck by a stick at time $t = 2.00$ s and moves in the opposite direction at 25 m/s. If the puck is in contact with the stick for 0.20 s, plot the momentum and the position of the puck, and the force acting on it as a function of time, from 0 to 5 s. Be sure to label the coordinate axes with reasonable numbers.

•7.94 Balls are sitting on a billiards table as shown in the figure. You are playing stripes and your opponent is playing solids. Your plan is to hit your target ball by bouncing the white ball off the table bumper.

a) At what angle relative to the normal does the cue ball need to hit the bumper for a purely elastic collision?

b) As an ever observant player, you determine that in fact collisions between billiard balls and bumpers have a coefficient of restitution of 0.6. What angle do you now choose to make the shot?

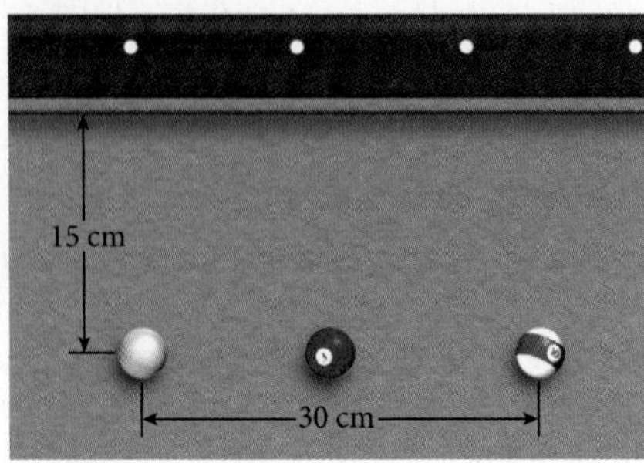


•7.95 You have dropped your cell phone behind a very long bookcase and cannot reach it from either the top or the sides. You decide to get the phone out by elastically colliding a set of keys with it so that both will slide out. If the mass of the cell phone is 0.111 kg, the key ring's mass is 0.020 kg, and each key's mass is 0.023 kg, what is the minimum number of keys you need on the ring so that both the keys and the cell phone will come out on the same side of the bookcase? If your key ring has five keys on it and the velocity is 1.21 m/s when it hits the cell phone, what are the final velocities of the cell phone and key ring? Assume the collision is one-dimensional and elastic, and neglect friction.

•7.96 After several large firecrackers have been inserted into its holes, a bowling ball is projected into the air using a homemade launcher and explodes in midair. During the launch, the 7-kg ball is shot into the air with an initial speed of 10.0 m/s at a 40° angle; it explodes at the peak of its trajectory, breaking into three pieces of equal mass. One piece travels straight up with a speed of 3.0 m/s. Another piece travels straight back with a speed of 2.0 m/s. What is the velocity of the third piece (speed and direction)?

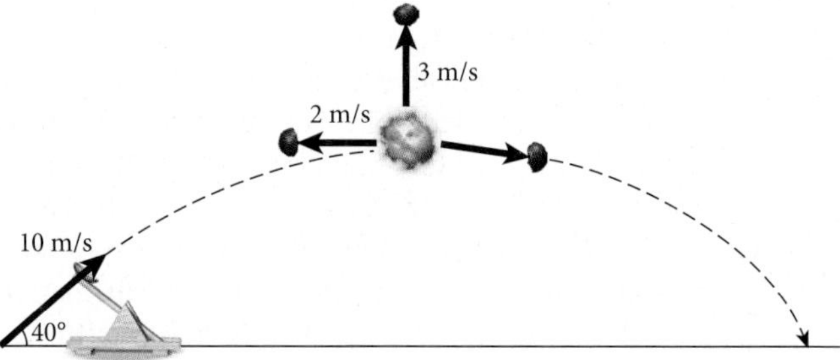


•7.97 In waterskiing, a "garage sale" occurs when a skier loses control and falls and waterskis fly in different directions. In one particular incident, a novice skier was skimming across the surface of the water at 22 m/s when he lost control. One ski, with a mass of 1.5 kg, flew off at an angle of 12° to the left of the initial direction of the skier with a speed of 25 m/s. The other identical ski flew from the crash at an angle of 5° to the right with a speed of 21 m/s. What was the velocity of the 61-kg skier? Give a speed and a direction relative to the initial velocity vector.

•7.98 An uncovered hopper car from a freight train rolls without friction or air resistance along a level track at a constant speed of 6.70 m/s in the positive $x$-direction. The mass of the car is $1.18 \cdot 10^5$ kg.

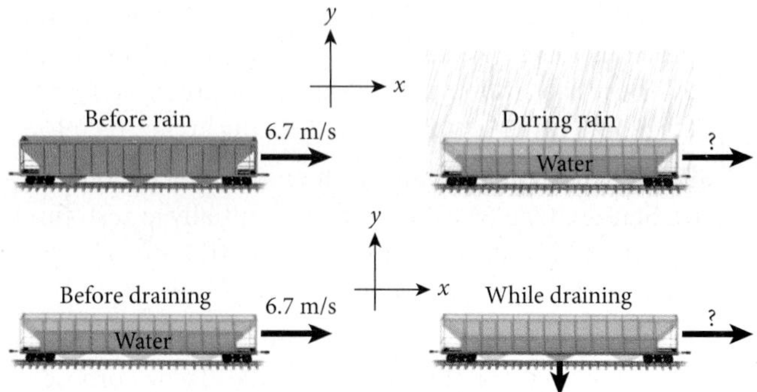

a) As the car rolls, a monsoon rainstorm begins, and the car begins to collect water in its hopper (see the figure). What is the speed of the car after $1.62 \cdot 10^4$ kg of water collects in the car's hopper? Assume that the rain is falling vertically in the negative $y$-direction.

b) The rain stops, and a valve at the bottom of the hopper is opened to release the water. The speed of the car when the valve is opened is again 6.70 m/s in the positive $x$-direction (see the figure).

The water drains out vertically in the negative $y$-direction. What is the speed of the car after all the water has drained out?

•**7.99** When a 99.5-g slice of bread is inserted into a toaster, the toaster's ejection spring is compressed by 7.50 cm. When the toaster ejects the toasted slice, the slice reaches a height 3.0 cm above its starting position. What is the average force that the ejection spring exerts on the toast? What is the time over which the ejection spring pushes on the toast?

•**7.100** A student with a mass of 60 kg jumps straight up in the air by using her legs to apply an average force of 770 N to the ground for 0.25 s. Assume that the initial momentum of the student and the Earth are zero. What is the momentum of the student immediately after this impulse? What is the momentum of the Earth after this impulse? What is the speed of the Earth after the impulse? What fraction of the total kinetic energy that the student produces with her legs goes to the Earth (the mass of the Earth is $5.98 \cdot 10^{24}$ kg)? Using conservation of energy, how high does the student jump?

•**7.101** A potato cannon is used to launch a potato on a frozen lake, as shown in the figure. The mass of the cannon, $m_c$, is 10.0 kg, and the mass of the potato, $m_p$, is 0.850 kg. The cannon's spring (with spring constant $k_c = 7.06 \cdot 10^3$ N/m) is compressed 2.00 m. Prior to launching the potato, the cannon is at rest. The potato leaves the cannon's muzzle moving horizontally to the right at a speed of $v_p = 175$ m/s. Neglect the effects of the potato spinning. Assume there is no friction between the cannon and the lake's ice or between the cannon barrel and the potato.

a) What are the direction and magnitude of the cannon's velocity, $v_c$, after the potato leaves the muzzle?

b) What is the total mechanical energy (potential and kinetic) of the potato/cannon system before and after the firing of the potato?

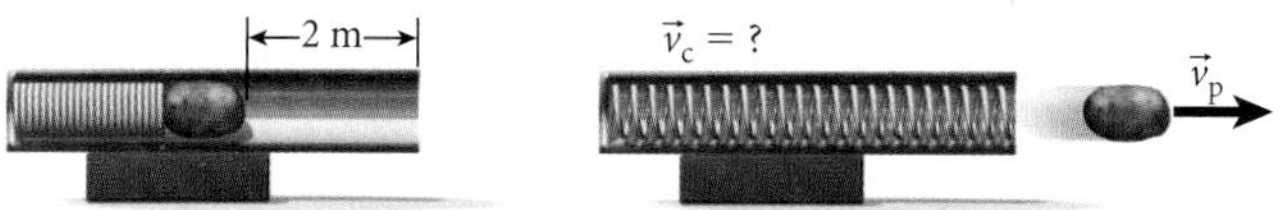


•**7.102** A potato cannon is used to launch a potato on a frozen lake, as in Problem 7.101. All quantities are the same as in that problem, except the potato has a large diameter and is very rough, causing friction between the cannon barrel and the potato. The rough potato leaves the cannon's muzzle moving horizontally to the right at a speed of $v_p = 165$ m/s. Neglect the effects of the potato spinning.

a) What are the direction and magnitude of the cannon's velocity, $v_c$, after the rough potato leaves the muzzle?

b) What is the total mechanical energy (potential and kinetic) of the potato/cannon system before and after the firing of the potato?

c) What is the work, $W_f$, done by the force of friction on the rough potato?

•**7.103** A particle ($M_1 = 1.00$ kg) moving at 30° downward from the horizontal with $v_1 = 2.50$ m/s hits a second particle ($M_2 = 2.00$ kg), which was at rest momentarily. After the collision, the speed of $M_1$ was reduced to 0.50 m/s, and it was moving at an angle of 32° downward with respect to the horizontal. Assume the collision is elastic.

a) What is the speed of $M_2$ after the collision?

b) What is the angle between the velocity vectors of $M_1$ and $M_2$ after the collision?

••**7.104** Many nuclear collisions are truly elastic. If a proton with kinetic energy $E_0$ collides elastically with another proton at rest and travels at an angle of 25° with respect to its initial path, what is its energy after the collision with respect to its original energy? What is the final energy of the proton that was originally at rest?

••**7.105** A method for determining the chemical composition of a material is Rutherford backscattering (RBS), named for the scientist who first discovered that an atom contains a high-density positively charged nucleus, rather than having positive charge distributed uniformly throughout (see Chapter 39). In RBS, alpha particles are shot straight at a target material, and the energy of the alpha particles that bounce directly back is measured. An alpha particle has a mass of $6.65 \cdot 10^{-27}$ kg. An alpha particle having an initial kinetic energy of 2.00 MeV collides elastically with atom X. If the backscattered alpha particle's kinetic energy is 1.59 MeV, what is the mass of atom X? Assume that atom X is initially at rest. You will need to find the square root of an expression, which will result in two possible answers (if $a = b^2$, then $b = \pm\sqrt{a}$). Since you know that atom X is more massive than the alpha particle, you can choose the correct root accordingly. What element is atom X? (Check a periodic table of elements, where atomic mass is listed as the mass in grams of 1 mol of atoms, which is $6.02 \cdot 10^{23}$ atoms.)

PART 2 EXTENDED OBJECTS, MATTER, AND CIRCULAR MOTION

# 8 Systems of Particles and Extended Objects

**FIGURE 8.1** The International Space Station photographed from the Space Shuttle *Discovery*.

## WHAT WE WILL LEARN

- The center of mass is the point at which we can imagine all the mass of an object to be concentrated.
- The position of the combined center of mass of two or more objects is found by taking the sum of their position vectors, weighted by their individual masses.
- The translational motion of the center of mass of an extended object can be described by Newtonian mechanics.
- The center-of-mass momentum is the sum of the linear momentum vectors of the parts of a system. Its time derivative is equal to the total net external force acting on the system, an extended formulation of Newton's Second Law.
- For systems of two particles, working in terms of center-of-mass momentum and relative momentum instead of the individual momentum vectors gives deeper insight into the physics of collisions and recoil phenomena.
- Analyses of rocket motion have to consider systems of varying mass. This variation leads to a logarithmic dependence of the velocity of the rocket on the ratio of initial to final mass.
- It is possible to calculate the location of the center of mass of an extended object by integrating its mass density over its entire volume, weighted by the coordinate vector, and then dividing by the total mass.
- If an object has a plane of symmetry, the center of mass lies in that plane. If the object has more than one symmetry plane, the center of mass lies on the line or point of intersection of the planes.

The International Space Station (ISS), shown in Figure 8.1, is a remarkable engineering achievement. It is scheduled to be completed in 2011, though it has been continuously inhabited since 2000. It orbits Earth at a speed of over 7.5 km/s, in an orbit ranging from 320 to 350 km above Earth's surface. When engineers track the ISS, they treat it as a point particle, even though it measures roughly 109 m by 73 m by 25 m. Presumably this point represents the center of the ISS, but how exactly do engineers determine where the center is?

Every object has a point where all the mass of the object can be considered to be concentrated. Sometimes this point, called the *center of mass,* is not even within the object. This chapter explains how to calculate the location of the center of mass and shows how to use it to simplify calculations involving collisions and conservation of momentum. We have been assuming in earlier chapters that objects could be treated as particles. This chapter shows why that assumption works.

This chapter also discusses changes in momentum for the situation where an object's mass varies as well as its velocity. This occurs with rocket propulsion, where the mass of fuel is often much greater than the mass of the rocket itself.

## 8.1 Center of Mass and Center of Gravity

So far, we have represented the location of an object by coordinates of a single point. However, a statement such as "a car is located at $x = 3.2$ m" surely does not mean that the entire car is located at that point. So, what does it mean to give the coordinate of one particular point to represent an extended object? Answers to this question depend on the particular application. In auto racing, for example, a car's location is represented by the coordinate of the frontmost part of the car. When this point crosses the finish line, the race is decided. On the other hand, in soccer, a goal is only counted if the entire ball has crossed the goal line; in this case, it makes sense to represent the soccer ball's location by the coordinates of the rearmost part of the ball. However, these examples are exceptions. In almost all situations, there is a natural choice of a point to represent the location of an extended object. This point is called the *center of mass.*

### Definition

The **center of mass** is the point at which we can imagine all the mass of an object to be concentrated.

Thus, the center of mass is also the point at which we can imagine the force of gravity acting on the entire object to be concentrated. If we can imagine all of the mass to be concentrated at this point when calculating the force due to gravity, it is legitimate to call this point the *center of gravity*, a term that can often be used interchangeably with *center of mass*. (To be precise, we should note that these two terms are only equivalent in situations where the gravitational force is constant everywhere throughout the object. In Chapter 12, we will see that this is not the case for very large objects.)

It is appropriate to mention here that if an object's mass density is constant, the center of mass (center of gravity) is located in the geometrical center of the object. Thus, for most objects in everyday experience, it is a reasonable first guess that the center of gravity is the middle of the object. The derivations in this chapter will bear out this conjecture.

## Combined Center of Mass for Two Objects

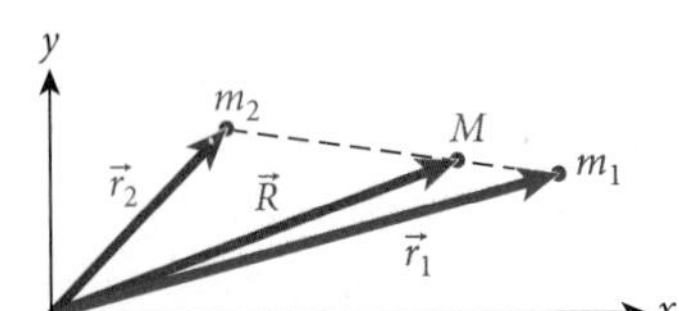


**FIGURE 8.2** Location of the center of mass for a system of two masses $m_1$ and $m_2$, where $M = m_1 + m_2$.

If we have two identical objects of equal mass and want to find the center of mass for the combination of the two, it is reasonable to assume from considerations of symmetry that the combined center of mass of this system lies exactly midway between the individual centers of mass of the two objects. If one of the two objects is more massive, then it is equally reasonable to assume that the center of mass for the combination is closer to that of the more massive one. Thus, we have a general formula for calculating the location of the center of mass, $\vec{R}$, for two masses $m_1$ and $m_2$ located at positions $\vec{r}_1$ and $\vec{r}_2$ to an arbitrary coordinate system (Figure 8.2):

$$\vec{R} = \frac{\vec{r}_1 m_1 + \vec{r}_2 m_2}{m_1 + m_2}. \tag{8.1}$$

This equation says that the center-of-mass position vector is an average of the position vectors of the individual objects, weighted by their mass. Such a definition is consistent with the empirical evidence we have just cited. For now, we will use this equation as an operating definition and gradually work out its consequences. Later in this chapter and in the following chapters, we will see additional reasons why this definition makes sense.

Note that we can immediately write vector equation 8.1 in Cartesian coordinates as follows:

$$X = \frac{x_1 m_1 + x_2 m_2}{m_1 + m_2}, \; Y = \frac{y_1 m_1 + y_2 m_2}{m_1 + m_2}, \; Z = \frac{z_1 m_1 + z_2 m_2}{m_1 + m_2}. \tag{8.2}$$

### 8.1 In-Class Exercise

In the case shown in Figure 8.2, what are the relative magnitudes of the two masses $m_1$ and $m_2$?

a) $m_1 < m_2$

b) $m_1 > m_2$

c) $m_1 = m_2$

d) Based solely on the information given in the figure, it is not possible to decide which of the two masses is larger.

In Figure 8.2, the location of the center of mass lies exactly on the straight (dashed black) line that connects the two masses. Is this a general result—does the center of mass always lie on this line? If yes, why? If no, what is the special condition that is needed for this to be the case? The answer is that this is a general result for all two-body systems: The center of mass of such a system always lies on the connecting line between the two objects. To see this, we can place the origin of the coordinate system at one of the two masses in Figure 8.2, say $m_1$. (As we know, we can always shift the origin of a coordinate system without changing the physics results.) Using equation 8.1, we then see that $\vec{R} = \vec{r}_2 m_2 / (m_1 + m_2)$, because with this choice of coordinate system, we define $\vec{r}_1$ as zero. Thus, the two vectors $\vec{R}$ and $\vec{r}_2$ point in the same direction, but $\vec{R}$ is shorter by a factor of $m_2 / (m_1 + m_2) < 1$. This shows that $\vec{R}$ always lies on the straight line that connects the two masses.

## SOLVED PROBLEM 8.1 Center of Mass of Earth and Moon

The Earth has a mass of $5.97 \cdot 10^{24}$ kg, and the Moon has a mass of $7.36 \cdot 10^{22}$ kg. The Moon orbits the Earth at a distance of 384,000 km; that is, the center of the Moon is a distance of 384,000 km from the center of Earth, as shown in Figure 8.3a.

**PROBLEM**

How far from the center of the Earth is the center of mass of the Earth-Moon system?

## SOLUTION

### THINK

The center of mass of the Earth-Moon system can be calculated by taking the center of the Earth to be located at $x = 0$ and the center of the Moon to be located at $x = 384{,}000$ km. The center of mass of the Earth-Moon system will lie along a line connecting the center of the Earth and the center of the Moon (as in Figure 8.3a).

### SKETCH

A sketch showing Earth and Moon to scale is presented in Figure 8.3b.

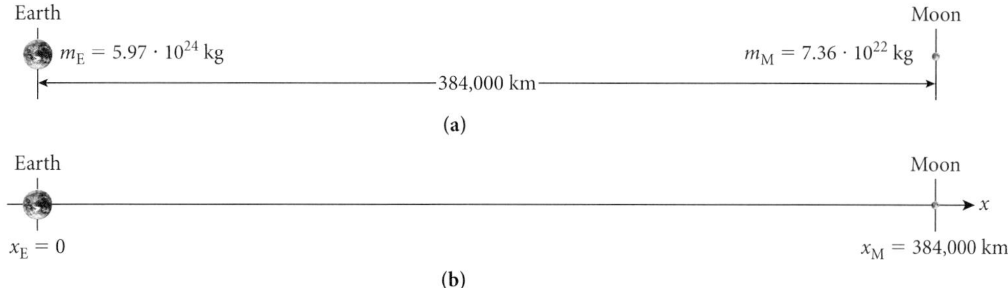


**FIGURE 8.3** (a) The Moon orbits the Earth at a distance of 384,000 km (drawing to scale). (b) A sketch showing the Earth at $x_E = 0$ and the Moon at $x_M = 384{,}000$ km.

### RESEARCH

We define an $x$-axis and place the Earth at $x_E = 0$ and the Moon at $x_M = 384{,}000$ km. We can use equation 8.2 to obtain an expression for the $x$-coordinate of the center of mass of the Earth-Moon system:

$$X = \frac{x_E m_E + x_M m_M}{m_E + m_M}.$$

### SIMPLIFY

Since we have put the origin of our coordinate system at the center of Earth, we set $x_E = 0$. This results in

$$X = \frac{x_M m_M}{m_E + m_M}.$$

### CALCULATE

Inserting the numerical values, we get the $x$-coordinate of the center of mass of the Earth-Moon system:

$$X = \frac{x_M m_M}{m_E + m_M} = \frac{(384{,}000 \text{ km})(7.36 \cdot 10^{22} \text{ kg})}{5.97 \cdot 10^{24} \text{ kg} + 7.36 \cdot 10^{22} \text{ kg}} = 4676.418 \text{ km}.$$

### ROUND

All of the numerical values were given to three significant figures, so we report our result as

$$X = 4680 \text{ km}.$$

### DOUBLE-CHECK

Our result is in kilometers, which is the correct unit for a position. The center of mass of the Earth-Moon system is close to the center of the Earth. This distance is small compared to the distance between the Earth and the Moon, which makes sense because the mass of the Earth is much larger than the mass of the Moon. In fact, this distance is less than the radius of the Earth, $R_E = 6370$ km. The Earth and the Moon actually each orbit the common center of mass. Thus, the Earth seems to wobble as the Moon orbits it.

## Combined Center of Mass for Several Objects

The definition of the center of mass in equation 8.1 can be generalized to a total of $n$ objects with different masses, $m_i$, located at different positions, $\vec{r}_i$. In this general case,

$$\vec{R} = \frac{\vec{r}_1 m_1 + \vec{r}_2 m_2 + \cdots + \vec{r}_n m_n}{m_1 + m_2 + \cdots + m_n} = \frac{\sum_{i=1}^{n} \vec{r}_i m_i}{\sum_{i=1}^{n} m_i} = \frac{1}{M}\sum_{i=1}^{n} \vec{r}_i m_i, \tag{8.3}$$

where $M$ represents the combined mass of all $n$ objects:

$$M = \sum_{i=1}^{n} m_i. \tag{8.4}$$

Writing equation 8.3 in Cartesian components, we obtain

$$X = \frac{1}{M}\sum_{i=1}^{n} x_i m_i;\ Y = \frac{1}{M}\sum_{i=1}^{n} y_i m_i;\ Z = \frac{1}{M}\sum_{i=1}^{n} z_i m_i. \tag{8.5}$$

The location of the center of mass is a fixed point relative to the object or system of objects and does not depend on the location of the coordinate system used to describe it. We can show this by taking the system of equation 8.3 and moving it by $\vec{r}_0$, resulting in a new center-of-mass position, $\vec{R} + \vec{R}_0$. Using equation 8.3, we find

$$\vec{R} + \vec{R}_0 = \frac{\sum_{i=1}^{n} (\vec{r}_0 + \vec{r}_i) m_i}{\sum_{i=1}^{n} m_i} = \vec{r}_0 + \frac{1}{M}\sum_{i=1}^{n} \vec{r}_i m_i.$$

Thus, $\vec{R}_0 = \vec{r}_0$, and the location of the center of mass does not change relative to the system.

Now we can determine the center of mass of a collection of objects in the following example.

### EXAMPLE 8.1 Shipping Containers

Large freight containers, which can be transported by truck, railroad, or ship, come in standard sizes. One of the most common sizes is the ISO 20' container, which has a length of 6.1 m, a width of 2.4 m, and a height of 2.6 m. This container is allowed to have a mass (including its contents, of course) of up to 30,400 kg.

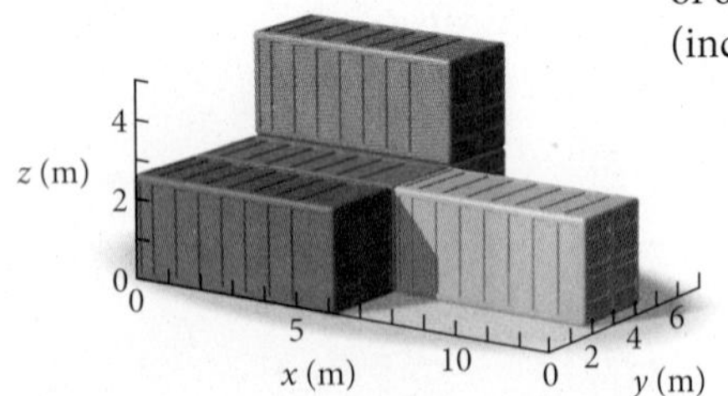


**FIGURE 8.4** Freight containers arranged on the deck of a container ship.

**PROBLEM**

The four freight containers shown in Figure 8.4 sit on the deck of a container ship. Each one has a mass of 9,000 kg, except for the red one, which has a mass of 18,000 kg. Assume that each of the containers has an individual center of mass at its geometric center. What are the $x$-coordinate and the $y$-coordinate of the containers' combined center of mass? Use the coordinate system shown in the figure to describe the location of this center of mass.

**SOLUTION**

We need to calculate the individual Cartesian components of the center of mass, so we'll use equation 8.5. There does not seem to be a shortcut we can utilize.

Let's call the length of each container $\ell$ (6.1 m), the width of each container $w$ (2.4 m), and the mass of the green container $m_0$ (9,000 kg). The mass of the red container is then $2m_0$, and all the others also have a mass of $m_0$.

First, we need to calculate the combined mass, $M$. According to equation 8.4, it is

$$\begin{aligned} M &= m_{\text{red}} + m_{\text{green}} + m_{\text{orange}} + m_{\text{blue}} + m_{\text{purple}} \\ &= 2m_0 + m_0 + m_0 + m_0 + m_0 \\ &= 6m_0. \end{aligned}$$

For the $x$-coordinate of the combined center of mass, we find

$$\begin{aligned} X &= \frac{x_{\text{red}} m_{\text{red}} + x_{\text{green}} m_{\text{green}} + x_{\text{orange}} m_{\text{orange}} + x_{\text{blue}} m_{\text{blue}} + x_{\text{purple}} m_{\text{purple}}}{M} \\ &= \frac{\frac{1}{2}\ell 2m_0 + \frac{1}{2}\ell m_0 + \frac{3}{2}\ell m_0 + \frac{1}{2}\ell m_0 + \frac{1}{2}\ell m_0}{6m_0} \\ &= \frac{\ell(1 + \frac{1}{2} + \frac{3}{2} + \frac{1}{2} + \frac{1}{2})}{6} \\ &= \tfrac{2}{3}\ell = 4.1 \text{ m}. \end{aligned}$$

In the last step, we substituted the value of 6.1 m for $\ell$.

In the same way, we can calculate the $y$-coordinate:

$$\begin{aligned} Y &= \frac{y_{\text{red}} m_{\text{red}} + y_{\text{green}} m_{\text{green}} + y_{\text{orange}} m_{\text{orange}} + y_{\text{blue}} m_{\text{blue}} + y_{\text{purple}} m_{\text{purple}}}{M} \\ &= \frac{\frac{1}{2}w 2m_0 + \frac{3}{2}w m_0 + \frac{3}{2}w m_0 + \frac{5}{2}w m_0 + \frac{5}{2}w m_0}{6m_0} \\ &= \frac{w(1 + \frac{3}{2} + \frac{3}{2} + \frac{5}{2} + \frac{5}{2})}{6} \\ &= \tfrac{3}{2}w = 3.6 \text{ m}. \end{aligned}$$

Here again we substituted the numerical value of 2.4 m in the last step. (Note that we rounded both center-of-mass coordinates to two significant figures to be consistent with the given values.)

### 8.1 Self-Test Opportunity

Determine the $z$-coordinate of the center of mass of the container arrangement in Figure 8.4.

## 8.2 Center-of-Mass Momentum

Now we can take the time derivative of the position vector of the center of mass to get $\vec{V}$, the velocity vector of the center of mass. We take the time derivative of equation 8.3:

$$\vec{V} \equiv \frac{d}{dt}\vec{R} = \frac{d}{dt}\left(\frac{1}{M}\sum_{i=1}^{n}\vec{r}_i m_i\right) = \frac{1}{M}\sum_{i=1}^{n} m_i \frac{d}{dt}\vec{r}_i = \frac{1}{M}\sum_{i=1}^{n} m_i \vec{v}_i = \frac{1}{M}\sum_{i=1}^{n}\vec{p}_i. \tag{8.6}$$

For now, we have assumed that the total mass, $M$, and the masses, $m_i$, of the individual objects remain constant. (Later in this chapter, we will give up this assumption and study the consequences for rocket motion.) Equation 8.6 is an expression for the velocity vector of the center of mass, $\vec{V}$. Multiplication of both sides of equation 8.6 by $M$ yields

$$\vec{P} = M\vec{V} = \sum_{i=1}^{n}\vec{p}_i. \tag{8.7}$$

We thus find that the center-of-mass momentum, $\vec{P}$, is the product of the total mass, $M$, and the center-of-mass velocity, $\vec{V}$, and is the sum of all the individual momentum vectors.

Taking the time derivative of both sides of equation 8.7 yields Newton's Second Law for the center of mass:

$$\frac{d}{dt}\vec{P} = \frac{d}{dt}(M\vec{V}) = \frac{d}{dt}\left(\sum_{i=1}^{n}\vec{p}_i\right) = \sum_{i=1}^{n}\frac{d}{dt}\vec{p}_i = \sum_{i=1}^{n}\vec{F}_i. \tag{8.8}$$

In the last step, we used the result from Chapter 7 that the time derivative of the momentum of particle $i$ is equal to the net force, $\vec{F}_i$, acting on it. Note that if the particles (objects) in a system exert forces on one another, those forces do not make a net contribution to the sum of these forces in equation 8.8. Why? According to Newton's Third Law, the forces that two objects exert on each other are equal in magnitude and opposite in direction. Therefore, adding them yields zero. Thus, we obtain Newton's Second Law for the center of mass:

$$\frac{d}{dt}\vec{P} = \vec{F}_{\text{net}}, \tag{8.9}$$

where $\vec{F}_{\text{net}}$ is the sum of all *external* forces acting on the system of particles.

The center of mass has the same relationships among position, velocity, momentum, force, and mass that have been established for point particles. It is thus possible to consider the center of mass of an extended object or a group of objects as a point particle. This conclusion justifies the approximation we used in earlier chapters that objects can be represented as points.

## Two-Body Collisions

One of the most interesting applications of center of mass arises with a frame of reference whose origin is placed at the center of mass of a system of interacting objects. Let's investigate the simplest example of this situation. Consider a system consisting of only two objects. In this case, the total momentum—the sum of individual momenta according to equation 8.7—is

$$\vec{P} = \vec{p}_1 + \vec{p}_2. \tag{8.10}$$

In Chapter 7, we saw that the relative velocity between two colliding objects plays a big role in two-body collisions. Thus, it is natural to define the relative momentum as half of the momentum difference:

$$\vec{p} = \tfrac{1}{2}(\vec{p}_1 - \vec{p}_2). \tag{8.11}$$

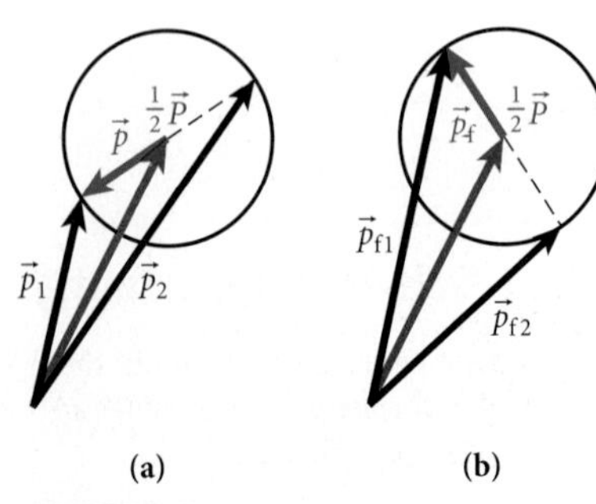


**FIGURE 8.5** Relationship between momentum vectors 1 and 2 (black), center-of-mass momentum (red), and relative momentum (blue) in some reference frame: (a) before an elastic collision; (b) after the elastic collision.

Why is the factor $\frac{1}{2}$ appropriate in this definition? The answer is that in a center-of-momentum reference frame—a frame in which the center of mass has zero momentum—the momentum of object 1 is $\vec{p}$ and that of object 2 is $-\vec{p}$. Let's see how this comes about.

Figure 8.5a illustrates the relationship between the center-of-mass momentum $\vec{P}$ (red arrow), the relative momentum $\vec{p}$ (blue arrow), and the momenta of objects 1 and 2 (black arrows). We can express the individual momenta in terms of the center-of-mass momentum and the relative momentum:

$$\begin{aligned}\vec{p}_1 &= \tfrac{1}{2}\vec{P} + \vec{p}\\ \vec{p}_2 &= \tfrac{1}{2}\vec{P} - \vec{p}.\end{aligned} \tag{8.12}$$

The biggest advantage of thinking in terms of center-of-mass momentum and relative momentum becomes clearer when we consider a collision between the two objects. During a collision, the dominant forces that act on the objects are the forces they exert on each other. These internal forces do not enter into the sum of forces in equation 8.8, so we obtain, for the collision of two objects:

$$\frac{d}{dt}\vec{P} = 0.$$

In other words, the center-of-mass momentum does not change; it remains the same during a two-body collision. This is true for elastic or totally inelastic or partially inelastic collisions.

For an inelastic collision, where the two objects stick together after colliding, we found in Chapter 7 that the velocity with which the combined mass moves is

$$\vec{v}_{\text{f}} = \frac{m_1\vec{v}_{\text{i}1} + m_2\vec{v}_{\text{i}2}}{m_1 + m_2}.$$

If we compare this equation with equation 8.6, we see that this velocity is just the center-of-mass velocity. In other words, in the case of a totally inelastic collision, the relative momentum after the collision is zero.

For elastic collisions, the total kinetic energy has to be conserved. If we compute the total kinetic energy in terms of the total momentum, $\vec{P}$, and the relative momentum, $\vec{p}$, the contribution from the total momentum has to remain constant, because $\vec{P}$ is constant. This finding implies that the kinetic energy contained in the relative motion also has to remain constant. Because this kinetic energy, in turn, is proportional to the square of the relative momentum vector, the length of the relative momentum vector has to remain unchanged during an elastic collision. Only the direction of this vector can change. As shown in Figure 8.5b, the new relative momentum vector after the elastic collision lies on the circumference of a circle, whose radius is equal to the length of the initial relative momentum vector and whose center is at the end point of the $\frac{1}{2}\vec{p}$. The situation depicted in Figure 8.5 implies that the motion is restricted to two spatial dimensions. For two-body collisions in three dimensions, the final relative momentum vector is located on the surface of a sphere instead of the perimeter of a circle.

In Figure 8.5, the momentum vectors for particle 1 and particle 2 are plotted in some arbitrary reference frame before and after the collision between the two particles. We can plot the same vectors in a frame that moves with the center of mass. (We have just shown that the center-of-mass momentum does not change in the collision!) The center-of-mass velocity in such a moving frame is zero, and, consequently, $\vec{P}=0$ in that frame. From equation 8.12, we can then see that in this case the initial momentum vectors of the two particles are $\vec{p}_1=+\vec{p}$ and $\vec{p}_2=-\vec{p}$. In the moving center-of-mass frame, the two-particle collision simply results in a rotation of the relative momentum vector about the origin, as shown in Figure 8.6, which automatically ensures that the conservation laws of momentum and kinetic energy (because this is an elastic collision!) are obeyed.

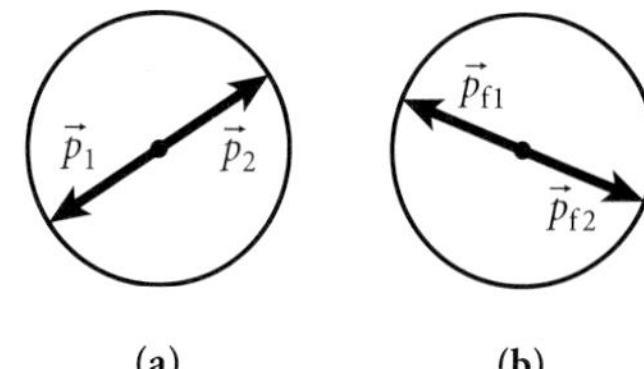

**FIGURE 8.6** Same collision as in Figure 8.5 but displayed in the center-of-mass frame.

## Recoil

When a bullet is fired from a gun, the gun **recoils;** that is, it moves in the direction opposite to that in which the bullet is fired. Another demonstration of the same physical principle occurs if you are sitting in a boat that is at rest and you throw an object off the boat: The boat moves in the direction opposite from that of the object. You also experience the same effect if you stand on a skateboard and toss a (reasonably heavy) ball. This well-known recoil effect can be understood using the framework we have just developed for two-body collisions. It is also a consequence of Newton's Third Law.

## SOLVED PROBLEM 8.2 Cannon Recoil

Suppose a cannonball of mass 13.7 kg is fired at a target that is 2.30 km away from the cannon, which has a mass of 249.0 kg. The distance 2.30 km is also the maximum range of the cannon. The target and cannon are at the same elevation, and the cannon is resting on a horizontal surface.

### PROBLEM

What is the velocity with which the cannon will recoil?

### SOLUTION

#### THINK

First, we realize that the cannon can recoil only in the horizontal direction, because the normal force exerted by the ground will prevent it from acquiring a downward velocity component. We use the fact that the $x$-component of the center-of-mass momentum of the system (cannon and cannonball) remains unchanged in the process of firing the cannon, because the explosion of the gunpowder inside the cannon, which sets the cannonball in motion, creates only forces internal to the system. No net external force component occurs in the horizontal direction because the two external forces (normal force and gravity) are both vertical. The $y$-component of the center-of-mass velocity changes because a net external force component does occur in the $y$-direction when the normal force increases to prevent the cannon from penetrating the ground. Because the cannonball and cannon are both initially at rest, the center-of-mass momentum of this system is initially zero, and its $x$-component remains zero after the firing of the cannon.

*Continued—*

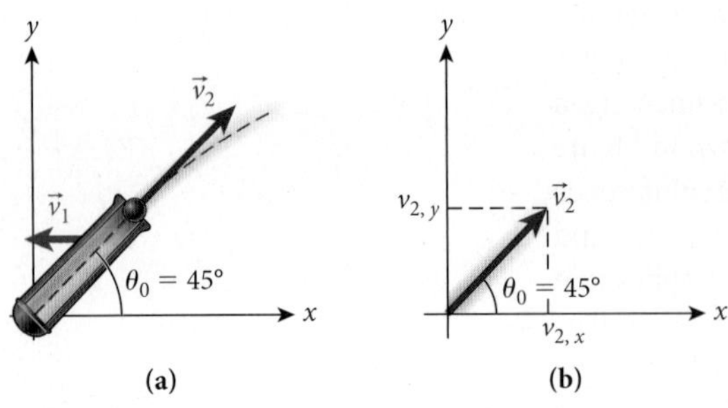


**FIGURE 8.7** (a) Cannonball being fired from a cannon. (b) The initial velocity vector of the cannonball.

### SKETCH

Figure 8.7a is a sketch of the cannon just as the cannonball is fired. Figure 8.7b shows the velocity vector of the cannonball, $\vec{v}_2$, including the $x$- and $y$-components.

### RESEARCH

Using equation 8.10 with index 1 for the cannon and index 2 for the cannonball, we obtain

$$\vec{P} = \vec{p}_1 + \vec{p}_2 = m_1\vec{v}_1 + m_2\vec{v}_2 = 0 \Rightarrow \vec{v}_1 = -\frac{m_2}{m_1}\vec{v}_2.$$

For the horizontal component of the velocity, we then have

$$v_{1,x} = -\frac{m_2}{m_1}v_{2,x}. \qquad \text{(i)}$$

We can obtain the horizontal component of the cannonball's initial velocity (at firing) from the fact that the range of the cannon is 2.30 km. In Chapter 3, we saw that the range of the cannon is related to the initial velocity via $R = (v_0^2/g)(\sin 2\theta_0)$. The maximum range is reached for $\theta_0 = 45°$ and is $R = v_0^2/g \Rightarrow v_0 = \sqrt{gR}$. For $\theta_0 = 45°$, the initial speed and horizontal velocity component are related via $v_{2,x} = v_0 \cos 45° = v_0/\sqrt{2}$. Combining these two results, we can relate the maximum range to the horizontal component of the initial velocity of the cannonball:

$$v_{2,x} = \frac{v_0}{\sqrt{2}} = \sqrt{\frac{gR}{2}}. \qquad \text{(ii)}$$

### SIMPLIFY

Substituting from equation (ii) into equation (i) gives us the result we are looking for:

$$v_{1,x} = -\frac{m_2}{m_1}v_{2,x} = -\frac{m_2}{m_1}\sqrt{\frac{gR}{2}}.$$

### CALCULATE

Inserting the numbers given in the problem statement, we obtain

$$v_{1,x} = -\frac{m_2}{m_1}\sqrt{\frac{gR}{2}} = -\frac{13.7 \text{ kg}}{249 \text{ kg}}\sqrt{\frac{(9.81 \text{ m/s}^2)(2.30 \cdot 10^3 \text{ m})}{2}} = -5.84392 \text{ m/s}.$$

### ROUND

Expressing our answer to three significant figures gives

$$v_{1,x} = -5.84 \text{ m/s}.$$

### DOUBLE-CHECK

The minus sign means that the cannon moves in the opposite direction to that of the cannonball, which is reasonable. The cannonball should have a much larger initial velocity than the cannon because the cannon is much more massive. The initial velocity of the cannonball was

$$v_0 = \sqrt{gR} = \sqrt{\left(9.81 \text{ m/s}^2\right)\left(2.3 \cdot 10^3 \text{ m}\right)} = 150 \text{ m/s}.$$

The fact that our answer for the velocity of the cannon is much less than the initial velocity of the cannonball also seems reasonable.

Mass can be ejected continuously from a system, producing a continuous recoil. As an example, let's consider the spraying of water from a fire hose.

### EXAMPLE 8.2 Fire Hose

**PROBLEM**

What is the magnitude of the force, $F$, that acts on a firefighter holding a fire hose that ejects 360 L of water per minute with a muzzle speed of $v = 39.0$ m/s, as shown in Figure 8.8?

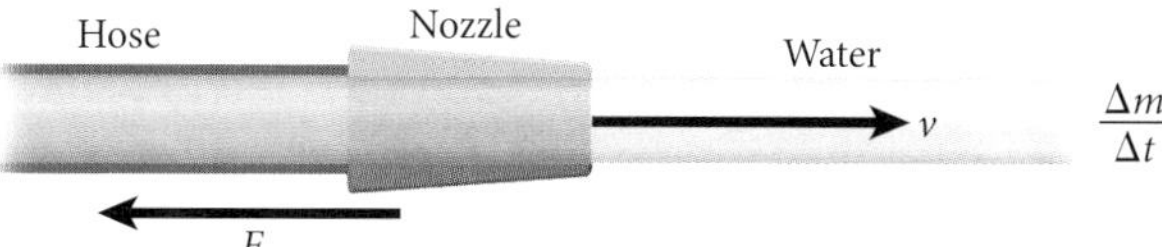


**FIGURE 8.8** A fire hose with water leaving at speed $v$.

**SOLUTION**

Let's first find the total mass of the water that is being ejected per minute. The mass density of water is $\rho = 1000 \text{ kg/m}^3 = 1.0 \text{ kg/L}$. Because $\Delta V = 360$ L, we get for the total mass of water ejected in a minute:

$$\Delta m = \Delta V \rho = (360 \text{ L})(1.0 \text{ kg/L}) = 360 \text{ kg}.$$

The momentum of the water is then $\Delta p = v\Delta m$, and, from the definition of the average force, $F = \Delta p/\Delta t$, we have:

$$F = \frac{v\Delta m}{\Delta t} = \frac{(39.0 \text{ m/s})(360 \text{ kg})}{60 \text{ s}} = 234 \text{ N}.$$

This force is sizable, which is why it is so dangerous for firefighters to let go of operating fire hoses: They would whip around, potentially causing injury.

### 8.2 In-Class Exercise

A garden hose is used to fill a 20-L bucket in 1 min. The velocity of the water leaving the hose is 1.05 m/s. What force is required to hold the hose in place?

a) 0.35 N

b) 2.1 N

c) 9.8 N

d) 12 N

e) 21 N

## General Motion of the Center of Mass

Extended solid objects can have motions that appear, at first sight, rather complicated. One example of such motion is high jumping. During the 1968 Olympic Games in Mexico City, the American track-and-field star Dick Fosbury won a gold medal using a new high-jump technique, which became known as the Fosbury flop (see Figure 8.9). Properly executed, the technique allows the athlete to cross over the bar while his or her center of mass remains below it, thus adding effective height to the jump.

**FIGURE 8.9** Dick Fosbury clears the high-jump bar during the finals of the Olympic Games in Mexico City on October 20, 1968.

Figure 8.10a shows a wrench twirling through the air, in a multiple-exposure series of images with equal time intervals between sequential frames. While this motion looks complicated, we can use what we know about the center of mass to perform a straightforward analysis of this motion. If we assume that all the mass of the wrench is concentrated at a point, then this point will move on a parabola through the air under the influence of gravity, as discussed in Chapter 3. Superimposed on this motion is a rotation of the wrench

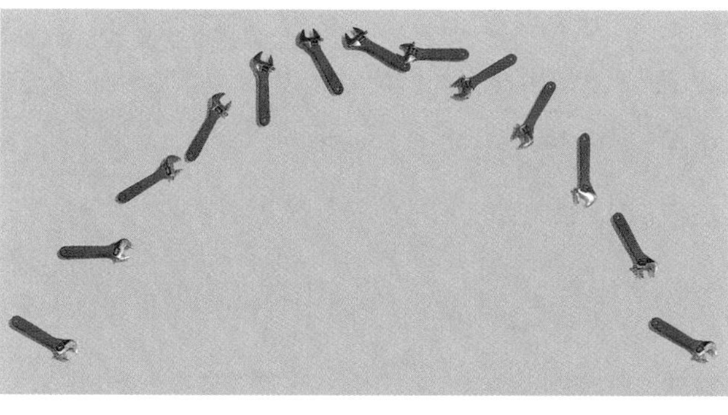

(a)

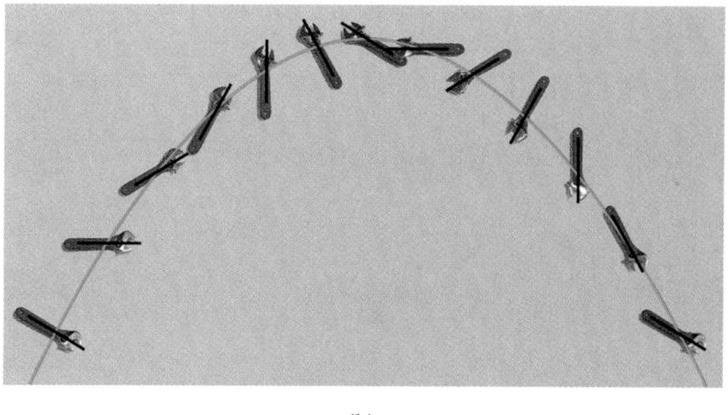

(b)

**FIGURE 8.10** (a) Digitally processed multiple-exposure series of images of a wrench tossed through the air. (b) Same series as in part (a), but with a parabola for the center-of-mass motion superimposed.

about its center of mass. You can see this parabolic trajectory clearly in Figure 8.10b, where a superimposed parabola (green) passes through the location of the center of mass of the wrench in each exposure. In addition, a superimposed black line rotates with a constant rate about the center of mass of the wrench. You can clearly see that the handle of the wrench is always aligned with the black line, indicating that the wrench rotates with constant rate about its center of mass (we will analyze such rotational motion in Chapter 10).

The techniques introduced here allow us to analyze many kinds of complicated problems involving moving solid objects in terms of superposition of the motion of the center of mass and a rotation of the object about the center of mass.

## 8.3 Rocket Motion

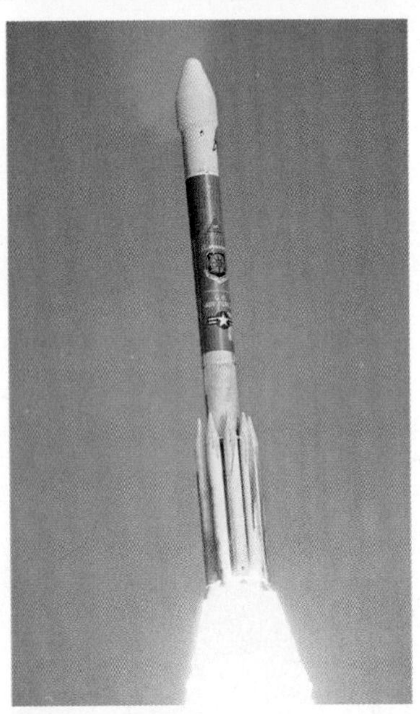

**FIGURE 8.11** A Delta II rocket lifting a GPS satellite into orbit.

Example 8.2 about the fire hose is the first situation we have examined that involves a change in momentum due to a change in mass rather than in velocity. Another important situation in which changing momentum is due to changing mass is rocket motion, where part of the mass of the rocket is ejected through a nozzle or nozzles at the rear (Figure 8.11). Rocket motion is an important case of the recoil effect discussed in Section 8.2. A rocket does not "push against" anything. Instead, its forward thrust is gained from ejecting its propellant from its rear, according to the law of conservation of total momentum.

In order to obtain an expression for the acceleration of a rocket, we'll first consider ejecting discrete amounts of mass out of the rocket. Then we can approach the continuum limit. Let's use a toy model of a rocket that moves in interstellar space, propelling itself forward by shooting cannonballs out its back end (Figure 8.12). (We specify that the rocket is in interstellar space so we can treat it and its components as an isolated system, for which we can neglect outside forces.) Initially, the rocket is at rest. All motion is in the $x$-direction, so we can use notation for one-dimensional motion, with the signs of the $x$-components of the velocities (which, for simplicity, we will refer to as velocities) indicating their direction. Each cannonball has a mass of $\Delta m$, and the initial mass of the rocket, including all cannonballs, is $m_0$. Each cannonball is fired with a velocity of $v_c$ relative to the rocket, resulting in a cannonball momentum of $v_c \Delta m$.

After the first cannonball is fired, the mass of the rocket is reduced to $m_0 - \Delta m$. Firing the cannonball does not change the center-of-mass momentum of the system (rocket plus cannonball). (Remember, this is an isolated system, on which no net external forces act.) Thus, the rocket receives a recoil momentum opposite to that of the cannonball. The momentum of the cannonball is

$$p_c = v_c \Delta m,$$

and the momentum of the rocket is

$$p_r = \left(m_0 - \Delta m\right) v_1,$$

where $v_1$ is the velocity of the rocket after the cannonball is fired. Because momentum is conserved, we can write $p_r + p_c = 0$, and then substitute $p_r$ and $p_c$ from the preceding two expressions:

$$\left(m_0 - \Delta m\right) v_1 + v_c \Delta m = 0.$$

We define the change in the velocity, $\Delta v_1$, of the rocket after firing one cannonball as

$$v_1 = v_0 + \Delta v = 0 + \Delta v = \Delta v_1,$$

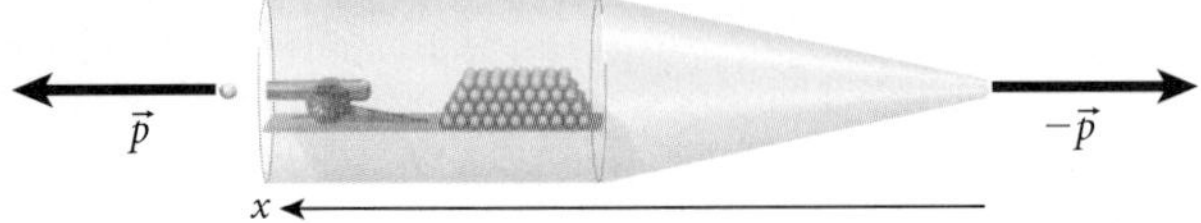


**FIGURE 8.12** Toy model for rocket propulsion: firing cannonballs.

where the assumption that the rocket was initially at rest means $v_0 = 0$. This gives us the recoil velocity of the rocket due to the firing of one cannonball:

$$\Delta v_1 = -\frac{v_c \Delta m}{m_0 - \Delta m}.$$

In the moving system of the rocket, we can then fire the second cannonball. Firing the second cannonball reduces the mass of the rocket from $m_0 - \Delta m$ to $m_0 - 2\Delta m$, which results in an additional recoil velocity of

$$\Delta v_2 = -\frac{v_c \Delta m}{m_0 - 2\Delta m}.$$

The total velocity of the rocket then increases to $v_2 = v_1 + \Delta v_2$. After firing the $n$th cannonball, the velocity change is:

$$\Delta v_n = -\frac{v_c \Delta m}{m_0 - n\Delta m}. \tag{8.13}$$

Thus, the velocity of the rocket after firing the $n$th cannonball is:

$$v_n = v_{n-1} + \Delta v_n.$$

This kind of equation, which defined the $n$th term of a sequence where each term is expressed as a function of the preceding terms, is called a *recursion relation.* It can be solved in a straightforward manner by using a computer. However, we can use a very helpful approximation for the case where the mass emitted per unit time is constant and small compared to $m$, the overall (time-dependent) mass of the rocket. In this limit, we obtain from equation 8.13

$$\Delta v = -\frac{v_c \Delta m}{m} \Rightarrow \frac{\Delta v}{\Delta m} = -\frac{v_c}{m}. \tag{8.14}$$

Here $v_c$ is the velocity with which the cannonball is ejected. In the limit $\Delta m \to 0$, we then obtain the derivative

$$\frac{dv}{dm} = -\frac{v_c}{m}. \tag{8.15}$$

The solution of this differential equation is

$$v(m) = -v_c \int_{m_0}^{m} \frac{1}{m'} dm' = -v_c \ln m \Big|_{m_0}^{m} = v_c \ln\left(\frac{m_0}{m}\right). \tag{8.16}$$

(You can verify that equation 8.16 is indeed the solution of equation 8.15 by taking the derivative of equation 8.16 with respect to $m$.)

If $m_i$ is the initial value for the total mass at some time $t_i$ and $m_f$ is the final mass at a later time, we can use equation 8.16 to obtain $v_i = v_c \ln(m_0/m_i)$ and $v_f = v_c \ln(m_0/m_f)$ for the initial and final velocities of the rocket. Then, using the property of logarithms, $\ln(a/b) = \ln a - \ln b$, we find the difference in those two velocities:

$$v_f - v_i = v_c \ln\left(\frac{m_0}{m_f}\right) - v_c \ln\left(\frac{m_0}{m_i}\right) = v_c \ln\left(\frac{m_i}{m_f}\right). \tag{8.17}$$

### EXAMPLE 8.3 Rocket Launch to Mars

One proposed scheme for sending astronauts to Mars involves assembling a spaceship in orbit around Earth, thus avoiding the need for the spaceship to overcome most of Earth's gravity at the start. Suppose such a spaceship has a payload of 50,000 kg, carries 2,000,000 kg of fuel, and is able to eject the propellant with a velocity of 23.5 km/s. (Current chemical rocket propellants yield a maximum velocity of approximately 5 km/s, but electromagnetic rocket propulsion is predicted to yield a velocity of perhaps 40 km/s.)

*Continued—*

**PROBLEM**

What is the final velocity that this spaceship can reach, relative to the velocity it initially had in its orbit around Earth?

**SOLUTION**

Using equation 8.17 and substituting the numbers given in this problem, we find

$$v_f - v_i = v_c \ln\left(\frac{m_i}{m_f}\right) = (23.5 \text{ km/s})\ln\left(\frac{2{,}050{,}000 \text{ kg}}{50{,}000 \text{ kg}}\right) = (23.5 \text{ km/s})(\ln 41) = 87.3 \text{ km/s}.$$

For comparison, the Saturn V multistage rocket that carried astronauts to the Moon in the late 1960s and early 1970s was able to reach a speed of only about 12 km/s.

However, even with advanced technology such as electromagnetic propulsion, it would still take several months for astronauts to reach Mars, even under the most favorable conditions. The *Mars Rover,* for example, took 207 days to travel from Earth to Mars. NASA estimates that astronauts on such a mission would receive approximately 10 to 20 times more radiation than the maximally allowable annual dose for radiation workers, leading to high probabilities of developing cancer and brain damage. No shielding mechanism has yet been proposed that could protect the astronauts from this danger.

Another and perhaps easier way to think of rocket motion is to go back to the definition of momentum as the product of mass and velocity and take the time derivative to obtain the force. However, now the mass of the object can change as well:

$$\vec{F}_{net} = \frac{d}{dt}\vec{p} = \frac{d}{dt}(m\vec{v}) = m\frac{d\vec{v}}{dt} + \vec{v}\frac{dm}{dt}.$$

(The last step in this equation represents the application of the product rule of differentiation from calculus.) If no external force is acting on an object ($\vec{F}_{net} = 0$), then we obtain

$$m\frac{d\vec{v}}{dt} = -\vec{v}\frac{dm}{dt}.$$

In the case of rocket motion (as illustrated in Figure 8.13), the outflow of propellant, *dm/dt,* is constant and creates the change in mass of the rocket. The propellant moves with a constant velocity, $\vec{v}_c$, relative to the rocket, so we obtain

$$m\frac{d\vec{v}}{dt} = m\vec{a} = -\vec{v}_c\frac{dm}{dt}.$$

The combination $v_c(dm/dt)$ is called the **thrust** of the rocket. It is a force and thus is measured in newtons:

$$\vec{F}_{thrust} = -\vec{v}_c\frac{dm}{dt}. \tag{8.18}$$

The thrust generated by space shuttle rocket engines and boosters is approximately 31.3 MN (31.3 meganewtons, or approximately 7.8 million pounds). The initial total mass of the Space Shuttle, including payload, fuel tanks, and rocket fuel, is slightly greater than 2.0 million kg; thus, the shuttle's rocket engines and boosters can produce an initial acceleration of

$$a = \frac{\vec{F}_{net}}{m} = \frac{3.13 \cdot 10^7 \text{ N}}{2.0 \cdot 10^6 \text{ kg}} = 16 \text{ m/s}^2.$$

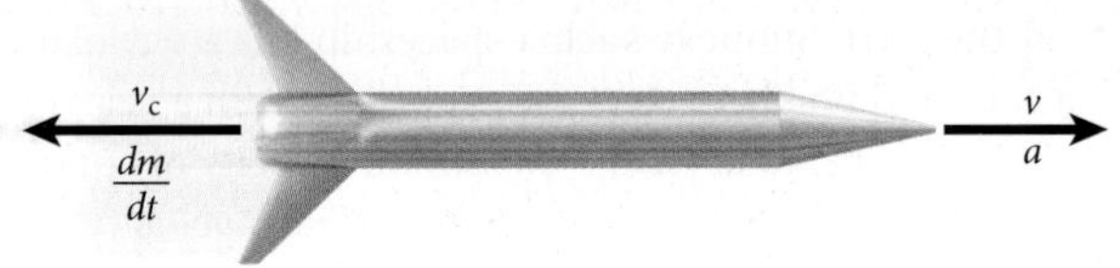


**FIGURE 8.13** Rocket motion.

This acceleration is sufficient to lift the shuttle off the launch pad against the acceleration of gravity ($-9.81 \text{ m/s}^2$). Once the shuttle rises and its mass decreases, it can generate a larger acceleration. As the fuel is expended, the main engines are throttled back to make sure that the acceleration does not exceed $3g$ (three times gravitational acceleration) in order to avoid damaging the cargo or injuring the astronauts.

## 8.4 Calculating the Center of Mass

So far, we have not addressed a key question: How do we calculate the location of the center of mass for an arbitrarily shaped object? To answer this question, let's find the location of the center of mass of the hammer shown in Figure 8.14. To do this, we can represent the hammer by small identical-sized cubes, as shown in the lower part of the figure. The centers of the cubes are their individual centers of mass, marked with red dots. The red arrows are the position vectors of the cubes. If we accept the collection of cubes as a good approximation for the hammer, we can use equation 8.3 to find the center of mass of the collection of cubes and thus that of the hammer.

**FIGURE 8.14** Calculating the center of mass for a hammer.

Note that not all the cubes have the same mass, because the densities of the wooden handle and the iron head are very different. The relationship between mass density ($\rho$), mass, and volume is given by

$$\rho = \frac{dm}{dV}. \tag{8.19}$$

If the mass density is uniform throughout an object, we simply have

$$\rho = \frac{M}{V} \quad \text{(for constant } \rho\text{)}. \tag{8.20}$$

We can then use the mass density and rewrite equation 8.3:

$$\vec{R} = \frac{1}{M}\sum_{i=1}^{n} \vec{r}_i m_i = \frac{1}{M}\sum_{i=1}^{n} \vec{r}_i \rho(\vec{r}_i) V.$$

Here we have assumed that the mass density of each small cube is uniform (but still possibly different from one cube to another) and that each cube has the same (small) volume, $V$.

We can obtain a better and better approximation by shrinking the volume of each cube and using a larger and larger number of cubes. This procedure should look very familiar to you, because it is exactly what is done in calculus to arrive at the limit for an integral. In this limit, we obtain for the location of the center of mass for an arbitrarily shaped object:

$$\vec{R} = \frac{1}{M}\int_V \vec{r}\rho(\vec{r})\, dV. \tag{8.21}$$

Here the three-dimensional volume integral extends over the entire volume of the object under consideration.

The next question that arises is what coordinate system to choose in order to evaluate this integral. You may have never seen a three-dimensional integral before and may have worked only with one-dimensional integrals of the form $\int f(x)dx$. However, all three-dimensional integrals that we will use in this chapter can be reduced to (at most) three successive one-dimensional integrals, most of which are very straightforward to evaluate, provided one selects an appropriate system of coordinates.

### Three-Dimensional Non-Cartesian Coordinate Systems

Chapter 1 introduced a three-dimensional orthogonal coordinate system, the Cartesian coordinate system, with coordinates $x$, $y$, and $z$. However, for some applications, it is mathematically simpler to represent the position vector in another coordinate system. This section briefly introduces two commonly used three-dimensional coordinate systems that can be used to specify a vector in three-dimensional space: spherical coordinates and cylindrical coordinates.

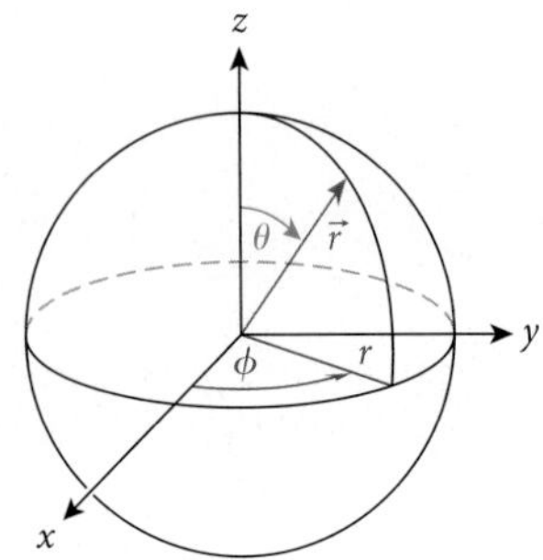


**FIGURE 8.15** Three-dimensional spherical coordinate system.

## Spherical Coordinates

In **spherical coordinates,** the position vector $\vec{r}$ is represented by giving its length, $r$; its polar angle relative to the positive $z$-axis, $\theta$; and the azimuthal angle of the vector's projection onto the $xy$-plane relative to the positive $x$-axis, $\phi$ (Figure 8.15).

We can obtain the Cartesian coordinates of the vector $\vec{r}$ from its spherical coordinates via the transformation

$$\begin{aligned} x &= r\cos\phi\sin\theta \\ y &= r\sin\phi\sin\theta \\ z &= r\cos\theta. \end{aligned} \tag{8.22}$$

The inverse transformation from Cartesian to spherical coordinates is

$$\begin{aligned} r &= \sqrt{x^2+y^2+z^2} \\ \theta &= \cos^{-1}\left(\frac{z}{\sqrt{x^2+y^2+z^2}}\right) \\ \phi &= \tan^{-1}\left(\frac{y}{x}\right). \end{aligned} \tag{8.23}$$

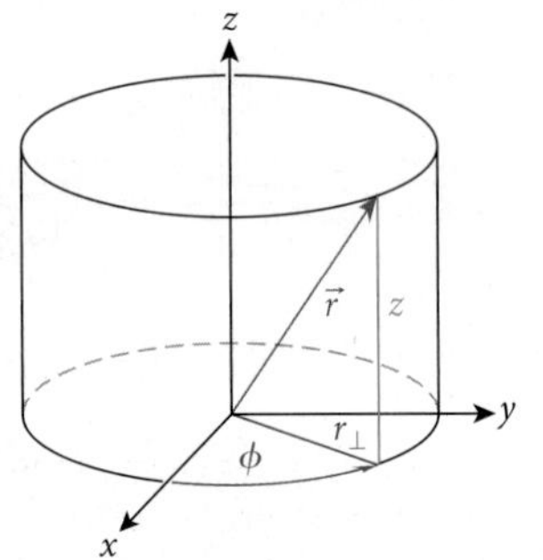


**FIGURE 8.16** Cylindrical coordinate system in three dimensions.

## Cylindrical Coordinates

**Cylindrical coordinates** can be thought of as an intermediate between Cartesian and spherical coordinate systems, in the sense that the Cartesian $z$-coordinate is retained, but the Cartesian coordinates $x$ and $y$ are replaced by the coordinates $r_\perp$ and $\phi$ (Figure 8.16). Here $r_\perp$ specifies the length of the projection of the position vector $\vec{r}$ onto the $xy$-plane, so it measures the perpendicular distance to the $z$-axis. Just as in spherical coordinates, $\phi$ is the angle of the vector's projection into the $xy$-plane relative to the positive $x$-axis.

We obtain the Cartesian coordinates from the cylindrical coordinates via

$$\begin{aligned} x &= r_\perp\cos\phi \\ y &= r_\perp\sin\phi \\ z &= z. \end{aligned} \tag{8.24}$$

The inverse transformation from Cartesian to cylindrical coordinates is

$$\begin{aligned} r_\perp &= \sqrt{x^2+y^2} \\ \phi &= \tan^{-1}\left(\frac{y}{x}\right) \\ z &= z. \end{aligned} \tag{8.25}$$

As a rule of thumb, you should use a Cartesian coordinate system in your first attempt to describe any physical situation. However, cylindrical and spherical coordinate systems are often preferable when working with objects that have symmetry about a point or a line. Later in this chapter, we will make use of a cylindrical coordinate system to perform a three-dimensional volume integral. Chapter 9 will discuss polar coordinates, which can be thought of as the two-dimensional equivalent of either cylindrical or spherical coordinates. Finally, in Chapter 10, we will again use spherical and cylindrical coordinates to solve slightly more complicated problems requiring integration.

## Mathematical Insert: Volume Integrals

Even though calculus is a prerequisite for physics, many universities allow students to take introductory physics and calculus courses concurrently. In general, this approach works well, but when students encounter multidimensional integrals in physics, it is often the first time they have seen this notation. Therefore, let's review the basic procedure for performing these integrations.

If we want to integrate any function over a three-dimensional volume, we need to find an expression for the volume element $dV$ in an appropriate set of coordinates. Unless there is an extremely important reason not to, you should always use orthogonal coordinate systems. The three commonly used three-dimensional orthogonal coordinate systems are the Cartesian, cylindrical, and spherical systems.

It is easiest by far to express the volume element $dV$ in Cartesian coordinates; it is simply the product of the three individual coordinate elements (Figure 8.17). The three-dimensional volume integral written in Cartesian coordinates is

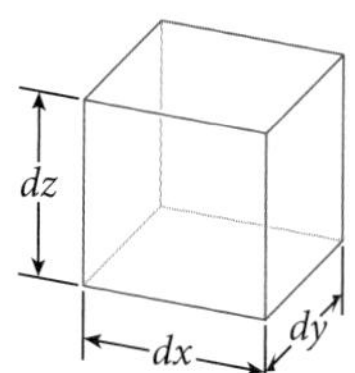

**FIGURE 8.17** Volume element in Cartesian coordinates.

$$\int_V f(\vec{r})dV = \int_{z_{\min}}^{z_{\max}} \left( \int_{y_{\min}}^{y_{\max}} \left( \int_{x_{\min}}^{x_{\max}} f(\vec{r})dx \right) dy \right) dz. \tag{8.26}$$

In this equation, $f(\vec{r})$ can be an arbitrary function of the position. The lower and upper boundaries for the individual coordinates are denoted by $x_{\min}, x_{\max}, \ldots$. The convention is to solve the innermost integral first and then work outward. For equation 8.26, this means that we first execute the integration over $x$, then the integration over $y$, and finally the integration over $z$. However, any other order is possible. An equally valid way of writing the integral in equation 8.26 is

$$\int_V f(\vec{r})dV = \int_{x_{\min}}^{x_{\max}} \left( \int_{y_{\min}}^{y_{\max}} \left( \int_{z_{\min}}^{z_{\max}} f(\vec{r})dz \right) dy \right) dx, \tag{8.27}$$

which implies that the order of integration is now $z, y, x$. Why might the order of integration make a difference? The only time the order of the integration matters is when the integration boundaries in a particular coordinate depend on one or both of the other coordinates. Example 8.4 will consider such a situation.

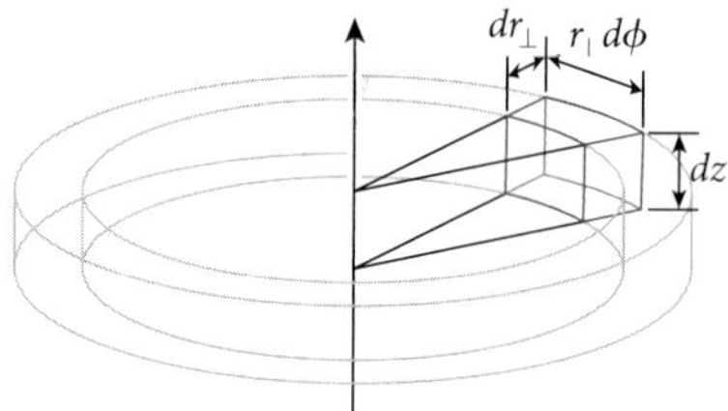

**FIGURE 8.18** Volume element in cylindrical coordinates.

Because the angle $\phi$ is one of the coordinates in the cylindrical coordinate system, the volume element is not cube-shaped. For a given differential angle, $d\phi$, the size of the volume element depends on how far away from the $z$-axis the volume element is located. This size increases linearly with the distance $r_\perp$ from the $z$-axis (Figure 8.18) and is given by

$$dV = r_\perp dr_\perp d\phi dz. \tag{8.28}$$

The volume integral is then

$$\int_V f(\vec{r})dV = \int_{z_{\min}}^{z_{\max}} \left( \int_{\phi_{\min}}^{\phi_{\max}} \left( \int_{r_{\perp\min}}^{r_{\perp\max}} f(\vec{r})r_\perp dr_\perp \right) d\phi \right) dz. \tag{8.29}$$

Again the order of integration can be chosen to make the task as simple as possible.

Finally, in spherical coordinates, we use two angular variables, $\theta$ and $\phi$ (Figure 8.19). Here the size of the volume element for a given value of the differential coordinates depends on the distance $r$ to the origin as well as the angle relative to the $\theta = 0$ axis (equivalent to the $z$-axis in Cartesian or cylindrical coordinates). The differential volume element in spherical coordinates is

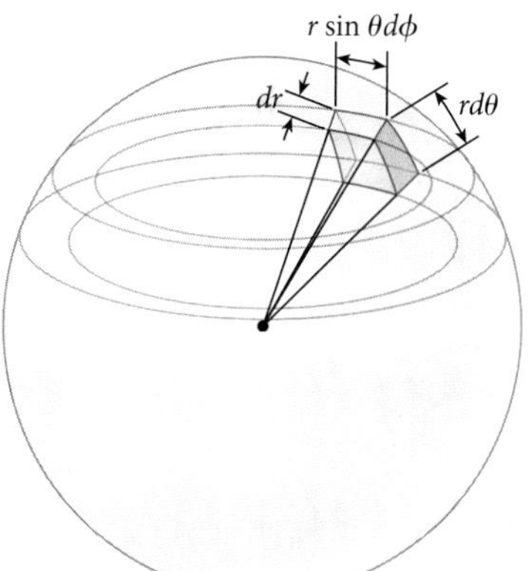

**FIGURE 8.19** Volume element in spherical coordinates.

$$dV = r^2 dr \sin\theta d\theta d\phi. \tag{8.30}$$

The volume integral in spherical coordinates is given by

$$\int_V f(\vec{r})dV = \int_{r_{\min}}^{r_{\max}} \left( \int_{\phi_{\min}}^{\phi_{\max}} \left( \int_{\theta_{\min}}^{\theta_{\max}} f(\vec{r})\sin\theta d\theta \right) d\phi \right) r^2 dr. \tag{8.31}$$

## EXAMPLE 8.4 Volume of a Cylinder

To illustrate why it may be simpler to use non-Cartesian coordinates in certain circumstances, let's use volume integrals to find the volume of a cylinder with radius $R$ and height $H$. We have to integrate the function $f(\vec{r}) = 1$ over the entire cylinder to obtain the volume.

*Continued—*

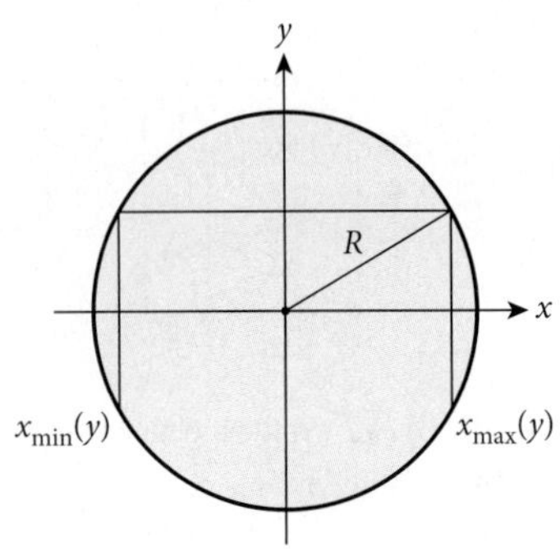


**FIGURE 8.20** Bottom surface of a right cylinder of radius $R$.

**PROBLEM**

Use a volume integral to find the volume of a right cylinder of height $H$ and radius $R$.

**SOLUTION**

In Cartesian coordinates, we place the origin of our coordinate system at the center of the cylinder's circular base (bottom surface), so the shape in the $xy$-plane that we have to integrate over is a circle with radius $R$ (Figure 8.20). The volume integral in Cartesian coordinates is then

$$\int_V dV = \int_0^H \left( \int_{y_{min}}^{y_{max}} \left( \int_{x_{min}(y)}^{x_{max}(y)} dx \right) dy \right) dz. \tag{i}$$

The innermost integral has to be done first and is straightforward:

$$\int_{x_{min}(y)}^{x_{max}(y)} dx = x_{max}(y) - x_{min}(y). \tag{ii}$$

The integration boundaries depend on $y$: $x_{max} = \sqrt{R^2 - y^2}$ and $x_{min} = -\sqrt{R^2 - y^2}$. Thus, the solution of equation (ii) is $x_{max}(y) - x_{min}(y) = 2\sqrt{R^2 - y^2}$. We insert this into equation (i) and obtain

$$\int_V dV = \int_0^H \left( \int_{-R}^{R} 2\sqrt{R^2 - y^2}\, dy \right) dz. \tag{iii}$$

The inner of these two remaining integrals evaluates to

$$\int_{-R}^{R} 2\sqrt{R^2 - y^2}\, dy = \left( y\sqrt{R^2 - y^2} + R^2 \tan^{-1}\left( \frac{y}{\sqrt{R^2 - y^2}} \right) \right)\Bigg|_{-R}^{R} = \pi R^2.$$

You can check this result by looking up the definite integral in an integral table. Inserting this result into equation (iii) finally yields our answer:

$$\int_V dV = \int_0^H \pi R^2 dz = \pi R^2 \int_0^H dz = \pi R^2 H.$$

As you can see, obtaining the volume of the cylinder was rather cumbersome in Cartesian coordinates. What about using cylindrical coordinates? According to equation 8.29, the volume integral is then

$$\int_V f(\vec{r})dV = \int_0^H \left( \int_0^{2\pi} \left( \int_0^R r_\perp dr_\perp \right) d\phi \right) dz = \int_0^H \left( \int_0^{2\pi} \left( \tfrac{1}{2}R^2 \right) d\phi \right) dz$$

$$= \tfrac{1}{2}R^2 \int_0^H \left( \int_0^{2\pi} d\phi \right) dz = \tfrac{1}{2}R^2 \int_0^H 2\pi dz = \pi R^2 \int_0^H dz = \pi R^2 H.$$

In this case, it was much easier to use cylindrical coordinates, a consequence of the geometry of the object over which we had to integrate.

## 8.2 Self-Test Opportunity

Using spherical coordinates, show that the volume $V$ of a sphere with radius $R$ is $V = \frac{4}{3}\pi R^3$.

Now we can return to the problem of calculating the location of an object's center of mass. For the Cartesian components of the position vector, we find, from equation 8.21:

$$X = \frac{1}{M}\int_V x\rho(\vec{r})dV, \quad Y = \frac{1}{M}\int_V y\rho(\vec{r})dV, \quad Z = \frac{1}{M}\int_V z\rho(\vec{r})dV. \tag{8.32}$$

If the mass density for the entire object is constant, $\rho(\vec{r}) \equiv \rho$, we can remove this constant factor from the integral and obtain a special case of equation 8.21 for constant mass density:

$$\vec{R} = \frac{\rho}{M}\int_V \vec{r}\,dV = \frac{1}{V}\int_V \vec{r}\,dV \quad \text{(for constant } \rho\text{)}, \tag{8.33}$$

where we have used equation 8.20 in the last step. Expressed in Cartesian components, we obtain for this case:

$$X = \frac{1}{V}\int_V x\,dV, \quad Y = \frac{1}{V}\int_V y\,dV, \quad Z = \frac{1}{V}\int_V z\,dV. \tag{8.34}$$

Equations 8.33 and 8.34 indicate that any object that has a symmetry plane has its center of mass located in that plane. An object having three mutually perpendicular symmetry planes (such as a cylinder, a rectangular solid, or a sphere) has its center of mass where these three planes intersect, which is the geometric center. Example 8.5 develops this idea further.

## EXAMPLE 8.5 Center of Mass for a Half-Sphere

### PROBLEM

Consider a solid half-sphere of constant mass density with radius $R_0$ (Figure 8.21a). Where is its center of mass?

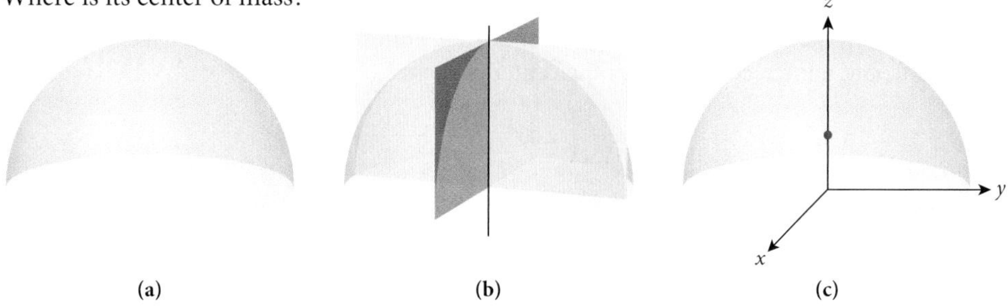


**FIGURE 8.21** Determination of the center of mass: (a) half-sphere; (b) symmetry planes and symmetry axis; (c) coordinate system, with location of center of mass marked by red dot.

### SOLUTION

As shown in Figure 8.21b, symmetry planes can divide this object into equal, mirror-image halves. Shown are two perpendicular planes in red and yellow, but any plane through the vertical symmetry axis (indicated by the thin black line) is a symmetry plane.

We now position the coordinate system so that one axis (the $z$-axis, in this case) coincides with this symmetry axis. We are then assured that the center of mass is located exactly on this axis. Because the mass distribution is symmetric and the integrands of equations 8.33 or 8.34 are odd powers of $\vec{r}$ the integral for $X$ or $Y$ has to have the value zero. Specifically,

$$\int_{-a}^{a} x\,dx = 0 \text{ for all values of the constant } a.$$

Positioning the coordinate system so that the $z$-axis is the symmetry axis ensures that $X = Y = 0$. This is shown in Figure 8.21c, where the origin of the coordinate system is positioned at the center of the half-sphere's circular bottom surface.

Now we have to find the value of the third integral in equation 8.34:

$$Z = \frac{1}{V}\int_V z\,dV.$$

The volume of a half-sphere is half the volume of a sphere, or

$$V = \frac{2\pi}{3}R_0^3. \tag{i}$$

*Continued—*

To evaluate the integral for $Z$, we use cylindrical coordinates, in which the differential volume element is given (see equation 8.28) as $dV = r_\perp dr_\perp d\phi\, dz$. The integral is then evaluated as follows:

$$\int_V z dV = \int_0^{R_0}\left(\int_0^{\sqrt{R_0^2-z^2}}\left(\int_0^{2\pi} z r_\perp d\phi\right) dr_\perp\right) dz$$

$$= \int_0^{R_0} z\left(\int_0^{\sqrt{R_0^2-z^2}} r_\perp\left(\int_0^{2\pi} d\phi\right) dr_\perp\right) dz$$

$$= 2\pi \int_0^{R_0} z\left(\int_0^{\sqrt{R_0^2-z^2}} r_\perp dr_\perp\right) dz$$

$$= \pi \int_0^{R_0} z(R_0^2 - z^2) dz$$

$$= \frac{\pi}{4} R_0^4.$$

Combining this result and the expression for the volume of a half-sphere from equation (i), we obtain the $z$-coordinate of the center of mass:

$$Z = \frac{1}{V}\int_V z dV = \frac{3}{2\pi R_0^3}\frac{\pi R_0^4}{4} = \frac{3}{8}R_0.$$

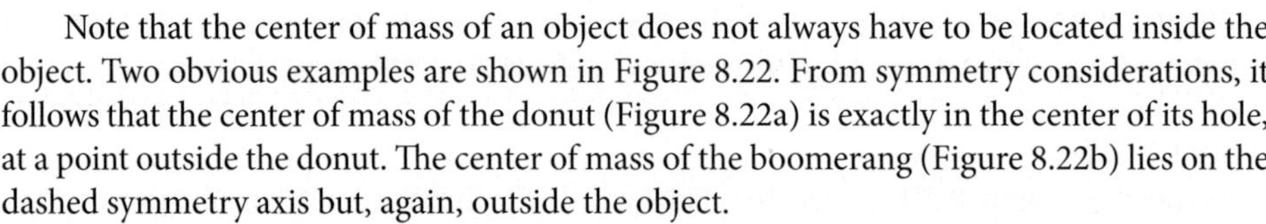

Note that the center of mass of an object does not always have to be located inside the object. Two obvious examples are shown in Figure 8.22. From symmetry considerations, it follows that the center of mass of the donut (Figure 8.22a) is exactly in the center of its hole, at a point outside the donut. The center of mass of the boomerang (Figure 8.22b) lies on the dashed symmetry axis but, again, outside the object.

(a)

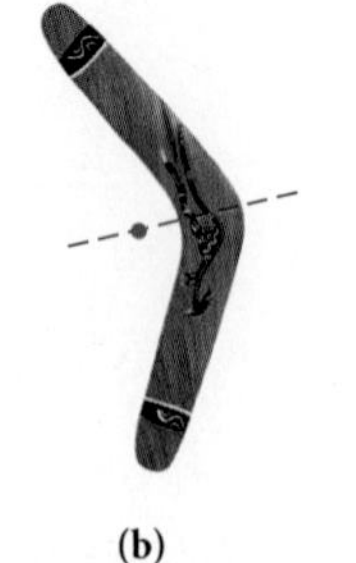

(b)

**FIGURE 8.22** Objects with a center of mass (indicated by the red dot) outside their mass distribution: (a) donut; (b) boomerang. The symmetry axis of the boomerang is shown by a dashed line.

## Center of Mass for One- and Two-Dimensional Objects

Not all problems involving calculation of the center of mass focus on three-dimensional objects. For example, you may want to calculate the center of mass of a two-dimensional object, such as a flat metal plate. We can write the equations for the center-of-mass coordinates of a two-dimensional object whose area mass density (or mass per unit area) is $\sigma(\vec{r})$ by modifying the expressions for $X$ and $Y$ given in equation 8.32:

$$X = \frac{1}{M}\int_A x\sigma(\vec{r})dA, \quad Y = \frac{1}{M}\int_A y\sigma(\vec{r})dA, \tag{8.35}$$

where the mass is

$$M = \int_A \sigma(\vec{r})dA. \tag{8.36}$$

If the area mass density of the object is constant, then $\sigma = M/A$, and we can rewrite equation 8.35 to give the coordinates of the center of mass of a two-dimensional object in terms of the area, $A$, and the coordinates $x$ and $y$:

$$X = \frac{1}{A}\int_A x dA, \quad Y = \frac{1}{A}\int_A y dA, \tag{8.37}$$

where the total area is obtained from

$$A = \int_A dA. \tag{8.38}$$

If the object is effectively one-dimensional, such as a long, thin rod with length $L$ and linear mass density (or mass per unit length) of $\lambda(x)$, the coordinate of the center of mass is given by

$$X = \frac{1}{M}\int_L x\lambda(x)dx, \tag{8.39}$$

where the mass is

$$M = \int_L \lambda(x)dx. \tag{8.40}$$

If the linear mass density of the rod is constant, then clearly the center of mass is located at the geometric center—the middle of the rod—and no further calculation is required.

## SOLVED PROBLEM 8.3 Center of Mass of a Long, Thin Rod

### PROBLEM

A long, thin rod lies along the $x$-axis. One end of the rod is located at $x = 1.00$ m, and the other end of the rod is located at $x = 3.00$ m. The linear mass density of the rod is given by $\lambda(x) = ax^2 + b$, where $a = 0.300$ kg/m$^3$ and $b = 0.600$ kg/m. What are the mass of the rod and the $x$-coordinate of its center of mass?

### SOLUTION

#### THINK

The linear mass density of the rod is not uniform but depends on the $x$-coordinate. Therefore, to get the mass, we must integrate the linear mass density over the length of the rod. To get the center of mass, we need to integrate the linear mass density, weighted by the distance in the $x$-direction, and then divide by the mass of the rod.

#### SKETCH

The long, thin rod oriented along the $x$-axis is shown in Figure 8.23.

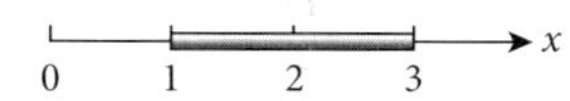


**FIGURE 8.23** A long, thin rod oriented along the $x$-axis.

#### RESEARCH

We obtain the mass of the rod by integrating the linear mass density, $\lambda$, over the rod from $x_1 = 1.00$ m to $x_2 = 3.00$ m (see equation 8.40):

$$M = \int_{x_1}^{x_2} \lambda(x)dx = \int_{x_1}^{x_2} \left(ax^2 + b\right)dx = \left[a\frac{x^3}{3} + bx\right]_{x_1}^{x_2}.$$

To find the $x$-coordinate of the center of mass of the rod, $X$, we evaluate the integral of the differential mass times $x$ and then divide by the mass, which we have just calculated (see equation 8.39):

$$X = \frac{1}{M}\int_{x_1}^{x_2} \lambda(x)x dx = \frac{1}{M}\int_{x_1}^{x_2} \left(ax^2 + b\right)x dx = \frac{1}{M}\int_{x_1}^{x_2} \left(ax^3 + bx\right)dx = \frac{1}{M}\left[a\frac{x^4}{4} + b\frac{x^2}{2}\right]_{x_1}^{x_2}.$$

#### SIMPLIFY

Inserting the upper and lower limits, $x_2$ and $x_1$, we get the mass of the rod:

$$M = \left[a\frac{x^3}{3} + bx\right]_{x_1}^{x_2} = \left(a\frac{x_2^3}{3} + bx_2\right) - \left(a\frac{x_1^3}{3} + bx_1\right) = \frac{a}{3}\left(x_2^3 - x_1^3\right) + b\left(x_2 - x_1\right).$$

*Continued—*

And, in the same way, we find the $x$-coordinate of the center of mass of the rod:

$$X=\frac{1}{M}\left[a\frac{x^4}{4}+b\frac{x^2}{2}\right]_{x_1}^{x_2}=\frac{1}{M}\left\{\left(a\frac{x_2^4}{4}+b\frac{x_2^2}{2}\right)-\left(a\frac{x_1^4}{4}+b\frac{x_1^2}{2}\right)\right\},$$

which we can further simplify to

$$X=\frac{1}{M}\left\{\frac{a}{4}\left(x_2^4-x_1^4\right)+\frac{b}{2}\left(x_2^2-x_1^2\right)\right\}.$$

**CALCULATE**

Substituting the given numerical values, we compute the mass of the rod:

$$M=\frac{0.300\text{ kg/m}^3}{3}\left[\left(3.00\text{ m}\right)^3-\left(1.00\text{ m}\right)^3\right]+\left(0.600\text{ kg/m}\right)\left(3.00\text{ m}-1.00\text{ m}\right)=3.8\text{ kg}.$$

With the numerical values, the $x$-coordinate of the rod is

$$X=\frac{1}{3.8\text{ kg}}\left\{\frac{0.300\text{ kg/m}^3}{4}\left[\left(3.00\text{ m}\right)^4-\left(1.00\text{ m}\right)^4\right]+\frac{0.600\text{ kg/m}}{2}\left[\left(3.00\text{ m}\right)^2-\left(1.00\text{ m}\right)^2\right]\right\}$$
$$=2.210526316\text{ m}.$$

**ROUND**

All of the numerical values in the problem statement were specified to three significant figures, so we report our results as

$$M=3.80\text{ kg}$$

and

$$X=2.21\text{ m}.$$

**DOUBLE-CHECK**

To double-check our answer for the mass of the rod, let's assume that the rod has a constant linear mass density equal to the linear mass density obtained by setting $x = 2$m (the middle of the rod) in the expression for $\lambda$ in the problem, that is,

$$\lambda=\left(0.3\cdot 4+0.6\right)\text{ kg/m}=1.8\text{ kg/m}.$$

The mass of the rod is then $m \approx 2\text{m}\cdot 1.8\text{ kg/m} = 3.6\text{ kg}$, which is reasonably close to our exact calculation of $M = 3.80$ kg.

To double-check the $x$-coordinate of the center of mass of the rod, we again assume that the linear mass density is constant. Then the center of mass will be located at the middle of the rod, or $X \approx 2$m. Our calculated answer is $X = 2.21$ m, which is slightly to the right of the middle of the rod. Looking at the function for the linear mass density, we see that the linear mass of the rod increases toward the right, which means that the center of mass of the rod must be to the right of the rod's geometric center. Our result is therefore reasonable.

### 8.3 Self-Test Opportunity

A plate with height $h$ is cut from a thin metal sheet with uniform mass density, as shown in the figure. The lower boundary of the plate is defined by $y = 2x^2$. Show that the center of mass of this plate is located at $x = 0$ and $y = \frac{3}{5}h$.

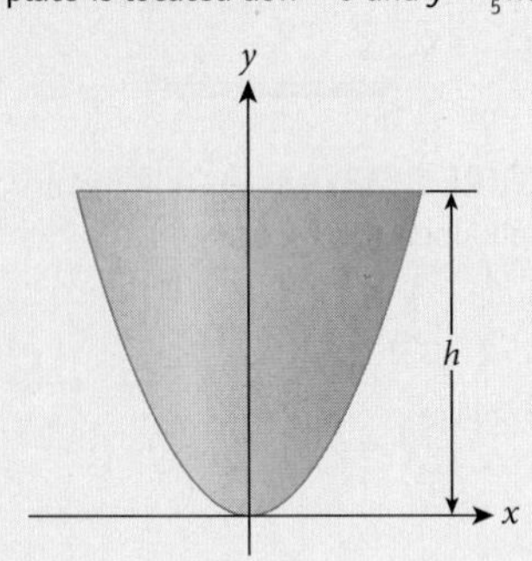


## WHAT WE HAVE LEARNED | EXAM STUDY GUIDE

- The center of mass is the point at which we can imagine all the mass of an object to be concentrated.
- The location of the center of mass for an arbitrarily shaped object is given by $\vec{R}=\frac{1}{M}\int_V \vec{r}\rho(\vec{r})dV$, where the mass density of the object is $\rho=\frac{dm}{dV}$, the integration extends over the entire volume $V$ of the object, and $M$ is its total mass.
- When the mass density is uniform throughout the object, that is, $\rho=\frac{M}{V}$, the center of mass is $\vec{R}=\frac{1}{V}\int_V \vec{r}\,dV.$
- If an object has a plane of symmetry, the location of the center of mass must be in that plane.
- The location of the center of mass for a combination of several objects can be found by taking the

mass-weighted average of the locations of the centers of mass of the individual objects:
$$\vec{R} = \frac{\vec{r}_1 m_1 + \vec{r}_2 m_2 + \cdots + \vec{r}_n m_n}{m_1 + m_2 + \cdots + m_n} = \frac{1}{M}\sum_{i=1}^{n} \vec{r}_i m_i .$$

- The motion of an extended rigid object can be described by the motion of its center of mass.
- The velocity of the center of mass is given by the derivative of its position vector: $\vec{V} \equiv \frac{d}{dt}\vec{R}$.
- The center-of-mass momentum for a combination of several objects is $\vec{P} = M\vec{V} = \sum_{i=1}^{n} \vec{p}_i$. This momentum obeys Newton's Second Law: $\frac{d}{dt}\vec{P} = \frac{d}{dt}(M\vec{V}) = \sum_{i=1}^{n} \vec{F}_i = \vec{F}_{\text{net}}$. Internal forces between the objects do not contribute to the sum that yields the net force (because they always come in action-reaction pairs adding up to zero) and thus do not change the center-of-mass momentum.
- For a system of two objects, the total momentum is $\vec{P} = \vec{p}_1 + \vec{p}_2$, and the relative momentum is $\vec{p} = \frac{1}{2}(\vec{p}_1 - \vec{p}_2)$. In collisions between two objects, the total momentum remains unchanged.
- Rocket motion is an example of motion during which the mass of the moving object is not constant. The equation of motion for a rocket in interstellar space is given by $\vec{F}_{\text{thrust}} = m\vec{a} = -\vec{v}_c \frac{dm}{dt}$, where $\vec{v}_c$ is the velocity of the propellant relative to the rocket and $\frac{dm}{dt}$ is the rate of change in mass due to outflow of propellant.
- The velocity of a rocket as a function of its mass is given by $v_f - v_i = v_c \ln(m_i/m_f)$, where the indices i and f indicate initial and final masses and velocities.

## KEY TERMS

center of mass, p. 247
recoil, p. 253
thrust, p. 258
spherical coordinates, p. 260
cylindrical coordinates, p. 260

## NEW SYMBOLS AND EQUATIONS

$\vec{R} = \frac{1}{M}\sum_{i=1}^{n} \vec{r}_i m_i$, combined center-of-mass position vector

$\vec{R} = \frac{1}{M}\int_V \vec{r}\rho(\vec{r})dV$, center of mass for an extended object

$dV = r_\perp dr_\perp d\phi\, dz$, volume element in cylindrical coordinates

$dV = r^2 dr \sin\theta\, d\theta\, d\phi$, volume element in spherical coordinates

$\vec{F}_{\text{thrust}}$, rocket thrust

## ANSWERS TO SELF-TEST OPPORTUNITIES

**8.1**

$$Z = \frac{z_{\text{red}} m_{\text{red}} + z_{\text{green}} m_{\text{green}} + z_{\text{orange}} m_{\text{orange}} + z_{\text{blue}} m_{\text{blue}} + z_{\text{purple}} m_{\text{purple}}}{M}$$
$$= \frac{\frac{1}{2}h2m_0 + \frac{1}{2}hm_0 + \frac{1}{2}hm_0 + \frac{1}{2}hm_0 + \frac{3}{2}hm_0}{6m_0}$$
$$= \frac{w1 + \frac{1}{2} + \frac{1}{2} + \frac{1}{2} + \frac{3}{2}}{6} = \frac{2}{3}h = 1.7 \text{ m}$$

**8.2** We use spherical coordinates and integrate the angle $\theta$ from 0 to $\pi$, the angle $\phi$ from 0 to $2\pi$, and the radial coordinate $r$ from 0 to $R$.

$$V = \int_0^R \left( \int_0^{2\pi} \left( \int_0^{\pi} \sin\theta\, d\theta \right) d\phi \right) r^2 dr$$

First, evaluate the integral over the azimuthal angle:

$$\int_0^{\pi} \sin\theta\, d\theta = \left[-\cos\theta\right]_0^{\pi} = -\left[\cos(\pi) - \cos(0)\right] = 2$$

$$V = 2\int_0^R \left( \int_0^{2\pi} d\phi \right) r^2 dr$$

Now evaluate the polar angle integral:

$$\int_0^{2\pi} d\phi = \left[\phi\right]_0^{2\pi} = 2\pi$$

*Continued—*

Finally:

$$V = 4\pi \int_0^R r^2 dr = 4\pi \left[\frac{r^3}{3}\right]_0^R = \frac{4}{3}\pi R^3$$

**8.3** $dA = x(y)dy;\ y = 2x^2 \Rightarrow x = \sqrt{y/2}$

$$x(y) = 2\sqrt{y/2} = \sqrt{2y} dA = \sqrt{2y} dy$$

$$Y = \frac{\int_0^h y\sqrt{2y}dy}{\int_0^h \sqrt{2y}dy} = \frac{\sqrt{2}\int_0^h y^{3/2}dy}{\sqrt{2}\int_0^h y^{1/2}dy} = \frac{\left[\frac{y^{5/2}}{5/2}\right]_0^h}{\left[\frac{y^{3/2}}{3/2}\right]_0^h}$$

$$Y = \tfrac{3}{5}h$$

## PROBLEM-SOLVING PRACTICE

### Problem-Solving Guidelines: Center of Mass

**1.** The first step in locating the center of mass of an object or a system of particles is to look for planes of symmetry. The center of mass must be located on the plane of symmetry, on the line of intersection of two planes of symmetry, or at the point of intersection of more than two planes.

**2.** For complicated shapes, break the object down into simpler geometric forms and locate the center of mass for each individual form. Then combine the separate centers of mass into one overall center of mass using the weighted average of distances and masses. Treat holes as objects of negative mass.

**3.** Any motion of an object can be treated as a superposition of motion of its center of mass (according to Newton's Second Law) and rotation of the object about the center of mass. Collisions can often be conveniently analyzed by considering a reference frame with the origin located at the center of mass.

**4.** Often, integration is unavoidable when you need to locate the center of mass. In such a case, it is always best to think carefully about the dimensionality of the situation and the choice of the coordinate system (Cartesian, cylindrical, or spherical).

### SOLVED PROBLEM 8.4 Thruster Firing

**PROBLEM**

Suppose a spacecraft has an initial mass of 1,850,000 kg. Without its propellant, the spacecraft has a mass of 50,000 kg. The rocket that powers the spacecraft is designed to eject the propellant with a speed of 25 km/s with respect to the rocket at a constant rate of 15,000 kg/s. The spacecraft is initially at rest in space and travels in a straight line. How far will the spacecraft travel before its rocket uses all the propellant and shuts down?

**SOLUTION**

**THINK**

The total mass of propellant is the total mass of the spacecraft minus the mass of the spacecraft after all the propellant is ejected. The rocket ejects the propellant at a fixed rate, so we can calculate the amount of time during which the rocket operates. As the propellant is used up, the mass of the spacecraft decreases and the speed of the spacecraft increases. If the spacecraft starts from rest, the speed $v(t)$ at any time while the rocket is operating can be obtained from equation 8.17, with the final mass of the spacecraft replaced by the mass of the spacecraft at that time. The distance traveled before all the propellant is used is given by the integral of the speed as a function of time.

**SKETCH**

The flight of the spacecraft is sketched in Figure 8.24.

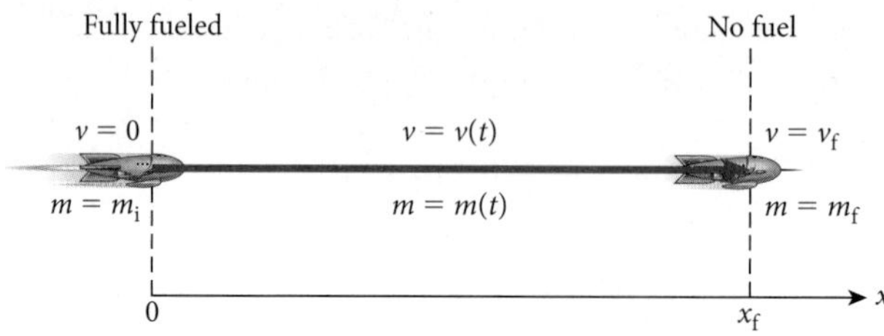


**FIGURE 8.24** The various parameters for the spacecraft as the rocket operates.

## RESEARCH

We symbolize the rate at which the propellant is ejected by $r_p$. The time $t_{max}$ during which the rocket will operate is then given by

$$t_{max} = \frac{(m_i - m_f)}{r_p},$$

where $m_i$ is the initial mass of the spacecraft and $m_f$ is the mass of the spacecraft after all the propellant is ejected. The total distance the spacecraft travels in this time interval is the integral of the speed over time:

$$x_f = \int_0^{t_{max}} v(t)dt. \quad \text{(i)}$$

While the rocket is operating, the mass of the spacecraft at a time $t$ is given by

$$m(t) = m_i - r_p t.$$

The speed of the spacecraft at any given time after the rocket starts to operate and before all the propellant is used up is given by (compare to equation 8.17)

$$v(t) = v_c \ln\left(\frac{m_i}{m(t)}\right) = v_c \ln\left(\frac{m_i}{m_i - r_p t}\right) = v_c \ln\left(\frac{1}{1 - r_p t / m_i}\right), \quad \text{(ii)}$$

where $v_c$ is the speed of the ejected propellant with respect to the rocket.

## SIMPLIFY

Now we substitute from equation (ii) for the time dependence of the speed of the spacecraft into equation (i) and obtain

$$x_f = \int_0^{t_{max}} v(t)dt = \int_0^{t_{max}} v_c \ln\left(\frac{1}{1 - r_p t / m_i}\right)dt = -v_c \int_0^{t_{max}} \ln\left(\frac{1 - r_p t}{m_i}\right)dt. \quad \text{(iii)}$$

Because $\int \ln(1-ax)dx = \frac{ax-1}{a}\ln(1-ax) - x$ (you can look up this result in an integral table), the integral evaluates to

$$\begin{aligned}
\int_0^{t_{max}} \ln(1 - r_p t / m_i)dt &= \left[\left(\frac{r_p t / m_i - 1}{r_p / m_i}\right)\ln(1 - r_p t / m_i) - t\right]_0^{t_{max}} \\
&= \left(\frac{r_p t_{max} / m_i - 1}{r_p / m_i}\right)\ln(1 - r_p t_{max} / m_i) - t_{max} \\
&= (t_{max} - m_i / r_p)\ln(1 - r_p t_{max} / m_i) - t_{max}.
\end{aligned}$$

The distance traveled is then

$$x_f = -v_c\left[(t_{max} - m_i / r_p)\ln(1 - r_p t_{max} / m)_i - t_{max}\right].$$

## CALCULATE

The time during which the rocket is operating is

$$t_{max} = \frac{m_i - m_f}{r_p} = \frac{1{,}850{,}000 \text{ kg} - 50{,}000 \text{ kg}}{15{,}000 \text{ kg/s}} = 120 \text{ s}.$$

Putting numerical values into the factor $1 - r_p t_{max} / m_i$ gives

$$1 - \frac{r_p t_{max}}{m_i} = 1 - \frac{15{,}000 \text{ kg/s} \cdot 120 \text{ s}}{1{,}850{,}000 \text{ kg}} = 0.027027.$$

*Continued—*

Thus, we find for the distance traveled

$$x_f = -\left(25\cdot 10^3 \text{ m/s}\right)\left[-(120 \text{ s}) + \{(120 \text{ s}) - \left(1.85\cdot 10^6 \text{ kg}\right)/\left(15\cdot 10^3 \text{ kg/s}\right)\}\ln(0.027027)\right]$$
$$= 2.69909\cdot 10^6 \text{ m}.$$

**ROUND**

Because the propellant speed was given to only two significant figures, we need to round to that accuracy:

$$x_f = 2.7\cdot 10^6 \text{ m}.$$

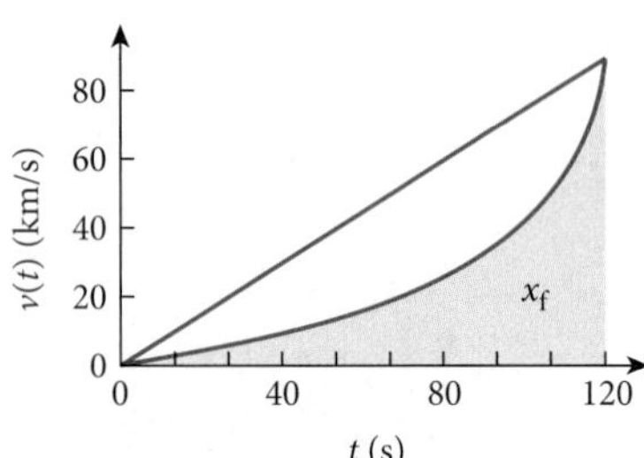


**FIGURE 8.25** Comparison of the exact solution for $v(t)$ (red curve) to one for constant acceleration (blue line).

**DOUBLE-CHECK**

To double-check our answer for the distance traveled, we use equation 8.17 to calculate the final velocity of the spacecraft:

$$v_f = v_c \ln\left(\frac{m_i}{m_f}\right) = \left(25 \text{ km/s}\right)\ln\left(\frac{1.85\cdot 10^6 \text{ kg}}{5\cdot 10^4 \text{ kg}}\right) = 90.3 \text{ km/s}.$$

If the spacecraft accelerated at a constant rate, the speed would increase linearly with time, as shown in Figure 8.25, and the average speed during the time the propellant was being ejected would be $\bar{v} = v_f/2$. Taking this average speed and multiplying by that time gives

$$x_{a\text{-const}} \approx \bar{v}t_{\max} = \left(v_f/2\right)t_{\max} = \left(90.3/2 \text{ km/s}\cdot 120 \text{ s}\right)/2 = 5.4\cdot 10^6 \text{ m}.$$

This approximate distance is bigger than our calculated answer, because in the calculation the velocity increases logarithmically in time until it reaches the value of 90.3 km/s. The approximation is about twice the calculated distance, giving us confidence that our answer at least has the right order of magnitude.

Figure 8.25 shows the exact solution for $v(t)$ (red curve). The distance traveled, $x_f$, is the area under the red curve. The blue line shows the case where constant acceleration leads to the same final velocity. As you can see, the area under the blue line is approximately twice that under the red curve. Since we just calculated the area under the blue line, $x_{a\text{-const}}$, and found it to be about twice as big as our calculated result, we gain confidence that we interpreted correctly.

## SOLVED PROBLEM 8.5 Center of Mass of a Disk with a Hole in It

**PROBLEM**

Where is the center of mass of a disk with a rectangular hole in it (Figure 8.26)? The height of the disk is $h = 11.0$ cm, and its radius is $R = 11.5$ cm. The rectangular hole has a width $w = 7.0$ cm and a depth $d = 8.0$ cm. The right side of the hole is located so that its midpoint coincides with the central axis of the disk.

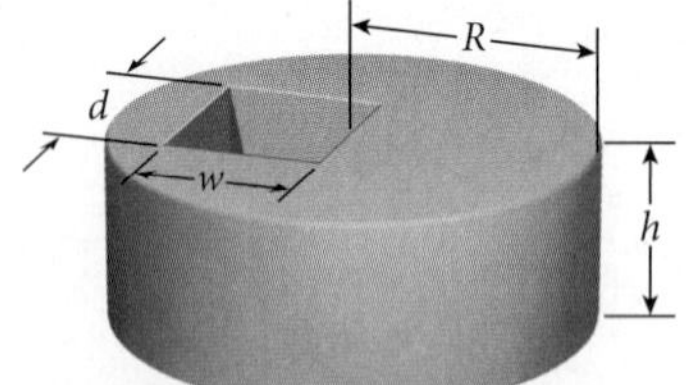


**FIGURE 8.26** Three-dimensional view of a disk with a rectangular hole in it.

**SOLUTION**

**THINK**

One way to approach this problem is to write mathematical formulas that describe the three-dimensional geometry of the disk with a hole in it and then integrate over that volume to obtain the coordinates of the center of mass. If we did that, we would be faced with several difficult integrals. A simpler way to approach this problem is to think of the disk with a hole in it as a solid disk *minus* a rectangular hole. That is, we treat the hole as a solid object with a negative mass. Using the symmetry of the solid disk and of the hole, we can specify the coordinates of the center of mass of the solid disk and of the center of mass of the hole. We can then combine these coordinates, using equation 8.1, to find the center of mass of the disk with a hole in it.

## SKETCH

Figure 8.27a shows a top view of the disk with a hole in it, with $x$- and $y$-axes assigned.

Figure 8.27b shows the two symmetry planes of the disk with a hole in it. One plane corresponds to the $x$-$y$ plane, and the second plane is a plane along the $x$-axis and perpendicular to the $x$-$y$ plane. The line where the two planes intersect is marked $A$.

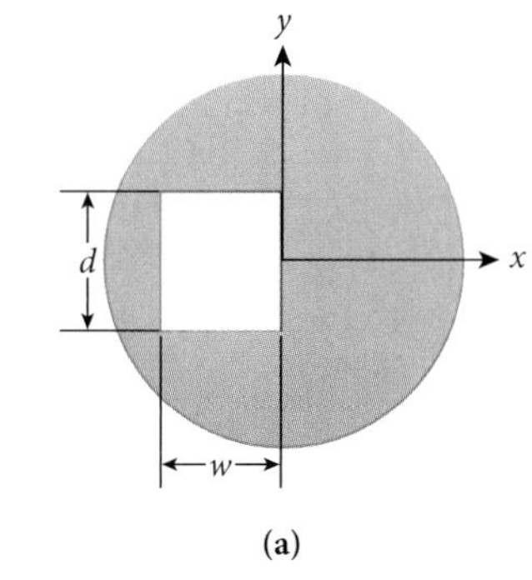

(a)

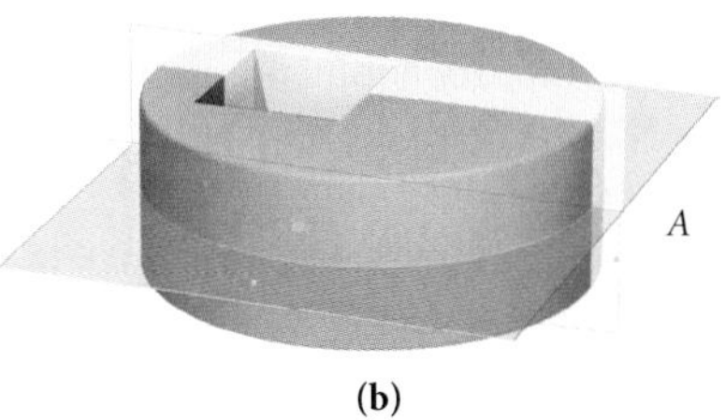

(b)

**FIGURE 8.27** (a) Top view of the disk with a hole in it with a coordinate system assigned. (b) Symmetry planes of the disk with a hole in it.

## RESEARCH

The center of mass must lie along the intersection of the two planes of symmetry. Therefore, we know that the center of mass can only be located along the $x$-axis. The center of mass for the disk without the hole is at the origin of the coordinate system, at $x_d = 0$, and the volume of the solid disk is $V_d = \pi R^2 h$. If the hole were a solid object with the same dimensions ($h = 11.0$ cm, $w = 7.0$ cm, and $d = 8.0$ cm), that object would have a volume of $V_h = hwd$. If this imagined solid object were located where the hole is, its center of mass would be in the middle of the hole, at $x_h = -3.5$ cm. We now multiply each of the volumes by $\rho$, the mass density of the material of the disk, to get the corresponding masses, and assign a negative mass to the hole. Then we use equation 8.1 to get the $x$-coordinate of the center of mass:

$$X = \frac{x_d V_d \rho - x_h V_h \rho}{V_d \rho - V_h \rho}. \qquad \text{(i)}$$

This method of treating a hole as an object of the same shape and then using its volume in calculations, but with negative mass (or charge), is very common in atomic and subatomic physics. We will encounter it again when we explore atomic physics (Chapter 37) and nuclear and particle physics (Chapters 39 and 40).

## SIMPLIFY

We can simplify equation (i) by realizing that $x_d = 0$ and that $\rho$ is a common factor:

$$X = \frac{-x_h V_h}{V_d - V_h}.$$

Substituting the expressions we obtained above for $V_d$ and $V_h$, we get

$$X = \frac{-x_h V_h}{V_d - V_h} = \frac{-x_h (hwd)}{\pi R^2 h - hwd} = \frac{-x_h wd}{\pi R^2 - wd}.$$

Defining the area of the disk in the $x$-$y$ plane to be $A_d = \pi R^2$ and the area of the hole in the $x$-$y$ plane to be $A_h = wd$, we can write

$$X = \frac{-x_h wd}{\pi R^2 - wd} = \frac{-x_h A_h}{A_d - A_h}.$$

## CALCULATE

Inserting the given numbers, we find that the area of the disk is

$$A_d = \pi R^2 = \pi (11.5 \text{ cm})^2 = 415.475 \text{ cm}^2,$$

and the area of the hole is

$$A_h = wd = (7.0 \text{ cm})(8.0 \text{ cm}) = 56 \text{ cm}^2.$$

Therefore, the location of the center of mass of the disk with the hole in it (remember that $x_h = -3.5$ cm ) is

$$X = \frac{-x_h A_h}{A_d - A_h} = \frac{-(-3.5 \text{ cm})(56 \text{ cm}^2)}{(415.475 \text{ cm}^2) - (56 \text{ cm}^2)} = 0.545239 \text{ cm}.$$

## ROUND

Expressing our answer with two significant figures, we report the $x$-coordinate of the center of mass of the disk with a hole in it as

$$X = 0.55 \text{ cm}.$$

*Continued—*

**DOUBLE-CHECK**

This point is slightly to the right of the center of the solid disk, by a distance that is a small fraction of the radius of the disk. This result seems reasonable because taking material out of the disk to the left of $x = 0$ should shift the center of gravity to the right, just as we calculated.

## MULTIPLE-CHOICE QUESTIONS

**8.1** A man standing on frictionless ice throws a boomerang, which returns to him. Choose the correct statement:

a) Since the momentum of the man-boomerang system is conserved, the man will come to rest holding the boomerang at the same location from which he threw it.

b) It is impossible for the man to throw a boomerang in this situation.

c) It is possible for the man to throw a boomerang, but because he is standing on frictionless ice when he throws it, the boomerang cannot return.

d) The total momentum of the man-boomerang system is not conserved, so the man will be sliding backward holding the boomerang after he catches it.

**8.2** When a bismuth-208 nucleus at rest decays, thallium-204 is produced, along with an alpha particle (helium-4 nucleus). The mass numbers of bismuth-208, thallium-204, and helium-4 are 208, 204, and 4, respectively. (The mass number represents the total number of protons and neutrons in the nucleus.) The kinetic energy of the thallium nucleus is

a) equal to that of the alpha particle.

b) less than that of the alpha particle.

c) greater than that of the alpha particle.

**8.3** Two objects with masses $m_1$ and $m_2$ are moving along the $x$-axis in the positive direction with speeds $v_1$ and $v_2$, respectively, where $v_1$ is less than $v_2$. The speed of the center of mass of this system of two bodies is

a) less than $v_1$.

b) equal to $v_1$.

c) equal to the average of $v_1$ and $v_2$.

d) greater than $v_1$ and less than $v_2$.

e) greater than $v_2$.

**8.4** An artillery shell is moving on a parabolic trajectory when it explodes in midair. The shell shatters into a very large number of fragments. Which of the following statements is true (select all that apply)?

a) The force of the explosion will increase the momentum of the system of fragments, and so the momentum of the shell is *not* conserved during the explosion.

b) The force of the explosion is an internal force and thus cannot alter the total momentum of the system.

c) The center of mass of the system of fragments will continue to move on the initial parabolic trajectory until the last fragment touches the ground.

d) The center of mass of the system of fragments will continue to move on the initial parabolic trajectory until the first fragment touches the ground.

e) The center of mass of the system of fragments will have a trajectory that depends on the number of fragments and their velocities right after the explosion.

**8.5** An 80-kg astronaut becomes separated from his spaceship. He is 15.0 m away from it and at rest relative to it. In an effort to get back, he throws a 500-g object with a speed of 8.0 m/s in a direction away from the ship. How long does it take him to get back to the ship?

a) 1 s
b) 10 s
c) 20 s
d) 200 s
e) 300 s

**8.6** You find yourself in the (realistic?) situation of being stuck on a 300-kg raft (including yourself) in the middle of a pond with nothing but a pile of 7-kg bowling balls and 55-g tennis balls. Using your knowledge of rocket propulsion, you decide to start throwing balls from the raft to move toward shore. Which of the following will allow you to reach the shore faster?

a) throwing the tennis balls at 35 m/s at a rate of 1 tennis ball per second

b) throwing the bowling balls at 0.5 m/s at a rate of 1 bowling ball every 3 s

c) throwing a tennis ball and a bowling ball simultaneously, with the tennis ball moving at 15 m/s and the bowling ball moving at 0.3 m/s, at a rate of 1 tennis ball and 1 bowling ball every 4 s

d) not enough information to decide

**8.7** The figures show a high jumper using different techniques to get over the crossbar. Which technique would allow the jumper to clear the highest setting of the bar?

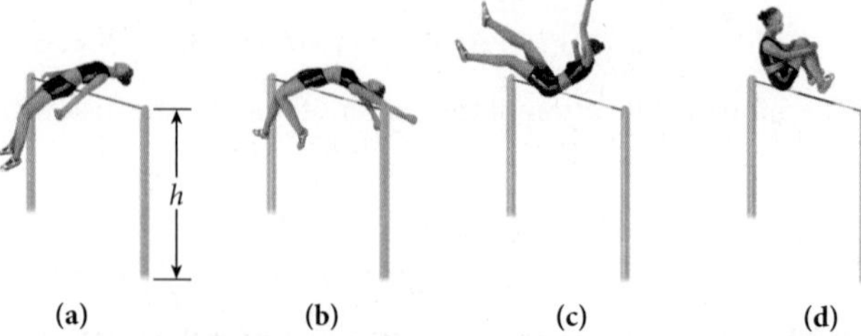


**8.8** The center of mass of an irregular rigid object is *always* located

a) at the geometrical center of the object.

b) somewhere within the object.

c) both of the above

d) none of the above

**8.9** A catapult on a level field tosses a 3-kg stone a horizontal distance of 100 m. A second 3-kg stone tossed in an identical fashion breaks apart in the air into 2 pieces, one with a mass of 1 kg and one with a mass of 2 kg. Both of the pieces hit the ground at the same time. If the 1-kg piece lands a distance of 180 m away from the catapult, how far away from the catapult does the 2-kg piece land? Ignore air resistance.

a) 20 m
b) 60 m
c) 100 m
d) 120 m
e) 180 m

**8.10** Two point masses are located in the same plane. The distance from mass 1 to the center of mass is 3.0 m. The distance from mass 2 to the center of mass is 1.0 m. What is $m_1/m_2$, the ratio of mass 1 to mass 2?

a) 3/4
b) 4/3
c) 4/7
d) 7/4
e) 1/3
f) 3/1

**8.11** A cylindrical bottle of oil-and-vinegar salad dressing whose volume is 1/3 vinegar ($\rho = 1.01$ g/cm$^3$) and 2/3 oil ($\rho = 0.910$ g/cm$^3$) is at rest on a table. Initially, the oil and the vinegar are separated, with the oil floating on top of the vinegar. The bottle is shaken so that the oil and vinegar mix uniformly, and the bottle is returned to the table. How has the height of the center of mass of the salad dressing changed as a result of the mixing?

a) It is higher.
b) It is lower.
c) It is the same.
d) There is not enough information to answer this question.

**8.12** A one-dimensional rod has a linear density that varies with position according to the relationship $\lambda(x) = cx$, where $c$ is a constant and $x = 0$ is the left end of the rod. Where do you expect the center of mass to be located?

a) the middle of the rod
b) to the left of the middle of the rod
c) to the right of the middle of the rod
d) at the right end of the rod
e) at the left end of the rod

## QUESTIONS

**8.13** A projectile is launched into the air. Part way through its flight, it explodes. How does the explosion affect the motion of the center of mass of the projectile?

**8.14** Find the center of mass of the arrangement of uniform identical cubes shown in the figure. The length of the sides of the each cube is $d$.

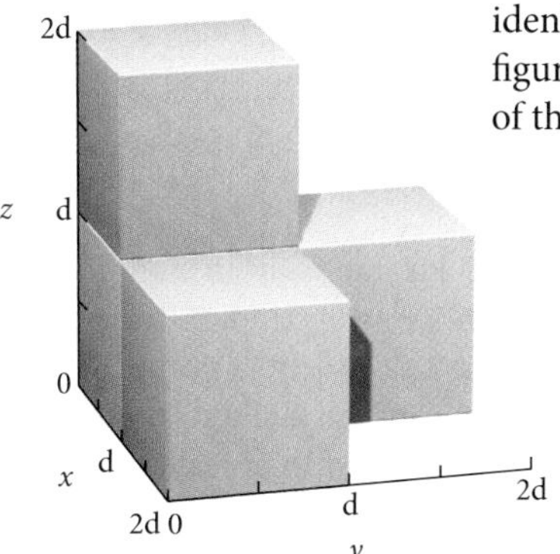


**8.15** A model rocket that has a horizontal range of 100 m is fired. A small explosion splits the rocket into two equal parts. What can you say about the points where the fragments land on the ground?

**8.16** Can the center of mass of an object be located at a point outside the object, that is, at a point in space where no part of the object is located? Explain.

**8.17** Is it possible for two masses to undergo a collision such that the system of two masses has more kinetic energy than the two separate masses had? Explain.

**8.18** Prove that the center of mass of a thin metal plate in the shape of an equilateral triangle is located at the intersection of the triangle's altitudes by direct calculation and by physical reasoning.

**8.19** A soda can of mass $m$ and height $L$ is filled with soda of mass $M$. A hole is punched in the bottom of the can to drain out the soda.

a) What is the center of mass of the system consisting of the can and the soda remaining in it when the level of soda in the can is $h$, where $0 < h < L$?

b) What is the minimum value of the center of mass as the soda drains out?

**8.20** An astronaut of mass $M$ is floating in space at a constant distance $D$ from his spaceship when his safety line breaks. He is carrying a toolbox of mass $M/2$ that contains a big sledgehammer of mass $M/4$, for a total mass of $3M/4$. He can throw the items with a speed $v$ relative to his final speed after each item is thrown. He wants to return to the spaceship as soon as possible.

a) To attain the maximum final speed, should the astronaut throw the two items together, or should he throw them one at a time? Explain.

b) To attain the maximum speed, is it best to throw the hammer first or the toolbox first, or does the order make no difference? Explain.

c) Find the maximum speed at which the astronaut can start moving toward the spaceship.

**8.21** A metal rod with a length density (mass per unit length) $\lambda$ is bent into a circular arc of radius $R$ and subtending a total angle of $\phi$, as shown in the figure. What is the distance of the center of mass of this arc from $O$ as a function of the angle $\phi$? Plot this center-of-mass coordinate as a function of $\phi$.

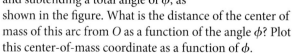

**8.22** The carton shown in the figure is filled with a dozen eggs, each of mass $m$. Initially, the center of mass of the eggs is at the center of the carton, which is the same point as the origin of the Cartesian coordinate system shown. Where is the center of mass of the remaining eggs, in terms of the egg-to-egg distance $d$, in each of the following situations? Neglect the mass of the carton.

Continued—

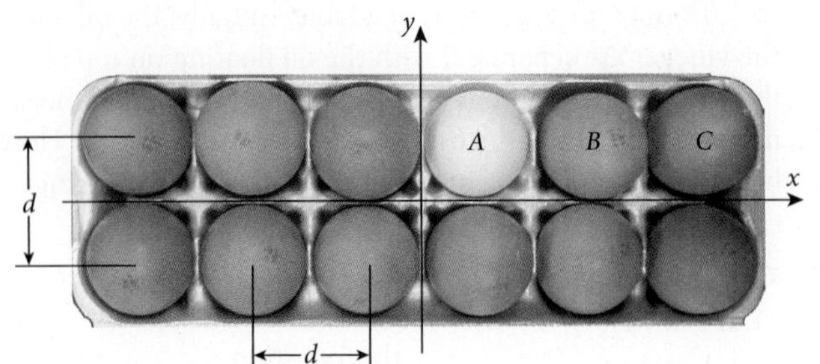


a) Only egg A is removed.
b) Only egg B is removed.
c) Only egg C is removed.
d) Eggs A, B and C are removed.

**8.23** A circular pizza of radius $R$ has a circular piece of radius $R/4$ removed from one side, as shown in the figure. Where is the center of mass of the pizza with the hole in it?

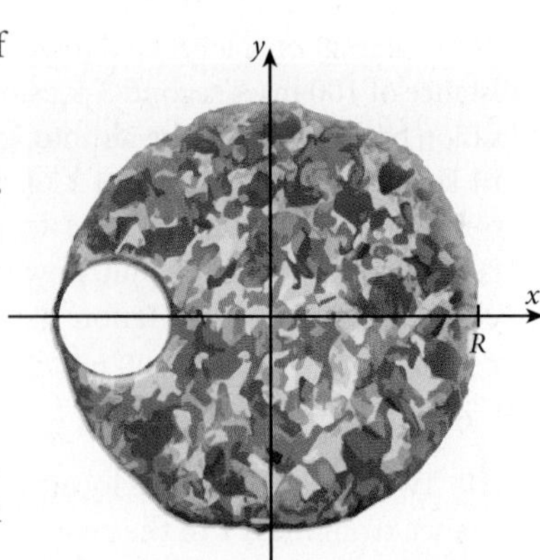


**8.24** Suppose you place an old-fashioned hourglass, with sand in the bottom, on a very sensitive analytical balance to determine its mass. You then turn it over (handling it with very clean gloves) and place it back on the balance. You want to predict whether the reading on the balance will be less than, greater than, or the same as before. What do you need to calculate to answer this question? Explain carefully what should be calculated and what the results would imply. You do not need to attempt the calculation.

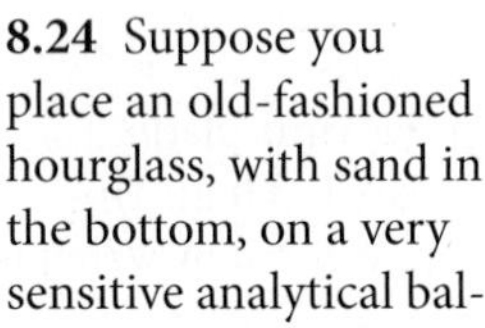

## PROBLEMS

A blue problem number indicates a worked-out solution is available in the Student Solutions Manual. One • and two •• indicate increasing level of problem difficulty.

### Section 8.1

8.25 Find the following center-of-mass information about objects in the Solar System. You can look up the necessary data on the Internet or in the tables in Chapter 12 of this book. Assume spherically symmetrical mass distributions for all objects under consideration.

a) Determine the distance from the center of mass of the Earth-Moon system to the geometric center of Earth.

b) Determine the distance from the center of mass of the Sun-Jupiter system to the geometric center of the Sun.

•**8.26** The coordinates of the center of mass for the extended object shown in the figure are $(L/4, -L/5)$. What are the coordinates of the 2-kg mass?

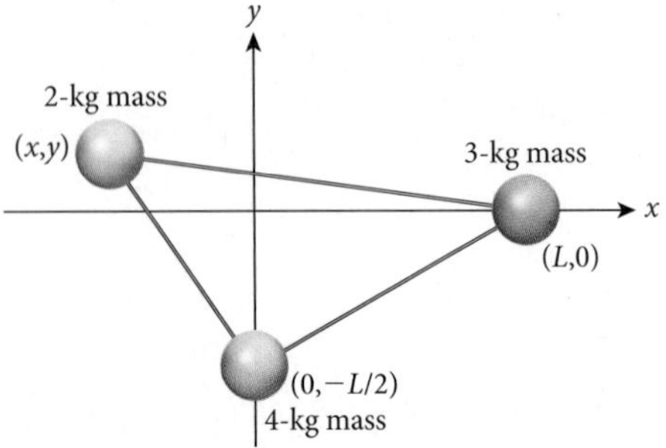


•**8.27** Young acrobats are standing still on a circular horizontal platform suspended at the center. The origin of the two-dimensional Cartesian coordinate system is assumed to be at the center of the platform. A 30-kg acrobat is located at (3m, 4m), and a 40-kg acrobat is located at (–2m, –2m). Assuming that the acrobats stand still in their positions, where must a 20-kg acrobat be located so that the center of mass of the system consisting of the three acrobats is at the origin and the platform is balanced?

### Section 8.2

**8.28** A man with a mass of 55 kg stands up in a 65-kg canoe of length 4.0 m floating on water. He walks from a point 0.75 m from the back of the canoe to a point 0.75 m from the front of the canoe. Assume negligible friction between the canoe and the water. How far does the canoe move?

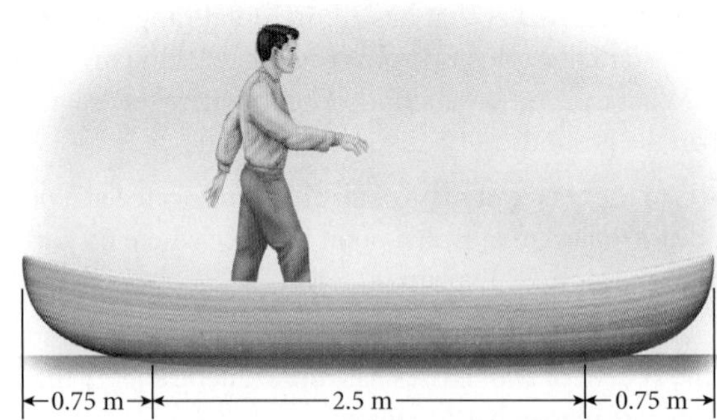


8.29 A toy car of mass 2.0 kg is stationary, and a child rolls a toy truck of mass 3.5 kg straight toward it with a speed of 4.0 m/s.

a) What is the velocity of the center of mass of the system consisting of the two toys?

b) What are the velocities of the truck and the car with respect to the center of mass of the system consisting of the two toys?

**8.30** A motorcycle stunt rider plans to start from one end of a railroad flatcar, accelerate toward the other end of the car, and jump from the flatcar to a platform. The motorcycle and rider have a mass of 350. kg and a length of 2.00 m. The flatcar has a mass of 1500. kg and a length of 20.0 m. Assume that there is negligible friction between the flatcar's wheels and the rails and that the motorcycle and rider can move through the air with negligible resistance. The flatcar is initially touching the platform. The promoters of the event have asked you how far the flatcar will be from the platform when the stunt rider reaches the end of the flatcar. What is your answer?

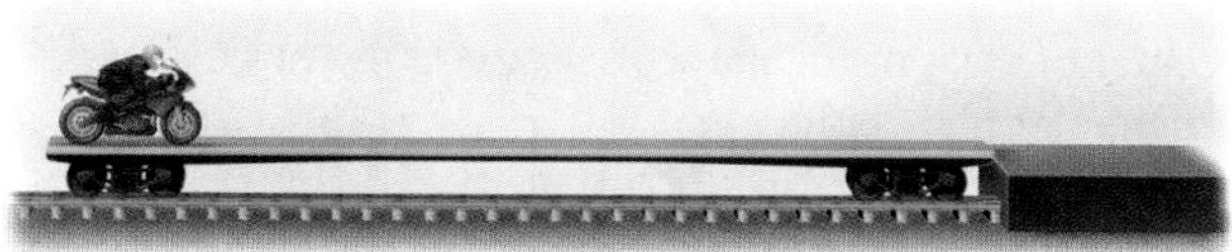

•**8.31** Starting at rest, two students stand on 10-kg sleds, which point away from each other on ice, and they pass a 5-kg medicine ball back and forth. The student on the left has a mass of 50 kg and can throw the ball with a relative speed of 10 m/s. The student on the right has a mass of 45 kg and can throw the ball with a relative speed of 12 m/s. (Assume there is no friction between the ice and the sleds and no air resistance.)

a) If the student on the left throws the ball horizontally to the student on the right, how fast is the student on the left moving right after the throw?

b) How fast is the student on the right moving right after catching the ball?

c) If the student on the right passes the ball back, how fast will the student on the left be moving after catching the pass from the student on the right?

(d) How fast is the student on the right moving after the pass?

•**8.32** Two skiers, Annie and Jack, start skiing from rest at different points on a hill at the same time. Jack, with mass 88 kg, skis from the top of the hill down a steeper section with an angle of inclination of 35°. Annie, with mass 64 kg, starts from a lower point and skis a less steep section, with an angle of inclination of 20°. The length of the steeper section is 100 m. Determine the acceleration, velocity, and position vectors of the combined center of mass for Annie and Jack as a function of time before Jack reaches the less steep section.

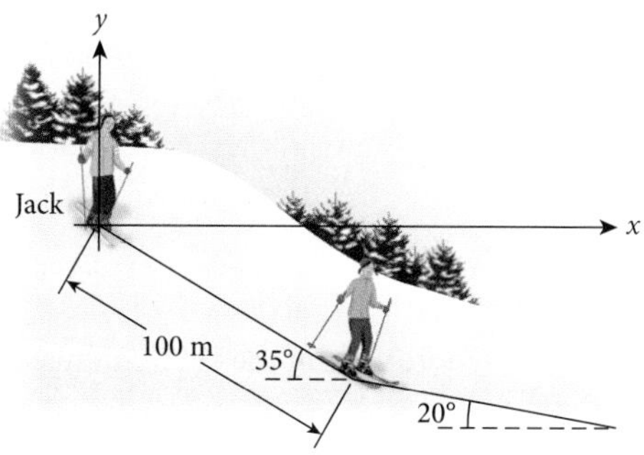


•8.33 Many nuclear collisions studied in laboratories are analyzed in a frame of reference relative to the laboratory. A proton, with a mass of $1.6605 \cdot 10^{-27}$ kg and traveling at a speed of 70% of the speed of light, $c$, collides with a tin-116 ($^{116}$Sn) nucleus with a mass of $1.9096 \cdot 10^{-25}$ kg. What is the speed of the center of mass with respect to the laboratory frame? Answer in terms of $c$, the speed of light.

•**8.34** A system consists of two particles. Particle 1 with mass 2.0 kg is located at (2.0 m, 6.0 m) and has a velocity of (4.0 m/s, 2.0 m/s). Particle 2 with mass 3.0 kg is located at (4.0 m, 1.0 m) and has a velocity of (0, 4.0 m/s).

a) Determine the position and the velocity of the center of mass of the system.

b) Sketch the position and velocity vectors for the individual particles and for the center of mass.

•**8.35** A fire hose 4.0 cm in diameter is capable of spraying water at a velocity of 10 m/s. For a continuous horizontal flow of water, what horizontal force should a fireman exert on the hose to keep it stationary?

••**8.36** A block of mass $m_b = 1.2$ kg slides to the right at a speed of 2.5 m/s on a frictionless horizontal surface, as shown in the figure. It "collides" with a wedge of mass $m_w$, which moves to the left at a speed of 1.1 m/s. The wedge is shaped so that the block slides seamlessly up the Teflon (frictionless!) surface, as the two come together. Relative to the horizontal surface, block and wedge are moving with a common velocity $v_{b+w}$ at the instant the block stops sliding up the wedge.

a) If the block's center of mass rises by a distance $h = 0.37$ m, what is the mass of the wedge?

b) What is $v_{b+w}$?

## Section 8.3

8.37 One important characteristic of rocket engines is the specific impulse, which is defined as the total impulse (time integral of the thrust) per unit ground weight of fuel/oxidizer expended. (The use of weight, instead of mass, in this definition is due to purely historical reasons.)

a) Consider a rocket engine operating in free space with an exhaust nozzle speed of $v$. Calculate the specific impulse of this engine.

b) A model rocket engine has a typical exhaust speed of $v_{toy} = 800$ m/s. The best chemical rocket engines have exhaust speeds of approximately $v_{chem} = 4.00$ km/s. Evaluate and compare the specific impulse values for these engines.

•**8.38** An astronaut is performing a space walk outside the International Space Station. The total mass of the astronaut with her space suit and all her gear is 115 kg. A small leak develops in her propulsion system and 7 g of gas are ejected each second into space with a speed of 800 m/s. She notices the leak 6 s after it starts. How much will the gas leak have caused her to move from her original location in space by that time?

•**8.39** A rocket in outer space has a payload of 5190.0 kg and $1.551 \cdot 10^5$ kg of fuel. The rocket can expel propellant at a speed of 5.600 km/s. Assume that the rocket starts from rest, accelerates to its final velocity, and then begins its trip. How long will it take the rocket to travel a distance of $3.82 \cdot 10^5$ km (approximately the distance between Earth and Moon)?

••**8.40** A uniform chain with a mass of 1.32 kg per meter of length is coiled on a table. One end is pulled upward at a constant rate of 0.47 m/s.

a) Calculate the net force acting on the chain.

b) At the instant when 0.15 m of the chain has been lifted off the table, how much force must be applied to the end being raised?

••**8.41** A spacecraft engine creates 53.2 MN of thrust with a propellant velocity of 4.78 km/s.

a) Find the rate ($dm/dt$) at which the propellant is expelled.

b) If the initial mass is $2.12 \cdot 10^6$ kg and the final mass is $7.04 \cdot 10^4$ kg, find the final speed of the spacecraft (assume the initial speed is zero and any gravitational fields are small enough to be ignored).

c) Find the average acceleration till burnout (the time at which the propellant is used up; assume the mass flow rate is constant until that time).

••**8.42** A cart running on frictionless air tracks is propelled by a stream of water expelled by a gas-powered pressure washer stationed on the cart. There is a 1.0-m$^3$ water tank on the cart to provide the water for the pressure washer. The mass of the cart, including the operator riding it, the pressure washer with its fuel, and the empty water tank, is 400. kg. The water can be directed, by switching a valve, either backward or forward. In both directions, the pressure washer ejects 200 L of water per min with a muzzle velocity of 25.0 m/s.

a) If the cart starts from rest, after what time should the valve be switched from backward (forward thrust) to forward (backward thrust) for the cart to end up at rest?

b) What is the mass of the cart at that time, and what is its velocity? (*Hint:* It is safe to neglect the decrease in mass due to the gas consumption of the gas-powered pressure washer!)

c) What is the thrust of this "rocket"?

d) What is the acceleration of the cart immediately before the valve is switched?

## Section 8.4

**8.43** A 32-cm-by-32-cm checkerboard has a mass of 100 g. There are four 20-g checkers located on the checkerboard, as shown in the figure. Relative to the origin located at the bottom left corner of the checkerboard, where is the center of mass of the checkerboard-checkers system?

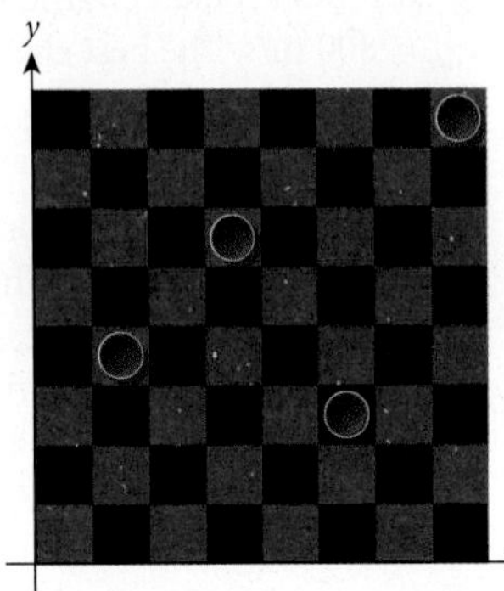


•**8.44** A uniform, square metal plate with side $L = 5.70$ cm and mass 0.205 kg is located with its lower left corner at $(x, y) = (0, 0)$, as shown in the figure. A square with side $L/4$ and its lower left edge located at $(x, y) = (0, 0)$ is removed from the plate. What is the distance from the origin of the center of mass of the remaining plate?

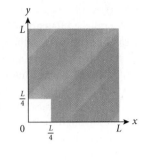

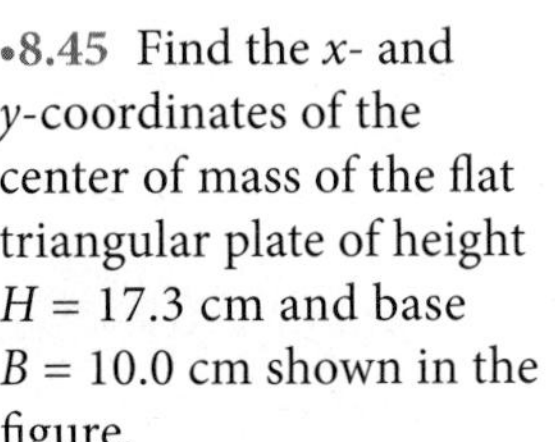

•**8.45** Find the $x$- and $y$-coordinates of the center of mass of the flat triangular plate of height $H = 17.3$ cm and base $B = 10.0$ cm shown in the figure.

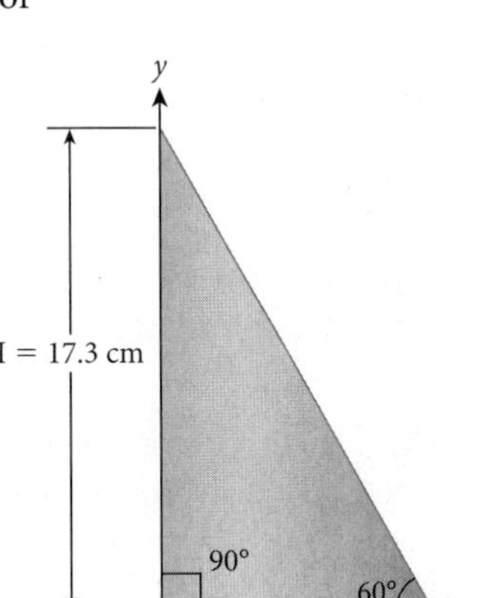


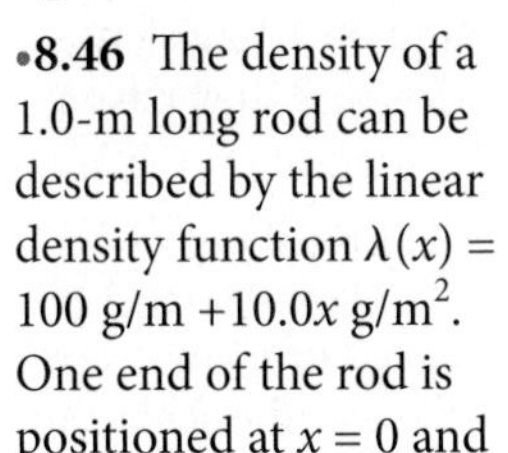

•**8.46** The density of a 1.0-m long rod can be described by the linear density function $\lambda(x) = 100 \text{ g/m} + 10.0x \text{ g/m}^2$. One end of the rod is positioned at $x = 0$ and the other at $x = 1$ m. Determine (a) the total mass of the rod, and (b) the center-of-mass coordinate.

•**8.47** A thin rectangular plate of uniform area density $\sigma_1 = 1.05$ kg/m$^2$ has a length $a = 0.600$ m and a width $b = 0.250$ m. The lower left corner is placed at the origin, $(x, y) = (0, 0)$. A circular hole of radius $r = 0.048$ m with center at $(x, y) = (0.068 \text{ m}, 0.068 \text{ m})$ is cut in the plate. The hole is plugged with a disk of the same radius that is composed of another material of uniform area density $\sigma_2 = 5.32$ kg/m$^2$. What is the distance from the origin of the resulting plate's center of mass?

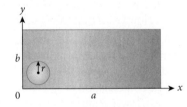

•**8.48** A uniform, square metal plate with side $L =$ 5.70 cm and mass 0.205 kg is located with its lower left corner at $(x, y) = (0, 0)$ as shown in the figure. Two squares with side length $L/4$ are removed from the plate.

a) What is the $x$-coordinate of the center of mass?

b) What is the $y$-coordinate of the center of mass?

••8.49 The linear mass density, $\lambda(x)$, for a one-dimensional object is plotted in the graph. What is the location of the center of mass for this object?

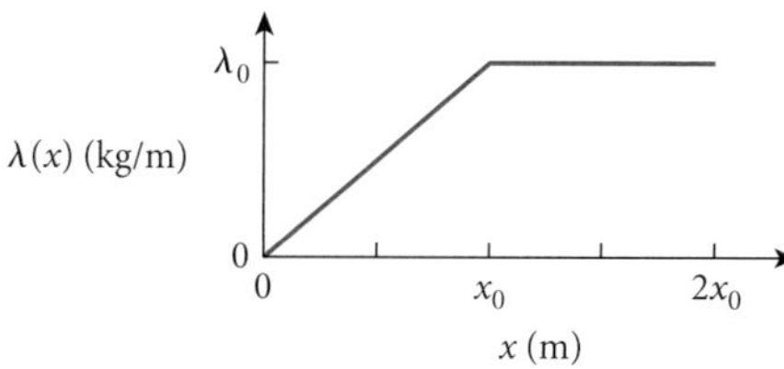


## Additional Problems

**8.50** A 750-kg cannon fires a 15-kg projectile with a speed of 250 m/s with respect to the muzzle. The cannon is on wheels and can recoil with negligible friction. Just after the cannon fires the projectile, what is the speed of the projectile with respect to the ground?

**8.51** The distance between a carbon atom ($m$ = 12 u) and an oxygen atom ($m$ = 16 u) in a carbon monoxide (CO) molecule is $1.13 \cdot 10^{-10}$ m. How far from the carbon atom is the center of mass of the molecule? (1 u = 1 atomic mass unit.)

**8.52** One method of detecting extrasolar planets involves looking for indirect evidence of a planet in the form of wobbling of its star about the star-planet system's center of mass. Assuming that the Solar System consisted mainly of the Sun and Jupiter, how much would the Sun wobble? That is, what back-and-forth distance would it move due to its rotation about the center of mass of the Sun-Jupiter system? How far from the center of the Sun is that center of mass?

8.53 The USS *Montana* is a massive battleship with a weight of 136,634,000 lb. It has twelve 16-inch guns, which are capable of firing 2700-lb projectiles at a speed of 2300 ft/s. If the battleship fires three of these guns (in the same direction), what is the recoil velocity of the ship?

**8.54** Three identical balls of mass $m$ are placed in the configuration shown in the figure. Find the location of the center of mass.

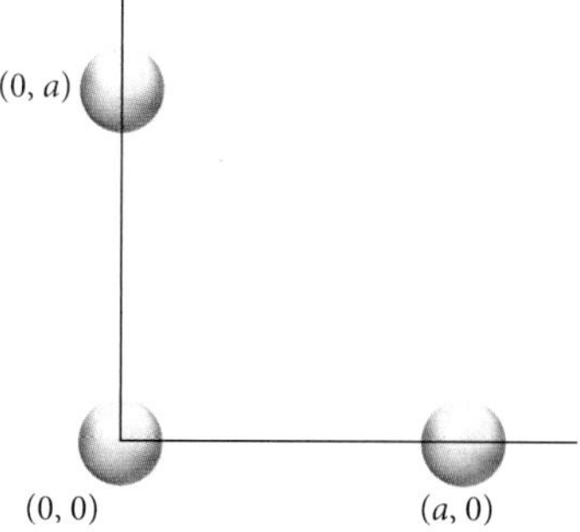


**8.55** Sam (61 kg) and Alice (44 kg) stand on an ice rink, providing them with a nearly frictionless surface to slide on. Sam gives Alice a push, causing her to slide away at a speed (with respect to the rink) of 1.20 m/s.

a) With what speed does Sam recoil?

b) Calculate the change in the kinetic energy of the Sam-Alice system.

c) Energy cannot be created or destroyed. What is the source of the final kinetic energy of this system?

**8.56** A baseball player uses a bat with mass $m_{\text{bat}}$ to hit a ball with mass $m_{\text{ball}}$. Right before he hits the ball, the bat's initial velocity is 35 m/s, and the ball's initial velocity is –30 m/s (the positive direction is along the positive $x$-axis). The bat and ball undergo a one-dimensional elastic collision. Find the speed of the ball after the collision. Assume that $m_{\text{bat}}$ is much greater than $m_{\text{ball}}$, so the center of mass of the two objects is essentially at the bat.

8.57 A student with a mass of 40 kg can throw a 5-kg ball with a relative speed of 10.0 m/s. The student is standing at rest on a cart of mass 10 kg that can move without friction. If the student throws the ball horizontally, what will the velocity of the ball with respect to the ground be?

•**8.58** Find the location of the center of mass of a two-dimensional sheet of constant density $\sigma$ that has the shape of an isosceles triangle (see the figure).

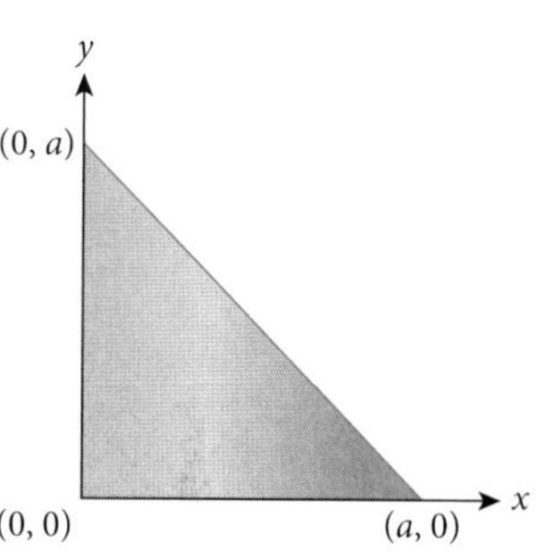


•**8.59** A rocket consists of a payload of 4390.0 kg and $1.761 \cdot 10^5$ kg of fuel. Assume that the rocket starts from rest in outer space, accelerates to its final velocity, and then begins its trip. What is the speed at which the propellant must be expelled to make the trip from the Earth to the Moon, a distance of $3.82 \cdot 10^5$ km, in 7.0 h?

•**8.60** A 350-kg cannon, sliding freely on a frictionless horizontal plane at a speed of 7.5 m/s, shoots a 15-kg cannonball at an angle of 55° above the horizontal. The velocity of the ball relative to the cannon is such that when the shot occurs, the cannon stops cold. What is the velocity of the ball relative to the cannon?

•8.61 The Saturn V rocket, which was used to launch the *Apollo* spacecraft on their way to the Moon, has an initial mass $M_0 = 2.8 \cdot 10^6$ kg and a final mass $M_1 = 0.8 \cdot 10^6$ kg and burns fuel at a constant rate for 160. s. The speed of the exhaust relative to the rocket is about $v$ = 2700. m/s.

a) Find the upward acceleration of the rocket, as it lifts off the launch pad (while its mass is the initial mass).

b) Find the upward acceleration of the rocket, just as it finishes burning its fuel (when its mass is the final mass).

c) If the same rocket were fired in deep space, where there is negligible gravitational force, what would be the net change in the speed of the rocket during the time it was burning fuel?

•**8.62** Find the location of the center of mass for a one-dimensional rod of length $L$ and of linear density $\lambda(x) = cx$, where $c$ is a constant. (*Hint:* You will need to calculate the mass in terms of $c$ and $L$.)

•**8.63** Find the center of mass of a rectangular plate of length 20 cm and width 10 cm. The mass density varies linearly along the length. At one end, it is 5 g/cm²; at the other end, it is 20 g/cm².

•**8.64** A uniform log of length 2.50 m has a mass of 91 kg and is floating in water. Standing on this log is a 72-kg man, located 22 cm from one end. On the other end is his daughter ($m = 20$ kg), standing 1 m from the end.

a) Find the center of mass of this system.

b) If the father jumps off the log backward away from his daughter ($v = 3.14$ m/s), what is the initial speed of log and child?

8.65 A sculptor has commissioned you to perform an engineering analysis of one of his works, which consists of regularly shaped metal plates of uniform thickness and density, welded together as shown in the figure. Using the intersection of the two axes shown as the origin of the coordinate system, determine the Cartesian coordinates of the center of mass of this piece.

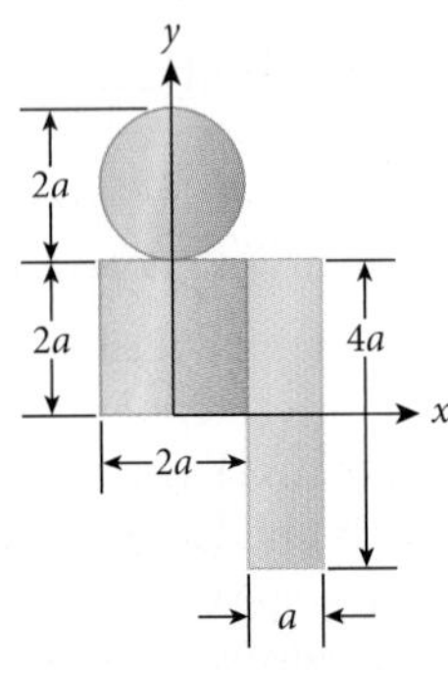


•**8.66** A jet aircraft is traveling at 223 m/s in horizontal flight. The engine takes in air at a rate of 80.0 kg/s and burns fuel at a rate of 3.00 kg/s. The exhaust gases are ejected at 600. m/s relative to the speed of the aircraft. Find the thrust of the jet engine.

•**8.67** A bucket is mounted on a skateboard, which rolls across a horizontal road with no friction. Rain is falling vertically into the bucket. The bucket is filled with water, and the total mass of the skateboard, bucket, and water is $M = 10$ kg. The rain enters the top of the bucket and simultaneously leaks out of a hole at the bottom of the bucket at equal rates of $\lambda = 0.10$ kg/s. Initially, bucket and skateboard are moving at a speed of $v_0$. How long will it take before the speed is reduced by half?

•**8.68** A 1000-kg cannon shoots a 30-kg shell at an angle of 25° above the horizontal and a speed of 500 m/s. What is the recoil velocity of the cannon?

•**8.69** Two masses, $m_1 = 2.0$ kg and $m_2 = 3.0$ kg, are moving in the $xy$-plane. The velocity of their center of mass and the velocity of mass 1 relative to mass 2 are given by the vectors $v_{cm} = (-1.0, +2.4)$ m/s and $v_{rel} = (+5.0, +1.0)$ m/s. Determine

a) the total momentum of the system

b) the momentum of mass 1, and

c) the momentum of mass 2.

••**8.70** You are piloting a spacecraft whose total mass is 1000 kg and attempting to dock with a space station in deep space. Assume for simplicity that the station is stationary, that your spacecraft is moving at 1.0 m/s toward the station, and that both are perfectly aligned for docking. Your spacecraft has a small retro-rocket at its front end to slow its approach, which can burn fuel at a rate of 1.0 kg/s and with an exhaust velocity of 100 m/s relative to the rocket. Assume that your spacecraft has only 20 kg of fuel left and sufficient distance for docking.

a) What is the initial thrust exerted on your spacecraft by the retro-rocket? What is the thrust's direction?

b) For safety in docking, NASA allows a maximum docking speed of 0.02 m/s. Assuming you fire the retro-rocket from time $t = 0$ in one sustained burst, how much fuel (in kilograms) has to be burned to slow your spacecraft to this speed relative to the space station?

c) How long should you sustain the firing of the retro-rocket?

d) If the space station's mass is 500,000 kg (close to the value for the ISS), what is the final velocity of the station after the docking of your spacecraft, which arrives with a speed of 0.02 m/s?

••**8.71** A chain whose mass is 3.0 kg and length is 5.0 m is held at one end so that the bottom end of the chain just touches the floor (see the figure). The top end of the chain is released. What is the force exerted by the chain on the floor just as the last link of the chain lands on the floor?

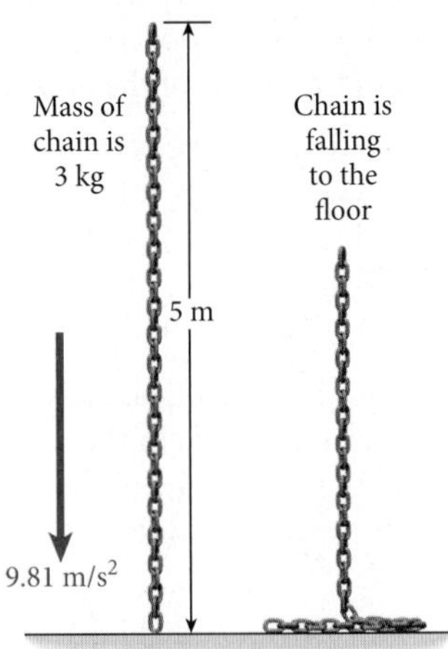

# Circular Motion

**FIGURE 9. 1** Two aircraft perform aerobatic maneuvers over Oshkosh, Wisconsin.

## WHAT WE WILL LEARN

- The motion of objects traveling in a circle rather than in a straight line can be described using coordinates based on radius and angle rather than Cartesian coordinates.
- There is a relationship between linear motion and circular motion.
- Circular motion can be described in terms of the angular coordinate, angular frequency, and period.
- An object undergoing circular motion can have angular velocity and angular acceleration.

Have you ever wondered why an airplane tilts, or banks, when it makes a turn in the air? An airplane has no road to fly on, so the force needed to make it turn in a circle cannot come from friction with a road surface. Instead, the angled orientation of the wings splits the normal force supporting the plane into components. One component continues to support the plane against gravity, while the other component acts horizontally and turns the plane in a circle, such as the aircraft shown in Figure 9.1. The same physics is applied in designing banked roads for cars or other vehicles.

In this chapter, we study circular motion and see how force is involved in making turns. This discussion relies on concepts of force, velocity, and acceleration presented in Chapters 3 and 4. Chapter 10 will combine these ideas with some of the concepts from Chapters 5 through 8, such as energy and momentum. Much of what you will learn in these two chapters on circular motion and rotation is analogous to earlier material on linear motion, force, and energy. Because most objects do not travel in perfectly straight lines, the concepts of circular motion will be applied many times in later chapters.

## 9.1 Polar Coordinates

**FIGURE 9.2** Circular motion in the horizontal and in the vertical plane.

In Chapter 3, we discussed motion in two dimensions. In this chapter, we examine a special case of motion in a two-dimensional plane: motion of an object along the circumference of a circle. To be precise, we will only study circular motion of objects that we can consider to be point particles. In Chapter 10 on rotation, we will relax this condition and also examine extended bodies.

Circular motion is surprisingly common. Riding on a carousel or many other amusement park rides, such as those shown in Figure 9.2, qualifies as circular motion. Indy-style car racing also involves circular motion, as the cars alternate between moving along straight sections and half-circle segments of the track. CD, DVD, and Blu-ray players also operate with circular motion, although this motion is usually hidden from the eye.

During an object's **circular motion,** its $x$- and $y$-coordinates change continuously, but the distance from the object to the center of the circular path stays the same. We can take advantage of this fact by using **polar coordinates** to study circular motion. Shown in Figure 9.3 is the position vector, $\vec{r}$, of an object in circular motion. This vector changes as a function of time, but its tip always moves on the circumference of a circle. We can specify $\vec{r}$ by giving its $x$- and $y$-components. However, we can specify the same vector by giving two other numbers: the angle of $\vec{r}$ relative to the $x$-axis, $\theta$, and the length of $\vec{r}$, $r = |\vec{r}|$ (Figure 9.3).

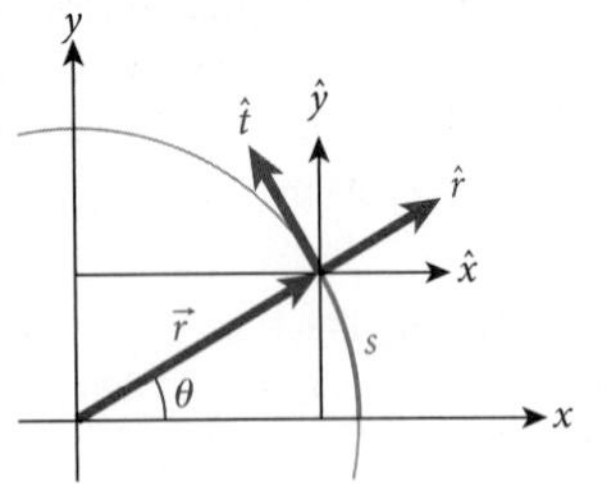


**FIGURE 9.3** Polar coordinate system for circular motion.

Trigonometry provides the relationship between the Cartesian coordinates $x$ and $y$ and the polar coordinates $\theta$ and $r$:

$$r = \sqrt{x^2 + y^2} \tag{9.1}$$

$$\theta = \tan^{-1}(y/x). \tag{9.2}$$

The inverse transformation from polar to Cartesian coordinates is given by (see Figure 9.4):

$$x = r\cos\theta \tag{9.3}$$

$$y = r\sin\theta. \tag{9.4}$$

The major advantage of using polar coordinates for analyzing circular motion is that $r$ never changes. It remains the same as long as the tip of the vector $\vec{r}$ moves along the circular path. Thus, we can reduce the description of two-dimensional motion on the circumference of a circle to a one-dimensional problem involving the angle $\theta$.

Figure 9.3 also shows the unit vectors in the radial and tangential directions, $\hat{r}$ and $\hat{t}$, respectively. The angle between $\hat{r}$ and $\hat{x}$ is again the angle $\theta$. Therefore, the Cartesian components of the radial unit vector can be written as follows (refer to Figure 9.4):

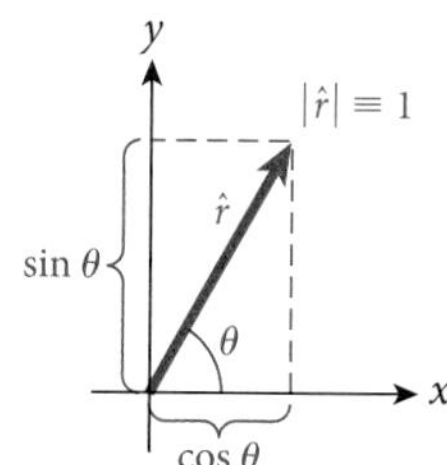


**FIGURE 9.4** Relationship between the radial unit vector and the sine and cosine of the angle.

$$\hat{r} = \frac{x}{r}\hat{x} + \frac{y}{r}\hat{y} = (\cos\theta)\hat{x} + (\sin\theta)\hat{y} \equiv (\cos\theta, \sin\theta). \tag{9.5}$$

Similarly, we obtain the Cartesian components of the tangential unit vector:

$$\hat{t} = \frac{-y}{r}\hat{x} + \frac{x}{r}\hat{y} = (-\sin\theta)\hat{x} + (\cos\theta)\hat{y} \equiv (-\sin\theta, \cos\theta). \tag{9.6}$$

(Note that the tangential unit vector is always denoted with a caret, $\hat{t}$, and can thus be readily distinguished from the time, $t$.) It is straightforward to verify that the radial and tangential unit vectors are perpendicular to each other by taking their scalar product:

$$\hat{r} \bullet \hat{t} = (\cos\theta, \sin\theta) \bullet (-\sin\theta, \cos\theta) = -\cos\theta\sin\theta + \sin\theta\cos\theta = 0.$$

We can also verify that these two unit vectors have a length of 1, as required:

$$\hat{r} \bullet \hat{r} = (\cos\theta, \sin\theta) \bullet (\cos\theta, \sin\theta) = \cos^2\theta + \sin^2\theta = 1$$
$$\hat{t} \bullet \hat{t} = (-\sin\theta, \cos\theta) \bullet (-\sin\theta, \cos\theta) = \sin^2\theta + \cos^2\theta = 1.$$

Finally, it is important to stress that, for circular motion, there is a major difference between the unit vectors for polar coordinates and those for Cartesian coordinates: The Cartesian unit vectors stay constant in time, while the radial and tangential unit vectors change their directions during the process of circular motion. This is because both of these vectors both depend on the angle $\theta$, which for circular motion depends on the time.

## 9.2 Angular Coordinates and Angular Displacement

Polar coordinates allow us to describe and analyze circular motion, where the distance to the origin, $r$, of the object in motion stays constant and the angle $\theta$ varies as a function of time, $\theta(t)$. As was already pointed out, the angle $\theta$ is measured relative to the positive $x$-axis. Any point on the positive $x$-axis has an angle $\theta = 0$. As the definition in equation 9.2 implies, a move in the counterclockwise direction away from the positive $x$-axis toward the positive $y$-axis results in positive values for the angle $\theta$. Conversely, a clockwise move away from the positive $x$-axis toward the negative $y$-axis results in negative values of $\theta$.

The two most commonly used units for angles are degrees (°) and radians (rad). These units are defined such that the angle measured by one complete circle is 360°, which corresponds to $2\pi$ rad. Thus, the unit conversion between the two angular measures is

$$\theta\text{ (degrees)}\frac{\pi}{180} = \theta\text{ (radians)} \Leftrightarrow \theta\text{ (radians)}\frac{180}{\pi} = \theta\text{ (degrees)}$$

$$1\text{ rad} = \frac{180°}{\pi} \approx 57.3°.$$

Like the linear position $x$, the angle $\theta$, can have positive and negative values. However, $\theta$ is periodic; a complete turn around the circle ($2\pi$ rad, or 360°) returns the coordinate for $\theta$ to the same point in space. Just as the linear displacement, $\Delta x$, is defined to be the difference between two positions, $x_2$ and $x_1$, the angular displacement, $\Delta\theta$, is the difference between two angles:

$$\Delta\theta = \theta_2 - \theta_1.$$

## EXAMPLE 9.1 Locating a Point with Cartesian and Polar Coordinates

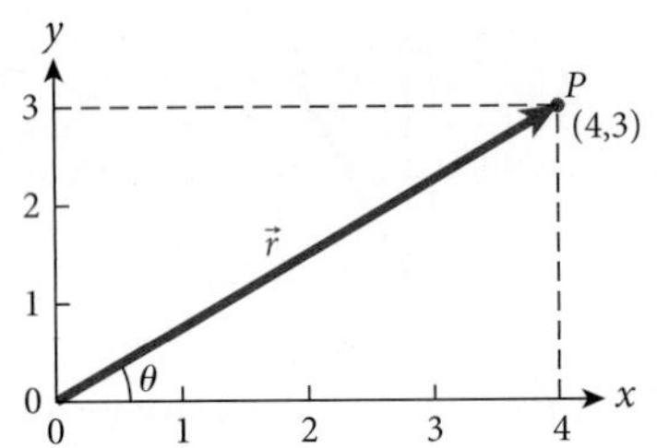


**FIGURE 9.5** A point located at (4,3) in a Cartesian coordinate system.

A point has a location given in Cartesian coordinates as (4,3), as shown in Figure 9.5.

### PROBLEM

How do we represent the position of this point in polar coordinates?

### SOLUTION

Using equation 9.1, we can calculate the radial coordinate:

$$r=\sqrt{x^2+y^2}=\sqrt{4^2+3^2}=5.$$

Using equation 9.2, we can calculate the angular coordinate:

$$\theta=\tan^{-1}(y/x)=\tan^{-1}(3/4)=0.64 \text{ rad}=37°.$$

Therefore, we can express the position of point $P$ in polar coordinates as $(r, \theta) = (5,0.64 \text{ rad}) = (5,37°)$. Note that we can specify the same position by adding (any integral multiple of) $2\pi$ rad, or 360°, to $\theta$:

$$(r,\theta)=(5,0.64 \text{ rad})=(5,37°)=(5,2\pi \text{ rad}+0.64 \text{ rad})=(5,360°+37°).$$

### 9.1 Self-Test Opportunity

Use polar coordinates and calculus to show that the circumference of a circle with radius $R$ is $2\pi R$.

### 9.2 Self-Test Opportunity

Use polar coordinates and calculus to show that the area of a circle with radius $R$ is $\pi R^2$.

## Arc Length

Figure 9.3 also shows (in green) the path on the circumference of the circle traveled by the tip of the vector $\vec{r}$ in going from an angle of zero to $\theta$. This path is called the *arc length, s*. It is related to the radius and angle via

$$s=r\theta. \tag{9.7}$$

For this relationship to work out numerically, the angle has to be measured in radians. The fact that the circumference of a circle is $2\pi r$ is a special case of equation 9.7 with $\theta = 2\pi$ rad, corresponding to one full turn around the circle. The arc length has the same unit as the radius.

For small angles, measuring a degree or less, the sine of the angle is approximately (to four significant figures) equal to the angle measured in radians. Because of this fact as well as the need to use equation 9.7 to solve problems, the preferred unit for angular coordinates is the radian. But the use of degrees is common, and this book uses both units.

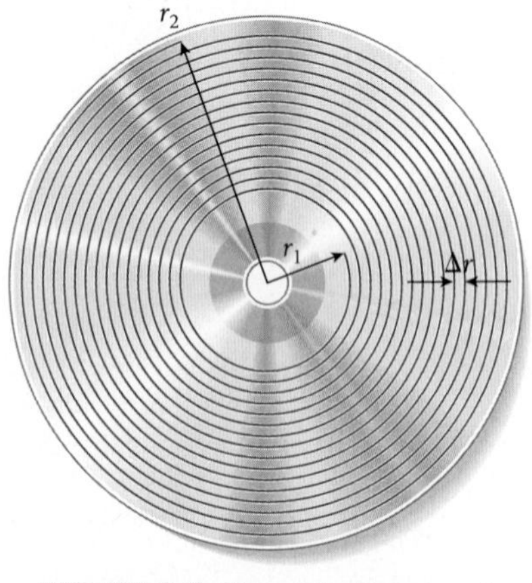

## EXAMPLE 9.2 CD Track

**FIGURE 9.6** The track of a compact disc.

The track on a compact disc (CD) is represented in Figure 9.6. The track is a spiral, originating at an inner radius of $r_1 = 25$ mm and terminating at an outer radius of $r_2 = 58$ mm. The spacing between successive loops of the track is a constant, $\Delta r = 1.6$ μm.

### PROBLEM

What is the total length of this track?

### SOLUTION 1, WITHOUT CALCULUS

At a given radius $r$ between $r_1$ and $r_2$, the track is almost perfectly circular. Since the track spacing is $\Delta r = 1.6$ μm, and the distance between the innermost and outermost parts is $r_2 - r_1 = (58 \text{ mm}) - (25 \text{ mm}) = 33$ mm, we find that the track winds around a total of

$$n=\frac{r_2-r_1}{\Delta r}=\frac{3.3\cdot 10^{-2}\text{ m}}{1.6\cdot 10^{-6}\text{ m}}=20{,}625 \text{ times}.$$

(Note that we did not round this intermediate result!) The radii of these 20,625 circles grow linearly from 25 mm to 58 mm; consequently, the circumferences ($c = 2\pi r$) also grow linearly. The average radius of the circles is then $\bar{r}=\frac{1}{2}(r_2+r_1)=41.5$ mm, and the average circumference is

$$\bar{c}=2\pi\bar{r}=2\pi\cdot 41.5 \text{ mm}=0.2608 \text{ m}.$$

Now we have to multiply this average circumference by the number of circles to obtain our result:

$$L = n\bar{c} = 20{,}625 \cdot 0.2608 \text{ m} = 5.4 \text{ km},$$

rounded to two significant figures. Thus, the length of a track on a CD is more than 3.3 mi!

### SOLUTION 2, WITH CALCULUS

The spacing between successive loops of the track is a constant, $\Delta r = 1.6\ \mu\text{m}$; therefore, the track density (that is, the number of times the track is crossed in the radial direction per unit length moving outward from the innermost point) is $1/\Delta r = 625{,}000\ \text{m}^{-1}$. At a given radius $r$ between $r_1$ and $r_2$, the track is almost perfectly circular. This segment of the spiral track has a length of $2\pi r$, but the loops' lengths grow steadily longer moving to the outside. We obtain the overall length of the track by integrating the length of each loop from $r_1$ to $r_2$, multiplied by the number of loops per unit length:

$$L = \frac{1}{\Delta r}\int_{r_1}^{r_2} 2\pi r\,dr = \frac{1}{\Delta r}\pi r^2\bigg|_{r_1}^{r_2} = \frac{1}{\Delta r}\pi\left(r_2^2 - r_1^2\right)$$

$$= (625{,}000\ \text{m}^{-1})\pi\left((0.058\ \text{m})^2 - (0.025\ \text{m})^2\right) = 5.4\ \text{km}.$$

Satisfyingly, this calculus-based solution agrees with the solution we obtained without the use of calculus. You may want to think about what conditions had to be met to avoid using calculus in this case. (For a DVD, by the way, the track density is higher by a factor of 2.2, resulting in a track that is almost 12 km long.)

Figure 9.7 shows a small portion of the tracks of two CDs, at a magnification of 500 times. Figure 9.7a shows a factory-pressed CD, with the individual aluminum bumps visible. In Figure 9.7b is a read-write CD, which is burned when a laser induces a phase change (Chapter 34 on wave optics will explain this process) in the continuous track. The lower right portion of this picture shows a portion of the CD on which nothing has yet been written.

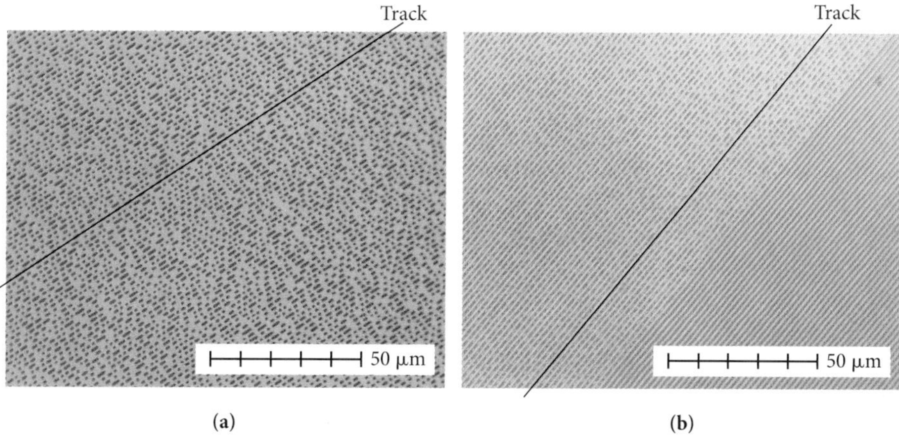


**FIGURE 9.7** Microscope pictures (magnification 500×) of (a) a factory-pressed CD and (b) a read-write CD. The solid lines indicate the direction of the tracks.

## 9.3 Angular Velocity, Angular Frequency, and Period

We have seen that the change of an object's linear coordinates in time is its velocity. Similarly, the change of an object's angular coordinate in time is its **angular velocity.** The average magnitude of the angular velocity is defined as

$$\bar{\omega} = \frac{\theta_2 - \theta_1}{t_2 - t_1} = \frac{\Delta\theta}{\Delta t}.$$

This definition uses the notation $\theta_1 \equiv \theta(t_1)$ and $\theta_2 \equiv \theta(t_2)$. The horizontal bar above the symbol $\omega$ for angular velocity again indicates a time average. By taking the limit of this expression as

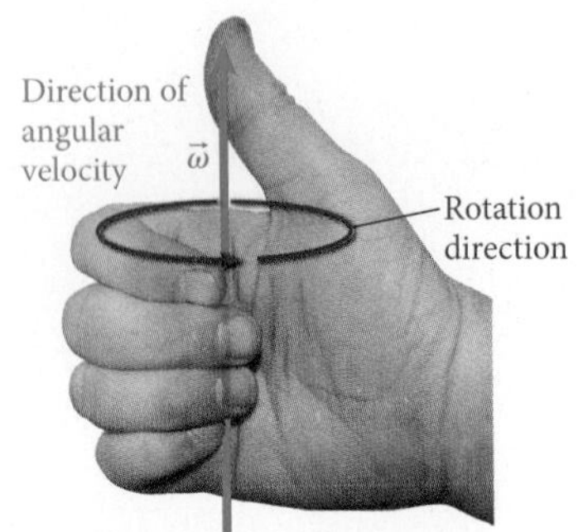


**FIGURE 9.8** The right-hand rule for determining the direction of the angular velocity vector.

**FIGURE 9.9** The tachometer of a car measures the frequency (in cycles per minute) of revolutions of the engine.

the time interval approaches zero, we find the instantaneous value of the magnitude of the angular velocity:

$$\omega = \lim_{\Delta t \to 0} \bar{\omega} = \lim_{\Delta t \to 0} \frac{\Delta \theta}{\Delta t} \equiv \frac{d\theta}{dt}. \tag{9.8}$$

The most common unit of angular velocity is radians per second (rad/s); degrees per second is not generally used.

Angular velocity is a vector. Its direction is that of an axis through the center of the circular path and perpendicular to the plane of the circle. (This axis is a rotation axis, as we will discuss in more detail in Chapter 10.) This definition allows for two possibilities for the direction in which the vector $\vec{\omega}$ can point: up or down, parallel or antiparallel to the axis of rotation. A right-hand rule helps us decide which is the correct direction: When the fingers point in the direction of rotation along the circle's circumference, the thumb points in the direction of $\vec{\omega}$, as shown in Figure 9.8.

Angular velocity measures how fast the angle $\theta$ changes in time. Another quantity also specifies how fast this angle changes in time—the **angular frequency,** or simply frequency, $f$. For example, the rpm number on the tachometer in your car indicates how many times per minute the engine cycles and thus specifies the frequency of engine revolution. Figure 9.9 shows a tachometer, with the units specified as "1/min × 1000"; the engine hits the red line at 6000 revolutions per minute. Thus, the frequency, $f$, measures cycles per unit time, instead of radians per unit time as the angular velocity does. The frequency is related to the magnitude of the angular velocity, $\omega$, by

$$f = \frac{\omega}{2\pi} \Leftrightarrow \omega = 2\pi f. \tag{9.9}$$

This relationship makes sense because one complete turn around a circle requires an angle change of $2\pi$ rad. (Be careful—both frequency and angular velocity have the same unit of inverse seconds and can be easily confused.)

Because the unit inverse second is used so widely, it was given its own name, the **hertz** (Hz), for the German physicist Heinrich Rudolf Hertz (1857–1894): 1 Hz = 1 s$^{-1}$. The **period of rotation,** $T$, is defined as the inverse of the frequency:

$$T = \frac{1}{f}. \tag{9.10}$$

The period measures the time interval between two successive instances where the angle has the same value; that is, the time it takes to pass once around the circle. The unit of the period is the same as that of time, the second (s). Given the relationships between period and frequency and between frequency and angular velocity, we also obtain

$$\omega = 2\pi f = \frac{2\pi}{T}. \tag{9.11}$$

## Angular Velocity and Linear Velocity

If we take the time derivative of the position vector, we obtain the linear velocity vector. To find the angular velocity, it is most convenient to write the radial position vector in Cartesian coordinates and take the time derivatives component by component:

$$\vec{r} = x\hat{x} + y\hat{y} = (x, y) = (r\cos\theta, r\sin\theta) = r(\cos\theta, \sin\theta) = r\hat{r} \Rightarrow$$

$$\vec{v} = \frac{d\vec{r}}{dt} = \frac{d}{dt}(r\cos\theta, r\sin\theta) = \left(\frac{d}{dt}(r\cos\theta), \frac{d}{dt}(r\sin\theta)\right).$$

Now we can use the fact that, for motion along a circle, the distance $r$ to the origin does not change in time but is constant. This results in

$$\begin{aligned}\vec{v} &= \left(\frac{d}{dt}(r\cos\theta), \frac{d}{dt}(r\sin\theta)\right) = \left(r\frac{d}{dt}(\cos\theta), r\frac{d}{dt}(\sin\theta)\right)\\ &= \left(-r\sin\theta\frac{d\theta}{dt}, r\cos\theta\frac{d\theta}{dt}\right)\\ &= (-\sin\theta, \cos\theta)r\frac{d\theta}{dt}.\end{aligned}$$

Here we used the chain rule of differentiation in the next to last step and then factored out the common factor $r\,d\theta/dt$. We already know that the time derivative of the angle is the angular velocity (see equation 9.8). In addition, we recognize the vector $(-\sin\theta, \cos\theta)$ as the tangential unit vector (see equation 9.6). Thus, we have the relationship between angular and linear velocities for circular motion:

$$\vec{v} = r\omega\hat{t}. \tag{9.12}$$

(Again, $\hat{t}$ is the symbol for the tangential unit vector and has no connection with the time, $t$!)

Because the velocity vector points in the direction of the tangent to the trajectory at any given time, it is always tangential to the circle's circumference, pointing in the direction of the motion, as shown in Figure 9.10. Thus, the velocity vector is always perpendicular to the position vector, which points in the radial direction. If two vectors are perpendicular to each other, their scalar product is zero. Thus, for circular motion, we always have

$$\vec{r}\bullet\vec{v} = (r\cos\theta, r\sin\theta)\bullet(-r\omega\sin\theta, r\omega\cos\theta) = 0.$$

If we take the absolute values of the left- and right-hand sides of equation 9.12, we obtain an important relationship between the magnitudes of the linear and angular speeds for circular motion:

$$v = r\omega. \tag{9.13}$$

Remember that this relationship holds only for the *magnitudes* of the linear and angular velocities. Their vectors point in different directions and, for uniform circular motion, are perpendicular to each other, with $\vec{\omega}$ pointing in the direction of the rotation axis, and $\vec{v}$ tangential to the circle.

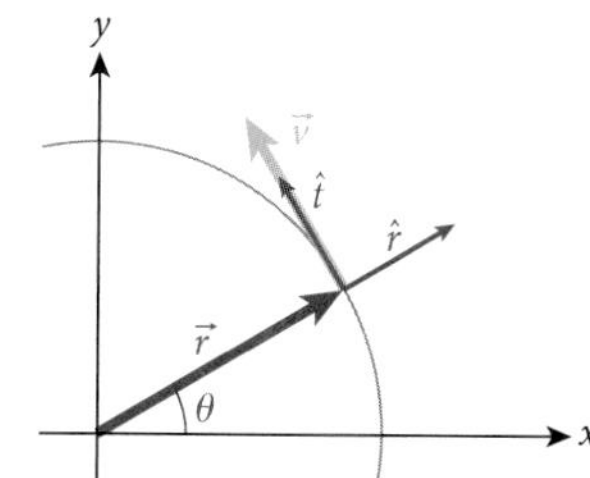


**FIGURE 9.10** Linear velocity and coordinate vectors.

### 9.1 In-Class Exercise

A bicycle has wheels with a radius of 33.0 cm. The bicycle is traveling at a speed of 6.5 m/s. What is the angular speed of the front tire?

a) 0.197 rad/s

b) 1.24 rad/s

c) 5.08 rad/s

d) 19.7 rad/s

e) 215 rad/s

## EXAMPLE 9.3 Revolution and Rotation of the Earth

### PROBLEM

The Earth orbits around the Sun and also rotates on its pole-to-pole axis. What are the angular velocities, frequencies, and linear speeds of these motions?

### SOLUTION

Any point on the surface of Earth moves in circular motion around the rotation axis (pole-to-pole), with a rotation period of 1 day. Expressed in seconds, this period is

$$T_{\text{Earth}} = 1\text{ day}\cdot\frac{24\text{ h}}{1\text{ day}}\cdot\frac{3600\text{ s}}{1\text{ h}} = 8.64\cdot10^4\text{ s}.$$

The Earth moves around the Sun on an elliptical path; the path is very close to circular, so let's treat the Earth's orbit as circular motion. The orbital period for the motion of the Earth around the Sun is 1 year. Expressing this period in seconds, we obtain

$$T_{\text{Sun}} = 1\text{ year}\cdot\frac{365\text{ days}}{1\text{ year}}\cdot\frac{24\text{ h}}{1\text{ day}}\cdot\frac{3600\text{ s}}{1\text{ h}} = 3.15\cdot10^7\text{ s}.$$

Both circular motions have constant angular velocity. Thus, we can use $T = 1/f$ and $\omega = 2\pi f$ to obtain the frequencies and angular velocities:

$$f_{\text{Earth}} = \frac{1}{T_{\text{Earth}}} = 1.16\cdot10^{-5}\text{ Hz};\ \omega_{\text{Earth}} = 2\pi f_{\text{Earth}} = 7.27\cdot10^{-5}\text{ rad/s}$$

$$f_{\text{Sun}} = \frac{1}{T_{\text{Sun}}} = 3.17\cdot10^{-8}\text{ Hz};\ \omega_{\text{Sun}} = 2\pi f_{\text{Sun}} = 1.99\cdot10^{-7}\text{ rad/s}.$$

Note that the 24-hour period that we used as the length of a day is how long it takes the Sun to reach the same position in the sky. If we wanted to specify the frequencies and angular velocities to greater precision, we would have to use the sidereal day: Because the Earth is also moving around the Sun during each day, the time it actually takes to complete one full rotation and bring the stars into the same position in the night sky is the sidereal day, which consists of 23 h, 56 min, and a little more than 4 s, or 86,164.09074 s =

$(1 - 1/365.2425) \cdot 86{,}400$ s. (We have used the fact that it takes a fraction of a day more than 365 to complete an orbit around the Sun—which is why leap years are necessary.)

Now let's find the linear velocity at which the Earth orbits the Sun. Because we are assuming circular motion, the relationship between the orbital speed and angular velocity is given by $v = r\omega$. To get our answer, we need to know the radius of the orbit. The radius of this orbit is the Earth-Sun distance, $r_{\text{Earth-Sun}} = 1.49 \cdot 10^{11}$ m. Therefore, the linear orbital speed—the speed with which the Earth moves around the Sun—is

$$v = r\omega = (1.49 \cdot 10^{11} \text{ m})(1.99 \cdot 10^{-7} \text{ s}^{-1}) = 2.97 \cdot 10^4 \text{ m/s}.$$

This is over 66,000 mph!

Now we want to find the speed of a point on the surface of the rotating Earth relative to the center of the Earth. We note that points at different latitudes have different distances to the axis of rotation, as shown in Figure 9.11. At the Equator, the radius of the orbit is $r = R_{\text{Earth}} = 6380$ km. Away from the Equator, the radius of rotation as a function of the latitude angle is $r = R_{\text{Earth}} \cos \vartheta$; see Figure 9.11. (The script theta, $\vartheta$, is used for the latitude angles here in order to avoid confusion with $\theta$, which is used for the motion in the circle.) In general, we obtain the following formula for the speed of rotation:

$$\begin{aligned} v &= \omega r = \omega R_{\text{Earth}} \cos \vartheta \\ &= (7.27 \cdot 10^{-5} \text{ s}^{-1})(6.38 \cdot 10^6 \text{ m})(\cos \vartheta) \\ &= (464 \text{ m/s})(\cos \vartheta). \end{aligned}$$

At the poles, where $\vartheta = 90°$, the rotation speed is zero; at the Equator, where $\vartheta = 0°$, the speed is 464 m/s. Seattle, with $\vartheta = 47.5°$, moves at $v = 313$ m/s, and Miami, with $\vartheta = 25.7°$, has a speed of $v = 418$ m/s.

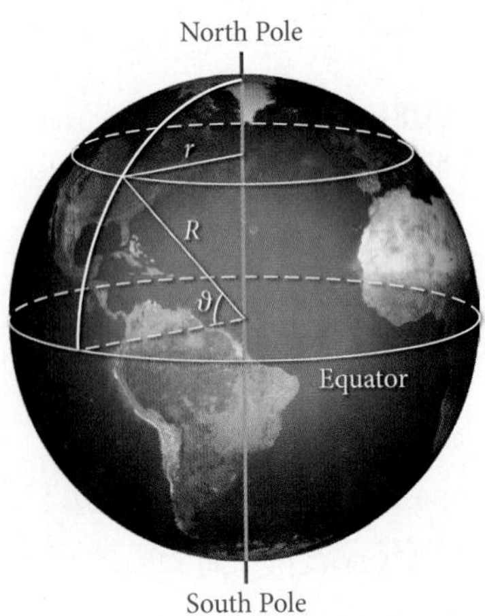


**FIGURE 9.11** Earth's rotation axis is indicated by the vertical line. Points at different latitudes on the Earth's surface move at different speeds.

## 9.4 Angular and Centripetal Acceleration

The rate of change of an object's angular velocity is its **angular acceleration,** denoted by the Greek letter $\alpha$. The definition of the magnitude of the angular acceleration is analogous to that for the linear acceleration. Its time average is defined as

$$\bar{\alpha} = \frac{\Delta\omega}{\Delta t}.$$

The instantaneous magnitude of the angular acceleration is obtained in the limit as the time interval approaches zero:

$$\alpha = \lim_{\Delta t \to 0} \bar{\alpha} = \lim_{\Delta t \to 0} \frac{\Delta\omega}{\Delta t} \equiv \frac{d\omega}{dt} = \frac{d^2\theta}{dt^2}. \tag{9.14}$$

Just as we related the linear velocity to the angular velocity, we can also relate the tangential acceleration to the angular acceleration. We start with the definition of the linear acceleration vector as the time derivative of the linear velocity vector. Then we substitute the expression for the linear velocity in circular motion from equation 9.12:

$$\vec{a}(t) = \frac{d}{dt}\vec{v}(t) = \frac{d}{dt}(v\hat{t}) = \left(\frac{dv}{dt}\right)\hat{t} + v\left(\frac{d\hat{t}}{dt}\right). \tag{9.15}$$

In the last step here, we used the product rule of differentiation. Thus, the acceleration in circular motion has two components. The first part arises from the change in the magnitude of the velocity; this is the **tangential acceleration.** The second part is due to the fact that the velocity vector always points in the tangential direction and thus has to change its direction continuously as the tip of the radial position vector moves around the circle; this is the **radial acceleration.**

Let's look at the two components individually. First, we can calculate the time derivative of the linear speed, $v$, using the relationship between linear speed and angular speed in equation 9.13 and again invoking the product rule:

$$\frac{dv}{dt}=\frac{d}{dt}(r\omega)=\left(\frac{dr}{dt}\right)\omega+r\frac{d\omega}{dt}.$$

Because $r$ is constant for circular motion, $dr/dt = 0$, and the first term in the sum on the right-hand side is zero. From equation 9.14, $d\omega/dt = \alpha$, and so the second term in the sum is equal to $r\alpha$. Thus, the change in speed is related to the angular acceleration by

$$\frac{dv}{dt}=r\alpha. \tag{9.16}$$

However, the acceleration vector of equation 9.15 also has a second component, which is proportional to the time derivative of the tangential unit vector. For this quantity, we find

$$\begin{aligned}\frac{d}{dt}\hat{t}&=\frac{d}{dt}(-\sin\theta,\cos\theta)=\left(\frac{d}{dt}(-\sin\theta),\frac{d}{dt}(\cos\theta)\right)\\&=\left(-\cos\theta\frac{d\theta}{dt},-\sin\theta\frac{d\theta}{dt}\right)=-\frac{d\theta}{dt}(\cos\theta,\sin\theta)\\&=-\omega\hat{r}.\end{aligned}$$

Therefore, we find that the time derivative of the tangential unit vector points in the direction opposite to that of the radial unit vector. With this result, we can finally write for the linear acceleration vector of equation 9.15

$$\vec{a}(t)=r\alpha\hat{t}-v\omega\hat{r}. \tag{9.17}$$

Again, for circular motion, the acceleration vector has two physical components (Figure 9.12): The first results from a change in speed and points in the tangential direction, and the second comes from the continuous change of direction of the velocity vector and points in the negative radial direction, toward the center of the circle. This second component is present even if the circular motion proceeds at constant speed. If the angular velocity is constant, the tangential angular acceleration is zero, but the velocity vector still changes direction continuously as the object moves in its circular path. The acceleration that changes the direction of the velocity vector without changing its magnitude is often called **centripetal acceleration** (*centripetal* means "center-seeking"), and it is directed in the inward radial direction. Thus, we can write equation 9.17 for the acceleration of an object in circular motion as the sum of the tangential acceleration and the centripetal acceleration:

$$\vec{a}=a_t\hat{t}-a_c\hat{r}. \tag{9.18}$$

The magnitude of the centripetal acceleration is

$$a_c=v\omega=\frac{v^2}{r}=\omega^2 r. \tag{9.19}$$

The first expression for the centripetal acceleration in equation 9.19 can be simply read off from equation 9.17 as the coefficient of the unit vector pointing in the negative radial direction. The second and third expressions for the centripetal acceleration then follow from the relationship between linear and angular speeds and the radius (equation 9.13).

For the magnitude of the acceleration in circular motion, we thus have, from equations 9.17 and 9.19,

$$a=\sqrt{a_t^2+a_c^2}=\sqrt{(r\alpha)^2+(r\omega^2)^2}=r\sqrt{\alpha^2+\omega^4}. \tag{9.20}$$

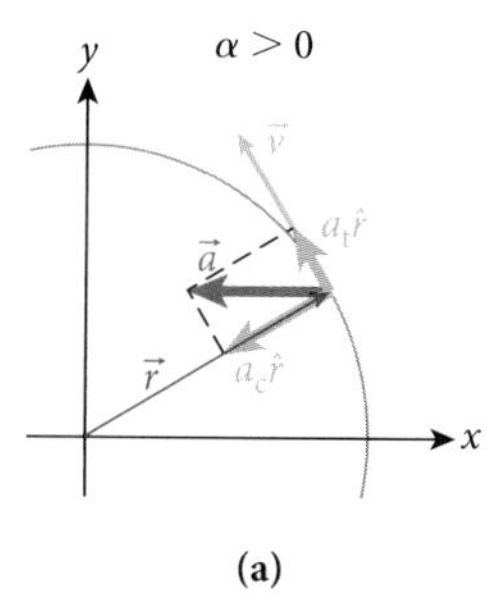

(a)

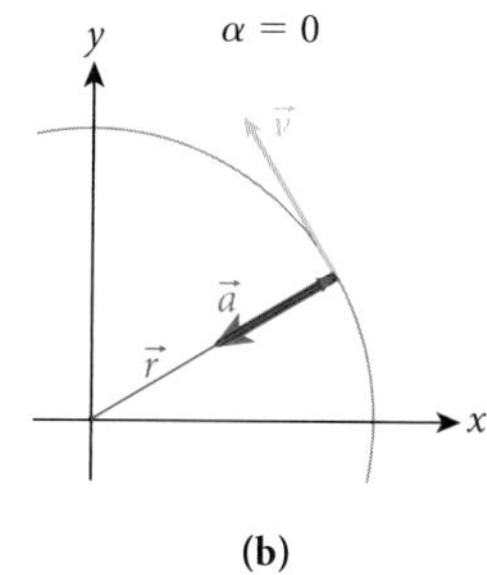

(b)

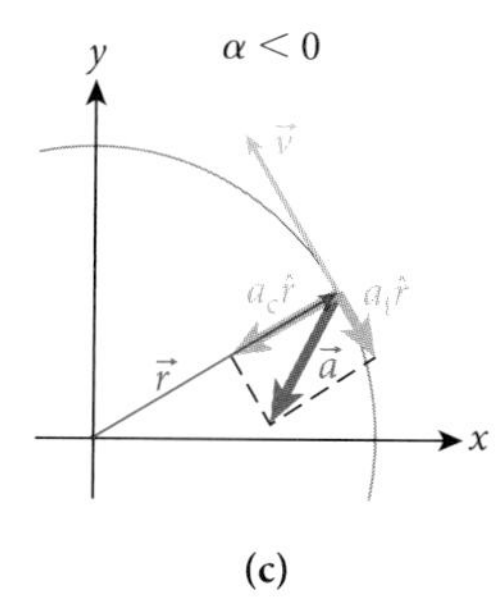

(c)

**FIGURE 9.12** Relationships among linear acceleration, centripetal acceleration, and angular acceleration for (a) increasing speed; (b) constant speed; and (c) decreasing speed.

## EXAMPLE 9.4 Ultracentrifuge

One of the most important pieces of equipment in biomedical labs is the ultracentrifuge (Figure 9.13). It is used for separation of compounds (such as colloids or proteins) consisting of *particles* of different masses through the process of sedimentation (more massive particles sink to the bottom). Instead of relying on the acceleration of gravity to accomplish sedimentation, an ultracentrifuge utilizes the centripetal acceleration from rapid rotation to speed up the process. Some ultracentrifuges can reach centripetal acceleration values of up to $10^6 g$ ($g = 9.81 \text{ m/s}^2$).

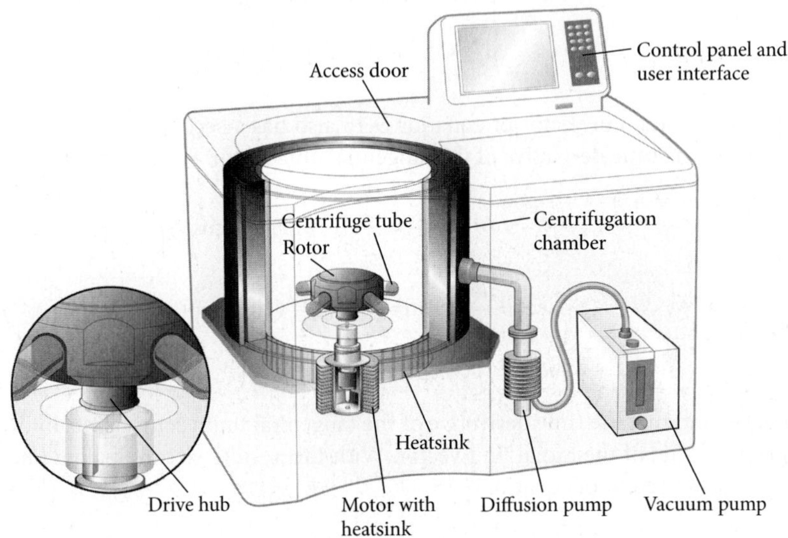


**FIGURE 9.13** Cutaway diagram of an ultracentrifuge.

### PROBLEM

If you want to generate 840,000$g$ of centripetal acceleration in a sample rotating at a distance of 23.5 cm from the ultracentrifuge's rotation axis, what is the frequency you have to enter into the controls? What is the linear speed with which the sample is then moving?

### SOLUTION

The centripetal acceleration is given by $a_c = \omega^2 r$, and the angular velocity is related to the frequency via equation 9.11: $\omega = 2\pi f$. Thus, the relationship between the frequency and the centripetal acceleration is $a_c = (2\pi f)^2 r$, or

$$f = \frac{1}{2\pi}\sqrt{\frac{a_c}{r}} = \frac{1}{2\pi}\sqrt{\frac{(840{,}000)(9.81 \text{ m/s}^2)}{0.235 \text{ m}}} = 942 \text{ s}^{-1} = 56{,}500 \text{ rpm}.$$

For the linear speed of the sample inside the centrifuge, we then find

$$v = r\omega = 2\pi r f = 2\pi(0.235 \text{ m})(942 \text{ s}^{-1}) = 1.39 \text{ km/s}.$$

Other types of centrifuges called *gas centrifuges* are used in the process of uranium enrichment. In this process, the isotopes $^{235}$U and $^{238}$U (discussed in Chapter 40), which differ in mass by only slightly more than 1%, are separated. Natural uranium contains more than 99% of the harmless $^{238}$U. But if uranium is enriched to contain more than 90% $^{235}$U, it can be used for nuclear weapons. The gas centrifuges used in the enrichment process have to spin at approximately 100,000 rpm, which puts an incredible strain on the mechanisms and materials of these centrifuges and makes them very hard to design and manufacture. In order to impede proliferation of nuclear weapons, the design of these centrifuges is a closely guarded secret.

## EXAMPLE 9.5 Centripetal Acceleration Due to Earth's Rotation

Because the Earth rotates, points on its surface move with a rotational velocity, and it is interesting to compute the corresponding centripetal acceleration. This acceleration can slightly alter the commonly stated value of the acceleration due to gravity on the surface of Earth.

We can insert the data for the Earth into equation 9.19 to find the magnitude of the centripetal acceleration:

$$\begin{aligned} a_c &= \omega^2 r = \omega^2 R_{\text{Earth}} \cos\vartheta \\ &= (7.27\cdot 10^{-5}\ \text{s}^{-1})^2 (6.38\cdot 10^6\ \text{m})(\cos\vartheta) \\ &= (0.034\ \text{m/s}^2)(\cos\vartheta). \end{aligned}$$

Here we have used the same notation as in Example 9.3, with $\vartheta$ indicating the latitude angle relative to the Equator. Our result shows that the centripetal acceleration due to the Earth's rotation changes the effective gravitational acceleration observed on the surface of Earth by a factor ranging between 0.33 percent (at the Equator) and zero (at the poles). Using Seattle and Miami as examples, we obtain a centripetal acceleration of 0.02 m/s$^2$ for Seattle and 0.03 m/s$^2$ for Miami. These values are relatively small compared to the quoted value for the acceleration of gravity, 9.81 m/s$^2$, but are not always negligible.

### 9.3 Self-Test Opportunity

The centripetal acceleration due to the Earth's rotation has a maximum value of approximately $g/300$. Can you determine the value of the centripetal acceleration due to the Earth's orbit around the Sun?

### 9.4 Self-Test Opportunity

You are standing on the surface of the Earth at the Equator. If the Earth stopped rotating on its axis, would you feel lighter, heavier, or the same?

### 9.2 In-Class Exercise

The period of rotation of the Earth on its axis is 24 h. At this angular velocity, the centripetal acceleration at the surface of the Earth is small compared with the acceleration of gravity. What would Earth's period of rotation have to be for the magnitude of the centripetal acceleration at the surface at the Equator to be equal to the magnitude of the acceleration of gravity? (With this period of rotation, you could levitate just above the Earth's surface!)

a) 0.043 h
b) 0.340 h
c) 0.841 h
d) 1.41 h
e) 3.89 h
f) 12.0 h

## EXAMPLE 9.6 CD Player

**PROBLEM**

In Example 9.2, we established that a CD track is 5.4 km long. A music CD can store 74 min of music. What are the angular velocity and the tangential acceleration of the disc as it spins inside a CD player, assuming a constant linear velocity?

**SOLUTION**

Since the track is 5.4 km long and has to pass the laser that reads it in a time interval of $\Delta t = 74$ min $= 4440$ s, the speed of the track passing by the reader has to be $v = (5.4\ \text{km})/(4440\ \text{s}) = 1.216$ m/s. From Example 9.2, the track is a spiral with 20,625 loops, starting at an inner radius $r_1 = 25$ mm and reaching an outer radius $r_2 = 58$ mm. At each value of the radius $r$, we can approximate the spiral track by a circle, as we did in Example 9.2. Then we can use the relationship between linear and angular speeds expressed in equation 9.13 to solve for the angular velocity as a function of the radius:

$$v = r\omega \Rightarrow \omega = \frac{v}{r}.$$

Inserting the values for $v$ and $r$, we obtain

$$\omega(r_1) = \frac{1.216\ \text{m/s}}{0.025\ \text{m}} = 48.64\ \text{s}^{-1}$$

$$\omega(r_2) = \frac{1.216\ \text{m/s}}{0.058\ \text{m}} = 20.97\ \text{s}^{-1}.$$

This means that a CD player has to slow the disc's rate of rotation during the course of playing it. The average angular acceleration during this process is

$$\alpha = \frac{\omega(r_2) - \omega(r_1)}{\Delta t} = \frac{20.97\ \text{s}^{-1} - 48.64\ \text{s}^{-1}}{4440\ \text{s}} = -6.23\cdot 10^{-3}\ \text{s}^{-2}.$$

### 9.5 Self-Test Opportunity

How does the centripetal acceleration of a CD rotating in a player compare to its tangential acceleration?

## 9.5 Centripetal Force

The centripetal force, $\vec{F}_c$, is not another fundamental force of nature but is simply the net inward force needed to provide the centripetal acceleration necessary for circular motion.

### 9.3 In-Class Exercise

You are sitting on a carousel, which is in motion. Where should you sit so that the largest possible centripetal force is acting on you?

a) close to the outer edge

b) close to the center

c) in the middle

d) The force is the same everywhere.

It has to point inward, toward the circle's center. Its magnitude is the product of the mass of the object and the centripetal acceleration required to force it onto a circular path:

$$F_c = ma_c = mv\omega = m\frac{v^2}{r} = m\omega^2 r. \quad (9.21)$$

To arrive at equation 9.21, we simply wrote the centripetal acceleration in terms of the linear velocity $v$, the angular velocity $\omega$, and the radius $r$, as in equation 9.19, and multiplied it by the mass of the object forced onto a circular path by the centripetal force.

Figure 9.14 shows a top view of a spinning table with three identical (except for color) poker chips on it. The black chip is located close to the center, the red chip close to the outer edge, and the blue chip in the middle between them. If we spin the table slowly as in part (a), all three chips are in circular motion. In this case, the static friction force between the table and the chips provides the centripetal force required to keep the chips in circular motion. In parts (b), (c), and (d), the table is spinning progressively faster. A higher angular velocity means a larger centripetal force, according to equation 9.21. The chips slide when the friction force is not large enough to provide the centripetal force needed. As you can see, the outermost chip slides off first, and the innermost chip last. This clearly indicates that, for a given angular velocity, the centripetal force increases with the distance from the center. Equation 9.21 in the form $F_c = m\omega^2 r$ can explain this observed behavior. All points on the surface of the spinning table have the same angular velocity, $\omega$, because all of them take the same time to complete one revolution. Thus, for the three poker chips, the centripetal force is proportional to the distance from the center, explaining why the red chip slides off first and the black chip slides off last.

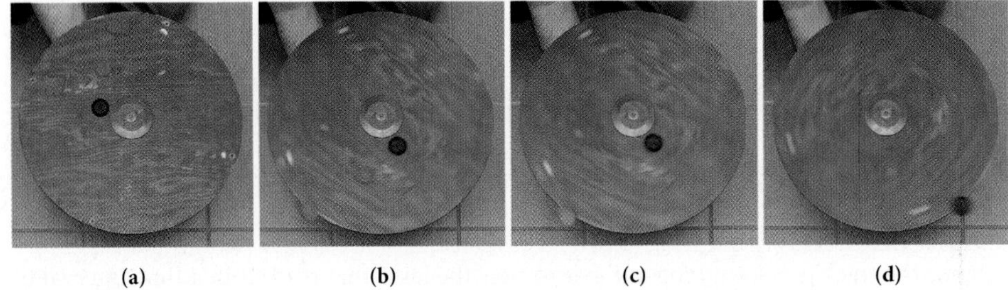


**FIGURE 9.14** Poker chips on a spinning table. Shown from left to right are the initial positions of the chips and the moments when the three chips slide off.

## Conical Pendulum

**FIGURE 9.15** Wave Swinger at an amusement park.

Figure 9.15 is a picture of the Wave Swinger, a ride at an amusement park. The riders sit in seats that are suspended from a solid disk by long chains. At the beginning of the ride, the chains hang straight down, but as the ride starts to rotate, the chains form an angle $\varphi$ with the vertical, as you can see. This angle is independent of the mass of the rider and depends only on the angular velocity of the circular motion. How can we find the value of this angle in terms of that velocity?

In order to gain this understanding, we consider a similar, but somewhat simpler situation: a mass suspended from the ceiling by a string of length $\ell$ and performing circular motion such that the angle between the string and the vertical is $\varphi$. The string outlines the surface of a cone, which is why this setup is called a *conical pendulum;* see Figure 9.16.

Figure 9.16c shows a free-body diagram for the mass. There are only two forces acting on it. Acting vertically downward is the force of gravity, $\vec{F}_g$, indicated by the red arrow in the

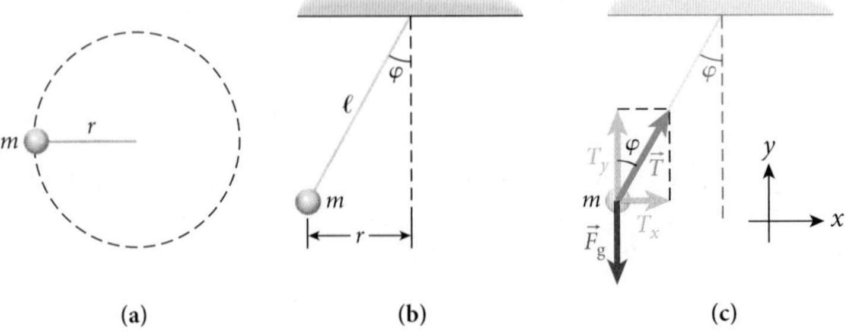


**FIGURE 9.16** Conical pendulum: (a) top view; (b) side view; (c) free-body diagram.

free-body diagram; as usual, its magnitude is $mg$. The only other force acting on the mass is the string tension, $\vec{T}$, which acts along the direction of the string at an angle $\varphi$ with the vertical. This string tension is broken down into its $x$- and $y$-components ($T_x = T\sin\varphi$, $T_y = T\cos\varphi$). There is no motion in the vertical direction; therefore, we have to have zero net force in that direction, $F_{\text{net},y} = T\cos\varphi - mg = 0$, leading to

$$T\cos\varphi = mg.$$

In the horizontal direction, the horizontal component of the string tension is the only force component; it provides the centripetal force. Newton's Second Law, $F_{\text{net},x} = F_c = ma_c$, then yields

$$T\sin\varphi = mr\omega^2.$$

As you can see from Figure 9.16b, the radius of the circular motion is given by $r = \ell\sin\varphi$. Using this relationship, we find the string tension in terms of the angular velocity:

$$T = m\ell\omega^2. \quad (9.22)$$

We have just seen that $T\cos\varphi = mg$, and we can substitute the expression for $T$ from equation 9.22 into this equation to get

$$(m\ell\omega^2)(\cos\varphi) = mg$$

$$\omega^2 = \frac{g}{\ell\cos\varphi}$$

$$\omega = \sqrt{\frac{g}{\ell\cos\varphi}}. \quad (9.23)$$

The mass cancels out, which explains why all the chains in Figure 9.15 have the same angle with the vertical. Clearly, there is a unique and interesting relationship between the angle of the conical pendulum and its angular velocity. As the angle $\varphi$ approaches zero, the angular velocity does not approach zero but rather some finite minimum value, $\sqrt{g/\ell}$. (We will gain a deeper understanding of this result in Chapter 14, when we study pendulum motion.) In the limit where the angle $\varphi$ approaches 90°, $\omega$ becomes infinite.

### 9.6 Self-Test Opportunity

Sketch a graph of the string tension as a function of the angle $\varphi$.

### 9.4 In-Class Exercise

A certain angular velocity, $\omega_0$, of a conical pendulum results in an angle $\varphi_0$. If the same conical pendulum was taken to the Moon, where the gravitational acceleration is a sixth of that on Earth, how would one have to adjust the angular velocity to obtain the same angle $\varphi_0$?

a) $\omega_{\text{Moon}} = 6\omega_0$

b) $\omega_{\text{Moon}} = \sqrt{6}\,\omega_0$

c) $\omega_{\text{Moon}} = \omega_0$

d) $\omega_{\text{Moon}} = \omega_0/\sqrt{6}$

e) $\omega_{\text{Moon}} = \omega_0/6$

## SOLVED PROBLEM 9.1 Analysis of a Roller Coaster

Perhaps the biggest thrill to be had at an amusement park is on a roller coaster with a vertical loop in it (Figure 9.17), where passengers feel almost weightless at the top of the loop.

### PROBLEM

Suppose the vertical loop has a radius of 5.00 m. What does the linear speed of the roller coaster have to be at the top of the loop for the passengers to feel weightless? (Assume that friction between roller coaster and rails can be neglected.)

### SOLUTION

#### THINK

A person feels weightless when there is no supporting force, from a seat or a restraint, acting to counter his or her weight. For a person to feel weightless at the top of the loop, no normal force can be acting on him or her at this point.

#### SKETCH

The free-body diagrams in Figure 9.18 may help to conceptualize the situation. The force of gravity and the normal force acting on a passenger in the roller coaster at the top of the loop are shown in Figure 9.18a. The sum of these two forces is the net force, which has to equal the centripetal force in circular motion. If the net force (centripetal force here) is equal to the gravitational force, then the normal force is zero, and the passenger feels weightless. This situation is illustrated in Figure 9.18b.

**FIGURE 9.17** Modern roller coaster with a vertical loop.

*Continued—*

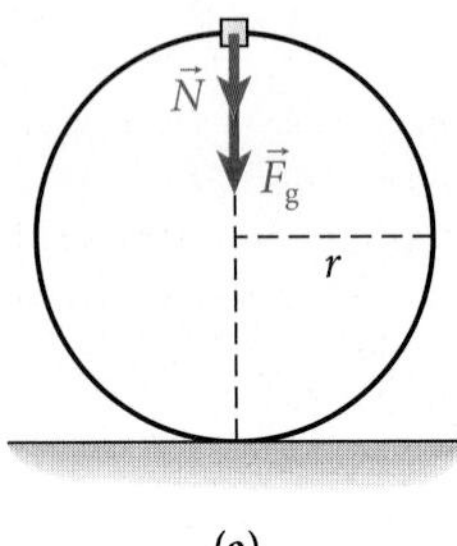


(a)

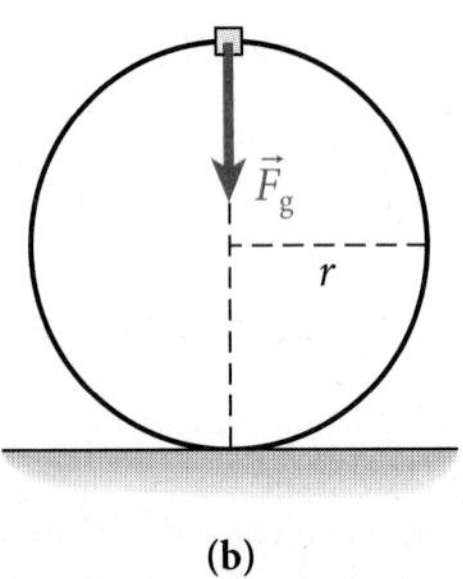


(b)

**FIGURE 9.18** (a) Free-body diagram for a passenger at the top of the vertical loop of a roller coaster. (b) Condition for the feeling of weightlessness.

**RESEARCH**

We have just stated that the net force is equal to the centripetal force and that the net force is the sum of the normal force and the force of gravity:

$$\vec{F}_c = \vec{F}_{net} = \vec{F}_g + \vec{N}.$$

For the feeling of weightlessness at the top of the loop, we need $\vec{N} = 0$, and thus

$$\vec{F}_c = \vec{F}_g \Rightarrow F_c = F_g. \tag{i}$$

As always, we have $F_g = mg$. For the magnitude of the centripetal force, use equation 9.21

$$F_c = ma_c = m\frac{v^2}{r}.$$

**SIMPLIFY**

After substituting the expressions for the centripetal and the gravitational forces into equation (i), we solve for the linear speed at the top of the loop:

$$F_c = F_g \Rightarrow m\frac{v_{top}^2}{r} = mg \Rightarrow v_{top} = \sqrt{rg}.$$

**CALCULATE**

Using $g = 9.81 \text{ m/s}^2$ and the value of 5.00 m given for the radius, we obtain

$$v_{top} = \sqrt{(5.00 \text{ m})(9.81 \text{ m/s}^2)} = 7.00357 \text{ m/s}.$$

**ROUND**

Rounding our result to three-digit precision, we have

$$v_{top} = 7.00 \text{ m/s}.$$

**DOUBLE-CHECK**

Obviously, our answer passes the simplest check in that the units are those of speed, meters per second. The formula for the linear speed at the top, $v_{top} = \sqrt{rg}$, indicates that a larger radius necessitates a higher speed, which seems plausible.

Is 7.00 m/s for the speed at the top reasonable? Converting this value gives 15.7 mph, which seems fairly slow for a ride that is typically experienced as tremendously fast. But keep in mind that this is the minimum speed needed at the top of the loop, and the operators of the rides do not want to get too close to this value.

Let's go a step further and calculate the velocity vectors at the 3 o'clock and 9 o'clock positions on the loop, assuming that the roller coaster moves counterclockwise around the loop. The directions of the velocity vectors for circular motion are always tangential to the circle, as shown in Figure 9.19.

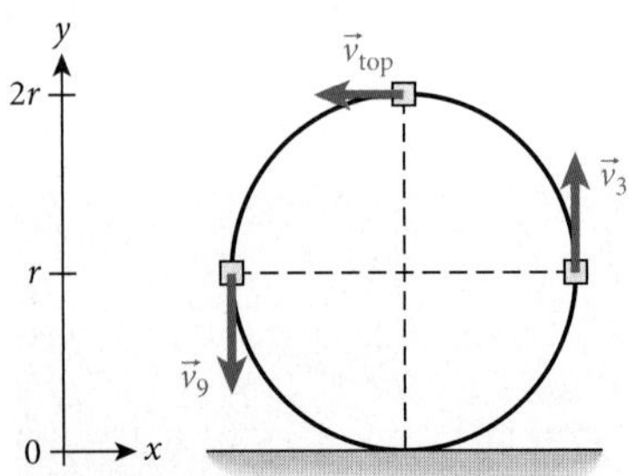


**FIGURE 9.19** Directions of the velocity vectors at several points along the vertical roller coaster loop.

How do we obtain the magnitudes of the velocities $v_3$ (at 3 o'clock) and $v_9$ (at 9 o'clock)? First, recall from Chapter 6, that the total energy is the sum of kinetic and potential energies, $E = K+U$, that the kinetic energy is $K = \frac{1}{2}mv^2$, and that the gravitational potential energy is proportional to the height above ground, $U = mgy$. In Figure 9.19, the coordinate system is placed so that the zero of the $y$-axis is at the bottom of the loop. We then can write the equation for the conservation of mechanical energy assuming no nonconservative forces are acting:

$$E = K_3 + U_3 = K_{top} + U_{top} = K_9 + U_9 \Rightarrow$$
$$\tfrac{1}{2}mv_3^2 + mgy_3 = \tfrac{1}{2}mv_{top}^2 + mgy_{top} = \tfrac{1}{2}mv_9^2 + mgy_9. \tag{ii}$$

We can see from the figure that the $y$-coordinates and therefore the potential energies are the same at the 3 o'clock and 9 o'clock positions; therefore, the kinetic energies at both points have to be the same. Consequently, the absolute values of the speeds at both points are the same: $v_3 = v_9$. Solving equation (ii) for $v_3$ we obtain

$$\tfrac{1}{2}mv_3^2 + mgy_3 = \tfrac{1}{2}mv_{\text{top}}^2 + mgy_{\text{top}} \Rightarrow$$
$$\tfrac{1}{2}v_3^2 + gy_3 = \tfrac{1}{2}v_{\text{top}}^2 + gy_{\text{top}} \Rightarrow$$
$$v_3 = \sqrt{v_{\text{top}}^2 + 2g(y_{\text{top}} - y_3)}.$$

Again, the mass cancels out. Further, only the difference in the $y$-coordinates enters into this formula for $v_3$; therefore, the choice of the origin of the coordinate system is irrelevant. The difference in the $y$-coordinates between the two points is $y_{\text{top}} - y_3 = r$. Inserting the given value of $r = 5.00$ m and the result $v_{\text{top}} = 7.00$ m/s that we found previously, we see that the speed at the 3 o'clock and at the 9 o'clock positions in the loop is

$$v_3 = \sqrt{(7.00 \text{ m/s})^2 + 2(9.81 \text{ m/s}^2)(5.00 \text{ m})} = 12.1 \text{ m/s}.$$

As you can see from this discussion, practically any force can act as the centripetal force. It was the force of static friction for the poker chips on the rotating table and the horizontal component of the tension in the string for the conical pendulum. But it can also be the gravitational force, which forces planets into (almost) circular orbits around the Sun (see Chapter 12), or the Coulomb force acting on the electrons in atoms (see Chapter 21).

## Is There a Centrifugal Force?

This is a good time to clarify an important point regarding the direction of the force responsible for circular motion. You often hear people talking about centrifugal (or "center-fleeing," in the radial outward direction) acceleration or centrifugal force (mass times acceleration). You can experience the sensation of seemingly being pulled outward on many rotating amusement park rides. This sensation is due to your body's inertia, which resists the centripetal acceleration toward the center. Thus, you feel a seemingly outward-pointing force—the centrifugal force. Keep in mind that this perception is due to your body moving in an accelerated reference frame; there is no centrifugal force. The real force that acts on your body and forces it to move on a circular path is the centripetal force, and it points inward.

You have also experienced a similar effect in straight-line motion. When you are sitting in your car at rest and then step on the gas pedal, you feel as if you are being pressed back into your seat. This sensation of a force that presses you in the backward direction also comes from the inertia of your body, which is accelerated forward by your car. Both of these sensations of forces acting on your body—the "centrifugal" force and the force "pushing" you into the seat—are the result of your body experiencing an acceleration in the opposite direction and putting up resistance, inertia, against this acceleration.

### 9.5 In-Class Exercise

In Solved Problem 9.1, at what speed does the roller coaster car have to enter the bottom of the loop in order to produce the feeling of weightlessness at the top?

a) 7.00 m/s

b) 12.1 m/s

c) 13.5 m/s

d) 15.7 m/s

e) 21.4 m/s

### 9.7 Self-Test Opportunity

What speed must the roller coaster of Solved Problem 9.1 have at the top of the loop to accomplish the same feeling of weightlessness if the radius of the loop is doubled?

### 9.6 In-Class Exercise

Figure 9.18a shows the general free-body diagram for the forces acting on a passenger of the roller coaster at the top of the loop, where a normal force acts and has a magnitude smaller than the magnitude of the gravitational force. Given that the speed of the roller coaster is 7.00 m/s, what does the radius of the loop have to be for that free-body diagram to be correct?

a) less than 5 m

b) 5 m

c) more than 5 m

## 9.6 Circular and Linear Motion

Table 9.1 summarizes the relationships between linear and angular quantities for circular motion. The relationships shown in the table relate the angular quantities ($\theta$, $\omega$, and $\alpha$) to the linear quantities ($s$, $v$, and $a$). The radius $r$ of the circular path is constant and provides the connection between the two sets of quantities. (In Chapter 10, the rotational counterparts to mass, kinetic energy, momentum, and force will be added to this list.)

As we have just seen, there is a formal correspondence between motion on a straight line with constant velocity and circular motion with constant angular velocity. However, there is one big difference. As we saw in Section 3.6 on relative motion, you cannot always distinguish between moving with constant velocity in a straight line and being at rest. This is because the origin of the coordinate system can be placed at any point—even a point that moves with constant velocity. The physics of translational motion does not change under this Galilean transformation. In contrast, in circular motion, you are always moving on a circular path with a well-defined center. Experiencing the "centrifugal" force is then a sure sign of circular motion, and the strength of that force is a measure of the magnitude of angular velocity. You may argue that you are in constant circular motion around the center of the Earth, around

**Table 9.1 Comparison of Kinematic Variables for Circular Motion**

| Quantity | Linear | Angular | Relationship |
|---|---|---|---|
| Displacement | $s$ | $\theta$ | $s = r\theta$ |
| Velocity | $v$ | $\omega$ | $v = r\omega$ |
| Acceleration | $a$ | $\alpha$ | $\vec{a} = r\alpha\hat{t} - r\omega^2\hat{r}$ |
| | | | $a_t = r\alpha$ |
| | | | $a_c = \omega^2 r$ |

the center of the Solar System, and around the center of the Milky Way Galaxy but you do not feel the effects of those circular motions. True, but the very small magnitudes of the angular velocities involved in these motions cause their perceived effects to be negligible.

## Constant Angular Acceleration

Chapter 2 discussed at some length the special case of constant acceleration. Under this assumption, we derived five equations that proved useful in solving all kinds of problems. For ease of reference, here are those five equations of linear motion with constant acceleration:

$$\text{(i)} \qquad x = x_0 + v_{x0}t + \tfrac{1}{2}a_x t^2$$

$$\text{(ii)} \qquad x = x_0 + \bar{v}_x t$$

$$\text{(iii)} \qquad v_x = v_{x0} + a_x t$$

$$\text{(iv)} \qquad \bar{v}_x = \tfrac{1}{2}(v_x + v_{x0})$$

$$\text{(v)} \qquad v_x^2 = v_{x0}^2 + 2a_x(x - x_0).$$

Now we'll take the same steps as in Chapter 2 to derive the equivalent equations for constant angular acceleration. We start with equation 9.14 and integrate, using the usual notation convention of $\omega_0 \equiv \omega(t_0)$:

$$\alpha(t) = \frac{d\omega}{dt} \Rightarrow$$

$$\int_{t_0}^{t} \alpha(t')dt' = \int_{t_0}^{t} \frac{d\omega(t')}{dt'}dt' = \omega(t) - \omega(t_0) \Rightarrow$$

$$\omega(t) = \omega_0 + \int_{t_0}^{t} \alpha(t')dt'.$$

This relationship is the inverse of equation 9.14 and holds in general. If we assume the angular acceleration, $\alpha$, is constant in time, we can evaluate the integral and obtain

$$\omega(t) = \omega_0 + \alpha\int_{0}^{t} dt' = \omega_0 + \alpha t. \tag{9.24}$$

For convenience, we have set $t_0 = 0$, just as we did in Chapter 2 at this juncture. Next, we use equation 9.8, expressing the angular velocity as the derivative of the angle with respect to time and with the notation $\theta_0 = \theta(t = 0)$:

$$\frac{d\theta(t)}{dt} = \omega(t) = \omega_0 + \alpha t \Rightarrow$$

$$\theta(t) = \theta_0 + \int_0^t \omega(t')dt' = \theta_0 + \int_0^t (\omega_0 + \alpha t')dt' \Rightarrow$$

$$= \theta_0 + \omega_0\int_0^t dt' + \alpha\int_0^t t'dt' \Rightarrow$$

$$\theta(t) = \theta_0 + \omega_0 t + \tfrac{1}{2}\alpha t^2. \tag{9.25}$$

Comparing equations 9.24 and 9.25 to equations (iii) and (i) for linear motion reveals that these two equations are the circular motion equivalents of the two kinematical equations for straight-line linear motion in one dimension. With the straightforward substitutions $x \to \theta$, $v_x \to \omega$, and $a_x \to \alpha$, we can write five kinematical equations for circular motion under constant angular acceleration:

$$\begin{aligned} &\text{(i)} && \theta = \theta_0 + \omega_0 t + \tfrac{1}{2}\alpha t^2 \\ &\text{(ii)} && \theta = \theta_0 + \bar{\omega} t \\ &\text{(iii)} && \omega = \omega_0 + \alpha t \\ &\text{(iv)} && \bar{\omega} = \tfrac{1}{2}(\omega + \omega_0) \\ &\text{(v)} && \omega^2 = \omega_0^2 + 2\alpha(\theta - \theta_0). \end{aligned} \tag{9.26}$$

### 9.8 Self-Test Opportunity

The derivation has been provided for two of these five kinematical equations for circular motion. Can you provide the derivation for the remaining equations? (*Hint:* The chain of reasoning proceeds along exactly the same lines as Derivation 2.1 in Chapter 2.)

## EXAMPLE 9.7 Hammer Throw

One of the most interesting events in track-and-field competitions is the hammer throw. The task is to throw the "hammer," a 12-cm-diameter iron ball attached to a grip by a steel cable, a maximum distance. The hammer's total length is 121.5 cm, and its total mass is 7.26 kg. The athlete has to accomplish the throw from within a circle of radius 2.135 m (7 ft), and the best way to throw the hammer is for the athlete to spin, allowing the hammer to move in a circle around him, before releasing it. At the 1988 Olympic Games in Seoul, the Russian thrower Sergey Litvinov won the gold medal with an Olympic record distance of 84.80 m. He took seven turns before releasing the hammer, and the period to complete each turn was obtained from examining the video recording frame by frame: 1.52 s, 1.08 s, 0.72 s, 0.56 s, 0.44 s, 0.40 s, and 0.36 s.

### PROBLEM 1

What was the average angular acceleration during the seven turns?

### SOLUTION 1

In order to find the average angular acceleration, we add all the time intervals for the seven turns, obtaining the total time:

$$t_{\text{all}} = 1.52 \text{ s} + 1.08 \text{ s} + 0.72 \text{ s} + 0.56 \text{ s} + 0.44 \text{ s} + 0.40 \text{ s} + 0.36 \text{ s} = 5.08 \text{ s}.$$

During this time, Litvinov completed seven full turns, resulting in a total angle of

$$\theta_{\text{all}} = 7(2\pi \text{ rad}) = 14\pi \text{ rad}.$$

Because we can assume constant acceleration, we can readily solve for the angular acceleration and insert the given numbers to obtain our answer:

$$\theta = \tfrac{1}{2}\alpha t^2 \Rightarrow \alpha = \frac{2\theta_{\text{all}}}{t^2} = 2\frac{14\pi \text{ rad}}{(5.08 \text{ s})^2} = 3.41 \text{ rad/s}^2.$$

### DISCUSSION

Because we know how long it took to complete each turn, we can generate a plot of the angle of the hammer in the horizontal plane as a function of time. This plot is shown in Figure 9.20, with the red dots representing the data points. The blue line that is fit to the data points in Figure 9.20 assumes a constant angular acceleration of $\alpha = 3.41 \text{ rad/s}^2$. As you can see, the assumption of constant angular acceleration is almost, but not quite justified.

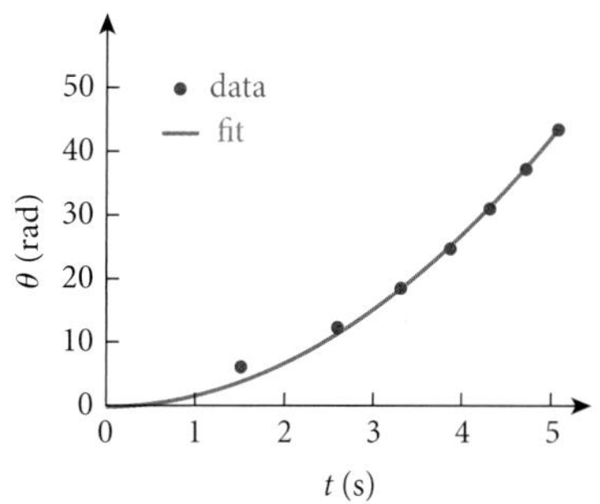


**FIGURE 9.20** Angle as a function of time for Sergey Litvinov's 1988 gold-medal-winning hammer throw.

### PROBLEM 2

Assuming that the radius of the circle on which the hammer moves is 1.67 m (the length of the hammer plus the arms of the athlete), what is the linear speed with which the hammer is released?

*Continued—*

**SOLUTION 2**

With constant angular acceleration from rest for a period of 5.08 s, the final angular velocity is

$$\omega = \alpha t = \left(3.41 \text{ rad/s}^2\right)\left(5.08 \text{ s}\right) = 17.3 \text{ rad/s}.$$

Using the relationship between linear and angular velocity, we obtain the linear speed at release:

$$v = r\omega = \left(1.67 \text{ m}\right)\left(17.3 \text{ rad/s}\right) = 28.9 \text{ m/s}.$$

**PROBLEM 3**

What is the centripetal force that the hammer thrower has to exert on the hammer right before he releases it?

**SOLUTION 3**

The centripetal acceleration right before release is given by

$$a_c = \omega^2 r = \left(17.3 \text{ rad/s}\right)^2 \left(1.67 \text{ m}\right) = 500. \text{ m/s}^2.$$

With a mass of 7.26 kg for the hammer, the centripetal force required is

$$F_c = ma_c = \left(7.26 \text{ kg}\right)\left(500 \text{ m/s}^2\right) = 3630 \text{ N}.$$

This is an astonishingly large force, equivalent to the weight of an object of mass 370 kg! This is why world-class hammer throwers have to be very strong.

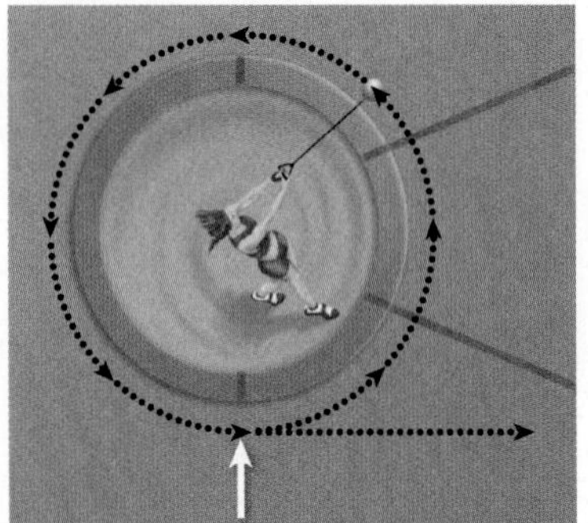

**FIGURE 9.21** Overhead view of the trajectory of the hammer (black dots, with the arrows indicating the direction of the velocity vector) during the time the athlete holds on to it (circular path) and after release (straight line). The white arrow marks the point of release.

**PROBLEM 4**

After release, what is the direction in which the hammer moves?

**SOLUTION 4**

It is a common misconception that the hammer "spirals" in some sort of circular motion with an ever increasing radius after release. This idea is wrong because there is no horizontal component of force once the athlete releases the hammer. Newton's Second Law tells us that there will be no horizontal component of acceleration and hence no centripetal acceleration. Instead, the hammer moves in a direction that is tangential to the circle at the point of release. If you were to look straight down on the stadium from a blimp, you would see that the hammer moves on a straight line, as shown in Figure 9.21. Viewed from the side, the shape of the hammer's trajectory is a parabola, as shown in Chapter 3.

## 9.7 More Examples for Circular Motion

Let's look at another example and solved problem that demonstrate how useful the concepts of circular motion we've just discussed are.

### EXAMPLE 9.8 Formula 1 Racing

If you watch a Formula 1 race, you can see that the race cars approach curves from the outside, cut through to the inside, and then drift again to the outside, as shown by the red path in Figure 9.22a. The blue path is shorter. Why don't the drivers follow the shortest path?

**PROBLEM**

Suppose that cars move through the U-turn shown in Figure 9.22a at constant speed and that the coefficient of static friction between the tires and the road is $\mu_s = 1.2$. (As was mentioned in Chapter 4, modern race car tires can have coefficients of friction that exceed 1 when they are heated to race temperature and thus are very sticky.) If the radius of the inner curve shown in the figure is $R_B = 10.3$ m and radius of the outer is $R_A = 32.2$ m and the cars move at their maximum speed, how much time will it take to move from point $A$ to $A'$ and from point $B$ to $B'$?

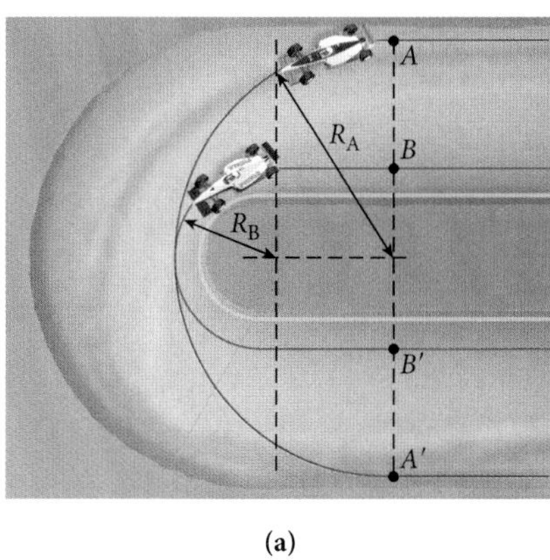


(a)

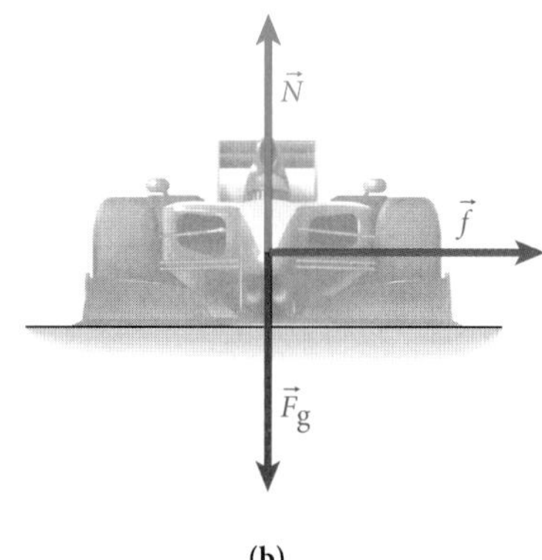


(b)

**FIGURE 9.22** (a) Paths of race cars negotiating a turn on an oval track in two ways. (b) Free-body diagram for a race car in a curve.

## SOLUTION

We start by drawing a free-body diagram, as shown in Figure 9.22b. The diagram shows all forces that act on the car with the force arrows originating at the center of mass of the car. The force of gravity, acting downward with magnitude $F_g = mg$, is shown in red. This force is exactly balanced by the normal force, which the road exerts on the car, shown in green. As a car makes the turn, a net force is required to change the car's velocity vector and act as the centripetal force that pushes the car onto a circular path. This net force is generated by the friction force (shown in blue) between the car's tires and the road. This force arrow points horizontally and inward, toward the center of the curve. As usual, the magnitude of the friction force is the product of the normal force and the coefficient of friction: $f_{max} = \mu_s mg$. (*Note:* In this case, we use the equals sign, because the race car drivers push their cars and tires to the limit and thus reach the maximum possible static friction force.) The arrow for the friction force is longer than that for the normal force by a factor of 1.2, because $\mu_s = 1.2$.

First, we need to calculate the maximum velocity that a race car can have on each trajectory. For each radius of curvature, $R$, the resulting centripetal force, $F_c = mv^2/R$, must be provided by the friction force, $f_{max} = \mu_s mg$:

$$m\mu_s g = m\frac{v^2}{R} \Rightarrow v = \sqrt{\mu_s gR}.$$

Therefore, for the red and blue curves, we get

$$v_{red} = \sqrt{\mu_s gR_A} = \sqrt{(1.2)(9.81 \text{ m/s}^2)(32.2 \text{ m})} = 19.5 \text{ m/s}$$
$$v_{blue} = \sqrt{\mu_s gR_B} = \sqrt{(1.2)(9.81 \text{ m/s}^2)(10.3 \text{ m})} = 11.0 \text{ m/s}.$$

These speeds are only about 43.6 mph and 24.6 mph, respectively! However, the curve shown is a very tight one, typically only encountered on a city course like the one in Monaco.

Even though a car can move much faster on the red curve, the blue curve is shorter than the red one. For the path length of the red curve, we simply have the distance along the semicircle, $\ell_{red} = \pi R_A = 101.$ m. For the path length of the blue curve, we have to add the two straight sections and the semicircular curve with the smaller radius:

$$\ell_{blue} = \pi R_B + 2(R_A - R_B) = 76.2 \text{ m}.$$

We then get for the time to go from $A$ to $A'$ on the red path:

$$t_{red} = \frac{\ell_{red}}{v_{red}} = \frac{101. \text{ m}}{19.5 \text{ m/s}} = 5.18 \text{ s}.$$

To travel along the blue curve from $B$ to $B'$ takes

$$t_{blue} = \frac{\ell_{blue}}{v_{blue}} = \frac{76.2 \text{ m}}{11.0 \text{ m/s}} = 6.92 \text{ s}.$$

*Continued—*

With the assumption that the cars have to use a constant speed, it is clearly a big advantage to cut through the curve, as shown by the red path.

### DISCUSSION

In a race situation, it is unreasonable to expect the car on the blue path to move along the straight-line segments with constant velocity. Instead, the driver will come to point $B$ with the maximum speed that allows him to slow down to 11.0 m/s when entering the circular segment. If we worked this out in detail, we would find that the blue path is still slower, but not by much. In addition, the car following the blue path can reach point $B$ with a slightly higher speed than that at which the red car can reach point $A$. In other words, we have a tradeoff to consider before declaring a winner in this situation. In real races, cars slow down as they approach a curve and then accelerate as they come out of the curve. The most advantageous path for cutting through a curve is not the red semicircle, but a path that looks closer to elliptical, starting on the outside, cutting to the extreme inside in the middle of the turn, and then drifting to the outside again while accelerating out of the turn.

Formula 1 racing generally involves flat tracks and tight turns. Indy-style and NASCAR races take place on tracks with bigger turning radii as well as banked curves. To study the forces involved in this type of racing situation, we must combine concepts of static equilibrium on an inclined plane with concepts of circular motion.

## SOLVED PROBLEM 9.2 NASCAR Racing

As a NASCAR racer moves through a banked curve, the banking helps the driver achieve higher speeds. Let's see how. Figure 9.23 shows a race car on a banked curve.

**FIGURE 9.23** Race car on a banked curve.

### PROBLEM

If the coefficient of static friction between the track surface and the car's tires is $\mu_s = 0.620$ and the radius of the turn is $R = 110.$ m, what is the maximum speed with which a driver can take a curve banked at $\theta = 21.1°$? (This is a fairly typical banking angle for NASCAR tracks. Indianapolis has only 9° banking, but there are some tracks with over 30° banking angles, including Daytona (31°), Talladega (33°), and Bristol (36°).)

### SOLUTION

#### THINK

The three forces acting on the race car are gravity, $\vec{F}_g$, the normal force, $\vec{N}$, and friction, $\vec{f}$. The curve is banked at an angle $\theta$, which is also the angle between the normal to the track surface and the gravity force vector, as shown in Figure 9.24a. To draw the vector for the force of friction, we have assumed that the car has entered the curve at high speed, so the direction of the force of friction is down the incline. In contrast to the situation of static equilibrium, these three forces do not add up to zero, but instead add up to a net force, $\vec{F}_{net}$, as shown in Figure 9.24b. This net force has to provide the centripetal force, $\vec{F}_c$, which forces the car to move in a circle. Thus, the net force has to act in the horizontal direction because this is the direction of the center of the circle in which the car is moving.

#### SKETCH

The free-body diagram for the race car on the banked curve, showing the $x$- and $y$-components of the forces, is presented in Figure 9.24c. The orientation for the coordinate system was selected to give a horizontal $x$-axis and a vertical $y$-axis.

#### RESEARCH

Like problems involving linear motion, we can solve problems involving circular motion by starting with the familiar Newton's Second Law: $\sum \vec{F} = m\vec{a}$. And just as in the linear case,

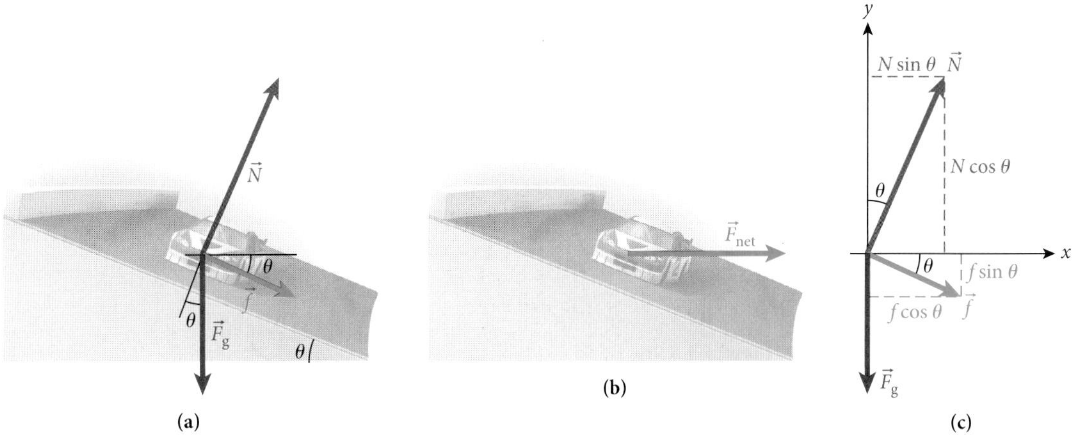


**FIGURE 9.24** (a) Forces on a race car going around a banked curve on a racetrack. (b) The net force, the sum of the three forces in part (a). (c) A free-body diagram showing the $x$- and $y$-components of the forces acting on the car.

we can generally work the problems in Cartesian components. From the free-body diagram in Figure 9.24c, we can see that the $x$-components of the forces acting on the race car are

$$N\sin\theta + f\cos\theta = F_{net}. \qquad \text{(i)}$$

Similarly, the forces acting in the $y$-direction are

$$N\cos\theta - F_g - f\sin\theta = 0. \qquad \text{(ii)}$$

As usual, the maximum friction force is given by the product of the coefficient of friction and the normal force: $f = \mu_s N$. The gravitational force is the product of mass and gravitational acceleration: $F_g = mg$.

The key to solving this problem is to realize that the net force has to be the force that causes the race car to move through the curve, that is, that provides the centripetal force. Therefore, using the expression for the centripetal force from equation 9.21, we have

$$F_{net} = F_c = m\frac{v^2}{R},$$

where $R$ is the radius of the curve.

## SIMPLIFY

We insert the expressions for the maximum friction force, the gravitational force, and the net force into equations (i) and (ii) for the $x$- and $y$-components of the forces:

$$N\sin\theta + \mu_s N\cos\theta = m\frac{v^2}{R} \Rightarrow N(\sin\theta + \mu_s\cos\theta) = m\frac{v^2}{R}$$
$$N\cos\theta - mg - \mu_s N\sin\theta = 0 \Rightarrow N(\cos\theta - \mu_s\sin\theta) = mg.$$

This is a system of two equations for two unknown quantities: the magnitude of the normal force, $N$, and the speed of the car, $v$. It is easy to eliminate $N$ by dividing the first equation above by the second:

$$\frac{\sin\theta + \mu_s\cos\theta}{\cos\theta - \mu_s\sin\theta} = \frac{v^2}{gR}.$$

We solve for $v$:

$$v = \sqrt{\frac{Rg(\sin\theta + \mu_s\cos\theta)}{\cos\theta - \mu_s\sin\theta}}. \qquad \text{(iii)}$$

Note that the mass of the car, $m$, canceled out. Thus, what matters in this situation is the coefficient of friction between the tires and the track surface, the radius of the turn, and the angle of banking.

*Continued—*

### CALCULATE

Putting in the numbers, we obtain

$$v = \sqrt{\frac{(110.\text{ m})(9.81\text{ m/s}^2)[\sin 21.1° + 0.620(\cos 21.1°)]}{\cos 21.1° - 0.620(\sin 21.1°)}} = 37.7726\text{ m/s}.$$

### ROUND

Expressing our result to three significant figures gives us

$$v = 37.8\text{ m/s}.$$

### DOUBLE-CHECK

To double-check our result, let's compare the speed for a banked curve to the maximum speed a race car can achieve on a curve with the same radius but without banking. Without banking, the only force keeping the race car on the circular path is the force of friction. Then our result reduces to the equation $v = \sqrt{\mu_s gR}$ that we found in Example 9.8, and we can insert the numerical values given here to obtain the maximum speed around a flat curve of the same radius:

$$v = \sqrt{\mu_s gR} = \sqrt{(0.620)(9.81\text{ m/s}^2)(110.\text{ m})} = 25.9\text{ m/s}.$$

Our result for the maximum speed around a banked curve, 37.8 m/s (84.6 mph) is considerably larger than this result for a flat curve, 25.9 m/s (57.9 mph), which seems reasonable.

Note that the vector for the force of friction in Figure 9.24 points along the surface of the track and toward the inside of the curve, just as the friction force vector for the nonbanked curve does in Figure 9.22b. However, you can see that as the banking angle increases, it reaches a value for which the denominator of the velocity formula, equation (iii), approaches zero. This occurs when $\cot\theta = \mu_s$. For the given value, $\mu_s = 0.620$, this angle is 58.2°. For larger angles, the friction force vector will point along the surface of the track and toward the outside of the curve. For these angles, the driver needs to maintain a minimum speed driving through the curve to prevent the car from sliding to the bottom of the incline.

## WHAT WE HAVE LEARNED | EXAM STUDY GUIDE

- The conversion between Cartesian coordinates, $x$ and $y$, and polar coordinates, $r$ and $\theta$, is given by

$$r = \sqrt{x^2 + y^2}$$
$$\theta = \tan^{-1}(y/x).$$

- The conversion between polar coordinates and Cartesian coordinates is given by

$$x = r\cos\theta$$
$$y = r\sin\theta.$$

- The linear displacement, $s$, is related to the angular displacement, $\theta$, by $s = r\theta$, where $r$ is the radius of the circular path and $\theta$ is measured in radians.
- The magnitude of the instantaneous angular velocity, $\omega$, is given by $\omega = \frac{d\theta}{dt}$.
- The magnitude of the angular velocity is related to the magnitude of the linear velocity, $v$, by $v = r\omega$.
- The magnitude of the instantaneous angular acceleration, $\alpha$, is given by

$$\alpha = \frac{d\omega}{dt} = \frac{d^2\theta}{dt^2}.$$

- The magnitude of the angular acceleration is related to the magnitude of the tangential acceleration, $a_t$, by $a_t = r\alpha$.
- The magnitude of the centripetal acceleration, $a_c$, required to keep an object moving in a circle with constant angular velocity is given by $a_c = \omega^2 r = \frac{v^2}{r}$.
- The magnitude of the total acceleration of an object in circular motion is $a = \sqrt{a_t^2 + a_c^2} = r\sqrt{\alpha^2 + \omega^4}$.

## KEY TERMS

circular motion, p. 280
polar coordinates, p. 280
angular velocity, p. 283
angular frequency, p. 284
hertz, p. 284
period of rotation, p. 284
angular acceleration, p. 286
tangential acceleration, p. 286
radial acceleration, p. 286
centripetal acceleration, p. 287

## NEW SYMBOLS AND EQUATIONS

$\bar{\omega} = \frac{\theta_2 - \theta_1}{t_2 - t_1} = \frac{\Delta\theta}{\Delta t}$, magnitude of the average angular velocity

$\omega = \frac{d\theta}{dt}$, magnitude of the instantaneous angular velocity

$f = \frac{\omega}{2\pi}$, angular frequency

$T = \frac{1}{f}$, period

$\bar{\alpha} = \frac{\Delta\omega}{\Delta t}$, magnitude of the average angular acceleration

$\alpha = \frac{d\omega}{dt} = \frac{d^2\theta}{dt^2}$, magnitude of the instantaneous angular acceleration

$a_c = \omega^2 r = \frac{v^2}{r}$, magnitude of the centripetal acceleration

## ANSWERS TO SELF-TEST OPPORTUNITIES

**9.1** Differential arc length is $Rd\theta$ for circle with radius $R$; integral arc length around the circle is the circumference $C$

$$C = \int_0^{2\pi} r\,d\theta = r\int_0^{2\pi} d\theta = r[\theta]_0^{2\pi} = 2\pi r.$$

**9.2** The differential area is shown in the sketch. The differential area is $dA = 2\pi dr$. Area of circle is

$$\int_0^R 2\pi r\,dr = 2\pi\int_0^R r\,dr$$

$$= 2\pi\left[\frac{r^2}{2}\right]_0^R = \pi R^2.$$

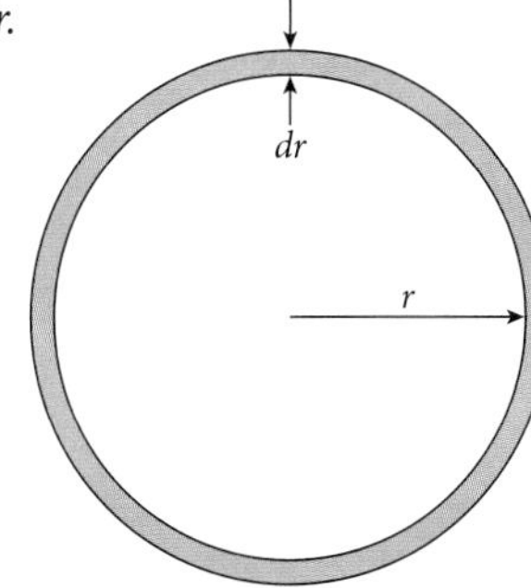


**9.3** $\omega^2 r = (2\pi/\text{yr})^2\,(1\text{ au}) = 5.9\cdot 10^{-4}\text{ m/s}^2 = g/1700.$

**9.4** Slightly heavier, but barely noticeable.

**9.5** At $r_1 = 25$ mm, the magnitude of the tangential acceleration is $a_t = \alpha r_1 = 1.56\cdot 10^{-4}\text{ m/s}^2$, and that of the centripetal acceleration is $a_c = v\omega(r_1) = 59.1\text{ m/s}^2$, larger by more than four orders of magnitude. At $r_2 = 58$ mm, the accelerations are $a_t = \alpha r_2 = 3.61\cdot 10^{-4}\text{ m/s}^2$ and $a_c = v\omega(r_2) = 25.5\text{ m/s}^2$.

**9.6** Substitute equation 9.23 into equation 9.22 and find $T = mg/\cos\varphi$.

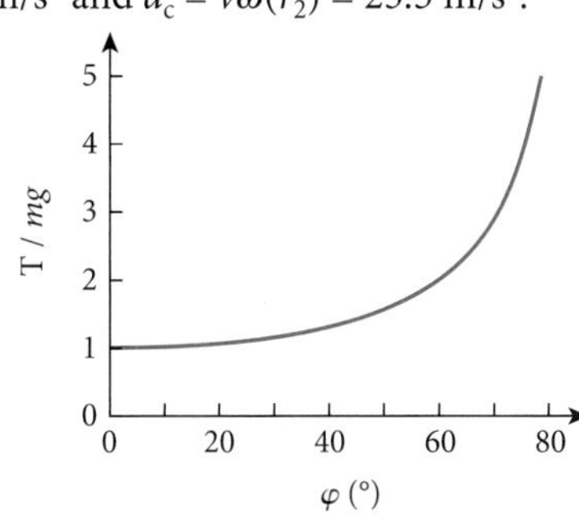


**9.7** If the radius is doubled, the speed at the top of the loop has to increase by a factor of $\sqrt{2}$. Thus, the required speed is $(7.00\text{ m/s})(\sqrt{2}) = 9.90\text{ m/s}$.

**9.8** (*iv*) $\bar{\omega} = \frac{1}{t}\int_0^t \omega(t')dt' = \frac{1}{t}\int_0^t (\omega_0 + at')dt'$

$$= \frac{\omega_0}{t}\int_0^t dt' + \frac{\alpha}{t}\int_0^t t'dt' = \omega_0 + \tfrac{1}{2}\alpha t$$

$$= \tfrac{1}{2}\omega_0 + \tfrac{1}{2}(\omega_0 + \alpha t)$$

$$= \tfrac{1}{2}(\omega_0 + \omega)$$

(*ii*) $\bar{\omega} = \omega_0 + \tfrac{1}{2}\alpha t$

$$\Rightarrow \bar{\omega}t = \omega_0 t + \tfrac{1}{2}\alpha t^2$$

$$\theta = \theta_0 + \theta_0 t + \tfrac{1}{2}\alpha t^2 = \theta_0 + \bar{\omega}t$$

(*v*) $\theta = \theta_0 + \omega_0 t + \tfrac{1}{2}\alpha t^2$

$$= \theta_0 + \theta_0\left(\frac{\omega - \omega_0}{\alpha}\right) + \tfrac{1}{2}\alpha\left(\frac{\omega - \omega_0}{\alpha}\right)^2$$

$$= \theta_0 + \frac{\omega\omega_0 - \omega_0^2}{\alpha} + \tfrac{1}{2}\frac{\omega^2 + \omega_0^2 - 2\omega\omega_0}{\alpha}$$

Now we subtract $\theta_0$ from both sides of the equation and then multiply by $\alpha$:

$$\alpha(\theta - \theta_0) = \omega\omega_0 - \omega_0^2 + \tfrac{1}{2}(\omega^2 + \omega_0^2 - 2\omega\omega_0)$$

$$\Rightarrow \alpha(\theta - \theta_0) = \tfrac{1}{2}\omega^2 - \tfrac{1}{2}\omega_0^2$$

$$\Rightarrow \omega^2 = \omega_0^2 + 2\alpha(\theta - \theta_0)$$

## PROBLEM-SOLVING PRACTICE

### Problem-Solving Guidelines: Circular Motion

**1.** Motion in a circle always requires centripetal force and centripetal acceleration. However, remember that centripetal force is not a new kind of force but is simply the net force causing the motion; it consists of the sum of whatever forces are acting on the moving object. This net force equals the mass times the centripetal acceleration; do not make the common mistake of counting mass times acceleration as a force to be added to the net force on one side of the equation of motion.

**2.** Be sure you note whether the situation involves angles in degrees or in radians. The radian is not a unit that necessarily has to be carried through a calculation, but check that your result makes sense in terms of the angular units.

**3.** The equations of motion with constant angular acceleration have the same form as the equations of motion with constant linear acceleration. However, neither set of equations applies if the acceleration is not constant.

### SOLVED PROBLEM 9.3 Carnival Ride

**FIGURE 9.25** A carnival ride consisting of a rotating cylinder.

**PROBLEM**

One of the rides found at carnivals is a rotating cylinder, as shown in Figure 9.25. The riders step inside the vertical cylinder and stand with their backs against the curved wall. The cylinder spins very rapidly, and at some angular velocity, the floor is pulled away. The thrill-seekers now hang like flies on the wall. (The cylinder in Figure 9.25, is lifted up and tilted after the ride reaches it operating angular velocity, but we will not deal with this additional complication.) If the radius of the cylinder is $r = 2.10$ m, the rotation axis of the cylinder remains vertical, and the coefficient of static friction between the people and the wall is $\mu_s = 0.390$, what is the minimum angular velocity, $\omega$, at which the floor can be withdrawn?

**SOLUTION**

**THINK**

When the floor drops away, the magnitude of the force of static friction between a rider and the wall of the rotating cylinder must equal the magnitude of the force of gravity acting on the rider. The static friction between the rider and the wall depends on the normal force being exerted on the rider and the coefficient of static friction. As the cylinder spins faster, the normal force (which acts as the centripetal force) being exerted on the rider increases. At a certain angular velocity, the maximum magnitude of the force of static friction will equal the magnitude of the force of gravity. That angular velocity is the minimum angular velocity at which the floor can be withdrawn.

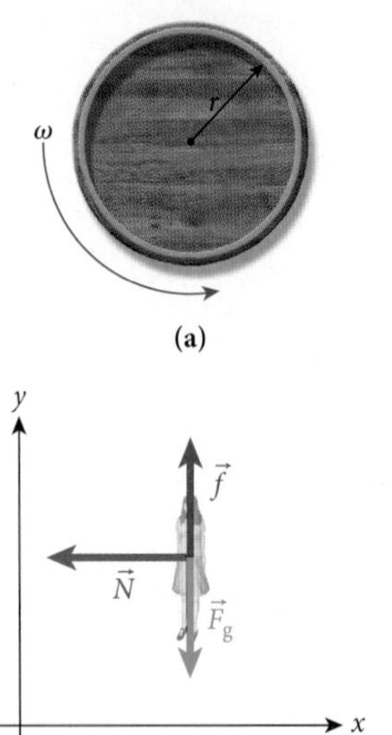


**FIGURE 9.26** (a) Top view of the rotating cylinder of a carnival ride. (b) Free-body diagram for one of the riders.

**SKETCH**

A top view of the rotating cylinder is shown in Figure 9.26a. The free-body diagram for one of the riders is shown in Figure 9.26b, where the rotation axis is assumed to be the $y$-axis. In this sketch, $\vec{f}$ is the force of static friction, $\vec{N}$ is the normal force exerted on the rider of mass $m$ by the wall of the cylinder, and $\vec{F}_g$ is the force of gravity acting on the rider.

**RESEARCH**

At the minimum angular velocity required to keep the rider from falling, the magnitude of the force of static friction between the rider and the wall is equal to the magnitude of the force of gravity acting on the rider. To analyze these forces, we start with the free-body diagram shown in Figure 9.26b. In the free-body diagram, the $x$-direction is along the radius of the cylinder, and the $y$-direction is vertical. In the $x$-direction, the normal force exerted by the wall on the rider provides the centripetal force that makes the rider move in a circle:

$$F_c = N. \tag{i}$$

In the $y$-direction, the thrill-seeking rider sticks to the wall only if the upward force of static friction between the rider and wall balances the downward force of gravity. The force of gravity on the rider is his or her weight, so we can write

$$f = F_g = mg. \tag{ii}$$

We know that the centripetal force is given by

$$F_c = mr\omega^2, \tag{iii}$$

and the force of static friction is given by

$$f \le f_{max} = \mu_s N. \tag{iv}$$

**SIMPLIFY**

We can combine equations (ii) and (iv) to obtain

$$mg \le \mu_s N. \tag{v}$$

Substituting for $F_c$ from equation (i) into (iii), we find

$$N = mr\omega^2. \tag{vi}$$

Combining equations (v) and (vi), we have

$$mg \le \mu_s mr\omega^2,$$

which we can solve for $\omega$:

$$\omega \ge \sqrt{\frac{g}{\mu_s r}}.$$

Thus, the minimum value of the angular velocity is given by

$$\omega_{min} = \sqrt{\frac{g}{\mu_s r}}.$$

Note that the mass of the rider canceled out. This is crucial since people of different masses want to ride at the same time!

**CALCULATE**

Putting in the numerical values, we find

$$\omega_{min} = \sqrt{\frac{g}{\mu_s r}} = \sqrt{\frac{9.81 \text{ m/s}^2}{(0.390)(2.10 \text{ m})}} = 3.46093 \text{ rad/s}.$$

**ROUND**

Expressing our result to three significant figures gives

$$\omega_{min} = 3.46 \text{ rad/s}.$$

**DOUBLE-CHECK**

To double-check, let's express our result for the angular velocity in revolutions per minute (rpm):

$$3.46 \frac{\text{rad}}{\text{s}} = \left(3.46 \frac{\text{rad}}{\text{s}}\right)\left(\frac{60 \text{ s}}{1 \text{ min}}\right)\left(\frac{1 \text{ rev}}{2\pi \text{ rad}}\right) = 33 \text{ rpm}.$$

An angular velocity of 33 rpm for the rotating cylinder seems reasonable, because this means that it completes almost one full turn every 2 s. If you have ever been on one of these rides or watched one, you know that the answer is in the right ballpark.

Note that the coefficient of friction, $\mu_s$, between the clothing of the rider and the wall is not identical in all cases. Our formula, $\omega_{min} = \sqrt{g/(\mu_s r)}$, indicates that a smaller coefficient of friction necessitates a larger angular velocity. Designers of this kind of ride need to make sure that they allow for the smallest coefficient of friction that can be expected to occur. Obviously, they want to have a somewhat sticky contact surface on the wall of the ride, just to make sure!

One last point to check: $\omega_{min} = \sqrt{g/(\mu_s r)}$ indicates that the minimum angular velocity required will decrease as a function of the radius of the cylinder. The example of the poker chips on the spinning table in Section 9.5 established that the centripetal force increases with radial distance, which is consistent with this result.

## SOLVED PROBLEM 9.4 Flywheel

### PROBLEM

The flywheel of a steam engine starts to rotate from rest with a constant angular acceleration of $\alpha = 1.43 \text{ rad/s}^2$. The flywheel undergoes this constant angular acceleration for $t = 25.9$ s and then continues to rotate at a constant angular velocity, $\omega$. After the flywheel has been rotating for 59.5 s, what is the total angle through which it has rotated since it started?

### SOLUTION

### THINK

Here we are trying to determine the total angular displacement, $\theta$. For the time interval when the flywheel is undergoing angular acceleration, we can use equation 9.26(i) with $\theta_0 = 0$ and $\omega_0 = 0$. When the flywheel is rotating at a constant angular velocity, we use equation 9.26(i)with $\theta_0 = 0$ and $\alpha = 0$. To get the total angular displacement, we add these two angular displacements.

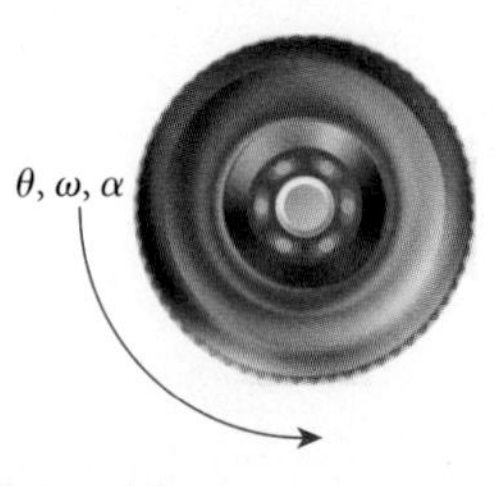


**FIGURE 9.27** Top view of the rotating flywheel.

### SKETCH

A top view of the rotating flywheel is shown in Figure 9.27.

### RESEARCH

Let's call the time during which the flywheel is undergoing angular acceleration $t_a$ and the total time the flywheel is rotating $t_b$. Thus, the flywheel rotates at a constant angular velocity for a time interval equal to $t_b - t_a$. The angular displacement, $\theta_a$, that occurs while the flywheel is undergoing angular acceleration is given by

$$\theta_a = \tfrac{1}{2}\alpha t_a^2. \qquad \text{(i)}$$

The angular displacement, $\theta_b$, that occurs while the flywheel is rotating at the constant angular velocity, $\omega$, is given by

$$\theta_b = \omega\left(t_b - t_a\right). \qquad \text{(ii)}$$

The angular velocity, $\omega$, reached by the flywheel after undergoing the angular acceleration $\alpha$ for a time $t_a$ is given by

$$\omega = \alpha t_a. \qquad \text{(iii)}$$

The total angular displacement is given by

$$\theta_{\text{total}} = \theta_a + \theta_b. \qquad \text{(iv)}$$

### SIMPLIFY

We can combine equations (ii) and (iii) to obtain the angular displacement while the flywheel is rotating at a constant angular velocity:

$$\theta_b = \left(\alpha t_a\right)\left(t_b - t_a\right) = \alpha t_a t_b - \alpha t_a^2. \qquad \text{(v)}$$

We can combine equations (v), (iv), and (i) to get the total angular displacement of the flywheel:

$$\theta_{\text{total}} = \theta_a + \theta_b = \tfrac{1}{2}\alpha t_a^2 + \left(\alpha t_a t_b - \alpha t_a^2\right) = \alpha t_a t_b - \tfrac{1}{2}\alpha t_a^2.$$

### CALCULATE

Putting in the numerical values gives us

$$\theta_{\text{total}} = \alpha t_a t_b - \tfrac{1}{2}\alpha t_a^2 = \left(1.43 \text{ rad/s}^2\right)\left(25.9 \text{ s}\right)\left(59.5 \text{ s}\right) - \tfrac{1}{2}\left(1.43 \text{ rad/s}^2\right)\left(25.9 \text{ s}\right)^2$$
$$= 1724.07 \text{ rad}.$$

### ROUND

Expressing our result to three significant figures, we have

$$\theta_{\text{total}} = 1720 \text{ rad}.$$

### DOUBLE-CHECK

It is comforting that our answer has the right unit, the rad. Our formula, $\theta_{\text{total}} = \alpha t_{\text{a}} t_{\text{b}} - \frac{1}{2}\alpha t_{\text{a}}^2 = \alpha t_{\text{a}}(t_{\text{b}} - \frac{1}{2}t_{\text{a}})$, gives a value that increases linearly with the value of the angular acceleration. It is also always larger than zero, as expected, because $t_{\text{b}} > t_{\text{a}}$.

To perform a further check, let's calculate the angular displacement in two steps. The first step is to calculate the angular displacement while the flywheel is accelerating

$$\theta_a = \tfrac{1}{2}\alpha t_{\text{a}}^2 = \tfrac{1}{2}\left(1.43 \text{ rad/s}^2\right)\left(25.9 \text{ s}\right)^2 = 480 \text{ rad}.$$

The angular velocity of the flywheel after the angular acceleration ends is

$$\omega = \alpha t_{\text{a}} = \left(1.43 \text{ rad/s}^2\right)\left(25.9 \text{ s}\right) = 37.0 \text{ rad/s}.$$

Next, we calculate the angular displacement while the flywheel is rotating at constant velocity:

$$\theta_{\text{b}} = \omega\left(t_{\text{b}} - t_{\text{a}}\right) = \left(37.0 \text{ rad/s}\right)\left(59.5 \text{ s} - 25.9 \text{ s}\right) = 1240 \text{ rad}.$$

The total angular displacement is then

$$\theta_{\text{total}} = \theta_{\text{a}} + \theta_{\text{b}} = 480 \text{ rad} + 1240 \text{ rad} = 1720 \text{ rad},$$

which agrees with our answer.

## MULTIPLE-CHOICE QUESTIONS

**9.1** An object is moving in a circular path. If the centripetal force is suddenly removed, how will the object move?

a) It will move radially outward.

b) It will move radially inward.

c) It will move vertically downward.

d) It will move in the direction in which its velocity vector points at the instant the centripetal force vanishes.

**9.2** The angular acceleration for an object undergoing circular motion is plotted versus time in the figure. If the object started from rest at $t = 0$ s, the net angular displacement of the object at $t = t_{\text{f}}$

a) is in the clockwise direction.

b) is in the counterclockwise direction.

c) is zero.

d) cannot be determined.

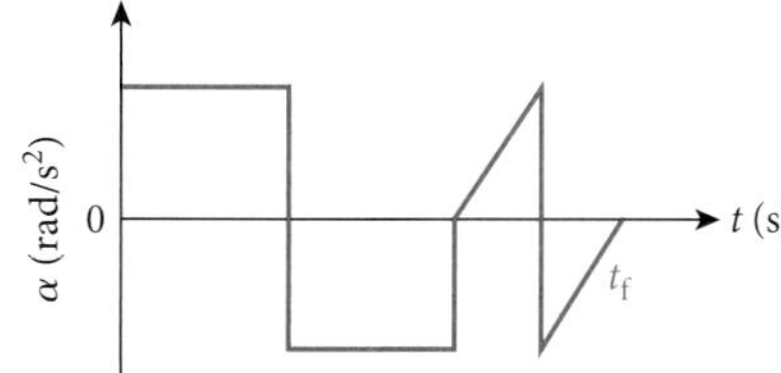


**9.3** The latitude of Lubbock, Texas (known as the Hub City of the South Plains), is 33° N. What is its rotational speed, assuming the radius of the Earth at the Equator to be 6380 km?

a) 464 m/s

b) 389 m/s

c) 253 m/s

d) 0.464 m/s

e) 0.389 m/s

**9.4** A rock attached to a string moves clockwise in uniform circular motion. In which direction from point *A* is the rock thrown off when the string is cut?

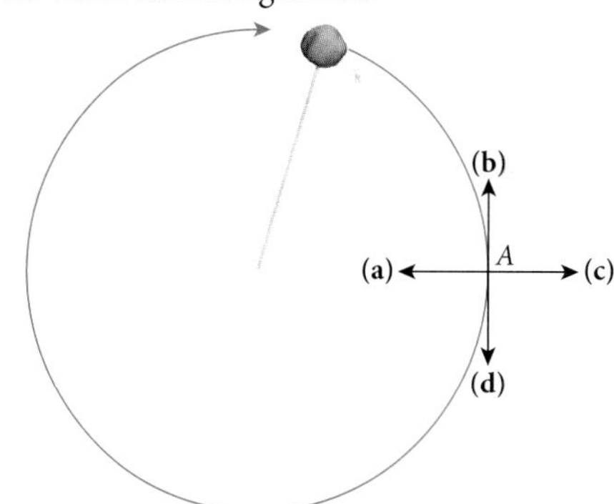


**9.5** A Ferris wheel rotates slowly about a horizontal axis. Passengers sit on seats that remain horizontal on the Ferris wheel as it rotates. Which type of force provides the centripetal acceleration on the passengers when they are at the top of the Ferris wheel?

a) centrifugal

b) normal

c) gravity

d) tension

**9.6** In a conical pendulum, a bob moves in a horizontal circle, as shown in the figure. The period of the pendulum (the time it takes for the bob to perform a complete revolution) is

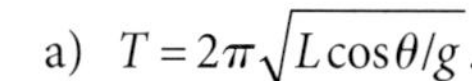

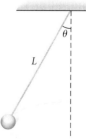

a) $T = 2\pi\sqrt{L\cos\theta/g}$.

b) $T = 2\pi\sqrt{g\cos\theta/L}$.

*Continued—*

c) $T = 2\pi\sqrt{Lg\sin\theta}$.

d) $T = 2\pi\sqrt{L\sin\theta/g}$.

e) $T = 2\pi\sqrt{L/g}$.

**9.7** A ball attached to the end of a string is swung around in a circular path of radius $r$. If the radius is doubled and the linear speed is kept constant, the centripetal acceleration

a) remains the same.
b) increases by a factor of 2.
c) decreases by a factor of 2.
d) increases by a factor of 4.
e) decreases by a factor of 4.

**9.8** The angular speed of the hour hand of a clock (in radians per second) is

a) $\frac{\pi}{7200}$
b) $\frac{\pi}{3600}$
c) $\frac{\pi}{1800}$
d) $\frac{\pi}{60}$
e) $\frac{\pi}{30}$
f) The correct value is not shown.

**9.9** You put three identical coins on a turntable at different distances from the center and then turn the motor on. As the turntable speeds up, the outermost coin slides off first, followed by the one at the middle distance, and, finally, when the turntable is going the fastest, the innermost one. Why is this?

a) For greater distances from the center the centripetal acceleration is higher, and so the force of friction becomes unable to hold the coin in place.

b) The weight of the coin causes the turntable to flex downward, so the coin nearest the edge falls off first.

c) Because of the way the turntable is made, the coefficient of static friction decreases with distance from the center.

d) For smaller distances from the center, the centripetal acceleration is higher.

**9.10** A point on a Blu-ray disc is a distance $R/4$ from the axis of rotation. How far from the axis of rotation is a second point that has, at any instant, a linear velocity twice that of the first point?

a) $R/16$
b) $R/8$
c) $R/2$
d) $R$

**9.11** The figure shows a rider stuck to the wall in the Barrel of Fun at a carnival. Which diagram correctly shows the forces acting on the rider?

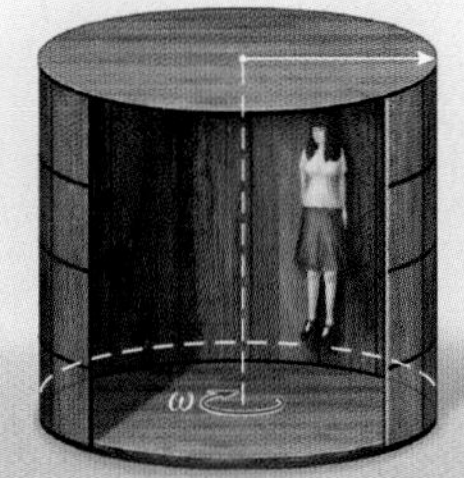


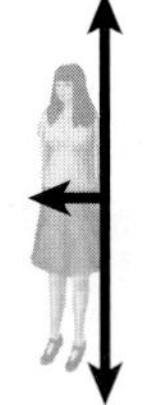
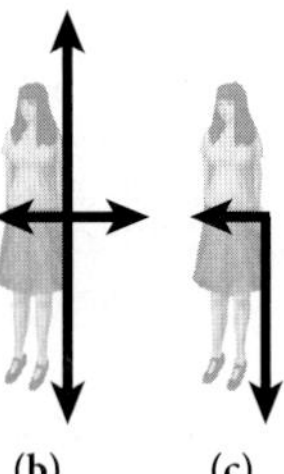
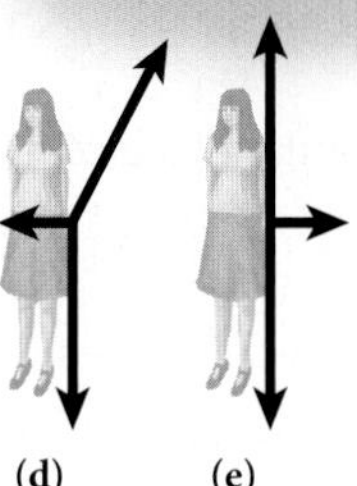

(a) (b) (c) (d) (e)

**9.12** A string is tied to a rock, and the rock is twirled around in a circle at a constant speed. If gravity is ignored and the period of the circular motion is doubled, the tension in the string is

a) reduced to $\frac{1}{4}$ of its original value.
b) reduced to $\frac{1}{2}$ of its original value.
c) unchanged.
d) increased to 2 times its original value.
e) increased to 4 times its original value.

## QUESTIONS

**9.13** A ceiling fan is rotating in clockwise direction (as viewed from below) but it is slowing down. What are the directions of $\omega$ and $\alpha$?

**9.14** A hook above a stage is rated to support 150 lb. A 3-lb rope is attached to the hook, and a 147-lb actor is going to attempt to swing across the stage on the rope. Will the hook hold the actor up during the swing?

**9.15** A popular carnival ride consists of seats attached to a central disk through cables, as shown in the figure. The passengers travel in uniform circular motion. The mass of one of the passengers (including the chair he is sitting on) is 65 kg; the mass of an empty chair on the opposite side of the central disk is 5 kg. If $\theta_1$ and $\theta_2$ are the angles that the cables attached to the two chairs make with respect to vertical, how do these two angles compare qualitatively? Is $\theta_2$ larger than, smaller than, or equal to $\theta_1$?

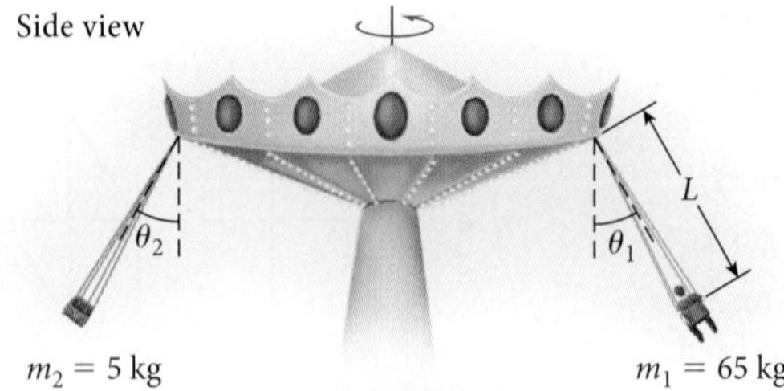


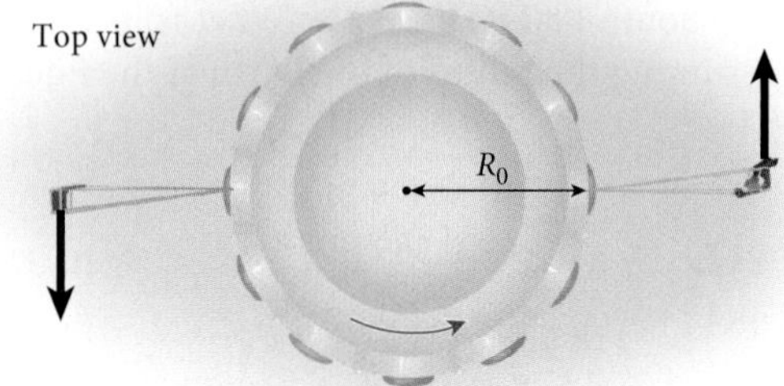


**9.16** A person rides on a Ferris wheel of radius $R$, which is rotating at a constant angular velocity $\omega$. Compare the normal force of the seat pushing up on the person at point $A$ to that at point $B$ in the figure. Which force is greater, or are they the same?

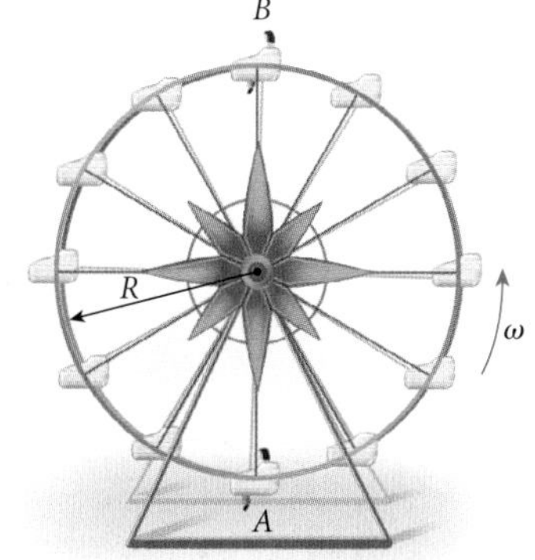


**9.17** Bicycle tires range in size from about 25 cm in diameter to about 70 cm in diameter. Why is it impractical to make tires much smaller than 25 cm in diameter? (You will learn why bicycle tires can't be too large in Chapter 10.)

**9.18** A CD starts from rest and speeds up to the operating angular frequency of the CD player. Compare the angular velocity and acceleration of a point on the edge of the CD to those of a point halfway between the center and the edge of the CD. Do the same for the linear velocity and acceleration.

**9.19** A car is traveling around an unbanked curve at a maximum speed. Which force(s) is(are) responsible for keeping it on the road?

**9.20** Two masses hang from two strings of equal length that are attached to the ceiling of a car. One mass is over the driver's seat; the other is over the passenger's seat. As the car makes a sharp turn, both masses swing away from the center of the turn. In their resulting positions, will they be farther apart, closer together, or the same distance apart as they were when the car wasn't turning?

**9.21** A point mass $m$ starts sliding from a height $h$ along the frictionless surface shown in the figure. What is the minimum value of $h$ in order for the mass to complete the loop of radius $R$?

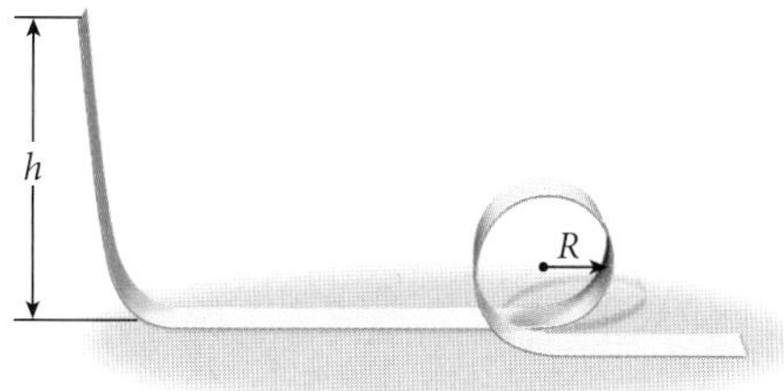


**9.22** In a conical pendulum, the bob attached to the string (which can be considered massless) moves in a horizontal circle at constant speed. The string sweeps out a cone as the bob rotates. What forces are acting on the bob?

**9.23** Is it possible to swing a mass attached to a string in a perfectly horizontal circle (with the mass and the string parallel to the ground)?

**9.24** A small ice block of mass $m$ starts from rest from the top of an inverted bowl in the shape of a hemisphere, as shown in the figure. The hemisphere is fixed to the ground, and the block slides without friction along the surface of the hemisphere. Find the normal force exerted by the block on the sphere when the line between the block and the center of the sphere makes an angle $\theta$ with the horizontal. Discuss the result.

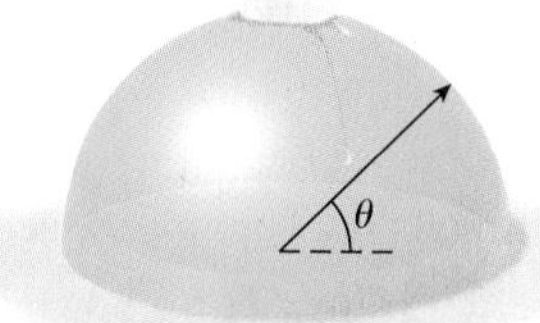


**9.25** Suppose you are riding on a roller coaster, which moves through a vertical circular loop. Show that the difference in your apparent weights at the top and the bottom of the loop is six times your weight and is independent of the size of the loop. Assume that friction is negligible.

**9.26** The following event actually occurred on the Sunshine Skyway Bridge near St. Petersburg, Florida, in 1997. Five daredevils tied a 55-m-long cable to the center of the bridge. They hoped to swing back and forth under the bridge at the end of this cable. The five people (total weight = $W$) attached themselves to the end of the cable, at the same level and 55 m away from where it was attached to the bridge and dropped straight down from the bridge, following the dashed circular path indicated in the figure. Unfortunately, the daredevils were not well versed in the laws of physics, and the cable broke (at the point it was linked to their seats) at the bottom of their swing. Determine how strong the cable (and all the links where the seats and the bridge are attached to it) would have to be in order to support the five people at the bottom of the swing. Express your result in terms of their total weight, $W$.

## PROBLEMS

A blue problem number indicates a worked-out solution is available in the Student Solutions Manual. One • and two •• indicate increasing level of problem difficulty.

### Section 9.2

9.27 What is the angle in radians that the Earth sweeps out in its orbit during winter?

**9.28** Assuming that the Earth is spherical and recalling that latitudes range from 0° at the Equator to 90° N at the North Pole, how far apart, measured on the Earth's surface, are Dubuque, Iowa (42.50° N latitude), and Guatemala City (14.62° N latitude)? The two cities lie on approximately the same longitude. Do *not* neglect the curvature of the Earth in determining this distance.

•**9.29** Refer to the information given in Problem 9.28. If one could burrow through the Earth and dig a straight-line tunnel from Dubuque to Guatemala City, how long would the tunnel be? From the point of view of the digger, at what angle below the horizontal would the tunnel be directed?

## Section 9.3

**9.30** A typical Major League fastball is thrown at approximately 88 mph and with a spin rate of 110 rpm. If the distance between the pitcher's point of release and the catcher's glove is exactly 60.5 ft, how many full turns does the ball make between release and catch? Neglect any effect of gravity or air resistance on the ball's flight.

**9.31** A vinyl record plays at 33.3 rpm. Assume it takes 5 s for it to reach this full speed, starting from rest.

a) What is its angular acceleration during the 5 s?

b) How many revolutions does the record make before reaching its final angular speed?

**9.32** At a county fair, a boy takes his teddy bear on the giant Ferris wheel. Unfortunately, at the top of the ride, he accidentally drops his stuffed buddy. The wheel has a diameter of 12.0 m, the bottom of the wheel is 2.0 m above the ground and its rim is moving at a speed of 1.0 m/s. How far from the base of the Ferris wheel will the teddy bear land?

•**9.33** Having developed a taste for experimentation, the boy in Problem 9.32 invites two friends to bring their teddy bears on the same Ferris wheel. The boys are seated in positions 45° from each other. When the wheel brings the second boy to the maximum height, they all drop their stuffed animals. How far apart will the three teddy bears land?

•**9.34** Mars orbits the Sun at a mean distance of 228 million km, in a period of 687 days. The Earth orbits at a mean distance of 149.6 million km, in a period of 365.26 days.

a) Suppose Earth and Mars are positioned such that Earth lies on a straight line between Mars and the Sun. Exactly 365.26 days later, when the Earth has completed one orbit, what is the angle between the Earth-Sun line and the Mars-Sun line?

b) The initial situation in part (a) is a closest approach of Mars to Earth. What is the time, in days, between two closest approaches? Assume constant speed and circular orbits for both Mars and Earth.

c) Another way of expressing the answer to part (b) is in terms of the angle between the lines drawn through the Sun, Earth, and Mars in the two closest approach situations. What is that angle?

••**9.35** Consider a large simple pendulum that is located at a latitude of 55° N and is swinging in a north-south direction with points *A* and *B* being the northernmost and the southernmost points of the swing, respectively. A stationary (with respect to the fixed stars) observer is looking directly down on the pendulum at the moment shown in the figure. The Earth is rotating once every 23 h and 56 min.

a) What are the directions (in terms of N, E, W, and S) and the magnitudes of the velocities of the surface of the Earth at points *A* and *B* as seen by the observer? *Note*: You will need to calculate answers to at least seven significant figures to see a difference.

b) What is the angular speed with which the 20.0-m diameter circle under the pendulum appears to rotate?

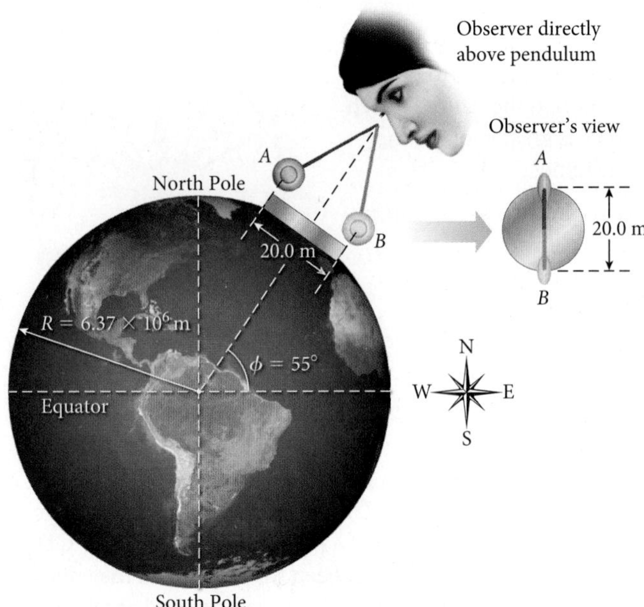


c) What is the period of this rotation?

d) What would happen to a pendulum swinging at the Equator?

## Section 9.4

**9.36** What is the centripetal acceleration of the Moon? The period of the Moon's orbit about the Earth is 27.3 days, measured with respect to the fixed stars. The radius of the Moon's orbit is $R_M = 3.85 \cdot 10^8$ m.

**9.37** You are holding the axle of a bicycle wheel with radius 35 cm and mass 1 kg. You get the wheel spinning at a rate of 75 rpm and then stop it by pressing the tire against the pavement. You notice that it takes 1.2 s for the wheel to come to a complete stop. What is the angular acceleration of the wheel?

**9.38** Life scientists use ultracentrifuges to separate biological components or to remove molecules from suspension. Samples in a symmetric array of containers are spun rapidly about a central axis. The centrifugal acceleration they experience in their moving reference frame acts as "artificial gravity" to effect a rapid separation. If the sample containers are 10 cm from the rotation axis, what rotation frequency is required to produce an acceleration of $1.00 \cdot 10^5 g$?

**9.39** A centrifuge in a medical laboratory rotates at an angular speed of 3600 rpm (revolutions per minute). When switched off, it rotates 60.0 times before coming to rest. Find the constant angular acceleration of the centrifuge.

**9.40** A discus thrower (with arm length of 1.2 m) starts from rest and begins to rotate counterclockwise with an angular acceleration of 2.5 rad/s$^2$.

a) How long does it take the discus thrower's speed to get to 4.7 rad/s?

b) How many revolutions does the thrower make to reach the speed of 4.7 rad/s?

c) What is the linear speed of the discus at 4.7 rad/s?

d) What is the linear acceleration of the discus thrower at this point?

e) What is the magnitude of the centripetal acceleration of the discus thrown?

f) What is the magnitude of the discus's total acceleration?

•9.41 In a department store toy display, a small disk (disk 1) of radius 0.1 m is driven by a motor and turns a larger disk (disk 2) of radius 0.5 m. Disk 2, in turn, drives disk 3, whose radius is 1.0 m. The three disks are in contact, and there is no slipping. Disk 3 is observed to sweep through one complete revolution every 30 s.

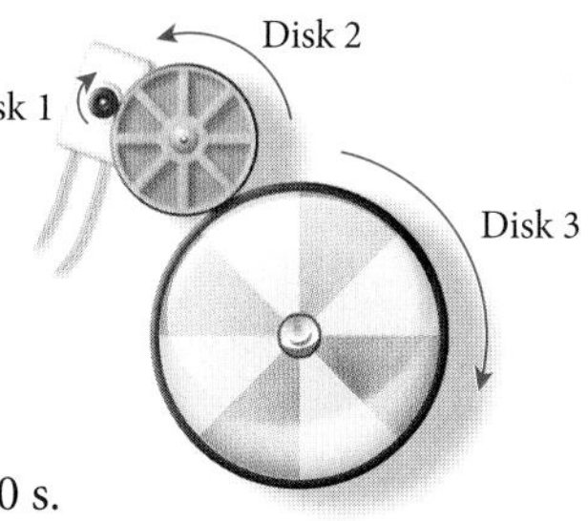


a) What is the angular speed of disk 3?

b) What is the ratio of the tangential velocities of the rims of the three disks?

c) What is the angular speed of disks 1 and 2?

d) If the motor malfunctions, resulting in an angular acceleration of 0.1 rad/s$^2$ for disk 1, what are disks 2 and 3's angular accelerations?

•9.42 A particle is moving clockwise in a circle of radius 1.00 m. At a certain instant, the magnitude of its acceleration is $a = |\vec{a}| = 25.0$ m/s$^2$, and the acceleration vector has an angle of $\theta = 50°$ with the position vector, as shown in the figure. At this instant, find the speed, $v = |\vec{v}|$, of this particle.

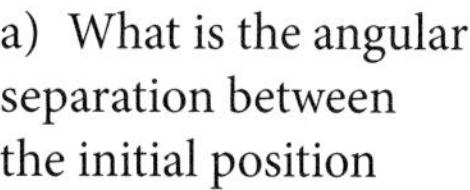

•9.43 In a tape recorder, the magnetic tape moves at a constant linear speed of 5.6 cm/s. To maintain this constant linear speed, the angular speed of the driving spool (the take-up spool) has to change accordingly.

a) What is the angular speed of the take-up spool when it is empty, with radius $r_1 = 0.80$ cm?

b) What is the angular speed when the spool is full, with radius $r_2 = 2.20$ cm?

c) If the total length of the tape is 100.80 m, what is the average angular acceleration of the take-up spool while the tape is being played?

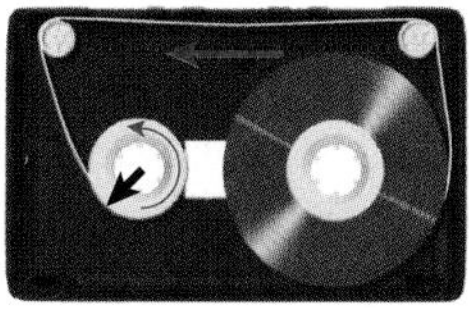

••9.44 A ring is fitted loosely (with no friction) around a long, smooth rod of length $L = 0.50$ m. The rod is fixed at one end, and the other end is spun in a horizontal circle at a constant angular velocity of $\omega = 4.0$ rad/s. The ring has zero radial velocity at its initial position, a distance of $r_0 = 0.30$ m from the fixed end. Determine the radial velocity of the ring as it reaches the moving end of the rod.

••9.45 A flywheel with a diameter of 1 m is initially at rest. Its angular acceleration versus time is graphed in the figure.

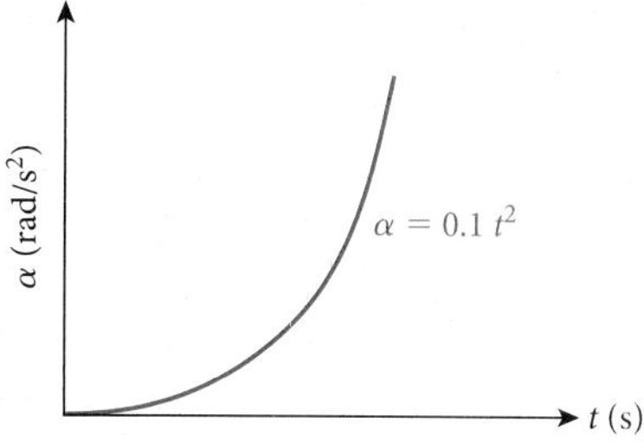


a) What is the angular separation between the initial position of a fixed point on the rim of the flywheel and the point's position 8 s after the wheel starts rotating?

b) The point starts its motion at $\theta = 0$. Calculate and sketch the linear position, velocity vector, and acceleration vector 8 s after the wheel starts rotating.

## Section 9.5

9.46 Calculate the centripetal force exerted on a vehicle of mass $m = 1500$ kg that is moving at a speed of 15 m/s around a curve of radius $R = 400$ m. Which force plays the role of the centripetal force in this case?

9.47 What is the apparent weight of a rider on the roller coaster of Solved Problem 9.1 at the *bottom* of the loop?

9.48 Two skaters, A and B, of equal mass are moving in clockwise uniform circular motion on the ice. Their motions have equal periods, but the radius of skater A's circle is half that of skater B's circle.

a) What is the ratio of the speeds of the skaters?

b) What is the ratio of the magnitudes of the forces acting on each skater?

•9.49 A small block of mass $m$ is in contact with the inner wall of a large hollow cylinder. Assume the coefficient of static friction between the object and the wall of the cylinder is $\mu$. Initially, the cylinder is at rest, and the block is held in place by a peg supporting its weight. The cylinder starts rotating about its center axis, as shown in the figure, with an angular acceleration of $\alpha$. Determine the minimum time interval after the cylinder begins to rotate before the peg can be removed without the block sliding against the wall.

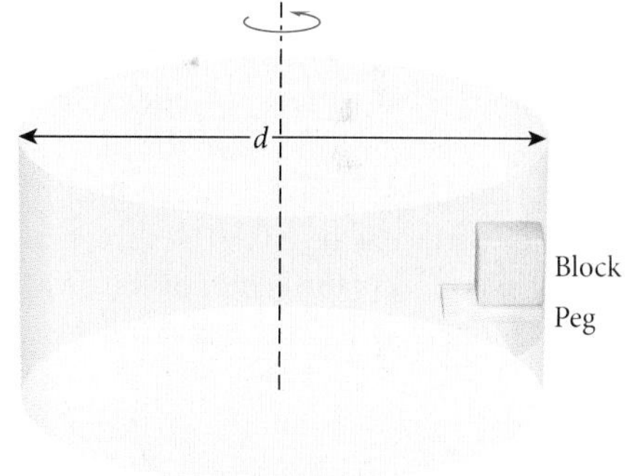


•9.50 A race car is making a U-turn at constant speed. The coefficient of friction between the tires and the track is $\mu_s = 1.2$. If the radius of the curve is 10 m, what is the maximum speed at which the car can turn without sliding? Assume that the car is performing uniform circular motion.

•9.51 A car speeds over the top of a hill. If the radius of curvature of the hill at the top is 9.0 m, how fast can the car be traveling and maintain constant contact with the ground?

•9.52 A ball of mass $m = 0.2$ kg is attached to a (massless) string of length $L = 1$ m and is undergoing circular motion in the horizontal plane, as shown in the figure.

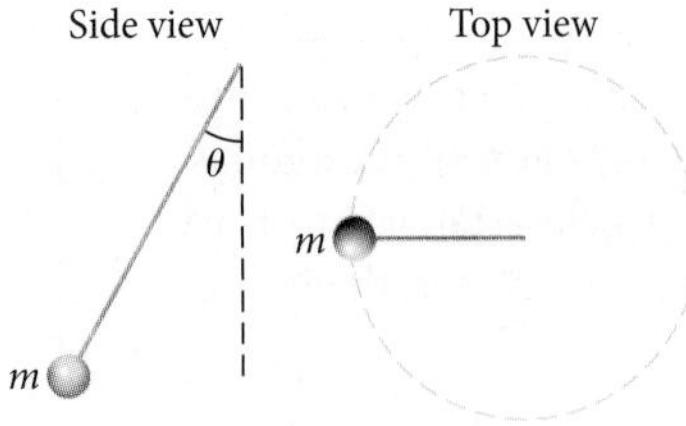


a) Draw a free-body diagram for the ball.

b) Which force plays the role of the centripetal force?

c) What should the speed of the mass be for $\theta$ to be 45°?

d) What is the tension in the string?

•**9.53** You are flying to Chicago for a weekend away from the books. In your last physics class, you learned that the airflow over the wings of the plane creates a *lift force,* which acts perpendicular to the wings. When the plane is flying level, the upward lift force exactly balances the downward *weight force*. Since O'Hare is one of the busiest airports in the world, you are not surprised when the captain announces that the flight is in a holding pattern due to the heavy traffic. He informs the passengers that the plane will be flying in a circle of radius 7 mi at a speed of 360 mph and an altitude of 20,000 ft. From the safety information card, you know that the total length of the wingspan of the plane is 275 ft. From this information, estimate the banking angle of the plane relative to the horizontal.

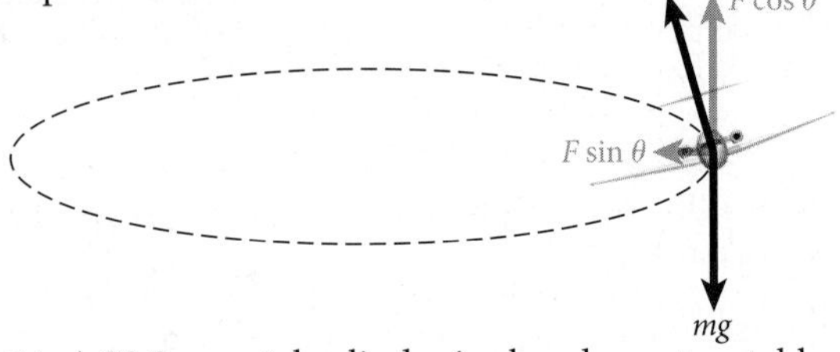


••**9.54** A 20.0-g metal cylinder is placed on a turntable, with its center 80 cm from the turntable's center. The coefficient of static friction between the cylinder and the turntable's surface is $\mu_s = 0.80$. A thin, massless string of length 80 cm connects the center of the turntable to the cylinder, and *initially*, the string has zero tension in it. Starting from rest, the turntable very slowly attains higher and higher angular velocities, but the turntable and the cylinder can be considered to have uniform circular motion at any instant. Calculate the tension in the string when the angular velocity of the turntable is 60 rpm (rotations per minute).

••**9.55** A speedway turn, with radius of curvature $R$, is banked at an angle $\theta$ above the horizontal.

a) What is the optimal speed at which to take the turn if the track's surface is iced over (that is, if there is very little friction between the tires and the track)?

b) If the track surface is ice-free and there is a coefficient of friction $\mu_s$ between the tires and the track, what are the maximum and minimum speeds at which this turn can be taken?

c) Evaluate the results of parts (a) and (b) for $R = 400$ m, $\theta = 45°$, and $\mu_s = 0.70$.

## Additional Problems

**9.56** A particular Ferris wheel takes riders in a vertical circle of radius 9.0 m once every 12.0 s.

a) Calculate the speed of the riders, assuming it to be constant.

b) Draw a free-body diagram for a rider at a time when she is at the bottom of the circle. Calculate the normal force exerted by the seat on the rider at that point in the ride.

c) Perform the same analysis as in part (b) for a point at the top of the ride.

**9.57** A boy is on a Ferris wheel, which takes him in a vertical circle of radius 9.0 m once every 12.0 s.

a) What is the angular speed of the Ferris wheel?

b) Suppose the wheel comes to a stop at a uniform rate during one quarter of a revolution. What is the angular acceleration of the wheel during this time?

c) Calculate the tangential acceleration of the boy during the time interval described in part (b).

**9.58** Consider a 53-cm-long lawn mower blade rotating about its center at 3400 rpm.

a) Calculate the linear speed of the tip of the blade.

b) If safety regulations require that the blade be stoppable within 3.0 s, what minimum angular acceleration will accomplish this? Assume that the angular acceleration is constant.

**9.59** A car accelerates uniformly from rest and reaches a speed of 22.0 m/s in 9.00 s. The diameter of a tire on this car is 58.0 cm.

a) Find the number of revolutions the tire makes during the car's motion, assuming that no slipping occurs.

b) What is the final angular speed of a tire in revolutions per second?

**9.60** Gear A, with a mass of 1 kg and a radius of 55 cm, is in contact with gear B, with a mass of 0.5 kg and a radius of 30 cm. The gears do not slip with respect to each other as they rotate. Gear A rotates at 120 rpm and slows to 60 rpm in 3 s. How many rotations does gear B undergo during this time interval?

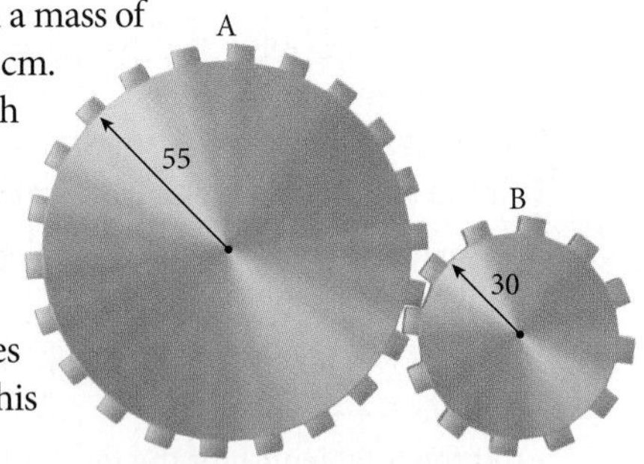


**9.61** A top spins for 10 min, beginning with an angular speed of 10.0 rev/s. Determine its angular acceleration, assuming it is constant, and its total angular displacement.

**9.62** A penny is sitting on the edge of an old phonograph disk that is spinning at 33 rpm and has a diameter of 12 inches. What is the minimum coefficient of static friction between the penny and the surface of the disk to ensure that the penny doesn't fly off?

**9.63** A vinyl record that is initially turning at $33\frac{1}{3}$ rpm slows uniformly to a stop in a time of 15 s. How many rotations are made by the record while stopping?

**9.64** Determine the linear and angular speeds and accelerations of a speck of dirt located 2.0 cm from the center of a CD rotating inside a CD player at 250 rpm.

9.65 What is the acceleration of the Earth in its orbit? (Assume the orbit is circular.)

9.66 A day on Mars is 24.6 Earth hours long. A year on Mars is 687 Earth days long. How do the angular velocities of Mars's rotation and orbit compare to the angular velocities of Earth's rotation and orbit?

9.67 A monster truck has tires with a diameter of 1.10 m and is traveling at 35.8 m/s. After the brakes are applied, the truck slows uniformly and is brought to rest after the tires rotate through 40.2 turns.

a) What is the initial angular speed of the tires?

b) What is the angular acceleration of the tires?

c) What distance does the truck travel before coming to rest?

•**9.68** The motor of a fan turns a small wheel of radius $r_m$ = 2.00 cm. This wheel turns a belt, which is attached to a wheel of radius $r_f$ = 3.00 cm that is mounted to the axle of the fan blades. Measured from the center of this axle, the tip of the fan blades are at a distance $r_b$ = 15.0 cm. When the fan is in operation, the motor spins at an angular speed of $\omega$ = 1200. rpm. What is the tangential speed of the tips of the fan blades?

•**9.69** A car with a mass of 1000. kg goes over a hill at a constant speed of 60 m/s. The top of the hill can be approximated as an arc length of a circle with a radius of curvature of 370 m. What force does the car exert on the hill as it passes over the top?

•**9.70** Unlike a ship, an airplane does not use its rudder to turn. It turns by banking its wings: The lift force, perpendicular to the wings, has a horizontal component, which provides the centripetal acceleration for the turn, and a vertical component, which supports the plane's weight. (The rudder counteracts yaw and thus it keeps the plane pointed in the direction it is moving.) The famous spy plane, the SR-71 Blackbird, flying at 4800 km/h, has a turning radius of 290. km. Find its banking angle.

•9.71 A 80-kg pilot in an aircraft moving at a constant speed of 500 m/s pulls out of a vertical dive along an arc of a circle of radius 4000 m.

a) Find the centripetal acceleration and the centripetal force acting on the pilot.

b) What is the pilot's apparent weight at the bottom of the dive?

•**9.72** A ball having a mass of 1 kg is attached to a string 1 m long and is whirled in a vertical circle at a constant speed of 10 m/s.

a) Determine the tension in the string when the ball is at the top of the circle.

b) Determine the tension in the string when the ball is at the bottom of the circle.

c) Consider the ball at some point other than the top or bottom. What can you say about the tension in the string at this point?

•**9.73** A car starts from rest and accelerates around a flat curve of radius $R$ = 36 m. The tangential component of the car's acceleration remains constant at $a_t$ = 3.3 m/s$^2$, while the centripetal acceleration increases to keep the car on the curve as long as possible. The coefficient of friction between the tires and the road is $\mu$ = 0.95. What distance does the car travel around the curve before it begins to skid? (Be sure to include both the tangential and centripetal components of the acceleration.)

•**9.74** A girl on a merry-go-round platform holds a pendulum in her hand. The pendulum is 6.0 m from the rotation axis of the platform. The rotational speed of the platform is 0.020 rev/s. It is found that the pendulum hangs at an angle $\theta$ to the vertical. Find $\theta$.

••9.75 A carousel at a carnival has a diameter of 6.0 m. The ride starts from rest and accelerates at a constant angular acceleration to an angular speed of 0.6 rev/s in 8.0 s.

a) What is the value of the angular acceleration?

b) What are the centripetal and angular accelerations of a seat on the carousel that is 2.75 m from the rotation axis?

c) What is the total acceleration, magnitude and direction, 8.0 s after the angular acceleration starts?

••**9.76** A car of weight $W$ = 10.0 kN makes a turn on a track that is banked at an angle of $\theta$ = 20.0°. Inside the car, hanging from a short string tied to the rear-view mirror, is an ornament. As the car turns, the ornament swings out at an angle of $\varphi$ = 30.0° measured from the vertical inside the car. What is the force of static friction between the car and the road?

••**9.77** A popular carnival ride consists of seats attached to a central disk through cables. The passengers travel in uniform circular motion. As shown in the figure, the radius of the central disk is $R_0$ = 3.00 m, and the length of the cable is $L$ = 3.20 m. The mass of one of the passengers (including the chair he is sitting on) is 65 kg.

a) If the angle $\theta$ that the cable makes with respect to the vertical is 30°, what is the speed, $v$, of this passenger?

b) What is the magnitude of the force exerted by the cable on the chair?

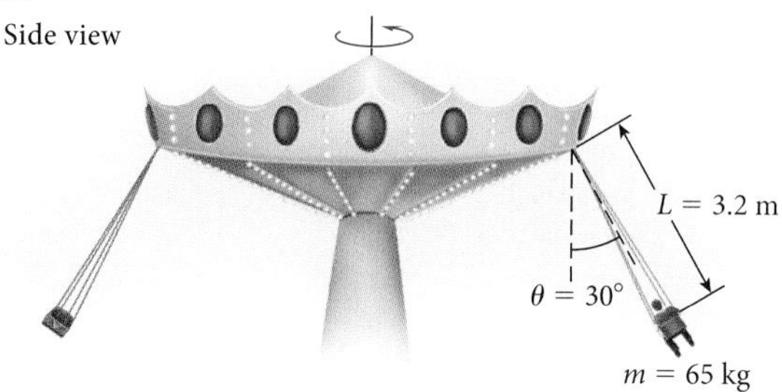


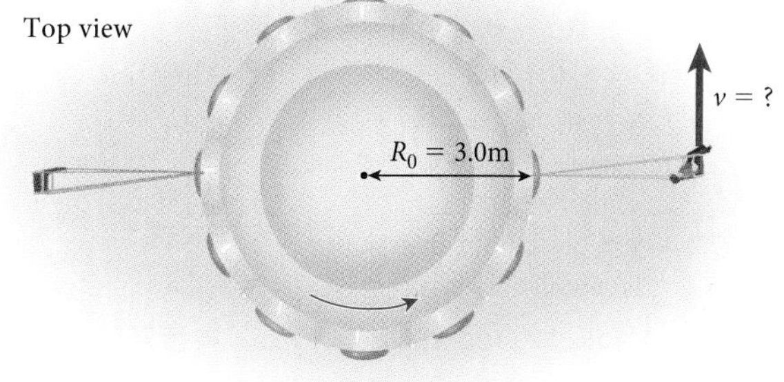

# 10 Rotation

**FIGURE 10.1** A modern jet engine.

## WHAT WE WILL LEARN

- The kinetic energy due to an object's rotational motion must be accounted for when considering energy conservation.
- For rotation of an object about an axis through its center of mass, the moment of inertia is proportional to the product of the object's mass and the square of the largest perpendicular distance from any part of the object to the axis of rotation. The proportionality constant has a value between zero and one and depends on the shape of the object.
- For rotation about an axis parallel to an axis through an object's center of mass, the moment of inertia equals the center-of-mass moment of inertia plus the product of the mass of the object and the square of the distance between the two axes.
- For rolling objects, the kinetic energies of rotation and translation are related.
- Torque is the vector product of the position vector and the force vector.
- Newton's Second Law also applies to rotational motion.
- Angular momentum is defined as the vector product of the position vector and the momentum vector.
- Relationships analogous to those for linear quantities exist among angular momentum, torque, moment of inertia, angular velocity, and angular acceleration.
- Another fundamental conservation law is the law of conservation of angular momentum.

The large fans at the front of modern jet engines, like the one shown in Figure 10.1, pull air into a compression chamber, where the air mixes with fuel and ignites. The explosion propels gases out the back of the engine, producing the thrust that moves the plane forward. These fans rotate at 7000–9000 rpm and must be inspected often—no one wants a broken fan blade at an altitude of 6 mi.

Almost all engines have rotating parts that transfer energy to the output device, which is often rotating as well. In fact, most objects in the universe rotate, from molecules to stars and galaxies. The laws that govern rotation are as fundamentally important as any other part of mechanics.

Chapter 9 introduced some basic concepts of circular motion, and this chapter uses some of those same ideas—angular velocity, angular acceleration, and axis of rotation. In this chapter, we complete our comparison of translational and rotational quantities and encounter another conservation law of basic importance: the law of conservation of angular momentum.

## 10.1 Kinetic Energy of Rotation

In Chapter 8, we saw that we can describe the motion of an extended object in terms of the path that its center of mass follows and the rotation of the object around its center of mass. However, even though we covered circular motion for point particles in Chapter 9, we have not yet considered the rotation of extended objects, such as those shown in Figure 10.2. Analyzing this motion is the purpose of this chapter.

(a)

(b)

**FIGURE 10.2** (a) Rotating air mass forming a hurricane. (b) Rotation on the largest scale—spiral galaxy M74.

### Point Particle in Circular Motion

Chapter 9 introduced the kinematic quantities of circular motion. Angular speed, $\omega$, and angular acceleration, $\alpha$, were defined in terms of the time derivatives of the angular displacement, $\theta$:

$$\omega = \frac{d\theta}{dt}$$

$$\alpha = \frac{d\omega}{dt} = \frac{d^2\theta}{dt^2}.$$

We saw that the angular quantities are related to the linear quantities as follows:

$$s = r\theta, \quad v = r\omega,$$

$$a_t = r\alpha, \quad a_c = \omega^2 r, \quad a = \sqrt{a_c^2 + a_t^2},$$

where $s$ is the arc length, $v$ is the linear speed of the center of mass, $a_t$ is the tangential acceleration, $a_c$ is the centripetal acceleration, and $a$ is the linear acceleration.

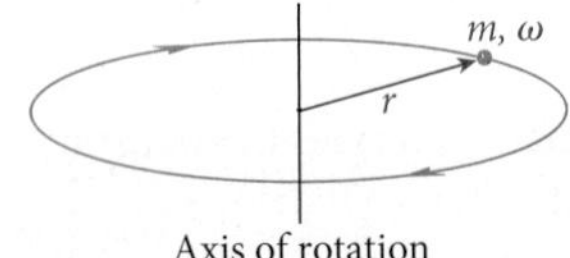


**FIGURE 10.3** A point particle moving in a circle about the axis of rotation.

The most straightforward way to introduce the physical quantities for the description of rotation is through the kinetic energy of rotation of an extended object. In Chapter 5 on work and energy, the kinetic energy of a moving object was defined as

$$K = \tfrac{1}{2}mv^2. \tag{10.1}$$

If the motion of this object is circular, we can use the relationship between linear and angular velocity to obtain

$$K = \tfrac{1}{2}mv^2 = \tfrac{1}{2}m(r\omega)^2 = \tfrac{1}{2}mr^2\omega^2, \tag{10.2}$$

which is the **kinetic energy of rotation** for a point particle's motion on the circumference of a circle of radius $r$ about a fixed axis, as illustrated in Figure 10.3.

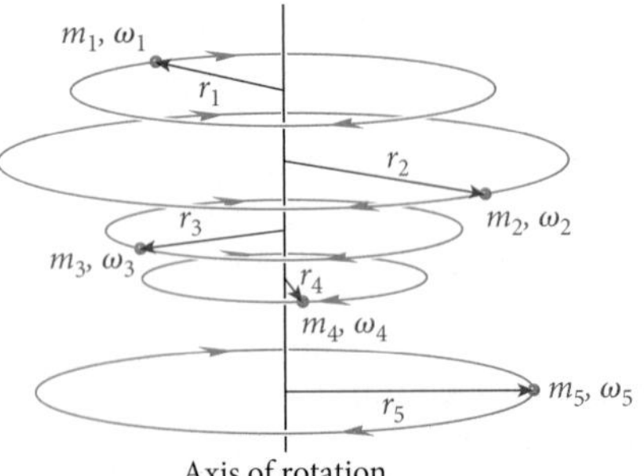


**FIGURE 10.4** Five point particles moving in circles about a common axis of rotation.

## Several Point Particles in Circular Motion

Just as we proceeded in Chapter 8 in finding the location of the center of mass of a system of particles, we start with a collection of individual rotating objects and then approach the continuous limit. The kinetic energy of a collection of rotating objects is given by

$$K = \sum_{i=1}^{n} K_i = \tfrac{1}{2}\sum_{i=1}^{n} m_i v_i^2 = \tfrac{1}{2}\sum_{i=1}^{n} m_i r_i^2 \omega_i^2.$$

This result is simply a consequence of using equation 10.2 for several point particles and writing the total kinetic energy as the sum of the individual kinetic energies. Here $\omega_i$ is the angular velocity of particle $i$ and $r_i$ is its perpendicular distance to a fixed axis. This fixed axis is the **axis of rotation** for these particles. An example of a system of five rotating point particles is shown in Figure 10.4.

Now we assume that all of the point particles whose kinetic energies we have summed keep their distances with respect to one another and with respect to the axis of rotation fixed. Then, all of the point particles in the system will undergo circular motion around the common axis of rotation with the same angular velocity. With this constraint, the sum of the particles' kinetic energies becomes

$$K = \tfrac{1}{2}\sum_{i=1}^{n} m_i r_i^2 \omega^2 = \tfrac{1}{2}\left(\sum_{i=1}^{n} m_i r_i^2\right)\omega^2 = \tfrac{1}{2}I\omega^2. \tag{10.3}$$

The quantity $I$ introduced in equation 10.3 is called the **moment of inertia,** also known as the *rotational inertia*. It depends only on the masses of the individual particles and their distances to the axis of rotation:

$$I = \sum_{i=1}^{n} m_i r_i^2. \tag{10.4}$$

In Chapter 9, we saw that all quantities associated with circular motion have equivalents in linear motion. The angular velocity $\omega$ and the linear velocity $v$ form such a pair. Comparing the expressions for the kinetic energy of rotation (equation 10.3) and the kinetic energy of linear motion (equation 10.1), we see that the moment of inertia $I$ plays the same role for circular motion as the mass $m$ does for linear motion.

### 10.1 In-Class Exercise

Consider two equal masses, $m$, connected by a thin, massless rod. As shown in the figures, the two masses spin in a horizontal plane around a vertical axis represented by the dashed line. Which system has the highest moment of inertia?

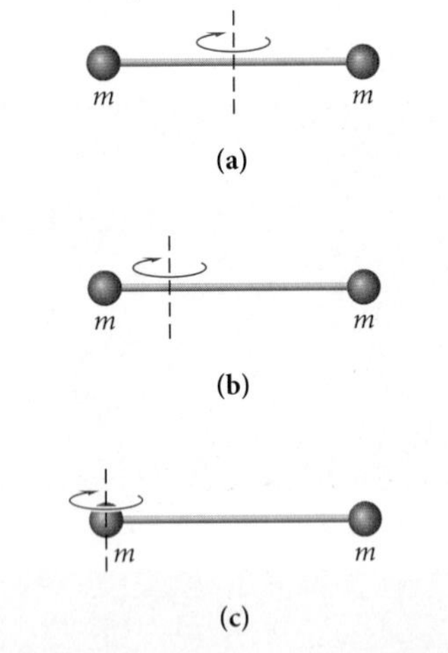


## 10.2 Calculation of Moment of Inertia

We can use the moment of inertia of several point particles, as given in equation 10.4, as a starting point to find the moment of inertia of an extended object. We'll proceed in the same way as we did in finding the center-of-mass location in Chapter 8. We again represent an extended object by a collection of small identical-sized cubes of volume $V$ and (possibly different) mass density $\rho$. Then equation 10.4 becomes

$$I = \sum_{i=1}^{n} \rho(\vec{r}_i) r_i^2 V. \tag{10.5}$$

Again, as in Chapter 8, we follow the conventional calculus approach, letting the volume of the cubes approach zero, $V \rightarrow 0$. In this limit, the sum in equation 10.5 approaches the integral, which gives an expression for the moment of inertia of an extended object:

$$I = \int_V r_\perp^2 \rho(\vec{r})dV. \tag{10.6}$$

The symbol $r_\perp$ represents the perpendicular distance of an infinitesimal volume element from the axis of rotation (Figure 10.5).

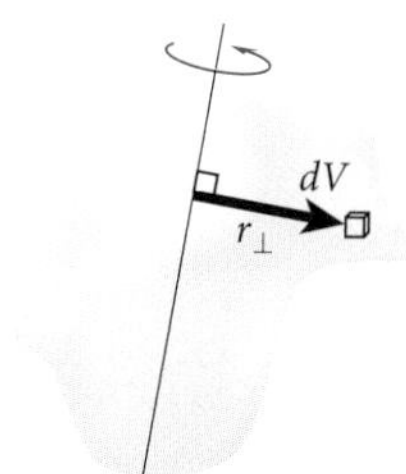


**FIGURE 10.5** Definition of $r_\perp$ as the perpendicular distance of an infinitesimal volume element from the axis of rotation.

We also know that the total mass of an object can be obtained by integrating the mass density over the total volume of the object:

$$M = \int_V \rho(\vec{r})dV. \tag{10.7}$$

Equations 10.6 and 10.7 are the most general expressions for the moment of inertia and the mass of an extended object. However, just as with the equations for the center of mass, some of the physically most interesting cases are those where the mass density is constant throughout the volume. In this case, equations 10.6 and 10.7 reduce to

$$I = \rho \int_V r_\perp^2 dV \quad \text{(for constant mass density, } \rho\text{)},$$

and

$$M = \rho \int_V dV = \rho V \quad \text{(for constant mass density, } \rho\text{)}.$$

Thus, the moment of inertia for an object with constant mass density is given by

$$I = \frac{M}{V} \int_V r_\perp^2 \, dV \quad \text{(for constant mass density, } \rho\text{)}. \tag{10.8}$$

We can now calculate moments of inertia for some objects with particular shapes. First we'll assume that the axis of rotation passes through the center of mass of the object. Then, we'll derive a theorem that connects this special case in a simple way to the general case, for which the axis of rotation does not pass through the center of mass.

## Rotation about an Axis through the Center of Mass

For any object with constant mass density, we can use equation 10.8 to calculate the moment of inertia with respect to rotation about a fixed axis that passes through the object's center of mass. For convenience, the location of the center of mass is usually chosen as the origin of the coordinate system. Because the integral in equation 10.8 is a three-dimensional volume integral, the choice of coordinate system is usually extremely important for evaluating the integral with as little computational work as possible.

In this section, we consider two cases: a hollow disk and a solid sphere. These cases represent the two most common classes of objects that can roll, and they illustrate the use of two different coordinate systems for the integral.

Figure 10.6a shows a hollow cylinder rotating about its symmetry axis. Its moment of inertia is

$$I = \tfrac{1}{2}M\left(R_1^2 + R_2^2\right) \quad \text{(hollow cylinder)}. \tag{10.9}$$

This is the general result for the moment of inertia of a hollow cylinder rotating about its symmetry axis, where $M$ is the cylinder's total mass, $R_1$ is its inner radius, and $R_2$ is its outer radius.

Using equation 10.9, we can obtain the moment of inertia for a solid cylinder rotating about its symmetry axis (see Figure 10.6b) by setting $R_1 = R$ and $R_2 = 0$:

$$I = \tfrac{1}{2}MR^2 \quad \text{(solid cylinder)}.$$

We can also obtain the limiting case of a thin cylindrical shell or hoop, for which all the mass is concentrated on the circumference, by setting $R_1 = R_2 = R$. In this case, the moment of inertia is

$$I = MR^2 \quad \text{(hoop or cylindrical thin shell)}.$$

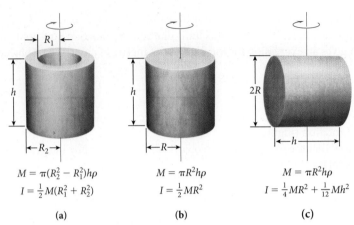

$M = \pi(R_2^2 - R_1^2)h\rho$

$I = \frac{1}{2}M(R_1^2 + R_2^2)$

(a)

$M = \pi R^2 h\rho$

$I = \frac{1}{2}MR^2$

(b)

$M = \pi R^2 h\rho$

$I = \frac{1}{4}MR^2 + \frac{1}{12}Mh^2$

(c)

**FIGURE 10.6** Moment of inertia for (a) a hollow cylinder and (b) a solid cylinder rotating about the axis of symmetry. (c) Moment of inertia for a cylinder rotating about an axis through its center of mass but perpendicular to its symmetry axis.

Finally, the moment of inertia for a solid cylinder of height $h$ rotating about an axis through its center of mass but perpendicular to its symmetry axis (see Figure 10.6c) is given by

$$I = \tfrac{1}{4}MR^2 + \tfrac{1}{12}Mh^2 \quad \text{(solid cylinder, perpendicular to axis of rotation).}$$

If the radius $R$ is very small compared to the height $h$, as is the case for a long thin rod, the moment of inertia in that limit is given by dropping the first term of the preceding equation:

$$I = \tfrac{1}{12}Mh^2 \qquad \text{(thin rod of length } h\text{, perpendicular to axis of rotation).}$$

We'll derive the formula for the moment of inertia of the hollow cylinder using a volume integral of the kind introduced in Section 8.4. This integral involves separate one-dimensional integrations over each of the three coordinates. This derivation and the next one show how these integrations are done for cylindrical and spherical coordinates. They are not essential for the physical concepts developed in this chapter, but they may be interesting to you.

## DERIVATION 10.1 Moment of Inertia for a Wheel

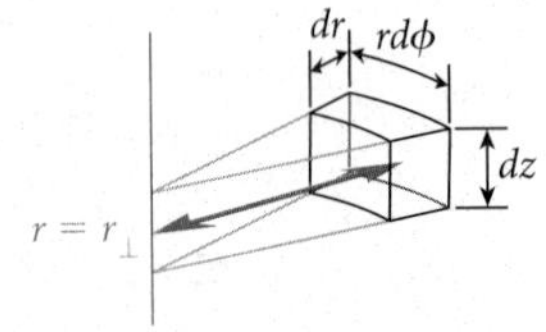


**FIGURE 10.7** Volume element in cylindrical coordinates.

In order to derive the moment of inertia of a hollow cylinder of constant density $\rho$, height $h$, inner radius $R_1$, and outer radius $R_2$, with its symmetry axis as the axis of rotation (see Figure 10.6a), we'll use cylindrical coordinates. For most problems involving cylinders or disks, the coordinate system chosen should generally be cylindrical coordinates. In this system (see the mathematical insert on volume integrals in Chapter 8), the volume element is given by (see Figure 10.7).

$$dV = r_\perp \, dr_\perp \, d\phi dz.$$

For cylindrical coordinates (and only for cylindrical coordinates!), the perpendicular distance, $r_\perp$, is the same as the radial coordinate, $r$. With this in mind, we can evaluate the integrals for the hollow cylinder. For the mass, we obtain

$$M = \rho \int_V dV = \rho \int_{R_1}^{R_2} \left( \int_0^{2\pi} \left( \int_{-h/2}^{h/2} dz \right) d\phi \right) r_\perp \, dr_\perp$$

$$= \rho h \int_{R_1}^{R_2} \left( \int_0^{2\pi} d\phi \right) r_\perp \, dr_\perp$$

$$
\begin{aligned}
&= \rho h 2\pi \int_{R_1}^{R_2} r_\perp dr_\perp \\
&= \rho h 2\pi \left( \tfrac{1}{2} R_2^2 - \tfrac{1}{2} R_1^2 \right) \\
&= \pi \left( R_2^2 - R_1^2 \right) h\rho.
\end{aligned}
$$

Alternatively, we can express the density as a function of the mass:

$$M = \pi \left( R_2^2 - R_1^2 \right) h\rho \Leftrightarrow \rho = \frac{M}{\pi \left( R_2^2 - R_1^2 \right) h}. \qquad \text{(i)}$$

The reason for performing this last step may not be entirely obvious right now, but it should become clear after we evaluate the integral for the moment of inertia:

$$
\begin{aligned}
I = \rho \int_V r_\perp^{\,2} dV &= \rho \int_{R_1}^{R_2} \left( \int_0^{2\pi} \left( \int_{-h/2}^{h/2} dz \right) d\phi \right) r_\perp^3 dr_\perp \\
&= \rho h \int_{R_1}^{R_2} \left( \int_0^{2\pi} d\phi \right) r_\perp^3 dr_\perp \\
&= \rho h 2\pi \int_{R_1}^{R_2} r_\perp^3 dr_\perp \\
&= \rho h 2\pi \left( \tfrac{1}{4} R_2^4 - \tfrac{1}{4} R_1^4 \right).
\end{aligned}
$$

Now we can substitute for the density from equation (i):

$$I = \tfrac{1}{2} \rho h \pi \left( R_2^4 - R_1^4 \right) = \frac{M}{\pi \left( R_2^2 - R_1^2 \right) h} \tfrac{1}{2} h \pi \left( R_2^4 - R_1^4 \right).$$

Finally, we make use of the identity $a^4 - b^4 = (a^2 - b^2)(a^2 + b^2)$ to obtain equation 10.9:

$$I = \tfrac{1}{2} M \left( R_1^2 + R_2^2 \right).$$

For objects other than disk-like ones, the use of cylindrical coordinates may not be advantageous. The most important such objects are spheres and rectangular blocks. The moment of inertia for a sphere rotating about any axis through its center of mass (see Figure 10.8a) is given by

$$I = \tfrac{2}{5} MR^2 \quad \text{(solid sphere).} \qquad \textbf{(10.10)}$$

The moment of inertia for a thin spherical shell rotating about any axis through its center of mass is

$$I = \tfrac{2}{3} MR^2 \quad \text{(thin spherical shell).}$$

The moment of inertia for a rectangular block with side lengths $a$, $b$, and $c$ rotating about an axis through the center of mass and parallel to side $c$ (see Figure 10.8b) is:

$$I = \tfrac{1}{12} M(a^2 + b^2) \quad \text{(rectangular block).}$$

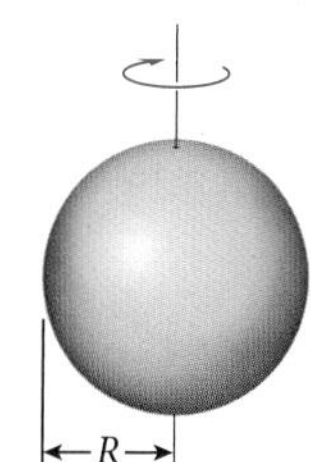


$M = \frac{4\pi}{3} R^3 \rho$

$I = \frac{2}{5} MR^2$

(a)

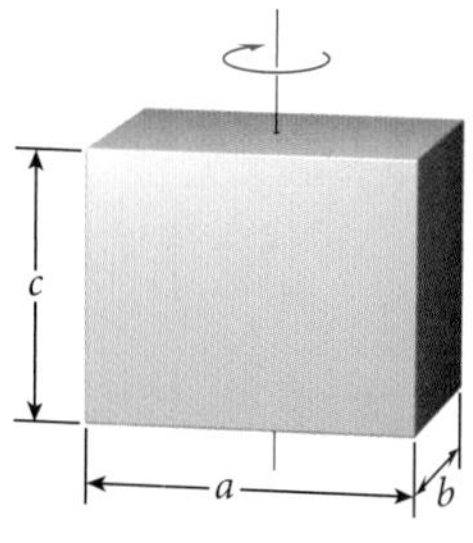


$M = abc\rho$

$I = \frac{1}{12} M(a^2 + b^2)$

(b)

**FIGURE 10.8** Moments of inertia of (a) a sphere and (b) a block.

Again, we'll derive the formula for the solid sphere only in order to illustrate working in a different coordinate system.

## DERIVATION 10.2 Moment of Inertia for a Sphere

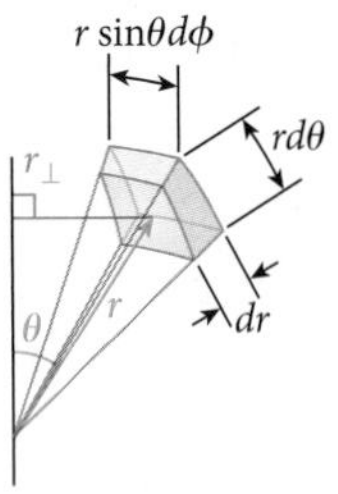

To calculate the moment of inertia for a solid sphere (Figure 10.8a) of constant mass density $\rho$ and radius $R$ rotating about an axis through its center, it is not appropriate to use cylindrical coordinates. Spherical coordinates are the best choice. In spherical coordinates, the volume element is given by (see Figure 10.9)

$$dV = r^2 \sin\theta \, dr \, d\theta \, d\phi.$$

Note that for spherical coordinates, the radial coordinate, $r$, and the perpendicular distance, $r_\perp$, are not identical. Instead, they are related via (see Figure 10.9)

$$r_\perp = r\sin\theta.$$

(It is a very common source of error in these kinds of calculations to omit the sine of the angle. You need to keep this in mind when you are using spherical coordinates.)

Again, we first evaluate the integral for the mass:

$$\begin{aligned} M = \rho\int_V dV &= \rho\int_0^R\left(\int_0^{2\pi}\left(\int_0^{\pi}\sin\theta\, d\theta\right)d\phi\right)r^2 dr \\ &= 2\rho\int_0^R\left(\int_0^{2\pi} d\phi\right)r^2 dr \\ &= 4\pi\rho\int_0^R r^2 dr \\ &= \frac{4\pi}{3}R^3\rho. \end{aligned}$$

Thus,

$$\rho = \frac{3M}{4\pi R^3}. \qquad \text{(i)}$$

Then we evaluate the integral for the moment of inertia in a similar way:

$$\begin{aligned} I = \rho\int_V r_\perp^2 dV &= \rho\int_V r^2\sin^2\theta\, dV \\ &= \rho\int_0^R\left(\int_0^{2\pi}\left(\int_0^{\pi}\sin^3\theta\, d\theta\right)d\phi\right)r^4 dr \\ &= \rho\frac{4}{3}\int_0^R\left(\int_0^{2\pi} d\phi\right)r^4 dr \\ &= \rho\frac{8\pi}{3}\int_0^R r^4 dr. \end{aligned}$$

Thus,

$$I = \rho\frac{8\pi}{15}R^5. \qquad \text{(ii)}$$



**FIGURE 10.9** Volume element in spherical coordinates.

Inserting the expression for the density from equation (i) into equation (ii) we obtain

$$I = \rho \frac{8\pi}{15} R^5 = \frac{3M}{4\pi R^3} \frac{8\pi}{15} R^5 = \tfrac{2}{5} MR^2 .$$

Finally, note this important general observation: If $R$ is the largest perpendicular distance of any part of the rotating object from the axis of rotation, then the moment of inertia is always related to the mass of an object by

$$I = cMR^2, \text{ with } 0 < c \leq 1. \tag{10.11}$$

The constant $c$ can be calculated from the geometrical configuration of the rotating object and always has a value between zero and one. The more the bulk of the mass is pushed toward the axis of rotation, the smaller the value of the constant $c$ will be. If all the mass is located on the outer edge of the object, as in a hoop, for example, then $c$ approaches the value 1. (Mathematically, this equation is a consequence of the mean value theorem, which you may have encountered in a calculus class.) For a cylinder rotating around its symmetry axis, $c_{\text{cyl}} = \frac{1}{2}$, and for a sphere, $c_{\text{sph}} = \frac{2}{5}$, as we have seen.

Various objects rotating about an axis that passes through the center of mass are shown in Figure 10.10. Table 10.1 gives the moment of inertia for each object as well as the constant $c$ from equation 10.11, where applicable.

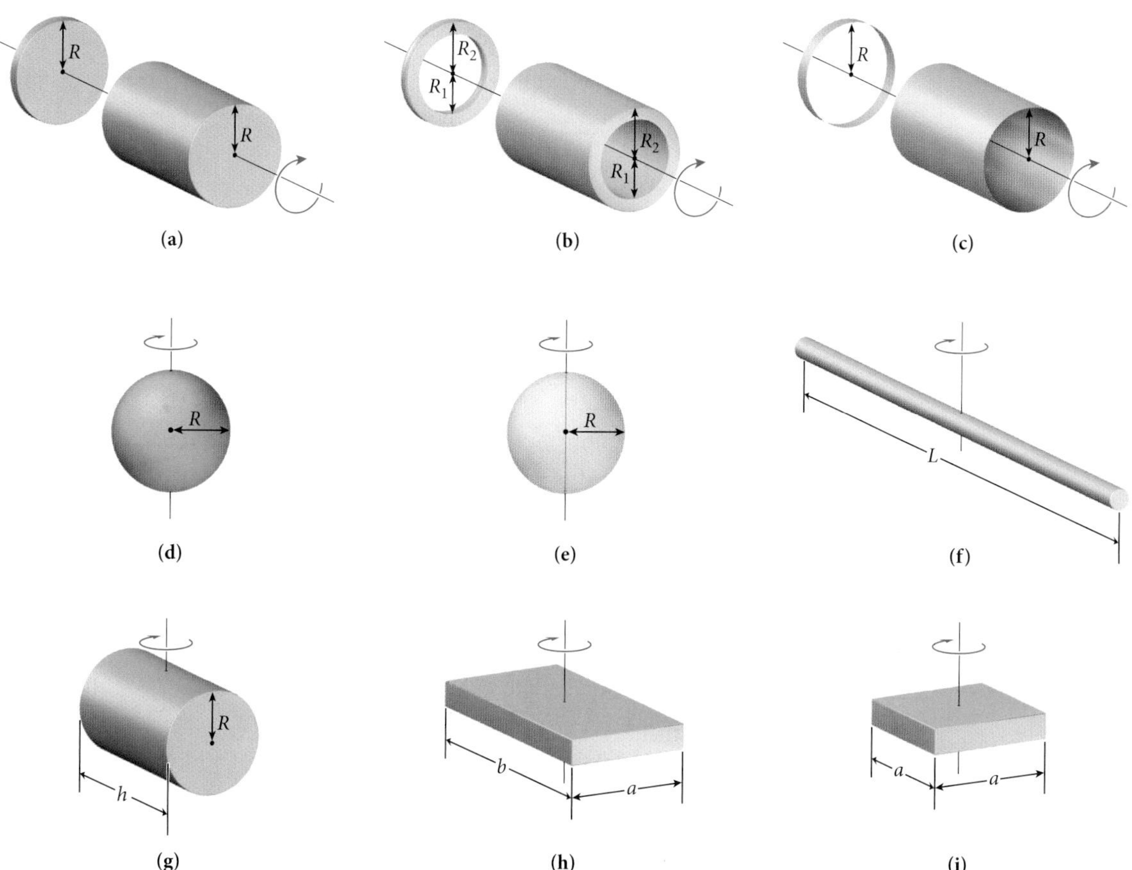


**FIGURE 10.10** Orientation of the axis of rotation passing through the center of mass and definition of dimensions for objects listed in Table 10.1.

**Table 10.1 The Moment of Inertia and Value of Constant $c$ for the Objects Shown in Figure 10.10. All Objects Have Mass $M$**

| Object | $I$ | $c$ |
|---|---|---|
| a) Solid cylinder or disk | $\frac{1}{2}MR^2$ | $\frac{1}{2}$ |
| b) Thick hollow cylinder or wheel | $\frac{1}{2}M(R_1^2+R_2^2)$ | |
| c) Hollow cylinder or hoop | $MR^2$ | 1 |
| d) Solid sphere | $\frac{2}{5}MR^2$ | $\frac{2}{5}$ |
| e) Hollow sphere | $\frac{2}{3}MR^2$ | $\frac{2}{3}$ |
| f) Thin rod | $\frac{1}{12}Mh^2$ | |
| g) Solid cylinder perpendicular to symmetry axis | $\frac{1}{4}MR^2+\frac{1}{12}Mh^2$ | |
| h) Flat rectangular plate | $\frac{1}{12}M(a^2+b^2)$ | |
| i) Flat square plate | $\frac{1}{6}Ma^2$ | |

## EXAMPLE 10.1 Rotational Kinetic Energy of Earth

Assume that the Earth is a solid sphere of constant density, with mass $5.98\cdot 10^{24}$ kg and radius 6370 km.

**PROBLEM**

What is the moment of inertia of the Earth with respect to rotation about its axis, and what is the kinetic energy of this rotation?

**SOLUTION**

Since the Earth is to be approximated by a sphere of constant density, its moment of inertia is

$$I=\tfrac{2}{5}MR^2.$$

Inserting the values for the mass and radius, we obtain

$$I=\tfrac{2}{5}MR^2=\tfrac{2}{5}(5.98\cdot 10^{24}\text{ kg})(6.37\cdot 10^6\text{ m})^2=9.71\cdot 10^{37}\text{ kg m}^2.$$

The angular frequency of Earth's rotation is

$$\omega=\frac{2\pi}{1\text{ day}}=\frac{2\pi}{86{,}164\text{ s}}=7.29\cdot 10^{-5}\text{ rad/s}.$$

(Note that we used the sidereal day here; see Example 9.3.)

With the results for the moment of inertia and the angular frequency, we can find the kinetic energy of the Earth's rotation:

$$K=\tfrac{1}{2}I\omega^2=0.5(9.71\cdot 10^{37}\text{ kg m}^2)(7.29\cdot 10^{-5}\text{ rad/s})^2=2.6\cdot 10^{29}\text{ J}.$$

Let's compare this to the kinetic energy of Earth's motion around the Sun. In Chapter 9, we calculated the orbital speed of Earth to be $v=2.97\cdot 10^4$ m/s. Thus, the kinetic energy of Earth's motion around the Sun is

$$K=\tfrac{1}{2}mv^2=0.5(5.98\cdot 10^{24}\text{ kg})(2.97\cdot 10^4\text{ m/s})^2=2.6\cdot 10^{33}\text{ J},$$

which is larger than the kinetic energy of rotation by a factor of 10,000.

## Parallel-Axis Theorem

We have determined the moment of inertia for a rotation axis through the center of mass of an object, but what is the moment of inertia for rotation about an axis that does not pass through the center of mass? The **parallel-axis theorem** answers this question. It states that the moment of inertia, $I_{\parallel}$, for rotation of an object of mass $M$ about an axis located a distance

$d$ away from the object's center of mass and parallel to an axis through the center of mass, for which the moment of inertia is $I_{cm}$, is given by

$$I_{\parallel} = I_{cm} + Md^2. \quad (10.12)$$

## DERIVATION 10.3 Parallel-Axis Theorem

For this derivation, consider the object in Figure 10.11. Suppose we have already calculated the moment of inertia of this object for rotation about the $z$-axis, which passes through the center of mass of this object. The origin of the $xyz$-coordinate system is at the center of mass, and the $z$-axis is the rotation axis. Any axis parallel to the rotation axis can then be described by a simple shift in the $xy$-plane indicated in the figure by the vector $\vec{d}$, with components $d_x$ and $d_y$.

If we shift the coordinate system in the $xy$-plane so that the new vertical axis, the $z'$-axis, coincides with the new rotation axis, then the transformation from the $xyz$-coordinates to the new $x'y'z'$-coordinates is given by

$$x' = x - d_x, \quad y' = y - d_y, \quad z' = z.$$

To calculate the moment of inertia of the object rotating about the new axis in the new coordinate system, we can simply use equation 10.6, the most general equation, which applies to the case where the mass density is not constant:

$$I_{\parallel} = \int_V (r_\perp')^2 \rho \, dV. \quad \text{(i)}$$

According to the coordinate transformation,

$$\begin{aligned}(r_\perp')^2 &= (x')^2 + (y')^2 = (x - d_x)^2 + (y - d_y)^2 \\ &= x^2 - 2xd_x + d_x^2 + y^2 - 2yd_y + d_y^2 \\ &= (x^2 + y^2) + (d_x^2 + d_y^2) - 2xd_x - 2yd_y \\ &= r_\perp^2 + d^2 - 2xd_x - 2yd_y.\end{aligned}$$

(Keep in mind that $\vec{r}_\perp$ lies in the $xy$-plane because of the way we constructed the coordinate system.) Now we substitute this expression for $(r_\perp')^2$ into equation (i) and obtain

$$\begin{aligned}I_{\parallel} &= \int_V (r_\perp')^2 \rho \, dV \\ &= \int_V r_\perp^2 \rho \, dV + d^2 \int_V \rho \, dV - 2d_x \int_V x\rho \, dV - 2d_y \int_V y\rho \, dV.\end{aligned} \quad \text{(ii)}$$

The first integral in equation (ii) gives the moment of inertia for rotation about the center of mass, which we already know. The second integral is simply equal to the mass (compare equation 10.7). The third and fourth integrals were introduced in Chapter 8 and give the locations of the $x$- and $y$-coordinates of the center of mass. However, by construction, they equal zero, because we put the origin of the $xyz$-coordinate system at the center of mass. Thus, we obtain the parallel-axis theorem:

$$I_{\parallel} = I_{cm} + d^2 M.$$

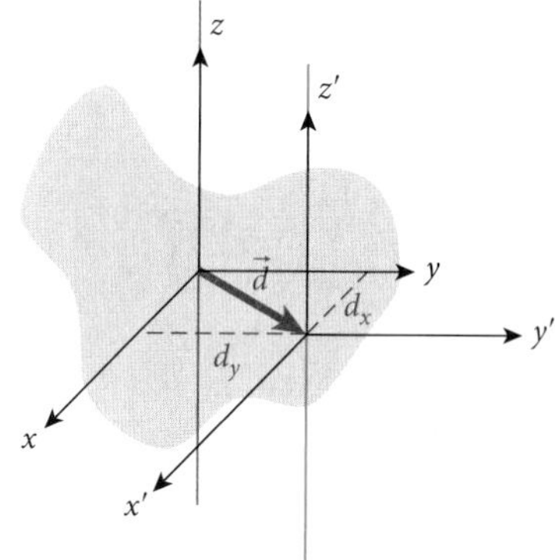


**FIGURE 10.11** Coordinates and distances for the parallel-axis theorem.

Note that, according to equations 10.11 and 10.12, the moment of inertia with respect to rotation about an arbitrary axis parallel to an axis through the center of mass can be written as

$$I = (cR^2 + d^2)M, \text{ with } 0 < c \leq 1.$$

Here $R$ is the maximum perpendicular distance of any part of the object from its axis of rotation through the center of mass, and $d$ is the distance of the rotation axis from a parallel axis through the center of mass.

### 10.1 Self-Test Opportunity

Show that the moment of inertia of a thin rod of mass $m$ and length $L$ rotating around one end is $I = \frac{1}{3}mL^2$.

## 10.3 Rolling without Slipping

**Rolling motion** is a special case of rotational motion that is performed by round objects of radius $R$ that move across a surface without slipping. For rolling motion, we can connect the linear and angular quantities by realizing that the linear distance moved by the center of mass is the same as the length of corresponding arc of the object's circumference. Thus, the relationship between the linear distance, $r$, traveled by the center of mass and the rotation angle is

$$r = R\theta.$$

Taking the time derivative and keeping in mind that the radius, $R$, remains constant, we obtain the relationships between linear and angular speeds and accelerations:

$$v = R\omega$$

and

$$a = R\alpha.$$

The total kinetic energy of an object in rolling motion is the sum of its translational and rotational kinetic energies:

$$K = K_{\text{trans}} + K_{\text{rot}} = \tfrac{1}{2}mv^2 + \tfrac{1}{2}I\omega^2. \tag{10.13}$$

We can now substitute for $\omega$ from $v = R\omega$ and for $I$ from equation 10.11:

$$\begin{aligned} K &= \tfrac{1}{2}mv^2 + \tfrac{1}{2}I\omega^2 \\ &= \tfrac{1}{2}mv^2 + \tfrac{1}{2}(cR^2m)\left(\frac{v}{R}\right)^2 \\ &= \tfrac{1}{2}mv^2 + \tfrac{1}{2}mv^2c \Rightarrow \end{aligned}$$

$$K = (1 + c)\tfrac{1}{2}mv^2, \tag{10.14}$$

where $0 < c \le 1$ is the constant introduced in equation 10.11. Equation 10.14 implies that the kinetic energy of a rolling object is always greater than that of an object that is sliding, provided they have the same mass and linear velocity.

With an expression for the kinetic energy that includes the contribution due to rotation, we can apply the concept of conservation of total mechanical energy (the sum of kinetic and potential energy) that we applied in Chapter 6.

### 10.2 In-Class Exercise

A bicycle is moving with a speed of 4.02 m/s. If the radius of the front wheel is 0.450 m, how long does it take for that wheel to make a complete revolution?

a) 0.703 s

b) 1.23 s

c) 2.34 s

d) 4.04 s

e) 6.78 s

### SOLVED PROBLEM 10.1 Sphere Rolling Down an Inclined Plane

**PROBLEM**

A solid sphere with a mass of 5.15 kg and a radius of 0.340 m starts from rest at a height of 2.10 m above the base of an inclined plane and rolls down without sliding under the influence of gravity. What is the linear speed of the center of mass of the sphere just as it leaves the incline and rolls onto a horizontal surface?

**SOLUTION**

**THINK**

At the top of the incline, the sphere is at rest. At that point, the sphere has gravitational potential energy and no kinetic energy. As the sphere starts to roll, it loses potential energy and gains kinetic energy of linear motion and kinetic energy of rotation. At the bottom of the inclined plane, all of the original potential energy is in the form of kinetic energy. The kinetic energy of linear motion is linked to the kinetic energy of rotation through the radius of the sphere.

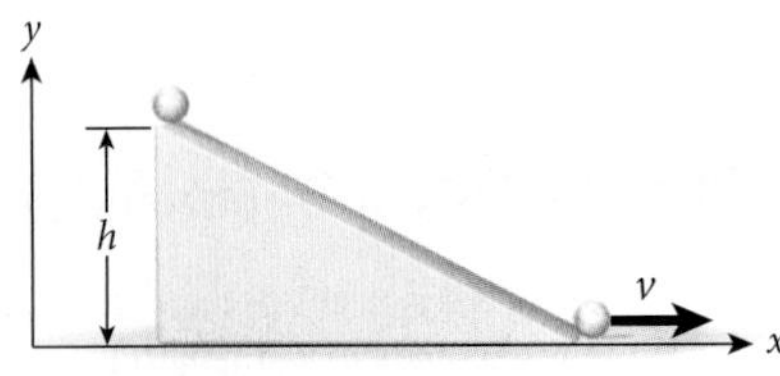


**FIGURE 10.12** Sphere rolling down an inclined plane.

**SKETCH**

A sketch of the problem situation is shown in Figure 10.12 with the zero of the $y$-coordinate at the bottom of the incline.

**RESEARCH**

At the top of the incline, the sphere is at rest and has zero kinetic energy. At the top, its energy is therefore its potential energy, $mgh$:

$$E_{\text{top}} = K_{\text{top}} + U_{\text{top}} = 0 + mgh = mgh,$$

where $m$ is the mass of the sphere, $h$ is the height of the sphere above the horizontal surface, and $g$ is the acceleration due to gravity. At the bottom of the incline, just as the sphere starts rolling onto the horizontal surface, the potential energy is zero. According to equation 10.14, the sphere has a total kinetic energy (sum of translational and rotational kinetic energy) of $(1 + c)\frac{1}{2}mv^2$. Thus, the total energy at the bottom of the incline is

$$E_{\text{bottom}} = K_{\text{bottom}} + U_{\text{bottom}} = (1+c)\tfrac{1}{2}mv^2 + 0 = (1+c)\tfrac{1}{2}mv^2.$$

Because the moment of inertia of a sphere is $I = \frac{2}{5}mR^2$ (see equation 10.10), the constant $c$ has the value $\frac{2}{5}$ in this case.

**SIMPLIFY**

Because of the conservation of energy, the energy at the top of the incline is equal to that at the bottom:

$$mgh = (1+c)\tfrac{1}{2}mv^2.$$

Solving for the linear velocity gives us

$$v = \sqrt{\frac{2gh}{1+c}}.$$

For a sphere, $c = \frac{2}{5}$, as noted above, and so the speed of the rolling object is in this case

$$v = \sqrt{\frac{2gh}{1+\frac{2}{5}}} = \sqrt{\frac{10}{7}gh}.$$

**CALCULATE**

Putting in the numerical values gives us

$$v = \sqrt{\frac{10}{7}(9.81 \text{ m/s}^2)(2.10 \text{ m})} = 5.42494 \text{ m/s}.$$

**ROUND**

Expressing our result with three significant figures leads to

$$v = 5.42 \text{ m/s}.$$

**DOUBLE-CHECK**

If the sphere did not roll, but instead slid down the inclined plane without friction, the final speed would be

$$v = \sqrt{2gh} = \sqrt{2(9.81 \text{ m/s}^2)(2.10 \text{ m})} = 6.42 \text{ m/s},$$

which is faster than the speed we found for the rolling sphere. It seems reasonable that the final linear velocity of the rolling sphere would be somewhat less than the final linear velocity of the sliding sphere, because some of the initial potential energy of the rolling sphere is transformed into rotational kinetic energy. This energy is then not available to become kinetic energy of linear motion of the center of mass of the sphere. Note that we did not need the mass or radius of the sphere in this calculation.

The formula derived in Solved Problem 10.1 for the speed of a rolling sphere at the bottom of the incline,

$$v = \sqrt{\frac{2gh}{1+c}}, \tag{10.15}$$

is a rather general result. It can be applied to various situations in which potential gravitational energy is converted into translational and rotational kinetic energy of a rolling object.

Galileo Galilei famously showed that the acceleration of an object in free fall is independent of its mass. This is also true for an object rolling down an incline, as we have just seen in Solved Problem 10.1, leading to equation 10.15. However, while the total mass of the rolling object does not matter, the distribution of mass within it does matter. Mathematically, this is reflected in equation 10.15 by the fact that the constant $c$ from equation 10.11, which is calculated from the geometric distribution of the mass, appears in the denominator. The following example demonstrates clearly that the distribution of mass in rolling objects matters.

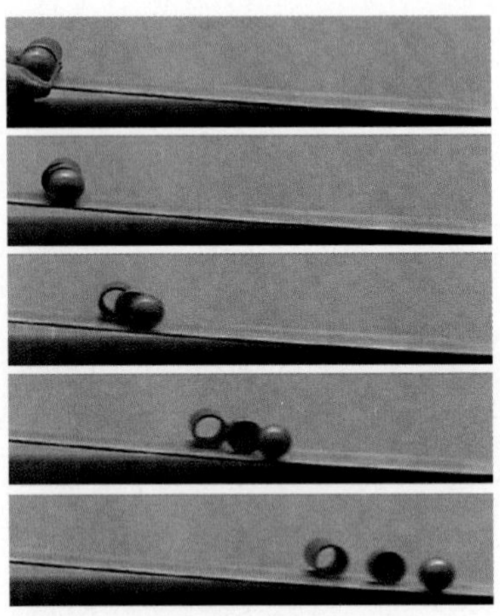

**FIGURE 10.13** Race of a sphere, a solid cylinder, and a hollow cylinder of the same mass and radius down an inclined plane. The frames were shot at intervals of 0.5 s.

### EXAMPLE 10.2 Race Down an Incline

**PROBLEM**

A solid sphere, a solid cylinder, and a hollow cylinder (a tube), all of the same mass $m$ and the same outer radius $R$, are released from rest at the top of an incline and start rolling without sliding. In which order do they arrive at the bottom of the incline?

**SOLUTION**

We can answer this question by using energy considerations only. Since the total mechanical energy is conserved for each of the three objects during its rolling motion, we can write for each object

$$E = K + U = K_0 + U_0.$$

The objects were released from rest, so $K_0 = 0$. For the potential energy, we again use $U = mgh$, and for the kinetic energy, we use equation 10.14. Therefore, we have

$$K_{\text{bottom}} = U_{\text{top}} \Rightarrow (1+c)\tfrac{1}{2}mv^2 = mgh \Rightarrow$$

$$v = \sqrt{\frac{2gh}{1+c}},$$

which is the same formula as equation 10.15.

We already noted that the mass of the object canceled out of this formula. However, we can make two additional important observations: (1) The radius of the rotating object does not appear in this expression for the linear velocity; and (2) the constant $c$ that is determined by the distribution of the mass appears in the denominator. We already know the value of $c$ for the three rolling objects: $c_{\text{sphere}} = \frac{2}{5}$, $c_{\text{cylinder}} = \frac{1}{2}$, and $c_{\text{tube}} \approx 1$. Since the constant for the sphere is the smallest, the sphere's velocity for any given height $h$ will be the largest, implying that the sphere will win the race. Physically, we see that since all three objects have the same mass and thus have the same change in potential energy, all final kinetic energies are equal. Thus, an object with a higher value of $c$ will have relatively more of its kinetic energy in rotation and therefore a lower translational kinetic energy and a lower linear speed. The solid cylinder will come in second in the race, followed by the tube. Figure 10.13 shows frames from a videotaped experiment that verify our conclusions.

### 10.3 In-Class Exercise

Suppose we repeat the race of Example 10.2, but let an unopened can of soda participate. In which place will the can finish?

a) first

b) second

c) third

d) fourth

### 10.2 Self-Test Opportunity

Can you explain why the can of soda finishes the race in the position determined in In-Class Exercise 10.3?

### SOLVED PROBLEM 10.2 Ball Rolling through a Loop

A solid sphere is released from rest and rolls down an incline and then into a circular loop of radius $R$ (see Figure 10.14).

**PROBLEM**

What is the minimum height $h$ from which the sphere has to be released so that it does not fall off the track when in the loop?

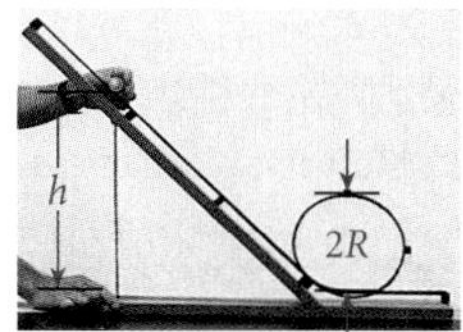

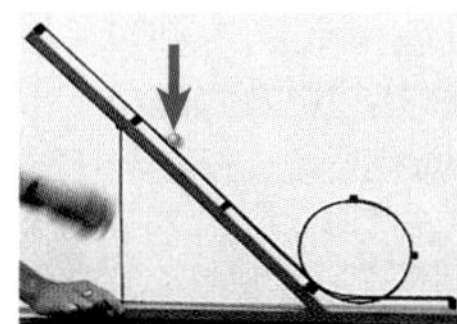
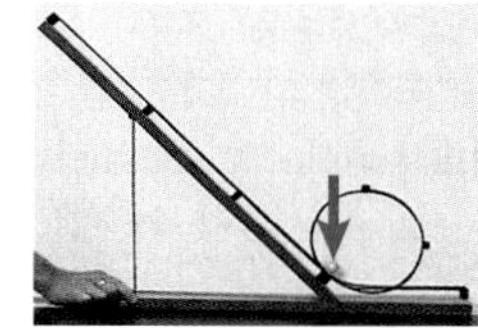
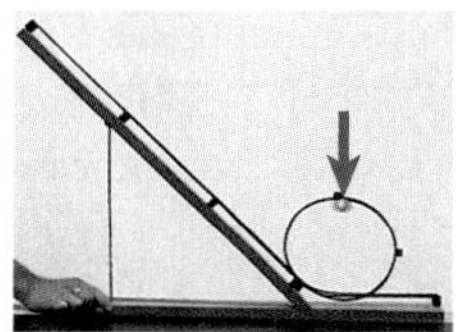
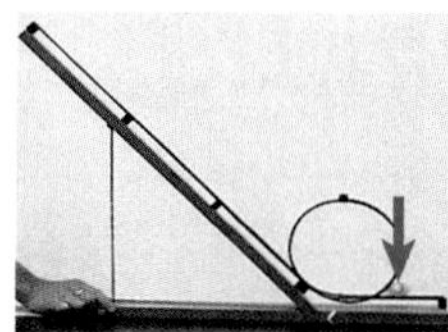

**FIGURE 10.14** Sphere rolling down an incline and into a loop. The frames were shot at intervals of 0.25 s.

## SOLUTION

### THINK

When released from rest at height $h$ on the incline, the sphere has gravitational potential energy but no kinetic energy. As the sphere rolls down the incline and around the loop, gravitational potential energy is converted to kinetic energy. At the top of the loop, the sphere has dropped a distance $h - 2R$. The key to solving this problem is to realize that at the top of the loop, the centripetal acceleration has to be equal to or larger than the acceleration due to gravity. (When these accelerations are equal, "weightlessness" of the object occurs, as shown in Solved Problem 9.1. When the centripetal acceleration is greater, there must be a downward supporting force, which the track can supply. When the centripetal acceleration is less, there must be an upward supporting force, which the track cannot supply, so the sphere falls off the track.) What is unknown here is the speed $v$ at the top of the loop. We can again employ energy conservation considerations to calculate the minimum velocity required and then find the minimum starting height required for the sphere to remain on the track.

### SKETCH

A sketch of the sphere rolling on the incline and loop is shown in Figure 10.15.

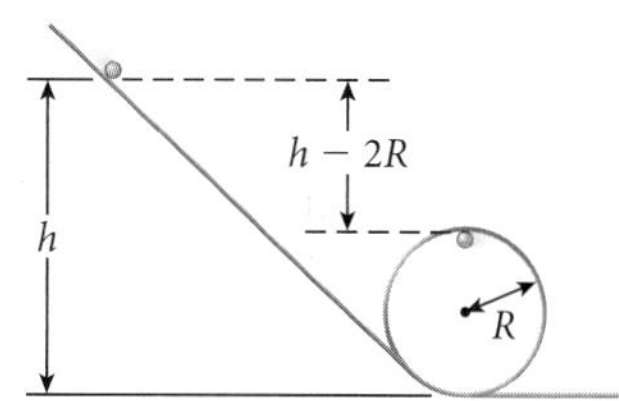


**FIGURE 10.15** Sphere rolling on an incline and a loop.

### RESEARCH

When the sphere is released, it has potential energy $U_0 = mgh$ and zero kinetic energy. At the top of the loop, the potential energy is $U = mg2R$, and the total kinetic energy is, according to equation 10.14, $K = (1 + c)\frac{1}{2}mv^2$, where $c_{\text{sphere}} = \frac{2}{5}$. Thus, conservation of total mechanical energy tells us that

$$E = K + U = (1+c)\tfrac{1}{2}mv^2 + mg2R = K_0 + U_0 = mgh. \qquad \text{(i)}$$

At the top of the loop, the centripetal acceleration, $a_c$, has to be equal to or larger than the acceleration due to gravity, $g$:

$$g \le a_c = \frac{v^2}{R}. \qquad \text{(ii)}$$

### SIMPLIFY

We can solve equation (i) for $v^2$:

$$v^2 = \frac{2g(h-2R)}{1+c}.$$

If we then substitute this expression for $v^2$ into equation (ii), we find

$$g \le \frac{2g(h-2R)}{R(1+c)}.$$

Multiplying both sides of this equation by the denominator of the fraction on the right side leads to

$$R(1+c) \le 2h - 4R.$$

Therefore, we have

$$h \ge \frac{5+c}{2}R.$$

*Continued—*

**CALCULATE**

This result is valid for any rolling object, with the constant $c$ determined from the object's geometry. In this problem, we have a solid sphere, so $c = \frac{2}{5}$. Thus, the result is

$$h \geq \frac{5+\frac{2}{5}}{2} R = \frac{27}{10} R.$$

**ROUND**

This problem situation was described in terms of variables rather than numerical values, so we can quote our result exactly as

$$h \geq 2.7R.$$

**DOUBLE-CHECK**

If the sphere were not rolling but sliding without friction, we could equate the kinetic energy of the sphere at the top of the loop with the change in gravitational potential energy:

$$\frac{1}{2}mv^2 = mg\left(h - 2R\right).$$

We could then express the inequality between gravitational and centripetal acceleration given in equation (ii) as follows:

$$g \leq a_c = \frac{v^2}{R} = \frac{2g(h-2R)}{R}.$$

Solving this equation for $h$ gives us

$$h \geq \frac{5}{2} R = 2.5R.$$

This required height for a sliding sphere is slightly lower than the height we found for a rolling sphere. We would expect that less energy, in the form of gravitational potential energy, would be required to keep the sphere on the track if it were not rolling because its kinetic energy would be entirely translational. Thus, our result for keeping a rolling sphere on the loop seems reasonable.

## 10.4 Torque

So far, in discussing forces, we have seen that a force can cause linear motion of an object, which can be described in terms of the motion of the center of mass of the object. However, we have not addressed one general question: Where are the force vectors acting on an extended object placed in a free-body diagram? A force can be exerted on an extended object at a point away from its center of mass, which can cause the object to rotate as well as move linearly.

### Moment Arm

Consider the hand attempting to use a wrench to loosen a bolt in Figure 10.16. It is clear that it will be easiest to turn the bolt in Figure 10.16c, quite a bit harder in Figure 10.16b, and

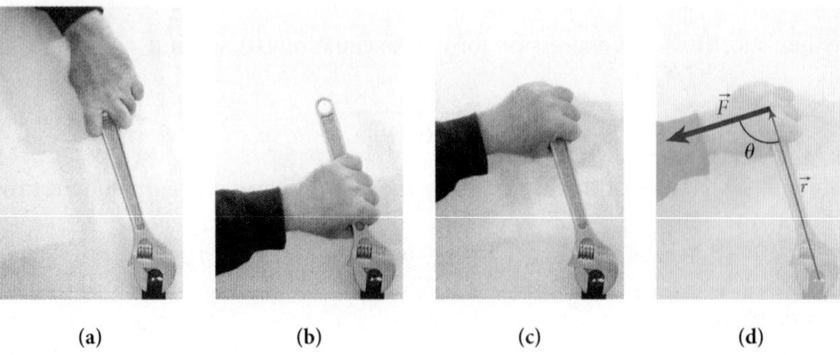


**FIGURE 10.16** (a–c) Three ways to use a wrench to loosen a bolt. (d) The force $\vec{F}$ and moment arm $r$, with the angle $\theta$ between them.

downright impossible in Figure 10.16a. This example shows that the magnitude of the force is not the only relevant quantity. The perpendicular distance from the line of action of the force to the axis of rotation, called the **moment arm,** is also important. In addition, the angle at which the force is applied, relative to the moment arm, matters as well. In parts (b) and (c) of Figure 10.16, this angle is 90°. (An angle of 270° would be just as effective, but then the force would act in the opposite direction.) An angle of 180° or 0° (Figure 10.16a) will not turn the bolt.

These considerations are quantified by the concept of torque, $\tau$. **Torque** (also called *moment*) is the vector product of the force $\vec{F}$ and the position vector $\vec{r}$:

$$\vec{\tau} = \vec{r} \times \vec{F}. \tag{10.16}$$

The position vector $\vec{r}$ is measured with the origin at the axis of rotation. The symbol $\times$ denotes the **vector product,** or *cross product.* In Section 5.4, we saw that we can multiply two vectors to yield a scalar quantity called the *scalar product,* denoted with the symbol •. Now we define a different multiplication of two vectors in such a way that the result is another vector.

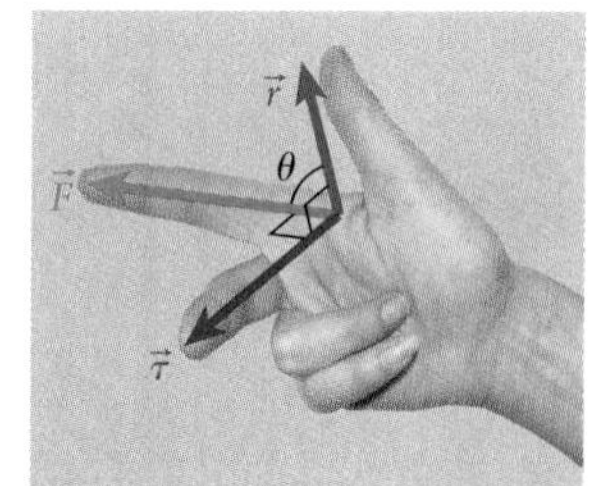

**FIGURE 10.17** Right-hand rule for the direction of the torque for a given force and position vector.

The SI unit of torque is N m, not to be confused with the unit of energy, which is the joule (J = N m)

$$[\tau] = [F] \cdot [r] = \text{N m}.$$

In English units, torque is often expressed in foot-pounds (ft-lb).

The magnitude of the torque is the product of the magnitude of the force and the distance to the axis of rotation (the magnitude of the position vector, or the moment arm) times the sine of the angle between the force vector and the position vector (see Figure 10.17):

$$\tau = rF \sin\theta. \tag{10.17}$$

Angular quantities can also be vectors, called *axial vectors.* (An axial vector is any vector that points along the rotation axis.) Torque is an example of an axial vector, and its magnitude is given by equation 10.17. The direction of the torque is given by a right-hand rule (Figure 10.17). The torque points in a direction perpendicular to the plane spanned by the force and position vectors. Thus, if the position vector points along the thumb and the force vector points along the index finger, then the direction of the axial torque vector is the direction of the middle finger, as shown in Figure 10.17. Note that the torque vector is perpendicular to both the force vector and the position vector.

## Mathematical Insert: Vector Product

The vector product (or cross product) between two vectors $\vec{A} = (A_x, A_y, A_z)$ and $\vec{B} = (B_x, B_y, B_z)$ is defined as

$$\vec{C} = \vec{A} \times \vec{B}$$
$$C_x = A_y B_z - A_z B_y$$
$$C_y = A_z B_x - A_x B_z$$
$$C_z = A_x B_y - A_y B_x.$$

In particular, for the vector products of the Cartesian unit vectors, this definition implies

$$\hat{x} \times \hat{y} = \hat{z}$$
$$\hat{y} \times \hat{z} = \hat{x}$$
$$\hat{z} \times \hat{x} = \hat{y}.$$

The absolute magnitude of the vector $\vec{C}$ is given by

$$\left|\vec{C}\right| = \left|\vec{A}\right|\left|\vec{B}\right| \sin\theta.$$

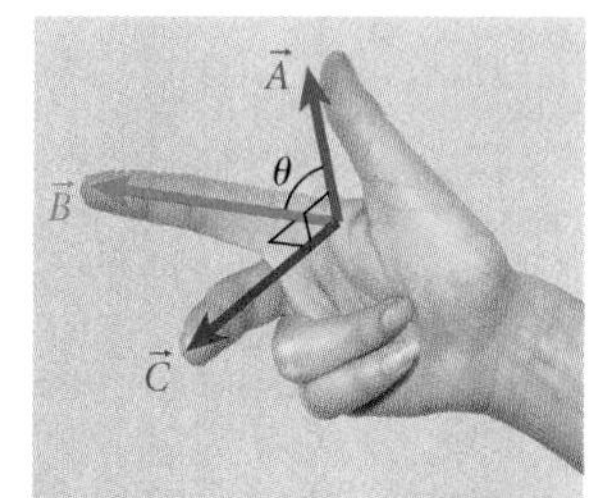

**FIGURE 10.18** Vector product.

Here $\theta$ is the angle between $\vec{A}$ and $\vec{B}$, as shown in Figure 10.18. This result implies that the magnitude of the vector product of two vectors is at its maximum when $\vec{A} \perp \vec{B}$ and is zero when $\vec{A} \parallel \vec{B}$. We can also interpret the right-hand side of this equation as either the product of the magnitude of vector $\vec{A}$ times the component of $\vec{B}$ perpendicular to $\vec{A}$ or the product of the magnitude of $\vec{B}$ times the component of $\vec{A}$ perpendicular to $\vec{B}$: $\left|\vec{C}\right| = \left|\vec{A}\right| B_{\perp A} = \left|\vec{B}\right| A_{\perp B}$. Either interpretation is valid.

The direction of the vector $\vec{C}$ can be found using the right-hand rule: If vector $\vec{A}$ points along the direction of the thumb and vector $\vec{B}$ points along the direction of the index finger, then the vector product is perpendicular to *both* vectors and points along the direction of the middle finger, as shown in Figure 10.18.

It is important to realize that for the vector product, the order of the factors matters:

$$\vec{B}\times\vec{A}=-\vec{A}\times\vec{B}.$$

Thus, the vector product differs from both regular multiplication of scalars and multiplication of vectors to form a scalar product.

We'll see immediately from the definition of the vector product that for any vector $\vec{A}$, the vector product with itself is always zero:

$$\vec{A}\times\vec{A}=0.$$

Finally, there is a handy rule for a double vector product involving three vectors: The vector product of the vector $\vec{A}$ with the vector product of the vectors $\vec{B}$ and $\vec{C}$ is the sum of two vectors, one pointing in the direction of the vector $\vec{B}$ and multiplied by the scalar product $\vec{A}\bullet\vec{C}$ and another one pointing in the direction of the vector $\vec{C}$ and multiplied by $-\vec{A}\bullet\vec{B}$:

$$\vec{A}\times(\vec{B}\times\vec{C})=\vec{B}(\vec{A}\bullet\vec{C})-\vec{C}(\vec{A}\bullet\vec{B}).$$

This *BAC-CAB rule* is straightforward to prove using the Cartesian components in the definitions of the vector product and the scalar product, but the proof is cumbersome and thus omitted here. However, the rule occasionally comes in handy, in particular when dealing with torque and angular momentum. And the mnemonic BAC-CAB makes it fairly easy to remember.

### 10.4 In-Class Exercise

Choose the combination of position vector, $\vec{r}$, and force vector, $\vec{F}$, that produces the torque of highest magnitude around the point indicated by the black dot.

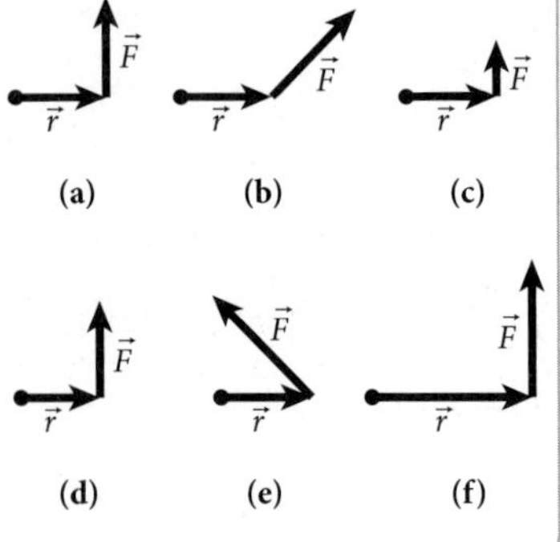


With the mathematical definition of the torque, its magnitude, and its relation to the force vector, the position vector, and their relative angle, we can understand why the approach shown in part (c) of Figure 10.16 yields maximum torque for a given magnitude of the force, whereas that in part (a) yields zero torque. We see that the magnitude of the torque is the decisive factor in determining how easy or hard it is to loosen (or tighten) a bolt.

Torques around any fixed axis of rotation can be clockwise or counterclockwise. As indicated by the force vector in Figure 10.16d, the torque generated by the hand pulling on the wrench would be counterclockwise. The **net torque** is defined as the difference between the sum of all clockwise torques and the sum of all counterclockwise torques:

$$\tau_{\text{net}}=\sum_i \tau_{\text{counterclockwise},i}-\sum_j \tau_{\text{clockwise},j}$$

## 10.5 Newton's Second Law for Rotation

In Section 10.1, we noted that moment of inertia $I$ is the rotational equivalent of mass. From Chapter 4, we know that the product of the mass and the linear acceleration is the net force acting on the object, as expressed by Newton's Second Law, $F_{\text{net}} = ma$. What is the equivalent of Newton's Second Law for rotational motion?

Let's start with a point particle of mass $M$ moving in a circle around an axis at a distance $R$ from the axis. If we multiply the moment of inertia for rotation about an axis parallel to the center of mass by the angular acceleration, we obtain

$$I\alpha=(R^2M)\alpha=RM(R\alpha)=RMa=RF_{\text{net}}.$$

To obtain this result, we first used equation 10.11 with $c = 1$, then the relationship between angular and linear acceleration for motion on a circle, and then Newton's Second Law. Thus, the product of the moment of inertia and the angular acceleration is proportional to the product of a distance quantity and a force quantity. In Section 10.4, we saw that this product is a torque, $\tau$. Thus, we can write the following form of Newton's Second Law for rotational motion:

$$\tau=I\alpha. \tag{10.18}$$

Combining equations 10.16 and 10.18 gives us

$$\vec{\tau}=\vec{r}\times\vec{F}_{\text{net}}=I\vec{\alpha}. \tag{10.19}$$

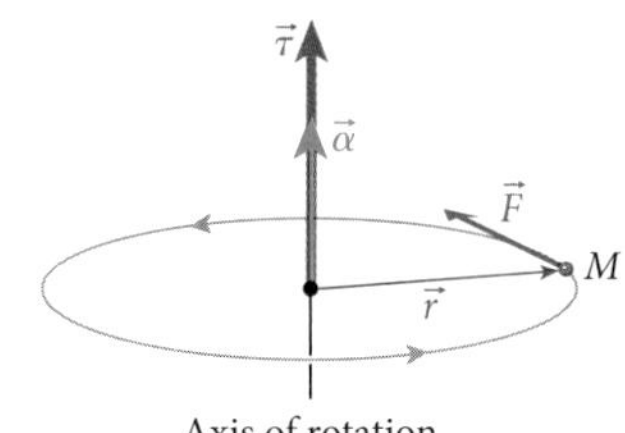

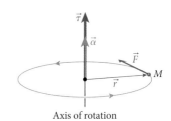

**FIGURE 10.19** A force exerted on a point particle creates a torque.

This equation for rotational motion is analogous to Newton's Second Law, $\vec{F} = m\vec{a}$, for linear motion. Figure 10.19 shows the relationship between force, position, torque, and angular acceleration for a point particle moving around an axis of rotation. Note that, strictly speaking, equation 10.19 holds only for a point particle in circular orbit. It seems plausible that this equation holds for a moment of inertia for an extended object in general, but we have not proven it. Later in this chapter, we will come back to this equation.

Newton's First Law stipulates that in the absence of a net force, an object experiences no acceleration and thus no change in velocity. The equivalent of Newton's First Law for rotational motion is that an object experiencing no net torque has no angular acceleration and thus no change in angular velocity. In particular, this means that for objects to remain stationary, the net torque on them has to be zero. We will return to this statement in Chapter 11, where we investigate static equilibrium.

Now, however, we can use the rotational form of Newton's Second Law (equation 10.19) to solve interesting problems of rotational motion, such as the following one.

## EXAMPLE 10.3 Toilet Paper

This may have happened to you: You are trying to put a new roll of toilet paper into its holder. However, you drop the roll, managing to hold onto just the first sheet. On its way to the floor, the toilet paper roll unwinds, as Figure 10.20a shows.

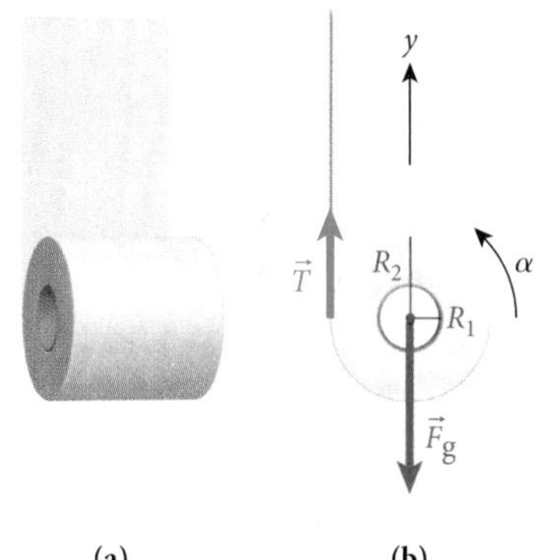


**FIGURE 10.20** (a) Unrolling toilet paper. (b) Free-body diagram of the roll of toilet paper.

### PROBLEM

How long does it take the roll of toilet paper to hit the floor, if it was released from a height of 0.73 m? The roll has an inner radius $R_1 = 2.7$ cm, an outer radius $R_2 = 6.1$ cm, and a mass of 274 g.

### SOLUTION

For the falling roll of toilet paper, the acceleration is $a_y = -g$, and in Chapter 2, we saw that for free fall from rest, the position as a function of time is given in general as $y = y_0 + v_0 t - \frac{1}{2}gt^2$. In this case, the initial velocity is zero; therefore, $y = y_0 - \frac{1}{2}gt^2$. If we put the origin at floor level, then we have to find the time at which $y = 0$. This implies $y_0 = \frac{1}{2}gt^2$ for the time it takes the roll to hit the floor. Therefore, the time it would take for the roll in free fall to hit the floor is

$$t_{\text{free}} = \sqrt{\frac{2y_0}{g}} = \sqrt{\frac{2(0.73 \text{ m})}{9.81 \text{ m/s}^2}} = 0.386 \text{ s.} \tag{i}$$

However, since you hold onto the first sheet and the toilet paper unwinds on the way down, the roll of toilet paper is rolling without slipping (in the sense in which this was defined earlier). Therefore, the acceleration will be different from the free-fall case. Once we know the value of this acceleration, we can use a formula relating the initial height and fall time similar to equation (i).

How do we calculate the acceleration that the toilet paper experiences? Again, we start with a free-body diagram. Figure 10.20b shows the toilet paper roll in side view and indicates the forces due to gravity, $\vec{F}_g = mg(-\hat{y})$, and to the tension from the sheet that is held by the hand, $\vec{T} = T\hat{y}$. Newton's Second Law then allows us to relate the net force acting on the toilet paper to the acceleration of the roll:

$$T - mg = ma_y. \tag{ii}$$

The tension and the acceleration are both unknown, so we need to find a second equation relating these quantities. This second equation can be obtained from the rotational motion of the roll, for which the net torque is the product of the moment of inertia and the angular acceleration: $\tau = I\alpha$. The moment of inertia of the roll of toilet paper is that of a hollow cylinder, $I = \frac{1}{2}m(R_1^2 + R_2^2)$, which is equation 10.9.

We can again relate the linear and angular accelerations via $a_y = R_2\alpha$, where $R_2$ is the outer radius of the roll of toilet paper. We need to specify which direction is positive

*Continued—*

for the angular acceleration—otherwise, we could get the sign wrong and thus obtain a false result. To be consistent with the choice of up as the positive $y$-direction, we need to select counterclockwise rotation as the positive angular direction, as indicated in Figure 10.20b.

For the torque about the axis of symmetry of the roll of toilet paper, we have $\tau = -R_2T$, with the sign convention for positive angular acceleration that we just established. The force of gravity does not contribute to the torque about the axis of symmetry, because its moment arm has a length of zero. Newton's Second Law for rotational motion then leads to

$$\tau = I\alpha$$

$$-R_2T = \left[\tfrac{1}{2}m(R_1^2 + R_2^2)\right]\frac{a_y}{R_2}$$

$$-T = \tfrac{1}{2}m\left(1+\frac{R_1^2}{R_2^2}\right)a_y. \qquad \text{(iii)}$$

Equations (ii) and (iii) form a set of two equations for the two unknown quantities $T$ and $a$. Adding them results in

$$-mg = \tfrac{1}{2}m\left(1+\frac{R_1^2}{R_2^2}\right)a_y + ma_y.$$

The mass of the roll of toilet paper cancels out, and we find for the acceleration

$$a_y = -\frac{g}{\frac{3}{2}+\frac{R_1^2}{2R_2^2}}.$$

Using the given values for the inner radius, $R_1 = 2.7$ cm and the outer radius, $R_2 = 6.1$ cm, we find the value of the acceleration:

$$a_y = -\frac{9.81\text{ m/s}^2}{\frac{3}{2}+\frac{(2.7\text{ cm})^2}{2(6.1\text{ cm})^2}} = -6.14\text{ m/s}^2.$$

Inserting this value for the acceleration into a formula for the fall time that is analogous to equation (i) gives our answer

$$t = \sqrt{\frac{2y_0}{-a_y}} = \sqrt{\frac{2(0.73\text{ m})}{6.14\text{ m/s}^2}} = 0.488\text{ s}.$$

This is approximately 0.1 s longer than the time for the free-falling roll of toilet paper released from the same height.

**DISCUSSION**

Note that we assumed that the outer radius of the roll of toilet paper did not change as the paper unwound. For the short distance of less than 1 m, this is justified. However, if we wanted to calculate how long it took for the toilet paper to unwind over a distance of, say, 10 m, we would need to take the change in outer radius into account. Of course, then we would also need to take into account the effect of air resistance!

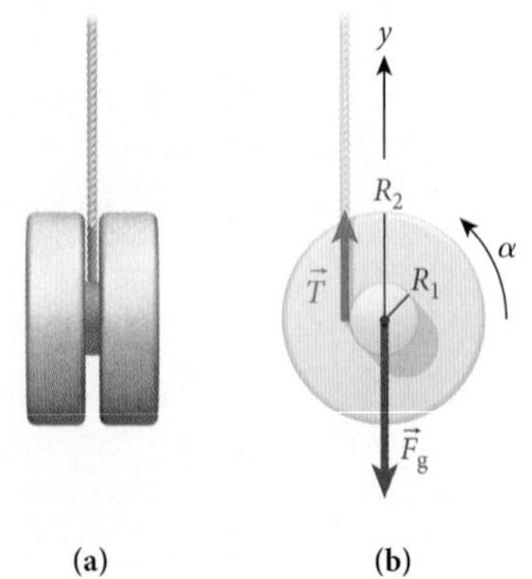


**FIGURE 10.21** (a) Yo-yo. (b) Free-body diagram of the yo-yo.

As an extension of Example 10.3, we can consider a yo-yo. A yo-yo consists of two solid disks of radius $R_2$, with a thin smaller disk of radius $R_1$ mounted between them, and a string wound around the smaller disk (see Figure 10.21). For the purpose of this analysis, we can consider the moment of inertia of the yo-yo to be that of a solid disk with radius $R_2$: $I = \frac{1}{2}mR_2^2$. The free-body diagrams of Figure 10.20b and Figure 10.21b are almost identical, except for one detail: For the yo-yo, the string tension acts on the surface at the inner radius

$R_1$, as opposed to the outer radius $R_2$, as was the case for the roll of toilet paper. This implies that the angular and linear accelerations for the yo-yo are proportional to each other, with proportionality constant $R_1$ (instead of $R_2$ as for the roll). Thus, the torque for the yo-yo rolling without slipping along the string is

$$\tau = I\alpha \Rightarrow -TR_1 = (\tfrac{1}{2}mR_2^2)\frac{a_y}{R_1}$$

$$-T = \tfrac{1}{2}m\frac{R_2^2}{R_1^2}a_y.$$

It is instructive to compare this equation to $-T = \frac{1}{2}m(1 + R_1^2/R_2^2)a_y$, which we derived in Example 10.3 for the roll of toilet paper. They look very similar, but the ratio of the radii is different. On the other hand, the equation derived from Newton's Second Law (for translation) is the same in both cases:

$$T - mg = ma_y.$$

Adding this equation and the equation for the torque of the yo-yo yields the acceleration of the yo-yo:

$$-mg = ma_y + \tfrac{1}{2}m\frac{R_2^2}{R_1^2}a_y \Rightarrow$$

$$a_y = -\frac{g}{1+\frac{1}{2}\frac{R_2^2}{R_1^2}} = -\frac{2R_1^2}{2R_1^2 + R_2^2}g.$$

For example, if $R_2 = 5R_1$, the acceleration of the yo-yo is

$$a_y = \frac{-g}{1+\frac{1}{2}(25)} = \frac{-g}{13.5} = -0.727 \text{ m/s}^2.$$

## Atwood Machine

Chapter 4 introduced an Atwood machine, which consists of two weights with masses $m_1$ and $m_2$ connected by a string or rope, which is guided over a pulley. The machines we analyzed in Chapter 4 were subject to the condition that the rope glides without friction over the pulley so that the pulley does not rotate (or, equivalently, that the pulley is massless). In such cases, the common acceleration of the two masses is $a = g(m_1 - m_2)/(m_1 + m_2)$. With the concepts of rotational dynamics, we can take another look at the Atwood machine and consider the case where friction occurs between the rope and the pulley, causing the rope to turn the pulley without slipping.

In Chapter 4, we saw that the magnitude of the tension, $T$, is the same everywhere in a string. However, now friction forces are involved to keep the string attached to the pulley, and we cannot assume that the tension is constant. Instead, the string tension is separately determined in each segment of the string from which one of the two masses hang. Thus, there are two different string tensions, $T_1$ and $T_2$, in the free-body diagrams for the two masses (as shown in Figure 10.22b). Just as in Chapter 4, applying Newton's Second Law individually to each free-body diagram leads to

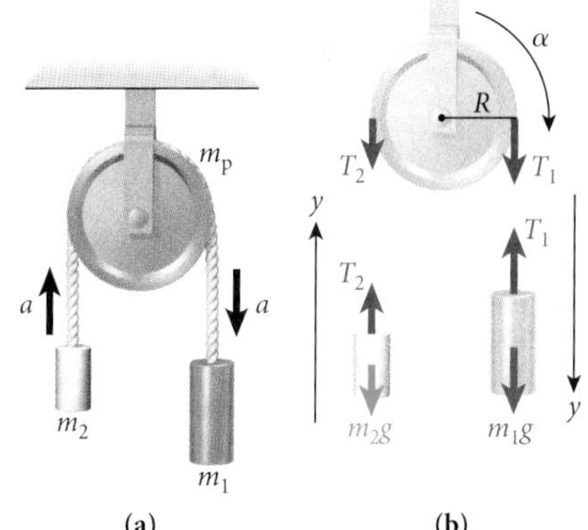


**FIGURE 10.22** Atwood machine: (a) physical setup; (b) free-body diagrams.

$$-T_1 + m_1g = m_1a \tag{10.20}$$

$$T_2 - m_2g = m_2a. \tag{10.21}$$

Here again we used the (arbitrary) sign convention that a positive acceleration ($a > 0$) is one for which $m_1$ moves downward and $m_2$ upward. This convention is indicated in the free-body diagrams by the direction of the positive $y$-axis.

Figure 10.22b also shows a free-body diagram for the pulley, but it includes only the forces that can cause a torque: the two string tensions, $T_1$ and $T_2$. The downward force of gravity and the upward force of the support structure on the pulley are not shown. The pulley does not have translational motion, so all forces acting on the pulley add up to zero.

However, a net torque does act on the pulley. According to equation 10.17, the magnitude of the torque due to the string tensions is given by

$$\tau = \tau_1 - \tau_2 = RT_1 \sin 90° - RT_2 \sin 90° = R(T_1 - T_2). \qquad (10.22)$$

These two torques have opposite signs, because one is acting clockwise and the other counterclockwise. According to equation 10.18, the net torque is related to the moment of inertia of the pulley and its angular acceleration by $\tau = I\alpha$. The moment of inertia of the pulley (mass of $m_p$) is that of a disk: $I = \frac{1}{2}m_pR^2$. Since the rope moves across the pulley without slipping, the acceleration of the rope (and masses $m_1$ and $m_2$) is related to the angular acceleration via $\alpha = a/R$, just like the correspondence established in Chapter 9 between linear and angular acceleration for a point particle moving on the circumference of a circle. Inserting the expressions for the moment of inertia and the angular acceleration results in $\tau = I\alpha = (\frac{1}{2}m_pR^2)(a/R)$. Substituting this expression for the torque into equation 10.22 we find

$$R(T_1 - T_2) = \tau = (\tfrac{1}{2}m_pR^2)\left(\frac{a}{R}\right) \Rightarrow$$

$$T_1 - T_2 = \tfrac{1}{2}m_pa. \qquad (10.23)$$

Equations 10.20, 10.21, and 10.23 form a set of three equations for three unknown quantities: the two values of the string tension, $T_1$ and $T_2$, and the acceleration, $a$. The easiest way to solve this system for the acceleration is to add the equations. We then find

$$m_1g - m_2g = (m_1 + m_2 + \tfrac{1}{2}m_p)a \Rightarrow$$

$$a = \frac{m_1 - m_2}{m_1 + m_2 + \frac{1}{2}m_p}g. \qquad (10.24)$$

Note that equation 10.24 matches the equation for the case of a massless pulley (or the case in which the rope slides over the pulley without friction), except for the additional term of $\frac{1}{2}m_p$ in the denominator, which represents the contribution of the pulley to the overall inertia of the system. The factor $\frac{1}{2}$ reflects the shape of the pulley, a disk, because $c = \frac{1}{2}$ for a disk in the relationship between moment of inertia, mass, and radius (equation 10.11).

Thus, we have answered the question of what happens when a force is exerted on an extended object some distance away from its center of mass: The force produces a torque as well as linear motion. This torque leads to rotation, which we left out of our original considerations of the result of exerting a force on an object because we assumed that all forces acted on the center of mass of the object.

## 10.6 Work Done by a Torque

In Chapter 5, we saw that the work $W$ done by a force $\vec{F}$ is given by the integral

$$W = \int_{x_0}^{x} F_x(x')dx'.$$

We can now consider the work done by a torque $\vec{\tau}$.

Torque is the angular equivalent of force. The angular equivalent of the linear displacement, $d\vec{r}$, is the angular displacement, $d\vec{\theta}$. Since both the torque and the angular displacement are axial vectors and point in the direction of the axis of rotation, we can write their scalar product as $\vec{\tau} \bullet d\vec{\theta} = \tau d\theta$. Therefore, the work done by a torque is

$$W = \int_{\theta_0}^{\theta} \tau(\theta')d\theta'. \qquad (10.25)$$

For the special case where the torque is constant and thus does not depend on $\theta$, the integral of equation 10.25 simply evaluates to

$$W = \tau(\theta - \theta_0). \qquad (10.26)$$

Chapter 5 also presented the first version of the work–kinetic energy theorem: $\Delta K \equiv K - K_0 = W$. The rotational equivalent of this work–kinetic energy relationship can be written with the aid of equation 10.3 as follows:

$$\Delta K \equiv K - K_0 = \tfrac{1}{2}I\omega^2 - \tfrac{1}{2}I\omega_0^2 = W. \tag{10.27}$$

For the case of a constant torque, we can use equation 10.26 and find the work–kinetic energy theorem for constant torque:

$$\tfrac{1}{2}I\omega^2 - \tfrac{1}{2}I\omega_0^2 = \tau(\theta - \theta_0). \tag{10.28}$$

## EXAMPLE 10.4 Tightening a Bolt

### PROBLEM

What is the total work required to completely tighten the bolt shown in Figure 10.23? The total number of turns is 30.5, the diameter of the bolt is 0.860 cm, and the friction force between the nut and the bolt is a constant 14.5 N.

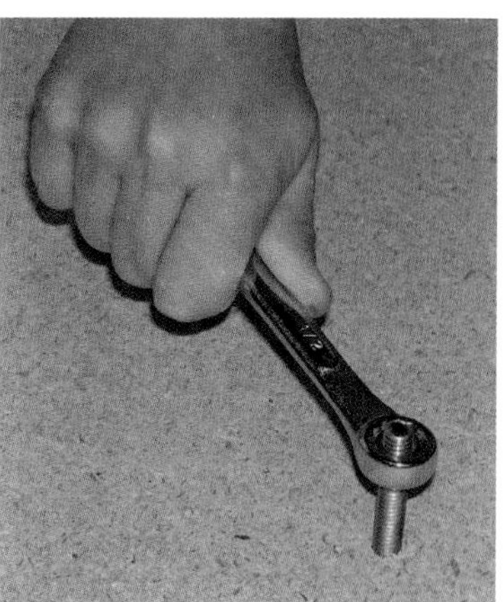

**FIGURE 10.23** Tightening a bolt.

### SOLUTION

Since the friction force is constant and the diameter of the bolt is constant, we can calculate directly the torque needed to turn the nut:

$$\tau = Fr = \tfrac{1}{2}Fd = \tfrac{1}{2}(14.5\text{ N})(0.860\text{ cm}) = 0.0623\text{ N m}.$$

In order to calculate the total work required to completely tighten the bolt, we need to figure out the total angle. Each turn corresponds to an angle of $2\pi$ rad, so the total angle in this case is $\Delta\theta = 30.5(2\pi) = 191.6$ rad.

The total work required is then obtained by making use of equation 10.26:

$$W = \tau\Delta\theta = (0.0623\text{ N m})(191.6) = 11.9\text{ J}.$$

As you can see, figuring out the work done is not very difficult with a constant torque. However, in many physical situations, the torque cannot be considered to be constant. The next example illustrates such a case.

## EXAMPLE 10.5 Driving a Screw

The friction force between a drywall screw and wood is proportional to the contact area between the screw and the wood. Since the drywall screw has a constant diameter, this means that the torque required for turning the screw increases linearly with the depth that the screw has penetrated into the wood.

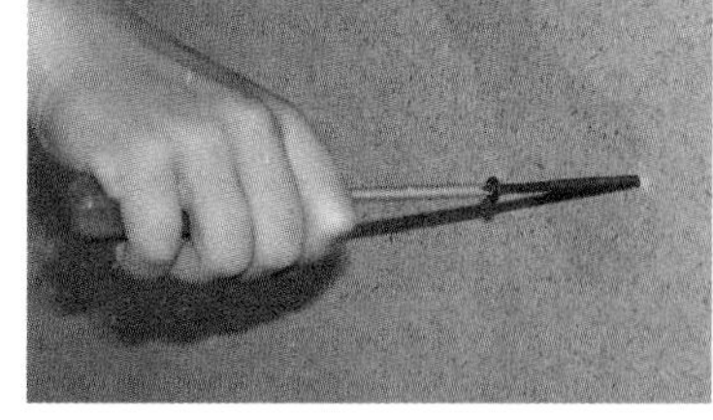

**FIGURE 10.24** Driving a drywall screw into a block of wood.

### PROBLEM

Suppose it takes 27.3 turns to screw a drywall screw completely into a block of wood (Figure 10.24). The torque needed to turn the screw increases linearly from zero at the beginning to a maximum of 12.4 N m at the end. What is the total work required to drive in the screw?

### SOLUTION

Clearly, the torque is a function of the angle in this situation and is not constant anymore. Thus, we have to use the integral of equation 10.25 to find our answer. First, let us calculate the total angle, $\theta_{\text{total}}$, that the screw turns through: $\theta_{\text{total}} = 27.3(2\pi) = 171.5$ rad. Now we need to find an expression for $\tau(\theta)$. A linear increase with $\theta$ from zero to 12.4 N m means

$$\tau(\theta) = \theta\frac{\tau_{\max}}{\theta_{\text{total}}} = \theta\frac{12.4\text{ N m}}{171.5} = \theta(0.0723\text{ N m}).$$

*Continued—*

### 10.5 In-Class Exercise

If you want to reduce the torque required to drive in a screw, you can rub soap on the thread beforehand. Suppose the soap reduces the coefficient of friction between the screw and the wood by a factor of 2 and therefore reduces the required torque by a factor of 2. How much does it change the total work required to turn the screw into the wood?

a) It leaves the work the same.

b) It reduces the work by a factor of 2.

c) It reduces the work by a factor of 4.

### 10.6 In-Class Exercise

If you get tired before you finish driving in the screw and you manage to screw it in only halfway, how does this change the total work you have done?

a) It leaves the work the same.

b) It reduces the work by a factor of 2.

c) It reduces the work by a factor of 4.

Now we can evaluate the integral as follows:

$$W = \int_0^{\theta_{\text{total}}} \tau(\theta')d\theta' = \int_0^{\theta_{\text{total}}} \theta' \frac{\tau_{\max}}{\theta_{\text{total}}} d\theta' = \frac{\tau_{\max}}{\theta_{\text{total}}} \int_0^{\theta_{\text{total}}} \theta' d\theta' = \frac{\tau_{\max}}{\theta_{\text{total}}} \frac{1}{2}\theta^2 \Big|_0^{\theta_{\text{total}}} = \tfrac{1}{2}\tau_{\max}\theta_{\text{total}}.$$

Inserting the numbers, we obtain

$$W = \tfrac{1}{2}\tau_{\max}\theta_{\text{total}} = \tfrac{1}{2}(12.4 \text{ N m})(171.5 \text{ rad}) = 1.06 \text{ kJ}.$$

## SOLVED PROBLEM 10.3 Atwood Machine

### PROBLEM

Two weights with masses $m_1 = 3.00$ kg and $m_2 = 1.40$ kg are connected by a very light rope that runs without sliding over a pulley (solid disk) of mass $m_p = 2.30$ kg. The two masses initially hang at the same height and are at rest. Once released, the heavier mass, $m_1$, descends and lifts up the lighter mass, $m_2$. What speed will $m_2$ have at height $h = 0.16$ m?

### SOLUTION

### THINK

We could try to calculate the acceleration of the two masses and then use kinematic equations to relate this acceleration to the vertical displacement. However, we can also use energy considerations, which will lead to a fairly direct solution. Initially, the two hanging masses and the pulley are at rest, so the total kinetic energy is zero. We can choose a coordinate system such that the initial potential energy is zero, and thus the total energy is zero. As one of the masses is lifted, it gains gravitational potential energy, and the other mass loses potential energy. Both masses gain translational kinetic energy, and the pulley gains rotational kinetic energy. Since the kinetic energy is proportional to the square of the speed, we can then use energy conservation to solve for the speed.

### SKETCH

Figure 10.25a shows the initial state of the Atwood machine with both hanging masses at the same height. We decide to set that height as the origin of the vertical axis, thus ensuring that the initial potential energy, and therefore the total initial energy, is zero. Figure 10.25b shows the Atwood machine with the masses displaced by $h$.

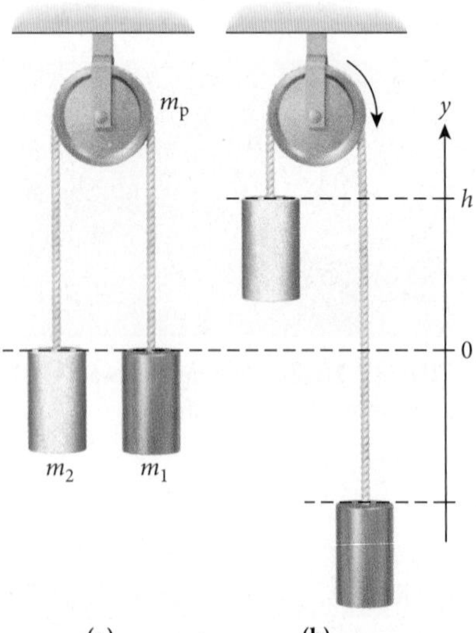


**FIGURE 10.25** One more Atwood machine: (a) initial positions; (b) positions after the weights move a distance $h$.

### RESEARCH

The gain in gravitational potential energy for $m_2$ is $U_2 = m_2gh$. At the same time, $m_1$ is lowered the same distance, so its potential energy is $U_1 = -m_1gh$. The kinetic energy of $m_1$ is $K_1 = \tfrac{1}{2}m_1v^2$, and the kinetic energy of $m_2$ is $K_2 = \tfrac{1}{2}m_2v^2$. Note that the same speed, $v$, is used in the energy expressions for the two masses. This equality is assured, because a rope connects them. (We assume that the rope does not stretch.)

What about the rotational kinetic energy of the pulley? The pulley is a solid disk with a moment of inertia given by $I = \tfrac{1}{2}m_pR^2$ and a rotational kinetic energy given by $K_r = \tfrac{1}{2}I\omega^2$. Since the rope runs over the pulley without slipping and also moves with the same speed as the two masses, points on the surface of the pulley also have to move with that same linear speed, $v$. Just as for a rolling solid disk, the linear speed is related to the angular speed via $\omega R = v$. Thus, the rotational kinetic energy of the pulley is

$$K_r = \tfrac{1}{2}I\omega^2 = \tfrac{1}{2}\left(\tfrac{1}{2}m_pR^2\right)\omega^2 = \tfrac{1}{4}m_pR^2\omega^2 = \tfrac{1}{4}m_pv^2.$$

Now we can write down the total energy as the sum of the potential and translational and rotational kinetic energies. This overall sum has to equal zero, because that was the initial value of the total energy and energy conservation applies:

$$\begin{aligned} 0 &= U_1 + U_2 + K_1 + K_2 + K_r \\ &= -m_1gh + m_2gh + \tfrac{1}{2}m_1v^2 + \tfrac{1}{2}m_2v^2 + \tfrac{1}{4}m_pv^2. \end{aligned}$$

**SIMPLIFY**

We can rearrange the preceding equation to isolate the speed, $v$:

$$(m_1 - m_2)gh = (\tfrac{1}{2}m_1 + \tfrac{1}{2}m_2 + \tfrac{1}{4}m_p)v^2 \Rightarrow$$

$$v = \sqrt{\frac{2(m_1 - m_2)gh}{m_1 + m_2 + \frac{1}{2}m_p}}. \qquad \text{(i)}$$

**CALCULATE**

Now we insert the numbers:

$$v = \sqrt{\frac{2(3.00\text{ kg} - 1.40\text{ kg})(9.81\text{ m/s}^2)(0.16\text{ m})}{3.00\text{ kg} + 1.40\text{ kg} + \frac{1}{2}(2.30\text{ kg})}} = 0.951312\text{ m/s}.$$

**ROUND**

The displacement $h$ was given with the least precision, to two digits, so we round our result to

$$v = 0.95\text{ m/s}.$$

**DOUBLE-CHECK**

The acceleration of the masses is given by equation 10.24, which we developed in Section 10.5 in discussing the Atwood machine:

$$a = \frac{m_1 - m_2}{m_1 + m_2 + \frac{1}{2}m_p}g. \qquad \text{(ii)}$$

Chapter 2 presented kinematic equations for one-dimensional linear motion. One of these, which relates the initial and final speeds, the displacement, and the acceleration, now comes in handy:

$$v^2 = v_0^2 + 2a(y - y_0).$$

Here $v_0 = 0$ and $y - y_0 = h$. Then, inserting the expression for $a$ from equation (ii) leads to our result

$$v^2 = 2ah = 2\frac{m_1 - m_2}{m_1 + m_2 + \frac{1}{2}m_p}gh \Rightarrow v = \sqrt{\frac{2(m_1 - m_2)gh}{m_1 + m_2 + \frac{1}{2}m_p}}.$$

This equation for the speed is identical to equation (i), which we obtained using energy considerations. The effort it took to develop the equation for the acceleration makes it clear that the energy method is the quicker way to proceed.

## 10.7 Angular Momentum

Although we have discussed the rotational equivalents of mass (moment of inertia), velocity (angular velocity), acceleration (angular acceleration), and force (torque), we have not yet encountered the rotational analogue to linear momentum. Since linear momentum is the product of an object's velocity and its mass, by analogy, the angular momentum should be the product of angular velocity and moment of inertia. In this section, we will find that this relationship is indeed true for an extended object with a fixed moment of inertia. However, to reach that conclusion, we need to start from a definition of angular momentum for a point particle and proceed from there.

### Point Particle

The **angular momentum,** $\vec{L}$, of a point particle is the vector product of its position and momentum vectors:

$$\vec{L} = \vec{r} \times \vec{p}. \qquad \textbf{(10.29)}$$

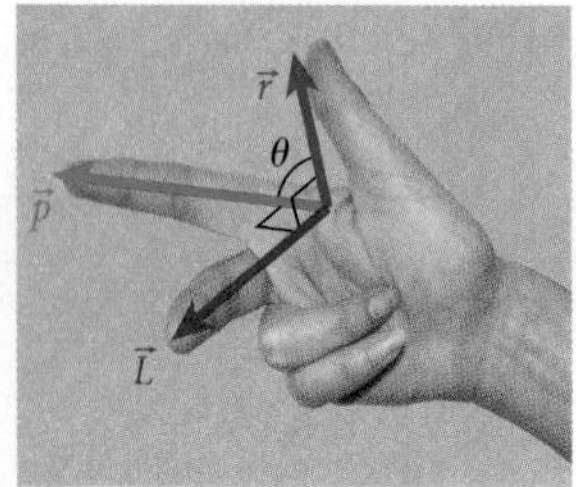


**FIGURE 10.26** Right-hand rule for the direction of the angular momentum vector: The thumb is aligned with the position vector and the index finger with the momentum vector; then the angular momentum vector points along the middle finger.

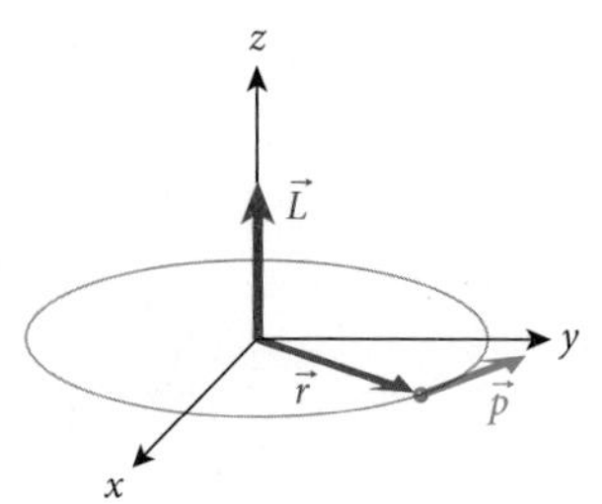


**FIGURE 10.27** The angular momentum of a point particle.

Because the angular momentum is defined as $\vec{L}=\vec{r}\times\vec{p}$ and the torque is defined as $\vec{\tau}=\vec{r}\times\vec{F}$, statements can be made about angular momentum that are similar to those made about torque in Section 10.4. For example, the magnitude of the angular momentum is given by

$$L = rp\sin\theta, \tag{10.30}$$

where $\theta$ is the angle between the position and momentum vectors. Also, just like the direction of the torque vector, the direction of the angular momentum vector is given by a right-hand rule. Let the thumb of the right hand point along the position vector, $\vec{r}$, of a point particle and the index finger point along the momentum vector, $\vec{p}$, then the middle finger will indicate the direction of the angular momentum vector $\vec{L}$ (Figure 10.26). As an example, the angular momentum vector of a point particle located in the $xy$-plane is illustrated in Figure 10.27.

With the definition of the angular momentum in equation 10.29, we can take the time derivative:

$$\frac{d}{dt}\vec{L}=\frac{d}{dt}(\vec{r}\times\vec{p})=\left[\left(\frac{d}{dt}\vec{r}\right)\times\vec{p}\right]+\left(\vec{r}\times\frac{d}{dt}\vec{p}\right)=(\vec{v}\times\vec{p})+(\vec{r}\times\vec{F}).$$

To take the derivative of the vector product, we apply the product rule of calculus. The term $\vec{v}\times\vec{p}$ is always zero, because $\vec{v}\parallel\vec{p}$. Also, from equation 10.16, we know that $\vec{r}\times\vec{F}=\vec{\tau}$. Thus, we obtain for the time derivative of the angular momentum vector:

$$\frac{d}{dt}\vec{L}=\vec{\tau}. \tag{10.31}$$

The time derivative of the angular momentum vector for a point particle is the torque vector acting on that point particle. This result is again analogous to the linear motion case, where the time derivative of the linear momentum vector is equal to the force vector.

The vector product allows us to revisit the relationship between the linear velocity vector, the coordinate vector, and the angular velocity vector, which was introduced in Chapter 9. For circular motion, the magnitudes of those vectors are related via $\omega = v/r$, and the direction of $\vec{\omega}$ is given by a right-hand rule. Using the definition of the vector product, we can write $\vec{\omega}$ as

$$\vec{\omega}=\frac{\vec{r}\times\vec{v}}{r^2}. \tag{10.32}$$

Comparing equations 10.29 and 10.32 reveals that the angular momentum and angular velocity vectors for the point particle are parallel, with

$$\vec{L}=\vec{\omega}\cdot(mr^2). \tag{10.33}$$

The quantity $mr^2$ is the moment of inertia of a point particle orbiting the axis of rotation at a distance $r$.

## System of Particles

It is straightforward to generalize the concept of angular momentum to a system of $n$ point particles. The total angular momentum of the system of particles is simply the sum of the angular momenta of the individual particles:

$$\vec{L}=\sum_{i=1}^{n}\vec{L}_i=\sum_{i=1}^{n}\vec{r}_i\times\vec{p}_i=\sum_{i=1}^{n}m_i\vec{r}_i\times\vec{v}_i. \tag{10.34}$$

Again, we take the time derivative of this sum of angular momenta in order to obtain the relationship between the total angular momentum of this system and the torque:

$$\frac{d}{dt}\vec{L}=\frac{d}{dt}\left(\sum_{i=1}^{n}\vec{L}_i\right)=\frac{d}{dt}\left(\sum_{i=1}^{n}\vec{r}_i\times\vec{p}_i\right)=\sum_{i=1}^{n}\frac{d}{dt}(\vec{r}_i\times\vec{p}_i)$$

$$=\sum_{i=1}^{n}\underbrace{\underbrace{\left(\frac{d}{dt}\vec{r}_i\right)}_{\text{Equals }\vec{v}_i}\times\vec{p}_i}_{\text{Equals zero, because }\vec{v}_i\parallel\vec{p}_i}+\vec{r}_i\times\underbrace{\left(\frac{d}{dt}\vec{p}_i\right)}_{\vec{F}_i}=\sum_{i=1}^{n}\vec{r}_i\times\vec{F}_i=\sum_{i=1}^{n}\vec{\tau}_i=\vec{\tau}_{\text{net}}.$$

As expected, we find that the time derivative of the total angular momentum for a system of particles is given by the total net external torque acting on the system. It is important to keep in mind that this is the net *external* torque due to the *external* forces, $\vec{F}_i$.

## Rigid Objects

A rigid object will rotate about a fixed axis with an angular velocity $\vec{\omega}$ that is the same for every part of the object. In this case, the angular momentum is proportional to the angular velocity, and the proportionality constant is the moment of inertia:

$$\vec{L} = I\vec{\omega}. \tag{10.35}$$

### 10.3 Self-Test Opportunity

Can you show that internal torques (those due to internal forces between particles in a system) do not contribute to the total net torque? (*Hint:* Use Newton's Third Law, $\vec{F}_{i\to j} = -\vec{F}_{j\to i}$.)

### DERIVATION 10.4 Angular Momentum of a Rigid Object

Representing the rigid object by a collection of point particles allows us to use the results of the preceding subsection as a starting point. In order for point particles to represent the rigid object, their relative distances from one another must remain constant (rigid). Then all of these point particles rotate with a constant angular velocity, $\vec{\omega}$, about the common rotation axis.

From equation 10.34, we obtain

$$\vec{L} = \sum_{i=1}^{n} \vec{L}_i = \sum_{i=1}^{n} m_i \vec{r}_i \times \vec{v}_i = \sum_{i=1}^{n} m_i r_{i\perp}^2 \vec{\omega}.$$

In the last step, we have used the relationship between the angular velocity and the vector product of the position and linear velocity vectors for point particles, equation 10.32, where $r_{i\perp}$ is the orbital radius of the point particle $i$. Note that the angular velocity vector is the same for all point particles in this rigid object. Therefore, we can move it out of the sum as a common factor:

$$\vec{L} = \vec{\omega} \sum_{i=1}^{n} m_i r_{i\perp}^2.$$

We can identify this sum as the moment of inertia of a collection of point particles; see equation 10.4. Thus, we finally have our result:

$$\vec{L} = I\vec{\omega}.$$

For rigid objects, just like for point particles, the direction of the angular momentum vector is the same as the direction of the angular velocity vector. Figure 10.28 shows the right-hand rule used to determine the direction of the angular momentum vector (arrow along the direction of the thumb) as a function of the sense of rotation (direction of the fingers).

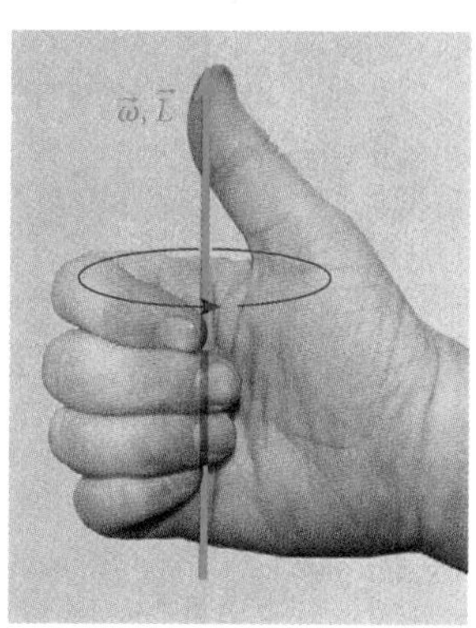


(a)

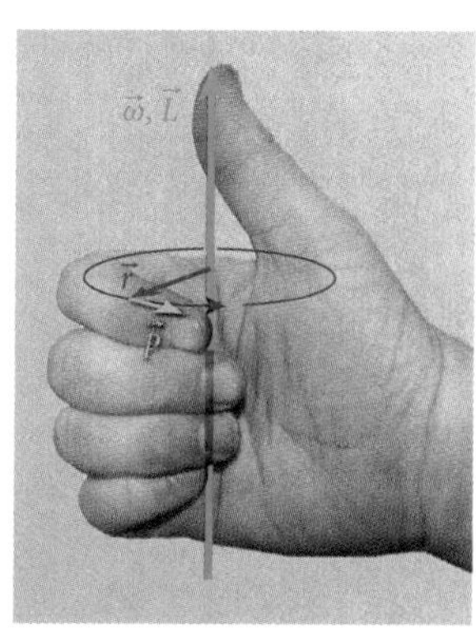


(b)

**FIGURE 10.28** (a) Right-hand rule for the direction of the angular momentum (along the thumb) as a function of the direction of rotation (along the fingers). (b) The momentum and position vectors of a point particle in circular motion.

### EXAMPLE 10.6 Golf Ball

**PROBLEM**

What is the magnitude of the angular momentum of a golf ball ($m = 4.59 \cdot 10^{-2}$ kg, $R = 2.13 \cdot 10^{-2}$m) spinning at 4250 rpm (revolutions per minute) after a good hit with a driver?

**SOLUTION**

First, we need to find the angular velocity of the golf ball, which involves use of the concepts introduced in Chapter 9.

$$\omega = 2\pi f = 2\pi(4250 \text{ min}^{-1}) = 2\pi(4250/60 \text{ s}^{-1}) = 445.0 \text{ rad/s}.$$

The moment of inertia of the golf ball is

$$I = \tfrac{2}{5}mR^2 = 0.4(4.59 \cdot 10^{-2} \text{ kg})(2.13 \cdot 10^{-2} \text{ m})^2 = 8.33 \cdot 10^{-6} \text{ kg m}^2.$$

*Continued—*

The magnitude of the angular momentum of the golf ball is then simply the product of these two numbers:

$$L=(8.33\cdot 10^{-6}\ \text{kg m}^2)(445.0\ \text{s}^{-1})=3.71\cdot 10^{-3}\ \text{kg m}^2\ \text{s}^{-1}.$$

Using equation 10.35 for the angular momentum of a rigid object, we can show that the relationship between the rate of change of the angular momentum and the torque is still valid. Taking the time derivative of equation 10.35, and assuming a rigid body whose moment of inertia is constant in time, we obtain

$$\frac{d}{dt}\vec{L}=\frac{d}{dt}(I\vec{\omega})=I\frac{d}{dt}\vec{\omega}=I\vec{\alpha}=\vec{\tau}_{\text{net}}. \tag{10.36}$$

Note the addition of the index "net" to the symbol for torque, indicating that this equation also holds if different torques are present. Earlier, equation 10.19 was said to be true only for a point particle. However, equation 10.36 clearly shows that equation 10.19 holds for any object with a fixed (constant in time) moment of inertia.

The time derivative of the angular momentum is equal to the torque, just as the time derivative of the linear momentum is equal to the force. Equation 10.31 is another formulation of Newton's Second Law for rotation and is more general than equation 10.19, because it also encompasses the case of a moment of inertia that is not constant in time.

## Conservation of Angular Momentum

If the net external torque on a system is zero, then according to equation 10.36, the time derivative of the angular momentum is also zero. However, if the time derivative of a quantity is zero, then the quantity is constant in time. Therefore, we can write the law of **conservation of angular momentum:**

$$\text{If } \vec{\tau}_{\text{net}}=0\Rightarrow\vec{L}=\text{constant}\Rightarrow\vec{L}(t)=\vec{L}(t_0)\equiv\vec{L}_0. \tag{10.37}$$

This is the third major conservation law that we have encountered—the first two applied to mechanical energy (Chapter 6) and linear momentum (Chapter 7). Like the other conservation laws, this one can be used to solve problems that would otherwise be very hard to attack.

If there are several objects present in a system with zero net external torque, the equation for conservation of angular momentum becomes

$$\sum_i \vec{L}_{i\,\text{initial}}=\sum_i \vec{L}_{i\,\text{final}}. \tag{10.38}$$

For the special case of a rigid body rotating around a fixed axis of rotation, we find (since, in this case, $\vec{L}=I\vec{\omega}$):

$$I\vec{\omega}=I_0\vec{\omega}_0\quad(\text{for }\vec{\tau}_{\text{net}}=0), \tag{10.39}$$

or, equivalently,

$$\frac{\omega}{\omega_0}=\frac{I_0}{I}\quad(\text{for }\vec{\tau}_{\text{net}}=0).$$

**FIGURE 10.29** Toy gyroscopes.

This conservation law is the basis for the functioning of gyroscopes (Figure 10.29). Gyroscopes are objects (usually disks) that spin around a symmetry axis at high angular velocities. The axis of rotation is able to turn on ball bearings, almost without friction, and the suspension system is able to rotate freely in all directions. This freedom of motion ensures that no net external torque can act on the gyroscope. Without torque, the angular momentum of the gyroscope remains constant and thus points in the same direction, no matter what the object carrying the gyroscope does. Both airplanes and satellites rely on gyroscopes for navigation. The Hubble Space Telescope, for example, is equipped with six gyroscopes, at least three of which must work to allow the telescope to orient itself in space.

Equation 10.39 is also of importance in many sports, most notably gymnastics, platform diving (Figure 10.30), and figure skating. In all three sports, the athletes rearrange their bodies and thus adjust their moments of inertia to manipulate their rotation frequencies. The

(a)

(b)

(c)

**FIGURE 10.30** Laura Wilkinson at the 2000 Olympic Games in Sydney, Australia. (a) She leaves the diving platform. (b) She holds the tucked position. (c) She stretches out before entering the water.

changing moment of inertia of a platform diver is illustrated in Figure 10.30. The platform diver starts her dive stretched out, as shown in Figure 10.30a. She then pulls her legs and arms into a tucked position, decreasing her moment of inertia, as shown in Figure 10.30b. She then completes several rotations as she falls. Before entering the water, she stretches out her arms and legs, increasing her moment of inertia and slowing her rotation, as shown in Figure 10.30c. Pulling her arms and legs into a tucked position reduces the moment of inertia of her body by some factor $I' = I/k$, where $k > 1$. Conservation of angular momentum then increases the angular velocity by the same factor $k$: $\omega' = k\omega$. Thus, the diver can control the rate of rotation. Tucking in the arms and legs can increase the rate of rotation relative to that in the stretched-out position by more than a factor of 2.

## EXAMPLE 10.7 Death of a Star

At the end of the life of a massive star more than five times as massive as the Sun, the core of the star consists almost entirely of the metal iron. Once this stage is reached, the core becomes unstable and collapses (as illustrated in Figure 10.31) in a process that lasts only about a second and is the initial phase of a supernova explosion. Among the most powerful energy-releasing events in the universe, supernova explosions are thought to be the source of most of the elements heavier than iron. The explosion throws off debris, including the heavier elements, into outer space, and may leave behind a neutron star, which consists of stellar material that is compressed to a density millions of times higher than the highest densities found on Earth.

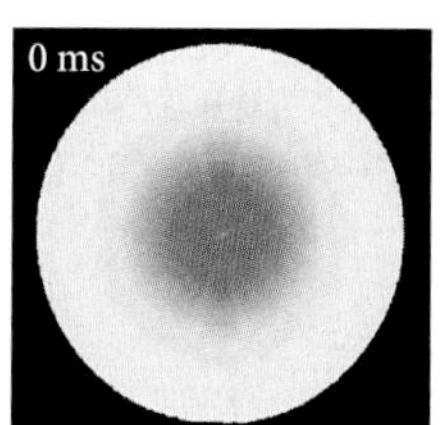


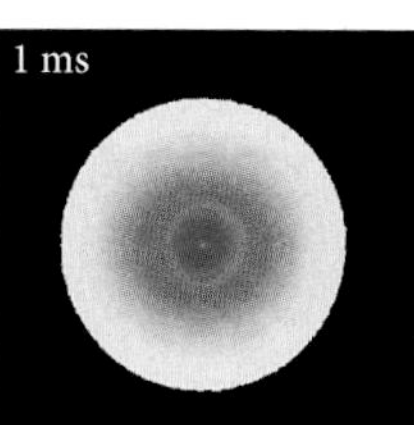


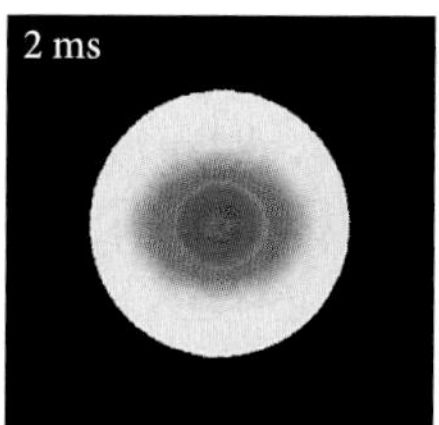


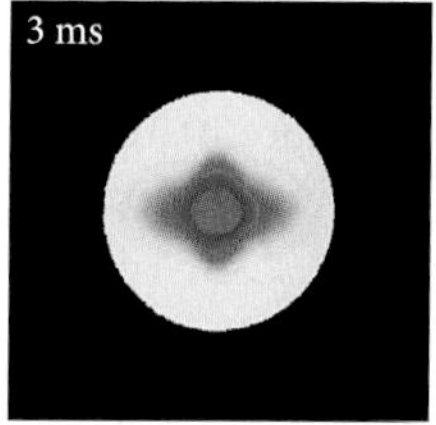


**FIGURE 10.31** Computer simulation of the initial stages of the collapse of the core of a massive star. The different colors represent the varying density of the star's core, increasing from yellow through green and blue to red.

### PROBLEM

If the iron core initially spins 9.00 revolutions per day and if its radius decreases during the collapse by a factor of 700, what is the angular velocity of the core at the end of the collapse? (The assumption that the iron core has a constant density is not really justified. Computer simulations show that it falls off exponentially in the radial direction. However, the same simulations show that the moment of inertia of the iron core is still approximately proportional to the square of its radius during the collapse process.)

*Continued—*

### SOLUTION

Because the collapse of the iron core occurs under the influence of its own gravitational pull, no net external torque acts on the core. Thus, according to equation 10.31, angular momentum is conserved. From equation 10.39 we then obtain

$$\frac{\omega}{\omega_0} = \frac{I_0}{I} = \frac{R_0^2}{R^2} = 700^2 = 4.90 \cdot 10^5.$$

With $\omega_0 = 2\pi f_0 = 2\pi[(9 \text{ rev})/(24 \cdot 3600 \text{ s})] = 6.55 \cdot 10^{-4}$ rad/s, we obtain the magnitude of the final angular velocity:

$$\omega = 4.90 \cdot 10^5 \, \omega_0 = 4.90 \cdot 10^5 (6.55 \cdot 10^{-4} \text{ rad/s}) = 321 \text{ rad/s}.$$

Thus, the neutron star that results from this collapse rotates with a rotation frequency of 51.1 rev/s.

### DISCUSSION

Astronomers can observe the rotation of neutron stars, which are called *pulsars*. It is estimated that the fastest a pulsar can rotate when formed from a single star supernova explosion is about 60 rev/s. One of the fastest known rotation frequencies known for a pulsar is

$$f = 716 \text{ rev/s},$$

which corresponds to an angular speed of

$$\omega = 2\pi f = 2\pi\left(716 \text{ s}^{-1}\right) = 4500 \text{ rad/s}.$$

The rotation frequencies of these pulsars are increased after their formation from a stellar collapse by taking matter from a companion star orbiting nearby.

The next example ends the section with a current cutting-edge engineering application, which ties together the concepts of moment of inertia, rotational kinetic energy, torque, and angular momentum.

## EXAMPLE 10.8 Flybrid

The process of braking to slow a car down decreases the car's kinetic energy and dissipates it through the action of the friction force between the brake pads and the drums. Gas-electric hybrid vehicles convert some or most of this kinetic energy into reusable electric energy stored in a large battery. However, there is a way to accomplish this energy storage without the use of a large battery by storing the energy temporarily in a flywheel (Figure 10.32). Interestingly, all Formula 1 race cars will be equipped with such an energy storage system, the *flybrid*, by 2013.

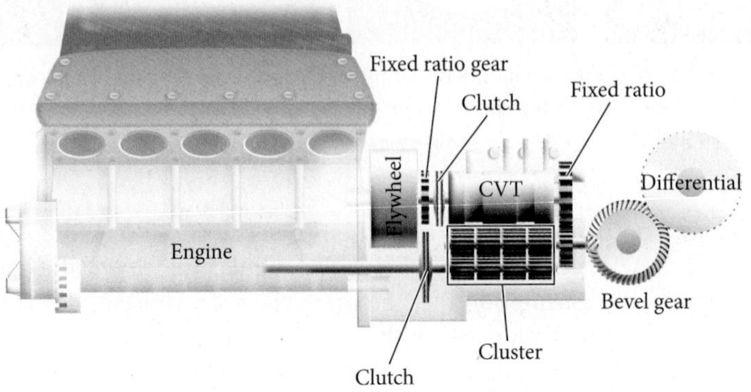


**FIGURE 10.32** Diagram of the integration of a flywheel into the drive train of a car.

**PROBLEM**

A flywheel made of carbon steel has a mass of 5.00 kg, an inner radius of 8.00 cm, and an outer radius of 14.2 cm. If it is supposed to store 400.0 kJ of rotational energy, how fast (in rpm) does it have to rotate? If the rotational energy can be stored or withdrawn in 6.67 s, how much average power and torque can this flywheel deliver during that time?

**SOLUTION**

The moment of inertia of the flywheel is given by equation 10.9: $I = \frac{1}{2}M(R_1^2 + R_2^2)$. The rotational kinetic energy (equation 10.3) is $K = \frac{1}{2}I\omega^2$. We solve this for the angular speed:

$$\omega = \sqrt{\frac{2K}{I}} = \sqrt{\frac{4K}{M\left(R_1^2 + R_2^2\right)}}.$$

So, for the rotation frequency, we find

$$f = \frac{\omega}{2\pi} = \sqrt{\frac{K}{\pi^2 M\left(R_1^2 + R_2^2\right)}}$$

$$= \sqrt{\frac{400.0 \text{ kJ}}{\pi^2 (5.00 \text{ kg})\left[(0.0800 \text{ m})^2 + (0.142 \text{ m})^2\right]}}$$

$$= 552 \text{ s}^{-1} = 33{,}100 \text{ rpm}.$$

Since the average power is given by the change in kinetic energy divided by the time (see Chapter 5), we have

$$P = \frac{\Delta K}{\Delta t} = \frac{400.0 \text{ kJ}}{6.67 \text{ s}} = 60.0 \text{ kW}.$$

We find the average torque from equation 10.36 and the fact that the average angular acceleration is the change in angular speed $\Delta\omega$, divided by the time interval, $\Delta t$:

$$\tau = I\alpha = I\frac{\Delta\omega}{\Delta t} = \frac{1}{2}M\left(R_1^2 + R_2^2\right)\frac{1}{\Delta t}\sqrt{\frac{4K}{M\left(R_1^2 + R_2^2\right)}} = \frac{1}{\Delta t}\sqrt{M\left(R_1^2 + R_2^2\right)K}$$

$$= \frac{1}{6.67 \text{ s}}\sqrt{(400.0 \text{ kJ})(5.00 \text{ kg})\left[(0.0800 \text{ m})^2 + (0.142 \text{ m})^2\right]}$$

$$= 34.6 \text{ N m}.$$

### 10.7 In-Class Exercise

The flybrid rotates fastest when the Formula 1 car is moving slowest, in the process of making a tight turn. Knowing that it takes torque to change an angular momentum vector, how would you orient the axis of rotation of the flywheel in order to have the least impact on steering the car through the curve?

a) The flywheel should be aligned with the main axis of the race car.

b) The flywheel should be vertical.

c) The flywheel should be aligned with the wheel axles.

d) It makes no difference; all three orientations are equally problematic.

e) Orientations (a) and (c) are both equally good and better than (b).

## 10.8 Precession

Spinning tops were popular toys when your parents or grandparents were kids. When put in rapid rotational motion, they stand upright without falling down. What's more, if tilted at an angle relative to the vertical, they still do not fall down. Instead, the rotation axis moves on the surface of a cone as a function of time (Figure 10.33). This motion is called **precession.** What causes it?

First, we note that a spinning top has an angular momentum vector, $\vec{L}$, which is aligned with its symmetry axis, pointing either up or down, depending on whether it is spinning clockwise or counterclockwise (Figure 10.34). Because the top is tilted, its center of mass (marked with a black dot in Figure 10.34) is not located above the contact point with the support surface. The gravitational force acting on the center of mass then results in a torque, $\vec{\tau}$, about the contact point, as indicated in the figure; in this case, the torque vector points straight out of the page. The position vector, $\vec{r}$, of the center of mass, which helps determine the torque, is exactly aligned with the angular momentum vector. The angle of the symmetry axis of the top with respect to the vertical is labeled $\phi$ in the figure. The angle between the gravitational force vector and the position vector is then $\pi - \phi$ (see Figure 10.34). Because $\sin(\pi - \phi) = \sin\phi$, we can write the magnitude of the torque as a function of the angle $\phi$:

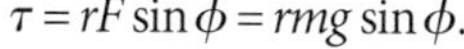

$$\tau = rF\sin\phi = rmg\sin\phi.$$

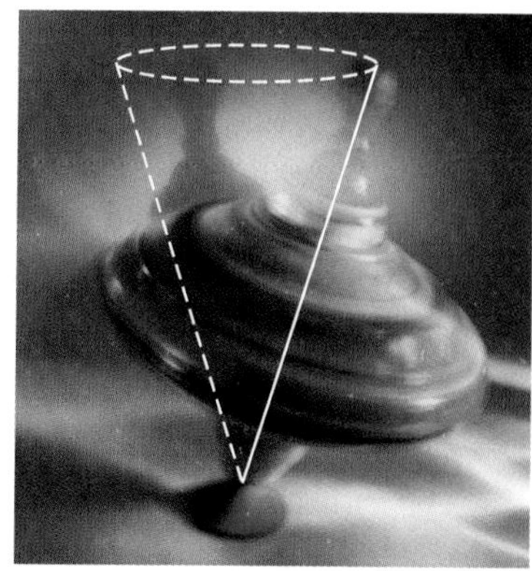

**FIGURE 10.33** A spinning top may tilt from the vertical but does not fall down.

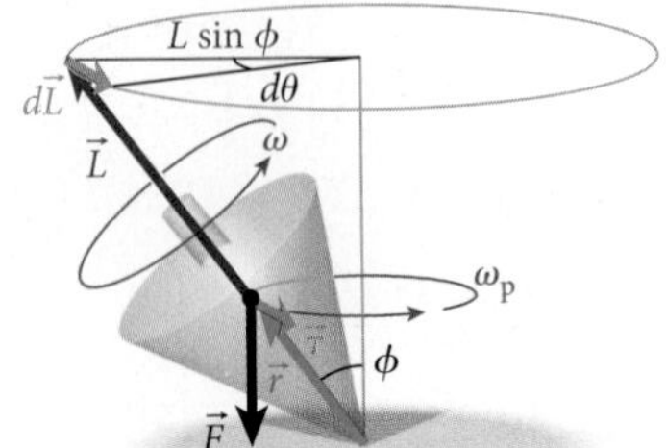


**FIGURE 10.34** Precession of a spinning top.

Since $d\vec{L}/dt = \vec{\tau}$, the change in the angular momentum vector, $d\vec{L}$, points in the same direction as the torque and is thus perpendicular to the angular momentum vector. This effect forces the angular momentum vector to sweep along the surface of a cone of angle $\phi$ as a function of time, with the tip of the angular momentum vector following a circle in the horizontal plane, shown in gray in Figure 10.34.

We can even calculate the magnitude of the angular velocity, $\omega_p$, for this precessional motion. Figure 10.34 indicates that the radius of the circle that the tip of the angular momentum vector sweeps out as a function of time is given by $L\sin\phi$. The magnitude of the differential change in angular momentum, $dL$, is the arc length of this circle, and it can be calculated as the product of the circle's radius and the differential angle swept out by the radius, $d\theta$:

$$dL = (L\sin\phi)d\theta.$$

Consequently, for the time derivative of the angular momentum, $dL/dt$, we find:

$$\frac{dL}{dt} = (L\sin\phi)\frac{d\theta}{dt}.$$

The time derivative of the deflection angle, $\theta$, is the angular velocity of precession, $\omega_p$. Since $dL/dt = \tau$, we use the preceding equation and the expression for the torque, $\tau = rmg\sin\phi$, to obtain

$$rmg\sin\phi = \tau = \frac{dL}{dt} = (L\sin\phi)\frac{d\theta}{dt} = (L\sin\phi)\omega_p \Rightarrow$$

$$\omega_p = \frac{rmg\sin\phi}{L\sin\phi}.$$

### 10.8 In-Class Exercise

Estimate the precessional angular speed of the wheel in Figure 10.35.

a) 0.01 rad/s
b) 0.6 rad/s
c) 5 rad/s
d) 10 rad/s

### 10.4 Self-Test Opportunity

The wheel shown in Figure 10.35 has a mass of 2.5 kg, almost all of it concentrated on its rim. It has a radius of 22 cm, and the distance between the suspension point and the center of mass is 5 cm. Estimate the angular speed with which it is spinning.

We see that the term $\sin\phi$ cancels out of the last expression, yielding $\omega_p = rmg/L$. The angular frequency of precession is the same for all values of $\phi$, the tilt angle of the rotation axis! This result may seem a bit surprising, but experiments verify that it is indeed the case. For the final step, we use the fact that the angular momentum for a rigid object, $L$, is the product of the moment of inertia, $I$, and the angular velocity, $\omega$. Thus, substituting $I\omega$ for $L$ in the expression for the precessional angular speed, $\omega_p$, gives us our final result:

$$\omega_p = \frac{rmg}{I\omega}. \qquad (10.40)$$

This formula reflects the interesting property that the precessional angular speed is inversely proportional to the angular speed of the spinning top. As the top slows down due to friction, its angular speed is gradually reduced, and therefore the precessional angular speed increases gradually. The faster and faster precession eventually causes the top to wobble and fall down.

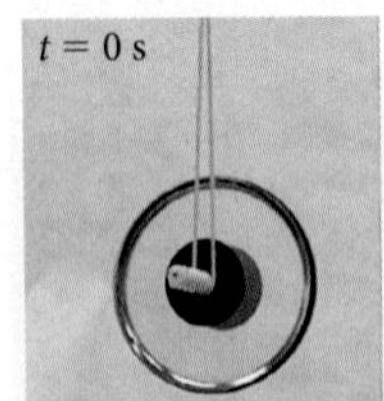

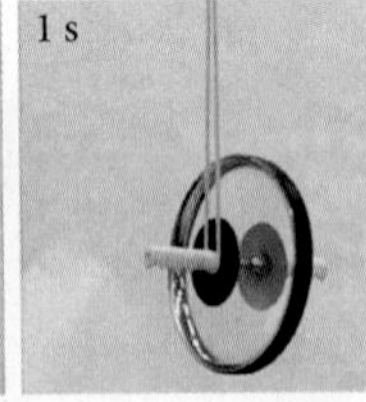

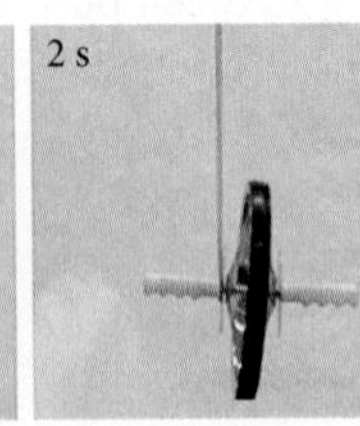

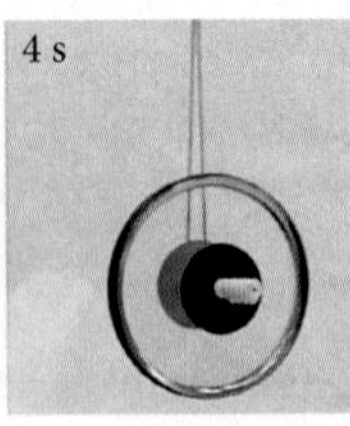

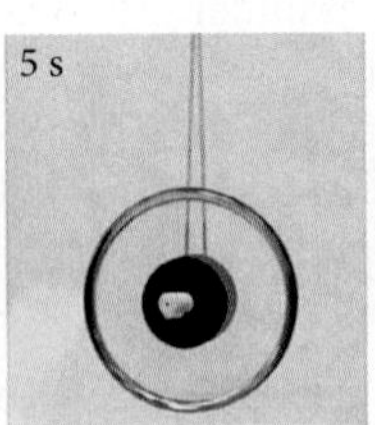

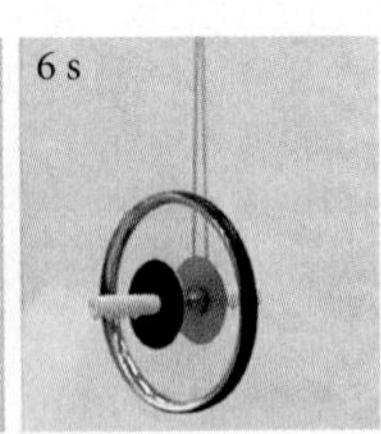


**FIGURE 10.35** Precession of a rapidly spinning wheel suspended from a rope.

Precession is impressively demonstrated in the photo sequence of Figure 10.35. Here a rapidly spinning wheel is suspended off-center from a string attached to the ceiling. As you can see, the wheel does not fall down, as one would expect a nonspinning wheel to do in the same situation, but slowly precesses around the suspension point.

## 10.9 Quantized Angular Momentum

To finish our discussion of angular momentum and rotation, let's consider the smallest quantity of angular momentum that an object can have. From the definition of the angular momentum of a point particle (equation 10.29), $\vec{L}=\vec{r}\times\vec{p}$ or $L = rp \sin\theta$, it would appear that there is no smallest amount of angular momentum, because either the distance to the rotation axis, $r$, or the momentum, $p$, can be reduced by a factor between 0 and 1, and the corresponding angular momentum will be reduced by the same factor.

However, for atoms and subatomic particles, the notion of a continuously variable angular momentum doesn't apply. Instead, a quantum of angular momentum is observed. This quantum of angular momentum is called **Planck's constant,** $h = 6.626\cdot 10^{-34}$ J s. Very often Planck's constant appears in equations divided by the factor $2\pi$, and physicists have given this ratio the symbol $\hbar$: $\hbar \equiv h/2\pi = 1.055\cdot 10^{-34}$ J s. Chapter 36 will provide a full discussion of the experimental observations that led to the introduction of this fundamental constant. Here we simply note an amazing fact: All elementary particles have an intrinsic angular momentum, often called *spin,* which is either an integer multiple ($0, 1\hbar, 2\hbar, \ldots$) or a half-integer multiple ($\frac{1}{2}\hbar, \frac{3}{2}\hbar, \ldots$) of Planck's quantum of angular momentum. Astonishingly, the integer- or half-integer spin values of particles make all the difference in the ways they interact with one another. Particles with integer-valued spins include photons, which are the elementary particles of light. Particles with half-integer values of spin include electrons, protons, and neutrons, the particles that constitute the building blocks of matter. We will return to the fundamental importance of angular momentum in atoms and subatomic particles in Chapters 37 through 40.

## WHAT WE HAVE LEARNED | EXAM STUDY GUIDE

- An object's kinetic energy of rotation is given by $K=\frac{1}{2}I\omega^2$. This relationship holds for point particles as well as for solid objects.
- The moment of inertia for rotation of an object about an axis through the center of mass is defined as $I=\int_V r_\perp^2 \rho(\vec{r})dV$, where $r_\perp$ is the perpendicular distance of the volume element $dV$ to the axis of rotation and $\rho(\vec{r})$ is the mass density.
- If the mass density is constant, the moment of inertia is $I=\frac{M}{V}\int_V r_\perp^2 dV$, where $M$ is the total mass of the rotating object and $V$ is its volume.
- The moment of inertia of all round objects is $I = cMR^2$, with $c \in [0,1]$.
- The parallel-axis theorem states that the moment of inertia, $I_\parallel$, for rotation about an axis parallel to one through the center of mass is given by $I_\parallel = I_{cm} + Md^2$, where $d$ is the distance between the two axes and $I_{cm}$ is the moment of inertia for rotation about the axis through the center of mass.
- For an object that is rolling without slipping, the center of mass coordinate, $r$, and the rotation angle, $\theta$, are related by $r = R\theta$, where $R$ is the radius of the object.
- The kinetic energy of a rolling object is the sum of its translational and rotational kinetic energies: $K = K_{trans} + K_{rot} = \frac{1}{2}mv_{cm}^2 + \frac{1}{2}I_{cm}\omega^2 = \frac{1}{2}(1+c)mv_{cm}^2$, with $c \in [0,1]$ and with $c$ depending on the shape of the object.
- Torque is defined as the vector product of the position vector and the force vector: $\vec{\tau}=\vec{r}\times\vec{F}$.
- The angular momentum of a point particle is defined as $\vec{L}=\vec{r}\times\vec{p}$.
- The rate of change of the angular momentum is equal to the torque: $\frac{d}{dt}\vec{L}=\vec{\tau}$. This is the rotational equivalent of Newton's Second Law.
- For rigid objects, the angular momentum is $\vec{L}=I\vec{\omega}$, and the torque is $\vec{\tau}=I\vec{\alpha}$.
- In the case of vanishing net external torque, angular momentum is conserved: $I\vec{\omega}=I_0\vec{\omega}_0$ (for $\vec{\tau}_{net}=0$).
- The equivalent quantities for linear and rotational motion are summarized in the table.

| Quantity | Linear | Circular | Relationship |
|---|---|---|---|
| Displacement | $\vec{s}$ | $\vec{\theta}$ | $\vec{s}=r\vec{\theta}$ |
| Velocity | $\vec{v}$ | $\vec{\omega}$ | $\vec{\omega}=\vec{r}\times\vec{v}/r^2$ |
| Acceleration | $\vec{a}$ | $\vec{\alpha}$ | $\vec{a}=r\alpha\,\hat{t}-r\omega^2\hat{r}$<br>$a_t=r\alpha$<br>$a_c=\omega^2 r$ |
| Momentum | $\vec{p}$ | $\vec{L}$ | $\vec{L}=\vec{r}\times\vec{p}$ |
| Mass/moment of inertia | $m$ | $I$ | |
| Kinetic energy | $\frac{1}{2}mv^2$ | $\frac{1}{2}I\omega^2$ | |
| Force/torque | $\vec{F}$ | $\vec{\tau}$ | $\vec{\tau}=\vec{r}\times\vec{F}$ |

## KEY TERMS



## NEW SYMBOLS AND EQUATIONS

$I=\sum_{i=1}^{n} m_i r_i^2$, moment of inertia of a system of particles

$I=\int_V r_\perp^2 \rho(\vec{r})dV$, moment of inertia of an extended object

$\vec{\tau}=\vec{r}\times\vec{F}$, torque

$\tau = rF\sin\theta$, magnitude of torque

$\vec{L}=\vec{r}\times\vec{p}$, angular momentum of a particle

$\vec{L}=I\vec{\omega}$, angular momentum of an extended object

## ANSWERS TO SELF-TEST OPPORTUNITIES

**10.1** $I_{\parallel}=\frac{1}{12}mL^2+m\left(\frac{L}{2}\right)^2=mL^2\left(\frac{1}{12}+\frac{1}{4}\right)=\frac{1}{3}mL^2.$

**10.2** The full can of soda is not a solid object and thus does not rotate as a solid cylinder. Instead, most of the liquid inside the can does not participate in the rotation even when the can reaches the bottom of the incline. The mass of the can itself is negligible compared to the mass of the liquid inside it. Therefore, a can of soda rolling down an incline approximates a mass sliding down an incline without friction. The constant $c$ used in equation 10.15 is then close to zero, and so the can wins the race.

**10.3** Newton's Third Law states that internal forces occur in equal and opposite pairs that act along the line of the separation of each pair of particles. Thus, the torque due to each pair of forces is zero. Summing up the torques from all the internal forces leads to zero net internal torque.

**10.4** Since the mass is concentrated on the rim of the wheel, the wheel's moment of inertia is $I = mR^2$. With the aid of equation 10.40, we then obtain

$$\omega_p=\frac{rmg}{mR^2\omega}=\frac{rg}{R^2\omega}=\frac{(0.05\text{ m})(9.81\text{ m/s}^2)}{(0.22\text{ m})^2(0.6\text{ rad/s})}=17\text{ rad/s}.$$

## PROBLEM-SOLVING PRACTICE

### Problem-Solving Guidelines: Rotational Motion

**1.** Newton's Second Law and the work–kinetic energy theorem are powerful and complementary tools for solving a wide variety of problems in rotational mechanics. As a general guideline, you should try an approach based on Newton's Second Law and free-body diagrams when the problem involves calculating an angular acceleration. An approach based on the work–kinetic energy theorem is more useful when you need to calculate an angular speed.

**2.** Many concepts of translational motion are equally valid for rotational motion. For example, conservation of linear momentum applies when no external forces are present; conservation of angular momentum applies when no external torques are present. Remember the correspondences between translational and rotational quantities.

**3.** It is crucial to remember that in situations involving rotational motion the shape of an object is important. Be sure to use the correct formula for moment of inertia, which depends on the location of the axis of rotation as well as on the geometry of the object. Torque also depends on the location of the axis of symmetry; be sure to be consistent in calculating clockwise and counterclockwise torques.

**4.** Many relationships for rotational motion depend on the geometry of the situation; for example, the relationship of the linear velocity of a hanging weight to the angular velocity of the rope moving over a pulley. Sometimes the geometry of a situation changes in a problem, for example, if there is a different rotational inertia between starting and ending points of a rotation. Be sure you understand what quantities change during the course of any rotational motion.

**5.** Many physical situations involve rotating objects that roll, with or without slipping. If rolling without slipping is occurring, you can relate linear and angular displacements, velocities, and accelerations to each other at points on the perimeter of the rolling object.

**6.** The law of conservation of angular momentum is just as important for problems involving circular or rotational motion as the law of conservation of linear momentum is for problems involving straight-line motion. Thinking of a problem situation in terms of conserved angular momentum often provides a straightforward path to a solution, which otherwise would be difficult to obtain. But keep in mind that angular momentum is only conserved if the net external torque is zero.

## SOLVED PROBLEM 10.4 Falling Horizontal Rod

A thin rod of length $L = 2.50$ m and mass $m = 3.50$ kg is suspended horizontally by a pair of vertical strings attached to the ends (Figure 10.36). The string supporting end $B$ is then cut.

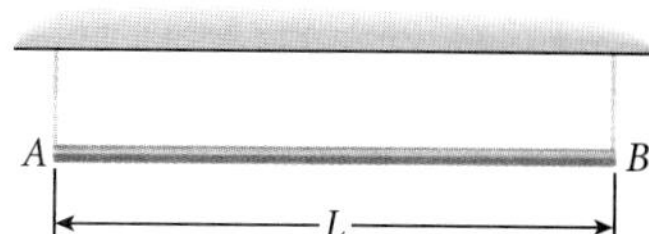


**FIGURE 10.36** A thin rod supported in a horizontal position by a vertical string at each end.

### PROBLEM

What is the linear acceleration of end $B$ of the rod just after the string is cut?

### SOLUTION

#### THINK

Before the string is cut, the rod is at rest. When the string supporting end $B$ is cut, a net torque acts on the rod, with a pivot point at end $A$. This torque is due to the force of gravity acting on the rod. We can consider the mass of the rod to be concentrated at its center of mass, which is located at $L/2$. The initial torque is then equal to the weight of the rod times the moment arm, which is $L/2$. The resulting initial angular acceleration can be related to the linear acceleration of end $B$ of the rod.

#### SKETCH

Figure 10.37 is a sketch of the rod after the string is cut.

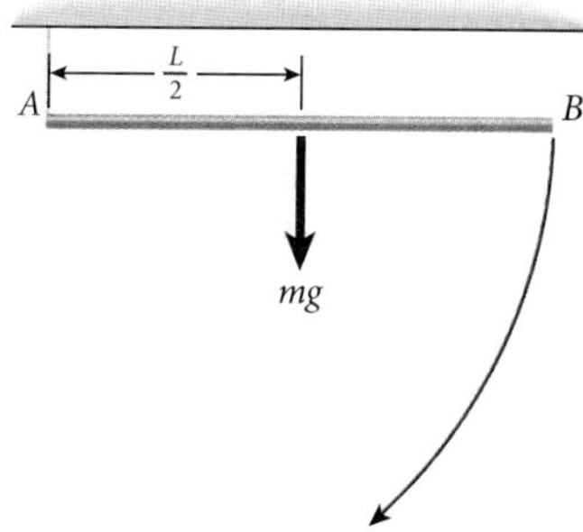


**FIGURE 10.37** The thin rod just after the string supporting end $B$ is cut.

#### RESEARCH

When the string supporting end $B$ is cut, the torque, $\tau$, on the rod is due to the force of gravity, $F_g$, acting on the rod times the moment arm, $r_\perp = L/2$:

$$\tau = r_\perp F_g = \left(\frac{L}{2}\right)(mg) = \frac{mgL}{2}. \quad \text{(i)}$$

The angular acceleration, $\alpha$, is given by

$$\tau = I\alpha, \quad \text{(ii)}$$

where the moment of inertia, $I$, of the thin rod rotating around end $A$ is given by

$$I = \tfrac{1}{3}mL^2. \quad \text{(iii)}$$

The linear acceleration, $a$, of end $B$ can be related to the angular acceleration through

$$a = L\alpha, \quad \text{(iv)}$$

because end $B$ of the rod is undergoing circular motion as the rod pivots around end $A$.

#### SIMPLIFY

We can combine equations (i) and (ii) to get

$$\tau = I\alpha = \frac{mgL}{2}. \quad \text{(v)}$$

Substituting for $I$ and $a$ from equations (iii) and (iv) into equation (v) gives us

$$I\alpha = \left(\frac{1}{3}mL^2\right)\left(\frac{a}{L}\right) = \frac{mgL}{2}.$$

Dividing out common factors, we obtain

$$\frac{a}{3} = \frac{g}{2}$$

or

$$a = 1.5g.$$

#### CALCULATE

Inserting the numerical value for the acceleration of gravity gives us

$$a = 1.5\left(9.81 \text{ m/s}^2\right) = 14.715 \text{ m/s}^2.$$

*Continued—*

**ROUND**

Expressing our result to three significant figures gives

$$a = 14.7 \text{ m/s}^2.$$

**DOUBLE-CHECK**

Perhaps this answer is somewhat surprising, because you may have assumed that the acceleration cannot exceed the free-fall acceleration, $g$. If both strings were cut at the same time, the acceleration of the entire rod would be $a = g$. Our result of an initial acceleration of end $B$ of $a = 1.5g$ seems reasonable because the full force of gravity is acting on the rod and end $A$ of the rod remains fixed. Therefore, the acceleration of the moving ends is not just that due to free fall—there is an additional acceleration due to the rotation of the rod.

## MULTIPLE-CHOICE QUESTIONS

**10.1** A circular object begins from rest and rolls without slipping down an incline, through a vertical distance of 4.0 m. When the object reaches the bottom, its translational velocity is 7.0 m/s. What is the constant $c$ relating the moment of inertia to the mass and radius (see equation 10.11) of this object?

a) 0.80
b) 0.60
c) 0.40
d) 0.20

**10.2** Two solid steel balls, one small and one large, are on an inclined plane. The large ball has a diameter twice as large as that of the small ball. Starting from rest, the two balls roll without slipping down the incline until their centers of mass are 1 m below their starting positions. What is the speed of the large ball ($v_L$) relative to that of the small ball ($v_S$) after rolling 1 m?

a) $v_L = 4v_S$
b) $v_L = 2v_S$
c) $v_L = v_S$
d) $v_L = 0.5v_S$
e) $v_L = 0.25v_S$

**10.3** A generator's flywheel, which is a homogeneous cylinder of radius $R$ and mass $M$, rotates about its longitudinal axis. The linear velocity of a point on the rim (side) of the flywheel is $v$. What is the kinetic energy of the flywheel?

a) $K = \frac{1}{2}Mv^2$
b) $K = \frac{1}{4}Mv^2$
c) $K = \frac{1}{2}Mv^2/R$
d) $K = \frac{1}{2}Mv^2R$
e) not given enough information to answer

**10.4** Four hollow spheres, each with a mass of 1 kg and a radius $R = 10$ cm, are connected with massless rods to form a square with sides of length $L = 50$ cm. In case 1, the masses rotate about an axis that bisects two sides of the square. In case 2, the masses rotate about an axis that passes through the diagonal of the square, as shown in the figure. Compute the ratio of the moments of inertia, $I_1/I_2$, for the two cases.

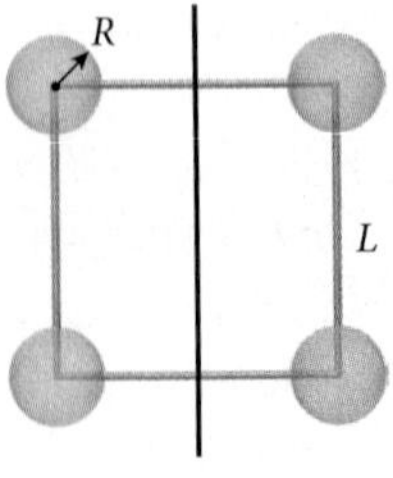

Case 1

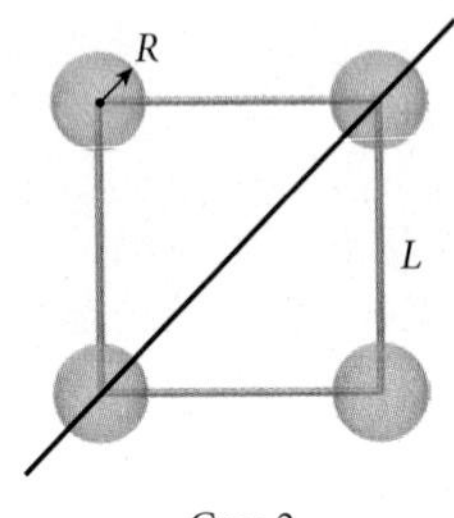

Case 2

a) $I_1/I_2 = 8$
b) $I_1/I_2 = 4$
c) $I_1/I_2 = 2$
d) $I_1/I_2 = 1$
e) $I_1/I_2 = 0.5$

**10.5** If the hollow spheres of Question 10.4 were replaced by solid spheres of the same mass and radius, the ratio of the moments of inertia for the two cases would

a) increase.
b) decrease.
c) stay the same.
d) be zero.

**10.6** An extended object consists of two point masses, $m_1$ and $m_2$, connected via a rigid massless rod of length $L$, as shown in the figure. The object is rotating at a constant angular velocity about an axis perpendicular to the page through the midpoint of the rod. Two time-varying tangential forces, $F_1$ and $F_2$, are applied to $m_1$ and $m_2$, respectively. After the forces have been applied, what will happen to the angular velocity of the object?

a) It will increase.
b) It will decrease.
c) It will remain unchanged.
d) Not enough information to make a determination

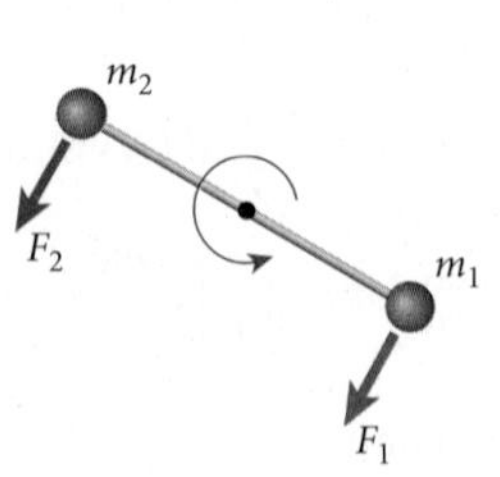


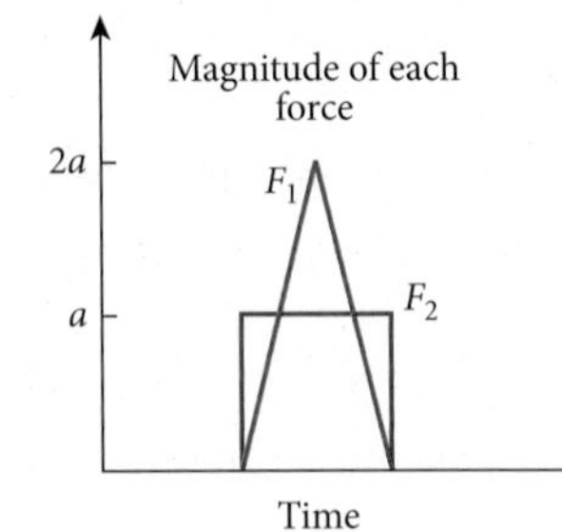


**10.7** Consider a cylinder and a hollow cylinder, rotating about an axis going through their centers of mass. If both objects have the same mass and the same radius, which object will have the larger moment of inertia?

a) The moment of inertia will be the same for both objects.

b) The solid cylinder will have the larger moment of inertia because its mass is uniformly distributed.

c) The hollow cylinder will have the larger moment of inertia because its mass is located away from the axis of rotation.

**10.8** A basketball of mass 610 g and circumference 76 cm is rolling without slipping across a gymnasium floor. Treating the ball as a hollow sphere, what fraction of its total kinetic energy is associated with its rotational motion?

a) 0.14
b) 0.19
c) 0.29
d) 0.40
e) 0.67

**10.9** A solid sphere rolls without slipping down an incline, starting from rest. At the same time, a box starts from rest at the same altitude and slides down the same incline, with negligible friction. Which object arrives at the bottom first?

a) The solid sphere arrives first.
b) The box arrives first.
c) Both arrive at the same time.
d) It is impossible to determine.

**10.10** A cylinder is rolling without slipping down a plane, which is inclined by an angle $\theta$ relative to the horizontal. What is the work done by the friction force while the cylinder travels a distance $s$ along the plane ($\mu_s$ is the coefficient of static friction between the plane and the cylinder)?

a) $+\mu_s mgs \sin\theta$
b) $-\mu_s mgs \sin\theta$
c) $+mgs \sin\theta$
d) $-mgs \sin\theta$
e) No work done.
f) Zero.

**10.11** A ball attached to the end of a string is swung in a vertical circle. The angular momentum of the ball at the top of the circular path is

a) greater than the angular momentum at the bottom of the circular path.
b) less than the angular momentum at the bottom of the circular path.
c) the same as the angular momentum at the bottom of the circular path.

**10.12** You are unwinding a large spool of cable. As you pull on the cable with a constant tension, what happens to the angular acceleration and angular velocity of the spool, assuming that the radius at which you are extracting the cable remains constant and there is no friction force?

a) Both increase as the spool unwinds.
b) Both decrease as the spool unwinds.
c) Angular acceleration increases, and angular velocity decreases.
d) Angular acceleration decreases, and angular velocity increases.
e) It is impossible to tell.

**10.13** A disk of clay is rotating with angular velocity $\omega$. A blob of clay is stuck to the outer rim of the disk, and it has a mass $\frac{1}{10}$ of that of the disk. If the blob detaches and flies off tangent to the outer rim of the disk, what is the angular velocity of the disk after the blob separates?

a) $\frac{5}{6}\omega$
b) $\frac{10}{11}\omega$
c) $\omega$
d) $\frac{11}{10}\omega$
e) $\frac{6}{5}\omega$

**10.14** An ice skater spins with her arms extended and then pulls her arms in and spins faster. Which statement is correct?

a) Her kinetic energy of rotation does not change because, by conservation of angular momentum, the fraction by which her angular velocity increases is the same as the fraction by which her rotational inertia decreases.
b) Her kinetic energy of rotation increases because of the work she does to pull her arms in.
c) Her kinetic energy of rotation decreases because of the decrease in her rotational inertia; she loses energy because she gradually gets tired.

**10.15** An ice skater rotating on frictionless ice brings her hands into her body so that she rotates faster. Which, if any, of the conservation laws hold?

a) conservation of mechanical energy and conservation of angular momentum
b) conservation of mechanical energy only
c) conservation of angular momentum only
d) neither conservation of mechanical energy nor conservation of angular momentum

**10.16** If the iron core of a collapsing star initially spins with a rotational frequency of $f_0 = 3.2\ \mathrm{s}^{-1}$, and if the core's radius decreases during the collapse by a factor of 22.7, what is the rotational frequency of the iron core at the end of the collapse?

a) 10.4 kHz
b) 1.66 kHz
c) 65.3 kHz
d) 0.46 kHz
e) 5.2 kHz

## QUESTIONS

**10.17** A uniform solid sphere of radius $R$, mass $M$, and moment of inertia $I = \frac{2}{5}MR^2$ is rolling without slipping along a horizontal surface. Its total kinetic energy is the sum of the energies associated with translation of the center of mass and rotation about the center of mass. Find the *fraction* of the sphere's total kinetic energy that is attributable to rotation.

**10.18** A thin ring, a solid sphere, a hollow spherical shell, and a disk of uniform thickness are placed side by side on a wide ramp of length $\ell$ and inclined at angle $\theta$ to the horizontal. At time $t = 0$, all four objects are released and roll without slipping on parallel paths down the ramp to the bottom. Friction and air resistance are negligible. Determine the order of finish of the race.

**10.19** In another race, a solid sphere and a thin ring roll without slipping from rest down a ramp that makes angle $\theta$ with the horizontal. Find the ratio of their accelerations, $a_{\text{ring}}/a_{\text{sphere}}$.

**10.20** A uniform solid sphere of mass $m$ and radius $r$ is placed on a ramp inclined at an angle $\theta$ to the horizontal. The coefficient of static friction between sphere and ramp is $\mu_s$. Find the maximum value of $\theta$ for which the sphere will roll without slipping, starting from rest, in terms of the other quantities.

**10.21** A round body of mass $M$, radius $R$, and moment of inertia $I$ about its center of mass is struck a sharp horizontal blow along a line at height $h$ above its center (with $0 \le h \le R$, of course). The body rolls away without slipping immediately after being struck. Calculate the ratio $I/(MR^2)$ for this body.

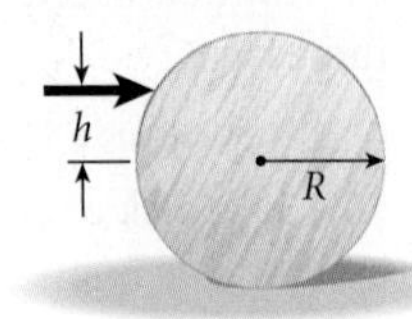


**10.22** A projectile of mass $m$ is launched from the origin at speed $v_0$ at angle $\theta_0$ above the horizontal. Air resistance is negligible.

a) Calculate the angular momentum of the projectile about the origin.

b) Calculate the rate of change of this angular momentum.

c) Calculate the torque acting on the projectile, about the origin, during its flight.

**10.23** A solid sphere of radius $R$ and mass $M$ is placed at a height $h_0$ on an inclined plane of slope $\theta$. When released, it rolls without slipping to the bottom of the incline. Next, a cylinder of same mass and radius is released on the same incline. From what height $h$ should it be released in order to have the same speed as the sphere at the bottom?

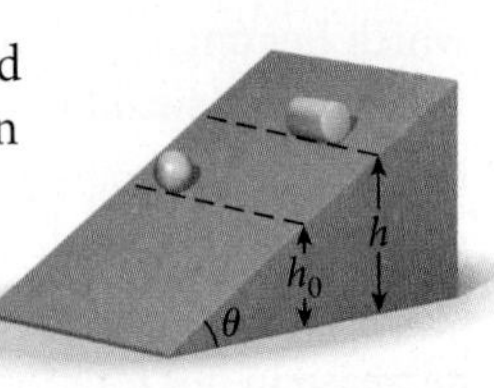


**10.24** It is harder to move a door if you lean against it (along the plane of the door) toward the hinge than if you lean against the door perpendicular to its plane. Why is this so?

**10.25** A figure skater draws her arms in during a final spin. Since angular momentum is conserved, her angular velocity will increase. Is her rotational kinetic energy conserved during this process? If not, where does the extra energy come from or go to?

**10.26** Does a particle traveling in a straight line have an angular momentum? Explain.

**10.27** A cylinder with mass $M$ and radius $R$ is rolling without slipping through a distance $s$ along an inclined plane that makes an angle $\theta$ with respect to the horizontal. Calculate the work done by (a) gravity, (b) the normal force, and (c) the frictional force.

**10.28** Using the conservation of mechanical energy, calculate the final speed and the acceleration of a cylindrical object of mass $M$ and radius $R$ after it rolls a distance $s$ without slipping along an inclined plane of angle $\theta$ with respect to the horizontal.

**10.29** A *couple* is a set of two forces of equal magnitude and opposite directions, whose lines of action are parallel but not identical. Prove that the net torque of a couple of forces is independent of the pivot point about which the torque is calculated and of the points along their lines of action where the two forces are applied.

**10.30** Why does a figure skater pull in her arms while increasing her angular velocity in a tight spin?

**10.31** To turn a motorcycle to the right, you do *not* turn the handlebars to the right, but instead slightly to the *left*. Explain, as precisely as you can, how this counter-steering turns the motorcycle in the desired direction. (*Hint:* The wheels of a motorcycle in motion have a great deal of angular momentum.)

**10.32** The Moon's tide-producing effect on the Earth is gradually slowing the rotation of the Earth, owing to tidal friction. Studies of corals from the Devonian Period indicate that the year was 400 days long in that period. What, if anything, does this indicate about the angular momentum of the Moon in the Devonian Period relative to its value at present?

**10.33** A light rope passes over a light, frictionless pulley. One end is fastened to a bunch of bananas of mass $M$, and a monkey of the same mass clings to the other end. The monkey climbs the rope in an attempt to reach the bananas. The radius of the pulley is $R$.

a) Treating the monkey, bananas, rope, and pulley as a system, evaluate the net torque about the pulley axis.

b) Using the result of part (a), determine the total angular momentum about the pulley axis as a function of time.

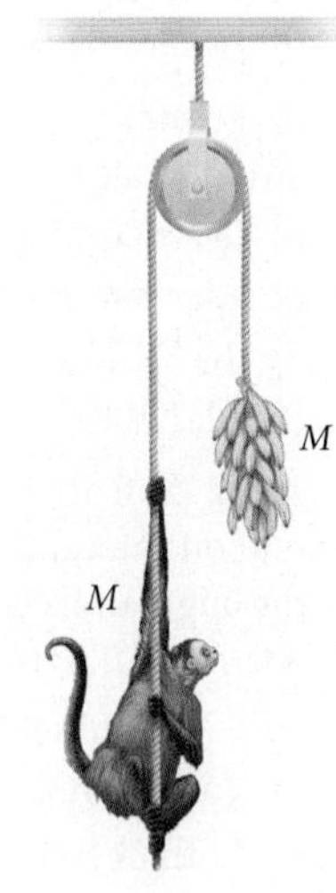


## PROBLEMS

A blue problem number indicates a worked-out solution is available in the Student Solutions Manual. One • and two •• indicate increasing level of problem difficulty.

### Sections 10.1 and 10.2

**10.34** A uniform solid cylinder of mass $M = 5$ kg is rolling without slipping along a horizontal surface. The velocity of its center of mass is 30 m/s. Calculate its energy.

**10.35** Determine the moment of inertia for three children weighing 60.0 lb, 45.0 lb and 80 lb sitting at different points on the edge of a rotating merry-go-round, which has a radius of 12 ft.

•**10.36** A 24-cm-long pen is tossed up in the air, reaching a maximum height of 1.2 m above its release point. On the way up, the pen makes 1.8 revolutions. Treating the pen as a thin uniform rod, calculate the ratio between the rotational kinetic energy and the translational kinetic energy at the instant the pen is released. Assume that the rotational speed does not change during the toss.

•**10.37** A solid ball and a hollow ball, each with a mass of 1 kg and radius of 0.1 m, start from rest and roll down a ramp of length 3 m at an incline of 35°. An ice cube of the same mass slides without friction down the same ramp.

a) Which ball will reach the bottom first? Explain!

b) Does the ice cube travel faster or slower than the solid ball at the base of the incline? Explain your reasoning.

c) What is the speed of the solid ball at the bottom of the incline?

•**10.38** A solid ball of mass $m$ and radius $r$ rolls without slipping through a loop of radius $R$, as shown in the figure. From what height $h$ should the ball be launched in order to make it through the loop without falling off the track?

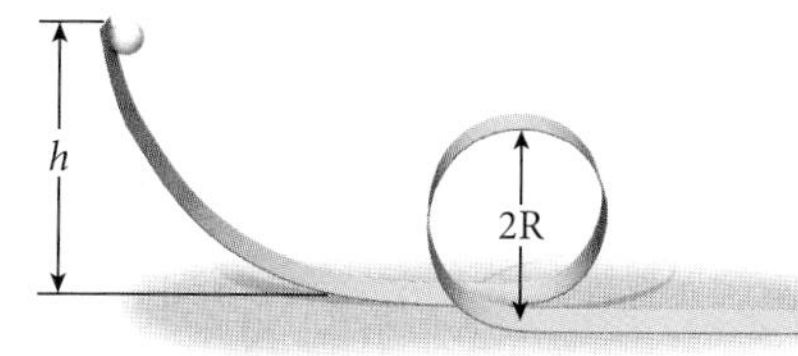


••**10.39** The Crab pulsar ($m \approx 2 \cdot 10^{30}$ kg) is a neutron star located in the Crab Nebula. The rotation rate of the Crab pulsar is currently about 30 rotations per second, or $60\pi$ rad/s. The rotation rate of the pulsar, however, is decreasing; each year, the rotation period decreases by $10^{-5}$ s. Justify the following statement: The loss in rotational energy of the pulsar is equivalent to 100,000 times the power output of the Sun. (The total power radiated by the Sun is about $4 \cdot 10^{26}$ W.)

••**10.40** A block of mass $m = 4$ kg is attached to a spring ($k = 32$ N/m) by a rope that hangs over a pulley of mass $M = 8$ kg and radius $R = 5$ cm, as shown in the figure. Treating the pulley as a solid homogeneous disk, neglecting friction at the axle of the pulley, and assuming the system starts from rest with the spring at its natural length, find (a) the speed of the block after it falls 1 m, and (b) the maximum extension of the spring.

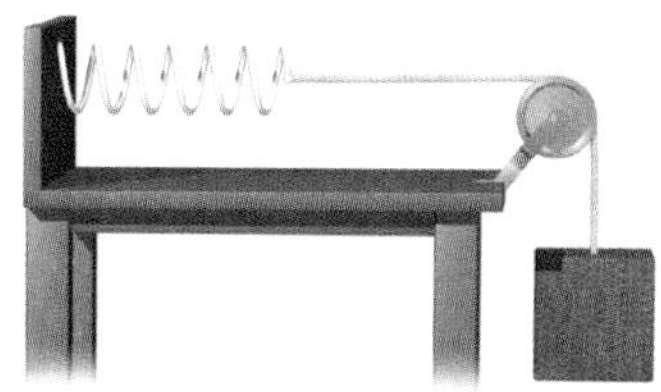

## Section 10.3

•**10.41** A small circular object with mass $m$ and radius $r$ has a moment of inertia given by $I = cmr^2$. The object rolls without slipping along the track shown in the figure. The track ends with a ramp of height $R = 2.5$ m that launches the object vertically. The object starts from a height $H = 6.0$ m. To what maximum height will it rise after leaving the ramp if $c = 0.40$?

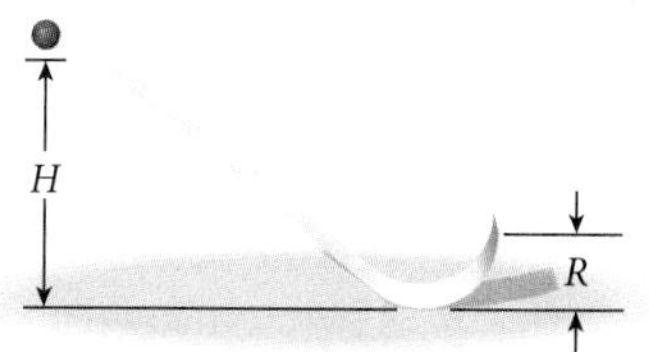


•**10.42** A uniform solid sphere of mass $M$ and radius $R$ is rolling without sliding along a level plane with a speed $v = 3.00$ m/s when it encounters a ramp that is at an angle $\theta = 23.0°$ above the horizontal. Find the maximum distance that the sphere travels up the ramp in each case:

a) The ramp is frictionless, so the sphere continues to rotate with its initial angular speed until it reaches its maximum height.

b) The ramp provides enough friction to prevent the sphere from sliding, so both the linear and rotational motion stop (instantaneously).

## Section 10.4

•**10.43** A disk with a mass of 30 kg and a radius of 40 cm is mounted on a frictionless horizontal axle. A string is wound many times around the disk and then attached to a 70-kg block, as shown in the figure. Find the acceleration of the block, assuming that the string does not slip.

•**10.44** A force, $\vec{F} = (2\hat{x} + 3\hat{y})$ N, is applied to an object at a point whose position vector with respect to the pivot point is $\vec{r} = (4\hat{x} + 4\hat{y} + 4\hat{z})$ m. Calculate the torque created by the force about that pivot point.

••**10.45** A disk with a mass of 14 kg, a diameter of 30 cm, and a thickness of 8 cm is mounted on a rough horizontal axle as shown on the left in the figure. (There is a friction force between the axle and the disk.) The disk is initially at rest. A constant force, $F = 70$ N, is applied to the edge of the disk at an angle of 37°, as shown on the right in the figure. After 2.0 s, the force is reduced to $F = 24$ N, and the disk spins with a constant angular velocity.

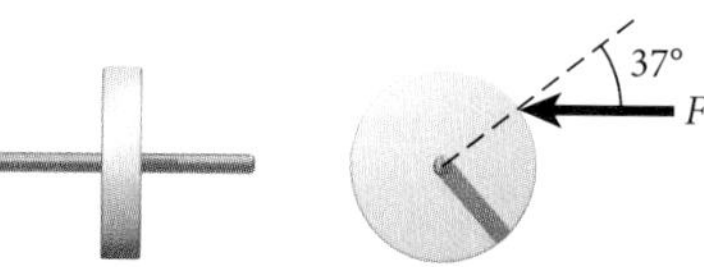


a) What is the magnitude of the torque due to friction between the disk and the axle?

b) What is the angular velocity of the disk after 2.0 s?

c) What is the kinetic energy of the disk after 2.0 s?

## Section 10.5

**10.46** A thin uniform rod (length = 1.0 m, mass = 2.0 kg) is pivoted about a horizontal frictionless pin through one of its ends. The moment of inertia of the rod through this axis is $\frac{1}{3}mL^2$. The rod is released when it is 60° below the horizontal. What is the angular acceleration of the rod at the instant it is released?

**10.47** An object made of two disk-shaped sections, A and B, as shown in the figure, is rotating about an axis through the center of disk A. The masses and the radii of disks A and B, respectively are, 2 kg and 0.2 kg and 25 cm and 2.5 cm.

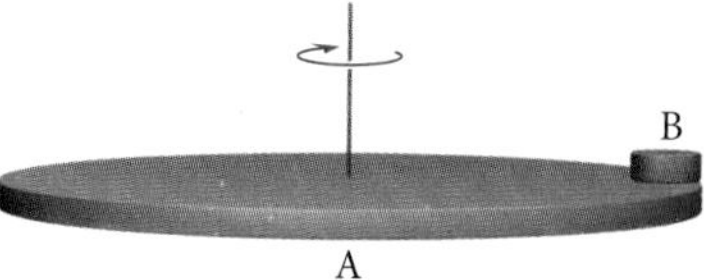


a) Calculate the moment of inertia of the object.

b) If the axial torque due to friction is 0.2 N m, how long will it take for the object to come to a stop if it is rotating with an initial angular velocity of $-2\pi$ rad/s?

•**10.48** You are the technical consultant for an action-adventure film in which a stunt calls for the hero to drop off a 20-m-tall building and land on the ground safely at a final vertical speed of 4 m/s. At the edge of the building's roof, there is a 100-kg drum that is wound with a sufficiently long rope (of negligible mass), has a radius of 0.5 m, and is free to rotate about its cylindrical axis with a moment of inertia $I_0$. The script calls for the 50-kg stuntman to tie the rope around his waist and walk off the roof.

a) Determine an expression for the stuntman's linear acceleration in terms of his mass $m$, the drum's radius $r$, and moment of inertia $I_0$.

b) Determine the required value of the stuntman's acceleration if he is to land safely at a speed of 4 m/s, and use this value to calculate the moment of inertia of the drum about its axis.

c) What is the angular acceleration of the drum?

d) How many revolutions does the drum make during the fall?

•**10.49** In a tire-throwing competition, a man holding a 23.5-kg car tire quickly swings the tire through three full turns and releases it, much like a discus thrower. The tire starts from rest and is then accelerated in a circular path. The orbital radius $r$ for the tire's center of mass is 1.10 m, and the path is horizontal to the ground. The figure shows a top view of the tire's circular path, and the dot at the center marks the rotation axis. The man applies a constant torque of 20.0 N m to accelerate the tire at a constant angular acceleration. Assume that all of the tire's mass is at a radius $R = 0.35$ m from its center.

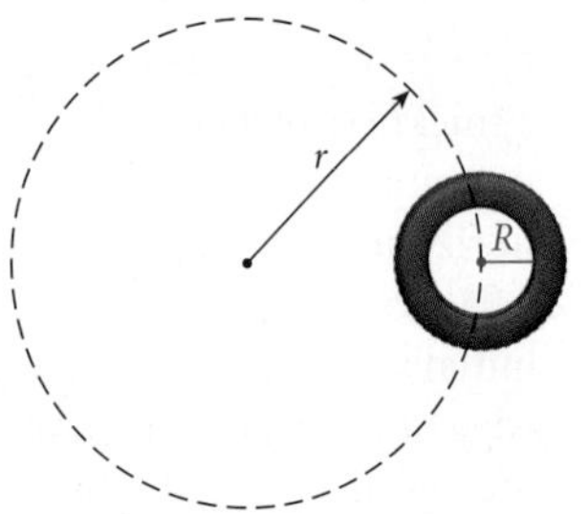


a) What is the time, $t_{\text{throw}}$, required for the tire to complete three full revolutions?

b) What is the final linear speed of the tire's center of mass (after three full revolutions)?

c) If, instead of assuming that all of the mass of the tire is at a distance 0.35 m from its center, you treat the tire as a hollow disk of inner radius 0.30 m and outer radius 0.40 m, how does this change your answers to parts (a) and (b)?

•**10.50** A uniform rod of mass $M = 250.0$ g and length $L = 50.0$ cm stands vertically on a horizontal table. It is released from rest to fall.

a) What forces are acting on the rod?

b) Calculate the angular speed of the rod, the vertical acceleration of the moving end of the rod, and the normal force exerted by the table on the rod as it makes an angle $\theta = 45.0°$ with respect to the vertical.

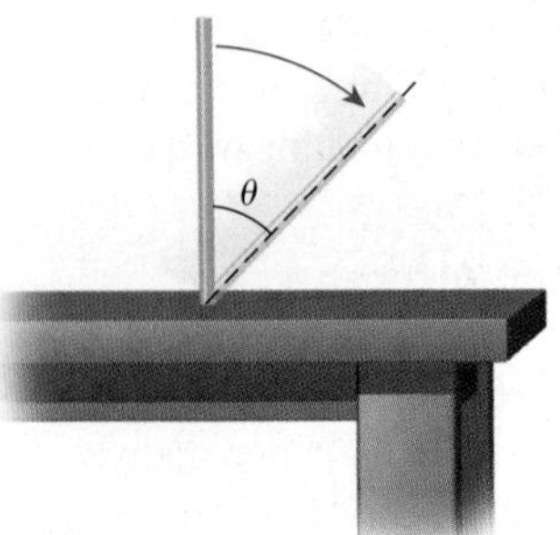


c) If the rod falls onto the table without slipping, find the linear acceleration of the end point of the rod when it hits the table and compare it with $g$.

•**10.51** A wheel with $c = \frac{4}{9}$, a mass of 40 kg, and a rim radius of 30 cm is mounted vertically on a horizontal axis. A 2-kg mass is suspended from the wheel by a rope wound around the rim. Find the angular acceleration of the wheel when the mass is released.

••**10.52** A 100-kg barrel with a radius of 50 cm has two ropes wrapped around it, as shown in the figure. The barrel is released from rest, causing the ropes to unwind and the barrel to fall spinning toward the ground. What is the speed of the barrel after it has fallen a distance of 10 m? What is the tension in each rope? Assume that the barrel's mass is uniformly distributed and that the barrel rotates as a solid cylinder.

••**10.53** A demonstration setup consists of a uniform board of length $L$, hinged at the bottom end and elevated at an angle $\theta$ by means of a support stick. A ball rests at the elevated end, and a light cup is attached to the board at the distance $d$ from the elevated end to catch the ball when the stick supporting the board is suddenly removed. You want to use a thin hinged board 1.00 m long and 10.0 cm wide, and you plan to have the vertical support stick located right at its elevated end.

a) How long should you make the support stick so that the ball has a chance to be caught?

b) Assume that you choose to use the longest possible support stick placed at the elevated end of the board. What distance $d$ from that end should the cup be located to ensure that the ball will be caught in the cup?

## Section 10.6

•**10.54** The flywheel of an old steam engine is a solid homogeneous metal disk of mass $M = 120$ kg and radius $R = 80$ cm. The engine rotates the wheel at 500 rpm. In an emergency, to bring the engine to a stop, the flywheel is disengaged from the engine and a brake pad is applied at the edge to provide a radially inward force $F = 100$ N. If the coefficient of kinetic friction between the pad and the flywheel is $\mu_k = 0.2$, how many revolutions does the flywheel make before coming to rest? How long does it take for the flywheel to come to rest? Calculate the work done by the torque during this time.

•**10.55** The turbine and associated rotating parts of a jet engine have a total moment of inertia of 25 kg m$^2$. The turbine is accelerated uniformly from rest to an angular speed of 150 rad/s in a time of 25 s. Find

a) the angular acceleration,

b) the net torque required,

c) the angle turned through in 25 s,

d) the work done by the net torque, and

e) the kinetic energy of the turbine at the end of the 25 s.

## Section 10.7

**10.56** Two small 6-kg masses are joined by a string, which can be assumed to be massless. The string has a tangle in it, as shown the figure. With the string tangled, the masses are separated by 1 m. The two masses are then made to rotate about their center of mass on a frictionless table at a rate of 5 rad/s. As they are rotating, the string untangles and lengthens to 1.4 m. What is the angular velocity of the masses after the string untangles?

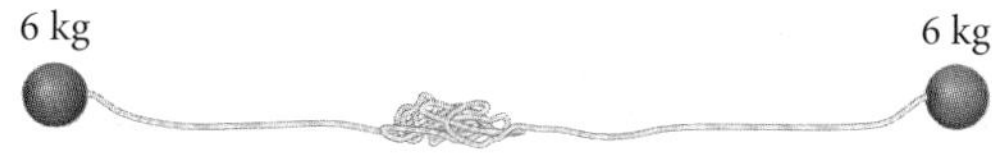


•**10.57** It is sometimes said that if the entire population of China stood on chairs and jumped off simultaneously, it would alter the rotation of the Earth. Fortunately, physics gives us the tools to investigate such speculations.

a) Calculate the moment of inertia of the Earth about its axis. For simplicity, treat the Earth as a uniform sphere of mass $m_E = 5.977 \cdot 10^{24}$ kg and radius 6371 km.

b) Calculate an upper limit for the contribution by the population of China to the Earth's moment of inertia, by assuming that the whole group is at the Equator. Take the population of China to be 1.3 billion people, of average mass 70. kg.

c) Calculate the change in the contribution in part (b) associated with a 1.0-m simultaneous change in the radial position of the entire group.

d) Determine the fractional change in the length of the day the change in part (c) would produce.

•**10.58** A bullet of mass $m_B = 0.01$ kg is moving with a speed of 100 m/s when it collides with a rod of mass $m_R = 5$ kg and length $L = 1$ m (shown in the figure). The rod is initially at rest, in a vertical position, and pivots about an axis going through its center of mass. The bullet imbeds itself in the rod at a distance $L/4$ from the pivot point. As a result, the bullet-rod system starts rotating.

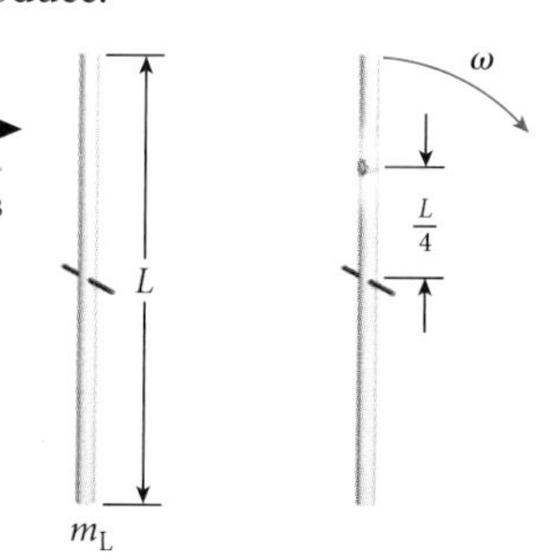


a) Find the angular velocity, $\omega$, of the bullet-rod system after the collision. You can neglect the width of the rod and can treat the bullet as a point mass.

b) How much kinetic energy is lost in the collision?

•**10.59** A sphere of radius $R$ and mass $M$ sits on a horizontal tabletop. A horizontally directed impulse with magnitude $J$ is delivered to a spot on the ball a vertical distance $h$ above the tabletop.

a) Determine the angular and translational velocity of the sphere just after the impulse is delivered.

b) Determine the distance $h_0$ at which the delivered impulse causes the ball to immediately roll without slipping.

•**10.60** A circular platform of radius $R_p = 4$ m and mass $M_p = 400$ kg rotates on frictionless air bearings about its vertical axis at 6 rpm. An 80-kg man standing at the very center of the platform starts walking (at $t = 0$) radially outward at a speed of 0.5 m/s with respect to the platform. Approximating the man by a vertical cylinder of radius $R_m = 0.2$ m, determine an equation (specific expression) for the angular velocity of the platform as a function of time. What is the angular velocity when the man reaches the edge of the platform?

•**10.61** A 25-kg boy stands 2 m from the center of a frictionless playground merry-go-round, which has a moment of inertia of 200 kg m$^2$. The boy begins to run in a circular path with a speed of 0.6 m/s relative to the ground.

a) Calculate the angular velocity of the merry-go-round.

b) Calculate the speed of the boy relative to the surface of the merry-go-round.

•**10.62** The Earth has an angular speed of $7.272 \cdot 10^{-5}$ rad/s in its rotation. Find the new angular speed if an asteroid ($m = 1.00 \cdot 10^{22}$ kg) hits the Earth while traveling at a speed of $1.40 \cdot 10^3$ m/s (assume the asteroid is a point mass compared to the radius of the Earth) in each of the following cases:

a) The asteroid hits the Earth dead center.

b) The asteroid hits the Earth nearly tangentially in the direction of Earth's rotation.

c) The asteroid hits the Earth nearly tangentially in the direction opposite to Earth's rotation.

## Section 10.8

**10.63** A demonstration gyroscope consists of a uniform disk with a 40-cm radius, mounted at the midpoint of a light 60-cm axle. The axle is supported at one end while in a horizontal position. How fast is the gyroscope precessing, in units of rad/s, if the disk is spinning around the axle at 30 rev/s?

## Additional Problems

**10.64** Most stars maintain an equilibrium size by balancing two forces—an inward gravitational force and an outward force resulting from the star's nuclear reactions. When the star's fuel is spent, there is no counterbalance to the gravitational force. Whatever material is remaining collapses in on itself. Stars about the same size as the Sun become white dwarfs, which glow from leftover heat. Stars that have about three times the mass of the Sun compact into neutron stars. And a star with a mass greater than three times the Sun's mass collapses into a single point, called a *black hole.* In most cases, protons and electrons are fused together to form neutrons—this is the reason for the name *neutron star.* Neutron stars rotate very fast because of the conservation of angular momentum. Imagine a star of mass $5 \cdot 10^{30}$ kg and radius $9.5 \cdot 10^8$ m that rotates once in 30 days. Suppose this star undergoes gravitational collapse to form a neutron star of radius 10 km. Determine its rotation period.

**10.65** In experiments at the Princeton Plasma Physics Laboratory, a plasma of hydrogen atoms is heated to over 500 million degrees Celsius (about 25 times hotter than the center of the Sun) and confined for tens of milliseconds by powerful magnetic fields (100,000 times greater than the Earth's magnetic field). For each experimental run, a huge amount of energy is required over a fraction of a second, which translates into a power requirement that would cause a blackout if electricity from the normal grid were to be used to power the experiment. Instead, kinetic energy is stored in a colossal flywheel, which is a spinning solid cylinder with a radius of 3.00 m and mass of $1.18 \cdot 10^6$ kg. Electrical energy from the power grid starts the flywheel spinning, and it takes 10 min to reach an angular speed of 1.95 rad/s. Once the flywheel reaches this angular speed, all of its energy can be drawn off very quickly to power an experimental run. What is the mechanical energy stored in the flywheel when it spins at 1.95 rad/s? What is the average torque required to accelerate the flywheel from rest to 1.95 rad/s in 10 min?

**10.66** A 2-kg thin hoop with a 50-cm radius rolls down a 30° slope without slipping. If the hoop starts from rest at the top of the slope, what is its translational velocity after it rolls 10 m along the slope?

**10.67** An oxygen molecule ($O_2$) rotates in the $xy$-plane about the $z$-axis. The axis of rotation passes through the center of the molecule, perpendicular to its length. The mass of each oxygen atom is $2.66 \cdot 10^{-26}$ kg, and the average separation between the two atoms is $d = 1.21 \cdot 10^{-10}$ m.

a) Calculate the moment of inertia of the molecule about the $z$-axis.

b) If the angular speed of the molecule about the $z$-axis is $4.60 \cdot 10^{12}$rad/s, what is its rotational kinetic energy?

**10.68** A 0.050-kg bead slides on a wire bent into a circle of radius 0.40 m. You pluck the bead with a force tangent to the circle. What force is needed to give the bead an angular acceleration of 6.0 rad/s$^2$?

**10.69** A professor doing a lecture demonstration stands at the center of a frictionless turntable, holding 5-kg masses in each hand with arms extended so that each mass is 1.2 m from his centerline. A (carefully selected!) student spins the professor up to a rotational speed of 1 rpm. If he then pulls his arms in by his sides so that each mass is 0.3 m from his centerline, what is his new rotation rate? Assume that his rotational inertia without the masses is 2.8 kg m/s, and neglect the effect on the rotational inertia of the position of his arms, since their mass is small compared to the mass of the body.

•**10.70** The system shown in the figure is held initially at rest. Calculate the angular acceleration of the system as soon as it is released. You can treat $M_A$ (1 kg) and $M_B$ (10 kg) as point masses located on either end of the rod of mass $M_C$ (20 kg) and length $L$ (5 m).

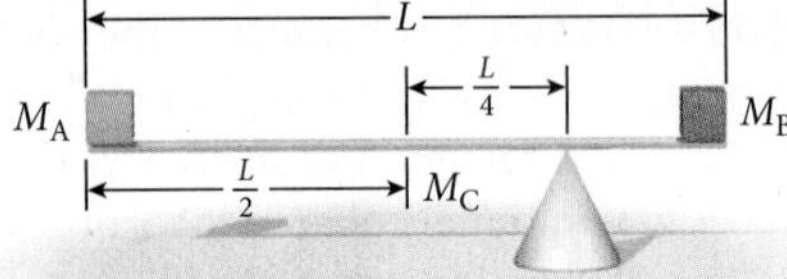


•**10.71** A child builds a simple cart consisting of a 0.60 m by 1.20 m sheet of plywood of mass 8.0 kg and four wheels, each 20.0 cm in diameter and with a mass of 2.00 kg. It is released from the top of a 15° incline that is 30 m long. Find the speed at the bottom. Assume that the wheels roll along the incline without slipping and that friction between the wheels and their axles can be neglected.

•**10.72** A CD has a mass of 15.0 g, an inner diameter of 1.5 cm, and an outer diameter of 11.9 cm. Suppose you toss it, causing it to spin at a rate of 4.3 revolutions per second.

a) Determine the moment of inertia of the CD, approximating its density as uniform.

b) If your fingers were in contact with the CD for 0.25 revolutions while it was acquiring its angular velocity and applied a constant torque to it, what was the magnitude of that torque?

•**10.73** A sheet of plywood 1.3 cm thick is used to make a cabinet door 55 cm wide by 79 cm tall, with hinges mounted on the vertical edge. A small 150-g handle is mounted 45 cm from the lower hinge at the same height as that hinge. If the density of the plywood is 550 kg/m$^3$, what is the moment of inertia of the door about the hinges? Neglect the contribution of hinge components to the moment of inertia.

•**10.74** A machine part is made from a uniform solid disk of radius $R$ and mass $M$. A hole of radius $R/2$ is drilled into the disk, with the center of the hole at a distance $R/2$ from the center of the disk (the diameter of the hole spans from the center of the disk to its outer edge). What is the moment of inertia of this machine part about the center of the disk in terms of $R$ and $M$?

•**10.75** A space station is to provide artificial gravity to support long-term stay of astronauts and cosmonauts. It is designed as a large wheel, with all the compartments in the rim, which is to rotate at a speed that will provide an acceleration similar to that of terrestrial gravity for the astronauts (their feet will be on the inside of the outer wall of the space station and their heads will be pointing toward the hub). After the space station is assembled in orbit, its rotation will be started by the firing of a rocket motor fixed to the outer rim, which fires tangentially to the rim. The radius of the space station is $R = 50$ m, and the mass is $M = 2.4 \cdot 10^5$ kg. If the thrust of the rocket motor is $F = 1.4 \cdot 10^2$ N, how long should the motor fire?

•**10.76** Many pulsars radiate radio frequency or other radiation in a periodic manner and are bound to a companion star in what is known as a *binary pulsar system.* In 2004, a double pulsar system, PSR J0737-3039A and J0737—3039B, was discovered by astronomers at the Jodrell Bank Observatory in the United Kingdom. In this system, *both* stars are pulsars. The pulsar with the faster rotation period rotates once every 0.023 s, while the other pulsar has a rotation period of 2.8 s. The faster pulsar has a mass 1.337 times that of the Sun, while the slower pulsar has a mass 1.250 times that of the Sun.

a) If each pulsar has a radius of 20.0 km, express the ratio of their rotational kinetic energies. Consider each star to be a uniform sphere with a fixed rotation period.

b) The orbits of the two pulsars about their common center of mass are rather eccentric (highly squashed ellipses), but an estimate of their average translational kinetic energy can be obtained by treating each orbit as circular with a radius equal to the mean distance from the system's center of mass. This radius is equal to $4.23 \cdot 10^8$ m for the larger star, and $4.54 \cdot 10^8$ m for the smaller star. If the orbital period is 2.4 h, calculate the ratio of rotational to translational kinetic energies for each star.

•**10.77** A student of mass 52 kg wants to measure the mass of a playground merry-go-round, which consists of a solid metal disk of radius $R = 1.5$ m that is mounted in a horizontal position on a low-friction axle. She tries an experiment: She runs with speed $v = 6.8$ m/s toward the outer rim of the merry-go-round and jumps on to the outer rim, as shown in the figure. The merry-go-round is initially at rest before the student jumps on and rotates at 1.3 rad/s immediately after she jumps on. You may assume that the student's mass is concentrated at a point.

a) What is the mass of the merry-go-round?

b) If it takes 35 s for the merry-go-round to come to a stop after the student has jumped on, what is the average torque due to friction in the axle?

c) How many times does the merry-go-round rotate before it stops, assuming that the torque due to friction is constant?

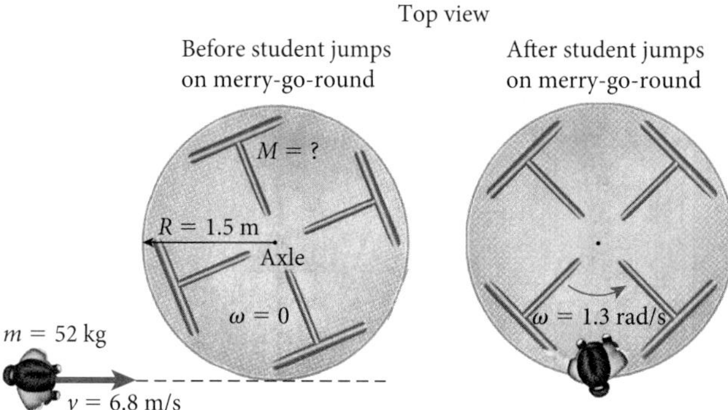


•**10.78** A ballistic pendulum consists of an arm of mass $M$ and length $L = 0.48$ m. One end of the arm is pivoted so that the arm rotates freely in a vertical plane. Initially, the arm is motionless and hangs vertically from the pivot point. A projectile of the same mass $M$ hits the lower end of the arm with a horizontal velocity of $V = 3.6$ m/s. The projectile remains stuck to the free end of the arm during their subsequent motion. Find the maximum angle to which the arm and attached mass will swing in each case:

a) The arm is treated as an ideal pendulum, with all of its mass concentrated as a point mass at the free end.

b) The arm is treated as a thin rigid rod, with its mass evenly distributed along its length.

••**10.79** A wagon wheel is made entirely of wood. Its components consist of a rim, 12 spokes, and a hub. The rim has mass 5.2 kg, outer radius 0.90 m, and inner radius 0.86 m. The hub is a solid cylinder with mass 3.4 kg and radius 0.12 m. The spokes are thin rods of mass 1.1 kg that extend from the hub to the inner side of the rim. Determine the constant $c = I/MR^2$ for this wagon wheel.

••**10.80** The figure shows a solid, homogeneous ball with radius $R$. Before falling to the floor, its center of mass is at rest, but it is spinning with angular velocity $\omega_0$ about a horizontal axis through its center. The lowest point of the ball is at a height $h$ above the floor. When released, the ball falls under the influence of gravity, and rebounds to a new height such that its lowest point is $ah$ above the floor. The deformation of the ball and the floor due to the impact can be considered negligible; the impact time, though, is finite. The mass of the ball is $m$, and the coefficient of kinetic friction between the ball and the floor is $\mu_k$. Ignore air resistance.

For the situation where the ball is slipping throughout the impact, find each of the following:

a) $\tan\theta$, where $\theta$ is the rebound angle indicated in the diagram

b) the horizontal distance traveled in flight between the first and second impacts

c) the minimum value of $\omega_0$ for this situation.

For the situation where the ball stops slipping before the impact ends, find each of the following:

d) $\tan\theta$

e) the horizontal distance traveled in flight between the first and second impacts.

Taking both of the situations into account, sketch the variation of $\tan\theta$ with $\omega_0$.

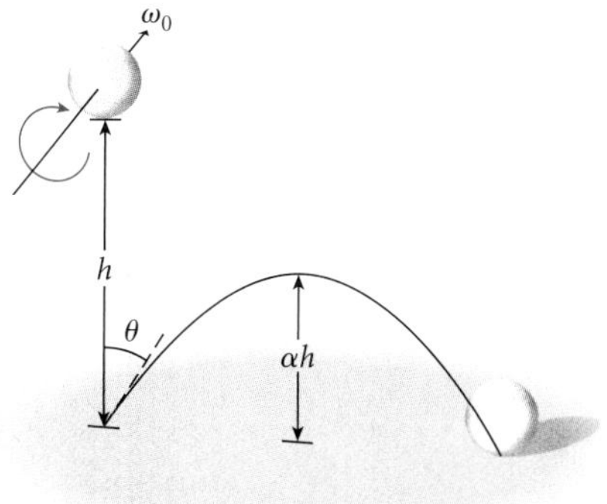

# 11 Static Equilibrium



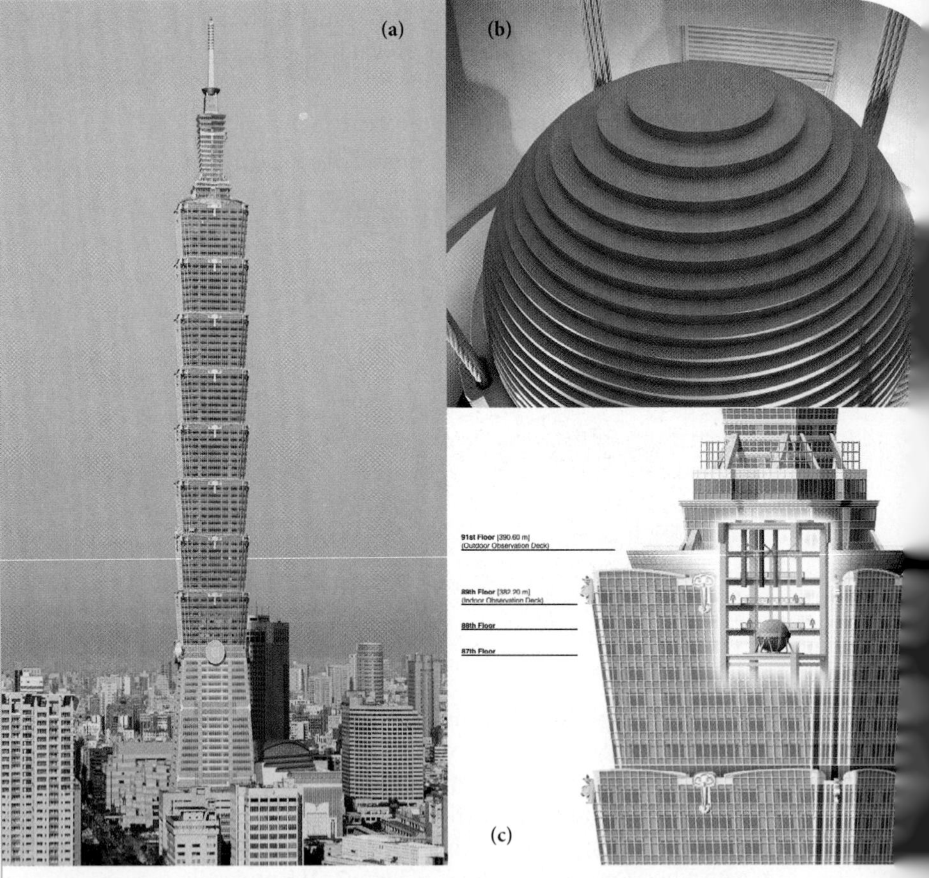


**FIGURE 11.1** The tallest building in the world as of 2008, Taipei 101 in Taiwan: (a) view of the tower; (b) view of the sway damper inside the tower; (c) cutaway drawing of the top of the tower showing the location of the damper.

## WHAT WE WILL LEARN

- Static equilibrium is defined as mechanical equilibrium for the special case of an object at rest.
- An object (or a collection of objects) can be in static equilibrium only if the net external force is zero and the net external torque is zero.
- A necessary condition for static equilibrium is that the first derivative of the potential energy function is zero at the equilibrium point.
- Stable equilibrium is achieved at points where the potential energy function has a minimum.
- Unstable equilibrium occurs at points where the potential energy function has a maximum.
- Neutral equilibrium (also called indifferent equilibrium or marginally stable equilibrium) exists at points where the first and second derivatives of the potential energy function are both zero.
- Equilibrium considerations are used to find otherwise unknown forces acting on an unmoving object or to find the forces required to prevent an object from moving.

The tallest building in the world as of 2008 was the Taipei 101 tower (Figure 11.1) in Taiwan, at 509 m (1670 ft) tall. Like any skyscraper, this building sways when winds near the top gust at high speeds. To minimize the motion, the Taipei 101 tower contains a mass damper between the 87th and 92nd floors, consisting of a steel ball built from 5-in-thick disks. The damper has a mass of 660 metric tons, enough to reduce the tower's motion by about 40%. Restaurants and observation decks surround the damper, making it a leading tourist attraction.

Stability and safety are of prime importance in the design and construction of any building. In this chapter, we examine the conditions for static equilibrium, which occurs when an object is at rest, and subject to zero net forces and torques. However, as we will see, a structure must be able to resist outside forces that tend to set it in motion. The long-term stability of a large structure—a building, a bridge, or a monument—depends on the builders' ability to judge how strong outside forces might be and to design the structure to withstand these forces.

## 11.1 Equilibrium Conditions

In Chapter 4, we saw that the necessary condition for static equilibrium is the absence of a net external force. In that case, Newton's First Law stipulates that an object stays at rest or moves with constant velocity. However, we often want to find the conditions necessary for a rigid object to stay at rest, in *static* equilibrium. An object (or collection of objects) is in **static equilibrium** if it is at rest and not experiencing translational or rotational motion. Figure 11.2 shows a famous example of a collection of objects in static equilibrium. Part of what makes this installation so amazing is that the eye does not want to accept that the configuration is stable.

**FIGURE 11.2** This 440-kg installation created by Alexander Calder hangs from the ceiling at the National Gallery of Art (Washington, DC) in perfect static equilibrium.

The requirement of no translational or rotational motion means that the linear and angular velocities of an object in static equilibrium are always zero. The fact that the linear and angular velocities do not change with time implies that the linear and angular accelerations are also zero at all times. In Chapter 4, we saw that Newton's Second Law,

$$\vec{F}_{\text{net}} = m\vec{a}, \tag{11.1}$$

implies that if the linear acceleration, $\vec{a}$, is zero, the external net force, $\vec{F}_{\text{net}}$, must be zero. Furthermore, we saw in Chapter 10 that Newton's Second Law for rotation,

$$\vec{\tau}_{\text{net}} = I\vec{\alpha}, \tag{11.2}$$

implies that if the angular acceleration, $\vec{\alpha}$, is zero, the external net torque, $\vec{\tau}_{\text{net}}$, must be zero. These facts lead to two conditions for static equilibrium.

### Static Equilibrium Condition 1

An object can stay in static equilibrium only if the net force acting on it is zero:

$$\vec{F}_{\text{net}} = 0. \tag{11.3}$$

*Continued—*

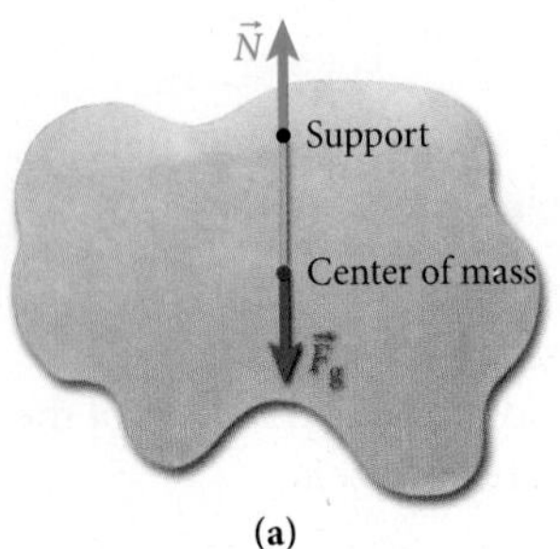


(a)

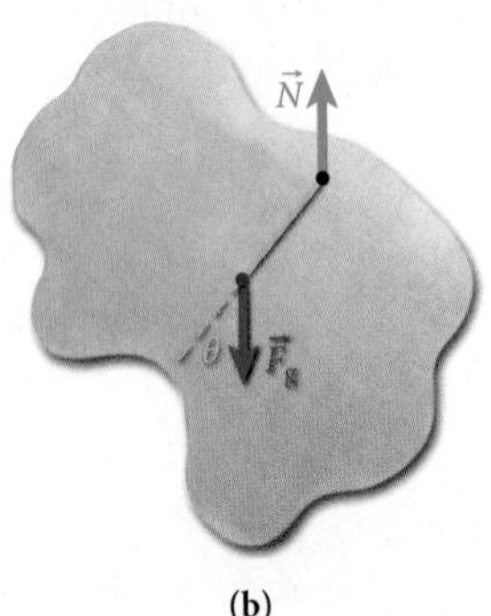


(b)

**FIGURE 11.3** (a) This object experiences zero net torque, because it is supported from a pin located exactly above the center of mass. (b) A net torque results when the center of mass of the same object is at a location not exactly below the support point.

### Static Equilibrium Condition 2

An object can stay in static equilibrium only if the net torque acting on it is zero:

$$\vec{\tau}_{\text{net}} = 0. \tag{11.4}$$

Even if Newton's First Law is satisfied (no net force acts on an object), and an object has no translational motion, it will still rotate if it experiences a net torque.

It is important to remember that torque is always defined with respect to a pivot point (the point where the axis of rotation intersects the plane defined by $\vec{F}$ and $\vec{r}$). When we compute the net torque, the pivot point must be the same for all forces involved in the calculation. If we try to solve a static equilibrium problem, with vanishing net torque, the net torque has to be zero for any pivot point chosen. Thus, we have the freedom to select a pivot point that best suits our purpose. A clever selection of a pivot point is often the key to a quick solution. For example, if an unknown force is present in the problem, we can select the point where the force acts as the pivot point. Then, that force will not enter into the torque equation because it has a moment arm of length zero.

If an object is supported from a pin located directly above its center of mass, as in Figure 11.3a (where the red dot marks the center of mass), then the object stays balanced; that is, it does not start to rotate. Why? Because in this case only two forces act on the object—the force of gravity, $\vec{F}_g$ (blue arrow), and the normal force $\vec{N}$ (green arrow), from the pin—and they lie on the same line (yellow line in Figure 11.3a). The two forces cancel each other out and produce no net torque, resulting in static equilibrium; the object is in balance.

On the other hand, if an object is supported in the same way from a pin but its center of mass is not below the support point, then the situation is that shown in Figure 11.3b. The normal and gravitational force vectors still point in opposite directions; however, a nonzero net torque now acts, because the angle $\theta$ between the gravitational force vector, $\vec{F}_g$, and the moment arm (directed along the yellow line) is not zero any more. This torque violates the condition that the net torque must be zero for static equilibrium. However, suspending an object from different points is a practical method for finding the center of mass of the object, even a strangely shaped object like the one in Figure 11.3.

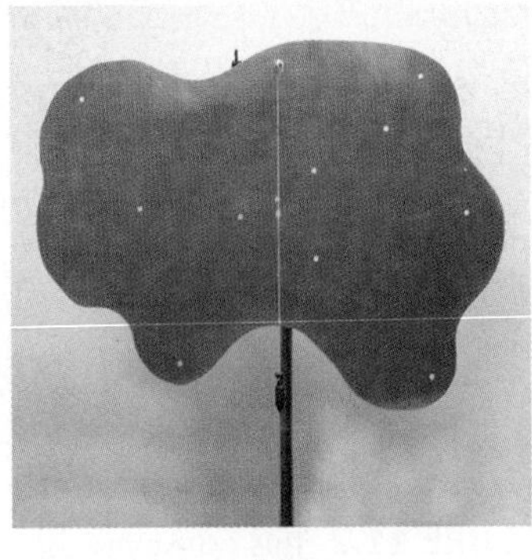

(a)

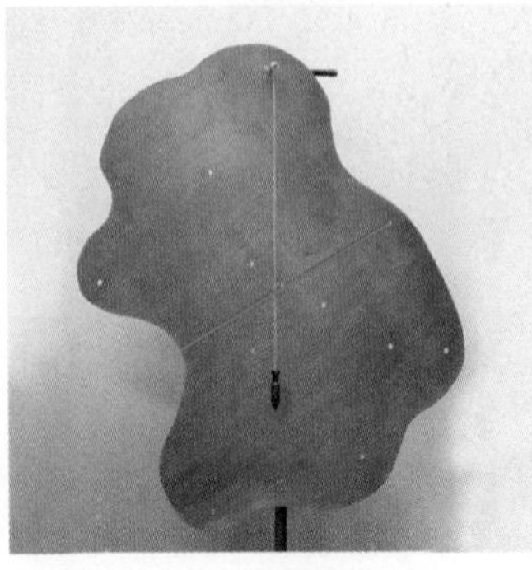

(b)

**FIGURE 11.4** Finding the center of mass for an arbitrarily shaped object.

## Experimentally Locating the Center of Mass

To locate an object's center of mass experimentally, we can support the object from a pin in such a way that it can rotate freely around the pin and then let it come to rest. Once the object has come to rest, its center of mass is located on the line directly below the pin. We hang a weight (a plumb bob in Figure 11.4) from the same pin used to support the object, and it identifies the line. We mark this line on the object. If we do this for two different support points, the intersection of the two lines will mark the precise location of the center of mass.

You can use another technique to determine the location of the center of mass for many objects (see Figure 11.5). You simply support the object on two fingers placed in such a way that the center of mass is located somewhere between them. (If this is not the case, you will know right away, because the object will fall.) Then slowly slide the fingers closer to each other. At the point where they meet, they are directly below the center of mass, and the object is balanced on top.

Why does this technique work? The finger that is closer to the center of mass exerts a larger normal force on the object. Thus, when moving, this finger exerts a larger friction force on the object than the finger that is farther away. Consequently, if the fingers slide toward each other, the finger that is closer to the center of mass will take the suspended object along with it. This continues until the other finger becomes closer to the center of mass, when the effect is reversed. In this way, the two fingers always keep the center of mass located between them. When the fingers are next to each other, the center of mass is located.

In Figure 4.6, showing a hand holding up a laptop computer, the force vector $\vec{N}$ exerted by the hand on the laptop acted at the laptop's center, just like the gravitational force vector but in the opposite direction. This placement is necessary. For a hand to hold up a laptop computer, it must be placed directly below the computer's center of mass. Otherwise, if the center of mass were not supported from directly below, the computer would tip over.

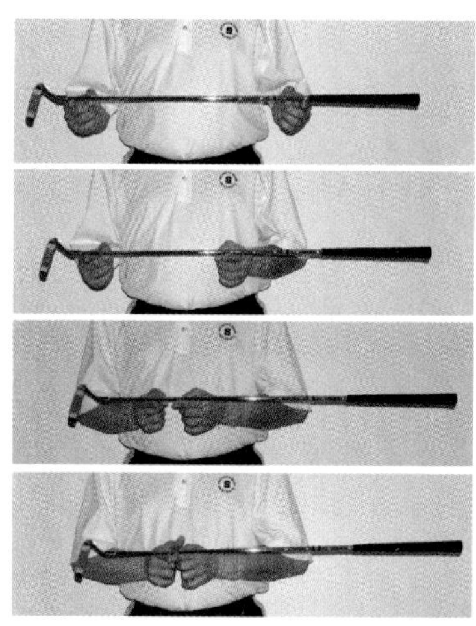

**FIGURE 11.5** Determining the center of mass of a golf club experimentally.

## Equilibrium Equations

With a qualitative understanding of the concepts and conditions for static equilibrium, we can formulate the equilibrium conditions for a more quantitative analysis. In Chapter 4, we found that the condition of zero net force translates into three independent equations in a three-dimensional space, one for each Cartesian component of the zero net force (refer to equation 11.3). In addition, the condition of zero net torque in three dimensions also implies three equations for the components of the net torque (refer to equation 11.4), representing independent rotations about the three possible axes of rotation, which are all perpendicular to each other. In this chapter, we will not deal with the three-dimensional situations (involving six equations), but rather will concentrate on problems of static equilibrium in two-dimensional space, that is, a plane. In a plane, there are two independent translational degrees of freedom for a rigid body (in the $x$- and the $y$-directions) and one possible rotation, either clockwise or counterclockwise around a rotation axis that is perpendicular to the plane. Thus, the two equations for the net force components are

$$F_{\text{net},x} = \sum_{i=1}^{n} F_{i,x} = F_{1,x} + F_{2,x} + \cdots + F_{n,x} = 0 \tag{11.5}$$

$$F_{\text{net},y} = \sum_{i=1}^{n} F_{i,y} = F_{1,y} + F_{2,y} + \cdots + F_{n,y} = 0. \tag{11.6}$$

In Chapter 10, the net torque about a fixed axis of rotation was defined as the difference between the sum of the counterclockwise torques and the sum of the clockwise torques. The static equilibrium condition of zero net torque about each axis of rotation can thus be written as

$$\tau_{\text{net}} = \sum_{i} \tau_{\text{counterclockwise},i} - \sum_{j} \tau_{\text{clockwise},j} = 0. \tag{11.7}$$

These three equations (11.5 through 11.7) form the basis for the quantitative analysis of static equilibrium in the problems in this chapter.

# 11.2 Examples Involving Static Equilibrium

The two conditions for static equilibrium (zero net force and zero net torque) are all we need to solve a very large class of problems involving static equilibrium. We do not need calculus to solve these problems; all the calculations use only algebra and trigonometry. Let's start with an example for which the answer seems obvious. This will provide practice with the method and show that it leads to the right answer.

### EXAMPLE 11.1 Seesaw

A playground seesaw consists of a pivot and a bar, of mass $M$, that is placed on the pivot so that the ends can move up and down freely (Figure 11.6a). If an object of mass $m_1$ is placed on one end of the bar at a distance $r_1$ from the pivot point, as shown in Figure 11.6b, that end goes down, simply because of the force and torque that the object exerts on it.

**PROBLEM 1**

Where do we have to place an object of mass $m_2$ (assumed to be equal to the mass of $m_1$) to get the seesaw to balance, so the bar is horizontal and neither end touches the ground?

**SOLUTION 1**

Figure 11.6b is a free-body diagram of the bar showing the forces acting on it and the points where they act. The force that $m_1$ exerts on the bar is simply $m_1g$, acting downward

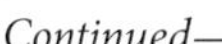

*Continued—*

(a)

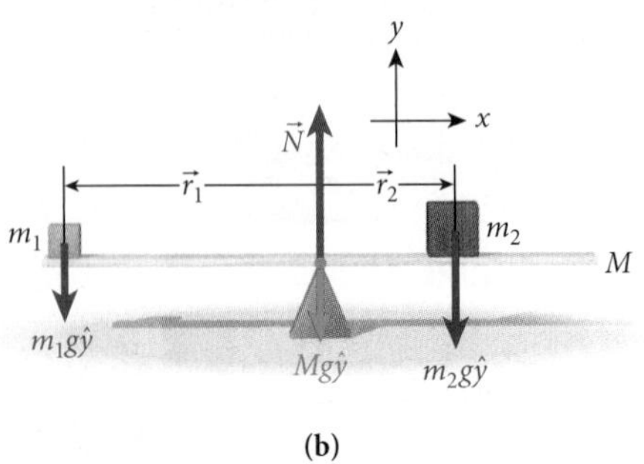


(b)

**FIGURE 11.6** (a) A playground seesaw; (b) free-body diagram showing forces and moment arms.

as shown in Figure 11.6b. The same is true for the force that $m_2$ exerts on the bar. In addition, because the bar has a mass $M$ of its own, it experiences a gravitational force, $Mg$. The gravitational force acts at the center of mass of the bar, right in the middle of the bar. The final force acting on the bar is the normal force, $N$, exerted by the bar's support. It acts exactly at the axle of the seesaw (marked with an orange dot).

The equilibrium equation for the $y$-components of the forces leads to an expression for the value of the normal force:

$$F_{\text{net},y} = \sum_i F_{i,y} = -m_1 g - m_2 g - Mg + N = 0$$
$$\Rightarrow N = g(m_1 + m_2 + M).$$

The signs in front of the individual force components indicate whether they act upward (positive) or downward (negative).

Because all forces act in the $y$-direction, it is not necessary to write equations for the net force components in the $x$- or $z$-directions.

Now we can consider the net torque. The selection of the proper pivot point can make our computations simple. For a seesaw, the natural selection is at the axle, the point marked with an orange dot in the center of the bar in Figure 11.6b. Because the normal force, $N$, and the weight of the bar, $Mg$, act exactly through this point, their moment arms have length zero. Thus, these two forces do not contribute to the torque equation if this is selected as the pivot point. The forces $F_1 = m_1 g$ and $F_2 = m_2 g$ are the only ones contributing torques: $F_1$ generates a counterclockwise torque, and $F_2$ a clockwise torque. The torque equation is then

$$\begin{aligned}\tau_{\text{net}} &= \sum_i \tau_{\text{clockwise},i} - \sum_j \tau_{\text{counterclockwise},j}\\ &= m_2 g r_2 \sin 90° - m_1 g r_1 \sin 90° = 0\\ &\Rightarrow m_2 r_2 = m_1 r_1 \qquad \text{(i)}\\ &\Rightarrow r_2 = r_1 \frac{m_1}{m_2}.\end{aligned}$$

Even though they equal 1 and thus have no effect, the factors sin90° are included above as a reminder that the angle between force and moment arm usually affects the calculation of the torques.

The question was where to put $m_2$ for the case that the two masses were the same, the answer is $r_2 = r_1$ in this case. This expected result shows that our systematic way of approaching the solution works in this easily verifiable case.

**PROBLEM 2**

How big does $m_2$ need to be to balance $m_1$ if $r_1 = 3r_2$, that is, if $m_2$ is three times closer to the pivot point than $m_1$?

**SOLUTION 2**

We use the same free-body diagram (Figure 11.6b) and arrive at the same general equation for the masses and distances. Solving equation (i) for $m_2$ gives

$$m_2 r_2 = m_1 r_1$$
$$\Rightarrow m_2 = m_1 \frac{r_1}{r_2}.$$

Using $r_1 = 3r_2$, we obtain

$$m_2 = m_1 \frac{r_1}{r_2} = m_1 \frac{3r_2}{r_2} = 3m_1.$$

For this case, we find that the mass of $m_2$ has to be three times that of $m_1$ to establish static equilibrium.

As Example 11.1 shows, a clever choice of a pivot point can often greatly simplify a solution. It is important, however, to realize that one can use *any* pivot point. If the torques are balanced about any pivot point, they are balanced about *all* pivot points. Thus, if we change the pivot point, it can make the calculations more complicated in certain situations, but the end result of the calculation will not change.

### 11.1 Self-Test Opportunity

Suppose that the pivot point for the seesaw in part 1 of Example 11.1 is placed instead below the center of mass of $m_2$. Show that this leads to the same result.

## EXAMPLE 11.2 Force on Biceps

Suppose you are holding a barbell in your hand, as shown in Figure 11.7a. Your biceps supports your forearm. The biceps is attached to the bone of the forearm at a distance $r_b$ = 2.0 cm from the elbow, as shown in Figure 11.7b. The mass of your forearm is 0.85 kg. The length of your forearm is 31 cm. Your forearm makes an angle $\theta$ = 75° with the vertical, as shown in Figure 11.7b. The barbell has a mass of 15 kg.

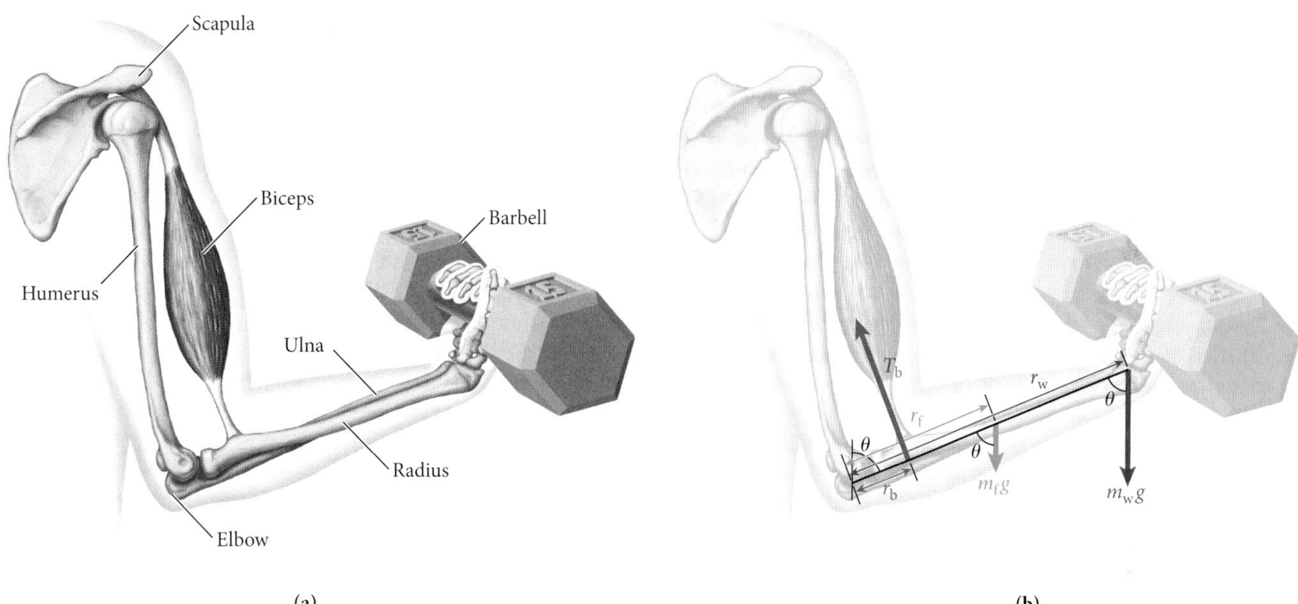


**FIGURE 11.7** (a) A human arm holding a barbell. (b) Forces and moment arms for a human arm holding a barbell.

### PROBLEM

What is the force that the biceps must exert to hold up your forearm and the barbell? Assume that the biceps exerts a force perpendicular to the forearm at the point of attachment.

### SOLUTION

The pivot point is the elbow. The net torque on your forearm must be zero, so the counterclockwise torque must equal the clockwise torque:

$$\sum_i \tau_{\text{counterclockwise},i} = \sum_j \tau_{\text{clockwise},j}.$$

The counterclockwise torque is provided by the biceps:

$$\sum_i \tau_{\text{counterclockwise},i} = T_b r_b \sin 90° = T_b r_b,$$

where $T_b$ is the force exerted by the biceps and $r_b$ is the moment arm for the force exerted by the biceps. The clockwise torque is the sum of the torque exerted by the weight of the forearm and the torque exerted by the barbell:

$$\sum_j \tau_{\text{clockwise},j} = m_f g r_f \sin\theta + m_w g r_w \sin\theta,$$

*Continued—*

where $m_f$ is the mass of the forearm, $r_f$ is the moment arm for the force exerted by the weight of the forearm, $m_w$ is the mass of the barbell, and $r_w$ is the moment arm of the force exerted by the weight of the barbell. We take the $r_w$ to be equal to the length of the forearm and $r_f$ to be half of that length, or $r_w/2$. Equating the counterclockwise torque and the clockwise torque gives us

$$T_b r_b = m_f g r_f \sin\theta + m_w g r_w \sin\theta.$$

Solving for the force exerted by the biceps, we obtain

$$T_b = \frac{m_f g r_f \sin\theta + m_w g r_w \sin\theta}{r_b} = g\sin\theta\left(\frac{m_f r_f + m_w r_w}{r_b}\right).$$

Putting in the given numbers gives the force exerted by the biceps

$$T_b = g\sin\theta\left(\frac{m_f r_f + m_w r_w}{r_b}\right)$$

$$= \left(9.81 \text{ m/s}^2\right)\left(\sin 75°\right)\left(\frac{(0.85 \text{ kg})\left(\frac{0.31 \text{ m}}{2}\right) + (15 \text{ kg})(0.31 \text{ m})}{0.020 \text{ m}}\right)$$

$$= 2300 \text{ N}.$$

You may wonder why evolution gave the biceps such a huge mechanical disadvantage. Apparently, it was more advantageous to be able to swing the arms a long distance while exerting a comparatively huge force than to be able to move them a short distance and exert a small force. This is in contrast, by the way, to the jaw muscles, which have evolved the ability to crunch tough food with huge force.

### 11.2 Self-Test Opportunity

In Example 11.2, suppose you hold the barbell so that your forearm makes an angle of 180° with the vertical. Why is it that you can still lift the barbell?

The following example for static equilibrium also shows an application of the formulas to compute the center of mass that we introduced in Chapter 8, and at the same time has a very surprising outcome.

## EXAMPLE 11.3 Stacking Blocks

**PROBLEM**

Consider a collection of identical blocks stacked at the edge of a table (Figure 11.8). How far out can we push the leading edge of the top block without the pile falling off?

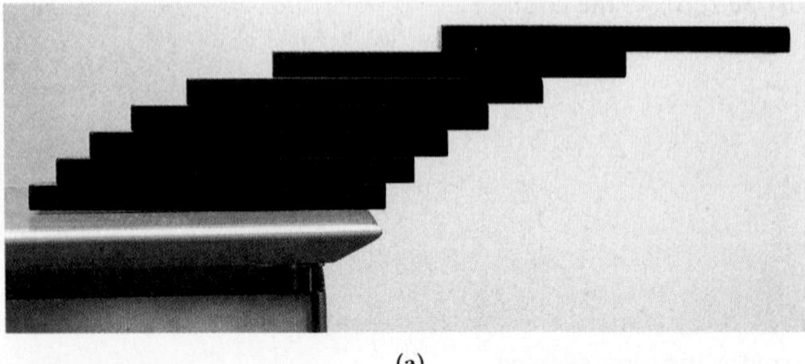

(a)

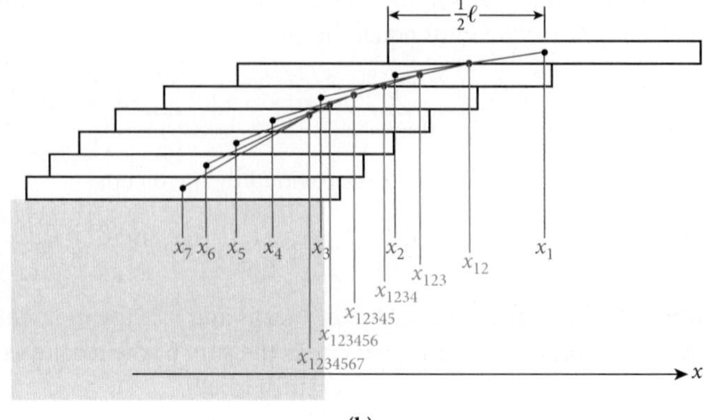


(b)

**FIGURE 11.8** (a): Stack of seven identical blocks piled on a table—note that the left edge of the top block is to the right of the right edge of the table. (b) Positions of the centers of mass of the individual blocks ($x_1$ through $x_7$) and locations of the combined centers of mass of the topmost blocks ($x_{12}$ through $x_{1234567}$).

**SOLUTION**

Let's start with one block. If the block has length $\ell$ and uniform mass density, then its center of mass is located at $\frac{1}{2}\ell$. Clearly, it can stay at rest as long as at least half of it is on the table, with its center of mass supported by the table from below. The block can stick out an infinitesimal amount less than $\frac{1}{2}\ell$ beyond the support, and it will remain at rest.

Next, we consider two identical blocks. If we call the $x$-coordinate of the center of mass of the upper block $x_1$ and that of the lower block $x_2$, we obtain for the $x$-coordinate of the center of mass of the combined system, according to Section 8.1,

$$x_{12}=\frac{x_1m_1+x_2m_2}{m_1+m_2}.$$

For identical blocks, $m_1 = m_2$, which simplifies the expression for $x_{12}$ to

$$x_{12}=\tfrac{1}{2}(x_1+x_2).$$

Since $x_1 = x_2 + \frac{1}{2}\ell$ in the limiting case that the center of mass of the first block is still supported from below by the second block, we obtain

$$x_{12}=\tfrac{1}{2}(x_1+x_2)=\tfrac{1}{2}((\tfrac{1}{2}\ell+x_2)+x_2)=x_2+\tfrac{1}{4}\ell. \qquad \text{(i)}$$

Now we can go to three blocks. The top two blocks will not topple if the combined center of mass, $x_{12}$, is supported from below. Shifting $x_{12}$ to the very edge of the third block, we obtain $x_{12} = x_3 + \frac{1}{2}\ell$. Combining this with equation (i), we have

$$x_{12}=x_2+\tfrac{1}{4}\ell=x_3+\tfrac{1}{2}\ell\Rightarrow x_2=x_3+\tfrac{1}{4}\ell.$$

Note that equation (i) is still valid after the shift because we have expressed $x_{12}$ in terms of $x_2$ and because $x_{12}$ and $x_2$ change by the same amount when the two blocks move together. We can now calculate the center of mass for the three blocks in the same way as before, by applying the same principle to find the new combined center of mass:

$$x_{123}=\frac{x_{12}(2m)+x_3m}{2m+m}=\tfrac{2}{3}x_{12}+\tfrac{1}{3}x_3=\tfrac{2}{3}(x_3+\tfrac{1}{2}\ell)+\tfrac{1}{3}x_3=x_3+\tfrac{1}{3}\ell. \qquad \text{(ii)}$$

Requiring that the top three blocks are supported by the fourth block from below results in $x_{123} = x_4 + \frac{1}{2}\ell$. Combined with equation (ii), this establishes

$$x_{123}=x_3+\tfrac{1}{3}\ell=x_4+\tfrac{1}{2}\ell\Rightarrow x_3=x_4+\tfrac{1}{6}\ell.$$

You can see how this series continues. If we have $n - 1$ blocks supported in this way by the $n$th block, then the coordinates of the $(n - 1)$st and $n$th block are related as follows:

$$x_{n-1}=x_n+\frac{\ell}{2n-2}.$$

We can now add up all the terms and find out how far away $x_1$ can be from the edge:

$$x_1=x_2+\tfrac{1}{2}\ell=x_3+\tfrac{1}{4}\ell+\tfrac{1}{2}\ell=x_4+\tfrac{1}{6}\ell+\tfrac{1}{4}\ell+\tfrac{1}{2}\ell=\cdots=x_{n+1}+\tfrac{1}{2}\ell\left(\sum_{i=1}^{n}\frac{1}{i}\right).$$

You may remember from calculus that the sum $\sum_{i=1}^{n} i^{-1}$ does not converge, that is, does not have an upper limit for $n \to \infty$. This gives the astonishing result that $x_1$ can move *infinitely* far away from the table's edge, provided there are enough blocks under it and that the edge of the table can support their weight without significant deformation! (But see Self-Test Opportunity 11.3 to put this *infinity* into perspective.) Figure 11.8a shows only 7 blocks stacked on a table, and the left edge of the top block is already to the right of the right edge of the table.

### 11.3 Self-Test Opportunity

Suppose you had 10,000 identical blocks of height 4.0 cm and length 15.0 cm. If you arranged them in the way demonstrated in Example 11.3, how far would the right edge of the top block stick out?

The following problem, involving a composite extended object, serves to review the calculation of the center of mass of such objects, which was introduced in Chapter 8.

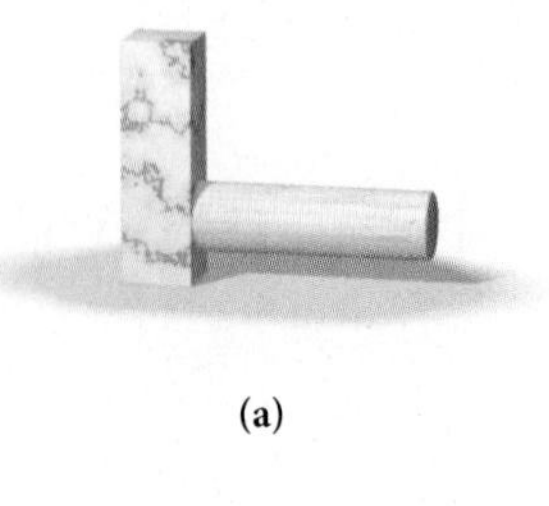

(a)

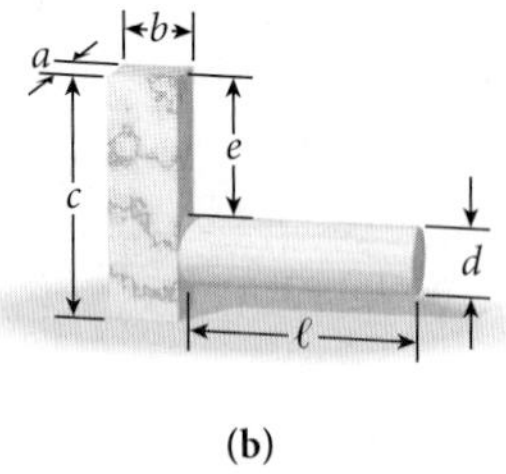


(b)

**FIGURE 11.9** (a) A wood and marble sculpture; (b) diagram of the sculpture with dimensions labeled.

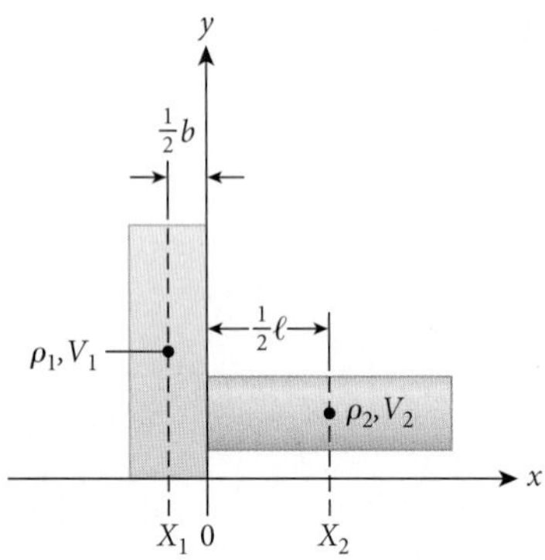


**FIGURE 11.10** Sketch for calculating the center of mass of the sculpture.

## SOLVED PROBLEM 11.1 An Abstract Sculpture

An alumnus of your university has donated a sculpture to be displayed in the atrium of the new physics building. The sculpture consists of a rectangular block of marble of dimensions $a = 0.71$ m, $b = 0.71$ m, and $c = 2.74$ m and a cylinder of wood with length $\ell = 2.84$ m and diameter $d = 0.71$ m, which is attached to the marble so that its upper edge is a distance $e = 1.47$ m from the top of the marble block (Figure 11.9).

### PROBLEM

If the mass density of the marble is $2.85 \cdot 10^3$ kg/m$^3$ and that of the wood is $4.40 \cdot 10^2$ kg/m$^3$, can the sculpture stand upright on the floor of the atrium, or does it need to be supported by some kind of bracing?

### SOLUTION

### THINK

Chapter 8 showed that a condition for stability of an object is that the object's center of mass needs to be directly supported from below. In order to decide if the sculpture can stand upright without additional support, we therefore need to determine the location of the center of mass of the sculpture and find out if it is located at a point inside the marble block. Since the floor supports the block from below, the sculpture will be able to stand upright if this is the case. If the center of mass of the sculpture is located outside the marble block, the sculpture will need bracing.

### SKETCH

We draw a side view of the sculpture (Figure 11.10), representing the marble block and the wooden cylinder by rectangles. The sketch also indicates a coordinate system with a horizontal $x$-axis and a vertical $y$-axis and the origin at the right edge of the marble block. (What about the $z$-coordinate? We can use symmetry arguments in the same way as we did in Chapter 8 and find that the $z$-component of the center of mass is located in the plane that divides block and cylinder into halves.)

### RESEARCH

With the chosen coordinate system, we also do not need to worry about calculating the $y$-coordinate of the center of mass of the sculpture. The condition for stability depends only on whether the $x$-coordinate of the center of mass is within the marble block; it does not depend on how high the center of mass is off the ground. Thus, the only task left is to calculate the $x$-component of the center of mass. According to the general principles for calculating center-of-mass coordinates, developed in Chapter 8, we can write

$$X = \frac{1}{M}\int_V x\rho(\vec{r})dV, \tag{i}$$

where $M$ is the mass of the entire sculpture and $V$ is its volume. Note that in this case the mass density is not homogeneous, since the marble and the wood have different densities.

In order to perform the integration, we split the volume $V$ into convenient parts: $V = V_1 + V_2$, where $V_1$ is the volume of the marble block and $V_2$ is the volume of the wooden cylinder. Then equation (i) becomes

$$\begin{aligned} X &= \frac{1}{M}\int_{V_1} x\rho(\vec{r})dV + \frac{1}{M}\int_{V_2} x\rho(\vec{r})dV \\ &= \frac{1}{M}\int_{V_1} x\rho_1 dV + \frac{1}{M}\int_{V_2} x\rho_2 dV. \end{aligned}$$

### SIMPLIFY

To calculate the location of the $x$-component of the center of mass of the marble block alone, we can use the equation for constant density from Chapter 8:

$$X_1 = \frac{1}{M_1}\int_{V_1} x\rho_1 dV = \frac{\rho_1}{M_1}\int_{V_1} x dV.$$

(Since the density is constant over this entire volume, we can move it out of the integral.) In the same way, we can find the $x$-component of the center of mass of the wooden cylinder:

$$X_2 = \frac{\rho_2}{M_2}\int_{V_2} x\,dV.$$

Therefore, the expression for the center of mass of the composite object—that is, the entire sculpture—is

$$\begin{aligned} X &= \frac{\rho_1}{M}\int_{V_1} x\,dV + \frac{\rho_2}{M}\int_{V_2} x\,dV \\ &= \frac{M_1}{M}\frac{\rho_1}{M_1}\int_{V_1} x\,dV + \frac{M_2}{M}\frac{\rho_2}{M_2}\int_{V_2} x\,dV \qquad \text{(ii)} \\ &= \frac{M_1}{M}X_1 + \frac{M_2}{M}X_2. \end{aligned}$$

This is a very important general result: Even for extended objects, the combined center of mass can be calculated in the same way as it is for point particles. Since the total mass of the sculpture is the combined mass of its two parts, $M = M_1 + M_2$, equation (ii) becomes

$$X = \frac{M_1}{M_1+M_2}X_1 + \frac{M_2}{M_1+M_2}X_2. \qquad \text{(iii)}$$

It is important to note that this relationship between the coordinate of combined center of mass of a composite object and the individual center-of-mass coordinates is true even in the case where the parts are separate objects. Further, it holds in the case when the density inside a given object is not constant. Formally, this result relies on the fact that a volume integral can always be split into a set of integrals over disjoint subvolumes that add up to the whole. That is, integration is linear, as it is merely addition.

Deriving equation (iii) has simplified a complicated problem greatly, because the center-of-mass coordinates of the two individual objects can be calculated easily. Since the density of each of them is constant, their center-of-mass locations are identical to their geometrical centers. One look at Figure 11.10 is enough to convince us that $X_2 = \frac{1}{2}\ell$ and $X_1 = -\frac{1}{2}b$. (Remember, we chose the origin of the coordinate system as the right edge of the marble block.)

All that is left now is to calculate the masses of the two objects. Since we know their densities, we only need to figure out each object's volume; then the mass is given by $M = \rho V$.

Since the marble block is rectangular, its volume is $V = abc$. Thus, we have

$$M_1 = \rho_1 abc.$$

The horizontal wooden part is a cylinder, so its mass is

$$M_2 = \frac{\rho_2 \ell \pi d^2}{4}.$$

## CALCULATE

Inserting the numbers given in the problem statement, we obtain for the individual masses

$$M_1 = (2850 \text{ kg/m}^3)(0.71 \text{ m})(0.71 \text{ m})(2.74 \text{ m}) = 3936.52 \text{ kg}$$

$$M_2 = \frac{(440 \text{ kg/m}^3)(2.84 \text{ m})\pi(0.71 \text{ m})^2}{4} = 494.741 \text{ kg}.$$

Thus, the combined mass is

$$M = M_1 + M_2 = 3936.52 \text{ kg} + 494.741 \text{ kg} = 4431.261 \text{ kg}.$$

The location of the $x$-component of the center of mass of the sculpture is then

$$X = \frac{3936.52 \text{ kg}}{4431.261 \text{ kg}}\left[(-0.5)(0.71 \text{ m})\right] + \frac{494.741 \text{ kg}}{4431.261 \text{ kg}}\left[(0.5)(2.84 \text{ m})\right] = -0.156825 \text{ m}.$$

*Continued—*

### ROUND

The densities were given to three significant figures, but the length dimensions were given to only two significant figures. Rounding thus results in

$$X = -0.16 \text{ m}.$$

Since this number is negative, the center of mass of the sculpture is located to the left of the right edge of the marble block. Thus, it is located above the base of the block and is supported from directly below. The sculpture is stable and can stand without bracing.

### DOUBLE-CHECK

From Figure 11.9a, it seems hardly possible that this sculpture wouldn't tip over. However, our eyes can deceive us, as the density of the sculpture isn't constant. The ratio of the densities of the two materials used in the sculpture is $\rho_1/\rho_2 = (2850 \text{ kg/m}^3)/(440 \text{ kg/m}^3) = 6.48$. Therefore, we would obtain the same location for the center of mass if the 2.84-m-long cylinder made of wood were replaced by a cylinder of the same length made of marble, with its central axis located at the same place as that of the wooden cylinder, but thinner by a factor of $\sqrt{6.48} = 2.55$. The sculpture shown in Figure 11.11 has the same center-of-mass location as the sculpture in Figure 11.9a. Figure 11.11 should convince you that the sculpture is able to stand in stable equilibrium without bracing.

**FIGURE 11.11** Sculpture made entirely of marble that has the same center-of-mass location as the marble-and-wood sculpture in Figure 11.9a.

### 11.1 In-Class Exercise

If a point mass were placed at the far right end of the wooden cylinder in the sculpture in Figure 11.9a, how big could this mass be before the sculpture would tip over?

a) 2.4 kg

b) 29.1 kg

c) 37.5 kg

d) 245 kg

e) 1210 kg

The next example considers a situation in which the force of static friction plays an essential role. Static friction forces help maintain many arrangements of objects in equilibrium.

## EXAMPLE 11.4 Person Standing on a Ladder

Typically, a ladder stands on a horizontal surface (the floor) and leans against a vertical surface (the wall). Suppose a ladder of length $\ell = 3.04$ m, with mass $m_l = 13.3$ kg, rests against a smooth wall at an angle of $\theta = 24.8°$. A student, who has a mass of $m_m = 62.0$ kg, stands on the ladder (Figure 11.12a). The student is standing on a rung that is $r = 1.43$ m along the ladder, measured from where the ladder touches the ground.

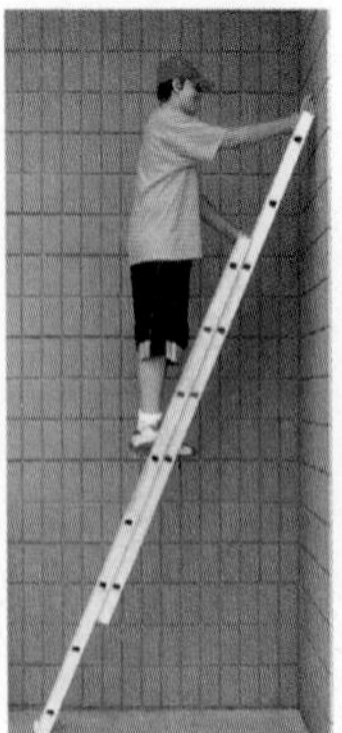

(a)

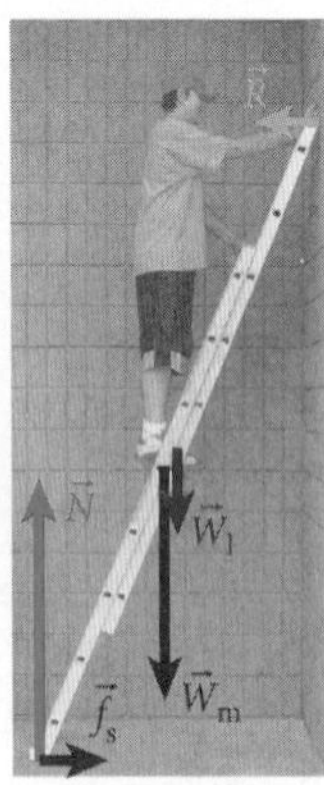



(b)

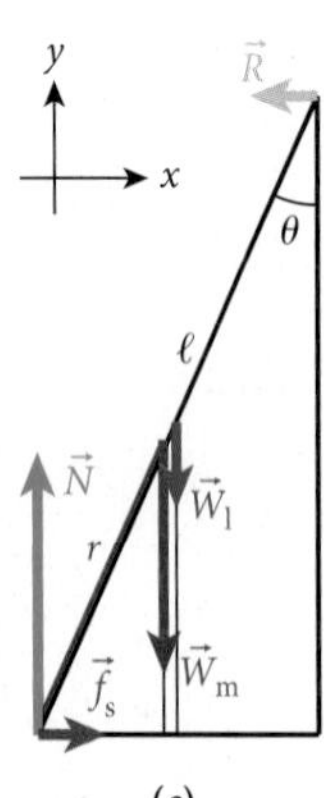



(c)

**FIGURE 11.12** (a) Student standing on a ladder. (b) Force vectors superimposed. (c) Free-body diagram of the ladder.

### PROBLEM 1

What friction force must act on the bottom of the ladder to keep it from slipping? Neglect the (small) force of friction between the smooth wall and the ladder.

### SOLUTION 1

Let's start with the free-body diagram shown in Figure 11.12c. Here $\vec{R} = -R\hat{x}$ is the normal force exerted by the wall on the ladder, $\vec{N} = N\hat{y}$ is the normal force exerted by the floor on the ladder, and $\vec{W}_m = -m_m g\hat{y}$ and $\vec{W}_l = -m_l g\hat{y}$ are the weights of the student and the ladder: $m_m g = (62.0 \text{ kg})(9.81 \text{ m/s}^2) = 608.$ N and $m_l g = (13.3 \text{ kg})(9.81 \text{ m/s}^2) = 130.$ N.

We'll let $\vec{f}_s = f_s\hat{x}$ be the force of static friction between the floor and the bottom of the ladder, which is the answer to the problem. Note that this force vector is directed in the positive $x$-direction (if the ladder slips, its bottom will slide in the negative $x$-direction, and the friction force must necessarily oppose that motion). As instructed, we neglect the force of friction between wall and ladder.

The ladder and the student are in translational and rotational equilibrium, so we have the three equilibrium conditions introduced in equations 11.5 through 11.7:

$$\sum_i F_{x,i} = 0, \quad \sum_i F_{y,i} = 0, \quad \sum_i \tau_i = 0.$$

Let's start with the equation for the force components in the horizontal direction:

$$\sum_i F_{x,i} = f_s - R = 0 \Rightarrow R = f_s.$$

From this equation, we learn that the force the wall exerts on the ladder and the friction force between the ladder and the floor have the same magnitude. Next, we write the equation for the force components in the vertical direction:

$$\sum_i F_{y,i} = N - m_m g - m_l g = 0 \Rightarrow N = g(m_m + m_l).$$

The normal force that the floor exerts on the ladder is exactly equal in magnitude to the sum of the weights of ladder and man: $N$ = 608. N +130. N = 738. N. (Again, we neglect the friction force between wall and ladder, which would otherwise have come in here.)

Now we sum the torques, assuming that the pivot point is where the ladder touches the ground. This assumption has the advantage of allowing us to ignore the forces acting at that point, because their moment arms are zero.

$$\sum_i \tau_i = (m_l g)\left(\frac{\ell}{2}\right)\sin\theta + (m_m g) r \sin\theta - R\ell\cos\theta = 0. \qquad \text{(i)}$$

Note that the torque from the wall's normal force acts counterclockwise, whereas the two torques from the weights of the student and the ladder act clockwise. Also, the angle between the normal force, $\vec{R}$, and its moment arm, $\vec{\ell}$, is 90° – $\theta$, and sin (90° – $\theta$) = cos $\theta$. Now we solve equation (i) for $R$:

$$R = \frac{\frac{1}{2}(m_l g)\ell\sin\theta + (m_m g) r\sin\theta}{\ell\cos\theta} = \left(\tfrac{1}{2}m_l g + m_m g\frac{r}{\ell}\right)\tan\theta.$$

Numerically, we obtain

$$R = \left(\tfrac{1}{2}(130.\text{ N}) + (608.\text{ N})\frac{1.43\text{ m}}{3.04\text{ m}}\right)(\tan 24.8°) = 162.\text{ N}.$$

However, we already found that $R = f_s$, so our answer is $f_s$ = 162. N.

**PROBLEM 2**

Suppose that the coefficient of static friction between ladder and floor is 0.31. Will the ladder slip?

**SOLUTION 2**

We found the normal force in the first part of this example: $N = g(m_m + m_l)$ = 738. N. It is related to the maximum static friction force via $f_{s,max} = \mu_s N$. So, the maximum static fraction force is 229. N, well above the 162. N that we just found necessary for static equilibrium. In other words, the ladder will not slip.

In general, the ladder will not slip as long as the force from the wall is smaller than the maximum force of static friction, leading to the condition

$$R = \left(\tfrac{1}{2}m_l + m_m\frac{r}{\ell}\right)g\tan\theta \le \mu_s(m_l + m_m)g. \qquad \text{(ii)}$$

*Continued—*

### 11.2 In-Class Exercise

What can the student in Example 11.4 do if he really has to get up just a bit higher than the maximum height allowed by equation (ii) for the given situation?

a) He can increase the angle $\theta$ between the wall and the ladder.

b) He can decrease the angle $\theta$ between the wall and the ladder.

c) Neither increasing nor decreasing the angle will make any difference.

**PROBLEM 3**

What happens as the student climbs higher on the ladder?

**SOLUTION 3**

From equation (ii), we see that $R$ grows larger with increasing $r$. Eventually, this force will overcome the maximum force of static friction, and the ladder will slip. You can now understand why it is not a good idea to climb too high on a ladder in this kind of situation.

## 11.3 Stability of Structures

**FIGURE 11.13** The bridge carrying Interstate 35W across the Mississippi River in Minneapolis collapsed on August 1, 2007, during rush hour.

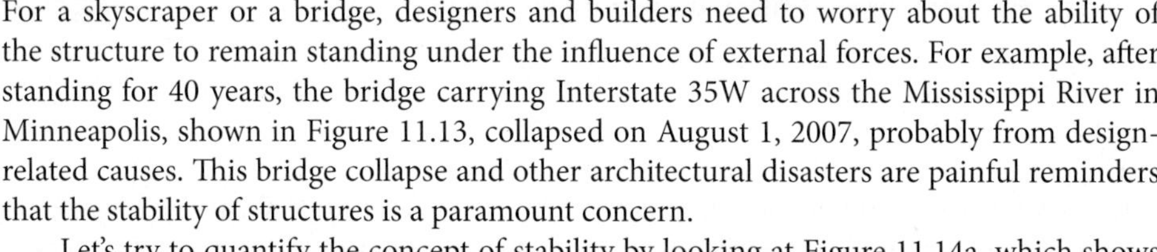

For a skyscraper or a bridge, designers and builders need to worry about the ability of the structure to remain standing under the influence of external forces. For example, after standing for 40 years, the bridge carrying Interstate 35W across the Mississippi River in Minneapolis, shown in Figure 11.13, collapsed on August 1, 2007, probably from design-related causes. This bridge collapse and other architectural disasters are painful reminders that the stability of structures is a paramount concern.

Let's try to quantify the concept of stability by looking at Figure 11.14a, which shows a box in static equilibrium, resting on a horizontal surface. Our experience tells us that if we use a finger to push with a small force in the way shown in the figure, the box remains in the same position. The small force we exert on the box is exactly balanced by the force of friction between the box and the supporting surface. The net force is zero, and there is no motion. If we steadily increase the magnitude of the force we apply, there are two possible outcomes: If the friction force is not sufficient to counterbalance the force exerted by the finger, the box begins to slide to the right. Or, if the torque of the friction force about the center of mass of the box is less than the torque due to the applied force, the box starts to tilt as shown in Figure 11.14b. Thus, the static equilibrium of the box is stable with respect to small external forces, but a sufficiently large external force destroys the equilibrium.

This simple example illustrates the characteristic of **stability.** Engineers need to be able to calculate the maximum external forces and torques that can be present without undermining the stability of a structure.

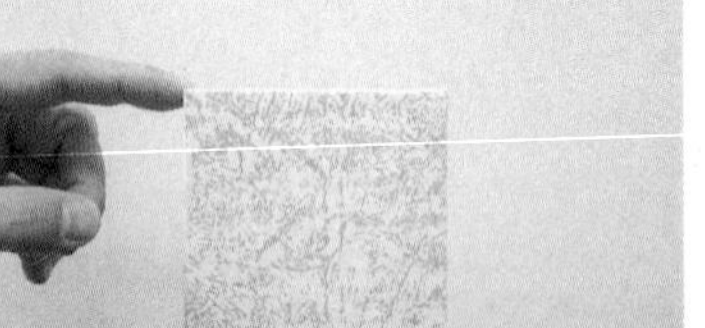

(a)

(b)

**FIGURE 11.14** (a) Pushing with a small force against the upper edge of a box. (b) Exerting a larger force on the box results in tilting it.

### Quantitative Condition for Stability

In order to be able to quantify the stability of an equilibrium situation, we start with the relationship between potential energy and force from Chapter 6:

$$\vec{F}(\vec{r}) = -\vec{\nabla}U(\vec{r}).$$

In one dimension, this is

$$F_x(x) = -\frac{dU(x)}{dx}.$$

A vanishing net force is one of the equilibrium conditions, which we can write as $\vec{\nabla}U(\vec{r}) \equiv \left( \frac{\partial U(\vec{r})}{\partial x}\hat{x} + \frac{\partial U(\vec{r})}{\partial y}\hat{y} + \frac{\partial U(\vec{r})}{\partial z}\hat{z} \right) = 0$, or as $\frac{dU(x)}{dx} = 0$ in one dimension, at a given point in space. So far, the condition of vanishing first derivative adds no new insight. However, we can use the second derivative of the potential energy function to distinguish three different cases, depending on the sign of the second derivative.

#### Case 1 Stable Equilibrium

$$\text{Stable equilibrium: } \left. \frac{d^2U(x)}{dx^2} \right|_{x=x_0} > 0. \tag{11.8}$$

If the second derivative of the potential energy function with respect to the coordinate is positive at a point, then the potential energy has a local minimum at that point. The system is in **stable equilibrium.** In this case, a small deviation from the equilibrium position creates a restoring force that drives the system back to the equilibrium point.

This situation is illustrated in Figure 11.15a: If the red dot is moved away from its equilibrium position at $x_0$ in either the positive or the negative direction and released, it will return to the equilibrium position.

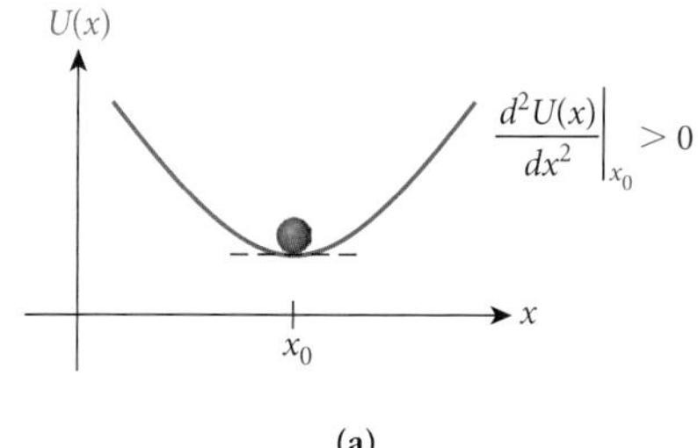


### Case 2 Unstable Equilibrium

$$\text{Unstable equilibrium: } \left.\frac{d^2U(x)}{dx^2}\right|_{x=x_0} < 0. \tag{11.9}$$

If the second derivative of the potential energy function with respect to the coordinate is negative at a point, then the potential energy has a local maximum at that point. The system is in **unstable equilibrium.** In this case, a small deviation from the equilibrium position creates a force that drives the system away from the equilibrium point. This situation is illustrated in Figure 11.15b: If the red dot is moved even slightly away from its equilibrium position at $x_0$ in either the positive or the negative direction and released, it will move away from the equilibrium position.

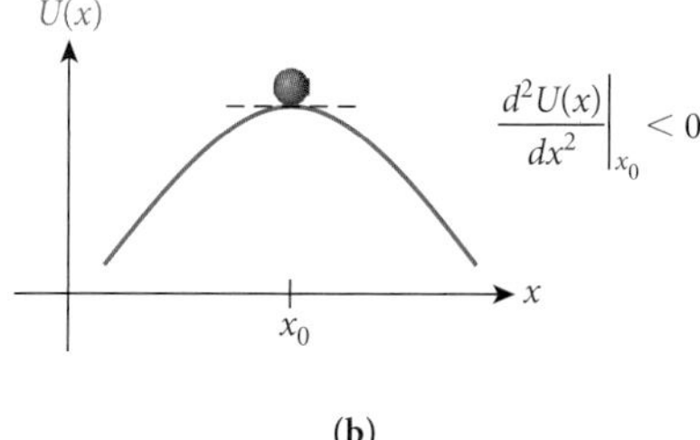


### Case 3 Neutral Equilibrium

$$\text{Neutral equilibrium: } \left.\frac{d^2U(x)}{dx^2}\right|_{x=x_0} = 0. \tag{11.10}$$

The case in which the sign of the second derivative of the potential energy function with respect to the coordinate is neither positive nor negative at a point is called **neutral equilibrium,** also referred to as *indifferent* or *marginally stable.* This situation is illustrated in Figure 11.15c: If the red dot is displaced by a small amount, it will neither return to nor move away from its original equilibrium position. Instead, it will simply stay in the new position, which is also an equilibrium position.

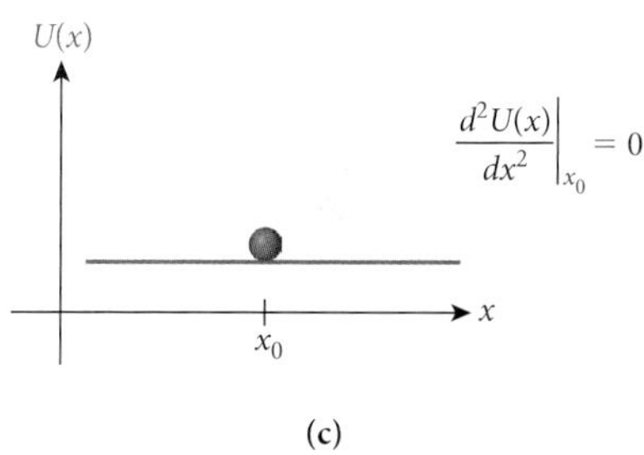


**FIGURE 11.15** Local shape of the potential energy function at an equilibrium point: (a) stable equilibrium; (b) unstable equilibrium; (c) neutral equilibrium.

## Multidimensional Surfaces and Saddle Points

The three cases just discussed cover all possible types of stability for one-dimensional systems. They can be generalized to two- and three-dimensional potential energy functions that depend on more than one coordinate. Instead of looking at only the derivative with respect to one coordinate, as in equations 11.8 through 11.10, we have to examine all partial derivatives. For the two-dimensional potential energy function $U(x,y)$, the equilibrium condition is that the first derivative with respect to each of the two coordinates is zero. In addition, stable equilibrium requires that at the point of equilibrium the second derivative of the potential energy function is positive for both coordinates, whereas unstable equilibrium implies that it is negative for both, and neutral equilibrium means that it is zero for both. Parts (a) through (c) of Figure 11.16 show these three cases, respectively.

However, in more than one spatial dimension, the possibility also exists that at one equilibrium point the second derivative with respect to one coordinate is positive, whereas it is negative for the other coordinate. These points are called *saddle points,* because the potential energy function is locally shaped like a saddle. Figure 11.16d shows such a saddle point, where one of the second partial derivatives is negative and one is positive. The equilibrium at this saddle point is stable with respect to small displacements in the $y$-direction, but unstable with respect to small displacements in the $x$-direction.

In a strict mathematical sense, the conditions noted above for the second derivative are sufficient for the existence of maxima and minima, but not necessary. Sometimes, the first derivative of the potential energy function is not continuous, but extrema can still exist, as the following example shows.

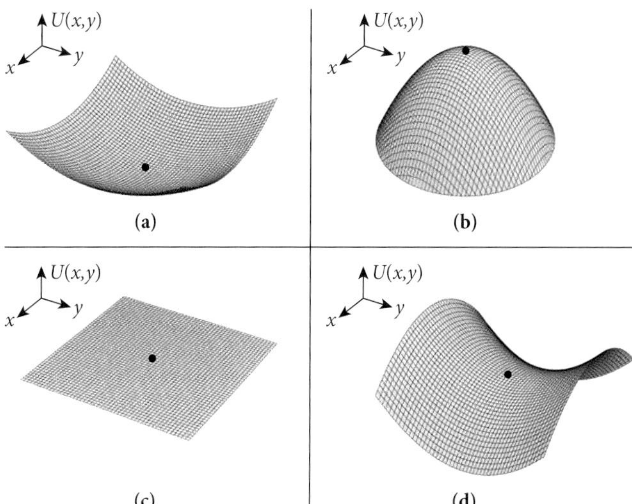


**FIGURE 11.16** Different types of equilibrium for a three-dimensional potential energy function.

### 11.3 In-Class Exercise

In Figure 11.16, which surface contains equilibrium points other than the one marked with the black dot?

a) c)

b) d)

## EXAMPLE 11.5 Pushing a Box

### PROBLEM 1

What is the force required to hold the box in Figure 11.14 in equilibrium at a given tilt angle?

### SOLUTION 1

Before the finger pushes on the box, the box is resting on a level surface. The only two forces acting on it are the force of gravity and a balancing normal force. There is no net force and no net torque; the box is in equilibrium (Figure 11.17a).

Once the finger starts pushing in the horizontal direction on the upper edge of the box and the box starts tilting, the normal force vector acts at the contact point (Figure 11.17b). The force of static friction acts at the same point but in the horizontal direction. Since the box does not slip, the friction force vector has exactly the same magnitude as the external force vector due to the pushing by the finger, but acts in the opposite direction.

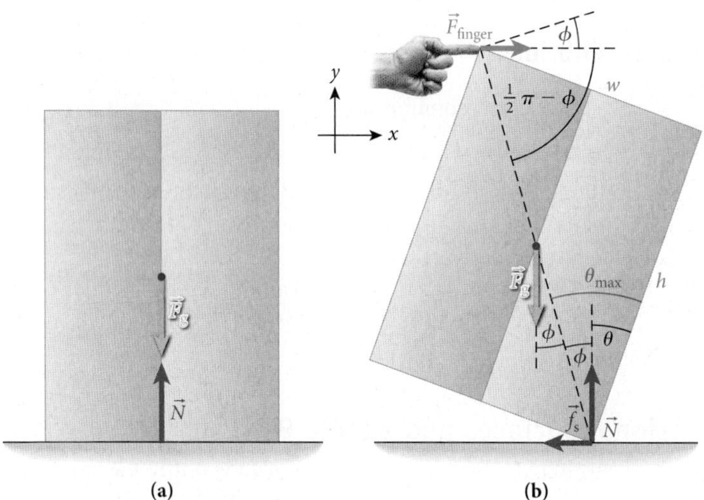


**FIGURE 11.17** Free-body diagrams for the box (a) resting on the level surface and (b) being tilted by a finger pushing in the horizontal direction.

We can now calculate the torques due to these forces and find the condition for equilibrium, that is, how much force it takes for the finger to hold the box at an angle $\theta$ with respect to the vertical. Figure 11.17b also indicates the angle $\theta_{max}$, which is a geometric property of the box that can be calculated from the ratio of the width $w$ and the height $h$: $\theta_{max} = \tan^{-1}(w/h)$. Of crucial importance is the angle $\phi$, which is the difference between these two angles (see Figure 11.17b): $\phi = \theta_{max} - \theta$. The angle $\phi$ decreases with increasing $\theta$ until $\theta = \theta_{max} \Rightarrow \phi = 0$, at which time the box falls over into the horizontal position.

Using equation 11.7, we can calculate the net torque. The natural pivot point in this situation is the contact point between the box and the support surface. The friction force and the normal force then have moment arms of zero length and thus do not contribute to the net torque. The only clockwise torque is due to the force from the finger, and the only counterclockwise torque results from the force of gravity. The length of the moment arm for the force from the finger is (see Figure 11.17b) $\ell = \sqrt{h^2 + w^2}$, and the length of the moment arm for the force of gravity is half of this value, or $\ell/2$. This means that equation 11.7 becomes

$$(F_g)(\tfrac{1}{2}\ell)\sin\phi - (F_{finger})(\ell)\sin(\tfrac{1}{2}\pi - \phi) = 0.$$

We can use $\sin(\frac{1}{2}\pi - \phi) = \cos\phi$ and $F_g = mg$ and then solve for the force the finger must provide to keep the box at equilibrium at a given angle:

$$F_{finger}(\theta) = \tfrac{1}{2}mg \tan\left[\tan^{-1}\left(\frac{w}{h}\right) - \theta\right]. \quad \text{(i)}$$

Figure 11.18 shows a plot of the force from the finger that is needed to hold the box at equilibrium at a given angle, from equation (i), for different ratios of box width to height. The curves reflect values of the angle $\theta$ between zero and $\theta_{max}$, which is the point where the force required from the finger is zero, and where the box tips over.

**PROBLEM 2**

Sketch the potential energy function for this box.

**SOLUTION 2**

Finding the solution to this part of the example is much more straightforward than for the first part. The potential energy is the gravitational potential energy, $U = mgy$, where $y$ is the vertical coordinate of the center of mass of the box. Figure 11.19, shows (red curve) the location of the center of mass of the box for different tipping angles. The curve traces out a segment of a circle with center at the lower right corner of the box. The dashed red line shows the same curve, but for angles $\theta > \theta_{max}$, for which the box tips over into the horizontal position without the finger exerting a force on it. You can clearly see that this potential energy function has a maximum at the point where the box stands on edge and its center of mass is exactly above the contact point with the surface. The curve for the location of the center of mass when the box is tilted to the left is shown in blue in Figure 11.19. You can see that the potential energy function has a minimum when the box rests flat on the table. *Note:* At this equilibrium point, the first derivative does not exist in the mathematical sense, but it is apparent from the figure that the function has a minimum, which is sufficient for stable equilibrium.

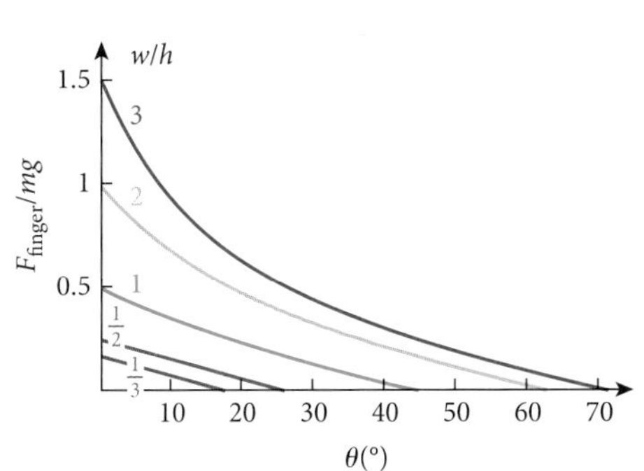


**FIGURE 11.18** Ratio of the force of the finger to the weight of the box needed to hold the box in equilibrium as a function of the angle, plotted for different representative values of the ratio of the width to the height of the box.

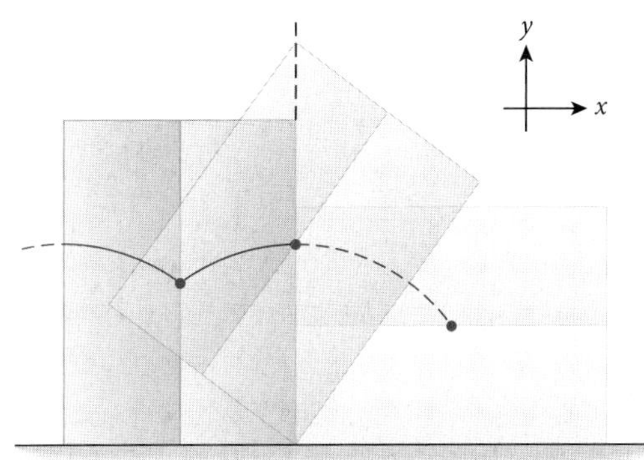


**FIGURE 11.19** Location of the center of mass of the box as a function of tipping angle.

## Dynamic Adjustments for Stability

How does the mass damper in the Taipei 101 tower provide stability to the structure? To answer this question, let's first look at how human beings stand up straight. When you stand up straight, your center of mass is located directly above your feet. The gravitational force then exerts no net torque on you, and you can remain standing up straight. If other forces act on you (a strong wind blowing, for example, or a load you have to lift) and provide additional torques, your brain senses this through nerves coupled to the fluid in your inner ears and provides corrective action through slight shifts in the body's mass distribution. You can get a demonstration of your brain's impressive ability to make these dynamic stability adjustments by holding out your backpack (loaded with books, laptop computer, etc.) with outstretched arms in front of your body. This action will not cause you to fall over. However, if you stand straight against a wall, with your heels touching the base of the wall, the same attempt to lift your backpack will cause you to fall forward. Why? Because the wall behind you prevents your brain from shifting your body's mass distribution in order to compensate for the torque due to the backpack's weight.

### 11.4 Self-Test Opportunity

What is the minimum value that the coefficient of static friction can have if the box in Example 11.5 is to tip over and the ratio of the width to the height of the box is 0.4?

**FIGURE 11.20** The Segway provides dynamic stability adjustments.

The same principles of dynamic stability adjustment are incorporated into the Segway Human Transport (Figure 11.20), an innovative system on two wheels, which is electric-powered and can reach speeds of up to 12 mph. Just like your brain, the Segway senses its orientation relative to the vertical. But it uses gyroscopes instead of the fluid in the inner ear. And just like the brain, it counterbalances a net torque by providing a compensating torque in the opposite direction. The Segway accomplishes this by slightly turning its wheels in a clockwise or counterclockwise direction.

Finally, the mass damper at the top of the Taipei 101 tower is used in a similar way, to provide a subtle shift in the building's mass distribution, thus contributing to stability in the presence of net external torques due to the forces of strong winds. But it also dampens the oscillation of the building due to these forces, a topic we will return to in Chapter 14.

## WHAT WE HAVE LEARNED | EXAM STUDY GUIDE

- Static equilibrium is mechanical equilibrium for the special case where the object in equilibrium is at rest.
- An object (or a collection of objects) can be in static equilibrium only if the net external force is zero and the net external torque is zero:

$$\vec{F}_{\text{net}} = \sum_{i=1}^{n} \vec{F}_i = \vec{F}_1 + \vec{F}_2 + \cdots + \vec{F}_n = 0$$

$$\vec{\tau}_{\text{net}} = \sum_i \vec{\tau}_i = 0$$

- The condition for static equilibrium can also be expressed as $\vec{\nabla} U(\vec{r})\big|_{\vec{r}_0} = 0$; that is, the first gradient derivative of the potential energy function with respect to the position vector is zero at the equilibrium point.
- The condition for stable equilibrium is that the potential energy function has a minimum at that point. A sufficient condition for stability is that the second derivative of the potential function with respect to the coordinate at the equilibrium point is positive.
- The condition for unstable equilibrium is that the potential energy function has a maximum at that point. A sufficient condition for instability is that the second derivative of the potential function with respect to the coordinate at the equilibrium point is negative.
- If the second derivative of the potential energy function with respect to the coordinate is zero at the equilibrium point, this type of equilibrium is called neutral (or indifferent or marginally stable).

## KEY TERMS

static equilibrium, p. 355
stability, p. 366
stable equilibrium, p. 366
unstable equilibrium, p. 367
neutral equilibrium, p. 367

## NEW SYMBOLS AND EQUATIONS

$\vec{F}_{\text{net}} = 0$, first condition for static equilibrium

$\vec{\tau}_{\text{net}} = 0$, second condition for static equilibrium

## ANSWERS TO SELF-TEST OPPORTUNITIES

**11.1** Chose pivot point to be at the location of $m_2$.

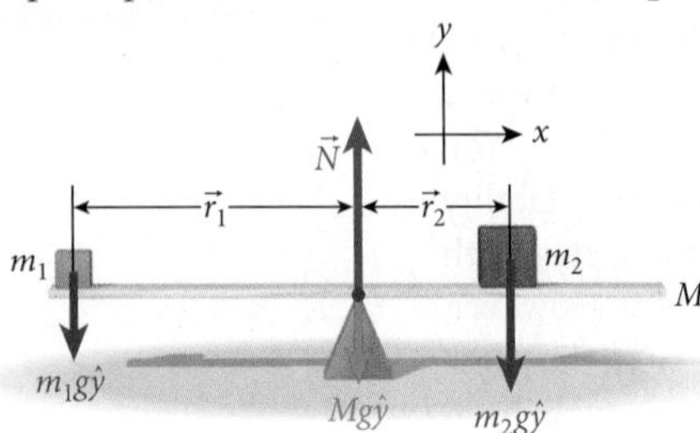


$$F_{\text{net},y} = \sum_i F_{i,y} = -m_1 g - m_2 g - Mg + N = 0$$

$$\Rightarrow N = g(m_1 + m_2 + M) = m_1 g + m_2 g + Mg$$

$$\tau_{\text{net}} = \sum_i \tau_{\text{clockwise},i} - \sum_j \tau_{\text{counterclockwise},j}$$

$$\tau_{\text{net}} = Nr_2 \sin 90° - m_1 g(r_1 + r_2)\sin 90° - Mgr_2 \sin 90° = 0$$

$$Nr_2 = m_1 g r_1 + m_1 g r_2 + Mgr_2$$

$$(m_1 g + m_2 g + Mg) r_2 = m_1 g r_2 + m_2 g r_2 + Mgr_2 = m_1 g r_1 + m_1 g r_2 + Mgr_2$$

$$\cancel{m_1 g r_2} + m_2 g r_2 + \cancel{Mgr_2} = m_1 g r_1 + \cancel{m_1 g r_2} + \cancel{Mgr_2}$$

$$m_2 \cancel{g} r_2 = m_1 \cancel{g} r_1$$
$$m_2 r_2 = m_1 r_1.$$

**11.2** The biceps still has a nonzero moment arm even when the arm is fully extended, because the tendon has to wrap around the elbow joint and is thus never parallel to the radius bone.

**11.3** The stack would be (10,000)(4 cm) = 400 m high, comparable to the tallest skyscrapers. And the top block's right edge would stick out by

$$\tfrac{1}{2}(15\text{ cm})\sum_{i=1}^{10,000}\frac{1}{i} = \tfrac{1}{2}(15\text{ cm})(9.78761) = 73.4\text{ cm}.$$

**11.4** From Figure 11.18, we can see that the force required is greatest at the beginning, for $\theta = 0$. If we set $\theta = 0$ in equation (i) of Example 11.5, we find the initial force that the finger needs to apply to displace the box from equilibrium:

$$F_{\text{finger}}(0) = \tfrac{1}{2}mg\tan\left[\tan^{-1}\left(\frac{w}{h}\right) - 0\right] = \tfrac{1}{2}mg\left(\frac{w}{h}\right).$$

A friction coefficient of at least $\mu_s = \frac{1}{2}(w/h)$ is needed to provide the matching friction force that prevents the box from sliding along the supporting surface: $\mu_s = \frac{1}{2}(0.4) = 0.2$.

## PROBLEM-SOLVING PRACTICE

### Problem-Solving Guidelines: Static Equilibrium

1. Almost all static equilibrium problems involve summing forces in the coordinate directions, summing torques, and setting the sums equal to zero. However, the right choices of coordinate axes and a pivot point for torques can make the difference between a hard solution and an easy one. Generally, choosing a pivot point that eliminates the moment arm for an unknown force (and often more than one force!) will simplify the equations so that you can solve for some force components.

2. The key step in writing correct equations for situations of static equilibrium is to draw a correct free-body diagram. Be careful about the locations where forces act; because torques are involved, you must represent objects as extended bodies, not point particles, and the point of application of a force makes a difference. Check each force to make sure that it is exerted *on* the object in equilibrium and not *by* the object.

## SOLVED PROBLEM 11.2 Hanging Storefront Sign

It is not uncommon for businesses to hang a sign over the sidewalk, suspended from a building's front wall. They often attach a post to the wall by a hinge and hold it horizontal by a cable that is also attached to the wall: the sign is then suspended from the post. Suppose the mass of the sign in Figure 11.21a is $M = 33.1$ kg, and the mass of the post is $m = 19.7$ kg. The length of the post is $l = 2.40$ m, and the sign is attached to the post as shown at a distance $r = 1.95$ m from the wall. The cable is attached to the wall a distance $d = 1.14$ m above the post.

### PROBLEM

What is the tension in the cable holding the post? What is the magnitude and direction of the force, $\vec{F}$, that the wall exerts on the post?

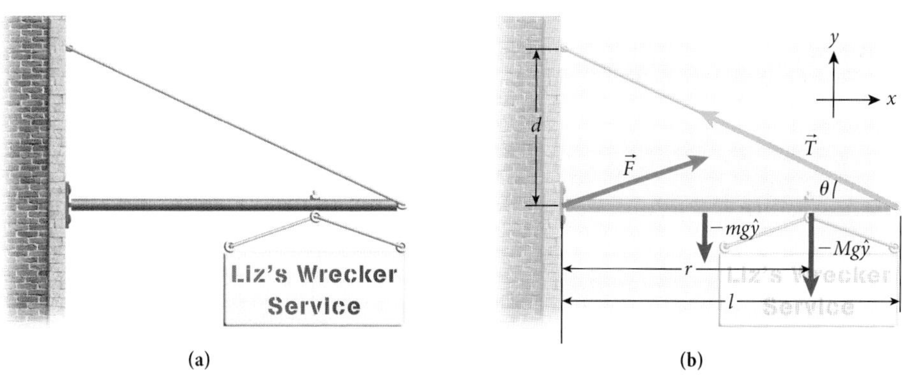


**FIGURE 11.21** (a) Hanging storefront sign; (b) free-body diagram for the post.

*Continued—*

## SOLUTION

### THINK

This problem involves static equilibrium, moment arms, and torques. Static equilibrium means vanishing net external force and torque. In order to determine the torques, we have to pick a pivot point. It seems natural to pick the point where the hinge attaches the post to the wall. Because a hinge is used, the post can rotate about this point. Picking this point also has the advantage that we do not need to pay attention to the force that the wall exerts on the post, because that force will act at the contact point (the hinge) and thus have a moment arm of zero and consequently no contribution to the torque.

### SKETCH

To calculate the net torque, we start with a free-body diagram that shows all the forces acting on the post (Figure 11.21b). We know that the weight of the sign (red arrow) acts at the point where the sign is suspended from the post. The gravitational force acting on the post is represented by the blue arrow, which points downward from the center of mass of the post. Finally, we know that the string tension, $\vec{T}$ (yellow arrow), is acting along the direction of the cable.

### RESEARCH

The angle $\theta$ between the cable and the post (see Figure 11.21b) can be found from the given data:

$$\theta = \tan^{-1}\left(\frac{d}{l}\right).$$

The equation for the torques about the point where the post touches the wall is then

$$mg\frac{l}{2}\sin 90° + Mgr\sin 90° - Tl\sin\theta = 0. \qquad \text{(i)}$$

Figure 11.21b also shows a green arrow for the force, $\vec{F}$, that the wall exerts on the post, but the direction and magnitude of this force vector are still to be determined. Equation (i) cannot be used to find this force, because the point where this force acts is the pivot point, and thus the corresponding moment arm has length zero.

On the other hand, once we have found the tension in the cable, we will have determined all the other forces involved in the problem situation, and we know from the condition of static equilibrium that the net force has to be zero. We can thus write separate equations for the horizontal and vertical force components. In the horizontal direction, we have only two force components, from the string tension and the force from the wall:

$$F_x - T\cos\theta = 0 \Rightarrow F_x = T\cos\theta.$$

In the vertical direction, we have the weights of the beam and the sign, in addition to the vertical components of the string tension and the force from the wall:

$$F_y + T\sin\theta - mg - Mg = 0 \Rightarrow F_y = (m+M)g - T\sin\theta.$$

### SIMPLIFY

We solve equation (i) for the tension:

$$T = \frac{(ml + 2Mr)g}{2l\sin\theta}. \qquad \text{(ii)}$$

For the magnitude of the force that the wall exerts on the post, we find

$$F = \sqrt{F_x^2 + F_y^2}.$$

The direction of this force is given by

$$\theta_F = \tan^{-1}\left(\frac{F_y}{F_x}\right).$$

**CALCULATE**

Inserting the numbers given in the problem statement, we find the angle $\theta$:

$$\theta = \tan^{-1}\left(\frac{1.14 \text{ m}}{2.40 \text{ m}}\right) = 25.4°.$$

We obtain the tension in the cable from equation (ii):

$$T = \frac{\left[(19.7 \text{ kg})(2.40 \text{ m}) + 2(33.1 \text{ kg})(1.95 \text{ m})\right](9.81 \text{ m/s}^2)}{2(2.40 \text{ m})(\sin 25.4°)} = 840.351 \text{ N}.$$

The magnitudes of the components of the force that the wall exerts on the post are

$$F_x = (840.351 \text{ N})(\cos 25.4°) = 759.119 \text{ N}$$

$$F_y = (19.7 \text{ kg} + 33.1 \text{ kg})(9.81 \text{ m/s}^2) - (840.351 \text{ N})(\sin 25.4°) = 157.512 \text{ N}.$$

Therefore, the magnitude and direction of that force are given by

$$F = \sqrt{(157.512 \text{ N})^2 + (759.119 \text{ N})^2} = 775.288 \text{ N}$$

$$\theta_F = \tan^{-1}\left(\frac{759.119}{157.512}\right) = 11.7°.$$

**ROUND**

All of the given quantities were specified to three significant figures, and so we round our final answers to three digits: $T = 840.$ N and $F = 775.$ N.

**DOUBLE-CHECK**

The two forces that we calculated have rather large magnitudes, considering that the combined weight of the post and attached sign is only

$$F_g = (m + M)g = (19.7 \text{ kg} + 33.1 \text{ kg})(9.81 \text{ m/s}^2) = 518 \text{ N}.$$

In fact, the sum of the magnitudes of the force from the cable on the post, $T$, and the force from the wall on the post, $F$, is larger than the combined weight of the post and the sign by more than a factor of 3. Does this make sense? Yes, because the two force vectors, $\vec{T}$ and $\vec{F}$, have rather large horizontal components that have to cancel each other. When we calculate the magnitudes of these forces, their horizontal components are also included. As the angle $\theta$ between the cable and the post approaches zero, the horizontal components of $\vec{T}$ and $\vec{F}$ become larger and larger. Thus, you can see that selecting a distance $d$ in Figure 11.21b that is too small relative to the length of the post will result in a huge tension in the cable and a very stressed suspension system.

### 11.4 In-Class Exercise

If all other parameters in Solved Problem 11.2 remain unchanged, but the sign is moved farther away from the wall toward the end of the post, what happens to the tension, $T$?

a) It decreases.

b) It stays the same.

c) It increases.

### 11.5 In-Class Exercise

If all other parameters in Solved Problem 11.2 remain unchanged, but the angle $\theta$ between the cable and the post is increased, what happens to the tension, $T$?

a) It decreases.

b) It stays the same.

c) It increases.

## MULTIPLE-CHOICE QUESTIONS

**11.1** A 3.0-kg broom is leaning against a coffee table. A woman lifts the broom handle with her arm fully stretched so that her hand is a distance of 0.45 m from her shoulder. What torque is produced on her shoulder if her arm is at an angle of 50° below the horizontal?

a) 7.0 N m  c) 8.5 N m
b) 5.8 N m  d) 10.1 N m

**11.2** A uniform beam of mass $M$ and length $L$ is held in static equilibrium, and so the magnitude of the net torque about its center of mass is zero. The magnitude of the net torque on this beam at one of its ends, a distance of $L/2$ from the center of mass, is

a) $MgL$.  c) zero.
b) $MgL/2$.  d) $2MgL$.

**11.3** A very light and rigid rod is pivoted at point $A$ and weights $m_1$ and $m_2$ are hanging from it, as shown in the figure. The ratio of the weight of $m_1$ to that of $m_2$ is 1:2. What is the ratio of $L_1$ to $L_2$, the distances from the pivot point to $m_1$ and $m_2$, respectively?

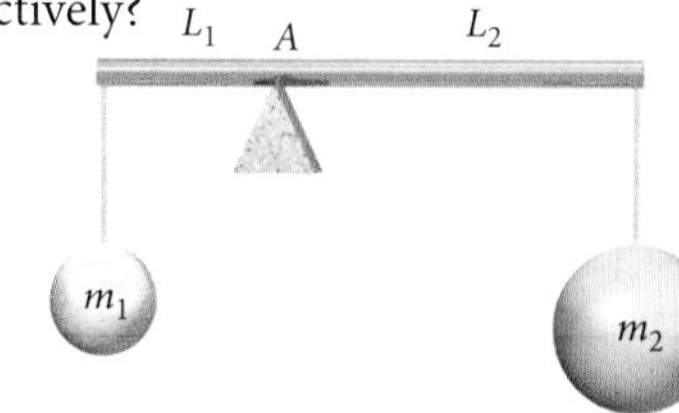


a) 1:2
b) 2:1
c) 1:1
d) not enough information to determine

**11.4** Which of the following are in static equilibrium?

a) a pendulum at the top of its swing
b) a merry-go-round spinning at constant angular velocity

c) a projectile at the top of its trajectory (with zero velocity)
d) all of the above
e) none of the above

**11.5** The object in the figure below is suspended at its center of mass—thus, it is balanced. If the object is cut in two pieces at its center of mass, what is the relation between the two resulting masses?

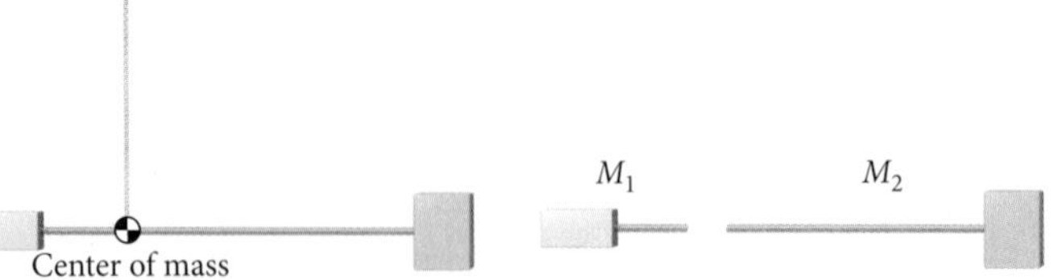


a) The masses are equal.
b) $M_1$ is less than $M_2$.
c) $M_2$ is less than $M_1$.
d) It is impossible to tell.

**11.6** As shown in the figure, two weights are hanging on a uniform wooden bar that is 60 cm long and has a mass of 100 g. Is this system in equilibrium?

a) yes
b) no
c) cannot be determined
d) depends on the value of the normal force

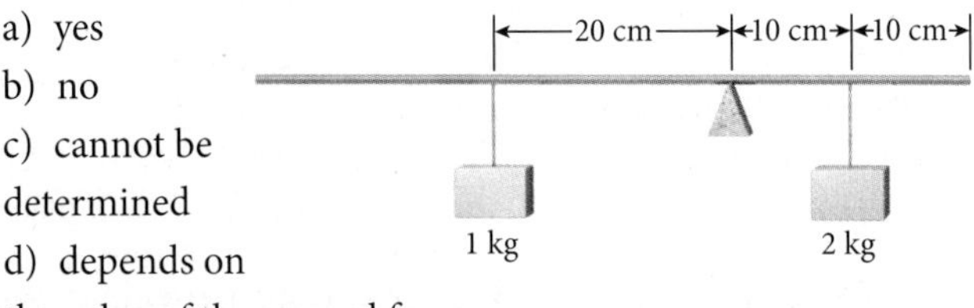


**11.7** A 15-kg child sits on a playground seesaw, 2.0 m from the pivot. A second child located 1.0 m on the other side of the pivot would have to have a mass of ________ to lift the first child off the ground.

a) greater than 30 kg
b) less than 30 kg
c) equal to 30 kg

**11.8** A mobile is constructed from a metal bar and two wooden blocks as shown in the figure. The metal bar has a mass of 1 kg and is 10 cm long. The metal bar has a 3-kg wooden block hanging from the left end and a string tied to it at a distance of 3 cm from the left end. What mass should the wooden block hanging from the right end of the bar have to keep the bar level?

a) 0.7 kg
b) 0.8 kg
c) 0.9 kg
d) 1.0 kg
e) 1.3 kg
f) 3.0 kg
g) 7.0 kg

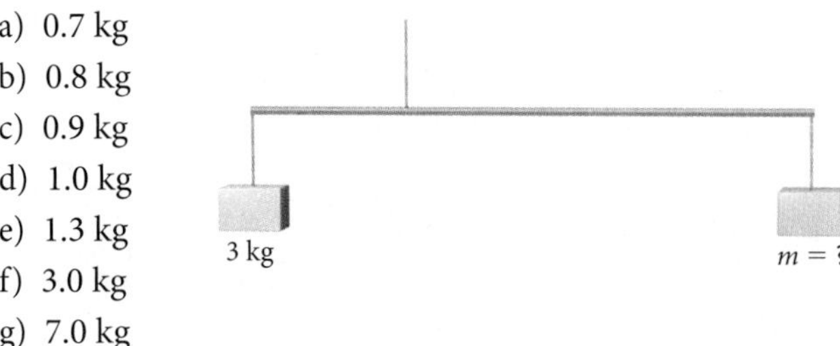


## QUESTIONS

**11.9** There are three sets of landing gear on an airplane: One main set is located under the centerline of each wing, and the third set is located beneath the nose of the plane. Each set of main landing gear has four tires, and the nose landing gear has two tires. If the load on all of the tires is the same when the plane is at rest, find the center of mass of the plane. Express your result as a fraction of the perpendicular distance between the centerline of the plane's wings to the plane's nose gear. (Assume that the landing gear struts are vertical when the plane is at rest and the dimensions of the landing gear are negligible compared to the dimensions of the plane.)

**11.10** A semicircular arch of radius $a$ stands on level ground as shown in the figure. The arch is uniform in cross section and density, with total weight $W$. By symmetry, at the top of the arch, each of the two legs exerts only horizontal forces on the other; ideally, the stress at that point is uniform compression over the cross section of the arch. What are the vertical and horizontal force components which must be supplied at the base of each leg to support the arch?

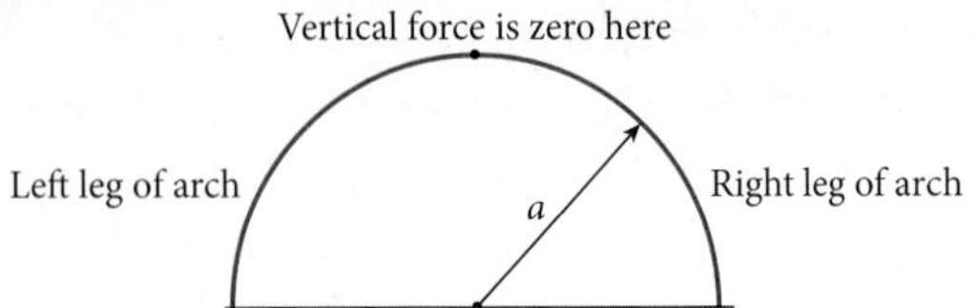


**11.11** In the absence of any symmetry or other constraints on the forces involved, how many unknown force components can be determined in a situation of static equilibrium in each of the following cases?

a) All forces and objects lie in a plane.
b) Forces and objects are in three dimensions.
c) Forces act in $n$ spatial dimensions.

**11.12** You have a meter stick that balances at the 55-cm mark. Is your meter stick homogeneous?

**11.13** You have a meter stick that balances at the 50-cm mark. Is it possible for your meter stick to be inhomogeneous?

**11.14** Why does a helicopter with a single main rotor generally have a second small rotor on its tail?

**11.15** The system shown in the figure consists of a uniform (homogeneous) rectangular board that is resting on two identical rotating cylinders. The two cylinders rotate in opposite directions at equal angular velocities. Initially, the board is placed perfectly symmetrically relative to the center point between the two cylinders. Is this an equilibrium position for the board? If yes, is it in stable or unstable equilibrium? What happens

if the board is given a very slight displacement out of the initial position?

**11.16** If the wind is blowing strongly from the east, stable equilibrium for an open umbrella is achieved if its shaft points west. Why is it relatively easy to hold the umbrella directly into the wind (in this case, easterly) but very difficult to hold it perpendicular to the wind?

**11.17** A sculptor and his assistant are carrying a wedge-shaped marble slab up a flight of stairs, as shown in the figure. The density of the marble is uniform. Both are lifting straight up as they hold the slab completely stationary for a moment. Does the sculptor have to exert more force than the assistant to keep the slab stationary? Explain.

**11.18** As shown in the figure, a thin rod of mass $M$ and length $L$ is suspended by two wires—one at the left end and one two-thirds of the distance from the left end to the right end.

a) What is the tension in each wire?

b) Determine the mass that an object hung by a string attached to the far right-hand end of the rod would have to have for the tension in the left-hand wire to be zero.

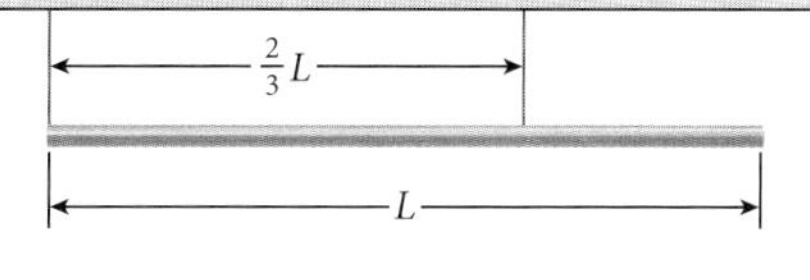

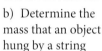

**11.19** A uniform disk of mass $M_1$ and radius $R_1$ has a circular hole of radius $R_2$ cut out as shown in the figure.

a) Find the center of mass of the resulting object.

b) How many equilibrium positions does this object have when resting on its edge? Which ones are stable, which neutral, and which unstable?

**11.20** Consider the system shown in the figure. If a pivot point is placed at a distance $L/2$ from the ends of the rod of length $L$ and mass $5M$, the system will rotate clockwise. Thus, in order for the system to not rotate, the pivot point should be placed away from the center of the rod. In which direction from the center of the rod should the pivot point be placed? How far from the center of the rod should the pivot point be placed in order for the system not to rotate? (Treat masses $M$ and $2M$ as point masses.)

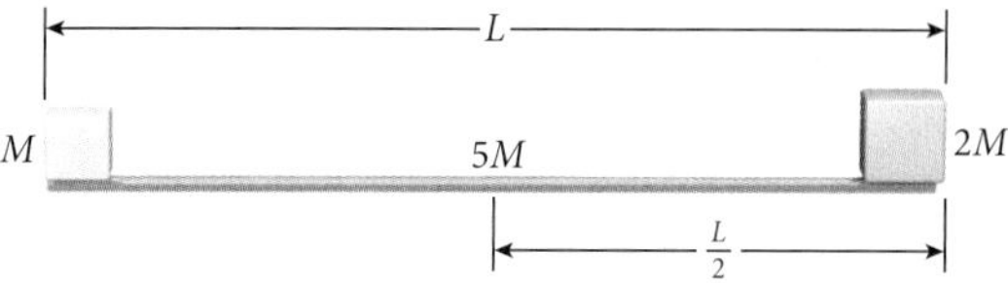


**11.21** A child has a set of blocks that are all made from the same type of wood. The blocks come in three shapes: a cube of side $L$, a piece the size of two cubes, and a piece equivalent in size to three of the cubes placed end to end. The child stacks three blocks as shown in the figure: a cube on the bottom, one of the longest blocks horizontally on top of that, and the medium-sized block placed vertically on top. The centers of each block are initially on a vertical line. How far can the top block be slid along the middle block before the middle block tips?

**11.22** Why didn't the ancient Egyptians build their pyramids upside-down? In other words, use force and center-of-mass principles to explain why it is more advantageous to construct buildings with broad bases and narrow tops than the other way around.

## PROBLEMS

A blue problem number indicates a worked-out solution is available in the Student Solutions Manual. One • and two •• indicate increasing level of problem difficulty.

### Section 11.1

**11.23** A 1000-N crate rests on a horizontal floor. It is being pulled up by two vertical ropes. The left rope has a tension of 400 N. Assuming the crate does not leave the floor, what can you say about the tension in the right rope?

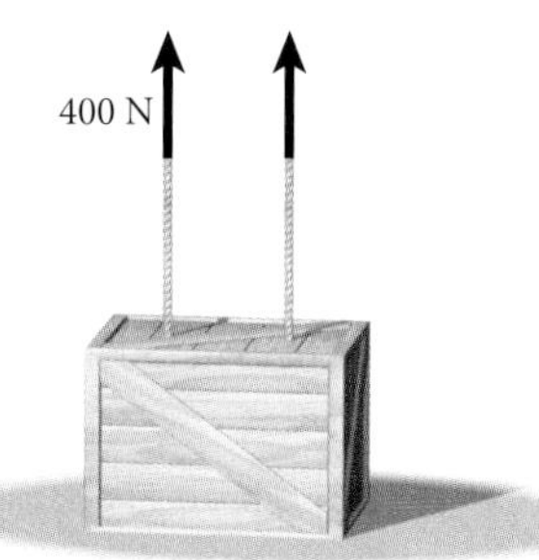


**11.24** In preparation for a demonstration on conservation of energy, a professor attaches a 5.0-kg bowling ball to a 4.0-m-long rope. He pulls the ball 20° away from the vertical and holds the ball while he discusses the physics principles involved. Assuming that the force he exerts on the ball is entirely in the horizontal direction, find the tension in the rope and the force the professor is exerting on the ball.

**11.25** A sculptor and his assistant stop for a break as they carry a marble slab of length $L = 2.00$ m and mass 75.0 kg up the steps, as shown in the figure. The mass of the slab is uniformly distributed along its length. As they rest, both the sculptor and his assistant are pulling *directly up* on each end of the slab, which is at an angle of 30° with respect to horizontal. What are the magnitudes of the forces that the sculptor and assistant must exert on the marble slab to keep it stationary during their break?

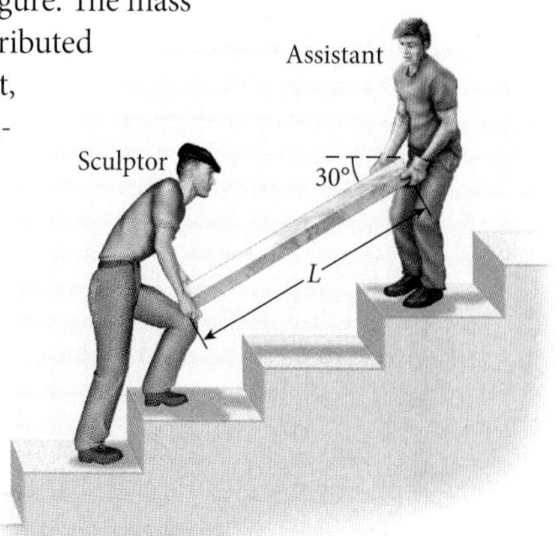

**11.26** During a picnic, you and two of your friends decide to have a three-way tug-of-war, with three ropes in the middle tied into a knot. Roberta pulls to the west with 420 N of force; Michael pulls to the south with 610 N. In what direction and with what magnitude of force should you pull to keep the knot from moving?

**11.27** The figure shows a photo of a typical merry-go-round, found at many playgrounds and a diagram giving a top view. Four children are standing on the ground and pulling on the merry-go-round as indicated by the force arrows. The four forces have the magnitudes $F_1 = 104.9$ N, $F_2 = 89.1$ N, $F_3 = 62.8$ N, and $F_4 = 120.7$ N. All the forces act in a tangential direction. With what force, $\vec{F}$, also in a tangential direction and acting at the black point, does a fifth child have to pull in order to prevent the merry-go-round from moving? Specify the magnitude of the force and state whether the force is acting counterclockwise or clockwise.

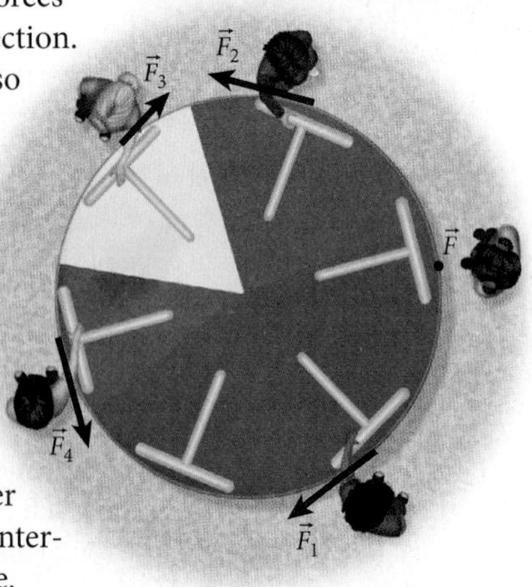

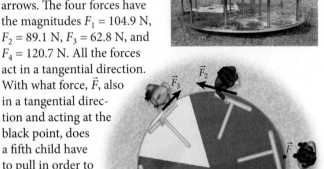

**11.28** A trapdoor on a stage has a mass of 19.2 kg and a width of 1.50 m (hinge side to handle side). The door can be treated as having uniform thickness and density. A small handle on the door is 1.41 m away from the hinge side. A rope is tied to the handle and used to raise the door. At one instant, the rope is horizontal, and the trapdoor has been partly opened so that the handle is 1.13 m above the floor. What is the tension, $T$, in the rope at this time?

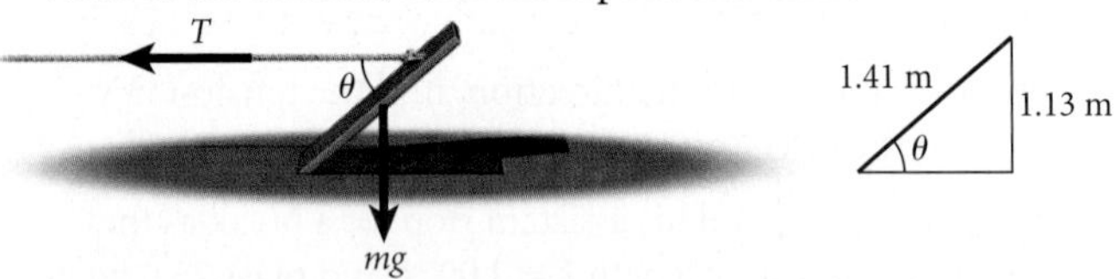


•**11.29** A rigid rod of mass $m_3$ is pivoted at point $A$, and masses $m_1$ and $m_2$ are hanging from it, as shown in the figure.

a) What is the normal force acting on the pivot point?

b) What is the ratio of $L_1$ to $L_2$, where these are the distances from the pivot point to $m_1$ and $m_2$, respectively? The ratio of the weights of $m_1$, $m_2$, and $m_3$ is 1:2:3.

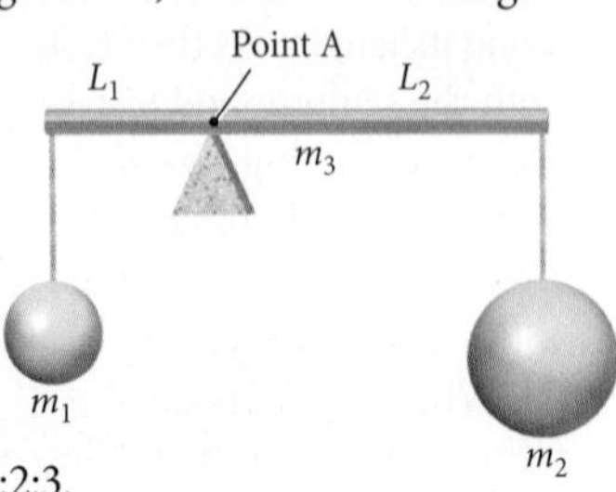


•**11.30** When only the front wheels of an automobile are on a platform scale, the scale balances at 8.0 kN; when only the rear wheels are on the scale, it balances at 6.0 kN. What is the weight of the automobile, and how far is its center of mass behind the front axle? The distance between the axles is 2.8 m.

•**11.31** By considering the torques about your shoulder, estimate the force your deltoid muscles (those on top of the shoulder) must exert on the bone of your upper arm, in order to keep your arm extended straight out at shoulder level. Then, estimate the force the muscles must exert to hold a 10-lb weight at arm's length. You'll need to estimate the distance from your shoulder pivot point to the point where your deltoid muscles connect to the bone of your upper arm in order to determine the necessary forces. Assume the deltoids are the only contributing muscles.

••**11.32** A uniform, equilateral triangle of side length 2.00 m and weight $4.00 \cdot 10^3$ N is placed across a gap. One point is on the north end of the gap, and the opposite side is on the south end. Find the force on each side.

## Section 11.2

**11.33** A 600.0-N bricklayer is 1.5 m from one end of a uniform scaffold that is 7.0 m long and weighs 800.0 N. A pile of bricks weighing 500.0 N is 3.0 m from the same end of the scaffold. If the scaffold is supported at both ends, calculate the force on each end.

**11.34** The uniform rod in the figure is supported by two strings. The string attached to the wall is horizontal, and the string attached to the ceiling makes an angle of $\phi$ with respect to the vertical. The rod itself is tilted from the vertical by an angle $\theta$. If $\phi = 30°$, what is the value of $\theta$?

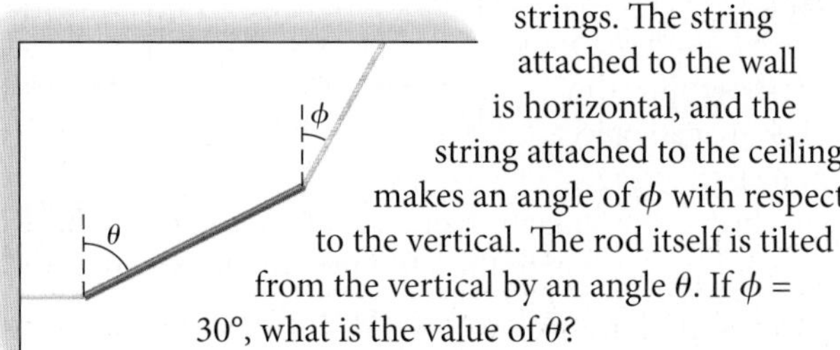


**11.35** A construction supervisor of mass $M = 92.1$ kg is standing on a board of mass $m = 27.5$ kg. Two sawhorses at a distance $\ell = 3.70$ m apart support the board. If the man stands a distance $x_1 = 1.07$ m away from the left-hand sawhorse as shown in the figure, what is the force that the board exerts on that sawhorse?

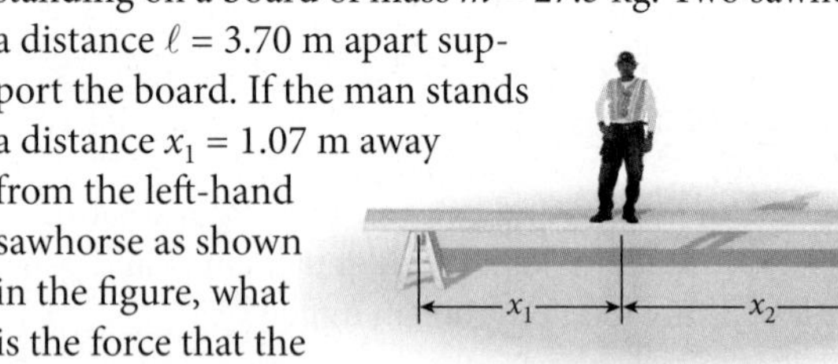


**11.36** In a butcher shop, a horizontal steel bar of mass 4.00 kg and length 1.20 m is supported by two vertical wires attached to its ends. The butcher hangs a sausage of mass 2.40 kg from a hook that is at a distance of 0.20 m from the left end of the bar. What are the tensions in the two wires?

•**11.37** Two uniform planks, each of mass $m$ and length $L$, are connected by a hinge at the top and by a chain of negligible mass attached at their centers, as shown in the figure. The assembly will stand upright, in the shape of an A, on a frictionless surface without

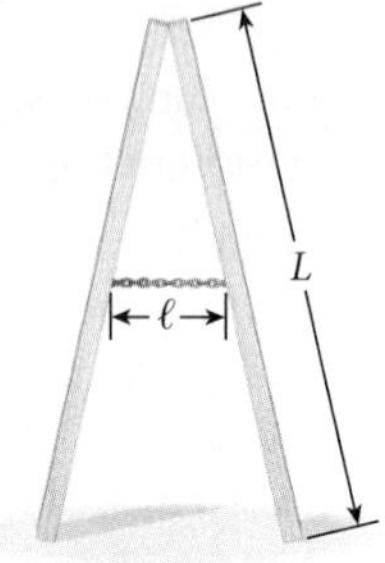

collapsing. As a function of the length of the chain, find each of the following:

a) the tension in the chain,

b) the force on the hinge of each plank, and

c) the force of the ground on each plank.

•**11.38** Three strings are tied together. They lie on top of a circular table, and the knot is exactly in the center of the table, as shown in the figure (top view). Each string hangs over the edge of the table, with a weight supported from it. The masses $m_1 = 4.30$ kg and $m_2 = 5.40$ kg are known. The angle $\alpha = 74°$ between strings 1 and 2 is also known. What is the angle $\beta$ between strings 1 and 3?

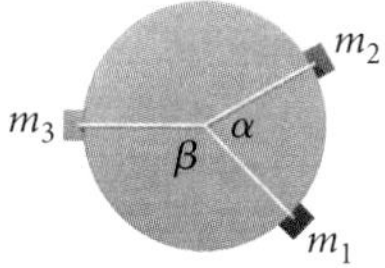


•11.39 A uniform ladder 10.0 m long is leaning against a frictionless wall at an angle of 60.0° above the horizontal. The weight of the ladder is 20.0 lb. A 61.0-lb boy climbs 4.00 m up the ladder. What is the magnitude of the frictional force exerted on the ladder by the floor?

•**11.40** Robin is making a mobile to hang over her baby sister's crib. She purchased four stuffed animals: a teddy bear (16 g), a lamb (18 g), a little pony (22 g) and a bird (15 g). She also purchased three small wooden dowels, each 15 cm long and of mass 5 g, and thread of negligible mass. She wants to hang the bear and the pony from the ends of one dowel and the lamb and the bird from the ends of the second dowel. Then, she wants to suspend the two dowels from the ends of the third dowel and hang the whole assembly from the ceiling. Explain where the thread should be attached on each dowel so that the entire assembly will hang level.

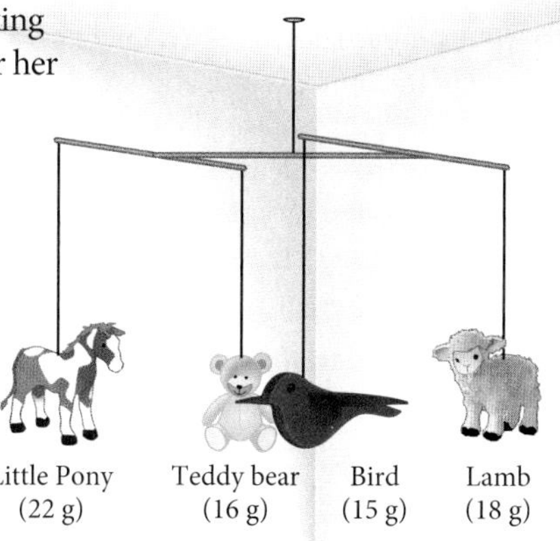


•**11.41** A door, essentially a uniform rectangle of height 2.00 m, width 0.80 m, and weight 100.0 N, is supported at one edge by two hinges, one 30.0 cm above the bottom of the door and one 170.0 cm above the bottom of the door. Calculate the horizontal components of the forces on the two hinges

••**11.42** The figure shows a 20-kg, uniform ladder of length $L$ hinged to a horizontal platform at point $P_1$ and anchored with a steel cable of the same length as the ladder attached at the ladder's midpoint. Calculate the tension in the cable and the forces in the hinge when an 80-kg person is standing three-quarters of the way up the ladder.

••11.43 A beam with a length of 8 m and a mass of 100 kg is attached by a large bolt to a support at a distance of 3 m from one end. The beam makes an angle $\theta = 30°$ with the horizontal, as shown in the figure. A mass $M = 500$ kg is attached with a rope to one end of the beam, and a second rope is attached at a right angle to the other end of the beam. Find the tension, $T$, in the second rope and the force exerted on the beam by the bolt.

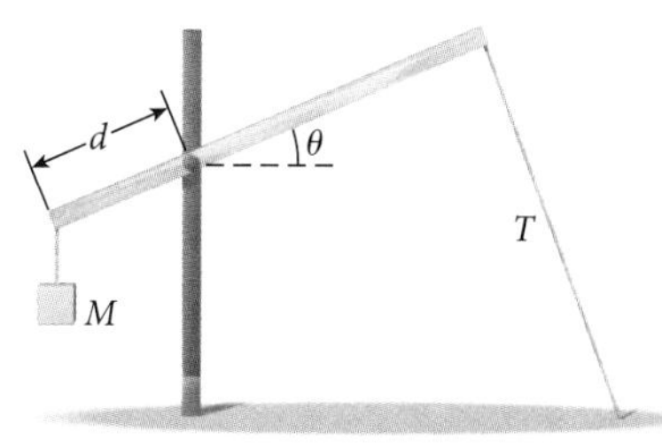


••**11.44** A ball of mass 15.49 kg rests on a table of height 0.72 m. The tabletop is a rectangular glass plate of mass 12.13 kg, which is supported at the corners by thin legs, as shown in the figure. The width of the tabletop is $w = 138.0$ cm, and its depth $d = 63.8$ cm. If the ball touches the tabletop at a point $(x,y) = (69.0\text{ cm}, 16.6\text{ cm})$ relative to corner 1, what is the force that the tabletop exerts on each leg?

••**11.45** A wooden bridge crossing a canyon consists of a plank with length density $\lambda = 2$ kg/m suspended at $h = 10.0$ m below a tree branch by two ropes of length $L = 2h$ and with a maximum rated tension of 2000 N, which are attached to the ends of the plank, as shown in the figure. A hiker steps onto the bridge from the left side, causing the bridge to tip to an angle of 25° with respect to the horizontal. What is the mass of the hiker?

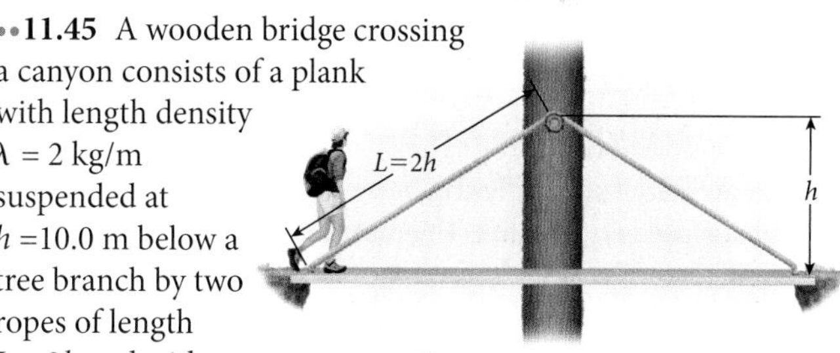


••**11.46** The famous Gateway Arch in St. Louis, Missouri, is approximately an *inverted catenary.* A simple example of such a curve is given by $y(x) = 2a - a\cosh(x/a)$, where $y$ is vertical height and $x$ is horizontal distance, measured from directly under the top of the curve; thus, $x$ varies from $-a\cosh^{-1}2$ to $+a\cosh^{-1}2$, with $a$ the height of the top of the curve (see

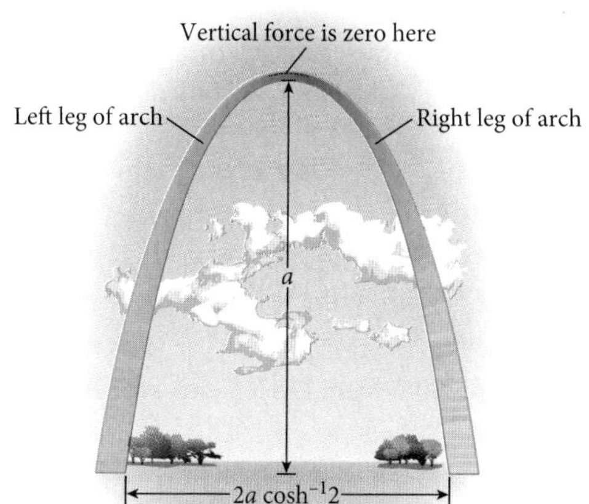

the figure). Suppose an arch of uniform cross section and density, with total weight $W$, has this shape. The two legs of the arch exert only horizontal forces on each other at the top; ideally, the stress there should be uniform compression across the cross section.

a) Calculate the vertical and horizontal force components that are acting at the base of each leg of this arch.

b) At what angle should the bottom face of the legs be oriented?

## Section 11.3

11.47 A uniform rectangular bookcase of height $H$ and width $W = H/2$ is to be pushed at a constant velocity across a level floor. The bookcase is pushed horizontally at its top edge, at the distance $H$ above the floor. What is the maximum value the coefficient of kinetic friction between the bookcase and the floor can have if the bookcase is not to tip over while being pushed?

11.48 The system shown in the figure is in static equilibrium. The rod of length $L$ and mass $M$ is held in an upright position. The top of the rod is tied to a fixed vertical surface by a string, and a force $F$ is applied at the midpoint of the rod. The coefficient of static friction between the rod and the horizontal surface is $\mu_s$. What is the maximum force, $F$, that can be applied and have the rod remain in static equilibrium?

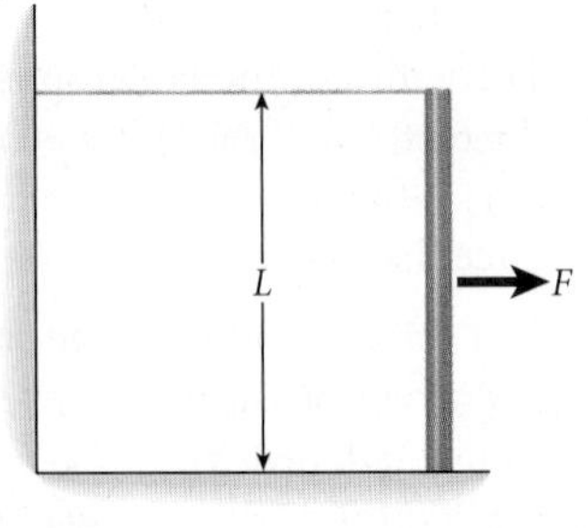


11.49 A ladder of mass 37.7 kg and length 3.07 m is leaning against a wall at an angle $\theta$. The coefficient of static friction between ladder and floor is 0.313; assume that the friction force between ladder and wall is zero. What is the maximum value that $\theta$ can have before the ladder starts slipping?

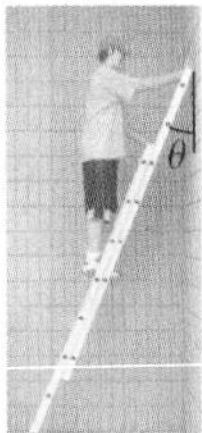


•11.50 A uniform rigid pole of length $L$ and mass $M$ is to be supported from a vertical wall in a horizontal position, as shown in the figure. The pole is not attached directly to the wall, so the coefficient of static friction, $\mu_s$, between the wall and the pole provides the only vertical force on one end of the pole. The other end of the pole is supported by a light rope that is attached to the wall at a point a distance $D$ directly above the point where the pole contacts the wall. Determine the minimum value of $\mu_s$, as a function of $L$ and $D$, that will keep the pole horizontal and not allow its end to slide down the wall.

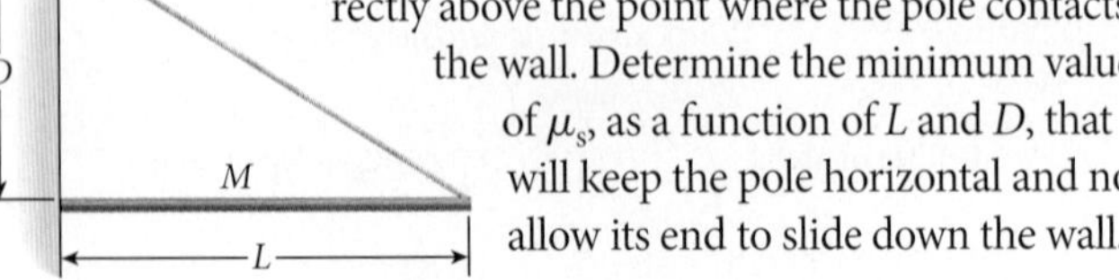


•11.51 A boy weighing 60.0 lb is playing on a plank. The plank weighs 30.0 lb, is uniform, is 8.00 ft long, and lies on two supports, one 2.00 ft from the left end and the other 2.00 ft from the right end.

a) If the boy is 3.00 ft from the left end, what force is exerted by each support?

b) The boy moves toward the right end. How far can he go before the plank will tip?

•11.52 A track has a height that is a function of horizontal position $x$, given by $h(x) = x^3 + 3x^2 - 24x + 16$. Find all the positions on the track where a marble will remain where it is placed. What kind of equilibrium exists at each of these positions?

•11.53 The figure shows a stack of seven identical aluminum blocks, each of length $l = 15.9$ cm and thickness $d = 2.2$ cm, stacked on a table.

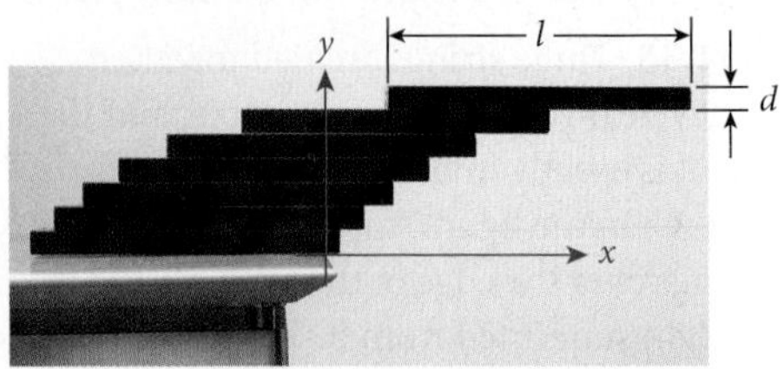


a) How far to the right of the edge of the table is it possible for the right edge of the top (seventh) block to extend?

b) What is the minimum height of a stack of these blocks for which the left edge of the top block is to the right of the right edge of the table?

•11.54 You are using a 5-m-long ladder to paint the exterior of your house. The point of contact between the ladder and the siding of the house is 4 m above the ground. The ladder has a mass of 20 kg. If you weigh 60 kg and stand three-quarters of the way up the ladder, determine (a) the forces exerted by the side wall and the ground on the ladder and (b) the coefficient of static friction between the ground and the base of the ladder that is necessary to keep the ladder stable.

•11.55 A ladder of mass $M$ and length $L = 4.0$ m is on a level floor leaning against a vertical wall. The coefficient of static friction between the ladder and the floor is $\mu_s = 0.60$, while the friction between the ladder and the wall is negligible. The ladder is at an angle of $\theta = 50°$ above the horizontal. A man of mass $3M$ starts to climb the ladder. To what distance up the ladder can the man climb before the ladder starts to slip on the floor?

•11.56 A machinist makes the object shown in the figure. The larger-diameter cylinder is made of brass (density of 8.6 g/cm$^3$); the smaller-diameter cylinder is made of aluminum (density of 2.7 g/cm$^3$). The dimensions are $r_1 = 2$ cm, $r_2 = 4$ cm, $d_1 = 20$ cm, and $d_2 = 4$ cm.

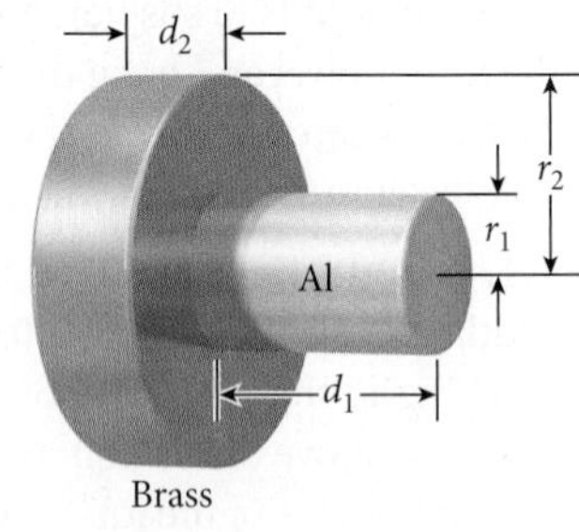


a) Find the location of the center of mass.

b) If the object is on its side, as shown in the figure, is it in equilibrium? If yes, is this stable equilibrium?

•11.57 An object is restricted to movement in one dimension. Its position is specified along the $x$-axis. The potential energy of the object as a function of its position is given by $U(x) = a(x^4 - 2b^2x^2)$, where $a$ and $b$ represent positive numbers. Determine the location(s) of any equilibrium point(s), and classify the equilibrium at each point as stable, unstable, or neutral.

•**11.58** A two-dimensional object with uniform mass density is in the shape of a thin square and has a mass of 2.00 kg. The sides of the square are each 20.0 cm long. The coordinate system has its origin at the center of the square. A point mass, $m$, of $2.00 \cdot 10^2$ g is placed at one corner of the square object, and the assembly is held in equilibrium by positioning the support at position $(x,y)$, as shown in the figure. Find the location of the support.

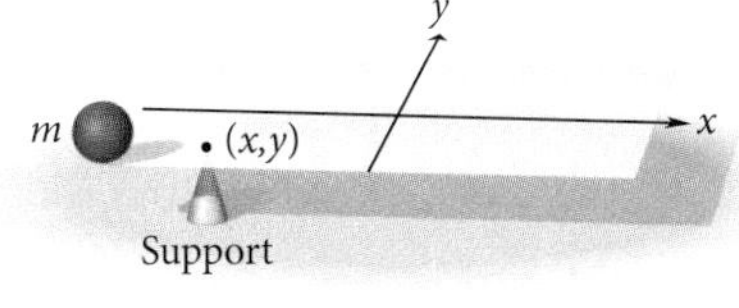


•**11.59** Persons $A$ and $B$ are standing on a board of uniform linear density that is balanced on two supports, as shown in the figure. What is the maximum distance $x$ from the right end of the board at which person $A$ can stand without tipping the board? Treat persons $A$ and $B$ as point masses. The mass of person $B$ is twice that of person $A$, and the mass of the board is half that of person $A$. Give your answer in terms of $L$, the length of the board.

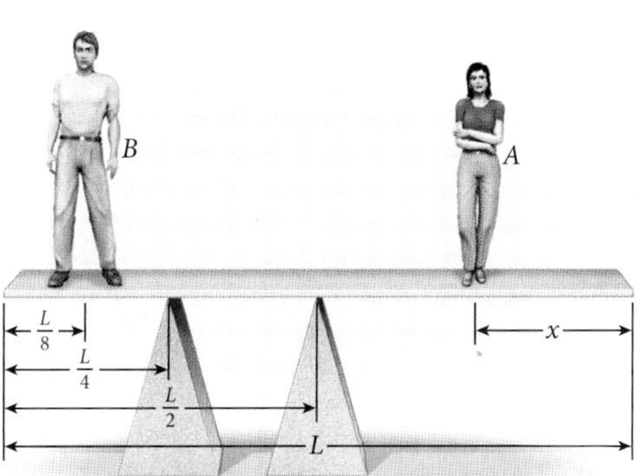


••**11.60** An SUV has a height $h$ and a wheelbase of length $b$. Its center of mass is midway between the wheels and at a distance $\alpha h$ above the ground, where $0 < \alpha < 1$. The SUV enters a turn at a dangerously high speed, $v$. The radius of the turn is $R$ ($R \gg b$), and the road is flat. The coefficient of static friction between the road and the properly inflated tires is $\mu_s$. After entering the turn, the SUV will either skid out of the turn or begin to tip.

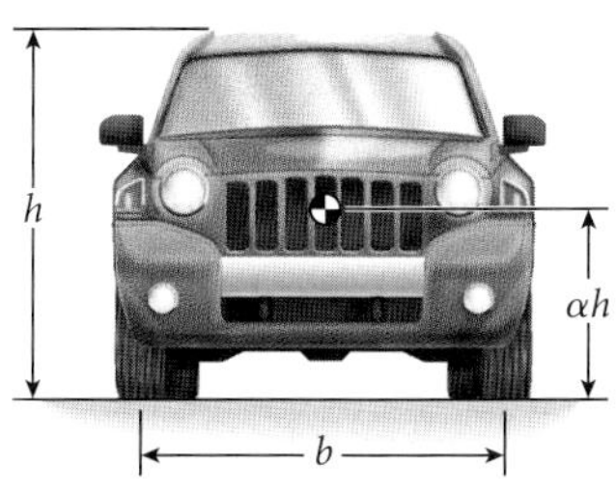


a) The SUV will skid out of the turn if the friction force reaches its maximum value, $F \rightarrow \mu_s N$. Determine the speed, $v_{skid}$, for which this will occur. Assume no tipping occurs.

b) The torque keeping the SUV from tipping acts on the outside wheel. The highest value this force can have is equal to the entire normal force. Determine the speed, $v_{tip}$, at which this will occur. Assume no skidding occurs.

c) It is safer if the SUV skids out before it tips. This will occur as long as $v_{skid} < v_{tip}$. Apply this condition, and determine the maximum value for $\alpha$ in terms of $b$, $h$ and $\mu_s$.

## Additional Problems

**11.61** A wooden plank with length $L = 8.0$ m and mass $M = 100$ kg is centered on a granite cube with side $S = 2.0$ m. A person of mass $m = 65$ kg begins walking from the center of the plank outward, as shown in the figure. How far from the center of the plank does the person get before the plank starts tipping?

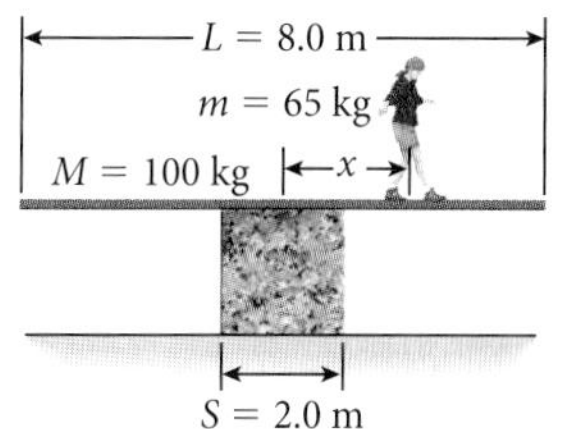


**11.62** A board, with a weight $mg = 120$ N and a length of 5 m, is supported by two vertical ropes, as shown in the figure. Rope $A$ is connected to one end of the board, and rope $B$ is connected at a distance $d = 1$ m from the other end of the board. A box with a weight $Mg = 20$ N is placed on the board with its center of mass at $d = 1$ m from rope $A$. What are the tensions in the two ropes?

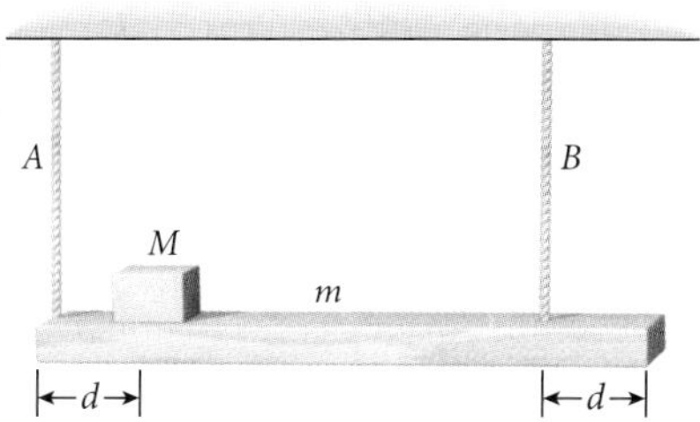


**11.63** In a car, which is accelerating at 5.00 m/s$^2$, an air freshener is hanging from the rear-view mirror, with the string maintaining a constant angle with respect to the vertical. What is this angle?

**11.64** Typical weight sets used for bodybuilding consist of disk-shaped weights with holes in the center that can slide onto 2.2-m-long barbells. A barbell is supported by racks located a fifth of its length from each end, as shown in the figure. What is the minimum mass $m$ of the barbell if a bodybuilder is to slide a weight with $M = 22$ kg onto the end without the barbell tipping off the rack? Assume that the barbell is a uniform rod.

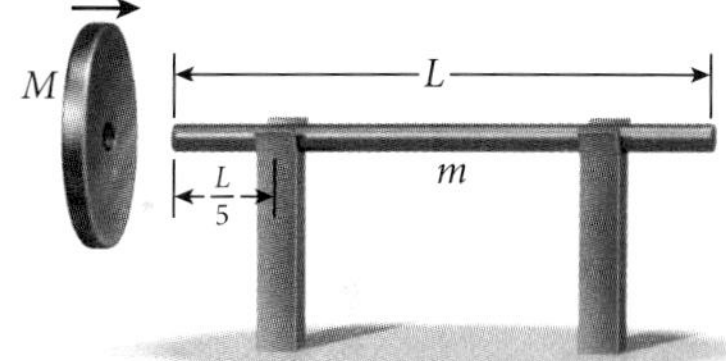


**11.65** A 5-m-long board of mass 50 kg is used as a seesaw. On the left end of the seesaw sits a 45-kg girl, and on the right end sits a 60-kg boy. Determine the position of the pivot point for static equilibrium.

**11.66** A mobile consists of two very lightweight rods of length $l = 0.400$ m connected to each other and the ceiling by vertical strings. (Neglect the masses of the rods and strings.) Three objects are suspended by strings from the rods. The masses of objects 1 and 3 are $m_1 = 6.40$ kg and $m_3 = 3.20$ kg. The distance $x$ shown in the figure is 0.160 m. What is the mass of $m_2$?

x
l
l
$m_1$
$m_3$
$m_2$

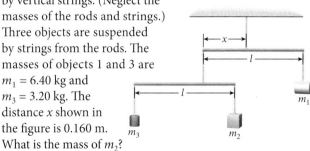

•**11.67** In the experimental setup shown in the figure, a beam, $B_1$, of unknown mass $M_1$ and length $L_1 = 1$ m is pivoted about its lowest point at $P_1$. A second beam, $B_2$, of mass $M_2 = 0.200$ kg and length $L_2 = 0.200$ m is suspended (pivoted) from $B_1$ at a point $P_2$, which is a horizontal distance

$d = 0.550$ m from $P_1$. To keep the system at equilibrium, a mass $m = 0.500$ kg has to be suspended from a massless string that runs horizontally from $P_3$, at the top of beam $B_1$, and passes over a frictionless pulley. The string runs at a vertical distance $y = 0.707$ m above the pivot point $P_1$. Calculate the mass of beam $B_1$.

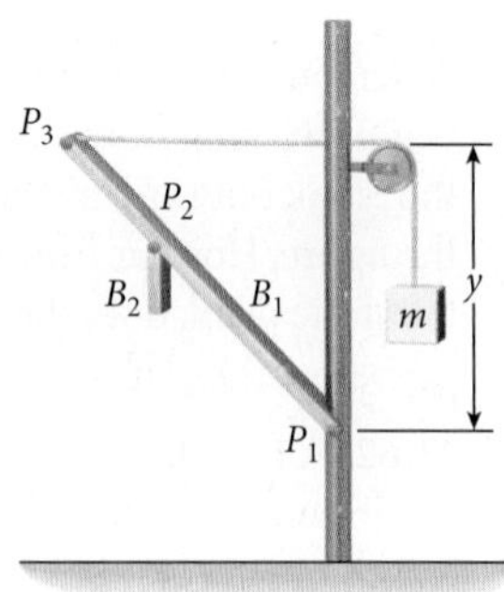


•**11.68** An important characteristic of the condition of static equilibrium is the fact that the net torque has to be zero irrespective of the choice of pivot point. For the setup in Problem 11.67, prove that the torque is indeed zero with respect to a pivot point at $P_1$, $P_2$, or $P_3$.

•**11.69** One end of a heavy beam of mass $M = 50.0$ kg is hinged to a vertical wall, and the other end is tied to a steel cable of length 3.0 m, as shown in the figure. The other end of the cable is also attached to the wall at a distance of 4.0 m above the hinge. A mass $m = 20.0$ kg is hung from one end of the beam by a rope.

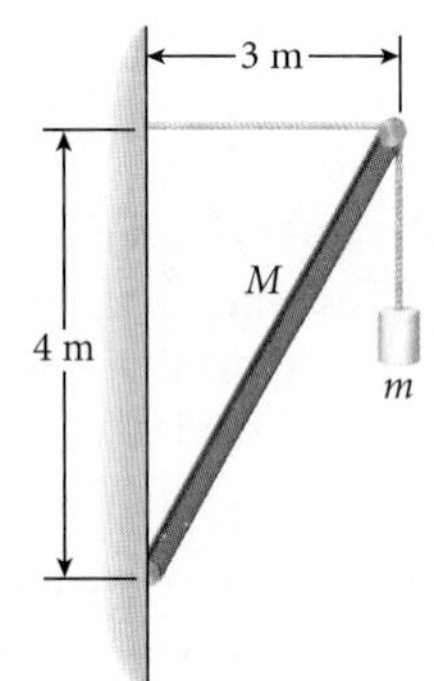


a) Determine the tensions in the cable and the rope.

b) Find the force the hinge exerts on the beam.

•**11.70** A 100-kg uniform bar of length $L = 5$ m is attached to a wall by a hinge at point $A$ and supported in a horizontal position by a light cable attached to its other end. The cable is attached to the wall at point $B$, at a distance $D = 2$ m above point $A$. Find:

a) the tension, $T$, on the cable and

b) the horizontal and vertical components of the force acting on the bar at point $A$.

•**11.71** A mobile over a baby's crib displays small colorful shapes. What values for $m_1$, $m_2$, and $m_3$ are needed to keep the mobile balanced (with all rods horizontal)?

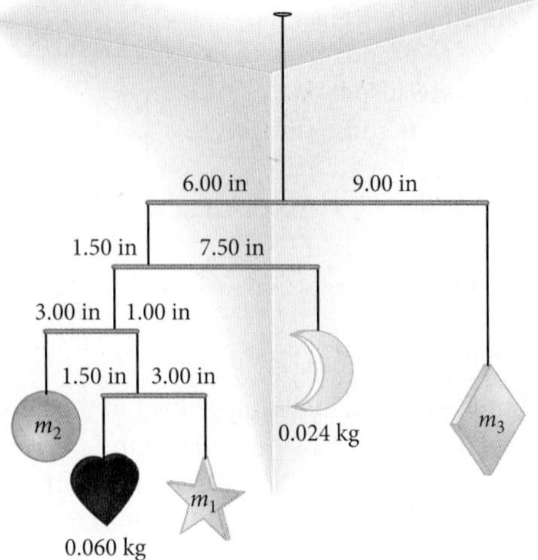


•**11.72** Consider the rod of length $L$ shown in the figure. The mass of the rod is $m = 2.00$ kg, and the pivot point is located at the left end (at $x = 0$). In order to prevent the rod from rotating, a variable force given by $F(x) = (15.0 \text{ N})(x/L)^4$ is applied to the rod. At what point $x$ on the rod should the force be applied in order to keep it from rotating?

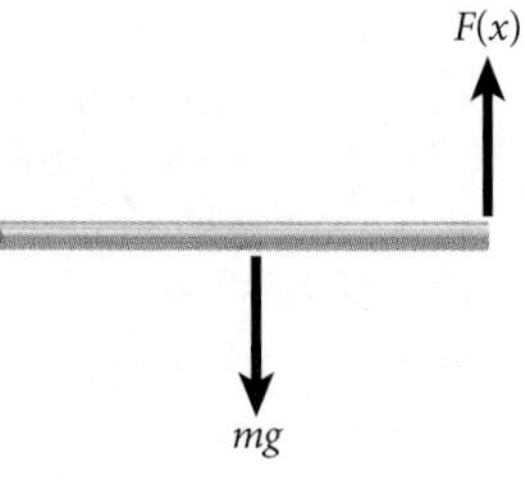


•**11.73** A pipe that is 2.20 m long and has a mass of 8.13 kg is suspended horizontally over a stage by two chains, each located 0.20 m from an end. Two 7.89-kg theater lights are clamped onto the pipe, one 0.65 m from the left end and the other 1.14 m from the left end. Calculate the tension in each chain.

•**11.74** A 2.00-m-long diving board of mass 12.0 kg is 3.00 m above the water. It has two attachments holding it in place. One is located at the very back end of the board, and the other is 25.0 cm away from that end.

a) Assuming that the board has uniform density, find the forces acting on each attachment (take the downward direction to be positive).

b) If a diver of mass 65.0 kg is standing on the front end, what are the forces acting on the two attachments?

•**11.75** A 20.0-kg box with a height of 80.0 cm and a width of 30.0 cm has a handle on the side that is 50.0 cm above the ground. The box is at rest, and the coefficient of static friction between the box and the floor is 0.28.

a) What is the minimum force, $F$, that can be applied to the handle so that the box will tip over without slipping?

b) In what direction should this force be applied?

••**11.76** The angular displacement of a torsional spring is proportional to the applied torque; that is $\tau = \kappa\theta$, where $\kappa$ is a constant. Suppose that such a spring is mounted to an arm that moves in a vertical plane. The mass of the arm is 45 g, and it is 12 cm long. The arm-spring system is at equilibrium with the arm at an angular displacement of 17° with respect to the horizontal. If a mass of 0.42 kg is hung from the arm 9 cm from the axle, what will be the angular displacement in the new equilibrium position (relative to that with the unloaded spring)?

# Gravitation 12

**FIGURE 12.1** Earthset photographed on November 7, 2007, by the Japanese satellite Kaguya orbiting the Moon. The Earth and the Moon orbit around each other and are kept together by their gravitational interaction.

## WHAT WE WILL LEARN

- The gravitational interaction between two point masses is proportional to the product of their masses and inversely proportional to the square of the distance separating them.
- The gravitational force on an object inside a homogeneous solid sphere rises linearly with the distance that the object is from the center of the solid sphere.
- At the surface of the Earth, it is a very good approximation to use a constant value for the acceleration due to gravity ($g$). The value of $g$ used for free-fall situations can be verified from the more general gravitational force law.
- A more general expression for the gravitational potential energy indicates that it is inversely proportional to the distance between two objects.
- Escape speed is the minimum speed with which a projectile must be launched to escape to infinity.
- Kepler's three laws of planetary motion state that planets move on elliptical orbits with the Sun at one focal point; that the radius vector connecting the Sun and a planet sweeps out equal areas in equal times; and that the square of the orbit's period for any planet is proportional to the cube of its semimajor axis.
- The kinetic, potential, and total energies of satellites in orbit have a fixed relationship with each other.
- There is evidence for a great amount of dark matter and dark energy in the universe.

Figure 12.1 shows the Earth setting over the horizon of the Moon, photographed by a satellite orbiting the Moon. We are so used to seeing the Moon in the sky that it's somewhat surprising to see the Earth in the sky. In fact, astronauts on the Moon do not see earthrise or earthset because the Moon always keeps the same face turned toward Earth. Only astronauts orbiting the Moon can see the Earth seeming to change position and rise or set. However, the image reminds us that, like all forces, the force of gravity is a mutual attraction between two objects—Earth pulls on the Moon, but the Moon also pulls on Earth.

We have examined forces in general terms in previous chapters. In this chapter, we focus on one particular force, the force of gravity, which is one of the four fundamental forces in nature. Gravity is the weakest of these four forces (inside atoms, for example, gravity is negligible relative to the electromagnetic forces), but it operates over all distances and is always a force of attraction between objects with mass (as opposed to the electromagnetic interaction, for which the charges come in positive and negative varieties so resulting forces can be attractive or repulsive and tend to sum to zero for most macroscopic objects). As a result, the gravitational force is of primary importance over the vast distances and for the enormous masses of astronomical studies.

**FIGURE 12.2** Center of our galaxy, the Milky Way, which contains a supermassive black hole. The size of the region shown here is 890 by 640 light-years. The Solar System is 26,000 light-years away from the center of the galaxy.

Figure 12.2 is an image of the center of our galaxy obtained with the space-based infrared Spitzer telescope. The galactic center contains a supermassive black hole, and knowledge of the gravitational interaction allows astronomers to calculate that the mass of this black hole is approximately 1 trillion times the mass of the Earth, or approximately 3.7 million times the mass of the Sun. (Example 12.4 shows how scientists arrived at this conclusion. A black hole is an object so massive and dense that nothing can escape from its surface, not even light.)

## 12.1 Newton's Law of Gravity

Up to now, we have encountered the gravitational force only in the form of a constant gravitational acceleration, $g = 9.81 \text{ m/s}^2$, multiplied by the mass of the object on which the force acts. However, from videos of astronauts running and jumping on the Moon (Figure 12.3), we know that the gravitational force is different there. Therefore, the approximation we have used of a constant gravitational force that depends only on the mass of the object the force acts on cannot be correct away from the surface of the Earth.

The general expression for the magnitude of the gravitational interaction between two point masses, $m_1$ and $m_2$, at a distance $r = |\vec{r}_2 - \vec{r}_1|$ from each other (Figure 12.4) is

$$F(r) = G\frac{m_1 m_2}{r^2}. \tag{12.1}$$

This relationship, known as **Newton's Law of Gravity,** is an empirical law, deduced from experiments and verified extensively. The proportionality constant $G$ is called the **universal gravitational constant** and has the value (to four significant figures)

$$G = 6.674 \cdot 10^{-11} \text{ N m}^2/\text{kg}^2. \tag{12.2}$$

Since 1 N = 1 kg m/s$^2$, we can also write this value as $G = 6.674 \cdot 10^{-11}\ \text{m}^3\text{kg}^{-1}\text{s}^{-2}$. Equation 12.1 says that the strength of the gravitational interaction is proportional to each of the two masses involved in the interaction and is inversely proportional to the square of the distance between them. For example, doubling one of the masses will double the strength of the interaction, whereas doubling the distance will reduce the strength of the interaction by a factor of 4.

Because a force is a vector, the direction of the gravitational force must be specified. The gravitational force $\vec{F}_{2\to1}$ acting from object 2 on object 1 always points toward object 2. We can express this concept in the form of an equation:

$$\vec{F}_{2\to1} = F(r)\hat{r}_{21} = F(r)\frac{\vec{r}_2 - \vec{r}_1}{\left|\vec{r}_2 - \vec{r}_1\right|}.$$

Combining this result with equation 12.1 results in

$$\vec{F}_{2\to1} = G\frac{m_1 m_2}{\left|\vec{r}_2 - \vec{r}_1\right|^3}\left(\vec{r}_2 - \vec{r}_1\right). \tag{12.3}$$

Equation 12.3 is the general form for the gravitational force acting on object 1 due to object 2. It is strictly valid for point particles, as well as for extended spherically symmetrical objects, in which case the position vector is the position of the center of mass. It is also a very good approximation for nonspherical extended objects, represented by their center-of-mass coordinates, provided that the separation between the two objects is large relative to their individual sizes. Note that the center of gravity is identical to the center of mass for spherically symmetrical objects.

Chapter 4 introduced Newton's Third Law: The force $\vec{F}_{1\to2}$ exerted on object 2 by object 1 has to be of the same magnitude as and in the opposite direction to the force $\vec{F}_{2\to1}$ exerted on object 1 by object 2:

$$\vec{F}_{1\to2} = -\vec{F}_{2\to1}.$$

You can see that the force described by equation 12.3 fulfills the requirements of Newton's Third Law by exchanging the indexes 1 and 2 on all variables ($F$, $m$, and $r$) and observing that the magnitude of the force remains the same, but the sign changes. Equation 12.3 governs the motion of the planets around the Sun, as well as of objects in free fall near the surface of the Earth.

**FIGURE 12.3** Apollo 16 Commander John Young jumps on the surface of the Moon and salutes the U.S. flag on April 20, 1972.

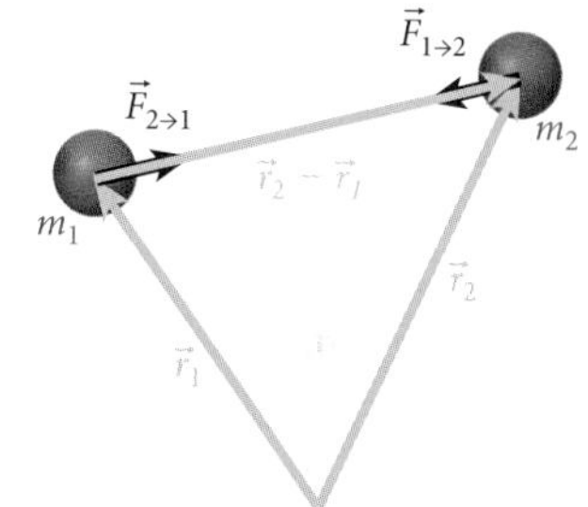


**FIGURE 12.4** The gravitational interaction of two point masses.

## Superposition of Gravitational Forces

If more than one object has a gravitational interaction with object 1, we can compute the total gravitational force on object 1 by using the **superposition principle,** which states that the vector sum of all gravitational forces on a specific object yields the total gravitational force on that object. That is, to find the total gravitational force acting on an object, we simply add the contributions from all other objects:

$$\vec{F}_1 = \vec{F}_{2\to1} + \vec{F}_{3\to1} + \cdots + \vec{F}_{n\to1} = \sum_{i=2}^{n} \vec{F}_{i\to1}. \tag{12.4}$$

The individual forces $\vec{F}_{i\to1}$ can be found from equation 12.3:

$$\vec{F}_{i\to1} = G\frac{m_i m_1}{\left|\vec{r}_i - \vec{r}_1\right|^3}\left(\vec{r}_i - \vec{r}_1\right).$$

Conversely, the total gravitational force on any one of $n$ objects experiencing mutual gravitational interaction can be written as

$$\vec{F}_j = \sum_{i=1, i\neq j}^{n} \vec{F}_{i\to j} = G\sum_{i=1, i\neq j}^{n} \frac{m_i m_j}{\left|\vec{r}_i - \vec{r}_j\right|^3}\left(\vec{r}_i - \vec{r}_j\right). \tag{12.5}$$

The notation "$i \neq j$" below the summation symbol indicates that the sum of forces does not include any object's interaction with itself.

The conceptualization of the superposition of forces is straightforward, but the solution of the resulting equations of motion can become complicated. Even in a system of three approximately equal masses that interact with one another, some initial conditions can lead to regular trajectories, whereas others lead to chaotic motion. The numerical investigation of this kind of system started to become possible only with the advent of computers. Over the last 10 years, the field of many-body physics has developed into one of the most interesting in all of physics, and many intriguing endeavors are likely to be undertaken, such as studying the origin of galaxies from small initial fluctuations in the density of the universe.

## DERIVATION 12.1 Gravitational Force from a Sphere

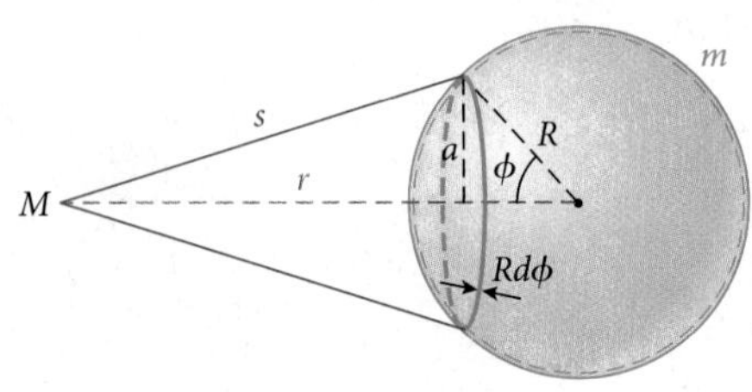

**FIGURE 12.5** Gravitational force on a particle of mass *M* exerted by a spherical shell (hollow sphere) of mass *m*.

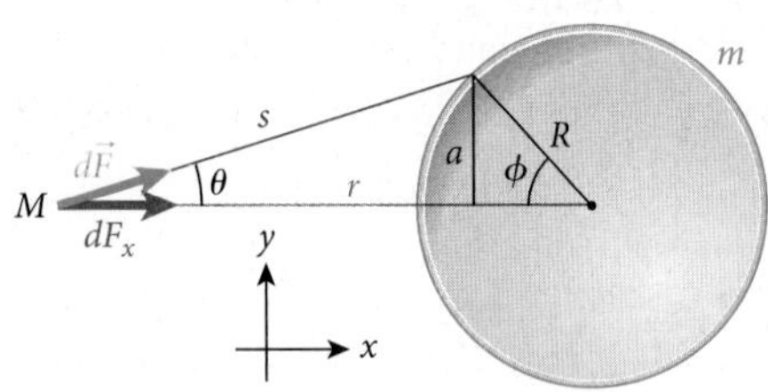

**FIGURE 12.6** Cross section through the center of the spherical shell of Figure 12.5.

Earlier, it was claimed that the gravitational interaction of an extended spherically symmetrical object could be treated like that of a point particle with the same mass located at the center of mass of the extended sphere. We can prove this statement with the aid of calculus and some elementary geometry.

To start, we treat a sphere as a collection of concentric and very thin spherical shells. If we can prove that a thin spherical shell has the same gravitational interaction as a point particle located at its center, then the superposition principle implies that a solid sphere does, too.

Figure 12.5 shows a point particle of mass $M$ located outside a spherical shell of mass $m$ at a distance $r$ from the center of the shell. We want to find the $x$-component of the force on the mass $M$ due to a ring of angular width $d\phi$. This ring has a radius of $a = R\sin\phi$ and thus a circumference of $2\pi R\sin\phi$. It has a width of $R\,d\phi$, as shown in Figure 12.5, and thus a total area of $2\pi R^2 \sin\phi\, d\phi$. Because the mass $m$ is homogeneously distributed over the spherical shell of area $4\pi R^2$, the differential mass of the ring is

$$dm = m\frac{2\pi R^2 \sin\phi d\phi}{4\pi R^2} = \tfrac{1}{2} m \sin\phi d\phi.$$

Because the ring is positioned symmetrically around the horizontal axis, there is no net force in the vertical direction from the ring acting on the mass $M$. The horizontal force component is (Figure 12.6)

$$dF_x = \cos\theta \left( G\frac{M\,dm}{s^2} \right) = \cos\theta \left( G\frac{Mm\sin\phi d\phi}{2s^2} \right). \quad \text{(i)}$$

Now we can relate $\cos\theta$ to $s$, $r$, and $R$ via the law of cosines:

$$\cos\theta = \frac{s^2 + r^2 - R^2}{2sr}.$$

In the same way,

$$\cos\phi = \frac{R^2 + r^2 - s^2}{2Rr}. \quad \text{(ii)}$$

If we differentiate both sides of equation (ii), we obtain

$$-\sin\phi d\phi = -\frac{s}{Rr}ds.$$

Inserting the expressions for $\sin\phi\, d\phi$ and $\cos\theta$ into equation (i) for the differential force component gives

$$\begin{aligned} dF_x &= \frac{s^2 + r^2 - R^2}{2sr} \left( G\frac{Mm}{2s^2} \right) \frac{s}{Rr} ds \\ &= G\frac{Mm}{r^2} \left( \frac{s^2 + r^2 - R^2}{4s^2 R} \right) ds. \end{aligned}$$

Now we can integrate over $ds$ from the minimum value of $s = r - R$ to the maximum value of $s = r + R$:

$$F_x = \int_{r-R}^{r+R} G\frac{Mm}{r^2}\frac{(s^2+r^2-R^2)}{4s^2R}ds = G\frac{Mm}{r^2}\underbrace{\int_{r-R}^{r+R}\frac{s^2+r^2-R^2}{4s^2R}ds}_{\text{Equals 1}} = G\frac{Mm}{r^2}.$$

(It is not obvious that the value of the integral is 1, but you can look this up in an integral table.) Thus, a spherical shell (and by the superposition principle a solid sphere also) exerts the same force on the mass $M$ as a point mass located at the center of the sphere, which is what we set out to prove.

## 12.1 Self-Test Opportunity

Derivation 12.1 assumes that $r > R$, which implies that the mass $M$ is located outside the spherical shell. What changes if $r < R$?

## The Solar System

The Solar System consists of the Sun, which contains the overwhelming majority of the total mass of the Solar System, the four Earthlike inner planets (Mercury, Venus, Earth, and Mars), the asteroid belt between the orbits of Mars and Jupiter, the four gas giants (Jupiter, Saturn, Uranus, and Neptune), a number of dwarf planets (including Ceres, Eris, Haumea, Makemake, and Pluto), and many other minor objects found in the Kuiper Belt. Figure 12.7 shows the orbits and relative sizes of the planets. Table 12.1 provides some physical data for the planets and the Sun.

Not listed as a planet is Pluto, and this omission deserves an explanation. The planet Neptune was discovered in 1846. This discovery had been predicted based on observed small irregularities in the orbit of Uranus, which hinted that another planet's gravitational interaction was the cause. Careful observations of the orbit of Neptune revealed more irregularities, which pointed toward the existence of yet another planet, and Pluto (mass = $1.3 \cdot 10^{22}$ kg) was discovered in 1930. After that, schoolchildren were taught that the Solar System has nine planets. However, in 2003, Sedna (mass = ~$5 \cdot 10^{21}$ kg) and, in 2005, Eris (mass = ~$2 \cdot 10^{22}$ kg) were discovered in the Kuiper Belt, a region in which various objects orbit the Sun at distances between 30 and 48 AU. When Eris's moon Dysnomia was discovered in 2006, it allowed astronomers to calculate that Eris is more massive than Pluto, which began the discussion about what defines a planet. The choice was either to give Sedna, Eris, Ceres (an asteroid that was actually classified as a planet from 1801 to about 1850), and many other Pluto-like objects in the Kuiper Belt the status of planet or to reclassify Pluto as a dwarf planet. In August 2006, the International Astronomical Union voted to remove full planetary status from Pluto.

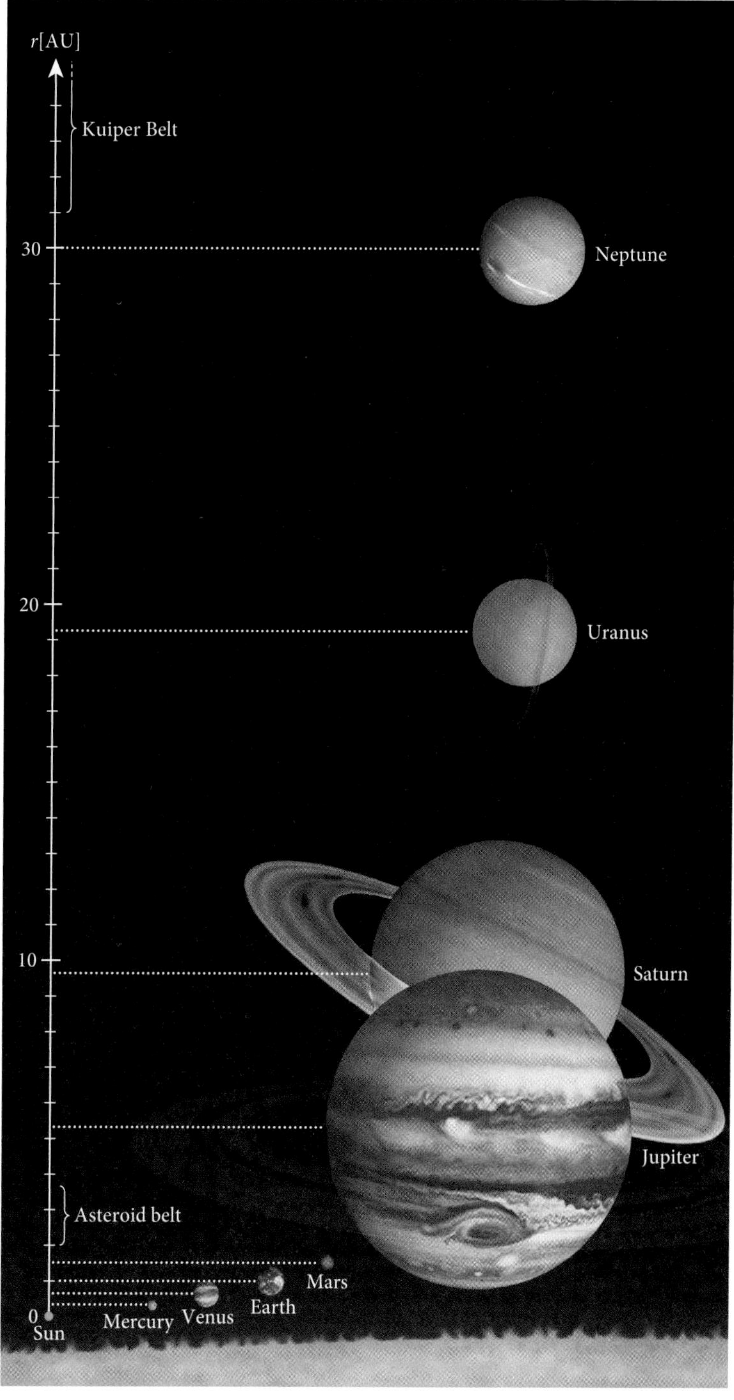


**FIGURE 12.7** The Solar System. On this scale, the sizes of the planets themselves would be too small to be seen. The small yellow dot at the origin of the axis represents the Sun, but it is 30 times bigger than the Sun would appear if it were drawn to scale. The pictures of the planets and the Sun (lower part of the figure) are all magnified by a factor of 30,000 relative to the scale for the orbits.

**Table 12.1 Selected Physical Data for the Solar System**

| Planet | Radius (km) | Mass ($10^{24}$ kg) | $g$ (m/s$^2$) | Escape Speed (km/s) | Mean Orbital Radius ($10^6$ km) | Eccentricity | Orbital Period (yr) |
|---|---|---|---|---|---|---|---|
| Mercury | 2,400 | 0.330 | 3.7 | 4.3 | 57.9 | 0.205 | 0.241 |
| Venus | 6,050 | 4.87 | 8.9 | 10.4 | 108.2 | 0.007 | 0.615 |
| Earth | 6,370 | 5.97 | 9.8 | 11.2 | 149.6 | 0.017 | 1 |
| Mars | 3,400 | 0.642 | 3.7 | 5.0 | 227.9 | 0.094 | 1.88 |
| Jupiter | 71,500 | 1,890 | 23.1 | 59.5 | 778.6 | 0.049 | 11.9 |
| Saturn | 60,300 | 568 | 9.0 | 35.5 | 1433 | 0.057 | 29.4 |
| Uranus | 25,600 | 86.8 | 8.7 | 21.3 | 2872 | 0.046 | 83.8 |
| Neptune | 24,800 | 102 | 11.0 | 23.5 | 4495 | 0.009 | 164 |
| Sun | 696,000 | 1,990,000 | 274 | 618.0 | — | — | — |

As the story of Pluto shows, the Solar System still holds the potential for many discoveries. For example, almost 400,000 asteroids have been identified, and about 5000 more are discovered each month. The total mass of all asteroids in the asteroid belt is less than 5% of the mass of the Moon. However, more than 200 of these asteroids are known to have a diameter of more than 100 km! Tracking them is very important, considering the damage one could cause if it hit the Earth. Another area of ongoing research is the investigation of the objects in the Kuiper Belt. Some models postulate that the combined mass of all Kuiper Belt objects is up to 30 times the mass of the Earth, but the observed mass so far is smaller than this value by a factor of about 1000.

## EXAMPLE 12.1 Influence of Celestial Objects

Astronomy is the science that focuses on planets, stars, galaxies, and the universe as a whole. The similarly named field of astrology has no scientific basis whatsoever. It may be fun to read the daily horoscope, but constellations of stars and/or alignments of planets have no influence on our lives. The only way stars and planets can interact with us is through the gravitational force. Let's calculate the force of gravity exerted on a person by the Moon, Mars, and the stars in the constellation Gemini.

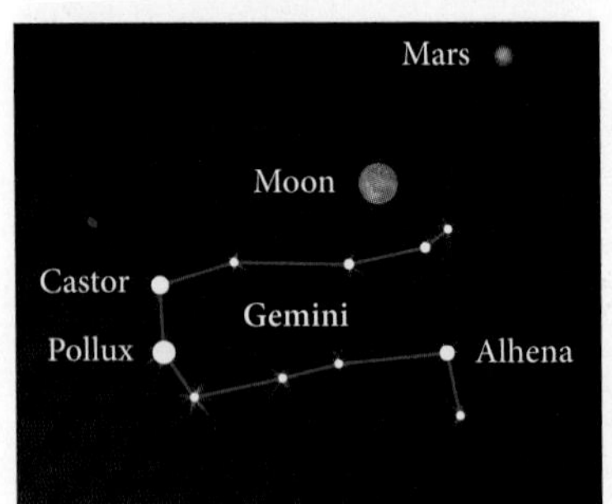


**FIGURE 12.8** Relative positions of the Moon, Mars, and the constellation Gemini in the sky over the Midwest on February 16, 2008.

### PROBLEM 1

Suppose you live in the Midwest of the United States. You go outside at 9:00 p.m. on February 16, 2008, and look up in the sky. You see the Moon, the planet Mars, and the constellation Gemini, as shown in Figure 12.8. Your mass is $m = 85$ kg. What is the force of gravity on you due to these celestial objects?

### SOLUTION 1

Let's start with the Moon. The mass of the Moon is $7.36 \cdot 10^{22}$ kg, and the distance from the Moon to you is $3.84 \cdot 10^8$ m. The gravitational force the Moon exerts on you is then

$$F_{\text{Moon}} = G\frac{M_{\text{Moon}}m}{r_{\text{Moon}}^2} = \left(6.67 \cdot 10^{-11} \text{ N m}^2/\text{kg}^2\right)\frac{\left(7.36 \cdot 10^{22} \text{ kg}\right)\left(85 \text{ kg}\right)}{\left(3.84 \cdot 10^8 \text{ m}\right)^2} = 0.0028 \text{ N}.$$

The distance between Mars and Earth on February 16, 2008, was 136 million km, and the mass of Mars is $M_{\text{Mars}} = 6.4 \cdot 10^{23}$ kg. Thus, the gravitational force that Mars exerts on you is

$$F_{\text{Mars}} = G\frac{M_{\text{Mars}}m}{r_{\text{Mars}}^2} = \left(6.67 \cdot 10^{-11} \text{ N m}^2/\text{kg}^2\right)\frac{\left(6.4 \cdot 10^{23} \text{ kg}\right)\left(85 \text{ kg}\right)}{\left(1.36 \cdot 10^{11} \text{ m}\right)^2} = 2.0 \cdot 10^{-7} \text{ N}.$$

We can estimate the gravitational force exerted by the constellation Gemini by calculating the force exerted by its three brightest objects: Castor, Pollux, and Alhena (see Figure 12.8). Castor is a triple binary star system with a mass 6.7 times the mass of the Sun and is located a distance of 51.5 light-years from you. Pollux is a star with a mass 1.7 times the mass of the Sun, located 33.7 light-years away. Alhena is a binary star system with a mass 3.8 times the mass of the Sun, and it is 105 light-years away. The mass of the Sun is $2.0 \cdot 10^{30}$ kg, and a light-year corresponds to $9.5 \cdot 10^{15}$ m. The gravitational force exerted on you by the constellation Gemini is then

$$F_{\text{Gemini}} = G\frac{M_{\text{Castor}}m}{r_{\text{Castor}}^2} + G\frac{M_{\text{Pollux}}m}{r_{\text{Pollux}}^2} + G\frac{M_{\text{Alhena}}m}{r_{\text{Alhena}}^2} = Gm\left(\frac{M_{\text{Castor}}}{r_{\text{Castor}}^2} + \frac{M_{\text{Pollux}}}{r_{\text{Pollux}}^2} + \frac{M_{\text{Alhena}}}{r_{\text{Alhena}}^2}\right)$$

$$= \left(6.67 \cdot 10^{-11} \text{ N m}^2/\text{kg}^2\right)\left(85 \text{ kg}\right)\frac{2.0 \cdot 10^{30} \text{ kg}}{\left(9.5 \cdot 10^{15} \text{ m}\right)^2}\left(\frac{6.7}{(51.5)^2} + \frac{1.7}{(33.7)^2} + \frac{3.8}{(105)^2}\right)$$

$$= 5.5 \cdot 10^{-13} \text{ N}.$$

The Moon exerts a measurable force on you, but Mars and Gemini exert only negligible forces.

### PROBLEM 2

When Mars and Earth are at their minimum separation distance, they are $r_{\text{M}} = 5.6 \cdot 10^{10}$ m apart. How far away from you does a truck of mass 16,000 kg have to be to have the same gravitational interaction with your body as Mars does at this minimum separation distance?

### SOLUTION

If the two gravitational forces are equal in magnitude, we can write

$$G\frac{M_{\text{M}}m}{r_{\text{M}}^2} = G\frac{m_{\text{T}}m}{r_{\text{T}}^2},$$

where $m$ is your mass, $m_{\text{T}}$ is the mass of the truck, and $r_{\text{T}}$ is the distance we want to find. Canceling out the mass and the universal gravitational constant, we have

$$r_{\text{T}} = r_{\text{M}}\sqrt{\frac{m_{\text{T}}}{M_{\text{M}}}}.$$

Inserting the given numerical values leads to

$$r_{\text{T}} = (5.6 \cdot 10^{10} \text{ m})\sqrt{\frac{1.6 \cdot 10^4 \text{ kg}}{6.4 \cdot 10^{23} \text{ kg}}} = 8.8 \text{ m}.$$

This result means that if you get closer to the truck than 8.8 m, it exerts a bigger gravitational pull on you than does Mars at its closest approach.

## 12.2 Gravitation near the Surface of the Earth

Now we can use the general expression for the gravitational interaction between two masses to reconsider the gravitational force due to the Earth on an object near the surface of the Earth. We can neglect the gravitational interaction of this object with any other objects, because the magnitude of the gravitational interaction with Earth is many orders of magnitudes larger, because of Earth's very large mass. Since we can represent an extended object by a point particle of the same mass located at the object's center of gravity, any object on the surface of Earth is experiencing a gravitational force directed toward the center of Earth. This corresponds to straight down anywhere on the surface of Earth, completely in accordance with empirical evidence.

It is more interesting to determine the magnitude of the gravitational force that an object experiences near the surface of the Earth. Inserting the mass of the Earth, $M_E$, as the mass of one object in equation 12.1 and expressing the altitude $h$ above the surface of the Earth as $h + R_E = r$, where $R_E$ is the radius of the Earth, we find for this special case that

$$F = G\frac{M_E m}{(R_E + h)^2}. \quad (12.6)$$

Because $R_E = 6370$ km, the altitude, $h$, of the object above the ground can be neglected for many applications. If we make this assumption, we find that $F = mg$, with

$$g = \frac{GM_E}{R_E^2} = \frac{\left(6.67\cdot 10^{-11}\ \text{m}^3\text{kg}^{-1}\text{s}^{-2}\right)\left(5.97\cdot 10^{24}\ \text{kg}\right)}{\left(6.37\cdot 10^6\ \text{m}\right)^2} = 9.81\ \text{m/s}^2. \quad (12.7)$$

As expected, near the surface of the Earth, the acceleration due to gravity can be approximated by the constant $g$ that was introduced in Chapter 2. We can insert the mass and radius of other planets (see Table 12.1), moons, or stars into equation 12.7 and find their surface gravity as well. For example, the gravitational acceleration at the surface of the Sun is approximately 28 times larger than that at the surface of the Earth.

If we want to find $g$ for altitudes where we cannot safely neglect $h$, we can start with equation 12.6, divide by the mass $m$ to find the acceleration of that mass,

$$g(h) = \frac{GM_E}{(R_E + h)^2} = \frac{GM_E}{R_E^2}\left(1 + \frac{h}{R_E}\right)^{-2} = g\left(1 + \frac{h}{R_E}\right)^{-2},$$

and then expand in powers of $h/R_E$ to obtain, to the first order,

$$g(h) \approx g\left(1 - 2\frac{h}{R_E} + \cdots\right). \quad (12.8)$$

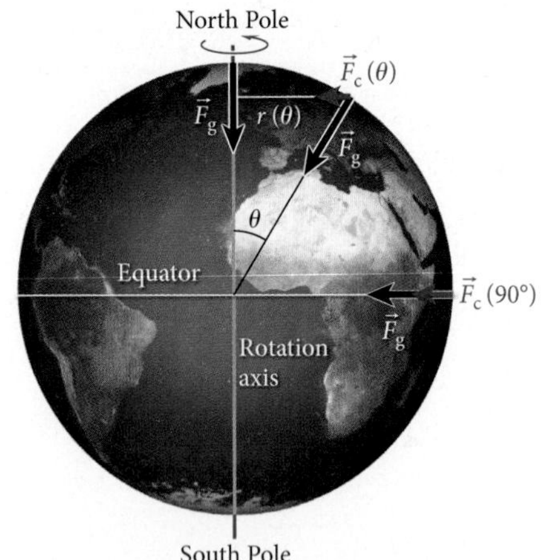


**FIGURE 12.9** Variation of the effective force of gravity due to the Earth's rotation. (The lengths of the red arrows representing the centripetal force have been scaled up by a factor of 200 relative to the black arrows representing the gravitational force.)

## 12.2 Self-Test Opportunity

What is the acceleration due to Earth's gravity at a distance $d = 3R_E$ from the center of the Earth, where $R_E$ is the radius of the Earth?

Equation 12.8 holds for all values of the altitude $h$ that are small compared to the radius of Earth, and it implies that the gravitational acceleration falls off approximately linearly as a function of the altitude above ground. At the top of Mount Everest, the Earth's highest peak, at an altitude of 8850 m, the gravitational acceleration is reduced by 0.27%, or by less than 0.03 m/s$^2$. The International Space Station is at an altitude of 365 km, where the gravitational acceleration is reduced by 11.4%, to a value of 8.7 m/s$^2$. For higher altitudes, the linear approximation of equation 12.8 should definitely not be used.

However, to obtain a more precise determination of the gravitational acceleration, we need to consider other effects. First, the Earth is not an exact sphere; it has a slightly larger radius at the Equator than at the poles. (The value of 6370 km in Table 12.1 is the mean radius of Earth; the radius varies from 6357 km at the poles to 6378 km at the Equator.) Second, the density of Earth is not uniform, and for a precise determination of the gravitational acceleration, the density of the ground right below the measurement makes a difference. Third, and perhaps most important, there is a systematic variation (as a function of the polar angle $\theta$; see Figure 12.9) of the apparent gravitational acceleration, due to the rotation of the Earth and the associated centripetal acceleration. From Chapter 9 on circular motion, we know that the centripetal acceleration is given by $a_c = \omega^2 r$, where $r$ is the radius of the circular motion. For the rotation of the Earth, this radius is the perpendicular distance to the rotation axis. At the poles, this distance is zero, and there is no contribution from the centripetal acceleration. At the equator, $r = R_E$, and the maximum value for $a_c$

$$a_{c,\max} = \omega^2 R_E = \left(7.29\cdot 10^{-5}\ \text{s}^{-1}\right)^2 (6378\ \text{km}) = 0.034\ \text{m/s}^2.$$

Thus, we find that the reduction of the apparent gravitational acceleration at the Equator due to the Earth's rotation is approximately equal to the reduction at the top of Mount Everest.

### EXAMPLE 12.2 Gravitational Tear from a Black Hole

A black hole is a very massive and extremely compact object that is so dense that light emitted from its surface cannot escape. (Thus, it appears black.)

**PROBLEM**

Suppose a black hole has a mass of $6.0 \cdot 10^{30}$ kg, three times the mass of the Sun. A spaceship of length $h = 85$ m approaches the black hole until the front of the ship is at a distance $R = 13{,}500$ km from the black hole. What is the difference in the acceleration due to gravity between the front and the back of the ship?

**SOLUTION**

We can determine the gravitational acceleration at the front of the spaceship due to the black hole via

$$g_{\text{bh}} = \frac{GM_{\text{bh}}}{R^2} = \frac{\left(6.67 \cdot 10^{-11}\ \text{m}^3\text{kg}^{-1}\text{s}^{-2}\right)\left(6.0 \cdot 10^{30}\ \text{kg}\right)}{\left(1.35 \cdot 10^7\ \text{m}\right)^2} = 2.2 \cdot 10^6\ \text{m/s}^2.$$

Now we can use the linear approximation of equation 12.8 to obtain

$$g_{\text{bh}}(h) - g_{\text{bh}}(0) = g_{\text{bh}}\left(1 - 2\frac{h}{R}\right) - g_{\text{bh}}(0) = -2g_{\text{bh}}\frac{h}{R},$$

where $h$ is the length of the spaceship. Inserting the numerical values, we find the difference in the gravitational acceleration at the front and the back:

$$g_{\text{bh}}(h) - g_{\text{bh}}(0) = -2(2.2 \cdot 10^6\ \text{m/s}^2)\frac{85\ \text{m}}{1.35 \cdot 10^7\ \text{m}} = -27.7\ \text{m/s}^2.$$

You can see that in the vicinity of the black hole, the differential acceleration between front and back is so large that the spaceship would have to have tremendous integral strength to avoid being torn apart! (Near a black hole, Newton's Law of Gravity needs to be modified, but this example has ignored that change. In Section 35.8, we will discuss black holes in more detail and see why light is affected by the gravitational interaction.)

## 12.3 Gravitation inside the Earth

Derivation 12.1 showed that the gravitational interaction of a mass $m$ with a spherically symmetrical mass distribution (where $m$ is located outside the sphere) is not affected if the sphere is replaced with a point particle with the same total mass, located at its center of mass (or center of gravity). Derivation 12.2 now shows that on the inside of a spherical shell of uniform density, the net gravitational force is zero.

### DERIVATION 12.2 Force of Gravity inside a Hollow Sphere

We want to show that the gravitational force acting on a point mass inside a hollow homogeneous spherical shell is zero everywhere inside the shell. To do this, we could use calculus and a mathematical law known as Gauss's Law. (Gauss's Law, discussed in Chapter 22, applies to electrostatic interaction and the Coulomb force, which is another force that falls off with $1/r^2$.) However, instead we'll use a geometrical argument that was presented by Newton in 1687 in his book *Philosophiae Naturalis Principia Mathematica,* usually known as the *Principia.*

Consider the (infinitesimally thin) spherical shell shown in Figure 12.10. We mark a point $P$ at an arbitrary location inside the shell and then draw a straight line through this point. This straight line intersects the shell at two points, and the distances between those two intersection points and $P$ are the radii $r_1$ and $r_2$. Now we draw cones with tips at $P$ and with small opening angle $\theta$ around the straight line. The areas where the two cones intersect the shell are

*Continued—*

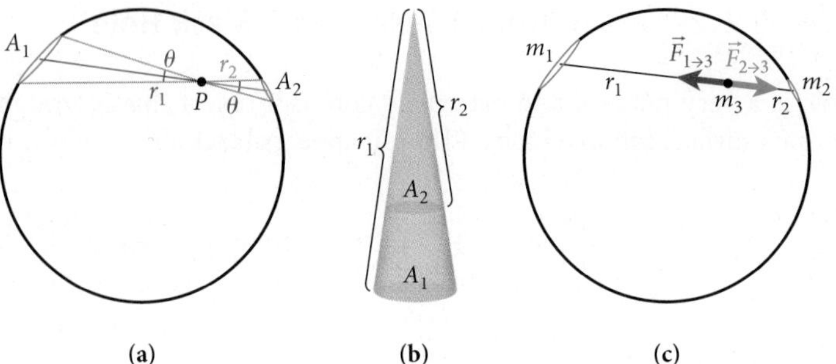


**FIGURE 12.10** Gravitational interaction of a point with the surface of a hollow sphere: (a) cones from point $P$ within the sphere to the sphere's surface; (b) details of the cones from $P$ to the surface; (c) the balance of gravitational forces acting on $P$ due to the opposite areas of the spherical shell.

called $A_1$ and $A_2$. These areas are proportional to the angle $\theta$, which is the same for both of them. Area $A_1$ is also proportional to $r_1^2$, and area $A_2$ is proportional to $r_2^2$ (see Figure 12.10b). Further, since the shell is homogeneous, the mass of any segment of it is proportional to the segment's area. Therefore, $m_1 = ar_1^2$ and $m_2 = ar_2^2$, with the same proportionality constant $a$.

The gravitational force $\vec{F}_{1\to3}$ that the mass $m_1$ of area $A_1$ exerts on the mass $m_3$ at point $P$ then points along $r_1$ toward the center of $A_1$. The gravitational force $\vec{F}_{2\to3}$ acting on $m_3$ points in exactly the opposite direction along $r_2$. We can also find the magnitudes of these two forces:

$$F_{1\to3} = \frac{Gm_1m_3}{r_1^2} = \frac{G(ar_1^2)m_3}{r_1^2} = Gam_3$$

$$F_{2\to3} = \frac{Gm_2m_3}{r_2^2} = \frac{G(ar_2^2)m_3}{r_2^2} = Gam_3.$$

Since the dependence on the distance cancels out, the magnitudes of the two forces are the same. Since their magnitudes are the same, and their directions are opposite, the forces $\vec{F}_{1\to3}$ and $\vec{F}_{2\to3}$ exactly cancel each other (see Figure 12.10c).

Since the location of the point and the orientation of the line drawn through it were arbitrary, the result is true for any point inside the spherical shell. The net force of gravity acting on a point mass inside this spherical shell is indeed zero.

We can now gain physical insight into the gravitational force acting inside the Earth. Think of the Earth as composed of many concentric thin spherical shells. Then the gravitational force at a point $P$ inside the Earth at a distance $r$ from the center is due to those shells with a radius less than $r$. All shells with a greater radius do not contribute to the gravitational force on $P$. Furthermore, the mass, $M(r)$, of all contributing shells can be imagined to be concentrated in the center, at a distance $r$ away from $P$. The gravitational force acting on an object of mass $m$ at a distance $r$ from the center of the Earth is then

$$F(r) = G\frac{M(r)m}{r^2}. \tag{12.9}$$

This is Newton's Law of Gravity (equation 12.1), with the mass of the contributing shells to be determined. In order to determine that mass, we make the simplifying assumption of a constant density, $\rho_E$, inside Earth. Then we obtain

$$M(r) = \rho_E V(r) = \rho_E \tfrac{4}{3}\pi r^3. \tag{12.10}$$

We can calculate the Earth's mass density from its total mass and radius:

$$\rho_E = \frac{M_E}{V_E} = \frac{M_E}{\frac{4}{3}\pi R_E^3} = \frac{\left(5.97\cdot 10^{24}\text{ kg}\right)}{\frac{4}{3}\pi\left(6.37\cdot 10^6\text{ m}\right)^3} = 5.5\cdot 10^3\text{ kg/m}^3.$$

Substituting the expression for the mass from equation 12.10 into equation 12.9, we obtain an equation for the radial dependence of the gravitational force inside Earth:

$$F(r)=G\frac{M(r)m}{r^2}=G\frac{\rho_E\frac{4}{3}\pi r^3 m}{r^2}=\tfrac{4}{3}\pi G\rho_E mr. \quad (12.11)$$

Equation 12.11 states that the gravitational force increases linearly with the distance from the center of the Earth. In particular, if the point is located exactly at the center of the Earth, zero gravitational force acts on it.

Now we can compare the radial dependence of the gravitational acceleration divided by $g$ (for an object acted on by just the force due to the Earth) inside and outside the Earth. Figure 12.11 shows the linear rise of this quantity inside the Earth, and the fall-off in an inverse square fashion outside. At the surface of the Earth, these curves intersect, and the gravitational acceleration has the value of $g$. Also shown in the figure is the linear approximation (equation 12.8) to the dependence of the gravitational acceleration as a function of the height above the Earth's surface: $g(h) \approx g(1 - 2h/R_E + \ldots) \Rightarrow g(r) \approx g(3 - 2r/R_E + \ldots)$, because $r \approx h + R_E$. You can clearly see that this approximation is valid to within a few percentage points for altitudes of a few hundred kilometers above the surface of the Earth.

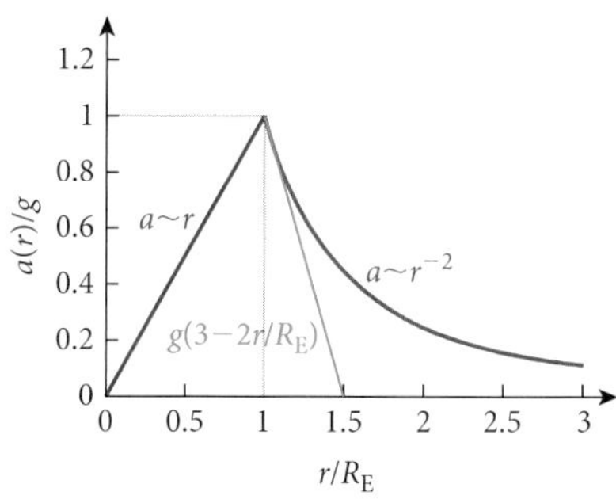


**FIGURE 12.11** Dependence of the gravitational acceleration on the radial distance from the center of the Earth.

Note that the functional form of the force in equation 12.11 is that of a spring force, with a restoring force that increases linearly as a function of displacement from equilibrium at $r = 0$. Equation 12.11 specifies the magnitude of the gravitational force. Because the force always points toward the center of the Earth, we can also write equation 12.11 as a one-dimensional vector equation in terms of $x$, the displacement from equilibrium:

$$F_x(x)=-\tfrac{4}{3}\pi G\rho_E mx=-\frac{mg}{R_E}x=-kx.$$

This result is Hooke's Law for a spring, which we encountered in Chapter 5. Therefore, the "spring constant" of the gravitational force is

$$k=\tfrac{4}{3}\pi G\rho_E m=\frac{mg}{R_E}.$$

Similar considerations also apply to the gravitational force inside other spherically symmetrical mass distributions, such as planets or stars.

### 12.1 In-Class Exercise

The Moon can be considered as a sphere of uniform density with mass $M_M$ and radius $R_M$. At the center of the Moon, the magnitude of the gravitational force acting on a mass $m$ due to the mass of the Moon is

a) $mGM_M/R_M^2$.

b) $\frac{1}{2}mGM_M/R_M^2$.

c) $\frac{3}{5}mGM_M/R_M^2$.

d) zero.

e) $2mGM_M/R_M^2$.

## 12.4 Gravitational Potential Energy

In Chapter 6, we saw that the gravitational potential energy is given by $U = mgh$, where $h$ is the distance in the $y$-direction, provided the gravitational force is written $\vec{F}=-mg\hat{y}$ (with the sign convention that positive is straight up). Using Newton's Law of Gravity, we can obtain a more general expression for gravitational potential energy. Integrating equation 12.1 yields an expression for the gravitational potential energy of a system of two masses $m_1$ and $m_2$ separated by a distance $r$:

$$U(r)-U(\infty)=-\int_\infty^r \vec{F}(\vec{r}\,')\bullet d\vec{r}\,'=\int_\infty^r F(r')dr'=\int_\infty^r G\frac{m_1m_2}{r'^2}dr'$$

$$=Gm_1m_2\int_\infty^r \frac{1}{r'^2}dr'=-Gm_1m_2\frac{1}{r'}\bigg|_\infty^r=-G\frac{m_1m_2}{r}.$$

The first part of this equation is the general relation between force and potential energy. For a gravitational interaction, the force depends only on the radial separation and points outward: $\vec{F}(\vec{r})=\vec{F}(r)$. The integration is equivalent to bringing the two masses together in the radial direction from an initial infinite separation to a final separation, $r$. Thus, $d\vec{r}$ points opposite to the force $\vec{F}(\vec{r})$, and so $\vec{F}(\vec{r})\bullet d\vec{r} = F(r)\,dr(\cos 180°) = -F(r)\,dr$.

Note that the equation describing gravitational potential energy only tells us the *difference* between the gravitational potential energy at separation $r$ and at separation infinity. We set $U(\infty) = 0$, which implies that the gravitational potential energy vanishes between two objects that are separated by an infinite distance. This choice gives us the following

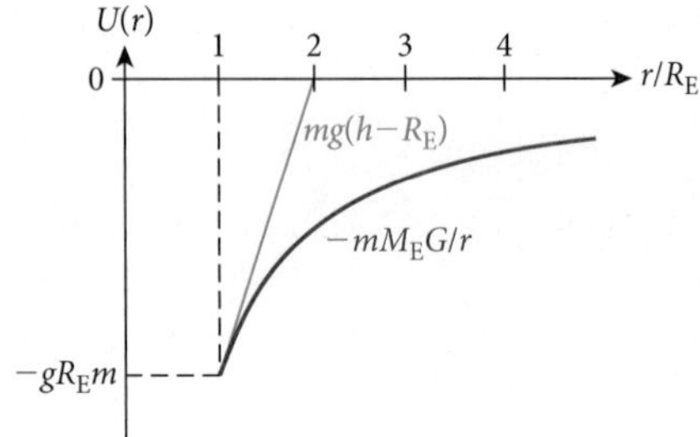


**FIGURE 12.12** Dependence of the gravitational potential energy on the distance to the center of the Earth, for distances larger than the radius of the Earth. The red curve represents the exact expression; the green line represents the linear approximation for values of $r$ not much larger than the radius of the Earth, $R_E$.

expression for the **gravitational potential energy** as a function of the separation of two masses:

$$U(r) = -G\frac{m_1 m_2}{r}. \tag{12.12}$$

Note that the gravitational potential energy is always less than zero with $U(\infty) = 0$. This dependence of the gravitational potential energy on $1/r$ is illustrated by the red curve in Figure 12.12 for an arbitrary mass near the surface of the Earth.

For interaction among more than two objects, we can write all pairwise interactions between two of them and integrate. The gravitational potential energies from these interactions are simply added to give the total gravitational potential energy. For three point particles, we find, for example:

$$U = U_{12} + U_{13} + U_{23} = -G\frac{m_1 m_2}{\left|\vec{r}_1 - \vec{r}_2\right|} - G\frac{m_1 m_3}{\left|\vec{r}_1 - \vec{r}_3\right|} - G\frac{m_2 m_3}{\left|\vec{r}_2 - \vec{r}_3\right|}.$$

An important special case occurs when one of the two interacting objects is the Earth. For altitudes $h$ that are small compared to the radius of the Earth, we expect to repeat the previous result that the gravitational potential energy is $mgh$. Because the Earth is many orders of magnitude more massive than any object on the surface of the Earth for which we might want to calculate the gravitational potential energy, the combined center of mass of the Earth and the object is practically identical to the center of mass of the Earth, which we then select as the origin of the coordinate system. Using equation 12.12, this results in

$$U(h) = -G\frac{M_E m}{R_E + h} = -G\frac{M_E m}{R_E}\left(1 + \frac{h}{R_E}\right)^{-1}$$
$$\approx -\frac{GM_E m}{R_E} + \frac{GM_E m}{R_E^2}h = -gmR_E + mgh.$$

In the second step here, we used the fact that $h \ll R_E$ and expanded. This result (plotted in green in Figure 12.12) looks almost like the expression $U = mgh$ from Chapter 6, except for the addition of the constant term $-gmR_E$. This constant term is a result of the choice of the integration constant in equation 12.12. However, as was stressed in Chapter 6, we can add any additive constant to the expression for potential energy without changing the physical results for the motion of objects. The only physically relevant quantity is the difference in potential energy between two different locations. Taking the difference between altitude $h$ and altitude zero results in

$$\Delta U = U(h) - U(0) = (-gmR_E + mgh) - (-gmR_E) = mgh.$$

As expected, the additive constant $-gmR_E$ cancels out, and we obtain the same result for low altitudes, $h$, that we had previously derived: $\Delta U = mgh$.

## Escape Speed

With an expression for the gravitational potential energy, we can determine the total mechanical energy for a system consisting of an object of mass $m_1$ and speed $v_1$, that has a gravitational interaction with another object of mass $m_2$ and speed $v_2$, if the two objects are separated by a distance $r = \left|\vec{r}_1 - \vec{r}_2\right|$:

$$E = K + U = \tfrac{1}{2}m_1 v_1^2 + \tfrac{1}{2}m_1 v_2^2 - \frac{Gm_1 m_2}{\left|\vec{r}_1 - \vec{r}_2\right|}. \tag{12.13}$$

Of particular interest is the case where one of the objects is the Earth ($m_1 \equiv M_E$). If we consider the reference frame to be that of the Earth ($v_1 = 0$), the Earth has no kinetic energy in this frame. Again, we put the origin of the coordinate system at the center of the Earth. Then the expression for the total energy of this system is just the kinetic energy of object 2 (but we have omitted the subscript 2), plus the gravitational potential energy:

$$E = \tfrac{1}{2}mv^2 - \frac{GM_E m}{R_E + h},$$

where, as before, $h$ is the altitude above Earth's surface of the object with mass $m$.

If we want to find out what initial speed a projectile must have to escape to an infinite distance from Earth, called the **escape speed,** $v_E$ (where E stands for Earth), we can use energy conservation. At infinite separation, the gravitational potential energy is zero, and the minimum kinetic energy is also zero. Thus, the total energy with which a projectile can barely escape to infinity from the Earth's gravitational pull is zero. Energy conservation then implies, starting from the Earth's surface:

$$E(h=0)=\tfrac{1}{2}mv_E^2-\frac{GM_Em}{R_E}=0.$$

Solving this for $v_E$ gives an expression for the minimum escape speed:

$$v_E=\sqrt{\frac{2GM_E}{R_E}}. \qquad (12.14)$$

Inserting the numerical values of these constants, we finally obtain

$$v_E=\sqrt{\frac{2\left(6.67\cdot10^{-11}\ \text{m}^3\text{kg}^{-1}\text{s}^{-2}\right)\left(5.97\cdot10^{24}\ \text{kg}\right)}{6.37\cdot10^6\ \text{m}}}=11.2\ \text{km/s}.$$

This escape speed is approximately equal to 25,000 mph. The same calculation can also be performed for other planets, moons, and stars by inserting the relevant constants (for example, see Table 12.1).

Note that the angle with which the projectile is launched into space does not enter into the expression for the escape speed. Therefore, it does not matter if the projectile is shot straight up or almost in a horizontal direction. However, we neglected air resistance, and the launch angle would make a difference if we accounted for its effect. An even bigger effect results from the Earth's rotation. Since the Earth rotates once around its axis each day, a point on the surface of the Earth located at the Equator has a speed of $v = 2\pi R_E/(1\ \text{day}) \approx 0.46$ km/s, which decreases to zero at the poles. The direction of the corresponding velocity vector points east, tangentially to the Earth's surface. Therefore, the launch angle matters most at the Equator. For a projectile fired in the eastern direction from any location on the Equator, the escape speed is reduced to approximately 10.7 km/s.

Can a projectile launched from the surface of the Earth with a speed of 11.2 km/s escape the Solar System? Doesn't the gravitational potential energy of the projectile due to its interaction with the Sun play a role? At first glance, it would appear not. After all, the gravitational force the Sun exerts on an object located near the surface of the Earth is negligible compared to the force the Earth exerts on that object. As proof, consider that if you jump up in the air, you land at the same place, independent of what time of day it is, that is, where the Sun is in the sky. Thus, we can indeed neglect the Sun's gravitational force near the surface of the Earth.

However, the gravitational force is far different from the gravitational potential energy. In contrast to the force, which falls off as $r^{-2}$, the potential energy falls off much more slowly, as it is proportional to $r^{-1}$. It is straightforward to generalize equation 12.14 for the escape speed from any planet or star with mass $M$, if the object is initially separated a distance $R$ from the center of that planet or star:

$$v=\sqrt{\frac{2GM}{R}}. \qquad (12.15)$$

Inserting the mass of the Sun and the size of the orbit of the Earth we find $v_S$, the speed needed for an object to escape from the gravitational influence of the Sun if it is initially a distance from it equal to the radius of the Earth's orbit:

$$v_S=\sqrt{\frac{2\left(6.67\cdot10^{-11}\ \text{m}^3\text{kg}^{-1}\text{s}^{-2}\right)\left(1.99\cdot10^{30}\ \text{kg}\right)}{1.49\cdot10^{11}\ \text{m}}}=42\ \text{km/s}.$$

This is quite an astonishing result: The escape speed needed to leave the Solar System from the orbit of the Earth is almost four times larger than the escape speed needed to get away from the gravitational attraction of the Earth.

### 12.2 In-Class Exercise

The escape speed from the surface of the Moon (mass = $7.35 \cdot 10^{22}$ kg and diameter = 3476 km) is

a) 2.38 km/s.
b) 1.68 km/s.
c) 11.2 km/s.
d) 5.41 km/s.

(a)

(b)

**FIGURE 12.13** Gravity assist technique: (a) trajectory of a spacecraft passing Jupiter, as seen in Jupiter's reference frame; (b) the same trajectory as seen in the reference frame of the Sun, in which Jupiter moves with a velocity of approximately 13 km/s.

Launching a projectile from the surface of the Earth with enough speed to leave the Solar System requires overcoming the combined gravitational potential energy of the Earth and the Sun. Since the two potential energies add, the combined escape speed is

$$v_{ES} = \sqrt{v_E^2 + v_S^2} = 43.5 \text{ km/s}.$$

We found that the rotation of the Earth has a nonnegligible, but still small, effect on the escape speed. However, a much bigger effect arises from the orbital motion of the Earth around the Sun. The Earth orbits the Sun with an orbital speed of $v_O = 2\pi R_{ES}/(1 \text{ yr}) = 30$ km/s, where $R_{ES}$ is the distance between Earth and Sun (149.6 or 150 million km, according to Table 12.1). A projectile launched in the direction of this orbital velocity vector needs a launch velocity of only $v_{ES,min} = (43.5 - 30)$ km/s $= 13.5$ km/s, whereas one launched in the opposite direction must have $v_{ES,max} = (43.5 + 30)$ km/s $= 73.5$ km/s. Other launch angles produce all values between these two extremes.

When NASA launches a probe to explore the outer planets or to leave the Solar System—for example, *Voyager 2*—the gravity assist technique is used to lower the required launch velocity. This technique is illustrated in Figure 12.13. Part (a) is a sketch of a flyby of Jupiter as seen by an observer at rest relative to Jupiter. Notice that the velocity vector of the spacecraft changes direction but has the same length at the same distance from Jupiter as the spacecraft approaches and as it leaves. This is a consequence of energy conservation. Figure 12.13b is a sketch of the spacecraft's trajectory as seen by an observer at rest relative to the Sun. In this reference frame, Jupiter moves with an orbital velocity of approximately 13 km/s. To transform from Jupiter's reference frame to the Sun's reference frame, we have to add the velocity of Jupiter to the velocities observed in Jupiter's frame (red arrows) to obtain the velocities in the Sun's frame (blue arrows). As you can see from the figure, in the Sun's frame, the length of the final velocity vector is significantly greater than that of the initial velocity vector. This means that the spacecraft has acquired significant additional kinetic energy (and Jupiter has lost this kinetic energy) during the flyby, allowing it to continue escaping the gravitational pull of the Sun.

## EXAMPLE 12.3 Asteroid Impact

One of the most likely causes of the extinction of dinosaurs at the end of the Cretaceous period, about 65 million years ago, was a large asteroid hitting the Earth. Let's look at the energy released during an asteroid impact.

### PROBLEM

Suppose a spherical asteroid, with a radius of 1.00 km and a mass density of 4750 kg/m$^3$, enters the Solar System with negligible speed and then collides with Earth in such a way that it hits Earth from a radial direction with respect to the Sun. What kinetic energy will this asteroid have in the Earth's reference frame just before its impact on Earth?

### SOLUTION

First, we calculate the mass of the asteroid:

$$m_a = V_a \rho_a = \tfrac{4}{3}\pi r_a^3 \rho_a = \tfrac{4}{3}\pi (1.00 \cdot 10^3 \text{ m})^3 (4750 \text{ kg/m}^3) = 1.99 \cdot 10^{13} \text{ kg}.$$

If the asteroid hits the Earth in a radial direction with respect to the Sun, the Earth's velocity vector will be perpendicular to that of the asteroid at impact, because the Earth moves tangentially around the Sun. Thus, there are three contributions to the kinetic energy of the asteroid as measured in Earth's reference frame: (1) conversion of the gravitational potential energy between Earth and asteroid, (2) conversion of the gravitational potential energy between Sun and asteroid, and (3) kinetic energy of the Earth's motion relative to that of the asteroid.

Because we have already calculated the escape speeds corresponding to the two gravitational potential energy terms and because the asteroid will arrive with the kinetic energy that corresponds to these escape speeds, we can simply write

$$K = \tfrac{1}{2} m_a (v_E^2 + v_S^2 + v_O^2).$$

Now we insert the numerical values:

$$K = 0.5(1.99 \cdot 10^{13}\ \text{kg})[(1.1 \cdot 10^4\ \text{m/s})^2 + (4.2 \cdot 10^4\ \text{m/s})^2 + (3.0 \cdot 10^4\ \text{m/s})^2]$$
$$= 2.8 \cdot 10^{22}\ \text{J}.$$

This value is equivalent to the energy released by approximately 300 million nuclear weapons of the magnitude of those used to destroy Hiroshima and Nagasaki in World War II. You can begin to understand the destructive power of an asteroid impact of this magnitude—an event like this could wipe out human life on Earth. A somewhat bigger asteroid, with a diameter of 6 to 10 km, hit Earth near the tip of the Yucatan peninsula in the Gulf of Mexico approximately 65 million years ago. It is believed to be responsible for the K-T (Cretaceous-Tertiary) extinction, which killed off the dinosaurs.

**DISCUSSION**

Figure 12.14 shows the approximately 1.5 km in diameter and almost 200 m deep Barringer impact crater, which was formed approximately 50,000 years ago when a meteorite of approximately 50-m diameter, with a mass of approximately 300,000 tons ($3 \cdot 10^8$ kg), hit Earth with a speed of approximately 12 km/s. This was a much smaller object than the asteroid described in this example, but the impact still had the destructive power of 150 atomic bombs of the Hiroshima/Nagasaki class.

**FIGURE 12.14** Barringer impact crater in central Arizona.

## Gravitational Potential

Equation 12.12 states that the gravitational potential energy of any object is proportional to that object's mass. When we employed energy conservation to calculate the escape speed, we saw that the mass of the object canceled out, because both kinetic energy and gravitational potential energy are proportional to an object's mass. Thus, the kinematics are independent of the object's mass. For example, let's consider the gravitational potential energy of a mass $m$ interacting with Earth, $U_E(r) = -GM_E m/r$. The Earth's gravitational potential $V_E(r)$ is defined as the ratio of the gravitational potential energy to the mass of the object, $V_E(r) = U_E(r)/m$, or

$$V_E(r) = -\frac{GM_E}{r}. \qquad (12.16)$$

This definition has the advantage of giving information on the gravitational interaction with Earth, independently of the other mass involved. (We will explore the concept of potentials in much more depth in Chapter 23 on electric potentials.)

# 12.5 Kepler's Laws and Planetary Motion

Johannes Kepler (1571–1630) used empirical observations, mainly from data gathered by Tycho Brahe, and sophisticated calculations to arrive at the famous **Kepler's laws of planetary motion,** published in 1609 and 1619. These laws were published decades before Isaac Newton was born in 1643, whose law of gravitation would eventually show *why* Kepler's laws were true. What is particularly significant about Kepler's laws is that they challenged the prevailing world view at the time, with the Earth in the center of the universe (a geocentric theory) and the Sun and all the planets and stars orbiting around it, just as the Moon does. Kepler and other pioneers, particularly Nicolaus Copernicus and Galileo Galilei, changed this geocentric view into a heliocentric (Sun-centered) cosmology. Today, space probes have provided direct observations from vantage points outside the Earth's atmosphere and verified that Copernicus, Kepler, and Galileo were correct. However, the simplicity with which the heliocentric model was able to explain astronomical observations won intelligent people over to their view long before external observations were possible.

### Kepler's First Law: Orbits

All planets move in elliptical orbits with the Sun at one focal point.

## Mathematical Insert: Ellipses

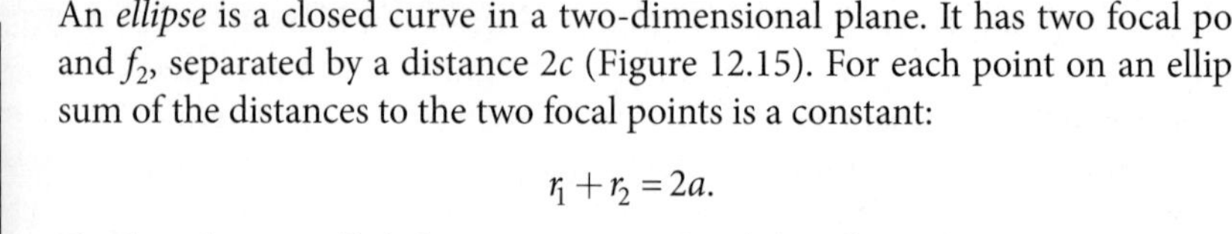

An *ellipse* is a closed curve in a two-dimensional plane. It has two focal points, $f_1$ and $f_2$, separated by a distance $2c$ (Figure 12.15). For each point on an ellipse, the sum of the distances to the two focal points is a constant:

$$r_1 + r_2 = 2a.$$

The length $a$ is called the semimajor axis of the ellipse (see Figure 12.15). (*Note:* Unfortunately, the standard notation for the semimajor axis of an ellipse uses the same letter, $a$, as is conventionally used to symbolize acceleration. You should be careful to avoid confusion.) The semiminor axis, $b$, is related to $a$ and $c$ via

$$b^2 \equiv a^2 - c^2.$$

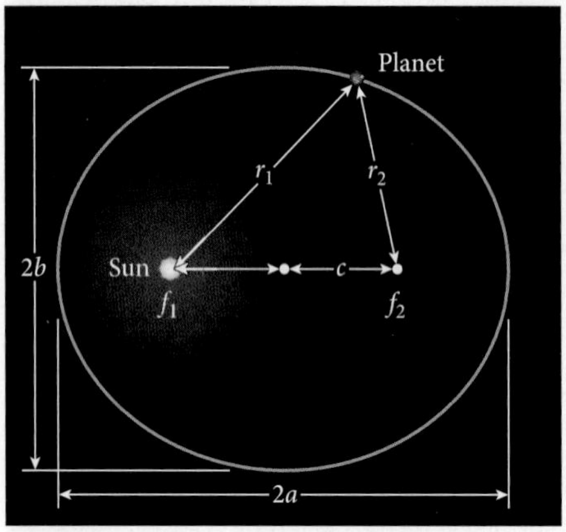

FIGURE 12.15 Parameters used in describing ellipses and elliptical orbits.

In terms of the Cartesian coordinates $x$ and $y$, the points on the ellipse satisfy the equation

$$\frac{x^2}{a^2} + \frac{y^2}{b^2} = 1,$$

where the origin of the coordinate system is at the center of the ellipse. If $a = b$, a circle (a special case of an ellipse) results.

It is useful to introduce the eccentricity, $e$, of an ellipse, defined as

$$e = \frac{c}{a} = \sqrt{1 - \frac{b^2}{a^2}}.$$

An eccentricity of zero, the smallest possible value, characterizes a circle. The ellipse shown in Figure 12.15 has an eccentricity of 0.6.

The eccentricity of the Earth's orbit around the Sun is only 0.017. If you were to plot an ellipse with this value of $e$, you could not distinguish it from a circle by visual inspection. The length of the semiminor axis of Earth's orbit is approximately 99.98% of the length of the semimajor axis. At its closest approach to the Sun, called the *perihelion*, the Earth is 147.1 million km away from the Sun. The *aphelion*, which is the farthest point from the Sun in Earth's orbit, is 152.6 million km.

It is important to note that the change in seasons is *not* caused primarily by the eccentricity of the Earth's orbit. (The point of closest approach to the Sun is reached in early January each year, in the middle of the cold season in the Northern Hemisphere.) Instead, the seasons are caused by the fact that the Earth's axis of rotation is tilted by an angle of 23.4° relative to the plane of the orbital ellipse. This tilt exposes the Northern Hemisphere to the Sun's rays for longer periods and at a more direct angle in the summer months.

Among the other planetary orbits, Mercury's has the largest eccentricity—0.205. (Pluto's orbital eccentricity is even larger, at 0.249, but Pluto has been declassified as a planet since August 2006.) Venus's orbit has the smallest eccentricity, 0.007, followed by Neptune with an eccentricity of 0.009.

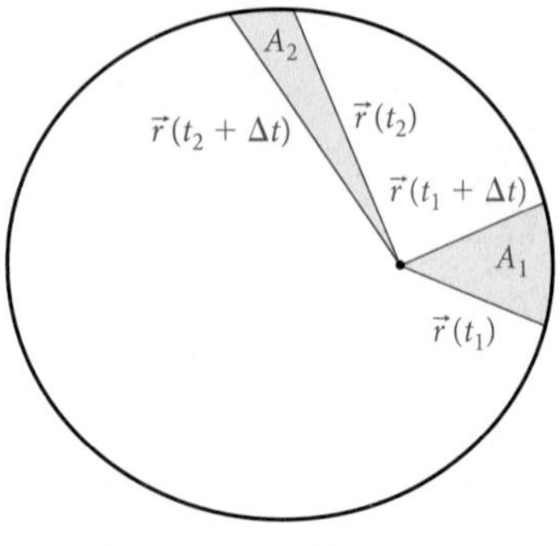

### Kepler's Second Law: Areas

A straight line connecting the center of the Sun and the center of any planet (Figure 12.16) sweeps out an equal area in any given time interval:

$$\frac{dA}{dt} = \text{constant}. \tag{12.17}$$

FIGURE 12.16 Kepler's Second Law states that equal areas are swept out in equal time periods, or $A_1 = A_2$.

### Kepler's Third Law: Periods

The square of the period of a planet's orbit is proportional to the cube of the semimajor axis of the orbit:

$$\frac{T^2}{a^3} = \text{constant}. \tag{12.18}$$

This proportionality constant can be expressed in terms of the mass of the Sun and the universal gravitational constant

$$\frac{T^2}{a^3} = \frac{4\pi^2}{GM}. \tag{12.19}$$

## DERIVATION 12.3 Kepler's Laws

The general proof of Kepler's laws uses Newton's Law of Gravity, equation 12.1, and the law of conservation of angular momentum. With quite a bit of algebra and calculus, it is then possible to prove all three of Kepler's laws. Here we will derive Kepler's Second and Third Laws for circular orbits with the Sun in the center, allowing us to put the Sun at the origin of the coordinate system and neglect the motion of the Sun around the common center of mass of the Sun-planet system.

First, we show that circular motion is indeed possible. From Chapter 9, we know that in order to obtain a closed circular orbit, the centripetal force needs to be equal to the gravitational force:

$$m\frac{v^2}{r} = G\frac{Mm}{r^2} \Rightarrow v = \sqrt{\frac{GM}{r}}.$$

This result establishes two important facts: First, the mass of the orbiting object has canceled out, so all objects can have the same orbit, provided that their mass is small compared to the Sun. Second, any given orbital radius, $r$, has a unique orbital velocity corresponding to it. Also, for a given orbital radius, we obtain a constant value of the angular velocity:

$$\omega = \frac{v}{r} = \sqrt{\frac{GM}{r^3}}. \tag{i}$$

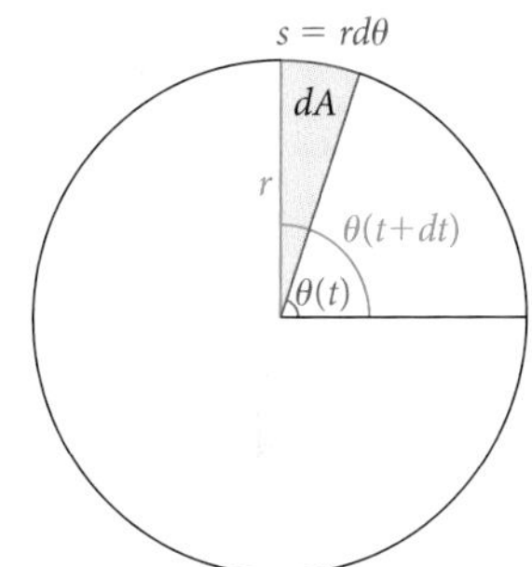


**FIGURE 12.17** Angle, arc length, and area as a function of time.

Next, we examine the area swept out by a radial vector connecting the Sun and the planet. As indicated in Figure 12.17, the area swept out is $dA = \frac{1}{2}rs = \frac{1}{2}d\theta$. Taking the time derivative results in

$$\frac{dA}{dt} = \frac{1}{2}r^2\frac{d\theta}{dt} = \frac{1}{2}r^2\omega.$$

Because, for a given orbit, $\omega$ and $r$ are constant, we have derived Kepler's Second Law, which states that $dA/dt = \text{constant}$.

Finally, for Kepler's Third Law, we use $T = 2\pi/\omega$ and substitute the expression for the angular velocity from equation (i). This results in

$$T = 2\pi\sqrt{\frac{r^3}{GM}}.$$

We can rearrange this equation and obtain

$$\frac{T^2}{r^3} = \frac{4\pi^2}{GM}.$$

This proves Kepler's Third Law and gives the value for the proportionality constant between the square of the orbital period and the cube of the orbital radius.

Again, keep in mind that Kepler's laws are valid for elliptic orbits in general, not just for circular orbits. Instead of referring to the radius of the circle, $r$ has to be taken as the semimajor axis of the ellipse for such orbits. Perhaps an even more useful formulation of Kepler's Third Law may be written as

$$\frac{T_1^2}{a_1^3} = \frac{T_2^2}{a_2^3}. \quad (12.20)$$

With this formula, we can easily find orbital periods and radii for two different orbiting objects.

## SOLVED PROBLEM 12.1 Orbital Period of Sedna

### PROBLEM

On November 14, 2003, astronomers discovered a previously unknown object in part of the Kuiper Belt beyond the orbit of Neptune. They named this object Sedna, after the Inuit goddess of the sea. The average distance of Sedna from the Sun is $78.7 \cdot 10^9$ km. How long does it take Sedna to complete one orbit around the Sun?

### SOLUTION

**THINK**

We can use Kepler's laws to relate Sedna's distance from the Sun to the period of Sedna's orbit around the Sun.

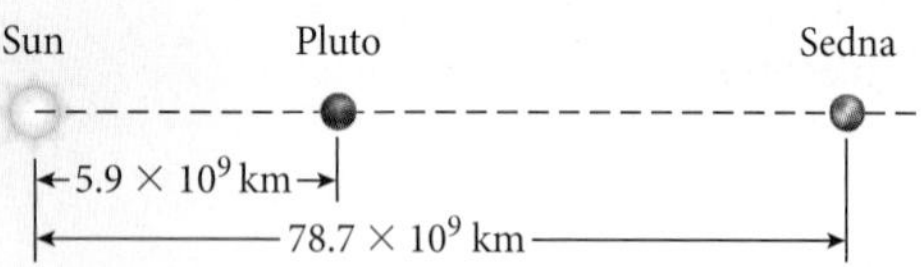


**FIGURE 12.18** The distance of Sedna from the Sun compared with the distance of Pluto from the Sun.

**SKETCH**

A sketch comparing the average distance of Sedna from the Sun with the average distance of Pluto from the Sun is shown in Figure 12.18.

**RESEARCH**

We can relate the orbit of Sedna to the known orbit of the Earth using equation 12.20 (a form of Kepler's Third Law):

$$\frac{T_{\text{Earth}}^2}{a_{\text{Earth}}^3} = \frac{T_{\text{Sedna}}^2}{a_{\text{Sedna}}^3}, \quad \text{(i)}$$

where $T_{\text{Earth}}$ is the period of Earth's orbit, $a_{\text{Earth}}$ is the radius of Earth's orbit, $T_{\text{Sedna}}$ is the period of Sedna's orbit, and $a_{\text{Sedna}}$ is the radius of Sedna's orbit.

**SIMPLIFY**

We can solve equation (i) for the period of Sedna's orbit:

$$T_{\text{Sedna}} = T_{\text{Earth}} \left( \frac{a_{\text{Sedna}}}{a_{\text{Earth}}} \right)^{3/2}.$$

**CALCULATE**

Putting in the numerical values, we get

$$T_{\text{Sedna}} = (1 \text{ yr}) \left( \frac{78.7 \cdot 10^9 \text{ km}}{0.150 \cdot 10^9 \text{ km}} \right)^{3/2} = 12{,}018 \text{ yr}.$$

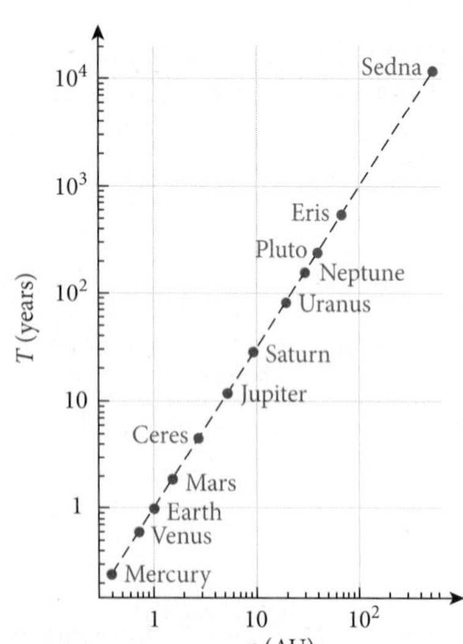


**FIGURE 12.19** Orbital period versus length of the semimajor axis of orbits of objects in the Solar System.

**ROUND**

We report our result to two significant figures:

$$T_{\text{Sedna}} = 1.21 \cdot 10^4 \text{ yr}.$$

**DOUBLE-CHECK**

We can compare our result for Sedna with the measured values for the semimajor axes of the orbits and orbital periods of the planets and several dwarf planets. As Figure 12.19 shows, our calculated result (dashed line, representing Kepler's Third Law) fits well with the extrapolation of the data from the planets (red dots) and dwarf planets (blue dots).

We can also use Kepler's Third Law to determine the mass of the Sun. We obtain this result by solving equation 12.19 for the mass of the Sun:

$$M = \frac{4\pi^2 a^3}{GT^2}. \tag{12.21}$$

Inserting the data for Earth's orbital period and radius gives

$$M = \frac{4\pi^2 (1.496 \cdot 10^{11}\ \text{m})^3}{(6.67 \cdot 10^{-11}\ \text{m}^3\text{kg}^{-1}\text{s}^{-2})(3.15 \cdot 10^7\ \text{s})^2} = 1.99 \cdot 10^{30}\ \text{kg}.$$

It is also possible to use Kepler's Third Law to determine the mass of the Earth from the period and radius of the Moon's orbit around Earth. In fact, astronomers can use this law to determine the mass of any astronomical object that has a satellite orbiting it if they know the radius and period of the orbit.

### 12.3 Self-Test Opportunity

Use the fact that the gravitational interaction between Earth and Sun provides the centripetal force that keeps Earth on its orbit to prove equation 12.21. (Assume a circular orbit.)

## EXAMPLE 12.4 Black Hole in the Center of the Milky Way

**PROBLEM**

There is a supermassive black hole in the center of the Milky Way. What is its mass?

**SOLUTION**

In June 2007, astronomers measured the mass of the center of the Milky Way. Seven stars orbiting near the galactic center had been tracked for 9 years, as shown in Figure 12.20. The periods and semimajor axes extracted by the astronomers are shown in Table 12.2. Using these data and Kepler's Third Law (equation 12.21), we can calculate the mass of the galactic center, indicated by a yellow star symbol in Figure 12.20. The resulting mass of the galactic center is shown in Table 12.2 for each set of star measurements. The average mass of the galactic center is $3.7 \cdot 10^6$ times the mass of the Sun. Thus, astronomers infer that there is a supermassive black hole at the center of the galaxy, because no star is visible at that point.

**DISCUSSION**

If there is a supermassive black hole in the center of the Milky Way, you may ask yourself why isn't Earth being pulled toward it? The answer is the same as the answer to the question of why the Earth does not fall into the Sun: The Earth orbits the Sun, and the Sun orbits the galactic center, a distance of 26,000 light-years away from the Solar System.

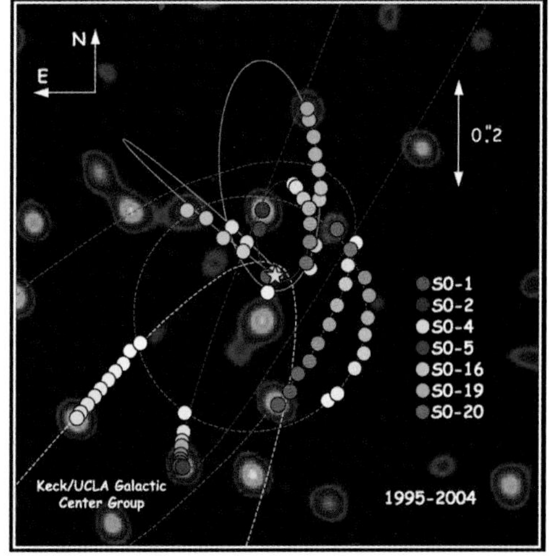


**FIGURE 12.20** The orbits of seven stars close to the center of the Milky Way as tracked by astronomers from the Keck/UCLA Galactic Center Group from 1995 to 2004. The measured positions, represented by colored dots, are superimposed on a picture of the stars taken at the start of the tracking. The lines represent fits to the measurements that were used to extract the periods and semimajor axes of the stars' orbits. The side of the image is a distance of approximately $\frac{1}{15}$ of a light-year.

**Table 12.2 Periods and Semimajor Axes for Stars Orbiting the Center of the Milky Way**

| Star | Period (yr) | Semimajor Axis (AU) | Period ($10^8$ s) | Semimajor Axis ($10^{14}$ m) | Mass of Galactic Center ($10^{36}$ kg) | Equivalent in Solar Masses ($10^6$) |
|---|---|---|---|---|---|---|
| S0-2 | 14.43 | 919 | 4.55 | 1.37 | 7.44 | 3.74 |
| S0-16 | 36 | 1680 | 113 | 2.51 | 7.31 | 3.67 |
| S0-19 | 37.2 | 1720 | 117 | 2.57 | 7.34 | 3.69 |
| S0-20 | 43 | 1900 | 135 | 2.84 | 7.41 | 3.72 |
| S0-1 | 190 | 5100 | 599 | 7.63 | 7.34 | 3.69 |
| S0-4 | 2600 | 30,000 | 819 | 44.9 | 7.98 | 4.01 |
| S0-5 | 9900 | 70,000 | 3120 | 105 | 6.99 | 3.51 |
| | Average | | | | 7.40 | 3.72 |

## 12.3 In-Class Exercise

The best estimate of the orbital period of the Solar System around the center of the Milky Way is between 220 and 250 million years. How much mass (in terms of solar masses) is enclosed by the 26,000 light-years ($1.7 \cdot 10^9$ AU) radius of the Solar System's orbit? (*Hint:* An orbital period of 1 yr for an orbit of radius 1 AU corresponds to 1 solar mass.)

a) 90 billion solar masses

b) 7.2 billion solar masses

c) 52 million solar masses

d) 3.7 million solar masses

e) 432,000 solar masses

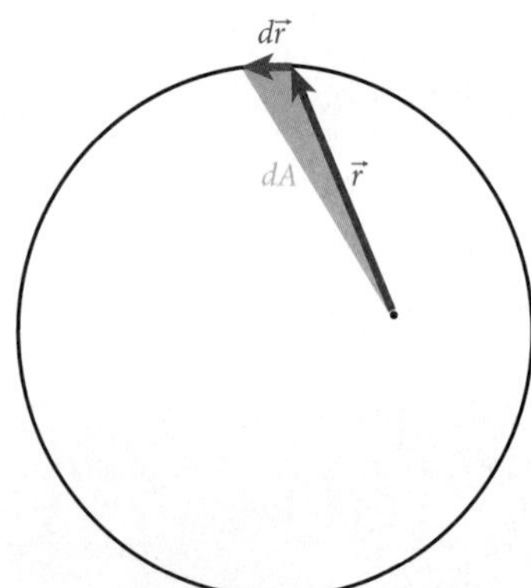


**FIGURE 12.21** Area swept out by the radius vector.

### Kepler's Second Law and Conservation of Angular Momentum

Chapter 10 (on rotation) stressed the importance of the concept of angular momentum, in particular, the importance of the conservation of angular momentum. It is quite straightforward to prove the law of conservation of angular momentum for planetary motion and, as a consequence, also derive Kepler's Second Law. Let's work through this proof.

First, we show that the angular momentum, $\vec{L} = \vec{r} \times \vec{p}$, of a point particle is conserved if the particle moves under the influence of a central force. A **central force** is a force that acts only in the radial direction, $\vec{F}_{\text{central}} = F\hat{r}$. To prove this statement, we take the time derivative of the angular momentum:

$$\frac{d\vec{L}}{dt} = \frac{d}{dt}(\vec{r} \times \vec{p}) = \frac{d\vec{r}}{dt} \times \vec{p} + \vec{r} \times \frac{d\vec{p}}{dt}.$$

For a point particle, the velocity vector, $\vec{v} = d\vec{r}/dt$, and the momentum vector, $\vec{p}$ are parallel; therefore, their vector product vanishes: $(d\vec{r}/dt) \times \vec{p} = 0$. This leaves only the term $\vec{r} \times (d\vec{p}/dt)$ in the preceding equation. Using Newton's Second Law, we find (see Chapter 7 on momentum) $d\vec{p}/dt = \vec{F}$. If this force is a central force, then it is parallel (or antiparallel) to the vector $\vec{r}$. Thus, for a central force, the vector product $\vec{r} \times (d\vec{p}/dt)$ also vanishes:

$$\frac{d\vec{L}}{dt} = \vec{r} \times \frac{d\vec{p}}{dt} = \vec{r} \times \vec{F}_{\text{central}} = \vec{r} \times F\hat{r} = 0.$$

Since $d\vec{L}/dt = 0$, we have shown that angular momentum is conserved for a central force. The force of gravity is such a central force, and therefore angular momentum is conserved for any planet moving on an orbit.

How does this general result help in deriving Kepler's Second Law? If we can show that the area $dA$ swept out by the radial vector, $\vec{r}$ during some infinitesimal time, $dt$, is proportional to the absolute value of the angular momentum, then we are done, because the angular momentum is conserved.

As you can see in Figure 12.21, the infinitesimal area $dA$ swept out by the vector $\vec{r}$ is the triangle spanned by that vector and the differential change in it, $d\vec{r}$:

$$dA = \tfrac{1}{2}\left|\vec{r} \times d\vec{r}\right| = \tfrac{1}{2}\left|\vec{r} \times \frac{d\vec{r}}{dt}dt\right| = \tfrac{1}{2}\left|\vec{r} \times \frac{1}{m}m\frac{d\vec{r}}{dt}dt\right| = \frac{dt}{2m}\left|\vec{r} \times \vec{p}\right| = \frac{dt}{2m}\left|\vec{L}\right|.$$

Therefore, the area swept out in each time interval, $dt$, is

$$\frac{dA}{dt} = \frac{\left|\vec{L}\right|}{2m} = \text{constant},$$

which is exactly what Kepler's Second Law states.

## 12.6 Satellite Orbits

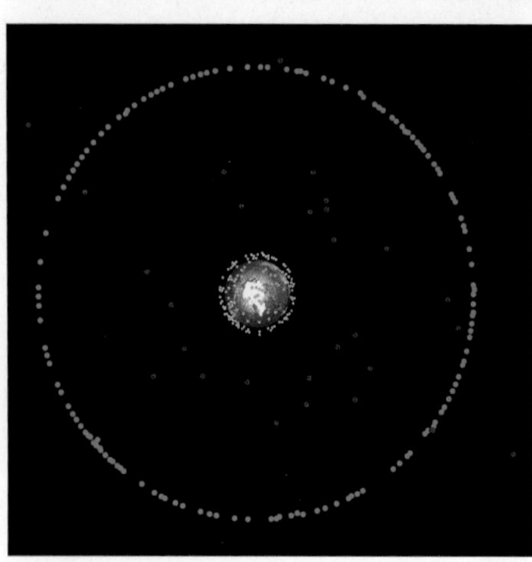

Figure 12.22 shows the positions of many of the several hundreds of satellites in orbit around Earth. Each dot represents the position of a satellite on the afternoon of June 23, 2004. In low orbits, only a few hundred kilometers above sea level, are communication satellites for phone systems, the International Space Station, the Hubble Space Telescope, and other applications (yellow dots). The perfect circle of satellites at a distance of approximately 5.6 Earth radii above the surface (green dots) is composed of **geostationary satellites,** which orbit at the same angular speed as Earth, and so remain above the same spot on the ground. The satellites in between the geostationary and the low-orbit satellites (red dots) are mainly those used for the Global Positioning System, but also those carrying research instruments.

**FIGURE 12.22** Positions of some of the satellites in orbit around Earth on June 23, 2004, looking down on the North Pole. This illustration was produced with data available from NASA.

## SOLVED PROBLEM 12.2 Satellite in Orbit

### PROBLEM

A satellite is in a circular orbit around the Earth. The orbit has a radius of 3.75 times the radius of the Earth. What is the linear speed of the satellite?

### SOLUTION

### THINK

The force of gravity provides the centripetal force that keeps the satellite in its circular orbit around the Earth. We can obtain the satellite's linear speed by equating the centripetal force expressed in terms of the linear speed with the force of gravity between the satellite and the Earth.

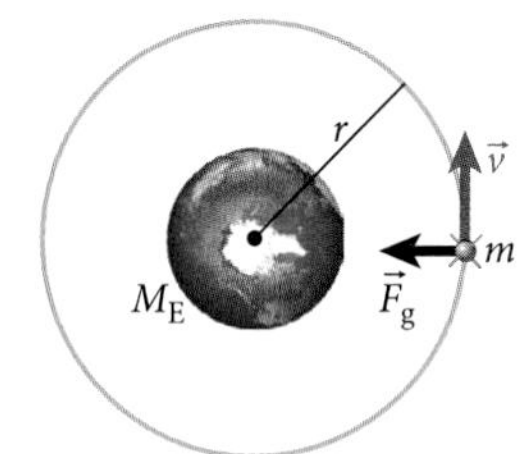


**FIGURE 12.23** Satellite in circular orbit around the Earth.

### SKETCH

A sketch of the problem situation is presented in Figure 12.23.

### RESEARCH

For a satellite with mass $m$ moving with linear speed $v$, the centripetal force required to keep the satellite moving in a circle with radius $r$ is

$$F_c = \frac{mv^2}{r}. \quad \text{(i)}$$

The gravitational force, $F_g$, between the satellite and the Earth is

$$F_g = G\frac{M_E m}{r^2}, \quad \text{(ii)}$$

where $G$ is the universal gravitational constant and $M_E$ is the mass of the Earth. Equating the forces described by equations (i) and (ii), we get

$$F_c = F_g \Rightarrow$$
$$\frac{mv^2}{r} = G\frac{M_E m}{r^2}.$$

### SIMPLIFY

The mass of the satellite cancels out; thus, the orbital speed of a satellite does not depend on its mass. We obtain

$$v^2 = G\frac{M_E}{r} \Rightarrow$$
$$v = \sqrt{\frac{GM_E}{r}}. \quad \text{(iii)}$$

### CALCULATE

The problem statement specified that the radius of the satellite's orbit is $r = 3.75R_E$, where $R_E$ is the radius of the Earth. Substituting for $r$ in equation (iii) and then inserting the known numerical values gives us

$$v = \sqrt{\frac{GM_E}{3.75R_E}} = \sqrt{\frac{\left(6.67 \cdot 10^{-11}\ \text{m}^3\text{kg}^{-1}\text{s}^{-2}\right)\left(5.97 \cdot 10^{24}\ \text{kg}\right)}{3.75(6.37 \cdot 10^6\ \text{m})}} = 4082.86\ \text{m/s}.$$

### ROUND

Expressing our result with three significant figures gives

$$v = 4080\ \text{m/s} = 4.08\ \text{km/s}.$$

*Continued—*

**DOUBLE-CHECK**

The time it takes this satellite to complete one orbit is

$$T = \frac{2\pi r}{v} = \frac{2\pi(3.75R_E)}{v} = \frac{2\pi\left[3.75(6.37 \cdot 10^6 \text{ m})\right]}{4080 \text{ m/s}} = 36{,}800 \text{ s} = 10.2 \text{ h},$$

which seems reasonable—communication satellites take 24 h but are at higher altitudes, and the Hubble Space Telescope takes 1.6 h at a lower altitude.

### 12.4 In-Class Exercise

The elliptical orbit of a small satellite orbiting a spherical planet is shown in the figure. At which point along the orbit is the linear speed of the satellite at a maximum?

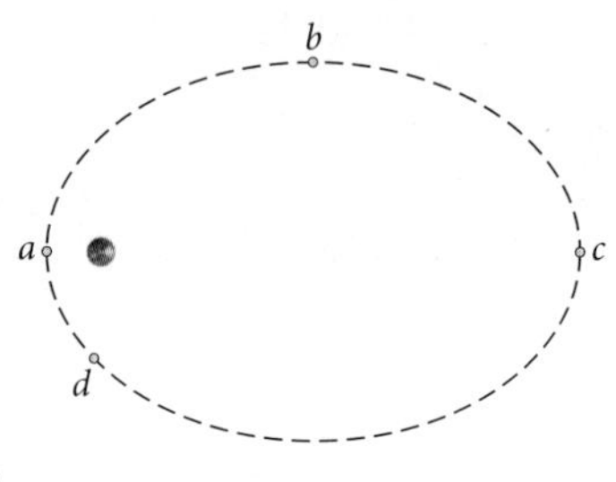


Combining the expression for the orbital speed from Solved Problem 12.2, $v = \sqrt{GM_E/r}$, with equation 12.15 for the escape speed, $v_{esc} = \sqrt{2GM_E/r}$, we find that the orbital velocity of a satellite is always

$$v(r) = \frac{1}{\sqrt{2}} v_{esc}(r). \tag{12.22}$$

The Earth is a satellite of the Sun, and, as we determined in Section 12.4, the escape speed from the Sun starting from the orbital radius of the Earth is 42 km/s. Using equation 12.22, we can predict that the orbital speed of the Earth moving around the Sun is $42/\sqrt{2}$ km/s, or approximately 30 km/s, which matches the value of the orbital speed we found earlier in Chapter 9.

## Energy of a Satellite

Having solved Sample Problem 12.2, we can readily obtain an expression for the kinetic energy of a satellite in orbit around Earth. Multiplying both sides of $mv^2/r = GM_E m/r^2$, which we found by equating the centripetal and gravitational forces, by $r/2$ yields

$$\tfrac{1}{2}mv^2 = \tfrac{1}{2}G\frac{M_E m}{r}.$$

The left-hand side of this equation is the kinetic energy of the satellite. Comparing the right-hand side to the expression for the gravitational potential energy, $U = -GM_E m/r$, we see that this side is equivalent to $-\tfrac{1}{2}U$. Thus, we obtain the kinetic energy of a satellite in circular orbit:

$$K = -\tfrac{1}{2}U. \tag{12.23}$$

The total mechanical energy of the satellite is then

$$E = K + U = -\tfrac{1}{2}U + U = \tfrac{1}{2}U = -\tfrac{1}{2}G\frac{M_E m}{r}. \tag{12.24}$$

Consequently, the total energy is exactly the negative of the satellite's kinetic energy:

$$E = -K. \tag{12.25}$$

It is important to note that equations 12.23 through 12.25 all hold for any orbital radius.

For an elliptical orbit with a semimajor axis $a$, obtaining the energy of the satellite requires a little more mathematics. The result is very similar to equation 12.24, with the radius $r$ of the circular orbit replaced by the semimajor axis $a$ of the elliptical orbit:

$$E = -\tfrac{1}{2}G\frac{M_E m}{a}.$$

## Orbit of Geostationary Satellites

For many applications, a satellite needs to remain at the same point in the sky. For example, satellite TV dishes always point at the same place in the sky, so we need a satellite to be located there, to be sure we can get reception of a signal. These satellites that are continuously at the same point in the sky are called geostationary.

What are the conditions that a satellite must fulfill to be geostationary? First, it has to move in a circle, because this is the only orbit that has a constant angular velocity. Second, the period of rotation must match that of the Earth's, exactly 1 day. And third, the axis of rotation of the satellite's orbit must be exactly aligned with that of the Earth's rotation. Because the center of the Earth must be at the center of a circular orbit for any satellite, the only possible geostationary orbit is one exactly above the Equator. These conditions leave only the radius of the orbit to be determined.

To find the radius, we use Kepler's Third Law in the form of equation 12.19 and solve for $r$:

$$\frac{T^2}{r^3} = \frac{4\pi^2}{GM} \Rightarrow r = \left( \frac{GMT^2}{4\pi^2} \right)^{1/3}. \tag{12.26}$$

The mass $M$ in this case is that of the Earth. Inserting the numerical values, we find

$$r = \left( \frac{(6.674 \cdot 10^{-11} \text{ m}^3\text{kg}^{-1}\text{s}^{-2})(5.9742 \cdot 10^{24} \text{ kg})(86{,}164 \text{ s})^2}{4\pi^2} \right)^{1/3} = 42{,}168 \text{ km}.$$

Note that we used the best available value of the mass of the Earth and the sidereal day as the correct period of the Earth's rotation (see Chapter 9). The distance of a geostationary satellite above sea level at the Equator is then 42,168 km $-R_\text{E}$. Taking into account that the Earth is not a perfect sphere, but slightly oblate, this distance is

$$d = r - R_\text{E} = 35{,}790 \text{ km}.$$

This distance is 5.61 times Earth's radius. This is why the geostationary satellites form an almost perfect circle with a radius of $6.61R_\text{E}$ in Figure 12.22.

Figure 12.24 shows a cross section through Earth and the location of a geostationary satellite, with the radius of its orbit drawn to scale. From this figure, the angle $\xi$ relative to the horizontal at which to orient a satellite dish for the best TV reception. Because any geostationary satellite is located in the plane of the Equator, a dish in the Northern Hemisphere should point in a southerly direction.

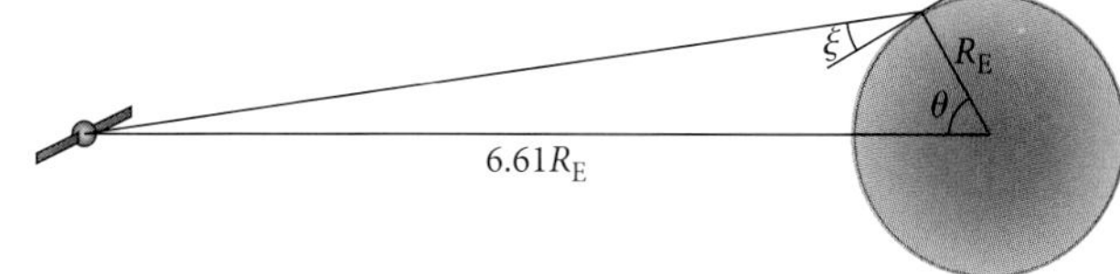


**FIGURE 12.24** Angle of a satellite dish, $\xi$, relative to the local horizontal as a function of the angle of latitude, $\theta$.

There are also geosynchronous satellites in orbit around the Earth. A geosynchronous satellite also has an orbital period of 1 day but does not need to remain at the same point in the sky as viewed from the surface of Earth. For example, NASA's Solar Dynamics Observatory (scheduled to launch in January 2010) will have a geosynchronous orbit that is inclined, which traces out a figure 8 in the sky, as viewed from the ground. A geostationary orbit is a special case of geosynchronous orbits.

## SOLVED PROBLEM 12.3 Satellite TV Dish

You just received your new television system, but the company cannot come out to install the dish for you immediately. You want to watch the big game tonight, so you decide to set up the satellite dish yourself.

### PROBLEM

Assuming that you live in a location at latitude 42.75° N and that the satellite TV company has a satellite aligned with your longitude, in what direction should you point the satellite dish?

### SOLUTION

### THINK

Satellite TV companies use geostationary satellites to broadcast the signals. Thus, we know you need to point the satellite dish southward, toward the Equator, but you also

*Continued—*

need to know the angle of inclination of the satellite dish with respect to the horizontal. In Figure 12.24, this is the angle $\xi$. To determine $\xi$, we can use the law of cosines, incorporating the distance of a satellite in geostationary orbit, the radius of the Earth, and the latitude of the location of the dish.

## SKETCH

Figure 12.25 is a sketch of the geometry of the location of the geostationary satellite and the point on the surface of the Earth where the dish is being set up. In this sketch, $R_E$ is the radius of the Earth, $R_S$ is the distance of the satellite from the center of the Earth, $d_S$ is the distance from the satellite to the point on the Earth's surface where the dish is located, $\theta$ is the angle of the latitude of the surface of that location, and $\phi$ is the angle between $d_S$ and $R_E$.

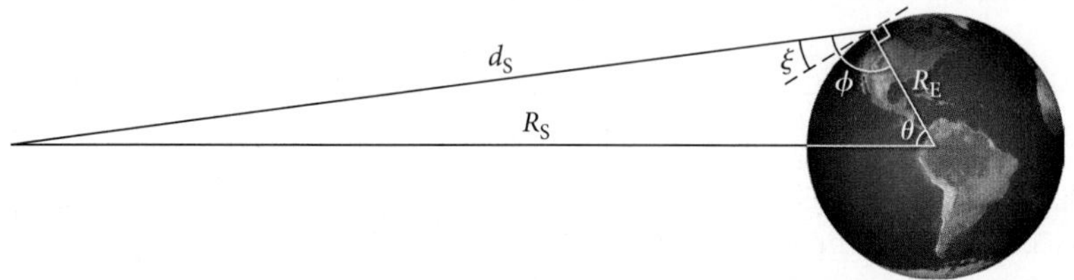


**FIGURE 12.25** Geometry of a geostationary satellite in orbit around the Earth.

## RESEARCH

To determine the angle $\xi$, we first need to determine the angle $\phi$. We can see from Figure 12.25 that $\xi = \phi - 90°$ because the dashed line is tangent to the surface of the Earth and thus is perpendicular to a line from the point to the center of the Earth. To determine $\phi$, we can apply the law of cosines to the triangle defined by $d_S$, $R_E$, and $R_S$. We will need to apply the law of cosines to this triangle twice. To use the law of cosines to determine $\phi$, we need to know the lengths of the sides $d_S$ and $R_E$. We know $R_E$ but not $d_S$. We can determine the length $d_S$ using the law of cosines, the angle $\theta$, and the lengths of the two known sides of $R_E$ and $R_S$:

$$d_S^2 = R_S^2 + R_E^2 - 2R_S R_E \cos\theta \qquad \text{(i)}$$

We can now get an equation for the angle $\phi$ using the law of cosines with the angle $\phi$ and the two known lengths $d_S$ and $R_E$:

$$R_S^2 = d_S^2 + R_E^2 - 2d_S R_E \cos\phi. \qquad \text{(ii)}$$

## SIMPLIFY

We know that $R_S = 6.61R_E$ for geostationary satellites. The angle $\theta$ corresponds to the latitude, $\theta = 42.75°$. We can substitute these quantities into equation (i):

$$d_S^2 = \left(6.61R_E\right)^2 + R_E^2 - 2\left(6.61R_E\right)R_E\left(\cos 42.75°\right).$$

We can now write an expression for $d_S$ in terms of $R_E$:

$$d_S^2 = R_E^2\left[6.61^2 + 1 - 2\left(6.61\right)\left(\cos 42.75°\right)\right] = 34.984R_E^2,$$

or

$$d_S = 5.915R_E.$$

We can solve equation (ii) for $\phi$:

$$\phi = \cos^{-1}\left(\frac{d_S^2 + R_E^2 - R_S^2}{2d_S R_E}\right). \qquad \text{(iii)}$$

**CALCULATE**

Inserting the values we have for $d_S$ and $R_S$ into equation (iii) gives

$$\phi = \cos^{-1}\left(\frac{34.984R_E^2 + R_E^2 - \left(6.61R_E\right)^2}{2\left(5.915R_E\right)R_E}\right) = 130.66°.$$

The angle at which you need to aim the satellite dish with respect to the horizontal is then

$$\xi = \phi - 90° = 130.66° - 90° = 40.66.$$

**ROUND**

Expressing our result with three significant figures gives

$$\xi = 40.7°.$$

**DOUBLE-CHECK**

If the satellite were very far away, the lines $d_S$ and $R_S$ would be parallel to each other, and the sketch of the geometry of the situation would be redrawn as shown in Figure 12.26. We can see from this sketch that $\phi = 180° - \theta$. Remembering that $\xi = \phi - 90°$, we can write

$$\xi = \left(180° - \theta\right) - 90° = 90° - \theta.$$

In this situation, $\theta = 42.75°$, so the estimated angle would be $\xi = 90° - 42.75° = 47.25°$, which is close to our result of $\xi = 40.7°$, but definitely larger, as required. Thus, our answer seems reasonable.

### 12.5 In-Class Exercise

If a permanent base is established on Mars, it will be necessary to have Martian-stationary satellites in orbit around Mars to facilitate communications. A day on Mars lasts 24 h, 39 min, and 35 s. What is the radius of the orbit for a Martian-stationary satellite?

a) 12,560 km
b) 15,230 km
c) 20,450 km
d) 29,320 km
e) 43,350 km

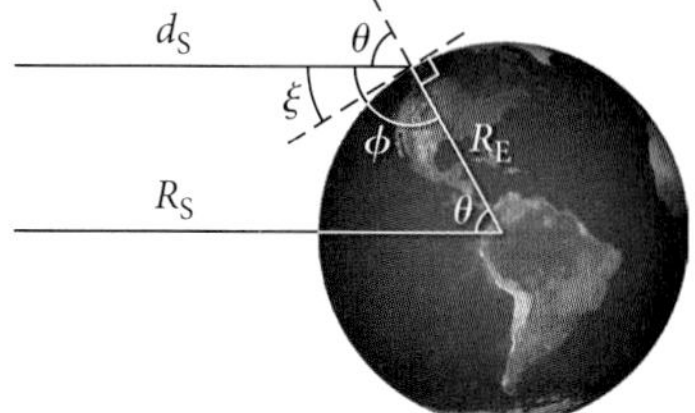


**FIGURE 12.26** Geometry for a satellite that is very far away.

## 12.7 Dark Matter

In the Solar System, almost all of the matter is concentrated in the Sun. The mass of the Sun is approximately 750 times greater than the mass of all planets combined (refer to Table 12.1). The Milky Way, our home galaxy, contains giant clouds of gas and dust, but their combined mass is only about a tenth of that contained in the galaxy's stars. Extrapolating from these facts, you might conclude that the entire universe is composed almost exclusively of luminous matter—that is, stars. Nonluminous matter in the form of dust, asteroids, moons, and planets should contribute only a small fraction of the mass of the universe.

Astronomers know the typical masses of stars and can estimate their numbers in galaxies. Thus, they have obtained fairly accurate estimates of the masses of galaxies and clusters of galaxies. Furthermore, they use modern X-ray telescopes, such as Chandra, to take images of the hot interstellar gas trapped within galaxy clusters. They can deduce the temperature of this gas from the X-ray emissions, and they can also determine how much gravitational force it takes to keep this hot gas from escaping the galaxy clusters. The surprise that has emerged from this research is that the mass contained in luminous matter is too small by a factor of approximately 3 to 5 to provide this gravitational force. This analysis leads to the conclusion that there must be other matter, referred to as *dark matter,* which provides the missing gravitational force.

Some of the observational evidence for dark matter has emerged from measurements of the orbital speeds of stars around the center of their galaxy as a function of their distance from the center, as shown in Figure 12.27 for M 31, the Andromeda galaxy. Even if we assume that there is a supermassive black hole, which is obviously nonluminous, at the center of the galaxy, we would expect star velocity to fall off toward zero for large distances from the center. Experimental evidence, however, indicates that this is not the case. Thus, it is likely that large quantities of dark matter, with a very large radial extent, are present.

Other evidence for dark matter has emerged from observations of gravitational lensing. In Figure 12.28a, the light blue shading represents the distribution of dark matter around a galaxy cluster, as calculated from the observed gravitational lensing. In Figure 12.28b, white

### 12.4 Self-Test Opportunity

One of the pieces of evidence for dark matter is the velocity curve of stars in rotating galaxies (see, for example, Figure 12.27). Astronomers observe that the velocity of stars in such a galaxy first increases and then remains constant as a function of distance from the center of the galaxy. What would you expect the velocity of stars as a function of distance from the center of the galaxy to be if the galaxy consisted of stars with equal mass distributed uniformly throughout a disk and no dark matter were present?

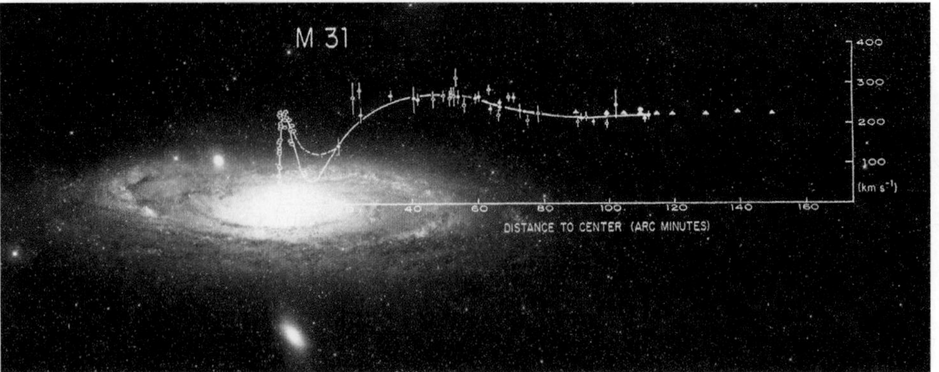


**FIGURE 12.27** Image of the Andromeda galaxy, with data on the orbital speeds of stars superimposed. (The triangles represent radio-telescope data from 1975, and the other symbols show optical-wavelength observations from 1970; the solid and dashed lines are simple fits to guide the eye.) The main feature of interest here is that the orbital speeds remain approximately constant well outside the luminous portion of Andromeda, indicating the presence of dark matter.

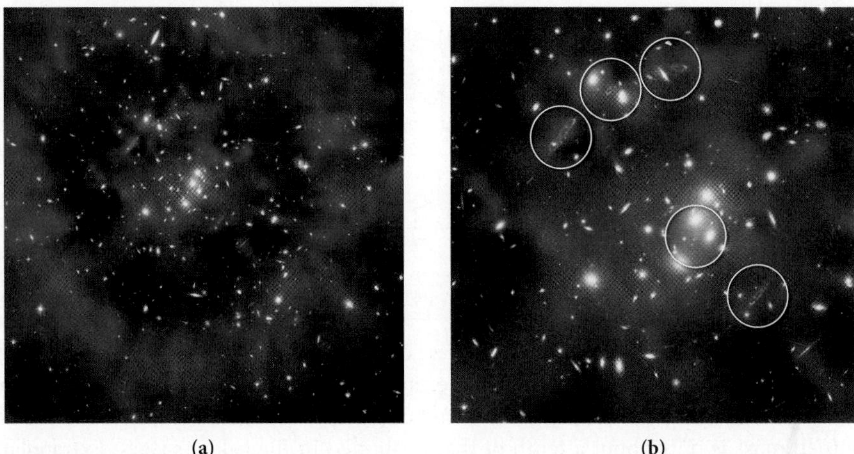


**FIGURE 12.28** An example of gravitational lensing by dark matter. (a) Photograph of the galaxy cluster Cl 0024+17 taken with the Hubble Space Telescope. The light blue shading represents the distribution of dark matter based on the observed gravitational lensing. (b) Expanded section of the center section of the photograph in part (a), with five images of the same galaxy produced by gravitational lensing marked by circles.

**FIGURE 12.29** Superposition of X-ray and optical images of the "bullet cluster," galaxy cluster 1E 0657-56, which contains direct empirical proof of the existence of dark matter.

circles mark the positions of five images of the same galaxy produced by the gravitational lensing from the unseen dark matter. (Gravitational lensing will be explained in Chapter 35 on relativity. For now, this observation is presented simply as one more empirical fact pointing toward the existence of dark matter.)

Supporting data have also emerged from the WMAP (Wilkinson Microwave Anisotropy Probe) mission, which measured the cosmic background radiation left over from the Big Bang. The best estimate based on this data is that 23% of the universe is composed of dark matter. In addition, the combination of images from the Hubble Space Telescope, the Chandra X-ray Observatory, and the Magellan telescope has yielded a direct empirical proof of the existence of dark matter in the "bullet cluster" (Figure 12.29). The measured temperature of the intergalactic gas in this galaxy cluster is too large for the gas to be contained within the cluster without the presence of dark matter.

There have been intense speculation and extensive theoretical investigation as to the nature of this dark matter during the last few years, and theories about it are still being modified as new observations emerge. However, all of the proposed theories require a fundamental rethinking of the standard model of the universe and quite possibly of the fundamental models for the interaction of particles. Whimsical names have been suggested for the possible constituents of dark matter, such as WIMP (Weakly Interacting Massive Particle) or MACHO (Massive Astrophysical Compact Halo Object). (The physical properties of these postulated constituents are being actively investigated.)

During the last few years, an even stranger phenomenon has been discovered: It seems that, in addition to dark matter, there is also *dark energy.* This dark energy seems to be responsible for an increasing acceleration in the expansion of the universe. A stunning 73% of the mass-energy of the universe is estimated to be dark energy. Together with the 23% estimated to be dark matter, this leaves only 4% of the universe for the stars, planets, moons, gas, and all other objects made of conventional matter.

These are very exciting new areas of research that are sure to change our picture of the universe in the coming decades.

## WHAT WE HAVE LEARNED | EXAM STUDY GUIDE

- The gravitational force between two point masses is proportional to the product of their masses and inversely proportional to the distance between them, $F(r) = G\frac{m_1 m_2}{r^2}$, with the proportionality constant $G = 6.674 \cdot 10^{-11}\ \text{m}^3\text{kg}^{-1}\text{s}^{-2}$, known as the universal gravitational constant.
- In vector form, the equation for the gravitational force can be written as $\vec{F}_{2\to 1} = G\frac{m_1 m_2}{|\vec{r}_2 - \vec{r}_1|^3}(\vec{r}_2 - \vec{r}_1)$. This is Newton's Law of Gravity.
- If more than two objects interact gravitationally, the resulting force on one object is given by the vector sum of the forces acting on it due to the other objects.
- Near the surface of the Earth, the gravitational acceleration can be approximated by the function $g(h) = g\left(1 - 2\frac{h}{R_\text{E}} + \cdots\right)$; that is, it falls off linearly with height above the surface.
- The gravitational acceleration at sea level can be derived from Newton's Law of Gravity: $g = \frac{GM_\text{E}}{R_\text{E}^2}$.
- No gravitational force acts on an object inside a massive spherical shell. Because of this, the gravitational force inside a uniform sphere increases linearly with radius: $F(r) = \frac{4}{3}\pi G\rho m r$.
- The gravitational potential energy between two objects is given by $U(r) = -G\frac{m_1 m_2}{r}$.
- The escape speed from the surface of the Earth is $v_\text{E} = \sqrt{\frac{2GM_\text{E}}{R_\text{E}}}$.
- Kepler's laws of planetary motion are as follows
  - All planets move in elliptical orbits with the Sun at one focal point.
  - A straight line connecting the center of the Sun and the center of any planet sweeps out an equal area in any given time interval: $\frac{dA}{dt} = \text{constant}$.
  - The square of the period of a planet's orbit is proportional to the cube of the semimajor axis of the orbit: $\frac{T^2}{r^3} = \text{constant}$.
- The relationships of the kinetic, potential, and total energy of a satellite in a circular orbit are $K = -\frac{1}{2}U$, $E = K + U = -\frac{1}{2}U + U = \frac{1}{2}U = -\frac{1}{2}G\frac{Mm}{r}$, $E = -K$.
- Geostationary satellites have an orbit that is circular, above the Equator, and with a radius of 42,168 km.
- Evidence points strongly to the existence of dark matter and dark energy, which make up the vast majority of the universe.

## KEY TERMS

Newton's Law of Gravity, p. 383
universal gravitational constant, p. 383
superposition principle, p. 383
gravitational potential energy, p. 392
escape speed, p. 393
Kepler's laws of planetary motion, p. 395
central force, p. 400
geostationary satellites, p. 400

## NEW SYMBOLS AND EQUATIONS

$F(r)=G\frac{m_1m_2}{r^2}$, Newton's Law of Gravity

$G=6.674\cdot10^{-11}\ \text{m}^3\text{kg}^{-1}\text{s}^{-2}$, universal gravitational constant

$U(r)=-G\frac{m_1m_2}{r}$, gravitational potential energy

$V_E(r)=-\frac{GM_E}{r}$, Earth's gravitational potential

$v_E=\sqrt{\frac{2GM_E}{R_E}}$, escape speed from Earth

## ANSWERS TO SELF-TEST OPPORTUNITIES

**12.1** The derivation is almost identical, but the integration from $R-r$ to $R+r$ yields zero force. Another proof is given in Derivation 12.2, based on geometry.

**12.2** $g_d = g/9 = 1.09\ \text{m/s}^2$.

**12.3** Gravitational attraction between Sun and Earth provides the centripetal force to keep Earth in orbit around the Sun.

$$\frac{m_E v^2}{r}=G\frac{m_E M}{r^2}$$

Now use $v=\frac{2\pi r}{T}$ and insert, then solve for $M$:

$$\frac{\left(\frac{2\pi r}{T}\right)^2}{r}=G\frac{M}{r^2}$$

$$M=\frac{(2\pi)^2 r^3}{GT^2}.$$

**12.4** The velocity would increase linearly until the distance of the star from the center of the galaxy, $r$, was equal to the radius of the disk-shaped galaxy. For stars outside the radius of the galaxy, the velocity would decrease proportional to $1/\sqrt{r}$.

## PROBLEM-SOLVING PRACTICE

### Problem-Solving Guidelines: Gravitation

**1.** This chapter introduced the general form of the equation for the gravitational force, and the approximation $F=mg$ is no longer valid in general. Be sure to keep in mind that in general the acceleration due to gravity is not constant either. And this means that you are not able to use the kinematic equations of Chapters 2 and 3 to solve problems.

**2.** Energy conservation is essential for many dynamic problems involving gravitation. Be sure to remember that the gravitational potential energy is not simply given by $U=mgh$, as presented in Chapter 6.

**3.** The principle of superposition of forces is important for situations involving interactions of more than two objects. It allows you to calculate the forces pairwise and then add them appropriately.

**4.** For planetary and satellite orbits, Kepler's laws are very useful computational tools, enabling you to connect orbital periods and orbital radii.

### SOLVED PROBLEM 12.4 Astronaut on a Small Moon

**PROBLEM**

A small spherical moon has a radius of $6.30\cdot10^4$ m and a mass of $8.00\cdot10^{18}$ kg. An astronaut standing on the surface of the moon throws a rock straight up. The rock reaches a maximum height of 2.20 km above the surface of the moon before it returns to the surface. What was the initial speed of the rock as it left the hand of the astronaut? (This moon is too small to have an atmosphere.)

**SOLUTION**

**THINK**

We know the mass and radius of the moon, allowing us to compute the gravitational potential. Knowing the height that the rock attains, we can use energy conservation to calculate the initial speed of the rock.

**SKETCH**

Figure 12.30 is a sketch showing the rock at its highest point.

y

h

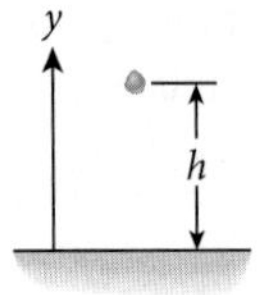

**FIGURE 12.30** The rock at the highest point on its straight-line trajectory.

**RESEARCH**

We denote the mass of the rock as $m$, the mass of the moon as $m_{\text{moon}}$, and the radius of the moon as $R_{\text{moon}}$. The potential energy of the rock on the surface of the moon is then

$$U(R_{\text{moon}}) = -G\frac{m_{\text{moon}}}{R_{\text{moon}}}m. \tag{i}$$

The potential energy at the top of the rock's trajectory is

$$U(R_{\text{moon}}+h) = -G\frac{m_{\text{moon}}}{R_{\text{moon}}+h}m. \tag{ii}$$

The kinetic energy of the rock as it leaves the hand of the astronaut depends on its mass and the initial speed $v$:

$$K(R_{\text{moon}}) = \tfrac{1}{2}mv^2. \tag{iii}$$

At the top of the rock's trajectory, $K(R_{\text{moon}} + h) = 0$. Conservation of total energy now helps us solve the problem:

$$U(R_{\text{moon}}) + K(R_{\text{moon}}) = U(R_{\text{moon}}+h) + K(R_{\text{moon}}+h). \tag{iv}$$

**SIMPLIFY**

We substitute the expressions from equations (i) through (iii) into (iv):

$$-G\frac{m_{\text{moon}}}{R_{\text{moon}}}m + \tfrac{1}{2}mv^2 = -G\frac{m_{\text{moon}}}{R_{\text{moon}}+h}m + 0 \Rightarrow$$

$$-G\frac{m_{\text{moon}}}{R_{\text{moon}}} + \tfrac{1}{2}v^2 = -G\frac{m_{\text{moon}}}{R_{\text{moon}}+h} \Rightarrow$$

$$v = \sqrt{2Gm_{\text{moon}}\left(\frac{1}{R_{\text{moon}}} - \frac{1}{R_{\text{moon}}+h}\right)}.$$

**CALCULATE**

Putting in the numerical values, we get

$$v = \sqrt{2(6.67\cdot 10^{-11}\ \text{N m}^2/\text{kg}^2)(8.00\cdot 10^{18}\ \text{kg})\left(\frac{1}{6.30\cdot 10^4\ \text{m}} - \frac{1}{(6.30\cdot 10^4\ \text{m})+(2.20\cdot 10^3\ \text{m})}\right)}$$
$$= 23.9078\ \text{m/s}.$$

**ROUND**

Rounding to three significant figures gives

$$v = 23.9\ \text{m/s}.$$

**DOUBLE-CHECK**

To double-check our result, let's make the simplifying assumption that the force of gravity on the small moon does not change with altitude. We can find the initial speed of the rock using conservation of energy:

$$mg_{\text{moon}}h = \tfrac{1}{2}mv^2, \tag{v}$$

where $h$ is the height attained by the rock, $v$ is the initial speed, $m$ is the mass of the rock, and $g_{\text{moon}}$ is the acceleration of gravity at the surface of the small moon. The acceleration of gravity is

$$g_{\text{moon}} = G\frac{m_{\text{moon}}}{R_{\text{moon}}^2} = \left(6.67\cdot 10^{-11}\ \text{N m}^2/\text{kg}^2\right)\frac{8.00\cdot 10^{18}\ \text{kg}}{\left(6.30\cdot 10^4\ \text{m}\right)^2} = 0.134\ \text{m/s}^2.$$

*Continued—*

Solving equation (v) for the initial speed of the rock gives us

$$v = \sqrt{2 g_{\text{moon}} h}.$$

Putting in the numerical values, we get

$$v = \sqrt{2\left(0.134 \text{ m/s}^2\right)\left(2200 \text{ m}\right)} = 24.3 \text{ m/s}.$$

This result is close to but larger than our answer, 23.9 m/s. The fact that the simple assumption of constant gravitational force leads to a larger initial speed than for the realistic case where the force decreases with altitude makes sense. The close agreement of the two results also makes sense because the altitude gained (2.20 km) is not too large compared to the radius of the small moon (63.0 km). Thus, our answer seems reasonable.

## MULTIPLE-CHOICE QUESTIONS

**12.1** A planet is in a circular orbit about a remote star, far from any other object in the universe. Which of the following statements is true?

a) There is only one force acting on the planet.

b) There are two forces acting on the planet and their resultant is zero.

c) There are two forces acting on the planet and their resultant is not zero.

d) None of the above statements are true.

**12.2** Two 30.0-kg masses are held at opposite corners of a square of sides 20.0 cm. If one of the masses is released and allowed to fall toward the other mass, what is the acceleration of the first mass just as it is released? Assume that the only force acting on the mass is the gravitational force of the other mass.

a) $1.5 \cdot 10^{-8}$ m/s$^2$
c) $7.5 \cdot 10^{-8}$ m/s$^2$
b) $2.5 \cdot 10^{-8}$ m/s$^2$
d) $3.7 \cdot 10^{-8}$ m/s$^2$

**12.3** With the usual assumption that the gravitational potential energy goes to zero at infinite distance, the gravitational potential energy due to the Earth at the center of Earth is

a) positive.
c) zero.
b) negative.
d) undetermined.

**12.4** A man inside a sturdy box is fired out of a cannon. Which of following statements regarding the weightless sensation for the man is correct?

a) The man senses weightlessness only when he and the box are traveling upward.

b) The man senses weightlessness only when he and the box are traveling downward.

c) The man senses weightlessness when he and the box are traveling both upward and downward.

d) The man does not sense weightlessness at any time of the flight.

**12.5** In a binary star system consisting of two stars of equal mass, where is the gravitational potential equal to zero?

a) exactly halfway between the stars

b) along a line bisecting the line connecting the stars

c) infinitely far from the stars

d) none of the above

**12.6** Two planets have the same mass, $M$, but one of them is much denser than the other. Identical objects of mass $m$ are placed on the surfaces of the planets. Which object will have the gravitational potential energy of larger magnitude?

a) Both objects will have the same gravitational potential energy.

b) The object on the surface of the denser planet will have the larger gravitational potential energy.

c) The object on the surface of the less dense planet will have the larger gravitational potential energy.

d) It is impossible to tell.

**12.7** Two planets have the same mass, $M$. Each planet has a constant density, but the density of planet 2 is twice as high as that of planet 1. Identical objects of mass $m$ are placed on the surfaces of the planets. What is the relationship of the gravitational potential energy, $U_1$, on planet 1 to $U_2$ on planet 2?

a) $U_1 = U_2$
d) $U_1 = 8U_2$
b) $U_1 = \frac{1}{2}U_2$
e) $U_1 = 0.794U_2$
c) $U_1 = 2U_2$

**12.8** For two identical satellites in circular motion around the Earth, which statement is true?

a) The one in the lower orbit has less total energy.

b) The one in the higher orbit has more kinetic energy.

c) The one in the lower orbit has more total energy.

d) Both have the same total energy.

**12.9** Which condition do all geostationary satellites orbiting the Earth have to fulfill?

a) They have to orbit above the Equator.

b) They have to orbit above the poles.

c) They have to have an orbital radius that locates them less than 30,000 km above the surface.

d) They have to have an orbital radius that locates them more than 42,000 km above the surface.

**12.10** An object is placed between the Earth and the Moon, along the straight line that joins them. Assume that the mass of the Moon is a sixth that of the Earth. About how far away from the Earth should the object be placed so that the net gravitational force on the object from the Earth and the Moon is zero? This point is known as the *L1 Point*, where the L stands for Lagrange, a famous French mathematician.

a) halfway to the Moon

b) 60% of the way to the Moon

c) 70% of the way to the Moon

d) 80% of the way to the Moon

e) 90% of the way to the Moon

**12.11** A man of mass 100 kg feels a gravitational force, $F_m$, from a woman of mass 50 kg sitting 1 m away. The gravitational force, $F_w$, experienced by the woman will be __________ that experienced by the man.

a) more than

b) less than

c) the same as

d) not enough information given

## QUESTIONS

**12.12** Can the expression for gravitational potential energy $U_g(y) = mgy$ be used to analyze high-altitude motion? Why or why not?

**12.13** Even though the Moon does not have an atmosphere, the trajectory of a projectile near its surface is only approximately a parabola. This is because the acceleration due to gravity near the surface of the Moon is only approximately constant. Describe as precisely as you can the *actual* shape of a projectile's path on the Moon, even one that travels a long distance over the surface of the Moon.

**12.14** A scientist working for a space agency noticed that a Russian satellite of mass 250 kg is on collision course with an American satellite of mass 600 kg orbiting at 1000 km above the surface. Both satellites are moving in circular orbits but in opposite directions. If the two satellites collide and stick together, will they continue to orbit or crash to the Earth? Explain.

**12.15** Three asteroids, located at points $P_1$, $P_2$, and $P_3$, which are not in a line, and having known masses $m_1$, $m_2$, and $m_3$, interact with one another through their mutual gravitational forces only; they are isolated in space and do not interact with any other bodies. Let $\sigma$ denote the axis going through the center of mass of the three asteroids, perpendicular to the triangle $P_1P_2P_3$. What conditions should the angular velocity $\omega$ of the system (around the axis $\sigma$) and the distances

$$P_1P_2 = a_{12}, \quad P_2P_3 = a_{23}, \quad P_1P_3 = a_{13},$$

fulfill to allow the shape and size of the triangle $P_1P_2P_3$ to remain unchanged during the motion of the system? That is, under what conditions does the system rotate around the axis $\sigma$ as a rigid body?

**12.16** The more powerful the gravitational force of a planet, the greater its escape speed, $v$, and the greater the gravitational acceleration, $g$, at its surface. However, in Table 12.1, the value for $v$ is much greater for Uranus than for Earth—but $g$ is smaller on Uranus than on Earth! How can this be?

**12.17** Is the orbital speed of the Earth when it is closest to the Sun greater than, less than, or equal to the orbital speed when it is farthest from the Sun? Explain.

**12.18** Point out any flaw in the following physics exam statement: "*Kepler's First Law states that all planets move in elliptical orbits with the Sun at one focal point. It follows that during one complete revolution around the Sun (1 year), the Earth will pass through a closest point to the Sun—the perihelion—as well as through a furthest point from the Sun—the aphelion. This is the main cause of the seasons (summer and winter) on Earth.*"

**12.19** A comet orbiting the Sun moves in an elliptical orbit. Where is its kinetic energy, and therefore its speed, at a maximum, at perihelion or aphelion? Where is its gravitational potential energy at a maximum?

**12.20** Where the International Space Station orbits, the gravitational acceleration is just 11.4% less than its value on the surface of the Earth. Nevertheless, astronauts in the space station float. Why is this so?

**12.21** Satellites in low orbit around the Earth lose energy from colliding with the gases of the upper atmosphere, causing them to slowly spiral inward. What happens to their kinetic energy as they fall inward?

**12.22** Compare the magnitudes of the gravitational force that the Earth exerts on the Moon and the gravitational force that the Moon exerts on the Earth. Which is larger?

**12.23** Imagine that two tunnels are bored completely through the Earth, passing through the center. Tunnel 1 is along the Earth's axis of rotation, and tunnel 2 is in the equatorial plane, with both ends at the Equator. Two identical balls, each with a mass

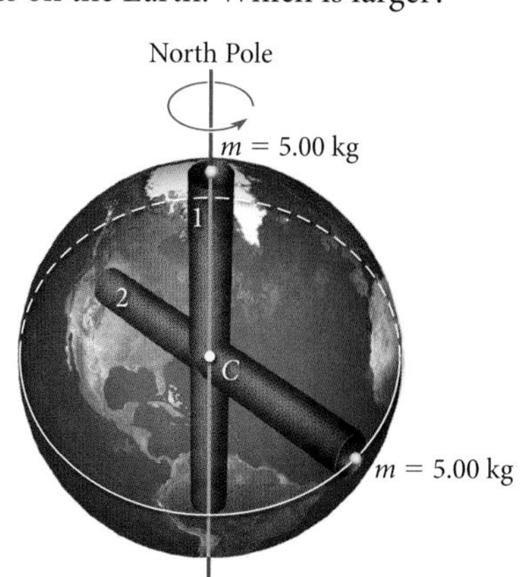

of 5.00 kg, are simultaneously dropped into both tunnels. Neglect air resistance and friction from the tunnel walls. Do the balls reach the center of the Earth (point $C$) at the same time? If not, which ball reaches the center of the Earth first?

**12.24** Imagine that a tunnel is bored in the Earth's equatorial plane, going completely through the center of the Earth with both ends at the Equator. A mass of 5.00 kg is dropped into the tunnel at one end, as shown in the figure. The tunnel has a radius that is slightly larger than that of the mass. The mass is dropped into the center of the tunnel. Neglect air resistance and friction from the tunnel wall. Does the mass ever touch the wall of the tunnel as it falls? If so, which side does it touch first, north, east, south, or west? (*Hint:* The angular momentum of the mass is conserved if the only forces acting on it are radial.)

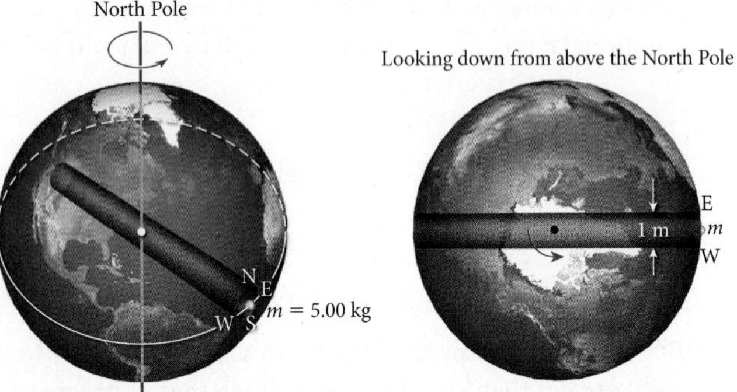


**12.25** A plumb bob located at latitude 55.0° N hangs motionlessly with respect to the ground beneath it. A straight line from the string supporting the bob does not go exactly through the Earth's center. Does this line intersect the Earth's axis of rotation south or north of the Earth's center?

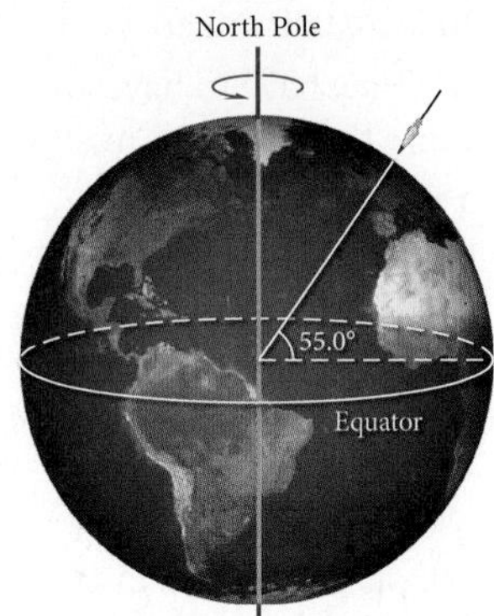


## PROBLEMS

A blue problem number indicates a worked-out solution is available in the Student Solutions Manual. One • and two •• indicate increasing level of problem difficulty.

### Section 12.1

**12.26** The Moon causes tides because the gravitational force it exerts differs between the side of the Earth nearest the Moon and that farthest from the Moon. Find the difference in the accelerations toward the Moon of objects on the nearest and farthest sides of the Earth.

12.27 After a spacewalk, a 1-kg tool is left 50 m from the center of gravity of a 20-metric ton space station, orbiting along with it. How much closer to the space station will the tool drift in an hour due to the gravitational attraction of the space station?

•**12.28** a) What is the total force on $m_1$ due to $m_2$, $m_3$, and $m_4$ if all four masses are located at the corners of a square of side $a$? Let $m_1 = m_2 = m_3 = m_4$.

b) Sketch all the forces acting on $m_1$.

•**12.29** A spaceship of mass $m$ is located between two planets of masses $M_1$ and $M_2$; the distance between the two planets is $L$, as shown in the figure. Assume that $L$ is much larger than the radius of either planet. What is the position of the spacecraft (given as a function of $L$, $M_1$, and $M_2$) if the net force on the spacecraft is zero?

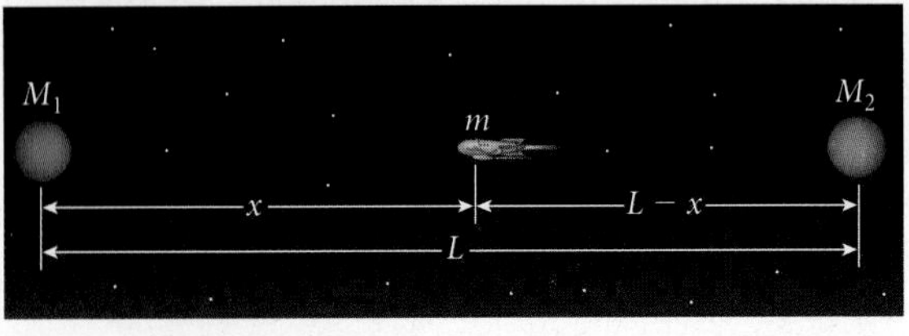


•**12.30** A carefully designed experiment can measure the gravitational force between masses of 1 kg. Given that the density of iron is 7860 kg/m$^3$, what is the gravitational force between two 1.00-kg iron spheres that are touching?

••**12.31** A uniform rod of mass 333 kg is in the shape of a semicircle of radius 5 m. Calculate the magnitude of the force on a 77-kg point mass placed at the center of the semicircle, as shown in the figure.

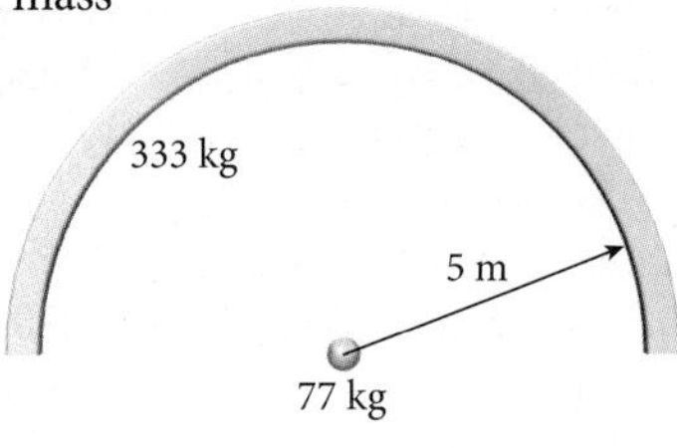


••**12.32** The figure shows a system of four masses. The center-to-center distance between any two adjacent masses is 10 cm. The base of the pyramid is in the $xz$-plane and the 20-kg mass is on the $y$ axis. What is the magnitude and direction of the gravitational force acting on the 10-kg mass? Give direction of the net force with respect to the $xyz$-coordinates shown.

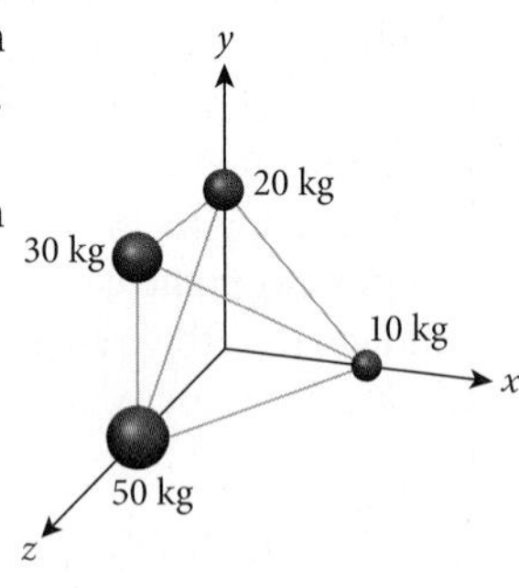


### Sections 12.2 and 12.3

**12.33** Suppose a new extrasolar planet is discovered. Its mass is double the mass of the Earth, but it has the same density and spherical shape as the Earth. How would the weight of an object at the new planet's surface differ from its weight on Earth?

**12.34** What is the magnitude of the free-fall acceleration of a ball (mass $m$) due to the Earth's gravity at an altitude of $2R$, where $R$ is the radius of the Earth?

12.35 Some of the deepest mines in the world are in South Africa and are roughly 3.5 km deep. Consider the Earth to be a uniform sphere of radius 6370 km.

a) How deep would a mine shaft have to be for the gravitational acceleration at the bottom to be reduced by a factor of 2 from its value on the Earth's surface?

b) What is the percentage difference in the gravitational acceleration at the bottom of the 3.5-km-deep shaft relative to that at the Earth's mean radius? That is, what is the value of $(a_{\text{surf}} - a_{3.5\text{km}})/a_{\text{surf}}$?

•**12.36** In an experiment performed at the bottom of a very deep vertical mine shaft, a ball is tossed vertically in the air with a known initial velocity of 10.0 m/s, and the maximum height the ball reaches (measured from its launch point) is determined to be 5.113 m. Knowing the radius of the Earth, $R_E = 6370$ km, and the gravitational acceleration at the surface of the Earth, $g(0) = 9.81$ m/s$^2$, calculate the depth of the shaft.

••**12.37** Careful measurements of local variations in the acceleration due to gravity can reveal the locations of oil deposits. Assume that the Earth is a uniform sphere of radius 6370 km and density 5500 kg/m$^3$, except that there is a spherical region of radius 1.0 km and density 900 kg/m$^3$, whose center is at a depth of 2.0 km. Suppose you are standing on the surface of the Earth directly above the anomaly with an instrument capable of measuring the acceleration due to gravity with great precision. What is the ratio of the acceleration due to gravity that you measure compared to what you would have measured had the density been 5500 kg/m$^3$ everywhere? (*Hint:* Think of this as a superposition problem involving two uniform spherical masses, one with a negative density.)

## Section 12.4

**12.38** A spaceship is launched from the Earth's surface with a speed $v$. The radius of the Earth is $R$. What will its speed be when it is very far from the Earth?

12.39 What is the ratio of the escape speed to the orbital speed of a satellite at the surface of the Moon, where the gravitational acceleration is about a sixth of that on Earth?

**12.40** Standing on the surface of a small spherical moon whose radius is $6.30 \cdot 10^4$ m and whose mass is $8.00 \cdot 10^{18}$ kg, an astronaut throws a rock of mass 2.00 kg straight upward with an initial speed 40.0 m/s. (This moon is too small to have an atmosphere.) What maximum height above the surface of the moon will the rock reach?

**12.41** An object of mass $m$ is launched from the surface of the Earth. Show that the minimum speed required to send the projectile to a height of $4R_E$ above the surface of the Earth is $v_{\min} = \sqrt{8GM_E/5R_E}$. $M_E$ is the mass of the Earth and $R_E$ is the radius of the Earth. Neglect air resistance.

**12.42** For the satellite in Solved Problem 12.2, orbiting the Earth at a distance of $3.75R_E$ with a speed of 4.08 km/s, with what speed would the satellite hit the Earth's surface if somehow it suddenly stopped and fell to Earth? Ignore air resistance.

•12.43 Estimate the radius of the largest asteroid from which you could escape by jumping. Assume spherical geometry and a uniform density equal to the Earth's average density.

•**12.44** Eris, the largest dwarf planet known in the Solar System, has a radius $R = 1200$ km and an acceleration due to gravity on its surface of magnitude $g = 0.77$ m/s$^2$.

a) Use these numbers to calculate the escape speed from the surface of Eris.

b) If an object is fired directly upward from the surface of Eris with half of this escape speed, to what maximum height above the surface will the object rise? (Assume that Eris has no atmosphere and negligible rotation.)

•**12.45** Two identical 20.0-kg spheres of radius 10 cm are 30.0 cm apart (center-to-center distance).

a) If they are released from rest and allowed to fall toward one another, what is their speed when they first make contact?

b) If the spheres are initially at rest and just touching, how much energy is required to separate them to 1.00 m apart? Assume that the only force acting on each mass is the gravitational force due to the other mass.

••**12.46** Imagine that a tunnel is bored completely through the Earth along its axis of rotation. A ball with a mass of 5.00 kg is dropped from rest into the tunnel at the North Pole, as shown in the figure. Neglect air resistance and friction from the tunnel wall. Calculate the potential energy of the ball as a function of its distance from the center of the Earth. What is the speed of the ball when it arrives at the center of the Earth (point $C$)?

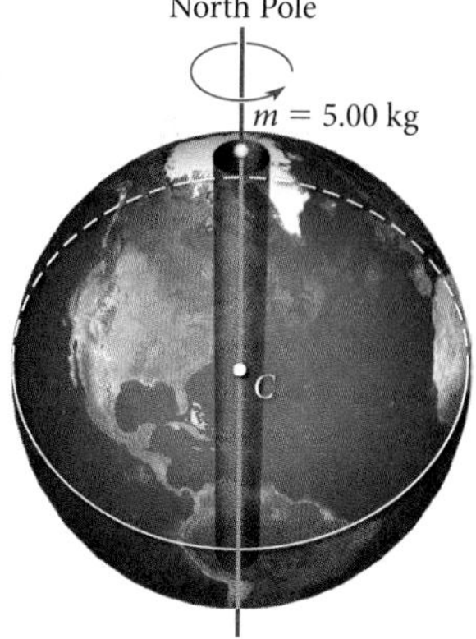


## Section 12.5

12.47 The Apollo 8 mission in 1968 included a circular orbit at an altitude of 111 km above the Moon's surface. What was the period of this orbit? (You need to look up the mass and radius of the Moon to answer this question!)

•**12.48** Halley's comet orbits the Sun with a period of 76.2 yr.

a) Find the semimajor axis of the orbit of Halley's comet in astronomical units (1 AU is equal to the semimajor axis of the Earth's orbit).

b) If Halley's comet is 0.56 AU from the Sun at perihelion, what is its maximum distance from the Sun, and what is the eccentricity of its orbit?

••**12.49** A satellite of mass $m$ is in an elliptical orbit (that satisfies Kepler's laws) about a body of mass $M$, with $m$ negligible compared to $M$.

a) Find the total energy of the satellite as a function of its speed, $v$, and distance, $r$, from the body it is orbiting.

b) At the maximum and minimum distance between the satellite and the body, and only there, the angular momentum is simply related to the speed and distance. Use this relationship and the result of part (a) to obtain a relationship between the extreme distances and the satellite's energy and angular momentum.

c) Solve the result of part (b) for the maximum and minimum radii of the orbit in terms of the energy and angular momentum per unit mass of the satellite.

d) Transform the results of part (c) into expressions for the semimajor axis, $a$, and eccentricity of the orbit, $e$, in terms of the energy and angular momentum per unit mass of the satellite.

••**12.50** Consider the Sun to be at the origin of an $xy$-coordinate system. A telescope spots an asteroid in the $xy$-plane at a position given by $(2.0 \cdot 10^{11}\text{ m}, 3.0 \cdot 10^{11}\text{ m})$ with a velocity given by $(-9.0 \cdot 10^3\text{ m/s}, -7.0 \cdot 10^3\text{ m/s})$. What will the asteroid's velocity and distance from the Sun be at closest approach?

## Section 12.6

12.51 A spy satellite was launched into a circular orbit with a height of 700 km above the surface of the Earth. Determine its orbital speed and period.

**12.52** Express algebraically the ratio of the gravitational force on the Moon due to the Earth to the gravitational force on the Moon due to the Sun. Why, since the ratio is so small, doesn't the Sun pull the Moon away from the Earth?

**12.53** A space shuttle is initially in a circular orbit at a radius of $r = 6.60 \cdot 10^6$ m from the center of the Earth. A retrorocket is fired forward, reducing the total energy of the space shuttle by 10% (that is, increasing the magnitude of the negative total energy by 10%), and the space shuttle moves to a new circular orbit with a radius that is smaller than $r$. Find the speed of the space shuttle (a) before and (b) after the retrorocket is fired.

•**12.54** A 200-kg satellite is in circular orbit around the Earth and moving at a speed of 5.00 km/s. How much work must be done to move the satellite into another circular orbit that is twice as high above the surface of the Earth?

•12.55 The radius of a black hole is the distance from the black hole's center at which the escape speed is the speed of light.

a) What is the radius of a black hole with a mass twice that of the Sun?

b) At what radius from the center of the black hole in part (a) would the orbital speed be equal to the speed of light?

c) What is the radius of a black hole with the same mass as that of the Earth?

•**12.56** A satellite is in a circular orbit around a planet. The ratio of the satellite's kinetic energy to its gravitational potential energy, $K/U_g$, is a constant whose value is independent of the masses of the satellite and planet, and of the radius and velocity of the orbit. Find the value of this constant. (Potential energy is taken to be zero at infinite separation.)

•**12.57** Determine the minimum amount of energy that a projectile of mass 100.0 kg must gain to reach a circular orbit 10.00 km above the Earth's surface if launched from (a) the North Pole or from (b) the Equator (keep answers to four significant figures). Do not be concerned about the direction of the launch or of the final orbit. Is there an advantage or disadvantage to launching from the Equator? If so, how significant is the difference? Do not neglect the rotation of the Earth when calculating the initial energies.

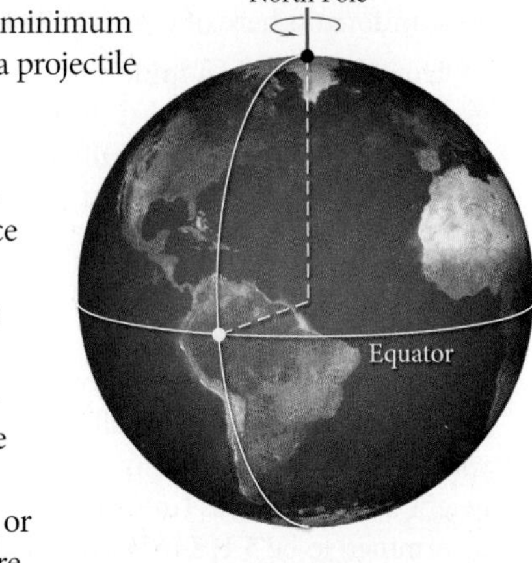


••**12.58** A rocket with mass $M = 12$ metric tons is moving around the Moon in a circular orbit at the height of $h = 100$ km. The braking engine is activated for a short time to lower the orbital height so that the rocket can make a lunar landing. The velocity of the ejected gases is $u = 10^4$ m/s. The Moon's radius is $R_M = 1.7 \cdot 10^3$ km; the acceleration of gravity near the Moon's surface is $g_M = 1.7\text{ m/s}^2$.

a) What amount of fuel will be used by the braking engine if it is activated at point $A$ of the orbit and the rocket lands on the Moon at point $B$ (see the left part of the figure)?

b) Suppose that, at point $A$, the rocket is given an impulse directed toward the center of the Moon, to put it on a trajectory that meets the Moon's surface at point $C$ (see the right part of the figure). What amount of fuel is needed in this case?

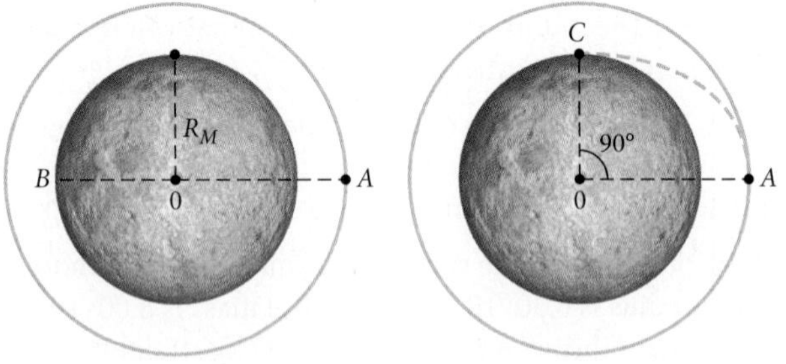


## Additional Problems

12.59 Calculate the magnitudes of the gravitational forces exerted on the Moon by the Sun and by the Earth when the two forces are in direct competition, that is, when the Sun, Moon, and Earth are aligned with the Moon between the Sun and the Earth. (This alignment corresponds to a solar eclipse.) Does the orbit of the Moon ever actually curve away from the Sun, toward the Earth?

**12.60** A projectile is shot from the surface of the Earth by means of a very powerful cannon. If the projectile reaches a height of 55.0 km above Earth's surface, what was the speed of the projectile when it left the cannon?

**12.61** Newton's Law of Gravity specifies the magnitude of the interaction force between two point masses, $m_1$ and $m_2$, separated by a distance $r$ as $F(r) = Gm_1m_2/r^2$. The gravitational constant $G$ can be determined by directly measuring the interaction force (gravitational attraction) between two sets of spheres by using the apparatus constructed in the late 18th century by the English scientist Henry Cavendish. This apparatus was a torsion balance consisting of a 6-ft wooden rod suspended from a torsion wire, with a lead sphere having a diameter of 2 in and a weight of 1.61 lb attached to each end. Two 12-in, 348-lb lead balls were located near the smaller balls, about 9 in away, and held in place with a separate suspension system. Today's accepted value for $G$ is $6.674 \cdot 10^{-11}\ \mathrm{m^3kg^{-1}s^{-2}}$. Determine the force of attraction between the larger and smaller balls that had to be measured by this balance. Compare this force to the weight of the small balls.

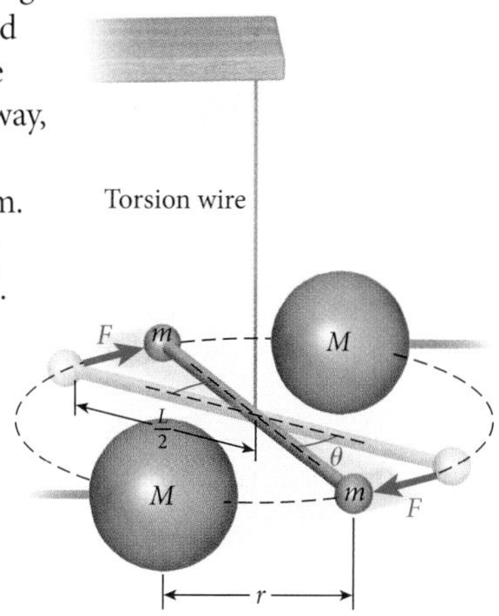


**12.62** Newton was holding an apple of mass 100 g and thinking about the gravitational forces exerted on the apple by himself and by the Sun. Calculate the magnitude of the gravitational force acting on the apple due to (a) Newton, (b) the Sun, and (c) the Earth, assuming that the distance from the apple to Newton's center of mass is 50 cm and Newton's mass is 80 kg.

12.63 A 1000-kg communications satellite is released from a space shuttle to initially orbit the Earth at a radius of $7 \cdot 10^6$ m. After being deployed, the satellite's rockets are fired to put it into a higher altitude orbit of radius $5 \cdot 10^7$ m. What is the minimum mechanical energy supplied by the rockets to effect this change in orbit?

**12.64** Consider a 0.3-kg apple (a) attached to a tree and (b) falling. Does the apple exert a gravitational force on the Earth? If so, what is the magnitude of this force?

**12.65** At what height $h$ above the Earth will a satellite moving in a circular orbit have half the period of the Earth's rotation about its own axis?

•**12.66** In the Earth-Moon system, there is a point where the gravitational forces balance. This point is known as the L1 point where the L stands for Lagrange, a famous French mathematician. Assume that the mass of the Moon is $\frac{1}{81}$ that of the Earth.

a) At what point, on a line between the Earth and the Moon, is the gravitational force exerted on an object by the Earth exactly balanced by the gravitational force exerted on the object by the Moon?

b) Is this point one of stable or unstable equilibrium?

c) Calculate the ratio of the force of gravity due to the Sun, acting on an object at this point, to the force of gravity due to Earth and, separately, to the force of gravity due to the Moon.

•12.67 Consider a particle on the surface of the Earth, at a position with an angle of latitude of $\lambda = 30°$ N as shown in the figure. Find (a) the magnitude, and (b) the direction of the effective gravitational force acting on the particle, taking into consideration the rotation of the Earth. (c) What angle $\lambda$ gives rise to the maximum deviation of the gravitational acceleration?

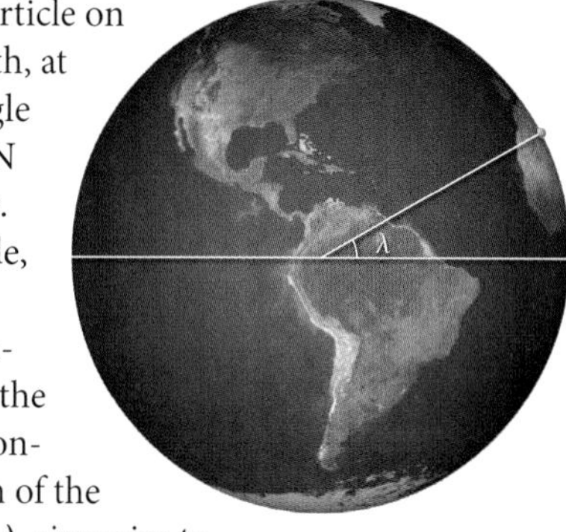


•**12.68** An asteroid is discovered to have a tiny moon that orbits it in a circular path at a distance of 100 km and with a period of 40 h. The asteroid is roughly spherical (unusual for such a small body) with a radius of 20 km.

a) Find the acceleration of gravity at the surface of the asteroid.

b) Find the escape velocity from the asteroid.

•**12.69** a) By what percentage does the gravitational potential energy of the Earth change between perihelion and aphelion? (Assume the Earth's potential energy would be zero if it is moved to a very large distance away from the Sun.)

b) By what percentage does the kinetic energy of the Earth change between perihelion and aphelion?

•**12.70** A planet with a mass of $7 \cdot 10^{21}$ kg is in a circular orbit around a star with a mass of $2 \cdot 10^{30}$ kg. The planet has an orbital radius of $3 \cdot 10^{10}$ m.

a) What is the linear orbital velocity of the planet?

b) What is the period of the planet's orbit?

c) What is the total mechanical energy of the planet?

•12.71 The astronomical unit (AU, equal to the mean radius of the Earth's orbit) is $1.4960 \cdot 10^{11}$ m, and a year is $3.1557 \cdot 10^7$ s. Newton's gravitational constant is $G = 6.6743 \cdot 10^{-11}\mathrm{m^3kg^{-1}s^{-2}}$. Calculate the mass of the Sun in kilograms. (Recalling or looking up the mass of the Sun does not constitute a solution to this problem.)

•**12.72** The distances from the Sun at perihelion and aphelion for Pluto are $4410 \cdot 10^6$ km and $7360 \cdot 10^6$ km, respectively. What is the ratio of Pluto's orbital speed around the Sun at perihelion to that at aphelion?

•**12.73** The weight of a star is usually balanced by two forces: the gravitational force, acting inward, and the force created by nuclear reactions, acting outward. Over a long period of time, the force due to nuclear reactions gets weaker, causing the gravitational collapse of the star and crushing atoms out of existence. Under such extreme conditions, protons and electrons are squeezed to form neutrons, giving birth to a neutron star. Neutron stars are massively heavy—a teaspoon of the substance of a neutron star would weigh 100 million metric tons on the Earth.

a) Consider a neutron star whose mass is twice the mass of the Sun and whose radius is 10 km. If it rotates with a period of 1 s, what is the speed of a point on the Equator of this star? Compare this speed with the speed of a point on the Earth's Equator.

b) What is the value of $g$ at the surface of this star?

c) Compare the weight of a 1-kg mass on the Earth with its weight on the neutron star.

d) If a satellite is to circle 10 km above the surface of such a neutron star, how many revolutions per minute will it make?

e) What is the radius of the geostationary orbit for this neutron star?

••**12.74** You have been sent in a small spacecraft to rendezvous with a space station that is in a circular orbit of radius $2.5000 \cdot 10^4$ km from the Earth's center. Due to a mishandling of units by a technician, you find yourself in the same orbit as the station but exactly halfway around the orbit from it! You do not apply forward thrust in an attempt to chase the station; that would be fatal folly. Instead, you apply a brief braking force against the direction of your motion, to put you into an elliptical orbit, whose highest point is your present position, and whose period is half that of your present orbit. Thus, you will return to your present position when the space station has come halfway around the circle to meet you. Is the minimum radius from the Earth's center—the low point—of your new elliptical orbit greater than the radius of the Earth (6370 km), or have you botched your last physics problem?

••**12.75** If you and the space station are initially in low Earth orbit—say, with a radius of 6720 km, approximately that of the orbit of the International Space Station—the maneuver of Problem 12.74 will fail unpleasantly. Keeping in mind that the life-support capabilities of your small spacecraft are limited and so time is of the essence, can you perform a similar maneuver that will enable you to rendezvous with the station? Find the radius and period of the transfer orbit you should use.

••**12.76** A satellite is placed between the Earth and the Moon, along a straight line that connects their centers of mass. The satellite has an orbital period around the Earth that is the same as that of the Moon, 27.3 days. How far away from the Earth should this satellite be placed?

# 13 Solids and Fluids

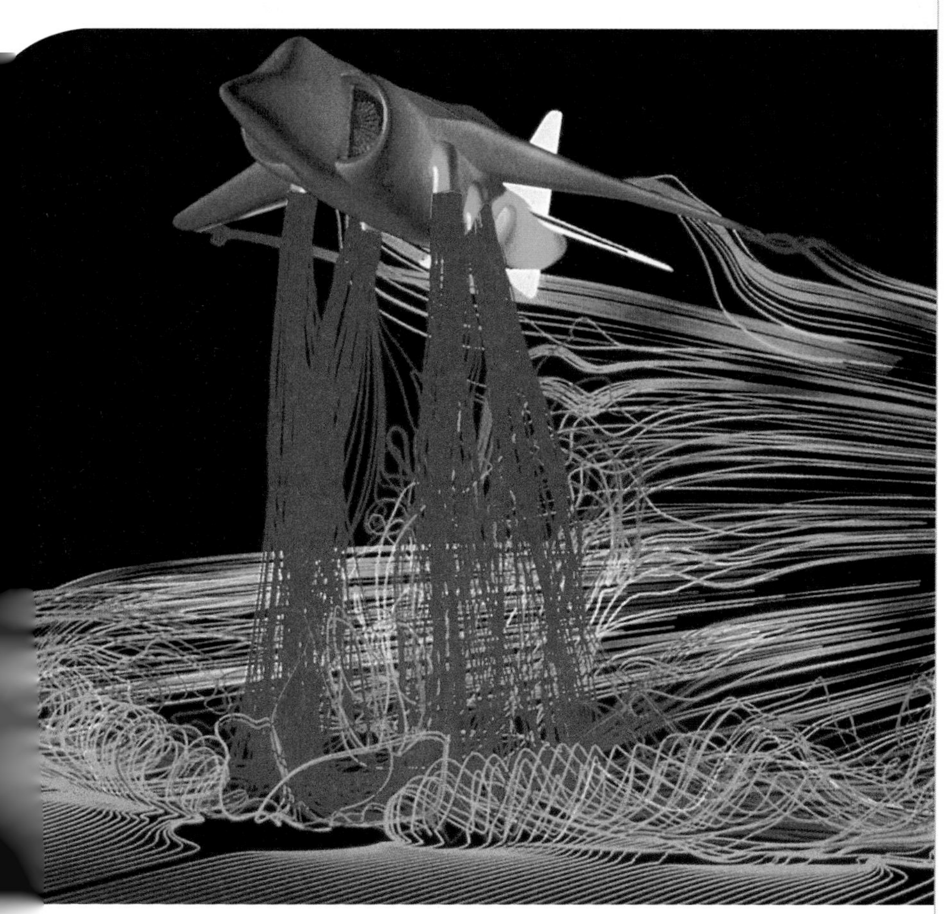

**FIGURE 13.1** NASA computer simulation of airflow streamlines from the propulsion systems of a Harrier jet in flight. The colors of the streamlines indicate the time elapsed since the exhaust was emitted.

## WHAT WE WILL LEARN

- The atom is the basic building block of macroscopic matter.
- The diameter of an atom is approximately $10^{-10}$ m.
- Matter can exist as a gas, a liquid, or a solid.
  - A gas is a system in which the atoms move freely through space.
  - A liquid is a system in which the atoms move freely but form a nearly incompressible substance.
  - A solid defines its own size and shape.
- Solids are nearly incompressible.
- Various forms of pressure, such as stretching, compression, and shearing, can deform solids. These deformations can be expressed in terms of linear relationships between the applied pressure and the resulting deformation.
- Pressure is force per unit area.
- The pressure of the Earth's atmosphere can be measured using a mercury barometer or a similar instrument.
- The pressure of gas can be measured using a mercury manometer.
- Pascal's Principle states that pressure applied to a confined fluid is transmitted to all parts of the fluid.
- Archimedes's Principle states that the buoyant force on an object in a fluid is equal to the weight of the fluid displaced by the object.
- Bernoulli's Principle states that the faster a fluid flows, the less pressure it exerts on its boundary.

The flow of air around a moving airplane is of great interest to airplane designers, builders, and users. However, the characteristics of moving fluids are not easy to describe mathematically. Often, scale models of new planes are built and placed in a wind tunnel to see how the air flows around them. In recent years, computer models have been developed that can show airflow and vortices in dramatic detail. Figure 13.1 illustrates airflow around a vertical-takeoff jet plane. The Boeing 777 was the first aircraft to be modeled entirely in the computer, without wind tunnel tests. Fluid dynamics is currently a major area of research, with applications to all areas of physics from astronomy to nuclear research.

So far, we have studied motion of idealized objects, ignoring factors such as the materials they are made of and the behavior of these materials in response to the forces exerted on them. We have usually ignored air resistance and other factors involving the medium through which an object moves. This chapter considers some of these factors, presenting an overview of the physical characteristics of solids, liquids, and gases.

## 13.1 Atoms and the Composition of Matter

During the evolution of physics, scientists have explored ever smaller dimensions, looking deeper into matter in order to examine its elementary building blocks. This general way of learning more about a system by studying its subsystems is called *reductionism* and has proven a fruitful guiding principle during the last four or five centuries of scientific advancements.

Today, we know that **atoms** are the elementary building blocks of matter, although they themselves are composite particles. The substructure of atoms, however, can be resolved only with accelerators and other tools of modern nuclear and particle physics. For the purposes of this chapter, it is reasonable to view atoms as the elementary building blocks. In fact, the word *atom* comes from the Greek *atomos*, which means "indivisible." The diameter of an atom is about $10^{-10}$ m = 0.1 nm. This distance is often called an angstrom Å.

The simplest atom is hydrogen, composed of a proton and an electron. Hydrogen is the most abundant element in the universe. The next most abundant element is helium. Helium has two protons and two neutrons in its nucleus, along with two electrons surrounding the nucleus. Another common atom is oxygen, with eight protons and eight electrons, as well as (usually) eight neutrons. The heaviest naturally occurring atom is uranium, with 92 protons, 92 electrons, and usually 146 neutrons. So far, 117 different elements have been recognized and classified in the periodic table of the elements.

Essentially all of the hydrogen in the universe, along with some helium, was produced in the Big Bang about 13.7 billion years ago. Heavier elements, up to iron, were and still are produced in stars. Elements with more protons and neutrons than iron (gold, mercury, lead, and so forth) are thought to have been produced by supernova explosions. Most of the atoms of the elements on Earth were produced more than 5 billion years ago, probably in a supernova explosion, and have been recycled ever since. Even our bodies are composed of atoms from the ashes of a dying star. (The production and recycling of the elements will be covered in detail in Chapters 39 and 40.)

Consider the number of atoms in 12 g of the most common isotope of carbon, $^{12}C$, where 12 represents the atomic mass number—that is, the total number of protons (six) plus neutrons (six) in the carbon nucleus. This number of atoms has been measured to be $6.022 \cdot 10^{23}$, which is called **Avogadro's number,** $N_A$:

$$N_A = 6.022 \cdot 10^{23}.$$

To get a feeling for how much carbon corresponds to $N_A$ atoms, consider two forms of carbon: diamond and graphite. Diamonds are composed of carbon atoms arranged in a crystal lattice, whereas the carbon atoms in graphite are arranged in two-dimensional layers (Figure 13.2). In the diamond structure, carbon atoms are bonded in an interlocking lattice, which makes diamond very hard and transparent to light. In graphite, carbon atoms are arranged in layers that can slide over one another, making graphite soft and slippery. It does not reflect light, but appears black instead. Graphite is used in pencils and for lubrication.

The density of diamond is 3.51 g/cm$^3$, and the density of graphite is 2.20 g/cm$^3$. Diamonds are often classified in terms of their mass in carats, where 1 carat = 200 mg. A 12-g diamond would have $N_A$ atoms, a mass of 60 carats, and a volume of about 3.4 cm$^3$, about 1.5 times bigger than the Hope Diamond. A typical wedding ring might have a 1-carat diamond containing approximately $10^{22}$ carbon atoms.

New structures composed entirely of carbon atoms have been produced; these include fullerenes and carbon nanotubes. Fullerenes are composed of sixty carbon atoms arranged in a truncated icosahedron, a shape like a soccer ball, as illustrated in Figure 13.3a. The 1996 Nobel Prize in Chemistry was awarded to Robert Curl, Harold Kroto, and Richard Smalley for their discovery of fullerenes. Fullerenes are also called *buckyballs* and were named after Buckminster Fuller (1895–1983), who invented the geodesic dome, whose geometry resembles that of the fullerenes. Figure 13.3b shows the structure of a carbon nanotube, consisting of interlocking hexagons of carbon atoms. A carbon nanotube can be thought of as a layer of graphite (Figure 13.2c) rolled up into a tube. Fullerenes and carbon nanotubes are early products of an emerging area of research termed *nanotechnology*, referring to the size of the objects under investigation. New materials constructed using nanotechnology may revolutionize materials science. For example, fullerene crystals are harder than diamonds, and fibers composed of carbon nanotubes are lightweight and stronger than steel.

One mole (mol) of a substance contains $N_A = 6.022 \cdot 10^{23}$ atoms or molecules. Because the masses of a proton and a neutron are about equal and the mass of either is far greater than that of an electron, the mass of 1 mol of a substance in grams is given by the atomic mass number. Thus, 1 mol of $^{12}C$ has a mass of 12 g, and 1 mol of $^4He$ has a mass of 4 g. The periodic table of the elements lists the atomic mass number for each element. This atomic mass number is

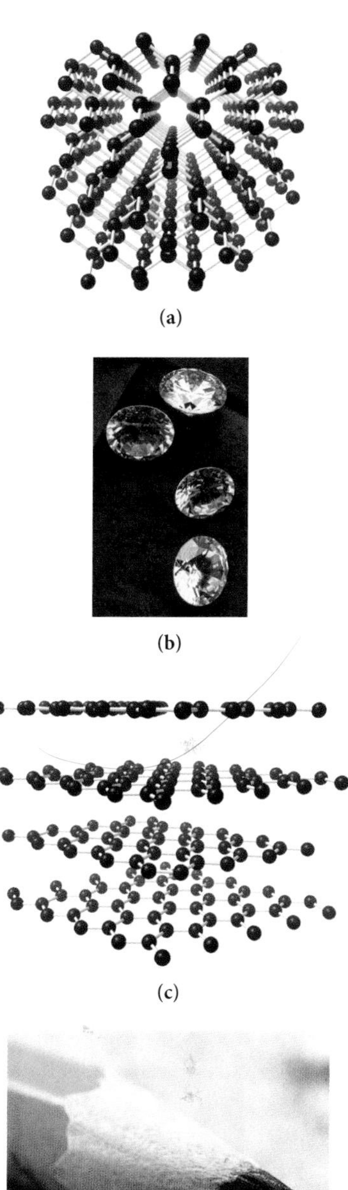


**FIGURE 13.2** Diamonds and graphite are composed of carbon atoms. (a) Structure of diamond consisting of carbon atoms bonded in a tetrahedral arrangement. (b) Diamonds. (c) Structure of graphite comprised of parallel layers of hexagonal structures. (d) A pencil showing the graphite lead.

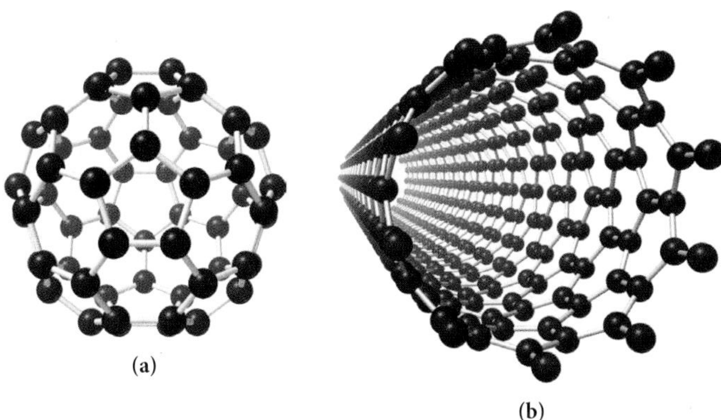


**FIGURE 13.3** Nanostructures consisting of arrangements of carbon atoms: (a) fullerene, or buckyball; (b) carbon nanotube.

### 13.1 Self-Test Opportunity

How many molecules of water are in a half-liter bottle of water? The mass of 1 mol of water is 18.02 g.

### 13.2 Self-Test Opportunity

One mole of any gas occupies a volume of 22.4 L at standard temperature and pressure (STP are $T = 0$ °C and $p = 1$ atm.) What are the densities of hydrogen gas and helium gas?

equal to the number of protons and neutrons contained in the nucleus of the atom; it is not an integer because it takes into account the natural isotopic abundances. (Isotopes of an element have varying numbers of neutrons in the nucleus. If a carbon nucleus has seven neutrons, it is the $^{13}C$ isotope.) For molecules, the molar mass is obtained by adding the mass numbers of all atoms in the molecule. Thus, 1 mol of water, $^1H_2{}^{16}O$, has a mass of 18.02 g. (It is not exactly 18 g, because 0.2% of oxygen atoms are the isotope $^{18}O$.)

Atoms are electrically neutral. They have the same number of positively charged protons as negatively charged electrons. The chemical properties of an atom are determined by its electronic structure. This structure allows bonding of certain atoms with other atoms to form molecules. For example, water is a molecule containing two hydrogen atoms and one oxygen atom. The electronic structures of atoms and molecules determine a substance's macroscopic properties, such as whether it exists as a gas, liquid, or solid at a given temperature and pressure.

## 13.2 States of Matter

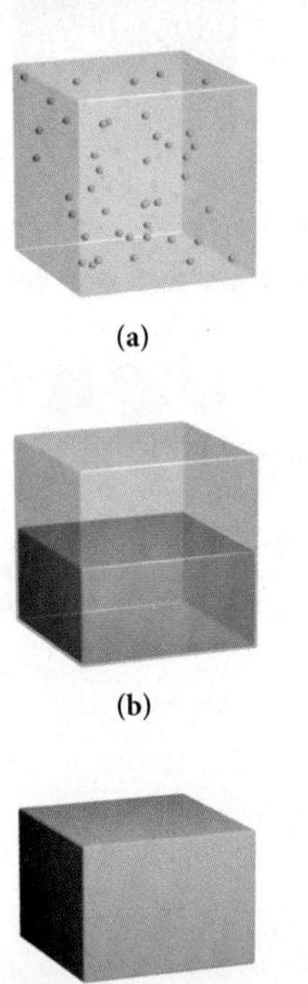

(a)

(b)

(c)

**FIGURE 13.4** (a) A cubical container filled with a gas; (b) the same container partially filled with a liquid; (c) a solid, which does not need a container.

A **gas** is a system in which each atom or molecule moves through space as a free particle. Occasionally, an atom or molecule collides with another atom or molecule or with the wall of the container. A gas can be treated as a **fluid** because it can flow and exert pressure on the walls of its container. A gas is compressible, which means that the volume of the container can be changed and the gas will still fill the volume, although the pressure it exerts on the walls of the container will change.

In contrast to gases, most liquids are nearly incompressible. If a gas is placed in a container, it will expand to fill the container (Figure 13.4a). When a **liquid** is placed in a container, it fills only the volume corresponding to its initial volume (Figure 13.4b). If the volume of the liquid is less than the volume of the container, the container is only partially filled.

A **solid** does not require a container but instead defines its own shape (Figure 13.4c). Like liquids, solids are nearly incompressible. However, solids can be compressed and deformed slightly.

The categorization of matter into solids, liquids, and gases does not cover the entire range of possibilities. Clearly, which state a certain substance is in depends on its temperature. Water, for example, can be ice (solid), water (liquid), or steam (gas). The same condition holds true for practically all other substances. However, there are states of matter that do not fit into the solid/liquid/gas classification. Matter in stars, for example, is not in any of those three states. Instead, it forms a *plasma,* a system of ionized atoms. On Earth, many beaches are made of sand, a prime example of *granular medium.* The grains of granular media are solids, but their macroscopic characteristics can be closer to those of liquids (Figure 13.5). For instance, sand can flow like a liquid. *Glasses* seem to be solids at first glance, because they do not change their shape. However, there is also some justification to the view that glass is a type of liquid with

(a)

(b)

**FIGURE 13.5** (a) Pouring liquid silver (a solid metal at room temperature); (b) pouring sand (a granular medium).

an extremely high viscosity. For the purposes of a classification of matter into states, a glass is neither a solid nor a liquid, but a separate state of matter. *Foams* and *gels* are yet other states of matter, which are currently receiving a lot of interest from researchers. In a foam, the material forms thin membranes around enclosed bubbles of gas of different sizes; thus, some foams are very rigid while having a very low mass density.

Less than two decades ago, the existence of a new form of matter called *Bose-Einstein condensates* was experimentally verified. An understanding of this new state of matter requires some basic concepts of quantum physics. However, in basic terms, at very low temperatures, a gas of certain kinds of atoms can assume an ordered state in which all of the atoms tend to have the same energy and momentum, very similar to the way that light assumes an ordered state in a laser (see Chapter 38).

Finally, the matter in our bodies and in most other biological organisms does not fit into any of these classifications. Biological tissue consists predominantly of water, yet it is able to keep or change its shape, depending on the environmental boundary conditions.

## 13.3 Tension, Compression, and Shear

Let's examine how solids respond to external forces.

### Elasticity of Solids

Many solids are composed of atoms arranged in a three-dimensional crystal lattice in which the atoms have a well-defined equilibrium distance from their neighbors. Atoms in a solid are held in place by interatomic forces that can be modeled as springs. The lattice is very rigid, which implies that the imaginary springs are very stiff. Macroscopic solid objects such as wrenches and spoons are composed of atoms arranged in such a rigid lattice. However, other solid objects, such as rubber balls, are composed of atoms arranged in long chains rather than in a well-defined lattice. Depending on their atomic or molecular structure, solids can be extremely rigid or more easily deformable.

All rigid objects are somewhat elastic, even though they do not appear to be. Compression, pulling, or twisting can deform a rigid object. If a rigid object is deformed a small amount, it will return to its original size and shape when the deforming force is removed. If a rigid object is deformed past a point called its **elastic limit,** it will not return to its original size and shape but will remain permanently deformed. If a rigid object is deformed too far beyond its elastic limit, it will break.

(a)

(b)

(c)

**FIGURE 13.6** Three examples of stress and strain: (a) the stretching of power lines; (b) the compression of the Hoover Dam; (c) shearing by scissors.

### Stress and Strain

Deformations of solids are usually classified into three types: stretching (or tension), compression, and shear. Examples of stretching, compression, and shear are shown in Figure 13.6. What these three deformations have in common is that a **stress,** or deforming force per unit area, produces a **strain,** or unit deformation. **Stretching,** or **tension,** is associated with tensile stress. **Compression** can be produced by hydrostatic stress. **Shear** is produced by shearing stress, sometimes also called *deviatory stress*. When a shear force is applied, planes of material parallel to the force and on either side of it remain parallel but shift relative to each other.

Although stress and strain take different forms for the three types of deformation, they are related linearly through a constant called the **modulus of elasticity:**

$$\text{stress} = \text{modulus of elasticity} \cdot \text{strain}. \quad (13.1)$$

This empirical relationship applies as long as the elastic limit of the material is not exceeded.

In the case of tension, a force $F$ is applied to opposite ends of an object of length $L$ and the object stretches to a new length, $L + \Delta L$ ( Figure 13.7). The stress for stretching or tension is defined as the force, $F$, per unit area, $A$, applied to the end of an object. The strain is defined as the fractional change in length of the object, $\Delta L/L$. The relationship between stress and strain up to the elastic limit is then

$$\frac{F}{A} = Y\frac{\Delta L}{L}, \quad (13.2)$$

| Table 13.1 Some Typical Values of Young's Modulus | |
|---|---|
| Material | Young's Modulus ($10^9$ N/m$^2$) |
| Aluminum | 70 |
| Bone | 10-20 |
| Concrete | 30-60 (compression) |
| Diamond | 1000-1200 |
| Glass | 70 |
| Polystyrene | 3 |
| Rubber | 0.01-0.1 |
| Steel | 200 |
| Titanium | 100-120 |
| Tungsten | 400 |
| Wood | 10-15 |

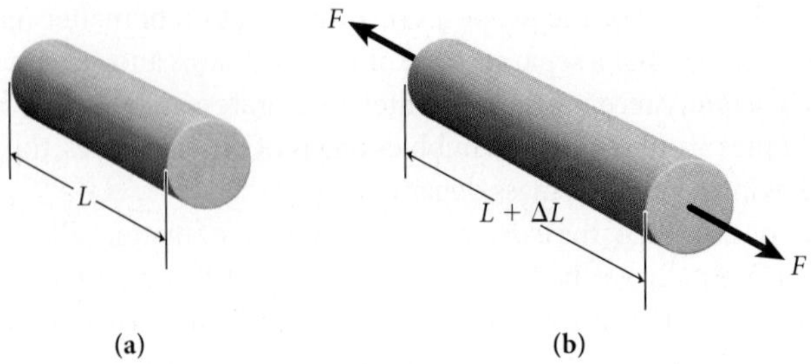


**FIGURE 13.7** Tension applied to opposite ends of an object by a pulling force. (a) Object before force is applied. (b) Object after force is applied. *Note:* Tension can also be applied by pushing, with a resulting negative change in length (not shown).

where $Y$ is called **Young's modulus** and depends only on the type of material and not on its size or shape . Some typical values of Young's modulus are given in Table 13.1.

Linear compression can be treated in a manner similar to stretching for most materials, within the elastic limits. However, many materials have different breaking points for stretching and compression. The most notable example is concrete, which resists compression much better than stretching, which is why steel rods are added to it in places where greater tolerance of stretching is required. A steel rod resists stretching much better than compression, under which it can buckle.

The stress related to volume compression is caused by a force per unit area applied to the entire surface area of an object for example, one submerged in a liquid (Figure 13.8). The resulting strain is the fractional change in the volume of the object, $\Delta V/V$. The modulus of elasticity in this case is the **bulk modulus,** $B$. We can thus write the equation relating stress and strain for volume compression as

$$\frac{F}{A} = B\frac{\Delta V}{V}. \tag{13.3}$$

| Table 13.2 Some Typical Values of the Bulk Modulus | |
|---|---|
| Material | Bulk Modulus ($10^9$ N/m$^2$) |
| Air | 0.000142 |
| Aluminum | 76 |
| Basalt rock | 50-80 |
| Gasoline | 1.5 |
| Granite rock | 10-50 |
| Mercury | 28.5 |
| Steel | 160 |
| Water | 2.2 |

Some typical values of the bulk modulus are given in Table 13.2. Note the extremely large jump in the bulk modulus from air, which is a gas and can be compressed rather easily, to liquids such as gasoline and water. Solids such as rocks and metals have values for the bulk modulus that are higher than those of liquids by a factor between 5 and 100.

In the case of shearing, the stress is again a force per unit area. However, for shearing, the force is parallel to the area rather than perpendicular to it (Figure 13.9). For shearing, stress is given by the force per unit area, $F/A$, exerted on the end of the object. The resulting strain is given by the fractional deflection of the object, $\Delta x/L$. The stress is related to the strain through the **shear modulus,** $G$:

$$\frac{F}{A} = G\frac{\Delta x}{L}. \tag{13.4}$$

Some typical values of the shear modulus are given in Table 13.3.

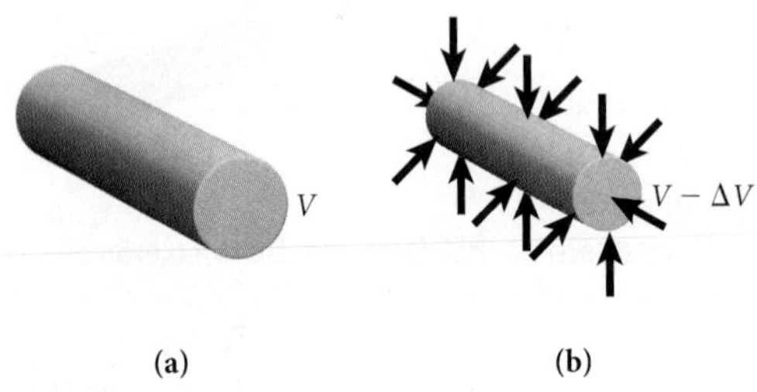


**FIGURE 13.8** Compression of an object by fluid pressure; (a) object before compression; (b) object after compression.

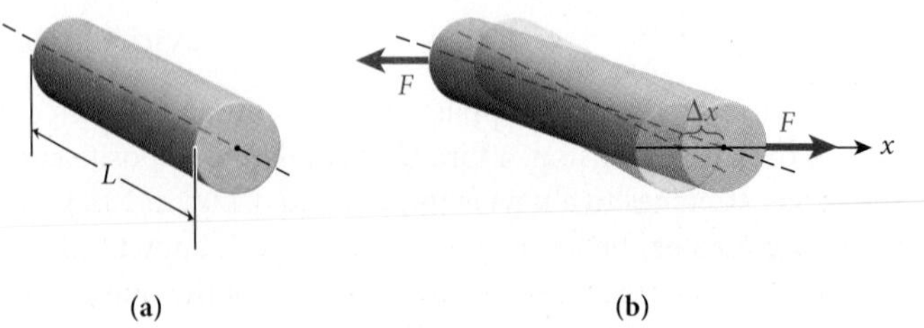


**FIGURE 13.9** Shearing of an object caused by a force parallel to the area of the end of the object. (a) Object before shear force is applied. (b) Object after shear force is applied.

## EXAMPLE 13.1 Wall Mount for a Flat-Panel TV

You've just bought a new flat-panel TV (Figure 13.10) and want to mount it on the wall with four bolts, each with a diameter each of 0.50 cm. You cannot mount the TV flush against the wall but have to leave a 10.0-cm gap between wall and TV for air circulation.

### PROBLEM 1

If the mass of your new TV is 42.8 kg, what is the shear stress on the bolts?

### SOLUTION 1

The combined cross-sectional area of the bolts is

$$A = 4\left(\frac{\pi d^2}{4}\right) = \pi(0.005 \text{ m})^2 = 7.85 \cdot 10^{-5} \text{ m}^2.$$

One force acting on the bolts is $\vec{F}_g$, the force of gravity on the TV, extected at one end of each bolt. This force is balanced by a force due to the wall, acting on the other end of the bolts. This wall force holds the TV in place; thus, it has exactly the same magnitude as the force of gravity but points in the opposite direction. Therefore, the force entering into equation 13.4 for the stress is

$$F = mg = (42.8 \text{ kg})(9.81 \text{ m/s}^2) = 420 \text{ N}.$$

Thus, we obtain the shear stress on the bolts:

$$\frac{F}{A} = \frac{420 \text{ N}}{7.85 \cdot 10^{-5} \text{ m}^2} = 5.35 \cdot 10^6 \text{ N/m}^2.$$

### PROBLEM 2

The shear modulus of the steel used in the bolts is $9.0 \cdot 10^{10}$ N/m$^2$. What is the resulting vertical deflection of the bolts?

### SOLUTION 2

Solving equation 13.4 for the deflection, $\Delta x$, we find

$$\Delta x = \left(\frac{F}{A}\right)\frac{L}{G} = (5.35 \cdot 10^6 \text{ N/m}^2)\frac{0.1 \text{ m}}{9.0 \cdot 10^{10} \text{ N/m}^2} = 5.94 \cdot 10^{-6} \text{ m}.$$

Even though the shear stress is more than 5 million N/m$^2$, the resulting sag of your flat-panel TV is only about 0.006 mm, a distance that is undetectable with the naked eye.

**Table 13.3 Some Typical Values of the Shear Modulus**

| Material | Shear Modulus ($10^9$ N/m$^2$) |
|---|---|
| Aluminum | 25 |
| Copper | 45 |
| Glass | 26 |
| Polyethylene | 0.12 |
| Rubber | 0.0003 |
| Titanium | 41 |
| Steel | 80–90 |

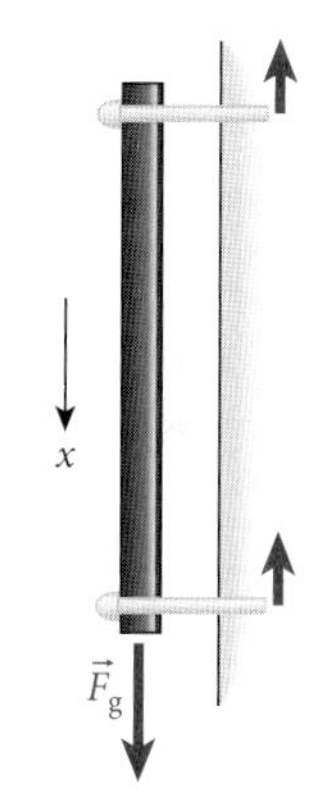


**FIGURE 13.10** Forces acting on a wall mount for a flat-panel TV.

## SOLVED PROBLEM 13.1 Stretched Wire

### PROBLEM

A 1.50-m-long steel wire with a diameter of 0.400 mm is hanging vertically. A spotlight of mass m = 1.50 kg is attached to the wire and released. How much does the wire stretch?

### SOLUTION

#### THINK

From the diameter of the wire, we can get its cross-sectional area. The weight of the spotlight provides a downward force. The stress on the wire is the weight divided by the cross-sectional area. The strain is the change in the length of the wire divided by its original length. The stress and strain are related through Young's modulus for steel.

#### SKETCH

Figure 13.11 shows the wire before and after the spotlight is attached.

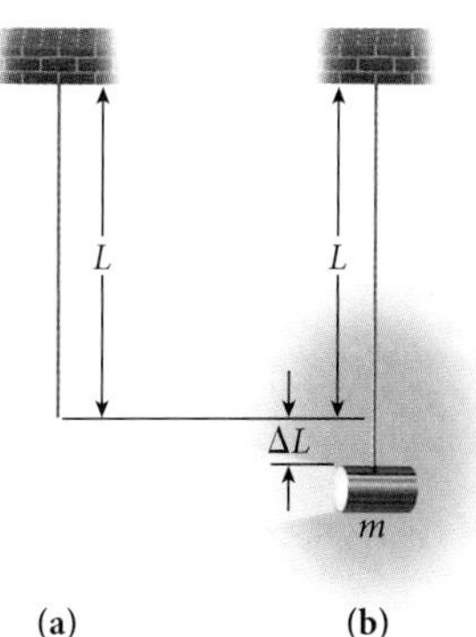


**FIGURE 13.11** A wire (a) before and (b) after a spotlight is attached to it.

Continued—

**RESEARCH**

The cross-sectional area of the wire, $A$, can be calculated from the diameter, $d$, of the wire:

$$A = \pi\left(\frac{d}{2}\right)^2 = \frac{\pi d^2}{4}. \qquad \text{(i)}$$

The force on the wire is the weight of the searchlight,

$$F = mg. \qquad \text{(ii)}$$

We can relate the stress and the strain on the wire through Young's modulus, $Y$, for steel:

$$\frac{F}{A} = Y\frac{\Delta L}{L}, \qquad \text{(iii)}$$

where $\Delta L$ is the change in the length of the wire and $L$ is its original length.

**SIMPLIFY**

We can combine equations (i), (ii), and (iii) to get

$$\frac{F}{A} = \frac{mg}{\left(\frac{1}{4}\pi d^2\right)} = \frac{4mg}{\pi d^2} = Y\frac{\Delta L}{L}.$$

Solving for the change in the wire's length, we obtain

$$\Delta L = \frac{4mgL}{Y\pi d^2}.$$

**CALCULATE**

Putting in the numerical values gives us

$$\Delta L = \frac{4mgL}{Y\pi d^2} = \frac{4 \cdot 1.50 \text{ kg} \cdot (9.81 \text{ m/s}^2) \cdot 1.50 \text{ m}}{\left(200 \cdot 10^9 \text{ N/m}^2\right)\pi\left(0.400 \cdot 10^{-3} \text{ m}\right)^2} = 0.00087824 \text{ m}.$$

**ROUND**

We report our result to three significant figures:

$$\Delta L = 0.000878 \text{ m} = 0.878 \text{ mm}.$$

**DOUBLE-CHECK**

An easy and important first check of a result is to confirm that the units are right. The change in length, which we have calculated, has the units of millimeters, which makes sense. Next, we examine the magnitude: The wire stretches just less than 1 mm compared with its original length of 1.50 m. This stretch is less than 0.1%, which seems reasonable. Any time a steel wire experiences noticeable stretch, breaking the wire becomes a concern. This is not the case here, which gives us some confidence that our result is of approximately the right order of magnitude.

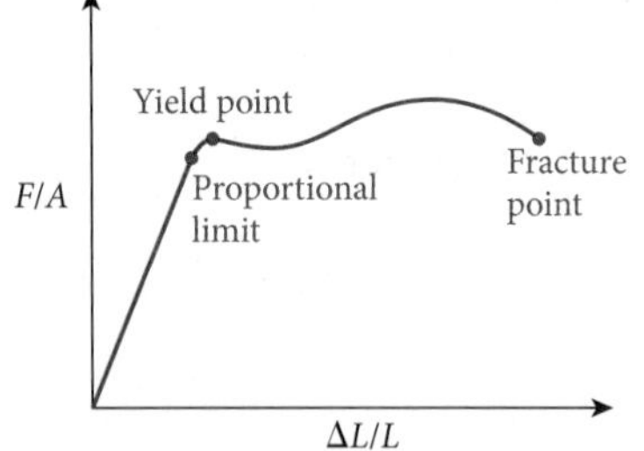


**FIGURE 13.12** A typical stress-strain diagram for a ductile metal under tension showing the proportional limit, the yield point, and the fracture point.

**Table 13.4 Breaking Stress for Common Materials**

| Material | Breaking Stress ($10^6$ N/m$^2$) |
|---|---|
| Aluminum | 455 |
| Brass | 550 |
| Copper | 220 |
| Steel | 400 |
| Bone | 130 |

Note that these values are approximate.

As noted earlier, the stress applied to an object is proportional to the strain as long as the elastic limit of the material is not exceeded. Figure 13.12 shows a typical stress-strain diagram for a ductile (easily drawn into a wire) metal under tension. Up to the proportional limit, the ductile metal responds linearly to stress. If the stress is removed, the material will return to its original length. If stress is applied past the proportional limit, the material will continue to lengthen until it reaches its yield point. If stress is applied between the proportional limit and the fracture point and then is removed, the material will not return to its original length but will be permanently deformed. The yield point is the point where the stress causes sudden deformation without any increase in force as can be seen from the flattening of the curve (see Figure 13.12). Additional stress will continue to stretch the material until it reaches its fracture point, where it breaks. This breaking stress is also called the *ultimate stress*, or in the case of tension, the *tensile strength*. Some approximate breaking stresses are given in Table 13.4.

## 13.4 Pressure

This section examines the properties of liquids and gases, together termed *fluids*. The properties of stress and strain, discussed in section 13.3, are most useful when studying solids, because substances that flow, such as liquids and gases, offer little resistance to shear or tension. In this section, we consider the properties of fluids at rest. Later in this chapter, we will discuss the properties of fluids in motion.

The **pressure,** $p$, is the force per unit area:

$$p = \frac{F}{A}. \tag{13.5}$$

Pressure is a scalar quantity. The SI unit of pressure is N/m$^2$, which has been named the **pascal,** abbreviated Pa:

$$1 \text{ Pa} \equiv \frac{1 \text{ N}}{1 \text{ m}^2}.$$

The average pressure of the Earth's atmosphere at sea level, 1 atm, is a commonly used non-SI unit that is expressed in other units as follows:

$$1 \text{ atm} = 1.01 \cdot 10^5 \text{ Pa} = 760 \text{ torr} = 14.7 \text{ lb/in}^2.$$

Gauges used to measure how much air has been removed from a vessel are often calibrated in *torr,* a unit named after the Italian physicist Evangelista Torricelli (1608–1647). Automobile tire pressures in the United States are often measured in pounds per square inch (lb/in$^2$, or psi).

### 13.1 In-Class Exercise

Suppose you pump the air out of a paint can, covered by a lid. The paint can is cylindrical, 22.4 cm tall and with a diameter of 16.0 cm. How much force does the atmosphere exert on the lid of the evacuated paint can?

a) 9.81 N

b) 511 N

c) 2030 N

d) 8120 N

### Pressure-Depth Relationship

Consider a tank of water open to the Earth's atmosphere, and imagine a cube of the water inside of it (shown in pink in Figure 13.13). Assume that the top surface of the cube is horizontal and at a depth of $y_1$ and that the bottom surface of the cube is horizontal and at a depth of $y_2$. The other sides of the cube are oriented vertically. The water pressure acting on the cube produces forces. However, by Newton's First Law, there must be no net force acting on this stationary cube of water. The forces acting on the vertical sides of the cube clearly cancel out. The vertical forces acting on the bottom and top sides of the cube must also add to zero:

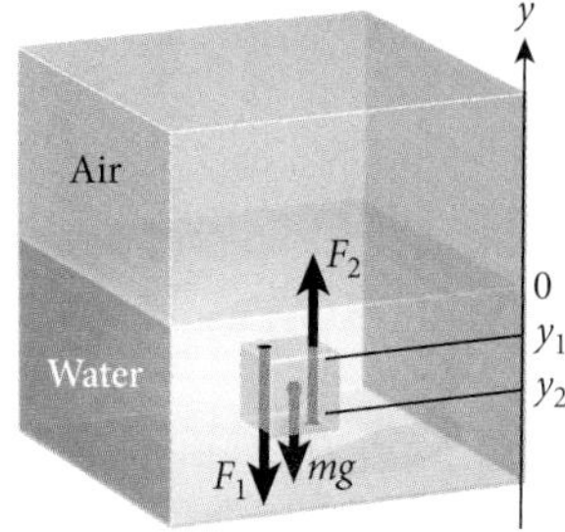


**FIGURE 13.13** Cube of water in a tank of water.

$$F_2 - F_1 - mg = 0, \tag{13.6}$$

where $F_1$ is the force downward on the top of the cube, $F_2$ is the force upward on the bottom of the cube, and $mg$ is the weight of the cube of water. The pressure at depth $y_1$ is $p_1$, and the pressure at depth $y_2$ is $p_2$. We can write the forces at these depths in terms of the pressures, assuming that the area of the top and bottom surfaces of the cube is $A$:

$$F_1 = p_1 A$$
$$F_2 = p_2 A.$$

We can also express the mass, $m$, of the water in terms of the density of water, $\rho$, assumed constant and the volume, $V$, of the cube, $m = \rho V$. Substituting for $F_1$, $F_2$, and $m$ in equation 13.6 gives

$$p_2 A - p_1 A - \rho V g = 0.$$

Rearranging this equation and substituting $A(y_1 - y_2)$ for $V$, we get

$$p_2 A = p_1 A + \rho A (y_1 - y_2) g.$$

Dividing out the area yields an expression for the pressure as a function of depth in a liquid of uniform density $\rho$:

$$p_2 = p_1 + \rho g (y_1 - y_2). \tag{13.7}$$

A common problem involves the pressure as a function of depth below the surface of a liquid. Starting with equation 13.7, we can define the pressure at the surface of the liquid

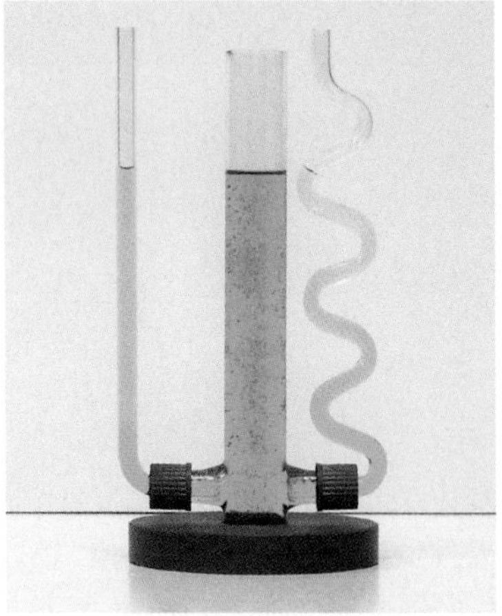

**FIGURE 13.14** Three connected columns—the fluid rises to the same height in each.

($y_1 = 0$) to be $p_0$ and the pressure at a given depth $h$ ($y_2 = -h$) to be $p$. These assumptions lead to the equation

$$p = p_0 + \rho g h \tag{13.8}$$

for the pressure at a given depth of a liquid with uniform density $\rho$.

Note that in deriving equation 13.8, we made use of the fact that the density of the fluid does not change as a function of depth. This assumption of incompressibility is essential for obtaining this result. Later in this section, relaxing this incompressibility requirement will lead to a different formula relating height and pressure for gases. Further, it is important to note that equation 13.8 specifies the pressure in a liquid at a vertical depth of $h$, without any dependence on horizontal position. Thus, the equation holds regardless of the shape of the vessel containing the liquid. Figure 13.14, for example, shows three connected columns containing a fluid. You can see that the fluid reaches the same height in each column, independent of the shape or cross-sectional area; this occurs because the bottoms are interconnected and thus have the same pressure.

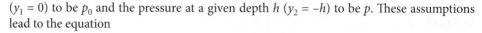

### EXAMPLE 13.2 Submarine

**FIGURE 13.15** Submarine surfacing.

A U.S. Navy submarine of the Los Angeles class is 110 m long and has a hull diameter of 10 m (Figure 13.15). Assume that the submarine has a flat top with an area of $A = 1100\ \text{m}^2$ and that the density of seawater is $1024\ \text{kg/m}^2$.

**PROBLEM**

What is the total force pushing down on the top of this submarine at a diving depth of 250 m?

**SOLUTION**

The pressure inside the submarine is normal atmospheric pressure, $p_0$. According to equation 13.8, the pressure at a depth of 250 m is given by $p = p_0 + \rho g h$. Therefore, the pressure difference between the inside and the outside of the submarine is

$$\Delta p = \rho g h = (1024\ \text{kg/m}^3)(9.81\ \text{m/s}^2)(250\ \text{m}) = 2.51 \cdot 10^6\ \text{N/m}^2 = 2.51\ \text{MPa},$$

or approximately 25 atm.

Multiplying the area times the pressure gives the total force, according to equation 13.5. Thus, we find a total force of

$$F = \Delta p A = (2.51 \cdot 10^6\ \text{N/m}^2)(1100\ \text{m}^2) = 2.8 \cdot 10^9\ \text{N}.$$

This number is astonishingly large and corresponds to the weight of a mass of ≈280,000 metric tons!

### 13.2 In-Class Exercise

A steel sphere with a diameter of 0.250 m is submerged in the ocean at a depth of 500.0 m. What is the percentage change in the volume of the sphere? The bulk modulus of steel is $160 \cdot 10^9$ Pa.

a) 0.0031%

b) 0.045%

c) 0.33%

d) 0.55%

e) 1.5%

## Gauge Pressure and Barometers

The pressure $p$ in equation 13.8 is an **absolute pressure,** which means it includes the pressure of the liquid as well as the pressure of the air above it. The difference between an absolute pressure and the atmospheric air pressure is called a **gauge pressure.** For example, a gauge used to measure the air pressure in a tire is calibrated so that the atmospheric pressure reads zero. When connected to the tire's compressed air, the gauge measures the additional pressure present in the tire. In equation 13.8, the gauge pressure is $\rho g h$.

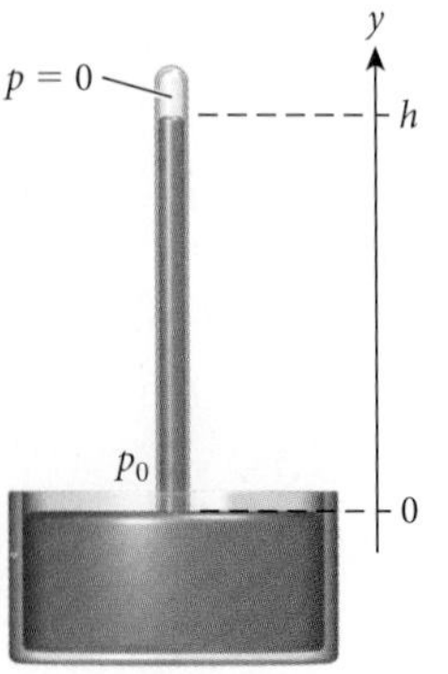


**FIGURE 13.16** A mercury barometer measures atmospheric pressure $p_0$.

A simple device used to measure atmospheric pressure is the mercury **barometer** (Figure 13.16). You can construct a mercury barometer by taking a long glass tube, closed at one end, filling it with mercury, and inverting it with the open end in a dish of mercury. The space above the mercury is a vacuum and thus has zero pressure. The difference in height between the top of the mercury in the tube and the top of the mercury in the dish, $h$, can be related to the atmospheric pressure, $p_0$, using equation 13.7 with $p_2 = p_1 + \rho g(y_1 - y_2)$, $y_2 - y_1 = h$, $p_2 = 0$, and $p_1 = p_0$:

$$p_0 = \rho gh.$$

Note that the measurement of atmospheric pressure using a mercury barometer depends on the local value of $g$. Atmospheric pressure is often expressed in millimeters of mercury (mmHg), corresponding to the height difference $h$. The torr is equivalent to 1 mmHg, so standard atmospheric pressure is 760 torr, or 29.92 in of mercury (101.325 kPa).

An open-tube **manometer** measures the gauge pressure of a gas. It consists of a U-shaped tube partially filled with a liquid such as mercury (Figure 13.17). The closed end of the manometer is connected to a vessel containing the gas whose gauge pressure, $p_g$, is being measured, and the other end is open and thus experiences atmospheric pressure, $p_0$. Using equation 13.7 with $y_1 = 0$, $p_1 = p_0$, $y_2 = -h$, and $p_2 = p$, where $p$ is the absolute pressure of the gas in the vessel, we obtain $p = p_0 + \rho gh$, the pressure-depth relationship derived earlier. The gauge pressure of the gas in the vessel is then

$$p_g = p - p_0 = \rho gh. \tag{13.9}$$

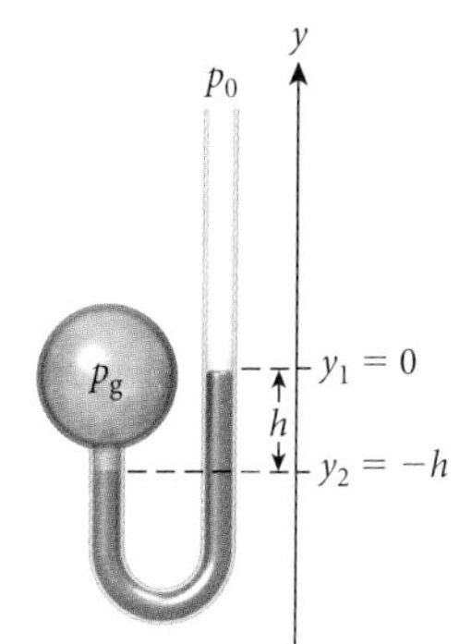

**FIGURE 13.17** An open-tube manometer measures the gauge pressure of a gas.

Note that gauge pressure can be positive or negative. The gauge pressure of the air in an inflated automobile tire is positive. The gauge pressure at the end of a straw being used by a person drinking a milkshake is negative.

## Barometric Altitude Relation for Gases

In deriving equation 13.8, we made use of the incompressibility of liquids. However, if the fluid is a gas, we cannot make this assumption. Let's start again with a thin layer of fluid in a column of the fluid. The pressure difference between the top and bottom surfaces is the negative of the weight of the thin layer of fluid divided by its area:

$$\Delta p = -\frac{F}{A} = -\frac{mg}{A} = -\frac{\rho Vg}{A} = -\frac{\rho(\Delta hA)g}{A} = -\rho g \Delta h. \tag{13.10}$$

The negative sign reflects the fact that the pressure decreases with increasing altitude ($h$), since the weight of the fluid column above the layer will be reduced. So far, nothing is different from the derivation for the incompressible fluid. However, for compressible fluids, the density is proportional to the pressure:

$$\frac{\rho}{\rho_0} = \frac{p}{p_0}. \tag{13.11}$$

Strictly speaking, this relationship is true only for an ideal gas at constant temperature, as we will see in Chapter 19. However, if we combine equations 13.10 and 13.11, we obtain

$$\frac{\Delta p}{\Delta h} = -\frac{g\rho_0}{p_0} p.$$

Taking the limit as $\Delta h \rightarrow 0$, we have

$$\frac{dp}{dh} = -\frac{g\rho_0}{p_0} p.$$

This is an example of a differential equation. We need to find a function whose derivative is proportional to the function itself, which leads us to an exponential function:

$$p(h) = p_0 e^{-h\rho_0 g/p_0}. \tag{13.12}$$

It is easy to convince yourself that equation 13.12 is indeed a solution to the preceding differential equation by simply taking the derivative with respect to $h$. Equation 13.12 is sometimes called the **barometric pressure formula,** and it relates the pressure to the altitude in gases. It applies only as long as the temperature does not change as a function of altitude and as long as the gravitational acceleration can be assumed to be constant. (We will consider the effect of temperature change when we discuss the Ideal Gas Law in Chapter 19.)

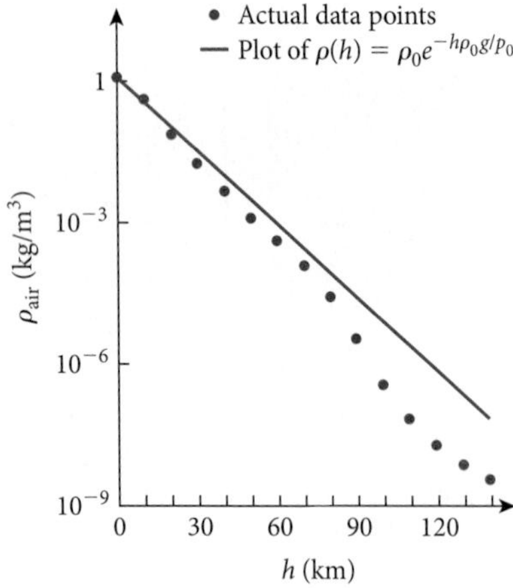

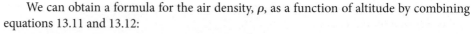

**FIGURE 13.18** Comparison of value for air density plotted from equation 13.13 (blue line) with actual data for the atmosphere (red dots).

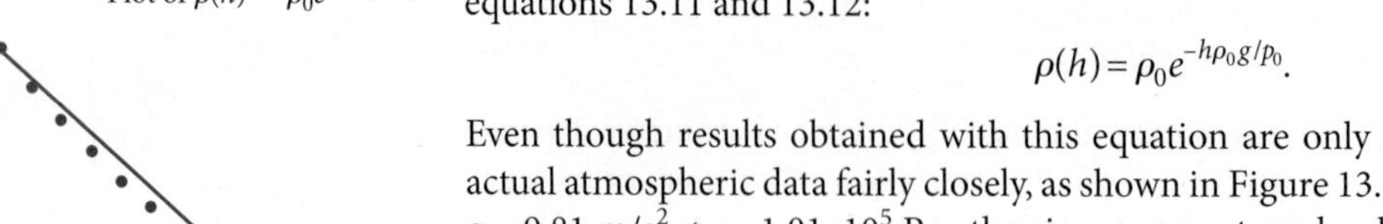

We can obtain a formula for the air density, $\rho$, as a function of altitude by combining equations 13.11 and 13.12:

$$\rho(h) = \rho_0 e^{-h\rho_0 g/p_0}. \tag{13.13}$$

Even though results obtained with this equation are only an approximation, they match actual atmospheric data fairly closely, as shown in Figure 13.18, where $\rho(h)$ was plotted with $g = 9.81$ m/s$^2$, $p_0 = 1.01 \cdot 10^5$ Pa , the air pressure at sea level ($h = 0$), and $\rho_0 = 1.229$ kg/m$^3$, the air density at sea level. As you can see, the agreement is very close up to the top of the stratosphere, approximately 50 km above ground.

## EXAMPLE 13.3 Air Pressure on Mount Everest

As climbers approach the peak of Earth's highest mountain, Mount Everest, they usually have to wear breathing equipment. The reason for this is that the air pressure is very low, too low for climbers' lungs, which are usually accustomed to near sea-level conditions.

### PROBLEM

What is the air pressure at the top of Mount Everest (Figure 13.19)?

**FIGURE 13.19** Mount Everest is Earth's highest peak, at 8850 m (29,035 ft).

### SOLUTION

This is a situation where we can use the barometric pressure formula (equation 13.12). We first note the constants: $p_0 = 1.01 \cdot 10^5$ Pa, the sea-level air pressure, and $\rho_0 = 1.229$ kg/m$^3$, the air density at sea level. Then we find the inverse of the constant part of the exponent in equation 13.12:

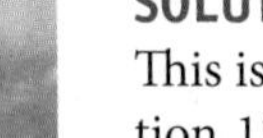

$$\frac{p_0}{\rho_0 g} = \frac{1.01 \cdot 10^5 \text{ Pa}}{(1.229 \text{ kg/m}^3)(9.81 \text{ m/s}^2)} = 8377 \text{ m}.$$

We can then rewrite equation 13.12 as

$$p(h) = p_0 e^{-h/(8377 \text{ m})}.$$

The height of Mount Everest is 8850 m. Therefore, we obtain

$$p(8850 \text{ m}) = p_0 e^{-8850/8377} = 0.348 p_0 = 35 \text{ kPa}.$$

The calculated air pressure at the top of Mount Everest is only 35% of the air pressure at sea level. (The actual pressure is slightly lower, mainly because of temperature effects.)

### 13.3 In-Class Exercise

If you descend into a mine shaft below sea level, the air pressure

a) decreases linearly.

b) decreases exponentially.

c) increases linearly.

d) increases exponentially.

You do not have to travel to the top of Mount Everest to appreciate the change in air pressure with altitude. You have probably experienced your ears "popping" while driving in the mountains. This physiological effect of feeling pressure on the eardrums results because of a lag in your body's adjusting the internal pressure to the change in external pressure due to the rapid change in altitude.

## Pascal's Principle

If pressure is exerted on a part of an incompressible fluid, that pressure will be transmitted to all parts of the fluid without loss. This is **Pascal's Principle,** which can be stated as follows:

> *When a change in pressure occurs at any point in a confined fluid, an equal change in pressure occurs at every point in the fluid.*

Pascal's Principle is the basis for many modern hydraulic devices, such as automobile brakes, large earth-moving machines, and car lifts.

Pascal's Principle can be demonstrated by taking a cylinder partially filled with water, placing a piston on top of the column of water, and placing a weight on top of the piston

(Figure 13.20). The air pressure and the weight exert a pressure, $p_t$, on top of the column of water. The pressure, $p$, at depth $h$ is given by

$$p = p_t + \rho g h.$$

Because water can be considered incompressible, if a second weight is added on top of the piston, the change in pressure, $\Delta p$, at depth $h$ is due solely to the change in the pressure, $\Delta p_t$, on top of the water. The density of the water does not change and the depth does not change, so we can write

$$\Delta p = \Delta p_t.$$

This result does not depend on $h$, so it must hold for all positions in the liquid.

Now consider two connected pistons in cylinders filled with oil, as shown in Figure 13.21. One piston has area $A_{in}$ and the other piston has area $A_{out}$, with $A_{in} < A_{out}$. A force, $F_{in}$, is exerted on the first piston, producing a change in pressure in the oil. This change in pressure is transmitted to all points in the oil, including points adjacent to the second piston. We can write

$$\Delta p = \frac{F_{in}}{A_{in}} = \frac{F_{out}}{A_{out}},$$

or

$$F_{out} = F_{in}\frac{A_{out}}{A_{in}}. \qquad (13.14)$$

Because $A_{out}$ is larger than $A_{in}$, $F_{out}$ is larger than $F_{in}$. Thus, the force applied to the first piston is magnified. This phenomenon is the basis of hydraulic devices that produce large output forces with small input forces.

The amount of work done on the first piston is the same as the amount of work done by the second piston. To calculate the work done, we need to calculate the distance over which the forces act. For both pistons, the volume, $V$, of incompressible oil that is moved is the same:

$$V = h_{in}A_{in} = h_{out}A_{out},$$

where $h_{in}$ is the distance the first piston moves and $h_{out}$ is the distance the second piston moves. We can see that

$$h_{out} = h_{in}\frac{A_{in}}{A_{out}}, \qquad (13.15)$$

which means that the second piston moves a smaller distance than the first piston because $A_{in} < A_{out}$. We can find the work done by using the fact that work is force times distance and using equations 13.14 and 13.15:

$$W = F_{in}h_{in} = \left(F_{out}\frac{A_{in}}{A_{out}}\right)\left(h_{out}\frac{A_{out}}{A_{in}}\right) = F_{out}h_{out}.$$

Thus, this hydraulic device transmits a larger force over a smaller distance. However, no additional work is done.

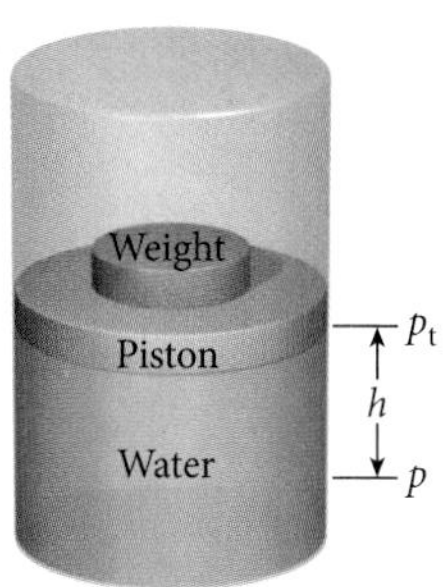


**FIGURE 13.20** Cylinder partially filled with water, with a piston placed on top of the water and a weight placed on the piston.

### 13.4 In-Class Exercise

A car with a mass of 1600 kg is supported by a hydraulic car lift, as illustrated in Figure 13.21. The large piston supporting the car has a diameter of 25.0 cm. The small piston has a diameter of 1.25 cm. How much force must be exterted on the small piston to support the car?

a) 1.43 N

b) 5.22 N

c) 10.2 N

d) 23.1 N

e) 39.2 N

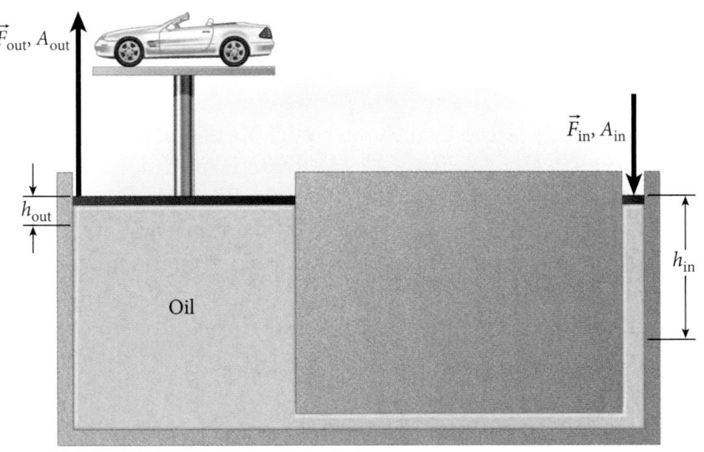


**FIGURE 13.21** Application of Pascal's Principle in a hydraulic lift. (The scale of the lift relative to the car is greatly out of proportion in order to show the essential details clearly.)

## 13.5 Archimedes' Principle

Archimedes (287-212 BC) of Syracuse, Sicily, was one of the greatest mathematicians of all time. The king, Hiero of Syracuse, ordered a new crown made and gave the goldsmith the exact amount of gold needed to create the crown. When the crown was finished, it had the correct weight, but Hiero suspected that the goldsmith had used some silver in the crown to replace the more valuable gold. Hiero could not prove this and went to Archimedes for help. According to legend, the answer occurred to Archimedes as he was about to take a bath and noticed the water level rising and his apparent weight diminishing when he got into the bathtub. He then ran naked through the streets of Syracuse to the palace, shouting "Eureka!" (Greek for "I have found it!") Using his discovery, he was able to demonstrate the silver-for-gold switch and thus prove the theft committed by the goldsmith. Later in this section, we'll examine Archimedes' method for solving this problem, after considering buoyant force and fluid displacement.

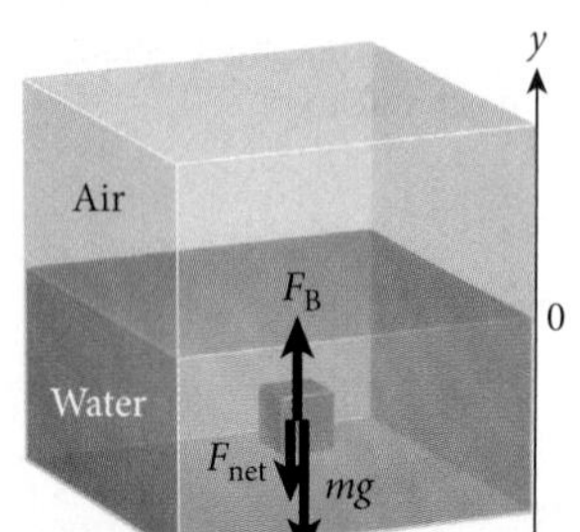


**FIGURE 13.22** The weight of a steel cube submerged in water is larger than the buoyant force acting on the cube.

### Buoyant Force

Figure 13.13, shows a cube of water within a larger volume of water. The weight of the cube of water is supported by the force resulting from the pressure difference between the top and bottom surfaces of the cube, as given by equation 13.6, which we can rewrite as

$$F_2 - F_1 = mg = F_B, \tag{13.16}$$

where $F_B$ is defined as the **buoyant force** acting on the cube of water. For the case of the cube of water, the buoyant force is equal to the weight of the water. In general, the buoyant force acting on a submerged object is given by the weight of the fluid displaced,

$$F_B = m_f g.$$

Now suppose the cube of water is replaced with a cube of steel (Figure 13.22). Because the steel cube has the same volume and is at the same depth as the cube of water, the buoyant force remains the same. However, the steel cube weighs more than the cube of water, so a net $y$-component of force acts on the steel cube, given by

$$F_{\text{net},y} = F_B - m_{\text{steel}} g < 0.$$

This net downward force causes the steel cube to sink.

If the cube of water is replaced by a cube of wood, the weight of the cube of wood is less than that of the cube of water, so the net force will be upward. The wooden cube will rise toward the surface. If an object that is less dense than water is placed in water, it will float. An object of mass $m_{\text{object}}$ will sink in water until the weight of the displaced water equals the weight of the object:

$$F_B = m_f g = m_{\text{object}} g.$$

(a)

(b)

**FIGURE 13.23** A ship in a lock: (a) low position; (b) high position.

A floating body displaces its own weight of fluid. This statement is true, independent of the amount of fluid present. To make this clearer, Figure 13.23 shows a ship in a lock. Part (a) shows the ship in the low position, and part (b) in the high position. In both positions, the ship floats, with the same fraction of the ship below the water level. What matters for the buoyant force is not the total amount of the water in the lock (light blue in the figure), but the amount of water displaced by the underwater portion of the ship (hatched blue and brown in the figure). Clearly, in Figure 13.23a, there is much less water left in the lock than the volume that has been displaced by the ship. The only thing that matters is the weight of the water that *would be where the ship is,* not the weight of the water still in the lock. (In fact, with a container of the right shape, even a single gallon of water could be "enough to float a battleship.")

If an object that has a higher density than water is placed under water, it will experience an upward buoyant force that is less than its weight. Its apparent weight is then given by

$$\text{Actual weight} - \text{buoyant force} = \text{apparent weight}. \tag{13.17}$$

## EXAMPLE 13.4 Floating Iceberg

Icebergs, such as the one shown in Figure 13.24a, pose grave dangers for ocean-going ships. Many ships, the *Titanic* most famously, have sunk after collisions with icebergs. The trouble is that a large fraction of an iceberg's volume is hidden below the waterline and thus practically invisible to sailors, as illustrated in Figure 13.24b.

(a)

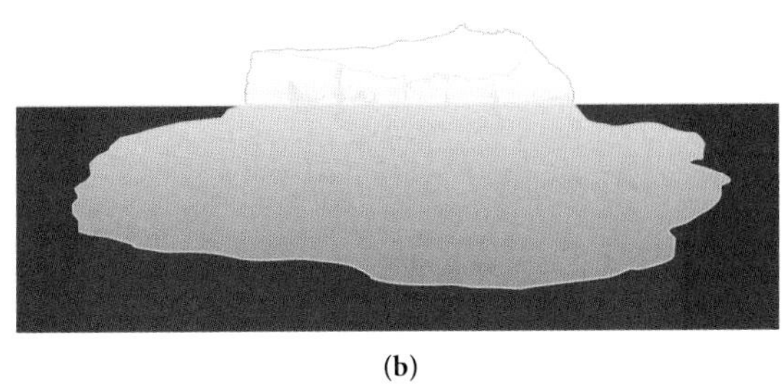

(b)

**FIGURE 13.24** (a) An iceberg floating in the ocean. (b) Illustration showing the fraction of the volume of the iceberg above and below the waterline.

### PROBLEM

What fraction of the volume of an iceberg floating in seawater is visible above the surface?

### SOLUTION

Let $V_t$ be the total volume of the iceberg and $V_s$ be the volume of the iceberg that is submerged. The fraction, $f$, above water is then

$$f = \frac{V_t - V_s}{V_t} = 1 - \frac{V_s}{V_t}.$$

Because the iceberg is floating, the submerged volume must displace a volume of seawater that has the same weight as the entire iceberg. The mass of the iceberg, $m_t$, can be calculated from the volume of the iceberg and the density of ice, $\rho_{ice} = 0.917\ g/cm^3$. The mass of the displaced seawater can be calculated from the submerged volume and the known density of seawater, $\rho_{seawater} = 1.024\ g/cm^3$. We equate the two weights:

$$\rho_{ice} V_t g = \rho_{seawater} V_s g,$$

or

$$\rho_{ice} V_t = \rho_{seawater} V_s.$$

We can rearrange this equation to get

$$\frac{V_s}{V_t} = \frac{\rho_{ice}}{\rho_{seawater}}.$$

We can now find the fraction above water:

$$f = 1 - \frac{V_s}{V_t} = 1 - \frac{\rho_{ice}}{\rho_{seawater}} = 1 - \frac{0.917\ g/cm^3}{1.024\ g/cm^3} = 0.104,$$

or about 10%. Figure 13.24b shows an iceberg with approximately 10% of its volume above the waterline.

An interesting experiment on buoyancy can be performed as follows. You pour water (with red food coloring) into a container (Figure 13.25a) and put a swimmer into it (Figure 13.25b). The swimmer is 90% submerged, so its average density is 90% of that of water, like the iceberg in Example 13.4. Next, you pour paint thinner on top of the water. Paint thinner does not mix with water and has a density equal to 80% of that of water. Thus, it rests on top of the water. If you put the swimmer into a container of paint thinner, the swimmer would sink to the bottom. So, what happens when you pour the paint thinner on top of the water with the swimmer in it? Will the swimmer rise, stay at the same level, or sink?

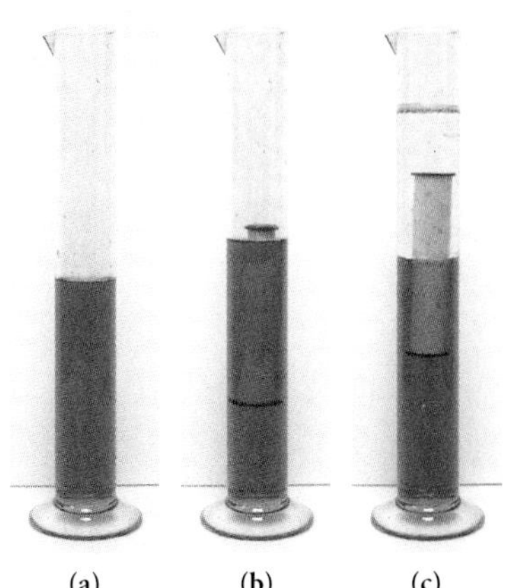

(a) (b) (c)

**FIGURE 13.25** Buoyancy experiment with two liquids: (a) water with red dye; (b) a swimmer placed in the water; (c) paint thinner added above the water.

The answer is shown in Figure 13.25c: The swimmer rises. Why? In the two liquids, two buoyant forces are present—one from the fraction of the swimmer's volume submerged in water, and the other from the fraction of the swimmer's volume submerged in the paint thinner. Therefore, to generate the same buoyant force, less of the volume of the swimmer needs to be submerged in the water when paint thinner is also present than when air and not paint thinner lies on top of the water.

## EXAMPLE 13.5 A Hot-Air Balloon

**FIGURE 13.26** Hot-air balloon in flight.

A typical hot-air balloon has a volume of 2200 m$^3$. The density of air at a temperature of 20 °C is 1.205 kg/m$^3$. The density of the hot air inside the balloon at a temperature of 100 °C is 0.946 kg/m$^3$.

**PROBLEM**

How much weight can the hot-air balloon shown in Figure 13.26 lift (counting the balloon itself)?

**SOLUTION**

The weight of the air at 20 °C that the balloon displaces is equal to the buoyant force:

$$F_B = \rho_{20} V g.$$

The weight of the hot air inside the balloon is

$$W_{\text{balloon}} = \rho_{100} V g.$$

The weight that can be lifted is

$$\begin{aligned} W &= F_B - W_{\text{balloon}} = \rho_{20} V g - \rho_{100} V g \\ &= V g\left(\rho_{20} - \rho_{100}\right) = \left(2200 \text{ m}^3\right)\left(9.81 \text{ m/s}^2\right)\left(1.205 \text{ kg/m}^3 - 0.946 \text{ kg/m}^3\right) \\ &= 5590 \text{ N}. \end{aligned}$$

Note that this weight must include the balloon envelope, the basket, and the fuel, as well as any payload such as the pilot and passengers.

### 13.3 Self-Test Opportunity

Suppose the balloon in Example 13.5 were filled with helium instead of hot air. How much weight could the helium-filled balloon lift? The density of helium is 0.164 kg/m$^3$.

## Determination of Density

Let's return to Archimedes' method of figuring out if the king's new crown was pure gold or a mixture of silver and gold. Since the density of gold is 19.3 g/cm$^3$ and the density of silver is only 10.5 g/cm$^3$, Archimedes needed to measure the density of the crown to find out what fraction of it was gold and what fraction was silver. However, how could he measure the density? He needed to submerge the crown in water and determine its volume from the rise in the water level. Weighing the crown gave him the mass, and dividing the mass by the volume yielded the density, the answer the king wanted.

## EXAMPLE 13.6 Finding the Density of an Object

The method shown in Figure 13.27 can be use to determine the density of an object just by weighing, without the need to measure water levels. This method is usually more precise and thus is preferred. We first weigh a beaker containing water (with green food coloring added) whose density is assumed to be 1000 kg/m$^3$, as shown in part (a). The mass of the beaker and water is determined to be $m_0 = 0.437$ kg. In part (b) we submerge a metal ball, suspended by a string, in the water, being careful not to let the ball touch the bottom of the beaker. The new total mass of this arrangement is determined in part (c) as $m_1 = 0.458$ kg. Finally, in part (d) we let the ball rest on the bottom of the beaker, without pulling on the string. Now the mass is determined to be $m_2 = 0.596$ kg.

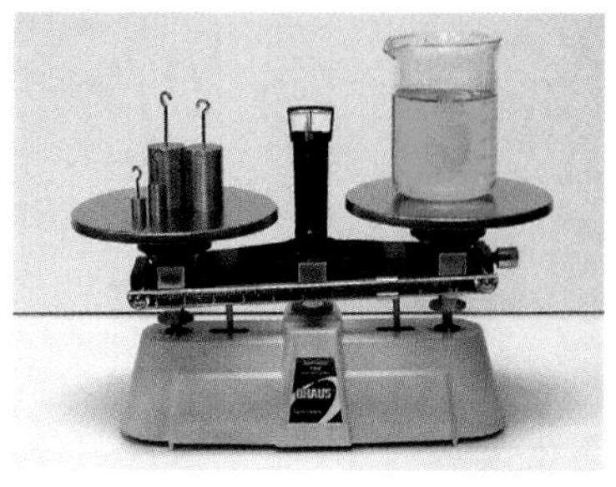

(a)

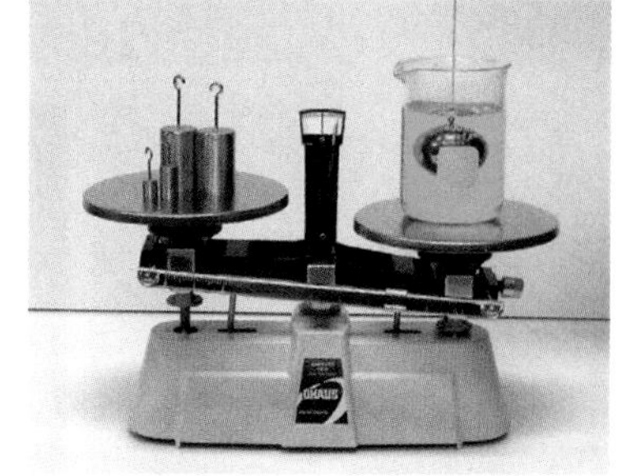

(b)

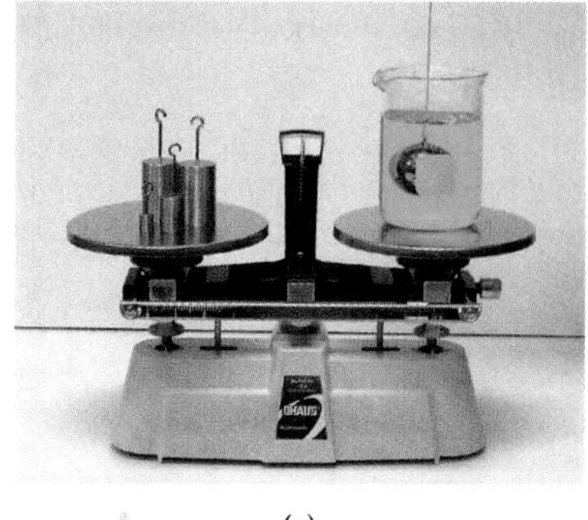

(c)

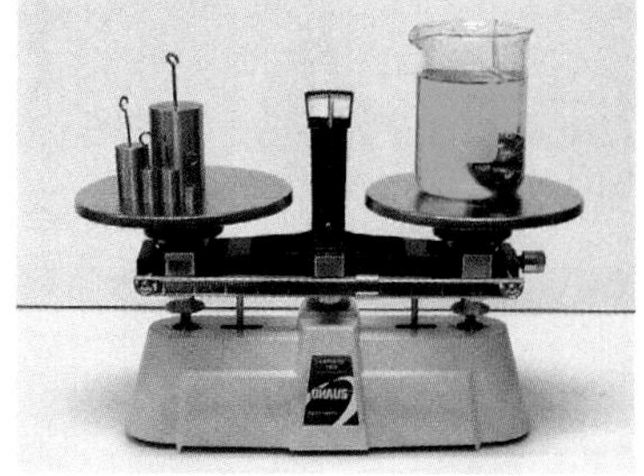

(d)

**FIGURE 13.27** Method for determining the density of a metal object.

**PROBLEM**

What is the density of the metal ball?

**SOLUTION**

Figure 13.28 shows free-body diagrams for the different parts of Figure 13.27. The leftmost diagram shows the ball suspended from the string in static equilibrium, with its weight $-m_b g\hat{y}$, balanced by the string tension, $\vec{T}$. The diagram second from the left in Figure 13.28 corresponds to Figure 13.27a and shows the only forces acting on the water-filled container, which are its weight, $-m_w g\hat{y}$, and the normal force, $m_0 g\hat{y}$. The value of $m_0$ was determined as the result of the first measurement. The diagram third from the left in Figure 13.28 corresponds to parts (b) and (c) in Figure 13.27 and shows the free-body diagram of the submerged ball as well as the free-body diagram of the container; it illustrates the effect of the buoyant force, $\vec{F}_B$, on the ball and on the liquid-filled container. Note that the ball experiences an upward buoyant force and thus the container must experience a force of equal magnitude and opposite direction (downward), or $-\vec{F}_B$, according to Newton's Third Law. The normal force required to keep the container in equilibrium in this situation is, $m_1 g\hat{y}$. The rightmost free-body diagram in Figure 13.28 corresponds to part (d) of Figure 13.27 and shows the normal force as $m_2 g\hat{y}$.

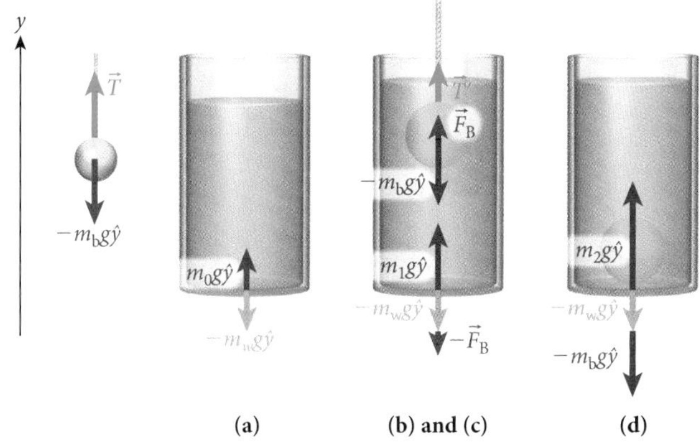


**FIGURE 13.28** Free-body diagrams for the different parts of the experiment.

The combined mass of water and beaker, $m_0$, was determined in part (a) of the method illustrated in Figure 13.28. In part (c), we measured in addition the mass of the water displaced by the (as yet unknown) volume $V$ of the metal ball. The mass of the displaced water is $m_w = \rho_w V_b$, and the total mass measured in part (c) is

$$m_1 = m_0 + \rho_w V_b. \qquad \text{(i)}$$

Finally, the mass measured in part (d) is the combined mass of the ball, the beaker, and the water. For the mass of the ball, we can again use the product of the density and the volume:

$$m_2 = m_0 + m_b = m_0 + \rho_b V_b. \qquad \text{(ii)}$$

Combining equations (i) and (ii), we have

$$\frac{m_2 - m_0}{m_1 - m_0} = \frac{\rho_b V_b}{\rho_w V_b} = \frac{\rho_b}{\rho_w}.$$

As you can see, the volume of the ball has canceled out, and we can obtain the density of the ball from the known density of water and our three measurements, $m_0$, $m_1$, and $m_2$:

$$\rho_b = \frac{m_2 - m_0}{m_1 - m_0}\rho_w = \frac{0.596 - 0.437}{0.458 - 0.437}(1000 \text{ kg/m}^3) = 7570 \text{ kg/m}^3.$$

### 13.5 In-Class Exercise

Suppose we use the same setup as in Example 13.6 and a ball of the same size, but half as dense: $\rho_b = 3785 \text{ kg/m}^3$. The measurement of $m_1$ (ball is held by a string so that it is under water, but not touching the bottom of the beaker) will yield

a) a lower value. b) the same value. c) a higher value.

The mass measurement of $m_2$ (ball is resting at the bottom of the beaker) will yield

a) a lower value. b) the same value. c) a higher value.

## 13.6 Ideal Fluid Motion

(a) (b) (c)

**FIGURE 13.29** (a) Laminar flow of the Firehole River at Yellowstone; (b) transition from laminar to turbulent flow in rising smoke; (c) turbulent flow at the Upper Falls on the Yellowstone River.

We have been examining the behavior of fluids at rest but now turn to fluids in motion. The motion of real-life fluids is complicated and difficult to describe. Numerical techniques and computers are often required to calculate the quantities related to fluid motion. In this section, we consider ideal fluids that can be treated more simply and still yield important and relevant results. To be considered ideal, a fluid must exhibit flow that is laminar, incompressible, nonviscous, and irrotational.

**Laminar flow** means that the velocity of the moving fluid relative to a fixed point in space does not change with time. A gently flowing river displays laminar flow (Figure 13.29a). Water in a waterfall exhibits nonlaminar flow, or **turbulent flow** (Figure 13.29c). Rising smoke exemplifies a transition from laminar to turbulent flow (Figure 13.29b). The warm smoke initially displays laminar flow as it rises. As the smoke continues to rise, its speed increases until turbulent flow sets in.

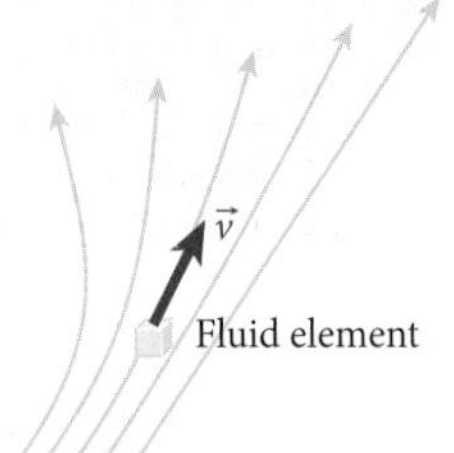


**FIGURE 13.30** Streamlines of a fluid in laminar flow.

**Incompressible flow** means that the density of the liquid does not change as the liquid flows. A **nonviscous fluid** flows completely freely. Some liquids that do not flow freely are pancake syrup and lava. The viscosity of a fluid has an effect analogous to friction. An object moving in a nonviscous liquid experiences no friction-like force, but the same object moving in a viscous liquid is subject to a drag force due to viscosity, a force similar to friction. A flowing viscous fluid loses kinetic energy of motion to thermal energy. We assume that ideal fluids do not lose energy as they flow.

**Irrotational flow** means that no part of the fluid rotates about its own center of mass. Rotational motion of a small part of the fluid would mean that the rotating part had rotational energy, which is assumed not to occur in ideal fluids.

Laminar flow can be described in terms of streamlines (Figure 13.30). A streamline represents the path that a small element of the fluid takes over time. The velocity, $\vec{v}$, of the small element is always tangent to the streamline. Note that streamlines never cross; if they did, the velocity of flow at the crossing point would have two values.

**FIGURE 13.31** Blowing air through the gap between two soft-drink cans.

### Bernoulli's Equation

What provides the lift that allows an airplane to fly through the air? To understand this phenomenon, you can do a simple demonstration with two empty soft-drink cans and five drinking straws. Place each empty can on two straws, with a gap of approximately 1 cm between them, as shown in Figure 13.31. In this arrangement, the cans are able to make lateral movements relatively easily. Now, using the fifth straw, blow air through the gap between the cans. What happens? (This question will be answered later in the section.)

To begin our studies of fluid motion, let us first introduce the equation of continuity. Consider an ideal fluid flowing with speed $v$ in a container or pipe with cross-sectional area $A$ (Figure 13.32). Then $\Delta V$ is the volume of fluid that flows through the pipe during time $\Delta t$ and is given by

$$\Delta V = A\Delta x = Av\Delta t.$$

We can write the volume of fluid passing a given point in the pipe per unit time as

$$\frac{\Delta V}{\Delta t} = Av.$$

Now consider an ideal fluid flowing in a pipe that has a cross-sectional area that changes (Figure 13.33). The fluid is initially flowing with speed $v_1$ through a part of the pipe with cross-sectional area $A_1$. Downstream, the fluid flows with speed $v_2$ in a part of the pipe with cross-sectional area $A_2$. The volume of ideal fluid that enters this section of the pipe per unit time must equal the volume of ideal fluid exiting the pipe per unit time because the fluid is incompressible and the pipe has no leaks. We can express the volume flowing in the first part of the pipe per unit time as

$$\frac{\Delta V}{\Delta t} = A_1 v_1$$

and the volume flowing in the second part of the pipe per unit time as

$$\frac{\Delta V}{\Delta t} = A_2 v_2.$$

The volume of fluid passing any point in the pipe per unit time must be the same in all parts of the pipe, or else fluid would somehow be created or destroyed. Thus, we have

$$A_1 v_1 = A_2 v_2. \tag{13.18}$$

This equation is called the **equation of continuity.**

We can express equation 13.18 as a constant volume flow rate, $R_V$:

$$R_V = Av.$$

Assuming an ideal fluid whose density does not change, we can also express a constant mass flow rate, $R_m$:

$$R_m = \rho Av.$$

The SI unit for mass flow rate is kilograms per second (kg/s).

Let's now consider what happens to the pressure in an ideal fluid flowing through a pipe at a steady rate (Figure 13.34). We start by applying conservation of energy to the ideal fluid flowing between the lower part and the upper part of the pipe. In the lower left part of the pipe, the flowing fluid of constant density, $\rho$, is characterized by pressure $p_1$, speed $v_1$, and elevation $y_1$. The same fluid flows through the transition region and into the upper right part of the pipe. Here the fluid is characterized by pressure $p_2$, speed $v_2$, and elevation $y_2$. We will show that the relationship between the pressures and velocities in this situation is given by

$$p_1 + \rho g y_1 + \tfrac{1}{2}\rho v_1^2 = p_2 + \rho g y_2 + \tfrac{1}{2}\rho v_2^2. \tag{13.19}$$

Another way of stating this relationship is

$$p + \rho g y + \tfrac{1}{2}\rho v^2 = \text{constant}. \tag{13.20}$$

This equation is **Bernoulli's Equation.** If there is no flow, $v = 0$ and equation 13.20 is equivalent to equation 13.8.

One major consequence of Bernoulli's Equation becomes evident if $y = 0$, which means that, at constant elevation,

$$p + \tfrac{1}{2}\rho v^2 = \text{constant}. \tag{13.21}$$

From equation 13.21, we can see that if the velocity of a moving fluid is increased, the pressure must decrease. The decrease of the pressure transverse to the fluid flow has many practical applications, including the measurement of the flow rate of a fluid and the creation of a partial vacuum.

Derivation 13.1 develops Bernoulli's Equation mathematically. The only two conditions that enter into this derivation are the law of conservation of energy and the equation of continuity, which simply restates the basic fact that the number of atoms in the ideal fluid is conserved.

### 13.6 In-Class Exercise

Another demonstration is done by holding two sheets of paper parallel to each other at a separation of approximately an inch and blowing air between them. What do you expect the motion of the sheets of paper will be?

a) They will move away from each other.

b) The will remain at approximately the same separation.

c) They will move toward each other.

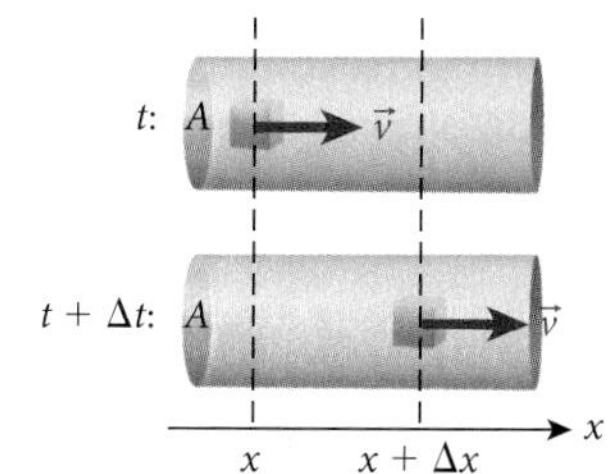


**FIGURE 13.32** Ideal fluid flowing through a pipe with a constant cross-sectional area.

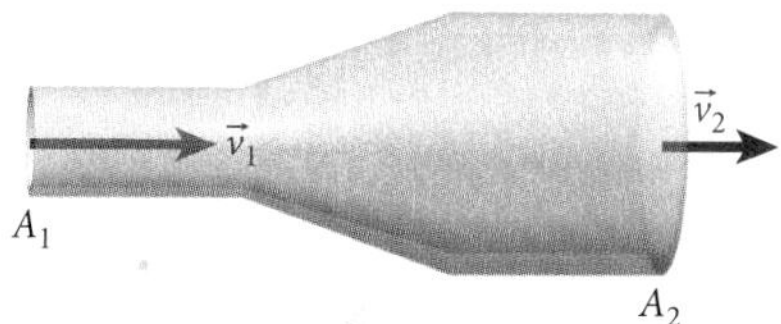


**FIGURE 13.33** Ideal fluid flowing through a pipe with a changing cross-sectional area.

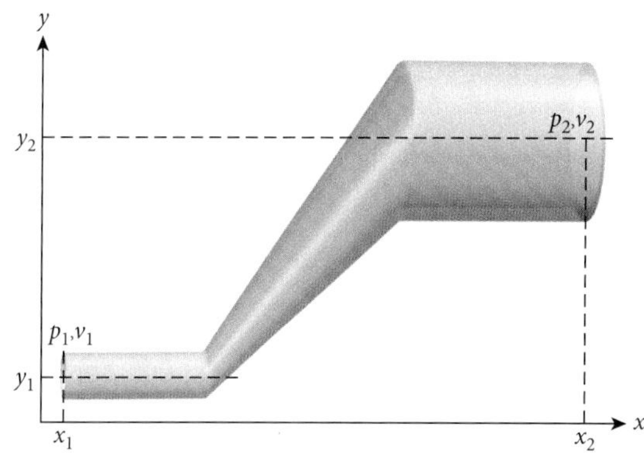


**FIGURE 13.34** Ideal fluid flowing through a pipe with changing cross-sectional area and elevation.

**DERIVATION 13.1 Bernoulli's Equation**

The net work done on the system, $W$, is equal to the change in kinetic energy, $\Delta K$, of the fluid flowing between the initial and final cross sections of the pipe in Figure 13.34:

$$W = \Delta K. \tag{i}$$

The change in kinetic energy is given by

$$\Delta K = \tfrac{1}{2}\Delta m v_2^2 - \tfrac{1}{2}\Delta m v_1^2, \tag{ii}$$

where $\Delta m$ is the amount of mass entering the lower part of the pipe and exiting the upper part of the pipe in time $\Delta t$. The flow of mass per unit time is the density of the fluid times the change in volume per unit time:

$$\frac{\Delta m}{\Delta t} = \rho \frac{\Delta V}{\Delta t}.$$

Thus, we can rewrite equation (ii) as

$$\Delta K = \tfrac{1}{2}\rho\Delta V\left(v_2^2 - v_1^2\right).$$

The work done by gravity on the flowing fluid is given by

$$W_g = -\Delta m g\left(y_2 - y_1\right), \tag{iii}$$

where the negative sign arises because negative work is being done by gravity on the fluid when $y_2 > y_1$. We can also rewrite equation (iii) in terms of the volume flow $\Delta V$ and the density of the fluid:

$$W_g = -\rho\Delta V g\left(y_2 - y_1\right).$$

The work done by a force, $F$, acting over a distance, $\Delta x$, is given by $W = F\Delta x$, which in this case we can express as

$$W = F\Delta x = \left(pA\right)\Delta x = p\Delta V,$$

because the force arises from the pressure in the fluid. We can then express the work done on the fluid by the pressure forcing the fluid to flow into the pipe as $p_1\Delta V$ and the work done on the exiting fluid as $-p_2\Delta V$, giving the work done as a result of pressure,

$$W_p = \left(p_1 - p_2\right)\Delta V.$$

$p_1\Delta V$ is positive because it arises from fluid to the left of flow exerting force on fluid entering the pipe. $p_2\Delta V$ is negative because it arises from fluid to the right of flow exerting a force on fluid exiting the pipe. Using equation (i) and $W = W_p + W_g$, we get

$$\left(p_1 - p_2\right)\Delta V - \rho\Delta V g\left(y_2 - y_1\right) = \tfrac{1}{2}\rho\Delta V\left(v_2^2 - v_1^2\right),$$

which we can simplify to

$$p_1 + \rho g y_1 + \tfrac{1}{2}\rho v_1^2 = p_2 + \rho g y_2 + \tfrac{1}{2}\rho v_2^2.$$

This is Bernoulli's Equation.

## Applications of Bernoulli's Equation

Now that we have derived Bernoulli's Equation, we can return to the demonstration of Figure 13.31. If you performed this demonstration, you found that the two soft-drink cans move closer to each other. This is the opposite of what most people expect, which is that blowing air in the gap will force the cans apart. Bernoulli's Equation explains this surprising result. Since $p + \tfrac{1}{2}\rho v^2 =$ constant, the large speed with which the air is moving between the cans causes the pressure to decrease. Thus, the air pressure between the cans is smaller than the air pressure on other parts of the cans, causing the cans to be pushed toward each other. This is the *Bernoulli effect*. Another way to show the same effect is to hold two sheets of paper parallel to each other about an inch apart and then blow air between them: The sheets are pushed together.

Truck drivers know about this effect. When two 18-wheelers with the typical rectangular box trailers drive in lanes next to each other at high speed, the drivers have to pay attention not to come too close to each other, because the Bernoulli effect can then push the trailers toward each other.

Race car designers also make use of the Bernoulli effect. The biggest limitation on the acceleration of a race car is the maximum friction force between the tires and the road. This friction force, in turn, is proportional to the normal force. The normal force can be increased by increasing the car's mass, but this defeats the purpose, because a larger mass means a smaller acceleration, according to Newton's Second Law. A much more efficient way to increase the normal force is to develop a large pressure difference between the upper and lower surfaces of the car. According to Bernoulli's Equation, this can be accomplished by causing the air to move faster across the bottom of the car than across the top. (Another way to accomplish this would be to use a wing to deflect the air up and thus create a downward force. However, wings have been ruled out in Formula 1 racing.)

Let's reconsider the question of what causes an airplane to fly. Figure 13.35 shows the forces acting on an airplane flying with constant velocity and at constant altitude. The thrust, $\vec{F}_t$, is generated by the jet engines taking in air from the front and expeling it out the back. This force is directed toward the front of the plane. The drag force, $\vec{F}_d$, due to air resistance (discussed in Chapter 4) is directed toward the rear of the plane. At constant velocity, these two forces cancel: $\vec{F}_t = \vec{F}_d$. The forces acting in the vertical direction are the weight, $\vec{F}_g$ and the lift, $\vec{F}_l$ which is provided almost exclusively by the wings. If the plane flies at constant altitude, these two forces cancel as well: $\vec{F}_l = \vec{F}_g$. Since a fully loaded and fueled Boeing 747 typically has a mass of 350 metric tons, its weight is 3.4 MN. Thus, the lift provided by the wings to keep the airplane flying must be very large. By comparison, the maximum thrust produced by all four engines on a 747 is 0.9 MN.

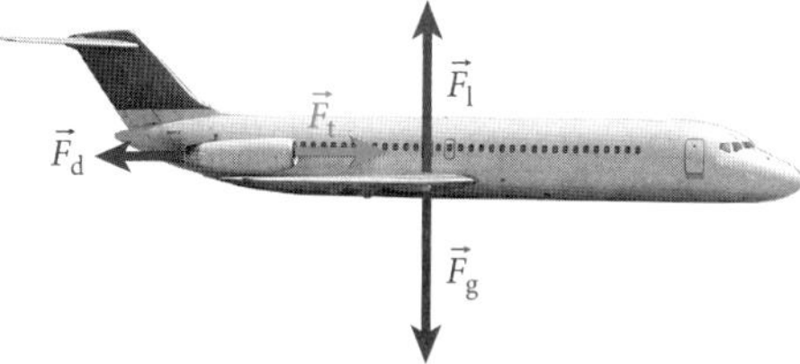


**FIGURE 13.35** Forces acting on an airplane in flight.

The most commonly held notion of what generates the lift that allows a plane to stay airborne is shown in Figure 13.36a: The wing moves through the air and forces the air streamlines that move above it onto a longer path. Thus, the air above the wing needs to move at a higher speed in order to reconnect with the air that moves below the wing. Bernoulli's Equation then implies that the pressure on the top side of the wing is lower than that on the bottom side. Since the combined surface area of the two wings of a Boeing 747 is 511. m$^2$, a pressure difference of $\Delta p = F/A = (3.4 \text{ MN})/(511. \text{ m}^2) = 6.6$ kPa (0.66% of the atmospheric pressure at sea level) between the bottom and top of the wings would be required for this notion to work out quantitatively.

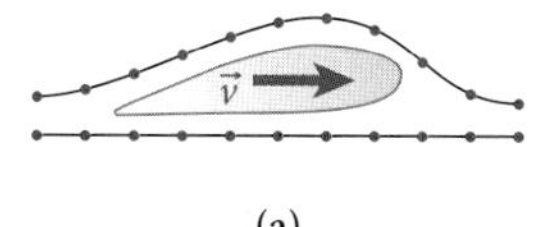


(a)

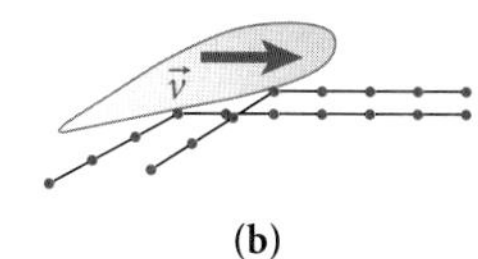


(b)

**FIGURE 13.36** Two extreme views of the process that creates lift on the wing of an airplane: (a) the Bernoulli effect; (b) Newton's Third Law.

An alternative physical interpretation is shown in Figure 13.36b. In this view, air molecules are deflected downward by the lower side of the wing, and according to Newton's Third Law, the wing then experiences an upward force that provides the lift. In order for this idea to work, the bottom of the wing has to have some nonzero angle (the angle of attack) relative to the horizontal. (To attain this position, the nose of the aircraft points slightly upward, so the engine thrust also contributes to the lift. Alternatively, the angle of attack can also be realized by changing the angle of the flaps on the wings.)

Both of these simple ideas contain elements of the truth, and the effects contribute to the lift to different degrees, depending on the type of aircraft and the flight phase. The Newtonian effect is more important during takeoff and landing, and during all flight phases for fighter planes (which have a smaller wing area). However, in general, air does not act as an ideal incompressible fluid as it streams around an airplane's wing. It is compressed at the front edge of the wing and decompressed at its back edge. This compression-decompression effect is stronger on the top surface than on the bottom surface, creating a higher net pressure on the bottom of the wing and providing the lift. The design of aircraft wings is still under intense study, and ongoing research seeks to develop new wing designs that can yield higher fuel efficiency and stability.

We can apply similar considerations to the flight of a curveball in baseball. Curveballs have a sideward spin, and if a left-handed pitcher gives a strong clockwise rotation (as

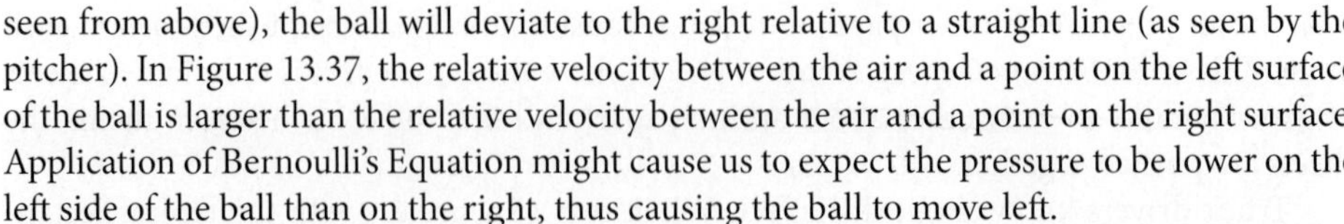

seen from above), the ball will deviate to the right relative to a straight line (as seen by the pitcher). In Figure 13.37, the relative velocity between the air and a point on the left surface of the ball is larger than the relative velocity between the air and a point on the right surface. Application of Bernoulli's Equation might cause us to expect the pressure to be lower on the left side of the ball than on the right, thus causing the ball to move left.

The interpretation using Newton's Third Law (Figure 13.37a) comes closer to explaining correctly why a clockwise spinning baseball experiences a clockwise deflection. The figure shows a top view of a clockwise spinning baseball, moving from the bottom of the page to the top, with the air molecules that the ball encounters represented by red arrows. As the air molecules collide with the surface of the ball, those on the side where the surface is rotating toward them (left side in Figure 13.37a) receive a stronger sideways hit, or impulse, than those on the other side. By Newton's Third Law, the net recoil force on the baseball thus deflects the baseball in the same direction as its rotation. While the Newtonian explanation gets the direction of the effect right, it is not quite the correct explanation, because the oncoming air molecules do not reach the surface of the baseball undisturbed, as is required for this model to work.

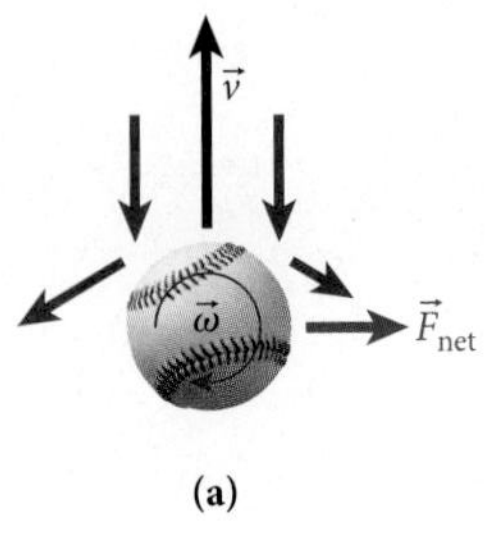


(a)

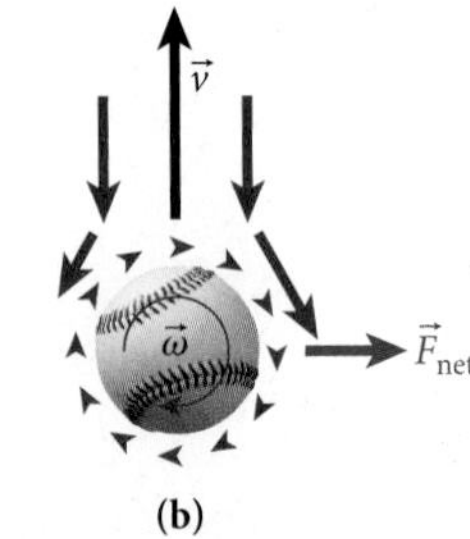


(b)

**FIGURE 13.37** Top view of clockwise spinning baseball moving from the bottom of the page to the top of the page: (a) interpretation using Newton's Third Law; (b) explanation involving a boundary layer.

Just like the lift of an airplane, the phenomenon of a curveball has a more complex explanation. A rotating sphere moving through air drags a boundary layer of air along with its surface. The air molecules that encounter this boundary layer get dragged along to some extent. This causes the air molecules on the right side of the ball in Figure 13.37b to be accelerated and those on the left side to be slowed down. The differentially higher speed of air on the right side implies a lower pressure due to the Bernoulli effect and thus causes a deflection to the right. This is known as the *Magnus effect*.

In tennis, topspin causes the ball to dip faster than a ball hit without spin; backspin causes the ball to sail longer. Both kinds of effect occur for the same reason as we just noted for the curveball in baseball. Golfers also use backspin to make their drives carry longer. A well driven golf ball will have backspin of approximately 4000 rpm. Sidespin in golf causes draws or hooks and fades or slices, depending on the severity and direction of the spin. Incidentally, the dimples on a golf ball are essential for its flight characteristics; they cause turbulence around the golf ball and thus reduce the air resistance. Even the best professionals would not be able to hit a golf ball without dimples more than 200 m.

## EXAMPLE 13.7 Spray Bottle

**FIGURE 13.38** Spray bottle for dispensing liquids in the form of a fine mist.

### PROBLEM

If you squeeze the handle of a spray bottle (Figure 13.38), you cause air to flow horizontally across the opening of a tube that extends down into the liquid almost to the bottom of the bottle. If the air is moving at 50.0 m/s, what is the pressure difference between the top of the tube and the atmosphere? Assume that the density of air is $\rho = 1.20 \text{ kg/m}^3$.

### SOLUTION

Before you squeeze the handle, the airflow speed is $v_0 = 0$. Using Bernoulli's Equation for a negligible height difference (equation 13.21), we find

$$p + \tfrac{1}{2}\rho v^2 = p_0 + \tfrac{1}{2}\rho v_0^2.$$

Solving this equation for the pressure difference, $p - p_0$, under the condition that $v_0 = 0$, we arrive at

$$p - p_0 = -\tfrac{1}{2}\rho v^2 = -\frac{\left(1.20 \text{ kg/m}^3\right)\left(50.0 \text{ m/s}\right)^2}{2} = -1500 \text{ Pa}.$$

Therefore, we see that the pressure is lowered by 1.50 kPa, causing the liquid to be pushed upward and broken up into small droplets in the air stream, to form a mist.

The same principle is used in old-fashioned carburetors, which mix the fuel with air in older cars. (In newer cars, the carburetor has been replaced by fuel injectors.)

## SOLVED PROBLEM 13.2 Venturi Tube

### PROBLEM

On some light aircraft, a device called a *Venturi tube* is used to create a pressure difference that can be used to drive gyroscope-based instruments for navigation. The Venturi tube is mounted on the outside of the fuselage in an area of free airflow. Suppose a Venturi tube has a circular opening with a diameter of 10.0 cm, narrowing down to a circular opening with a diameter of 2.50 cm, and then opening back up to the original diameter of 10.0 cm. What is the pressure difference between the 10-cm opening and the narrowest region of the Venturi tube, assuming that the aircraft is flying at a constant speed of 38.0 m/s at a low altitude where the density of air can be assumed that at sea level ($\rho = 1.30\ \text{kg/m}^3$) at 5 °C?

### SOLUTION

### THINK

The equation of continuity (equation 13.18) tells us that the product of the area and the velocity of the flow through the Venturi tube is constant. We can then relate the area of the opening, the area of the narrowest region, the velocity of the air entering the Venturi tube, and the velocity of the air in the narrowest region of the tube. Using Bernoulli's Equation, we can then relate the pressure at the opening to the pressure in the narrowest region.

### SKETCH

A Venturi tube is sketched in Figure 13.39.

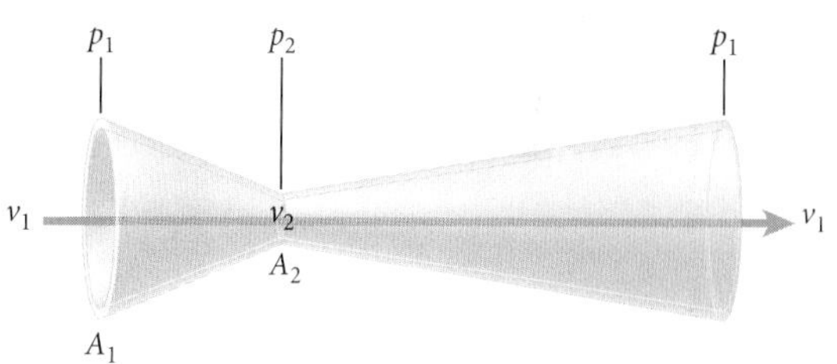


**FIGURE 13.39** A Venturi tube with air flowing through it.

### RESEARCH

The equation of continuity is

$$A_1 v_1 = A_2 v_2.$$

Bernoulli's Equation tells us that

$$p_1 + \tfrac{1}{2}\rho v_1^2 = p_2 + \tfrac{1}{2}\rho v_2^2.$$

### SIMPLIFY

The pressure difference between the opening of the Venturi tube and the narrowest area is found by rearranging the Bernoulli Equation:

$$p_1 - p_2 = \Delta p = \tfrac{1}{2}\rho v_2^2 - \tfrac{1}{2}\rho v_1^2 = \tfrac{1}{2}\rho\left(v_2^2 - v_1^2\right).$$

Solving the equation of continuity for $v_2$ and substituting that result into the rearranged Bernoulli Equation gives us the pressure difference, $\Delta p$:

$$\Delta p = \tfrac{1}{2}\rho\left(v_2^2 - v_1^2\right) = \tfrac{1}{2}\rho\left[\left(\frac{A_1}{A_2}v_1\right)^2 - v_1^2\right] = \tfrac{1}{2}\rho v_1^2\left(\frac{A_1^2}{A_2^2} - 1\right).$$

Since both areas are circular, we have $A_1 = \pi r_1^2$ and $A_2 = \pi r_2^2$, and thus $(A_1/A_2)^2 = (\pi r_1^2/\pi r_2^2)^2 = (r_1/r_2)^4$, which leads to

$$\Delta p = \tfrac{1}{2}\rho v_1^2\left[\left(\frac{r_1}{r_2}\right)^4 - 1\right].$$

### CALCULATE

Putting in the numerical values gives us

$$\Delta p = \tfrac{1}{2}\left(1.30\ \text{kg/m}^3\right)\left(38.0\ \text{m/s}\right)^2\left[\left(\frac{10.0/2\ \text{cm}}{2.5/2\ \text{cm}}\right)^4 - 1\right] = 239343.0\ \text{Pa},$$

where we have used $\rho = 1.30\ \text{kg/m}^3$ as the density of air.

### ROUND

We report our result to three significant figures:

$$\Delta p = 239.\ \text{kPa}.$$

*Continued—*

**DOUBLE-CHECK**

The pressure difference between the interior of the airplane and the narrowest part of the Venturi tube is 239 kPa, or more than twice normal atmospheric pressure. Connecting a hose between the narrowest part of the tube and the interior of the plane thus allows for a steady stream of air through the tube, which can be used to drive a rapidly rotating gyroscope.

## Draining a Tank

Let's analyze another simple experiment. A large container with a small hole at the bottom is filled with water and measurements are made of how long it takes to drain. Figure 13.40 illustrates this experiment. We can analyze this process quantitatively and arrive at a description of the height of the fluid column as a function of time. Since the bottle is open at the top, the atmospheric pressure is the same at the upper surface of the fluid column as it is at the small hole. We thus obtain from Bernoulli's Equation (equation 13.19)

$$\rho g y_1 + \tfrac{1}{2}\rho v_1^2 = \rho g y_2 + \tfrac{1}{2}\rho v_2^2.$$

Canceling out the density and reordering the terms, we obtain

$$v_2^2 - v_1^2 = 2g(y_1 - y_2) = 2gh,$$

where $h = y_1 - y_2$ is the height of the fluid column. We obtained this same result in Chapter 2 for a particle in free fall! Thus, the streaming of an ideal fluid under the influence of gravity proceeds in the same way as free fall of a point particle.

The equation of continuity (equation 13.18) relates the two speeds, $v_1$ and $v_2$, to each other via the ratio of their corresponding cross-sectional areas: $A_1v_1 = A_2v_2$. Thus, we find the speed with which the fluid flows from the container as a function of the height of the fluid column above the hole:

$$v_1 = v_2\frac{A_2}{A_1} \Rightarrow v_2^2 - v_1^2 = v_2^2\left(1 - \frac{A_2^2}{A_1^2}\right) \Rightarrow$$

$$v_2^2 = \frac{2gh}{1 - \dfrac{A_2^2}{A_1^2}} = \frac{2A_1^2 g}{A_1^2 - A_2^2}h.$$

If $A_2$ is small compared to $A_1$, this result simplifies to

$$v_2 = \sqrt{2gh}. \tag{13.22}$$

The speed with which the fluid streams from the container is sometimes called the **speed of efflux,** and equation 13.22 is often called Torricelli's Theorem.

How does the height of the fluid column change as a function of time? To answer this, we note that the speed $v_1$ is the negative of the time derivative of the height of the fluid column, $h$: $v_1 = -dh/dt$, since the height decreases with time. From the equation of continuity, we then find

$$A_1v_1 = -A_1\frac{dh}{dt} = A_2v_2 = A_2\sqrt{2gh} \Rightarrow$$

$$\frac{dh}{dt} = -\frac{A_2}{A_1}\sqrt{2gh}. \tag{13.23}$$

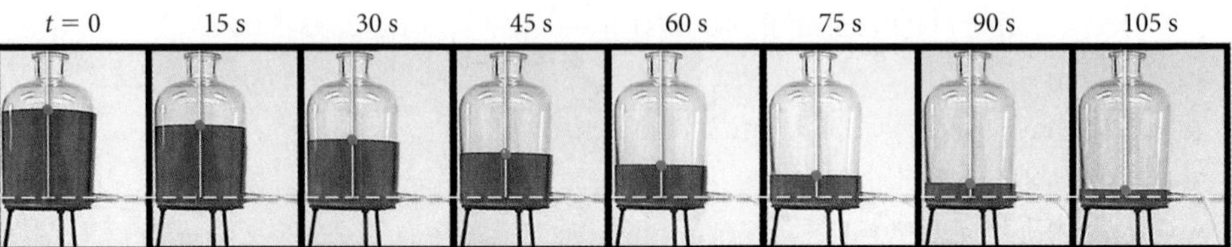


**FIGURE 13.40** Draining a bottle of water through a small hole at the base.

Because equation 13.23 relates the height to its derivative, it is a differential equation. Its solution is

$$h(t)=h_0-\frac{A_2}{A_1}\sqrt{2gh_0}t+\frac{g}{2}\frac{A_2^2}{A_1^2}t^2, \tag{13.24}$$

where $h_0$ is the initial height of the fluid colomn. You can convince yourself that equation 13.24 is really the solution by taking its derivative and inserting it into equation 13.23. The derivative is

$$\frac{dh}{dt}=-\frac{A_2}{A_1}\sqrt{2gh_0}+g\frac{A_2^2}{A_1^2}t.$$

The point in time at which this derivative reaches zero is the time at which the draining process has finished:

$$\frac{dh}{dt}=0\Rightarrow t_{\text{f}}=\frac{A_1}{A_2}\sqrt{\frac{2h_0}{g}}. \tag{13.25}$$

If you substitute this expression for $t$ into equation 13.24, you see that $t_{\text{f}}$ is also the time at which the height of the fluid column reaches zero, which is as expected. Because, according to equation 13.23, the time derivative of the height is proportional to the square root of the height, the height has to be zero when the derivative is zero.

Equation 13.25 also tells us that the time it takes to drain a cylindrical container is proportional to the cross-sectional area of the container and inversely proportional to the area of the hole the liquid is draining through. However, the time is also proportional to the square root of the initial height of the fluid. Thus, if two containers hold the same volume of liquid and drain through holes of the same size, the container that is taller and with a smaller cross-sectional area will drain faster.

Finally, keep in mind that equation 13.25 holds only for an ideal fluid in a container with constant cross-sectional area as a function of height and when the cross-sectional area of the hole is small compared to that of the container.

## EXAMPLE 13.8 Draining a Bottle

Figure 13.40 shows a large cylindrical bottle of cross-sectional area $A_1 = 0.100 \text{ m}^2$. The liquid (water with red food coloring) drained through a small hole of radius 7.40 mm, or of area $A_2 = 1.72 \cdot 10^{-4} \text{ m}^2$. The frames in the figure represent times at 15-s intervals. The initial height of the fluid column above the hole was $h_0 = 0.300$ m.

### PROBLEM

How long did it take to drain this bottle?

### SOLUTION

We can simply use equation 13.25 and insert the given values:

$$t_{\text{f}}=\frac{A_1}{A_2}\sqrt{\frac{2h_0}{g}}=\frac{0.100 \text{ m}^2}{1.72\cdot 10^{-4} \text{ m}^2}\sqrt{\frac{2(0.300 \text{ m})}{9.81 \text{ m/s}^2}}=144 \text{ s}.$$

While the solution to this problem is straightforward, it is instructive to plot the height as a function of time, using equation 13.24, and compare the plot to the experimental data. In Figure 13.40, the level of the hole is marked with a dashed horizontal line, and the height of the fluid column in each frame is marked with a green dot. Figure 13.41 shows the comparison between the calculated solution (the red line) and the data obtained from the experiment. You can see that the agreement is well within the measurement uncertainty, represented by the size of the green dots.

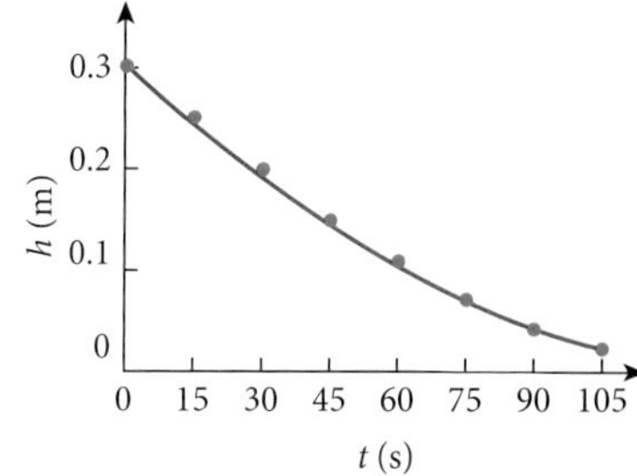


**FIGURE 13.41** Height of fluid column as a function of time: Green dots are observed data points; the red line is the calculated solution.

### 13.7 In-Class Exercise

If the bottle and initial fluid level remain the same as in Example 13.8 but the diameter of the hole is halved, the time required to drain all the fluid will be approximately

a) 36 s.

b) 72 s.

c) 144 s.

d) 288 s.

e) 578 s.

## 13.7 Viscosity

(a)

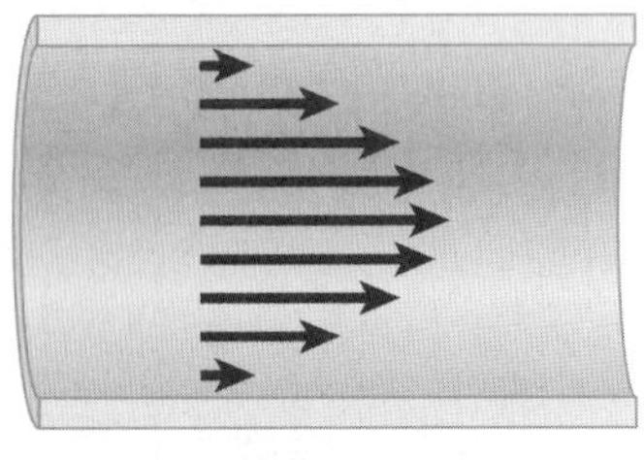

(b)

**FIGURE 13.42** Velocity profiles in a cylindrical tube: (a) nonviscous ideal fluid flow; (b) viscous flow.

If you have ever drifted in a boat on a gentle river, you may have noticed that your boat moved faster in the middle of the river than very close to the banks. Why would this happen? If the water in the river were an ideal fluid in laminar motion, it should make no difference how far away from shore you are. However, water is not quite an ideal fluid. Instead, it has some degree of "stickiness," called **viscosity.** For water, the viscosity is quite low; for heavy motor oil, it is significantly higher, and it is even higher yet for substances like honey, which flow very slowly. Viscosity causes the fluid streamlines at the surface of a river to partially stick to the boundary and neighboring streamlines to partially stick to one another.

The velocity profile for the streamlines in viscous flow in a tube is sketched in Figure 13.42b. The profile is parabolic, with the velocity approaching zero at the walls and reaching its maximum value in the center. This flow is still laminar, with the streamlines all flowing parallel to one another.

How is the viscosity of a fluid measured? The standard procedure is to use two parallel plates of area $A$ and fill the gap of width $h$ between them with the fluid. Then one of the plates is dragged across the other and the force $F$ that is required to do so is measured. The resulting velocity profile of the fluid flow is linear (Figure 13.43). The viscosity, $\eta$, is defined as the ratio of the force per unit area divided by the velocity difference between the top and bottom plates over the distance between the plates:

$$\eta = \frac{F/A}{\Delta v/h} = \frac{Fh}{A\Delta v}. \tag{13.26}$$

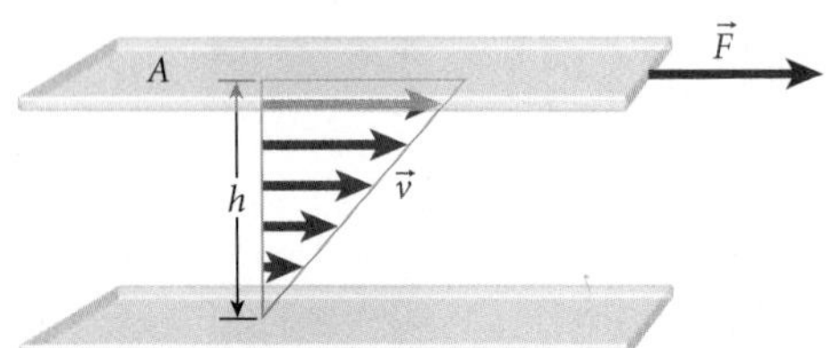


**FIGURE 13.43** Measuring the viscosity of a liquid with two parallel plates.

The unit of viscosity represents pressure (force per unit area) multiplied by time, or pascal seconds (Pa s). This unit is also called a *poiseuille* (Pl). (Care must be taken to avoid confusing this SI unit with the cgi unit poise (P), because 1 P = 0.1 Pa s.)

It is important to realize that the viscosity of any fluid depends strongly on temperature. You can see an example of this temperature dependence in the kitchen. If you store olive oil in the refrigerator and then pour it from the bottle, you can see how slowly it flows. Heat the same olive oil in a pan, and it flows almost as readily as water. Temperature dependence is of great concern for motor oils, and the goal is to have a small temperature dependence. Table 13.5 lists some typical viscosity values for different fluids. All values are those at room temperature (20 °C = 68 °F) except that of blood, whose value is given for the physiologically relevant temperature of human body temperature (37 °C = 98.6 °F). Incidentally, the viscosity of blood increases by about 20% during a human's lifetime, and the average value for men is slightly higher than that for women ($4.7 \cdot 10^{-3}$ Pa s vs. $4.3 \cdot 10^{-3}$ Pa s).

The viscosity of a fluid is important in determining how much fluid can flow through a pipe of given radius $r$ and length $\ell$. Gotthilf Heinrich Ludwig Hagen (in 1839) and Jean Louis Marie Poiseuille (in 1840) found independently that $R_v$, the volume of fluid that can flow per unit time, is

$$R_v = \frac{\pi r^4 \Delta p}{8\eta\ell}. \tag{13.27}$$

**Table 13.5 Some Typical Values of Viscosity at Room Temperature**

| Material | Viscosity (Pa s) |
|---|---|
| Air | $1.8 \cdot 10^{-5}$ |
| Alcohol (ethanol) | $1.1 \cdot 10^{-3}$ |
| Blood (at body temperature) | $4 \cdot 10^{-3}$ |
| Honey | 10 |
| Mercury | $1.5 \cdot 10^{-3}$ |
| Motor oil (SAE10 to SAE40) | 0.06 to 0.7 |
| Olive oil | 0.08 |
| Water | $1.0 \cdot 10^{-3}$ |

Here $\Delta p$ is the pressure difference between the two ends of the pipe. As expected, the flow is inversely proportional to the viscosity and the length of the pipe. Most significantly, though, it is proportional to the fourth power of the radius of the pipe. If we consider a blood vessel as a pipe, this relationship helps us understand the problem associated with clogging of the arteries. If cholesterol-induced deposits reduce the diameter of a blood vessel by 50%, then the blood flow through the vessel is reduced to $1/2^4 = 1/16$ or 6.25% of the original rate—a reduction of 93.75%.

## EXAMPLE 13.9 Hypodermic Needle

For many people, the scariest part of a visit to the doctor is an injection. Learning about the fluid mechanics of the hypodermic needle (Figure 13.44) won't change that, but is interesting nonetheless.

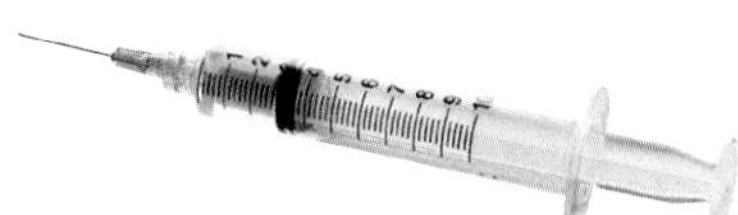

**FIGURE 13.44** A hypodermic needle illustrates fluid flow with viscosity.

### PROBLEM 1

If 2.0 cm$^3$ of water is to be pushed out of a 1.0-cm-diameter syringe through a 3.5-cm-long 15 gauge needle (interior needle diameter = 1.37 mm) in 0.4 s, what force must be applied to the plunger of the syringe?

### SOLUTION 1

The Hagen-Poiseuille Law (equation 13.27) relates fluid flow to the pressure difference causing the flow. We can solve that equation for the pressure difference, $\Delta p$, between the tip of the needle and the end that is connected to the syringe:

$$\Delta p = \frac{8\eta \ell R_{\text{V}}}{\pi r^4}.$$

The viscosity of water can be obtained from Table 13.5: $\eta = 1.0 \cdot 10^{-3}$ Pa s. The flow rate, $R_{\text{V}}$, is just the ratio of volume to time:

$$R_{\text{V}} = \frac{\Delta V}{\Delta t} = \frac{2.0 \cdot 10^{-6} \text{ m}^3}{0.4 \text{ s}} = 5.0 \cdot 10^{-6} \text{ m}^3\text{/s}.$$

The geometric dimensions of the syringe are specified in the problem statement, so we obtain

$$\Delta p = \frac{8\left(1.0 \cdot 10^{-3} \text{ Pa s}\right)\left(0.035 \text{ m}\right)\left(5.0 \cdot 10^{-6} \text{ m}^3\text{/s}\right)}{\pi\left[0.5\left(1.37 \cdot 10^{-3} \text{ m}\right)\right]^4} = 2020 \text{ Pa}.$$

Because pressure is force per unit area, we can obtain the required force by multiplying the pressure difference we just calculated by the appropriate area. What is this area? The required force is provided by pushing on the plunger, so we use the area of the plunger:

$$A = \pi R^2 = \pi\left[0.5\left(0.01 \text{ m}\right)\right]^2 = 7.8 \cdot 10^{-5} \text{ m}^2.$$

Thus, the force required to push out 2.0 cm$^3$ of water in 0.4 s is only

$$F = A\Delta p = (7.8 \cdot 10^{-5} \text{ m}^2)(2020 \text{ Pa}) = 0.16 \text{ N}.$$

### PROBLEM 2

What is the speed with which the water emerges from the needle of the syringe?

### SOLUTION 2

We saw in Section 13.6 that the speed of fluid flow is related to the volume rate of flow by $R_{\text{V}} = A_{\text{v}}$, where $A$ is the cross-sectional area—in this case, the area of the needle tip whose the diameter is 1.37 mm. Solving this equation for the speed, we find

$$v = \frac{R_{\text{V}}}{A} = \frac{5.0 \cdot 10^{-6} \text{ m}^3\text{/s}}{\pi\left[0.5\left(1.37 \cdot 10^{-3} \text{ m}\right)\right]^2} = 3.4 \text{ m/s}.$$

*Continued—*

**DISCUSSION**

If you have ever pushed the plunger of a syringe, you know it takes more force than 0.16 N. (A force of 0.16 N is equivalent to the weight of a cheap ballpoint pen.) What gives rise to the higher force requirement? Remember that the plunger has to provide a tight seal with the wall of the syringe. Thus, the main effort in pushing the plunger results from overcoming the friction force between the syringe wall and the plunger. It would be an altogether different story, though, if the syringe were filled with honey. Note also that the 15 gauge needle used in this example is larger than the needle typically use, to make an injection, which is a 24 or 25 gauge needle.

## 13.8 Turbulence and Research Frontiers in Fluid Flow

**FIGURE 13.45** Extreme turbulent flow—a tornado is a vortex.

In laminar flow, the streamlines of a fluid follow smooth paths. In contrast, for a fluid in turbulent flow, vortices form, detach, and propagate (Figure 13.45). We have already seen that ideal laminar flow or viscous laminar flow transitions into turbulent flow when the velocity of flow exceeds a certain value. This transition is clearly illustrated in Figure 13.29b, which shows how rising cigarette smoke undergoes a transition from laminar to turbulent flow. What is the criterion that determines whether flow is laminar or turbulent?

The answer lies in the **Reynolds number,** Re, which is the ratio of the typical inertial force to the viscous force and thus is a pure dimensionless number. The inertial force has to be proportional to the density, $\rho$, and the typical velocity of the fluid, $\bar{v}$, because $F = dp/dt$, according to Newton's Second Law. The viscous force is proportional to the viscosity, $\eta$, and inversely proportional to the characteristic length scale, $L$, over which the flow varies. For flow through a pipe with a circular cross section, this length scale is the diameter of the pipe, $L = 2r$. Thus, the formula for calculating the Reynolds number is

$$\mathrm{Re} = \frac{\rho \bar{v} L}{\eta}. \tag{13.28}$$

As a rule of thumb, a Reynolds number less than 2000 means laminar flow, and one higher than 4000 means turbulent flow. For Reynolds numbers between 2000 and 4000, the character of the flow depends on many fine details of the exact configuration. Engineers try hard to avoid this interval because of this essential unpredictability.

The true power of the Reynolds number lies in the fact that fluid flows in systems with the same geometry and the same Reynolds number behave similarly. This allows engineers to reduce typical length scales or velocity scales, build scale models of boats or airplanes, and test their performance in water tanks or wind tunnels of relatively modest scale (Figure 13.46).

Rather than scale models, modern research on fluid flow and turbulence relies on computer models (Figure 13.1). Hydrodynamic modeling is employed for studying applications of an incredible variety of physical systems, such as the performance and aero-

(a) (b)

**FIGURE 13.46** (a) Wind-tunnel testing of a scale model of a wing. (b) Scale model of fighter jet with interchangeable parts for wind-tunnel testing.

dynamics of cars, airplanes, rockets, and boats. However, hydrodynamic modeling is also utilized in studying the collisions of atomic nuclei at the highest energies attainable in modern accelerators and in the modeling of supernova explosions (Figure 13.47). In 2005, an experimental group working at the Relativistic Heavy Ion Collider in Brookhaven, New York, discovered that gold nuclei show the characteristics of a perfect nonviscous fluid when they smash into each other at the highest attainable energies. (One of the authors (Westfall) had the privilege to announce this discovery at the 2006 annual meeting of the American Physical Society.) Exciting research results on fluid motion will continue to emerge over the next decades, as this is one of the most interesting interdisciplinary areas in the physical sciences.

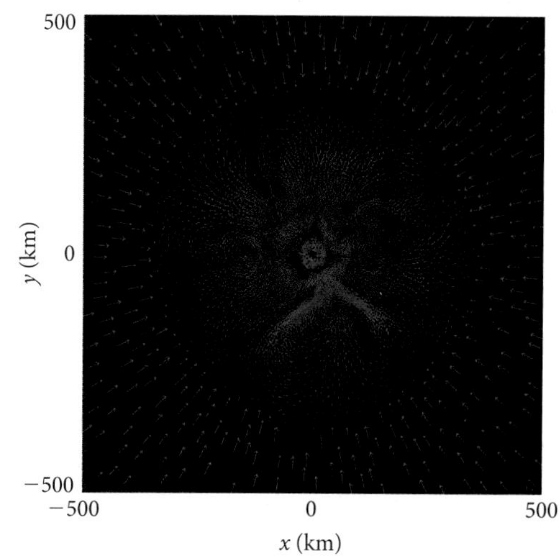


**FIGURE 13.47** Hydrodynamic modeling of the collapse of a supernova core. The arrows indicate the flow directions of the fluid elements, and the color indicates the temperature.

## WHAT WE HAVE LEARNED | EXAM STUDY GUIDE

- One mole of a material has $N_A = 6.022 \cdot 10^{23}$ atoms or molecules. The mass of 1 mol of a material in grams is given by the sum of the atomic mass numbers of the atoms that make up the material.
- For a solid, stress = modulus·strain, where stress is force per unit area and strain is a unit deformation. There are three types of stress and strain, each with its own modulus:
  - Tension or linear compression leads to a positive or negative change in length: $\frac{F}{A} = Y\frac{\Delta L}{L}$, where $Y$ is Young's modulus.
  - Volume compression leads to a change in volume: $p = B\frac{\Delta V}{V}$, where $B$ is the bulk modulus.
  - Shear leads to bending: $\frac{F}{A} = G\frac{\Delta x}{L}$, where $G$ is the shear modulus.
- Pressure is defined as force per unit area: $p = \frac{F}{A}$.
- The absolute pressure, $p$, at a depth, $h$, in a liquid with density, $\rho$, and pressure, $p_0$, on the surface of the liquid is $p = p_0 + \rho gh$.
- Gauge pressure is the difference in pressure between the gas in a container and the pressure of the Earth's atmosphere.
- Pascal's Principle states that when a change in pressure occurs at any point in a confined incompressible fluid, an equal change in pressure occurs at every point in the fluid.
- The buoyant force on an object immersed in a fluid is equal to the weight of the fluid displaced: $F_B = m_f g$.
- An ideal fluid is assumed to exhibit laminar flow, incompressible flow, nonviscous flow, and irrotational flow.
- The flow of an ideal gas follows streamlines.
- The equation of continuity for a flowing ideal fluid relates the velocity and area of the fluid flowing through a container or pipe: $A_1v_1 = A_2v_2$.
- Bernoulli's Equation relates the pressure, height, and velocity of an ideal fluid flowing through a container or pipe: $p + \rho gy + \frac{1}{2}\rho v^2 = \text{constant}$.
- For viscous fluids, the viscosity, $\eta$, is the ratio of the force per unit area to the velocity difference per unit length: $\eta = \frac{F/A}{\Delta v/h} = \frac{Fh}{A\Delta v}$.
- For viscous fluids, the volume flow rate through a cylindrical pipe of radius $r$ and length $\ell$ is given by: $R_V = \frac{\pi r^4 \Delta p}{8\eta\ell}$, where $\Delta p$ is the pressure difference between the two ends of the pipe.
- The Reynolds number determines the ratio of inertial force to viscous force and is defined as $\text{Re} = \frac{\rho\bar{v}L}{\eta}$, where $\bar{v}$ is the average fluid velocity and $L$ is the characteristic length scale over which the flow changes. A Reynolds number less than 2000 means laminar flow, and one greater than 4000 means turbulent flow.

## KEY TERMS

atoms, p. 418
Avogadro's number, p. 419
gas, p. 420
fluid, p. 420
liquid, p. 420
solid, p. 420
elastic limit, p. 421
stress, p. 421
strain, p. 421
stretching/tension, p. 421
compression, p. 421
shear, p. 421
modulus of elasticity, p. 421
Young's modulus, p. 422
bulk modulus, p. 422
shear modulus, p. 422
pressure, p. 425
pascal, p. 425
absolute pressure, p. 426
gauge pressure, p. 426
barometer, p. 426
manometer, p. 427
barometric pressure formula, p. 427
Pascal's Principle, p. 428
buoyant force, p. 430
laminar flow, p. 434
turbulent flow, p. 434
incompressible flow, p. 434
nonviscous fluid, p. 434
irrotational flow, p. 434
equation of continuity, p. 435
Bernoulli's Equation, p. 435
speed of efflux, p. 440
viscosity, p. 442
Reynolds number, p. 444

## NEW SYMBOLS AND EQUATIONS

$N_A = 6.022 \cdot 10^{23}$ atoms, Avogadro's number

Stress = modulus · strain, stress-strain relationship

$p = p_0 + \rho g h$, absolute pressure

$F_B = m_f g$, buoyant force

$A_1 v_1 = A_2 v_2$, continuity equation

$p + \rho g y + \frac{1}{2}\rho v^2 =$ constant, Bernoulli's Equation

$\eta = \dfrac{F/A}{\Delta v/h} = \dfrac{Fh}{A\Delta v}$, viscosity

$R_V = \dfrac{\pi r^4 \Delta p}{8\eta \ell}$, volume flow rate

$\text{Re} = \dfrac{\rho \bar{v} L}{\eta}$, Reynolds number

## ANSWERS TO SELF-TEST OPPORTUNITIES

**13.1** The volume of the water in the bottle is 500 cm$^3$. The density of water is 1 g/cm$^3$. Thus, the bottle contains 500 g of water. The mass of 1 mol of water is 18.02 g, so $n = (500\text{ g})/(18.02\text{ g/mol}) = 27.7$ mol. Thus, $N = n \cdot N_A = 27.7(6.02 \cdot 10^{23}\text{ mol}^{-1}) = 1.67 \cdot 10^{25}$ molecules.

**13.2** $\rho_H = \dfrac{2\text{ g}}{22.4\text{ L}} = 0.089\text{ g/L} = 0.089\text{ kg/m}^3$

$\rho_{He} = \dfrac{4\text{ g}}{22.4\text{ L}} = 0.18\text{ g/L} = 0.18\text{ kg/m}^3$

**13.3** $W = F_B - W_{\text{balloon}} = \rho_{20} V g - \rho_{He} V g$

$= Vg(\rho_{20} - \rho_{100}) = (2200\text{ m}^3)(9.81\text{ m/s}^2)(1.205\text{ kg/m}^3 - 0.164\text{ kg}$

$= 22{,}500$ N, or 5050 lb$_f$.

## PROBLEM-SOLVING PRACTICE

### Problem-Solving Guidelines: Solids and Fluids

**1.** The three types of stress all have the same relationship with strain: The ratio of stress to strain equals a constant for the material, which can be Young's modulus, the bulk modulus, or the shear modulus. Be sure you understand what kind of stress is acting in a particular situation.

**2.** Remember that the buoyant force is exerted by a fluid on an object floating or submerged in it. The net force depends on the density of the object as well as the density of the fluid; you will often need to calculate mass and volume separately and take their ratio to obtain a value for the density.

**3.** Bernoulli's Equation is derived from the conservation of energy, and the main problem-solving guidelines for energy problems apply to flow problems as well. In particular, be sure to clearly identify where point 1 and point 2 are located for applying Bernoulli's Equation and to list known values for pressure, height, and velocity of the fluid at each point.

### SOLVED PROBLEM 13.3 Finding the Density of an Unknown Liquid

**PROBLEM**

A solid aluminum sphere (density = 2700. kg/m$^3$) is suspended in air from a scale (Figure 13.48a). The scale reads 13.06 N. The sphere is then submerged in a liquid with unknown density (Figure 13.48b). The scale then reads 8.706 N. What is the density of the liquid?

## SOLUTION

### THINK

From Newton's Third Law, the buoyant force exerted by the unknown liquid on the aluminum sphere is the difference between the scale reading when the sphere is suspended in air and the scale reading when the sphere is submerged in the liquid.

### SKETCH

Free-body diagrams of the sphere in each situation are shown in Figure 13.49.

### RESEARCH

The buoyant force, $F_B$, exerted by the unknown liquid is

$$F_B = m_l g, \tag{i}$$

where the mass of the displaced liquid, $m_l$, can be expressed in terms of the volume of the sphere, $V$, and the density of the liquid, $\rho_l$:

$$m_l = \rho_l V. \tag{ii}$$

We can obtain the buoyant force exerted by the unknown liquid by subtracting the measured weight of the sphere when it is submerged in the unknown liquid, $F_{sub}$, from the measured weight when it is suspended in air, $F_{air}$:

$$F_B = F_{air} - F_{sub}. \tag{iii}$$

The mass of the sphere can be obtained from its weight in air:

$$F_{air} = m_s g. \tag{iv}$$

Using the known density of aluminum, $\rho_{Al}$, we can express the volume of the aluminum sphere as

$$V = \frac{m_s}{\rho_{Al}}. \tag{v}$$

### SIMPLIFY

We can combine equations (i), (ii), and (iii) to get

$$F_B = m_l g = \rho_l V g = F_{air} - F_{sub}. \tag{vi}$$

We can use equations (iv) and (v) to get an expression for the volume of the sphere in terms of known quantities:

$$V = \frac{m_s}{\rho_{Al}} = \frac{\left(\dfrac{F_{air}}{g}\right)}{\rho_{Al}} = \frac{F_{air}}{\rho_{Al} g}.$$

Substituting this expression for $V$ into equation (vi) gives us

$$\rho_l V g = \rho_l \left(\frac{F_{air}}{\rho_{Al} g}\right) g = F_{air} \frac{\rho_l}{\rho_{Al}} = F_{air} - F_{sub}.$$

Now, we solve for the density of the unknown liquid:

$$\rho_l = \rho_{Al} \frac{(F_{air} - F_{sub})}{F_{air}}.$$

### CALCULATE

Putting in the numerical values gives us the density of the unknown liquid:

$$\rho_l = \rho_{Al} \frac{(F_{air} - F_{sub})}{F_{air}} = \left(2700.\ \text{kg/m}^3\right) \frac{13.06\ \text{N} - 8.706\ \text{N}}{13.06\ \text{N}} = 900.138\ \text{kg/m}^3.$$

### ROUND

We report our result to four significant figures:

$$\rho_l = 900.1\ \text{kg/m}^3.$$

*Continued—*

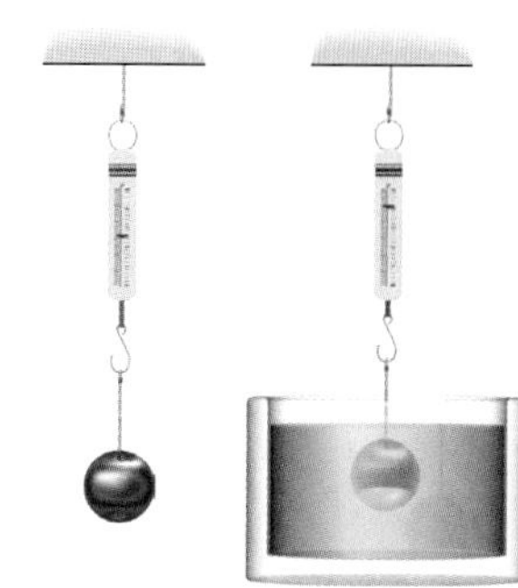

**FIGURE 13.48** (a) An aluminum sphere suspended in air. (b) The aluminum sphere submerged in an unknown liquid.

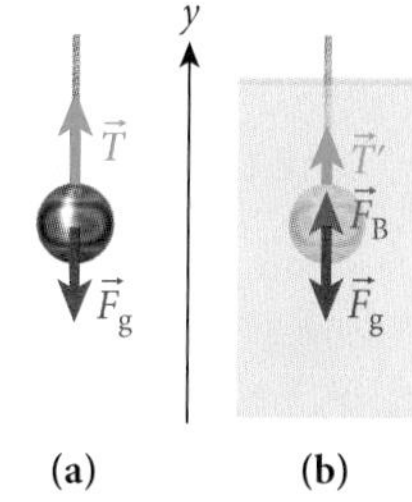


**FIGURE 13.49** Free-body diagrams for the sphere (a) suspended in air and (b) submerged in the liquid.

### DOUBLE-CHECK

The calculated density of the unknown liquid is 90% of the density of water. Many liquids have such densities, so our answer seems reasonable.

## SOLVED PROBLEM 13.4 Diving Bell

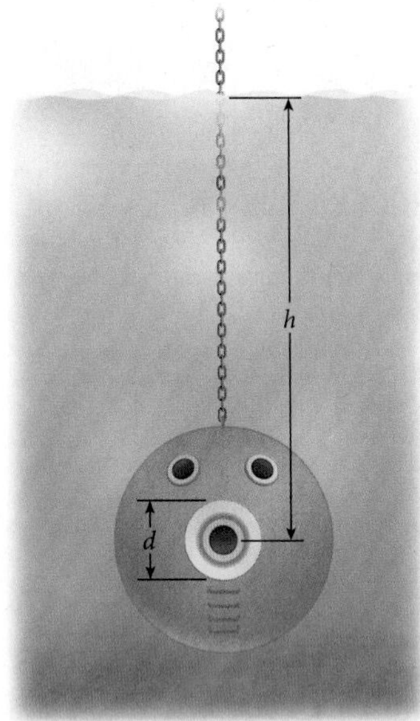


**FIGURE 13.50** Diving bell submerged in Lake Michigan.

### PROBLEM

A diving bell with interior air pressure equal to atmospheric pressure is submerged in Lake Michigan at a depth of 185 m. The diving bell has a flat, transparent, circular viewing port with a diameter of 20.0 cm. What is the magnitude of the net force on the viewing port?

### SOLUTION

### THINK

The pressure on the viewing port depends on the depth of the diving bell. The pressure on the port is the total force divided by the area of the port.

### SKETCH

A sketch of the diving bell submerged in Lake Michigan is presented in Figure 13.50.

### RESEARCH

The pressure $p$ at a depth $h$ is given by

$$p = p_0 + \rho g h,$$

where $\rho$ is the density of water and $p_0$ is the atmospheric pressure at the surface of Lake Michigan. The pressure inside the diving bell is atmospheric pressure. Therefore, the pressure difference between the outside and the inside of the viewing port is

$$p = \rho g h. \tag{i}$$

We can obtain the net force, $F$, on the viewing port using the definition of pressure:

$$p = \frac{F}{A}, \tag{ii}$$

where $A$ is the area of the viewing port.

### SIMPLIFY

We can combine equations (i) and (ii) to get

$$p = \rho g h = \frac{F}{A},$$

which we can solve for the net force on the viewing port:

$$F = \rho g h A.$$

### CALCULATE

Putting in the numerical values gives us the net force on the viewing port:

$$F = \rho g h A = \left(1000 \text{ kg/m}^3\right)\left(9.81 \text{ m/s}^2\right)\left(185 \text{ m}\right)\left[\pi\left(\frac{0.200}{2}\right)^2 \text{ m}^2\right] = 57015.2 \text{ N}.$$

### ROUND

We report our result to three significant figures:

$$F = 5.70 \cdot 10^4 \text{ N}.$$

### DOUBLE-CHECK

The net force of $5.70 \cdot 10^4$ N corresponds to 12,800 lb. This viewing port would have to be constructed of very thick glass or quartz. In addition, the size of such a viewing port is limited, which is why the problem specified a relatively small diameter of 20 cm (approximately 8 in).

## MULTIPLE-CHOICE QUESTIONS

**13.1** Salt water has a greater density than freshwater. A boat floats in both freshwater and salt water. The buoyant force on the boat in salt water is ________ that in freshwater.

a) equal to  b) smaller than  c) larger than

**13.2** You fill a tall glass with ice and then add water to the level of the glass's rim, so some fraction of the ice floats above the rim. When the ice melts, what happens to the water level? (Neglect evaporation, and assume that the ice and water remain at 0 °C during the melting process.)

a) The water overflows the rim.
b) The water level drops below the rim.
c) The water level stays at the top of the rim.
d) It depends on the difference in density between water and ice.

**13.3** The figure shows four identical open-top tanks filled to the brim with water and sitting on a scale. Balls float in tanks (2) and (3), but an object sinks to the bottom in tank (4). Which of the following correctly ranks the weights shown on the scales?

(1)

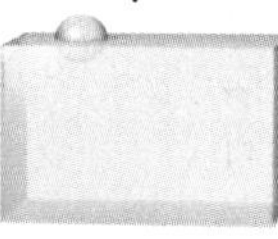
(2)

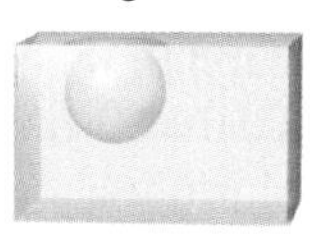
(3)

(4)

a) (1) < (2) < (3) < (4)
b) (1) < (2) = (3) < (4)
c) (1) < (2) = (3) = (4)
d) (1) = (2) = (3) < (4)

**13.4** You are in a boat filled with large rocks in the middle of a small pond. You begin to drop the rocks into the water. What happens to the water level of the pond?

a) It rises.
b) It falls.
c) It doesn't change.
d) It rises momentarily and then falls when the rocks hit bottom.
e) There is not enough information to say.

**13.5** Rank in order, from largest to smallest, the magnitudes of the forces $F_1$, $F_2$, and $F_3$ required for balancing the masses shown in the figure.

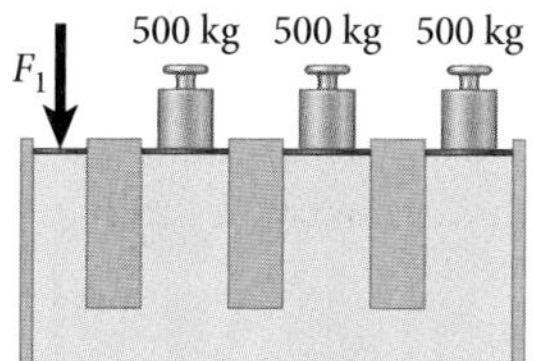


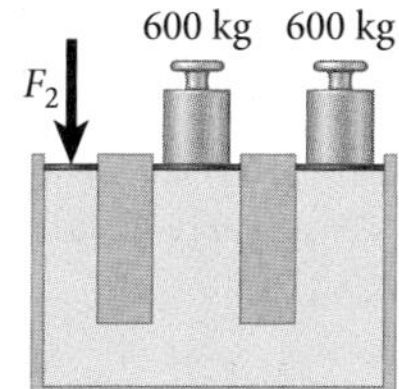


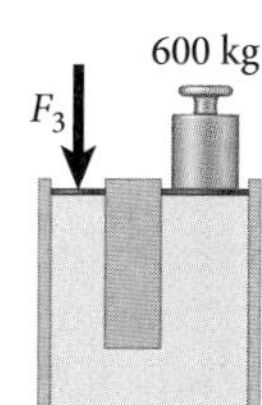


**13.6** In a horizontal water pipe that narrows to a smaller radius, the velocity of the water in the section with the smaller radius will be larger. What happens to the pressure?

a) The pressure will be the same in both the wider and narrower sections of the pipe.
b) The pressure will be higher in the narrower section of the pipe.
c) The pressure will be higher in the wider section of the pipe.
d) It is impossible to tell.

**13.7** In one of the *Star Wars* ™© movies, four of the heroes are trapped in a trash compactor on the Death Star. The compactor's walls begin to close in, and the heroes need to pick an object from the trash to place between the closing walls to stop them. All the objects are the same length and have a circular cross section, but their diameters and compositions are different. Assume that each object is oriented horizontally and does not bend. They have the time and strength to hold up only one object between the walls. Which of the objects shown in the figure will work best—that is, will withstand the greatest force per unit of compression?

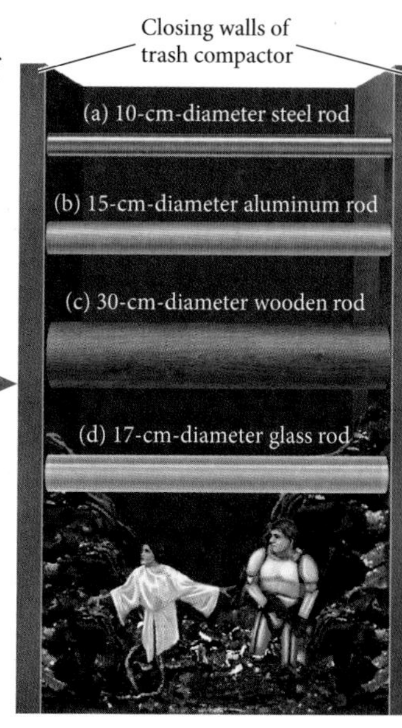


**13.8** Many altimeters determine altitude changes by measuring changes in the air pressure. An altimeter that is designed to be able to detect altitude changes of 100 m near sea level should be able to detect pressure changes of

a) approximately 1 Pa.
b) approximately 10 Pa.
c) approximately 100 Pa.
d) approximately 1 kPa.
e) approximately 10 kPa.

**13.9** Which of the following assumptions is *not* made in the derivation of Bernoulli's Equation?

a) Streamlines do not cross.
b) There is negligible viscosity.
c) There is negligible friction.
d) There is no turbulence.
e) There is negligible gravity.

**13.10** A beaker is filled with water to the rim. Gently placing a plastic toy duck in the beaker causes some of the water to spill out. The weight of the beaker with the duck floating in it is

a) greater than the weight before adding the duck.
b) less than the weight before adding the duck.
c) the same as the weight before adding the duck.
d) greater or less than the weight before the duck was added, depending on the weight of the duck.

**13.11** A piece of cork (density = 0.33 g/cm$^3$) with a mass of 10 g is held in place under water by a string, as shown in the figure. What is the tension, $T$, in the string?

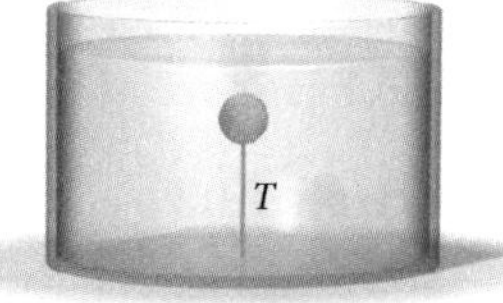


a) 0.10 N
b) 0.20 N
c) 0.30 N
d) 100 N
e) 200 N
f) 300 N

## QUESTIONS

**13.12** You know from experience that if a car you are riding in suddenly stops, heavy objects in the rear of the car move toward the front. Why does a helium-filled balloon in such a situation move, instead, toward the rear of the car?

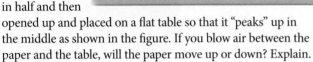

**13.13** A piece of paper is folded in half and then opened up and placed on a flat table so that it "peaks" up in the middle as shown in the figure. If you blow air between the paper and the table, will the paper move up or down? Explain.

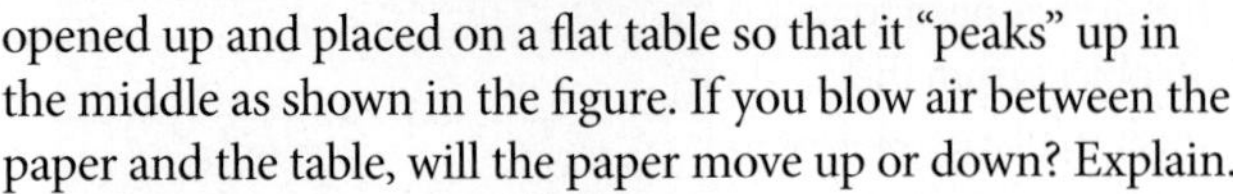

**13.14** In what direction does a force due to water flowing from a showerhead act on a shower curtain, inward toward the shower or outward? Explain.

**13.15** Point out and discuss any flaws in the following statement: *The hydraulic car lift is a device that operates on the basis of Pascal's Principle. Such a device can produce large output forces with small input forces. Thus, with a small amount of work done by the input force, a much larger amount of work is produced by the output force, and the heavy weight of a car can be lifted.*

**13.16** Given two springs of identical size and shape, one made of steel and the other made of aluminum, which has the higher spring constant? Why? Does the difference depend more on the shear modulus or the bulk modulus of the material?

**13.17** One material has a higher density than another. Are the individual atoms or molecules of the first material necessarily more massive than those of the second?

**13.18** Analytic balances are calibrated to give correct mass values for such items as steel objects of density $\rho_s = 8000.00\ \text{kg/m}^3$. The calibration compensates for the buoyant force arising because the measurements are made in air, of density $\rho_a = 1.205\ \text{kg/m}^3$. What compensation must be made to measure the masses of objects of a different material, of density $\rho$? Does the buoyant force of air matter?

**13.19** If you turn on the faucet in the bathroom sink, you will observe that the stream seems to narrow from the point at which it leaves the spigot to the point at which it hits the bottom of the sink. Why does this occur?

**13.20** In many problems involving application of Newton's Second Law to the motion of solid objects, friction is neglected for the sake of making the solution easier. The counterpart of friction between solids is viscosity of liquids. Do problems involving fluid flow become simpler if viscosity is neglected? Explain.

**13.21** You have two identical silver spheres and two unknown fluids, A and B. You place one sphere in fluid A, and it sinks; you place the other sphere in fluid B, and it floats. What can you conclude about the buoyant force of fluid A versus that of fluid B?

**13.22** Water flows from a circular faucet opening of radius $r_0$, directed vertically downward, at speed $v_0$. As the stream of water falls, it narrows. Find an expression for the radius of the stream as a function of distance fallen, $r(y)$, where $y$ is measured downward from the opening. Neglect the eventual breakup of the stream into droplets, and any resistance due to drag or viscosity.

## PROBLEMS

A blue problem number indicates a worked-out solution is available in the Student Solutions Manual. One • and two •• indicate increasing level of problem difficulty.

### Sections 13.1 and 13.2

13.23 Air consists of molecules of several types, with an average molar mass of 28.95 g. An adult who inhales 0.50 L of air at sea level takes in about how many molecules?

•**13.24** Ordinary table salt (NaCl) consists of sodium and chloride ions arranged in a *face-centered cubic crystal lattice*. That is, a sodium chloride crystal consists of cubic unit cells with a sodium ion on each corner and at the center of each face, and a chloride ion at the center of the cube and at the midpoint of each edge. The density of sodium chloride is $2.165 \cdot 10^3\ \text{kg/m}^3$. Calculate the spacing between adjacent sodium and chloride ions in the crystal.

### Section 13.3

**13.25** A 20-kg chandelier is suspended from the ceiling by four vertical steel wires. Each wire has an unloaded length of 1 m and a diameter of 2 mm, and each bears an equal load. When the chandelier is hung, how far do the wires stretch?

**13.26** Find the minimum diameter of a 50-m-long nylon string that will stretch no more than 1 cm when a load of 70 kg is suspended from its lower end. Assume that $Y_{\text{nylon}} = 3.51 \cdot 10^8\ \text{N/m}^2$.

13.27 A 2.0-m-long steel wire in a musical instrument has a radius of 0.03 mm. When the wire is under a tension of 90 N, how much does its length change?

•**13.28** A rod of length $L$ is attached to a wall. The load on the rod increases linearly (as shown by the arrows in the figure) from zero at the left end to $W$ newtons per unit length at the right end. Find the shear force at

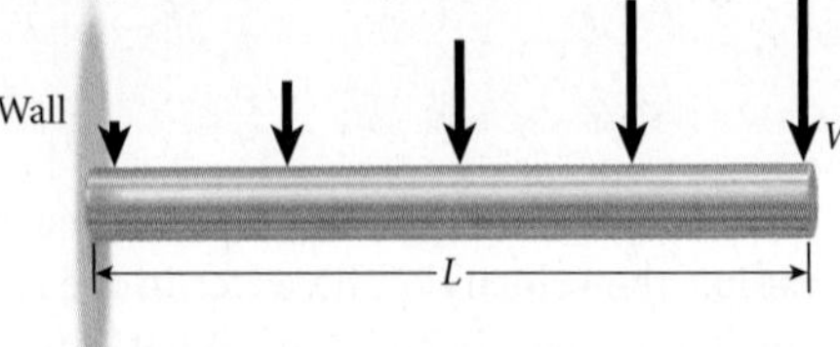


a) the right end, b) the center, and c) the left end.

•**13.29** Challenger Deep in the Marianas Trench of the Pacific Ocean is the deepest known spot in the Earth's oceans, at 10.922 km below sea level. Taking density of seawater at atmospheric pressure ($p_0 = 101.3$ kPa) to be $1024\ \text{kg/m}^3$ and its bulk modulus to be $B(p) = B_0 + 6.67(p - p_0)$, with $B_0 = 2.19 \cdot 10^9$ Pa, calculate the pressure and the density of the seawater at the bottom of Challenger Deep. Disregard variations in water temperature and salinity with depth. Is it a good approximation to treat the density of seawater as essentially constant?

## Section 13.4

**13.30** How much force is the air exerting on the front cover of your textbook? What mass has a weight equivalent to this force?

13.31 Blood pressure is usually reported in millimeters of mercury (mmHg) or the height of a column of mercury producing the same pressure value. Typical values for an adult human are 130/80; the first value is the *systolic* pressure, during the contraction of the ventricles of the heart, and the second is the *diastolic* pressure, during the contraction of the auricles of the heart. The head of an adult male giraffe is 6.0 m above the ground; the giraffe's heart is 2.0 m above the ground. What is the minimum systolic pressure (in mmHg) required at the heart to drive blood to the head (neglect the additional pressure required to overcome the effects of viscosity)? The density of giraffe blood is 1.00 g/cm$^3$, and that of mercury is 13.6 g/cm$^3$.

**13.32** A scuba diver must decompress after a deep dive to allow excess nitrogen to exit safely from his bloodstream. The length of time required for decompression depends on the total change in pressure that the diver experienced. Find this total change in pressure for a diver who starts at a depth of $d = 20.0$ m in the ocean (density of seawater = 1024 kg/m$^3$) and then travels aboard a small plane (with an unpressurized cabin) that rises to an altitude of $h = 5000$ m above sea level.

•**13.33** A child loses his balloon, which rises slowly into the sky. If the balloon is 20 cm in diameter when the child loses it, what is its diameter at an altitude of (a) 1000 m, (b) 2000 m, and (c) 5000 m? Assume that the balloon is very flexible and so surface tension can be neglected.

•**13.34** The atmosphere of Mars exerts a pressure of only 600 Pa on the surface and has a density of only 0.02 kg/m$^3$.

a) What is the thickness of the Martian atmosphere, assuming the boundary between atmosphere and outer space to be the point where atmospheric pressure drops to 0.01% of its value at surface level?

b) What is the atmospheric pressure at the bottom of Mars's Hellas Planitia canyon, at a depth of 7 km?

c) What is the atmospheric pressure at the top of Mars's Olympus Mons volcano, at a height of 27 km?

d) Compare the relative change in air pressure, $\Delta p/p$, between these two points on Mars and between the equivalent extremes on Earth—the Dead Sea shore, at 400 m below sea level, and Mount Everest, at an altitude of 8850 m.

•13.35 The air density at the top of Mount Everest ($h_{\text{Everest}}$ = 8850 m) is 35.00% of the air density at sea level. If the air *pressure* at the top of Mount McKinley in Alaska is 47.74% of the air pressure at sea level, calculate the height of Mount McKinley, using only the information given here.

•**13.36** A sealed vertical cylinder of radius $R$ and height $h = 0.60$ m is initially filled halfway with water, and the upper half is filled with air. The air is initially at standard atmospheric pressure, $p_0 = 1.01 \cdot 10^5$ Pa. A small valve at the bottom of the cylinder is opened, and water flows out of the cylinder until the reduced pressure of the air in the upper part of the cylinder prevents any further water from escaping. By what distance is the depth of the water lowered? (Assume that the temperature of water and air do not change and that no air leaks into the cylinder.)

••**13.37** A square pool with 100-m-long sides is created in a concrete parking lot. The walls are concrete 50 cm thick and have a density of 2.5 g/cm$^3$. The coefficient of static friction between the walls and the parking lot is 0.45. What is the maximum possible depth of the pool?

••**13.38** The calculation of atmospheric pressure at the summit of Mount Everest carried out in Example 13.3 used the model known as the *isothermal atmosphere*, in which gas pressure is proportional to density: $p = \gamma\rho$, with $\gamma$ constant. Consider a spherical cloud of gas supporting itself under *its own* gravitation and following this model.

a) Write the equation of hydrostatic equilibrium for the cloud, in terms of the gas density as a function of radius, $\rho(r)$.

b) Show that $\rho(r) = A/r^2$ is a solution of this equation, for an appropriate choice of constant $A$. Explain why this solution is not suitable as a model of a star.

## Section 13.5

13.39 A racquetball with a diameter of 5.6 cm and a mass of 42 g is cut in half to make a boat for American pennies made after 1982. The mass and volume of an American penny made after 1982 are 2.5 g and 0.36 cm$^3$. How many pennies can be placed in the racquetball boat without sinking it?

**13.40** A supertanker filled with oil has a total mass of $10.2 \cdot 10^8$ kg. If the dimensions of the ship are those of a rectangular box 250 m long, 80 m wide, and 80 m high, determine how far the bottom of the ship is below sea level ($\rho_{\text{sea}} = 1020$ kg/m$^3$).

**13.41** A box with a volume $V = 0.05$ m$^3$ lies at the bottom of a lake whose water has a density of $10^3$ kg/m$^3$. How much force is required to lift the box, if the mass of the box is (a) 1000 kg, (b) 100 kg, and (c) 55 kg?

•**13.42** A man of mass 64 kg and density 970 kg/m$^3$ stands in a shallow pool with 32% of the volume of his body below water. Calculate the normal force the bottom of the pool exerts on his feet.

•13.43 A block of cherry wood that is 20 cm long, 10 cm wide, and 2 cm thick has a density of 800 kg/m$^3$. What is the volume of a piece of iron that, if glued to the bottom of the block makes the block float in water with its top just at the surface of the water? The density of iron is 7860 kg/m$^3$, and the density of water is 1000 kg/m$^3$.

•**13.44** The average density of the human body is 985 kg/m$^3$, and the typical density of seawater is about 1020 kg/m$^3$.

a) Draw a free-body diagram of a human body floating in seawater and determine what percentage of the body's volume is submerged.

b) The average density of the human body, after maximum inhalation of air, changes to 945 kg/m$^3$. As a person floating in seawater inhales and exhales slowly, what percentage of his volume moves up out of and down into the water?

c) The Dead Sea (a saltwater lake between Israel and Jordan) is the world's saltiest large body of water. Its average salt content is more than six times that of typical seawater, which explains why there is no plant and animal life in it. Two-thirds of the volume of the body of a person floating in the Dead Sea is observed to be submerged. Determine the density (in kg/m$^3$) of the seawater in the Dead Sea.

•**13.45** A tourist of mass 60 kg notices a chest with a short chain attached to it at the bottom of the ocean. Imagining the riches it could contain, he decides to dive for the chest. He inhales fully, thus setting his average body density to 945 kg/m$^3$, jumps into the ocean (with saltwater density = 1020 kg/m$^3$), grabs the chain, and tries to pull the chest to the surface. Unfortunately, the chest is too heavy and will not move. Assume that the man does not touch the bottom.

a) Draw the man's free-body diagram, and determine the tension on the chain.

b) What mass (in kg) has a weight that is equivalent to the tension force in part (a)?

c) After realizing he cannot free the chest, the tourist releases the chain. What is his upward acceleration (assuming that he simply allows the buoyant force to lift him up to the surface)?

•**13.46** A very large balloon with mass $M = 10$ kg is inflated to a volume of 20 m$^3$ using a gas of density $\rho_{gas} = 0.2$ kg/m$^3$. What is the maximum mass $m$ that can be tied to the balloon using a 2 kg piece of rope without the balloon falling to the ground? (Assume that the density of air is 1.3 kg/m$^3$ and that the volume of the gas is equal to the volume of the inflated balloon).

•**13.47** The *Hindenburg*, the German zeppelin that caught fire in 1937 while docking in Lakehurst, New Jersey, was a rigid duralumin-frame balloon filled with 200,000 m$^3$ of hydrogen. The *Hindenburg's useful* lift (beyond the weight of the zeppelin structure itself) is reported to have been $1.099 \cdot 10^6$ N (or 247,000 lb). Use $\rho_{air} = 1.205$ kg/m$^3$, $\rho_H = 0.08988$ kg/m$^3$ and $\rho_{He} = 0.1786$ kg/m$^3$.

a) Calculate the weight of the zeppelin structure (without the hydrogen gas).

b) Compare the useful lift of the (highly flammable) hydrogen-filled *Hindenburg* with the useful lift the *Hindenburg* would have had had it been filled with (nonflammable) helium, as originally planned.

••**13.48** Brass weights are used to weigh an aluminum object on an analytical balance. The weighing is done one time in dry air and another time in humid air, with a water vapor pressure of $P_h = 2 \cdot 10^3$ Pa. The total atmospheric pressure ($P = 10^5$ Pa) and the temperature ($T = 20$ °C) are the same in both cases. What should the mass of the object be to be able to notice a difference in the balance readings, provided the balance's sensitivity is $m_0 = 0.10$ mg? (The density of aluminum is $\rho_A = 2700$ kg/m$^3$; the density of brass is $\rho_B = 8500$ kg/m$^3$.)

## Section 13.6

**13.49** A fountain sends water to a height of 100 m. What is the difference between the pressure of the water just before it is released upward and the atmospheric pressure?

**13.50** Water enters a horizontal pipe with a rectangular cross section at a speed of 1 m/s. The width of the pipe remains constant but the height decreases. Twenty meters from the entrance, the height is half of what it is at the entrance. If the water pressure at the entrance is 3000 Pa, what is the pressure 20 m downstream?

•**13.51** Water at room temperature flows with a constant speed of 8 m/s through a nozzle with a square cross section, as shown in the figure. Water enters the nozzle at point $A$ and exits the nozzle at point $B$. The lengths of the sides of the square cross section at $A$ and $B$ are 50 cm and 20 cm, respectively.

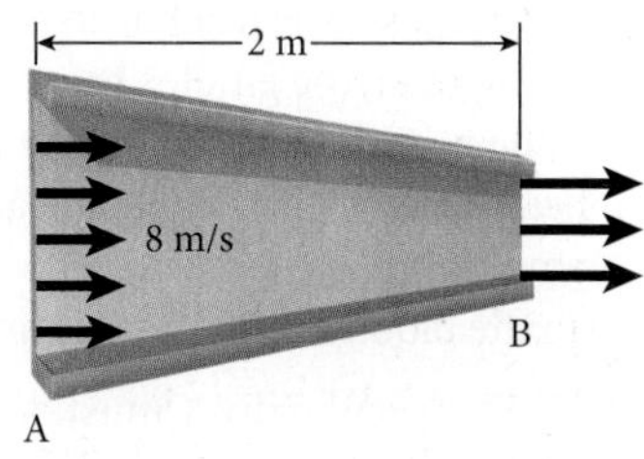


a) What is the volume flow rate at the exit?

b) What is the acceleration at the exit? The length of the nozzle is 2 m.

c) If the volume flow rate through the nozzle is increased to 6 m$^3$/s, what is the acceleration of the fluid at the exit?

•**13.52** Water is flowing in a pipe as depicted in the figure. What pressure is indicated on the upper pressure gauge?

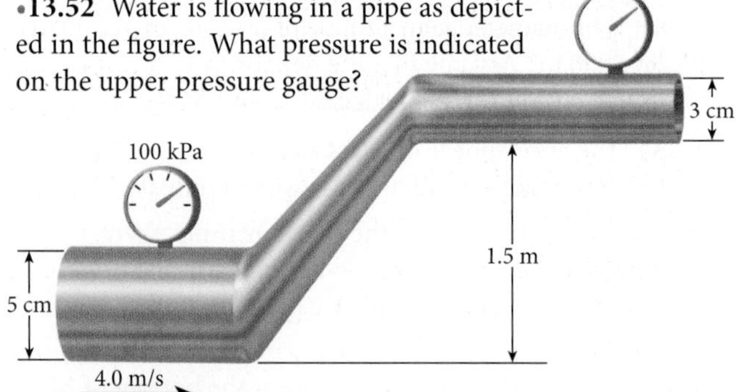


•**13.53** An open-topped tank completely filled with water has a release valve near its bottom. The valve is 1.0 m below the water surface. Water is released from the valve to power a turbine, which generates electricity. The area of the top of the tank, $A_T$, is 10 times the cross-sectional area, $A_V$, of the valve opening. Calculate the speed of the water as it exits the valve. Neglect friction and viscosity. In addition, calculate the speed of a drop of water released from rest at $h = 1.0$ m when it reaches the elevation of the valve. Compare the two speeds.

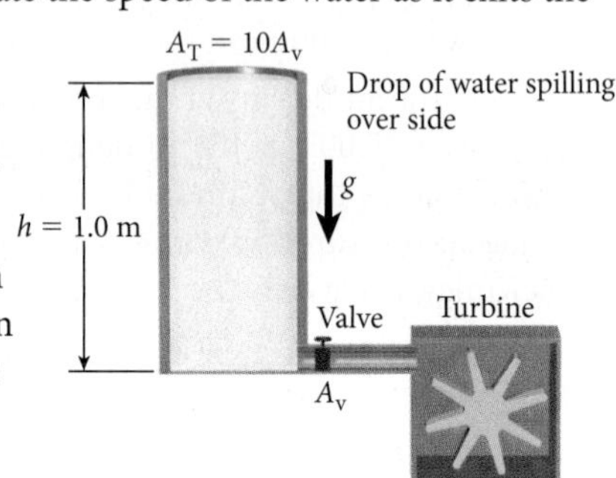

•**13.54** A tank of height $H$ is filled with water and sits on the ground, as shown in the figure. Water squirts from a hole at a height $y$ above the ground and has a range $R$. For two $y$ values, 0 and $H$, $R$ is zero. Determine the value of $y$ for which the range will be a maximum.

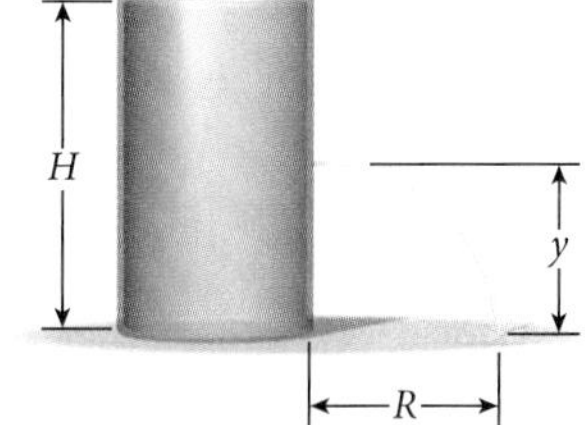


••13.55 A water-powered backup sump pump uses tap water at a pressure of 3 atm ($p_1 = 3p_{\text{atm}} = 3.03 \cdot 10^5$ Pa) to pump water out of a well, as shown in the figure ($p_{\text{well}} = p_{\text{atm}}$). This system allows water to be pumped out of a basement sump well when the electric pump stops working during an electrical power outage. Using water to pump water may sound strange at first, but these pumps are quite efficient, typically pumping out 2 L of well water for every 1 L of pressurized tap water. The supply water moves to the right in a large pipe with cross-sectional area $A_1$ at a speed $v_1 = 2.05$ m/s. The water then flows into a pipe of smaller diameter with a cross-sectional area that is ten times smaller ($A_2 = A_1/10$).

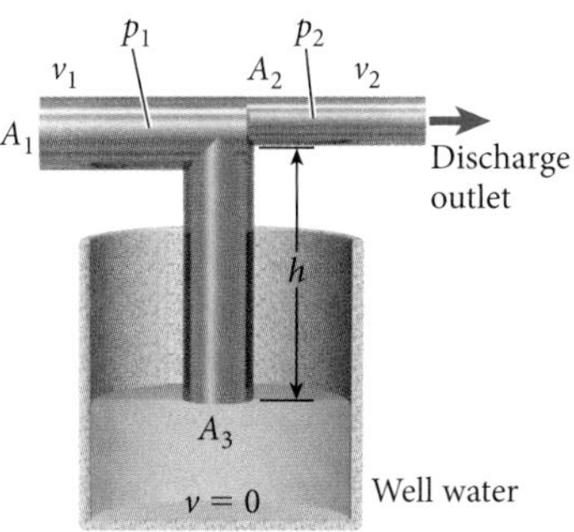


a) What is the speed $v_2$ of the water in the smaller pipe, with area $A_2$?

b) What is the pressure $p_2$ of the water in the smaller pipe, with area $A_2$?

c) The pump is designed so that the vertical pipe, with cross-sectional area $A_3$, that leads to the well water also has a pressure of $p_2$ at its top. What is the maximum height, $h$, of the column of water that the pump can support (and therefore act on) in the vertical pipe?

## Section 13.7

**13.56** A basketball of circumference 75.5 cm and mass 598 g is forced to the bottom of a swimming pool and then released. After initially accelerating upward, it rises at a constant velocity.

a) Calculate the buoyant force on the basketball.

b) Calculate the drag force the basketball experiences while it is moving upward at constant velocity.

•**13.57** The cylindrical container shown in the figure has a radius of 1 m and contains motor oil with a viscosity of 0.3 Pa s and a density of 670 kg m$^{-3}$. Oil flows out of the 20 cm long, 0.2 cm diameter tube at the bottom of the container. How much oil flows out of the tube in a period of 10 s if the container is originally filled to a height of 0.5 m?

## Additional Problems

**13.58** Estimate the atmospheric pressure inside Hurricane Rita. The wind speed was 290 km/h.

13.59 The following data are obtained for a car: The tire pressure is measured as 28.0 psi, and the width and length of contact surface of each tire are 7.50 in and 8.75 in, respectively. What is the approximate weight of the car?

**13.60** Calculate the ratio of the lifting powers of helium (He) gas and hydrogen ($H_2$) gas under identical circumstances. Assume that the molar mass of air is 29.5 g/mol.

**13.61** Water is poured into a large barrel whose height is 2.0 m and which has a cylindrical, 3.0-cm-diameter cork stuck into the side at a height of 0.50 m above the ground. As the water level in the barrel just reaches maximum height, the cork flies out of the barrel.

a) What was the magnitude of the static friction force between the barrel and the cork?

b) If the barrel were filled with seawater, would the cork have flown out before the barrel's full capacity was reached?

**13.62** In a hydraulic lift, the maximum gauge pressure is 17.00 atm. If the diameter of the output line is 22.5 cm, what is the heaviest vehicle that can be lifted?

13.63 A water pipe narrows from a radius of $r_1 = 5$ cm to a radius of $r_2 = 2$ cm. If the speed of the water in the wider part of the pipe is 2 m/s, what is the speed of the water in the narrower part?

**13.64** Donald Duck and his nephews manage to sink Uncle Scrooge's yacht ($m = 4500$ kg), which is made of steel ($\rho = 7800$ kg/m$^3$). In typical comic-book fashion, they decide to raise the yacht by filling it with ping-pong balls. A ping-pong ball has a mass of 2.7 g and a volume of $3.35 \cdot 10^{-5}$ m$^3$.

a) What is the buoyant force on one ping-pong ball in water?

b) How many balls are required to float the ship?

**13.65** A wooden block floating in seawater has two thirds of its volume submerged. When the block is placed in mineral oil, 80% of its volume is submerged. Find the density of the (a) wooden block, and (b) the mineral oil.

**13.66** An approximately round tendon that has an average diameter of 8.5 mm and is 15 cm long is found to stretch 3.7 mm when acted on by a force of 13.4 N. Calculate Young's modulus for the tendon.

•13.67 The Jovian moon Europa may have oceans (covered by ice, which can be ignored). What would the pressure be 1 km below the surface of a Europan ocean? The surface gravity of Europa is 13.5% that of the Earth's.

•**13.68** Two balls of the same volume are released inside a tank filled with water as shown in the figure. The densities of balls A and B are 0.90 g/cm$^3$ and 0.80 g/cm$^3$. Find the acceleration of (a) ball A, and of (b) ball B. (c) Which ball wins the race to the top?

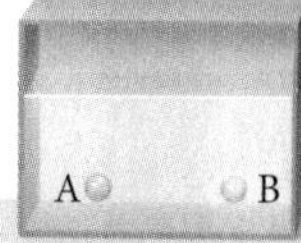

•**13.69** An airplane is moving through the air at a velocity $v = 200$ m/s. Streamlines just over the top of the wing are compressed to 80% of their original area, and those under the wing are not compressed at all.

a) Determine the velocity of the air just over the wing.

b) Find the difference in the pressure between the air just over the wing, $P$, and that under the wing, $P'$.

c) Find the net upward force on both wings due to the pressure difference, if the area of the wing is 40 m$^2$ and the density of the air is 1.3 kg/m$^3$.

•**13.70** A cylindrical buoy with hemispherical ends is dropped in seawater of density 1027 kg/m$^3$, as shown in the figure. The mass of the buoy is 75.0 kg, the radius of the cylinder and the hemispherical caps is $R = 0.200$ m, and the length of cylindrical center section of the buoy is $L = 0.600$ m. The lower part of the buoy is weighted so that the buoy is upright in the water as shown in the figure. In calm water, how high (distance $h$) will the top of the buoy be above the waterline?

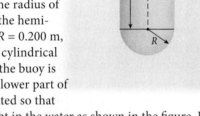

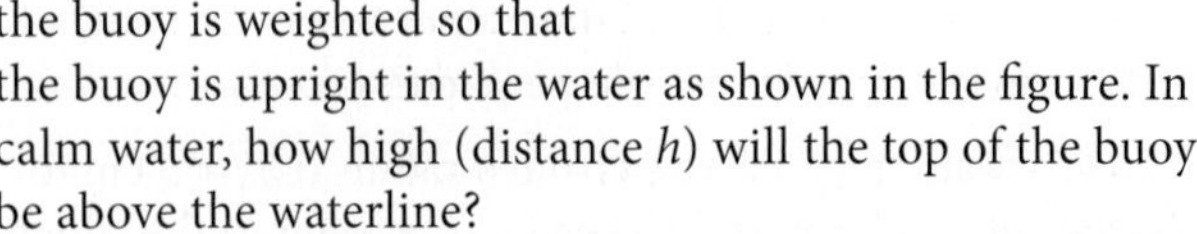


•**13.71** Water of density 998.2 kg/m$^3$ is moving at negligible speed under a pressure of 101.3 kPa but is then accelerated to high speed by the blades of a spinning propeller. The vapor pressure of the water at the initial temperature of 20.0 °C is 2.3388 kPa. At what flow speed will the water begin to boil? This effect, known as *cavitation*, limits the performance of propellers in water.

•**13.72** An astronaut wishes to measure the atmospheric pressure on Mars using a mercury barometer like that shown in the figure. The calibration of the barometer is the standard calibration for Earth: 760 mmHg corresponds to a pressure due to Earth's atmosphere, 1 atm or 101.325 kPa. How does the barometer need to be recalibrated for use in the atmosphere of Mars—by what factor does the barometer's scale need to be "stretched"? In her handy table of planetary masses and radii, the astronaut finds that Mars has an average radius of $3.37 \cdot 10^6$ m and a mass of $6.42 \cdot 10^{23}$ kg.

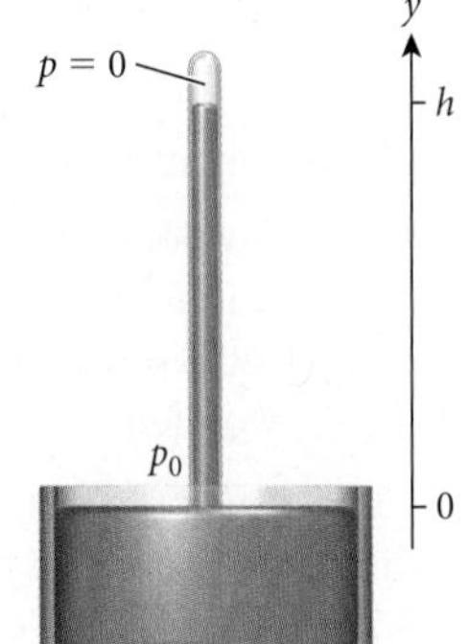


•**13.73** In many locations, such as Lake Washington in Seattle, floating bridges are preferable to conventional bridges. Such a bridge can be constructed out of concrete pontoons, which are essentially concrete boxes filled with air, Styrofoam, or another extremely low-density material. Suppose a floating bridge pontoon is constructed out of concrete and Styrofoam, which have densities of 2200 kg/m$^3$ and 50.0 kg/m$^3$. What must the volume ratio of concrete to Styrofoam be if the pontoon is to float with 35.0% of its overall volume above water?

•**13.74** A 1.0-g balloon is filled with helium gas. When a mass of 4.0 g is attached to the balloon, the combined mass hangs in static equilibrium in midair. Assuming that the balloon is spherical, what is its diameter?

•**13.75** A large water tank has an inlet pipe and an outlet pipe. The inlet pipe has a diameter of 2 cm and is 1 m above the bottom of the tank. The outlet pipe has a diameter of 5 cm and is 6 m above the bottom of the tank. A volume of 0.3 m$^3$ of water enters the tank every minute at a gauge pressure of 1 atm.

a) What is the velocity of the water in the outlet pipe?

b) What is the gauge pressure in the outlet pipe?

•**13.76** Two spheres of the same diameter (in air), one made of a lead alloy the other one made of steel are submerged at a depth $h = 2000$ m below the surface of the ocean. The ratio of the volumes of the two spheres at this depth is $V_{\text{Steel}}(h)/V_{\text{Lead}}(h) = 1.001206$. Knowing the density of ocean water $\rho = 1024$ kg/m$^3$ and the bulk modulus of steel, $B_{\text{Steel}} = 160 \cdot 10^9$ N/m$^2$, calculate the bulk modulus of the lead alloy used in the sphere.

PART 3 OSCILLATIONS AND WAVES

# 14 Oscillations

**FIGURE 14.1** Harmonic oscillations used for keeping time. (a) Multiple-exposure picture of an old-fashioned pendulum clock. (b) The world's smallest atomic clock from the National Institute of Standards and Technology, in which cesium atoms perform 9.2 billion oscillations each second. The clock is the size of a grain of rice and is accurate to 1 part in 10 billion—or off by less than 1 s over a period of 300 yr.

## WHAT WE WILL LEARN

- The spring force leads to a sinusoidal oscillation in time referred to as simple harmonic motion.
- A similar force law and time oscillation can be found for a pendulum swinging through small angles.
- Oscillations can be represented as a projection of circular motion onto one of the two Cartesian coordinate axes.
- In the presence of damping, oscillations slow down exponentially over time. Depending on the strength of the damping, it is possible that no oscillations will occur.
- Periodic external driving of an oscillator leads to sinusoidal motion at the driving frequency, with a maximum amplitude close to the resonant angular speed.
- Plotting the motion of oscillators in terms of the velocity and position shows that the motion of undamped oscillators follows an ellipse and that of damped oscillators spirals in.
- A damped and driven oscillator can exhibit chaotic motion in which the trajectory in time depends sensitively on the initial conditions.

Even when an object appears to be perfectly at rest, its atoms and molecules are rapidly vibrating. Sometimes these vibrations can be put to use; for example, the atoms in a quartz crystal vibrate with a very steady frequency if the crystal is subjected to a periodic electric field. This vibration is used to keep track of time in modern quartz-crystal clocks and wrist watches. Vibrations of cesium atoms are used in atomic clocks (Figure 14.1).

In this chapter, we examine the nature of oscillatory motion. Most of the situations we'll consider involve springs or pendulums, but these are just the simplest examples of oscillators. Later in the book, we will study other kinds of vibrating systems, which can be modeled as a spring or a pendulum for the purpose of analyzing the motion. In this chapter, we also investigate the concept of resonance, which is an important property of all oscillating systems, from the atomic level to bridges and skyscrapers. Chapters 15 and 16 will apply the concepts of oscillations to analyze the nature of waves and sound.

## 14.1 Simple Harmonic Motion

Repetitive motion, usually called **periodic motion,** is important in science, engineering, and daily life. Common examples of objects in periodic motion are windshield wipers on a car and the pendulum on a grandfather clock. However, periodic motion is also involved in the alternating current that powers the electronic grid of modern cities, atomic vibrations in molecules, and your own heartbeat and circulatory system.

Simple harmonic motion is a particular type of repetitive motion, which is displayed by a pendulum or by a massive object on a spring. Chapter 5 introduced the spring force, which is described by Hooke's Law: The spring force is proportional to the displacement of the spring from equilibrium. The spring force is a restoring force, always pointing toward the equilibrium position, and is thus opposite in direction to the displacement vector:

$$F_x = -kx.$$

The proportionality constant $k$ is called the *spring constant.*

We have encountered the spring force repeatedly in other chapters. The main reason why forces that depend linearly on displacement are so important in many branches of physics was emphasized in Section 6.4, where we saw that for a system at equilibrium, a small displacement from the equilibrium position results in a springlike force with linear dependence on the displacement from the equilibrium position.

Now let's consider the situation in which an object of mass $m$ is attached to a spring that is then stretched or compressed out of its equilibrium position. When the object is released, it oscillates back and forth. This motion is called **simple harmonic motion (SHM),** and it occurs whenever the restoring force is proportional to the displacement. (As we just noted, a linear restoring force is present in all systems close to a stable equilibrium point, so simple harmonic motion is seen in many physical systems.) Figure 14.2 shows frames of a video-

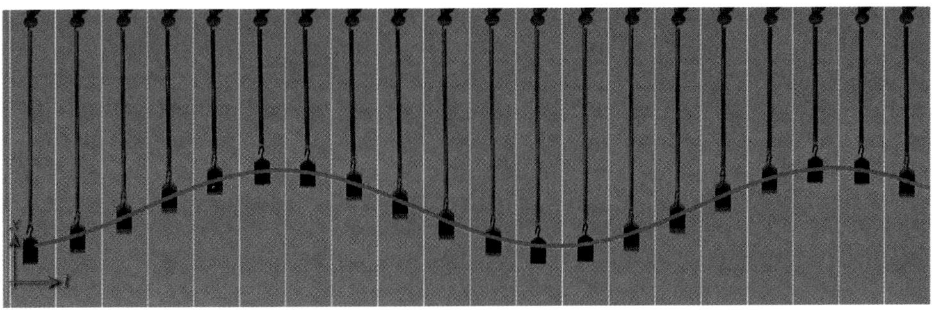

**FIGURE 14.2** Consecutive video images of a weight hanging from a spring undergoing simple harmonic motion. A coordinate system and graph of the position as a function of time are superimposed on the images.

tape of the vertical oscillation of a weight on a spring. A vertical $x$-axis is superimposed, and a horizontal axis that represents time, with each frame 0.06 s from the neighboring ones. The red curve running through this sequence is a sine function.

With the insight gained from Figure 14.2, we can describe this type of motion mathematically. We start with the force law for the spring force, $F_x = -kx$, and use Newton's Second Law, $F_x = ma$, to obtain

$$ma = -kx.$$

We know that acceleration is the second time derivative of the position: $a = d^2x/dt^2$. Substituting this expression for $a$ into the preceding equation obtained with Newton's Second Law gives us

$$m\frac{d^2x}{dt^2} = -kx,$$

or

$$\frac{d^2x}{dt^2} + \frac{k}{m}x = 0. \tag{14.1}$$

Equation 14.1 includes both the position, $x$, and its second derivative with respect to time. Both are functions of the time, $t$. This type of equation is called a *differential equation.* The solution to this particular differential equation is the mathematical description of simple harmonic motion.

From the curve in Figure 14.2, we can see that the solution to the differential equation should be a sine or a cosine function. Let's see if the following will work:

$$x = A\sin(\omega_0 t).$$

The constants $A$ and $\omega_0$ are called the **amplitude** of the oscillation and its **angular speed,** respectively. The amplitude is the maximum displacement away from the equilibrium position, as you learned in Chapter 5. This sinusoidal function works for any value of the amplitude, $A$. However, the amplitude cannot be arbitrarily large or the spring will be overstretched. On the other hand, we'll see that *not* all values of $\omega_0$ produce a solution.

Taking the second derivative of the trial sine function results in

$$x = A\sin(\omega_0 t) \Rightarrow$$
$$\frac{dx}{dt} = \omega_0 A\cos(\omega_0 t) \Rightarrow$$
$$\frac{d^2x}{dt^2} = -\omega_0^2 A\sin(\omega_0 t).$$

Inserting this result and the sinusoidal expression for $x$ into equation 14.1 yields

$$\frac{d^2x}{dt^2} + \frac{k}{m}x = -\omega_0^2 A\sin(\omega_0 t) + \frac{k}{m}A\sin(\omega_0 t) = 0.$$

This condition is fulfilled if $\omega_0^2 = k/m$, or

$$\omega_0 = \sqrt{\frac{k}{m}}.$$

### 14.1 Self-Test Opportunity

Show that $x(t) = A\sin(\omega_0 t + \theta_0)$ is a solution to equation 14.1.

### 14.2 Self-Test Opportunity

Show that $A\sin(\omega_0 t + \theta_0) = B\sin(\omega_0 t) + C\cos(\omega_0 t)$, where the relationships between the constants are given in equations 14.4 and 14.5.

We have thus found a valid solution to the differential equation (equation 14.1). In the same way, we can show that the cosine function leads to a solution as well, also with arbitrary amplitude and with the same angular speed. Thus, the complete solution for constants $B$ and $C$ is

$$x(t) = B\sin(\omega_0 t) + C\cos(\omega_0 t), \quad \text{with } \omega_0 = \sqrt{\frac{k}{m}}. \tag{14.2}$$

The units of $\omega_0$ are radians per second (rad/s). When applying equation 14.2, $\omega_0 t$ must be expressed in radians, not degrees.

Here is another useful form of equation 14.2:

$$x(t) = A\sin(\omega_0 t + \theta_0), \quad \text{with } \omega_0 = \sqrt{\frac{k}{m}}. \tag{14.3}$$

This form allows you to see more readily that the motion is sinusoidal. Instead of having two amplitudes for the sine and cosine functions, equation 14.3 has one amplitude, $A$, and a *phase angle*, $\theta_0$. These two constants are related to the constants $B$ and $C$ of equation 14.2 via

$$A = \sqrt{B^2 + C^2} \tag{14.4}$$

and

$$\theta_0 = \tan^{-1}\left(\frac{C}{B}\right). \tag{14.5}$$

## Initial Conditions

How do we determine the values to use for the constants $B$ and $C$, the amplitudes of the sine and cosine functions in equation 14.2? The answer is that we need two pieces of information, usually given in the form of the initial position, $x_0 = x(t = 0)$, and initial velocity, $v_0 = v(t = 0) = (dx/dt)|_{t=0}$, as in the following example.

### EXAMPLE 14.1 Initial Conditions

**PROBLEM 1**

A spring with spring constant $k = 56.0$ N/m has a lead weight of mass 1.00 kg attached to its end (Figure 14.3). The weight is pulled +5.5 cm from its equilibrium position and then pushed so that it receives an initial velocity of −0.32 m/s. What is the equation of motion for the resulting oscillation?

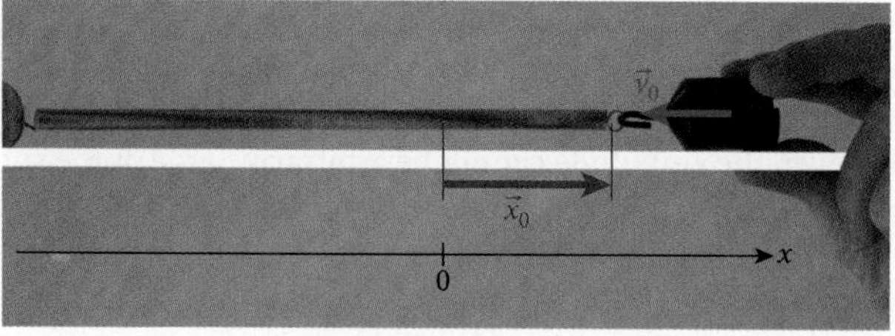


**FIGURE 14.3** A weight attached to a spring, with the initial position and velocity vectors shown.

**SOLUTION 1**

The general equation of motion for this situation is equation 14.2 for simple harmonic motion:

$$x(t) = B\sin(\omega_0 t) + C\cos(\omega_0 t), \quad \text{with } \omega_0 = \sqrt{\frac{k}{m}}.$$

From the data given in the problem statement, we can calculate the angular speed:

$$\omega_0 = \sqrt{\frac{k}{m}} = \sqrt{\frac{56.0 \text{ N/m}}{1.00 \text{ kg}}} = 7.48 \text{ s}^{-1}.$$

Now we must determine the values of the constants $B$ and $C$.

We take the first derivative of the general equation of motion:

$$x(t) = B\sin(\omega_0 t) + C\cos(\omega_0 t)$$
$$\Rightarrow v(t) = \omega_0 B\cos(\omega_0 t) - \omega_0 C\sin(\omega_0 t).$$

At time $t = 0$, $\sin(0) = 0$ and $\cos(0) = 1$, so these equations reduce to

$$x_0 = x(t=0) = C$$
$$v_0 = v(t=0) = \omega_0 B.$$

We were given the initial conditions for the position, $x_0 = 0.055$ m, and velocity, $v_0 = -0.32$ m/s. Thus, we find that $C = x_0 = 0.055$ m and $B = -0.043$ m.

**PROBLEM 2**

What is the amplitude of this oscillation? What is the phase angle?

**SOLUTION 2**

With the values of the constants $B$ and $C$, we can calculate the amplitude, $A$, from equation 14.4:

$$A = \sqrt{B^2 + C^2} = \sqrt{(0.043\ \text{m})^2 + (0.055\ \text{m})^2} = 0.070\ \text{m}.$$

Therefore, the amplitude of this oscillation is 7.0 cm. Note that, as a consequence of the nonzero initial velocity, the amplitude is *not* 5.5 cm, the value of the initial elongation of the spring.

The phase angle is found by straightforward application of equation 14.5:

$$\theta_0 = \tan^{-1}\left(\frac{C}{B}\right) = \tan^{-1}\left(-\frac{0.055}{0.043}\right) = -0.907\ \text{rad}.$$

Expressed in degrees, this phase angle is $\theta_0 = -52.0°$.

## Position, Velocity, and Acceleration

Let's look again at the relationships of position, velocity, and acceleration. In a form that describes oscillatory motion in terms of amplitude, $A$, and a phase angle determined by $\theta_0$, they are

$$\begin{aligned} x(t) &= A\sin(\omega_0 t + \theta_0) \\ v(t) &= \omega_0 A\cos(\omega_0 t + \theta_0) \\ a(t) &= -\omega_0^2 A\sin(\omega_0 t + \theta_0). \end{aligned} \tag{14.6}$$

Here the velocity and acceleration are obtained from the position vector by taking successive time derivatives. These equations suggest that the velocity and acceleration vectors have the same phase angle as the position vector, determined by $\theta_0$, but they have an additional phase angle of $\pi/2$ (for the velocity) and $\pi$ (for the acceleration), which correspond to the phase difference between the sine and cosine functions and between the sine and the negative sine functions, respectively.

The position, velocity, and acceleration, as given by equations 14.6, are plotted in Figure 14.4, with $\omega_0 = 1.25\ \text{s}^{-1}$ and $\theta_0 = -0.5$ rad. Indicated in Figure 14.4 is the phase angle, as well as the three amplitudes of the oscillations: $A$ is the amplitude of the oscillation of the position vector, $\omega_0 A$ (or $1.25A$, in this case) is the amplitude of the oscillation of the velocity vector, and $\omega_0^2 A$ [or $(1.25)^2 A$ here] is the amplitude of the oscillation of the acceleration vector. You can see that whenever the position vector passes through zero, the value of the velocity vector is at a maximum or a minimum, and vice versa. You can also observe that the acceleration (just like the force) is always in the opposite direction to the position vector. When the position passes through zero, so does the acceleration.

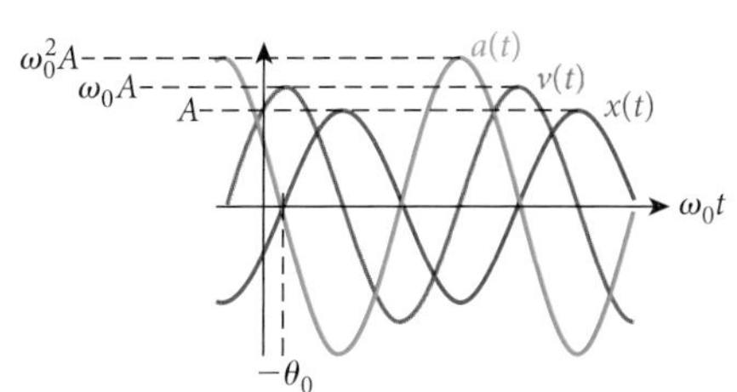


**FIGURE 14.4** Graphs of position, velocity, and acceleration for simple harmonic motion as a function of time.

Figure 14.5 shows a block connected to a spring and undergoing simple harmonic motion by sliding on a frictionless surface. The velocity and acceleration vectors of the block at eight different positions are shown. In Figure 14.5a, the block is released from $x = A$. The block accelerates to the left as indicated. The block has reached $x = A/\sqrt{2}$ in Figure 14.5b. At this point, the velocity of the block and the acceleration of the block are directed toward the left. In Figure 14.5c, the block has reached the equilibrium position. The block at this

**FIGURE 14.5** A block on a spring undergoing simple harmonic motion. The velocity and acceleration vectors are shown at different points in the oscillation: (a) $x = A$; (b) $x = A/\sqrt{2}$; (c) $x = 0$; (d) $x = -A/\sqrt{2}$; (e) $x = -A$; (f) $x = -A/\sqrt{2}$; (g) $x = 0$; (h) $x = A/\sqrt{2}$.

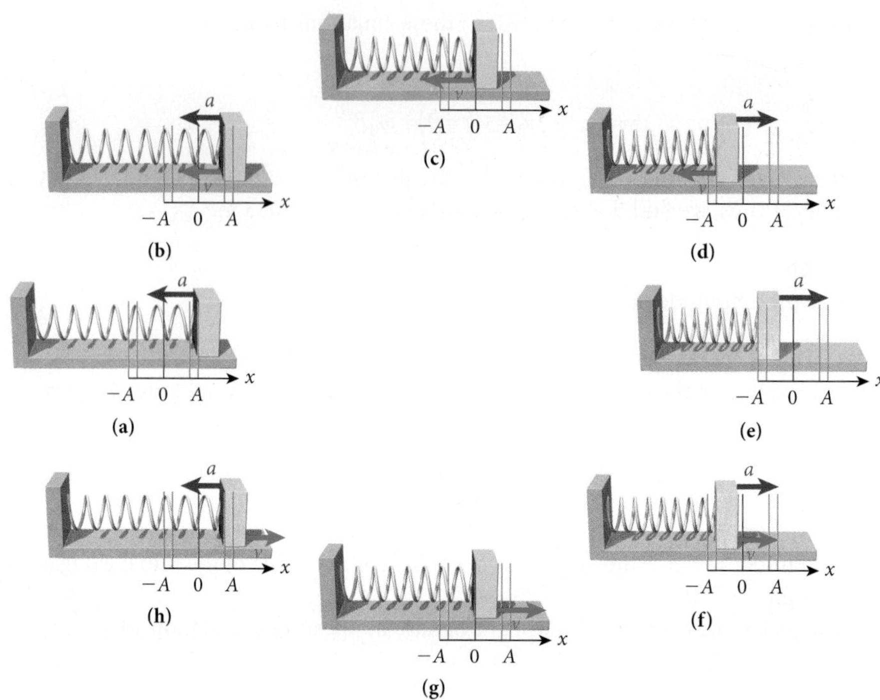


### 14.1 In-Class Exercise

A spring with $k = 12.0$ N/m has a weight of mass 3.00 kg attached to its end. The weight is pulled +10.0 cm from its equilibrium position and released from rest. What is the velocity of the weight as it passes the equilibrium position?

a) −0.125 m/s
b) +0.750 m/s
c) −0.200 m/s
d) +0.500 m/s
e) −0.633 m/s

position has zero acceleration and its maximum velocity to the left. The block continues through the equilibrium position of the spring and begins to slow down. In Figure 14.5d, the block is located at $x = -A/\sqrt{2}$. The acceleration of the block is now directed toward the right, although the block is still moving to the left. The block reaches $x = -A$ in Figure 14.5e. At this position, the velocity of the block is zero, and the acceleration of the block is directed toward the right. In Figure 14.5f, the block is again located at $x = -A/\sqrt{2}$, but now the velocity of the block and the acceleration are directed toward the right. The block again reaches the equilibrium position in Figure 14.5g with its velocity vector pointing to the right. The block continues through the equilibrium position and reaches $x = A/\sqrt{2}$ in Figure 14.5h, where the velocity is still toward the right but the acceleration is now toward the left. The block returns to its original configuration in Figure 14.5a, and the cycle continues.

## Period and Frequency

As you know, the sine and cosine functions are periodic, with a period of $2\pi$. The position, velocity, and acceleration for the oscillations of simple harmonic motion are described by a sine or cosine function, and adding a multiple of $2\pi$ to the argument of such a function does not change its value:

$$\sin(\omega t) = \sin(2\pi + \omega t) = \sin\left[\omega\left(\frac{2\pi}{\omega} + t\right)\right].$$

(To obtain the right-hand side of this equation, we rewrote the expression in the middle by multiplying and dividing $2\pi$ by $\omega$ and then factoring out the common factor $\omega$.) We have dropped the index 0 on $\omega$, because we are deriving a universal relationship that is valid for all angular speeds, not just for the particular situation of a mass on a spring.

The time interval over which a sinusoidal function repeats itself is the **period,** denoted by $T$. From the preceding equation for the periodicity of the sine function, we can see that

$$T = \frac{2\pi}{\omega}, \tag{14.7}$$

because $\sin(\omega t) = \sin[\omega(T + t)]$. The same argument works for the cosine function. In other words, replacing $t$ by $t + T$ yields the same position, velocity, and acceleration vectors, as demanded by the definition of the period of simple harmonic motion.

The inverse of the period is the **frequency,** $f$:

$$f = \frac{1}{T}, \tag{14.8}$$

where $f$ is the number of complete oscillations per unit time. For example, if $T = 0.2$ s, then 5 oscillations occur in 1 s, and $f = 1/T = 1/(0.2 \text{ s}) = 5.0 \text{ s}^{-1} = 5.0$ Hz. Substituting for $T$ from equation 14.7 into equation 14.8 yields an expression for the angular speed in terms of the frequency:

$$f = \frac{1}{T} = \frac{1}{2\pi/\omega} = \frac{\omega}{2\pi}$$

or

$$\omega = 2\pi f. \tag{14.9}$$

For a mass on a spring, we have the following for the period and the frequency:

$$T = \frac{2\pi}{\omega_0} = \frac{2\pi}{\sqrt{k/m}} = 2\pi\sqrt{\frac{m}{k}} \tag{14.10}$$

and

$$f = \frac{\omega_0}{2\pi} = \frac{1}{2\pi}\sqrt{\frac{k}{m}}. \tag{14.11}$$

Interestingly, the period does not depend on the amplitude of the motion.

Figure 14.6 illustrates the effect of changing the values of the variables affecting the simple harmonic motion of an object on a spring. The simple harmonic motion is described by equation 14.3 with $\theta_0 = 0$:

$$x = A\sin\left(\sqrt{\frac{k}{m}}t\right).$$

From Figure 14.6a, you can see that increasing the mass, $m$, increases the period of the oscillations. Figure 14.6b shows that increasing the spring constant, $k$, decreases the period of the oscillations. Figure 14.6c reinforces the earlier conclusion that increasing the amplitude, $A$, does not change the period of the oscillations.

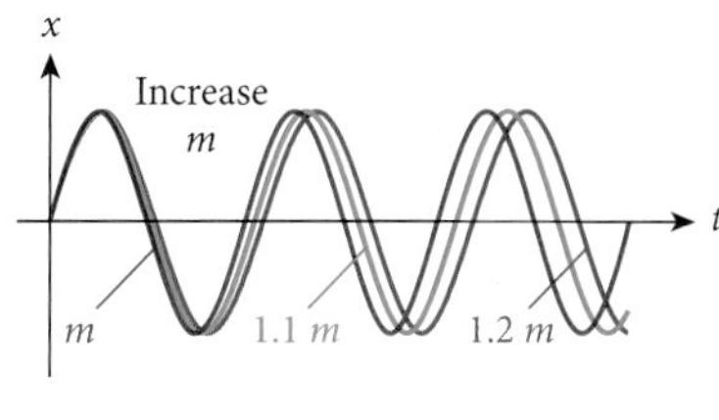


(a)

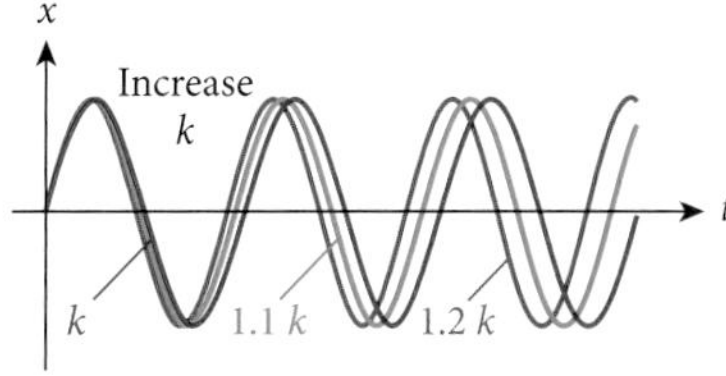


(b)

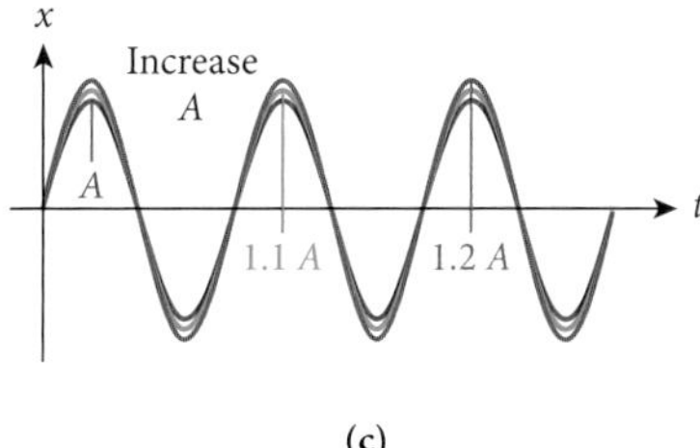


(c)

**FIGURE 14.6** The effect on the simple harmonic motion of an object connected to a spring resulting from increasing (a) the mass, $m$; (b) the spring constant, $k$; and (c) the amplitude, $A$.

## EXAMPLE 14.2 Tunnel through the Moon

Suppose we could drill a tunnel straight through the center of the Moon, from one side to the other. (The facts that the Moon has no atmosphere and that it is composed of solid rock make this scenario slightly less fantastic than drilling a tunnel through the center of Earth.)

### PROBLEM

If we released a steel ball of mass 5.0 kg from rest at one end of this tunnel, what would its motion be like?

### SOLUTION

From Chapter 12, the magnitude of the gravitational force inside a spherical mass distribution of constant density is $F_g = mgr/R$, where $r$ is the distance from the center of the Earth and $R$ is the radius of the Earth, with $r < R$. This force points toward the center of the Earth, that is, in the direction opposite to the displacement. In other words, the gravitational force inside a homogeneous spherical mass distribution follows Hooke's Law, $F(x) = -kx$, with a "spring constant" of $k = mg/R$, where $g$ is the gravitational acceleration experienced at the surface.

First, we need to calculate the gravitational acceleration at the surface on the Moon. Since the mass of the Moon is $7.35 \cdot 10^{22}$ kg (1.2% of the mass of Earth) and its radius is $1.735 \cdot 10^6$ m (27% of the radius of Earth), we find (see Chapter 12):

$$g_M = \frac{GM_M}{R_M^2} = \frac{\left(6.67 \cdot 10^{-11} \text{ m}^3\text{kg}^{-1}\text{s}^{-2}\right)\left(7.35 \cdot 10^{22} \text{ kg}\right)}{\left(1.735 \cdot 10^6 \text{ m}\right)^2} = 1.63 \text{ m/s}^2.$$

*Continued—*

The acceleration due to gravity at the Moon's surface is approximately a sixth of what it is on the surface of the Earth.

The appropriate equation of motion is equation 14.2:

$$x(t) = B\sin(\omega_0 t) + C\cos(\omega_0 t).$$

Releasing the ball from the surface of the Moon at time $t = 0$ implies that $x(0) = R_M = B \sin(0) + C\cos(0)$, or $C = R_M$. To determine the other initial condition, we use the velocity equation from Example 14.1: $v(t) = \omega_0 B\cos(\omega_0 t) - \omega_0 C\sin(\omega_0 t)$. The ball was released from rest, so $v(0) = 0 = \omega_0 B\cos(0) - \omega_0 C\sin(0) = \omega_0 B$, giving us $B = 0$. Thus, the equation of motion in this case becomes

$$x(t) = R_M \cos(\omega_0 t).$$

The angular speed of the oscillation is

$$\omega_0 = \sqrt{\frac{k}{m}} = \sqrt{\frac{g_M}{R_M}} = \sqrt{\frac{1.63 \text{ m/s}^2}{1.735 \cdot 10^6 \text{ m}}} = 9.69 \cdot 10^{-4} \text{ s}^{-1}.$$

Note that the mass of the steel ball turns out to be irrelevant. The period of the oscillation is

$$T = \frac{2\pi}{\omega_0} = 6485 \text{ s}.$$

The steel ball would arrive at the surface on the other side of the Moon 3242 s after it was released, and then oscillate back. Passing through the entire Moon in a little less than an hour would characterize an extremely efficient mode of transportation, especially since no power supply would be needed.

The velocity of the steel ball during its oscillation would be

$$v(t) = \frac{dx}{dt} = -\omega_0 R_M \sin(\omega_0 t).$$

The maximum velocity would be reached as the ball crosses the center of the Moon and would have the numerical value

$$v_{max} = \omega_0 R_M = (9.69 \cdot 10^{-4} \text{ s}^{-1})(1.735 \cdot 10^6 \text{ m}) = 1680 \text{ m/s} = 3760 \text{ mph}.$$

If the tunnel were big enough, the same motion could be achieved by a vehicle holding one or more people, providing a very efficient means of transportation to the other side of the Moon, without the need for propulsion. During the entire journey, the people inside the vehicle would feel absolute weightlessness, because they would experience no supporting force from the vehicle! In fact, it would not even be necessary to use a vehicle to make this trip—wearing a space suit, you could just jump into the tunnel.

## SOLVED PROBLEM 14.1 Block on a Spring

### PROBLEM

A 1.55-kg block sliding on a frictionless horizontal plane is connected to a horizontal spring with spring constant $k = 2.55$ N/m. The block is pulled to the right a distance $d = 5.75$ cm and released from rest. What is the block's velocity 1.50 s after it is released?

### SOLUTION

#### THINK

The block will undergo simple harmonic motion. We can use the given initial conditions to determine the parameters of the motion. With those parameters, we can calculate the velocity of the block at the specified time.

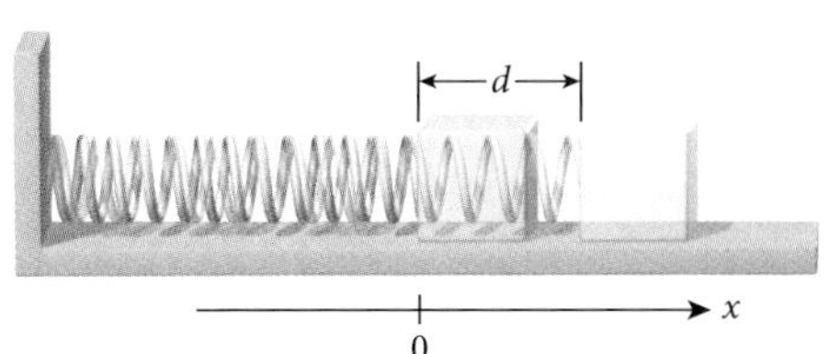


**FIGURE 14.7** A block attached to a spring and displaced a distance $d$ from the equilibrium position.

**SKETCH**

Figure 14.7 shows the block attached to a spring and displaced a distance $d$ from the equilibrium position.

**RESEARCH**

The first initial condition is that at $t = 0$, the position is $x = d$. Thus, we can write

$$x(t=0) = d = A\sin\left[(\omega_0 \cdot 0) + \theta_0\right] = A\sin\theta_0. \qquad \text{(i)}$$

We have one equation with two unknowns. To get a second equation, we use the second initial condition: at $t = 0$, the velocity is zero. This leads to

$$v(t=0) = 0 = \omega_0 A\cos\left[(\omega_0 \cdot 0) + \theta_0\right] = \omega_0 A\cos\theta_0. \qquad \text{(ii)}$$

We now have two equations and two unknowns.

**SIMPLIFY**

We can simplify equation (ii) to obtain $\cos\theta_0 = 0$, from which we get the phase angle, $\theta_0 = \pi/2$. Substituting this result into equation (i) gives us

$$d = A\sin\theta_0 = A\sin\left(\frac{\pi}{2}\right) = A.$$

Thus, we can write the velocity as a function of time as

$$v(t) = \omega_0 d\cos\left[\omega_0 t + \left(\frac{\pi}{2}\right)\right] = -\omega_0 d\sin(\omega_0 t).$$

Since the angular speed is given by $\omega_0 = \sqrt{k/m}$, we obtain

$$v(t) = -\sqrt{\frac{k}{m}}\,d\sin\left(\sqrt{\frac{k}{m}}\,t\right).$$

**CALCULATE**

Putting in the numerical values gives us

$$v(t = 1.50\text{ s}) = -\sqrt{\frac{2.55\text{ N/m}}{1.55\text{ kg}}}\,(0.0575\text{ m})\sin\left[\sqrt{\frac{2.55\text{ N/m}}{1.55\text{ kg}}}\,(1.50\text{ s})\right] = -0.06920005\text{ m/s}.$$

**ROUND**

We report our result to three significant figures:

$$v = -0.0692\text{ m/s} = -6.92\text{ cm/s}.$$

**DOUBLE-CHECK**

As usual, it is a good idea to verify that the answer has appropriate units. This is the case here, because m/s is a velocity unit. The maximum speed that the block can attain is $v = \omega_0 d = 7.38$ cm/s. The magnitude of our result is less than that maximum speed, so it seems reasonable.

## Relationship of Simple Harmonic Motion to Circular Motion

In Chapter 9, we analyzed circular motion with constant angular velocity, $\omega$, along a path with constant radius, $r$. We saw that the $x$- and $y$-coordinates of such motion are given by the equations $x(t) = r\cos(\omega t + \theta_0)$ and $y(t) = r\sin(\omega t + \theta_0)$. Figure 14.8a shows how the vector $\vec{r}(t)$ performs circular motion with constant angular velocity as a function of time with an initial angle of $\theta_0 = 0$. The red arc segment shows the path of the tip of this radius vector. Figure 14.8b shows the projection of the radius vector onto

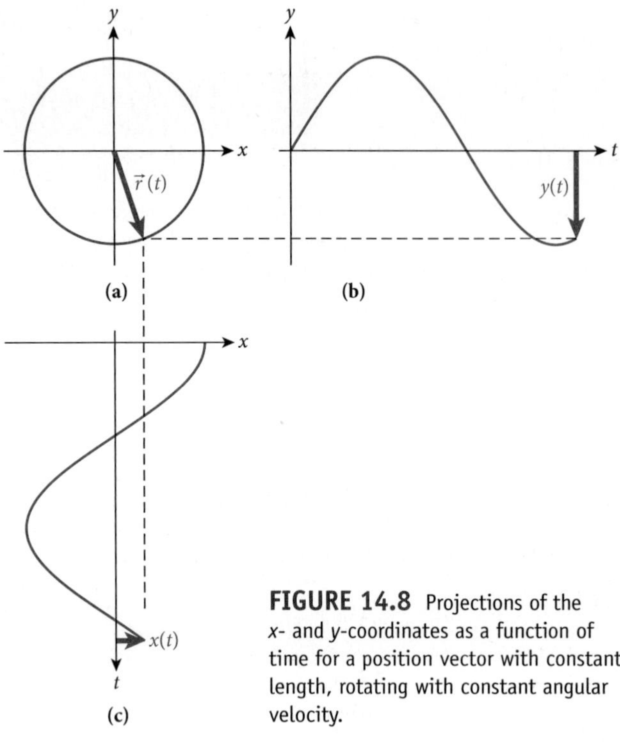


**FIGURE 14.8** Projections of the $x$- and $y$-coordinates as a function of time for a position vector with constant length, rotating with constant angular velocity.

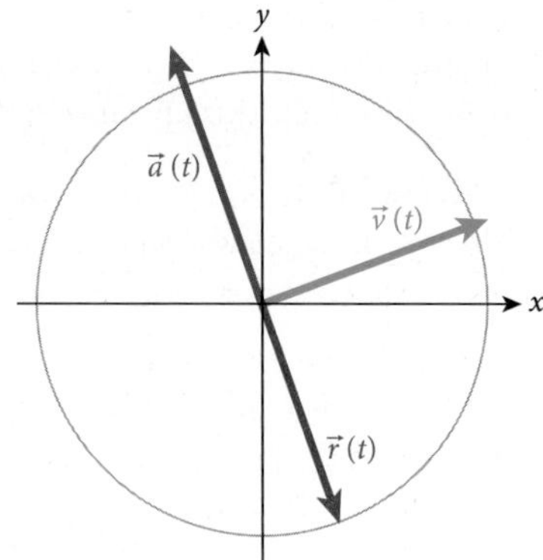


**FIGURE 14.9** The position vector, linear velocity vector, and acceleration vector for circular motion.

the $y$-coordinate. You can clearly see that the motion of the $y$-component of the radius vector traces out a sine function. The motion of the $x$-component as a function of time is shown in Figure 14.8c, and it traces out a cosine function. These two projections of circular motion of constant angular velocity exhibit simple harmonic oscillations. This observation makes it clear that the frequency, angular speed, and period defined here for oscillatory motion are identical to the quantities introduced in Chapter 9 for circular motion.

In Figure 14.8, the position vector, $\vec{r}(t)$, originates at $(x,y) = (0,0)$ and rotates with angular velocity $\omega$. In Chapter 9, we saw that for circular motion the linear velocity, $\vec{v}(t)$, is tangent to the circle and the linear acceleration, $\vec{a}(t)$, always points toward the center of the circle. The vectors $\vec{v}(t)$ and $\vec{a}(t)$ can be moved so that they originate at $(x,y) = (0,0)$, as shown in Figure 14.9. Thus, Figure 14.9 shows that, for circular motion, the three vectors $\vec{r}(t)$, $\vec{v}(t)$, and $\vec{a}(t)$ rotate together with angular velocity $\omega$. The linear velocity vector is always 90° out of phase with the position vector, and the linear acceleration vector is always 180° out of phase with the position vector. The projections of the linear velocity and linear acceleration vectors on the $x$- and $y$-axes correspond to the velocity and acceleration of an object undergoing simple harmonic motion.

## 14.2 Pendulum Motion

We are all familiar with another common oscillating system: the **pendulum.** In its ideal form, a pendulum consists of a thin string attached to a massive object that swings back and forth. The string is assumed to be massless, that is, of such small mass that the mass can be neglected. This assumption is, by the way, a very good approximation to the situation of a person on a swing.

Let's determine the equation of motion for any pendulum-like object. Shown in Figure 14.10 is a ball at the end of a rope of length $\ell$ at an angle $\theta$ relative to the vertical. For small angles $\theta$, the differential equation for the motion of the pendulum (which is derived in Derivation 14.1) is

$$\frac{d^2\theta}{dt^2} + \frac{g}{\ell}\theta = 0. \qquad (14.12)$$

Equation 14.12 has the solution

$$\theta(t) = B\sin(\omega_0 t) + C\cos(\omega_0 t), \quad \text{with } \omega_0 = \sqrt{\frac{g}{\ell}}. \tag{14.13}$$

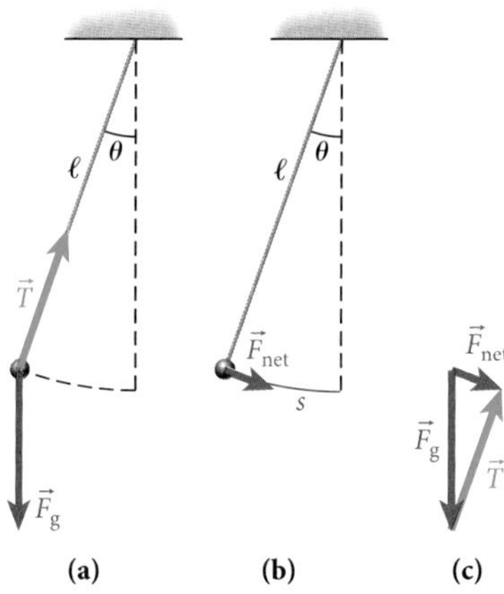


**FIGURE 14.10** A pendulum (a) with the force vectors due to gravity and string tension and (b) with the net force vector. (c) Vector construction of the net force.

### DERIVATION 14.1 Pendulum Motion

The displacement, $s$, of the pendulum in Figure 14.10 is measured along the circumference of a circle with radius $\ell$. This displacement can be obtained from the length of the string and the angle: $s = \ell\theta$. Because the length does not change with time, we can write the second derivative of the displacement as

$$\frac{d^2s}{dt^2} = \frac{d^2(\ell\theta)}{dt^2} = \ell\frac{d^2\theta}{dt^2}.$$

The next task is to find the angular acceleration, $d^2\theta/dt^2$, as a function of time. To do this, we need to determine the force that causes the acceleration. Two forces act on the ball: the force of gravity, $\vec{F}_g$, acting downward, and the force of the tension, $\vec{T}$, acting along the rope. Since the rope stays taut and does not stretch, the tension must equal the component of the gravitational force along the rope: $T = mg\cos\theta$. The vector sum of the forces gives a net force of magnitude $F_{net} = mg\sin\theta$, directed as shown in Figure 14.10. The net force is always in the direction opposite to the displacement, $s$.

Using $F_{net} = ma$, we then obtain

$$m\frac{d^2s}{dt^2} = -mg\sin\theta \Rightarrow$$

$$\frac{d^2s}{dt^2} = -g\sin\theta \Rightarrow$$

$$\ell\frac{d^2\theta}{dt^2} + g\sin\theta = 0 \Rightarrow$$

$$\frac{d^2\theta}{dt^2} + \frac{g}{\ell}\sin\theta = 0.$$

This equation is difficult to solve without using the *small-angle approximation*, $\sin\theta \approx \theta$ (in radians). Doing so yields the desired differential equation 14.12. Figure 14.11 shows that the small-angle approximation introduces little error for $\theta < 0.5$ rad (approximately 30°). For these small angles, the motion of a pendulum is approximately simple harmonic motion because the restoring force is approximately proportional to $\theta$.

To solve equation 14.12, we could go through exactly the same steps that led to the solution of the differential equation for the spring. However, since equations 14.1 and 14.12 are identical in form, we simply take the solution for spring motion and perform the appropriate substitutions: The angle $\theta$ takes the place of $x$, and $g/\ell$ takes the place of $k/m$. In this manner we arrive at our solution without deriving our previous result a second time.

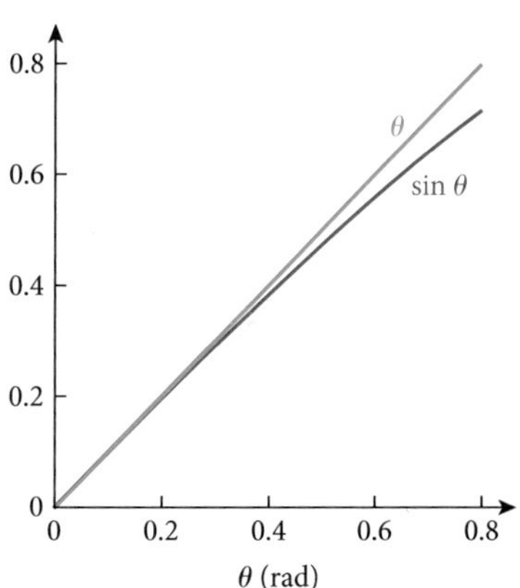


**FIGURE 14.11** Graph indicating the error incurred using the small-angle approximation, $\sin\theta \approx \theta$, where $\theta$ is measured in radians.

### 14.2 In-Class Exercise

A pendulum is released from an angle of 6.0° relative to the vertical and performs simple harmonic oscillations with period $T$. If the initial angle is doubled, to 12.0°, the pendulum will oscillate with period

a) $T/2$.

b) $T/\sqrt{2}$.

c) $T$.

d) $\sqrt{2}T$.

e) $2T$.

## Period and Frequency of a Pendulum

The period and frequency of a pendulum are related to the angular speed just as for a mass on a spring, but with the angular speed given by $\omega_0 = \sqrt{g/\ell}$:

$$T = \frac{2\pi}{\omega_0} = 2\pi\sqrt{\frac{\ell}{g}} \tag{14.14}$$

$$f = \frac{1}{2\pi}\sqrt{\frac{g}{\ell}}. \tag{14.15}$$

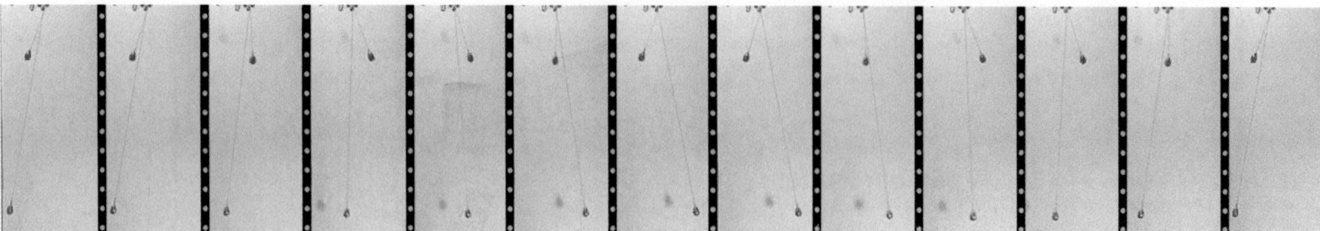

FIGURE 14.12 Video sequence of the oscillation of two pendulums whose lengths have the ratio 4:1.

### 14.3 Self-Test Opportunity

You have a pendulum that has the period $T$ on Earth. You take this pendulum to the Moon. What is the period of this pendulum on the Moon in terms of its period on Earth?

Thus, the solution of the equation of motion for the pendulum leads to harmonic motion, just as for the case of a mass on a spring. However, for the pendulum—*unlike* the spring—the frequency is independent of the mass of the oscillating object. This result means that two otherwise identical pendulums with different mass have the same period. The only way to change the period of a pendulum—other than by taking it to another planet or the Moon, where the gravitational acceleration is different—is by varying its length. (Of course, the gravitational acceleration also has small variations on Earth, for example, depending on altitude; so a pendulum's period is not *exactly* the same everywhere on Earth.)

To shorten the period of a pendulum by a factor of 2 requires shortening the length of the string by a factor of 4. This effect is illustrated in Figure 14.12, which shows the motion of two pendulums, with one four times longer than the other. In this sequence, the shorter pendulum completes two oscillations in the same time in which the longer pendulum completes one.

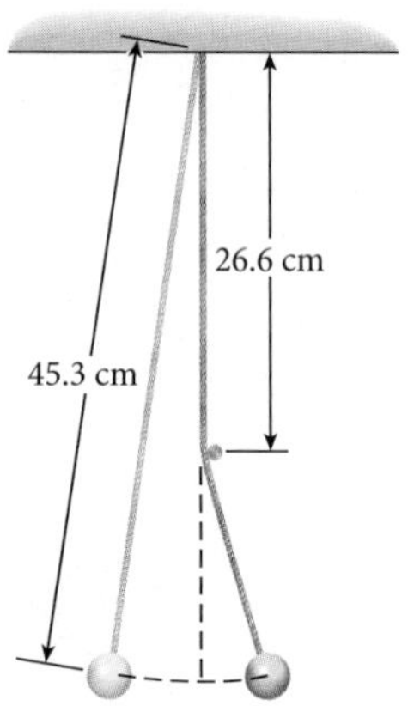


FIGURE 14.13 Restricted pendulum.

### EXAMPLE 14.3 Restricted Pendulum

**PROBLEM**

A pendulum of length 45.3 cm is hanging from the ceiling. Its motion is restricted by a peg that is sticking out of the wall 26.6 cm directly below the pivot point (Figure 14.13). What is the period of oscillation? (Note that specifying the mass is not necessary.)

**SOLUTION**

We must solve this problem separately for motion on the left side and on the right side of the peg. On the left side, the pendulum oscillates with its full length, $\ell_1 = 45.3$ cm. On the right side, the pendulum oscillates with a reduced length, $\ell_2 = 45.3 \text{ cm} - 26.6 \text{ cm} = 18.7$ cm. On each side, it performs exactly $\frac{1}{2}$ of a full oscillation. Thus, the overall period of the pendulum is $\frac{1}{2}$ of the sum of the two periods calculated with the different lengths:

$$T = \frac{1}{2}(T_1 + T_2) = \frac{2\pi}{2}\left(\sqrt{\frac{\ell_1}{g}} + \sqrt{\frac{\ell_2}{g}}\right) = \frac{\pi}{\sqrt{g}}\left(\sqrt{\ell_1} + \sqrt{\ell_2}\right)$$

$$= \frac{\pi}{\sqrt{9.81 \text{ m/s}^2}}\left(\sqrt{0.453 \text{ m}} + \sqrt{0.187 \text{ m}}\right)$$

$$= 1.11 \text{ s}.$$

### 14.3 In-Class Exercise

A grandfather clock keeps time using a pendulum consisting of a light rod connected to a small heavy mass. What should the length of the rod be for the period of the oscillations to be 1.00 s?

a) 0.0150 m

b) 0.145 m

c) 0.248 m

d) 0.439 m

e) 0.750 m

## 14.3 Work and Energy in Harmonic Oscillations

The concepts of work and kinetic energy were introduced in Chapter 5. In Chapter 6, we found the potential energy due to the spring force. In this section, we analyze the energies associated with motion of a mass on a spring. Then we'll see that almost all those results are applicable to a pendulum as well, but that we need to correct for the small-angle approximation used in solving the differential equation for pendulum motion.

## Mass on a Spring

In Chapter 6, we did not have the equation for the displacement as a function of time at our disposal. However, we derived the potential energy stored in a spring, $U_s$, as

$$U_s = \tfrac{1}{2}kx^2,$$

where $k$ is the spring constant and $x$ is the displacement from the equilibrium position. We also saw that the total mechanical energy of a mass on a spring undergoing oscillations of amplitude, $A$, is given by

$$E = \tfrac{1}{2}kA^2.$$

Conservation of total mechanical energy means that this expression gives the value of the energy for any point in the oscillation. Using the law of energy conservation, we can write

$$\tfrac{1}{2}kA^2 = \tfrac{1}{2}mv^2 + \tfrac{1}{2}kx^2.$$

Then we can solve for the velocity as a function of the position:

$$v = \sqrt{(A^2 - x^2)\frac{k}{m}}. \qquad (14.16)$$

The functions $v(t)$ and $x(t)$ given in equations 14.6 describe the oscillation over time of a mass on a spring. We can use these functions to verify the relationship between position and velocity expressed in equation 14.16, which we obtained from energy conservation. In this way, we can directly test whether this new result is consistent with what we found previously. First, we evaluate $A^2 - x^2$:

$$\begin{aligned} A^2 - x^2 &= A^2 - A^2\sin^2(\omega_0 t + \theta_0) \\ &= A^2\left[1 - \sin^2(\omega_0 t + \theta_0)\right] \\ &= A^2\cos^2(\omega_0 t + \theta_0). \end{aligned}$$

Multiplying both sides by $\omega_0^2 = k/m$, we obtain

$$\frac{k}{m}\left(A^2 - x^2\right) = \frac{k}{m}A^2\cos^2(\omega_0 t + \theta_0) = v^2.$$

By taking the square root of both sides of this equation, we obtain the same equation for the velocity that we got by applying energy considerations, equation 14.16.

Figure 14.14 illustrates the oscillations of the kinetic and potential energies as a function of time for a mass oscillating on a spring. As you can see, even though the potential and kinetic energies oscillate in time, their sum—the total mechanical energy—is constant. The kinetic energy always reaches its maximum wherever the displacement passes through zero, and the potential energy reaches its maximum for maximum elongation of the spring from the equilibrium position.

The position, velocity, and acceleration vectors at several positions for a block on a spring undergoing simple harmonic motion were shown in Figure 14.5. Figure 14.15 reproduces that figure but also shows the potential and kinetic energies ($U$ and $K$) at each position. In Figure 14.15a, the block is released from rest at position $x = A$. At this point, all the energy in the system exists in the form of potential energy stored in the spring, and the block has no kinetic energy. The block then accelerates to the left. At $x = A/\sqrt{2}$, shown in Figure 14.15b, the potential energy stored in the spring and the kinetic energy of the block are equal. In Figure 14.15c, the block reaches the equilibrium position of the spring, $x = 0$, at which point the potential energy stored in the spring is zero and the block has its maximum kinetic energy. The block continues moving, through the equilibrium position and in Figure 14.15d, it is located at $x = -A/\sqrt{2}$. Again the potential energy in the spring and kinetic energy of the block are equal. In Figure 14.15e, the block is located at $x = -A$,

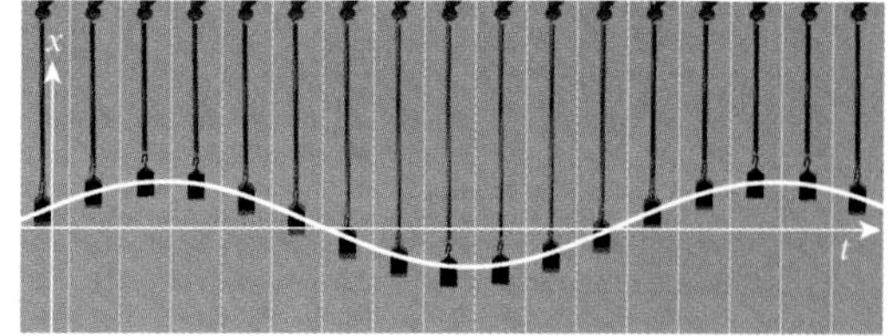


(a)

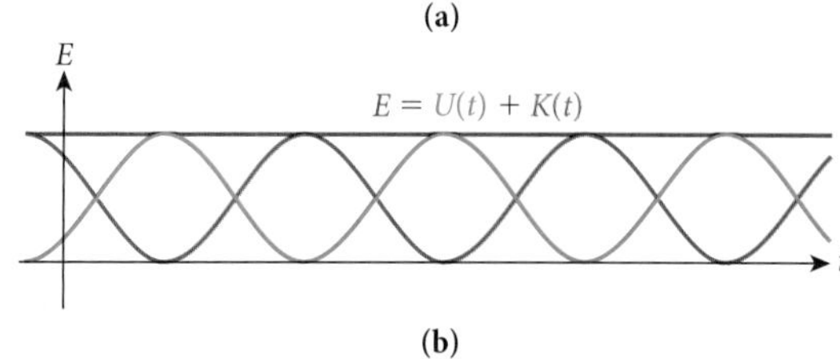


(b)

**FIGURE 14.14** Harmonic oscillation of a mass on a spring: (a) displacement as a function of time (same as Figure 14.2); (b) potential and kinetic energies as a function of time on the same time scale.

**FIGURE 14.15** Position, velocity, and acceleration vectors for a mass on a spring (as in Figure 14.5), with the potential and kinetic energies at each position.

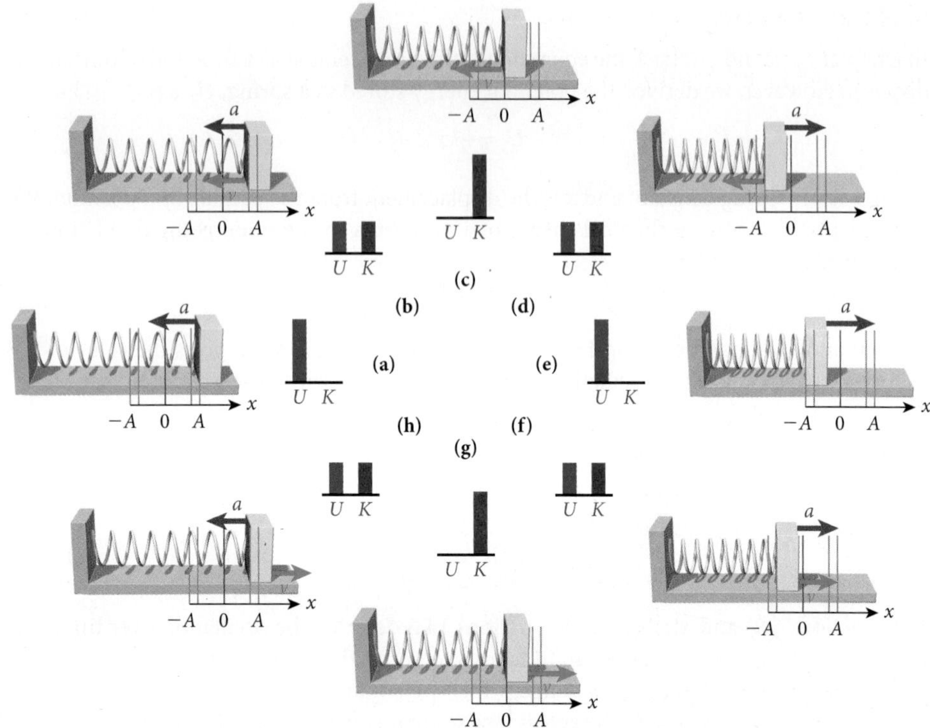


where the potential energy stored in the spring is at its maximum and the kinetic energy of the block is zero. The block has returned to $x = -A/\sqrt{2}$ in Figure 14.15f, and the potential energy in the spring is again equal to the kinetic energy of the block. In Figure 14.15g, the block is located at $x = 0$, where the potential energy in the spring is zero and the kinetic energy of the block is at its maximum. The block continues through that equilibrium position and reaches $x = A/\sqrt{2}$ in Figure 14.15h, and again the potential energy in the spring is equal to the kinetic energy of the block. The block returns to its original position in Figure 14.15a.

## Energy of a Pendulum

In Section 14.2, we saw that the time dependence of the deflection angle of a pendulum is $\theta(t) = \theta_0 \cos\left(\sqrt{g/\ell}\, t\right)$, if the initial conditions of maximum deflection and zero speed at time zero, $\theta(t=0) = \theta_0$ are known. We can then find the linear velocity at each point in time by taking the derivative, $d\theta/dt = \omega$, and multiplying this angular velocity by the radius of the circle, $\ell$:

$$v = \ell \frac{d\theta(t)}{dt} = -\theta_0 \sqrt{g\ell}\, \sin\left(\sqrt{\frac{g}{\ell}} t\right).$$

Because $\sin\alpha = \sqrt{1-\cos^2\alpha}$, we can insert the expression for the deflection angle as a function of time into this equation for the velocity and obtain the speed of the pendulum as a function of the angle:

$$\begin{aligned} |v| &= |\theta_0| \sqrt{g\ell} \left| \sin\left(\sqrt{\frac{g}{\ell}} t\right) \right| \\ &= |\theta_0| \sqrt{g\ell} \sqrt{1 - \cos^2\left(\sqrt{\frac{g}{\ell}} t\right)} \\ &= |\theta_0| \sqrt{g\ell} \sqrt{1 - \frac{\theta^2}{\theta_0^2}} \Rightarrow \end{aligned}$$

$$|v| = \sqrt{g\ell\left(\theta_0^2 - \theta^2\right)}. \tag{14.17}$$

What is the energy of the pendulum? At time zero, the pendulum has only gravitational potential energy. We can assign the potential energy a value of zero at the lowest point of the arc. From Figure 14.16, the potential energy at maximum deflection, $\theta_0$, is

$$E = K + U = 0 + U = mg\ell\left(1-\cos\theta_0\right).$$

This also gives the value of the total mechanical energy, because by definition the pendulum has zero kinetic energy at the point of maximum deflection, just as a spring does. For any other deflection, the energy is the sum of kinetic and potential energies:

$$E = mg\ell\left(1-\cos\theta\right)+\tfrac{1}{2}mv^2.$$

Combining the preceding two equations for $E$ (because the total energy is conserved) leads to

$$mg\ell\left(1-\cos\theta_0\right) = mg\ell\left(1-\cos\theta\right)+\tfrac{1}{2}mv^2 \Rightarrow$$
$$mg\ell\left(\cos\theta-\cos\theta_0\right)=\tfrac{1}{2}mv^2.$$

Solving this equation for the speed (absolute value of the velocity), we obtain

$$|v| = \sqrt{2g\ell\left(\cos\theta-\cos\theta_0\right)}. \quad (14.18)$$

Equation 14.18 is the exact expression for the speed of the pendulum at any angle $\theta$, which we obtained quite straightforwardly and without the need to solve a differential equation. This equation, however, does not match equation 14.17, which we obtained from the solution of a differential equation. However, remember that we used the small-angle approximation earlier for solving a differential equation. For small angles, we can approximate $\cos\theta \approx 1 - \frac{1}{2}\theta^2 + \ldots$, and so equation 14.17 is a special case of equation 14.18. Our derivation of equation 14.18 using energy considerations did not make use of the small-angle approximation, and thus this equation is valid for all deflection angles. However, the differences between the values obtained with the two equations are quite small, as you can see from Figure 14.17.

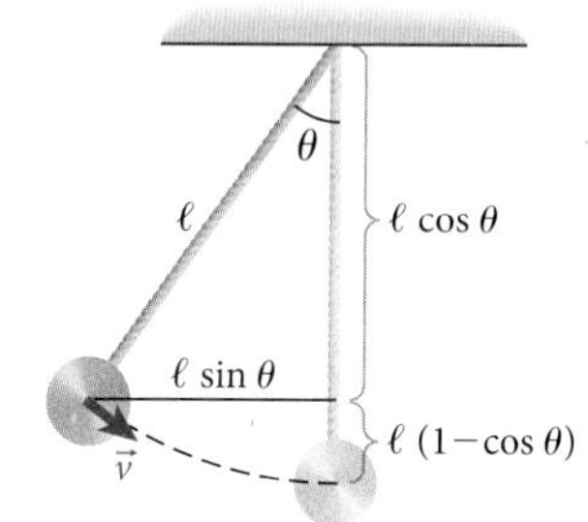


**FIGURE 14.16** Geometry of a pendulum.

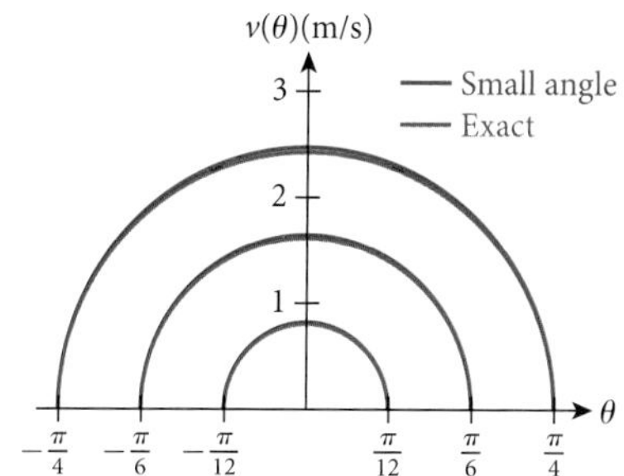


**FIGURE 14.17** Comparison of the speeds of a pendulum (of length 1 m) as a function of the deflection angle, calculated with the small-angle approximation (red curves) and the exact formula based on conservation of energy (blue curves). The three sets of curves represent initial angles of 15°, 30°, and 45° (from inside to outside).

## EXAMPLE 14.4 Speed on a Trapeze

### PROBLEM

A circus trapeze artist starts her motion on the trapeze from rest with the rope at an angle of 45° relative to the vertical. The rope has a length of 5.00 m. What is her speed at the lowest point in her trajectory?

### SOLUTION

The initial condition is $\theta_0 = 45° = \pi/4$ rad. We are interested in finding $v(\theta = 0)$. Applying conservation of energy, we use equation 14.18:

$$v(\theta) = \sqrt{2g\ell\left(\cos\theta-\cos\theta_0\right)}.$$

Inserting the numbers, we obtain

$$v\left(\frac{\pi}{4}\right) = \sqrt{2\left(9.81 \text{ m/s}^2\right)\left(5.00 \text{ m}\right)\left(1-\sqrt{\tfrac{1}{2}}\right)} = 5.36 \text{ m/s}.$$

With the small-angle approximation, for comparison, we get $v(\pi/4) = \theta_0\sqrt{g\ell} = 5.50$ m/s. This is close to the exact result, but for many applications it is not sufficiently precise.

## 14.4 Damped Harmonic Motion

Springs and pendulums do not go on oscillating forever; after some time interval, they come to rest. Thus, some force must be present that slows them down. This speed-diminishing effect is called **damping.** An example is shown in Figure 14.18, which shows a mass on a spring oscillating in water, which provides resistance to the motion of the mass and damps it. The center of the mass follows the red curve superimposed on the video frames in Figure 14.18b. The red curve is the product of a plot of simple harmonic motion (orange curve) and an exponentially falling function (yellow curve).

In order to quantify the effects of damping, we need to consider the damping force. As we have just seen, a force like the spring force, $F_s$, which depends linearly on position, does not accomplish damping. However, a force that depends on velocity does. For speeds that are not too large, the damping is given by a drag force of the form $F_\gamma = -bv$, where $b$ is a constant, known as the damping constant, and $v = dx/dt$ is the velocity. With this damping force, $F_\gamma$, we can write the differential equation describing damped harmonic motion:

$$ma = F_\gamma + F_s$$

$$m\frac{d^2x}{dt^2} = -b\frac{dx}{dt} - kx \Rightarrow$$

$$\frac{d^2x}{dt^2} + \frac{b}{m}\frac{dx}{dt} + \frac{k}{m}x = 0. \tag{14.19}$$

The solution of this equation depends on how large the damping force is relative to the linear restoring force responsible for the harmonic motion. This ratio establishes three general cases: small damping, large damping, and the intermediate case.

### Small Damping

For small values of the damping constant $b$ ("small" will be specified below), the solution of equation 14.19 is

$$x(t) = Be^{-\omega_\gamma t}\cos(\omega' t) + Ce^{-\omega_\gamma t}\sin(\omega' t). \tag{14.20}$$

The coefficients $B$ and $C$ are determined by the initial conditions, that is, the position, $x_0$, and velocity, $v_0$, at time $t = 0$:

$$B = x_0 \text{ and } C = \frac{v_0 + x_0\omega_\gamma}{\omega'}.$$

The angular speeds in this solution are given by

$$\omega_\gamma = \frac{b}{2m}$$

$$\omega' = \sqrt{\omega_0^2 - \omega_\gamma^2} = \sqrt{\frac{k}{m} - \left(\frac{b}{2m}\right)^2}.$$

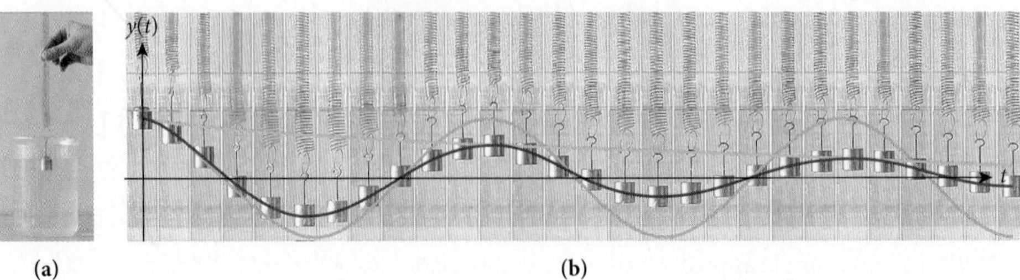


(a) (b)

**FIGURE 14.18** Example of damped harmonic motion: mass on a spring oscillating in water. (a) Initial conditions. (b) Video frames showing the motion of the mass after it is released.

This solution is valid for all values of the damping constant, $b$, for which the argument of the square root that determines $\omega'$ remains positive:

$$b < 2\sqrt{mk}. \tag{14.21}$$

This is the condition for small damping, also known as underdamping.

## DERIVATION 14.2 Small Damping

We can show that equation 14.20 satisfies the differential equation for damped harmonic motion (equation 14.19) in the limit of small damping. We'll start with an assumed solution, which mathematicians call an *Ansatz* (the German word for "attempt"). In order to derive the result from the *Ansatz,* no further knowledge of differential equation is needed; all that is required is taking derivatives.

*Ansatz:*

$$x(t) = Be^{-\omega_\gamma t}\cos(\omega' t)$$

$$\Rightarrow \frac{dx}{dt} = -\omega_\gamma Be^{-\omega_\gamma t}\cos(\omega' t) - \omega' Be^{-\omega_\gamma t}\sin(\omega' t)$$

$$\Rightarrow \frac{d^2x}{dt^2} = \left[\omega_\gamma^2 - (\omega')^2\right]Be^{-\omega_\gamma t}\cos(\omega' t) + 2\omega_\gamma\omega' Be^{-\omega_\gamma t}\sin(\omega' t).$$

We insert these expressions into equation 14.19:

$$\left[\omega_\gamma^2 - (\omega')^2\right]Be^{-\omega_\gamma t}\cos(\omega' t) + 2\omega_\gamma\omega' Be^{-\omega_\gamma t}\sin(\omega' t)$$
$$+\frac{b}{m}\left[-\omega_\gamma Be^{-\omega_\gamma t}\cos(\omega' t) - \omega' Be^{-\omega_\gamma t}\sin(\omega' t)\right]$$
$$+\frac{k}{m}\left[Be^{-\omega_\gamma t}\cos(\omega' t)\right] = 0.$$

Now we rearrange terms:

$$\left[\omega_\gamma^2 - (\omega')^2 - \frac{b}{m}\omega_\gamma + \frac{k}{m}\right]Be^{-\omega_\gamma t}\cos(\omega' t) +$$
$$\left(2\omega_\gamma\omega' - \frac{b}{m}\omega'\right)Be^{-\omega_\gamma t}\sin(\omega' t) = 0.$$

This equation can hold for all times $t$ only if the coefficients in front of the sine and the cosine functions are zero. We thus have two conditions:

$$2\omega_\gamma\omega' - \frac{b}{m}\omega' = 0 \Rightarrow \omega_\gamma = \frac{b}{2m}$$

and

$$\omega_\gamma^2 - (\omega')^2 - \frac{b}{m}\omega_\gamma + \frac{k}{m} = 0.$$

To simplify the second condition, we use the first condition, $\omega_\gamma = b/(2m)$, and the expression for the angular speed we obtained earlier for simple harmonic motion, $\omega_0 = \sqrt{k/m}$. We then obtain for the second condition:

$$\left[-(\omega')^2 - \omega_\gamma^2 + \omega_0^2\right] = 0 \Rightarrow \omega' = \sqrt{\omega_0^2 - \omega_\gamma^2}.$$

We could go through the same steps to show that $Ce^{-\omega_\gamma t}\sin(\omega' t)$ is also a valid solution and could further show that these two solutions are the only possible solutions (but we will skip these demonstrations).

In the derivations in this chapter that involve differential equations we have proceeded in the same manner. This method is a general approach to the solution of problems involving differential equations: Choose a trial solution, and then adjust parameters and make other changes based on the results obtained by inserting the trial solution into the differential equation.

Just as for the case of undamped motion, we can also write the solution in terms of one amplitude, $A$, and a phase shift determined by $\theta_0$ instead of using the coefficients $B$ and $C$ for the sine and cosine functions as in equation 14.20:

$$x(t) = Ae^{-\omega_\gamma t}\sin(\omega' t + \theta_0). \tag{14.22}$$

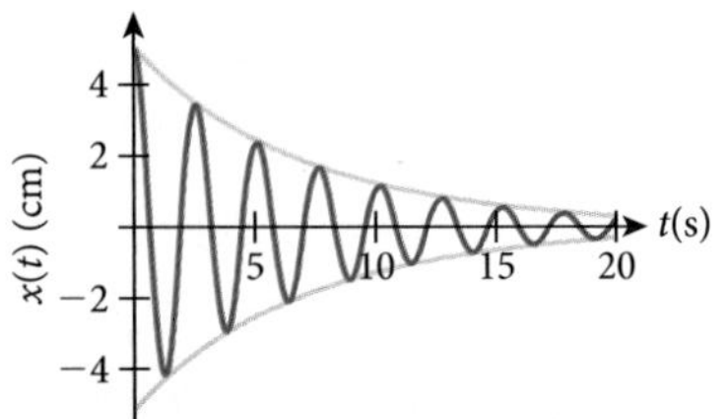


**FIGURE 14.19** Position versus time for a weakly damped harmonic oscillator.

Now let's look at a plot of equation 14.20, which describes weakly damped harmonic motion (Figure 14.19). The sine and cosine functions describe the oscillating behavior. Their combination results in another sine function, with a phase shift. The exponential functions that multiply them can be thought of as reducing the amplitude in time. Thus, the oscillations show exponentially decaying amplitude. The angular speed, $\omega'$, of the oscillation is reduced relative to the angular speed of oscillation without damping, $\omega_0$. The graph in Figure 14.19 was generated with $k = 11.00$ N/m, $m = 1.800$ kg, and $b = 0.500$ kg/s. These parameters result in an angular speed $\omega' = 2.468\ \text{s}^{-1}$, compared to an angular speed $\omega_0 = 2.472\ \text{s}^{-1}$ for the case without damping. The amplitude is $A = 5$ cm, and $\theta_0 = 1.6$. The dark blue curve is the plot of the function, and the two light blue curves show the exponential envelope, within which the amplitude is reduced as a function of time.

## EXAMPLE 14.5 Bungee Jumping

A bridge over a deep valley is ideal for bungee jumping. The first part of a bungee jump consists of a free fall of the same length as the unstretched rope. Suppose the height of the bridge is 50 m. A 30-m-long bungee rope is used and is stretched 5 m by the weight of a 70-kg person. Thus, the equilibrium length of the bungee rope is 35 m. This bungee rope has been found to have a damping angular speed of $0.3\ \text{s}^{-1}$.

### PROBLEM

Describe the vertical motion of the bungee jumper as a function of time.

### SOLUTION

From the top of the bridge, the jumper experiences free fall for the first 30 m of her descent, shown by the red part of the trajectory in Figure 14.20. Once she falls the length of the rope, 30 m, she reaches a velocity of $v_0 = -\sqrt{2g\ell} = -24.26$ m/s. This part of the trajectory takes time $t = \sqrt{2\ell/g} = 2.47$ s.

She then enters a damped oscillation $y(t)$ about the equilibrium position, $y_e = 50\text{ m} - 35\text{ m} = 15$ m, with the initial displacement $y_0 = 5$m. We can calculate the angular speed of her vertical motion from knowing that the rope is stretched by 5 m due to the 70-kg mass and using $mg = k(5\text{ m})$:

$$\omega_0 = \sqrt{\frac{k}{m}} = \sqrt{\frac{g}{\ell_0 - \ell}} = \sqrt{\frac{9.8\text{ m/s}^2}{5\text{ m}}} = 1.4\ \text{s}^{-1}.$$

Because $\omega_\gamma = 0.3\ \text{s}^{-1}$, was specified, we have $\omega' = \sqrt{\omega_0^2 - \omega_\gamma^2} = 1.367\ \text{s}^{-1}$. The bungee jumper oscillates according to equation 14.20, with coefficients $B = y_0 = 5$ m and $C = (v_0 + y_0\,\omega_\gamma)/\omega' = -16.65$ m/s. This motion is represented by the green part of the curve in Figure 14.20.

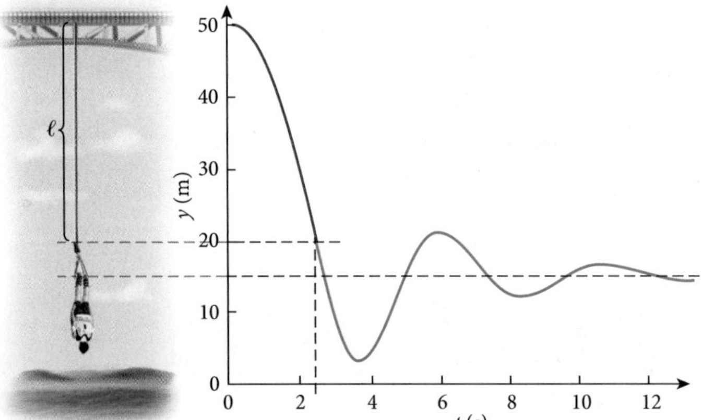


**FIGURE 14.20** Idealized vertical motion as a function of time of a bungee jump.

**DISCUSSION**

The approximation of damped harmonic motion for the final part of this bungee jump is not quite accurate. If you examine Figure 14.20 carefully, you see that the jumper rises above 20 m for approximately 1 s, starting at approximately 5.5 s. During this interval, the bungee cord is not stretched and so this part of the motion is not sinusoidal. Instead, the bungee jumper is again in free-fall motion. However, for the present qualitative discussion, the approximation of damped harmonic motion is acceptable. If you plan to bungee jump (strongly discouraged in light of past fatalities!), you should always test the setup first with an object of weight equal to or greater than yours.

## Large Damping

What happens when the condition for small damping, $b<2\sqrt{mk}$, is no longer fulfilled? The argument of the square root that determines $\omega'$ will be smaller than zero, and so the *Ansatz* we used for small damping does not work any more. The solution of the differential equation for damped harmonic oscillations (equation 14.19) for the case where $b>2\sqrt{mk}$, a situation called **overdamping,** is

$$x(t)=Be^{-\left(\omega_\gamma+\sqrt{\omega_\gamma^2-\omega_0^2}\right)t}+Ce^{-\left(\omega_\gamma-\sqrt{\omega_\gamma^2-\omega_0^2}\right)t}, \quad \text{for } b>2\sqrt{mk}, \tag{14.23}$$

with the amplitudes given by

$$B=\tfrac{1}{2}x_0-\frac{x_0\omega_\gamma+v_0}{2\sqrt{\omega_\gamma^2-\omega_0^2}}, \quad C=\tfrac{1}{2}x_0+\frac{x_0\omega_\gamma+v_0}{2\sqrt{\omega_\gamma^2-\omega_0^2}}.$$

Again, we have for the angular speeds

$$\omega_\gamma=\frac{b}{2m}, \quad \omega_0=\sqrt{\frac{k}{m}}.$$

The coefficients $B$ and $C$ are determined by the initial conditions.

The solution given by equation 14.23 has no oscillations but instead consists of two exponential terms. The term with the argument $-\omega_\gamma t+\sqrt{\omega_\gamma^2-\omega_0^2}\,t$ governs the long-time behavior of the system because it decays more slowly than the other term.

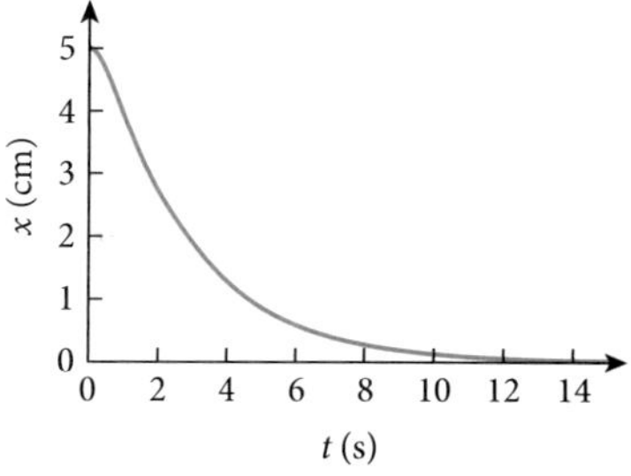

**FIGURE 14.21** Position versus time for an overdamped oscillator.

An example of the motion of an overdamped oscillator is graphed in Figure 14.21. The graph was generated for $k$ = 1.0 N/m, $m$ = 1.0 kg, and $b$ = 3.0 kg/s, such that $b(=3.0 \text{ kg/s})>2\sqrt{mk}\,(=2 \text{ kg/s})$. These parameters result in an angular speed of $\omega_\gamma = 1.5 \text{ s}^{-1}$, compared to an angular speed of $\omega_0$ = 1.0 s$^{-1}$ for the case without damping. The initial displacement was $x_0$ = 5.0 cm, and the system was assumed to be released from rest. The displacement goes to zero without any oscillations.

The coefficients $B$ and $C$ can have opposite signs, so the value of $x$ determined from equation 14.23 can change sign one time at most. Physically, this sign change corresponds to the situation in which the oscillator receives a large initial velocity toward the equilibrium position. In that case, the oscillator can overshoot the equilibrium position and approach it from the other side.

## Critical Damping

Because we have already covered the solutions of equation 14.19 for $b<2\sqrt{mk}$ and $b>2\sqrt{mk}$, you might be tempted to think that we can obtain the solution for $b=2\sqrt{mk}$ via some limit process or some interpolation between the two solutions. This case, however, is not quite so straightforward, and we have to employ a slightly different *Ansatz*. The solution for the case where $b=2\sqrt{mk}$, the solution known as **critical damping,** is given by

$$x(t)=Be^{-\omega_\gamma t}+tCe^{-\omega_\gamma t}, \quad \text{for } b=2\sqrt{mk}, \tag{14.24}$$

where the amplitude coefficients have the values

$$B = x_0,\ C = v_0 + x_0\omega_\gamma.$$

The angular speed in this solution is still $\omega_\gamma = b/2m$.

Let's look at an example in which the coefficients $B$ and $C$ are determined from the initial conditions for the damped harmonic motion.

## EXAMPLE 14.6 Damped Harmonic Motion

### PROBLEM

A spring with the spring constant $k = 1.00$ N/m has an object of mass $m = 1.00$ kg attached to it, which moves in a medium with damping constant $b = 2.00$ kg/s. The object is released from rest at $x = +5$ cm from the equilibrium position. Where will it be after 1.75 s?

### SOLUTION

First, we have to decide which of the three types of damping applies in this situation. To find out, we calculate $2\sqrt{mk} = 2\sqrt{(1\text{ kg})(1\text{ N/m})} = 2$ kg/s, which happens to be exactly equal to the value that was given for the damping constant, $b$. Therefore, this motion is critically damped. We can also calculate the damping angular speed: $\omega_\gamma = b/2m = 1.00\text{ s}^{-1}$.

The problem statement gives the initial conditions: $x_0 = +5$ cm and $v_0 = 0$. We determine the constants $B$ and $C$ from these initial conditions (expressions for these constants). We use equation 14.24 for critical damping:

$$x(t) = Be^{-\omega_\gamma t} + tCe^{-\omega_\gamma t}$$
$$\Rightarrow x(0) = Be^{-\omega_\gamma 0} + 0 \cdot Ce^{-\omega_\gamma 0} = B.$$

Next, we take the time derivative:

$$v = \frac{dx}{dt} = -\omega_\gamma Be^{-\omega_\gamma t} + C(1 - \omega_\gamma t)e^{-\omega_\gamma t}$$
$$\Rightarrow v(0) = -\omega_\gamma Be^{-\omega_\gamma 0} + C\left[1 - (\omega_\gamma \cdot 0)\right]e^{-\omega_\gamma 0} = -\omega_\gamma B + C.$$

According to the problem statement, $v_0 = 0$, which yields $C = \omega_\gamma B$, and $x_0 = +5$ cm, which determines $B$. We have already calculated $\omega_\gamma = 1.00\text{ s}^{-1}$. Thus, we have

$$x(t) = (5\text{ cm})(1 + \omega_\gamma t)e^{-\omega_\gamma t}$$
$$= (5\text{ cm})\left(1 + (1.00\text{ s}^{-1})t\right)e^{-(1.00/\text{s})t}.$$

We now calculate the position of the mass after 1.75 s:

$$x(1.75\text{ s}) = (5\text{ cm})(1 + 1.75)e^{-1.75} = 2.39\text{ cm}.$$

Figure 14.22 shows a plot of the displacement of this critically damped oscillator as a function of time. It also includes the curve from Figure 14.21 for the overdamped oscillator, which has the same mass and spring constant but a larger damping constant, $b = 3.0$ kg/s. Note that the critically damped oscillator reaches zero displacement more quickly than the overdamped oscillator does.

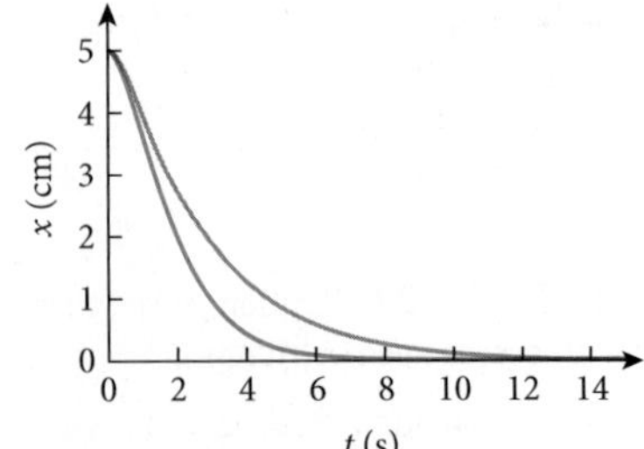


**FIGURE 14.22** The displacement of a critically damped oscillator versus time. The gray line reproduces the curve for the overdamped oscillator from Figure 14.21.

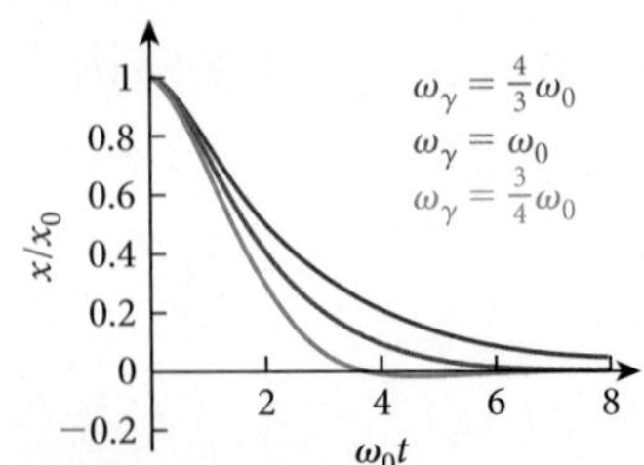


**FIGURE 14.23** Position versus time for underdamped motion (green line), overdamped motion (red line), and critically damped motion (blue line).

To compare underdamped, overdamped, and critically damped motion, Figure 14.23 graphs position against angular speed times time for each type of motion. In all three cases, the oscillator started from rest at time $t = 0$. The green line represents the underdamped case; you can see that oscillation continues. The red line reflects large damping above the critical value, and the blue line reflects critical damping.

An engineering solution providing a damped oscillation that returns as fast as possible to the equilibrium position and has no oscillations needs to use the condition of critical damping. Examples of such an application are the shock absorbers in cars, motorcycles (Figure 14.24), and bicycles. For maximum performance, they need to work at the critical damping limit or just slightly below. As they get worn, the damping effect becomes weaker, and the shock absorbers provide only small damping. This results in a "bouncing" sensation for riders and indicates that it is time to change the vehicle's shock absorbers.

**FIGURE 14.24** Shock absorbers on the front wheel of a motorcycle.

## Energy Loss in Damped Oscillations

Damped oscillations are the result of the velocity-dependent (and thus nonconservative) damping force, $F_\gamma = -bv$, and are described by the differential equation (compare to equation 14.19)

$$\frac{d^2x}{dt^2} + 2\omega_\gamma \frac{dx}{dt} + \omega_0^2 x = 0, \tag{14.25}$$

where $\omega_0 = \sqrt{k/m}$ ($k$ is the spring constant and $m$ is the mass) is the angular speed of the harmonic oscillation without damping and $\omega_\gamma = b/2m$ is the damping angular speed. The nonconservative damping force must result in a loss of total mechanical energy. We'll examine the energy loss for small damping only, because this case is by far of the greatest technological importance. However, large and critical damping can be treated in an analogous manner.

For small damping, we saw that the displacement as a function of time can be written in terms of a single sinusoidal function with a phase shift (compare to equation 14.22):

$$x(t) = Ae^{-\omega_\gamma t} \sin\left(\omega' t + \theta_0\right). \tag{14.26}$$

This solution can be understood as a sinusoidal oscillation in time with angular speed $\omega' = \sqrt{\omega_0^2 - \omega_\gamma^2}$ and with exponentially decreasing amplitude.

We know that the damping force is nonconservative, and so we can be sure that mechanical energy is lost over time. Since the energy of harmonic oscillation is proportional to the square of the amplitude, $E = \frac{1}{2}kA^2$, we might be led to conclude that the energy decreases smoothly and exponentially for a damped oscillation. However, this is not quite the case, which we can demonstrate as follows.

First, we can find the velocity as a function of time by taking the time derivative of equation 14.26:

$$v\left(t\right) = \omega' Ae^{-\omega_\gamma t} \cos\left(\omega' t + \theta_0\right) - \omega_\gamma Ae^{-\omega_\gamma t} \sin\left(\omega' t + \theta_0\right).$$

Next, we know the potential energy stored in the spring as a function of time, $U = \frac{1}{2}kx^2(t)$, as well as the kinetic energy due to the motion of the mass, $K = \frac{1}{2}mv^2(t)$. We can plot these energies and the displacement and velocity as functions of time, as shown in the graphs in Figure 14.25, where $k = 11.00$ N/m, $m = 1.800$ kg, and $b = 0.500$ kg/s. These parameters result in an angular speed of $\omega' = 2.468\ \text{s}^{-1}$, compared to an angular speed of $\omega_0 = 2.472\ \text{s}^{-1}$ for the case without damping, and a damping angular speed of $\omega_\gamma = 0.139\ \text{s}^{-1}$. The amplitude is $A = 5$ cm, and $\theta_0 = 1.6$.

Figure 14.25 shows that the total energy does not decrease in a smooth exponential way but instead declines in a stepwise fashion. This energy loss can be represented in a straightforward way by writing an expression for the power, that is, the rate of energy loss:

$$\frac{dE}{dt} = -bv^2 = vF_\gamma.$$

Thus, the rate of energy loss is largest wherever the velocity has the largest absolute value, which happens as the mass passes through the equilibrium position. When the mass reaches the points where the velocity is zero and it is about to reverse direction, all energy is potential energy and there is no energy loss at these points.

### 14.4 Self-Test Opportunity

Consider a pendulum whose motion is underdamped by a term given by the product of a damping constant, $\alpha$, and the speed of the moving mass. Making the small-angle approximation, we can write the differential equation

$$\frac{d^2\theta}{dt^2} + \alpha \frac{d\theta}{dt} + \frac{g}{\ell}\theta = 0.$$

By analogy with a mass on a spring, give expressions for the angular speeds characterizing this angular motion: $\omega_\gamma$, $\omega'$, and $\omega_0$.

### 14.4 In-Class Exercise

An automobile with a mass of 1640 kg is lifted into the air. When the car is lifted, the suspension spring on each wheel lengthens by 30.0 cm. What damping constant is required for the shock absorber on each wheel to produce critical damping?

a) 101 kg/s

b) 234 kg/s

c) 1230 kg/s

d) 2310 kg/s

e) 4690 kg/s

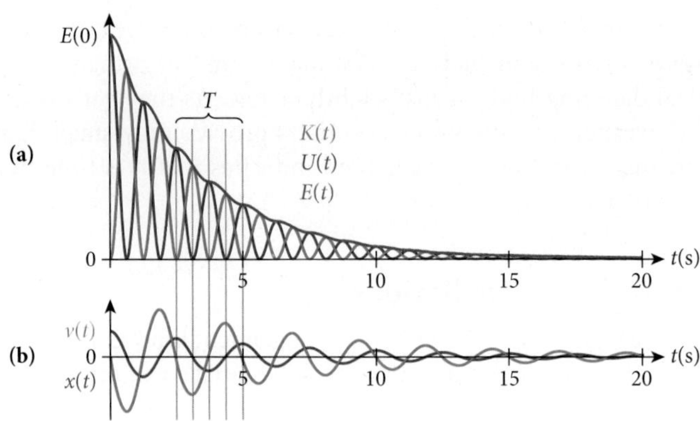


**FIGURE 14.25** (a) Potential energy (red), kinetic energy (green), and total energy (blue). (b) Displacement (red) and velocity (green) for a weakly damped harmonic oscillator. The five thin vertical lines mark one complete oscillation period, *T*.

## DERIVATION 14.3 Energy Loss

We can derive the formula for the energy loss in damped oscillation by taking the time derivatives of the expressions for kinetic and potential energies:

$$\frac{dE}{dt}=\frac{dK}{dt}+\frac{dU}{dt}=\frac{d}{dt}\left(\frac{1}{2}mv^2\right)+\frac{d}{dt}\left(\frac{1}{2}kx^2\right)=mv\frac{dv}{dt}+kx\frac{dx}{dt}.$$

Now we use $v = dx/dt$ and factor out the common factor, $m\,dx/dt$:

$$\frac{dE}{dt}=m\frac{dx}{dt}\frac{d^2x}{dt^2}+kx\frac{dx}{dt}=m\frac{dx}{dt}\left(\frac{d^2x}{dt^2}+\frac{k}{m}x\right).$$

Now we can use $\omega_0^2 = k/m$ to obtain

$$\frac{dE}{dt}=m\frac{dx}{dt}\left(\frac{d^2x}{dt^2}+\omega_0^2x\right). \qquad \text{(i)}$$

We can reorder equation 14.25 by moving the damping term to the right-hand side:

$$\frac{d^2x}{dt^2}+\omega_0^2x=-2\omega_\gamma\frac{dx}{dt}=-\frac{b}{m}\frac{dx}{dt}.$$

Because the left-hand side of this differential equation is equal to the term inside the parentheses in equation (i) for the rate of energy loss, we obtain

$$\frac{dE}{dt}=m\frac{dx}{dt}\left(-\frac{b}{m}\frac{dx}{dt}\right)=-b\left(\frac{dx}{dt}\right)^2=-bv^2.$$

The damping force is $F_\gamma = -bv$, and so we have shown that the rate of energy loss is indeed $dE/dt = vF_\gamma$.

Practical applications require consideration of the **quality** of the oscillator, *Q*, which specifies the ratio of total energy, *E*, to the energy loss, $\Delta E$, over one complete oscillation period, *T*:

$$Q=2\pi\frac{E}{|\Delta E|}. \qquad (14.27)$$

In the limit of zero damping, the oscillator experiences no energy loss, and $Q \rightarrow \infty$. In the limit of small damping, the quality of the oscillator can be approximated by

$$Q\approx\frac{\omega_0}{2\omega_\gamma}. \qquad (14.28)$$

Combining these two results provides a handy formula for the energy loss during a complete oscillation period of weakly damped motion:

$$|\Delta E| = E\frac{2\pi}{Q} = E\frac{4\pi\omega_\gamma}{\omega_0}. \tag{14.29}$$

For the harmonic oscillation graphed in Figure 14.25, equation 14.28 results in $Q = 2.472\ \text{s}^{-1}/2(0.139)\ \text{s}^{-1} = 8.89$. Substituting this value of $Q$ into equation 14.29 gives an energy loss of $|\Delta E| = 0.71E$ during each oscillation period. We can compare this to the exact result for Figure 14.25 by measuring the energy at the beginning and end of the interval indicated in the figure. We find for the energy difference divided by the average energy a value of 0.68, in fairly good agreement with the value of 0.71 obtained by using the approximation of equation 14.28.

## 14.5 Forced Harmonic Motion and Resonance

If you are pushing someone sitting on a swing, you give the person periodic pushes, with the general goal of making him or her swing higher—that is, of increasing the amplitude of the oscillation. This situation is an example of **forced harmonic motion.** Periodically forced or driven oscillations arise in many kinds of problems in various areas of physics, including mechanics, acoustics, optics, and electromagnetism. To represent the periodic driving force, we use

$$F(t) = F_\text{d}\cos(\omega_\text{d}t), \tag{14.30}$$

where $F_\text{d}$ and $\omega_\text{d}$ are constants.

To analyze forced harmonic motion, we start by considering the case of no damping. Using the driving force, $F(t)$, the differential equation for this situation is

$$m\frac{d^2x}{dt^2} = -kx + F_\text{d}\cos(\omega_\text{d}t) \Rightarrow$$

$$\frac{d^2x}{dt^2} + \frac{k}{m}x - \frac{F_\text{d}}{m}\cos(\omega_\text{d}t) = 0.$$

The solution to this differential equation is

$$x(t) = B\sin(\omega_0 t) + C\cos(\omega_0 t) + A_\text{d}\cos(\omega_\text{d}t).$$

The coefficients $B$ and $C$ for the part of the motion that oscillates with the intrinsic angular speed, $\omega_0 = \sqrt{k/m}$, can be determined from the initial conditions. However, much more interesting is the part of the motion that oscillates with the driving angular speed. It can be shown that the amplitude of this forced oscillation is given by

$$A_\text{d} = \frac{F_\text{d}}{m\left(\omega_0^2 - \omega_\text{d}^2\right)}.$$

Thus, the closer the driving angular speed, $\omega_\text{d}$, is to the intrinsic angular speed, $\omega_0$, the larger the amplitude becomes. In the situation of pushing a person on a swing, this amplification is apparent. You can increase the amplitude of the swing's motion only when you push at approximately the same frequency with which the swing is already oscillating. If you push with that frequency, the person on the swing will go higher and higher.

If the driving angular speed is exactly equal to the intrinsic angular speed of the oscillator, the equation $A_\text{d} = F_\text{d}/\left[m(\omega_0^2 - \omega_\text{d}^2)\right]$ predicts that the amplitude will grow infinitely big. In real life, this infinite growth does not occur. Some damping is always present in the system. In the presence of damping, the differential equation to solve becomes

$$\frac{d^2x}{dt^2} + \frac{b}{m}\frac{dx}{dt} + \frac{k}{m}x - \frac{F_\text{d}}{m}\cos(\omega_\text{d}t) = 0. \tag{14.31}$$

This equation has the steady-state solution

$$x(t) = A_\gamma\cos(\omega_\text{d}t - \theta_\gamma), \tag{14.32}$$

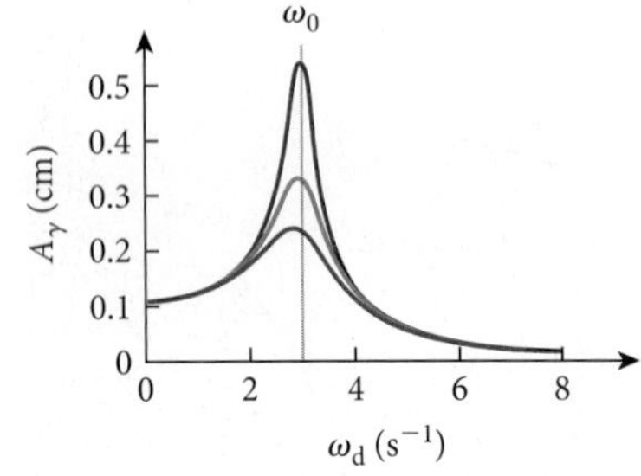


**FIGURE 14.26** The amplitude of forced oscillations as a function of the angular speed of the driving force. The three curves represent different damping angular speeds: $\omega_\gamma = 0.3\ \text{s}^{-1}$ (red curve), $\omega_\gamma = 0.5\ \text{s}^{-1}$ (green curve), and $\omega_\gamma = 0.7\ \text{s}^{-1}$ (blue curve).

where $x$ oscillates at the driving angular speed and the amplitude is

$$A_\gamma = \frac{F_d}{m\sqrt{\left(\omega_0^2 - \omega_d^2\right)^2 + 4\omega_d^2\omega_\gamma^2}}. \tag{14.33}$$

As you can see, the amplitude given by equation 14.33 cannot be infinite, because even at $\omega_d = \omega_0$, it still has a finite value, $F_d/(2m\omega_d\omega_\gamma)$, which is also *approximately* its maximum. The driving frequency corresponding to the maximum amplitude can be calculated by taking the derivative of $A_\gamma$ with respect to $\omega_d$, setting that derivative equal to 0, and solving for $\omega_d$, which gives the result that $A_\gamma$ is maximum when $\omega_d = \sqrt{\omega_0^2 - 2\omega_\gamma^2}$. Note that for $\omega_\gamma \ll \omega_0$, $\omega_d \approx \omega_0$. The shape of the curve of the amplitude $A_\gamma$, as a function of the driving angular speed, $\omega_d$, is called a **resonance shape,** and it is characteristic of all resonance phenomena. When $\omega_d = \sqrt{\omega_0^2 - 2\omega_\gamma^2}$, the driving angular speed is called the **resonant angular speed;** $A_\gamma$ is close to a maximum at this angular speed. In Figure 14.26, $A_\gamma$ is plotted from equation 14.33 for $\omega_0 = 3\ \text{s}^{-1}$ and three values of the damping angular speed, $\omega_\gamma$. As the damping gets weaker, the resonance shape becomes sharper. You can see that in every case, the amplitude curve reaches its maximum at a value of $\omega_d$ slightly below $\omega_0 = 3\ \text{s}^{-1}$.

The phase angle $\theta_\gamma$ in equation 14.32 depends on the driving angular speed as well as on the damping and intrinsic angular speeds:

$$\theta_\gamma = \frac{\pi}{2} - \tan^{-1}\left(\frac{\omega_0^2 - \omega_d^2}{2\omega_d\omega_\gamma}\right). \tag{14.34}$$

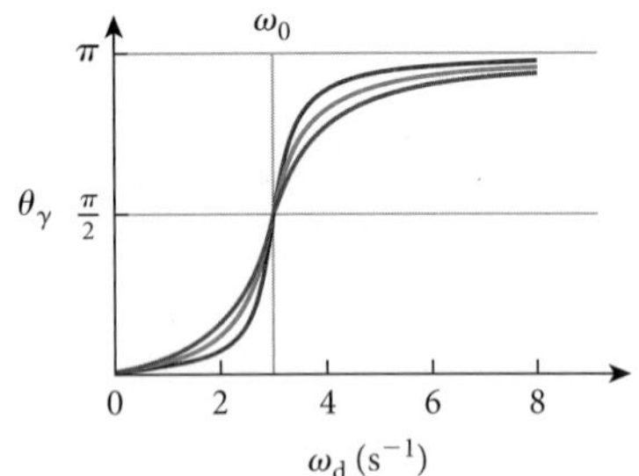


**FIGURE 14.27** Phase shift as a function of driving angular speed with the same parameters as in Figure 14.26. The three curves represent different damping angular speeds: $\omega_\gamma = 0.3\ \text{s}^{-1}$ (red curve), $\omega_\gamma = 0.5\ \text{s}^{-1}$ (green curve), and $\omega_\gamma = 0.7\ \text{s}^{-1}$ (blue curve).

The phase shift as a function of the driving angular speed is graphed in Figure 14.27 with the same parameters used in Figure 14.26. You can see that for small driving angular speeds, the forced oscillation follows the driving in phase. As the driving angular speed increases, the phase difference begins to grow and reaches $\pi/2$ at $\omega_d = \omega_0$. For driving angular speeds much larger than the resonant angular speed, the phase shift approaches $\pi$. The lower the damping angular speed, $\omega_\gamma$, the steeper the transition from 0 to $\pi$ phase shift becomes.

A practical note is in order at this point: The solution presented in equation 14.32 is usually only reached after some time. The full solution for the differential equation for damped and driven oscillation (equation 14.31) is given by a combination of equations 14.20 and 14.32. During the transition time, the effect of the particular initial conditions given to the system is damped out, and the solution given by equation 14.32 is approached asymptotically.

Figure 14.28 shows motion of a system that is driven with a driving angular speed of $\omega_d = 1.2\ \text{s}^{-1}$ and an acceleration of $F/m = 0.6\ \text{m/s}^2$. The system has an intrinsic angular speed of $\omega_0 = 2.2\ \text{s}^{-1}$ and a damping angular speed of $\omega_\gamma = 0.4\ \text{s}^{-1}$ and starts at $x_0 = 0$ with a positive initial velocity. The unforced motion of the system follows equation 14.20 and is shown by the green dashed curve. The motion according to equation 14.32 is shown in red. The full solution (blue curve) is the sum of the two. After some complicated-looking oscillations, during which the initial conditions are damped out, the full solution approaches the simple harmonic motion given by equation 14.32.

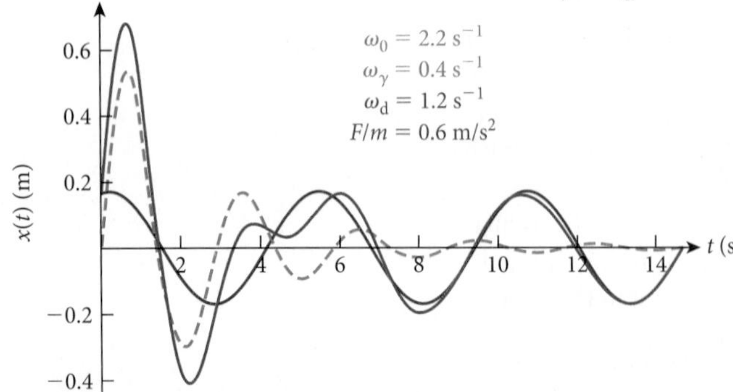


**FIGURE 14.28** Position as a function of time for damped and driven harmonic oscillation: The red curve plots equation 14.32, and the green dashed curve shows the motion with damping (from equation 14.20). The blue curve is the sum of the red and green curves, giving the complete picture.

Driving mechanical systems to resonance has important technical consequences, not all of them desirable. Most worrisome are resonance frequencies in architectural structures. Builders have to be very careful, for example, not to construct high-rises in such a way that earthquakes can drive them into oscillations near resonance. Perhaps the most infamous example of a structure that was destroyed by a resonance phenomenon was the Tacoma Narrows Bridge, known as "Galloping Gertie," in the state of Washington. It collapsed on November 7, 1940, only a few months after construction was finished (Figure 14.29). A 40-mph wind was able to drive the oscillations of the bridge into resonance, and catastrophic mechanical failure occurred.

### 14.5 In-Class Exercise

The system of Figure 14.28 is driven at four different driving angular speeds, and position versus time for all four cases is plotted in the figures. Which of the four cases is closest to resonance?

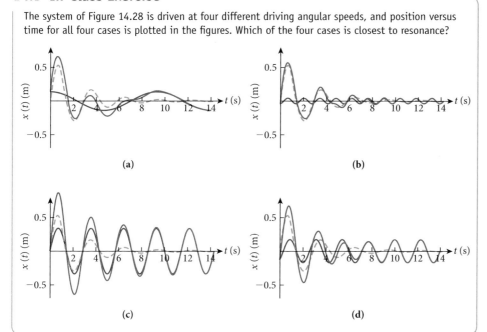

(a)

(b)

(c)

(d)

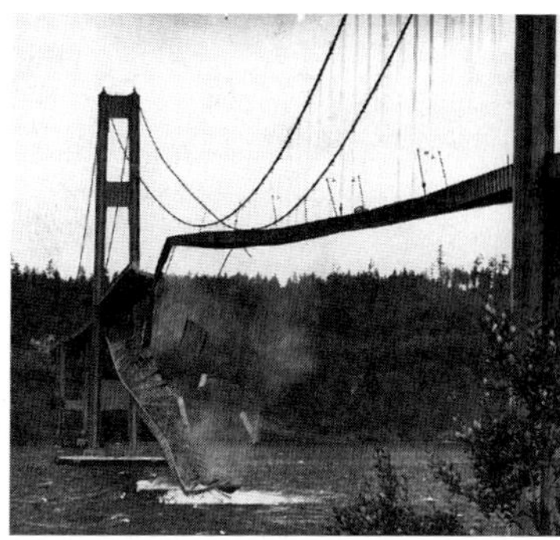

**FIGURE 14.29** Collapse of the midsection of the Tacoma Narrows Bridge on November 7, 1940.

Columns of soldiers marching across a bridge are often told to fall out of lockstep to avoid creating a resonance due to the periodic stomping of their feet. However, a large number of walkers can get locked into phase through a feedback mechanism. On June 12, 2000, the London Millennium Bridge on the Thames River had to be closed after only three days of operation, because 2000 pedestrians walking over the bridge drove it into resonance and earned it the nickname "wobbly bridge." The lateral swaying of the bridge was such that it provided feedback and locked the pedestrians into a stepping rhythm that amplified the oscillations.

## 14.6 Phase Space

We will return repeatedly to the physics of oscillations during the course of this book. As mentioned earlier, the reason for this emphasis is that many systems have restoring forces that drive them back to equilibrium if they are nudged only a small distance from equilibrium. Many problem situations in physics can therefore be modeled in terms of a pendulum or a spring. Before we leave the topic of oscillations for now, let's consider one more interesting way of displaying the motion of oscillating systems. Instead of plotting displacement, velocity, or acceleration as a function of time, we can plot velocity versus displacement. A display of velocity versus displacement is called a **phase space.** Looking at physical motion in this way provides interesting insights.

Let's start with a harmonic oscillation without damping. Plotting an oscillation with time dependence given by $x(t) = A\sin(\omega t)$, amplitude of $A = 4$ cm, and angular speed $\omega = 0.7\ \text{s}^{-1}$ in a phase space gives an ellipse, the red curve in Figure 14.30. Other values of the amplitude, between 3 and 7 in steps of 1, yield different ellipses, shown in blue in Figure 14.30.

Consider another example: Figure 14.31 shows a plot in a phase space of the motion of a weakly damped oscillator (with $k = 11$ N/m, $m = 1.8$ kg, $b = 0.5$ kg/s, $A = 5$ cm, and $\theta_0 = 1.6$), the same oscillator whose position was plotted in Figure 14.19 as a function of time. You can see the starting point of the oscillatory motion,

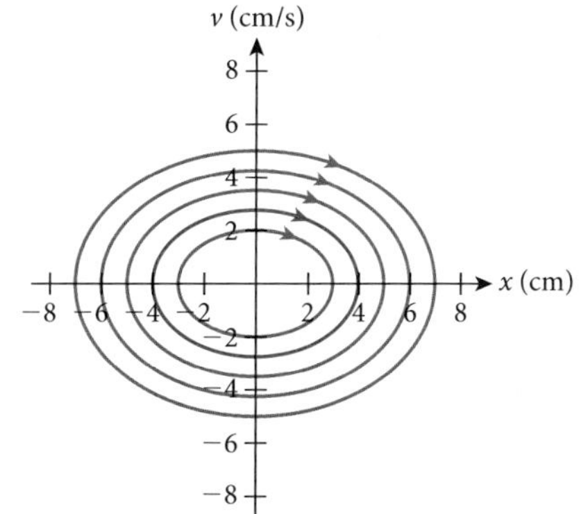


**FIGURE 14.30** Velocity versus displacement for different amplitudes of simple harmonic motion.

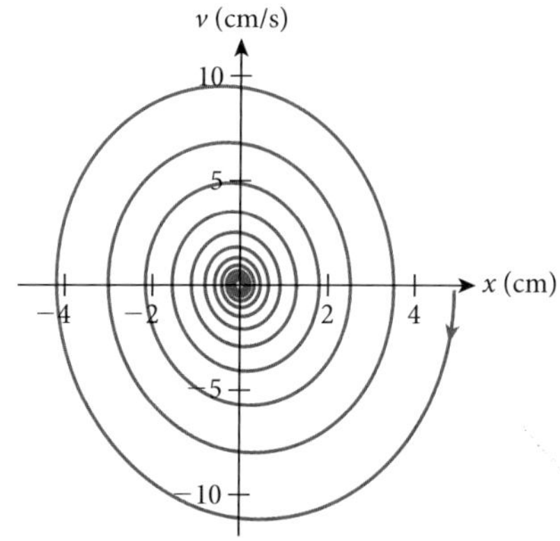


**FIGURE 14.31** Velocity versus displacement for a damped oscillator.

at $x(t = 0) = (5 \text{ cm})\sin(1.6)e^{-0}$ and $v(t = 0) = -1.05$ cm/s. The trajectory in the phase space spirals inward toward the point (0,0). It looks as though the trajectory is attracted to that point. If the oscillation had different initial conditions, the plot would have spiraled inward to the same point, but on a different trajectory. A point to which all trajectories eventually converge is called a *point attractor.*

## 14.7 Chaos

One field of research in physics has emerged only since the advent of computers. This field of **nonlinear dynamics,** including those of **chaos,** has yielded interesting and relevant results during the last decade. If we look hard enough and under certain conditions, we can discover signs of chaotic motion even in the oscillators we have been dealing with in this chapter. In Chapter 7, we looked at chaos and billiards, noting the sensitivity of a system's motion to initial conditions; here we continue this examination in greater detail.

When we analyzed the motion of a pendulum, we used the small-angle approximation (Section 14.2). However, suppose we want to study large deflection angles. We can do this by using a pendulum whose string has been replaced with a thin solid rod. Such a pendulum can move through more than 180°, even through a full 360°. In this case, the approximation $\sin\theta \approx \theta$ does not work, and the sine function must be included in the differential equation. If a periodic external driving force as well as damping is present for this pendulum, we cannot solve the equation of motion in an analytic fashion. However, we can use numerical integration procedures with a computer to obtain the solution. The solution depends on the given initial conditions. Based on what you have learned so far about oscillators, you might expect that similar initial conditions give similar patterns of motion over time. However, for some combinations of driving angular speed and amplitude, damping constant, and pendulum length, this expectation is not borne out.

Figure 14.32 shows plots of the pendulum's deflection angle as a function of time using the same equation of motion and the same value for the initial velocity. However, for the red curve, the initial angle is exactly zero radians. For the green curve, the initial angle is $10^{-3}$ rad. You can see that these two curves stay close to one another for a few oscillations, but then rapidly diverge. If the initial angle is $10^{-5}$ rad (blue curve), the curves do not stay close for much longer. Since the blue curve starts 100 times closer to the red curve, it might be expected to stay close to the red curve longer in time by a factor of 100. This is not the case. This behavior is called *sensitive dependence on initial conditions.* If a system is sensitively dependent on the initial conditions, a long-term prediction of its motion is impossible. This sensitivity is one of the hallmarks of chaotic motion.

Precise long-range weather forecasts are impossible for reasons of sensitive dependence on initial conditions. With the advent of computers in the 1950s, meteorologists thought they would be able to describe the future weather accurately if only they generated enough measurements of all the relevant variables—such as temperature and wind direction—at as many points as possible across the country. However, by the 1960s, it had become obvious that the basic differential equations for air masses moving in the atmosphere show sensitive dependence on initial conditions. Thus, the dream of precise long-range weather forecasting was ended.

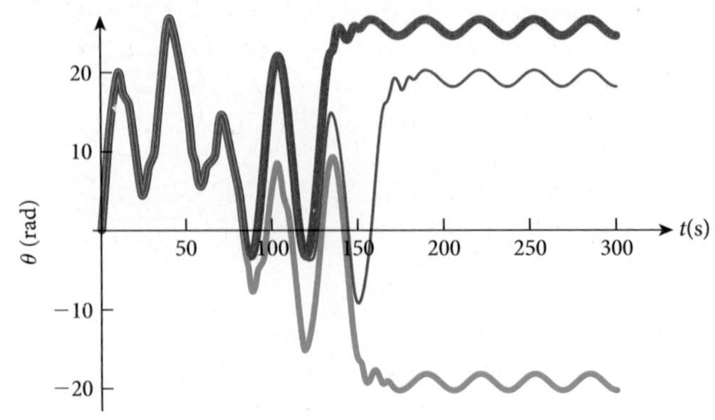


**FIGURE 14.32** Deflection angle as a function of time for a driven pendulum with three different initial conditions: The red line resulted from an initial angle of zero. The green line resulted from an initial angle of $10^{-3}$ rad. The blue line resulted from an initial angle of $10^{-5}$ rad.

## WHAT WE HAVE LEARNED | EXAM STUDY GUIDE

- An object acting under Hooke's Law, $F = -kx$, undergoes simple harmonic motion, with a restoring force that acts in the opposite direction to the displacement from equilibrium.
- The equation of motion for a mass on a spring (with no damping) is $x(t) = B\sin(\omega_0 t) + C\cos(\omega_0 t)$, where $\omega_0 = \sqrt{k/m}$, or, alternatively, $x(t) = A\sin(\omega_0 t + \theta_0)$
- The equation of motion for a pendulum (with no damping) is $\theta(t) = B\sin(\omega_0 t) + C\cos(\omega_0 t)$ with $\omega_0 = \sqrt{g/\ell}$.
- Velocity and acceleration for simple harmonic oscillations:
$$x(t) = A\sin(\omega_0 t + \theta_0)$$
$$\Rightarrow v(t) = \omega_0 A\cos(\omega_0 t + \theta_0)$$
$$\Rightarrow a(t) = -\omega_0^2 A\sin(\omega_0 t + \theta_0).$$
- The period of an oscillation is $T = 2\pi/\omega$.
- The frequency of an oscillation is $f = 1/T = \omega/2\pi$, or $\omega = 2\pi f$.
- The equation of motion for a mass on a spring with small damping $\left(b < 2\sqrt{mk}\right)$ is $x(t) = Be^{-\omega_\gamma t}\cos(\omega' t) + Ce^{-\omega_\gamma t}\sin(\omega' t)$, with $\omega_\gamma = b/2m$, $\omega' = \sqrt{k/m - (b/2m)^2} = \sqrt{\omega_0^2 - \omega_\gamma^2}$, and $\omega_0 = \sqrt{k/m}$.
- The equation of motion for a mass on a spring with large damping $\left(b > 2\sqrt{mk}\right)$ is
$$x(t) = Be^{-\left(\omega_\gamma + \sqrt{\omega_\gamma^2 - \omega_0^2}\right)t} + Ce^{-\left(\omega_\gamma - \sqrt{\omega_\gamma^2 - \omega_0^2}\right)t}.$$
- The equation of motion for a mass on a spring with critical damping $\left(b = 2\sqrt{mk}\right)$ is $x(t) = (B + tC)e^{-\omega_\gamma t}$.
- For a damped harmonic oscillator, the rate of energy loss is $dE/dt = -bv^2 = vF_\gamma$.
- The quality of an oscillator is related to the energy loss during one complete oscillation period, $Q = 2\pi E/|\Delta E|$.
- If a periodic external driving force, $F(t) = F_d\cos(\omega_d t)$, is applied to an oscillating system, the equation of motion becomes (after some transient time) $x(t) = A_\gamma\cos(\omega_d t - \theta_\gamma)$, with amplitude $A_\gamma = \dfrac{F_d}{m\sqrt{\left(\omega_0^2 - \omega_d^2\right)^2 + 4\omega_d^2\omega_\gamma^2}}$. The graph of this amplitude displays a resonance shape in which the amplitude is a maximum when the driving angular speed approximately equals the intrinsic angular speed ($\omega_d = \omega_0$).
- A phase space displays a plot of the velocity of an object versus its position. Simple harmonic motion yields an ellipse in such a phase space. A plot of damped harmonic motion spirals in toward the point attractor at (0,0).
- With the right choice of parameters, a damped and driven rod pendulum can show chaotic motion which is sensitively dependent on the initial conditions. In this case, the system's long-term behavior is not predictable.

## KEY TERMS

periodic motion, p. 456
simple harmonic motion (SHM), p. 456
amplitude, p. 457
angular speed, p. 457
period, p. 460
frequency, p. 461
pendulum, p. 464
damping, p. 470
overdamping, p. 473
critical damping, p. 473
quality, p. 476
forced harmonic motion, p. 477
resonance shape, p. 478
resonant angular speed, p. 478
phase space, p. 479
nonlinear dynamics, p. 480
chaos, p. 480

## NEW SYMBOLS AND EQUATIONS

$\omega_0 = \sqrt{\dfrac{k}{m}}$, angular speed of simple harmonic motion

$x(t) = B\sin(\omega_0 t) + C\cos(\omega_0 t)$, displacement function of oscillation

$T = \dfrac{2\pi}{\omega}$, period of oscillation

$f = \dfrac{1}{T}$, frequency of oscillation

$\omega_0 = \sqrt{\dfrac{g}{\ell}}$, angular speed for a pendulum

$b = 2\sqrt{mk}$, damping constant for critical damping

$Q = 2\pi\dfrac{E}{|\Delta E|}$, quality of an oscillator

## ANSWERS TO SELF-TEST OPPORTUNITIES

**14.1** $\frac{d^2x}{dt^2}+\frac{k}{m}x=0$. Use the trial function $x(t)=A\sin(\omega_0 t+\theta_0)$ and take derivatives:

$$\frac{dx}{dt}=A\omega_0\cos(\omega_0 t+\theta_0);$$

$$\frac{d^2x}{dt^2}=-A\omega_0^2\sin(\omega_0 t+\theta_0).$$

Now insert back into the differential equation with $\omega_0^2=\frac{k}{m}$. We find:

$$-A\omega_0^2\sin(\omega_0 t+\theta_0)+A\omega_0^2\sin(\omega_0 t+\theta_0)=0.$$

**14.2** Use the trig identity $\sin(\alpha+\beta)=\sin(\alpha)\cos(\beta)+\cos(\alpha)\sin(\beta)$

$A\sin(\omega_0 t+\theta_0)=A\sin(\omega_0 t)\cos(\theta_0)+A\cos(\omega_0 t)\sin(\theta_0)$

$A\sin(\omega_0 t+\theta_0)=(A\cos(\theta_0))\sin(\omega_0 t)+(A\sin(\theta_0))\cos(\omega_0 t)=B\sin(\omega_0 t)+C\cos(\omega_0 t)$

$B=A\cos(\theta_0)$

$C=A\sin(\theta_0)$

square and add these two equations

$B^2+C^2=A^2\sin^2(\theta_0)+A^2\cos^2(\theta_0)=A^2(\sin^2(\theta_0)+\cos^2(\theta_0))=A^2$

$A=\sqrt{B^2+C^2}$.

Now divide these same two equations

$$\frac{C}{B}=\frac{A\sin(\theta_0)}{A\cos(\theta_0)}=\tan(\theta_0)$$

$$\theta_0=\tan^{-1}\left(\frac{C}{B}\right).$$

**14.3** $T_{\text{Moon}}=\sqrt{\frac{g_{\text{Earth}}}{g_{\text{Moon}}}}T=\sqrt{\frac{9.81\text{ m/s}^2}{1.63\text{ m/s}^2}}T=2.45T.$

**14.4** $\omega_\gamma=\frac{\alpha}{2};\ \omega_0=\sqrt{\frac{g}{\ell}};\ \omega'=\sqrt{\omega_0^2-\omega_\gamma^2}.$

## PROBLEM-SOLVING PRACTICE

### Problem-Solving Guidelines

**1.** In oscillatory motion, some quantities are characteristic of the motion: position, $x$; velocity, $v$; acceleration, $a$; energy, $E$; phase angle, $\theta_0$; amplitude, $A$; and maximum velocity, $v_{max}$. Other quantities are characteristic of the particular system: mass, $m$; spring constant, $k$; pendulum length, $\ell$; period, $T$; frequency, $f$; and angular frequency, $\omega$. It is helpful to list the quantities you are looking for to solve a problem and the quantities you know or need to determine to find those unknowns.

**2.** Some motion starts out as motion along a line, sometimes with constant acceleration (for example, free fall), and then changes to oscillatory motion. In such cases, you need to identify where or when the change in motion takes place and which equations you need to use for each kind of motion.

**3.** In problems involving damping, it is essential to first determine whether the system is weakly damped, critically damped, or overdamped, because the form of the equation of motion you should use depends on this.

### SOLVED PROBLEM 14.2 Determining the Damping Constant

**PROBLEM**

A 1.75-kg block is connected to a vertical spring with spring constant $k = 3.50$ N/m. The block is pulled upward a distance $d = 7.50$ cm and released from rest. After 113 complete oscillations, the amplitude of the oscillation is half the original amplitude. The damping of the block's motion is proportional to the speed. What is the damping constant, $b$?

**SOLUTION**

**THINK**

Because the block executes 113 oscillations and the amplitude decreases by a factor of 2, we know we are dealing with small damping. After an integral number of oscillations, the cosine term will be 1 and the sine term will be zero in equation 14.20. Knowing that the amplitude has decreased by a factor of 2 after 113 oscillations, we can get an expression for $\omega_\gamma$ in terms of the time required to complete 113 oscillations, from which we can obtain the damping constant.

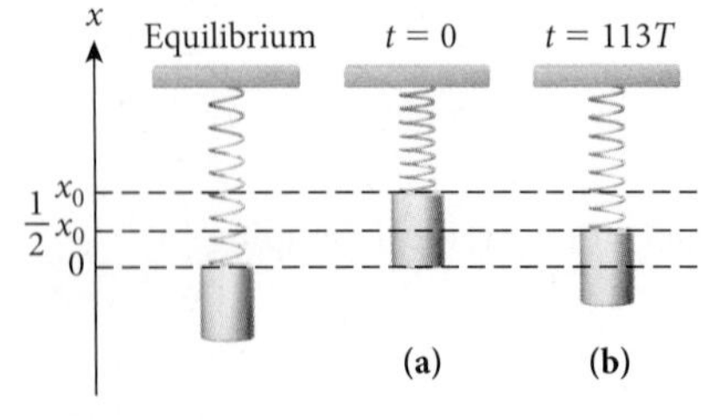


**FIGURE 14.33** Position of the oscillating block at (a) $t = 0$ and (b) $t = 113T$.

**SKETCH**

Figure 14.33 shows the position of the block at $t = 0$ and at $t = 113T$, where $T$ is the period of oscillation.

**RESEARCH**

Because we know that we are dealing with small damping, we can apply equation 14.20:

$$x(t) = Be^{-\omega_\gamma t}\cos(\omega' t) + Ce^{-\omega_\gamma t}\sin(\omega' t).$$

At $t = 0$, we know that $x = x_0$ and so $B = x_0$. After 113 complete oscillations, the cosine term will be 1, the sine term will be zero, and $x = x_0/2$. Thus, we can write

$$\frac{x_0}{2} = \left(x_0 e^{-\omega_\gamma(113T)} \cdot 1\right) + 0, \qquad \text{(i)}$$

where $T = 2\pi/\omega'$ is the period of oscillation. We can express equation (i) as

$$0.5 = e^{-\omega_\gamma\left[113(2\pi/\omega')\right]}. \qquad \text{(ii)}$$

Taking the natural logarithm of both sides of equation (ii) gives us

$$\ln 0.5 = -\omega_\gamma\left[113\left(\frac{2\pi}{\omega'}\right)\right].$$

Remembering that $\omega' = \sqrt{\omega_0^2 - \omega_\gamma^2}$, where $\omega_0 = \sqrt{k/m}$ ($k$ is the spring constant and $m$ is the mass), we can write

$$-\frac{\ln 0.5}{113(2\pi)} = \frac{\omega_\gamma}{\omega'} = \frac{\omega_\gamma}{\sqrt{\omega_0^2 - \omega_\gamma^2}}. \qquad \text{(iii)}$$

We can square equation (iii) to obtain

$$\frac{\omega_\gamma^2}{\omega_0^2 - \omega_\gamma^2} = \left(-\frac{\ln 0.5}{113(2\pi)}\right)^2. \qquad \text{(iv)}$$

**SIMPLIFY**

In order to save a little effort in writing the constants on the right-hand side of equation (iv), we use the abbreviation

$$c^2 = \left(-\frac{\ln 0.5}{113(2\pi)}\right)^2.$$

Then we can write

$$\omega_\gamma^2 = \left(\omega_0^2 - \omega_\gamma^2\right)c^2 = \omega_0^2 c^2 - \omega_\gamma^2 c^2.$$

We can now solve for $\omega_\gamma$:

$$\omega_\gamma = \sqrt{\frac{\omega_0^2 c^2}{1 + c^2}}.$$

Remembering that $\omega_\gamma = b/2m$, we can write an expression for the damping constant, $b$:

$$b = 2m\sqrt{\frac{\omega_0^2 c^2}{1 + c^2}}.$$

Substituting $\omega_0 = \sqrt{k/m}$, we obtain

$$b = 2m\sqrt{\frac{(k/m)c^2}{1 + c^2}} = 2\sqrt{\frac{mkc^2}{1 + c^2}}.$$

**CALCULATE**

We first put in the numerical values to find $c^2$:

$$c^2 = \left(-\frac{\ln 0.5}{113(2\pi)}\right)^2 = 9.53091 \cdot 10^{-7}.$$

*Continued—*

We can now calculate the damping constant:

$$b = 2\sqrt{\frac{(1.75 \text{ kg})(3.50 \text{ N/m})(9.53091 \cdot 10^{-7})}{1 + 9.53091 \cdot 10^{-7}}} = 0.00483226 \text{ kg/s}.$$

**ROUND**

We report our result to three significant figures:

$$b = 4.83 \cdot 10^{-3} \text{ kg/s}.$$

**DOUBLE-CHECK**

We can check that the calculated damping constant is consistent with small damping by verifying that $b < \sqrt{mk}$:

$$4.83 \cdot 10^{-3} \text{ kg/s} < \sqrt{(1.75 \text{ kg})(3.50 \text{ N/m})} = 2.47 \text{ kg/s}.$$

Thus, the damping constant is consistent with small damping.

## SOLVED PROBLEM 14.3 Oscillation of a Thin Rod

**PROBLEM**

A long, thin rod swings about a frictionless pivot at one end. The rod has a mass of 2.50 kg and a length of 1.25 m. The bottom of the rod is pulled to the right until the rod makes an angle $\theta = 20.0°$ with respect to the vertical. The rod is then released from rest and oscillates in simple harmonic motion. What is the period of this motion?

**SOLUTION**

**THINK**

We cannot apply the standard analysis to the motion of this pendulum because we must take into account the moment of inertia of the thin rod. This type of pendulum is called a *physical pendulum*. We can relate the torque due to the weight of the rod and acting at its center of gravity to the angular acceleration. From this relationship, we can get a differential equation that has a form similar to that of the equation we derived for a simple pendulum. By analogy to the simple pendulum, we can find the period of the oscillating thin rod.

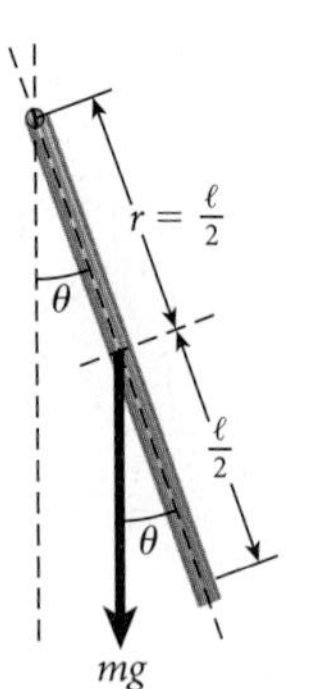

**FIGURE 14.34** A physical pendulum consisting of a thin rod swinging from one end.

**SKETCH**

Figure 14.34 shows the oscillating thin rod.

**RESEARCH**

Looking at Figure 14.34, we can see that the weight, $mg$, of the thin rod acts as a force producing a torque of magnitude $\tau$, given by

$$\tau = I\alpha = -mgr\sin\theta, \quad \text{(i)}$$

where $I$ is the moment of inertia, $\alpha$ is the resulting angular acceleration, $r$ is the moment arm, and $\theta$ is the angle between the force and the moment arm. The problem states that the initial angular displacement is 20°, so we can use the small-angle approximation, $\sin\theta \approx \theta$. We can then rewrite equation (i) as

$$I\alpha + mgr\theta = 0.$$

We now replace the angular acceleration with the second derivative of the angular displacement and divide through by the moment of inertia, which gives us

$$\frac{d^2\theta}{dt^2} + \frac{mgr}{I}\theta = 0.$$

This differential equation has the same form as equation 14.12; therefore, we can write its solution as

$$\theta(t) = B\sin(\omega_0 t) + C\cos(\omega_0 t), \quad \text{with } \omega_0 = \sqrt{\frac{mgr}{I}}.$$

**SIMPLIFY**

We can write the period of oscillation of the thin rod:

$$T = \frac{2\pi}{\omega_0} = 2\pi\sqrt{\frac{I}{mgr}}.$$

This result holds in general for the period of a physical pendulum that has moment of inertia $I$ and whose center of gravity is a distance $r$ from the pivot point. In this case, $r = \ell/2$ and $I = \frac{1}{3}m\ell^2$, so we have

$$T = 2\pi\sqrt{\frac{\frac{1}{3}m\ell^2}{mg(\ell/2)}} = 2\pi\sqrt{\frac{2\ell}{3g}}.$$

**CALCULATE**

Putting in the numerical values gives us

$$T = 2\pi\sqrt{\frac{2(1.25\text{ m})}{3(9.81\text{ m/s})^2}} = 1.83128\text{ s.}$$

**ROUND**

We report our result to three significant figures:

$$T = 1.83\text{ s.}$$

**DOUBLE-CHECK**

If the mass $m$ of the rod were concentrated at a point a distance $\ell$ from the pivot point, it would be a simple pendulum and the period would be

$$T = 2\pi\sqrt{\frac{\ell}{g}} = 2.24\text{ s.}$$

The period of the thin rod is less than this value, which is reasonable because the mass of the thin rod is spread out over its length rather than being concentrated at the end.

As a second check, we start with $T = 2\pi\sqrt{I/mgr}$ and substitute the expression for the moment of inertia of a point mass moving around a point at a distance $r$ away: $I = mr^2$. Taking $r = \ell$, we get

$$T = 2\pi\sqrt{\frac{m\ell^2}{mg\ell}} = 2\pi\sqrt{\frac{\ell}{g}},$$

which is the result for the period of a simple pendulum.

## MULTIPLE-CHOICE QUESTIONS

**14.1** Two children are on adjacent playground swings of the same height. They are given pushes by an adult and then left to swing. Assuming that each child on a swing can be treated as a simple pendulum and that friction is negligible, which child takes the longer time for one complete swing (has a longer period)?

a) the bigger child
b) the lighter child
c) neither child
d) the child given the biggest push

**14.2** Identical blocks oscillate on the end of a vertical spring, one on Earth and one on the Moon. Where is the period of the oscillations greater?

a) on Earth
b) on the Moon
c) same on both Earth and Moon
d) cannot be determined from the information given

**14.3** A mass that can oscillate without friction on a horizontal surface is attached to a horizontal spring that is pulled to the right 10.0 cm and is released from rest. The period of oscillation for the mass is 5.60 s. What is the speed of the mass at $t = 2.50$ s?

a) $-2.61 \cdot 10^{-1}$m/s
b) $-1.06 \cdot 10^{-1}$m/s
c) $-3.71 \cdot 10^{-1}$m/s
d) $-2.01 \cdot 10^{-1}$m/s

**14.4** The spring constant for a spring-mass system undergoing simple harmonic motion is doubled. If the total energy remains unchanged, what will happen to the maximum amplitude of the oscillation? Assume that the system is underdamped.

a) It will remain unchanged.
b) It will be multiplied by 2.
c) It will be multiplied by $\frac{1}{2}$.
d) It will be multiplied by $1/\sqrt{2}$.

**14.5** With the right choice of parameters, a damped and driven physical pendulum can show chaotic motion, which is sensitively dependent on the initial conditions. Which statement about such a pendulum is true?

a) Its long-term behavior can be predicted.

b) Its long-term behavior is not predictable.

c) Its long-term behavior is like that of a simple pendulum of equivalent length.

d) Its long-term behavior is like that of a conical pendulum.

e) None of the above is true.

**14.6** A spring is hanging from the ceiling with a mass attached to it. The mass is pulled downward, causing it to oscillate vertically with simple harmonic motion. Which of the following will increase the frequency of oscillation?

a) adding a second, identical spring with one end attached to the mass and the other to the ceiling

b) adding a second, identical spring with one end attached to the mass and the other to the floor

c) increasing the mass

d) adding both springs, as described in (a) and (b)

**14.7** A child of mass $M$ is swinging on a swing of length $L$ to a maximum deflection angle of $\theta$. A man of mass $4M$ is swinging on a similar swing of length $L$ to a maximum angle of $2\theta$. Each swing can be treated as a simple pendulum undergoing simple harmonic motion. If the period for the child's motion is $T$, then the period for the man's motion is

a) $T$. b) $2T$. c) $T/2$. d) $T/4$.

**14.8** Object A is four times heavier than object B. Each object is attached to a spring, and the springs have equal spring constants. The two objects are then pulled from their equilibrium positions and released from rest. What is the ratio of the periods of the two oscillators if the amplitude of A is half that of B?

a) $T_A:T_B = 1:4$

b) $T_A:T_B = 4:1$

c) $T_A:T_B = 2:1$

d) $T_A:T_B = 1:2$

**14.9** Rank the simple harmonic oscillators shown in the figure in order by their intrinsic frequencies, from highest to lowest. All the springs have identical spring constants, and all the blocks have identical masses.

(a)
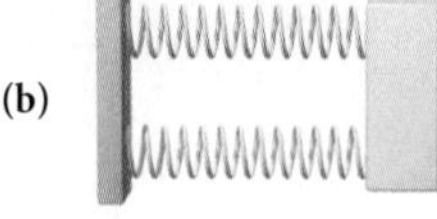

(b)
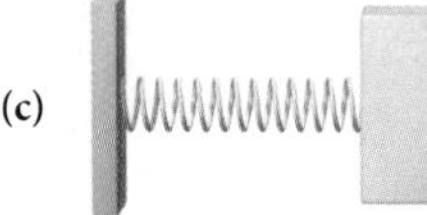

(c)

(d)
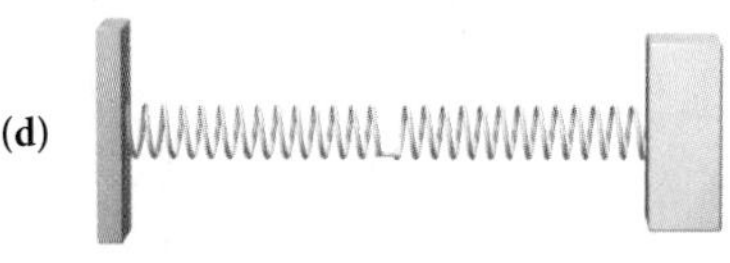

(e)
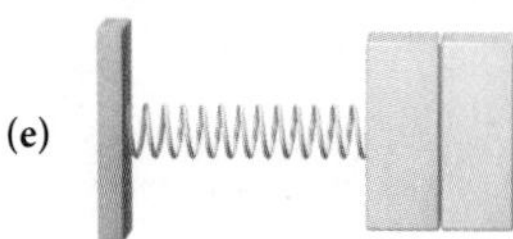

**14.10** A pendulum is suspended from the ceiling of an elevator. When the elevator is at rest, the period of the pendulum is $T$. The elevator accelerates upward, and the period of the pendulum is then

a) still $T$. b) less than $T$. c) greater than $T$.

**14.11** A 36-kg mass is placed on a horizontal frictionless surface and then connected to walls by two springs with spring constants $k_1 = 3$ N/m and $k_2 = 4$ N/m, as shown in the figure. What is the period of oscillation for the 36-kg mass if it is displaced slightly to one side?

a) 11 s

b) 14 s

c) 17 s

d) 20 s

e) 32 s

f) 38 s

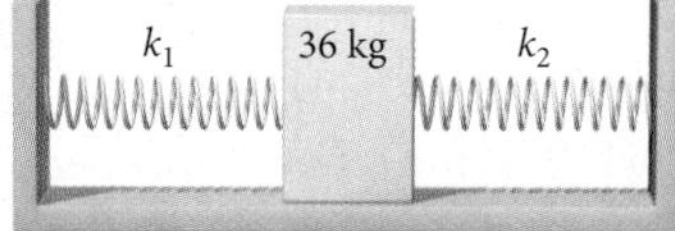


## QUESTIONS

**14.12** You sit in your SUV in a traffic jam. Out of boredom, you rock your body from side to side and then sit still. You notice that your SUV continues rocking for about 2 s and makes three side-to-side oscillations, before subsiding. You estimate that the amplitude of the SUV's motion must be less than 5% of the amplitude of your initial rocking. Knowing that your SUV has a mass of 2200 kg (including the gas, your stuff, and yourself), what can you say about the effective spring constant and damping of your SUV's suspension?

**14.13** An astronaut, taking his first flight on a space shuttle, brings along his favorite miniature grandfather clock. At launch, the clock and his digital wristwatch are synchronized, and the grandfather clock is pointed in the direction of the shuttle's nose. During the boost phase, the shuttle has an upward acceleration whose magnitude is several times the value of the gravitational acceleration at the Earth's surface. As the shuttle reaches a constant cruising speed after completion of the boost phase, the astronaut compares his grandfather clock's time to that of his watch. Are the timepieces still synchronized? If not, which is ahead of the other? Explain.

**14.14** A small cylinder of mass $m$ can slide without friction on a shaft that is attached to a turntable, as shown in the figure. The shaft also passes through the coils of a spring with spring constant $k$, which is attached to the turntable at one end, and to the cylinder at the other end. The equilibrium length of the spring (unstretched and uncompressed) matches the radius of the turntable; thus, when the turntable is not

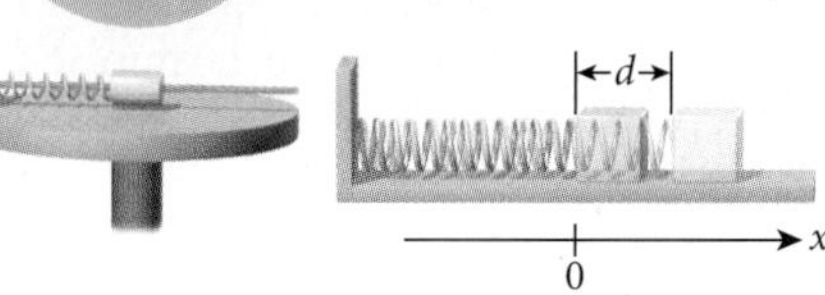

rotating, the cylinder is at equilibrium at the center of the turntable. The cylinder is given a small initial displacement and, at the same time, the turntable is set into uniform circular motion with angular speed $\omega$. Calculate the period of the oscillations performed by the cylinder. Discuss the result.

14.15 A door closer on a door allows the door to close by itself, as shown in the figure. The door closer consists of a return spring attached to an oil-filled damping piston. As the spring pulls the piston to the right, teeth on the piston rod engage a gear, which rotates and causes the door to close. Should the system be underdamped, critically damped, or overdamped for the best performance (where the door closes quickly without slamming into the frame)?

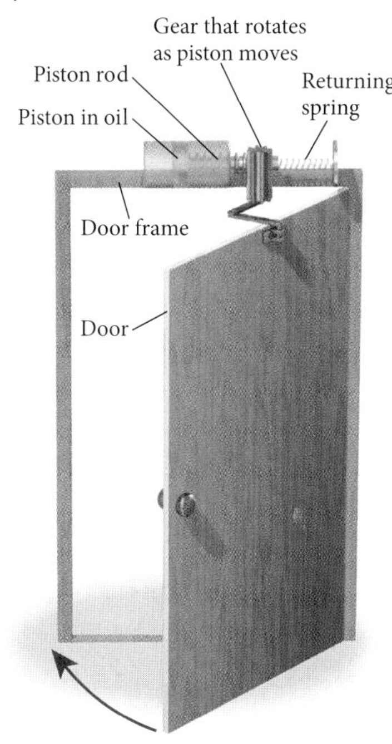


14.16 When the amplitude of the oscillation of a mass on a stretched string is increased, why doesn't the period of oscillation also increase?

14.17 Mass-spring systems and pendulum systems can both be used in mechanical timing devices. What are the advantages of using one type of system rather than the other in a device designed to generate reproducible time measurements over an extended period of time?

14.18 You have a linear (following Hooke's Law) spring with an unknown spring constant, a standard mass, and a timer. Explain carefully how you could most practically use these to measure masses *in the absence of gravity*. Be as quantitative as you can. Regard the mass of the spring as negligible.

14.19 Pendulum A has a bob of mass $m$ hung from a string of length $L$; pendulum B is identical to A except its bob has mass $2m$. Compare the frequencies of small oscillations of the two pendulums.

14.20 A sharp spike in a plotted curve of frequency can be represented as a sum of sinusoidal functions of all possible frequencies, with equal amplitudes. A bell struck with a hammer rings at its natural frequency, that is, the frequency at which it vibrates as a free oscillator. Explain why, as clearly and concisely as you can.

## PROBLEMS

A blue problem number indicates a worked-out solution is available in the Student Solutions Manual. One • and two •• indicate increasing level of problem difficulty.

### Section 14.1

14.21 A mass $m = 5$ kg is suspended from a spring and oscillates according to the equation of motion $x(t) = 0.5\cos(5t + \pi/4)$. What is the spring constant?

14.22 Determine the frequency of oscillation of a 200-g block that is connected by a spring to a wall and sliding on a frictionless surface for the three conditions shown in the figure.

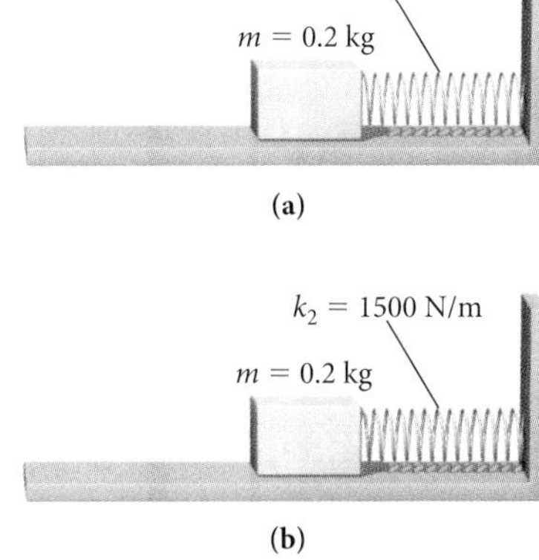


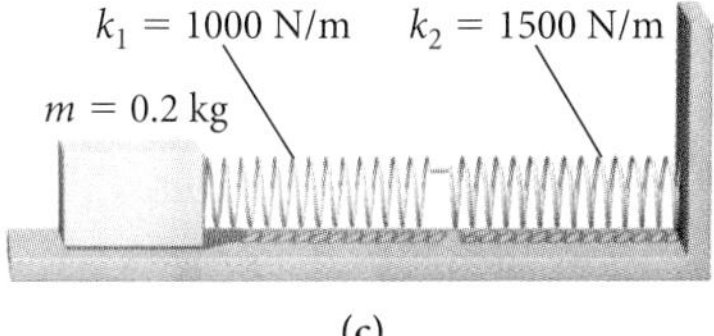


•14.23 A mass of 10.0 kg is hanging by a steel wire 1.00 m long and 1.00 mm in diameter. If the mass is pulled down slightly and released, what will be the *frequency* of the resulting oscillations? Young's modulus for steel is $2.0 \cdot 10^{11}$ N/m$^2$.

•14.24 A 100-g block hangs from a spring with $k = 5.0$ N/m. At $t = 0$ s, the block is 20.0 cm below the equilibrium position and moving upward with a speed of 200 cm/s. What is the block's speed when the displacement from equilibrium is 30.0 cm?

•14.25 A block of wood of mass 55 g floats in a swimming pool, oscillating up and down in simple harmonic motion with a frequency of 3 Hz.

a) What is the value of the effective spring constant of the water?

b) A partially filled water bottle of almost the same size and shape as the block of wood but with mass 250 g is placed on the water's surface. At what frequency will the bottle bob up and down?

•14.26 When a mass is attached to a vertical spring, the spring is stretched a distance $d$. The mass is then pulled down from this position and released. It undergoes 50 oscillations in 30.0 s. What was the distance $d$?

•14.27 A cube of density $\rho_c$ floats in a liquid of density $\rho_l$ as shown in the figure. At rest, an amount $h$ of the cube's height is submerged in liquid. If the cube is pushed down, it bobs up and down like a spring and oscillates about its equilibrium position. Show that the frequency of its oscillations is given by $f = (2\pi)^{-1}\sqrt{g/h}$.

•**14.28** A U-shaped glass tube with a uniform cross-sectional area, $A$, is partly filled with fluid of density $\rho$. Increased pressure is applied to one of the arms resulting in a difference in elevation of $L$ between the two arms of the tube, as shown in the figure. The increased pressure is then removed, and the fluid oscillates in the tube. Determine the period of the oscillations of the fluid column. (You have to determine what the known quantities are.)

•**14.29** The figure shows a mass $m_2 = 20$ g resting on top of a mass $m_1 = 20$ g which is attached to a spring with $k = 10$ N/m. The coefficient of static friction between the two masses is 0.6. The masses are oscillating with simple harmonic motion on a frictionless surface. What is the maximum amplitude the oscillation can have without $m_2$ slipping off $m_1$?

••**14.30** Consider two identical oscillators, each with spring constant $k$ and mass $m$, in simple harmonic motion. One oscillator is started with initial conditions $x_0$ and $v_0$; the other starts with slightly different conditions, $x_0 + \delta x$ and $v_0 + \delta v$.

a) Find the difference in the oscillators' positions, $x_1(t) - x_2(t)$, for all $t$.

b) This difference is *bounded;* that is, there exists a constant $C$, independent of time, for which $|x_1(t) - x_2(t)| \leq C$ holds for all $t$. Find an expression for $C$. What is the *best* bound, that is, the smallest value of $C$ that works? (*Note:* An important characteristic of *chaotic* systems is exponential sensitivity to initial conditions; the difference in position of two such systems with slightly different initial conditions grows exponentially with time. You have just shown that an oscillator in simple harmonic motion is not a chaotic system.)

## Section 14.2

**14.31** What is the period of a simple pendulum that is 1 m long in each situation?

a) in the physics lab

b) in an elevator accelerating at 2.1 m/s$^2$ upward

c) in an elevator accelerating 2.1 m/s$^2$ downward

d) in an elevator that is in free fall

**14.32** Suppose a simple pendulum is used to measure the acceleration due to gravity at various points on the Earth. If $g$ varies by 0.16% over the various locations where it is sampled, what is the corresponding variation in the period of the pendulum? Assume that the length of the pendulum does not change from one site to another.

•**14.33** A 1-kg mass is connected to a 2-kg mass by a massless rod 30 cm long, as shown in the figure. A hole is then drilled in the rod 10 cm away from the 1-kg mass, and the rod and masses are free to rotate about this pivot point, $P$. Calculate the period of oscillation for the masses if they are displaced slightly from the stable equilibrium position.

•**14.34** A torsional pendulum is a vertically suspended wire with a mass attached to one end that is free to rotate. The wire resists twisting as well as elongation and compression, so it is subject to a rotational equivalent of Hooke's Law: $\tau = -\kappa\theta$. Use this information to find the frequency of a torsional pendulum.

•**14.35** A physical pendulum consists of a uniform rod of mass $M$ and length $L$. The pendulum is pivoted at a point that is a distance $x$ from the center of the rod, so the period for oscillation of the pendulum depends on $x$: $T(x)$.

a) What value of $x$ gives the maximum value for $T$?

b) What value of $x$ gives the minimum value for $T$?

•**14.36** Shown in the figure are four different pendulums: (a) a rod has a mass $M$ and a length $L$; (b) a rod has a mass $2M$ and a length $L$; (c) a mass $M$ is attached to a massless string of a length $L$; and (d) a mass $M$ is attached to a massless string of length $\frac{1}{2}L$. Find the period of each pendulum when it is pulled 20° to the right and then released.

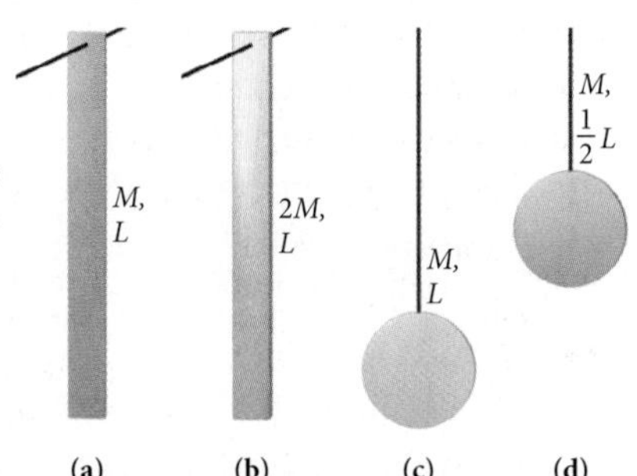


•**14.37** A grandfather clock uses a physical pendulum to keep time. The pendulum consists of a uniform thin rod of mass $M$ and length $L$ that is pivoted freely about one end, with a solid sphere of the same mass, $M$, and a radius of $L/2$ centered about the free end of the rod.

a) Obtain an expression for the moment of inertia of the pendulum about its pivot point as a function of $M$ and $L$.

b) Obtain an expression for the period of the pendulum for small oscillations.

c) Determine the length $L$ that gives a period of $T = 2.0$ s.

•**14.38** A standard CD has a diameter of 12 cm, and a hole that is centered on the axis of symmetry and has a diameter of 1.5 cm. The CD's thickness is 2.0 mm. If a (very thin) pin extending through the hole of the CD suspends it in a vertical orientation, as shown in the figure, the CD may oscillate about an axis parallel to the pin, rocking back and forth. Calculate the oscillation frequency.

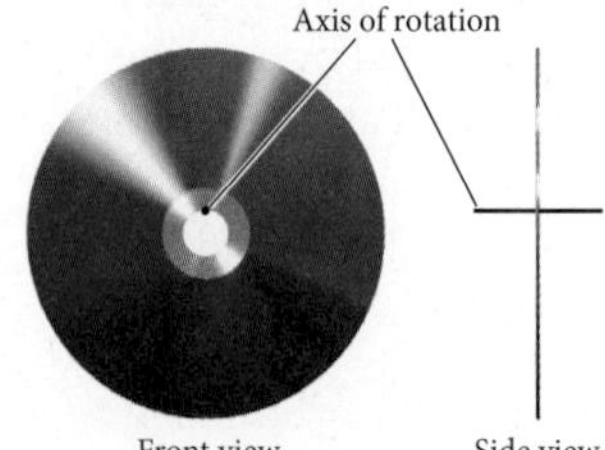


## Section 14.3

**14.39** A massive object of $m = 5$ kg oscillates with simple harmonic motion. Its position as a function of time varies according to the equation $x(t) = 2\sin(\pi/2t + \pi/6)$.

a) What is the position, velocity, and acceleration of the object at $t = 0$ s?

b) What is the kinetic energy of the object as a function of time?

c) At which time after $t = 0$ s is the kinetic energy first at a maximum?

**14.40** The mass $m$ shown in the figure is displaced a distance $x$ to the right from its equilibrium position.

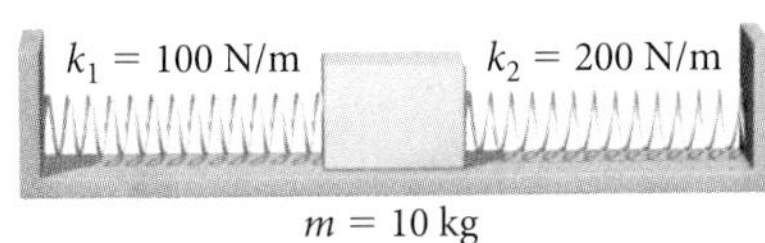


a) What is the net force acting on the mass, and what is the effective spring constant?

b) If $k_1 = 100$ N/m and $k_2 = 200$ N/m, what will the frequency of the oscillation be when the mass is released?

c) If $x = 0.1$ m, what is the total energy of the mass-spring system after the mass is released, and what is the maximum velocity of the mass?

•**14.41** A 2.0-kg mass attached to a spring is displaced 8.0 cm from the equilibrium position. It is released and then oscillates with a frequency of 4.0 Hz.

a) What is the energy of the motion when the mass passes through the equilibrium position?

b) What is the speed of the mass when it is 2.0 cm from the equilibrium position?

•**14.42** A Foucault pendulum is designed to demonstrate the effect of the Earth's rotation. A Foucault pendulum displayed in a museum is typically quite long, making the effect easier to see. Consider a Foucault pendulum of length 15 m with a 110-kg brass bob. It is set to swing with an amplitude of 3.5°.

a) What is the period of the pendulum?

b) What is the maximum kinetic energy of the pendulum?

c) What is the maximum speed of the pendulum?

••**14.43** A mass, $m_1 = 8$ kg, is at rest on a frictionless horizontal surface and connected to a wall by a spring with $k =$ 70 N/m, as shown in the figure. A second mass, $m_2 = 5$ kg, is moving to the right at $v_0 = 17$ m/s. The two masses collide and stick together.

a) What is the maximum compression of the spring?

b) How long will it take after the collision to reach this maximum compression?

••**14.44** A mass $M = 0.460$ kg moves with an initial speed $v = 3.20$ m/s on a level frictionless air track. The mass is initially a distance $D = 0.250$ m away from a spring with $k =$ 840 N/m, which is mounted rigidly at one end of the air track. The mass compresses the spring a maximum distance $d$, before reversing direction. After bouncing off the spring, the mass travels with the same speed $v$, but in the opposite direction.

a) Determine the maximum distance that the spring is compressed.

b) Find the total elapsed time until the mass returns to its starting point. (*Hint:* The mass undergoes a partial cycle of simple harmonic motion while in contact with the spring.)

••**14.45** The relative motion of two atoms in a molecule can be described as the motion of a single body of mass $m$ moving in one dimension, with potential energy $U(r) = A/r^{12} - B/r^6$, where $r$ is the separation between the atoms and $A$ and $B$ are positive constants.

a) Find the equilibrium separation, $r_0$, of the atoms, in terms of the constants $A$ and $B$.

b) If moved slightly, the atoms will oscillate about their equilibrium separation. Find the angular frequency of this oscillation, in terms of $A$, $B$, and $m$.

## Section 14.4

**14.46** A 3-kg mass attached to a spring with $k = 140$ N/m is oscillating in a vat of oil, which damps the oscillations.

a) If the damping constant of the oil is $b = 10$ kg/s, how long will it take the amplitude of the oscillations to decrease to 1% of its original value?

b) What should the damping constant be to reduce the amplitude of the oscillations by 99% in 1 s?

**14.47** A vertical spring with a spring constant of 2.00 N/m has a 0.30-kg mass attached to it, and the mass moves in a medium with a damping constant of 0.025 kg/s. The mass is released from rest at a position 5.0 cm from the equilibrium position. How long will it take for the amplitude to decrease to 2.5 m?

**14.48** A mass of 0.404 kg is attached to a spring with a spring constant of 206.9 N/m. Its oscillation is damped, with damping constant $b = 14.5$ kg/s. What is the frequency of this damped oscillation?

•**14.49** Cars have shock absorbers to damp the oscillations that would otherwise occur when the springs that attach the wheels to the car's frame are compressed or stretched. Ideally, the shock absorbers provide critical damping. If the shock absorbers fail, they provide less damping, resulting in an underdamped motion. You can perform a simple test of your shock absorbers by pushing down on one corner of your car and then quickly releasing it. If this results in an up-and-down oscillation of the car, you know that your shock absorbers need changing. The spring on each wheel of a car has a spring constant of 4005 N/m, and the car has a mass of 851 kg, equally distributed over all four wheels. Its shock absorbers have gone bad and provide only 60.7% of the damping they were initially designed to provide. What will the period of the underdamped oscillation of this car be if the pushing-down shock absorber test is performed?

•**14.50** In a lab, a student measures the unstretched length of a spring as 11.2 cm. When a 100.0-g mass is hung from the spring, its length is 20.7 cm. The mass-spring system is set into oscillatory motion, and the student observes that the amplitude of the oscillation decreases by about a factor of 2 after five complete cycles.

a) Calculate the period of oscillation for this system, assuming no damping.

b) If the student can measure the period to the nearest 0.05 s, will she be able to detect the difference between the period with no damping and the period with damping?

•**14.51** An 80-kg bungee jumper is enjoying an afternoon of jumps. The jumper's first oscillation has an amplitude of 10 m and a period of 5 s. Treating the bungee cord as a spring with no damping, calculate each of the following:

a) the spring constant of the bungee cord,

b) the bungee jumper's maximum speed during the oscillation, and

c) the time for the amplitude to decrease to 2 m (with air resistance providing the damping of the oscillations at 7.5 kg/s).

••**14.52** A small mass, $m = 50.0$ g, is attached to the end of a massless rod that is hanging from the ceiling and is free to swing. The rod has length $L = 1.00$ m. The rod is displaced 10.0° from the vertical and released at time $t = 0$. Neglect air resistance. What is the period of the rod's oscillation? Now suppose the entire system is immersed in a fluid with a small damping constant, $b = 0.0100$ kg/s, and the rod is again released from an initial displacement angle of 10.0°. What is the time for the amplitude of the oscillation to reduce to 5.00°? Assume that the damping is small. Also note that since the amplitude of the oscillation is small and all the mass of the pendulum is at the end of the rod, the motion of the mass can be treated as strictly linear, and you can use the substitution $R\theta(t) = x(t)$, where $R = 1.0$ m is the length of the pendulum rod.

## Section 14.5

**14.53** A 3.0-kg mass is vibrating on a spring. It has a resonant angular speed of 2.4 rad/s and a damping angular speed of 0.14 rad/s. If the driving force is 2.0 N, find the maximum amplitude if the driving angular speed is (a) 1.2 rad/s, (b) 2.4 rad/s, and (c) 4.8 rad/s.

•**14.54** A mass of 0.404 kg is attached to a spring with a spring constant of 204.7 N/m and hangs at rest. The spring hangs from a piston. The piston then moves up and down, driven by a force given by (29.4 N) cos [(17.1 Hz)$t$].

a) What is the maximum displacement from its equilibrium position that the mass can reach?

b) What is the maximum speed that the mass can attain in this motion?

••**14.55** A mass, $M = 1.6$ kg, is attached to a wall by a spring with $k = 578$ N/m. The mass slides on a frictionless floor. The spring and mass are immersed in a fluid with a damping constant of 6.4 kg/s. A horizontal force, $F(t) = F_d \cos(\omega_d t)$, where $F_d = 52$ N, is applied to the mass through a knob, causing the mass to oscillate back and forth. Neglect the mass of the spring and of the knob and rod. At approximately what frequency will the amplitude of the mass's oscillation be greatest, and what is the maximum amplitude? If the driving frequency is reduced slightly (but the driving amplitude remains the same), at what frequency will the amplitude of the mass's oscillation be half of the maximum amplitude?

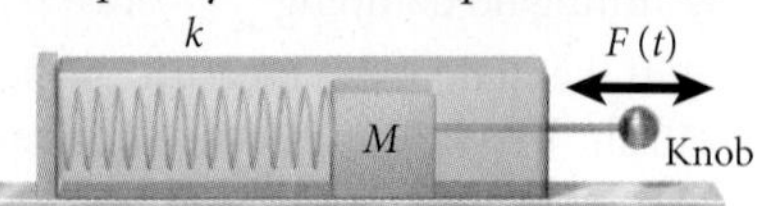


## Additional Problems

**14.56** When the displacement of a mass on a spring is half of the amplitude of its oscillation, what fraction of the mass's energy is kinetic energy?

**14.57** A mass $m$ is attached to a spring with a spring constant of $k$ and set into simple harmonic motion. When the mass has half of its maximum kinetic energy, how far away from its equilibrium position is it, expressed as a fraction of its maximum displacement?

**14.58** If you kick a harmonic oscillator sharply, you impart to it an initial velocity but no initial displacement. For a weakly damped oscillator with mass $m$, spring constant $k$, and damping force $F_\gamma = -bv$, find $x(t)$, if the total impulse delivered by the kick is $J_0$.

**14.59** A mass $m = 1$ kg in a spring-mass system with $k = 1$ N/m is observed to be moving to the right, past its equilibrium position with a speed of 1 m/s at time $t = 0$.

a) Ignoring all damping, determine the equation of motion.

b) Suppose the initial conditions are such that at time $t = 0$, the mass is at $x = 0.5$ m and moving to the right with a speed of 1 m/s. Determine the new equation of motion. Assume the same spring constant and mass.

**14.60** The motion of a block-spring system is described by $x(t) = A\sin(\omega t)$. Find $\omega$, if the potential energy equals the kinetic energy at $t = 1.00$ s.

**14.61** A hydrogen gas molecule can be thought of as a pair of protons bound together by a spring. If the mass of a proton is $1.7 \cdot 10^{-27}$ kg, and the period of oscillation is $8.0 \cdot 10^{-15}$ s, what is the effective spring constant for the bond in a hydrogen molecule?

**14.62** A shock absorber that provides critical damping with $\omega_\gamma = 72.4$ Hz is compressed by 6.41 cm. How far from the equilibrium position is it after 0.0247 s?

**14.63** Imagine you are an astronaut who has landed on another planet and wants to determine the free-fall acceleration on that planet. In one of the experiments you decide to conduct, you use a pendulum 0.50 m long and find that the period of oscillation for this pendulum is 1.50 s. What is the acceleration due to gravity on that planet?

**14.64** A horizontal tree branch is directly above another horizontal tree branch. The elevation of the higher branch is 9.65 m above the ground, and the elevation of the lower branch is 5.99 m above the ground. Some children decide to use the two branches to hold a tire swing. One end of the tire swing's rope is tied to the higher tree branch so that the bottom of the tire swing is 0.47 m above the ground. This swing is thus a restricted pendulum. Starting with the complete length of the rope at an initial angle of 14.2° with respect to the vertical, how long does it take a child of mass 29.9 kg to complete one swing back and forth?

•14.65 Two pendulums of identical length of 1 m are suspended from the ceiling and begin swinging at the same time. One is at Manila, in the Philippines, where $g = 9.784\text{ m/s}^2$, and the other is at Oslo, Norway, where $g = 9.819\text{ m/s}^2$. After how many oscillations of the Manila pendulum will the two pendulums be in phase again? How long will it take for them to be in phase again?

•**14.66** Two springs, each with $k = 125\text{ N/m}$, are hung vertically, and 1-kg masses are attached to their ends. One spring is pulled down 5 cm and released at $t = 0$; the other is pulled down 4 cm and released at $t = 0.3$ s. Find the phase difference, in degrees, between the oscillations of the two masses and the equations for the vertical displacements of the masses, taking upward to be the positive direction.

•**14.67** A piston that has a small toy car sitting on it is undergoing simple harmonic motion vertically, with amplitude of 5 cm, as shown in the figure. When the frequency of oscillation is low, the car stays on the piston. However, when the frequency is increased sufficiently, the car leaves the piston. What is the maximum frequency at which the car will remain in place on the piston?

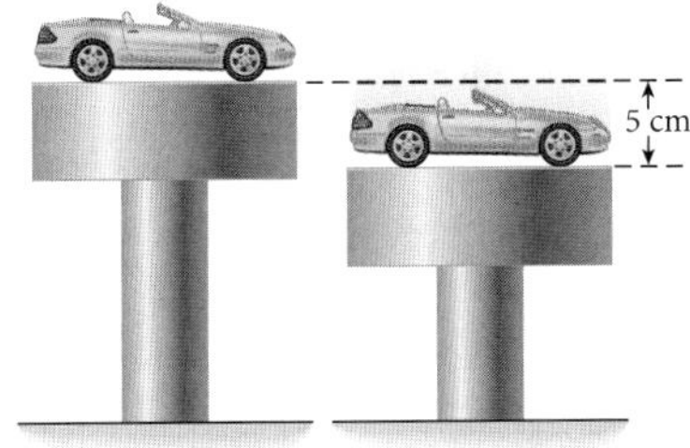


•**14.68** The period of a pendulum is 0.24 s on Earth. The period of the same pendulum is found to be 0.48 s on planet X, whose mass is equal to that of Earth. (a) Calculate the gravitational acceleration at the surface of planet X. (b) Find the radius of planet X in terms of that of Earth.

•14.69 A grandfather clock uses a pendulum and a weight. The pendulum has a period of 2.0 s, and the mass of the bob is 250 g. The weight slowly falls, providing the energy to overcome the damping of the pendulum due to friction. The weight has a mass of 1.0 kg, and it moves down 25 cm every day. Find $Q$ for this clock. Assume that the amplitude of the oscillation of the pendulum is 10°.

•**14.70** A cylindrical can of diameter 10.0 cm contains some ballast so that it floats vertically in water. The mass of can and ballast is 800.0 g, and the density of water is $1.00\text{ g/cm}^3$. The can is lifted 1.00 cm from its equilibrium position and released at $t = 0$. Find its vertical displacement from equilibrium as a function of time. Determine the period of the motion. Ignore the damping effect due to the viscosity of the water.

•**14.71** The period of oscillation of an object in a frictionless tunnel running through the center of the Moon is $T = 2\pi/\omega_0 = 6485$ s, as shown in Example 14.2. What is the period of oscillation of an object in a similar tunnel through the Earth ($R_E = 6.37 \cdot 10^6$ m; $R_M = 1.74 \cdot 10^6$ m; $M_E = 5.98 \cdot 10^{24}$ kg; $M_M = 7.36 \cdot 10^{22}$ kg)?

•**14.72** The motion of a planet in a circular orbit about a star obeys the equations of simple harmonic motion. If the orbit is observed edge-on, so the planet's motion appears to be one-dimensional, the analogy is quite direct: The motion of the planet looks just like the motion of an object on a spring.

a) Use Kepler's Third Law of planetary motion to determine the "spring constant" for a planet in circular orbit around a star with period $T$.

b) When the planet is at the extremes of its motion observed edge-on, the analogous "spring" is extended to its largest displacement. Using the "spring" analogy, determine the orbital velocity of the planet.

••14.73 An object in simple harmonic motion is *isochronous,* meaning that the period of its oscillations is independent of their amplitude. (Contrary to a common assertion, the operation of a pendulum clock is not based on this principle. A pendulum clock operates at fixed, finite amplitude. The gearing of the clock compensates for the anharmonicity of the pendulum.) Consider an oscillator of mass $m$ in one-dimensional motion, with a restoring force $F(x) = -cx^3$, where $x$ is the displacement from equilibrium and $c$ is a constant with appropriate units. The motion of this oscillator is periodic but not isochronous.

a) Write an expression for the period of the undamped oscillations of this oscillator. If your expression involves an integral, it should be a definite integral. You do not need to evaluate the expression.

b) Using the expression of part (a), determine the dependence of the period of oscillation on the amplitude.

c) Generalize the results of parts (a) and (b) to an oscillator of mass $m$ in one-dimensional motion with a restoring force corresponding to the potential energy $U(x) = \gamma|x|^\alpha/\alpha$, where $\alpha$ is any positive value and $\gamma$ is a constant.

# 15 Waves

**FIGURE 15.1** Surfer riding a huge wave in Australia.

## WHAT WE WILL LEARN

- Waves are disturbances (or excitations) that propagate through space or some medium as a function of time. A wave can be understood as a series of individual oscillators coupled to their nearest neighbors.
- Waves can be either longitudinal or transverse.
- The wave equation governs the general motion of waves.
- Waves transport energy across distances; however, even though a wave is propagated by vibrating atoms in a medium, it does not generally transport matter with it.
- The superposition principle states a key property of waves: Two waveforms can be added to form another one.
- Waves can interfere with each other constructively or destructively. Under certain conditions, interference patterns form and can be analyzed.
- A standing wave can be formed by superposition of two traveling waves.
- Boundary conditions on strings determine the possible wavelengths and frequencies of standing waves.

When most people hear the word *waves,* the first thing they think of is ocean waves. Everybody has seen pictures of surfers riding enormous ocean waves in Hawaii or Australia (Figure 15.1), or of the towering waves created by a fierce storm on the open sea. These walls of water are certainly waves, and so are gentle breakers that wash onto a beach. However, waves are far more widespread than disturbances in water. Light and sound are both waves, and wave motion is involved in earthquakes and tsunamis as well as radio and television broadcasts. In fact, later in this book, we'll see that the smallest, most fundamental particles of matter share many properties with waves. In a sense, we are all made of waves as much as particles.

In this chapter, we examine the properties and behavior of waves, building directly on Chapter 14's coverage of oscillations. We'll return to the subject of wave motion when we study light and when we consider quantum mechanics and atomic structure. Whole areas of physics are devoted to wave studies; for example, optics is the science of light waves and acoustics is the study of sound waves (see Chapter 16). Waves are also of fundamental importance in astronomy and in electronics. The concepts in this chapter are just the first steps toward understanding a large part of the physical world around us.

## 15.1 Wave Motion

If you throw a stone in a still pond, it creates circular ripples that travel along the surface of the water, radially outward (Figure 15.2). If an object is floating on the surface of the pond (a stick, a leaf, or a rubber duck), you can observe that the object moves up and down as the ripples move outward underneath it but does not move noticeably outward with the waves.

**FIGURE 15.2** Circular ripples on the surface of water.

If you tie a rope to a wall, hold one end and stretch the rope out, and then rapidly move your arm up and down, this motion will create a wave crest that travels down the rope, as illustrated in Figure 15.3. Again this wave moves, but the material that the wave travels through (the rope) moves up and down but otherwise remains in place.

These two examples illustrate a major characteristic of waves: A **wave** is an excitation that propagates through space or some medium as a function of time but does not generally transport matter along with it. In this fundamental way, wave motion is completely different from the motion we have studied so far. Electromagnetic and gravitational waves do not need a medium in which to propagate; they can move through empty space. Other waves do propagate in a medium; for example, sound waves can travel through a gas (air) or a liquid (water) or a solid (steel rails), but not through a vacuum.

Many other examples of waves surround us in everyday life, most of them with oscillations that are too small for us to observe. Light is an electromagnetic wave, as are the radio waves that FM and AM radios receive and the waves that carry TV signals into our houses—either through coaxial cables or through the air via broadcasts from antenna towers or geostationary satellites. We'll focus on electromagnetic waves in Chapter 31 (after we have studied electricity and magnetism in Chapters 21 through 30).

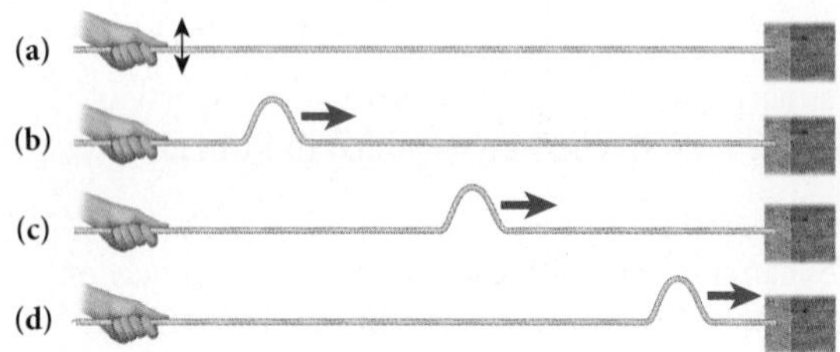

**FIGURE 15.3** A rope held by a hand at one end and attached firmly to a wall at the other end. (a) The end of the rope is moved up and down by moving the hand. In (b), (c), and (d), a wave travels along the rope toward the wall.

In this chapter, we study the motion of waves through media. Mechanical vibrations fall in this category, as do sound waves, earthquake waves, and water waves. Since sound is such an important part of human communication, Chapter 16 is devoted exclusively to it.

The stadium wave—called simply "the wave"—is a part of most major sports events: Spectators stand up and raise their arms as soon as they see their neighbor doing it, and then they sit down again. Thus, the wave travels around the stadium, but each spectator stays in place. How fast does this wave travel? Suppose that each person has a time delay comparable to his or her reaction time, 0.1 s, before imitating the adjacent spectator and getting up from the seat. Typically, seats are approximately 0.6 m (2 ft) wide. This means that the wave travels 0.6 m in 0.1 s. From simple kinematics, we know that this corresponds to a velocity of $v = \Delta x/dt = (0.6 \text{ m})/(0.1\text{s}) = 6 \text{ m/s}$. Since a typical football stadium is an oval with a circumference of approximately 400 m, we can estimate that it takes about 1 min for "the wave" to make it around the stadium once. This rather crude estimate is approximately right, within a factor of 2. Empirically, we find that "the wave" moves with a speed of approximately 12 m/s. The reason for this higher speed is that people anticipate the arrival of the wave and jump up when they see the person two or three seats ahead of them jump up.

## 15.2 Coupled Oscillators

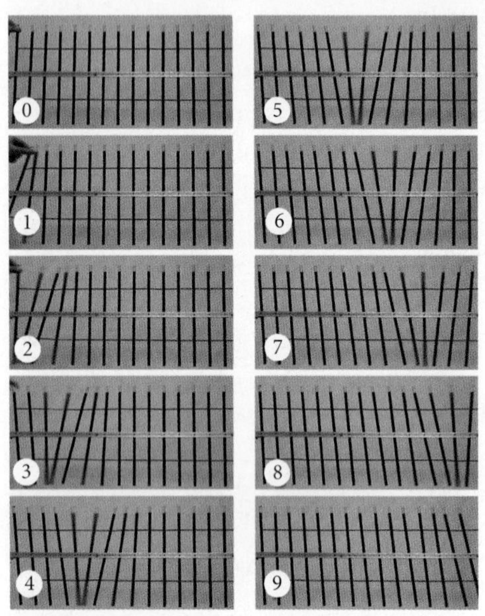

**FIGURE 15.4** Series of identical rods coupled to their nearest neighbors by springs. The numbers on the frames indicate the time sequence. Each pair of frames is separated by a time interval of 0.133 s.

Let's begin our quantitative discussion of waves by examining a series of coupled oscillators—for example, a string of identical massive objects coupled by identical springs. One physical realization is shown in Figure 15.4, where a series of identical metal rods is held in place by a central horizontal bar. The rods can pivot freely about their centers and are connected to their nearest neighbors by identical springs. When the first rod is pushed, this excitation is transmitted from each rod to its neighbor, and the pulse generated in this way travels down the row of rods. Each rod is an oscillator and moves back and forth in the plane of the paper. This setup represents a longitudinal coupling between the oscillators, in the sense that the direction of motion and the direction of coupling to the nearest neighbors are the same.

A physical realization of a transverse coupling is shown in Figure 15.5. In part (a) of the figure, all rods are attached to a tension strip in their horizontal equilibrium position; in part (b), rod $i$ is deflected from the horizontal by an angle $\theta_i$. A deflection of this kind has two effects: First, the tension strip provides a restoring force that tries to pull rod $i$ back to the equilibrium position. Second, since the tension strip is now twisted on both sides of rod $i$, the deflection of that rod at an angle $\theta_i$ creates a force that acts to deflect rods $i-1$ and $i+1$ toward $\theta_i$. If those rods are allowed to move freely, they will be deflected in that direction, and each rod in turn will generate a force acting on its two nearest neighbors.

Figure 15.6 shows a side view of the arrangement of parallel rods in Figure 15.5. Each circle represents the end of a rod, and the motion of this circle can be characterized by a coordinate $y_i$. Then $y_i$ is related to the deflection angle, $\theta_i$, via $y_i = \frac{1}{2}l \sin \theta_i$, where $l$ is the length of the rod. The line connecting the circles represents the edge of the tension strip. In Figure 15.6a, all rods are in equilibrium, which implies that all deflections are zero. In Figure 15.6b, rod $i$ is experiencing a deflection, the same as is depicted in Figure 15.5b. The force acting on rod $i$ due to the tension in the strip is a type

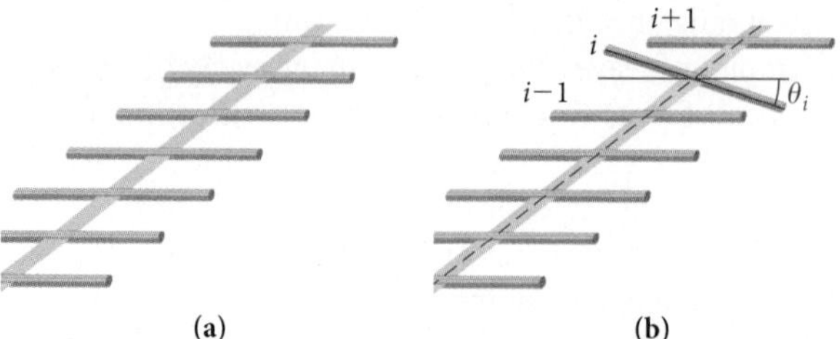

**FIGURE 15.5** Parallel rods supported by a tension strip as a model for transversely coupled oscillators: (a) initial position, all rods parallel; (b) rod $i$ is deflected by an angle $\theta_i$.

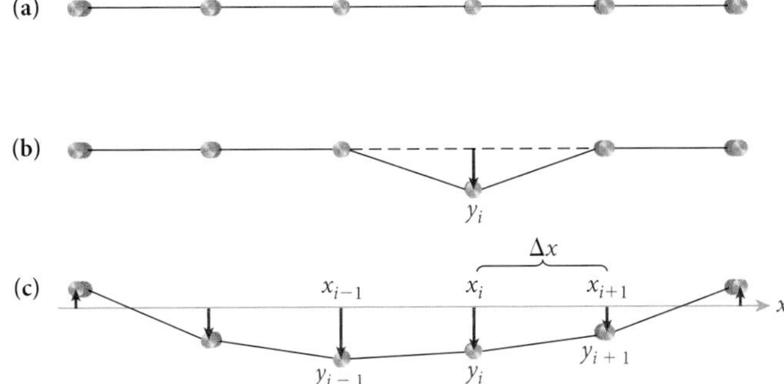


**FIGURE 15.6** Side views of coupled rods of Figure 15.5: (a) equilibrium position; (b) one rod is deflected; (c) all rods are allowed to move and become deflected.

of spring force given by $F_i = -ky_i$, where the effective spring constant is determined from the properties of the tension strip. Finally, in Figure 15.6c, all rods are allowed to move. The net force acting on rod $i$ is then the sum of the forces exerted on it by rods $i+1$ and $i-1$, as mediated by the tension strip connecting the rods. These forces depend on the difference between the $y$-coordinates at the beginning and the end of the section of tension strip that connects the three rods:

$$F_+ = -k(y_i - y_{i+1})$$

and

$$F_- = -k(y_i - y_{i-1}).$$

Thus, the net force acting on rod $i$ is the sum of these two forces:

$$F_i = F_+ + F_- = -k(y_i - y_{i+1}) - k(y_i - y_{i-1}) = k(y_{i+1} - 2y_i + y_{i-1}).$$

From Newton's Second Law, $F = ma$, we then find the acceleration, $a_i$, of rod $i$:

$$ma_i = k(y_{i+1} - 2y_i + y_{i-1}). \tag{15.1}$$

With initial conditions for the position and velocity of each oscillator, we could solve the resulting set of equations on a computer. However, we can actually derive a wave equation using equation 15.1, and we'll do so in Section 15.4.

### Transverse and Longitudinal Waves

We need to emphasize an essential difference between the two systems of coupled oscillators shown in Figure 15.4 and Figure 15.5. In the first case, the oscillators are allowed to move only in the direction of their nearest neighbors, whereas in the second case, the oscillators are restricted to moving in a direction perpendicular to the direction of their nearest neighbors (Figure 15.7). In general, a wave that propagates along the direction in which the oscillators move is called a **longitudinal wave.** In Chapter 16, we'll see that sound waves are prototypical longitudinal waves. A wave that moves in a direction perpendicular to the direction in which the individual oscillators move is called a **transverse wave.** Light waves are transverse waves, as we will discuss in Chapter 31 on electromagnetic waves.

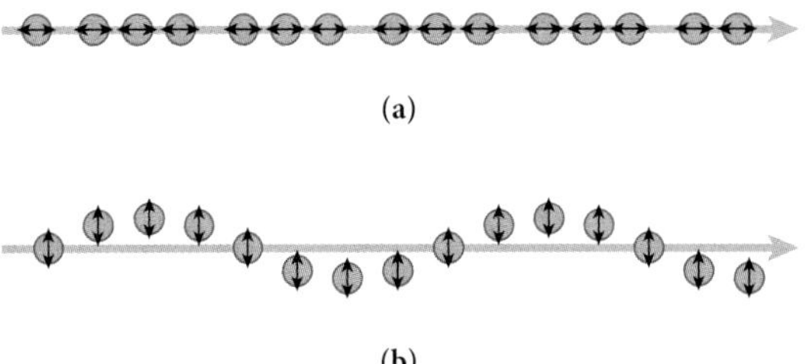


**FIGURE 15.7** Representations of (a) longitudinal wave and (b) transverse wave propagating horizontally to the right.

Seismic waves created by earthquakes exist as longitudinal (or compression) waves and as transverse waves. These waves can travel along the surface of the Earth as well as through the Earth and be detected far from the epicenter of the earthquake. The study of seismic waves has revealed detailed information about the composition of the interior of the Earth, as we'll see in Example 15.2.

## 15.3 Mathematical Description of Waves

So far, we have examined single wave pulses, such as the stadium wave and the wave that results from pushing once on a chain of coupled oscillators. A much more common class of wave phenomena involves a periodic excitation—in particular, a sinusoidal excitation. In the chain of oscillators of Figure 15.4 or Figure 15.5, a continuous wave can be generated

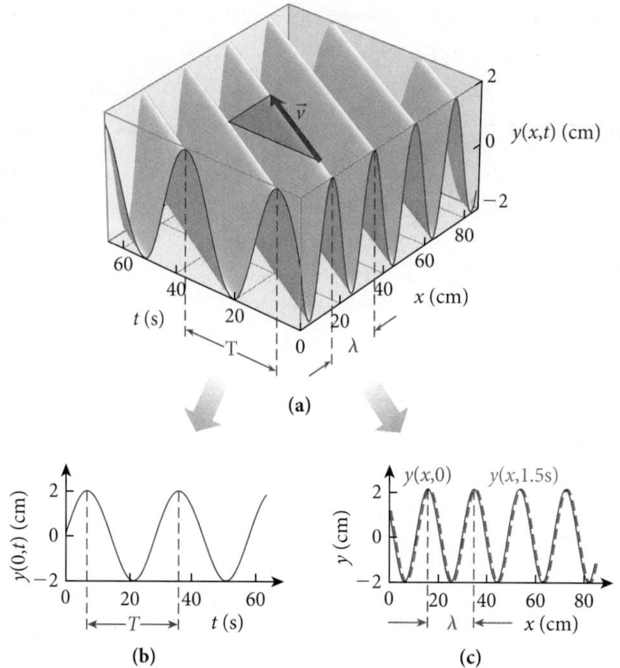


**FIGURE 15.8** Sinusoidal wave: (a) the wave as a function of space and time coordinates; (b) dependence of the wave on time at position $x = 0$; (c) dependence of the wave on position at time $t = 0$ (blue curve) and at a later time $t = 1.5$ s (gray dashed curve).

by moving the first rod back and forth periodically. This sinusoidal oscillation also travels down the chain of oscillators, just like a single wave pulse.

## Period, Wavelength, Velocity

Figure 15.8a shows the result of a sinusoidal excitation as a function of both time and the horizontal coordinate, in the limit where a very large number of coupled oscillators are very close to each other—that is, in the continuum limit. First, let's examine this wave along the time axis for $x = 0$. This projection is shown in Figure 15.8b, which is a plot of the sinusoidal oscillation of the first oscillator in the chain:

$$y(x=0,t) = A\sin(\omega t + \theta_0) = A\sin(2\pi ft + \theta_0), \tag{15.2}$$

where $\omega$ is the angular frequency (or angular velocity), $f$ is the frequency, $A$ is the amplitude of the oscillation, and $\theta_0$ gives the phase shift. (The values used in Figure 15.8 are $A = 2.0$ cm, $\omega = 0.2\ \text{s}^{-1}$, and $\theta_0 = 0.5$.) As for all oscillators, the period $T$ is defined as the time interval between two successive maxima and is related to the frequency by

$$T = \frac{1}{f}.$$

Figure 15.8c shows that for any given time, $t$, the dependence on the horizontal coordinate, $x$, of the oscillations is also sinusoidal. This dependence is shown for $t = 0$ (blue curve) and $t = 1.5$ s (gray dashed curve). The distance in space between two consecutive maxima is defined as the **wavelength,** $\lambda$.

As you can see in Figure 15.8a, diagonal lines of equal height describe the wave motion in the $xt$-plane. These lines are called **wave fronts.** What is most important to realize about the wave motion depicted in Figure 15.8 and wave motion in general is that during one period, any given wave front advances by one wavelength.

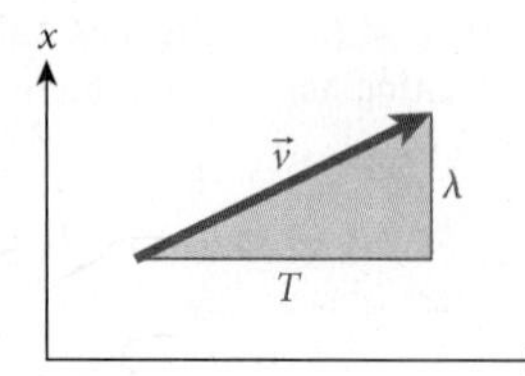


**FIGURE 15.9** View of the $xt$-plane of Figure 15.8 showing the relationship of the wave velocity to the period and the wavelength.

Figure 15.9 presents the light purple triangle shown on the top surface of the wave in Figure 15.8a in the $xt$-plane to further illustrate the relationship among the wave velocity, the period, and the wavelength. Now you can see that the speed of propagation of the wave is in general given by

$$v = \frac{\lambda}{T},$$

or, using the relationship between period and frequency, $T = 1/f$, by

$$v = \lambda f. \tag{15.3}$$

This relationship among wavelength, frequency, and velocity holds for all types of waves and is one of the most important results in this chapter.

## 15.1 Self-Test Opportunity

You are sitting in a boat in the middle of Lake Michigan. The water waves hitting the boat are traveling at 3.0 m/s and the crests are 7.5 m apart. What is the frequency with which the wave crests strike the boat?

## Sinusoidal Waveform, Wave Number, and Phase

How do we obtain a mathematical description of a sinusoidal wave as a function of space and time? We start with the motion of the oscillator at $x = 0$ (see equation 15.2), and realize that it takes a time $t = x/v$ for the excitation to move a distance $x$. Thus, we can simply replace $t$ in equation 15.2 by $t - x/v$ to find the displacement $y(x,t)$ away from the equilibrium position as a function of space and time:

$$y(x,t) = A\sin\left(\omega\left(t - \frac{x}{v}\right) + \theta_0\right).$$

We can then use equation 15.3 for the wave velocity, $v = \lambda f$, and obtain

$$\begin{aligned} y(x,t) &= A\sin\left(2\pi f\left(t - \frac{x}{v}\right) + \theta_0\right) \\ &= A\sin\left(2\pi f t - \frac{2\pi f x}{\lambda f} + \theta_0\right) \\ &= A\sin\left(\frac{2\pi}{T}t - \frac{2\pi}{\lambda}x + \theta_0\right). \end{aligned}$$

Similar to $\omega = 2\pi/T$, the expression for the **wave number,** $\kappa$, is

$$\kappa = \frac{2\pi}{\lambda}. \quad (15.4)$$

Substituting $\omega$ and $\kappa$ into the above expression for $y(t)$ then yields the descriptive function we want:

$$y(x,t) = A\sin(\omega t - \kappa x + \theta_0). \quad (15.5)$$

A sinusoidal waveform of this kind is almost universally applicable when there is only one spatial dimension in which the wave can propagate. Later in this chapter, we'll generalize this waveform to more than one spatial coordinate.

We can also start with a given sinusoidal waveform at some point in space and ask where this waveform will be at a later time. From this perspective, we can write the expression for the waveform as a function of space and time as

$$y(x,t) = A\sin(\kappa x - \omega t + \phi_0). \quad (15.6)$$

Equations 15.5 and 15.6 are equivalent to each other, because $\sin(-x) = -\sin(x) = \sin(\pi + x)$. The phase shift, $\phi_0$, in equation 15.6 is related to the phase angle, $\theta_0$, in equation 15.5 via $\phi_0 = \pi - \theta_0$. We'll use equation 15.6, which is the more common form, from here on.

The argument of the sine function in equation 15.6 is called the wave's **phase,** $\phi$:

$$\phi = \kappa x - \omega t + \phi_0. \quad (15.7)$$

In considering more than one wave simultaneously, the relationship of the phases of the waves plays an essential role in the analysis. Later in this chapter, we'll return to this point.

Finally, using $\kappa = 2\pi/\lambda$ and $\omega = 2\pi/T$, we can rewrite the equation for the velocity of the wave front (15.3) in terms of the wave number and the angular frequency:

$$v = \frac{\omega}{\kappa}. \quad (15.8)$$

The angular frequency, $\omega$, counts the number of oscillations in each time interval of $2\pi$ s: $\omega = 2\pi/T$, where $T$ is the period. Similarly, the wave number, $\kappa$, counts the number of wavelengths, $\lambda$, that fit into a distance of $2\pi$ m: $\kappa = 2\pi/\lambda$. For example, $\kappa = 0.33\ \text{cm}^{-1}$ in the plot of Figure 15.8.

## 15.2 Self-Test Opportunity

The figure shows a plot at $t = 0$ of a wave of the form $y(x,t) = A\sin(\kappa x - \omega t + \phi_0)$. Determine the amplitude, $A$, the wave number, $\kappa$, and the phase shift, $\phi_0$, of the wave.

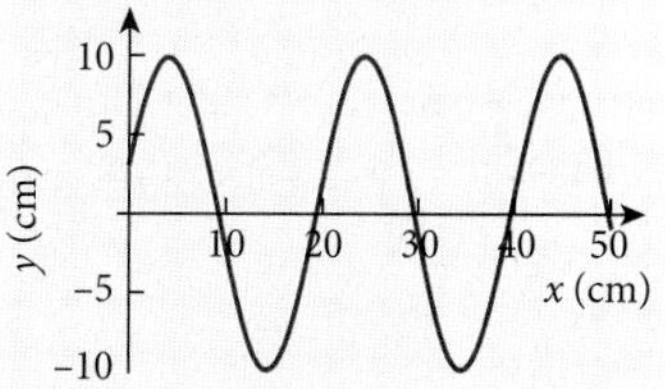


A traveling wave at the instant when $t = 0$.

## 15.4 Derivation of the Wave Equation

In this section, we derive a general equation of motion for waves, called simply the **wave equation:**

$$\frac{\partial^2}{\partial t^2}y(x,t)-v^2\frac{\partial^2}{\partial x^2}y(x,t)=0, \tag{15.9}$$

where $y(x,t)$ is the displacement from the equilibrium position as a function of a single space coordinate, $x$, and a time coordinate, $t$, and $v$ is the speed of propagation of the wave. This wave equation describes all undamped wave motion in one dimension.

### DERIVATION 15.1 Wave Equation

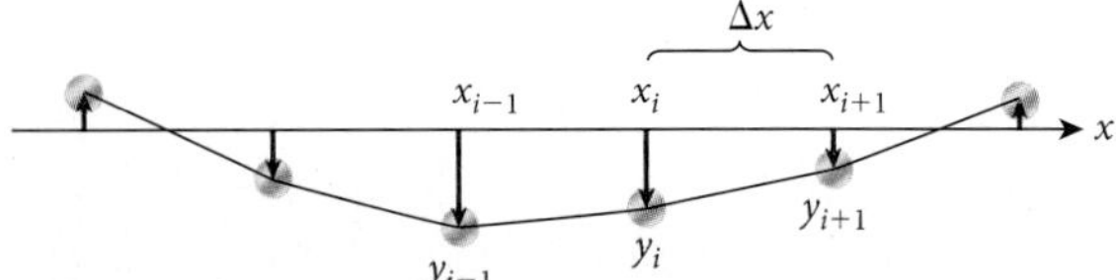


**FIGURE 15.10** Side views of coupled rods where all rods are allowed to deflect.

Figure 15.10 reproduces Figure 15.6c. The $x$-axis, pointing in the conventional horizontal direction and to the right, is (as usual) perpendicular to the $y$-axis, along which the motion of each individual oscillator occurs. The horizontal distance between neighboring oscillators is $\Delta x$. The displacements of the oscillators, given by $y(x,t)$, are then functions of time and of position along the $x$-axis, with $y_i = y(x_i,t)$. The acceleration is defined as the second derivative of the position vector, $y(x,t)$, with respect to time. Since the position vector depends on more than one variable in this case (time as well as the horizontal coordinate), we need to use a partial derivative to express this relationship:

$$a=\frac{\partial^2}{\partial t^2}y(x,t). \tag{i}$$

If we go back to the original definition of the derivative as the limit of the ratio of differences, we can approximate a partial derivative with respect to the $x$-coordinate as

$$\frac{\partial}{\partial x}y(x,t)=\frac{\Delta y}{\Delta x}=\frac{y_{i+1}-y_i}{\Delta x}.$$

We can then approximate the second derivative as

$$\frac{\partial^2}{\partial x^2}y(x,t)=\frac{\Delta\left(\frac{\Delta y}{\Delta x}\right)}{\Delta x}=\frac{\frac{(y_{i+1}-y_i)}{\Delta x}-\frac{(y_i-y_{i-1})}{\Delta x}}{\Delta x}=\frac{y_{i+1}-2y_i+y_{i-1}}{(\Delta x)^2}.$$

Both the right-hand side of this equation and the right-hand side of equation 15.1 contain $y_{i+1}-2y_i+y_{i-1}$. Thus, we can combine these two equations:

$$ma_i=k(\Delta x)^2\frac{\partial^2}{\partial x^2}y(x,t).$$

Substituting the second partial derivative of position with respect to time from equation (i) for the acceleration, we find

$$m\left[\frac{\partial^2}{\partial t^2}y(x,t)\right]=k(\Delta x)^2\frac{\partial^2}{\partial x^2}y(x,t).$$

Finally, we divide by the mass of the oscillator and obtain

$$\frac{\partial^2}{\partial t^2}y(x,t)-\frac{k}{m}(\Delta x)^2\frac{\partial^2}{\partial x^2}y(x,t)=0. \tag{ii}$$

This equation relating the second partial derivatives with respect to time and position is a form of the wave equation. The ratio of the spring constant to the mass of the oscillator is equal to the square of the angular frequency, as we saw in Chapter 14. This is also the angular frequency of the traveling wave. Thus, the product of the angular frequency, $\omega$, times the horizontal displacement, $\Delta x$, is the velocity of the wave (how fast the wave moves in space),

$$\frac{k}{m}(\Delta x)^2 = \omega^2(\Delta x)^2 = v^2. \tag{iii}$$

Thus, equation (ii) becomes the wave equation given by equation 15.9:

$$\frac{\partial^2}{\partial t^2}y(x,t) - v^2\frac{\partial^2}{\partial x^2}y(x,t) = 0.$$

In the limit where the oscillators get very close to one another—that is, as $\Delta x \to 0$—the solution to the wave equation 15.9 provides the exact solution to the difference equation 15.1. Equation 15.9 is of general importance, and its applicability far exceeds the problem of coupled oscillators. The basic form given in equation 15.9 holds for the propagation of sound waves, light waves, or waves on a string. The difference in physical systems exhibiting wave motion manifests itself only in the way $v$, the speed of propagation of the wave, is calculated.

In mathematical terms, equation 15.9 is a partial differential equation. As you may know, some semester-long courses are entirely devoted to partial differential equations, and you will encounter many more partial differential equations if you major in the physical sciences or engineering. However, this wave equation will be the only partial differential equation that you'll encounter in this book, and it will be sufficient to treat the partial derivatives just like conventional derivatives of functions in one variable.

## Solutions of the Wave Equation

In Section 15.3, we developed equation 15.6, describing a sinusoidal waveform. Let's see if this is a solution of the general wave equation 15.9. To do this, we have to take the partial derivatives of equation 15.6 with respect to the $x$-coordinate and with respect to time:

$$\begin{aligned}
\frac{\partial}{\partial t}y(x,t) &= \frac{\partial}{\partial t}\left(A\sin\left(\kappa x - \omega t + \phi_0\right)\right) = -A\omega\cos\left(\kappa x - \omega t + \phi_0\right)\\
\frac{\partial^2}{\partial t^2}y(x,t) &= \frac{\partial}{\partial t}\left(-A\omega\cos\left(\kappa x - \omega t + \phi_0\right)\right) = -A\omega^2\sin\left(\kappa x - \omega t + \phi_0\right)\\
\frac{\partial}{\partial x}y(x,t) &= \frac{\partial}{\partial x}\left(A\sin\left(\kappa x - \omega t + \phi_0\right)\right) = A\kappa\cos\left(\kappa x - \omega t + \phi_0\right)\\
\frac{\partial^2}{\partial x^2}y(x,t) &= \frac{\partial}{\partial x}\left(A\kappa\cos\left(\kappa x - \omega t + \phi_0\right)\right) = -A\kappa^2\sin\left(\kappa x - \omega t + \phi_0\right).
\end{aligned}$$

Inserting these partial derivatives into equation 15.9, we find

$$\begin{aligned}
&\frac{\partial^2}{\partial t^2}y(x,t) - v^2\frac{\partial^2}{\partial x^2}y(x,t) =\\
&\left[-A\omega^2\sin\left(\kappa x - \omega t + \phi_0\right)\right] - v^2\left[-A\kappa^2\sin\left(\kappa x - \omega t + \phi_0\right)\right] =\\
&-A\sin\left(\kappa x - \omega t + \phi_0\right)\left(\omega^2 - v^2\kappa^2\right) = 0.
\end{aligned} \tag{15.10}$$

Thus, we see that the function given by equation 15.6 is a solution of the wave equation 15.9, provided that $\omega^2 - v^2\kappa^2 = 0$. However, we already know from equation 15.8 that this is the case, so we have established that equation 15.6 is truly a solution of the wave equation.

Is there a larger class of solutions that includes the one we have already found? Yes, any arbitrary, continuously differentiable function $Y$ with an argument containing the same linear combination of $x$ and $t$ as in equation 15.6, $\kappa x - \omega t$, is a solution to the wave equation. The constants $\omega$, $\kappa$, and $\phi_0$ can have arbitrary values. However, the convention generally followed is that the angular frequency and the wave number are both positive numbers. Thus, there are two general solutions to the wave equation:

$$Y(\kappa x - \omega t + \phi_0) \text{ for a wave moving in the positive } x\text{-direction} \quad (15.11)$$

and

$$Y(\kappa x + \omega t + \phi_0) \text{ for a wave moving in the negative } x\text{-direction}. \quad (15.12)$$

This result is perhaps more obvious if we write these functions in terms of arguments that involve the wave speed, $v = \omega/\kappa$. We then see that $Y(x - vt + \phi_0/\kappa)$ is a solution for an arbitrary waveform moving in the positive $x$-direction and $Y(x + vt + \phi_0/\kappa)$ is a solution for a waveform moving in the negative $x$-direction.

As previously stated, the functional form of $Y$ does not matter at all. We can see this by looking at the chain rule for differentiation, $df(g(x))/dx = (df/dg)(dg/dx)$. The partial derivatives in space and time contain the common factor $df/dg$, which cancels out in the same way that the sine function factored out in equation 15.10. Thus, the derivatives of the function $g = \kappa x \pm \omega t + \phi_0$ always lead to the same condition, $\omega^2 - v^2\kappa^2 = 0$, as applied to the solution for the sinusoidal waveform.

## Waves on a String

String instruments form a large class of musical instruments. Guitars (Figure 15.11), cellos, violins, mandolins, harps, and others fall into this category. These instruments produce musical sounds when vibrations are induced on their strings.

Suppose a string has a mass $M$ and a length $L$. In order to find the wave speed, $v$, on this string, we can think of it as composed of many small individual segments of length $\Delta x$ and mass $m$, where

$$m = \frac{M\Delta x}{L} = \mu \Delta x.$$

Here the linear mass density, $\mu$, of the string is its mass per unit length (assumed to be constant):

$$\mu = \frac{M}{L}.$$

We can then treat the motion of a wave on a string as that of coupled oscillators. The restoring force is provided by the string tension, $T = k\Delta x$. Substituting from $m = \mu\Delta x$ into the equation for the wave velocity for coupled oscillator systems, $k(\Delta x)^2/m = v^2$ [equation (iii) in Derivation 15.1] after solving for $v$, we find the wave velocity on a string:

$$v = \sqrt{\frac{k}{m}}\Delta x = \sqrt{\frac{k}{\mu \Delta x}}\Delta x = \sqrt{\frac{k\Delta x}{\mu}} = \sqrt{\frac{T}{\mu}}. \quad (15.13)$$

Note that the letter $T$ is used to represent both the string tension and the oscillation period, in accordance with common convention. Make sure you understand which quantity applies to a given situation.

What does equation 15.13 imply? There are two immediate consequences. First, increasing the linear mass density of the string (that is, using a heavier string) reduces the wave velocity on the string. Second, increasing the string tension (for example, by tightening the tuning knobs at the top of the electric guitar shown in Figure 15.11) increases the wave velocity. If you play a string instrument, you know that tightening the tension increases the pitch of the sound produced by a string. We'll return to the relationship of tension to sound later in this chapter.

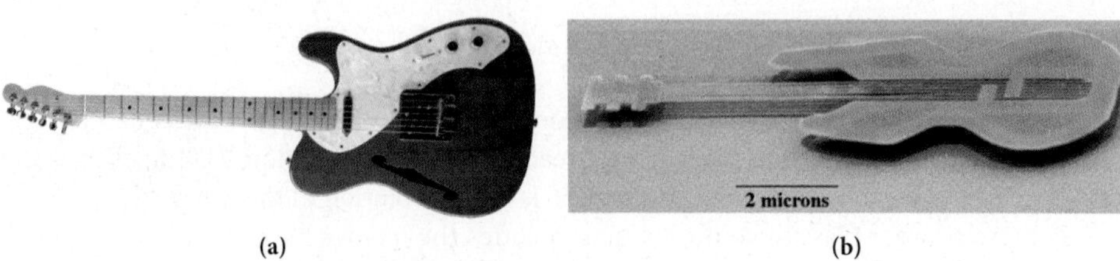


(a) (b)

**FIGURE 15.11** (a) Typical electric guitar used by a rock band. (b) World's smallest guitar, produced in 2004 in the Cornell NanoScale Science and Technology Facility. It is smaller than the guitar in part (a) by a factor of $10^5$ and has a length of 10 μm, approximately $\frac{1}{20}$ of the width of a human hair.

### EXAMPLE 15.1 Elevator Cable

An elevator repairman (mass = 73 kg) sits on top of an elevator cabin of mass 655 kg inside a shaft in a skyscraper. A 61-m-long steel cable of mass 38 kg suspends the cabin. The man sends a signal to his colleague at the top of the elevator shaft by tapping the cable with his hammer.

**PROBLEM**

How long will it take for the wave pulse generated by the hammer to travel up the cable?

**SOLUTION**

The cable is under tension from the combined weight of the elevator cabin and the repairman:

$$T = mg = (73\text{ kg} + 655\text{ kg})(9.81\text{ m/s}^2) = 7142\text{ N}.$$

The linear mass density of the steel cable is

$$\mu = \frac{M}{L} = \frac{38\text{ kg}}{61\text{ m}} = 0.623\text{ kg/m}.$$

Therefore, according to equation 15.13, the wave speed on this cable is

$$v = \sqrt{\frac{T}{\mu}} = 107\text{ m/s}.$$

To travel up the 61-m-long cable, the pulse needs a time of

$$t = \frac{L}{v} = \frac{61\text{ m}}{107\text{ m/s}} = 0.57\text{ s}.$$

**DISCUSSION**

This solution implies that the string tension is the same everywhere in the cable. Is this true? Not quite! We cannot really neglect the weight of the cable itself in determining the tension. The speed we calculated is correct for the lowest point of the cable, where it attaches to the cabin. However, the tension at the upper end of the cable also has to include the complete weight of the cable. There the tension is

$$T = mg = (73\text{ kg} + 655\text{ kg} + 38\text{ kg})(9.81\text{ m/s}^2) = 7514\text{ N},$$

which yields a wave speed of 110 m/s, about 3% greater than what we calculated. If we had to take this effect into account, we would have to calculate how the wave speed depends on the height and then perform the proper integration. However, since the maximum correction is 3%, we can assume that the average correction is about 1.5%, use an average wave speed of 108.6 m/s, and obtain a signal propagation time of 0.56 s. Thus, taking the mass of the cable into account changes the result by only 0.01 s.

## Reflection of Waves

What happens when a wave traveling on a string encounters a boundary? The end of a string tied to a support on the wall can be considered a boundary. In Figure 15.3, a wave was initiated on a rope by moving one end of the rope up and down. The wave then traveled along the rope. When the wave reaches the end of the rope that is tied to a fixed support on the wall, the wave reflects as shown in Figure 15.12. Because the end of the rope is fixed, the reflected wave has its phase shifted by 180° and returns in the opposite direction with a negative height. The amplitude of the wave at the fixed end point is zero.

If the end of the rope is attached via a frictionless movable connection, such as the one shown in Figure 15.13, the wave is reflected with no change in phase, and the returning wave has a positive amplitude. The maximum amplitude of the wave at the movable end point is twice the initial amplitude of the wave.

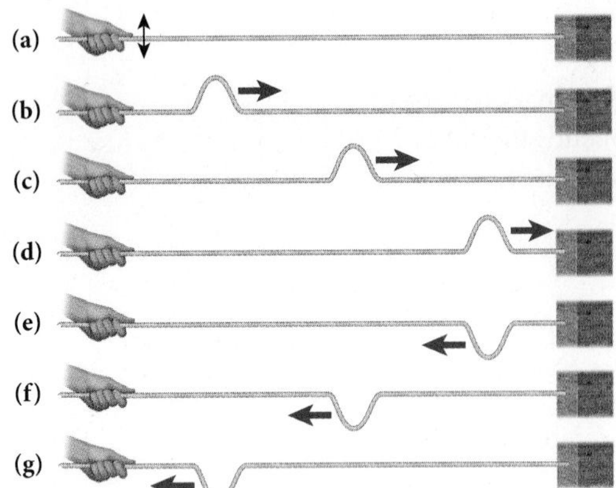


**FIGURE 15.12** A rope held by a hand at one end and attached firmly to a wall at the other end. (a) The end of the rope is moved up and down. In (b), (c), and (d), a wave travels along the rope toward the right. In (e), (f), and (g), the wave travels along the rope toward the left after reflecting from the fixed end point at the wall.

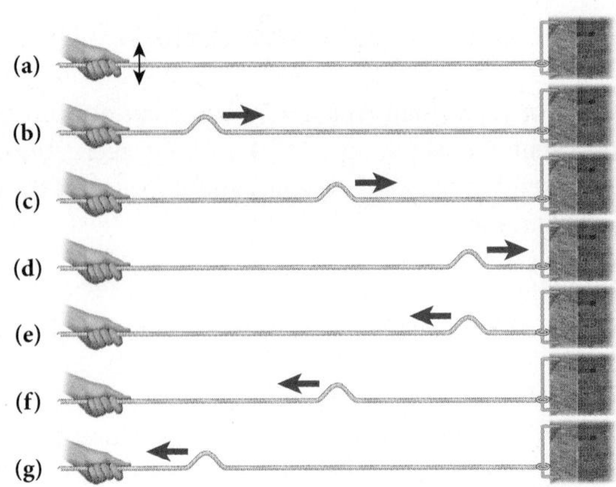


**FIGURE 15.13** A rope held by a hand at one end and attached with a frictionless movable connection to a wall at the other end. (a) The end of the rope is moved up and down. In (b), (c), and (d), a wave travels along the rope toward the right. In (e), (f), and (g), the wave travels along the rope toward the left after reflecting from the movable end point.

In Chapter 34, we'll see that a light wave can be reflected at a boundary between two optical media with or without a phase change, depending on the properties of the two optical media.

## 15.5 Waves in Two- and Three-Dimensional Spaces

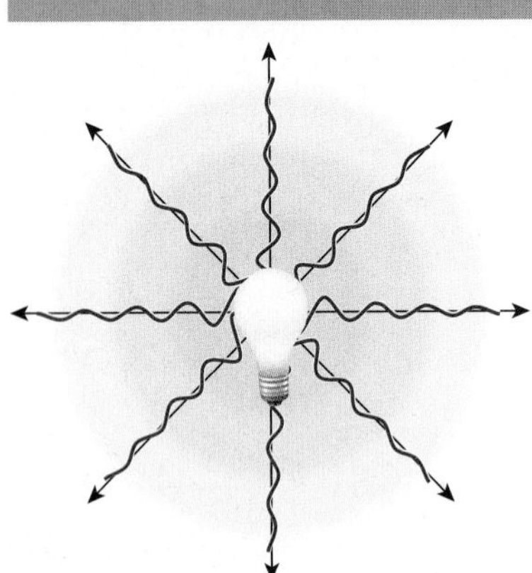

**FIGURE 15.14** Waves spreading radially outward from a point source (a light bulb, in this case). The inner and outer surfaces of each concentric colored band are 1 wavelength apart.

So far, we have only described one-dimensional waves, those moving on a string and on a linear chain of coupled oscillators. However, waves on the surface of water, like those in Figure 15.2, spread in a two-dimensional space—the plane of the water surface. A light bulb and a loudspeaker also emit waves, and these waves spread in three-dimensions. Thus, a more complete mathematical description of wave motion is needed. A completely general treatment exceeds the scope of this text, but we can examine two special cases that cover most of the important wave phenomena that occur in nature.

### Spherical Waves

If you examine Figure 15.2 and Figure 15.14, you see waves generated from one point and spreading as concentric circles. The mathematical idealization of waves emitted from point sources describes the waves as moving outward in concentric circles in two spatial dimensions or in concentric spheres in three spatial dimensions. Sinusoidal waveforms that spread in this way are described mathematically by

$$y(\vec{r},t) = A(r)\sin\left(\kappa r - \omega t + \phi_0\right). \tag{15.14}$$

Here $\vec{r}$ is the position vector (in two or three dimensions), and $r = |\vec{r}|$ is its absolute value. You can convince yourself that this function indeed describes waveforms in concentric circles (in two dimensions) or spheres (in three dimensions) if you remember that the circumference of a circle or surface of a sphere has a constant distance to the origin and thus $r$ has a constant value. A constant value of $r$ at some time $t$ implies a constant phase, $\phi = \kappa r - \omega t + \phi_0$. Since all of the points on a given circle or on the surface of a given sphere have the same phase, they have the same state of oscillation. Therefore, equation 15.14 indeed describes circular or spherical waves.

A waveform described by equation 15.14 does not have a constant amplitude, $A$, but one that depends on the radial distance to the origin of the wave (Figure 15.15). When we talk about energy transport by waves in Section 15.6, it will become clear that radial dependence of the amplitude is necessary for two- and three-dimensional waves and that the amplitude has to be a monotonically decreasing function of the distance from the source.

We'll use energy conservation to find out exactly how the amplitude varies with distance, but for now we'll use $A(r) = C/r$ (where $C$ is a constant) for a three-dimensional spherical wave. For such a wave, equation 15.14 becomes

$$y(\vec{r},t) = \frac{C}{r}\sin\left(\kappa r - \omega t + \phi_0\right). \tag{15.15}$$

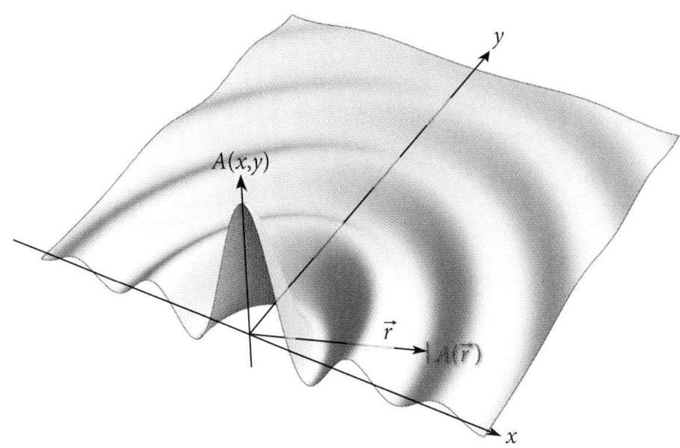


**FIGURE 15.15** A spherical wave as a function of two spatial coordinates.

## Plane Waves

Spherical waves originating from a point source can be approximated as plane waves at points that are very far away from the source. As shown in Figure 15.16, **plane waves** are waves for which the surfaces of constant phase are planes. For example, the light coming from the Sun is in the form of plane waves when it hits the Earth. Why can we use the plane-wave approximation? If we calculate the curvature of a sphere as a function of its radius and let that radius approach infinity, the curvature approaches zero. This means that any surface segment of this sphere locally looks like a plane. For example, the Earth is basically a sphere, but it is so large that locally the surface looks like a plane to us.

For a mathematical description of a plane wave, we can use the same functional form as for a one-dimensional wave:

$$y(\vec{r},t) = A\sin\left(\kappa x - \omega t + \phi_0\right), \tag{15.16}$$

where the $x$-axis is locally defined as the axis perpendicular to the plane of the wave. You may wonder why equation 15.16 again includes a constant amplitude. The answer is that very far away from the origin, $A/r$ varies very slowly as a function of $r$, so using a constant amplitude gives the same degree of approximation as approximating the curvature of a sphere as zero. For a perfectly flat planar wave, the amplitude is indeed constant.

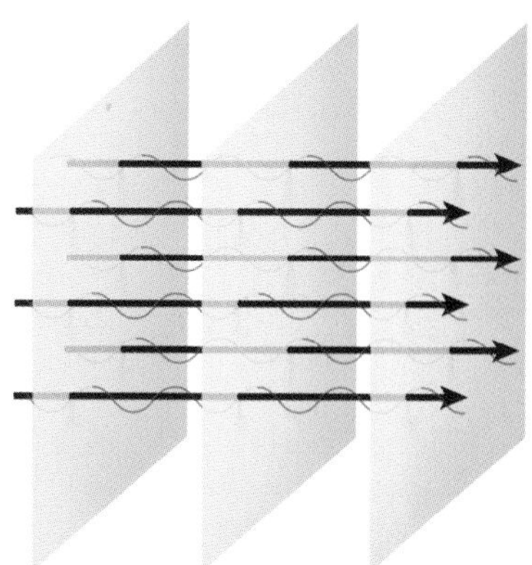

**FIGURE 15.16** Plane waves. Each parallel plane represents a surface of constant phase. Adjacent planes are two wavelengths apart.

## Surface Waves

On December 26, 2004, an earthquake of magnitude 9.0 under the Indian Ocean off the coast of Sumatra triggered a *tsunami* (a Japanese word meaning "harbor wave"). This tsunami caused incredible destruction in the Aceh province of Indonesia, and it also raced across the Indian Ocean with the speed of a jetliner and caused many casualties in India and Sri Lanka. A computer simulation of the propagation of the tsunami is shown in Figure 15.17. What is the physical basis of the motion of a tsunami?

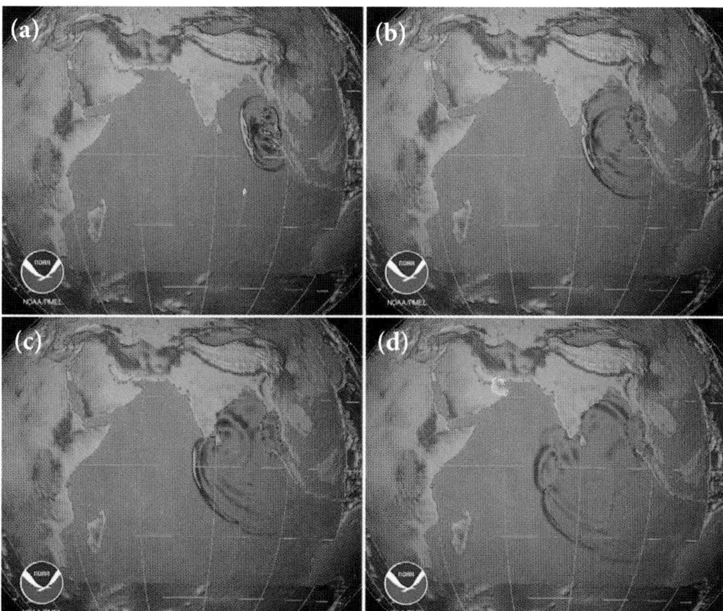


**FIGURE 15.17** A computer simulation of the propagation of the tsunami near Sumatra that occurred on December 26, 2004. The four frames cover a time period of several hours.

When an underwater earthquake occurs, the seafloor is often locally displaced and shifted upward as two plates of the Earth's crust move over each other. Thus, the water in the ocean is displaced, causing a pressure wave. This is one of the rare instances of a wave that carries matter (in this case, water) along with it; water does crash on the shore. However, the water reaching shore did not originate at the site of the earthquake, so no long-distance transportation of water takes place in a tsunami. We'll consider pressure waves in more detail in Chapter 16. For now, you only need to know that the pressure waves in the ocean have very long wavelengths, up to 100 km or even more. Since this wavelength is much larger than the typical depth of the ocean, which is approximately 4 km, the pressure waves resulting from earthquakes cannot spread spherically, but only as surface waves. For these surface waves, the local wave speed is not constant, but depends on the depth of the ocean:

$$v_{\text{surface}} = \sqrt{gd}, \tag{15.17}$$

where $g = 9.81\ \text{m/s}^2$ is the acceleration due to gravity and $d$ is the depth of the ocean, in meters. For a depth of 4000 m, this results in a wave speed of 200 m/s, or approximately 450 mph, as fast as a jetliner. However, when the leading edge of a tsunami reaches the shallow coastal waters, the wave speed decreases according to equation 15.17. This causes a pileup, as water farther out in the ocean still has a larger speed and catches up with the water that forms the leading edge of the tsunami. Thus, a tsunami wave crest has a height on the order of only a foot in the open ocean and can pass unnoticed under ships, while near shore it creates a wall of water up to 100 ft high that sweeps away everything in its path.

### 15.1 In-Class Exercise

On a golf course, your ball is right at the edge of a 120.0-m-diameter circular water hazard that is 1.00 m deep. Instead of hitting over the hazard as planned, you hit your ball exactly into the middle of it. How long does it take for the water wave created by your golf ball to arrive at your feet?

a) 19.2 s

b) 34.7 s

c) 56.1 s

d) 67.1 s

e) 89.3 s

## Seismic Waves

Earthquakes also generate waves that travel through the Earth. Two kinds of seismic waves generated by earthquakes are compression waves and transverse waves. Seismic compression waves are often called *P (or primary) waves,* and seismic transverse waves are often termed *S (or secondary) waves.* P waves can travel in solids and liquids, but S waves can travel only in solids. Liquids such as molten iron cannot sustain S waves.

The Earth has a variable composition as a function of distance from the center, as shown in Figure 15.18a. At the center of the Earth is an inner core composed mainly of solid iron. Next comes the outer core, which is composed mostly of molten iron. The inner and outer mantles are composed of lighter elements and exist in a solid, plastic state. The crust is hard and brittle. Most of the mass of the Earth is concentrated in the mantle.

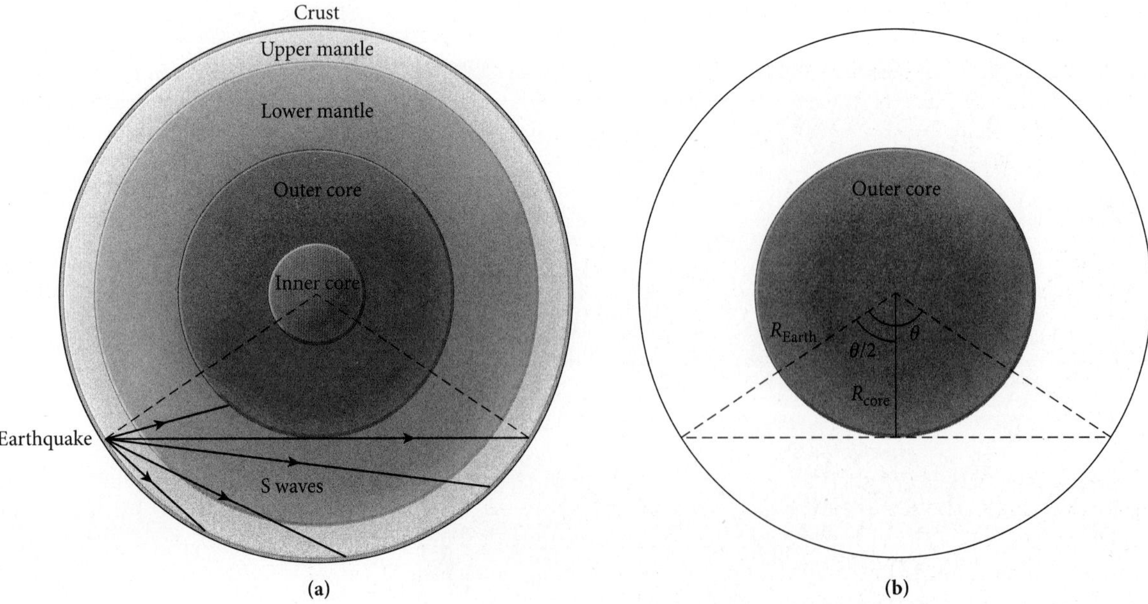


**FIGURE 15.18** (a) The composition of the interior of the Earth. An earthquake takes place, creating S waves. The S waves can travel in the crust, the upper mantle, and the lower mantle but cannot travel in the molten outer core. (b) No S waves are observed at points around the Earth at angles greater than $\theta$ from the earthquake because the S waves cannot transit the molten outer core.

Seismic waves travel from the epicenter of the earthquake through the Earth and can be detected on the surface far away. Seismic compression (P) waves can traverse the entire volume of the Earth, while seismic transverse (S) waves cannot penetrate the molten core. Thus, the core limits the extent of observation for S waves, as illustrated in Figure 15.18.

### EXAMPLE 15.2 Measuring the Size of the Earth's Molten Core

Scientists have observed that the maximum angle around the surface of the Earth where S waves from an earthquake are observed is $\theta = 105°$ (see Figure 15.18b). This effect is due to the fact that S waves cannot travel in the Earth's outer core because molten iron cannot support shear forces. The radius of the Earth is 6370 km.

**PROBLEM**

What is the radius of the molten core of the Earth?

**SOLUTION**

Looking at Figure 15.18b, we see that we can relate the radius of the molten core, $R_{\text{core}}$, the radius of the Earth, $R_{\text{Earth}}$, and the maximum angle of observation of S waves at the surface of the Earth:

$$\cos\left(\frac{\theta}{2}\right) = \frac{R_{\text{core}}}{R_{\text{Earth}}}.$$

We can then solve for the radius of the molten core:

$$R_{\text{core}} = R_{\text{Earth}} \cos\left(\frac{\theta}{2}\right) = (6370 \text{ km})\cos\left(\frac{105°}{2}\right) = 3880 \text{ km}.$$

Note that this result is only approximate. Seismic transverse waves do not travel in exactly straight lines because of variations in density as a function of depth in the Earth. However, the calculated result is reasonably close to the measured value, $R_{\text{core}} = 3480$ km, which has been obtained by combining several different methods.

Earthquakes are not the only source of seismic waves. Such waves can also be generated through the underground use of explosives such as dynamite or TNT or even nuclear bombs. Thus, the monitoring of seismic waves is used to verify nuclear test bans and nonproliferation of nuclear weapons. Other means of generating seismic waves are thumper-trucks, which lift a large mass approximately 3 m off the ground and drop it, air guns, plasma sound sources, and hydraulic or electromechanical shakers. Why would anyone want to generate seismic waves? The waves can be used similar to the way S waves were used in Example 15.2 to look for underground deposits of different density. This basic technique of reflection seismology is widely utilized in oil and gas exploration and is thus of great current research interest.

## 15.6 Energy, Power, and Intensity of Waves

Section 15.1 pointed out that waves do not generally transport matter. However, they do transport energy, and in this section, we will determine how much.

### Energy of a Wave

In Chapter 14, we saw that the energy of a mass oscillating on a spring is proportional to the spring constant and to the square of the oscillation amplitude: $E = \frac{1}{2}kA^2$. Since we can envision a wave as the oscillations of many coupled oscillators, each small volume of the medium through which the wave passes has this energy per oscillator. Using (again) $k/m = \omega^2$, or $k = m\omega^2$, the energy of the wave is

$$E = \tfrac{1}{2}m\omega^2\left[A(r)\right]^2, \quad (15.18)$$

where the amplitude of the wave has a radial dependence, as explained in the general discussion of waves in two and three dimensions. We see that the energy of a wave is proportional to the square of the amplitude and the square of the frequency.

If the wave moves through an elastic medium, the mass, $m$, that appears in equation 15.18 can be expressed as $m = \rho V$, where $\rho$ is the density and $V$ is the volume through which the wave passes. This volume is the product of the cross-sectional area, $A_\perp$ (the area perpendicular to the velocity vector of the wave), that the wave passes through and the distance that the wave travels in time $t$, or $l = vt$. (*Note:* The symbol $A_\perp$ indicates the area, whereas the symbol $A$, without the subscript $_\perp$, is used for the wave amplitude!) Thus, we find for the energy in this case

$$E = \tfrac{1}{2}\rho V\omega^2\left[A(r)\right]^2 = \tfrac{1}{2}\rho A_\perp l\omega^2\left[A(r)\right]^2 = \tfrac{1}{2}\rho A_\perp vt\omega^2\left[A(r)\right]^2. \qquad (15.19)$$

If the wave is a spherical wave radiating outward from a point source and if no energy is lost to damping, we can determine the radial dependence of the amplitude from equation 15.19. In such a case the energy is a constant at a fixed value of $r$, as are the density of the medium, the angular frequency, and the wave velocity. Equation 15.19 then yields the condition $A_\perp[A(r)]^2 = \text{constant}$. For a spherical wave in three dimensions, the area through which the wave passes is the surface of a sphere, as shown in Figure 15.14. Thus, in this case, $A_\perp = 4\pi r^2$, and we obtain $r^2\,[A(r)]^2 = \text{constant} \Rightarrow A(r) \propto 1/r$. This is exactly the result for the radial dependence of the amplitude that was used in equation 15.15.

If the wave is spreading strictly in a two-dimensional space, as is true of surface waves on water, then the cross-sectional area is proportional to the circumference of a circle and is thus proportional to $r$, not $r^2$. In this case, we obtain for the radial dependence of the wave amplitude $r[A(r)]^2 = \text{constant} \Rightarrow A(r) \propto 1/\sqrt{r}$.

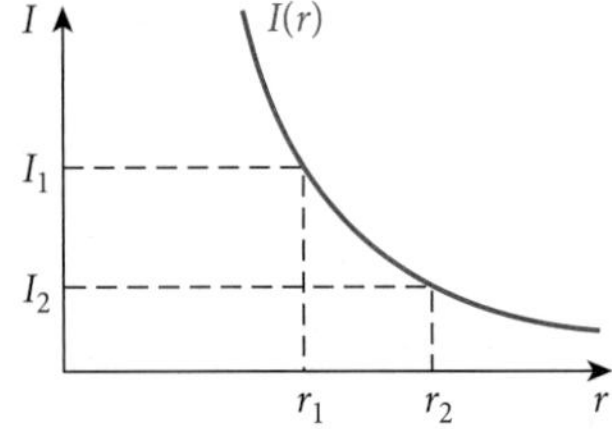

**FIGURE 15.19** Inverse square law for the intensities of a spherical wave at two different distances from the source.

## Power and Intensity of a Wave

In Chapter 5, we saw that the average power is the rate of energy transfer per time. With the energy contained in a wave as given by equation 15.19, we then have

$$\overline{P} = \frac{E}{t} = \tfrac{1}{2}\rho A_\perp v\omega^2\left[A(r)\right]^2. \qquad (15.20)$$

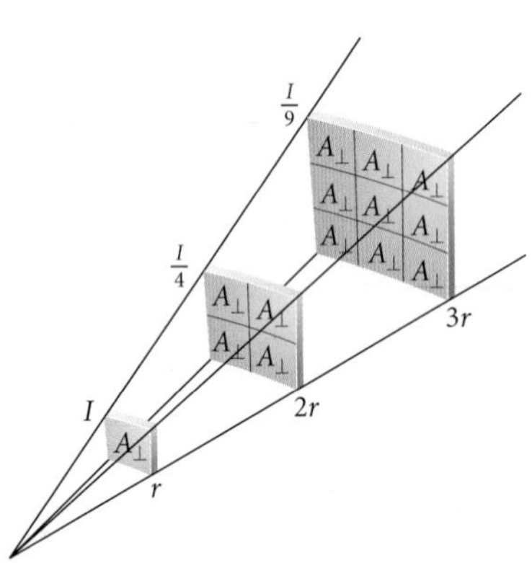

**FIGURE 15.20** The intensity of a wave at a distance $r$ is $I$. At a distance of $2r$, the intensity is $I/4$. At a distance of $3r$, the intensity is $I/9$.

Since all quantities on the right-hand side of equation 15.20 are independent of time, the power radiated by a wave is constant in time.

The intensity, $I$, of a wave is defined as the power radiated per unit cross-sectional area:

$$I = \frac{\overline{P}}{A_\perp}.$$

Substituting for $\overline{P}$ from equation 15.20 into this definition leads to

$$I = \tfrac{1}{2}\rho v\omega^2\left[A(r)\right]^2. \qquad (15.21)$$

As you can see from this result, the intensity depends on the square of the amplitude, and so the radial dependence of the intensity is due to the dependence of the amplitude on the radial distance from the point of propagation. For a spherical wave in three dimensions, $A(r) \propto 1/r$, and thus

$$I \propto \frac{1}{r^2}.$$

Thus, for two points that are at distances of $r_1$ and $r_2$ from the point of propagation of a spherical wave, the intensities $I_1 = I(r_1)$ and $I_2 = I(r_2)$ measured at these two locations are related via

$$\frac{I_1}{I_2} = \left(\frac{r_2}{r_1}\right)^2, \qquad (15.22)$$

as shown in Figure 15.19.

Figure 15.20 illustrates the dependence of the intensity of a wave on the distance from the source. If the intensity of a wave at a distance $r$ is $I$, then the intensity of the wave at a distance of $2r$ is $I/4$. You can see why this relationship is true in Figure 15.20 because the

### 15.3 Self-Test Opportunity

The intensity of electromagnetic waves from the Sun at the Earth is approximately 1400 W/m$^2$. What is the intensity of electromagnetic waves from the Sun at Mars? The distance of the Earth from the Sun is $149.6 \cdot 10^6$ km, and the distance from Mars to the Sun is $227.9 \cdot 10^6$ km.

same amount of power is passing through an area four times the size of $A_\perp$. Similarly, at a distance of $3r$, the intensity is $I/9$ because the same amount of power is passing through nine times the area.

### 15.2 In-Class Exercise

The intensity of electromagnetic waves from the Sun at the Earth is 1400 W/m$^2$. How much total power does the Sun generate? The distance from the Earth to the Sun is $149.6 \cdot 10^6$ km.

a) $1.21 \cdot 10^{20}$ W

b) $2.43 \cdot 10^{24}$ W

c) $3.94 \cdot 10^{26}$ W

d) $2.11 \cdot 10^{28}$ W

e) $9.11 \cdot 10^{30}$ W

## SOLVED PROBLEM 15.1 Power Transmitted by a Wave on a String

### PROBLEM

The transverse displacement from equilibrium of a particular stretched string as a function of position and time is given by $y(x,t) = (0.130 \text{ m}) \cos [(9.00 \text{ m}^{-1})x + (72.0 \text{ s}^{-1})t]$. The linear mass density of the string is 0.00677 kg/m. What is the average power transmitted by the string?

### SOLUTION

#### THINK

We start with equation 15.20 for the average power transmitted by a wave. We can extract the amplitude, $A(r)$, the wave number, $\kappa$, and the angular frequency, $\omega$, from the given function $y(x,t)$ by comparing it with the solution to the wave equation for a wave traveling in the negative $x$-direction (compare equation 15.6). The product $\rho A_\perp$ is given by the linear mass density, $\mu$.

#### SKETCH

Figure 15.21 identifies the relevant factors of the wave function.

$$y(x,t) = \overbrace{(0.130 \text{ m})}^{A}\cos(\overbrace{(9.00 \text{ m}^{-1})}^{\kappa}x + \overbrace{(72.0 \text{ s}^{-1})}^{\omega}t)$$

**FIGURE 15.21** The wave function with the relevant factors identified.

#### RESEARCH

The power transmitted by a wave is given by equation 15.20:

$$\bar{P} = \tfrac{1}{2}\rho A_\perp v\omega^2 \left[A(r)\right]^2.$$

For a wave moving on an elastic string, the product of the density, $\rho$, and the cross-sectional area, $A_\perp$, can be written as

$$\rho A_\perp = \frac{m}{V}A_\perp = \frac{m}{L \cdot A_\perp}A_\perp = \frac{m}{L} = \mu,$$

where $\mu = m/L$ is the linear mass density of the string. For a wave moving on an elastic string, $A(r)$ is constant, and so we can express the transmitted power as

$$\bar{P} = \tfrac{1}{2}\mu v\omega^2 \left[A(r)\right]^2.$$

We can use equation 15.8 to obtain the speed of the wave:

$$v = \frac{\omega}{\kappa}.$$

#### SIMPLIFY

From the wave function specified in the problem statement, we see that $A(r) = 0.130$ m, $\kappa = 9.00 \text{ m}^{-1}$, and $\omega = 72.0 \text{ s}^{-1}$. By convention, $\omega$ is a positive number, and the specified wave function represents a wave traveling in the negative $x$-direction. We can then calculate $v$:

$$v = \frac{\omega}{\kappa} = \frac{72.0 \text{ s}^{-1}}{9.00 \text{ m}^{-1}} = 8.00 \text{ m/s}.$$

#### CALCULATE

Putting the numerical values into equation 15.20, we get

$$\bar{P} = \tfrac{1}{2}\mu v\omega^2 \left[A(r)\right]^2 = \tfrac{1}{2}\left(0.00677 \text{ kg/m}\right)\left(8.00 \text{ m/s}\right)\left(72.0 \text{ s}^{-1}\right)^2\left(0.130 \text{ m}\right)^2$$
$$= 2.37247 \text{ W}.$$

#### ROUND

We report our result to three significant figures:

$$\bar{P} = 2.37 \text{ W}.$$

Continued—

**DOUBLE-CHECK**

To double-check our answer, let's assume that the energy of the wave corresponds to the energy of an object whose mass is equal to the mass of a one wavelength-long section of string. This object is moving in the negative $x$-direction at the speed of the traveling wave. The power is then the energy transmitted in one period. From the values for $\kappa$ and $\omega$ extracted from the given wave function, we get the wavelength and the period:

$$\lambda = \frac{2\pi}{\kappa} = 0.698 \text{ m}$$

$$T = \frac{2\pi}{\omega} = 0.0873 \text{ s.}$$

We can write the kinetic energy as

$$K = \tfrac{1}{2}(\mu\lambda)v^2 = 0.151 \text{ J.}$$

The power is then

$$P = \frac{K}{T} = \frac{0.151 \text{ J}}{0.0873 \text{ s}} = 1.73 \text{ W.}$$

This result is similar to our result for the power transmitted along the string. Thus, our answer seems to be reasonable.

## 15.7 Superposition Principle and Interference

Wave equations like equation 15.9 have a very important property: They are linear. What does this mean? If you find two different solutions, $y_1(x,t)$ and $y_2(x,t)$, to a linear differential equation, then any linear combination of those solutions, such as

$$y(x,t) = ay_1(x,t) + by_2(x,t), \qquad (15.23)$$

where $a$ and $b$ are arbitrary constants, is also a solution to the same linear differential equation. You can see this linear property in the differential equation, because it contains the function $y(x,t)$ to the first power only [there is no term containing $y(x,t)^2$ or $\sqrt{y(x,t)}$ or any other power of the function].

In physical terms, the linear property means that wave solutions can be added, subtracted, or combined in any other linear combination, and the result is again a wave solution. This physical property is the mathematical basis of the superposition principle:

**Two or more wave solutions can be added, resulting in another wave solution.**

$$y(x,t) = y_1(x,t) + y_2(x,t). \qquad (15.24)$$

Equation 15.24 is a special case of equation 15.23 with $a = b = 1$. Let's look at an example of the superposition of two wave pulses with different amplitudes and different velocities.

### EXAMPLE 15.3 Superposition of Wave Pulses

Two wave pulses, $y_1(x,t) = A_1e^{-(x-v_1t)^2}$ and $y_2(x,t) = A_2e^{-(x+v_2t)^2}$, with $A_2 = 1.7A_1$ and $v_1 = 1.6v_2$, are at some distance from each other. Because of their mathematical form, these are called *Gaussian wave packets*. We assume that the two waves are of the same type, such as waves on a rope.

**PROBLEM**

What is the amplitude of the resulting wave at maximum overlap of the two pulses, and at what time does this occur?

### SOLUTION

Both of these Gaussian wave packets are of the functional form $y(x,t) = Y(x \pm vt)$ and are therefore valid solutions of the one-dimensional wave equation 15.9. Thus, the superposition principle (equation 15.23) holds, and we can simply add the two wave functions to obtain the function for the resulting wave. The two Gaussian wave packets have maximum overlap when their centers are at the same position in coordinate space. The center of a Gaussian wave packet is at the location where the exponent has a value of zero. As you can see from the given functions, both wave packets will be centered at $x = 0$ at time $t = 0$, which is the answer to the question as to when the maximum overlap occurs. The amplitude at maximum overlap is simply $A = A_1 + A_2 = 2.7A_1$ (because $A_2 = 1.7A_1$, as specified in the problem statement).

### DISCUSSION

It is instructive to look at plots illustrating these waves and their superposition. The two Gaussian wave packets start at time $t = -3$, in units of the width of a wave packet divided by the speed $v_1$, and propagate. Figure 15.22a is a three-dimensional plot of the wave function $y(x,t) = y_1(x,t) + y_2(x,t)$ as a function of coordinate $x$ and time $t$. The velocity vectors $\vec{v}_1$ and $\vec{v}_2$ are shown. Figure 15.22b shows coordinate space plots of the waves at different points in time. The blue curve represents $y_1(x,t)$, the red curve represents $y_2(x,t)$, and the black curve is their sum. At the middle point in time, the two wave packets achieve maximum overlap.

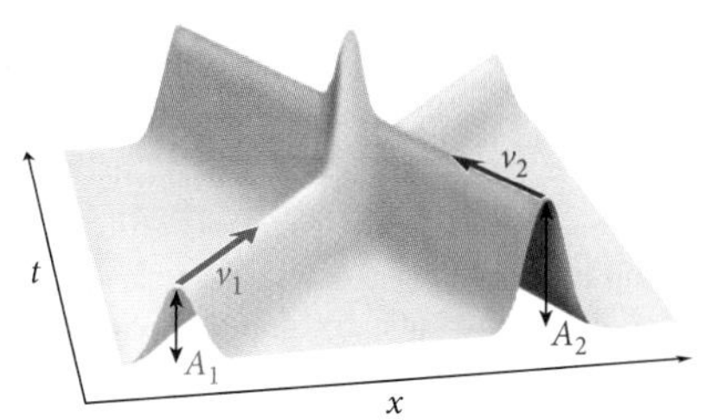

(a)

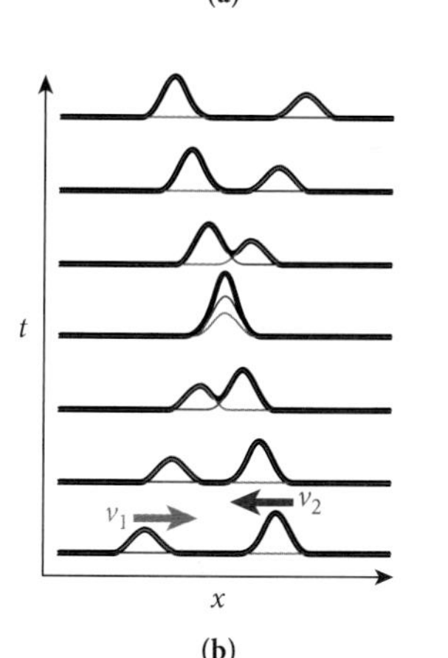

(b)

**FIGURE 15.22** Superposition of two Gaussian wave packets: (a) three-dimensional representation; (b) coordinate space plots for different times.

What is essential to remember from the superposition principle is that waves can penetrate each other without changing their frequency, amplitude, speed, or direction. Modern communications technology is absolutely dependent on this fact. In any city, several TV and radio stations broadcast their signals at different frequencies; there are also cell phone and satellite TV transmissions, as well as light waves and sound waves. All of these waves have to be able to penetrate each other without changing, or everyday communication would be impossible.

## Interference of Waves

Interference is one consequence of the superposition principle. If waves pass through each other, their displacements simply add, according to equation 15.24. Interesting cases arise when the wavelengths and frequencies of the two waves are the same, or at least close to each other. Here we'll look at the case of two waves with identical wavelengths and frequencies. The case of waves whose frequencies are close to each other will be discussed in Chapter 16 on sound, when we discuss beats.

First, let's consider two one-dimensional waves with identical amplitude, $A$, wave number, $\kappa$, and angular frequency, $\omega$. The phase shift $\phi_0$ is set at zero for wave 1 but allowed to vary for wave 2. The sum of the waveforms is then

$$y(x,t) = A\sin(\kappa x - \omega t) + A\sin(\kappa x - \omega t + \phi_0). \tag{15.25}$$

For $\phi_0 = 0$, the two sine functions in equation 15.25 have identical arguments, so the sum is simply $y(x,t) = 2A \sin(\kappa x - \omega t)$. In this situation, the two waves add to the maximum extent possible, which is called **constructive interference.**

If we shift the argument of a sine or cosine function by $\pi$, we obtain the negative of that function: $\sin(\theta + \pi) = -\sin\theta$. For this reason, the two terms in equation 15.25 add up to exactly zero when $\phi_0 = \pi$. This situation is called **destructive interference.** The interference patterns for these two values of $\phi_0$, along with three others, are shown in Figure 15.23 for $t = 0$.

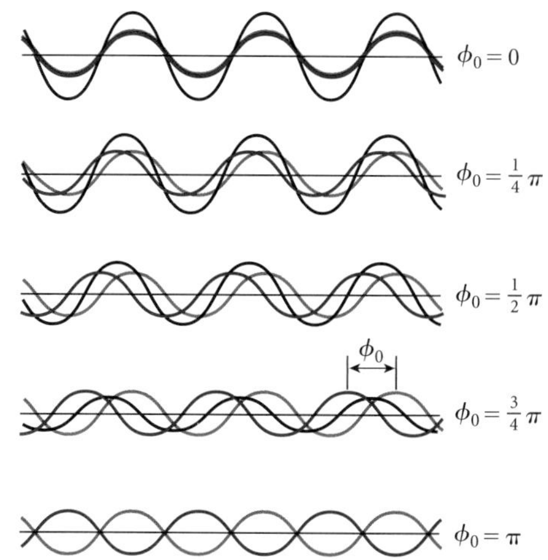


**FIGURE 15.23** Interference of one-dimensional waves as a function of their relative phase shifts at $t = 0$. At later times, the patterns move rigidly to the right. Wave 1 is red, wave 2 is blue, and black is the sum of the two waves.

Interesting interference patterns are obtained with two identical spherical waves whose sources are some distance from each other. Figure 15.24 shows the schematic interference pattern for two identical waves with $\kappa = 1\ \text{m}^{-1}$ and the same values of $\omega$ and $A$. One of the waves was shifted by $\Delta x$ to the right and the other by $-\Delta x$ to the left. The figure shows ten different values of $\Delta x$, which should give you a good idea of how the interference pattern develops as a function of the separation of the centers of the waves. If you have ever dropped two stones at a time into a pond, you will recognize these interference patterns as similar to the surface ripples that resulted.

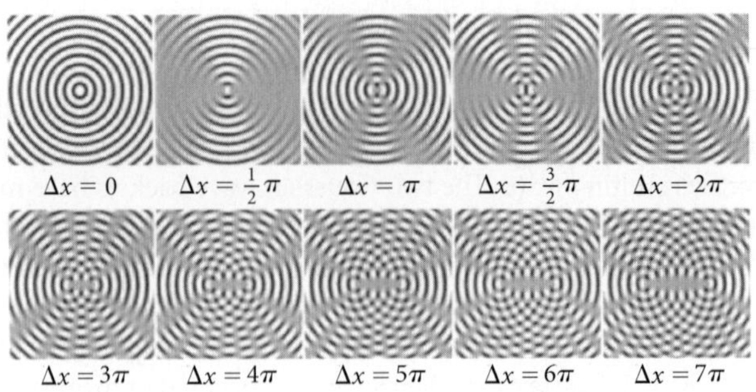


**FIGURE 15.24** Schematic interference patterns of two identical two-dimensional periodic waves shifted a distance $\Delta x$ to either side of the origin, where the unit of $\Delta x$ is meters. White and black circles mark minima and maxima; gray indicates points where the sum of the waves equals zero.

## 15.8 Standing Waves and Resonance

A special type of superposition occurs for two traveling waves if they are identical except for opposite signs for $\omega$: $y_1(x,t) = A\sin(\kappa x + \omega t)$, $y_2(x,t) = A\sin(\kappa x - \omega t)$. Let's first look at the mathematical result of this superposition (after which we'll gain physical insight):

$$y(x,t) = y_1(x,t) + y_2(x,t)$$
$$= A\sin(\kappa x + \omega t) + A\sin(\kappa x - \omega t) \Rightarrow$$
$$y(x,t) = 2A\sin(\kappa x)\cos(\omega t). \tag{15.26}$$

In the last step to obtain equation 15.26, we used the trigonometric addition formula, $\sin(\alpha \pm \beta) = \sin\alpha\cos\beta \pm \cos\alpha\sin\beta$. Thus, for the superposition of two traveling waves with the same amplitude, same wave number, and same speed but opposite direction of propagation, the dependence on the spatial coordinate and the dependence on the time separate (or factorize) into a function of just $x$ times a function of just $t$. The result of this superposition is a wave that has **nodes** (where $y = 0$) and **antinodes** (where $y$ can reach its maximum value) at particular points along the $x$-axis. Each antinode is located midway between neighboring nodes.

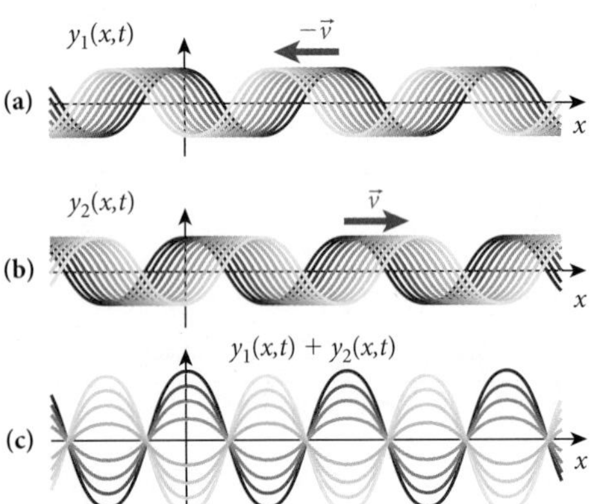


**FIGURE 15.25** (a) and (b) Two traveling waves with velocity vectors in opposite directions. (c) Superposition of the two waves results in a standing wave.

This superposition is much easier to visualize in graphical form. Figure 15.25a shows a wave given by $y_1(x,t) = A\sin(\kappa x + \omega t)$, and Figure 15.25b shows a wave given by $y_2(x,t) = A\sin(\kappa x - \omega t)$. Figure 15.25c shows the addition of the two waves. In all three plots, the waveform is shown as a function of the $x$-coordinate. Each plot also shows the wave at 11 different instants in time, with a time difference of $\pi/10$ between times as if viewing a time-lapse photograph. To compare the same instants of time in each plot, the curves are color-coded starting from red and progressing through orange to yellow. You can see that the traveling waves move toward the left and the right, respectively, but the wave resulting from their superposition oscillates in place, with the nodes and the antinodes remaining at fixed locations on the $x$-axis. This interference wave is known as a **standing wave** and results from the factorization of the time dependence and the spatial dependence in equation 15.26. The two traveling waves are always out of phase at the nodes.

The nodes of a standing wave always have zero amplitude. According to equation 15.18, the energy contained in a wave is proportional to the square of the amplitude. Thus, no energy is transported across a node by a standing wave. Wave energy gets trapped between the nodes of a standing wave and remains localized there. Also note that even though a standing wave is not moving, the relationship between wavelength, frequency, and wave speed, $v = \lambda f$, still holds. Here, $v$ is the speed of the two traveling waves, which make up the standing wave.

### Standing Waves on a String

The basis for the creation of musical sounds by string instruments is the generation of standing waves on strings that are put under tension. Section 15.4 discussed how the velocity of a wave on a string depends on the string tension and the linear mass density of the string (see equation 15.13). In this section, we discuss the basic physics of one-dimensional standing waves.

Let's start with a demonstration. Shown in Figure 15.26 is a string tied to an anchor on the left side and to a piston on the right. The piston oscillates up and down, in sinusoidal fashion, with a variable frequency, $f$. The up-and-down motion of the piston has only a very small amplitude, so we can consider the string to be fixed at both ends. The frequency of the piston's oscillation is varied, and at a certain frequency, $f_0$, the string develops a large-amplitude motion with a central antinode (Figure 15.26a). Clearly, this excitation of the string produces a standing wave. This is a resonant excitation of the string in the sense that the amplitude gets large only at a well-defined **resonance frequency.** If the frequency of the piston's motion is increased by 10%, as shown in Figure 15.26b, this resonant oscillation of the string vanishes. Lowering the piston's frequency by 10% has the same effect. Increasing the frequency to twice the value of $f_0$ results in another resonant excitation, one that has a single node in the center and antinodes at $\frac{1}{4}$ and $\frac{3}{4}$ of the length of the string (Figure 15.26c). Increasing the frequency to triple the value of $f_0$ results in a standing wave with two nodes and three antinodes (Figure 15.26d). The continuation of this series is clear: A frequency of $nf_0$ will result in a standing wave with $n$ antinodes and $n-1$ nodes (plus the two nodes at both ends of the string), which are equally spaced along the length of the string.

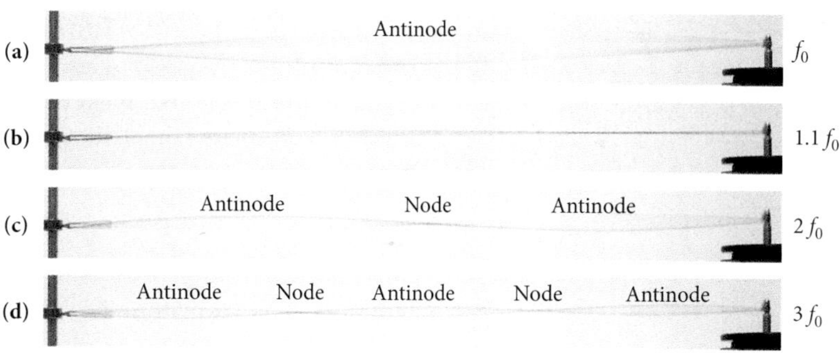


**FIGURE 15.26** Generation of standing waves on a string.

Figure 15.27 shows that the condition for a standing wave is that an integer multiple, $n$, of half wavelengths fits exactly into the length of the string $L$. We give these special wavelengths the index $n$ and write

$$n\frac{\lambda_n}{2}=L,\quad n=1,2,3,\ldots.$$

Solving for the wavelengths, we obtain

$$\lambda_n=\frac{2L}{n},\quad n=1,2,3,\ldots. \tag{15.27}$$

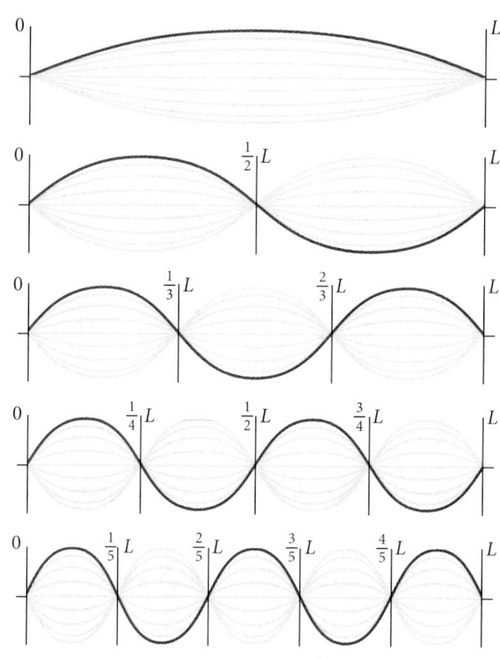


**FIGURE 15.27** Lowest five standing-wave excitations of a string.

The index $n$ on the wavelength (or frequency) indicates the **harmonic.** That is, $n = 1$ identifies the first harmonic (also called the *fundamental frequency*), $n = 2$ labels the second harmonic, $n = 3$ the third, and so on.

As mentioned before, $v = \lambda f$ still holds for a standing wave. Using this relationship, we can find the resonance frequencies of a string from equation 15.27:

$$f_n=\frac{v}{\lambda_n}=n\frac{v}{2L},\quad n=1,2,3,\ldots.$$

Finally, using equation 15.13 for the wave velocity on a string with linear mass density $\mu$ under tension $T$, $v=\sqrt{T/\mu}$, we have

$$f_n=\frac{v}{\lambda_n}=n\frac{\sqrt{T}}{2L\sqrt{\mu}}=n\sqrt{\frac{T}{4L^2\mu}}. \tag{15.28}$$

Equation 15.28 reveals several interesting facts about the construction of string instruments. First, the longer the string, the lower the resonance frequencies. This is the basic reason why a cello, which produces lower notes, is longer than a violin. Second, the resonance frequencies are proportional to the square root of the tension on the string. If an instrument's frequency is too low (it sounds "flat"), you need to increase the tension. Third, the higher the linear mass density, $\mu$, of the string, the lower the frequency. "Fatter" strings produce bass notes. Fourth, the second harmonic is twice as high as the fundamental frequency, the third is three times as high, and so on. When we discuss sound in Chapter 16, we'll see that this fact is the basis of the definition of the octave. For now, you can see that you can produce the second harmonic on the same string that is used for the first harmonic by putting a finger on the exact middle of the string and thus forcing a node to be located there.

### 15.3 In-Class Exercise

A string of a given length is clamped at both ends and given a certain tension. Which of the following statements about a standing wave on this string is true?

a) The higher the frequency of a standing wave on the string, the closer together are the nodes.

b) For any frequency of the standing wave, the nodes are always the same distance apart.

c) For a standing wave on a string at a given tension, only one frequency is ever possible.

d) The lower the frequency of a standing wave on the string, the closer together are the nodes.

## EXAMPLE 15.4 Tuning a Piano

A piano tuner's job is to make sure that all keys on the instrument produce the proper tones. On one piano, the string for the middle-A key is under a tension of 2900 N and has a mass of 0.006000 kg and a length of 0.6300 m.

### PROBLEM

By what fraction does the tuner have to change the tension on this string in order to obtain the proper frequency of 440.0 Hz?

### SOLUTION

First, we calculate the fundamental frequency (or first harmonic), using $n = 1$ in equation 15.28 and the given tension, length, and mass:

$$f_1 = \sqrt{\frac{2900 \text{ N}}{4(0.6300 \text{ m})(0.006000 \text{ kg})}} = 438 \text{ Hz}.$$

This frequency is off by 2 Hz, or too low by 0.5%, so the tension has to be increased. The proper tension can be found by solving equation 15.28 for $T$ and then using the fundamental frequency of 440.0 Hz, the proper frequency for middle A:

$$f_n = n\sqrt{\frac{T}{4L^2(m/L)}} = n\sqrt{\frac{T}{4Lm}} \Rightarrow T = 4Lm\left(\frac{f_n}{n}\right)^2.$$

Inserting the numbers, we find

$$T = 4(0.6300 \text{ m})(0.006000 \text{ kg})\left(440.0 \text{ Hz}\right)^2 = 2927 \text{ N}.$$

The tension needs to be increased by 27 N, which is 0.93% of 2900 N. The piano tuner needs to increase the tension on the string by 0.93% for it to produce the proper middle-A frequency.

### DISCUSSION

We could also have found the required change in the tension by arguing in the following way: The frequency is proportional to the square root of the tension, and $\sqrt{1-x} \approx 1 - \frac{1}{2}x$ for small $x$. Since a 0.5% change in the frequency is needed, the tension has to be changed by twice that amount, or 1%.

By the way, a piano tuner can detect a frequency that is off by 2 Hz easily, by listening for beats, which we'll discuss in Chapter 16.

### 15.4 In-Class Exercise

A guitar string is 0.750 m long and has a mass of 5.00 g. The string is tuned to E (660 Hz) when it vibrates at its fundamental frequency. What is the required tension on the string?

a) $2.90 \cdot 10^3$ N

b) $4.84 \cdot 10^3$ N

c) $6.53 \cdot 10^3$ N

d) $8.11 \cdot 10^3$ N

e) $1.23 \cdot 10^4$ N

### 15.4 Self-Test Opportunity

Researchers at the Cornell NanoScale Science and Technology Facility, where the guitar in Figure 15.11b was built, have also made an even smaller guitar string, from a single carbon nanotube only 1 to 4 nm in diameter. It is suspended over a trench of width $W = 1.5$ μm. In their 2004 paper in *Nature*, the Cornell researchers reported a resonance frequency of the carbon nanotube at 55 MHz. What is the speed of a wave on this carbon nanotube?

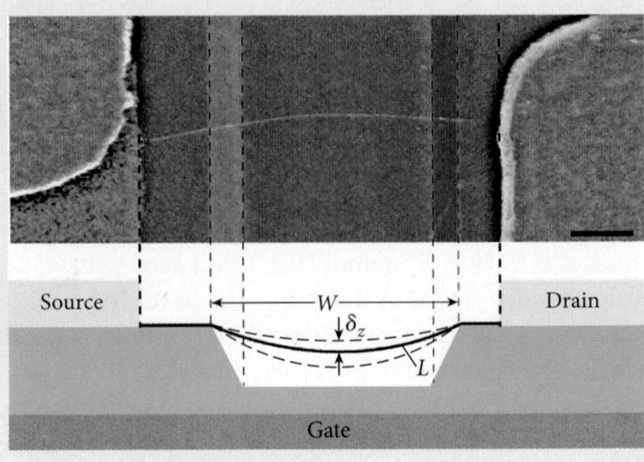

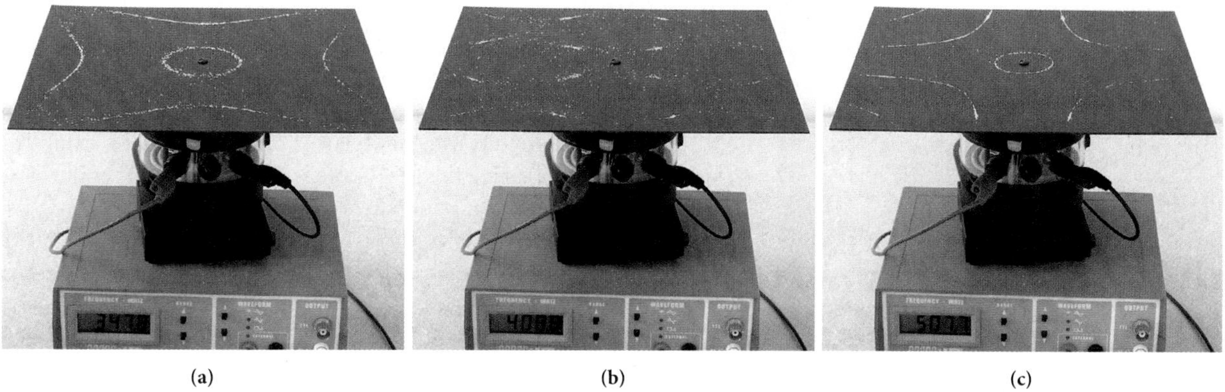

**FIGURE 15.28** Standing-wave nodal patterns on a square plate driven at frequencies of (a) 348 Hz, (b) 409 Hz, and (c) 508 Hz.

Standing waves can also be created in two- and three-dimensions. While we will not treat these cases in any mathematical detail, it is interesting to see the wave patterns that can emerge. To visualize standing wave patterns on two-dimensional sheets, the sheets are driven with a tunable oscillator from below and salt grains are scattered evenly on top. The salt grains gather at the wave nodes, where they do not receive large kicks from the oscillator below. Figure 15.28 shows the result for a square plate driven at three different frequencies. You can see that the standing waves have nodal lines, which have different shapes at different frequencies. We won't discuss these nodal lines in quantitative detail, but it is important to realize that the higher the frequency, the closer the nodal lines are to each other. This is in complete analogy to the case of a one-dimensional standing wave on a string.

### 15.5 In-Class Exercise

The circular plate shown in the figure is driven at three different frequencies. From examining the three parts, rank the three frequencies from lowest to highest.

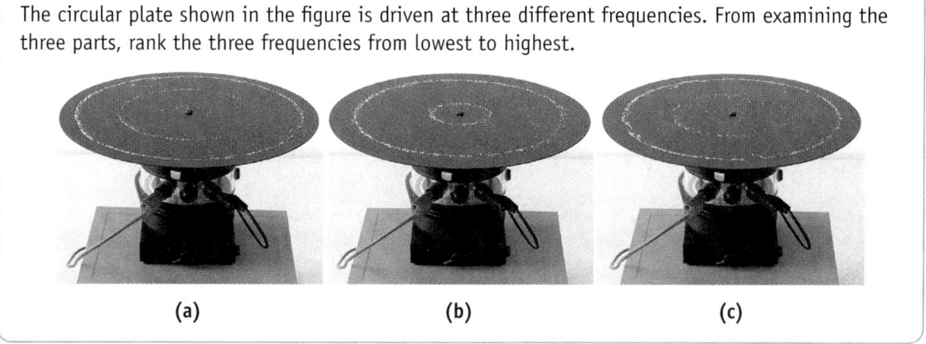

(a) (b) (c)

## 15.9 Research on Waves

This chapter gives only a brief overview of wave phenomena. All physics majors and most engineering majors will take another complete course on waves that goes into more detail. Several other chapters in this book deal with waves. Chapter 16 is devoted to the longitudinal pressure waves called *sound waves.* Chapter 31 focuses on electromagnetic waves, Chapter 34 examines wave optics, and Chapter 36 explores a wave description of matter. This wave character of matter led directly to the development of quantum mechanics, the basis for most of modern physics, including nanoscience and nanotechnology.

Scientists and engineers continue to discover new aspects of wave physics, from fundamental concepts to applications. On the applied side, engineers are studying ways to guide microwaves through different geometries with the aid of finite element methods (which are computational methods that break a large object into a large number of small objects) and very time-intensive computer programs. If these geometries are not spatially symmetric—that is, if they are not exactly round or exactly rectangular—some surprising results can occur. For example, physicists can study quantum chaos with the aid of microwaves that are sent into a metal cavity in the shape of a stadium (a shape that has semicircles at both ends, connected by straight sections). Figure 15.29 shows such a stadium geometry, not for microwaves but

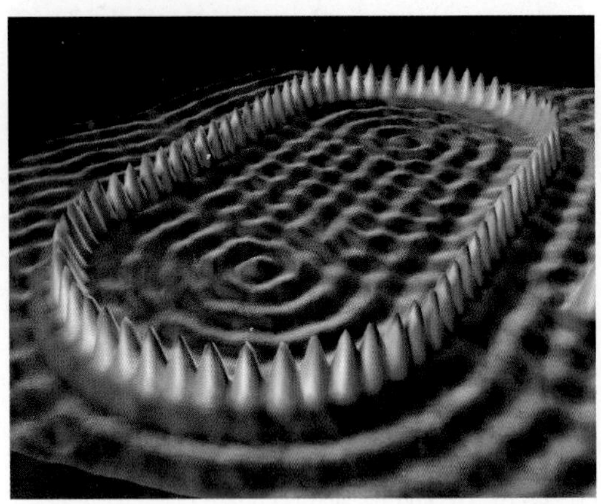

**FIGURE 15.29** Iron atoms on a copper surface form a stadium-shaped cavity in which physicists can study electron matter waves.

for a physical system—a collection of individual iron atoms arranged on a copper surface by using a scanning tunneling microscope. The wave ripples in the interior of the stadium are standing electron matter waves.

On the most advanced level of investigation, particle theorists and mathematicians have come up with new classes of theories called *string theories.* The mathematics of string theories exceeds the scope of this book. However, it is interesting to note that string theorists obtain their physical insights from considering systems similar to those we have considered in this chapter—in particular, standing waves on a string. We'll touch on this subject again in Chapter 39 on particle physics.

Perhaps one of the most interesting experimental investigations into wave physics is the search for gravitational waves. Gravitational waves were predicted by Einstein's theory of general relativity (see Chapter 35) but are very weak and hard to detect unless generated by an extremely massive object. Astrophysicists have postulated that some astronomical objects emit gravitational waves of such great intensity that they should be observable on Earth. Indirect evidence of the existence of such gravitational waves has come from the study of orbital frequencies of neutron star pairs, since the decrease of measured rotational energy in the pair can be explained by the radiation of gravitational waves. However, no direct observation of gravitational waves has yet been made. Groups of American and international physicists have constructed or are in the process of constructing several gravitational wave detectors, such as LIGO (Laser Interferometer Gravitational-Wave Observatory). These detectors all work by using the principle of interference of light waves, which we will cover in more detail in Chapter 34. The basic idea is to suspend in a vacuum by very thin wires several heavy test masses with mirrored sides, placed at the ends of two perpendicular arms of the instrument. As a gravitational wave passes by, it should cause alternating stretching and compressing of space, which will be out of phase for the two arms and should be observable as minute movements of the mirrors. In principle, bouncing a laser light off the mirrors and then recombining the beams would reveal interference of light due to the motion, detectable as an interference pattern. These gravitational wave detectors require very long tunnels for the lasers. LIGO consists of two observatories, one in Hanford, Washington, and the other in Livingston, Louisiana (see Figure 15.30), with a

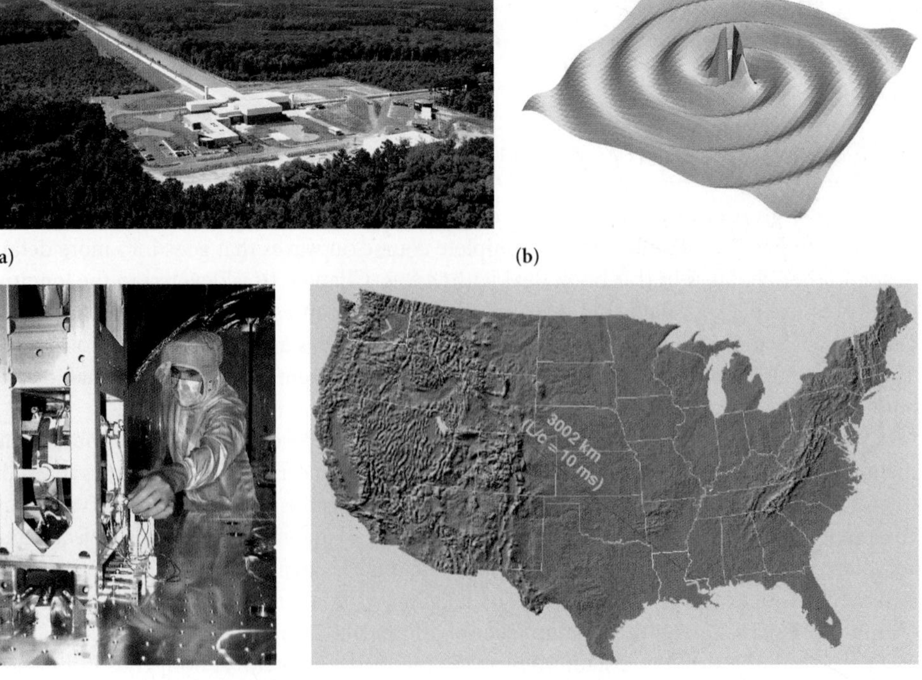


**FIGURE 15.30** (a) Aerial view of the Livingston site of LIGO (Laser Interferometer Gravitational-Wave Observatory); (b) visualization of a gravitational wave; (c) a scientist adjusts the interferometer; (d) map of LIGO.

distance of 3002 km between them. Each observatory has two tunnels, each 4 km long, arranged in an L shape. These experiments push the envelope of what is technically possible and measurable. For example, the expected movement of the test masses due to a gravitational wave is on the order of only $10^{-18}$ m, which is smaller than the diameter of a hydrogen atom by a factor of 100 million, or $10^8$. Measuring a deflection that small is a huge technological challenge.

Other laboratories in Europe (Virgo and GEO-600) and Japan (TAMA) are also trying to measure the effect of gravitational waves. Plans also exist for a space-based gravitational wave observatory, named LISA (Laser Interferometer Space Antenna), which will consist of three spacecraft arranged in an equilateral triangle with side length $5 \cdot 10^6$ km. Scientists have recently used supercomputers to simulate gravitational waves emitted from the collision of two massive black holes, as illustrated in Figure 15.31 in a calculation done by a group from NASA. Such waves might be intense enough to be detected on Earth.

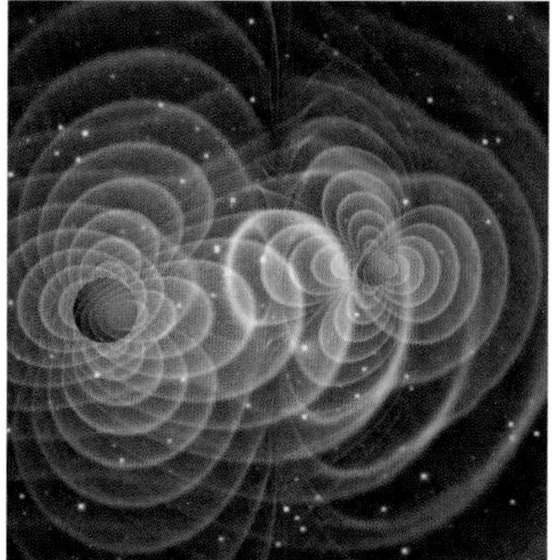

**FIGURE 15.31** Supercomputer simulation of the collision of two massive black holes, creating gravitational waves that may be observable.

## WHAT WE HAVE LEARNED | EXAM STUDY GUIDE

- The wave equation describes the displacement $y(x,t)$ from the equilibrium position for any wave motion: $\frac{\partial^2}{\partial t^2} y(x,t) - v^2 \frac{\partial^2}{\partial x^2} y(x,t) = 0.$
- For any wave, $v$ is the wave velocity, $\lambda$ is the wavelength, and $f$ is the frequency, and they are related by $v = \lambda f$.
- The wave number is defined as $\kappa = 2\pi/\lambda$, just as the angular frequency is related to the period by $\omega = 2\pi/T$.
- Any function of the form $Y(\kappa x - \omega t + \phi_0)$ or $Y(\kappa x + \omega t + \phi_0)$ is a solution to the wave equation. The first function describes a wave traveling to the right, and the second function describes a wave traveling to the left.
- The wave speed on a string is $v = \sqrt{T/\mu}$, where $\mu = m/L$ is the linear mass density of the string and $T$ is the string tension.
- A one-dimensional sinusoidal wave is given by $y(x,t) = A\sin(\kappa x - \omega t + \phi_0)$, where the argument of the sine function is the phase of the wave and $A$ is the wave amplitude.
- A three-dimensional spherical wave is described by $y(\vec{r},t) = \frac{A}{r}\sin(\kappa r - \omega t + \phi_0)$, and a plane wave is described by $y(\vec{r},t) = A\sin(\kappa x - \omega t + \phi_0)$.
- The energy contained in a wave is $E = \frac{1}{2}\rho V\omega^2[A(r)]^2 = \frac{1}{2}\rho(A_\perp l)\omega^2[A(r)]^2 = \frac{1}{2}\rho(A_\perp vt)\omega^2[A(r)]^2$, where $A_\perp$ is the perpendicular cross-sectional area through which the wave passes.
- The power transmitted by a wave is $\overline{P} = \frac{E}{t} = \frac{1}{2}\rho A_\perp v\omega^2 [A(r)]^2$, and its intensity is $I = \frac{1}{2}\rho v\omega^2[A(r)]^2$.
- For a spherical three-dimensional wave, the intensity is inversely proportional to the square of the distance to the source.
- The superposition principle holds for waves: Adding two solutions of the wave equation results in another valid solution.
- Waves can interfere in space and time, constructively or destructively, depending on their relative phases.
- Adding two traveling waves that are identical except for having velocity vectors that point in opposite directions yields a standing wave such as $y(x,t) = 2A\sin(\kappa x)\cos(\omega t)$, for which the dependence on space and time factorizes.
- The resonance frequencies (or harmonics) for waves on a string are given by $f_n = \frac{v}{\lambda_n} = n\frac{\sqrt{T}}{2L\sqrt{\mu}}$, for $n = 1, 2, 3, \ldots$. The index $n$ indicates the harmonic (for example, $n = 4$ corresponds to the fourth harmonic).

## KEY TERMS

## NEW SYMBOLS AND EQUATIONS

$\frac{\partial^2}{\partial t^2} y(x,t) - v^2 \frac{\partial^2}{\partial x^2} y(x,t) = 0$, wave equation

$v = \lambda f$, velocity, wavelength, and frequency of a wave

$\kappa = 2\pi/\lambda$, wave number

$v = \sqrt{T/\mu}$, speed of a wave on a string

$y(x,t) = A \sin(\kappa x - \omega t + \phi_0)$, one-dimensional traveling wave

$y(x,t) = 2A \sin(\kappa x) \cos(\omega t)$, one-dimensional standing wave

$f_n = \frac{v}{\lambda_n} = n \frac{\sqrt{T}}{2L\sqrt{\mu}}$, resonance frequencies of a wave on a string

## ANSWERS TO SELF-TEST OPPORTUNITIES

**15.1** $f = 0.40$ Hz.

**15.2** $A = 10$ cm, $\kappa = 0.31\ \text{cm}^{-1}$, $\phi_0 = 0.30$.

**15.3**

$$\frac{I_E}{I_M} = \left(\frac{r_M}{r_E}\right)^2$$

$$I_M = I_E \left(\frac{r_E}{r_M}\right)^2 = \left(1400\ \text{W/m}^2\right)\left(\frac{149.6 \cdot 10^6\ \text{km}}{227.9 \cdot 10^6\ \text{km}}\right)^2 = 603.3\ \text{W/m}^2.$$

**15.4** Lowest resonance frequency: $f_1 = \frac{v}{\lambda_1} = \frac{v}{2L}$

Wave speed: $v = 2Lf_1 = 2(1.5 \cdot 10^{-6}\ \text{m})(5.5 \cdot 10^7\ \text{s}^{-1}) = 165$ m/s.

## PROBLEM-SOLVING PRACTICE

### Problem-Solving Guidelines

**1.** If a wave fits the description of a one-dimensional traveling wave, you can write a mathematical equation describing it if you know the amplitude, $A$, of the motion and any two of these three quantities: velocity, $v$, wavelength, $\lambda$, and frequency, $f$. (Alternatively, you need to know two of the three quantities velocity, $v$, wave number, $\kappa$, and angular frequency, $\omega$.)

**2.** If a wave is a one-dimensional standing wave on a string, its resonance frequencies are determined by the length, $L$, of the string and the speed, $v$, of the component waves. The speed of the wave, in turn, is determined by the tension, $T$, in the string and the linear mass density, $\mu$, of the string.

### SOLVED PROBLEM 15.2 Traveling Wave

**PROBLEM**

A wave on a string is given by the function

$$y(x,t) = (0.00200\ \text{m})\sin\left[\left(78.8\ \text{m}^{-1}\right)x + \left(346\ \text{s}^{-1}\right)t\right].$$

What is the wavelength of this wave? What is its period? What is its velocity?

**SOLUTION**

**THINK**

From the given wave function, we can identify the wave number and the angular frequency, from which we can extract the wavelength and period. We can then calculate the velocity of the traveling wave from the ratio of the angular frequency to the wave number.

**SKETCH**

Figure 15.32 shows plots of the wave function at $t = 0$ (part a) and at $x = 0$ (part b).

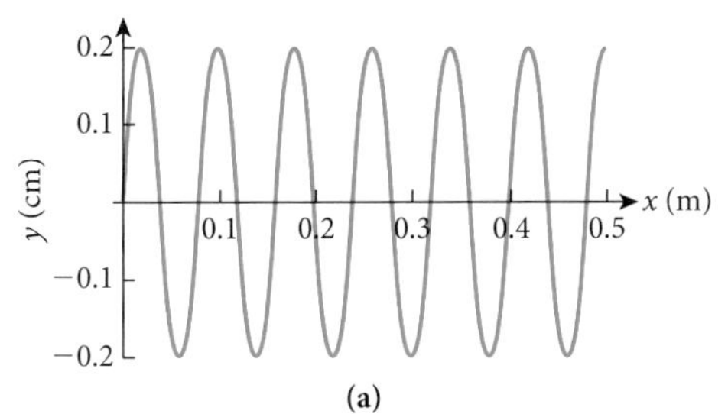


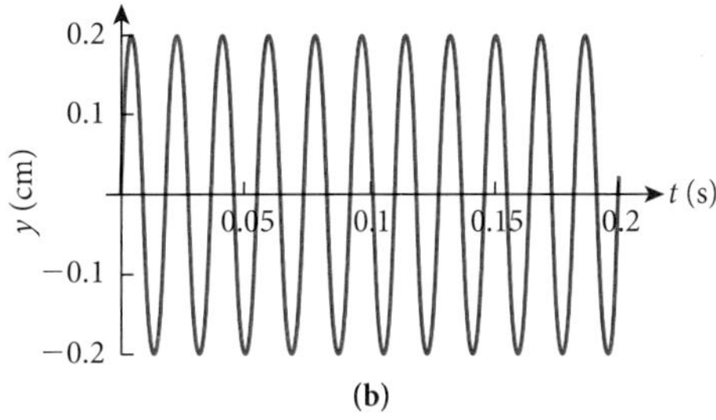


**FIGURE 15.32** (a) The wave function at $t = 0$. (b) The wave function as a function of time at $x = 0$.

## RESEARCH

Looking at the wave function given in the problem statement, $y(x,t) = (0.00200 \text{ m}) \sin\left[\left(78.8 \text{ m}^{-1}\right)x + \left(346 \text{ s}^{-1}\right)t\right]$, and comparing it with the solution to the wave equation for a wave traveling in the negative $x$-direction (see equation 15.6),

$$y(x,t) = A\sin(\kappa x + \omega t + \phi_0),$$

we see that $\kappa = 78.8 \text{ m}^{-1}$ and $\omega = 346 \text{ s}^{-1}$.

## SIMPLIFY

We can obtain the wavelength from the wave number,

$$\lambda = \frac{2\pi}{\kappa},$$

and the period from the angular frequency,

$$T = \frac{2\pi}{\omega}.$$

We can then calculate the velocity of the wave from

$$v = \frac{\omega}{\kappa}.$$

## CALCULATE

Putting in the numerical values, we get the wavelength and the period:

$$\lambda = \frac{2\pi}{78.8 \text{ m}^{-1}} = 0.079736 \text{ m}$$

and

$$T = \frac{2\pi}{346 \text{ s}^{-1}} = 0.018159 \text{ s}.$$

We now calculate the velocity of the traveling wave:

$$v = \frac{346 \text{ s}^{-1}}{78.8 \text{ m}^{-1}} = 4.39086 \text{ m/s}.$$

## ROUND

We report our results to three significant figures:

$$\lambda = 0.0797 \text{ m}$$
$$T = 0.0182 \text{ s}$$
$$v = 4.39 \text{ m/s}.$$

## DOUBLE-CHECK

To double-check our result for the wavelength, we look at Figure 15.32a. We can see that the wavelength (the distance required to complete one oscillation) is approximately 0.08 m, in agreement with our result. Looking at Figure 15.32b, we can see that the period (the time required to complete one oscillation) is approximately 0.018 s, also in agreement with our result.

## SOLVED PROBLEM 15.3 Standing Wave on a String

### PROBLEM

A mechanical driver is used to set up a standing wave on an elastic string, as shown in Figure 15.33a. Tension is put on the string by running it over a frictionless pulley and hanging a metal block from it (Figure 15.33b). The length of the string from the top of the pulley to the driver is 1.25 m, and the linear mass density of the string is 5.00 g/m. The frequency of the driver is 45.0 Hz. What is the mass of the metal block?

### SOLUTION

**THINK**

The weight of the metal block is equal to the tension placed on the elastic string. We can see from Figure 15.33a that the string is vibrating at its third harmonic because three antinodes are visible. We can thus use equation 15.28 to relate the tension on the string, the harmonic, the linear mass density of the string, and the frequency. Once we have determined the tension on the string, we can calculate the mass of the metal block.

**SKETCH**

Figure 15.33b shows the elastic string under tension due to a metal block hanging from it. The mechanical driver sets up a standing wave on the string. Here $L$ is the length of the string from the pulley to the wave driver, $\mu$ is the linear mass density of the string, and $m$ is the mass of the metal block. Figure 15.33c shows the free-body diagram of the hanging metal block, where $T$ is the tension on the string and $mg$ is the weight of the metal block.

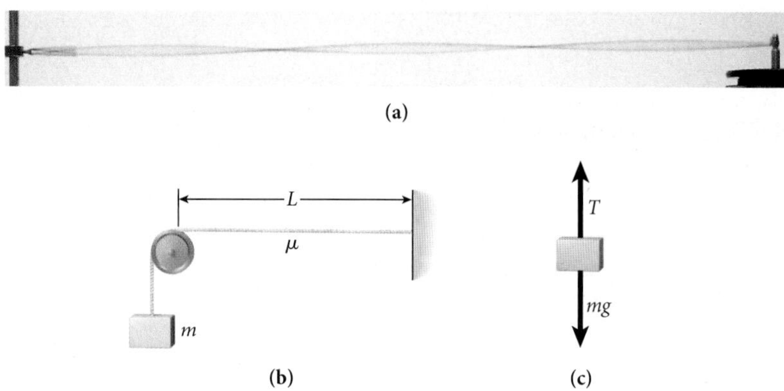


**FIGURE 15.33** (a) A standing wave on an elastic string driven by a mechanical driver. (b) The string is put under tension by a hanging mass. (c) Free-body diagram for the hanging block.

**RESEARCH**

A standing wave with harmonic $n$ and frequency $f_n$ on an elastic string of length $L$ and linear mass density $\mu$ satisfies equation 15.28:

$$f_n = n\sqrt{\frac{T}{4L^2\mu}}.$$

From the free-body diagram in Figure 15.33c, and because the metal block does not move, we can write

$$T - mg = 0.$$

Therefore, the tension on the string is $T = mg$.

**SIMPLIFY**

We solve equation 15.28 for the tension on the string:

$$T = 4L^2\mu\frac{f_n^2}{n^2}.$$

Substituting $mg$ for $T$ and rearranging terms, we obtain

$$m = \frac{\mu}{g}\left(\frac{2Lf_n}{n}\right)^2.$$

**CALCULATE**

Putting in the numerical values, we get

$$m=\frac{\mu}{g}\left(\frac{2Lf_n}{n}\right)^2=\frac{0.00500\text{ kg/m}}{9.81\text{ m/s}^2}\left(\frac{2(1.25\text{ m})(45.0\text{ Hz})}{3}\right)^2$$
$$=0.716743\text{ kg}.$$

**ROUND**

We report our result to three significant figures:

$$m=0.717\text{ kg}.$$

**DOUBLE-CHECK**

As a double-check of our result, we use the fact that a half-liter bottle of water has a mass of about 0.5 kg. Even a light string could support this mass without breaking. Thus, our answer seems reasonable.

## MULTIPLE-CHOICE QUESTIONS

**15.1** Fans at a local football stadium are so excited that their team is winning that they start "the wave" in celebration. Which of the following four statements is (are) true?

I. This wave is a traveling wave.

II. This wave is a transverse wave.

III. This wave is a longitudinal wave.

IV. This wave is a combination of a longitudinal wave and a transverse wave.

a) I and II
b) II only
c) III only
d) I and IV
e) I and III

**15.2** You wish to decrease the speed of a wave traveling on a string to half its current value by changing the tension in the string. By what factor must you decrease the tension in the string?

a) 1
b) $\sqrt{2}$
c) 2
d) 4
e) none of the above

**15.3** Suppose that the tension is doubled for a string on which a standing wave is propagated. How will the velocity of the standing wave change?

a) It will double.
b) It will quadruple.
c) It will be multiplied by $\sqrt{2}$.
d) It will be multiplied by $\frac{1}{2}$.

**15.4** Which of the following transverse waves has the greatest power?

a) a wave with velocity $v$, amplitude $A$, and frequency $f$
b) a wave of velocity $v$, amplitude $2A$, and frequency $f/2$
c) a wave of velocity $2v$, amplitude $A/2$, and frequency $f$
d) a wave of velocity $2v$, amplitude $A$, and frequency $f/2$
e) a wave of velocity $v$, amplitude $A/2$, and frequency $2f$

**15.5** The speed of light waves in air is greater than the speed of sound in air by *about* a factor of a million. Given a sound wave and a light wave of the same wavelength, both traveling through air, which statement about their frequencies is true?

a) The frequency of the sound wave will be about a million times greater than that of the light wave.
b) The frequency of the sound wave will be about a thousand times greater than that of the light wave.
c) The frequency of the light wave will be about a thousand times greater than that of the sound wave.
d) The frequency of the light wave will be about a million times greater than that of the sound wave.
e) There is insufficient information to determine the relationship between the two frequencies.

**15.6** A string is made to oscillate, and a standing wave with three antinodes is created. If the tension in the string is increased by a factor of 4,

a) the number of antinodes increases.
b) the number of antinodes remains the same.
c) the number of antinodes decreases.
d) the number of antinodes will equal the number of nodes.

**15.7** The different colors of light we perceive are a result of the varying frequencies (and wavelengths) of the electromagnetic radiation. Infrared radiation has lower frequencies than does visible light, and ultraviolet radiation has higher frequencies than visible light does. The primary colors are red (R), yellow (Y), and blue (B). Order these colors by their wavelength, shortest to longest.

a) B, Y, R
b) B, R, Y
c) R, Y, B
d) R, B, Y

**15.8** If transverse waves on a string travel with a velocity of 50 m/s when the string is under a tension of 20 N, what tension on the string is required for the waves to travel with a velocity of 30 m/s?

a) 7.2 N
b) 12 N
c) 33 N
d) 40 N
e) 45 N
f) 56 N

## QUESTIONS

**15.9** You and a friend are holding the two ends of a Slinky stretched out between you. How would you move your end of the Slinky to create (a) transverse waves or (b) longitudinal waves?

**15.10** A steel cable consists of two sections with different cross-sectional areas, $A_1$ and $A_2$. A sinusoidal traveling wave is sent down this cable from the thin end of the cable. What happens to the wave on encountering the $A_1/A_2$ boundary? How do the speed, frequency and wavelength of the wave change?

**15.11** Noise results from the superposition of a very large number of sound waves of various frequencies (usually in a continuous spectrum), amplitudes, and phases. Can interference arise with noise produced by two sources?

**15.12** The $1/R^2$ dependency for intensity can be thought of to be due to the fact that the same power is being spread out over the surface of a larger and larger sphere. What happens to the intensity of a sound wave inside an enclosed space, say a long hallway?

**15.13** If two traveling waves have the same wavelength, frequency, and amplitude and are added appropriately, the result is a standing wave. Is it possible to combine two standing waves in some way to give a traveling wave?

**15.14** A ping-pong ball is floating in the middle of a lake and waves begin to propagate on the surface. Can you think of a situation in which the ball remains stationary? Can you think of a situation involving a single wave on the lake in which the ball remains stationary?

**15.15** Why do circular water waves on the surface of a pond decrease in amplitude as they travel away from the source?

**15.16** Consider a monochromatic wave on a string, with amplitude $A$ and wavelength $\lambda$, traveling in one direction. Find the relationship between the maximum speed of any portion of string, $v_{max}$, and the wave speed, $v$.

## PROBLEMS

A blue problem number indicates a worked-out solution is available in the Student Solutions Manual. One • and two •• indicate increasing level of problem difficulty.

### Sections 15.1 through 15.3

15.17 One of the main things allowing humans to determine whether a sound is coming from the left or the right is the fact that the sound will reach one ear before the other. Given that the speed of sound in air is 343 m/s and that human ears are typically 20 cm apart, what is the maximum time resolution for human hearing that allows sounds coming from the left to be distinguished from sounds coming from the right? Why is it impossible for a diver to be able to tell from which direction the sound of a motor boat is coming? The speed of sound in water is 1500 m/s.

**15.18** Hiking in the mountains, you shout "hey," wait 2 s and shout again. What is the distance between the sound waves you cause? If you hear the first echo after 5 s, what is the distance between you and the point where your voice hit a mountain?

**15.19** The displacement from equilibrium caused by a wave on a string is given by $y(x,t) = (-0.00200\text{ m})\sin[(40.0\text{ m}^{-1})x - (800.\text{ s}^{-1})t]$. For this wave, what are the (a) amplitude, (b) number of waves in 1 m, (c) number of complete cycles in 1 s, (d) wavelength, and (e) speed?

**•15.20** A traveling wave propagating on a string is described by the following equation:

$$y(x,t) = (5\text{ mm})\left(\sin(157.08\text{ m}^{-1})x - (314.16\text{ s}^{-1})t + 0.7854\right)$$

a) Determine the minimum separation, $\Delta x_{min}$, between two points on the string that oscillate in perfect opposition of phases (move in opposite directions *at all times*).

b) Determine the separation, $\Delta x_{AB}$, between two points $A$ and $B$ on the string, if point $B$ oscillates with a phase difference of 0.7854 rad compared to point $A$.

c) Find the number of crests of the wave that pass through point $A$ in a time interval $\Delta t = 10$ s and the number of troughs that pass through point $B$ in the same interval.

d) At what point along its trajectory should a linear driver connected to one end of the string at $x = 0$ start its oscillation to generate this sinusoidal traveling wave on the string?

••15.21 Consider a linear array of $n$ masses, each equal to $m$, connected by $n + 1$ springs, all massless and having spring constant $k$, with the outer ends of the first and last springs fixed. The masses can move without friction in the linear dimension of the array.

a) Write the equations of motion for the masses.

b) Configurations of motion for which all parts of a system oscillate with the same angular frequency are called *normal modes* of the system; the corresponding angular frequencies are the system's *normal-mode angular frequencies*. Find the normal-mode angular frequencies of this array.

### Section 15.4

**15.22** Show that the function $D = A\ln(x + vt)$ is a solution of the wave equation (equation 15.9).

**15.23** A wave travels along a string in the positive $x$-direction at 30 m/s. The frequency of the wave is 50 Hz. At $x = 0$ and $t = 0$, the wave velocity is 2.5 m/s and the vertical displacement is $y = 4$ mm. Write the function $y(x,t)$ for the wave.

**15.24** A wave on a string has a wave function given by
$y(x,t) = (0.0200 \text{ m})\sin\left[\left(6.35 \text{ m}^{-1}\right)x + \left(2.63 \text{ s}^{-1}\right)t\right]$.

a) What is the amplitude of the wave?
b) What is the period of the wave?
c) What is the wavelength of the wave?
d) What is the speed of the wave?
e) In which direction does the wave travel?

•**15.25** A sinusoidal wave traveling in the positive $x$-direction has a wavelength of 12 cm, a frequency of 10 Hz, and an amplitude of 10 cm. The part of the wave that is at the origin at $t = 0$ has a vertical displacement of 5 cm. For this wave, determine the

a) wave number,
b) period,
c) angular frequency,
d) speed,
e) phase angle, and
f) equation of motion.

•**15.26** A mass $m$ hangs on a string that is connected to the ceiling. You pluck the string just above the mass, and a wave pulse travels up to the ceiling, reflects off the ceiling, and travels back to the mass. Compare the round-trip time for this wave pulse to that of a similar wave pulse on the same string if the attached mass is increased to $3m$. (Assume that the string does not stretch in either case and the contribution of the mass of the string to the tension is negligible.)

•**15.27** Point $A$ in the figure is 30 cm below the ceiling. Determine how much longer it will take for a wave pulse to travel along wire 1 than along wire 2.

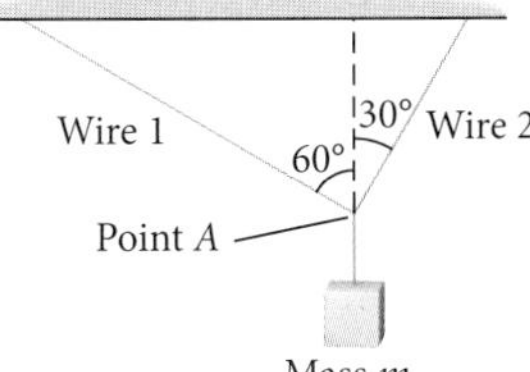


•**15.28** A particular steel guitar string has mass per unit length of 1.93 g/m.

a) If the tension on this string is 62.2 N, what is the wave speed on the string?

b) For the wave speed to be increased by 1.0%, how much should the tension be changed?

•**15.29** Bob is talking to Alice using a tin can telephone, which consists of two steel cans connected by a 20.0-m-long taut steel wire (see the figure). The wire has a linear density of 6.13 g/m, and the tension on the wire is 25.0 N. The sound waves leave Bob's mouth, are collected by the can on the left, and then create vibrations in the wire, which travel to Alice's can and are transformed back into sound waves in air. Alice hears both the sound waves that have traveled through the wire (wave 1) and those that have traveled through the air (wave 2), bypassing the wire. Do these two kinds of waves reach her at the same time? If not, which wave arrives sooner and by how much? The speed of sound in air is 343 m/s.

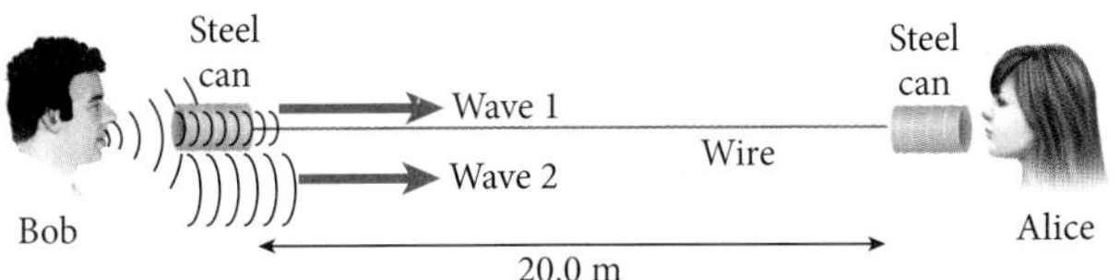


••**15.30** A wire of uniform linear mass density hangs from the ceiling. It takes 1.00 s for a wave pulse to travel the length of the wire. How long is the wire?

••**15.31** a) Starting from the general wave equation (equation 15.9), prove through direct derivation that the Gaussian wave packet described by the equation $y(x,t) = (5m)e^{-0.1(x-5t)^2}$ is indeed a traveling wave (that it satisfies the differential wave equation).

b) If $x$ is specified in meters and $t$ in seconds, determine the speed of this wave. On a single graph, plot this wave as a function of $x$ at $t = 0$, $t = 1$ s, $t = 2$ s, and $t = 3$ s.

c) More generally, prove that any function $f(x,t)$ that depends on $x$ and $t$ through a combined variable $x \pm vt$ is a solution of the wave equation, irrespective of the specific form of the function $f$.

## Section 15.5

•**15.32** Suppose the slope of a beach underneath the ocean is 12 cm of dropoff for every 1.0 m of horizontal distance. A wave is moving inland, slowing down as it enters shallower water. What is its acceleration when it is 10 m from the shoreline?

•**15.33** An earthquake generates three kinds of waves: surface waves (L waves), which are the slowest and weakest; shear (S) waves, which are transverse waves and carry most of the energy; and pressure (P) waves, which are longitudinal waves and travel the fastest. The speed of P waves is approximately 7.0 km/s, and that of S waves is about 4.0 km/s. Animals seem to feel the P waves. If a dog senses the arrival of P waves and starts barking 30.0 s before an earthquake is felt by humans, approximately how far is the dog from the earthquake's epicenter?

## Section 15.6

**15.34** A string with a mass of 30 g and a length of 2 m is stretched under a tension of 70 N. How much power must be supplied to the string to generate a traveling wave that has a frequency of 50 Hz and an amplitude of 4 cm?

**15.35** A string with linear mass density of 0.1 kg/m is under a tension of 100 N. How much power must be supplied to the string to generate a sinusoidal wave of amplitude 2 cm and frequency 120 Hz?

•**15.36** A sinusoidal wave on a string is described by the equation $y = (0.10 \text{ m}) \sin (0.75x - 40t)$, where $x$ and $y$ are in meters and $t$ is in seconds. If the linear mass density of the string is 10 g/m, determine (a) the phase constant, (b) the phase of the wave at $x = 2$ cm and $t = 0.1$ s, (c) the speed of the wave, (d) the wavelength, (e) the frequency, and (f) the power transmitted by the wave.

## Sections 15.7 and 15.8

**15.37** In an acoustics experiment, a piano string with a mass of 5.0 g and a length of 70 cm is held under tension by running the string over a frictionless pulley and hanging a 250-kg weight from it. The whole system is placed in an elevator.

a) What is the fundamental frequency of oscillation for the string when the elevator is at rest?

b) With what acceleration and in what direction (up or down) should the elevator move for the string to produce the proper frequency of 440 Hz, corresponding to middle A?

**15.38** A string is 35.0 cm long and has a mass per unit length of $5.51 \cdot 10^{-4}$ kg/m. What tension must be applied to the string so that it vibrates at the fundamental frequency of 660 Hz?

**15.39** A 2.00-m-long string of mass 10.0 g is clamped at both ends. The tension in the string is 150 N.

a) What is the speed of a wave on this string?

b) The string is plucked so that it oscillates. What is the wavelength and frequency of the resulting wave if it produces a standing wave with two antinodes?

•**15.40** Write the equation for a standing wave that has three antinodes of amplitude 2 cm on a 3-m-long string that is fixed at both ends and vibrates 15 times a second. The time $t = 0$ is chosen to be an instant when the string is flat. If a wave pulse were propagated along this string, how fast would it travel?

•**15.41** A 3-m-long string, fixed at both ends, has a mass of 6 g. If you want to set up a standing wave in this string having a frequency of 300 Hz and three antinodes, what tension should you put the string under?

•**15.42** A cowboy walks at a pace of about two steps per second, holding a glass of diameter 10 cm that contains milk. The milk sloshes higher and higher in the glass until it eventually starts to spill over the top. Determine the maximum speed of the waves in the milk.

•**15.43** Students in a lab produce standing waves on stretched strings connected to vibration generators. One such wave is described by the wave function $y(x,t) = (2.00 \text{ cm})\sin\left[(20.0 \text{ m}^{-1})x\right]\cos\left[(150. \text{ s}^{-1})t\right]$, where $y$ is the transverse displacement of the string, $x$ is the position along the string, and $t$ is time. Rewrite this wave function in the form for a right-moving and a left-moving wave: $y(x,t) = f(x - vt) + g(x + vt)$; that is, find the functions $f$ and $g$ and the speed, $v$.

•**15.44** An array of wave emitters, as shown in the figure, emits a wave of wavelength $\lambda$ that is to be detected at a distance $L$ directly above the rightmost emitter. The distance between adjacent wave emitters is $d$.

a) Show that when $L \gg d$, the wave from the $n$th emitter (counting from right to left with $n = 0$ being the rightmost emitter) has to travel an extra distance of $\Delta s = n^2 (d^2/2L)$.

b) If $\lambda = d^2/2L$, will the interference at the detector be constructive or destructive?

c) If $\lambda = d^2/2L = 10^{-3}$ m and $L = 1000$ m, what is $d$, the distance between adjacent emitters?

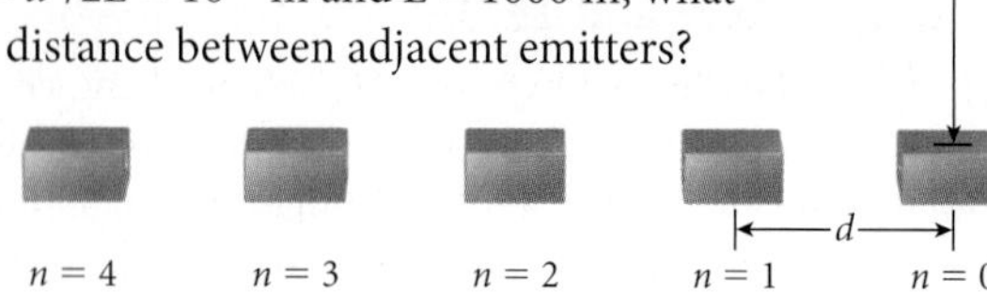


••**15.45** A small ball floats in the center of a circular pool that has a radius of 5 m. Three wave generators are placed at the edge of the pool, separated by 120°. The first wave generator operates at a frequency of 2 Hz. The second wave generator operates at a frequency of 3 Hz. The third wave generator operates at a frequency of 4 Hz. If the speed of each water wave is 5 m/s, and the amplitude of the waves is the same, sketch the height of the ball as a function of time from $t = 0$ to $t = 2$ s, assuming that the water surface is at zero height. Assume that all the wave generators impart a phase shift of zero. How would your answer change if one of the wave generators was moved to a different location at the edge of the pool?

••**15.46** A string with linear mass density $\mu = 0.0250$ kg/m under a tension of $T = 250.$ N is oriented in the $x$-direction. Two transverse waves of equal amplitude and with a phase angle of zero (at $t = 0$) but with different frequencies ($\omega = 3000.$ rad/s and $\omega/3 = 1000.$ rad/s) are created in the string by an oscillator located at $x = 0$. The resulting waves, which travel in the positive $x$-direction, are reflected at a distant point, so there is a similar pair of waves traveling in the negative $x$-direction. Find the values of $x$ at which the first two nodes in the standing wave are produced by these four waves.

••**15.47** The equation for a standing wave on a string with mass density $\mu$ is $y(x,t) = 2A\cos(\omega t)\sin(\kappa x)$. Show that the average kinetic energy and potential energy over time for this wave per unit length are given by $K_{\text{ave}}(x) = \mu\omega^2 A^2 \sin^2 \kappa x$ and $U_{\text{ave}}(x) = T(\kappa A)^2 (\cos^2 \kappa x)$.

## Additional Problems

**15.48** A sinusoidal wave traveling on a string is moving in the positive $x$-direction. The wave has a wavelength of 4 m, a frequency of 50 Hz, and an amplitude of 3 cm. What is the wave function for this wave?

**15.49** A guitar string with a mass of 10.0 g is 1.00 m long and attached to the guitar at two points separated by 65.0 cm.

a) What is the frequency of the first harmonic of this string when it is placed under a tension of 81.0 N?

b) If the guitar string is replaced by a heavier one that has a mass of 16.0 g and is 1.00 m long, what is the frequency of the replacement string's first harmonic?

**15.50** Write the equation for a sinusoidal wave propagating in the negative $x$-direction with a speed of 120 m/s, if a particle in the medium in which the wave is moving is observed to swing back and forth through a 6-cm range in 4 s. Assume that $t = 0$ is taken to be the instant when the particle is at $y = 0$ and that the particle moves in the positive $y$-direction immediately after $t = 0$.

**15.51** Shown in the figure is a plot of the displacement, $y$, due a sinusoidal wave traveling along a string as a function of time, $t$. What are the (a) period, (b) maximum speed, and (c) maximum acceleration of this wave perpendicular to the direction that the wave is traveling?

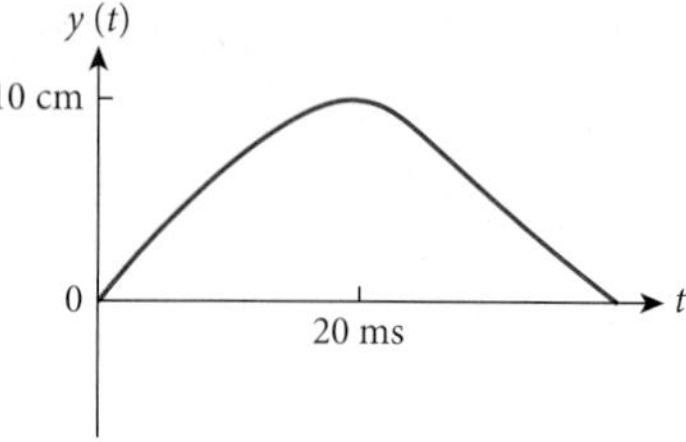

**15.52** A 50-cm-long wire with a mass of 10.0 g is under a tension of 50.0 N. Both ends of the wire are held rigidly while it is plucked.

a) What is the speed of the waves on the wire?

b) What is the fundamental frequency of the standing wave?

c) What is the frequency of the third harmonic?

15.53 What is the wave speed along a brass wire with a radius of 0.500 mm stretched at a tension of 125 N? The density of brass is $8.60 \cdot 10^3$ kg/m$^3$.

**15.54** Two steel wires are stretched under the same tension. The first wire has a diameter of 0.500 mm, and the second wire has a diameter of 1.00 mm. If the speed of waves traveling along the first wire is 50.0 m/s, what is the speed of waves traveling along the second wire?

**15.55** The middle-C key (key 52) on a piano corresponds to a fundamental frequency of about 262 Hz, and the soprano-C key (key 64) corresponds to a fundamental frequency of 1046.5 Hz. If the strings used for both keys are identical in density and length, determine the ratio of the tensions in the two strings.

**15.56** A sinusoidal wave travels along a stretched string. A point along the string has a maximum velocity of 1 m/s and a maximum displacement of 2 cm. What is the maximum acceleration at that point?

•15.57 As shown in the figure, a sinusoidal wave travels to the right at a speed of $v_1$ along string 1, which has linear mass density $\mu_1$. This wave has frequency $f_1$ and wavelength $\lambda_1$. Since string 1 is attached to string 2 (which has linear mass density $\mu_2 = 3\mu_1$), the first wave will excite a new wave in string 2, which will also move to the right. What is the frequency, $f_2$ of the wave produced in string 2? What is the speed, $v_2$, of the wave produced in string 2? What is the wavelength, $\lambda_2$, of the wave produced in string 2? Write all answers in terms of $f_1$, $v_1$, and $\lambda_1$.

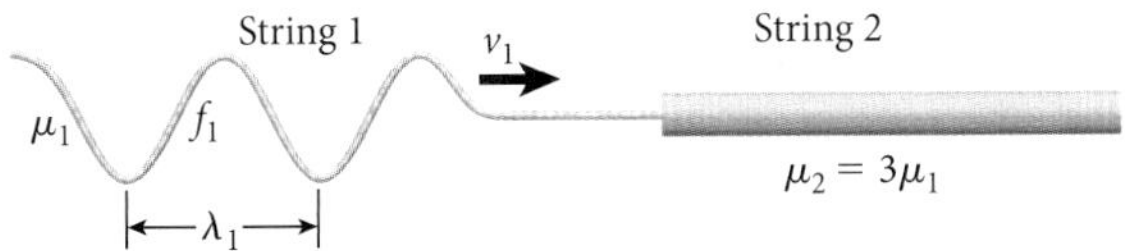


**•15.58** The tension in a 2.7-m-long, 1.0-cm-diameter steel cable ($\rho = 7800$ kg/m$^3$) is 840 N. What is the fundamental frequency of vibration of the cable?

**•15.59** A wave traveling on a string has the equation of motion $y(x,t) = 0.02 \sin (5x - 8t)$.

a) Calculate the wavelength and the frequency of the wave.

b) Calculate its velocity.

c) If the linear mass density of the string is $\mu = 0.10$ kg/m, what is the tension on the string?

**•15.60** Calvin sloshes back and forth in his bathtub, producing a standing wave. What is the frequency of such a wave if the bathtub is 150 cm long and 80 cm wide and contains water that is 38 cm deep?

•15.61 Consider a guitar string stretching 80.0 cm between its anchored ends. The string is tuned to play middle C, with a frequency of 256 Hz, when oscillating in its fundamental mode, that is, with one antinode between the ends. If the string is displaced 2.00 mm at its midpoint and released to produce this note, what are the wave speed, $v$, and the maximum speed, $V_{max}$, of the midpoint of the string?

**•15.62** The largest tension that can be sustained by a stretched string of linear mass density $\mu$, even in principle, is given by $\tau = \mu c^2$, where $c$ is the speed of light in vacuum. (This is an enormous value. The breaking tensions of all ordinary materials are about 12 orders of magnitude less than this.)

a) What is the speed of a traveling wave on a string under such tension?

b) If a 1.000-m-long guitar string, stretched between anchored ends, were made of this hypothetical material, what frequency would its first harmonic have?

c) If that guitar string were plucked at its midpoint and given a displacement of 2 mm there to produce the fundamental frequency, what would be the maximum speed attained by the midpoint of the string?

**•15.63** A rubber band of mass 0.21 g is stretched between two fingers, putting it under a tension of 2.8 N. The overall stretched length of the band is 21.3 cm. One side of the band is plucked, setting up a vibration in 8.7 cm of the band's stretched length. What is the lowest frequency of vibration that can be set up on this part of the rubber band? Assume that the band stretches uniformly.

**15.64** Two waves traveling in opposite directions along a string fixed at both ends create a standing wave described by $y(x,t) = 0.01 \sin (25x) \cos (1200t)$. The string has a linear mass density of 0.01 kg/m, and the tension in the string is supplied by a mass hanging from one end. If the string vibrates in its third harmonic, calculate (a) the length of the string, (b) the velocity of the waves, and (c) the mass of the hanging mass.

15.65 A sinusoidal transverse wave of wavelength 20.0 cm and frequency 500 Hz travels along a string in the positive $z$-direction. The wave oscillations take place in the $xz$-plane and have an amplitude of 3.00 cm. At time $t = 0$, the displacement of the string at $x = 0$ is $z = 3.00$ cm.

a) A photo of the wave is taken at $t = 0$. Make a simple sketch (including axes) of the string at this time.

b) Determine the speed of the wave.

c) Determine the wave's wave number.

d) If the linear mass density of the string is 30.0 g/m, what is the tension in the string?

e) Determine the function $D(z,t)$ that describes the displacement $x$ that is produced in the string by this wave.

**••15.66** A heavy cable of total mass $M$ and length $L = 5.0$ m has one end attached to a rigid support and the other end hanging free. A small transverse displacement is initiated at the bottom of the cable. How long does it take for the displacement to travel to the top of the cable?

# 16 Sound

**FIGURE 16.1** Jerry Garcia of the Grateful Dead in the process of producing sound waves.

## WHAT WE WILL LEARN

- Sound consists of longitudinal pressure waves and needs a medium in which to propagate.
- The speed of sound in solids is usually higher than in liquids, and that in liquids is higher than in gases.
- The speed of sound in air depends on the temperature; it is approximately 343 m/s at normal atmospheric pressure and a temperature of 20 °C.
- The intensity of sound detectable by the human ear spans a large range and is usually expressed on a logarithmic scale in terms of decibels (dB).
- Sound waves from two or more sources can interfere in space and time, resulting in destructive or constructive interference.
- Interference of sound waves in time produces beats, which occur with a specific beat frequency.
- The Doppler effect is the shift in the observed frequency of a sound due to the fact that the source moves (is approaching or receding) relative to the observer.
- If a sound source moves with a speed greater than that of sound, a shock wave, or Mach cone, develops.
- Standing waves in open or closed pipes can be generated only at discrete wavelengths.

Recognizing sounds is one of the most important ways in which we learn about the world around us. As we'll see in this chapter, human hearing is very sensitive and can distinguish a wide range of frequencies and degrees of loudness. However, although interpreting sound probably developed in animals from earliest times, as it warned of predators or provided help in hunting prey, sound has also been a powerful source of cultural ritual and entertainment for humans (Figure 16.1). Music has been part of human life for as long as society has existed.

Sound is a type of wave, so this chapter's study of sound builds directly on the concepts about waves presented in Chapter 15. Some characteristics of sound waves also pertain to light waves and will be useful when we study those waves in later chapters. This chapter also examines sources of musical sounds and applications of sound in a wide range of other areas, from medicine to geography.

## 16.1 Longitudinal Pressure Waves

**Sound** is a pressure variation that propagates through some medium. In air, the pressure variation causes abnormal motion of air molecules in the direction of propagation; thus, a sound wave is longitudinal. When we hear a sound, our eardrums are set in vibration by the air next to them. If this air has a pressure variation that repeats with a certain frequency, the eardrums vibrate at this frequency.

Sound requires a medium for propagation. If the air is pumped out of a glass jar that contains a ringing bell, the sound ceases as the air is evacuated from the jar, even though the hammer clearly still hits the bell. Thus, sound waves originate from a source, or emitter, and need a medium to travel through. In spite of movie scenes that include a huge roar when a spaceship flies by or when a star or planet explodes, interstellar space is a silent place because it is a vacuum.

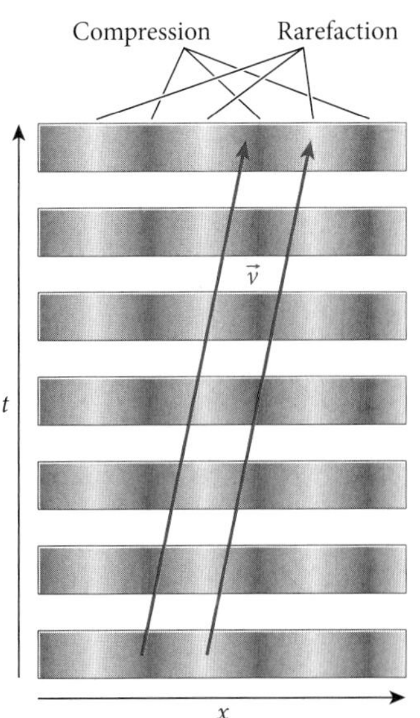


**FIGURE 16.2** Propagation of a longitudinal pressure wave along an $x$-axis as a function of time.

Figure 16.2 shows a continuous pressure wave composed of alternating variations of air pressure—a pressure excess (compression) followed by a pressure reduction (rarefaction). Plotting these variations along an $x$-axis for various times, as in Figure 16.2, allows us to deduce the wave speed.

Although sound has to propagate through a medium, that medium does not have to be air. When you are under water, you can still hear sounds, so you know that sound also propagates through liquids. In addition, you may have put your ear to a train track to obtain advance information on an approaching train (not recommended, by the way, because the train may be closer than you think). Thus, you know that sound propagates through solids. In fact, sound propagates faster and with less loss through metals than through air. If this were not so, putting your ear to the train track would be pointless.

**FIGURE 16.3** You see fireworks before you hear their explosive sounds.

## Sound Velocity

When you watch fireworks on the Fourth of July, you see the explosions of the rockets before you hear the sounds (Figure 16.3). The reason is that the light emitted from the explosions reaches your eyes almost instantaneously, because the speed of light is approximately 300,000 km/s. However, the speed of sound is much slower, and so the sound waves from the explosions reach you some time after the light waves have already arrived.

How fast does sound propagate—in other words, what is the speed of sound? For a wave on a string, we saw in the Chapter 15 that the wave speed is $v=\sqrt{T/\mu}$, where $T$ is the string tension (a force) and $\mu$ is the linear mass density of the string. From this equation, you can think of $v$ as the square root of the ratio of the restoring force to the inertial response. If the wave propagates through a three-dimensional medium instead of a one-dimensional string, the inertial response originates from the density of the medium, $\rho$. Chapter 13 discussed the elasticity of solids, and introduced the elastic modulus known as Young's modulus, $Y$. This modulus determines the fractional change in length of a thin rod as a function of the force per unit area applied to the rod. Analysis reveals that Young's modulus is the appropriate force term for a wave propagating along a thin solid rod; so we obtain

$$v=\sqrt{\frac{Y}{\rho}} \tag{16.1}$$

for the speed of sound in a thin solid rod.

The speed of sound in fluids—both liquids and gases—is similarly related to the bulk modulus, $B$, defined in Chapter 13, as determining a material's volume change in response to external pressure. Thus, the speed of sound in a gas or liquid is given by,

$$v=\sqrt{\frac{B}{\rho}}. \tag{16.2}$$

Although, these dimensional arguments show that the speed of sound is proportional to the square root of the modulus divided by the appropriate density, a more fundamental analysis such as that in the following derivation shows that the proportionality constants have a value of exactly 1 in equations 16.1 and 16.2.

### DERIVATION 16.1 Speed of Sound

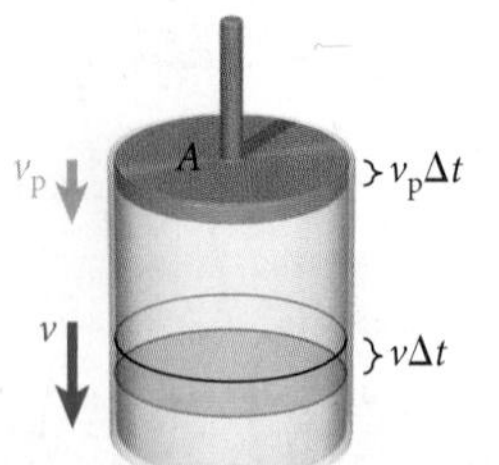


**FIGURE 16.4** Piston compressing a fluid.

Suppose a fluid is contained in a cylinder, with a movable piston on one end (Figure 16.4). If this piston moves with speed $v_p$ into the fluid, it will compress the fluid element in front of it. This fluid element, in turn, will move as a result of the pressure change; its leading edge will move with the speed of sound, $v$, which, by definition, is the speed of the pressure waves in the medium.

Pushing the piston into the fluid with a force, $F$, causes a pressure change, $p = F/A$ in the fluid (see Chapter 13), where $A$ is the cross-sectional area of the cylinder and also of the piston. The force exerted on the fluid element of mass $m$ causes an acceleration given by $\Delta v/\Delta t$:

$$F=m\frac{\Delta v}{\Delta t}=m\frac{v_p}{\Delta t}\Rightarrow p=\frac{F}{A}=\frac{m}{A}\frac{v_p}{\Delta t}.$$

Because the mass is the density of the fluid times its volume, $m=\rho V$, and the volume is that of the cylinder of base area $A$ and length $l$, we can find the mass of the fluid element:

$$m=\rho V=\rho A l\Rightarrow p=\frac{m}{A}\frac{v_p}{\Delta t}=\frac{\rho A l}{A}\frac{v_p}{\Delta t}=\frac{\rho l v_p}{\Delta t}.$$

Since this fluid element responds to the compression by moving with speed $v$ during the time interval $\Delta t$, the length of the fluid element that has experienced the pressure wave is $l = v\Delta t$, which finally yields the pressure difference:

$$p = \frac{\rho l v_\text{p}}{\Delta t} = \frac{\rho (v\Delta t) v_\text{p}}{\Delta t} = \rho v v_\text{p}. \quad \text{(i)}$$

From the definition of the bulk modulus (see Chapter 13), $p = B\Delta V/V$. We can combine this expression for the pressure and that obtained in equation (i):

$$p = \rho v v_\text{p} = B\frac{\Delta V}{V}. \quad \text{(ii)}$$

Again, the volume of the moving fluid, $V$, that has experienced the pressure wave is proportional to $v$ since $V = Al = Av\Delta t$. In addition, the volume change in the fluid caused by pushing the piston into the cylinder is $\Delta V = Av_\text{p}\Delta t$. Thus, the ratio $\Delta V/V$ is equivalent to $v_\text{p}/v$, the ratio of the piston's speed to the speed of sound. Inserting this result into equation (ii) yields

$$\rho v v_\text{p} = B\frac{\Delta V}{V} = B\frac{v_\text{p}}{v} \Rightarrow$$
$$\rho v^2 = B \Rightarrow$$
$$v = \sqrt{\frac{B}{\rho}}.$$

As you can see, the speed with which the piston is pushed into the fluid cancels out. Therefore, it does not matter what the speed of the excitation is; the sound always propagates with the same speed in the medium.

Equations 16.1 and 16.2 both state that the speed of sound in a given state of matter (gas, liquid, or solid) is inversely proportional to the square root of the density, which means that in two different gases, the speed of sound is higher in the one with the lower density. However, the values of Young's modulus for solids are much larger than the values of the bulk modulus for liquids, which are larger than those for gases. This difference is more important than the density dependence. From equations 16.1 and 16.2, it then follows that for the speeds of sound in solids, liquids, and gases,

$$v_\text{solid} > v_\text{liquid} > v_\text{gas}. \quad \textbf{(16.3)}$$

Representative values for the speed of sound in different materials under standard conditions of pressure (1 atm) and temperature (20 °C ) are listed in Table 16.1.

We are most interested in the speed of sound in air, because this is the most important medium for sound propagation in everyday life. At normal atmospheric pressure and 20 °C, this speed of sound is

$$v_\text{air} = 343 \text{ m/s}. \quad \textbf{(16.4)}$$

Knowing the speed of sound in air explains the 5-second rule for thunderstorms: If 5 s or less pass between the instant you see a lightning strike (Figure 16.5) and the instant you hear the thunder, the lightning strike is 1 mi or less away. Since sound travels approximately 340 m/s, it travels 1700 m, or approximately 1 mi, in 5 s. The speed of light is approximately 1 million times greater than the speed of sound, so the visual perception of the lightning strike occurs with essentially no delay. Countries that use the metric system have a 3-second rule: A 3-s delay between lightning and thunder correspond to 1 km of distance to the lightning strike (Figure 16.6).

The speed of sound in air depends (weakly) on the air temperature, $T$. The following linear dependence is obtained experimentally:

$$v(T) = \left(331 + 0.6T/°\text{C}\right) \text{ m/s}. \quad \textbf{(16.5)}$$

### 16.1 In-Class Exercise

You are in the middle of a concert hall that is 120.0 m deep. What is the time difference between the arrival of the sound directly from the orchestra at your position and the arrival of the sound reflected from the back of the concert hall?

a) 0.010 s

b) 0.056 s

c) 0.11 s

d) 0.35 s

e) 0.77 s

**FIGURE 16.5** The 5-second rule for lightning strikes is due to the difference in speed between light and sound.

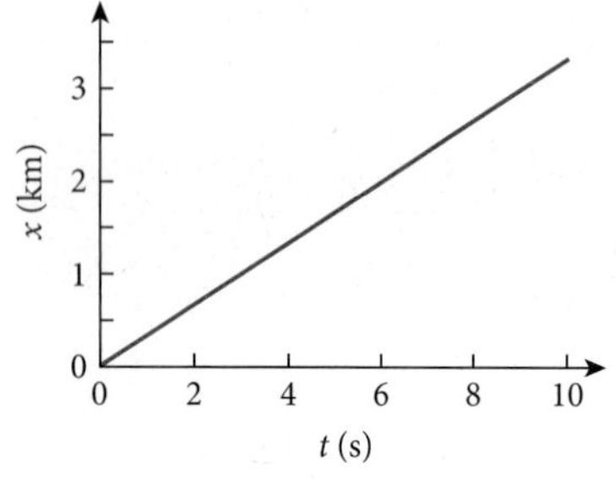


**FIGURE 16.6** Distance $x$ to the source (lightning strike) of the sound as a function of the time delay $t$ in hearing the sound (thunder).

**Table 16.1 Speed of Sound in Some Common Substances**

| | Substance | Speed of Sound (m/s) |
|---|---|---|
| Gases | Krypton | 220 |
| | Carbon Dioxide | 260 |
| | Air | 343 |
| | Helium | 960 |
| | Hydrogen | 1280 |
| Liquids | Methanol | 1143 |
| | Mercury | 1451 |
| | Water | 1480 |
| | Seawater | 1520 |
| Solids | Lead | 2160 |
| | Concrete | 3200 |
| | Hardwood | 4000 |
| | Steel | 5800 |
| | Aluminum | 6400 |
| | Diamond | 12,000 |

## EXAMPLE 16.1 Football Cheers

A student's apartment is located exactly 3.75 km from the football stadium. He is watching the game live on TV and sees the home team score a touchdown. According to his clock, 11.2 s pass after he hears the roar of the crowd on TV until he hears it again from outside.

### PROBLEM

What is the temperature at game time?

### SOLUTION

The TV signal moves with the speed of light and thus arrives at the student's TV at essentially the same instant as the action is happening in the stadium. Thus, the delay in the arrival of the roar at the student's apartment can be attributed entirely to the finite speed of sound. We find this speed of sound from the given data:

$$v = \frac{\Delta x}{\Delta t} = \frac{3750 \text{ m}}{11.2 \text{ s}} = 334.8 \text{ m/s}.$$

Using equation 16.5, we find the temperature at game time:

$$T = \frac{v(T)/(\text{m/s}) - 331}{0.6} \text{ °C} = \frac{334.8 - 331}{0.6} \text{ °C} = \frac{3.8}{0.6} \text{ °C} = 6.4 \text{ °C}$$

### DISCUSSION

A word of caution is in order: Several uncertainties are inherent in this result. First, if we redid the calculation for a time delay of 11.1 s, we would find the temperature to be 11.4 °C, or 5 °C higher than what we found using 11.2 s as the delay. Second, if 0.1 s makes such a big difference in the determined temperature, we need to ask if the TV signal really gets to the student's apartment instantaneously. The answer is no. If the student watches satellite TV, it takes about 0.2 s for the signal to make it up to the geostationary satellite and back down to the student's dish. In addition, short intentional delays are often inserted in TV broadcasts. Thus, our calculation is not very useful for practical purposes.

## Sound Reflection

You can measure the distance to a distant large object by measuring the time between producing a short, loud sound and hearing that sound again after it has traveled to the object, reflected off the object, and returned to you. For example, if you were standing in Yosemite Valley and shouted in the direction of the flat face of Half Dome, 1.0 km away, the sound of your voice would carry across the valley, reflect off Half Dome, and return to you, making a round trip of 2.0 km. The speed of the sound of your voice is 343 m/s, so the time delay between shouting and hearing your reflected shout would be $t = (2000 \text{ m})/(343 \text{ m/s}) = 5.8 \text{ s}$, a time interval that you could easily measure with a wristwatch.

This principle is used in ultrasound imaging for medical diagnostic purposes. Ultrasound waves have a frequency much higher than a human can hear, between 2 and 15 MHz. The frequency is chosen to provide detailed images and to penetrate deeply into human tissue. When the ultrasound waves encounter a change in density of the tissue, some of them are reflected back. By measuring the time the ultrasound waves take to travel from the emitter to the receiver and recording how much of the emission is reflected as well as the direction of the original waves, an image can be formed. The average value of the speed of ultrasound waves used for imaging human tissue is 1540 m/s. The time for these ultrasound waves to travel 2.5 cm and return is $t = (0.050 \text{ m})/(1540 \text{ m/s}) = 32\ \mu\text{s}$, and the time to travel 10.0 cm and return is 130 $\mu$s. Thus, the ultrasound device needs to be able to measure accurately times in the range from 30 to 130 $\mu$s. A typical image of a fetus produced by ultrasound imaging is shown in Figure 16.7.

Bats and dolphins navigate using sound reflection (Figure 16.8). They emit sound waves in a frequency range from 14,000 Hz to well over 100,000 Hz in a specific direction and determine information about their surroundings from the reflected sound. This process of echolocation allows bats to navigate in darkness.

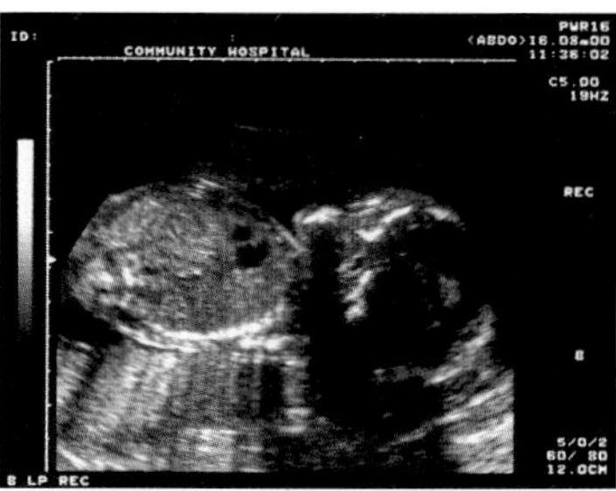


**FIGURE 16.7** An image of a fetus produced by the reflection of ultrasound waves.

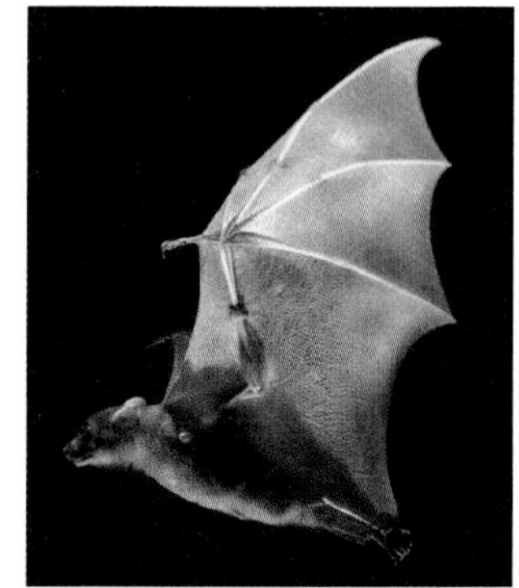

**FIGURE 16.8** Bat flying in darkness, relying on echolocation for navigation.

## 16.2 Sound Intensity

In Chapter 15, the intensity of a wave was defined as the power per unit area. We saw that for spherical waves the intensity falls as the second power of the distance to the source, $I \propto r^{-2}$, giving the ratio

$$\frac{I(r_1)}{I(r_2)} = \left(\frac{r_2}{r_1}\right)^2. \tag{16.6}$$

This relationship also holds for sound waves. Since the intensity is the power per unit area, its physical units are watts per square meter (W/m$^2$).

Sound waves that can be detected by the human ear have a very large range of intensities, from whispers as low as $10^{-12}$ W/m$^2$ to the output of a jet engine or a rock band at close distance, which can reach 1 W/m$^2$. The oscillations in pressure for even the loudest sounds, at the pain threshold of 10 W/m$^2$, are on the order of only tens of micropascals ($\mu$Pa). Normal atmospheric air pressure, by comparison, is $10^5$ Pa. Thus, you can see that the air pressure varies by only 1 part in 10 billion, even for the loudest sounds. And the variation is several orders of magnitude less for the quietest sounds you can hear. This might give you a new appreciation for the capabilities of your ears.

Since human ears can register sounds over many orders of magnitude of intensity, a logarithmic scale is used to measure sound intensities. The unit of this scale is the bel (B), named for Alexander Graham Bell, but much more commonly used is the **decibel** (dB): 1 dB = 0.1 bel. The Greek letter $\beta$ symbolizes the sound level measured on this decibel scale and is defined as

$$\beta = 10 \log \frac{I}{I_0}. \tag{16.7}$$

Here $I_0 = 10^{-12}$ W/m$^2$, which corresponds approximately to the minimum intensity that a human ear can hear. The notation "log" refers to the base-10 logarithm. (See Appendix A for a refresher on logarithms.) Thus, a sound intensity 1000 times the reference intensity, $I_0$, gives $\beta = 10 \log 1000 \text{ dB} = 10 \cdot 3 \text{ dB} = 30 \text{ dB}$. This sound level corresponds to a whisper not very far from your ear (Table 16.2).

### 16.1 Self-Test Opportunity

You hear a sound with level 80.0 dB, and you are located 10.0 m from the sound source. What is the power being emitted by the sound source?

**Table 16.2 Levels of Common Sounds**

| Sound | Sound Level (dB) |
|---|---|
| Quietest sound heard | 0 |
| Background sound in library | 30 |
| Golf course | 40–50 |
| Street traffic | 60–70 |
| Train at railroad crossing | 90 |
| Dance club | 110 |
| Jackhammer | 120 |
| Jet taking off from aircraft carrier | 130–150 |

## Relative Intensity and Dynamic Range

The relative sound level, $\Delta\beta$, is the difference between two sound levels:

$$\begin{aligned}\Delta\beta &= \beta_2 - \beta_1\\ &= 10\log\frac{I_2}{I_0} - 10\log\frac{I_1}{I_0}\\ &= 10(\log I_2 - \log I_0) - 10(\log I_1 - \log I_0)\\ &= 10\log I_2 - 10\log I_1\end{aligned}$$

$$\Delta\beta = 10\log\frac{I_2}{I_1}. \quad (16.8)$$

The *dynamic range* is a measure of the relative sound levels of the loudest and the quietest sounds produced by a source. The dynamic range of a compact disc is approximately 90 dB, whereas the range was about 70 dB for vinyl records. (Vinyl records, 8-tracks, and audiocassettes are all old technologies that you may find in your parents' or grandparents' house.) Manufacturers list the dynamic range for all high-end loudspeakers and headphones as well. In general, the higher the dynamic range, the better the sound quality. However, the price usually increases with the dynamic range.

### SOLVED PROBLEM 16.1 Relative Sound Levels at a Rock Concert

Two friends attend a rock concert and bring along a sound meter. With this device, one of the friends measures a sound level of $\beta_1 = 105.0$ dB, whereas the other, who sits 4 rows (2.8 m) closer to the stage, measures $\beta_2 = 108.0$ dB.

**PROBLEM**

How far away are the two friends from the loudspeakers on stage?

**SOLUTION**

**THINK**

We express the intensities at the two seats using equation 16.6 and the relative sound levels at the two seats using equation 16.8. We can combine these two equations to obtain an expression for relative sound levels at the two seats. Knowing the distance between the two seats, we can then calculate the distance of the seats to the loudspeakers on the stage.

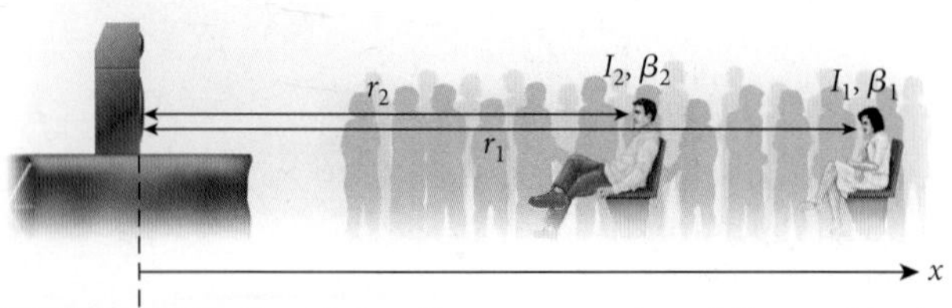


**FIGURE 16.9** The relative distances, $r_1$ and $r_2$, of the two friends from the loudspeakers at a concert.

**SKETCH**

Figure 16.9 shows the relative positions of the two friends.

**RESEARCH**

The distance of the first friend from the loudspeakers is $r_1$, and the distance of the second friend from the loudspeakers is $r_2$. The intensity of the sound at $r_1$ is $I_1$, and the intensity of the sound at $r_2$ is $I_2$. The sound level at $r_1$ is $\beta_1$, and the sound level at $r_2$ is $\beta_2$. We can express the two intensities in terms of the two distances using equation 16.6:

$$\frac{I_1}{I_2}=\left(\frac{r_2}{r_1}\right)^2. \qquad \text{(i)}$$

We can then use equation 16.8 to relate the sound levels at $r_1$ and $r_2$ to the sound intensities:

$$\beta_2-\beta_1=10\log\frac{I_2}{I_1}. \qquad \text{(ii)}$$

Combining equations (i) and (ii), we get

$$\beta_2-\beta_1=\Delta\beta=10\log\left(\frac{r_1}{r_2}\right)^2=20\log\left(\frac{r_1}{r_2}\right). \qquad \text{(iii)}$$

The relative sound level, $\Delta\beta$, is specified in the problem and we know that

$$r_2=r_1-2.8 \text{ m}.$$

Thus, we can solve for the distance $r_1$ and then obtain $r_2$.

**SIMPLIFY**

Substituting the relationship between the two distances into equation (iii) gives us

$$\Delta\beta=20\log\left(\frac{r_1}{r_1-2.8 \text{ m}}\right)\Rightarrow$$

$$10^{\Delta\beta/20}=\frac{r_1}{r_1-2.8 \text{ m}}\Rightarrow$$

$$\left(r_1-2.8 \text{ m}\right)10^{\Delta\beta/20}=r_1 10^{\Delta\beta/20}-\left(2.8 \text{ m}\right)10^{\Delta\beta/20}=r_1\Rightarrow$$

$$r_1=\frac{\left(2.8 \text{ m}\right)10^{\Delta\beta/20}}{10^{\Delta\beta/20}-1}.$$

**CALCULATE**

Putting in the numerical values, we obtain the distance of the first friend from the loudspeakers:

$$r_1=\frac{\left(2.8 \text{ m}\right)10^{(108.0 \text{ dB}-105.0 \text{ dB})/20}}{10^{(108.0 \text{ dB}-105.0 \text{ dB})/20}-1}=9.58726 \text{ m}.$$

The distance of the second friend is:

$$r_2=(9.58726 \text{ m})-(2.8 \text{ m})=6.78726 \text{ m}.$$

**ROUND**

We report our results to two significant figures:

$$r_1=9.6 \text{ m} \quad \text{and} \quad r_2=6.8 \text{ m}.$$

**DOUBLE-CHECK**

To double-check, we substitute our results for the distances back into equation (iii) to verify that we get the specified sound level difference:

$$\Delta\beta=(108.0 \text{ dB})-(105.0 \text{ dB})=3.0 \text{ dB}=20\log\left(\frac{9.6 \text{ m}}{6.8 \text{ m}}\right)=3.0 \text{ dB}.$$

Therefore, our answer seems reasonable, or at least consistent with what was originally stated in the problem.

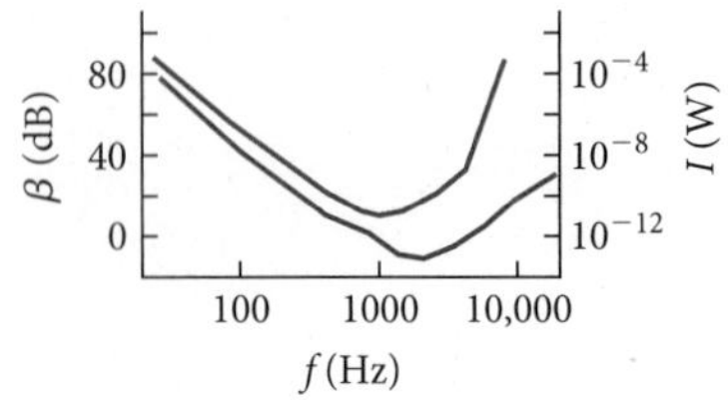


**FIGURE 16.10** Frequency dependence of the human hearing threshold. The red curve represents a typical teenager; the blue curve represents someone of retirement age.

## Limits of Human Hearing

Human ears can detect sound waves with frequencies between approximately 20 Hz and 20,000 Hz. (Dogs, by the way, can detect even higher frequencies of sound.) As mentioned earlier, the threshold of human hearing was used as the anchor for the decibel scale. However, the ability of the human ear to detect sound depends strongly on the frequency. Figure 16.10 is a graph of this dependence of the human hearing threshold on the frequency of the sound. In addition to a dependence on frequency, the hearing threshold also has a strong dependence on age. The red curve in Figure 16.10 represents the values for a typical teenager, and the blue curve those for a person at retirement age. Teenagers have no trouble hearing frequencies of 10,000 Hz, but most retirees cannot hear them at all. One example of such a high-pitched sound is the buzzing of cicadas in the summer, which many older people cannot hear anymore.

The sounds that we can hear best have frequencies around 1000 Hz. When you listen to music, the spectrum of sound typically extends over the entire range of human hearing. This occurs because musical instruments produce a characteristic mixture of the base frequencies of the notes played and several overtones at higher frequencies. If you play music on a stereo system and turn down the volume, you reduce the sound intensity in all frequency ranges approximately uniformly. Depending on how far down you turn the volume, you can approach sound intensities in the very high and very low frequencies that are very close to or below your hearing threshold. As a result, the music sounds flat. In order to correct for this perceived narrowing of the spectrum of the sound, many stereo systems have a loudness switch, which artificially enhances the power output of the very low and very high frequencies relative to the rest of the spectrum, giving the music a fuller sound at low volume settings.

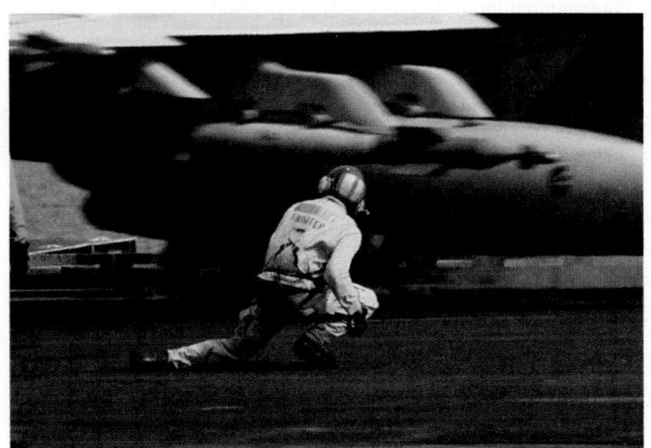

**FIGURE 16.11** A Navy launch officer ducks under the wing of an F/A-18F being launched off the USS *John C. Stennis.* Ear protection is required in this environment of high-intensity sound.

A sound level above 130 dB will cause pain, and sound levels above 150 dB can rupture an eardrum. In addition, long-term exposure to sound levels above 120 dB causes loss of sensitivity of the human ear. For this reason, it is advisable to avoid very loud music played over long times. In addition, where loud noises are part of the work environment, for example, on the deck of an aircraft carrier (Figure 16.11), it is necessary to wear ear protection to avoid ear damage.

### EXAMPLE 16.2 Wavelength Range of Human Hearing

The range of frequencies for sounds that the human ear can detect corresponds to a range of wavelengths.

**PROBLEM**

What is the range of wavelengths for the sounds to which the human ear is sensitive?

**SOLUTION**

The range of frequencies the ear can detect is 20–20,000 Hz. At room temperature, the speed of sound is 343 m/s. Wavelength, speed, and frequency are related by

$$v = \lambda f \Rightarrow \lambda = \frac{v}{f}.$$

For the lowest detectable frequency, we then obtain

$$\lambda_{\max} = \frac{v}{f_{\min}} = \frac{343 \text{ m/s}}{20 \text{ Hz}} = 17 \text{ m}.$$

Since the highest detectable frequency is 1000 times the lowest, the shortest wavelength of audible sound is 0.001 times the value we just calculated: $\lambda_{\min} = 0.017$ m.

### 16.2 In-Class Exercise

A teenager is using a new cell phone ring tone with a frequency of 17 kHz. The hearing threshold for a typical teenager at this frequency is 30 dB, and that for an adult might be 100 dB. Thus, the teenager's cell phone ring is audible only to teenagers and not to adults at distances of several meters. How close would an adult need to be to the cell phone to hear it ring if the teenager can hear it at a distance of 10.0 m?

a) 0.32 cm

b) 4.5 cm

c) 8.9 cm

d) 25 cm

e) 3.0 m

## 16.3 Sound Interference

Like all three-dimensional waves, sound waves from two or more sources can interfere in space and time. Let's first consider spatial interference of sound waves emitted by two **coherent sources**—that is, two sources that produce sound waves that have the same frequency and are in phase.

Figure 16.12 shows two loudspeakers emitting identical sinusoidal pressure fluctuations in phase. The circular arcs represent the maxima of the sound waves at some given instant in time. The horizontal line to the right of each speaker anchors a sine curve to show that an arc occurs wherever the sine has a maximum. The distance between neighboring arcs emanating from one source is exactly one wavelength, $\lambda$. You can clearly see that the arcs from the different loudspeakers intersect. Examples of such intersections are points $A$ and $C$. If you count the maxima, you see that both $A$ and $C$ are exactly $8\lambda$ away from the lower loudspeaker, while $A$ is $5\lambda$ away from the upper speaker and $C$ is $6\lambda$ away. Thus, the path length difference, $\Delta r = r_2 - r_1$, is an integer multiple of the wavelength. This relationship is a general condition for constructive interference at a given spatial point:

$$\Delta r = n\lambda, \quad \text{for all } n = 0, \pm 1, \pm 2, \pm 3, \ldots \text{ (constructive interference).} \qquad (16.9)$$

Approximately halfway between points $A$ and $C$ in Figure 16.12 is point $B$. Just as for the other two points, the distance between $B$ and the lower speaker is $8\lambda$. However, this point is $5.5\lambda$ away from the upper speaker. As a result, at point $B$, the maximum of the sound wave from the lower speaker falls on a minimum of the sound wave from the upper speaker, and they cancel out. This means that destructive interference occurs at this point. You have probably already noticed that the path length difference is an odd number of half wavelengths. This is the general condition for destructive interference:

$$\Delta r = (n + \tfrac{1}{2})\lambda, \quad \text{for all } n = 0, \pm 1, \pm 2, \pm 3, \ldots \text{ (destructive interference).} \qquad (16.10)$$

In fact, all points of constructive interference fall on lines, shown in green in Figure 16.13. These lines are approximately straight lines at sufficient distance from the sources, although they are actually hyperbolas. Lines of destructive interference are shown in red in Figure 16.13. These lines of constructive and destructive interference remain constant in time because the path length difference remains the same at any fixed point. If your ear were

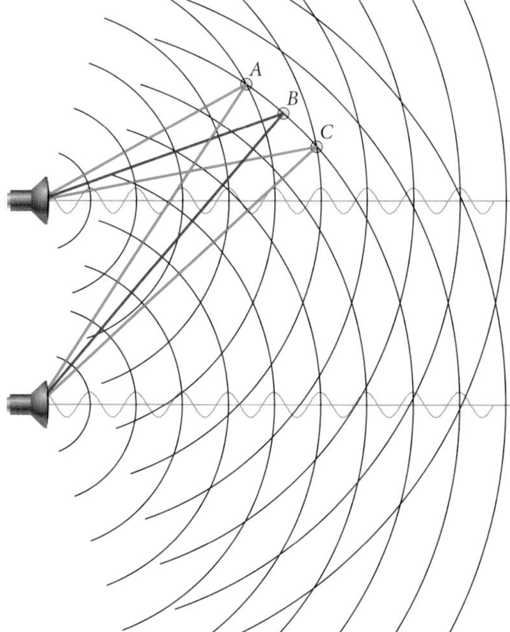


**FIGURE 16.12** Interference of two identical sets of sinusoidal sound waves.

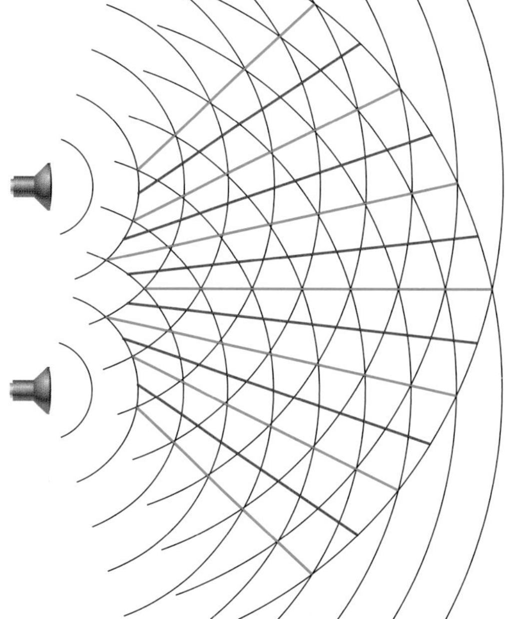

**FIGURE 16.13** Interference of identical sinusoidal sound waves from two sources, with the lines of constructive interference indicated in green and the lines of destructive interference in red.

positioned on any of the red lines, you would hear nothing of the sound emitted from both speakers; the two sound sources would cancel each other out (assuming an ideal situation where there are no sound reflections).

Why do we not detect these dead zones due to the lines of detstructive interference in front of our stereo speakers when we listen to music at home? The answer is that these lines of interference depend on the wavelength and thus on the frequency. Sounds of different frequencies cancel each other out and add up maximally along different lines. Any note played by any instrument has a rich mixture of different frequencies; so, even if one particular frequency cancels, we can still hear all or most of the rest. Generally, we do not detect what is missing. In addition, all objects in a room scatter and reflect the sound emitted from stereo speakers to some degree, which makes detection of a dead zone even less likely. However, for pure sinusoidal tones, the existence of dead zones can be easily detected by ear.

## Beats

It is also possible to have interference of two waves in time. To consider this effect, suppose an observer is located at some arbitrary point, $x_0$, in space, and two sound waves with slightly different frequencies and the same amplitude are emitted:

$$y_1(x_0,t) = A\sin(\kappa_1 x_0 + \omega_1 t + \phi_1) = A\sin(\omega_1 t + \tilde{\phi}_1)$$
$$y_2(x_0,t) = A\sin(\kappa_2 x_0 + \omega_2 t + \phi_2) = A\sin(\omega_2 t + \tilde{\phi}_2).$$

To obtain the right-hand side of each equation, we used the fact that the product of the wave number and the position is a constant for a given point in space and then just added that constant to the phase shift, which in this case is also a constant. We wrote these wave functions as one-dimensional waves, but the outcome is the same in three dimensions. For the next step, we simply set the two phase constants, $\tilde{\phi}_1$ and $\tilde{\phi}_2$, to zero, because they only cause a phase shift, but otherwise have no essential role. Therefore, the two time-dependent oscillations that form the starting waves are

$$y_1(x_0,t) = A\sin(\omega_1 t)$$
$$y_2(x_0,t) = A\sin(\omega_2 t).$$

If we do this experiment using two angular frequencies, $\omega_1$ and $\omega_2$, that are close to each other, we can hear a sound that oscillates in volume as a function of time. Why does this happen?

In order to answer this question, we add the two sine waves. To add two sine functions, we can use an addition theorem for trigonometric functions:

$$\sin\alpha + \sin\beta = 2\cos\left[\tfrac{1}{2}(\alpha - \beta)\right]\sin\left[\tfrac{1}{2}(\alpha + \beta)\right].$$

Thus, we obtain for the two sine waves

$$\begin{aligned} y(x_0,t) &= y_1(x_0,t) + y_2(x_0,t) \\ &= A\sin(\omega_1 t) + A\sin(\omega_2 t) \\ &= 2A\cos\left[\tfrac{1}{2}(\omega_1 - \omega_2)t\right]\sin\left[\tfrac{1}{2}(\omega_1 + \omega_2)t\right]. \end{aligned}$$

It is more common to write this result in terms of frequencies instead of angular speeds:

$$y(x_0,t) = 2A\cos\left[2\pi\tfrac{1}{2}(f_1 - f_2)t\right]\sin\left[2\pi\tfrac{1}{2}(f_1 + f_2)t\right]. \quad \textbf{(16.11)}$$

The term $\frac{1}{2}(f_1 + f_2)$ is simply the average of the two individual frequencies:

$$\bar{f} = \tfrac{1}{2}(f_1 + f_2). \quad \textbf{(16.12)}$$

The factor $2A\cos(2\pi\frac{1}{2}(f_1 - f_2)t)$ in equation 16.11 can be thought of as a slowly varying amplitude of a rapidly varying function $\sin(2\pi\frac{1}{2}(f_1 + f_2)t)$ when $|f_1 - f_2|$ is small.

When the cosine term has a maximum or minimum value (+1 or −1), a **beat** occurs; $f_1 - f_2$ is the beat frequency:

$$f_b = |f_1 - f_2|. \quad \textbf{(16.13)}$$

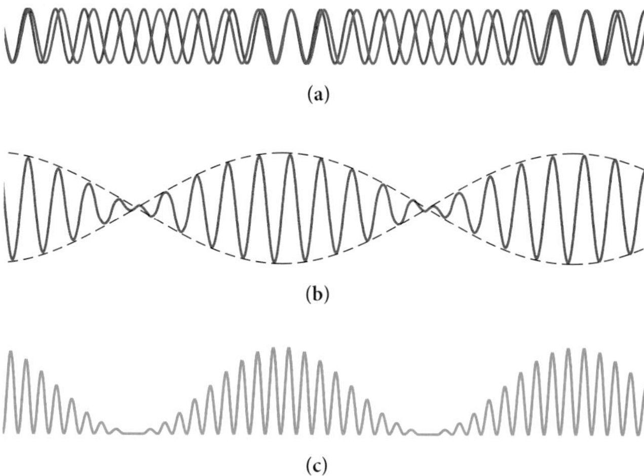


**FIGURE 16.14** Beats of two sinusoidal tones. (a) The blue and red sine curves have frequencies that differ by 10%; (b) addition of the two curves; (c) square of the sum of the functions.

Two questions now arise: First, what happened to the factor $\frac{1}{2}$ in the cosine function of equation 16.11, and second, why is the beat frequency defined in terms of an absolute value? Use of the absolute value ensures that the beat frequency is positive, independent of which of the two frequencies is larger. For the cosine function, use of the absolute value does not matter at all, because $\cos|\alpha| = \cos\alpha = \cos(-\alpha)$. For the answer to why there is no factor $\frac{1}{2}$, take a look at Figure 16.14: Part (a) plots two sinusoidal tones as a function of time; part (b) shows the addition of the two functions, with the dashed envelope given $\pm 2A\cos(2\pi\frac{1}{2}(f_1 - f_2)t)$. You can see that the amplitude of the addition of the two functions reaches a maximum twice for a given complete oscillation of the cosine function. This is perhaps more apparent from Figure 16.14c, the plot of the square of the function (equation 16.11), which is proportional to the sound intensity. Thus, the beat frequency, the frequency at which the volume oscillates up and down, is properly defined as shown in equation 16.13, omitting the factor $\frac{1}{2}$.

For example, striking two bars on a xylophone—one that emits at the proper frequency for middle A, 440 Hz, and another that emits a frequency of 438 Hz—will produce two discernable volume oscillations per second. This is because the beat frequency is $f_b = |440\text{ Hz} - 438\text{ Hz}| = 2\text{ Hz}$, and the period of oscillation of the volume is therefore $T_b = 1/f_b = \frac{1}{2}$ s. From this example, you can understand how useful beats can be in tuning one musical instrument relative to another.

## Active Noise Cancellation

We all know that noise can be muffled or attenuated and that some materials do this better than others. For example, if noise bothers you at night, you can use your pillow to cover your ears. Construction workers and other workers who have to spend extended periods in very loud environments wear ear-protecting headphones. Such ear protection reduces the amplitude of all sounds.

If you want to reduce background noise while listening to music, a technique to accomplish this is based on the principle of **active noise cancellation,** which relies on sound interference. An external sinusoidal sound wave arrives at the headphone and is recorded by a microphone (Figure 16.15). A processor inverts the phase of this sound wave and sends out the sound wave with the same frequency and amplitude, but opposite phase. The two sine waves add up (superposition principle), interfere destructively, and cancel out completely. At the same time, the speaker inside the headphones emits the music you want to hear, and the result is a listening experience that is free of background noise.

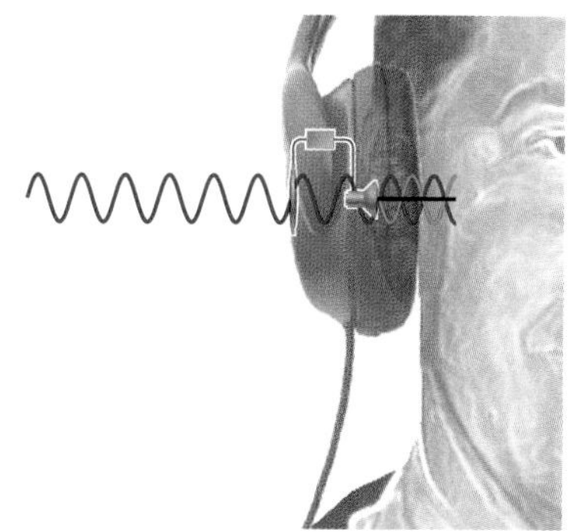

**FIGURE 16.15** Active noise cancellation.

In practice, background noise never consists of a pure sinusoidal sound wave. Instead, many sounds of many different frequencies are mixed. In particular, the presense of high-frequency sounds constitutes a problem for active noise cancellation. However, this method works very well for periodic low-frequency sounds, such as the engine noise from airliners. Another application is in some luxury automobiles, where active noise cancellation techniques reduce wind and tire noise.

## 16.4 Doppler Effect

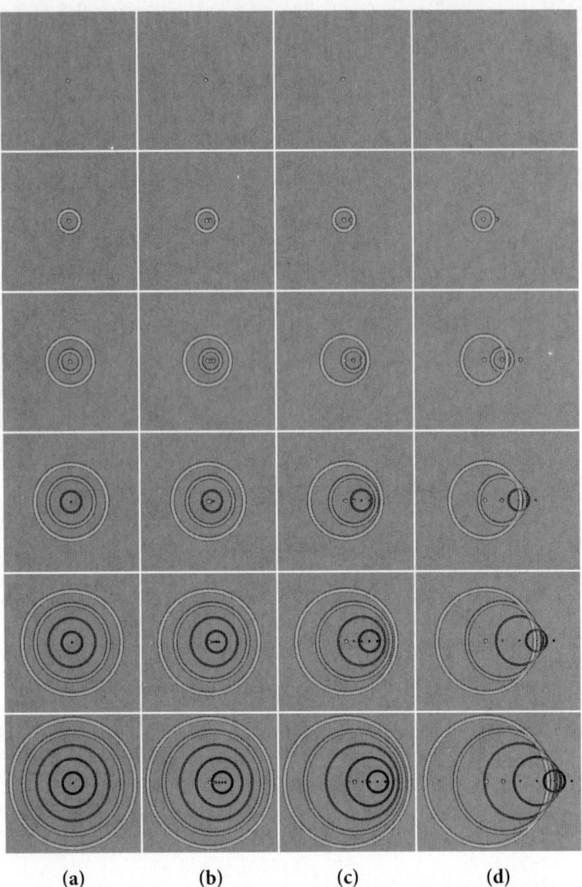


**FIGURE 16.16** The Doppler effect for sound waves emitted at six equally spaced points in time: (a) stationary source; (b) source moving to the right; (c) source moving faster to the right; (d) source moving to the right faster than the speed of sound.

We've all had the experience when a train approaching a railroad crossing sounds its warning horn and then, as it moves past our position, the sound pitch changes from a higher frequency to a lower one. This change in frequency is called the **Doppler effect.**

In order to gain a qualitative understanding of the Doppler effect, consider Figure 16.16. In part (a), the leftmost column, a stationary sound source emits spherical waves that travel outward. The time evolution of these radial waves is depicted at six equally spaced points in time, from top to bottom. The maxima of the five sine waves radiate outward as circles, color coded in each frame from yellow to red.

When the source moves, the same sound waves are emitted from different points in space. However, from each of these points, the maxima again travel outward as circles, as shown in parts (b) and (c) Figure 16.16. The only difference between these two columns is the speed with which the source of the sound waves moves from left to right. In each case, you can see a "crowding" of the sound waves in front of the emitter, on the right side of it in this context. This crowding is greater with higher source velocity and is the entire key to understanding the Doppler effect. An observer located to the right of the source—that is to say, with the source moving toward him or her—experiences more wave fronts per unit time and thus hears a higher frequency. An observer located to the left of the moving source, with the source moving away from him or her, experiences fewer wave fronts per unit time and thus hears a reduced frequency.

Quantitatively, the observed frequency, $f_o$, is given as

$$f_o = f\left(\frac{v_{\text{sound}}}{v_{\text{sound}} \pm v_{\text{source}}}\right), \tag{16.14}$$

where $f$ is the frequency of the sound as emitted by the source, and $v_{\text{sound}}$ and $v_{\text{source}}$ are the speeds of the sound and of the source, respectively. The upper sign (+) applies when the source moves away from the observer, and the lower sign (–) applies when the source moves toward the observer.

The Doppler effect also occurs if the source is stationary and the observer moves. In this case, the observed frequency is given by

$$f_o = f\left(\frac{v_{\text{sound}} \mp v_{\text{observer}}}{v_{\text{sound}}}\right) = f\left(1 \mp \frac{v_{\text{observer}}}{v_{\text{sound}}}\right). \tag{16.15}$$

The upper sign (–) applies when the observer moves away from the source, and the lower sign (+) applies when the observer moves toward the source.

In each case, moving source or moving observer, the observed frequency is lower than the source frequency when observer and source are moving away from each other and is higher than the source frequency when they are moving toward each other.

### DERIVATION 16.2 Doppler Shift

Let's begin with an observer at rest and a sound source moving toward the observer. If the sound source is at rest and emits a sound with frequency $f$, then the wavelength is $\lambda = v_{\text{sound}}/f$, which is also the distance between two successive wave crests. If the sound source moves toward the observer with speed $v_{\text{source}}$, then the distance between two successive crests as perceived by the observer is reduced to

$$\lambda_o = \frac{v_{\text{sound}} - v_{\text{source}}}{f}.$$

Using the relationship between wavelength, frequency, and (fixed!) speed of sound, $v_{sound} = \lambda f$, we find the frequency of the sound as detected by the observer:

$$f_o = \frac{v_{sound}}{\lambda_o} = f\left(\frac{v_{sound}}{v_{sound} - v_{source}}\right).$$

This result proves equation 16.14 with the minus sign in the denominator. The plus sign is obtained by reversing the sign of the source velocity vector.

You may want to try the derivation for the case in which the observer is moving as an exercise.

### 16.3 In-Class Exercise

Consider the case of a stationary observer and a source moving with constant velocity (less than the speed of sound) and emitting a sound of frequency $f$. If the source moves toward the observer, equation 16.14 predicts a higher observed frequency, $f_+$. If the source moves away from the observer, a lower observed frequency, $f_-$, results. Let's call the difference between the higher frequency and the original frequency $\Delta_+$ and the difference between the original frequency and the lower frequency $\Delta_-$: $\Delta_+ = f_+ - f$ and $\Delta_- = f - f_-$. Which statement about the frequency differences is correct?

a) $\Delta_+ > \Delta_-$
b) $\Delta_+ = \Delta_-$
c) $\Delta_+ < \Delta_-$
d) It depends on the original frequency, $f$.
e) It depends on the source velocity, $v$.

### 16.4 In-Class Exercise

Now consider the case of a stationary source emitting a sound of frequency $f$ and an observer moving with constant velocity (less than the speed of sound). If the observer moves toward the source, equation 16.15 predicts a higher observed frequency, $f_+$. If the observer moves away from the source, a lower observed frequency, $f_-$, results. Let's call the difference between the higher frequency and the original frequency $\Delta_+$ and the difference between the original frequency and the lower frequency $\Delta_-$: $\Delta_+ = f_+ - f$ and $\Delta_- = f - f_-$. Which statement about the frequency differences is correct?

a) $\Delta_+ > \Delta_-$
b) $\Delta_+ = \Delta_-$
c) $\Delta_+ < \Delta_-$
d) It depends on the original frequency, $f$.
e) It depends on the source velocity, $v$.

Finally, what happens if both the source and the observer move? The answer is that the expressions for the Doppler effect for moving source and moving observer simply combine, and the observed frequency is given by

$$f_o = f\left(\frac{v_{sound} \mp v_{observer}}{v_{sound}}\right)\left(\frac{v_{sound}}{v_{sound} \pm v_{source}}\right) = f\left(\frac{v_{sound} \mp v_{observer}}{v_{sound} \pm v_{source}}\right). \quad (16.16)$$

Here the upper signs in the numerator and the denominator apply when observer or source is moving away from the other, and the lower signs apply when observer or source is moving toward the other.

### 16.5 In-Class Exercise

Suppose a source emitting a sound with frequency $f$ moves to the right (in the positive $x$-direction) with a speed of 30 m/s. An observer is located to the right of the source and also moves to the right (in the positive $x$-direction) with a speed of 50 m/s. The observed frequency, $f_o$, will be _______ the original frequency, $f$.

a) lower than
b) the same as
c) higher than

### 16.6 In-Class Exercise

Suppose a source emitting a sound with frequency $f$ moves to the right (in the positive $x$-direction) with a speed of 30 m/s. An observer is located to the right of the source and also moves to the right (in the positive $x$-direction) with a speed of 50 m/s. Which of the following is the correct expression for the observed frequency, $f_o$?

a) $f_o = f\left(\frac{v_{sound} - 50 \text{ m/s}}{v_{sound} - 30 \text{ m/s}}\right)$

b) $f_o = f\left(\frac{v_{sound} - 50 \text{ m/s}}{v_{sound} + 30 \text{ m/s}}\right)$

c) $f_o = f\left(\frac{v_{sound} + 50 \text{ m/s}}{v_{sound} - 30 \text{ m/s}}\right)$

d) $f_o = f\left(\frac{v_{sound} + 50 \text{ m/s}}{v_{sound} + 30 \text{ m/s}}\right)$

## Doppler Effect in Two- and Three-Dimensional Space

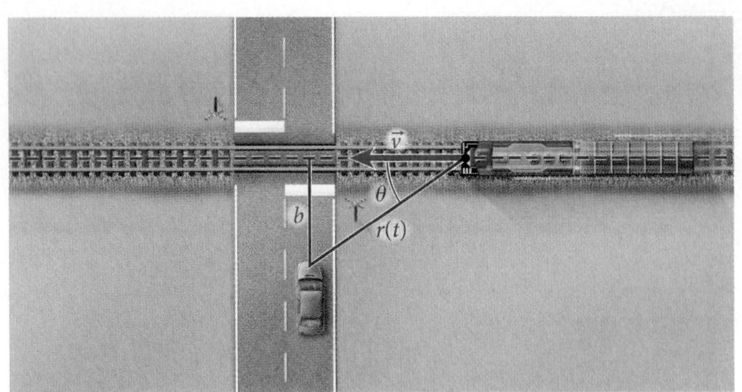


**FIGURE 16.17** Doppler effect in a two-dimensional space—a car at a railroad crossing.

So far, our discussion of the Doppler effect has assumed that the source (or observer) moves with constant velocity on a straight line, which passes right through the location of the observer (or source). However, if you are waiting at a railroad crossing, for example, a train does not pass directly through your location but has a perpendicular distance of closest approach, $b$, as shown in Figure 16.17. In this situation, what you hear is not an instantaneous switch from the higher frequency of the approaching train to the lower frequency of the receding train that equation 16.14 implies, but instead a gradual and smooth change from the higher to the lower frequency.

To obtain a quantitative expression of the Doppler effect for this case, we let $t = 0$ to be the instant when the train is closest to the car, that is, as it crosses the road. For $t < 0$, the train is then moving toward the observer, and for $t > 0$, it is moving away from the observer. The origin of the spatial coordinate system is chosen as the location of the observer, the driver of the car. Then the distance of the train to the origin is given by $r(t) = \sqrt{b^2 + v^2t^2}$, where $v$ is the (constant) speed of the train (see Figure 16.17). The angle $\theta$ between the train's velocity vector and the direction from which the sound is emitted relative to the car is given by

$$\cos\theta(t) = \frac{vt}{r(t)} = \frac{vt}{\sqrt{b^2 + v^2t^2}}.$$

The velocity of the sound source as observed in the car as a function of time is the projection of the train's velocity vector onto the radial direction:

$$v_{\text{source}}(t) = v\cos\theta(t) = \frac{v^2t}{\sqrt{b^2 + v^2t^2}}.$$

We can now substitute for $v_{\text{source}}$ in equation 16.14 and obtain

$$f_o(t) = f\frac{v_{\text{sound}}}{v_{\text{sound}} + v_{\text{source}}(t)} = f\frac{v_{\text{sound}}}{v_{\text{sound}} + \dfrac{v^2t}{\sqrt{b^2 + v^2t^2}}}. \qquad (16.17)$$

### 16.2 Self-Test Opportunity

Show that the time derivative of the function determining the observed frequency (equation 16.17) at $t = 0$ is given by

$$\left.\frac{df_o}{dt}\right|_{t=0} = -\frac{fv^2}{bv_{\text{sound}}}.$$

In equation 16.17, the source velocity is negative for $t < 0$, corresponding to an approaching source, while the source velocity is positive for $t > 0$, corresponding to a receding source.

The formula of equation 16.17 may look complicated, but it is actually a smoothly varying function, as shown in Figure 16.18. Assume that the moving train emits a sound of frequency $f = 440$ Hz and that the sound speed is 343 m/s. Figure 16.18a shows three plots of the frequency observed by a stationary observer in a car 10 m from the railroad crossing when a train moving at 20, 30, and 40 m/s passes through. Figure 16.18b shows plots of the frequencies observed by three stationary observers in cars parked at distances of 10, 40, and 70 m from the railroad crossing as a train passes at a speed of 40 m/s. These three curves converge

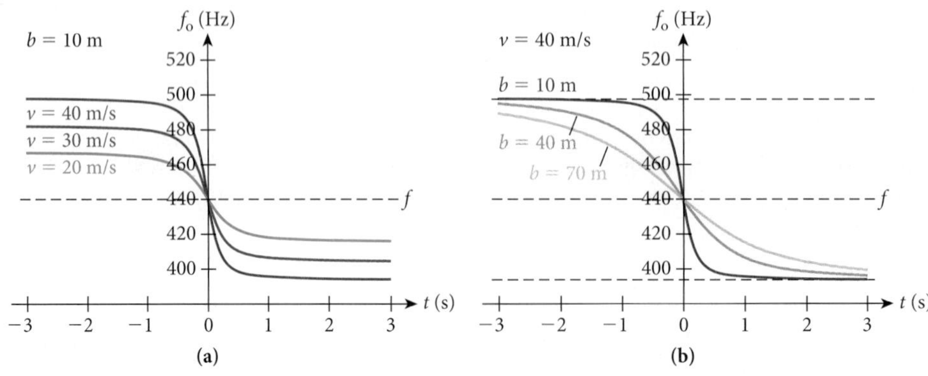


**FIGURE 16.18** Time dependence of the frequency observed by a stationary observer: (a) source velocities of 20 m/s, 30 m/s, and 40 m/s and distance of closest approach of 10 m; (b) source velocity of 40 m/s and distances of closest approach of 10 m, 40 m, and 70 m.

**FIGURE 16.19** Demonstration of the Doppler effect.

toward the asymptotic frequencies of $f = 498$ Hz for the approaching train and $f = 394$ Hz for the receding train, which are the values predicted by equation 16.14. However, the closer the observer is to the railroad crossing, the more sudden the transition from the higher frequency to the lower one becomes.

It is fairly easy to set up a demonstration of the Doppler effect. Just drive down a road at constant speed, continuously honking your horn, and have a friend videotape you. Then you can analyze the video clip with a frequency analyzer and see how the frequencies of the car horn change as you pass by the camera. In Figure 16.19, the car drove along a country highway with a speed of 26.5 m/s and passed by the camera at a closest distance of 14.0 m. The six dominant frequencies of the car's horn, given by $f = n(441\text{ Hz})$, $n = 1, 2, \ldots, 6$, are clearly visible in this figure as the narrow red bands representing high decibel values, and you can see the frequency shift as the car passes by. The superimposed dashed gray lines are the results of the calculations using equation 16.17 with a distance of closest approach $b = 14.0$ m and a source speed $v_{\text{source}} = 26.5$ m/s. You can see that they are in complete agreement with the experimental measurements.

### 16.3 Self-Test Opportunity

Derive a formula similar to equation 16.17 for an observer moving on a straight line and passing a stationary source at some nonzero closest approach distance.

## Applications of the Doppler Effect

In Section 16.1, we examined the application of ultrasound waves to the imaging of human tissues. The Doppler effect for ultrasound waves can be used to measure the speed of blood flowing in an artery, which can be an important diagnostic tool for cardiac disease. An example of the measurement of blood flow in the carotid artery is shown in Figure 16.20. This kind of measurement is implemented by transmitting ultrasound waves toward the flowing blood. The ultrasound waves reflect off the moving blood cells back toward the ultrasound device, where they are detected.

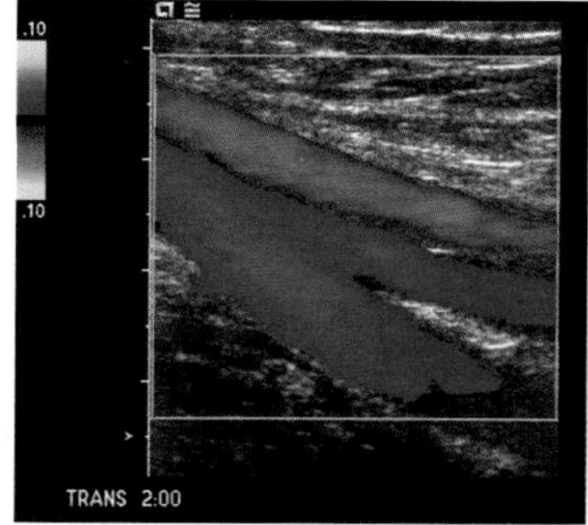


**FIGURE 16.20** Doppler ultrasound image of blood flowing in the carotid artery. Red and blue indicate the speed of the blood flow.

### EXAMPLE 16.3 Doppler Ultrasound Measurement of Blood Flow

**PROBLEM**

What is the typical frequency change for ultrasound waves reflecting off blood flowing in an artery?

**SOLUTION**

Ultrasound waves have a typical frequency of $f = 2.0$ MHz. In an artery, blood flows with a speed of $v_{\text{blood}} = 1.0$ m/s. The speed of ultrasound waves in human tissue is $v_{\text{sound}} = 1540$ m/s.

*Continued—*

The blood cells can be thought of as moving observers for the ultrasound waves. Thus, if a blood cell is moving toward the source of the ultrasound, the frequency observed by the blood cell, $f_1$, is given by equation 16.15 with a plus sign

$$f_1 = f\left(1+\frac{v_{\text{blood}}}{v_{\text{sound}}}\right). \quad \text{(i)}$$

The reflected ultrasound waves with frequency $f_1$ constitute a moving source. The stationary Doppler ultrasound device will then observe reflected ultrasound waves with a frequency, $f_2$, given by equation 16.14 with a minus sign

$$f_2 = f_1\left(\frac{v_{\text{sound}}}{v_{\text{sound}} - v_{\text{blood}}}\right). \quad \text{(ii)}$$

Thus, the frequency observed by the Doppler ultrasound device is obtained by combining equations (i) and (ii)

$$f_2 = f\left(1+\frac{v_{\text{blood}}}{v_{\text{sound}}}\right)\left(\frac{v_{\text{sound}}}{v_{\text{sound}} - v_{\text{blood}}}\right).$$

Note that this equation is just a special case of equation 16.16, the expression for the Doppler effect with a moving source and a moving observer; here the speed of the observer and that of the source are both that of the blood cell. Putting in the numbers gives us

$$f_2 = (2.0 \text{ MHz})\left(1+\frac{1.0 \text{ m/s}}{1540 \text{ m/s}}\right)\left(\frac{1540 \text{ m/s}}{1540 \text{ m/s} - 1.0 \text{ m/s}}\right) = 2.0026 \text{ MHz}.$$

The frequency difference, $\Delta f$, is then

$$\Delta f = f_2 - f = 2.6 \text{ kHz}.$$

This frequency change is the maximum observed when the blood is flowing as a result of a heart pulse. Between heart pulses, the blood slows down and almost stops. Combining the original ultrasound waves with frequency $f$ and the reflected ultrasound waves with frequency $f_2$ yields a beat frequency, $f_{\text{beat}} = f_2 - f$, which varies from 2.6 kHz to zero as the heart pulses. These frequencies can be perceived by human ears, which means the beat frequency can simply be amplified and heard as a pulse monitor. This Doppler ultrasound effect is the basis of fetal heart probes.

Note that this example assumes that the blood is flowing directly toward the Doppler ultrasound device. In general, there is an angle between the direction of the transmitted ultrasound waves and the direction of the blood flow, which decreases the frequency change. Doppler ultrasound devices take this angle into account using information on the orientation of the artery.

The Doppler radar that you hear about on TV weather reports uses the Doppler effect for electromagnetic waves. Moving raindrops change the reflected radar waves to different frequencies and thus are detected by the Doppler radar. (However, the Doppler shift for electromagnetic waves such as radar is governed by a different formula than is used for sound. Electromagnetic waves will be discussed in Chapter 31.) Doppler radar is also used to detect rotation (and possible tornados) in a thunderstorm by finding regions with motion both toward and away from the observer. Another common use of the Doppler effect is radar speed detectors employed by police forces around the world. And similar techniques are used by astronomers to measure red shifts of galaxies (more on this topic in Chapter 35 on relativity).

## Mach Cone

We have yet to examine Figure 16.16d. In this column, the source speed exceeds the speed of sound. As you can see, in this case the origin of a subsequent wave crest lies outside the circle formed by the previous wave crest.

Figure 16.21a shows the successive wave crests emitted by a source whose speed is greater than the speed of sound, called a **supersonic source.** You can see that all of these circles have a common tangent line that make an angle $\theta_M$, called the **Mach angle,** with the direction of the velocity vector (in this case, the horizontal). This accumulation of wave fronts produces a large, abrupt, conical-shaped wave called a **shock wave,** or **Mach cone.** The Mach angle of the cone is given by

$$\theta_M = \sin^{-1}\left(\frac{v_{\text{sound}}}{v_{\text{source}}}\right). \quad (16.18)$$

This formula is valid only for source speeds that exceed the speed of sound, and in the limit where the source speed is equal to the speed of sound, $\theta_M$ approaches 90°. The greater the source speed, the smaller is the Mach angle.

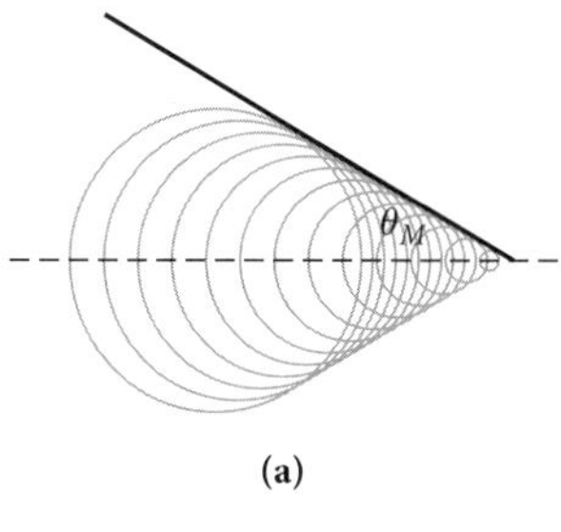


(a)

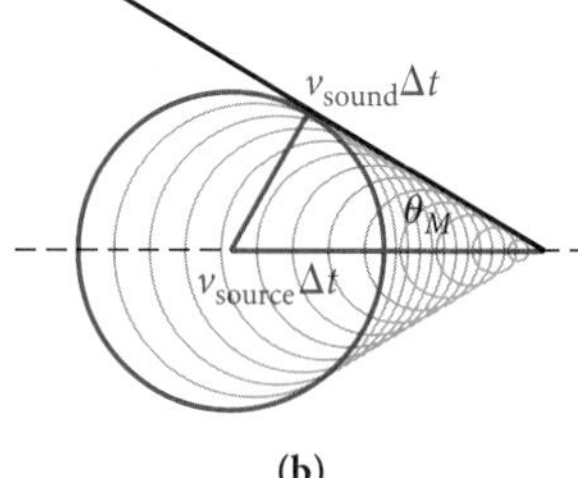


(b)

**FIGURE 16.21** (a) The Mach cone, or shock wave. (b) Geometry for Derivation 16.3.

## DERIVATION 16.3 Mach Angle

The derivation of equation 16.18 is quite straightforward if we examine Figure 16.21b, which shows the circle of the wave crest that was formed at some initial time, $t = 0$, but is now at a later time, $\Delta t$. The radius of this circle at time $\Delta t$ is $v_{\text{sound}}\Delta t$ and is shown by the blue straight line. Note that the radius makes a 90° angle with the tangent line, indicated in black. However, during the same time, $\Delta t$, the source has moved a distance $v_{\text{source}}\Delta t$, which is indicated by the red line in the figure. The red, blue, and black lines in the figure form a right triangle. Then, by definition of the sine function, the sine of the Mach angle, $\theta_M$, is

$$\sin\theta_M = \frac{v_{\text{sound}}\Delta t}{v_{\text{source}}\Delta t}.$$

Canceling out the common factor $\Delta t$ and taking the inverse sine on both sides of this equation yields equation 16.18, as desired.

Interestingly, shock waves can be detected in all kinds of physical systems involving all kinds of waves. For example, surface waves on shallow water have relatively low wave speeds, as discussed in Chapter 15. Therefore, motorboats can create shock waves relatively easily by moving across the water surface at high speeds. From these triangular bow waves, you can determine the speed of the surface waves on a lake, if you know the speed of the boat.

The speed of sound in atomic nuclei is approximately a third of the speed of light. Since particle accelerators can collide nuclei at speeds very close to the speed of light, shock waves should be created even in this environment. Some computer simulations have shown that this effect is present in collisions of atomic nuclei at very high energies, but definite experimental proof has been difficult to obtain. Several nuclear physics laboratories around the world are actively investigating this phenomenon.

Finally, it would seem unlikely that shock waves involving light could be observed, since the speed of light in vacuum is the maximum speed possible in nature, and thus the source velocity cannot exceed the speed of light in vacuum. When the Russian physicist Pavel Cherenkov (also sometimes spelled Čerenkov) proposed in the 1960s that this effect did occur, it was dismissed as impossible. However, so-called *Cherenkov radiation* (Figure 16.22) can be emitted when a source such as a high-energy proton or electron moving at close to the speed of light enters a medium, like water, where the speed of light is significantly lower. Then the source does move faster than the speed of light in that medium, and a Mach cone can form. Modern particle detectors make use of this Cherenkov radiation; measuring the emission angle allows the velocity of the particle that emitted the radiation to be calculated.

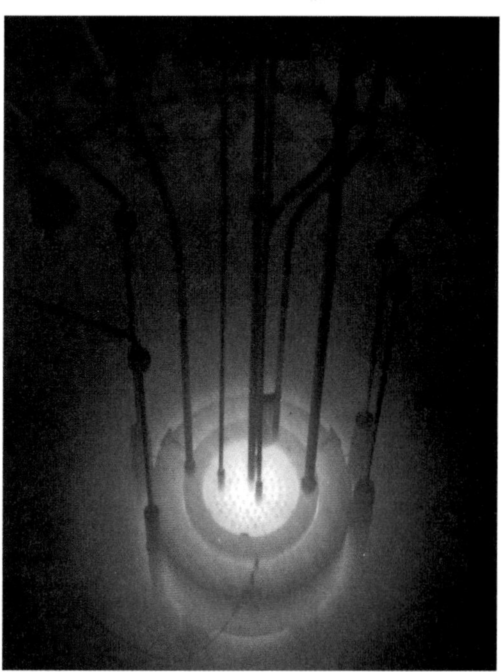

**FIGURE 16.22** Cherenkov radiation (bluish glow) from the core of a nuclear reactor.

**FIGURE 16.23** The supersonic Concorde jetliner taking off. Regular commercial service on the Concorde began in 1976 and ended in 2003.

### 16.7 In-Class Exercise

What is the maximum cruise speed of the Concorde at 60,000 ft?

a) 665 mph

b) 834 mph

c) 1350 mph

d) 2130 mph

e) 3450 mph

### EXAMPLE 16.4 The Concorde

The speed of a supersonic aircraft is often given as a Mach number, $M$. A speed of Mach 1 ($M = 1$) means that the aircraft is traveling at the speed of sound. A speed of Mach 2 ($M = 2$) means that the aircraft is traveling at twice the speed of sound. The Concorde supersonic airliner (Figure 16.23) cruised at 60,000 feet, where the speed of sound is 295 m/s (661 mph). The maximum cruise speed of the Concorde was Mach 2.04 ($M = 2.04$).

**PROBLEM**

At this speed, what was the angle of the Mach cone produced by the Concorde?

**SOLUTION**

The angle of the Mach cone is given by equation 16.18:

$$\theta_M = \sin^{-1}\left(\frac{v_{\text{sound}}}{v_{\text{source}}}\right).$$

The speed of the source in this case is that of the Concorde, which is traveling at a speed of

$$v_{\text{source}} = Mv_{\text{sound}} = 2.04v_{\text{sound}}.$$

Thus, we can write the Mach angle as

$$\theta_M = \sin^{-1}\left(\frac{v_{\text{sound}}}{Mv_{\text{sound}}}\right) = \sin^{-1}\left(\frac{1}{M}\right).$$

For the Concorde, we then have

$$\theta_M = \sin^{-1}\left(\frac{1}{2.04}\right) = 0.512 \text{ rad} = 29.4°.$$

## 16.5 Resonance and Music

Basically all musical instruments rely on the excitation of resonances to generate sound waves of discrete, reproducible, and predetermined frequencies. No discussion of sound could possibly be complete without a discussion of musical tones.

### Tones

Which frequencies correspond to which musical tones? The answer to this question is not simple and has changed over time. The current tonal scale dates back to 1722, the year of publication of Johann Sebastian Bach's work *Das wohltemperierte Klavier (The Well-Tempered Clavier)*. Let's review these accepted values.

The tones associated with the white keys of the piano are called A, B, C, D, E, F, and G. The C closest to the left end of the keyboard is designated C1. The note below it, oddly enough, is B0. Thus the sequence of tones for the white keys, from left to right, on a piano is A0, B0, C1, D1, E1, F1, G1, A1, B1, C2, D2, ... , A7, B7, C8. The scale is anchored with middle A (A4) at precisely 440 Hz. The next higher A (A5) is an octave higher, which means that it has exactly twice the frequency, or 880 Hz.

Between these two A's are 11 halftones: A-sharp/B-flat, B, C, C-sharp/D-flat, D, D-sharp/E-flat, E, F, F-sharp/G-flat, G, and G-sharp/A-flat. Ever since the work of J. S. Bach, these 11 halftones divide the octave into exactly 12 equal intervals, all of which differ by the same factor, a factor that will give the number 2 when multiplied with itself 12 times. This factor is: $2^{1/12} = 1.0595$. The frequencies of all the tones can be found through successive multiplication with this factor. For example, D (D5) is 5 steps up from middle A: $(1.0595^5)(440 \text{ Hz}) = 587.4$ Hz. Frequencies of notes in other octaves are found by multiplication or division by a factor of 2. For example, the next higher D (D6) has a frequency of 2(587.4 Hz) = 1174.9 Hz.

**Table 16.3 Frequency Ranges for Classifications of Human Singers**

| Classification | Lowest Note | Highest Note |
|---|---|---|
| Bass | E2 (82 Hz) | G4 (392 Hz) |
| Baritone | A2 (110 Hz) | A4 (440 Hz) |
| Tenor | D3 (147 Hz) | B4 (494 Hz) |
| Alto | G3 (196 Hz) | F5 (698 Hz) |
| Soprano | C4 (262 Hz) | C6 (1046 Hz) |

Human voices have classifications based on the average ranges shown in Table 16.3. Thus, basically all songs sung by humans exhibit a range of frequencies of one order of magnitude, between 100 and 1000 Hz. No human voice and practically no musical instrument generates a pure single-frequency tone. Instead, musical tones have a rich admixture of overtones that gives each instrument and each human voice its own characteristics. For string instruments, the hollow wooden body of the instrument, which serves to amplify the sound of the strings, influences this admixture.

The touch-tone phones that have almost completely replaced dial phones in the United States rely on sounds of preassigned frequencies. For each key that you press on the phone, two relatively pure tones are generated. Figure 16.24 shows the arrangement of these tones. Each key in a given row sets off the tone displayed to its right, and each key in a column triggers the tone displayed under it. If you press the 2 key, for example, the frequencies 697 Hz and 1336 Hz are produced.

**FIGURE 16.24** The frequencies generated by the keys of a touch-tone phone.

## Half-Open and Open Pipes

We discussed standing waves on strings in Chapter 15. These waves form the basis for the sounds produced by all string instruments. It is important to remember that standing waves can be excited on strings only at discrete resonance frequencies. Most percussion instruments, like drums, also work on the principle of exciting discrete resonances. However, the waveforms produced by these instruments are typically two-dimensional and are much more complicated than those produced on strings.

Wind instruments use half-open or open pipes to generate sounds. A *half-open pipe* is a pipe that is open at only one end and closed at the other end (such as a clarinet or trumpet); an open pipe is open on both ends (such as a flute). You know that you can generate a sound by blowing air over the opening of a bottle, as shown in Figure 16.25. You may remember that the sound's pitch is higher if the bottle is fuller. Also, an empty liter bottle produces a lower sound than an empty half-liter bottle, like the one used in Figure 16.25. A bottle is only an approximation (because of its noncylindrical shape) for a half-open pipe. The air molecules at the bottom of such a pipe are in contact with the wall and thus do not vibrate. Therefore, the closed end of the pipe establishes a node in the sound wave. The other end of the pipe is open, and the air molecules can vibrate freely. Blowing air over the end of the pipe creates a resonant sound wave that has an antinode at the open end of the pipe and a node at the closed end.

**FIGURE 16.25** Exciting a resonant sound from an approximation of a half-open pipe.

Figure 16.26a shows some of the possible standing waves in a half-open pipe. In the top pipe is the standing wave with the largest wavelength, for which the first antinode coincides with the length of the pipe. Since the distance between a node and the first antinode is a fourth of a wavelength, this resonance condition corresponds to $L = \frac{1}{4}\lambda$, where $L$ is the length of the pipe.The middle and bottom pipes in Figure 16.26a have standing waves with smaller possible wavelengths, for which the second or third antinodes fall on the pipe opening, and $L = \frac{3}{4}\lambda$ and $L = \frac{5}{4}\lambda$, respectively. In general, the condition for a resonant standing wave in a half-open pipe is thus

$$L = \frac{2n-1}{4}\lambda, \quad \text{for} \quad n = 1, 2, 3, \ldots .$$

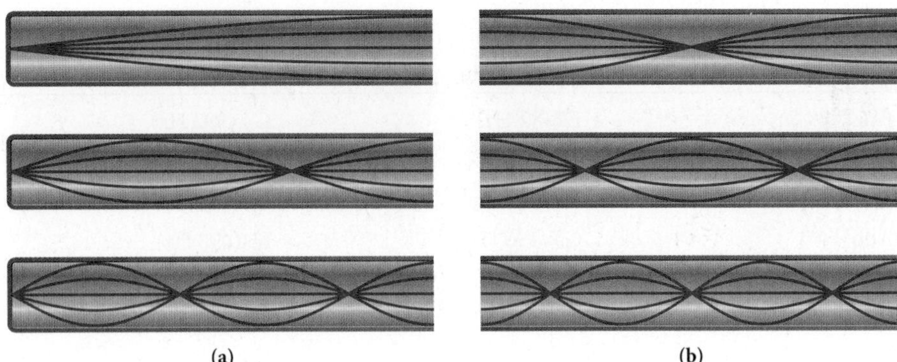


**FIGURE 16.26** (a) Standing waves in a half-open pipe; (b) standing waves in an open pipe. The red lines represent the sound amplitude at various times.

Solving this equation for the possible wavelengths results in

$$\lambda_n = \frac{4L}{2n-1}, \quad \text{for} \quad n = 1,2,3,\ldots. \tag{16.19}$$

Using $v = \lambda f$, we obtain the possible frequencies:

$$f_n = (2n-1)\frac{v}{4L}, \quad \text{for} \quad n = 1,2,3,\ldots, \tag{16.20}$$

where $n$ correspondes to the number of nodes.

Figure 16.26b displays the possible standing waves in an open pipe. In this case, there are antinodes at both ends of the pipe, and the condition for a resonant standing wave is

$$L = \frac{n}{2}\lambda, \quad n = 1,2,3,\ldots.$$

This relationship results in wavelengths and frequencies of

$$\lambda_n = \frac{2L}{n}, \quad n = 1,2,3,\ldots. \tag{16.21}$$

$$f_n = n\frac{v}{2L}, \quad n = 1,2,3,\ldots. \tag{16.22}$$

where again $n$ is the number of nodes.

For both half-open and open pipes, the fundamental frequency is obtained with $n = 1$. For a half-open pipe, the first harmonic ($n = 2$) is three times as high as the fundamental frequency. However, for an open pipe, the first harmonic ($n = 2$) is twice as high as the fundamental frequency, or one octave higher.

### 16.4 Self-Test Opportunity

Estimate the fundamental frequency of the resonant sound induced by blowing on the open end of a half-liter bottle, as illustrated in Figure 16.25.

### 16.8 In-Class Exercise

The half-liter bottle in Figure 16.25 contains air above the liquid. How would the frequency change if the air in the bottle were replaced with krypton gas? (*Hint:* Table 16.1 may help.)

a) It would be lower.

b) It would stay the same.

c) It would be higher.

## EXAMPLE 16.5 A Pipe Organ

Church organs operate on the principle of creating standing waves in pipes. If you have ever been in an old cathedral, you have very likely seen an impressive collection of organ pipes (Figure 16.27).

**FIGURE 16.27** Pipes of a church organ.

### PROBLEM

If you wanted to build an organ whose fundamental frequencies cover the frequency range of a piano, from A0 to C8, what is the range of lengths of the pipes you would need to use?

**SOLUTION**

The frequencies for A0 and C8 are $f_{A0} = 27.5$ Hz, $f_{C8} = 4186$ Hz. According to equations 16.20 and 16.22, the fundamental frequencies for a half-open pipe and an open pipe are

$$f_{1,\text{half-open}} = \frac{v}{4L} \quad \text{and} \quad f_{1,\text{open}} = \frac{v}{2L}.$$

This means that the open pipes have to be twice as long as the corresponding half-open pipes for the same frequency. The lengths of the half-open pipes needed for the highest and lowest frequencies are

$$L_{\text{half-open,C8}} = \frac{v}{4f_{C8}} = \frac{343 \text{ m/s}}{16,744 \text{ Hz}} = 0.0205 \text{ m}$$

$$L_{\text{half-open,A0}} = \frac{v}{4f_{A0}} = \frac{343 \text{ m/s}}{110 \text{ Hz}} = 3.118 \text{ m}.$$

**COMMENT**

Real organ pipes can indeed be as long as 10 ft. The part that produces the standing wave can be as short as 0.5 in; however, the base of a pipe is approximately 0.5 ft in length, so the shortest pipes typically observed in a church organ are at least that long.

## WHAT WE HAVE LEARNED | EXAM STUDY GUIDE

- Sound is a longitudinal pressure wave that requires a medium in which to propagate.
- The speed of sound in air at normal pressure and temperature (1 atm, 20 °C) is 343 m/s.
- In general, the speed of sound in a solid is given by $v = \sqrt{Y/\rho}$, and that in a liquid or a gas is given by $v = \sqrt{B/\rho}$.
- A logarithmic scale is used to measure sound intensities. The unit of this scale is the decibel (dB). The sound level, $\beta$, in this decibel scale is defined as $\beta = 10\log\frac{I}{I_0}$, where $I$ is the intensity of the sound waves and $I_0 = 10^{-12}$ W/m$^2$, which corresponds approximately to the minimum intensity that a human ear can hear.
- Sound waves emitted by two coherent sources constructively interfere if the path length difference is $\Delta r = n\lambda$; for all $n = 0, \pm1, \pm2, \pm3, \ldots$ and will destructively interfere if the path length difference is $\Delta r = (n + \frac{1}{2})\lambda$; for all $n = 0, \pm1, \pm2, \pm3, \ldots$.
- Two sine waves with equal amplitudes and slightly different frequencies produce beats with a beat frequency of $f_b = |f_1 - f_2|$.
- As a result of the Doppler effect, the frequency of sound from a moving source as perceived by a stationary observer is $f_o = f\left(\frac{v_{\text{sound}}}{v_{\text{sound}} \pm v_{\text{source}}}\right)$, where $f$ is the frequency of the sound as emitted by the source and $v_{\text{sound}}$ and $v_{\text{source}}$ are the speeds of sound and of the source, respectively. The upper sign (+) applies when the source moves away from the observer, and the lower sign (−) applies when the source moves toward the observer.
- The source of sound can be stationary while the observer moves. In this case, the observed frequency of the sound is given by $f_o = f\left(\frac{v_{\text{sound}} \mp v_{\text{observer}}}{v_{\text{sound}}}\right) = f\left(1 \mp \frac{v_{\text{observer}}}{v_{\text{sound}}}\right)$. Here the upper sign (−) applies when the observer moves away from the source, and the lower sign (+) applies when the observer moves toward the source.
- The Mach angle of the shock wave formed by a sound source moving at supersonic speeds is $\theta_M = \sin^{-1}\left(\frac{v_{\text{sound}}}{v_{\text{source}}}\right)$.
- The relationship between the length of a half-open pipe and the wavelengths of possible standing waves inside it is $L = \frac{2n-1}{4}\lambda$, for $n = 1, 2, \ldots$; for open pipes, the relationship is $L = \frac{n}{2}\lambda$ for $n = 1, 2, \ldots$.

## KEY TERMS

sound, p. 525
decibel, p. 529
coherent sources, p. 533
beats, p. 534
active noise cancellation, p. 535
Doppler effect, p. 536
supersonic source, p. 541
Mach angle, p. 541
shock wave (or Mach cone), p. 541

## NEW SYMBOLS AND EQUATIONS

$v_{\text{sound}}$, speed of sound

$\beta = 10\log\frac{I}{I_0}$, sound level in decibels

$f_b = |f_1 - f_2|$, beat frequency

$v_{\text{source}}$, speed with which sound source is moving

$v_{\text{observer}}$, speed with which the observer is moving

$f_o$, sound frequency detected by observer

$\theta_M = \sin^{-1}\left(\frac{v_{\text{sound}}}{v_{\text{source}}}\right)$, Mach angle of shock wave

## ANSWERS TO SELF-TEST OPPORTUNITIES

**16.1** $\beta = 10\log\frac{I}{I_0} \Rightarrow I = I_0\left(10^{\beta/10}\right)$

$$P = IA = I\left(4\pi r^2\right) = I_0\left(10^{\beta/10}\right)\left(4\pi r^2\right)$$

$$P = 4\pi\left(10^{-12}\text{ W/m}^2\right)10^{(80\text{ dB})/10}\left(10.0\text{ m}\right)^2 = 0.126\text{ W}.$$

**16.2** $f_o(t) = f\dfrac{v_{\text{sound}}}{v_{\text{sound}} + \dfrac{v^2 t}{\sqrt{b^2 + v^2t^2}}}$

$$\frac{df_o(t)}{dt} = -fv_{\text{sound}}\frac{\left(-\dfrac{t^2v^4}{\left(b^2+v^2t^2\right)^{3/2}} + \dfrac{v^2}{\sqrt{b^2+v^2t^2}}\right)}{\left(v_{\text{sound}} + \dfrac{v^2t^2}{\sqrt{b^2+v^2t^2}}\right)^2}$$

$$\left.\frac{df_o(t)}{dt}\right|_{t=0} = -fv_{\text{sound}}\frac{\left(\dfrac{v^2}{\sqrt{b^2}}\right)}{\left(v_{\text{sound}}\right)^2} = -f\frac{v^2}{bv_{\text{sound}}}.$$

**16.3** $v_{\text{observer}}(t) = \dfrac{v^2t}{\sqrt{b^2+v^2t^2}}$

$$f_o(t) = f\frac{v_{\text{sound}} \mp v_{\text{observer}}(t)}{v_{\text{sound}}}$$

$$f_o(t) = f\frac{v_{\text{sound}} - \dfrac{v^2t}{\sqrt{b^2+v^2t^2}}}{v_{\text{sound}}}.$$

**16.4** $f_n = (2n-1)\dfrac{v}{4L}$

$$f_1 = \frac{v}{4L} = \frac{343\text{ m/s}}{4(0.215\text{ m})} = 399\text{ Hz}$$

estimate 400 Hz.

## PROBLEM-SOLVING PRACTICE

### Problem-Solving Guidelines

**1.** Be sure you're familiar with the properties of logarithms to solve problems involving the intensity of sound. In particular, remember that log $(A/B)$ = log $A$ – log $B$ and that log $x^n = n$ log $x$.

**2.** Situations involving the Doppler effect can be tricky. Sometimes you need to break the path of the sound wave into segments and treat each segment separately, depending on whether motion of the source or motion of the observer is occurring. Often a sketch is useful to identify which expression for the Doppler effect pertains to which part of the situation.

**3.** You don't need to memorize formulas for the harmonic frequencies of open pipes or half-open pipes. Instead, remember that the closed end of a pipe is the location of a node and the open end of a pipe is the location of an antinode. Then you can reconstruct the standing waves from that point.

### SOLVED PROBLEM 16.2 Doppler Shift for a Moving Observer

**PROBLEM**

A person in a parked car sounds the horn. The frequency of the horn's sound is 290.0 Hz. A driver in an approaching car measures the frequency of the sound coming from the parked car to be 316.0 Hz. What is the speed of the approaching car?

## SOLUTION

### THINK

The frequency of the sound measured in the approaching car has been Doppler shifted. Knowing the emitted frequency, $f$, the observed frequency, $f_o$, and the speed of sound, $v_{sound}$, we can calculate the speed of the approaching car.

### SKETCH

Figure 16.28 shows the vehicle containing the observer moving toward the stationary sound-emitting vehicle.

**FIGURE 16.28** A vehicle approaching another that is sounding its horn.

### RESEARCH

The frequency measured by a moving observer, $f_o$, resulting from a sound of frequency $f$ being emitted by a stationary source is given by

$$f_o = f\left(1 + \frac{v_{observer}}{v_{sound}}\right), \tag{i}$$

where $v_{observer}$ is the speed with which the observer approaches the sound-emitting source and $v_{sound}$ is the speed of sound in air.

### SIMPLIFY

We can rearrange equation (i) to obtain

$$\frac{f_o}{f} - 1 = \frac{v_{observer}}{v_{sound}}.$$

From this equation, we can solve for the speed of the moving observer:

$$v_{observer} = v_{sound}\left(\frac{f_o}{f} - 1\right).$$

### CALCULATE

Putting in the numerical values, we get

$$v_{observer} = (343 \text{ m/s})\left(\frac{316.0 \text{ Hz}}{290.0 \text{ Hz}} - 1\right) = 30.7517 \text{ m/s}.$$

### ROUND

We report our result to three significant figures:

$$v_{observer} = 30.8 \text{ m/s}.$$

### DOUBLE-CHECK

A useful first check is that the units work out correctly. Units of meters per second, which we obtained for our answer, are certainly appropriate for a speed. Next, we can check the order of magnitude. A speed of 30.8 m/s (68.9 mph) is reasonable for a car.

## SOLVED PROBLEM 16.3 Standing Wave in a Pipe

### PROBLEM

A standing sound wave is set up in a pipe of length 0.410 m (Figure 16.29) containing air at normal pressure and temperature. What is the frequency of this sound?

**FIGURE 16.29** A standing sound wave in a pipe.

### SOLUTION

### THINK

First, we recognize that the standing sound wave shown in Figure 16.29 corresponds to one in a pipe that is closed on the left end and open on the right end. Thus, we are dealing with a standing wave in a half-open pipe. Four nodes are visible. Combining these observations and knowing the speed of sound, we can calculate the frequency of the standing wave.

### SKETCH

The standing wave with the pipe drawn in and the nodes marked is shown in Figure 16.30.

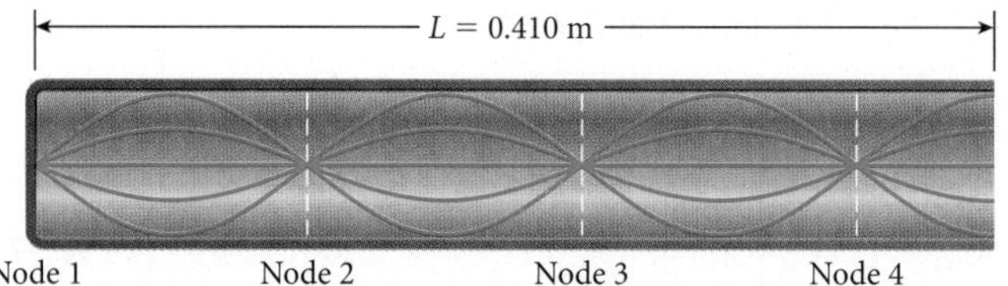


**FIGURE 16.30** A standing sound wave with four nodes in a half-open pipe.

### RESEARCH

The frequency of this standing wave in a half-open pipe is given by

$$f_n = (2n-1)\frac{v}{4L}, \quad \text{(i)}$$

where $n = 4$ is the number of nodes. The speed of sound in air is $v = 343$ m/s, and the length of the pipe is $L = 0.410$ m.

### SIMPLIFY

For $n = 4$, we can write equation (i) for the frequency of the standing wave as

$$f_4 = \left(2\left(4\right)-1\right)\frac{v}{4L} = \frac{7v}{4L}.$$

### CALCULATE

Putting in the numerical values, we get

$$f = \frac{7(343 \text{ m/s})}{4(0.410 \text{ m})} = 1464.02 \text{ Hz}.$$

### ROUND

We report our result to three significant figures:

$$f = 1460 \text{ Hz}.$$

### DOUBLE-CHECK

This frequency is well within the range of human hearing and just above the range of frequencies produced by the human voice. Therefore, our answer seems reasonable.

## SOLVED PROBLEM 16.4 Doppler Shift of Reflected Ambulance Siren

### PROBLEM

An ambulance has picked up an injured rock climber and is heading directly away from the canyon wall (where the climber was injured) at a speed of 31.3 m/s (70 mph). The ambulance's siren has a frequency of 400.0 Hz. After the ambulance turns off its siren, the injured rock climber can hear the reflected sound from the canyon wall for a few seconds. The velocity of sound in air is $v_{\text{sound}} = 343$ m/s. What is the frequency of the reflected sound from the ambulance's siren as heard by the injured rock climber in the ambulance?

### SOLUTION

**THINK**

The ambulance can be considered to be a moving source with speed $v_{\text{ambulance}}$ emitting a sound with frequency $f$. The frequency, $f_1$, of the siren at the fixed wall of the canyon can be calculated using the sound velocity, $v_{\text{sound}}$. The canyon wall reflects the sound of the siren back toward the moving ambulance. The climber in the ambulance can be thought of as a moving observer of the reflected sound of the siren. Combining the two Doppler shifts gives the frequency, $f_2$, of the reflected siren as heard on the moving ambulance.

**SKETCH**

Figure 16.31 shows the ambulance moving away from the canyon wall with speed $v_{\text{ambulance}}$ while emitting a sound from its siren of frequency $f$.

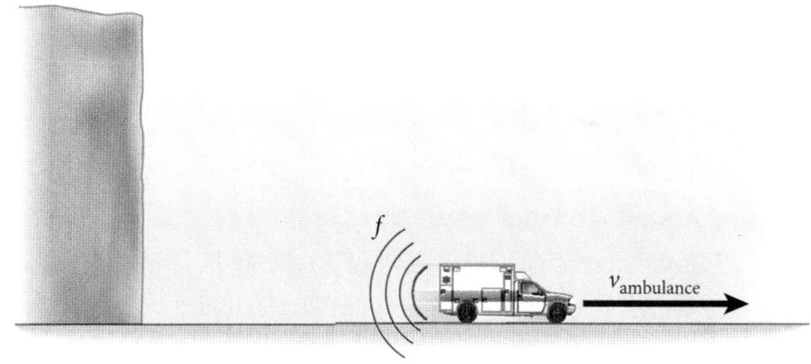


**FIGURE 16.31** An ambulance moves away from a canyon wall with its siren operating.

**RESEARCH**

The frequency, $f_1$, of the sound measured by a stationary observer standing at the canyon wall resulting from the sound of frequency $f$ emitted by the ambulance's siren moving with speed $v_{\text{ambulance}}$ is given by equation 16.14:

$$f_1 = f\left(\frac{v_{\text{sound}}}{v_{\text{sound}} + v_{\text{ambulance}}}\right), \quad \text{(i)}$$

where the plus sign is applicable because the ambulance is moving away from the stationary observer. The sound is reflected from the canyon wall back toward the moving ambulance. Now the injured rock climber in the ambulance is a moving observer. We can calculate the frequency observed by the rock climber, $f_2$, using equation 16.15:

$$f_2 = f_1\left(1 - \frac{v_{\text{ambulance}}}{v_{\text{sound}}}\right), \quad \text{(ii)}$$

where the minus sign is used because the observer is moving away from the source.

**SIMPLIFY**

We can combine equations (i) and (ii) to obtain

$$f_2 = f\left(\frac{v_{\text{sound}}}{v_{\text{sound}} + v_{\text{ambulance}}}\right)\left(1 - \frac{v_{\text{ambulance}}}{v_{\text{sound}}}\right).$$

You can see that this result is a special case of equation 16.16, where the speeds of observer and source are both the speed of the ambulance and both observer and source are moving away from each other.

**CALCULATE**

Putting in the numerical values, we get

$$f_2 = (400\text{ Hz})\left(\frac{343\text{ m/s}}{343\text{ m/s} + 31.3\text{ m/s}}\right)\left(1 - \frac{31.3\text{ m/s}}{343\text{ m/s}}\right) = 333.1018\text{ Hz}.$$

*Continued—*

**ROUND**

We report the result to three significant figures:

$$f_2 = 333 \text{ Hz}.$$

**DOUBLE-CHECK**

The frequency of the reflected sound observed by the injured rock climber is approximately 20% lower than the original frequency of the ambulance's siren. The speed of the ambulance is about 10% of the speed of sound. The two Doppler shifts, one for the sound emitted from the moving ambulance and one for the reflected sound observed in the moving ambulance, should each shift the frequency by approximately 10%. So the observed frequency seems reasonable.

## MULTIPLE-CHOICE QUESTIONS

**16.1** You stand on the curb waiting to cross a street. Suddenly you hear the sound of a horn from a car approaching you at a constant speed. You hear a frequency of 80 Hz. After the car passes by, you hear a frequency of 72 Hz. At what speed was the car traveling?

a) 17 m/s
b) 18 m/s
c) 19 m/s
d) 20 m/s

**16.2** A sound level of 50 decibels is

a) 2.5 times as intense as a sound of 20 decibels.
b) 6.25 times as intense as a sound of 20 decibels.
c) 10 times as intense as a sound of 20 decibels.
d) 100 times as intense as a sound of 20 decibels.
e) 1000 times as intense as a sound of 20 decibels.

**16.3** A police car is moving in your direction, constantly accelerating, with its siren on. As it gets closer, the sound you hear will

a) stay at the same frequency.
b) drop in frequency.
c) increase in frequency.
d) More information is needed.

**16.4** You are creating a sound wave by shaking a paddle in a liquid medium. How can you increase the speed of the resulting sound wave?

a) You shake the paddle harder to give the medium more kinetic energy.
b) You vibrate the paddle more rapidly to increase the frequency of the wave.
c) You create a resonance with a faster-moving wave in air.
d) All of these will work.
e) None of these will work.
f) Only (a) and (b) will work.
g) Only (a) and (c) will work.
h) Only (b) and (c) will work.

**16.5** Standing on the sidewalk, you listen to the horn of a passing car. As the car passes, the frequency of the sound changes from high to low in a continuous manner; that is, there is no abrupt change in the perceived frequency. This occurs because

a) the pitch of the sound of the horn changes continuously.
b) the intensity of the observed sound changes continuously.
c) you are not standing directly in the path of the moving car.
d) of all of the above reasons.

**16.6** A sound meter placed 3 m from a speaker registers a sound level of 80 dB. If the volume on the speaker is then turned down so that the power is reduced by a factor of 25, what will the sound meter read?

a) 3.2 dB
b) 11 dB
c) 32 dB
d) 55 dB
e) 66 dB

**16.7** What has the greatest effect on the speed of sound in air?

a) temperature of the air
b) frequency of the sound
c) wavelength of the sound
d) pressure of the atmosphere

**16.8** A driver sits in her car at a railroad crossing. Three trains go by at different (constant) speeds, each emitting the same sound. Bored, the driver makes a recording of the sounds. Later, at home, she plots the frequency as a function of time on her computer, resulting in the plots shown in the figure. Which of the three trains had the highest speed?

a) the one represented by the solid line
b) the one represented by the dashed line
c) the one represented by the dotted line
d) impossible to tell

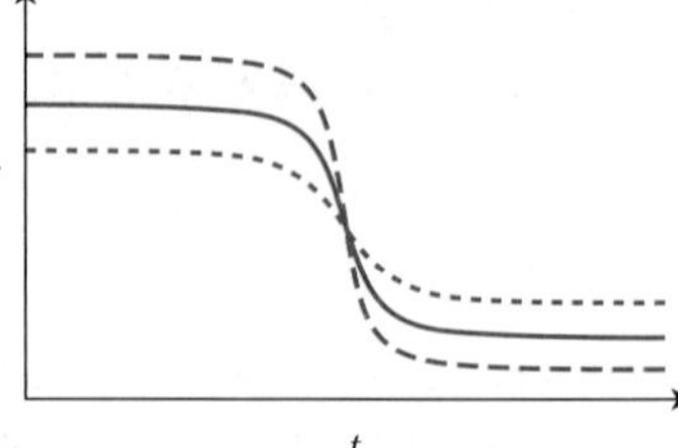


**16.9** Three physics faculty members sit in three different cars at a railroad crossing. Three trains go by at different (constant) speeds, each emitting the same sound. Each faculty member uses a cell phone to record the sound of a different train. On the next day, during a faculty meeting, they plot the frequency as a function of time on a computer, resulting in the plots shown in the figure. Which of the three trains had the highest speed?

a) the one represented by the solid line

b) the one represented by the dashed line

c) the one represented by the dotted line

d) impossible to tell

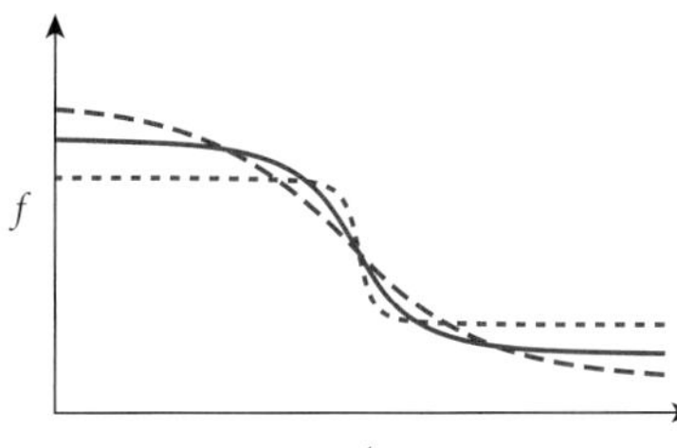


16.10 In Question 16.9, which of the three faculty members was closest to the train tracks?

a) the one who recorded the sound represented by the solid line

b) the one who recorded the sound represented by the dashed line

c) the one who recorded the sound represented by the dotted line

d) impossible to tell

## QUESTIONS

**16.11** A (somewhat risky) way of telling if a train that cannot be seen or heard is approaching is by placing your ear on the rail. Explain why this works.

**16.12** A classic demonstration of the physics of sound involves an alarm clock in a bell vacuum jar. The demonstration starts with air in the vacuum jar at normal atmospheric pressure and then the jar is evacuated to lower and lower pressures. Describe the expected outcome.

**16.13** You are sitting near the back of a twin-engine commercial jet, which has the engines mounted on the fuselage near the tail. The nominal rotation speed for the turbofan in each engine is 5200 revolutions per minute, which is also the frequency of the dominant sound emitted by the engine. What audio clue would suggest that the engines are not perfectly synchronized and that the turbofan in one of them is rotating approximately 1% faster than that in the other? How would you measure this effect if you had only your wristwatch (could measure time intervals only in seconds)? To hear this effect, go to http://qbx6.ltu.edu/s_schneider/physlets/main/beats.shtml. This simulation works best with integer natural frequencies, so round your frequencies to the ones place.

**16.14** You determine the direction from which a sound is coming by subconsciously judging the difference in time it takes to reach the right and left ears. Sound directly in front (or back) of you arrives at both ears at the same time; sound from your left arrives at your left ear before your right ear. What happens to this ability to determine the location of a sound if you are underwater? Will sounds appear to be located more in front or more to the side of where they actually are?

**16.15** On a windy day, a child standing outside a school hears the school bell. If the wind is blowing toward the child from the direction of the bell, will it alter the frequency, the wavelength, or the velocity of the sound heard by the child?

**16.16** A police siren contains at least two frequencies, producing the wavering sound (beats). Explain how the siren sound changes as a police car approaches, passes, and moves away from a pedestrian.

**16.17** The Moon has no atmosphere. Is it possible to generate sound waves on the Moon?

**16.18** When two pure tones with similar frequencies combine to produce beats, the result is a train of wave packets. That is, the sinusoidal waves are partially localized into packets. Suppose two sinusoidal waves of equal amplitude $A$, traveling in the same direction, have wave numbers $\kappa$ and $\kappa + \Delta\kappa$ and angular frequencies $\omega$ and $\omega + \Delta\omega$, respectively. Let $\Delta x$ be the length of a wave packet, that is, the distance between two nodes of the envelope of the combined sine functions. What is the value of the product $\Delta x \Delta\kappa$?

**16.19** As you walk down the street, a convertible drives by on the other side of the street. You hear the deep bass speakers in the car thumping out a groovy but annoyingly loud beat. Estimate the intensity of the sound you perceive, and then calculate the minimum intensity that the poor ears of the convertible's driver and passengers are subject to.

**16.20** If you blow air across the mouth of an empty soda bottle, you hear a tone. Why is it that if you put some water in the bottle, the pitch of the tone increases?

## PROBLEMS

A blue problem number indicates a worked-out solution is available in the Student Solutions Manual. One • and two •• indicate increasing level of problem difficulty.

### Section 16.1

**16.21** Standing on a sidewalk, you notice that you are halfway between a clock tower and a large building. When the clock strikes the hour, you hear an echo of the bell 0.5 s after hearing it directly from the tower. How far apart are the clock tower and the building?

**16.22** Two farmers are standing on opposite sides of a very large empty field that is 510 m across. One farmer yells out some instructions, and 1.5 s pass until the sound reaches the other farmer. What is the temperature of the air?

**16.23** The density of a sample of air is 1.205 kg/m$^3$, and the bulk modulus is $1.42 \cdot 10^5$ N/m$^2$.

a) Find the speed of sound in the air sample.

b) Find the temperature of the air sample.

•**16.24** You drop a stone down a well that is 9.50 m deep. How long is it before you hear the splash? The speed of sound in air is 343 m/s.

**16.25** Electromagnetic radiation (light) consists of waves. More than a century ago, scientists thought that light, like other waves, required a medium (called the *ether*) to support its transmission. Glass, having a typical mass density of $\rho = 2500 \text{ kg/m}^3$, also supports the transmission of light. What would the elastic modulus of glass have to be to support the transmission of light waves at a speed of $v = 2.0 \cdot 10^8$ m/s? Compare this to the actual elastic modulus of window glass, which is $5 \cdot 10^{10} \text{ N/m}^2$.

## Section 16.2

**16.26** Compare the intensity of sound at the pain level, 120 dB, with that at the whisper level, 20 dB.

**16.27** The sound level in decibels is typically expressed as $\beta = 10 \log (I/I_0)$, but since sound is a pressure wave, the sound level can be expressed in terms of a pressure difference. Intensity depends on the amplitude squared, so the expression is $\beta = 20 \log (P/P_0)$, where $P_0$ is the smallest pressure difference noticeable by the ear: $P_0 = 2 \cdot 10^{-5}$ Pa. A loud rock concert has a sound level of 110 dB, find the amplitude of the pressure wave generated by this concert.

**16.28** At a state championship high school football game, the intensity level of the shout of a single person in the stands at the center of the field is about 50 dB. What would be the intensity level at the center of the field if all 10,000 fans at the game shouted from roughly the same distance away from that center point?

•**16.29** Two people are talking at a distance of 3.0 m from where you are, and you measure the sound intensity as $1.1 \cdot 10^{-7} \text{ W/m}^2$. Another student is 4.0 m away from the talkers. What sound intensity does the other student measure?

•**16.30** An excited child in a campground lets out a yell. Her parent, 1.2 m away, hears it as a 90.0-dB sound. A climber sits 850 m away from the child, on top of a nearby mountain. How loud does the yell sound to the climber?

•**16.31** While enjoyable, rock concerts can damage people's hearing. In the front row at a rock concert, 5 m from the speaker system, the sound intensity is 145 dB. How far back would you have to sit for the sound intensity to drop below the recommended safe level of 90 dB?

## Section 16.3

**16.32** Two sources, A and B, emit a sound of a certain wavelength. The sound emitted from both sources is detected at a point away from the sources. The sound from source A is a distance $d$ from the observation point, whereas the sound from source B has to travel a distance of $3\lambda$. What is the largest value of the wavelength, in terms of $d$, for the maximum sound intensity to be detected at the observation point? If $d = 10.0$ m and the speed of sound is 340 m/s, what is the frequency of the emitted sound?

**16.33** A college student is at a concert and really wants to hear the music, so she sits between two in-phase loudspeakers, which point toward each other and are 50.0 m apart. The speakers emit sound at a frequency of 490. Hz. At the midpoint between the speakers, there will be constructive interference, and the music will be at its loudest. At what distance closest to the midpoint could she also sit to experience the loudest sound?

**16.34** A string of a violin produces 2 beats per second when sounded along with a standard fork of frequency 400. Hz. The beat frequency increases when the string is tightened.

a) What was the frequency of the violin at first?

b) What should be done to tune the violin?

•**16.35** You are standing against a wall opposite two speakers that are separated by 3.0 m, as shown in the figure. The two speakers begin emitting a 343-Hz tone in phase. Where along the far wall should you stand so that the sound from the speakers is as soft as possible? Be specific; how far away from a spot centered between the speakers will you be? The far wall is 120 m from the wall that has the speakers. (Assume the walls are good absorbers, and therefore, the contribution of reflections to the perceived sound is negligible.)

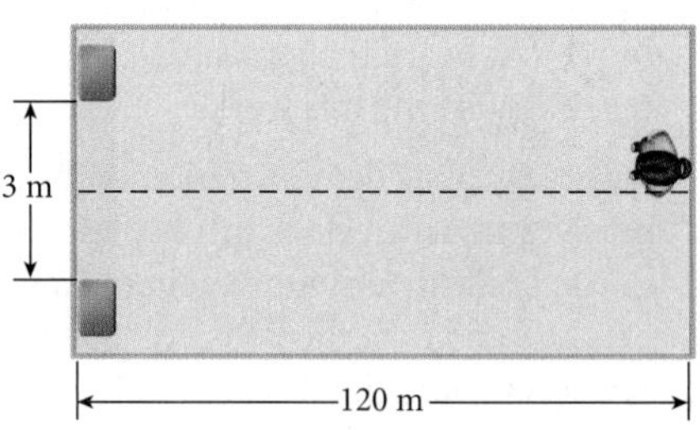


•**16.36** You are playing a note that has a fundamental frequency of 400 Hz on a guitar string of length 50 cm. At the same time, your friend plays a fundamental note on an open organ pipe, and 4 beats per seconds are heard. The mass per unit length of the string is 2 g/m. Assume the velocity of sound is 343 m/s.

a) What are the possible frequencies of the open organ pipe?

b) When the guitar string is tightened, the beat frequency decreases. Find the original tension in the string.

c) What is the length of the organ pipe?

•**16.37** Two 100.0-W speakers, A and B, are separated by a distance $D = 3.6$ m. The speakers emit in-phase sound

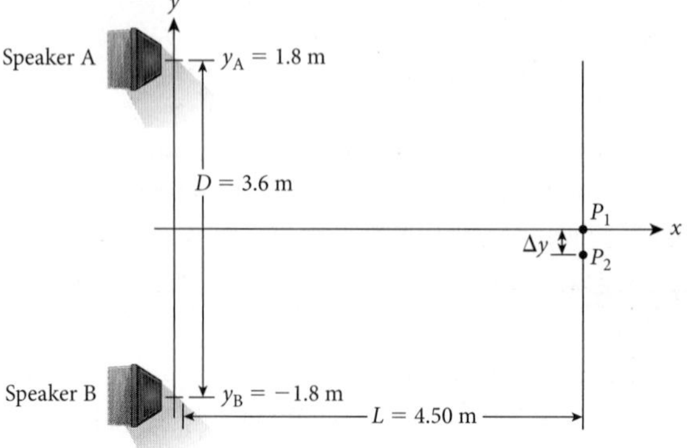

waves at a frequency $f = 10{,}000.0$ Hz. Point $P_1$ is located at $x_1 = 4.50$ m and $y_1 = 0$ m; point $P_2$ is located at $x_2 = 4.50$ m and $y_2 = -\Delta y$. Neglecting speaker B, what is the intensity, $I_{A1}$ (in W/m$^2$), of the sound at point $P_1$ due to speaker A? Assume that the sound from the speaker is emitted uniformly in all directions. What is this intensity in terms of decibels (sound level, $\beta_{A1}$)? When both speakers are turned on, there is a maximum in their combined intensities at $P_1$. As one moves toward $P_2$, this intensity reaches a single minimum and then becomes maximized again at $P_2$. How far is $P_2$ from $P_1$, that is, what is $\Delta y$? You may assume that $L \gg \Delta y$ and that $D \gg \Delta y$, which will allow you to simplify the algebra by using $\sqrt{a \pm b} \approx a^{1/2} \pm \frac{b}{2a^{1/2}}$ when $a \gg b$.

## Section 16.4

**16.38** A policeman with a very good ear and a good understanding of the Doppler effect stands on the shoulder of a freeway assisting a crew in a 40-mph work zone. He notices a car approaching that is honking its horn. As the car gets closer, the policeman hears the sound of the horn as a distinct B4 tone (494 Hz). The instant the car passes by, he hears the sound as a distinct A4 tone (440 Hz). He immediately jumps on his motorcycle, stops the car, and gives the motorist a speeding ticket. Explain his reasoning.

**16.39** A meteorite hits the surface of the ocean at a speed of 8800 m/s. What are the shock wave angles it produces (a) in the air just before hitting the ocean surface, and (b) in the ocean just after entering? Assume the speed of sound in air and in water is 343 m/s and 1560 m/s, respectively.

**16.40** A train whistle emits a sound at a frequency $f = 3000$ Hz when stationary. You are standing near the tracks when the train goes by at a speed of $v = 30$ m/s. What is the magnitude of the change in the frequency ($|\Delta f|$) of the whistle as the train passes? (Assume that the speed of sound is $v = 343$ m/s.)

•**16.41** You are driving along a highway at 30.0 m/s when you hear a siren. You look in the rear-view mirror and see a police car approaching you from behind with a constant speed. The frequency of the siren that you hear is 1300 Hz. Right after the police car passes you, the frequency of the siren that you hear is 1280 Hz.

a) How fast was the police car moving?

b) You are so nervous after the police car passes you that you pull off the road and stop. Then you hear another siren, this time from an ambulance approaching from behind. The frequency of its siren that you hear is 1400 Hz. Once it passes, the frequency is 1200 Hz. What is the actual frequency of the ambulance's siren?

•**16.42** A bat flying toward a wall at a speed of 7.0 m/s emits an ultrasound wave with a frequency of 30.0 kHz. What frequency does the reflected wave have when it reaches the flying bat?

•**16.43** A plane flies at Mach 1.3, and its shock wave reaches a man on the ground 50 s after the plane passes directly overhead.

a) What is the Mach angle?

b) What is the altitude of the plane? The speed of the plane is 300 m/s.

•**16.44** You are traveling in a car toward a hill at a speed of 40.0 mph. The car's horn emits sound waves of frequency 250 Hz, which move with a speed of 340 m/s.

a) Determine the frequency with which the waves strike the hill.

b) What is the frequency of the reflected sound waves you hear?

c) What is the beat frequency produced by the direct and the reflected sounds at your ears?

•**16.45** A car is parked at a railroad crossing. A train passes, and the driver of the car records the time dependence of the frequency of the sound emitted by the train, as shown in the figure.

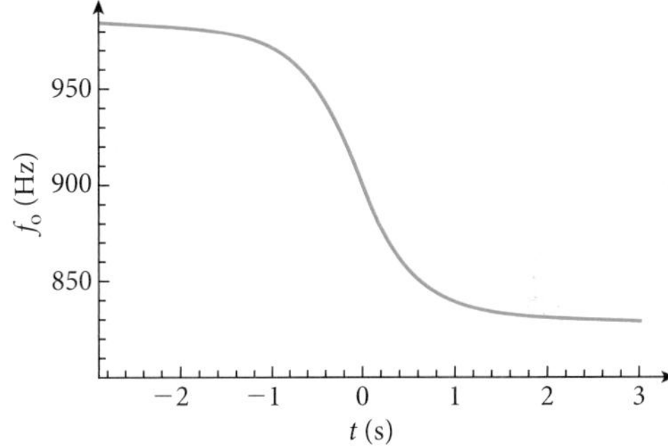


a) What is the frequency heard by someone riding on the train?

b) How fast is the train moving?

c) How far away from the train tracks is the driver of the car? $\left( Hint: \left.\frac{df_o}{dt}\right|_{t=0} = -\frac{fv^2}{b\,v_{sound}}. \right)$

••**16.46** Many towns have tornado sirens, large elevated sirens used to warn locals of imminent tornados. In one small town, a siren is elevated 100 m off the ground. A car is being driven at 100 km/h directly away from this siren while it is emitting a 440-Hz sound. What is the frequency of the sound heard by the driver as a function of the distance from the siren at which he starts? Plot this frequency as a function of the car's position up to 1000 m. Explain this plot in terms of the Doppler effect.

## Section 16.5

**16.47** A standing wave in a pipe with both ends open has a frequency of 440 Hz. The next higher harmonic has a frequency of 660 Hz.

a) Determine the fundamental frequency.

b) How long is the pipe?

**16.48** A bugle can be represented by a cylindrical pipe of length $L = 1.35$ m. Since the ends are open, the standing waves produced in the bugle have antinodes at the open

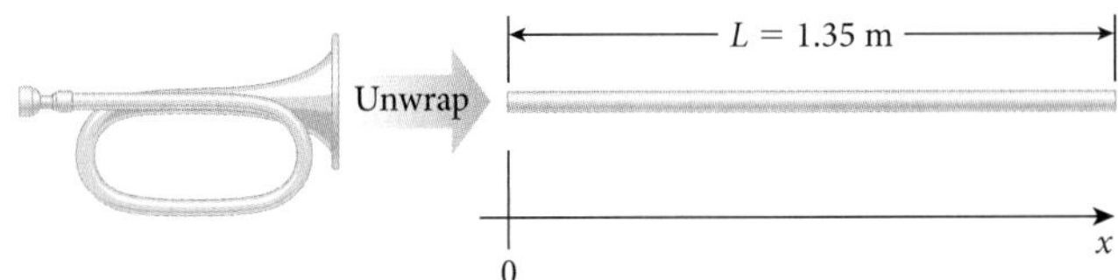

ends, where the air molecules move back and forth the most. Calculate the longest three wavelengths of standing waves inside the bugle. Also calculate the three lowest frequencies and the three longest wavelengths of the sound that is produced in the air around the bugle.

16.49 A soprano sings the note C6 (1046 Hz) across the mouth of a soda bottle. To excite a fundamental frequency in the soda bottle equal to this note, describe how far the top of the liquid must be below the top of the bottle.

**16.50** A thin aluminum rod of length $L = 2.0$ m is clamped at its center. The speed of sound in aluminum is 5000 m/s. Find the lowest resonance frequency for vibrations in this rod.

•**16.51** Find the resonance frequency of the ear canal. Treat it as a half-open pipe of diameter 8.0 mm and length 25 cm. Assume that the temperature inside the ear canal is body temperature (37 °C).

•**16.52** A half-open pipe is constructed to produce a fundamental frequency of 262 Hz when the air temperature is 22 °C. It is used in an overheated building when the temperature is 35 °C. Neglecting thermal expansion in the pipe, what frequency will be heard?

## Additional Problems

16.53 A car horn emits a sound of frequency 400.0 Hz. The car is traveling at 20.0 m/s toward a stationary pedestrian when the driver sounds the horn. What frequency does the pedestrian hear?

**16.54** An observer stands between two sound sources. Source A is moving away from the observer, and source B is moving toward the observer. Both sources emit sound of the same frequency. If both sources are moving with a speed, $v_{sound}/2$, what is the ratio of the frequencies detected by the observer?

**16.55** An F16 jet takes off from the deck of an aircraft carrier. A diver 1 km away from the ship is floating in the water with one ear below the surface and one ear above the surface. How much time elapses between the time he initially hears the jet's engines in one ear and the time he initially hears them in the other ear?

**16.56** A train has a horn that produces a sound with a frequency of 311 Hz. Suppose you are standing next to a track when the train, horn blaring, approaches you at a speed of 22.3 m/s. By how much will the frequency of the sound you hear shift as the train passes you?

16.57 You have embarked on a project to build a five-pipe wind chime. The notes you have selected for your open pipes are presented in the table.

| Note | Frequency (Hz) | Length (m) |
|---|---|---|
| G4 | 392 | |
| A4 | 440 | |
| B4 | 494 | |
| F5 | 698 | |
| C6 | 1046 | |

Calculate the length for each of the five pipes to achieve the desired frequency, and complete the table.

**16.58** Two trains are traveling toward each other in still air at 25.0 m/s relative to the ground. One train is blowing a whistle at 300. Hz. The speed of sound is 343 m/s.

a) What frequency is heard by a man on the ground facing the whistle-blowing train?

b) What frequency is heard by a man on the other train?

•**16.59** At a distance of 20 m from a sound source, the intensity of the sound is 60 dB. What is the intensity (in dB) at a point 2.0 m from the source? Assume that the sound radiates equally in all directions from the source.

•**16.60** Two vehicles carrying speakers that produce a tone of frequency 1000.0 Hz are moving directly toward each other. Vehicle A is moving at 10.00 m/s and vehicle B is moving at 20.00 m/s. Assume the speed of sound in air is 343.0 m/s, and find the frequencies that the driver of each vehicle hears.

•16.61 You are standing between two speakers that are separated by 80 m. Both speakers are playing a pure tone of 286 Hz. You begin running directly toward one of the speakers, and you measure a beat frequency of 10 Hz. How fast are you running?

•**16.62** In a sound interference experiment, two identical loudspeakers are placed 4 m apart, facing in a direction perpendicular to the line that connects them. A microphone attached to a carrier sliding on a rail picks up the sound from the speakers at a distance of 400 m, as shown in the figure. The two speakers are driven in phase by the same signal generator at a frequency of 3400 Hz. Assume that the speed of sound in air is 340 m/s.

a) At what point(s) on the rail should the microphone be located for the sound reaching it to have maximum intensity?

b) At what point(s) should it be located for the sound reaching it to be zero?

c) What is the separation between two points of maximum intensity?

d) What is the separation between two points of zero intensity?

e) How would things change if the two loudspeakers produced sounds of the same frequency but different intensities?

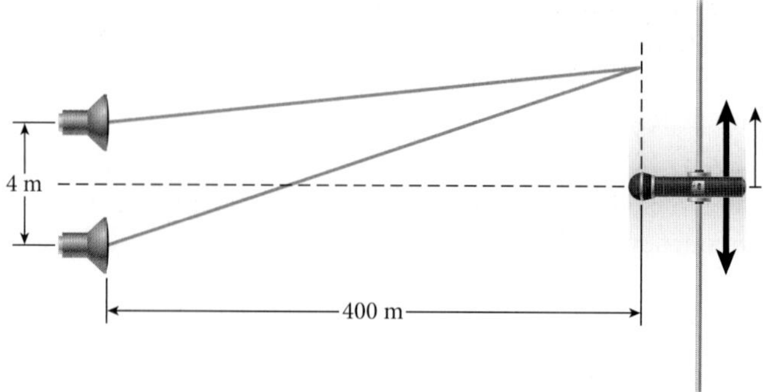

•**16.63** A car traveling at 25 m/s honks its horn as it directly approaches the side of a large building. The horn produces a long sustained note of frequency $f_0 = 230$ Hz. The sound is reflected off the building back to the car's driver. The sound wave from the original note and that reflected off the building combine to create a beat frequency. What is the beat frequency that the driver hears (which tells him that he had better hit the brakes!)?

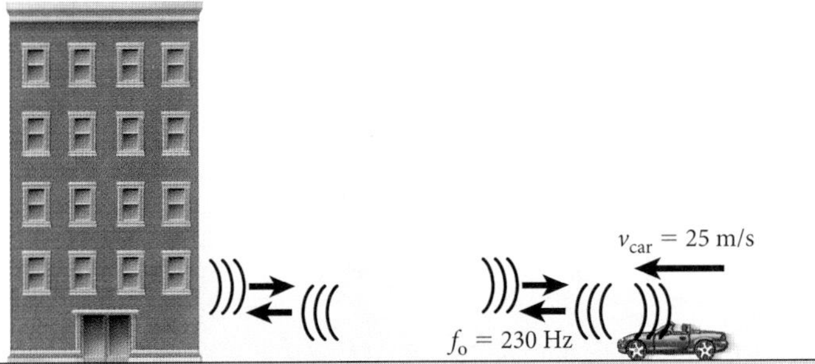


•**16.64** Two identical half-open pipes each have a fundamental frequency of 500 Hz. What percentage change in the length of one of the pipes will cause a beat frequency of 10 Hz when they are sounded simultaneously?

•**16.65** A source traveling to the right at a speed of 10.00 m/s emits a sound wave at a frequency of 100.0 Hz. The sound wave bounces off of a reflector, which is traveling to the left at a speed of 5.00 m/s. What is the frequency of the reflected sound wave detected by a listener back at the source?

•**16.66** In a suspense-thriller movie, two submarines, X and Y, approach each other, traveling at 10 m/s and 15 m/s, respectively. Submarine X "pings" submarine Y by sending a sonar wave of frequency 2000 Hz. Assume that the sound travels at 1500 m/s in the water.

a) Determine the frequency of the sonar wave detected by submarine Y.

b) What is the frequency detected by submarine X for the sonar wave reflected off submarine Y?

c) Suppose the submarines barely miss each other and begin to move away from each other. What frequency does submarine Y detect from the pings sent by X? How much is the Doppler shift?

••**16.67** Consider a sound wave (that is, a longitudinal displacement wave) in an elastic medium with Young's modulus $Y$ (solid) or bulk modulus $B$ (fluid) and unperturbed density $\rho_0$. Suppose this wave is described by the wave function $\delta x(x,t)$, where $\delta x$ denotes the displacement of a point in the medium from its equilibrium position, $x$ is position along the path of the wave, and $t$ is time. The wave can also be regarded as a pressure wave, described by wave function $\delta p(x,t)$, where $\delta p$ denotes the change of pressure in the medium from its equilibrium value.

a) Find the relationship between $\delta p(x,t)$ and $\delta x(x,t)$, in general.

b) If the displacement wave is a pure sinusoidal function, with amplitude $A$, wave number $\kappa$, and angular frequency $\omega$, given by $\delta x(x,t) = A\cos(\kappa x - \omega t)$, what is the corresponding pressure wave function, $\delta p(x,t)$? What is the amplitude of the pressure wave?

••**16.68** Consider the sound wave of Problem 16.67.

a) Find the intensity, $I$, of the general wave, in terms of the wave functions $\delta x(x,t)$ and $\delta p(x,t)$.

b) Find the intensity of the sinusoidal wave of part (b), in terms of the displacement and amplitude of the pressure wave.

•**16.69** Using the results of Problems 16.67 and 16.68, determine the displacement and amplitude of the pressure wave corresponding to a pure tone of frequency $f = 1.000$ kHz in air (density = 1.20 kg/m$^3$, speed of sound 343. m/s), at the threshold of hearing ($\beta = 0.00$ dB) and at the threshold of pain ($\beta = 120.$ dB).

17

PART 4 THERMAL PHYSICS

# Temperature



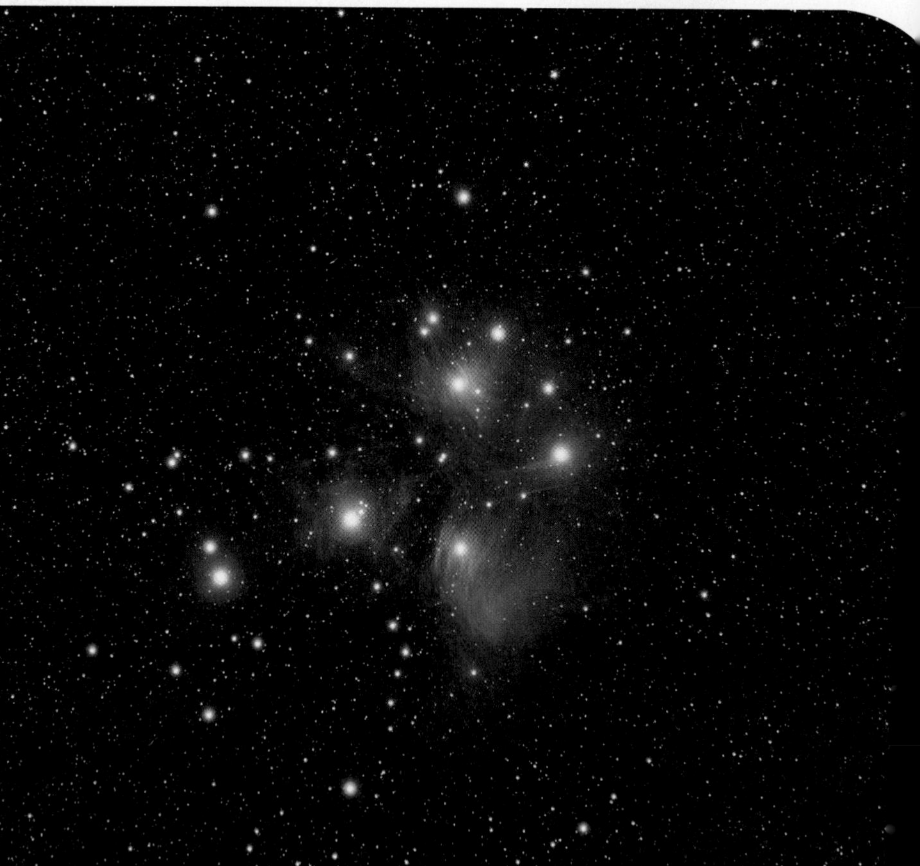

**FIGURE 17.1** The star cluster known as the Pleiades, or Seven Sisters.

## WHAT WE WILL LEARN

- Temperature is measured using any of several different physical properties of certain materials.
- The Fahrenheit temperature scale sets the temperature of the freezing point of water at 32 °F and that of the boiling point of water at 212 °F.
- The Celsius temperature scale sets the temperature of the freezing point of water at 0 °C and that of the boiling point of water at 100 °C.
- The Kelvin temperature scale is defined in terms of an absolute zero temperature, or the lowest temperature at which any object could theoretically exist. In the Kelvin temperature scale, the freezing point of water is 273.15 K, and the boiling point of water is 373.15 K.
- Heating a long, thin metal rod causes its length to increase linearly with the temperature, measured in K.
- Heating a liquid generally causes its volume to increase linearly with the temperature, measured in K.
- The Earth's average surface temperature was 14.4 °C in 2005 and increased by 1 °C in the preceding 155 years.
- Analysis of the cosmic microwave background radiation shows the temperature of the universe to be 2.725 K.

The Pleiades star cluster, shown in Figure 17.1, can be seen with the naked eye and was known to the ancient Greeks. What they could not know is that these stars exhibit some of the highest temperatures that occur in nature. Surface temperatures of these stars range from 4,000 to 10,000 °C, depending on size and other factors. However, interior temperatures can reach over 10 million °C, hot enough to vaporize any substance. On the other end of the temperature range, space itself, far from any stars, registers a temperature of roughly −270 °C.

This chapter begins our study of thermodynamics, including the concepts of temperature, heat, and entropy. In its widest sense, thermodynamics is the physics of energy and energy transfer—how energy is stored, how it is transformed from one kind to another, and how it can be made available to do work. We'll examine energy at the atomic and molecular level as well as at the macroscopic level of engines and machines.

This chapter looks at temperature—how it is defined and measured and how changes in temperature can affect objects. We'll consider various scales with which to quantify temperature as well as the ranges of temperatures observed in nature and in the lab.

In practical terms, it is almost impossible to describe temperature without also discussing heat, which is the subject of Chapter 18. Be sure you keep in mind the differences between these concepts as you study the next few chapters.

## 17.1 Definition of Temperature

Temperature is a concept we all understand from experience. We hear weather forecasters tell us that the temperature will be 72 °F today. We hear doctors tell us that our body temperature is 98.6 °F. When we touch an object, we can tell whether it is hot or cold. If we put a hot object in contact with a cold object, the hot object will cool off and the cold object will warm up. If we measure the temperatures of the two objects after some time has passed, they will be equal. The two objects are then in **thermal equilibrium.**

**Heat** is the transfer of a type of energy. This energy, sometimes called **thermal energy,** is in the form of random motion of the atoms and molecules making up the matter being studied. Chapter 18 will quantify the concept of heat as thermal energy that is transferred because of a temperature difference.

The **temperature** of an object is related to the tendency of the object to transfer heat to or from its surroundings. Heat will transfer from the object to its surroundings if the object's temperature is higher than that of its surroundings. Heat will transfer to the object if its temperature is less than its surroundings. Note that cold is simply the absence of heat; there is no such thing as a flow of "coldness" between an object and its surroundings. If an object feels cold to your touch, this is simply a consequence of heat being transferred from your fingers to the object. (This is the macroscopic definition of temperature; we'll see in

Chapter 19 that on a microscopic level, temperature is proportional to the kinetic energy of random motion of particles.)

Measuring temperature relies on the fact that if two objects are in thermal equilibrium with a third object, they are in thermal equilibrium with each other. This third object could be a **thermometer,** which measures temperature. This idea, often called the **Zeroth Law of Thermodynamics,** defines the concept of temperature and underlies the ability to measure temperature. That is, in order to find out if two objects have the same temperature, you do not need to bring them into thermal contact and monitor whether thermal energy transfers (which may be hard or even impossible in some instances). Instead, you can use a thermometer and measure each object's temperature separately; if your readings are the same, you know that the objects have the same temperature.

Temperature measurements can be taken using any of several common scales. Let's examine these.

## Temperature Scales

### Fahrenheit Scale

Several systems have been proposed and used to quantify temperature; the most widely used are the Fahrenheit, Celsius, and Kelvin scales. The **Fahrenheit temperature scale** was proposed in 1724 by Gabriel Fahrenheit, a German-born scientist living in Amsterdam. Fahrenheit also invented the mercury-expansion thermometer. The Fahrenheit scale went through several iterations. Fahrenheit finally defined the unit of the Fahrenheit scale (°F) by fixing 0 °F for the temperature of an ice-salt bath, 32 °F for the freezing point of water, and 96 °F for the temperature of the human body as measured under the arm. Later, other scientists defined the boiling point of water as 212 °F. This temperature scale is used widely in the United States.

### Celsius Scale

Anders Celsius, a Swedish astronomer, proposed the **Celsius temperature scale,** often called the *centigrade scale,* in 1742. Several iterations of this scale resulted in the Celsius scale unit (°C) being determined by setting the freezing point of water at 0 °C and the boiling point of water at 100 °C (at normal atmospheric pressure). This temperature scale is used worldwide, except in the United States.

Ice water
0 °C

(a)

Boiling water
100 °C

(b)

**FIGURE 17.2** A hollow aluminum sphere fitted with a pressure gauge and filled with nitrogen gas. (a) The container is held at 0 °C by placing it in ice water. (b) The container is held at 100 °C by placing it in boiling water.

### Kelvin Scale

In 1848, William Thomson (Lord Kelvin), a British physicist, proposed another temperature scale, which is now called the **Kelvin temperature scale.** This scale is based on the existence of **absolute zero,** the minimum possible temperature.

The behavior of the pressure of gases at fixed volume as a function of temperature was studied, and the observed behavior was extrapolated to zero pressure to establish this absolute zero temperature. To see how this works, suppose we have a fixed volume of nitrogen gas in a hollow spherical aluminum container connected to a pressure gauge (Figure 17.2). There is enough nitrogen gas in the container that when the container is immersed in an ice-water bath, the pressure gauge reads 0.200 atm. We then place the container in boiling water. The pressure gauge now reads 0.273 atm. Therefore, the pressure has increased while the volume has stayed constant. Suppose we repeat this procedure with different starting pressures. Figure 17.3 summarizes the results, with four solid lines representing the different starting pressures. You can see that the pressure of the gas goes down as the temperature decreases. Conversely, lowering the pressure of the gas must lower its temperature. Theoretically, the lowest temperature of the gas can be determined by extrapolating the measured behavior until the pressure becomes zero. The dashed lines in Figure 17.3 shows the extrapolations.

The relationship among pressure, volume, and temperature for gas is the focus of Chapter 19. However, for now, you only need to understand the following observation: The data from the four sets of measurements starting at different initial pressures extrapolate to the same temperature at zero pressure. This temperature is called *absolute zero* and corresponds to −273.15 °C. When researchers attempt to lower the temperature of real nitrogen gas to

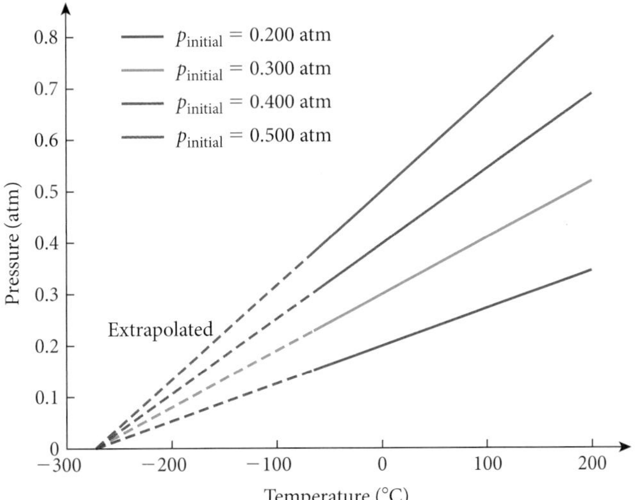


**FIGURE 17.3** Solid lines: Pressures of a gas measured at fixed volume and different temperatures. Each line represents an experiment starting at a different initial pressure. Dashed lines: Extrapolation of the pressure of nitrogen gas in a fixed volume as the temperature is lowered.

very low temperatures, the linear relationship breaks down because the gaseous nitrogen liquefies, bringing subsequent interactions between the nitrogen molecules. In addition, different gases show slightly different behavior at very low temperatures. However, for low pressures and relatively high temperatures, this general result stands.

Absolute zero is the lowest temperature at which matter could theoretically exist. (Experimentally, it is impossible to reach absolute zero, just as it is impossible to build a perpetual motion machine.) We'll see in Chapter 19 that temperature corresponds to motion at the atomic and molecular level, so making an object reach absolute zero would imply that all motion in the atoms and molecules of the object ceases. However, in Chapters 36 and 37, we'll see that some motion of this type is required by quantum mechanics. This is sometimes called the **Third Law of Thermodynamics,** and means that absolute zero is never actually attainable.

Kelvin used the size of the Celsius degree (°C) as the size of the unit of his temperature scale, now called the **kelvin** (K). On the Kelvin scale, the freezing point of water is 273.15 K and the boiling point of water is 373.15 K. This temperature scale is used in many scientific calculations, as we'll see in the next few chapters. Because of these considerations, the kelvin is the standard SI unit for temperature. To achieve greater consistency, scientists have proposed defining the kelvin in terms of other fundamental constants, rather than in terms of the properties of water. These new definitions are scheduled to take effect by the year 2011.

### 17.1 In-Class Exercise

Which of the following temperatures is the coldest?

a) 10 °C

b) 10 °F

c) 10 K

### 17.2 In-Class Exercise

Which of the following temperatures is the warmest?

a) 300 °C

b) 300 °F

c) 300 K

## 17.2 Temperature Ranges

Temperature measurements span a huge range (shown in Figure 17.4, using a logarithmic scale), from the highest measured temperatures ($2 \cdot 10^{12}$ K), observed in relativistic heavy ion collisions (RHIC), to the lowest measured temperatures ($1 \cdot 10^{-10}$ K), observed in spin systems of rhodium atoms. The temperature in the center of the Sun is estimated to be $15 \cdot 10^{6}$ K, and at the surface of the Sun, it has been measured to be 5780 K. The lowest measured air temperature on the Earth's surface is 183.9 K (−89.2 °C) in Antarctica; the highest measured air temperature on the Earth's surface is 330.8 K (57.7 °C) in the Sahara desert in Libya. You can see from Figure 17.4 that the observed temperature range on the surface of the Earth covers only a very small fraction of the range of observed temperatures. The cosmic microwave background radiation left over from the Big Bang 13.7 billion years ago has a temperature of 2.73 K (which is the temperature of "empty" intergalactic space), an observation that will be explained in more detail in later chapters. From Figure 17.4, you can see that temperatures attained in relativistic heavy ion collisions are 300 million times higher than the surface temperature of the Sun and that temperatures of atoms measured in ion traps are more than a billion times lower than the temperature of intergalactic space.

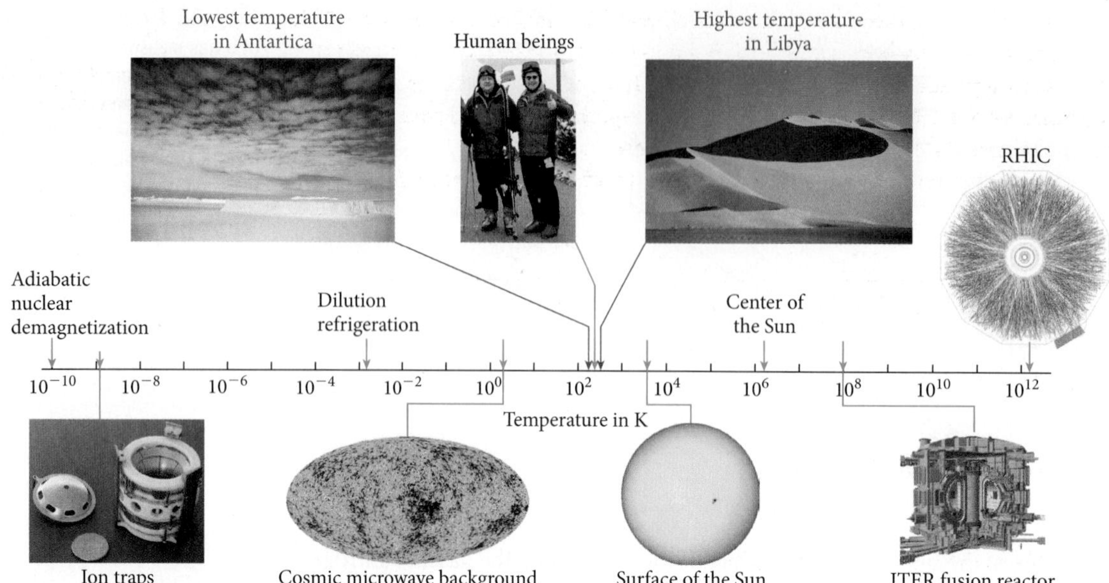


**FIGURE 17.4** Range of observed temperatures, plotted on a logarithmic scale.

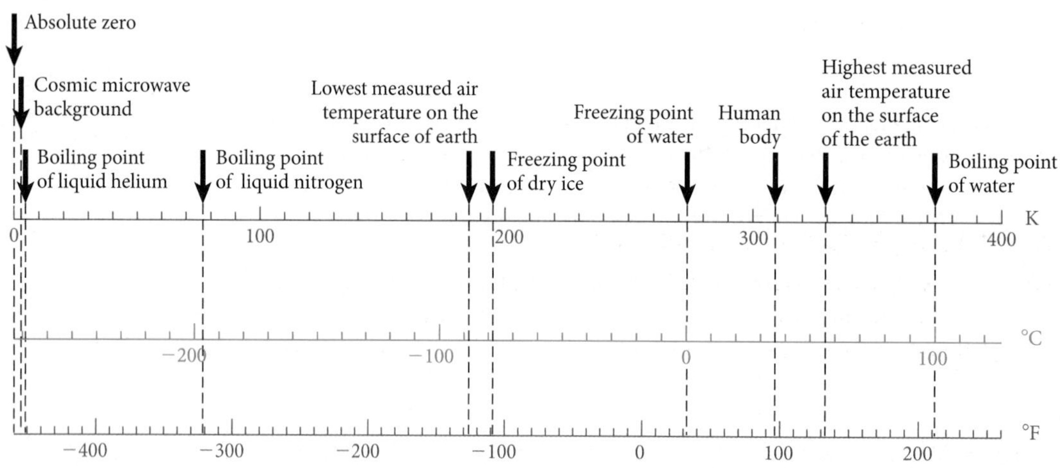


**FIGURE 17.5** Representative temperatures, expressed in the three common temperature scales.

## 17.1 Self-Test Opportunity

At what temperature do the Fahrenheit and Celsius scales have the same numerical value?

Figure 17.5 shows some representative temperatures above absolute zero and below 400 K, expressed in the Fahrenheit, Celsius, and Kelvin scales. This temperature range is shown on a linear scale.

In the following formulas for converting between the various temperature scales, the Fahrenheit temperature is $T_F$, measured in °F; the Celsius temperature is $T_C$, measured in °C; and the Kelvin temperature is $T_K$, measured in K.

Fahrenheit to Celsius:

$$T_C = \frac{5}{9}\left(T_F - 32\ ^\circ\text{F}\right). \tag{17.1}$$

Celsius to Fahrenheit:

$$T_F = \frac{9}{5}T_C + 32\ ^\circ\text{C}. \tag{17.2}$$

Celsius to Kelvin:

$$T_K = T_C + 273.15\ ^\circ\text{C}. \tag{17.3}$$

Kelvin to Celsius:

$$T_C = T_K - 273.15\ \text{K}. \tag{17.4}$$

The mean human body temperature taken orally is 36.8 °C, which corresponds to 98.2 °F. The often quoted mean oral temperature of 98.6 °F corresponds to a 19th-century measurement of 37 °C. This temperature is highlighted on the oral mercury thermometer in Figure 17.6. The difference between 98.6 °F and 98.2 °F thus corresponds to the error due to rounding the Celsius scale measurement of human body temperature to two digits. Table 17.1 lists human body temperature and other representative temperatures, expressed in the three temperature scales.

**FIGURE 17.6** An oral, mercury-expansion thermometer.

**EXAMPLE 17.1 Room Temperature**

Room temperature is often quoted as 72.0 °F.

**PROBLEM**

What is room temperature in the Celsius and Kelvin scales?

**SOLUTION**

Using equation 17.1, we can convert room temperature from degrees Fahrenheit to degrees Celsius:

$$T_C = \tfrac{5}{9}\left(72\ ^\circ\text{F} - 32\ ^\circ\text{F}\right) = 22.2\ ^\circ\text{C}.$$

Using equation 17.3, we can then express room temperature in kelvins as

$$T_K = 22\ ^\circ\text{C} + 273.15\ ^\circ\text{C} = 295.\ \text{K}.$$

## Research at the Low-Temperature Frontier

How are very low temperatures attained in a lab? Researchers start by taking advantage of the properties of gases related to their pressure, volume, and temperature (again, Chapter 19 will go into these properties in much greater depth). In addition, phase

**Table 17.1 Various Temperatures, Expressed in the Three Most Commonly Used Temperature Scales**

| | Fahrenheit (°F) | Celsius (°C) | Kelvin (K) |
|---|---|---|---|
| Absolute zero | –459.67 | –273.15 | 0 |
| Water freezing point | 32 | 0 | 273.15 |
| Water boiling point | 212 | 100 | 373.15 |
| Typical human body temperature | 98.2 | 36.8 | 310 |
| Lowest measured air temperature | –129 | –89.2 | 184 |
| Highest measured air temperature | 136 | 57.8 | 331 |
| Lowest temperature ever measured in a lab | –459.67 | –273.15 | $1.0 \cdot 10^{-10}$ |
| Highest temperature ever measured in a lab | $3.6 \cdot 10^{12}$ | $2 \cdot 10^{12}$ | $2 \cdot 10^{12}$ |
| Cosmic background microwave radiation | –454.76 | –270.42 | 2.73 |
| Liquid nitrogen boiling point | –321 | –196 | 77.3 |
| Liquid helium boiling point | –452 | –269 | 4.2 |
| Temperature at the surface of the Sun | 11,000 | 6,000 | 6,300 |
| Temperature at the center of the Sun | $27 \cdot 10^{6}$ | $15 \cdot 10^{6}$ | $15 \cdot 10^{6}$ |
| Average temperature of the surface of the Earth | 59 | 15 | 288 |
| Temperature at the center of the Earth | 12,000 | 6,700 | 7,000 |

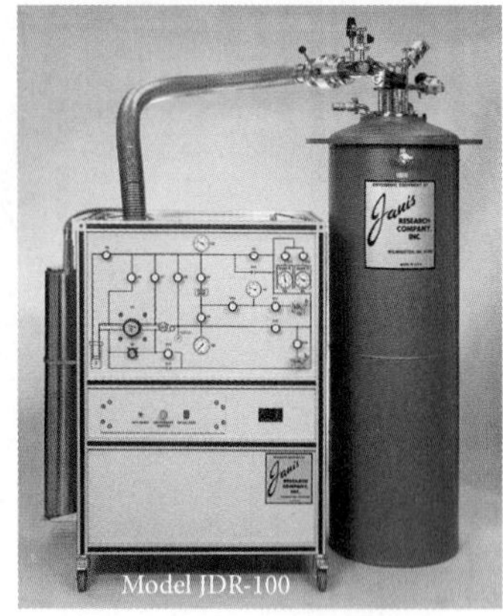


(a)

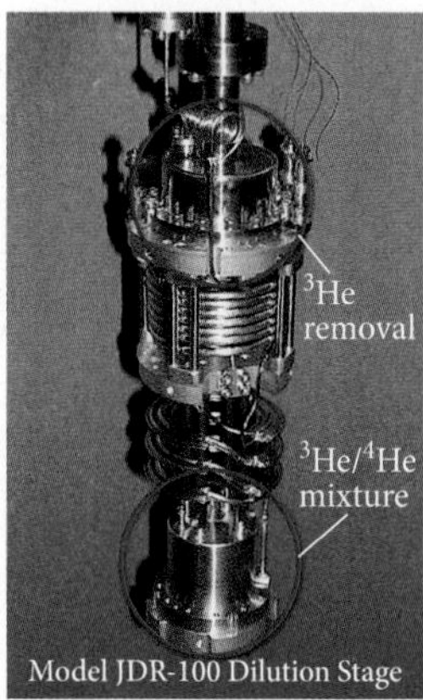


(b)

**FIGURE 17.7** (a) A commercial dilution refrigerator. (b) Interior of the dilution refrigerator with the $^3$He/$^4$He mixture container (mixer) and the $^3$He removal container (still).

**FIGURE 17.8** A laser-cooling device for cooling trapped sodium atoms.

changes (discussed in Chapter 18) from liquid to gas and back to liquid are crucial in producing low temperatures in the lab. For now, all you need to know is that when a liquid evaporates, the energy required to change the liquid to a gas is taken from the liquid—therefore, the liquid cools. For example, air is liquefied through a multistep process in which the air is compressed and cooled by removing the heat produced in the compression stage. Then, the fact that different gases have different temperatures at which they liquefy allows liquid nitrogen and liquid oxygen to be separated. At normal atmospheric pressure, liquid nitrogen has a boiling point of 77 K, while liquid oxygen has a boiling point of 90 K. Liquid nitrogen has many applications, not the least of which is in classroom lecture demonstrations.

Even lower temperatures are achieved using the compression process, with liquid nitrogen extracting the heat as helium is compressed. Liquid helium boils at 4.2 K. Starting with liquid helium and reducing the pressure so that it evaporates, researchers can cool the remaining liquid down to about 1.2 K. For the isotope helium-3 ($^3$He), a temperature as low as 0.3 K can be reached with this evaporative-cooling technique.

To reach still lower temperatures, researchers use an apparatus called a *dilution refrigerator* (see Figure 17.7). A dilution refrigerator uses two isotopes of helium: $^3$He (two protons and one neutron) and $^4$He (two protons and two neutrons). (Isotopes are discussed in Chapter 40 on nuclear physics.) When the temperature of a mixture of liquid $^3$He and liquid $^4$He is reduced below 0.7 K, the two gases separate into a $^3$He-poor phase and a $^4$He-poor phase. Energy is required to move $^3$He atoms into the $^3$He-poor phase. If the atoms can be forced to cross the boundary between the two phases, the mixture can be cooled in a way similar to evaporative cooling. A dilution refrigerator can cool to temperatures around 10 mK, and the best commercial models can reach 2 mK.

Still lower temperatures can be reached with a gas of atoms confined to a trap. To attain these temperatures, researchers use techniques such as laser cooling. Laser cooling takes advantage of the electronic structure of certain atoms. (Atoms, transitions between their electron energy levels, and lasers will be covered in detail in Chapter 38.) Transitions between specific electron energy levels in an atom can cause photons with a wavelength close to visible light to be emitted. The inverse process can also occur, in which the atoms absorb photons. (Light absorption and emission are explained in Chapters 36 through 38.) To cool atoms using laser cooling, researchers use a laser that emits light with a very specific wavelength that is longer than the wavelength of light emitted during an atomic transition. Thus, any atom moving toward the laser experiences a slightly shorter wavelength due to the Doppler effect (see Chapter 16 and Chapter 31), while any atom moving away from the laser experiences a longer wavelength. Atoms moving toward the laser absorb photons, but atoms moving away from the laser are not affected. Atoms that absorb photons re-emit them in random directions, effectively cooling the atoms. Temperatures approaching $10^{-9}$ K have been achieved using this method (Figure 17.8). Steven Chu, Claude Cohen-Tannoudji, and William D. Phillips were awarded the 1997 Nobel Prize in Physics for their work with laser cooling.

The coldest temperatures achieved in a lab have been accomplished using a technique called *adiabatic nuclear demagnetization.* The sample, typically a piece of rhodium metal, is first cooled using a dilution refrigerator. A strong magnetic field is then applied. (Magnetic fields will be discussed in Chapters 27 and 28.) A small amount of heat is produced in the rhodium metal, which is removed by the dilution refrigerator. Then the magnetic field is slowly turned off. As the magnetic field is reduced, the rhodium metal cools further, reaching temperatures as low as $10^{-10}$ K.

## Research at the High-Temperature Frontier

How are high temperatures attained in the lab? The most common methods are by burning fuels or by creating explosions. For example, the yellow part of a candle flame has a temperature of 1470 K.

The ITER fusion reactor to be constructed in Cadarache, France, by 2016 (Figure 17.9) is designed to fuse the hydrogen isotopes deuterium ($^2$H) and tritium ($^3$H) to yield helium ($^4$He) plus a neutron, releasing energy. This process of nuclear fusion (to be discussed in Chapter 40) is similar to the process that the Sun uses to fuse hydrogen into helium and thereby produce energy. In the Sun, the huge gravitational force compresses and heats the hydrogen nuclei to produce fusion. In ITER, magnetic confinement will be used to hold ionized hydrogen in the form of a plasma. A *plasma* is a state of matter in which electrons and nuclei move separately. A plasma cannot be contained in a physical container, because it is so hot that any contact with it would vaporize the container. In the fusion reactor, plasma is heated by flowing a current through it. In addition, the plasma is compressed and further heated by the applied magnetic field. At high temperatures, up to $9.9 \cdot 10^7$ K, and densities, ITER will produce usable energy from the fusion of hydrogen into helium.

It is possible to achieve even higher temperatures in particle accelerators. The highest temperature has been reached by colliding gold nuclei accelerated by the Relativistic Heavy Ion Collider at Brookhaven National Laboratory and by the Large Hadron Collider at the European CERN laboratory. When two gold nuclei collide, a very hot system with a temperature of $2 \cdot 10^{12}$ K is created; this system is also very small ($\sim 10^{-14}$ m) and exists for very short periods of time ($\sim 10^{-22}$ s).

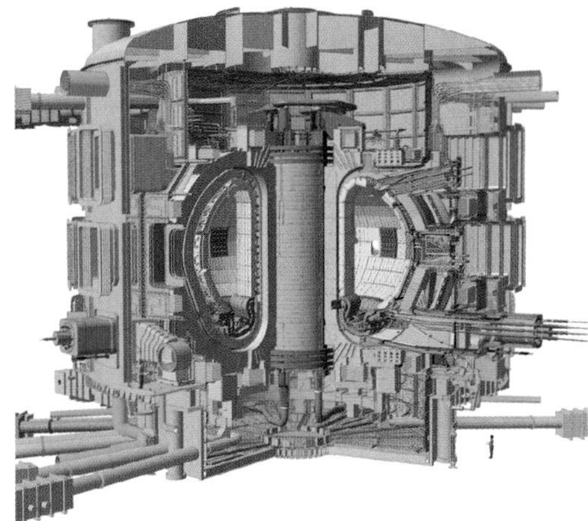

**FIGURE 17.9** Cutaway drawing of the central core of ITER, the plasma fusion reactor to be built in France by the year 2016. For size comparison, a person is shown standing at the bottom.

## 17.3 Measuring Temperature

How are temperatures measured? A device that measures temperature is called a *thermometer.* Any thermometer that can be calibrated directly by using a physical property is called a *primary thermometer.* A primary thermometer does not need calibration to external standard temperatures. An example of a primary thermometer is one based on the speed of sound in a gas. A *secondary thermometer* is one that requires external calibration to standard temperature references. Often, secondary thermometers are more sensitive than primary thermometers.

A common thermometer is the mercury-expansion thermometer, which is a secondary thermometer. This type of thermometer takes advantage of the thermal expansion of mercury (discussed in Section 17.4). Other types of thermometers include bimetallic, thermocouple, chemoluminescence, and thermistor thermometers. It is also possible to measure the temperature of a system by studying the distribution of the speeds of the molecules inside the constituent material.

To measure the temperature of an object or a system using a thermometer, the thermometer must be placed in thermal contact with the object or system. (*Thermal contact* is physical contact that allows relatively fast transfer of heat.) Heat will then transfer to or from the object or system to or from the thermometer, until they have the same temperature. A good thermometer should require as little thermal energy transfer as possible to reach thermal equilibrium, so the temperature measurement does not significantly change the object's temperature. The thermometer should also be easily calibrated so that anyone making the same measurement will get the same temperature.

Calibrating a thermometer requires reproducible conditions. It is difficult to reproduce the freezing point of water exactly, so scientists use a condition called the *triple point of water.* Solid ice, liquid water, and gaseous water vapor can coexist at only one temperature and pressure. By international agreement, the temperature of the triple point of water has been assigned the temperature 273.16 K (and a pressure of 611.73 Pa) for the calibration of thermometers.

### 17.2 Self-Test Opportunity

You have an uncalibrated thermometer, which is intended to be used to measure air temperatures. How would you calibrate the thermometer?

## 17.4 Thermal Expansion

Most of us are familiar in some way with **thermal expansion.** Perhaps you know that you can loosen a metal lid on a glass jar by heating the lid. You may have seen that bridge spans contain gaps in the roadway to allow for the expansion of sections of the bridge in warm weather. Or you may have observed that power lines sag in warm weather.

The thermal expansion of liquids and solids can be put to practical use. Bimetallic strips, which are often used in room thermostats, meat thermometers, and thermal protection devices

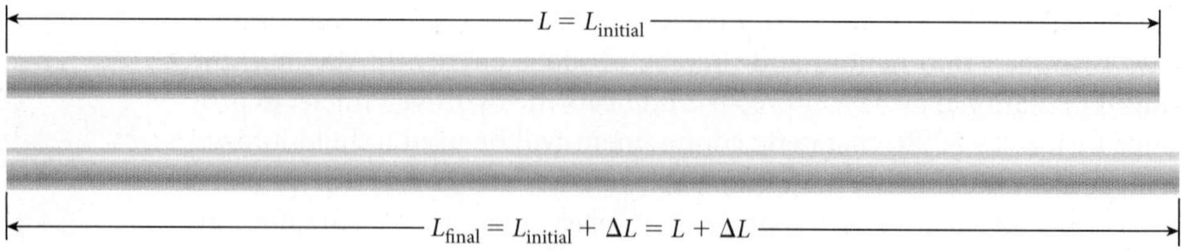

$L = L_{\text{initial}}$

$L_{\text{final}} = L_{\text{initial}} + \Delta L = L + \Delta L$

**FIGURE 17.10** Thermal expansion of a rod with initial length $L$. (The thermally expanded rod at the bottom has been shifted so that the left edges coincide.)

**Table 17.2 The Linear Expansion Coefficients of Some Common Materials**

| Material | $\alpha(10^{-6}\ ^\circ\text{C}^{-1})$ |
|---|---|
| Aluminum | 22 |
| Brass | 19 |
| Concrete | 15 |
| Copper | 17 |
| Diamond | 1 |
| Gold | 14 |
| Lead | 29 |
| Plate glass | 9 |
| Rubber | 77 |
| Steel | 13 |
| Tungsten | 4.5 |

in electrical equipment, take advantage of linear thermal expansion. (A bimetallic strip consists of two thin long strips of different metals, which are welded together.) A mercury thermometer uses volume expansion to provide precise measurements of temperature. Thermal expansion can occur as linear expansion, area expansion, or volume expansion; all three classifications describe the same phenomena.

## Linear Expansion

Let's consider a metal rod of length $L$ (Figure 17.10). If we raise the temperature of the rod by $\Delta T = T_{\text{final}} - T_{\text{initial}}$, the length of the rod increases by the amount $\Delta L = L_{\text{final}} - L_{\text{initial}}$, given by

$$\Delta L = \alpha L \Delta T, \tag{17.5}$$

where $\alpha$ is the **linear expansion coefficient** of the metal from which the rod is constructed and the temperature difference is expressed in degrees Celsius or kelvins. The linear expansion coefficient is a constant for a given material within normal temperature ranges. Some typical linear expansion coefficients are listed in Table 17.2.

### EXAMPLE 17.2 Thermal Expansion of the Mackinac Bridge

The main span of the Mackinac Bridge (Figure 17.11) has a length of 1158 m. The bridge is built of steel. Assume that the lowest possible temperature experienced by the bridge is −50 °C and the highest possible temperature is 50 °C.

**FIGURE 17.11** The Mackinac Bridge across the Straits of Mackinac in Michigan is the third-longest suspension bridge in the United States.

**PROBLEM**

How much room must be made available for thermal expansion of the center span of the Mackinac Bridge?

**SOLUTION**

The linear expansion coefficient of steel is $\alpha = 13 \cdot 10^{-6}\ ^\circ\text{C}^{-1}$. Thus, the total linear expansion of the center span of the bridge that must be allowed for is given by

$$\Delta L = \alpha L \Delta T = \left(13 \cdot 10^{-6}\ ^\circ\text{C}^{-1}\right)\left(1158\ \text{m}\right)\left[50\ ^\circ\text{C} - \left(-50\ ^\circ\text{C}\right)\right] = 1.5\ \text{m}.$$

**DISCUSSION**

A change in length of 1.5 m is fairly large. How is this length change accommodated in practice? (Obviously, we cannot have gaps in the road surface.) The answer lies in expansion joints, which are metal connectors between bridge segments whose parts can move relative to each other. A popular type of expansion joint is the finger joint (see Figure 17.12). The Mackinac Bridge has two large finger joints at the towers to accommodate the expansion of the suspended parts of the roadway and eleven smaller finger joints and five sliding joints across the main span of the bridge.

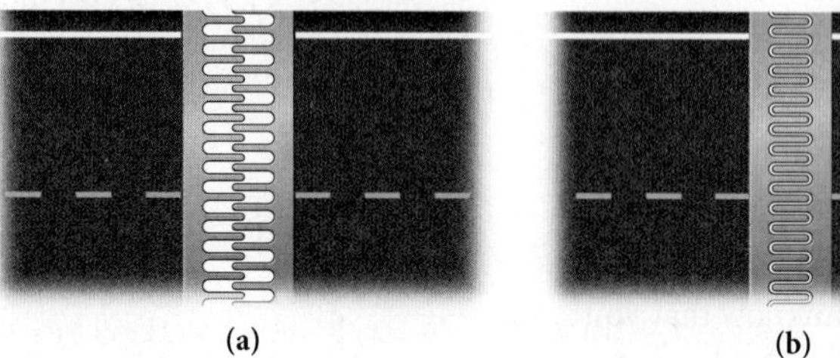


**FIGURE 17.12** Finger joints between road segments: (a) open and (b) closed.

From Table 17.2, you can see that various materials, such as brass and steel, have different linear expansion coefficients. This makes them useful in bimetallic strips. The following solved problem considers the result of heating a bimetallic strip.

### 17.3 In-Class Exercise

A concrete section of a bridge has a length of 10.0 m at 10.0 °C. By how much does the length of the concrete section increase if its temperature increases to 40.0 °C?

a) 0.025 cm

b) 0.051 cm

c) 0.075 cm

d) 0.22 cm

e) 0.45 cm

## SOLVED PROBLEM 17.1 Bimetallic Strip

A straight bimetallic strip consists of a steel strip and a brass strip, each 1.25 cm wide and 30.5 cm long, welded together (see Figure 17.13a). Each strip is $t = 0.500$ mm thick. The bimetallic strip is heated uniformly along its length, as shown in Figure 17.13c. (It doesn't matter that the flame is on the right; the heating is uniform throughout the strip. If the flame were on the left, the strip would bend in the same direction!) The strip curves such that the radius of curvature is $R = 36.9$ cm.

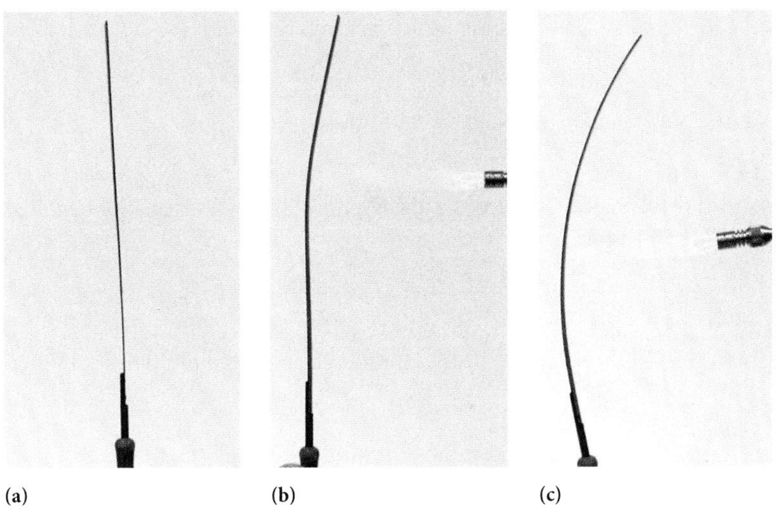


**FIGURE 17.13** A bimetallic strip. (a) The bimetallic strip at room temperature. (b) The bimetallic strip as it begins to be heated by a gas torch (on the right edge of the frame). (c) The bimetallic strip heated to a uniform temperature over its full length.

### PROBLEM

What is the temperature of the bimetallic strip after it is heated?

### SOLUTION

#### THINK

The bimetal strip is constructed from two materials, steel and brass, that have different linear expansion coefficients, listed in Table 17.2. As the temperature of the bimetallic strip increases, the brass expands more than the steel does, so the strip curves toward the steel side. When the bimetallic strip is heated uniformly, both the steel and brass strips lay along the arc of a circle, with the brass strip on the outside and the steel strip on the inside. The ends of the bimetallic strip subtend the same angle, measured from the center of the circle. The arc length of each metal strip then equals the length of the bimetallic strip at room temperature plus the length due to linear thermal expansion. Equating the angle subtended by the steel strip to the angle subtended by the brass strip allows the temperature to be calculated.

#### SKETCH

Figure 17.14 shows the bimetallic strip after it is heated. The angle subtended by the two ends of the strip is $\theta$, and the radius of the inner strip is $r_1$. A portion of a circle with radius $r_1 = 36.9$ cm is superimposed on the curved strip.

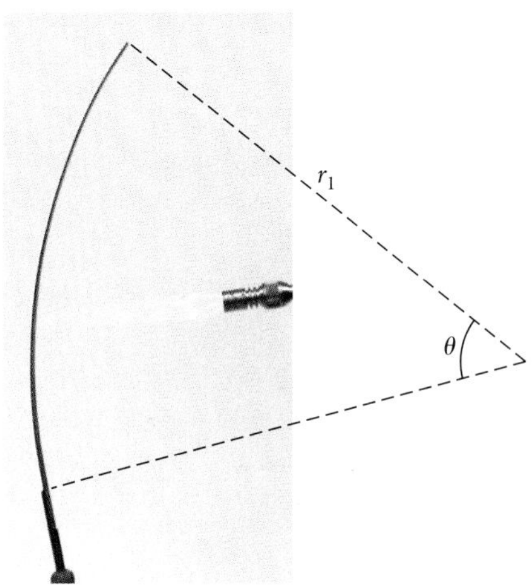


**FIGURE 17.14** The bimetallic strip after it is heated, showing the angle subtended by the two ends of the strip.

#### RESEARCH

The arc length, $s_1$, of the heated steel strip is $s_1 = r_1\theta$, where $r_1$ is the radius of the circle along which the steel strip lies, and $\theta$ is the angle subtended by the steel strip. Also, the arc length, $s_2$, of the heated brass strip is $s_2 = r_2\theta$, where $r_2$ is the radius of

*Continued—*

the circle along which the brass strip lies and $\theta$ is the *same angle* as that subtended by the steel strip. The two radii differ by the thickness, $t$, of the steel strip:

$$r_2 = r_1 + t \Leftrightarrow t = r_2 - r_1. \quad \text{(i)}$$

We can equate the expressions that are equal to the angle subtended by the two strips:

$$\theta = \frac{s_1}{r_1} = \frac{s_2}{r_2}. \quad \text{(ii)}$$

The arc length, $s_1$, of the steel strip after it is heated is given by

$$s_1 = s + \Delta s_1 = s + \alpha_1 s \Delta T = s(1 + \alpha_1 \Delta T),$$

where $s$ is the original length of the bimetallic strip. The factor $\alpha_1$ is the linear expansion coefficient of steel from Table 17.2, and $\Delta T$ is the temperature difference between room temperature and the final temperature of the bimetallic strip. Correspondingly, the arc length, $s_2$, of the brass strip after being heated is given by

$$s_2 = s + \Delta s_2 = s + \alpha_2 s \Delta T = s(1 + \alpha_2 \Delta T),$$

where $\alpha_2$ is the linear expansion coefficient of brass given in Table 17.2.

**SIMPLIFY**

We can substitute the expressions for the arc lengths of the two strips after heating, $s_1$ and $s_2$, into equation (ii) to get

$$\frac{s(1+\alpha_1 \Delta T)}{r_1} = \frac{s(1+\alpha_2 \Delta T)}{r_2}.$$

Dividing both sides of this equation by the common factor $s$ and multiplying by $r_1 r_2$ gives

$$r_2 + r_2 \alpha_1 \Delta T = r_1 + r_1 \alpha_2 \Delta T.$$

We can rearrange this equation and gather common terms to obtain

$$r_2 - r_1 = r_1 \alpha_2 \Delta T - r_2 \alpha_1 \Delta T.$$

Solving this equation for the temperature difference gives

$$\Delta T = \frac{r_2 - r_1}{r_1 \alpha_2 - r_2 \alpha_1}.$$

Using the relationship between the two radii from equation (i) leads to

$$\Delta T = \frac{t}{r_1(\alpha_2 - \alpha_1) - t\alpha_1}. \quad \text{(iii)}$$

**CALCULATE**

From Table 17.2, the linear expansion coefficient for steel is $\alpha_1 = 13 \cdot 10^{-6}\,°\text{C}^{-1}$, and the linear expansion coefficient for brass is $\alpha_2 = 19 \cdot 10^{-6}\,°\text{C}^{-1}$. Inserting the numerical values gives us

$$\Delta T = \frac{0.500 \cdot 10^{-3}\text{ m}}{(0.369\text{ m})(19 \cdot 10^{-6}\,°\text{C}^{-1} - 13 \cdot 10^{-6}\,°\text{C}^{-1}) - (0.500 \cdot 10^{-3}\text{ m})(13 \cdot 10^{-6}\,°\text{C}^{-1})} = 226.5\,°\text{C}.$$

**ROUND**

Taking room temperature to be 20 °C and reporting our result to two significant figures gives us the temperature of the bimetal strip after heating:

$$T = 20\,°\text{C} + \Delta T = 250\,°\text{C}.$$

**DOUBLE-CHECK**

First, we check that the magnitude of the calculated temperature is reasonable. Our answer of 250 °C is well below the melting points of brass (900 °C) and steel (1450 °C), which is important because Figure 17.13c shows that the strip does not melt. Our answer is also significantly above room temperature, which is important because Figure 17.13a shows that the bimetal strip is straight at room temperature.

We can further check that the steel and brass strips do subtend the same angle. The angle subtended by the steel strip is

$$\theta_1=\frac{s_1}{r_1}=\frac{30.5\text{ cm}+(30.5\text{ cm})(13\cdot 10^{-6}\text{ °C}^{-1})(226.5\text{ °C})}{36.9\text{ cm}}=0.829\text{ rads}\equiv 47.5°.$$

The angle subtended by the brass strip is

$$\theta_2=\frac{s_2}{r_2}=\frac{30.5\text{ cm}+(30.5\text{ cm})(19\cdot 10^{-6}\text{ °C}^{-1})(226.5\text{ °C})}{36.9\text{ cm}+0.05\text{ cm}}=0.829\text{ rads}\equiv 47.5°.$$

Note that because the thickness of the strips is small compared with the radius of the circle, we can rewrite equation (iii) as

$$\Delta T\approx\frac{t}{r_1(\alpha_2-\alpha_1)}=\frac{0.500\cdot 10^{-3}\text{ m}}{(0.369\text{ m})(19\cdot 10^{-6}\text{ °C}^{-1}-13\cdot 10^{-6}\text{ °C}^{-1})}=226\text{ °C},$$

which agrees within rounding error with our calculated result. Thus, our answer seems reasonable.

### 17.4 In-Class Exercise

Suppose the bimetallic strip in Figure 17.13 were made of aluminum on the right side and copper on the left side. Which way would the strip bend if it were heated in the same way as shown in the figure? (Consult Table 17.2 for the linear expansion coefficients of the two metals.)

a) It would bend to the right.

b) It would stay straight.

c) It would bend to the left.

## Area Expansion

The effect of a change in temperature on the area of an object is analogous to using a copy machine to enlarge or reduce a picture. Each dimension of the object will change linearly with the change in temperature. Quantitatively, for a square object with side $L$ (Figure 17.15), the area is given by $A = L^2$. Taking the differential of both sides of this equation, we get $dA = 2L\,dL$. If we make the approximations that $\Delta A = dA$ and $\Delta L = dL$, we can write $\Delta A = 2L\Delta L$. Using equation 17.5, we then obtain

$$\Delta A = 2L(\alpha L\Delta T) = 2\alpha A\Delta T. \tag{17.6}$$

Although a square was used to derive equation 17.6, it holds for a change in area of any shape.

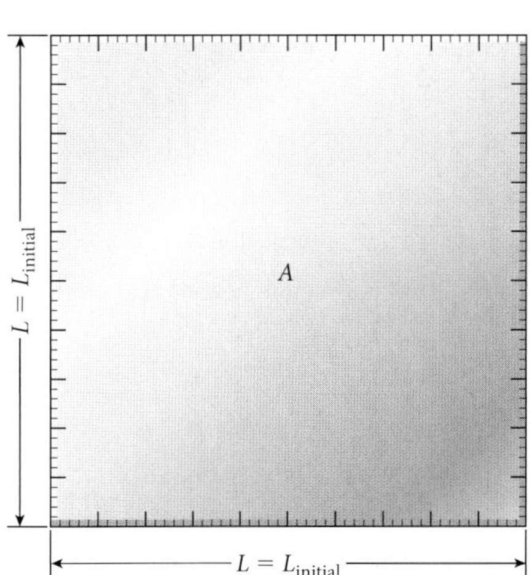


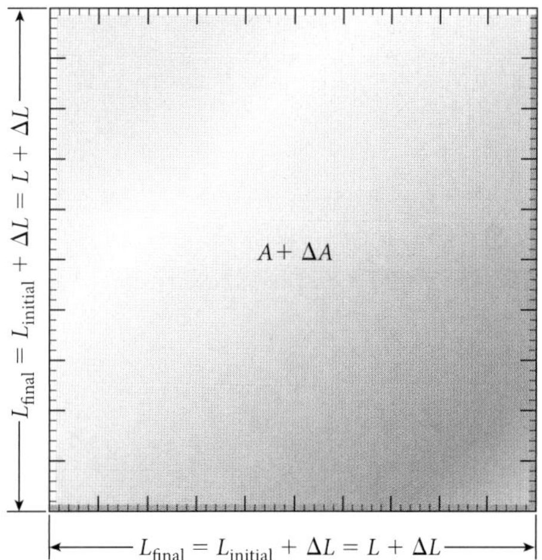


**FIGURE 17.15** Thermal expansion of a square plate with side $L$.

### SOLVED PROBLEM 17.2 Expansion of a Plate with a Hole in It

A brass plate has a hole in it (Figure 17.16a); the hole has a diameter $d = 2.54$ cm. The hole is too small to allow a brass sphere to pass through (Figure 17.6b). However, when the plate is heated from 20.0 °C to 220.0 °C, the brass sphere passes through the hole in the plate.

*Continued—*

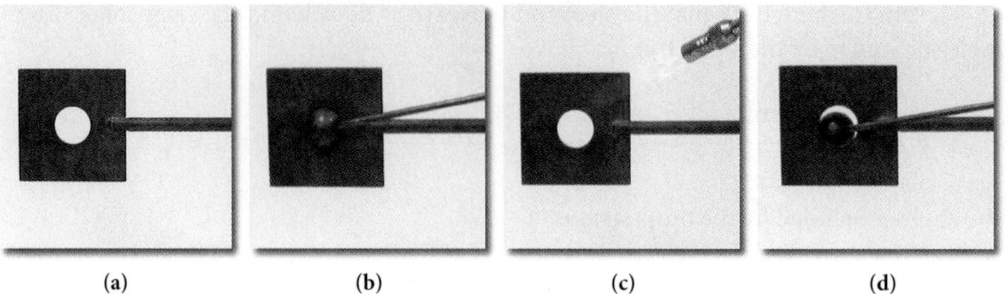

**FIGURE 17.16** (a) The plate before it is heated. (b) A brass sphere will not pass through the hole in the unheated plate. (c) The plate is heated. (d) The same brass sphere passes through the hole in the heated brass plate.

## PROBLEM

How much does the area of the hole in the brass plate increase as a result of heating?

## SOLUTION

### THINK

The area of the brass plate increases as the temperature of the plate increases. Correspondingly, the area of the hole in the plate also increases. We can calculate the increase in the area of the hole using equation 17.6.

### SKETCH

Figure 17.17a shows the brass plate before it is heated, and Figure 17.17b shows the plate after it is heated.

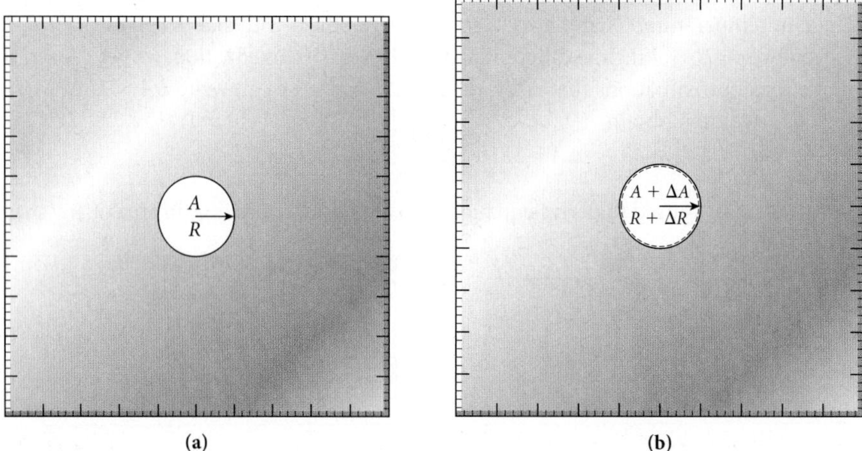


**FIGURE 17.17** The thermal expansion of a plate with a hole in it: (a) before heating; (b) after heating.

### RESEARCH

The area of the plate increases as the temperature increases, as given by equation 17.6. The area of the hole will increase proportionally. This increase in the area of the hole might seem surprising. However, you can convince yourself that the area of the hole will increase when the plate undergoes thermal expansion by looking at Figure 17.17. The plate with a hole at $T = 20$ °C is shown in part (a). The same plate scaled up by 5% in all dimensions is shown in part (b). The dashed circle in the hole in the plate in (b) is the same size as the hole in the original plate. Clearly, the hole in (b) is larger than the hole in (a). The area of the hole at $T = 20.0$ °C is $A = \pi R^2$. Equation 17.6 gives the change in the area of the hole:

$$\Delta A = 2\alpha A \Delta T,$$

where $\alpha$ is the linear expansion coefficient for brass and $\Delta T$ is the change in temperature of the brass plate.

**SIMPLIFY**

Using $A = \pi R^2$ in equation 17.6, we have the change in the area of the hole

$$\Delta A = 2\alpha\left(\pi R^2\right)\Delta T.$$

Remembering that $R = d/2$, we get

$$\Delta A = 2\alpha\pi\left(\tfrac{1}{2}d\right)^2 \Delta T = \frac{\pi\alpha d^2 \Delta T}{2}.$$

**CALCULATE**

From Table 17.2, the linear expansion coefficient of brass is $\alpha = 19 \cdot 10^{-6}\ ^\circ\mathrm{C}^{-1}$. Thus, the change in the area of the hole is

$$\Delta A = \frac{\pi\left(19 \cdot 10^{-6}\ ^\circ\mathrm{C}^{-1}\right)\left(0.0254\ \mathrm{m}\right)^2\left(220.0\ ^\circ\mathrm{C} - 20.0\ ^\circ\mathrm{C}\right)}{2} = 3.85098 \cdot 10^{-6}\ \mathrm{m}^2.$$

**ROUND**

We report our result to two significant figures:

$$\Delta A = 3.9 \cdot 10^{-6}\ \mathrm{m}^2.$$

**DOUBLE-CHECK**

From everyday experience with objects that are heated and cooled, we know that the relative changes in area are not very big. Since the original area of the hole is $A = \pi d^2/4 = \pi(0.0254\ \mathrm{m})^2/4 = 5.07 \cdot 10^{-4}\ \mathrm{m}^2$, we obtain for the fractional change $\Delta A/A = (3.9 \cdot 10^{-6}\ \mathrm{m}^2)/(5.07 \cdot 10^{-4}\ \mathrm{m}^2) = 7.7 \cdot 10^{-3}$, or less than 0.8%. Thus, the magnitude of our answer seems in line with physical intuition.

The change in radius of the hole as the temperature increases is given by

$$\Delta R = \alpha R \Delta T = \left(19 \cdot 10^{-6}\ ^\circ\mathrm{C}^{-1}\right)\left(\frac{0.0254\ \mathrm{m}}{2}\right)\left(200\ ^\circ\mathrm{C}\right) = 4.83 \cdot 10^{-5}\ \mathrm{m}.$$

For that change in radius, the increase in the area of the hole is

$$\Delta A = \Delta\left(\pi R^2\right) = 2\pi R \Delta R = 2\pi\left(\frac{0.0254\ \mathrm{m}}{2}\right)\left(4.8 \cdot 10^{-5}\ \mathrm{m}\right) = 3.85 \cdot 10^{-6}\ \mathrm{m}^2,$$

which agrees within rounding error with our result. Thus, our answer seems reasonable.

## Volume Expansion

Now let's consider the change in volume of an object with a change in temperature. For a cube with side $L$, the volume is given by $V = L^3$. Taking the differential of both sides of this equation, we get $dV = 3L^2 dL$. Making the approximations that $\Delta V = dV$ and $\Delta L = dL$, we can write $\Delta V = 3L^2 \Delta L$. Then, using equation 17.5, we obtain

$$\Delta V = 3L^2\left(\alpha L \Delta T\right) = 3\alpha V \Delta T. \quad (17.7)$$

Because the change in volume with a change in temperature is often of interest, it is convenient to define the **volume expansion coefficient:**

$$\beta = 3\alpha. \quad (17.8)$$

Thus, we can rewrite equation 17.7 as

$$\Delta V = \beta V \Delta T. \quad (17.9)$$

Although a cube was used to derive equation 17.9, it can be generally applied to a change in the volume of any shape. Some typical volume expansion coefficients are listed in Table 17.3.

Equation 17.9 applies to the thermal expansion of most solids and liquids. However, it does not describe the thermal expansion of water. Between 0 °C and about 4 °C, water

**Table 17.3 Volume Expansion Coefficient for Some Common Liquids**

| Material | $\beta(10^{-6}\ ^\circ\mathrm{C}^{-1})$ |
|---|---|
| Mercury | 181 |
| Gasoline | 950 |
| Kerosene | 990 |
| Ethyl alcohol | 1120 |
| Water (1 °C) | −47.8 |
| Water (4 °C) | 0 |
| Water (7 °C) | 45.3 |
| Water (10 °C) | 87.5 |
| Water (15 °C) | 151 |
| Water (20 °C) | 207 |

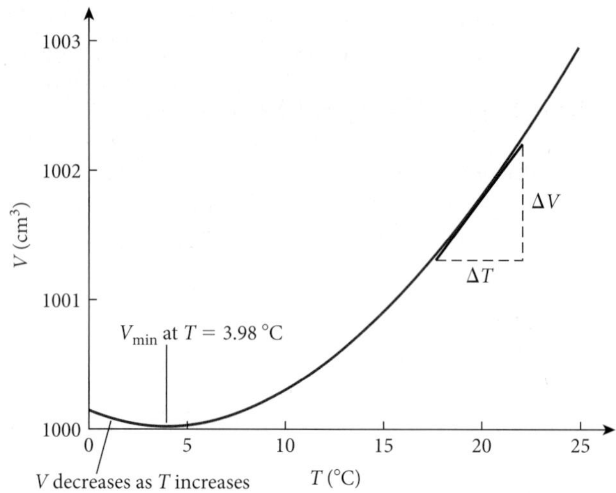


**FIGURE 17.18** The dependence of the volume of 1 kg of water on temperature.

contracts as the temperature increases (Figure 17.18). Water with a temperature above 4 °C is denser than water with a temperature just below 4 °C. This property of water has a dramatic effect on the way a lake freezes in the winter. As the air temperature drops from warm summer to cold winter temperatures, the water in a lake cools from the surface down. The cooler, denser water sinks to the bottom of the lake. However, as the temperature of the water on the surface falls below 4 °C, the downward motion ceases, and the cooler water remains at the surface of the lake, with the denser, warmer water below. The top layer eventually cools to 0 °C and then freezes. Ice is less dense than water, so the ice floats on the water. This newly formed layer of ice acts as insulation, which slows the freezing of the rest of the water in the lake. If water had the same thermal expansion properties as other common materials, instead of freezing from the top down, the lake would freeze from the bottom up, with warmer water remaining at the surface of the lake and cooler water sinking to the bottom. This would mean that lakes would freeze solid more often, and any forms of life in them that could not exist in ice would not survive the winter.

In addition, you can see from Figure 17.18 that the volume of a given amount of water never depends linearly on the temperature. However, the linear dependence of the volume of water on temperature can be approximated by considering a small temperature interval. The slope of the volume/temperature curve is $\Delta V/\Delta T$, so we can extract an effective volume expansion coefficient for small temperature changes. For example, the volume expansion coefficient for water at six different temperatures is given in Table 17.3; note that at 1 °C, $\beta = -47.8 \cdot 10^{-6}\ °C^{-1}$, which means that the volume of water will decrease as the temperature is increased.

### 17.5 In-Class Exercise

You have a metal cube, which you heat. After heating, the area of one of the cube's surfaces has increased by 0.02%. Which statement about the volume of the cube after heating is correct?

a) It has decreased by 0.02%.

b) It has increased by 0.02%.

c) It has increased by 0.01%.

d) It has increased by 0.03%.

e) Not enough information is given to determine the volume change.

### EXAMPLE 17.3 Thermal Expansion of Gasoline

You pull your car into a gas station on a hot summer day, when the air temperature is 40 °C. You fill your empty 55-L gas tank with gasoline that comes from an underground storage tank where the temperature is 12 °C. After paying for the gas, you decide to walk to the restaurant next door and have lunch. Two hours later, you come back to your car and discover that gasoline has spilled out of the gas tank onto the ground.

**PROBLEM**

How much gasoline has spilled?

**SOLUTION**

We know the following: The temperature of the gasoline you put in the tank starts out at 12 °C. The gasoline warms up to the outside air temperature of 40 °C. The volume expansion coefficient of gasoline is $950 \cdot 10^{-6}\ °C^{-1}$.

While you were gone, the temperature of the gasoline changed from 12 °C to 40 °C. Using equation 17.9, we can find the change in volume of the gasoline as the temperature increases:

$$\Delta V = \beta V \Delta T = \left(950 \cdot 10^{-6}\ ^\circ\text{C}^{-1}\right)\left(55\ \text{L}\right)\left(40\ ^\circ\text{C} - 12\ ^\circ\text{C}\right) = 1.5\ \text{L}.$$

Thus, the volume of the gasoline increases by 1.5 liter as the temperature of the gasoline is raised from 12 °C to 40 °C. The gas tank was full when the temperature of the gasoline was 12 °C, so this excess volume spills out of the tank onto the ground.

## 17.5 Surface Temperature of the Earth

A report on daily surface temperatures is part of every weather report in newspapers and TV and radio newscasts. It is clear that it is usually colder at night than during the day, colder in the winter than in the summer, and hotter close to the Equator than near the poles. A current topic of intense discussion is whether the temperature of Earth is rising. A conclusive answer to this question requires data giving appropriate averages. The first average that is useful is over time. Figure 17.19 is a plot of the surface temperature of the Earth, time-averaged over one month (June 1992).

The time-averaged values of the temperature over the entire surface of Earth are obtained by systematically taking temperature measurements over the surface of the Earth, including the oceans. These measurements are then corrected for any systematic biases, such as the fact that many temperature-measuring stations are located in populated areas and many sparsely populated areas have few temperature measurements. After all of the corrections are taken into account, the result is the average surface temperature of Earth in a given year. The current year-round average surface temperature of Earth is approximately 287.5 K (14.4 °C). In Figure 17.20, this average global temperature is plotted for the years 1880 to 2005. You can see that since around 1900, the temperature seems to be increasing with time, indicating global warming. The blue horizontal line in the graph represents the average global temperature for the 20th century, 13.9 °C.

Several models predict that the average global surface temperature of Earth will continue to increase. Although the magnitude of the increase in average global temperature over the past 155 years is around 1 °C, which does not seem like a large increase, combined with predicted future increases, it is sufficient to cause observable effects, such as the raising of ocean water levels, the disappearance of the Arctic ice cover in summers, climate shifts, and increased severity of storms and droughts around the world.

Figure 17.21 shows a record of the difference between the current average annual surface temperature in Antarctica and the average annual surface temperature over the last 420,000 years, which was determined from ice cores. Note that past temperatures were evaluated from measurements of carbon dioxide in the ice cores, that their values may be somewhat model-dependent, and that the

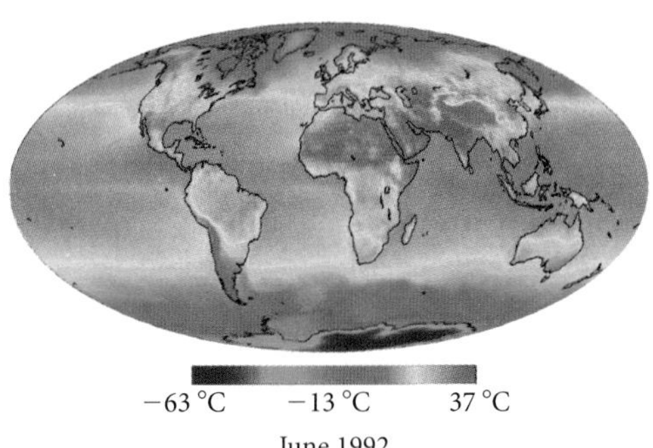


**FIGURE 17.19** Time-averaged surface temperature of the Earth in June 1992. The colors represent a range of temperatures from −63 °C to +37 °C.

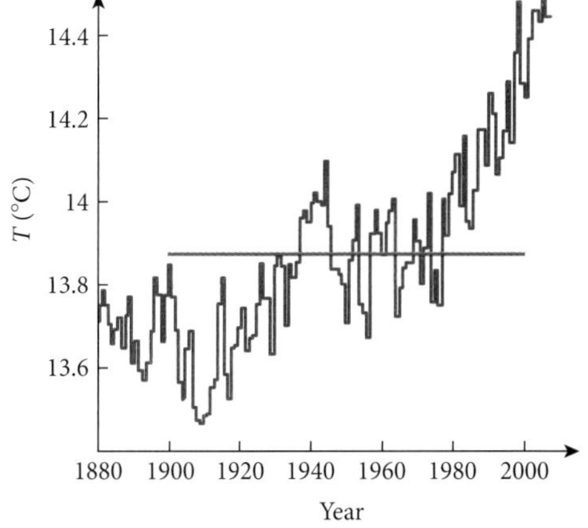


**FIGURE 17.20** Annual average global surface temperature from 1880 to 2005 as measured by thermometers on land and in the ocean (red histogram). The blue horizontal line represents the average global temperature for the 20th century, 13.9 °C.

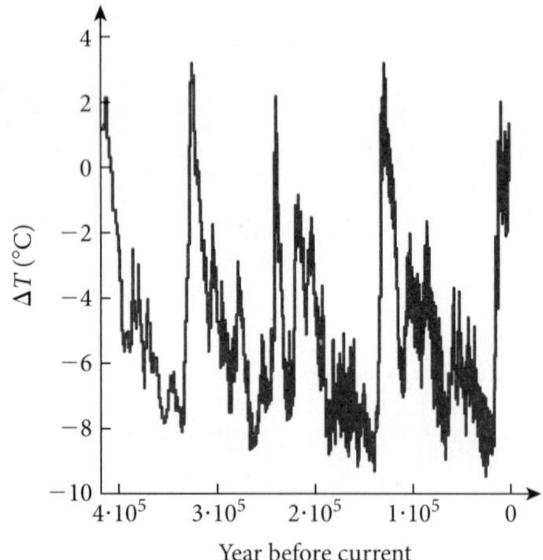


**FIGURE 17.21** Average annual surface temperature of Antarctica in the past, extracted from carbon dioxide content of ice cores, relative to the present value.

resulting temperature differences may be significantly larger than the corresponding global temperature differences. Several distinct periods are apparent in Figure 17.21. An interval of time when the temperature difference is around –7 °C corresponds to a period in which ice sheets covered parts of North America and Europe and is termed a *glacial period.* The last glacial period ended about 10,000 years ago. The warmer periods between glacial periods, called *interglacial periods,* correspond to temperature differences of around zero. In Figure 17.21, four glacial periods are visible, going back 400,000 years. Attempts have been made to relate these temperature differences to differences in the heat received from the Sun due to variations in the Earth's orbit and the orientation of its rotational axis known as the Milankovitch Hypothesis. However, these variations cannot account for all of the observed temperature differences.

The current warm, interglacial period began about 10,000 years ago and seems to be slightly cooler than previous interglacial periods. Previous interglacial periods have lasted from 10,000 to 25,000 years. However, human activities, such as the burning of fossil fuels and the resulting greenhouse effect (more on this in Chapter 18), are influencing the average global temperature. Models predict that the effect of these activities will be to warm the Earth, at least for the next several hundred years.

One effect of the warming of Earth's surface is a rise in sea level. Sea level has risen 120 m since the peak of the last glacial period, about 20,000 years ago, as a result of the melting of the glaciers that covered large areas of land. The melting of large amounts of ice resting on solid ground is the largest potential contributor to a further rise in sea level. For example, if all the ice in Antarctica melted, sea level would rise 61 m. If all the ice in Greenland melted, the rise in sea level would be 7 m. However, it would take several centuries for these large deposits of ice to melt completely, even if pessimistic predictions of climate models are correct. The rise in sea level due to thermal expansion is small compared with that due to the melting of large glaciers. The current rate of the rise in sea level is 2.8 mm/yr, as measured by the TOPEX/Poseidon satellite.

### 17.3 Self-Test Opportunity

Identify the years corresponding to glacial and interglacial periods in Figure 17.21.

### EXAMPLE 17.4 Rise in Sea Level Due to Thermal Expansion of Water

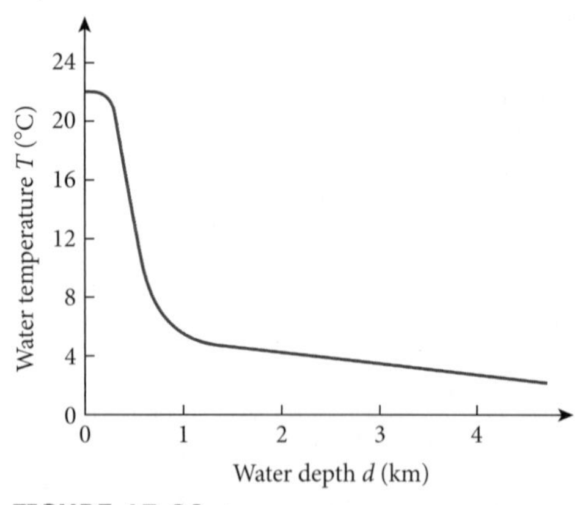


**FIGURE 17.22** Average ocean water temperature as a function of depth below the surface.

The rise in the level of the Earth's oceans is of current concern. Oceans cover $3.6 \cdot 10^8$ km$^2$, slightly more than 70% of Earth's surface area. The average ocean depth is 3700 m. The surface ocean temperature varies widely, between 35 °C in the summer in the Persian Gulf and –2 °C in the Arctic and Antarctic regions. However, even if the ocean surface temperature exceeds 20 °C, the water temperature rapidly falls off as a function of depth and reaches 4 °C at a depth of 1000 m (Figure 17.22). The global average temperature of all seawater is approximately 3 °C. Table 17.3 lists a volume expansion coefficient of zero for water at a temperature of 4 °C. Thus, it is safe to assume that the volume of ocean water changes very little at a depth greater than 1000 m. For the top 1000 m of ocean water, let's assume a global average temperature of 10.0 °C and calculate the effect of thermal expansion.

**PROBLEM**

By how much would sea level change, solely as a result of the thermal expansion of water, if the water temperature of all the oceans increased by $\Delta T = 1.0$ °C?

**SOLUTION**

The volume expansion coefficient of water at 10.0 °C is $\beta = 87.5 \cdot 10^{-6}$ °C$^{-1}$ (from Table 17.3), and the volume change of the oceans is given by equation 17.9, $\Delta V = \beta V \Delta T$, or

$$\frac{\Delta V}{V} = \beta \Delta T. \qquad \text{(i)}$$

We can express the total surface area of the oceans as $A = (0.7)4\pi R^2$, where $R$ is the radius of Earth and the factor 0.7 reflects the fact that about 70% of the surface of the sphere is covered by water. We assume that the surface area of the oceans increases only minutely from the water moving up the shores and neglect the change in surface area due to this effect. Then, essentially all of the change of the oceans' volume will result from the change in the depth, and we can write

$$\frac{\Delta V}{V} = \frac{\Delta d \cdot A}{d \cdot A} = \frac{\Delta d}{d}. \quad \text{(ii)}$$

Combining equations (i) and (ii), we obtain an expression for the change in depth:

$$\frac{\Delta d}{d} = \beta \Delta T \Rightarrow \Delta d = \beta d \Delta T.$$

Inserting the numerical values, $d = 1000$ m, $\Delta T = 1.0$ °C, and $\beta = 87.5 \cdot 10^{-6}$ °C$^{-1}$, we obtain

$$\Delta d = (1000 \text{ m})(87.5 \cdot 10^{-6} \text{ °C}^{-1})(1.0 \text{ °C}) = 9 \text{ cm}.$$

So, for every increase of the average ocean temperature by 1 °C, the sea level will rise by 9 cm (almost 4 in). This rise is smaller than the anticipated rise due to the melting of the ice cover on Greenland or Antarctica but will contribute to the problem of coastal flooding.

## 17.6 Temperature of the Universe

In 1965, while working on an early radio telescope, Arno Penzias and Robert Wilson discovered the **cosmic microwave background radiation.** They detected "noise," or "static," that seemed to come from all directions in the sky. Penzias and Wilson figured out what was producing this noise (which earned them the 1978 Nobel Prize): It was electromagnetic radiation left over from the Big Bang, which occurred 13.7 billion years ago. (We'll discuss electromagnetic radiation in Chapter 31.) It is astonishing to realize that an "echo" of the Big Bang still reverberates in "empty" intergalactic space after such a long time. The wavelength of the cosmic background radiation is similar to the wavelength of the electromagnetic radiation used in a microwave oven. An analysis of the distribution of wavelengths of this radiation led to the deduction that the background temperature of the universe is 2.725 K (exactly how this analysis was done is described in Chapter 36 on quantum physics). George Gamov had already predicted a cosmic background temperature of 2.7 K in 1948, when it was not clear at all yet that the Big Bang was a scientifically established fact.

In 2001, the Wilkinson Microwave Anisotropy Probe (WMAP) satellite measured variations in the background temperature of the universe. This mission followed the successful Cosmic Background Explorer (COBE) satellite, launched in 1989, which resulted in the 2006 Nobel Prize for Physics being awarded to John Mather and George Smoot. The COBE and WMAP missions found that the cosmic microwave background radiation was very uniform in all directions, but small differences in the temperature were superimposed on the smooth background. The WMAP results for the background temperature in all directions are shown in Figure 17.23. The effects of the Milky Way galaxy have been subtracted. You can see that the variation in the background temperature in the universe is very small, since $\pm 200$ μK/2.725 K = $\pm 7.3 \cdot 10^{-5}$. From interpretation of these results and other observations, scientists deduced that the age of the universe is 13.7 billion years, with a margin of error of less than 1%. In addition, scientists were able to deduce that the universe is

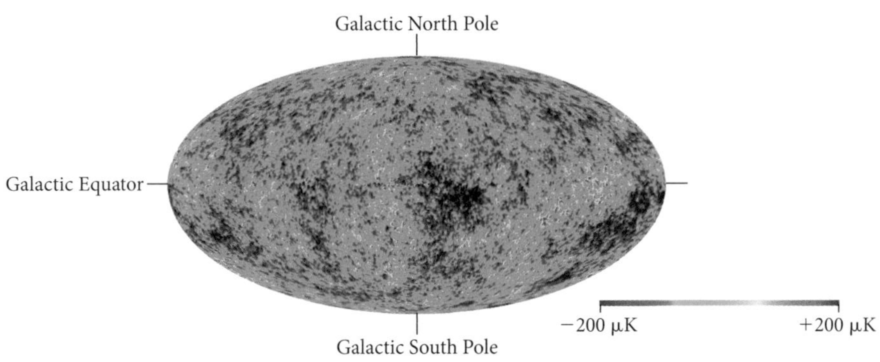


**FIGURE 17.23** The temperature of the cosmic microwave background radiation everywhere in the universe. The colors represent a range of temperatures from 200 μK below to 200 μK above the average temperature of the cosmic microwave background radiation, which has an average temperature of 2.725 K.

composed of 4% ordinary matter, 23% dark matter, and 73% dark energy. Dark matter is matter that exerts an observable gravitational pull but seems to be invisible, as discussed in Chapter 12. Dark energy seems to be causing the expansion of the universe to speed up. (Chapter 39, on particle physics and cosmology, covers these phenomena in detail.) Both dark matter and dark energy are under intense current investigation, and understanding of them should improve in the next decade.

## WHAT WE HAVE LEARNED | EXAM STUDY GUIDE

- The three commonly used temperature scales are the Fahrenheit scale, the Celsius scale, and the Kelvin scale.
- The Fahrenheit scale is used only in the United States and defines the freezing point of water as 32 °F and the boiling point of water as 212 °F.
- The Celsius scale defines the freezing point of water as 0 °C and the boiling point of water as 100 °C.
- The Kelvin scale defines 0 K as absolute zero and the freezing point of water as 273.15 K. The size of the kelvin and the Celsius degree are the same.
- The conversion from °F to °C is given by $T_C = \frac{5}{9}(T_F - 32\ °F)$.
- The conversion from °C to °F is given by $T_F = \frac{9}{5}T_C + 32\ °C$.
- The conversion from °C to K is given by $T_K = T_C + 273.15\ °C$.
- The conversion from K to °C is given by $T_C = T_K - 273.15\ K$.
- The change in length, $\Delta L$, of an object of length $L$ as the temperature changes by $\Delta T$ is given by $\Delta L = \alpha L \Delta T$, where $\alpha$ is the linear expansion coefficient.
- The change in volume, $\Delta V$, of an object with volume $V$ as the temperature changes by $\Delta T$ is given by $\Delta V = \beta V \Delta T$, where $\beta$ is the volume expansion coefficient.

## KEY TERMS

thermal equilibrium, p. 557
heat, p. 557
thermal energy, p. 557
temperature, p. 557
thermometer, p. 558
Zeroth Law of Thermodynamics, p. 558
Fahrenheit temperature scale, p. 558
Celsius temperature scale, p. 558
Kelvin temperature scale, p. 558
absolute zero, p. 558
Third Law of Thermodynamics, p. 559
kelvin, p. 559
thermal expansion, p. 563
linear expansion coefficient, p. 564
volume expansion coefficient, p. 569
cosmic microwave background radiation, p. 573

## NEW SYMBOLS

$T$, temperature

$\alpha$, linear expansion coefficient

$\beta$, volume expansion coefficient

## ANSWERS TO SELF-TEST OPPORTUNITIES

**17.1** Take $T = T_F = T_C$, and solve for $T$:

$$T = \tfrac{9}{5}T + 32$$
$$-\tfrac{4}{5}T = 32$$
$$T = -40\ °C \text{ or } °F.$$

**17.2** Use an ice-and-water mixture, which is at 0 °C, and boiling water, which is at 100 °C. Take the measurements and mark the corresponding places on the uncalibrated thermometer.

**17.3** The glacial periods are in years ago:

12,000 to 120,000
150,000 to 230,000
250,000 to 310,000
330,000 to 400,000.

The interglacial periods are in years ago:

0 to 12,000
110,000 to 130,000
230,000 to 250,000
310,000 to 330,000
400,000 to 410,000.

## PROBLEM-SOLVING PRACTICE

### Problem-Solving Guidelines

**1.** Be consistent in your use of Celsius, Kelvin, or Fahrenheit temperatures. For calculating a temperature change, a Celsius degree is equal to a kelvin. If you need to find a temperature value, however, remember which temperature scale is asked for.

**2.** Thermal expansion has an effect similar to enlarging a photograph: All parts of the picture get expanded in the same way. Remember that a hole in an object expands with increasing temperature just as the object itself does. Sometimes you can simplify a problem involving volume expansion to one dimension and then use linear expansion coefficients. Be alert for these kinds of situations.

### SOLVED PROBLEM 17.3 Linear Expansion of a Steel Bar and a Brass Bar

At 20.0 °C, a steel bar is 3.0000 m long and a brass bar is 2.9970 m long.

**PROBLEM**

At what temperature will the two bars be the same length?

**SOLUTION**

**THINK**

The coefficient of linear expansion for steel is less than that for brass. Thus, as the two bars are heated, the brass bar will expand more. To obtain the temperature at which the two bars have the same length, we equate the expressions for the final lengths of the bars, given in terms of the initial lengths, the expansion coefficients, and the temperature increase.

**SKETCH**

Figure 17.24 shows the two bars before and after heating.

The length of the steel bar before heating is $L_s$, its change in length is $\Delta L_s$, and the linear expansion coefficient of steel is $\alpha_s = 13 \cdot 10^{-6}\ °\text{C}^{-1}$. The length of the brass bar before heating is $L_b$, its change in length is $\Delta L_b$, and the linear expansion coefficient of brass is $\alpha_b = 19 \cdot 10^{-6}\ °\text{C}^{-1}$. The change in temperature is $\Delta T$.

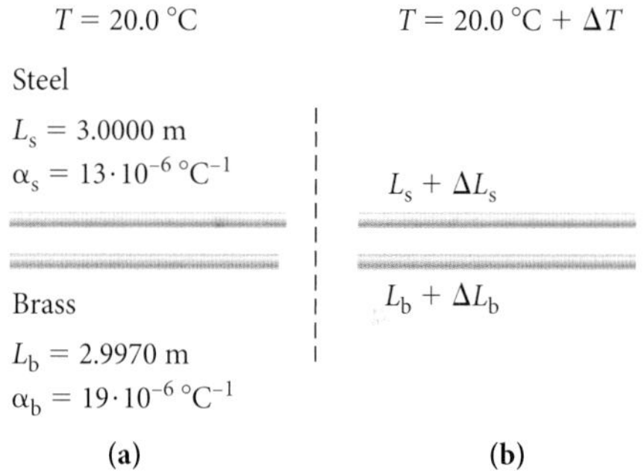

**FIGURE 17.24** A steel bar and a brass bar: (a) before heating; (b) after heating.

**RESEARCH**

The change in length of the steel bar is given by

$$\Delta L_s = \alpha_s L_s \Delta T.$$

The change in length of the brass bar is given by

$$\Delta L_b = \alpha_b L_b \Delta T.$$

When the two bars have the same length,

$$L_s + \Delta L_s = L_b + \Delta L_b.$$

**SIMPLIFY**

We can combine the preceding three equations to obtain

$$L_s + \alpha_s L_s \Delta T = L_b + \alpha_b L_b \Delta T.$$

Rearranging and solving for the temperature difference, we get

$$\alpha_s L_s \Delta T - \alpha_b L_b \Delta T = L_b - L_s \Rightarrow$$

$$\Delta T = \frac{L_b - L_s}{\alpha_s L_s - \alpha_b L_b}.$$

**CALCULATE**

Putting in the numerical values gives

$$\Delta T = \frac{(2.9970\text{ m}) - (3.0000\text{ m})}{(13 \cdot 10^{-6}\ °\text{C}^{-1})(3.0000\text{ m}) - (19 \cdot 10^{-6}\ °\text{C}^{-1})(2.9970\text{ m})} = 167.1961\ °\text{C}.$$

*Continued—*

**ROUND**

We report our result to three significant figures:

$$T = 20.0\ °\text{C} + 167.1961\ °\text{C} = 187.\ °\text{C}.$$

**DOUBLE-CHECK**

To double-check our result, we calculate the final length of both bars. For the steel bar, we have

$$L_s(1+\alpha_s\Delta T) = (3.0000\ \text{m})\left[1+(13\cdot 10^{-6}\ °\text{C}^{-1})(167.1961\ °\text{C})\right] = 3.00652\ \text{m}.$$

For the brass bar, we have

$$L_b(1+\alpha_b\Delta T) = (2.9970\ \text{m})\left[1+(19\cdot 10^{-6}\ °\text{C}^{-1})(167.1961\ °\text{C})\right] = 3.00652\ \text{m}.$$

Thus, our answer seems reasonable.

## MULTIPLE-CHOICE QUESTIONS

**17.1** Two mercury-expansion thermometers have identical reservoirs and cylindrical tubes made of the same glass but of different diameters. Which of the two thermometers can be graduated to a better resolution?

a) The thermometer with the smaller diameter tube will have better resolution.

b) The thermometer with the larger diameter tube will have better resolution.

c) The diameter of the tube is irrelevant; it is only the volume expansion coefficient of mercury that matters.

d) Not enough information is given to tell.

**17.2** For a class demonstration, your physics instructor uniformly heats a bimetallic strip that is held in a horizontal orientation. As a result, the bimetallic strip bends upward. This tells you that the coefficient of linear thermal expansion for metal T, on the top is _____ that of metal B, on the bottom.

a) smaller than b) larger than c) equal to

**17.3** Two solid objects, A and B, are in contact. In which case will thermal energy transfer from A to B?

a) A is at 20 °C, and B is at 27 °C.

b) A is at 15 °C, and B is at 15 °C.

c) A is at 0 °C, and B is at –10 °C.

**17.4** Which of the following bimetallic strips will exhibit the greatest sensitivity to temperature changes? That is, which one will bend the most as temperature increases?

a) copper and steel
b) steel and aluminum
c) copper and aluminum
d) aluminum and brass
e) copper and brass

**17.5** The background temperature of the universe is

a) 6000 K.
b) 288 K.
c) 3 K.
d) 2.73 K.
e) 0 K.

**17.6** Which air temperature feels coldest?

a) –40 °C
b) –40 °F
c) 233 K
d) All three are equal.

**17.7** At what temperature do the Celsius and Fahrenheit temperature scales have the same numeric value?

a) –40 degrees
b) 0 degrees
c) 40 degrees
d) 100 degrees

**17.8** The city of Yellowknife in the Northwest Territories of Canada is on the shore of Great Slave Lake. The average high in July is 21 °C and the average low in January is –31 °C. Great Slave Lake has a volume of 2090 $\text{km}^3$ and is the deepest lake in North America, with a depth of 614 m. What is the temperature of the water at the bottom of Great Slave Lake in January?

a) –31 °C
b) –10 °C
c) 0 °C
d) 4 °C
e) 32 °C

**17.9** Which object has the higher temperature after being left outside for an entire winter night: a metal door knob or a rug?

a) The metal door knob has the higher temperature.

b) The rug has the higher temperature.

c) Both have the same temperature.

d) It depends on the outside temperature.

## QUESTIONS

**17.10** A common way of opening a tight lid on a glass jar is to place it under warm water. The thermal expansion of the metal lid is greater than that of the glass jar; thus, the space between the two expands and it is easier to open the jar. Will this work for a metal lid on a container of the same kind of metal?

**17.11** Would it be possible to have a temperature scale defined in such a way that the hotter an object or system got, the lower (less positive or more negative) its temperature was?

**17.12** The solar corona has a temperature of about $1\cdot 10^6$ K. However, a spaceship flying in the corona will not be burned up. Why is this?

**17.13** Explain why it might be difficult to weld aluminum to steel or to weld any two unlike metals together.

**17.14** Two solid objects are made of different materials. Their volumes and volume expansion coefficients are $V_1$ and $V_2$ and $\beta_1$ and $\beta_2$, respectively. It is observed that during a temperature change of $\Delta T$, the volume of each object changes by the same amount. If $V_1 = 2V_2$ what is the ratio of the volume expansion coefficients?

**17.15** Some textbooks use the unit $K^{-1}$ rather than $°C^{-1}$ for values of the linear expansion coefficient; see Table 17.2. How will the numerical values of the coefficient differ if expressed in $K^{-1}$?

**17.16** You are outside on a hot day, with the air temperature at $T_o$. Your sports drink is at a temperature $T_d$ in a sealed plastic bottle. There are a few remaining ice cubes in the sports drink, which are at a temperature $T_i$, but they are melting fast.

a) Write an inequality expressing the relationship among the three temperatures.

b) Give reasonable values for the three temperatures in degrees Celsius.

**17.17** The Rankine temperature scale is an absolute temperature scale that uses Fahrenheit degrees; that is, temperatures are measured in Fahrenheit degrees, starting at absolute zero. Find the relationships between temperature values on the Rankine scale and those on the Fahrenheit, Kelvin, and Celsius scales.

**17.18** The Zeroth Law of Thermodynamics forms the basis for the definition of temperature with regard to thermal energy. But the concept of temperature is used in other areas of physics. In a system with energy levels, such as electrons in an atom or protons in a magnetic field, the population of a level with energy $E$ is proportional to the factor $e^{-E/k_BT}$, where $T$ is the absolute temperature of the system and $k_B = 1.381 \cdot 10^{-23}$ J/K is Boltzmann's constant. In a two-level system with the levels' energies differing by $\Delta E$, the ratio of the populations of the higher-energy and lower-energy levels is $p_{high}/p_{low} = e^{-\Delta E/k_BT}$. Such a system can have an infinite or even negative absolute temperature. Explain the meaning of such temperatures.

**17.19** Suppose a bimetallic strip is constructed of two strips of metals with linear expansion coefficients $\alpha_1$ and $\alpha_2$, where $\alpha_1 > \alpha_2$.

a) If the temperature of the bimetallic strip is reduced by $\Delta T$, what way will the strip bend (toward the side made of metal 1 or the side made of metal 2)? Briefly explain.

b) If the temperature is increased by $\Delta T$, which way will the strip bend?

**17.20** For food storage, what is the advantage of placing a metal lid on a glass jar? (*Hint:* Why does running the metal lid under hot water for a minute help you open such a jar?)

**17.21** A solid cylinder and a cylindrical shell, of identical radius and length and made of the same material, experience the same temperature increase $\Delta T$. Which of the two will expand to a larger outer radius?

## PROBLEMS

A blue problem number indicates a worked-out solution is available in the Student Solutions Manual. One • and two •• indicate increasing level of problem difficulty.

### Sections 17.1 and 17.2

**17.22** Express each of the following temperatures in degrees Celsius and in kelvins.

a) −19 °F b) 98.6 °F c) 52 °F

**17.23** One thermometer is calibrated in degrees Celsius, and another in degrees Fahrenheit. At what temperature is the reading on the thermometer calibrated in degrees Celsius three times the reading on the other thermometer?

**17.24** During the summer of 2007, temperatures as high as 47 °C were recorded in Southern Europe. The highest temperature ever recorded in the United States was 134 °F (at Death Valley, California, in 1913). What is the difference between these two temperatures, in degrees Celsius?

**17.25** The lowest air temperature recorded on Earth is −129 °F in Antarctica. Convert this temperature to the Celsius scale.

**17.26** What air temperature will feel twice as warm as 0 °F?

**17.27** A piece of dry ice (solid carbon dioxide) sitting in a classroom has a temperature of approximately −79 °C.

a) What is this temperature in kelvins?

b) What is this temperature in degrees Fahrenheit?

**17.28** In 1742, the Swedish astronomer Anders Celsius proposed a temperature scale on which water boils at 0 degrees and freezes at 100 degrees. In 1745, after the death of Celsius, Carolus Linnaeus (another Swedish scientist) reversed those standards, yielding the scale that is most commonly used today. Find room temperature (77 °F) on Celsius's original temperature scale.

**17.29** At what temperature do the Kelvin and Fahrenheit scales have the same numeric value?

### Section 17.4

**17.30** How does the density of copper that is just above its melting temperature of 1356 K compare to that of copper at room temperature?

**17.31** The density of steel is 7800.0 kg/m³ at 20.0 °C. Find the density at 100.0 °C.

**17.32** Two cubes with sides of length 100.0 mm fit in a space that is 201.0 mm wide, as shown in the figure. One cube is made of aluminum, and the other is made of brass. What temperature increase is required for the cubes to completely fill the gap?

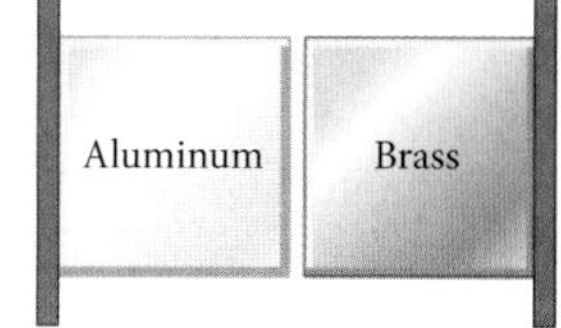


**17.33** A brass piston ring is to be fitted onto a piston by first heating up the ring and then slipping it over the piston. The piston ring has an inner diameter of 10.0 cm and an outer diameter of 10.2 cm. The piston has an outer diameter of 10.1 cm, and a groove for the piston ring has an outer diameter of 10.0 cm. To what temperature must the piston ring be heated so that it will slip onto the piston?

**17.34** An aluminum sphere of radius 10.0 cm is heated from 100.0 °F to 200.0 °F. Find (a) the volume change and (b) the radius change.

**17.35** Steel rails for a train track are laid in a region subject to extremes of temperature. The distance from one juncture to the next is 5.2 m, and the cross-sectional area of the rails is 60 cm$^2$. If the rails touch each other without buckling at the maximum temperature, 50 °C, how much space will there be between the rails at −10 °C?

**17.36** Even though steel has a relatively low linear expansion coefficient ($\alpha_{steel} = 13 \cdot 10^{-6}$ °C$^{-1}$), the expansion of steel railroad tracks can potentially create significant problems on very hot summer days. To accommodate for the thermal expansion, a gap is left between consecutive sections of the track. If each section is 25.0 m long at 20.0 °C and the gap between sections is 10.0 mm wide, what is the highest temperature the tracks can take before the expansion creates compressive forces between sections?

**17.37** A medical device used for handling tissue samples has two metal screws, one 20 cm long and made from brass ($\alpha_b = 18.9 \cdot 10^{-6}$ °C$^{-1}$) and the other 30 cm long and made from aluminum ($\alpha_a = 23.0 \cdot 10^{-6}$ °C$^{-1}$). A gap of 1 mm exists between the ends of the screws at 22 °C. At what temperature will the two screws touch?

•**17.38** You are designing a precision mercury thermometer based on the thermal expansion of mercury ($\beta = 1.8 \cdot 10^{-4}$ °C$^{-1}$), which causes the mercury to expand up a thin capillary as the temperature increases. The equation for the change in volume of the mercury as a function of temperature is $\Delta V = \beta V_0 \Delta T$, where $V_0$ is the initial volume of the mercury and $\Delta V$ is the change in volume due to a change in temperature, $\Delta T$. In response to a temperature change of 1.0 °C, the column of mercury in your precision thermometer should move a distance $D = 1.0$ cm up a cylindrical capillary of radius $r = 0.10$ mm. Determine the initial volume of mercury that allows this change. Then find the radius of a spherical bulb that contains this volume of mercury.

•**17.39** On a hot summer day, a cubical swimming pool is filled to within 1.0 cm of the top with water at 21 °C. When the water warms to 37 °C, the pool overflows. What is the depth of the pool?

•**17.40** A steel rod of length 1 m is welded to the end of an aluminum rod of length 2 m (lengths measured at 22 °C). The combination rod is then heated to 200 °C. What is the length of the combination rod at 200 °C?

•**17.41** A clock based on a simple pendulum is situated outdoors in Anchorage, Alaska. The pendulum consists of a mass of 1.00 kg that is hanging from a thin brass rod that is 2.000 m long. The clock is calibrated perfectly during a summer day with an average temperature of 25.0 °C. During the winter, when the average temperature over one 24-h period is −20.0 °C, find the time elapsed for that period according to the simple pendulum clock.

•**17.42** In a thermometer manufacturing plant, a type of mercury thermometer is built at room temperature (20 °C) to measure temperatures in the 20 °C to 70 °C range, with a 1-cm$^3$ spherical reservoir at the bottom and a 0.5-mm inner diameter expansion tube. The wall thickness of the reservoir and tube is negligible, and the 20 °C mark is at the junction between the spherical reservoir and the tube. The tubes and reservoirs are made of fused silica, a transparent *glass* form of $SiO_2$ that has a very low linear expansion coefficient ($\alpha = 0.4 \cdot 10^{-6}$ °C$^{-1}$). By mistake, the material used for one batch of thermometers was *quartz*, a transparent *crystalline* form of $SiO_2$ with a much higher linear expansion coefficient ($\alpha = 12.3 \cdot 10^{-6}$ °C$^{-1}$). Will the manufacturer have to scrap the batch, or will the thermometers work fine, within the expected uncertainty of 5% in reading the temperature? The volume expansion coefficient of mercury is $\beta = 181 \cdot 10^{-6}$ °C$^{-1}$.

•**17.43** The ends of the two rods shown in the figure are separated by 5 mm at 25 °C. The left-hand rod is brass and 1 m long; the right-hand rod is steel and 1 m long. Assuming that the outside ends of both rods rest firmly against rigid supports, at what temperature will the ends of the rods that face each other just touch?

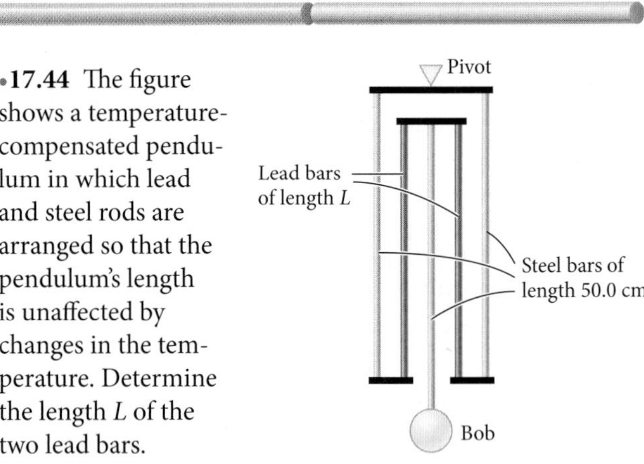


•**17.44** The figure shows a temperature-compensated pendulum in which lead and steel rods are arranged so that the pendulum's length is unaffected by changes in the temperature. Determine the length $L$ of the two lead bars.

•**17.45** Consider a bimetallic strip consisting of a 0.5-mm-thick brass upper strip welded to a 0.5-mm-thick steel lower strip. When the temperature of the bimetallic strip is increased by 20 K, the unattached tip deflects by 3 mm from its original straight position, as shown in the figure. What is the length of the strip at its original position?

•**17.46** Thermal expansion seems like a small effect, but it can engender tremendous, often damaging, forces. For example, steel has a linear expansion coefficient of $\alpha = 1.2 \cdot 10^{-5}$ °C$^{-1}$ and a bulk modulus of $B = 160$ GPa. Calculate the pressure engendered in steel by a 1.0 °C temperature increase.

•17.47 At room temperature, an iron horseshoe, when dunked into a cylindrical tank of water (radius of 10.0 cm) causes the water level to rise 0.25 cm above the level without the horseshoe in the tank. When heated in the blacksmith's stove from room temperature to a temperature of $7.00 \cdot 10^2$ K, worked into its final shape, and then dunked back into the water, how much does the water level rise above the "no horseshoe" level (ignore any water that evaporates as the horseshoe enters the water)? *Note:* The linear expansion coefficient for iron is roughly that of steel: $11 \cdot 10^{-6}$ °C$^{-1}$.

•**17.48** A clock has an aluminum pendulum with a period of 1.000 s at 20.0 °C. Suppose the clock is moved to a location where the average temperature is 30.0 °C. Determine (a) the new period of the clock's pendulum and (b) how much time the clock will lose or gain in 1 week.

•**17.49** Using techniques similar to those that were originally developed for miniaturized semiconductor electronics, scientists and engineers are creating Micro-Electro-Mechanical Systems (MEMS). An example is an electrothermal actuator that is driven by heating its different parts using an electrical current. The device is used to position 125-μm-diameter optical fibers with submicron resolution and consists of thin and thick silicon arms connected in the shape of a U, as shown in the figure. The arms are not attached to the substrate under the device but are free to move, whereas the electrical contacts (marked + and – in the figure) are attached to the substrate and cannot move. The thin arm is $3.0 \cdot 10^1$ μm wide, and the thick arm is 130 μm wide. Both arms are 1800 μm long. Electrical current flows through the arms, causing them to heat up. Although the same current flows through both arms, the thin arm has a greater electrical resistance than the thick arm and therefore dissipates more electrical power and gets substantially hotter. When current is made to flow through the beams, the thin beam reaches a temperature of $4.0 \cdot 10^2$ °C, and the thick beam reaches a temperature of $2.0 \cdot 10^2$ °C. Assume that the temperature in each beam is constant along the entire length of that beam (strictly speaking, this is not the case) and that the two arms remain parallel and bend only in the plane of the paper at higher temperatures. How much and in which direction will the tip move? The linear expansion coefficient for silicon is $3.2 \cdot 10^{-6}$ °C$^{-1}$.

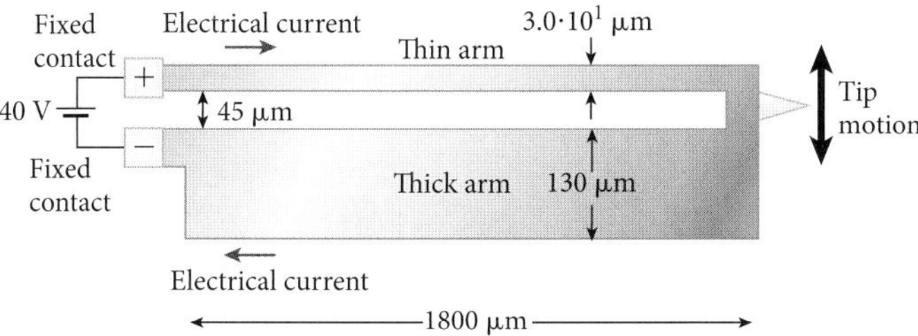


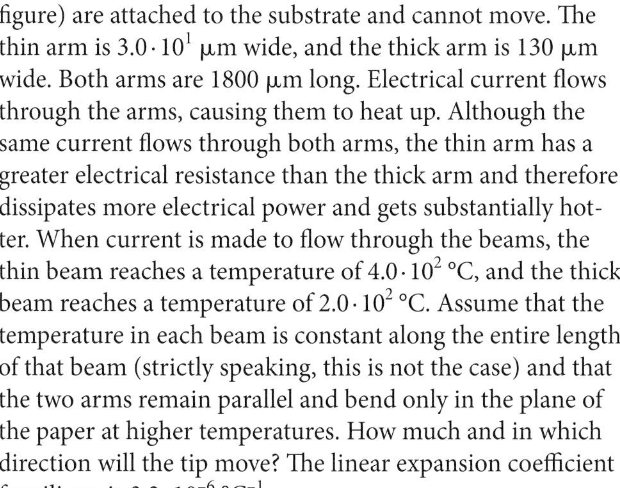


•**17.50** Another MEMS device, used for the same purpose as the one in Problem 17.49, has a different design. This electrothermal actuator consists of a thin V-shaped silicon beam, as shown in the figure. The beam is not attached to the substrate under the device but is free to move, whereas the electrical contacts (marked + and – in the figure) are attached to the substrate and cannot move. The beam spans a gap between the electrical contacts that is 1800 μm wide, and the two halves of the beam slant up from horizontal by 0.10 rad. Electrical current flows through the beam, causing it to heat up. When current is made to flow across the beam, the beam reaches a temperature of 500 °C. Assume that the temperature is constant along the entire length of the beam (strictly speaking, this is not the case). How much and in which direction will the tip move? The linear expansion coefficient for silicon is $3.2 \cdot 10^{-6}$ °C$^{-1}$.

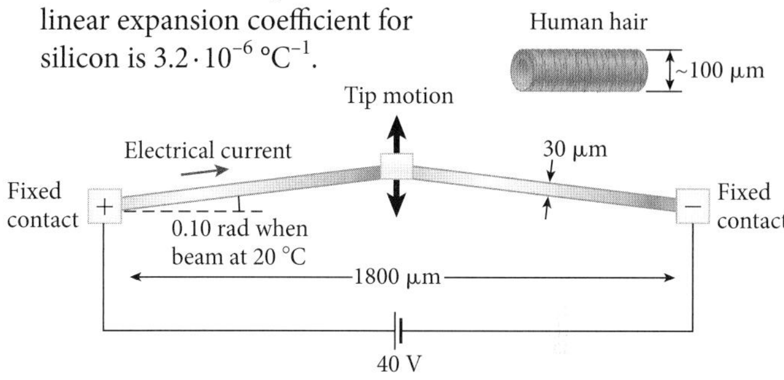


•17.51 The volume of 1 kg of liquid water over the temperature range from 0 °C to 50 °C fits reasonably well to the polynomial function $V = 1.00016 - (4.52 \cdot 10^{-5})T + (5.68 \cdot 10^{-6})T^2$, where the volume is measured in cubic meters and $T$ is the temperature in degrees Celsius.

a) Use this information to calculate the volume expansion coefficient for liquid water as a function of temperature.

b) Evaluate your expression at 20.0 °C, and compare the value to that listed in Table 17.3.

••**17.52** a) Suppose a bimetallic strip is constructed of copper and steel strips of thickness 1.0 mm and length 25 mm, and the temperature of the strip is reduced by 5.0 K. Determine the radius of curvature of the cooled strip (the radius of curvature of the interface between the two strips).

b) If the strip is 25 mm long, how far is the maximum deviation of the strip from the straight orientation?

## Additional Problems

**17.53** A copper cube of side length 40 cm is heated from 20 °C to 120 °C. What is the change in the volume of the cube? The linear expansion coefficient of copper is $17 \cdot 10^{-6}$ °C$^{-1}$.

**17.54** When a 50-m-long metal pipe is heated from 10 °C to 40 °C, it lengthens by 2.85 cm.

a) Determine the linear expansion coefficient.

b) What type of metal is the pipe made of?

17.55 On a cool morning, with the temperature at 15 °C, a painter fills a 5-gal aluminum container to the brim with turpentine. When the temperature reaches 27 °C, how much fluid spills out of the container? The volume expansion coefficient for this brand of turpentine is $9.0 \cdot 10^{-4}$ °C$^{-1}$.

**17.56** A building having a steel infrastructure is $6.00 \cdot 10^2$ m high on a day when the temperature is 0.00 °C. How much taller is the building on a day when the temperature is 45.0 °C? The linear expansion coefficient of steel is $1.30 \cdot 10^{-5}$ °C$^{-1}$.

**17.57** In order to create a tight fit between two metal parts, machinists sometimes make the interior part larger than the hole into which it will fit and then either cool the interior part or heat the exterior part until they fit together. Suppose an aluminum rod with diameter $D_1$ (at $2.0 \cdot 10^1$ °C) is to be fit into a hole in a brass plate that has a diameter $D_2 = 10.000$ mm (at $2.0 \cdot 10^1$ °C). The machinists can cool the rod to 77.0 K by immersing it in liquid nitrogen. What is the largest possible diameter that the rod can have at $2.0 \cdot 10^1$ °C and just fit into the hole if the rod is cooled to 77.0 K and the brass plate is left at $2.0 \cdot 10^1$ °C? The linear expansion coefficients for aluminum and brass are $22 \cdot 10^{-6}$ °C$^{-1}$ and $19 \cdot 10^{-6}$ °C$^{-1}$, respectively.

$D_1$

$D_2 = 10.000$ mm

**17.58** A military vehicle is fueled with gasoline in the United States, preparatory to being shipped overseas. Its fuel tank has a capacity of 213 L. When it is fueled, the temperature is 57 °F. At its destination, it may be required to operate in temperatures as high as 120 °F. What is the maximum volume of gasoline that should be put in its tank?

**17.59** A mercury thermometer contains 8.0 mL of mercury. If the tube of the thermometer has a cross-sectional area of 1.0 mm$^2$, what should the spacing between the °C marks be?

**17.60** A 14-gal container is filled with gasoline. Neglect the change in volume of the container and find how many gallons are lost if the temperature increases by 27 °F. The volume expansion coefficient of gasoline is $9.6 \cdot 10^{-4}$ °C$^{-1}$.

**17.61** A highway of concrete slabs is to be built in the Libyan desert, where the highest air temperature recorded is 57.8 °C. The temperature is 20.0 °C during construction of the highway. The slabs are measured to be 12.0 m long at this temperature. How wide should the expansion cracks between the slabs be (at 20.0 °C) in order to prevent buckling at the highest temperatures?

**17.62** An aluminum vessel with a volume capacity of 500 cm$^3$ is filled with water to the brim at 20 °C. The vessel and contents are heated to 50 °C. During the heating process, will the water spill over the top, will there be more room for water to be added, or will the water level remain the same? Calculate the volume of water that will spill over or that could be added if either is the case.

**17.63** By how much does the temperature of a given mass of kerosene need to change in order for its volume to increase by 1.0%?

**17.64** A plastic-epoxy sheet has uniform holes of radius 1.99 cm. The holes are intended to allow solid ball bearings with an outer radius of 2.00 cm to just go through. Over what temperature rise must the plastic-epoxy sheet be heated so that the ball bearings will go through the holes? The linear expansion coefficient of plastic-epoxy is about $1.3 \cdot 10^{-4}$ °C$^{-1}$.

•**17.65** A uniform brass disk of radius $R$ and mass $M$ with a moment of inertia $I$ about its cylindrical axis of symmetry is at a temperature $T = 20$ °C. Determine the fractional change in its moment of inertia if it is heated to a temperature of 100 °C.

•**17.66** A 25.01-mm-diameter brass ball sits at room temperature on a 25.00-mm-diameter hole made in an aluminum plate. The ball and plate are heated uniformly in a furnace, so both are at the same temperature at all times. At what temperature will the ball fall through the plate?

•**17.67** In a pickup basketball game, your friend cracked one of his teeth in a collision with another player while attempting to make a basket. To correct the problem, his dentist placed a steel band of initial internal diameter 4.4 mm, and a cross-sectional area of width 3.5 mm, and thickness 0.45 mm on the tooth. Before placing the band on the tooth, he heated the band to 70 °C. What will be the tension in the band once it cools down to the temperature in your friend's mouth (37 °C)?

•**17.68** Your physics teacher assigned a thermometer-building project. He gave you a glass tube with an inside diameter of 1 mm and a receptacle at one end. He also gave you 8.63 cm$^3$ of mercury to pour into the tube, which filled the receptacle and some of the tube. You are to add marks indicating degrees Celsius on the glass tube. At what increments should the marks be put on? You know the volume expansion coefficient of mercury is $1.82 \cdot 10^{-4}$ °C$^{-1}$.

•**17.69** You are building a device for monitoring ultracold environments. Because the device will be used in environments where its temperature will change by 200 °C in 3 s, it must have the ability to withstand thermal shock (rapid temperature changes). The volume of the device is 0.00005 m$^3$, and if the volume changes by 0.0000001 m$^3$ in a time interval of 5 s, the device will crack and be rendered useless. What is the maximum volume expansion coefficient that the material you use to build the device can have?

•**17.70** A steel rod of length 1.0000 m and cross-sectional area $5.00 \cdot 10^{-4}$ m$^2$ is placed snugly against two immobile end points. The rod is initially placed when the temperature is 0 °C. Find the stress in the rod when the temperature rises to 40 °C.

•**17.71** A brass bugle can be thought of as a tube that is open on both ends (the actual physics is complicated by the interaction of the bugler's mouth and the mouthpiece and the bell at the end). The overall length if the bugle is stretched out is 183.0 cm (at 20.0 °C). A bugle is played on a hot summer day (41.0 °C). Find the fundamental frequency if

a) only the change in air temperature is considered,

b) only the change in the length of the bugle is considered, and

c) both effects in parts (a) and (b) are taken into account.

# Heat and the First Law of Thermodynamics

# 18

**FIGURE 18.1** A thunderstorm.

## WHAT WE WILL LEARN

- Heat is thermal energy that transfers between a system and its environment or between two systems, as a result of a temperature difference between them.
- The First Law of Thermodynamics states that the change in internal energy of a closed system is equal to the thermal energy absorbed by the system minus the work done by the system.
- Adding thermal energy to an object increases its temperature. This temperature increase is proportional to the heat capacity, $C$, of the object.
- The thermal energy added to an object of mass $m$ is equal to the product of the specific heat of the object, $c$, and $m$, and the temperature increase of the object.
- If thermal energy is added to a solid object, its temperature increases until it reaches the melting point. If thermal energy then continues to be added, the temperature of the object remains the same until the entire object melts to a liquid. The heat required to melt an object at its melting point, divided by its mass, is the latent heat of fusion.
- If thermal energy is added to a liquid, its temperature increases until it reaches the boiling point. If thermal energy then continues to be added, the temperature of the liquid remains the same until all of the liquid vaporizes to a gas. The heat required to vaporize a liquid at its boiling point, divided by its mass, is the latent heat of vaporization.
- The three main modes of transferring thermal energy are conduction, convection, and radiation.

Earth's weather is driven by thermal energy in the atmosphere. The equatorial regions receive more of the Sun's radiation than the polar regions do; so warm air moves north and south from the Equator toward the poles to distribute the thermal energy more evenly. This transfer of thermal energy, called *convection,* sets up wind currents around the world, carrying clouds and rain as well as air. In extreme cases, rising warm air and sinking cooler air form dramatic storms, like the thunderstorm shown in Figure 18.1. Circular storms—tornados and hurricanes—are also driven by the violent collision of warm air with cooler air.

This chapter examines the nature of heat and the mechanisms of thermal energy transfer. Heat is a form of energy that is transferred into or out of a system. Thus, heat is governed by a more general form of the law of conservation of energy, known as the First Law of Thermodynamics. We'll focus on this law in this chapter, along with some of its applications to thermodynamic processes and to changes in heat and temperature.

Heat is essential to life processes; no life could exist on Earth without heat from the Sun or from the Earth's interior. However, heat can also cause problems with the operation of electrical circuits, motors, and other mechanical devices. Every branch of science and engineering has to deal with heat in one way or another, and thus the concepts in this chapter are important for all areas of research, design, and development.

## 18.1 Definition of Heat

Heat is one of the most common forms of energy in the universe, and we all experience it every day. Yet many people have misconceptions about heat that often cause confusion. For instance, a burning object such as a candle flame does not "have" heat that it emits when it becomes hot enough. Instead, the candle flame transmits energy, in the form of heat, to the air around it. To clarify these ideas, we need to start with clear and precise definitions of heat and the units used to measure it.

If you pour cold water into a glass and then put the glass on the kitchen table, the water will slowly warm until it reaches the temperature of the air in the room. Similarly, if you pour hot water into a glass and place it on the kitchen table, the water will slowly cool until it reaches the temperature of the air in the room. The warming or cooling takes place rapidly at first and then more slowly as the water comes into thermal equilibrium with the air in the kitchen. At thermal equilibrium, the water, the glass, and the air in the room are all at the same temperature.

The water in the glass is a **system** with temperature $T_s$, and the air in the kitchen is an **environment** with temperature $T_e$. If $T_s \neq T_e$, then the temperature of the system changes until it is equal to the temperature of the environment. A system can be simple or complicated; it is just any object or collection of objects we wish to examine. The difference between the environment and the system is that the environment is large compared to the system. The temperature of the system affects the environment, but we'll assume that the environment is so large that any changes in its temperature are imperceptible.

The change in temperature of the system is due to a transfer of energy between the system and its environment. This type of energy is **thermal energy,** an internal energy related to the motion of the atoms, molecules, and electrons that make up the system or the environment. Thermal energy in the process of being transferred from one body to another is called **heat** and is symbolized by $Q$. If thermal energy is transferred into a system, then $Q > 0$ (Figure 18.2a). That is, the system gains thermal energy when it receives heat from its environment. If thermal energy is transferred from the system to its environment, then $Q < 0$ (Figure 18.2c). If the system and its environment have the same temperature (Figure 18.2b), then $Q = 0$. Thermal energy flow causes a system to gain or lose thermal energy.

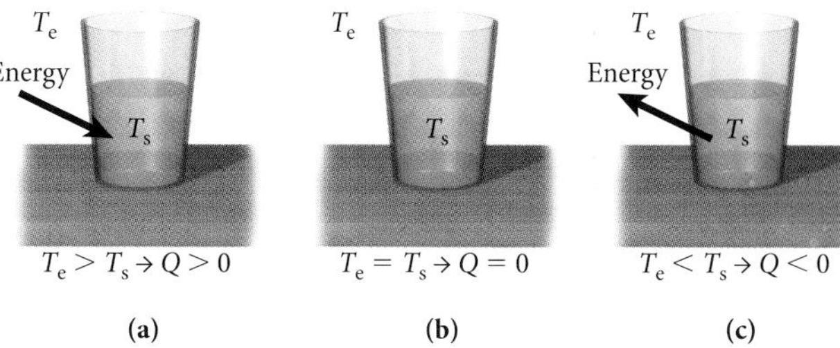


**FIGURE 18.2** (a) A system embedded in an environment that has a higher temperature. (b) A system embedded in an environment that has the same temperature. (c) A system embedded in an environment that has a lower temperature.

### Definition

Heat, $Q$, is the energy transferred between a system and its environment (or between two systems) because of a temperature difference between them. When energy flows into the system, $Q > 0$; when energy flows out of the system, $Q < 0$.

## 18.2 Mechanical Equivalent of Heat

Recall from Chapter 5 that energy can also be transferred between a system and its environment as work done by a force acting on a system or by a system. The concepts of heat and work, discussed in Section 18.3, can be defined in terms of the transfer of energy to or from a system. We can refer to the internal energy of a system, but not to the heat contained in a system. If we observe hot water in a glass, we do not know whether thermal energy was transferred to the water or work was done on the water.

Heat is transferred energy and can be quantified using the SI unit of energy, the joule. Originally, heat was measured in terms of its ability to raise the temperature of water. The **calorie** (cal) was defined as the amount of heat required to raise the temperature of 1 gram of water by 1 °C. Another common measure of heat, still used in the United States, is the British thermal unit (BTU), defined as the amount of heat required to raise the temperature of 1 pound of water by 1 °F. However, the change in temperature of water as a function of the amount of thermal energy transferred to it depends on the original temperature of the water. Reproducible definitions of the calorie and the British thermal unit required that the measurements be made at a specified starting temperature.

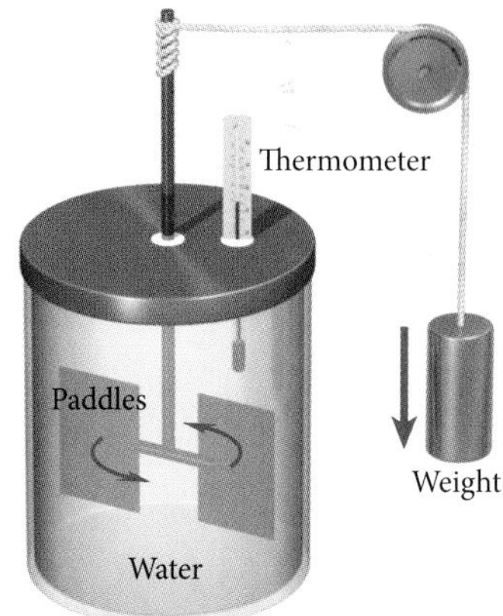


**FIGURE 18.3** Apparatus for Joule's experiment to determine the mechanical equivalent of heat.

In a classic experiment performed in 1843, the English physicist James Prescott Joule showed that the mechanical energy of an object could be converted into thermal energy. The apparatus Joule used consisted of a large mass supported by a rope that was run over a pulley and wound around an axle (Figure 18.3). As the mass descended, the unwinding rope turned a pair of large paddles in a container of water. Joule showed that the rise in temperature of the water was directly related to the mechanical work done by the falling object. Thus, Joule demonstrated that mechanical work could be turned into thermal energy, and he obtained a relationship between the calorie and the joule (the energy unit named in his honor).

The modern definition of the calorie is based on the joule. The calorie is defined to be exactly 4.186 J, with no reference to the change in temperature of water. The following are some conversion factors for energy units:

$$\begin{aligned} 1 \text{ cal} &= 4.186 \text{ J} \\ 1 \text{ BTU} &= 1055 \text{ J} \\ 1 \text{ kW h} &= 3.60 \cdot 10^6 \text{ J} \\ 1 \text{ kW h} &= 3412 \text{ BTU}. \end{aligned} \quad (18.1)$$

The energy content of food is usually expressed in terms of calories. A food calorie, often called a *Calorie* or a *kilocalorie,* is not equal to the calorie just defined; a food calorie is equivalent to 1000 cal. The cost of electrical energy is usually stated in cents per kilowatt-hour (kWh).

### EXAMPLE 18.1 Energy Content of a Candy Bar

**PROBLEM**

The label of a candy bar states that it has 275 Calories. What is the energy content of this candy bar in joules?

**SOLUTION**

Food calories are kilocalories and 1 kcal = 4186 J. The candy bar has 275 kcal or

$$275 \text{ kcal}\left(\frac{4186 \text{ J}}{1 \text{ kcal}}\right) = 1.15 \cdot 10^6 \text{ J}.$$

**DISCUSSION**

Note that this energy is enough to raise a small truck with a weight of 22.2 kN (mass of 5000 lb) a distance of 52 m. A 73-kg (160-lb) person would have to walk an hour at 1.6 m/s (3.5 mph) to burn off the calories from consuming this candy bar.

## 18.3 Heat and Work

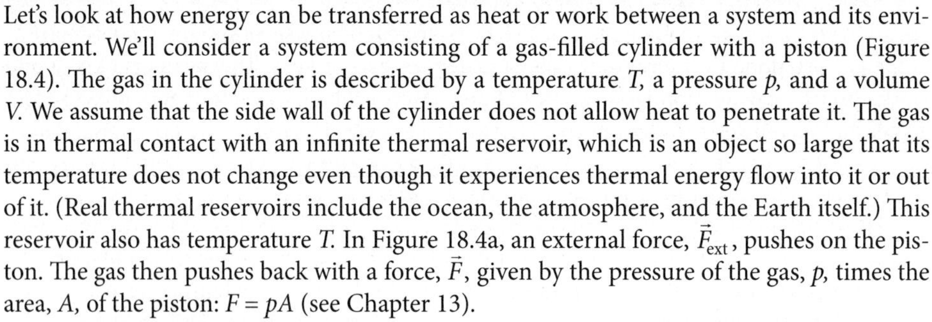

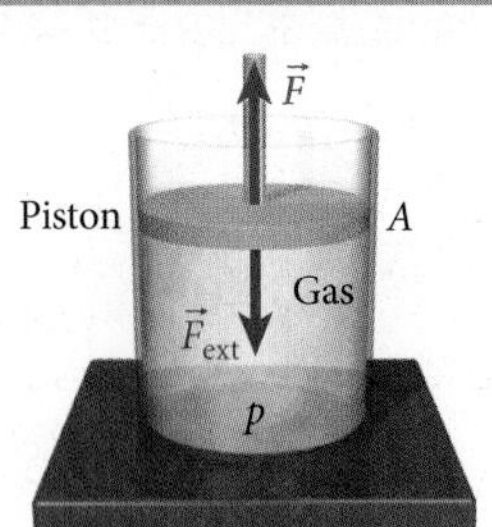


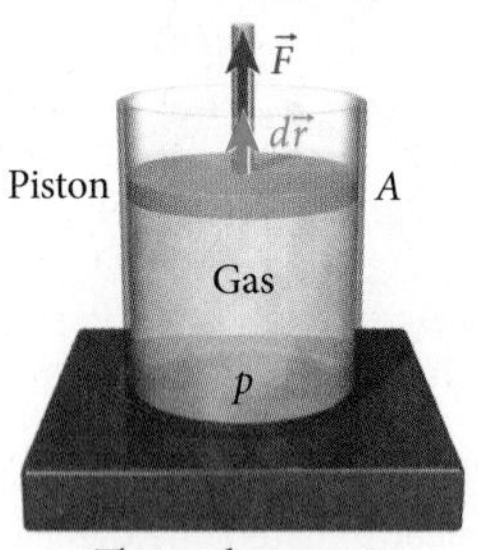


**FIGURE 18.4** A gas-filled cylinder with a piston. The gas is in thermal contact with an infinite thermal reservoir. (a) An external force pushes on the piston, creating a pressure in the gas. (b) The external force is removed, allowing the gas to push the piston out.

Let's look at how energy can be transferred as heat or work between a system and its environment. We'll consider a system consisting of a gas-filled cylinder with a piston (Figure 18.4). The gas in the cylinder is described by a temperature $T$, a pressure $p$, and a volume $V$. We assume that the side wall of the cylinder does not allow heat to penetrate it. The gas is in thermal contact with an infinite thermal reservoir, which is an object so large that its temperature does not change even though it experiences thermal energy flow into it or out of it. (Real thermal reservoirs include the ocean, the atmosphere, and the Earth itself.) This reservoir also has temperature $T$. In Figure 18.4a, an external force, $\vec{F}_{ext}$, pushes on the piston. The gas then pushes back with a force, $\vec{F}$, given by the pressure of the gas, $p$, times the area, $A$, of the piston: $F = pA$ (see Chapter 13).

To describe the behavior of this system, we consider the change from an initial state specified by pressure $p_i$, volume $V_i$, and temperature $T_i$ and a final state specified by pressure $p_f$, volume $V_f$, and temperature $T_f$. Imagine changes made slowly so that the gas remains close to equilibrium, allowing measurements of $p$, $V$, and $T$. The progression from the initial state to the final state is a **thermodynamic process.** During this process, thermal energy may be transferred into the system (positive heat), or it may be transferred out of the system (negative heat).

When the external force is removed (Figure 18.4b), the gas in the cylinder pushes the piston out a distance $dr$. The work done by the system in this process is

$$dW = \vec{F} \cdot d\vec{r} = (pA)(dr) = p(A\,dr) = p\,dV,$$

where $dV = A\,dr$ is the change in the volume of the system. Thus, the work done by the system in going from the initial configuration to the final configuration is given by

$$W = \int dW = \int_{V_i}^{V_f} p\,dV. \tag{18.2}$$

Note that during this change in volume, the pressure may also change. To evaluate this integral, we need to know the relationship between pressure and volume for the process. For example, if the pressure remains constant, we obtain

$$W = \int_{V_i}^{V_f} p\,dV = p\int_{V_i}^{V_f} dV = p(V_f - V_i) \quad \text{(for constant pressure).} \tag{18.3}$$

Equation 18.3 indicates that, for constant pressure, a negative change in the volume corresponds to negative work done by the system.

Figure 18.5 shows graphs of pressure versus volume, sometimes called ***pV*-diagrams.** The three parts of the figure show different paths, or ways to change the pressure and volume of a system from an initial condition to a final condition. Figure 18.5a illustrates a process that starts at an initial point $i$ and proceeds to a final point $f$ in such a way that the pressure decreases as the volume increases. The work done by the system is given by equation 18.2. The integral can be represented by the area under the curve, shown by the green shading in Figure 18.5a. In this case, the work done by the system is positive because the volume of the system increases.

Figure 18.5b illustrates a process that starts at an initial point $i$ and proceeds to a final point $f$ through an intermediate point $m$. The first step involves increasing the volume of the system while holding the pressure constant. One way to complete this step is to increase the temperature of the system as the volume increases, maintaining a constant pressure. The second step consists of decreasing the pressure while holding the volume constant. One way to accomplish this task is to lower the temperature by taking thermal energy away from the system. Again, the work done by the system can be represented by the area under the curve, shown by the green shading in Figure 18.5b. The work done by the system is positive because the volume of the system increases. However, the work done by the system originates only in the first step. In the second step, the system does no work because the volume does not change.

Figure 18.5c illustrates another process that starts at an initial point $i$ and proceeds to a final point $f$ through an intermediate point $m$. The first step involves decreasing the pressure of the system while holding the volume constant. The second step consists of increasing the volume while holding the pressure constant. Again, the work done by the system can be represented by the area under the curve, shown by the green shading in Figure 18.5c. The work done by the system is again positive because the volume of the system increases. However, the work done by the system originates only in the second step. In the first step, the system does no work because the volume does not change. The net work done in this process is less than the work done by the process in Figure 18.5b, as the green area is smaller. The thermal energy absorbed must also be less because the initial and final states are the same in these two processes, so the change in internal energy is the same. (This connection between work, heat, and the change in internal energy is discussed in more detail in Section 18.4.)

Thus, the work done by a system and the thermal energy transferred to the system depend on the manner in which the system moves from an initial point to a final point in a $pV$-diagram. This type of process is thus referred to as a **path-dependent process.**

The $pV$-diagrams in Figure 18.6 reverse the processes shown in Figure 18.5 by starting at the final points in Figure 18.5 and proceeding along the same path to the initial points. In each of these cases, the area under the curve represents the negative of the work done by the system in moving from an initial point $i$ to a final point $f$. In all three cases, the work done by the system is negative because the volume of the system decreases.

Suppose a process starts at some initial point on a $pV$-diagram, follows some path, and returns to the original point. A path that returns to its starting point is called a **closed path.** Two examples of closed paths are illustrated in Figure 18.7. In Figure 18.7a, a process starts at an initial point $i$ and proceeds to an intermediate point $m_1$; the volume is increased while the pressure is held constant. From $m_1$, the path goes to intermediate point $m_2$; the volume is held

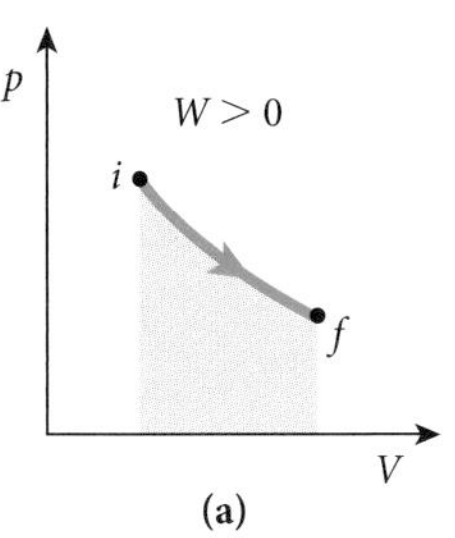


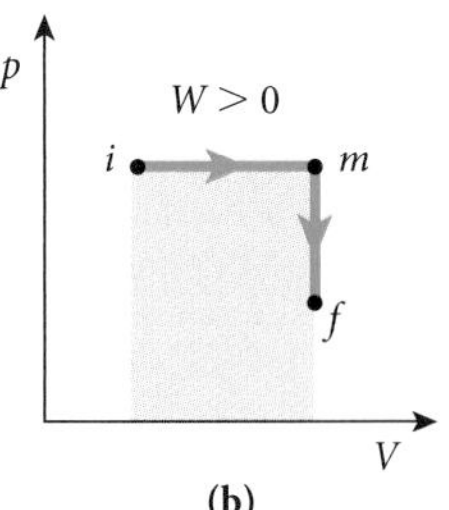


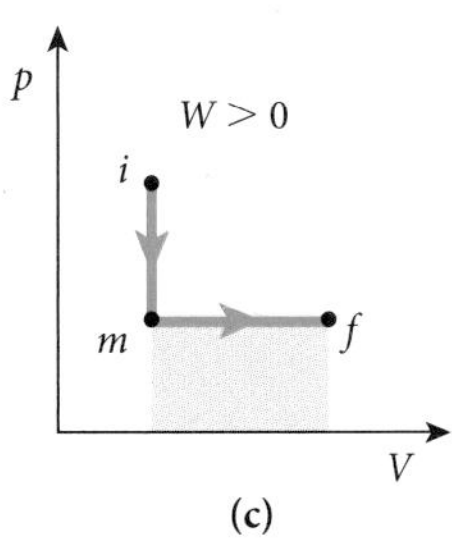


**FIGURE 18.5** The paths of three different processes in $pV$-diagrams. (a) A process in which the pressure decreases as the volume increases. (b) A two-step process in which the first step consists of increasing the volume while keeping the pressure constant and the second step consists of decreasing the pressure while holding the volume constant. (c) Another two-step process, in which the first step consists of keeping the volume constant and decreasing the pressure and the second step consists of holding the pressure constant while increasing the volume. In all three cases, the green area below the curve represents the work done during the process.

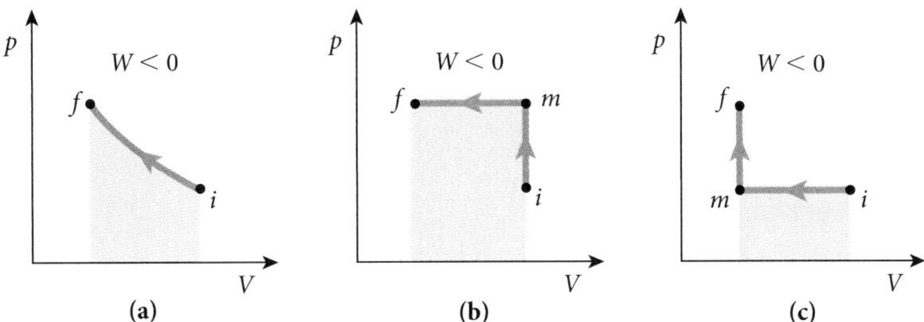


**FIGURE 18.6** The paths of three processes in $pV$-diagrams. These are the reverse of the processes shown in Figure 18.5.

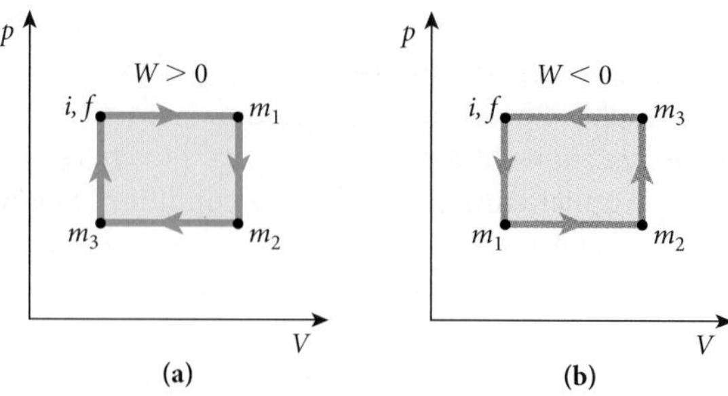


**FIGURE 18.7** Two closed-path processes in $pV$-diagrams.

### 18.1 Self-Test Opportunity

Consider the process shown in the $pV$-diagram. The path goes from point $i$ to point $f$ and back to point $i$. Is the work done by the system negative, zero, or positive?

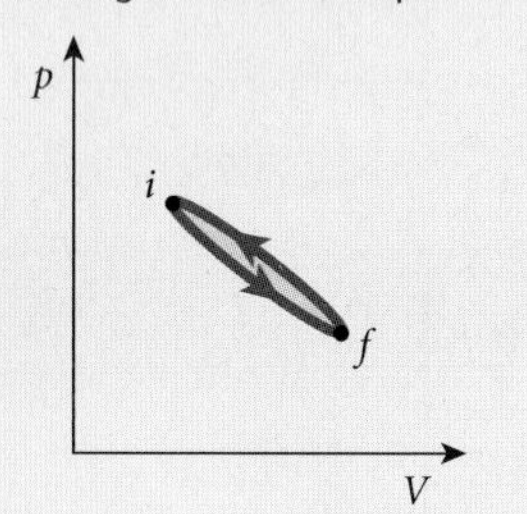


constant while the pressure decreases. From point $m_2$, the path goes to intermediate point $m_3$, with the pressure constant and the volume decreasing. Finally, the process moves from point $m_3$ to the final point, $f$, which is the same as the initial point, $i$. The area under the path from $i$ to $m_1$ represents positive work done by the system (the volume increases), while the area under the path from $m_2$ to $m_3$ represents negative work done on the system (volume decreases). The addition of the positive and the negative work yields net positive work being done by the system, as depicted by the area of the green rectangle in Figure 18.7a. In this case, heat is positive.

Figure 18.7b shows the same path on a $pV$-diagram as in Figure 18.7a, but the process occurs in the opposite direction. The area under the path from intermediate point $m_1$ to intermediate point $m_2$ corresponds to positive work done by the system. The area under the path from intermediate point $m_3$ to final point $f$ corresponds to negative work done on the system. The net work done by the system in this case is negative, and its magnitude is represented by the area of the orange rectangle in Figure 18.7b. In this case, the heat is negative.

The amount of work done by the system and the thermal energy absorbed by the system depend on the path taken on a $pV$-diagram, as well as on the direction in which the path is taken.

## 18.4 First Law of Thermodynamics

A **closed system** is a system into or out of which thermal energy can be transferred but from which no constituents can escape and to which no additional constituents are added. Combining several of the concepts already covered in this chapter allows us to express the change in internal energy of a closed system in terms of heat and work as

$$\Delta E_{\text{int}} = E_{\text{int,f}} - E_{\text{int,i}} = Q - W. \tag{18.4}$$

This equation is known as the **First Law of Thermodynamics.** It can be stated as follows:

> The change in the internal energy of a closed system is equal to the heat acquired by the system minus the work done by the system.

In other words, energy is conserved. Heat and work can be transformed into internal energy, but no energy is lost. Note that here the work is done by the system, not done on the system. Essentially, the First Law of Thermodynamics extends the conservation of energy (first encountered in Chapter 6) beyond mechanical energy to include heat as well as work. (In other contexts, such as chemical reactions, work is defined as the work done on the system, which leads to a different sign convention for the work. The sign convention can be assigned in either way, as long as it is consistent.) Note also that the change in internal energy is path-independent, while changes in heat and work are path-dependent.

**FIGURE 18.8** A weightlifter competes in the 2008 Olympic Games.

### EXAMPLE 18.2 Weightlifter

#### PROBLEM

A weightlifter snatches a barbell with mass $m = 180.0$ kg and moves it a distance $h = 1.25$ m vertically, as illustrated in Figure 18.8. If we consider the weightlifter to be a thermodynamic system, how much heat must he give off if his internal energy decreases by 4000.0 J?

**SOLUTION**

We start with the First Law of Thermodynamics (equation 18.4):

$$\Delta E_{\text{int}} = Q - W.$$

The work is the mechanical work done on the weight by the weightlifter:

$$W = mgh.$$

The heat is then given by

$$Q = \Delta E_{\text{int}} + W = \Delta E_{\text{int}} + mgh = -4000 \text{ J} + (180.0 \text{ kg})(9.81 \text{ m/s}^2)(1.25 \text{ m})$$
$$= -1790 \text{ J} = -0.428 \text{ kcal}.$$

The weightlifter cannot turn internal energy into useful work without giving off heat. Note that the decrease in internal energy of the weightlifter is less than 1 food calorie: (4000 J)/(4186 J/kcal) = 0.956 kcal. The heat output is only 0.428 kcal. This seemingly small amount of internal energy and heat associated with the large effort required to lift the weight is analogous to the amount of exercise required to burn off the calories in a candy bar (see Example 18.1).

## EXAMPLE 18.3 Truck Sliding to a Stop

**PROBLEM**

The brakes of a moving truck with a mass $m$ = 3000.0 kg lock up. The truck skids to a stop on a horizontal road surface in a distance $L$ = 83.2 m. The coefficient of kinetic friction between the truck tires and the surface of the road is $\mu_{\text{k}}$ = 0.600. What happens to the internal energy of the truck?

**SOLUTION**

The First Law of Thermodynamics (equation 18.4) relates the internal energy, the heat, and the work done on the system:

$$\Delta E_{\text{int}} = Q - W.$$

In this case, no thermal energy is transferred to or from the truck because the process of sliding to a complete stop is sufficiently fast, so that there is no time to transfer thermal energy in appreciable quantity. Thus, $Q = 0$. Work is done by the force of kinetic friction, $F_{\text{f}}$, to slow and stop the truck. The magnitude of the work done on the truck is (see Chapter 5)

$$W = F_{\text{f}} L = \mu_{\text{k}} mgL,$$

where $mg$ is the normal force exerted on the road by the truck. Because the work is done on the truck, it is negative, and we have

$$\Delta E_{\text{int}} = 0 - (-\mu_{\text{k}} mgL) = \mu_{\text{k}} mgL.$$

Putting in the numerical values gives

$$\Delta E_{\text{int}} = (0.600)(3000.0 \text{ kg})(9.81 \text{ m/s}^2)(83.2 \text{ m}) = 1.47 \text{ MJ}.$$

This increase in internal energy can warm the truck's tires. Chapter 6 discussed conservation of energy for conservative and nonconservative forces. Here we see that energy is conserved because mechanical work can be converted to internal energy.

**DISCUSSION**

Note that we have assumed that the truck is a closed system. However, when the truck begins to skid, the tires may leave skid marks on the road, removing matter and energy from

*Continued—*

the system. The assumption that no thermal energy is transferred to or from the truck in this process is also not exactly valid. In addition, the internal energy of the road surface will also increase as a result of the friction between the tires and the road, taking some of the total energy available. So, the 1.47 MJ added to the internal energy of the truck should be considered an upper limit. However, the basic lesson from this example is that the mechanical energy lost due to the action of nonconservative forces is transformed to internal energy of parts or all of the system, and the total energy is conserved.

## 18.5 First Law for Special Processes

The First Law of Thermodynamics—that is, the basic conservation of energy—holds for all kinds of processes with a closed system, but energy can be transformed and thermal energy transported in some special ways in which only a single or a few of the variables that characterize the state of the system change. Some special processes, which occur often in physical situations, can be described using the First Law of Thermodynamics. These special processes are also usually the only ones for which we can compute numerical values for heat and work. For this reason, they will be discussed several times in the following chapters. Keep in mind that these processes are simplifications or idealizations, but they often approximate real-world situations fairly closely.

### Adiabatic Processes

An **adiabatic process** is one in which no heat flows when the state of a system changes. This can happen, for example, if a process occurs quickly and there is not enough time for heat to be exchanged. Adiabatic processes are common because many physical processes do occur quickly enough that no thermal energy transfer takes place. For adiabatic processes, $Q = 0$ in equation 18.4, so

$$\Delta E_{\text{int}} = -W \quad \text{(for an adiabatic process).} \tag{18.5}$$

Another situation in which an adiabatic process can occur is if the system is thermally isolated from its environment while pressure and volume changes occur. An example is compressing a gas in an insulated container or pumping air into a bicycle tire using a manual tire pump. The change in the internal energy of the gas is due only to the work done on the gas.

### Constant-Volume Processes

Processes that occur at constant volume are called **isochoric processes.** For a process in which the volume is held constant, the system can do no work, so $W = 0$ in equation 18.4, giving

$$\Delta E_{\text{int}} = Q \quad \text{(for a process at constant volume).} \tag{18.6}$$

An example of a constant-volume process is warming a gas in a rigid, closed container that is in contact with other bodies. No work can be done because the volume of the gas cannot change. The change in the internal energy of the gas occurs because of heat flowing into or out of the gas as a result of contact between the container and the other bodies. Cooking food in a pressure cooker is an isochoric process.

### Closed-Path Processes

In a **closed-path process,** the system returns to the same state at which it started. Regardless of how the system reached that point, the internal energy must be the same as at the start, so $\Delta E_{\text{int}} = 0$ in equation 18.4. This gives

$$Q = W \quad \text{(for a closed-path process).} \tag{18.7}$$

Thus, the net work done by a system during a closed-path process is equal to the thermal energy transferred into the system. Such cyclical processes form the basis of many kinds of heat engines (discussed in Chapter 20).

### Free Expansion

If the thermally insulated (so $Q = 0$) container for a gas suddenly increases in size, the gas will expand to fill the new volume. During this **free expansion,** the system does no work and no heat is absorbed. That is, $W = 0$ and $Q = 0$, and equation 18.4 becomes

$$\Delta E_{\text{int}} = 0 \quad \text{(for free expansion of a gas).} \tag{18.8}$$

To illustrate this situation, consider a box with a barrier in the center (Figure 18.9). A gas is confined in the left half of the box. When the barrier between the two halves is removed, the gas fills the new volume. However, the gas performs no work. This last statement requires a bit of explanation: In its free expansion, the gas does not move a piston or any other material device; thus, it does no work on anything. During the expansion, the gas particles move freely until they encounter the walls of the expanded container. The gas is not in equilibrium while it is expanding. For this system, we can plot the initial state and the final state on a $pV$-diagram, but not the intermediate state.

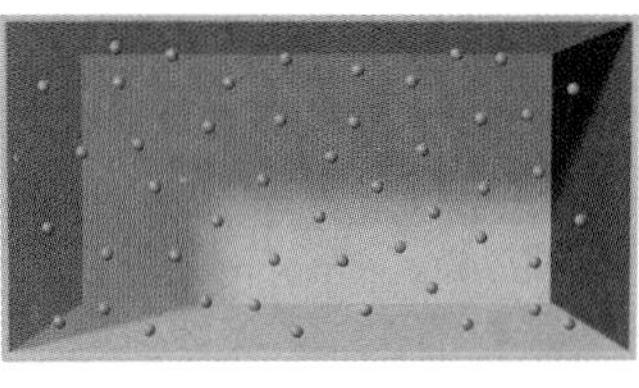

(a)

(b)

**FIGURE 18.9** (a) A gas is confined to half the volume of a box. (b) The barrier separating the two halves is removed, and the gas expands to fill the volume.

### Constant-Pressure Processes

Constant-pressure processes are called **isobaric processes.** Such processes are common in the study of the specific heat of gases. In an isobaric process, the volume can change, allowing the system to do work. Since the pressure is kept constant, $W = p(V_f - V_i) = p\Delta V$, according to equation 18.3. Thus, equation 18.4 can be written as follows:

$$\Delta E_{\text{int}} = Q - p\Delta V \quad \text{(for a process at constant pressure).} \tag{18.9}$$

An example of an isobaric process is the slow warming of a gas in a cylinder fitted with a frictionless piston, which can move to keep the pressure constant. The path of an isobaric process on a $pV$-diagram is a horizontal straight line. If the system moves in the positive-volume direction, the system is expanding. If the system moves in the negative-volume direction, the system is contracting. Cooking food in an open saucepan is another example of an isobaric process.

### Constant-Temperature Processes

Constant-temperature processes are called **isothermal processes.** The temperature of the system is held constant through contact with an external thermal reservoir. Isothermal processes take place slowly enough that heat can be exchanged with the external reservoir to maintain a constant temperature. For example, heat can flow from a warm reservoir to the system, allowing the system to do work. The path of an isothermal process in a $pV$-diagram is called an **isotherm.** As we'll see in Chapter 19, the product of the pressure and the volume is constant for an ideal gas undergoing an isothermal process, giving the isotherm the form of a hyperbola. In addition, as we'll see in Chapter 20, isothermal processes play an important part in the analysis of devices that produce useful work from sources of heat.

## 18.6 Specific Heats of Solids and Fluids

Suppose a block of aluminum is at room temperature. If heat, $Q$, is then transferred to the block, the temperature of the block goes up proportionally to the amount of heat. The proportionality constant between the temperature difference and the heat is the **heat capacity,** $C$, of an object. Thus,

$$Q = C\Delta T, \tag{18.10}$$

where $\Delta T$ is the change in temperature.

The term *heat capacity* does not imply that an object contains a certain amount of heat. Rather, it tells how much heat is required to raise the temperature of the object by a given amount. The SI units for heat capacity are joules per kelvin (J/K).

The temperature change of an object due to heat can be described by the **specific heat,** $c$, which is defined as the heat capacity per unit mass, $m$:

$$c = \frac{C}{m}. \tag{18.11}$$

**Table 18.1 Specific Heats for Selected Substances**

| Material | Specific Heat, $c$ kJ/(kg K) | cal/(g K) |
|---|---|---|
| Lead | 0.129 | 0.0308 |
| Copper | 0.386 | 0.0922 |
| Steel | 0.448 | 0.107 |
| Aluminum | 0.900 | 0.215 |
| Glass | 0.840 | 0.20 |
| Ice | 2.06 | 0.500 |
| Water | 4.19 | 1.00 |
| Steam | 2.01 | 0.48 |

With this definition, the relationship between temperature change and heat can be written as

$$Q = cm\Delta T. \tag{18.12}$$

The SI units for the specific heat are J/(kg K). In practical applications, specific heat is also often in cal/(g K). The units J/(kg K) and J/(kg °C) can be used interchangeably for specific heat since it is defined in terms of $\Delta T$. The specific heats of various materials are given in Table 18.1.

Note that specific heat and heat capacity are usually measured in two ways. For most substances, they are measured under constant pressure (as in Table 18.1) and denoted by $c_p$ and $C_p$. However, for fluids (gases and liquids), the specific heat and heat capacity can also be measured at constant volume and denoted by $c_V$ and $C_V$. In general, measurements under constant pressure produce larger values, because mechanical work has to be performed in the process, and the difference is particularly large for gases. We'll discuss this in greater detail in Chapter 19.

The specific heat of a substance can also be defined in terms of the number of moles of a material, rather than its mass. This type of specific heat is called the *molar specific heat,* and is also discussed in Chapter 19.

The effect of differences in specific heats of substances can be observed readily at the seashore, where the sun transfers energy to the land and to the water roughly equally during the day. The specific heat of the land is about five times lower than the specific heat of the water. Thus, the land warms up more quickly than the water, and it warms the air above it more than the water warms the air above it. This temperature difference creates an onshore breeze during the day. The high specific heat of water helps to moderate climates around oceans and large lakes.

## EXAMPLE 18.4 Energy Required to Warm Water

**PROBLEM**

You have 2.00 L of water at a temperature of 20.0 °C. How much energy is required to raise the temperature of that water to 95.0 °C? Assuming you use electricity to warm the water, how much will it cost at 10.0 cents per kilowatt-hour?

**SOLUTION**

The mass of 1.00 L of water is 1.00 kg. From Table 18.1, $c_{\text{water}} = 4.19$ kJ/(kg K). Therefore, the energy required to warm 2.00 kg of water from 20.0 °C to 95.0 °C is

$$Q = c_{\text{water}} m_{\text{water}} \Delta T = \left[4.19 \text{ kJ/}\left(\text{kg K}\right)\right]\left(2.00 \text{ kg}\right)\left(95.0\text{ °C} - 20.0\text{ °C}\right)$$
$$= 629{,}000 \text{ J}.$$

Using the conversion factor for converting joules to kilowatt-hours (equation 18.1), we calculate the cost of warming the water as

$$\text{Cost} = (629{,}000\text{ J})\left(\frac{10.0\text{ cents}}{1\text{ kW h}}\right)\left(\frac{1\text{ kW h}}{3.60\cdot 10^{6}\text{ J}}\right) = 1.75\text{ cents.}$$

### 18.1 In-Class Exercise

How much energy is needed to raise the temperature of a copper block with a mass of 3.00 kg from 25.0 °C to 125 °C?

a) 116 kJ

b) 278 kJ

c) 421 kJ

d) 576 kJ

e) 761 kJ

## Calorimetry

A **calorimeter** is a device used to study internal energy changes by measuring thermal energy transfers. The thermal energy transfer and internal energy change may result from a chemical reaction, a physical change, or differences in temperature and specific heat. A simple calorimeter consists of an insulated container and a thermometer. For simple measurements, a Styrofoam cup and an ordinary alcohol thermometer will do. We'll assume that no heat is lost or gained in the calorimeter or the thermometer.

When two materials with different temperatures and specific heats are placed in a calorimeter, heat will flow from the warmer substance to the cooler substance until the temperatures of the two substances are the same. No heat will flow into or out of the calorimeter. The heat lost by the warmer substance will be equal to the heat gained by the cooler substance.

A substance such as water with a high specific heat, $c = 4.19$ kJ/(kg K), requires more heat to raise its temperature by the same amount than does a substance with a low specific heat, like steel, with $c = 0.488$ kJ/(kg K). Solved Problem 18.1 illustrates the concepts of calorimetry.

### SOLVED PROBLEM 18.1 Water and Lead

**PROBLEM**

A metalsmith pours 3.00 kg of lead shot (which is the material used to fill shotgun shells) at a temperature of 94.7 °C into 1.00 kg of water at 27.5 °C in an insulated container, which acts as a calorimeter. What is the final temperature of the mixture?

**SOLUTION**

**THINK**

The lead shot will give up heat and the water will absorb heat until both are at the same temperature. (Something that may not occur to you is that lead shot has a melting temperature, but it is significantly above 94.7 °C; so we do not have to deal with a phase change in this situation. The water will stay liquid, and the lead shot will stay solid. Section 18.7 addresses situations involving phase changes.)

**SKETCH**

Figure 18.10 shows the problem situation before and after the lead shot is added to the water.

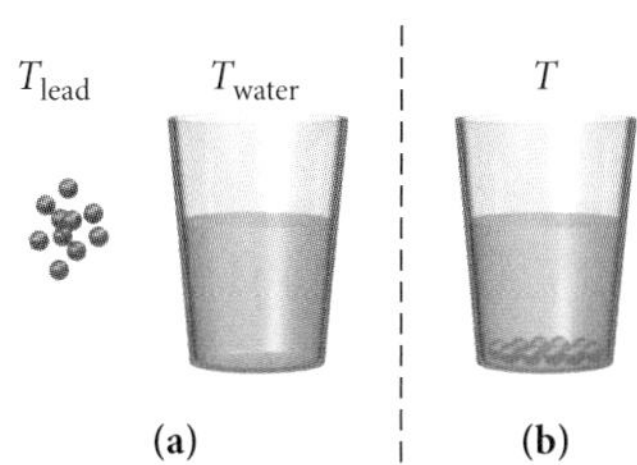


**FIGURE 18.10** Lead shot and water (a) before and (b) after the lead shot is added to the water.

**RESEARCH**

The heat lost by the lead shot, $Q_{lead}$, to its environment is given by $Q_{lead} = m_{lead}c_{lead}(T - T_{lead})$, where $c_{lead}$ is the specific heat of lead, $m_{lead}$ is the mass of the lead shot, $T_{lead}$ is the original temperature of the lead shot, and $T$ is the final equilibrium temperature.

The heat gained by the water, $Q_{water}$, is given by $Q_{water} = m_{water}c_{water}(T - T_{water})$, where $c_{water}$ is the specific heat of water, $m_{water}$ is the mass of the water, and $T_{water}$ is the original temperature of the water.

The sum of the heat lost by the lead shot and the heat gained by the water is zero, because the process took place in an insulated container and because the total energy is conserved, a consequence of the First Law of Thermodynamics. So we can write

$$Q_{lead} + Q_{water} = 0 = m_{lead}c_{lead}\left(T - T_{lead}\right) + m_{water}c_{water}\left(T - T_{water}\right).$$

*Continued—*

**SIMPLIFY**

We multiply through on both sides and reorder so that all of the terms containing the unknown temperature are on the left side of the equation:

$$m_{\text{lead}}c_{\text{lead}}T_{\text{lead}} + m_{\text{water}}c_{\text{water}}T_{\text{water}} = m_{\text{water}}c_{\text{water}}T + m_{\text{lead}}c_{\text{lead}}T.$$

We can solve this equation for $T$ by dividing both sides by $m_{\text{lead}}c_{\text{lead}} + m_{\text{water}}c_{\text{water}}$:

$$T = \frac{m_{\text{lead}}c_{\text{lead}}T_{\text{lead}} + m_{\text{water}}c_{\text{water}}T_{\text{water}}}{m_{\text{lead}}c_{\text{lead}} + m_{\text{water}}c_{\text{water}}}.$$

**CALCULATE**

Putting in the numerical values gives

$$T = \frac{(3.00\text{ kg})[0.129\text{ kJ/(kg K)}](94.7\text{ °C}) + (1.00\text{ kg})[4.19\text{ kJ/(kg K)}](27.5\text{ °C})}{(3.00\text{ kg})[0.129\text{ kJ/(kg K)}] + (1.00\text{ kg})[4.19\text{ kJ/(kg K)}]}$$
$$= 33.182\text{ °C}.$$

**ROUND**

We report our result to three significant figures:

$$T = 33.2\text{ °C}.$$

**DOUBLE-CHECK**

The final temperature of the mixture of lead shot and water is only 5.7 °C higher than the original temperature of the water. The masses of the lead shot and water differ by a factor of 3, but the specific heat of lead is much lower than the specific heat of water. Thus, it is reasonable that the lead had a large change in temperature and the water had a small change in temperature. To double-check, we calculate the heat lost by the lead,

$$Q_{\text{lead}} = m_{\text{lead}}c_{\text{lead}}(T - T_{\text{lead}}) = (3.00\text{ kg})[0.129\text{ kJ/(kg K)}](33.2\text{ °C} - 94.7\text{ °C})$$
$$= -23.8\text{ kJ},$$

and compare the result to the heat gained by the water,

$$Q_{\text{water}} = m_{\text{water}}c_{\text{water}}(T - T_{\text{water}}) = (1.00\text{ kg})[4.19\text{ kJ/(kg K)}](33.2\text{ °C} - 27.5\text{ °C})$$
$$= 23.9\text{ kJ}.$$

These results add up to 0 within rounding error, as required.

## 18.7 Latent Heat and Phase Transitions

As noted in Chapter 13, the three common **states of matter** (sometimes also called **phases**) are solid, liquid, and gas. We have been considering objects for which the change in temperature is proportional to the amount of heat that is added. This linear relationship between heat and temperature is, strictly speaking, an approximation, but it has high accuracy for solids and liquids. For a gas, adding heat will raise the temperature but may also change the pressure or volume, depending on whether and how the gas is contained. Substances can have different specific heats depending on whether they are in a solid, liquid, or gaseous state.

If enough heat is added to a solid, it melts into a liquid. If enough heat is added to a liquid, it vaporizes into a gas. These are examples of **phase changes,** or **phase transitions** (Figure 18.11). During a phase change, the temperature of an object remains constant. The heat that is required to melt a solid, divided by its mass, is called the **latent heat of fusion,** $L_{\text{fusion}}$. Melting changes a substance from a solid to a liquid. The heat that is required to vaporize a liquid, divided by its mass, is called the **latent heat of vaporization,** $L_{\text{vaporization}}$. Vaporizing changes a substance from a liquid to a gas.

The temperature at which a solid melts to a liquid is the melting point, $T_{\text{melting}}$. The temperature at which a liquid vaporizes to a gas is the boiling point, $T_{\text{boiling}}$. The relationship

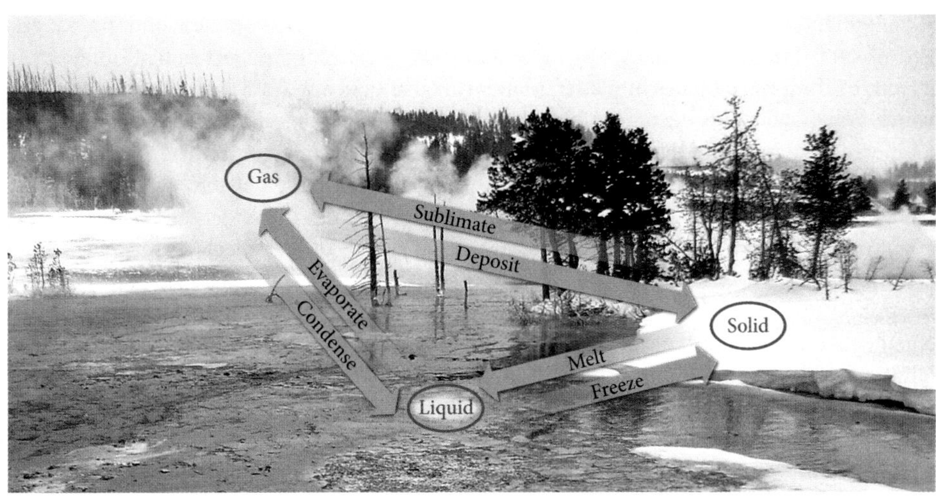


**FIGURE 18.11** Phase changes involving three states of water, which are simultaneously present in this picture taken at Yellowstone National Park.

between the mass of an object at its melting point and the heat needed to change the object from a solid to a liquid is given by

$$Q = mL_{\text{fusion}} \quad \left(\text{for } T = T_{\text{melting}}\right). \tag{18.13}$$

Similarly, the relationship between an object at its boiling point and the heat needed to change the object from a liquid to a gas is given by

$$Q = mL_{\text{vaporization}} \quad \left(\text{for } T = T_{\text{boiling}}\right). \tag{18.14}$$

The SI units for latent heats of fusion and vaporization are joules per kilogram (J/kg); units of calories per gram (cal/g) are also often used. The latent heat of fusion for a given substance is different from the latent heat of vaporization for that substance. Representative values for melting point, latent heat of fusion, boiling point, and latent heat of vaporization are listed in Table 18.2.

It is also possible for a substance to change directly from a solid to a gas. This process is called **sublimation.** For example, sublimation occurs when dry ice, which is solid (frozen) carbon dioxide, changes directly to gaseous carbon dioxide without passing through the liquid state. When a comet approaches the Sun, some of its frozen carbon dioxide sublimates, helping to produce the comet's visible tail.

If we continue to heat a gas, it will become ionized, which means that some or all of the electrons in the atoms of the gas are removed. An ionized gas and its free electrons form a state of matter called a **plasma.** Plasmas are very common in the universe; in fact, as much as 99% of the mass of the Solar System exists in the form of a plasma. The dusty cloud of gas known as the Omega/Swan Nebula (M17), shown in Figure 18.12, is also a plasma.

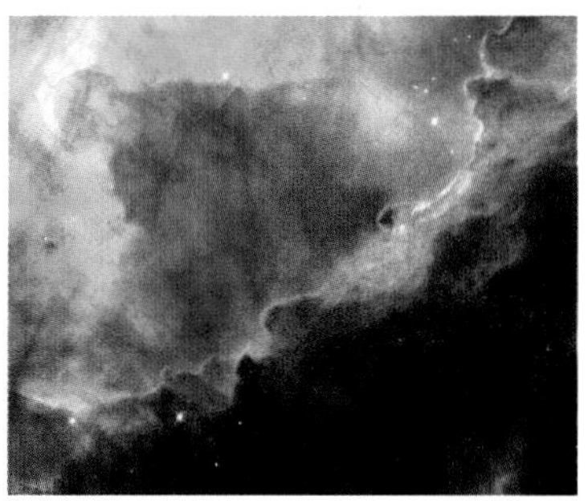

**FIGURE 18.12** Image of the Omega/Swan Nebula (M17), taken by the Hubble Space Telescope, shows an immense region of gas ionized by radiation from young stars.

**Table 18.2 Some Representative Melting Points, Boiling Points, Latent Heats of Fusion, and Latent Heats of Vaporization**

| | Melting Point | Latent Heat of Fusion, $L_{\text{fusion}}$ | | Boiling Point | Latent Heat of Vaporization, $L_{\text{vaporization}}$ | |
|---|---|---|---|---|---|---|
| **Material** | **(K)** | **(kJ/kg)** | **(cal/g)** | **(K)** | **(kJ/kg)** | **(cal/g)** |
| Hydrogen | 13.8 | 58.6 | 14.0 | 20.3 | 452 | 108 |
| Ethyl Alcohol | 156 | 104 | 24.9 | 351 | 858 | 205 |
| Mercury | 234 | 11.3 | 2.70 | 630 | 293 | 70.0 |
| Water | 273 | 334 | 79.7 | 373 | 2260 | 539 |
| Aluminum | 932 | 396 | 94.5 | 2740 | 10,500 | 2500 |
| Copper | 1359 | 205 | 49.0 | 2840 | 4730 | 1130 |

As mentioned in Chapter 13, there are other states of matter besides solid, liquid, gas, and plasma. For example, matter in a granular state has specific properties that differentiate it from the four major states of matter. Matter can also exist as a Bose-Einstein condensate in which individual atoms become indistinguishable, under very specific conditions at very low temperatures,, as we'll see in Chapter 36 on quantum mechanics. American physicists Eric Cornell and Carl Wieman, along with German physicist Wolfgang Ketterle, were awarded the Nobel Prize in Physics in 2001 for their studies of a Bose-Einstein condensate.

Cooling an object means reducing the internal energy of the object. As thermal energy is removed from a substance in the gaseous state, the temperature of the gas decreases, in relation to the specific heat of the gas, until the gas begins to condense into a liquid. This change takes place at a temperature called the *condensation point,* which is the same temperature as the boiling point of the substance. To convert all the gas to liquid requires the removal of an amount of heat corresponding to the latent heat of vaporization times the mass of the gas. If thermal energy continues to be removed, the temperature of the liquid will go down, as determined by the specific heat of the liquid, until the temperature reaches the freezing point, which is the same temperature as the melting point of the substance. To convert all of the liquid to a solid requires removing an amount of heat corresponding to the latent heat of fusion times the mass. If heat then continues to be removed from the solid, its temperature will decrease in relation to the specific heat of the solid.

### 18.2 In-Class Exercise

How much energy is needed to melt a copper block with a mass of 3.00 kg that is initially at a temperature of 1359 K?

a) 101 kJ
b) 221 kJ
c) 390 kJ
d) 615 kJ
e) 792 kJ

## EXAMPLE 18.5 Warming Ice to Water and Water to Steam

### PROBLEM

How much heat, $Q$, is required to convert 0.500 kg of ice (frozen water) at a temperature of –30 °C to steam at 140 °C?

### SOLUTION

We solve this problem in steps, with each step corresponding to either a rise in temperature or a phase change. We first calculate how much heat is required to raise the temperature of the ice from –30 °C to 0 °C. The specific heat of ice is 2.06 kJ/(kg K), so the heat required is

$$Q_1 = cm\Delta T = \left[2.06 \text{ kJ/(kg K)}\right](0.500 \text{ kg})(30 \text{ K}) = 30.9 \text{ kJ}.$$

We continue to add heat to the ice until it melts. The temperature remains at 0 °C until all the ice is melted. The latent heat of fusion for ice is 334 kJ/kg, so the heat required to melt all of the ice is

$$Q_2 = mL_{\text{fusion}} = (334 \text{ kJ/kg})(0.500 \text{ kg}) = 167 \text{ kJ}.$$

Once all the ice has melted to water, we continue to add heat until the water reaches the boiling point, at 100 °C. The required heat for this step is

$$Q_3 = cm\Delta T = \left[4.19 \text{ kJ/(kg K)}\right](0.500 \text{ kg})(100 \text{ K}) = 209.5 \text{ kJ}.$$

We continue to add heat to the water until it vaporizes. The heat required for vaporization is

$$Q_4 = mL_{\text{vaporization}} = (2260 \text{ kJ/kg})(0.500 \text{ kg}) = 1130 \text{ kJ}.$$

We now warm the steam and raise its temperature from 100 °C to 140 °C. The heat necessary for this step is

$$Q_5 = cm\Delta T = \left[2.01 \text{ kJ/(kg K)}\right](0.500 \text{ kg})(40 \text{ K}) = 40.2 \text{ kJ}.$$

Thus, the total heat required is

$$\begin{aligned} Q &= Q_1 + Q_2 + Q_3 + Q_4 + Q_5 \\ &= 30.9 \text{ kJ} + 167 \text{ kJ} + 209.5 \text{ kJ} + 1130 \text{ kJ} + 40.2 \text{ kJ} = 1580 \text{ kJ}. \end{aligned}$$

Figure 18.13 shows a graph of the temperature of the water as a function of the heat added. In region (a), the temperature of the ice increases from –30 °C to 0 °C. In region

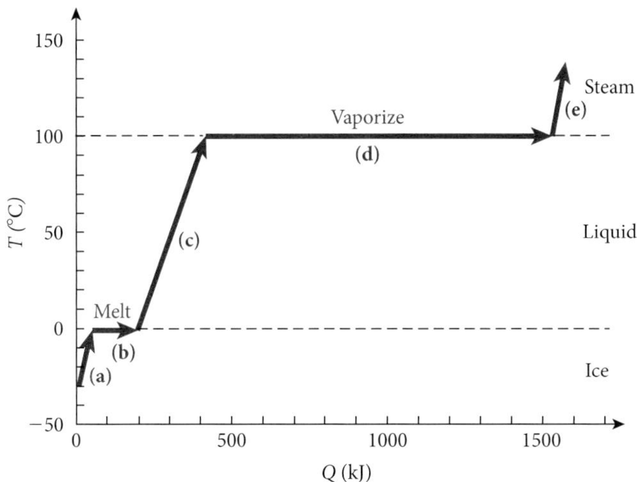


**FIGURE 18.13** Plot of temperature versus heat added to change 0.500 kg of ice starting at a temperature of −30 °C to steam at 140 °C.

(b), the ice melts to water while the temperature remains at 0 °C. In region (c), the water warms from 0 °C to 100 °C. In region (d), the water boils and changes to steam while the temperature remains at 100 °C. In region (e), the temperature of the steam increases from 100 °C to 140 °C.

Note that almost two thirds of the total heat in this entire process of warming the sample from −30 °C to 140 °C has to be spent to convert the liquid water to steam in the process of vaporizing.

### 18.3 In-Class Exercise

You have a block of ice of mass $m$ at a temperature of −3 °C in a thermally insulated container and you add the same mass $m$ of liquid water at a temperature of 6 °C and let the mixture come to equilibrium. What is the temperature of the mixture?

a) −3 °C

b) 0 °C

c) +3 °C

d) +4.5 °C

e) +6 °C

## EXAMPLE 18.6 Work Done Vaporizing Water

Suppose 10.0 g of water at a temperature of 100.0 °C is in an insulated cylinder equipped with a piston to maintain a constant pressure of $p$ = 101.3 kPa. Enough heat is added to the water to vaporize it to steam at a temperature of 100.0 °C. The volume of the water is $V_{\text{water}} = 10.0\ \text{cm}^3$, and the volume of the steam is $V_{\text{steam}} = 16{,}900\ \text{cm}^3$.

### PROBLEM

What is the work done by the water as it vaporizes? What is the change in internal energy of the water?

### SOLUTION

This process is carried out at a constant pressure. The work done by the vaporizing water is given by equation 18.3:

$$W = \int_{V_i}^{V_f} p\,dV = p\int_{V_i}^{V_f} dV = p\left(V_{\text{steam}} - V_{\text{water}}\right).$$

Putting in the numerical values gives the work done by the water as it increases in volume at a constant pressure:

$$W = \left(101.3\cdot 10^3\ \text{Pa}\right)\left(16{,}900\cdot 10^{-6}\ \text{m}^3 - 10.0\cdot 10^{-6}\ \text{m}^3\right) = 1710\ \text{J}.$$

The change in the internal energy of the water is given by the First Law of Thermodynamics (equation 18.4)

$$\Delta E_{\text{int}} = Q - W.$$

*Continued—*

In this case, the thermal energy transferred is the heat required to vaporize the water. From Table 18.2, the latent heat of vaporization of water is $L_{\text{vaporization}} = 2260$ kJ/kg. Thus, the thermal energy transferred to the water is

$$Q = mL_{\text{vaporization}} = \left(10.0 \cdot 10^{-3}\ \text{kg}\right)\left(2.260 \cdot 10^{6}\ \text{J/kg}\right) = 22{,}600\ \text{J}.$$

The change in the internal energy of the water is then

$$\Delta E_{\text{int}} = 22{,}600\ \text{J} - 1710\ \text{J} = 20{,}900\ \text{J}.$$

Most of the added energy remains in the water as increased internal energy. This internal energy is related to the phase change of water to steam. The energy is used to overcome the attractive forces between the water molecules in the liquid state in converting them to the gaseous state.

## 18.8 Modes of Thermal Energy Transfer

**FIGURE 18.14** A campfire illustrates the three main modes of thermal energy transfer: conduction, convection, and radiation.

The three main modes of thermal energy transfer are conduction, convection, and radiation. All three are illustrated by the campfire shown in Figure 18.14. **Radiation** is the transfer of thermal energy via electromagnetic waves. You can feel heat radiating from a campfire when you sit near it. **Convection** involves the physical movement of a substance (such as water or air) from thermal contact with one system to thermal contact with another system. The moving substance carries internal energy. You can see the thermal energy rising in the form of flames and hot air above the flames in the campfire. **Conduction** involves the transfer of thermal energy within an object (such as thermal energy transfer along a fire poker whose tip is hot) or the transfer of heat between two (or more) objects in thermal contact. Heat is conducted through a substance by the vibration of atoms and molecules and by motion of electrons. The water and food in the pots on the campfire are heated by conduction. The pots themselves are warmed by convection and radiation.

### Conduction

Consider a bar of material with cross-sectional area $A$ and length $L$ (Figure 18.15a). This bar is put in physical contact with a thermal reservoir at a hotter temperature, $T_h$, and a thermal reservoir at colder temperature $T_c$, which means $T_h > T_c$ (Figure 18.15b). Heat then flows from the reservoir at higher temperature to the reservoir at lower temperature. The thermal energy transferred per unit time, $P_{\text{cond}}$, by the bar connecting the two heat reservoirs has been found to be given by

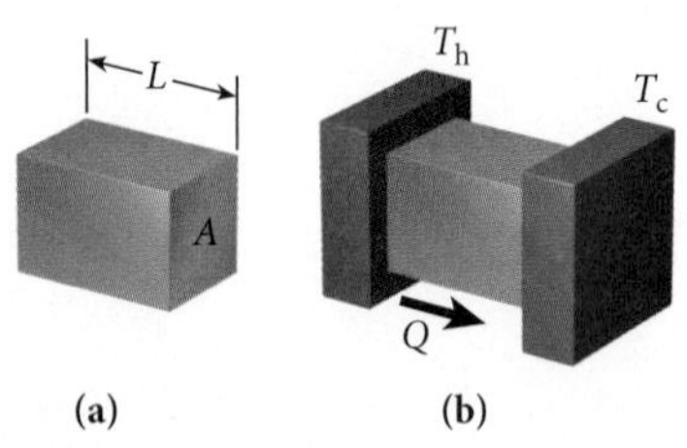


**FIGURE 18.15** (a) A bar with cross-sectional area $A$ and length $L$. (b) The bar is placed between two thermal reservoirs with temperatures $T_h$ and $T_c$.

$$P_{\text{cond}} = \frac{Q}{t} = kA\frac{T_h - T_c}{L}, \tag{18.15}$$

where $k$ is the **thermal conductivity** of the material of the bar. The SI units of thermal conductivity are W/(m K). Some typical values of thermal conductivity are listed in Table 18.3.

We can rearrange equation 18.15:

$$P_{\text{cond}} = \frac{Q}{t} = A\frac{T_h - T_c}{L/k} = A\frac{T_h - T_c}{R}, \tag{18.16}$$

where the **thermal resistance,** $R$, of the bar is defined to be

$$R = \frac{L}{k}. \tag{18.17}$$

The SI units of $R$ are $\text{m}^2$ K/W. A higher $R$ value means a lower rate of thermal energy transfer. A good insulator has a high $R$ value. In the United States, thermal resistance is often specified by an $R$ factor, which has the units $\text{ft}^2$ °F h/BTU. A common insulating material marked with an $R$ factor of R-30 is shown in Figure 18.16. To convert an $R$ factor value from $\text{ft}^2$ °F h/BTU to $\text{m}^2$ K/W, divide the $R$ factor by 5.678.

(a)

(b)

**FIGURE 18.16** (a) A common insulating material with an $R$ factor of R-30. (b) Insulation being installed in an attic.

**Table 18.3 Some Representative Thermal Conductivities**

| Material | $k$[W/(m k)] |
|---|---|
| Copper | 386 |
| Aluminum | 220 |
| Concrete | 0.8 |
| Glass | 0.8 |
| Rubber | 0.16 |
| Wood | 0.16 |

### 18.2 Self-Test Opportunity

Show that the conversion factor for changing from ft$^2$ °F h/BTU to m$^2$ K/W is 5.678.

## EXAMPLE 18.7 Roof Insulation

Suppose you insulate above the ceiling of a room with an insulation material having an $R$ factor of R-30. The room is 5.00 m by 5.00 m. The temperature inside the room is 21.0 °C, and the temperature above the insulation is 40.0 °C.

**PROBLEM**

How much heat enters the room through the ceiling in a day if the room is maintained at a temperature of 21.0 °C?

**SOLUTION**

We solve equation 18.16 for the heat:

$$Q = At\frac{T_h - T_c}{R}.$$

Putting in the numerical values, we obtain (1 day has 86400 s):

$$Q = (5.00\text{ m} \cdot 5.00\text{ m})(86400\text{ s})\frac{313\text{ K} - 294\text{ K}}{(30/5.678)\text{ m}^2\text{ K/W}} = 7.77 \cdot 10^6\text{ J}.$$

## SOLVED PROBLEM 18.2 Cost of Warming a House in Winter

You build a small house with four rooms (Figure 18.17a). Each room is 10.0 ft by 10.0 ft, and the ceiling is 8.00 ft high. The exterior walls are insulated with a material with an $R$ factor of R-19, and the floor and ceiling are insulated with a material with an $R$ factor of R-30. During the winter, the average temperature inside the house is 20.0 °C, and the average temperature outside the house is 0.00 °C. You warm the house for 6 months in winter using electricity that costs 9.5 cents per kilowatt-hour.

**PROBLEM**

How much do you pay to warm your house for the winter?

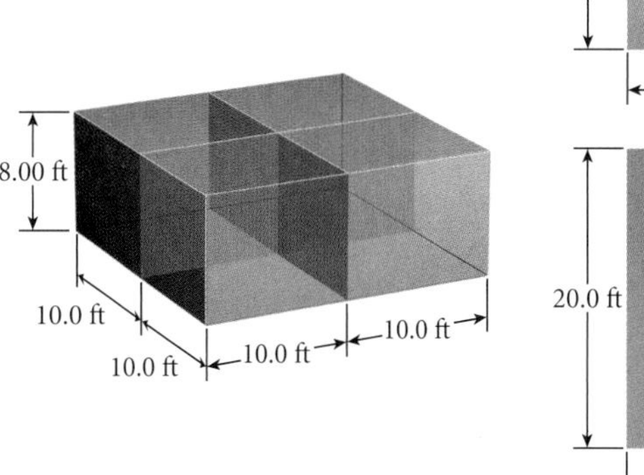

(a)

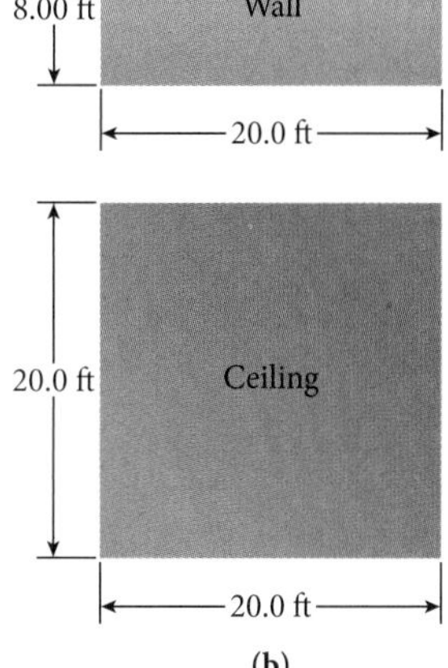

(b)

**FIGURE 18.17** (a) An insulated four-room house. (b) One wall and the ceiling of the house.

*Continued—*

## SOLUTION

### THINK

Using equation 18.16, we can calculate the total heat lost over 6 months through the walls (R-19) and the floor and ceiling (R-30). We can then calculate how much it will cost to add this amount of heat to the house.

### SKETCH

The dimensions of one wall and the ceiling are shown in Figure 18.17b.

### RESEARCH

Each of the four exterior walls, as shown in Figure 18.17b, has an area of $A_{\text{wall}}$. The ceiling has an area of $A_{\text{ceiling}}$, shown in Figure 18.17b. The floor and ceiling have the same area. The thermal resistance of the walls is given by R-19; the thermal resistance of the floor and ceiling is given by R-30. Thus, the thermal resistance of the walls in SI units is

$$R_{\text{wall}} = \frac{19}{5.678}\ \text{m}^2\ \text{K/W} = 3.346\ \text{m}^2\ \text{K/W},$$

and the thermal resistance of the floor and ceiling is

$$R_{\text{ceiling}} = \frac{30}{5.678}\ \text{m}^2\ \text{K/W} = 5.283\ \text{m}^2\ \text{K/W}.$$

### SIMPLIFY

We can calculate the heat lost per unit time using equation 18.16, taking the total area of the walls to be four times the area of one wall and the total area of the floor and ceiling as twice the area of the ceiling:

$$\frac{Q}{t} = 4A_{\text{wall}}\left(\frac{T_{\text{h}} - T_{\text{c}}}{R_{\text{wall}}}\right) + 2A_{\text{ceiling}}\left(\frac{T_{\text{h}} - T_{\text{c}}}{R_{\text{ceiling}}}\right) = 2\left(T_{\text{h}} - T_{\text{c}}\right)\left(\frac{2A_{\text{wall}}}{R_{\text{wall}}} + \frac{A_{\text{ceiling}}}{R_{\text{ceiling}}}\right).$$

### CALCULATE

The area of each exterior wall is $A_{\text{wall}} = (8.00\ \text{ft})(20.0\ \text{ft}) = 160.0\ \text{ft}^2 = 14.864\ \text{m}^2$. The area of the ceiling is $A_{\text{ceiling}} = (20.0\ \text{ft})(20.0\ \text{ft}) = 400.0\ \text{ft}^2 = 37.161\ \text{m}^2$. The number of seconds in 6 months is

$$t = (6\ \text{months})(30\ \text{days/month})(24\ \text{h/day})(3600\ \text{s/h}) = 1.5552 \cdot 10^7\ \text{s}.$$

The amount of heat lost in 6 months is then

$$\begin{aligned} Q &= 2t\left(T_{\text{h}} - T_{\text{c}}\right)\left(\frac{2A_{\text{wall}}}{R_{\text{wall}}} + \frac{A_{\text{ceiling}}}{R_{\text{ceiling}}}\right) \\ &= 2\left(1.5552 \cdot 10^7\ \text{s}\right)\left(293\ \text{K} - 273\ \text{K}\right)\left(\frac{2\left(14.864\ \text{m}^2\right)}{3.346\ \text{m}^2\ \text{K/W}} + \frac{37.161\ \text{m}^2}{5.283\ \text{m}^2\ \text{K/W}}\right) \\ &= 9.9027 \cdot 10^9\ \text{J}. \end{aligned}$$

Calculating the total cost for 6 months of electricity for warming, we get

$$\text{Cost} = \left(\frac{\$0.095}{1\ \text{kW h}}\right)\left(\frac{1\ \text{kW h}}{3.60 \cdot 10^6\ \text{J}}\right)\left(9.9027 \cdot 10^9\ \text{J}\right) = \$261.32125.$$

### ROUND

Since the input data were given to two significant figures, we report our result to two significant figures as well:

$$\text{Cost} = \$260.$$

(However, rounding down the amount on a bill is not acceptable to the utility company, and you would have to pay \$261.32.)

**DOUBLE-CHECK**

To double-check, let's use the $R$ values the way a contractor in the United States might do it. The total area of the walls is 4(160 ft$^2$) and the total area of the floor and ceiling is 2(400 ft$^2$). The heat lost per hour through the walls is

$$\frac{Q}{t}=4\left(160\text{ ft}^2\right)\frac{20\left(\frac{9}{5}\right)\,°\text{F}}{19\text{ ft}^2\;°\text{F h/BTU}}=1213\text{ BTU/h}.$$

The heat lost through the ceiling and floor is

$$\frac{Q}{t}=2\left(400\text{ ft}^2\right)\frac{20\left(\frac{9}{5}\right)\,°\text{F}}{30\text{ ft}^2\;°\text{F h/BTU}}=960\text{ BTU/h}.$$

The total heat lost in 6 months (4320 h) is

$$\begin{aligned}Q&=(4320\text{ h})(1213+960)\text{ BTU/h}\\&=9{,}387{,}360\text{ BTU}.\end{aligned}$$

The cost is then

$$\text{Cost}=\left(\frac{\$0.095}{1\text{ kW h}}\right)\left(\frac{1\text{ kW h}}{3412\text{ BTU}}\right)(9{,}387{,}360\text{ BTU})=\$261.37.$$

This is close to our answer and seems reasonable.

Note that the cost of warming the small house for the winter seems small compared with real-world experience. The reduction arises because the house has no doors or windows and is well insulated.

Thermal insulation is a key component of spacecraft that have to reenter the Earth's atmosphere. The reentry process creates thermal energy from friction with the air molecules. Due to the high speed, a shock wave forms in front of the spacecraft, which deflects most of the thermal energy created in the process. However, excellent thermal insulation is still required to prevent this heat from being conducted to the frame of the spacecraft, which is basically made of aluminum and cannot stand temperatures significantly higher than 180 °C. The thermal protection system has to be of very low mass, like all parts of a spacecraft, and be able to protect the spacecraft from very high temperatures, up to 1650 °C. In the spacecraft for Apollo missions, in which astronauts performed the Moon landings in the late 1960s and 1970s, and in planetary probes like the *Mars Rover,* the heat protection is simply an ablative heat shield, which burns off during entry into the atmosphere. For a reusable spacecraft like the Space Shuttle, such a shield is not an option, because it would require prohibitively expensive maintenance after each trip. The nose and wing edges of the Space Shuttle are instead covered with reinforced carbon, able to withstand the high reentry temperatures. The underside of the Space Shuttle is covered with passive heat protection, consisting of over 20,000 ceramic tiles made of 10% silica fibers and 90% empty space. They are such good thermal insulators that after being warmed to a temperature of 1260 °C (which is the maximum temperature the underside of the Space Shuttle encounters during reentry), they can be held by unprotected hands (see Figure 18.18).

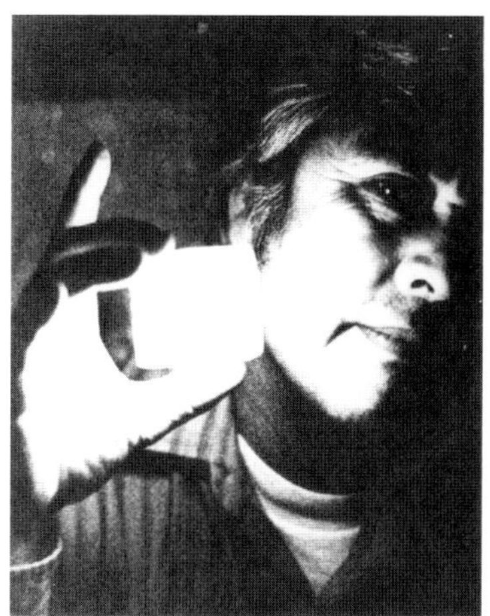

**FIGURE 18.18** White-hot (1260 °C) ceramic tile used for the thermal protection system of the Space Shuttle is held by an unprotected hand.

## Convection

If you hold your hand above a burning candle, you can feel the thermal energy transferred from the flame. The warmed air is less dense than the surrounding air and rises. The rising air carries thermal energy upward from the candle flame. This type of thermal energy transfer is called *convection.* Figure 18.19 shows a candle flame on Earth and on the orbiting Space Shuttle. You can see that the thermal energy travels upward from the candle burning on Earth but expands almost spherically from the candle aboard the Space Shuttle. The air in the Space Shuttle (actually in a small container on the Space Shuttle) has the same density in all directions, so the warmed air does not have a preferential direction in which to travel. The bottom of the flame on the candle aboard the Space Shuffle is extinguished because the wick carries thermal energy

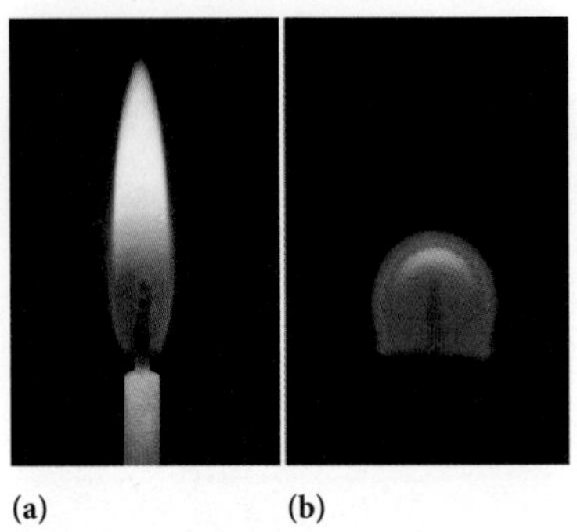

(a) (b)

**FIGURE 18.19** Candle flames under (a) terrestrial conditions and (b) microgravity conditions.

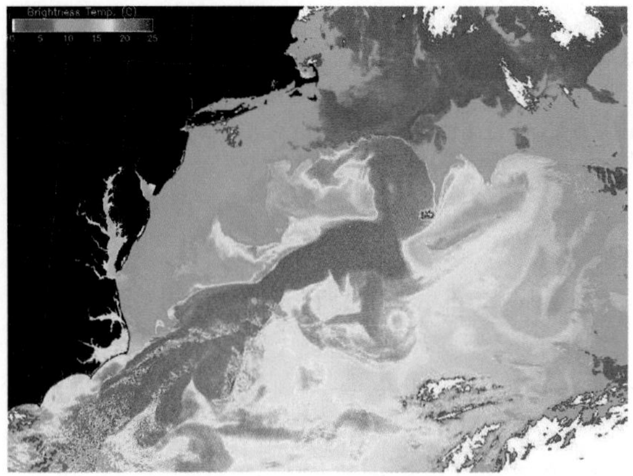

**FIGURE 18.20** A satellite image taken by the NASA TERRA satellite on May 5, 2001, shows the temperature of the water in the North Atlantic Ocean. The false colors represent the temperature range of the water. The warm Gulf Stream is visible as red and the East Coast of the United States is shown in black.

away. (In addition to convection involving a large-scale flow of mass, convection can occur via individual particles in a process called *diffusion,* which is not covered in this chapter.)

Most houses and office buildings in the United States have forced-air heating; that is, warm air is blown through heating ducts into rooms. This is an excellent example of thermal energy transfer via convection. The same air ducts are used in the summer for cooling the same structures by blowing cooler air through the ducts and into the rooms, which is another example of convective thermal energy transfer. (But, of course, then the heat has the opposite sign, because the temperature in the room is lowered.)

Large amounts of energy are transferred by convection in the Earth's atmosphere and in the oceans. For example, the Gulf Stream carries warm water from the Gulf of Mexico northward through the Straits of Florida, and up the east coast of the United States. The warmer temperature of the water in the Gulf Stream is revealed in the NASA satellite image in Figure 18.20. The Gulf Stream has a temperature around 20 °C as it flows with a speed of approximately 2 m/s up the east coast of the United States into the North Atlantic. The Gulf Stream then splits. One part continues to flow toward Britain and Western Europe, while the other part turns south along the African coast. The average temperature of Britain and Western Europe is approximately 5 °C higher than it would be without the thermal energy carried by the warmer waters of the Gulf Stream.

Some climate models predict that global warming may possibly threaten the Gulf Stream because of the melting of ice at the North Pole. The extra freshwater from the melting polar ice cap reduces the salinity of the water at the northern end of the Gulf Stream, which interferes with the mechanism that allows the cooled water from the Gulf Stream to sink and return to the south. Paradoxically, global warming could make Northern Europe colder.

## SOLVED PROBLEM 18.3 Gulf Stream

Let's assume that a rectangular pipe of water 100 km wide and 500 m deep can approximate the Gulf Stream. The water in this pipe is moving with a speed of 2.0 m/s. The temperature of the water is 5.0 °C warmer than the surrounding water.

### PROBLEM

Estimate how much power the Gulf Stream is transporting to the North Atlantic Ocean.

### SOLUTION

#### THINK

We can calculate the volume flow rate by taking the product of the speed of the flow and the cross-sectional area of the pipe. Using the density of water, we can then calculate the mass flow rate. Using the specific heat of water and the temperature difference between the Gulf Stream water and the surrounding water, we can calculate the power transported by the Gulf Stream to the North Atlantic.

### SKETCH

Figure 18.21 shows the idealized Gulf Stream flowing to the northeast in the North Atlantic.

**FIGURE 18.21** Idealized Gulf Stream flowing in the North Atlantic along the eastern coastline of the United States.

### RESEARCH

We assume that the Gulf Stream has a rectangular cross section of width $w = 100$ km and depth $d = 500$ m. The area of the Gulf Stream flow is then

$$A = wd.$$

The speed of the flow of the Gulf Stream is assumed to be $v = 2.0$ m/s. The volume flow rate is given by

$$R_V = vA.$$

The density of seawater is $\rho = 1025\ \text{kg/m}^3$. We can express the mass flow rate as

$$R_m = \rho R_V.$$

The specific heat of water is $c = 4186$ J/(kg K). The heat required to raise the temperature of a mass $m$ by $\Delta T$ is given by

$$Q = cm\Delta T.$$

The power carried by the Gulf Stream is equal to the heat per unit time:

$$\frac{Q}{t} = P = \frac{cm\Delta T}{t} = cR_m \Delta T.$$

(*Note:* The capital $P$ symbolizes power; do not confuse this with the lowercase $p$ used for the pressure.)

### SIMPLIFY

The power carried by the Gulf Stream is given by

$$P = cR_m\Delta T = c\rho R_V \Delta T = c\rho vwd\Delta T.$$

### CALCULATE

The temperature difference is $\Delta T = 5\ °\text{C} = 5$ K. Putting in the numerical values gives

$$P = \left[4186\ \text{J}/\left(\text{kg K}\right)\right]\left(1025\ \text{kg/m}^3\right)\left(2.0\ \text{m/s}\right)\left(100 \cdot 10^3\ \text{m}\right)\left(500\ \text{m}\right)\left(5.0\ \text{K}\right)$$
$$= 2.1453 \cdot 10^{15}\ \text{W}.$$

### ROUND

We report our result to two significant figures:

$$P = 2.1 \cdot 10^{15}\ \text{W} = 2.1\ \text{PW}\ \ (1\ \text{PW} = 1\ \text{petawatt} = 10^{15}\ \text{W}).$$

### DOUBLE-CHECK

To double-check this result, let's calculate how much power is incident on the Earth from the Sun. This total power is given by the cross-sectional area of Earth times the power incident on Earth per unit area:

$$P_{\text{total}} = \pi\left(6.4 \cdot 10^6\ \text{m}\right)^2\left(1400\ \text{W/m}^2\right) = 180\ \text{PW}.$$

Let's calculate how much of this power could be absorbed by the Gulf of Mexico. The intensity of sunlight at a distance from the Sun corresponding to the radius of Earth's orbit around the Sun is $S = 1400\ \text{W/m}^2$. We can estimate the area of the Gulf of Mexico as

*Continued—*

$1.0 \cdot 10^6 \text{ km}^2 = 1.0 \cdot 10^{12} \text{ m}^2$. If the water of the Gulf of Mexico absorbed all the energy incident from the Sun for half of each day, then the power available from the Gulf would be

$$P = \frac{\left(1.0 \cdot 10^{12} \text{ m}^2\right)\left(1400 \text{ W/m}^2\right)}{2} = 7.0 \cdot 10^{14} \text{ W} = 0.7 \text{ PW},$$

which is less than our estimate of the power transported by the Gulf Stream. Thus, more of the Atlantic Ocean must be involved in providing energy for the Gulf Stream than just the Gulf of Mexico. In fact, the Gulf Stream gets its energy from a large fraction of the Atlantic Ocean. The Gulf Stream is part of a network of currents flowing in the Earth's oceans, induced by prevailing winds, temperature differences, and the Earth's topology and rotation.

### 18.3 Self-Test Opportunity

List some other convection phenomena encountered in everyday life.

## Radiation

Radiation occurs via the transmission of electromagnetic waves. (In Chapter 31, we'll see how electromagnetic waves carry energy; for now, you can just accept this as an empirical fact.) Further, unlike mechanical and sound waves (covered in Chapters 15 and 16), these waves need no medium to support them. Thus, electromagnetic waves can carry energy from one site to another without any matter having to be present between the two sites. One example of the transport of energy via electromagnetic radiation occurs during cell phone conversation; a cell phone acts as both a transmitter of radiation to the nearest cell tower and a receiver of radiation originating from that tower.

All objects emit electromagnetic radiation. The temperature of the object determines the radiated power of the object, $P_{\text{radiated}}$, which is given by the **Stefan-Boltzmann equation:**

$$P_{\text{radiated}} = \sigma \epsilon A T^4, \tag{18.18}$$

where $\sigma = 5.67 \cdot 10^{-8} \text{ W/K}^4\text{m}^2$ is called the **Stefan-Boltzmann constant,** $\epsilon$ is the **emissivity,** which has no units, and $A$ is the surface (radiating) area. The temperature in equation 18.18 must be in kelvins, and it is assumed to be constant. The emissivity varies between 0 and 1, with 1 being the emissivity of an idealized object called a **blackbody.** A blackbody radiates 100% of its energy and absorbs 100% of any radiation incident on it. Although some real-world objects are close to being a blackbody, no perfect blackbodies exist; thus, the emissivity is always less than 1.

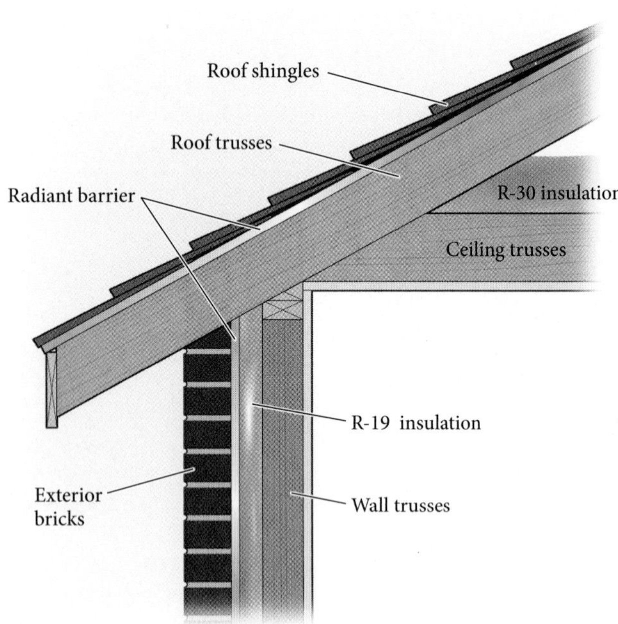


**FIGURE 18.22** A schematic drawing of the corner of a house, showing part of the roof, part of the ceiling, and part of a wall. The roof consists of an outer layer of shingles, a radiant barrier, and trusses supporting the roof. The ceiling consists of ceiling trusses supporting insulation with an $R$ factor of R-30. The wall consists of exterior bricks, a radiant barrier, insulation with an $R$ factor of R-19, and supporting wall trusses.

The earlier subsection on conduction discussed how the insulating ability of various materials is quantified by the $R$ value. The heat loss of a house in winter or the heat gain in summer depends not only on conduction but also on radiation. New building techniques have aimed at increasing the efficiency of house insulation by using radiant barriers. A *radiant barrier* is a layer of material that effectively reflects electromagnetic waves, especially infrared radiation (which is the radiation we usually feel as warmth). The use of radiant barriers in house insulation is illustrated in Figure 18.22.

A radiant barrier is constructed of a reflective substance, usually aluminum. A typical commercial radiant barrier is shown in Figure 18.23. Its material is aluminum-coated polyolefin, which reflects 97% of infrared radiation. Tests by Oak Ridge National Laboratory of houses in Florida with and without radiant barriers have shown that summer heat gains of ceilings with R-19 insulation can be reduced by 16% to 42%, resulting in the reduction of air-conditioning costs by as much as 17%.

The house illustrated in Figure 18.22 is designed to prevent heat from entering or leaving by conduction through its insulating layers, which have high $R$ factors. The radiant barriers block heat from entering the house in the form of radiation. Unfortunately, this type of barrier also prevents the house from being warmed by the Sun in the winter. Heat gain or loss by convection is reduced by the dead space between the ceiling and the roof. Thus, the house is designed to reduce heat gain or loss by any of the three modes of thermal energy transfer: conduction, convection, or radiation.

**FIGURE 18.23** One type of radiant barrier material, ARMA foil, produced by Energy Efficient Solutions.

### EXAMPLE 18.8 Earth as a Blackbody

Suppose that the Earth absorbed 100% of the incident energy from the Sun and then radiated all the energy back into space, just as a blackbody would.

**PROBLEM**

What would the temperature of the surface of Earth be?

**SOLUTION**

The intensity of sunlight arriving at the Earth is approximately $S = 1400\ \text{W/m}^2$. The Earth absorbs energy as a disk with the radius of the Earth, $R$, whereas it radiates energy from its entire surface. At equilibrium, the energy absorbed equals the energy emitted:

$$(S)\left(\pi R^2\right) = (\sigma)(1)\left(4\pi R^2\right)T^4.$$

Solving for the temperature, we get

$$T = \sqrt[4]{\frac{S}{4\sigma}} = \sqrt[4]{\frac{1400\ \text{W/m}^2}{4(5.67)\left(10^{-8}\ \text{W/K}^4\text{m}^2\right)}} = 280\ \text{K}.$$

This simple calculation gives a result close to the actual value of the average temperature of the surface of the Earth, which is about 288 K.

## Global Warming

The difference between the temperature calculated for Earth as a blackbody in Example 18.8 and the actual temperature of the Earth's surface is partly due to the Earth's atmosphere, as illustrated in Figure 18.24.

Clouds in the Earth's atmosphere reflect 20% and absorb 19% of the Sun's energy. The atmosphere reflects 6% of the Sun's energy, and 4% is reflected by the surface of the Earth. The Earth's atmosphere transmits 51% of the energy from the Sun to the surface of the

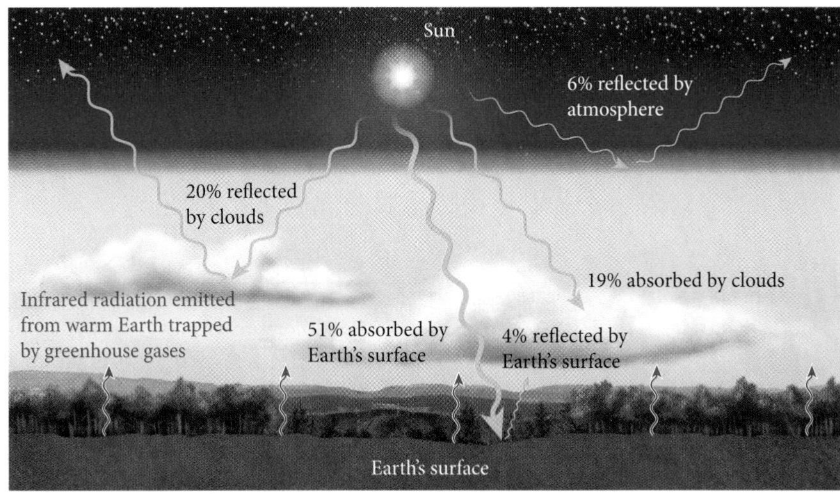


**FIGURE 18.24** The Earth's atmosphere strongly affects the amount of energy absorbed by the Earth from the Sun.

(a)

(b)

**FIGURE 18.25** (a) A NASA scientist with his arm inside a black plastic bag, illuminated with visible light. (b) The same NASA scientist photographed with an infrared camera. The scientist radiated infrared radiation that passed through the black plastic bag.

Earth. This solar energy is absorbed by the surface of the Earth and warms it, causing the surface to give off infrared radiation. Certain gases in the atmosphere—notably water vapor and carbon dioxide, in addition to other gases—absorb some of this infrared radiation, thereby trapping a fraction of the energy that would otherwise be radiated back into space. This trapping of thermal energy is called the **greenhouse effect.** The greenhouse effect keeps the Earth warmer than it would otherwise be and minimizes day-to-night temperature variations.

The two photographs in Figure 18.25 illustrate how radiation of different wavelengths can penetrate materials differently. In Figure 18.25a, a NASA scientist has his arm inside a black plastic bag. The camera was sensitive to visible light, and his arm is not visible inside the bag. In Figure 18.25b, the same person was photographed with the lights off, using a camera sensitive to infrared radiation. The human body emits infrared radiation because its metabolism produces heat. This radiation passes through the black plastic that blocks visible light and the previously concealed arm is visible.

The burning of fossil fuels and other human activities have increased the amount of carbon dioxide in the Earth's atmosphere and increase the surface temperature by trapping infrared radiation that would otherwise be emitted into space. The concentration of carbon dioxide in the Earth's atmosphere as a function of time is graphed in Figure 18.26. In Figure 18.26a, the concentration of carbon dioxide in the Earth's atmosphere is shown for the years 1832 through 2004. The concentration of carbon dioxide has been increasing for the past 150 years from around 284 ppmv (parts per million by volume) in 1832 to 377 ppmv in 2004.

Figure 18.26b shows the concentration of carbon dioxide in the air for the past 420,000 years. Visible in this plot, are glacial periods with relatively low carbon dioxide concentrations around 200 ppmv and interglacial periods with relatively high carbon dioxide concentrations around 275 ppmv. The combination of the direct measurements of the last 50 years and the inferred concentrations from the ice cores indicates that the present-day concentration of carbon dioxide in the atmosphere is higher than at any time in the past 420,000 years. Some researchers estimate that the current concentration of carbon dioxide is at its highest level in the past 20 million years. Models of the composition of the Earth's atmosphere based on current trends predict that the concentration of carbon dioxide will continue to increase in the next 100 years. This increase in the atmospheric concentration of carbon dioxide contributes to the observed global warming described in Chapter 17.

Worldwide, governments are reacting to these observations and predictions in many ways, including the Kyoto Protocol, which went into effect in 2005. In the Kyoto Protocol, signing nations agreed to make substantial cuts in their greenhouse gas emissions by the year 2012. However, several emerging nations were not required to reduce their greenhouse emissions.

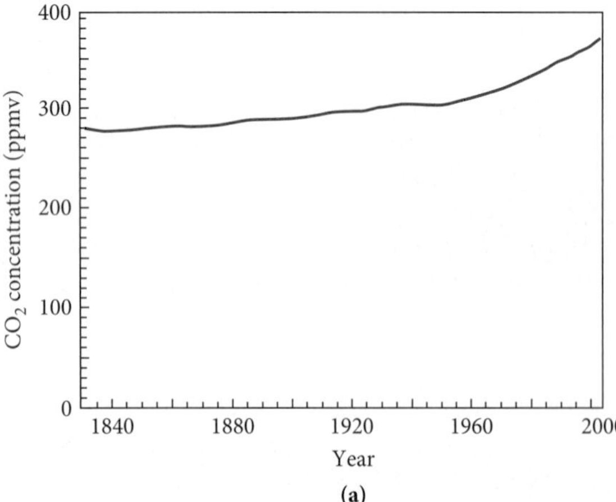


(a)

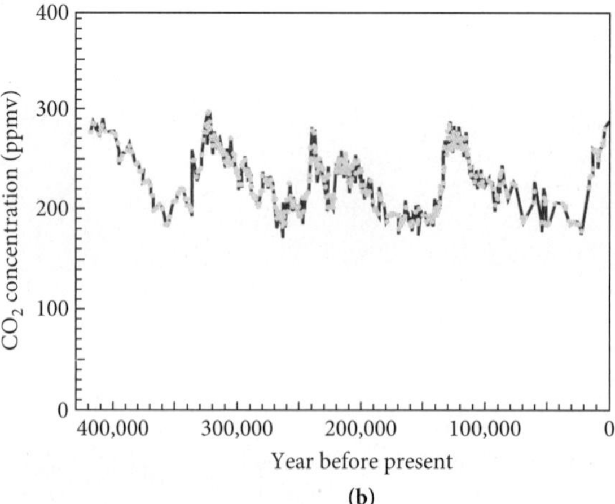


(b)

**FIGURE 18.26** Concentration of carbon dioxide ($CO_2$) in the Earth's atmosphere in parts per million by volume (ppmv). (a) Concentration of carbon dioxide in the atmosphere from 1832 to 2004. The measurements from 1832 to 1978 were done using ice cores in Antarctica and the measurements from 1959 to 2004 were carried out in the atmosphere at Mauna Loa in Hawaii. (b) Carbon dioxide concentrations for the past 420,000 years, extracted from ice cores in Antarctica.

## 18.4 In-Class Exercise

Indicate whether each of the following statements is true (T) or false (F).

1. A cold object will not radiate energy.
2. When heat is added to a system, the temperature must rise.
3. When the Celsius temperature doubles, the Fahrenheit temperature also doubles.
4. The temperature of the melting point is the same as the temperature of the freezing point.

## Heat in Computers

Thinking about thermal energy transfer may not bring your computer to mind right away. But cooling a computer is a major engineering problem. A typical desktop computer consumes between 100 and 150 W of electrical power, and a laptop uses between 25 and 70 W. As a general guideline, the higher the computer chips' frequency, the higher the power consumption. Most of this electrical power is converted into thermal energy and has to be removed from the computer. To accomplish this, computers have passive heat sinks, which consist of pieces of metal with large surface areas attached to the parts of the computer in need of cooling, mainly CPU and graphics chips (Figure 18.27). Passive heat sinks apply conduction to move the thermal energy away from the computer parts and then radiation to transfer the thermal energy to the surrounding air. An active heat sink includes a small fan to move more air past the metal surfaces to increase the thermal energy transfer. Of course, the fan also consumes electrical power and thus reduces the charge period of the battery in laptop computers.

**FIGURE 18.27** Active and passive heat sink technology mounted on a computer CPU.

While it may be only a slight annoyance to have a laptop computer warming your lap, cooling large computer clusters in data centers is very expensive. If a server farm contains 10,000 individual CPUs, its electrical power consumption is on the order of 1 MW, most of which is transformed to heat that has to be removed by very large air-conditioning systems. Thus, more efficient computer components, from power supplies to CPU chips, and more efficient cooling methods are currently the focus of much research.

## WHAT WE HAVE LEARNED | EXAM STUDY GUIDE

- Heat is energy transferred between a system and its environment or between two systems because of a temperature difference between them.
- A calorie is defined in terms of the joule: 1 cal = 4.186 J.
- The work done by a system in going from an initial volume $V_i$ to a final volume $V_f$ is $W = \int dW = \int_{V_i}^{V_f} p dV$.
- The First Law of Thermodynamics states that the change in internal energy of an isolated system is equal to the heat flowing into a system minus the work done by the system, or $\Delta E_{int} = Q - W$. The First Law of Thermodynamics states that energy in a closed system is conserved.
- An adiabatic process is one in which $Q = 0$.
- In a constant-volume process, $W = 0$.
- In a closed-loop process, $Q = W$.
- In an adiabatic free expansion, $Q = W = \Delta E_{int} = 0$.
- If heat, $Q$, is added to an object, its change in temperature, $\Delta T$, is given by $\Delta T = \frac{Q}{C}$, where $C$ is the heat capacity of the object.
- If heat, $Q$, is added to an object with mass $m$, its change in temperature, $\Delta T$, is given by $\Delta T = \frac{Q}{cm}$, where $c$ is the specific heat of the object.
- The energy required to melt a solid to a liquid, divided by its mass, is the latent heat of fusion, $L_{fusion}$. During the melting process, the temperature of the system remains at the melting point, $T = T_{melting}$.
- The energy required to vaporize a liquid to a gas, divided by its mass, is the latent heat of vaporization, $L_{vaporization}$. During the vaporization process, the temperature of the system remains at the boiling point, $T = T_{boiling}$.

- If a bar of cross-sectional area $A$ is placed between a thermal reservoir with temperature $T_h$ and a thermal reservoir with temperature $T_c$, where $T_h > T_c$, the rate of heat through the bar is $P_{cond} = \frac{Q}{t} = A\frac{T_h - T_c}{R}$, where $R$ is the thermal resistance of the bar.
- The radiated power from an object with temperature $T$ and surface area $A$ is given by the Stefan-Boltzmann equation: $P_{radiated} = \sigma\epsilon AT^4$, where $\sigma = 5.67 \cdot 10^{-8}$ W/K$^4$m$^2$ is the Stefan-Boltzmann constant and $\epsilon$ is the emissivity.

## KEY TERMS

system, p. 582
environment, p. 582
thermal energy, p. 583
heat, p. 583
calorie, p. 583
thermodynamic process, p. 584
$pV$-diagram, p. 585
path-dependent process, p. 585
closed path, p. 585
closed system, p. 586
First Law of Thermodynamics, p. 586
adiabatic process, p. 588
isochoric (constant-volume) process, p. 588
closed-path process, p. 588
free expansion, p. 589
isobaric (constant-pressure) process, p. 589
isothermal process, p. 589
isotherm, p. 589
heat capacity, p. 589
specific heat, p. 589
calorimeter, p. 591
state of matter, p. 592
phase, p. 592
phase changes (phase transitions), p. 592
latent heat of fusion, p. 592
latent heat of vaporization, p. 592
sublimation, p. 593
plasma, p. 593
radiation, p. 596
convection, p. 596
conduction, p. 596
thermal conductivity, p. 596
thermal resistance, p. 596
Stefan-Boltzmann equation, p. 602
Stefan-Boltzmann constant, p. 602
emissivity, p. 602
blackbody, p. 602
greenhouse effect, p. 604

## NEW SYMBOLS AND EQUATIONS

$Q$, heat

$\Delta E_{int} = Q - W$, change in internal energy of a system, First Law of Thermodynamics

$C$, heat capacity in J/K

$c$, specific heat in J/(kg K)

$k$, thermal conductivity of a material in W/(m K)

$R$, thermal resistance of a slab of material in m$^2$ K/W

$L_{fusion}$, latent heat of fusion

$L_{vaporization}$, latent heat of vaporization

$T_{melting}$, melting point of a substance

$T_{boiling}$, boiling point of a substance

## ANSWERS TO SELF-TEST OPPORTUNITIES

**18.1** The work is negative.

**18.2**
$$\left(\frac{1\text{ m}^2\text{ K}}{1\text{ W}}\right)\left(\frac{(3.281)^2\text{ ft}^2}{\text{m}^2}\right)\left(\frac{\frac{9}{5}\,{}^\circ\text{F}}{1\text{ K}}\right)\left(\frac{1\text{ W s}}{1\text{ J}}\right)\left(\frac{1055\text{ J}}{1\text{ BTU}}\right)\left(\frac{1\text{ h}}{3600\text{ s}}\right) = 5.678\,\frac{\text{ft}^2\,{}^\circ\text{F h}}{\text{BTU}}.$$

**18.3** thunderstorms, jet stream, heating water in a pan, heating a house

## PROBLEM-SOLVING PRACTICE

### Problem-Solving Guidelines

**1.** When using the First Law of Thermodynamics, always check the signs of work and heat. In this book, work done on a system is positive, and work done by a system is negative; heat added to a system is positive, and heat given off by a system is negative. Some books define the signs of work and heat differently; be sure you know what sign convention applies for a particular problem.

**2.** Work and heat are path-dependent quantities, but a system's change in internal energy is path-independent. Thus, to calculate a change in internal energy, you can use any processes that start from the same initial position and end at the same final position in a $pV$-diagram. Work and heat may vary, depending on the path in the diagram, but their difference, $Q - W$, will remain the same. For these kinds of problems, be sure you have a clearly defined system and you know what the initial and final conditions are for each step of a process.

**3.** For calorimetry problems, conservation of energy demands that transfers of energy sum up to zero. In other words, heat gained by one system must equal heat lost by the surroundings or some other system. This fact sets up the basic equation describing any transfer of heat between objects.

**4.** For calculating amounts of heat and corresponding temperature changes, remember that specific heat refers to a heat change per unit mass of a material, corresponding to a temperature change; for an object of known mass, you need to use the heat capacity. Also be aware of the possibility of a phase change. If a phase change is possible, break up the heat transfer process into steps, calculating heat corresponding to temperature changes and latent heat corresponding to phase changes. Remember that temperature changes are always final temperature minus initial temperature.

**5.** Be sure to check calorimetry calculations against reality. For example, if a final temperature is higher than an initial temperature, even though heat was extracted from a system, you may have overlooked a phase change.

## SOLVED PROBLEM 18.4 Thermal Energy Flow through a Copper/Aluminum Bar

### PROBLEM

A copper (Cu) bar of length $L = 90.0$ cm and cross-sectional area $A = 3.00$ cm$^2$ is in thermal contact at one end with a thermal reservoir at a temperature of 100.0 °C. The other end of the copper bar is in thermal contact with an aluminum (Al) bar of the same cross-sectional area and a length of 10.0 cm. The other end of the aluminum bar is in thermal contact with a thermal reservoir at a temperature of 1.00 °C. What is the thermal energy flow through the composite bar?

### SOLUTION

#### THINK

The thermal energy flow depends on the temperature difference between the two ends of the bar, the length and cross-sectional area of the bar, and the thermal conductivity of the materials. All the heat that flows from the high-temperature end must flow through both the copper and aluminum segments.

#### SKETCH

Figure 18.28 is a sketch of the copper/aluminum bar.

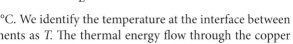

**FIGURE 18.28** Copper/aluminum bar held at 100 °C at one end and 1 °C at the other end.

#### RESEARCH

We can use equation 18.15 to describe the thermal energy flow through a bar of length $L$ and cross-sectional area $A$:

$$P_{cond} = kA\frac{T_h - T_c}{L}.$$

We have $T_h = 100$ °C and $T_c = 1$ °C. We identify the temperature at the interface between the copper and aluminum segments as $T$. The thermal energy flow through the copper segment is then

$$P_{Cu} = k_{Cu}A\frac{T_h - T}{L_{Cu}}.$$

The thermal energy flow through the aluminum segment is

$$P_{Al} = k_{Al}A\frac{T - T_c}{L_{Al}}.$$

The thermal energy flow through the copper segment must equal the thermal energy flow through the aluminum segment, so we have

$$P_{Cu} = P_{Al}$$

$$= k_{Cu}A\frac{T_h - T}{L_{Cu}} = k_{Al}A\frac{T - T_c}{L_{Al}}. \quad \text{(i)}$$

*Continued—*

**SIMPLIFY**

We can solve this equation for $T$. First we divide through by $A$ and then multiply both sides with $L_{\text{Cu}}L_{\text{Al}}$

$$k_{\text{Cu}}L_{\text{Al}}\left(T_{\text{h}}-T\right)=k_{\text{Al}}L_{\text{Cu}}\left(T-T_{\text{c}}\right).$$

Now we multiply through on both sides and collect all terms with $T$ on one side:

$$k_{\text{Cu}}L_{\text{Al}}T_{\text{h}}-k_{\text{Cu}}L_{\text{Al}}T=k_{\text{Al}}L_{\text{Cu}}T-k_{\text{Al}}L_{\text{Cu}}T_{\text{c}}$$
$$k_{\text{Al}}L_{\text{Cu}}T+k_{\text{Cu}}L_{\text{Al}}T=k_{\text{Cu}}L_{\text{Al}}T_{\text{h}}+k_{\text{Al}}L_{\text{Cu}}T_{\text{c}}$$
$$T(k_{\text{Al}}L_{\text{Cu}}+k_{\text{Cu}}L_{\text{Al}})=k_{\text{Cu}}L_{\text{Al}}T_{\text{h}}+k_{\text{Al}}L_{\text{Cu}}T_{\text{c}}$$
$$T=\frac{k_{\text{Cu}}L_{\text{Al}}T_{\text{h}}+k_{\text{Al}}L_{\text{Cu}}T_{\text{c}}}{k_{\text{Cu}}L_{\text{Al}}+k_{\text{Al}}L_{\text{Cu}}}.$$

Substituting this expression for $T$ into equation (i) gives us the thermal energy flow through the copper/aluminum bar.

**CALCULATE**

Putting in the numerical values gives us

$$\begin{aligned}T&=\frac{k_{\text{Cu}}L_{\text{Al}}T_{\text{h}}+k_{\text{Al}}L_{\text{Cu}}T_{\text{c}}}{k_{\text{Cu}}L_{\text{Al}}+k_{\text{Al}}L_{\text{Cu}}}\\&=\frac{\left[386\text{ W/(m K)}\right](0.100\text{ m})(373\text{ K})+\left[220\text{ W/(m K)}\right](0.900\text{ m})(274\text{ K})}{\left[386\text{ W/(m K)}\right](0.100\text{ m})+\left[220\text{ W/(m K)}\right](0.900\text{ m})}\\&=290.1513\text{ K}.\end{aligned}$$

Putting this result for $T$ into equation (i) gives us the thermal energy flow through the copper segment

$$\begin{aligned}P_{\text{Cu}}&=k_{\text{Cu}}A\frac{T_{\text{h}}-T}{L_{\text{Cu}}}\\&=\left[386\text{ W/}\left(\text{m K}\right)\right]\left(3.00\cdot10^{-4}\text{ m}^2\right)\frac{373\text{ K}-290.1513\text{ K}}{0.900\text{ m}}\\&=10.6599\text{ W}.\end{aligned}$$

**ROUND**

We report our result to three significant figures:

$$P_{\text{Cu}}=10.7\text{ W}.$$

**DOUBLE-CHECK**

To double-check, let's calculate the thermal energy flow through the aluminum segment:

$$P_{\text{Al}}=k_{\text{Al}}A\frac{T-T_{\text{c}}}{L_{\text{Al}}}=\left[220\text{ W/}\left(\text{m K}\right)\right]\left(3.00\cdot10^{-4}\text{ m}^2\right)\frac{290.1513\text{ K}-274\text{ K}}{0.100\text{ m}}=10.7\text{ W}.$$

This agrees with our result for the copper segment.

## MULTIPLE-CHOICE QUESTIONS

**18.1** A 2.0-kg metal object with a temperature of 90 °C is submerged in 1.0 kg of water at 20 °C. The water-metal system reaches equilibrium at 32 °C. What is the specific heat of the metal?

a) 0.840 kJ/kg K
b) 0.129 kJ/kg K
c) 0.512 kJ/kg K
d) 0.433 kJ/kg K

**18.2** A gas enclosed in a cylinder by means of a piston that can move without friction is warmed, and 1000 J of heat enters the gas. Assuming that the volume of the gas is constant, the change in the internal energy of the gas is

a) 0. b) 1000 J. c) −1000 J. d) none of the above.

**18.3** In the isothermal compression of a gas, the volume occupied by the gas is decreasing, but the temperature of the gas remains constant. In order for this to happen,

a) heat must enter the gas.
b) heat must exit the gas.
c) no heat exchange should take place between the gas and the surroundings.

**18.4** Which surface should you set a pot on to keep it hotter for a longer time?

a) a smooth glass surface
b) a smooth steel surface
c) a smooth wood surface
d) a rough wood surface

**18.5** Assuming the severity of a burn increases as the amount of energy put into the skin increases, which of the following would cause the most severe burn (assume equal masses)?

a) water at 90 °C
b) copper at 110 °C
c) steam at 180 °C
d) aluminum at 100 °C
e) lead at 100 °C

**18.6** In which type of process is no work done on a gas?

a) isothermal
b) isochoric
c) isobaric
d) none of the above

**18.7** An aluminum block of mass $M_{Al} = 2.0$ kg and specific heat $C_{Al} = 910$ J/(kg K) is at an initial temperature of 1000 °C and is dropped into a bucket of water. The water has mass $M_{H_2O} = 12$ kg and specific heat $C_{H_2O} = 4190$ J/(kg K) and is at room temperature (25 °C). What is the approximate final temperature of the system when it reaches thermal equilibrium? (Neglect heat loss out of the system.)

a) 50 °C
b) 60 °C
c) 70 °C
d) 80 °C

**18.8** A material has mass density $\rho$, volume $V$, and specific heat $c$. Which of the following is a correct expression for the heat exchange that occurs when the material's temperature changes by $\Delta T$ in degrees Celsius?

a) $(\rho c/V)\Delta T$
b) $(\rho c V)(\Delta T + 273.15)$
c) $(\rho c V)/\Delta T$
d) $\rho c V \Delta T$

**18.9** Which of the following *does not* radiate heat?

a) ice cube
b) liquid nitrogen
c) liquid helium
d) a device at $T = 0.010$ K
e) all of the above
f) none of the above

**18.10** Which of the following statements is (are) true?

a) When a system does work, its internal energy always decreases.

b) Work done on a system always decreases its internal energy.

c) When a system does work on its surroundings, the sign of the work is always positive.

d) Positive work done on a system is always equal to the system's gain in internal energy.

e) If you push on the piston of a gas-filled cylinder, the energy of the gas in the cylinder will increase.

## QUESTIONS

**18.11** Estimate the power radiated by an average person. (Approximate the human body as a cylindrical blackbody.)

**18.12** Several days after the end of a snowstorm, the roof of one house is still completely covered with snow, and another house's roof has no snow cover. Which house is most likely better insulated?

**18.13** Why does tile feel so much colder to your feet after a bath than a bath rug? Why is this effect more striking when your feet are cold?

**18.14** Can you think of a way to make a blackbody, a material that absorbs essentially all of the radiant energy falling in it, if you only have a material that reflects half the radiant energy that falls on it?

**18.15** In 1883, the volcano on Krakatau Island in the Pacific erupted violently in the largest explosion in Earth's recorded history, destroying much of the island in the process. Global temperature measurements indicate that this explosion reduced the average temperature of Earth by about 1 °C during the next two decades. Why?

**18.16** Fire walking is practiced in parts of the world for various reasons and is also a tourist attraction at some seaside resorts. How can a person walk across hot coals at a temperature well over 500 °F without burning his or her feet?

**18.17** Why is a dry, fluffy coat a better insulator than the same coat when it is wet?

**18.18** It has been proposed that global warming could be offset by dispersing large quantities of dust in the upper atmosphere. Why would this work, and how?

**18.19** A thermos bottle fitted with a piston is filled with a gas. Since the thermos bottle is well insulated, no heat can enter or leave it. The piston is pushed in, compressing the gas.

a) What happens to the pressure of the gas? Does it increase, decrease, or stay the same?

b) What happens to the temperature of the gas? Does it increase, decrease, or stay the same?

c) Do any other properties of the gas change?

**18.20** How would the rate of heat transfer between a thermal reservoir at a higher temperature and one at a lower temperature differ if the reservoirs were in contact with a 10-cm-long glass rod instead of a 10-m-long aluminum rod having an identical cross-sectional area?

**18.21** Why might a hiker prefer a plastic bottle to an old-fashioned aluminum canteen for carrying his drinking water?

**18.22** A girl has discovered a very old U.S. silver dollar and is holding it tightly in her little hands. Suppose that she put the silver dollar on the wooden (insulating) surface of a table, and then a friend came in from outside and placed on top of the silver dollar an equally old penny that she just found in the snow, where it had been left all night. Estimate the final equilibrium temperature of the system of the two coins in thermal contact.

# PROBLEMS

A blue problem number indicates a worked-out solution is available in the Student Solutions Manual. One • and two •• indicate increasing level of problem difficulty.

## Sections 18.2 and 18.3

18.23 You are going to lift an elephant (mass = $5.0 \cdot 10^3$ kg) over your head (2.0 m vertical displacement).

a) Calculate the work required to do this. You will lift the elephant slowly (no tossing of the elephant allowed!). If you want, you can use a pulley system. (As you saw in Chapter 5, this does not change the energy required to lift the elephant, but it definitely reduces the force required to do so.)

b) How many doughnuts (250 food calories each) must you metabolize to supply the energy for this feat?

**18.24** A gas has an initial volume of 2.00 $m^3$. It is expanded to three times its original volume through a process for which $P = \alpha V^3$, with $\alpha = 4.00\ \text{N/m}^{11}$. How much work is done by the expanding gas?

**18.25** How much work is done per cycle by a gas following the path shown on the $pV$-diagram?

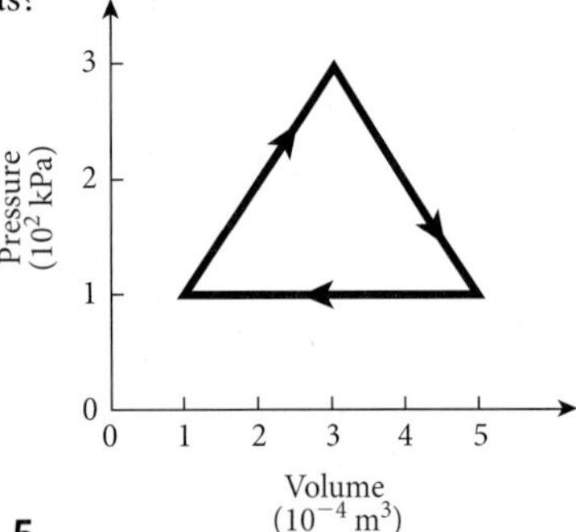


## Sections 18.4 and 18.5

**18.26** The internal energy of a gas is 500 J. The gas is compressed adiabatically, and its volume decreases by 100 $cm^3$. If the pressure applied on the gas during compression is 3 atm, what is the internal energy of the gas after the adiabatic compression?

## Section 18.6

18.27 You have 1 $cm^3$ of each of the materials listed in Table 18.1, all at room temperature, 22 °C. Which material has the highest temperature after 1 J of thermal energy is added to each sample? Which has the lowest temperature? What are these temperatures?

**18.28** Suppose you mix 7.0 L of water at $2.0 \cdot 10^1$ °C with 3.0 L of water at 32 °C; the water is insulated so that no energy can flow into it or out of it. (You can achieve this, approximately, by mixing the two fluids in a foam cooler of the kind used to keep drinks cool for picnics.) The 10 L of water will come to some final temperature. What is this final temperature?

**18.29** A 25-g piece of aluminum at 85 °C is dropped in 1.0 L of water at $1.0 \cdot 10^1$ °C, which is in an insulated beaker. Assuming that there is negligible heat loss to the surroundings, determine the equilibrium temperature of the system.

**18.30** A 12-g lead bullet is shot with a speed of 250 m/s into a wooden wall. Assuming that 75% of the kinetic energy is absorbed by the bullet as heat (and 25% by the wall), what is the final temperature of the bullet?

•18.31 A 1.00-kg block of copper at 80.0 °C is dropped into a container with 2.00 L of water at 10.0 °C. Compare the magnitude of the change in energy of the copper to the magnitude of the change in energy of the water. Which value is larger?

•**18.32** A 1.19-kg aluminum pot contains 2.31 L of water. Both pot and water are initially at 19.7 °C. How much heat must flow into the pot and the water to bring their temperature up to 95.0 °C? Assume that the effect of water evaporation during the heating process can be neglected and that the temperature remains uniform throughout the pot and the water.

•**18.33** A metal brick found in an excavation was sent to a testing lab for nondestructive identification. The lab weighed the sample brick and found its mass to be 3.0 kg. The brick was heated to a temperature of $3.0 \cdot 10^2$ °C and dropped into an insulated copper calorimeter of mass 1.5 kg containing 2.0 kg of water at $2.0 \cdot 10^1$ °C. The final temperature at equilibrium was noted to be 31.7 °C. By calculating the specific heat of the sample from this data, can you identify the brick's material?

•**18.34** A $2.0 \cdot 10^2$ g piece of copper at a temperature of 450 K and a $1.0 \cdot 10^2$ g piece of aluminum at a temperature of $2.0 \cdot 10^2$ K are dropped into an insulated bucket containing $5.0 \cdot 10^2$ g of water at 280 K. What is the equilibrium temperature of the mixture?

••18.35 When an immersion glass thermometer is used to measure the temperature of a liquid, the temperature reading will be affected by an error due to heat transfer between the liquid and the thermometer. Suppose you want to measure the temperature of 6 mL of water in a Pyrex glass vial thermally insulated from the environment. The empty vial has a mass of 5.0 g. The thermometer you use is made of Pyrex glass as well and has a mass of 15 g, of which 4.0 g is the mercury inside the thermometer. The thermometer is initially at room temperature (20.0 °C). You place the thermometer in the water in the vial and, after a while, you read an equilibrium temperature of 29 °C. What was the actual temperature of the water in the vial *before* the temperature was measured? The specific heat capacity of Pyrex glass around room temperature is 800 J/(kg K) and that of liquid mercury at room temperature is 140 J/(kg K).

## Section 18.7

•**18.36** Suppose 400 g of water at 30 °C is poured over a 60-g cube of ice with a temperature of –5 °C. If all the ice melts, what is the final temperature of the water? If all of the ice does not melt, how much ice remains when the water-ice mixture reaches equilibrium?

**18.37** A person gave off 180 kcal of heat in the evaporation of water from the skin in a workout session. How much water did the person lose, assuming that the heat given off was used only to evaporate the water?

**18.38** A 1.3-kg block of aluminum at 21 °C is to be melted and reshaped. How much heat must flow into the block in order to melt it?

18.39 The latent heat of vaporization of liquid nitrogen is about 200 kJ/kg. Suppose you have 1 kg of liquid nitrogen boiling at 77 K. If you supply heat at a constant rate of 10 W via an electric heater immersed in the liquid nitrogen, how long will it take to vaporize all of it? What is the time for 1 kg of liquid helium, whose heat of vaporization is 20.9 kJ/kg?

•**18.40** Suppose 0.010 kg of steam (at 100.00 °C) is added to 0.10 kg of water (initially at 19.0 °C). The water is inside an aluminum cup of mass 35 g. The cup is inside a perfectly insulated calorimetry container that prevents heat flow with the outside environment. Find the final temperature of the water after equilibrium is reached.

•**18.41** Suppose $1.0 \cdot 10^2$ g of molten aluminum at 932 K are dropped into 1.00 L of water at room temperature, 22 °C.

a) How much water will boil away?

b) How much aluminum will solidify?

c) What will be the final temperature of the water-aluminum system?

d) Suppose the aluminum were initially at 1150 K. Could you still do this problem using just the information given in this problem? What would be the result?

•**18.42** In one of your rigorous workout sessions, you lost 150 g of water through evaporation. Assume that the amount of work done by your body was $1.80 \cdot 10^5$ J and that the heat required to evaporate the water came from your body.

a) Find the loss in internal energy of your body, assuming the latent heat of vaporization is $2.42 \cdot 10^6$ J/kg.

b) Determine the minimum number of food calories that must be consumed to replace the internal energy lost (1 food calorie = 4186 J).

••18.43 Knife blades are often made of hardened carbon steel. The hardening process is a heat treatment in which the blade is first heated to a temperature of 1346 °F and then cooled down rapidly by immersing it in a bath of water. To achieve the desired hardness, after heating to 1346 °F, a blade needs to be brought to a temperature below $5.00 \cdot 10^2$ °F. If the blade has a mass of 0.500 kg and the water is in an open copper container of mass 2.000 kg and sufficiently large volume, what is the minimum quantity of water that needs to be in the container for this hardening process to be successful? Assume the blade is not in direct mechanical (and thus thermal) contact with the container, and neglect cooling through radiation into the air. Assume no water boils but reaches 100 °C. The heat capacity of copper around room temperature is $c_{\text{copper}} = 386$ J/(kg K). Use the data in the table below for the heat capacity of carbon steel.

| Temperature Range (°C) | Heat Capacity (J/kg K) |
|---|---|
| 150 to 200 | 519 |
| 200 to 250 | 536 |
| 250 to 350 | 553 |
| 350 to 450 | 595 |
| 450 to 550 | 662 |
| 550 to 650 | 754 |
| 650 to 750 | 846 |

••**18.44** It has been postulated that ethanol "snow" falls near the poles of the planets Jupiter, Saturn, Uranus, and Neptune. If the polar regions of Uranus, defined to be north of latitude 75° N and south of latitude 75° S, experience 1 ft of ethanol snow, what is the minimum amount of energy lost to the atmosphere to produce this much snow from ethanol vapor? Assume that solid ethanol has a density of 1 g/cm$^3$ and that ethanol snow—which is fluffy like Earth snow—is about 90% empty space. The specific heat capacity is 1.3 J/(g K) for ethanol vapor, 2.44 J/(g K) for ethanol liquid, and 1.2 J/(g K) for solid ethanol. How much power is dissipated if 1 ft of ethanol snow falls in one Earth day?

## Section 18.8

**18.45** A 100 mm by 100 mm by 5 mm block of ice at 0 °C is placed on its flat face on a 10-mm-thick metal disk that covers a pot of boiling water at normal atmospheric pressure. The time needed for the entire ice block to melt is measured to be 0.4 s. The density of ice is 920 kg/m$^3$. Use the data in Table 18.3 to determine the metal the disk is most likely made of.

**18.46** A copper sheet of thickness 2.00 mm is bonded to a steel sheet of thickness 1.00 mm. The outside surface of the copper sheet is held at a temperature of 100.0 °C and the steel sheet at 25.0 °C.

a) Determine the temperature of the copper-steel interface.

b) How much heat is conducted through 1 m$^2$ of the combined sheets per second?

18.47 The Sun is approximately a sphere of radius $6.963 \cdot 10^5$ km, at a mean distance $a = 1.496 \cdot 10^8$ km from the Earth. The *solar constant,* the intensity of solar radiation at the outer edge of Earth's atmosphere, is 1370. W/m$^2$. Assuming the Sun radiates as a blackbody, calculate its surface temperature.

•**18.48** An air-cooled motorcycle engine loses a significant amount of heat through thermal radiation according to the Stefan-Boltzmann equation. Assume that the ambient temperature is $T_0 = 27$ °C (300 K). Suppose the engine generates 15 hp (11 kW) of power and, due to several deep surface fins, has a surface area of $A = 0.50$ m$^2$. A shiny engine has an emissivity $e = 0.050$, whereas an engine that is painted black has $e = 0.95$. Determine the equilibrium temperatures for the black engine and the shiny engine. (Assume that radiation is the only mode by which heat is dissipated from the engine.)

•**18.49** One summer day, you decide to make a popsicle. You place a popsicle stick into an 8-oz glass of orange juice, which is at room temperature (71 °F). You then place the glass into the freezer which is at –15 °F and has a cooling power of 4000 BTU/h. How long does it take your popsicle to freeze?

•**18.50** An ice cube at 0 °C measures 10 cm on a side. It sits on top of a copper block with a square cross section 10 cm on a side and a length of 20 cm. The block is partially immersed in a large pool of water at 90 °C. How long does it take the ice cube to melt? Assume that only the part in contact with the copper liquefies; that is, the cube gets shorter as it melts. The density of ice is 0.96 g/cm$^3$.

•**18.51** A single-pane window is a poor insulator. On a cold day, the temperature of the inside surface of the window is often much less than the room air temperature. Likewise, the outside surface of the window is likely to be much warmer than the outdoor air. The actual surface temperatures are strongly dependent on convection effects. For instance, suppose the air temperatures are 21.5 °C inside and –3.0 °C outside, the inner surface of the window is at 8.5 °C, and the outer surface is at 4.1 °C. At what rate will heat flow through the window? Take the thickness of the window to be 0.32 cm, the height to be 1.2 m, and the width to be 1.4 m.

•**18.52** A cryogenic storage container holds liquid helium, which boils at 4.2 K. Suppose a student painted the outer shell of the container black, turning it into a pseudo-blackbody, and that the shell has an effective area of 0.50 $m^2$ and is at $3.0 \cdot 10^2$ K.

a) Determine the rate of heat loss due to radiation.

b) What is the rate at which the volume of the liquid helium in the container decreases as a result of boiling off? The latent heat of vaporization of liquid helium is 20.9 kJ/kg. The density of liquid helium is 0.125 kg/l.

•**18.53** Mars is 1.52 times farther away from the Sun than the Earth is and has a diameter 0.532 times that of the Earth.

a) What is the intensity of solar radiation (in W/$m^2$) on the surface of Mars?

b) Estimate the temperature on the surface of Mars.

••**18.54** Two thermal reservoirs are connected by a solid copper bar. The bar is 2 m long, and the temperatures of the reservoirs are 80.0 °C and 20.0 °C.

a) Suppose the bar has a constant rectangular cross section, 10 cm on a side. What is the rate of heat flow through the bar?

b) Suppose the bar has a rectangular cross section that gradually widens from the colder reservoir to the warmer reservoir. The area $A$ is determined by $A = (0.010 \text{ m}^2)[1.0 + x/(2.0 \text{ m})]$, where $x$ is the distance along the bar from the colder reservoir to the warmer one. Find the heat flow and the rate of change of temperature with distance at the colder end, at the warmer end, and at the middle of the bar.

••**18.55** The radiation emitted by a blackbody at temperature $T$ has a frequency distribution given by the Planck spectrum:

$$\epsilon_T(f) = \frac{2\pi h}{c^2}\left(\frac{f^3}{e^{hf/k_B T} - 1}\right),$$

where $\epsilon_T(f)$ is the energy density of the radiation per unit increment of frequency, $v$ (for example, in watts per square meter per hertz), $h = 6.626 \cdot 10^{-34}$ J s is Planck's constant, $k_B = 1.38 \cdot 10^{-23}$ $m^2$ kg $s^{-2}$ $K^{-1}$ is the Boltzmann constant, and $c$ is the speed of light in vacuum. (We'll derive this distribution in Chapter 36 as a consequence of the quantum hypothesis of light, but here it can reveal something about radiation. Remarkably, the most accurately and precisely measured example of this energy distribution in nature is the cosmic microwave background radiation.) This distribution goes to zero in the limits $f \to 0$ and $f \to \infty$ with a single peak in between those limits. As the temperature is increased, the energy density at each frequency value increases, and the peak shifts to a higher frequency value.

a) Find the frequency corresponding to the peak of the Planck spectrum, as a function of temperature.

b) Evaluate the peak frequency at temperature $T = 6.00 \cdot 10^3$ K, approximately the temperature of the photosphere (surface) of the Sun.

c) Evaluate the peak frequency at temperature $T = 2.735$ K, the temperature of the cosmic background microwave radiation.

d) Evaluate the peak frequency at temperature $T = 300$ K, which is approximately the surface temperature of Earth.

## Additional Problems

**18.56** How much energy is required to warm 0.30 kg of aluminum from 20.0 °C to 100.0 °C?

**18.57** The thermal conductivity of fiberglass batting, which is 4.0 in thick, is $8.0 \cdot 10^{-6}$ BTU/(ft °F s). What is the $R$ value (in $ft^2$ °F h/BTU)?

**18.58** Water is an excellent coolant as a result of its very high heat capacity. Calculate the amount of heat that is required to change the temperature of 10 kg of water by 10 K. Now calculate the kinetic energy of a car with $m = 1000$ kg moving at a speed of 27 m/s (60 mph). Compare the two quantities.

**18.59** Approximately 95% of the energy developed by the filament in a spherical $1.0 \cdot 10^2$ W light bulb is dissipated through the glass bulb. If the thickness of the bulb is 0.50 mm and its radius is 3.0 cm, calculate the temperature difference between the inner and outer surfaces of the bulb.

**18.60** The label on a soft drink states that 12 fl. oz (355 g) provides 150 kcal. The drink is cooled to 10.0 °C before it is consumed. It then reaches body temperature of 37 °C. Find the net energy content of the drink. (*Hint:* You can treat the soft drink as being identical to water in terms of heat capacity.)

**18.61** The human body transports heat from the interior tissues, at temperature 37.0 °C, to the skin surface, at temperature 27.0 °C, at a rate of 100. W. If the skin area is 1.5 $m^2$ and its thickness is 3.0 mm, what is the effective thermal conductivity, $\kappa$, of skin?

•**18.62** It has been said that sometimes lead bullets melt upon impact. Assume that a bullet receives 75% of the work done on it by a wall on impact as an increase in internal energy.

a) What is the minimum speed with which a 15-g lead bullet would have to hit a surface (assuming the bullet stops completely and all the kinetic energy is absorbed by it) in order to begin melting?

b) What is the minimum impact speed required for the bullet to melt completely?

•18.63 Solar radiation at the Earth's surface has an intensity of about 1.4 kW/m$^2$. Assuming that the Earth and Mars are blackbodies, calculate the intensity of sunlight at the surface of Mars.

•**18.64** You were lost while hiking outside wearing only a bathing suit.

a) Calculate the power radiated from your body, assuming that your body's surface area is about 2.0 m$^2$ and your skin temperature is about 33 °C. Also, assume that your body has an emissivity of 1.0.

b) Calculate the net radiated power from your body when you were inside a shelter at 20 °C.

c) Calculate the net radiated power from your body when your skin temperature dropped to 27 °C.

•**18.65** A 10-g ice cube at –10° C is dropped into 40 g of water at 30 °C.

a) After enough time has passed to allow the ice cube and water to come into equilibrium, what is the temperature of the water?

b) If a second ice cube is added, what will the temperature be?

•**18.66** Arthur Clarke wrote an interesting short story called "A Slight Case of Sunstroke." Disgruntled football fans came to the stadium one day equipped with mirrors and were ready to barbecue the referee if he favored one team over the other. Imagine the referee to be a cylinder filled with water of mass 60 kg at 35 °C. Also imagine that this cylinder absorbs all the light reflected on it from 50,000 mirrors. If the heat capacity of water is 4200 J/(kg °C), how long will it take to raise the temperature of the water to 100 °C? Assume that the Sun gives out 1000 W/m$^2$, the dimensions of each mirror are 25 cm by 25 cm, and the mirrors are held at an angle of 45°.

•18.67 If the average temperature of the North Atlantic is 12 °C and the Gulf Stream temperature averages 17 °C, estimate the net amount of heat the Gulf Stream radiates to the surrounding ocean. Use the details of Solved Problem 18.3 (the length is about 8000 km) and assume that $e = 0.93$. Don't forget to include the heat absorbed by the Gulf Stream.

•**18.68** For a class demonstration, your physics instructor pours 1.00 kg of steam at 100.0 °C over 4.00 kg of ice at 0.00 °C and allows the system to reach equilibrium. He is then going to measure the temperature of the system. While the system reaches equilibrium, you are given the latent heats of ice and steam and the specific heat of water: $L_{\text{ice}} = 3.33 \cdot 10^5$ J/kg, $L_{\text{steam}} = 2.26 \cdot 10^6$ J/kg, $c_{\text{water}} = 4186$ J/(kg °C). You are asked to calculate the final equilibrium temperature of the system. What value do you find?

•**18.69** Determine the ratio of the heat flow into a six-pack of aluminum soda cans to the heat flow into a 2-L plastic bottle of soda when both are taken out of the same refrigerator, that is, have the same initial temperature difference with the air in the room. Assume that each soda can has a diameter of 6 cm, a height of 12 cm, and a thickness of 0.1 cm. Use 205 W/(m K) as the thermal conductivity of aluminum. Assume that the 2-L bottle of soda has a diameter of 10 cm, a height of 25 cm, and a thickness of 0.1 cm. Use 0.10 W/(mK) as the thermal conductivity of plastic.

•**18.70** The $R$ factor for housing insulation gives the thermal resistance in units of ft$^2$ °F h/BTU. A good wall for harsh climates, corresponding to about 10 in of fiberglass, has $R = 40$ ft$^2$ °F h/BTU.

a) Determine the thermal resistance in SI units.

b) Find the heat flow per square meter through a wall that has insulation with an $R$ factor of 40, with an outside temperature of –22 °C and an inside temperature of 23 °C.

•18.71 Suppose you have an attic room measuring 5.0 m by 5.0 m and maintained at 21 °C when the outside temperature is 4.0 °C.

a) If you used R-19 insulation instead of R-30 insulation, how much more heat will exit this room in 1 day?

b) If electrical energy for heating the room costs 12 cents per kilowatt-hour, how much more will it cost you for heating for a period of 3 months with the R-19 insulation?

••**18.72** A thermal window consists of two panes of glass separated by an air gap. Each pane of glass is 3.00 mm thick, and the air gap is 1.00 cm thick. Window glass has a thermal conductivity of 1.00 W/(m K), and air has a thermal conductivity of 0.0260 W/(m K). Suppose a thermal window separates a room at temperature 20.00 °C from the outside at 0.00 °C.

a) What is the temperature at each of the four air-glass interfaces?

b) At what rate is heat lost from the room, per square meter of window?

c) Suppose the window had no air gap but consisted of a single layer of glass 6.00 mm thick. What would the rate of heat loss per square meter be then, under the same temperature conditions?

d) Heat conduction through the thermal window could be reduced essentially to zero by evacuating the space between the glass panes. Why is this not done?

# 19 Ideal Gases

**FIGURE 19.1** A scuba diver breathes compressed air under water.

## WHAT WE WILL LEARN

- A gas is a substance that, when placed in a container, expands to fill the container.
- The physical properties of a gas are pressure, volume, temperature, and number of molecules.
- An ideal gas is one in which the molecules of the gas are treated as point particles that do not interact with one another.
- The Ideal Gas Law gives the relationship among pressure, volume, temperature, and number of molecules of an ideal gas.
- The work done by an ideal gas is proportional to the change in volume of the gas if the pressure is constant.
- Dalton's Law says that the total pressure exerted by a mixture of gases is equal to the sum of the partial pressures exerted by each gas in the mixture.
- The specific heat of a gas can be calculated for both constant-volume and constant-pressure processes, and gases composed of monatomic, diatomic, or polyatomic molecules have different specific heats.
- Kinetic theory, describing the motion of the constituents of an ideal gas, accounts for its macroscopic properties, such as temperature and pressure.
- The temperature of a gas is proportional to the average kinetic energy of its molecules.
- The distribution of speeds of the molecules of a gas is described by the Maxwell speed distribution, and the distribution of kinetic energies of the molecules of a gas is described by the Maxwell kinetic energy distribution.
- The mean free path of a molecule in a gas is the average distance the molecule travels before encountering another molecule.

Scuba diving (*scuba* was originally an acronym for "self-contained underwater breathing apparatus") is a popular activity in locations with warm seawater. However, it is also a great application of the physics of gases. The scuba diver shown in Figure 19.1 relies on a tank of compressed air for breathing underwater. The air is typically kept at a pressure of 200 atm (~3000 psi, or ~20 MPa). However, to breathe this air, its pressure must be modified so that it is about the same as the pressure surrounding the diver's body in the water, which increases by about 1 atm for every 10 m below the surface (see section 13.4).

In this chapter, we study the physics of gases. The results are based on an ideal gas, which doesn't actually exist, but many real gases behave approximately like an ideal gas in many situations. We first examine the properties of gases based on observations, including laws first stated by pioneers of hot-air balloon navigation, who had a very practical interest in the behavior of gases at high altitudes. Then we gain additional insights from the kinetic theory of an ideal gas, which applies mathematical analysis to gas particles under several assumed conditions.

The key properties of gases are the thermodynamic quantities of temperature, pressure, and volume, which is why this chapter is with the others on thermodynamics. However, the physics of gases has applications to many different areas of science, from astronomy to meteorology, from chemistry to biology. Many of these concepts will be important in later chapters.

## 19.1 Empirical Gas Laws

Chapter 13 presented an overview of the states of matter. It introduced pressure as a physical quantity and stated that a gas is a special case of a fluid. With the concept of temperature, introduced in Chapter 17, we can now look at how gases respond to changes in temperature, pressure, and volume. The gas laws introduced in this section provide the empirical evidence that will lead to the derivation of the Ideal Gas Law in the next section. We'll see that all of the empirical gas laws are special cases of the Ideal Gas Law.

A **gas** is a substance that expands to fill the container in which it is placed. Thus, the volume of a gas is the volume of its container. This chapter uses the term *gas molecules* to refer to the constituents of a gas, although a gas may consist of atoms or molecules or may be a combination of atoms and molecules.

In Chapter 17, the temperature of a substance was defined in terms of its tendency to give off heat to its surroundings or to absorb heat from its surroundings. The pressure of a gas is defined as the force per unit area exerted by the gas molecules on the walls of a container. For many applications, standard temperature and pressure (STP) is defined as 0 °C (273.15 K) and 100 kPa.

Another characteristic of a gas, in addition to volume, temperature, and pressure, is the number of gas molecules in the volume of gas in a container. This number is expressed in terms of moles: 1 **mole** of a gas is defined to have $6.022 \cdot 10^{23}$ molecules. This number is known as *Avogadro's number,* introduced in Chapter 13 and will be discussed further in connection with Avogadro's Law later in this section.

Several simple relationships exist among the four properties of a gas—pressure, volume, temperature, and number of molecules. This section covers four simple gas laws that relate these properties. All of these gas laws are named after their discoverers (Boyle, Charles, Gay-Lussac, Avogadro) and are empirical—that is, they were found from performing a series of measurements and not derived from some more fundamental theory. In the next section, we'll combine these four laws to form the Ideal Gas Law, relating all the macroscopic characteristics of gases. If you understand the Ideal Gas Law, all of the empirical gas laws follow immediately. Why not skip them, then? The answer is that these empirical gas laws form the basis from which the Ideal Gas Law is derived.

## Boyle's Law

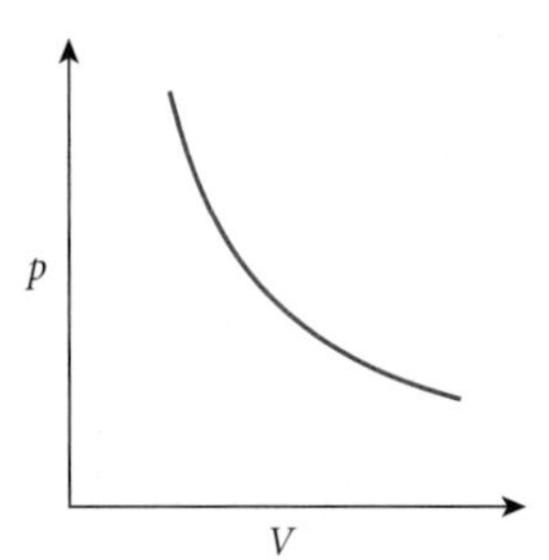

**FIGURE 19.2** Relationship between pressure and volume as expressed by Boyle's Law.

The first empirical gas law is Boyle's Law, which is also known in Europe as Mariotte's Law. English scientist Robert Boyle published this law in 1662; French scientist Edme Mariotte published a similar result in 1676. **Boyle's Law** states that the product of a gas's pressure, $p$, and its volume, $V$, at constant temperature is a constant (Figure 19.2). Mathematically, Boyle's Law is expressed as as $pV = \text{constant}$ (at constant temperature).

Another way to express Boyle's Law is to state that the product of the pressure, $p_1$, and volume, $V_1$, of a gas at time $t_1$ is equal to the product of the pressure, $p_2$, and the volume, $V_2$, of the same gas at the same temperature at another time, $t_2$:

$$p_1V_1 = p_2V_2 \quad \text{(at constant temperature).} \tag{19.1}$$

An everyday example of the application of Boyle's Law is breathing. When you take in a breath, your diaphragm expands, and the expansion produces a larger volume in your chest cavity. According to Boyle's Law, the air pressure in your lungs is reduced relative to the normal atmospheric pressure around you. The higher pressure outside your body then forces air into your lungs to equalize the pressure. To breathe out, your diaphragm contracts, reducing the volume of your chest cavity. This reduction in volume produces a higher pressure, which forces air out of your lungs.

## Charles's Law

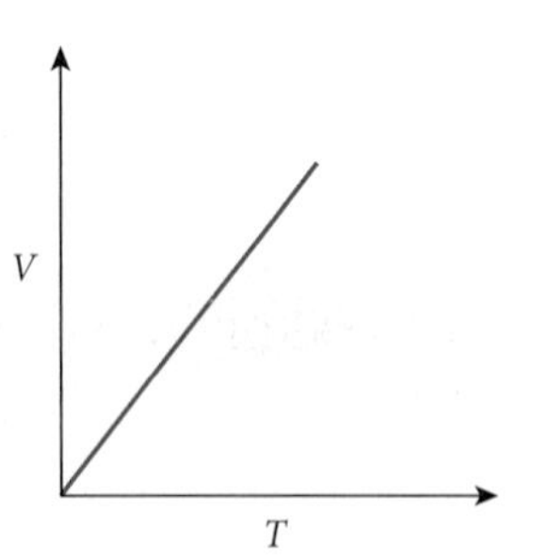

**FIGURE 19.3** Relationship between volume and temperature as expressed by Charles's Law.

The second empirical gas law is **Charles's Law,** which states that for a gas kept at constant pressure, the volume of the gas, $V$, divided by its temperature, $T$, is constant (Figure 19.3). The French physicist (and pioneering high-altitude balloonist) Jacques Charles proposed this law in 1787. Mathematically, Charles's Law is $V/T = \text{constant}$ (at constant pressure).

Another way to state Charles's Law is to state that the ratio of the temperature, $T_1$, and volume, $V_1$, of a gas at a given time, $t_1$, is equal to the ratio of the temperature, $T_2$, and the volume, $V_2$, of the same gas at the same pressure at another time, $t_2$:

$$\frac{V_1}{T_1} = \frac{V_2}{T_2} \Leftrightarrow \frac{V_1}{V_2} = \frac{T_1}{T_2} \text{(at constant pressure).} \tag{19.2}$$

Note that the temperatures must be expressed in kelvins (K).

Because the density, $\rho$, of a given mass, $m$, of gas is $\rho = m/V$, Charles's Law can also be written as $\rho T = \text{constant}$ (at constant pressure). As a corollary to equation 19.2, we then have

$$\rho_1 T_1 = \rho_2 T_2 \Leftrightarrow \frac{\rho_1}{\rho_2} = \frac{T_2}{T_1} \quad \text{(at constant pressure).} \tag{19.3}$$

## Gay-Lussac's Law

A third empirical gas law is **Gay-Lussac's Law,** which states that ratio of the pressure, $p$, of a gas to its temperature, $T$, at the same volume is constant (Figure 19.4). This law was presented in 1802 by the French chemist Joseph Louis Gay-Lussac (another avid high-altitude balloonist, by the way). Mathematically, the Gay-Lussac Law is expressed as $p/T$ = constant (at constant volume).

Another way to express Gay-Lussac's Law is to state that the ratio of the pressure of a gas, $p_1$, and its temperature, $T_1$, at a given time, $t_1$, is equal to the ratio of the pressure, $p_2$, and the temperature, $T_2$, of the same gas at the same volume at another time, $t_2$:

$$\frac{p_1}{T_1} = \frac{p_2}{T_2} \text{ (at constant pressure).} \tag{19.4}$$

Again, the temperatures must be in kelvins (K).

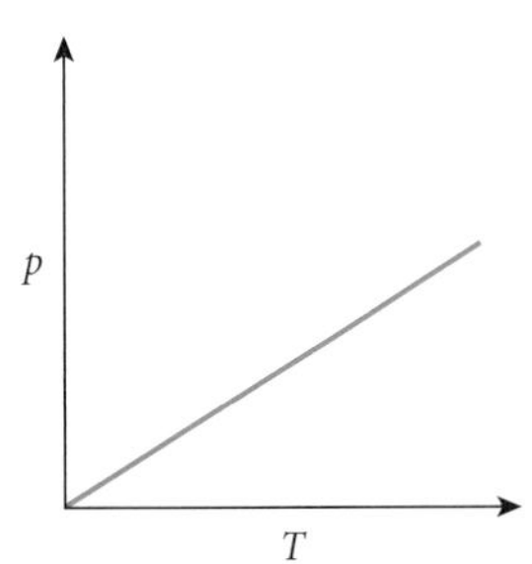


**FIGURE 19.4** Relationship between pressure and temperature as expressed in Gay-Lussac's Law.

## Avogadro's Law

The fourth empirical gas law deals with the quantity of a gas. **Avogadro's Law** states that the ratio of the volume of a gas, $V$, to the number of gas molecules, $N$, in that volume is constant if the pressure and temperature are held constant. This law was introduced in 1811 by the Italian chemist Amadeo Avogadro. Mathematically, Avogadro's Law is expressed as $V/N$ = constant (at constant pressure and temperature).

Another way to express Avogadro's Law is to state that the ratio of the volume, $V_1$, and the number of molecules, $N_1$, of a gas at a given time, $t_1$, is equal to the ratio of the volume, $V_2$, and the number of molecules, $N_2$, of the same gas at the same pressure and temperature at another time, $t_2$:

$$\frac{V_1}{N_1} = \frac{V_2}{N_2} \text{ (at constant pressure and temperature).} \tag{19.5}$$

It has been found that a volume of 22.4 L (1 L = $10^{-3}$ $m^3$) of a gas at standard temperature and pressure contains $6.022 \cdot 10^{23}$ molecules. This number of molecules is called **Avogadro's number,** $N_A$. The currently accepted value for Avogadro's number is

$$N_A = (6.02214179 \pm 0.00000030) \cdot 10^{23}.$$

One mole of any gas will have Avogadro's number of molecules. The number of moles is usually symbolized by $n$. Thus, the number of molecules, $N$, and the number of moles, $n$, of a gas are related by Avogadro's number:

$$N = nN_A. \tag{19.6}$$

Therefore, the mass of one mole of a gas is equal to the atomic mass or molecular mass of the constituent particles in grams. For example, nitrogen gas consists of molecules composed of two nitrogen atoms. Each nitrogen atom has an atomic mass number of 14. Thus, the nitrogen molecule has a molecular mass of 28, and a 22.4-L volume of nitrogen gas at standard temperature and pressure has a mass of 28 g.

# 19.2 Ideal Gas Law

We can combine the empirical gas laws described in Section 19.1 to obtain a more general law relating the properties of gases, called the **Ideal Gas Law:**

$$pV = nRT, \tag{19.7}$$

where $p$, $V$, and $T$ are the pressure, volume, and temperature, respectively, of $n$ moles of a gas and $R$ is the **universal gas constant.** The value of $R$ is determined experimentally and is given by

$$R = (8.314472 \pm 0.000015) \text{ J/(mol K)}.$$

The constant $R$ can also be expressed in other units, which changes its numerical value, for example, $R = 0.08205746$ L atm/(mol K).

The Ideal Gas Law was first stated by the French scientist Benoit Paul Émile Clapeyron in 1834. It is the main result of this chapter.

## DERIVATION 19.1 Ideal Gas Law

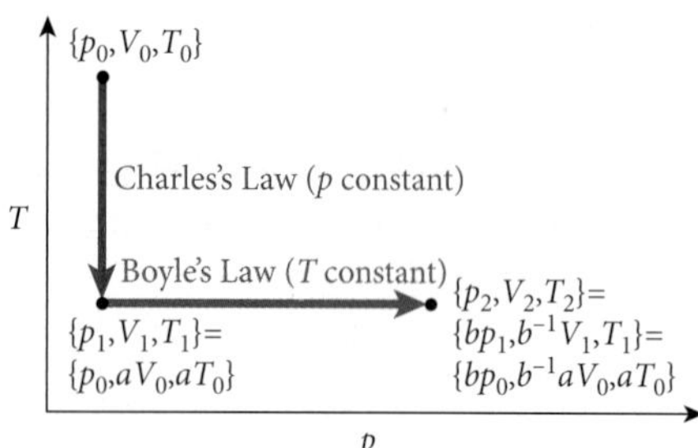


**FIGURE 19.5** Path of the state of a gas in a plot of temperature versus pressure (*Tp*-diagram). The first part of the path takes place at a constant pressure and is described by Charles's Law. The second part of the path takes place at constant temperature and is described by Boyle's Law.

To derive the Ideal Gas Law, we start with one mole of an ideal gas, described by a pressure $p_0$, a volume $V_0$, and a temperature $T_0$, which can be indicated using the shorthand notation $\{p_0, V_0, T_0\}$. (Remember, an ideal gas has no interactions among its molecules.) This particular state of the gas is shown on a plot of temperature versus pressure (a $Tp$-diagram ) in Figure 19.5.

Next, the state of the gas is changed by varying its volume and temperature while holding the pressure constant. In Figure 19.5, the gas's state goes from $\{p_0, V_0, T_0\}$ to $\{p_1, V_1, T_1\}$ on the $Tp$-diagram (red arrow). Using Charles's Law (equation 19.2), we can write

$$\frac{V_0}{T_0} = \frac{aV_0}{aT_0} = \frac{V_1}{T_1},$$

where we have multiplied the numerator and denominator of the ratio $V_0/T_0$ by a constant, $a$. We can describe the gas in this new state with

$$\{p_1, V_1, T_1\} = \{p_0, aV_0, aT_0\}. \quad \text{(i)}$$

From this point in the $Tp$-diagram, the pressure and volume of the gas change while the temperature is held constant. In Figure 19.5, the path goes from $\{p_1, V_1, T_1\}$ to $\{p_2, V_2, T_2\}$, as indicated by the blue arrow. Using Boyle's Law (equation 19.1), we can write

$$p_1V_1 = \frac{b}{b}p_1V_1 = (bp_1)(b^{-1}V_1) = p_2V_2,$$

where the product $p_1V_1$ is multiplied and divided by a constant, $b$. Now we can describe the state of the gas with

$$\{p_2, V_2, T_2\} = \{bp_1, b^{-1}V_1, T_1\}, \quad \text{(ii)}$$

Combining equations (i) and (ii) gives us

$$\{p_2, V_2, T_2\} = \{bp_0, b^{-1}aV_0, aT_0\}.$$

We can now write the ratio

$$\frac{p_2V_2}{T_2} = \frac{(bp_0)(b^{-1}aV_0)}{aT_0} = \frac{p_0V_0}{T_0}.$$

This implies that $p_0V_0/T_0$ is a constant, which we can call $R$, the universal gas constant.

This derivation was done for 1 mole of gas. For $n$ moles of gas, Avogadro's Law says that with constant pressure and temperature, the volume of $n$ moles of gas will be equal to the number of moles times the volume of one mole of gas. We can rearrange $p_0V_0/T_0 = R$ and multiply by $n$ to get

$$np_0V_0 = nRT_0.$$

We now write the Ideal Gas Law as $pV/T = nR$, or, more commonly, as

$$pV = nRT,$$

where $p = p_0$, $V = nV_0$, and $T = T_0$ are the pressure, volume, and temperature, respectively, of $n$ moles of gas.

The Ideal Gas Law can also be written in terms of the number of gas molecules instead of the number of moles of gas. This is sometimes useful for examining relationships at the molecular level of matter. In this form, the Ideal Gas Law is

$$pV = Nk_BT, \tag{19.8}$$

where $N$ is the number of atoms or molecules and $k_B$ is the **Boltzmann constant,** given by

$$k_B = \frac{R}{N_A}.$$

The currently accepted value of the Boltzmann constant is

$$k_B = (1.38106504 \pm 0.0000024) \cdot 10^{-23} \text{ J/K}.$$

The Boltzmann constant is an important fundamental physical constant that often arises in relationships based on atomic or molecular behavior. It will be used again later in this chapter and again in discussions of solid-state electronics and quantum mechanics. You may have noticed by now that the letter $k$ is often used in mathematics and physics to denote a constant (the German word for *constant* is *Konstante,* which is where the use of $k$ originated). However, the Boltzmann constant is of such basic significance that the subscript B is used to distinguish it from other physical constants denoted by $k$.

Another way to express the Ideal Gas Law for a constant number of moles of gas is

$$\frac{p_1V_1}{T_1} = \frac{p_2V_2}{T_2}, \tag{19.9}$$

where $p_1$, $V_1$, and $T_1$ are the pressure, volume, and temperature, respectively, at time 1 and $p_2$, $V_2$, and $T_2$ are the pressure, volume, and temperature, respectively, at time 2. The advantage of this formulation is that you do not have to know the numerical values of the universal gas constant or the Boltzmann constant to relate the pressures, volumes, and temperatures at two different times.

It is straightforward to confirm that Boyle's Law, Charles's Law, and Gay-Lussac's Law are embodied in the Ideal Gas Law. In Figure 19.6, a three-dimensional surface represents the relationship among the pressure, volume, and temperature of an ideal gas. Projecting this surface onto the constant-pressure axis yields a curve reflecting Charles's Law. Projecting onto the constant-temperature axis yields a curve reflecting Boyle's Law. Projecting onto the constant-volume axis yields a curve reflecting Gay-Lussac's Law.

Note that the Ideal Gas Law has been obtained in this discussion by combining the empirical gas laws. The next section will analyze the behavior of ideal gas molecules in a container and thereby provide insight into the microscopic physical basis for the Ideal Gas Law. Let's first explore the consequences and predictive powers of the Ideal Gas Law with a few illustrative examples.

### 19.1 In-Class Exercise

The pressure inside a scuba tank is 205 atm at 22.0 °C. Suppose the tank is left out in the sun and the temperature of the compressed air inside the tank rises to 40.0 °C. What will the pressure inside the tank be?

a) 205 atm

b) 218 atm

c) 254 atm

d) 321 atm

e) 373 atm

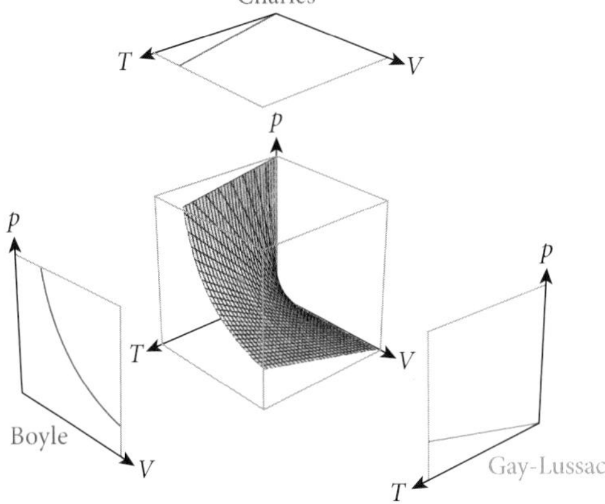


**FIGURE 19.6** The relationship among the Ideal Gas Law and Charles's Law (constant pressure), Boyle's Law (constant temperature), and Gay-Lussac's Law (constant volume).

## EXAMPLE 19.1 Gas in a Cylinder

Consider a gas at a pressure of 24.9 kPa in a cylinder with a volume of 0.100 m$^3$ and a piston. This pressure and volume corresponds to point 1 in Figure 19.7. The pressure of this gas has to be decreased to allow a manufacturing process to work efficiently. The piston is designed to increase the volume of the cylinder from 0.100 m$^3$ to 0.900 m$^3$ while keeping the temperature constant.

### PROBLEM

What is the pressure of the gas at a volume of 0.900 m$^3$?

### SOLUTION

Point 2 in Figure 19.7 represents the pressure of the gas at a volume of 0.900 m$^3$. You can see that the pressure decreases as the volume increases.

*Continued—*

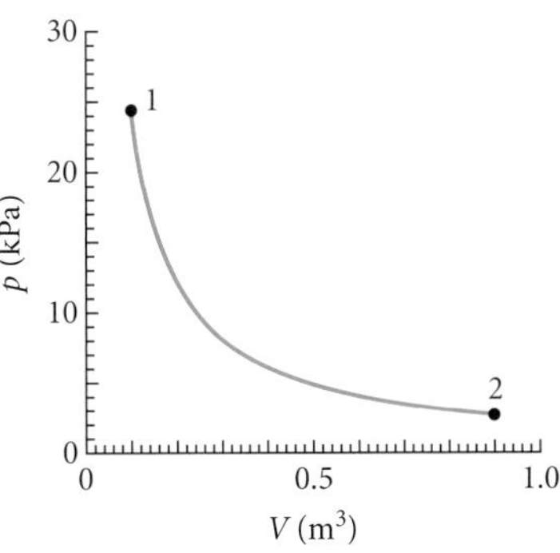


**FIGURE 19.7** Plot of pressure versus volume for a gas described by Boyle's Law.

Since the temperature is constant ($T_1 = T_2$) in this situation, the Ideal Gas Law (in the form in equation 19.9), $p_1V_1/T_1 = p_2V_2/T_2$, reduces to $p_1V_1 = p_2V_2$. (Section 19.1 identified this special case of the Ideal Gas Law as Boyle's Law.) We can use this to calculate the new pressure:

$$p_2 = \frac{p_1V_1}{V_2} = \frac{(24.9 \text{ kPa})(0.100 \text{ m}^3)}{0.900 \text{ m}^3} = 2.77 \text{ kPa}.$$

There are, of course, many applications where the temperature of a gas changes. Example 19.2 considers one.

## EXAMPLE 19.2 Cooling a Balloon

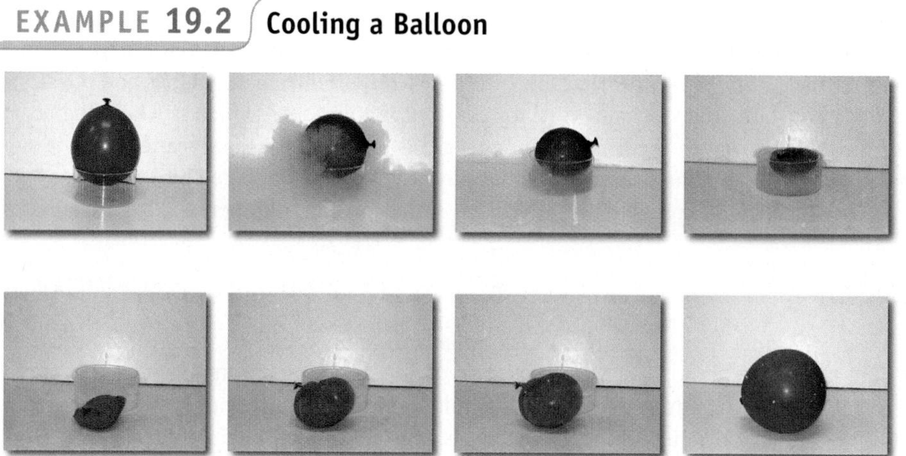

**FIGURE 19.8** The top row shows a balloon blown up at room temperature and then placed in liquid nitrogen. In the bottom row, the cold balloon is removed from the liquid nitrogen and allowed to warm to room temperature. The time between subsequent frames in each sequence is 20 s. Note that the liquid nitrogen causes a phase change in the air inside the balloon, causing it to contract more than the Ideal Gas Law alone predicts.

A balloon is blown up at room temperature and then placed in liquid nitrogen (Figure 19.8). The balloon is allowed to cool to liquid-nitrogen temperature and it shrinks dramatically. The cold balloon is then removed from the liquid nitrogen and allowed to warm back to room temperature. The balloon returns to its original volume.

### PROBLEM

By what factor does the volume of the balloon decrease as the temperature of the air inside the balloon goes from room temperature to the temperature of liquid nitrogen?

### SOLUTION

The gas inside the balloon is kept at constant pressure ($p_1 = p_2$), because it is always at atmospheric pressure, which is the pressure outside the balloon. The Ideal Gas Law (equation 19.9), $p_1V_1/T_1 = p_2V_2/T_2$, in this case becomes $V_1/T_1 = V_2/T_2$. (In Section 19.1, this special case of the Ideal Gas Law was identified as Charles's Law.) Assume room temperature is 22 °C or $T_1 = 295$ K, and the temperature of liquid nitrogen is $T_2 = 77.2$ K. We can calculate the fractional change in volume of the balloon:

$$\frac{V_2}{V_1} = \frac{T_2}{T_1} = \frac{77.2 \text{ K}}{295 \text{ K}} = 0.262.$$

### DISCUSSION

The balloon in Figure 19.8 shrank by much more than the factor calculated from Charles's Law. (A ratio of cold volume to warm volume of 25% implies that the radius of the cold balloon should be 63% of the radius of the room-temperature balloon.) There are several reasons for the smaller than expected size of the cold balloon. First, the water vapor in the air in the balloon freezes out as ice particles. Second, some of the oxygen and the nitrogen in the air inside the balloon condense into liquids, to which the Ideal Gas Law does not apply.

Example 19.3 deals with the relationship of air temperature and air pressure.

### EXAMPLE 19.3 Heat on the Golf Course

The 2007 PGA Championship was played in August in Oklahoma at the Southern Hills Country Club, at the highest average temperature of any major golf championship in history, reaching approximately 101 °F (38.3 °C = 311.5 K). Thus, the air density was lower than at cooler temperatures, causing the golf balls to fly farther. The players had to correct for this effect, just as they have to allow for longer flights when they play at higher elevations.

**FIGURE 19.9** Teeing off at the Old Course in St. Andrews, Scotland, at sea level and in late winter. At lower elevations or lower temperatures, the ball does not carry as far.

**PROBLEM**

The Southern Hills Country Club is at an elevation of 213 m (700 ft) above sea level. At what elevation would the Old Course at St. Andrews, Scotland, shown in Figure 19.9, have to be located, if the air temperature was 48 °F (8.9 °C = 282.0 K), for the golf balls to have the same increase in flight length as at Southern Hills?

**SOLUTION**

First, we calculate how the temperature influences the air density. We can use the relationship between the density and temperature for constant pressure, $\rho_1/\rho_2 = T_2/T_1$, equation 19.3. Putting in the known values (remember that we need to use kelvins for air temperature values), we obtain

$$\frac{\rho_1}{\rho_2} = \frac{T_2}{T_1} = \frac{282.0 \text{ K}}{311.5 \text{ K}} = 0.9053.$$

In other words, the air at 101 °F has only 90.53% of the density of air at 48 °F, at the same pressure. Now we have to compute the altitude that corresponds to the same density ratio. In Section 13.4, we derived a formula that relates the density to the height,

$$\rho(h) = \rho_0 e^{-h\rho_0 g/p_0},$$

where $h$ is the height above sea level, $\rho_0$ is the density of air at sea level, and $p_0$ is the air pressure at sea level. The constants in the formula relating density to height can be written as

$$\frac{p_0}{\rho_0 g} = \frac{1.01\cdot 10^5 \text{ Pa}}{\left(1.229 \text{ kg/m}^3\right)\left(9.81 \text{ m/s}^2\right)} = 8377 \text{ m}.$$

The elevation at Southern Hills is $h_1 = 213$ m and the air density is

$$\rho(h_1) = \rho_0 e^{-h_1/(8377 \text{ m})}.$$

The air density at the Old Course at an elevation $h_2$ would be

$$\rho(h_2) = \rho_0 e^{-h_2/(8377 \text{ m})}.$$

Taking the ratio of these two densities gives

$$\frac{\rho(h_2)}{\rho(h_1)} = \frac{\rho_0 e^{-h_2/(8377 \text{ m})}}{\rho_0 e^{-h_1/(8377 \text{ m})}} = e^{-(h_2-h_1)/(8377 \text{ m})}.$$

Setting this ratio equal to the ratio of the densities from the temperature difference gives

$$e^{-(h_2-h_1)/(8377 \text{ m})} = 0.9053.$$

Taking the natural log of both sides of this equation leads to

$$-(h_2 - h_1)/(8377 \text{ m}) = \ln(0.9053),$$

which can be solved for $h_2$:

$$h_2 = h_1 - (8377 \text{ m})\ln(0.9053) = (213 \text{ m}) - (8377 \text{ m})\ln(0.9053) = 1046 \text{ m}.$$

*Continued—*

Thus, at the temperature of 101 °F, the course at Southern Hills played as long as the Old Course would at an elevation of 1046 m (3432 ft) and a temperature of 48 °F.

**DISCUSSION**

Our solution neglected the effect of the high humidity at Southern Hills, which reduced the air density by an additional 2% relative to the density of dry air, causing the course to play like one with dry air and a temperature of 48 °F at an altitude over 4000 ft above sea level. (The effect of water vapor in the air on the air density is discussed later, in the subsection on Dalton's Law.)

## Work Done by an Ideal Gas at Constant Temperature

Suppose we have an ideal gas at a constant temperature in a closed container whose volume can be changed, such as a cylinder with a piston. This setup allows us to perform an isothermal process, described in Chapter 18. For an isothermal process, the Ideal Gas Law says that the pressure is equal to a constant times the inverse of the volume: $p = nRT/V$. If the volume of the container is changed from an initial volume, $V_i$, to a final volume, $V_f$, the work done by the gas (see Chapter 18) is given by

$$W = \int_{V_i}^{V_f} p dV.$$

Substituting the expression for the pressure into this integral gives

$$W = \int_{V_i}^{V_f} p dV = (nRT)\int_{V_i}^{V_f} \frac{dV}{V} = nRT\left[\ln V\right]_{V_i}^{V_f},$$

which evaluates to

$$W = nRT \ln\left(\frac{V_f}{V_i}\right). \tag{19.10}$$

Equation 19.10 indicates that the work done by the gas is positive if $V_f > V_i$ and is negative if $V_f < V_i$.

We can compare this result to those found in Chapter 18 for the work done by the gas under other assumptions. For example, if the volume is held constant rather than the temperature, the gas can do no work: $W = 0$. If the pressure is held constant, the work done by the gas is given by (see Chapter 18)

$$W = \int_{V_i}^{V_f} p dV = p\int_{V_i}^{V_f} dV = p(V_f - V_i) = p\Delta V.$$

### 19.2 In-Class Exercise

Suppose an ideal gas has pressure $p$, volume $V$, and temperature $T$. If the volume is doubled at constant pressure, the gas does work $W_{p=\text{const}}$. If instead the volume is doubled at constant temperature, the gas does work $W_{T=\text{const}}$. What is the relationship between these two amounts of work done by the gas?

a) $W_{p=\text{const}} < W_{T=\text{const}}$

b) $W_{p=\text{const}} = W_{T=\text{const}}$

c) $W_{p=\text{const}} > W_{T=\text{const}}$

d) The relationship cannot be determined from the information given.

## Dalton's Law

How do ideal gas considerations change, if there is more than one type of gas in a given volume, as in the Earth's atmosphere, for example? **Dalton's Law** states that the pressure of a gas composed of a homogeneous mixture of different gases is equal to the sum of the partial pressure of each of the gases. **Partial pressure** is defined as the pressure the gas would exert if the other gases were not present. Dalton's Law means that each gas is unaffected by the presence of other gases, as long as there is no interaction between the gas molecules. The law is named for John Dalton, a British chemist, who published it in 1801. Dalton's Law gives the total pressure, $p_{\text{total}}$, exerted by a mixture of $m$ gases, each with partial pressure $p_i$:

$$p_{\text{total}} = p_1 + p_2 + p_3 + \cdots + p_m = \sum_{i=1}^{m} p_i. \tag{19.11}$$

Avogadro's Law can be extended to state that the total number of moles of gas, $n_{\text{total}}$, contained in a mixture of $m$ gases is equal to the sum of the numbers of moles of each gas, $n_i$:

$$n_{\text{total}} = n_1 + n_2 + n_3 + \cdots + n_m = \sum_{i=1}^{m} n_i.$$

**Table 19.1 Major Gases Making up the Earth's Atmosphere**

| Gas | Percentage by Volume | Chemical Symbol | Molecular Mass |
|---|---|---|---|
| Nitrogen | 78.08 | $N_2$ | 28.0 |
| Oxygen | 20.95 | $O_2$ | 32.0 |
| Argon | 0.93 | Ar | 40.0 |
| Carbon dioxide | 0.038 | $CO_2$ | 44.0 |
| Neon | 0.0018 | Ne | 20.2 |
| Helium | 0.0005 | He | 4.00 |
| Methane | 0.0002 | $CH_4$ | 16.0 |
| Krypton | 0.0001 | Kr | 83.8 |

Then, the **mole fraction,** $r_i$, for each gas in a mixture is the number of moles of that gas divided by the total number of moles of gas:

$$r_i = \frac{n_i}{n_{\text{total}}}. \tag{19.12}$$

The sum of the mole fractions is equal to 1: $\sum_{i=1}^{m} r_i = 1$. If the number of moles of gas increases, with the temperature constant, the pressure must go up. We can then express each partial pressure as

$$p_i = r_i p_{\text{total}} = n_i \frac{p_{\text{total}}}{n_{\text{total}}}. \tag{19.13}$$

This means that the partial pressures in a mixture of gases are simply proportional to the mole fractions of those gases present in the mixture.

### 19.3 In-Class Exercise

What is the mass of 22.4 L of dry air at standard temperature and pressure?

a) 14.20 g

b) 28.00 g

c) 28.95 g

d) 32.22 g

e) 60.00 g

### Earth's Atmosphere

The atmosphere of the Earth has a mass of approximately $5.1 \cdot 10^{18}$ kg (5 quadrillion metric tons). It is a mixture of several gases; see Table 19.1. The table lists only the components of dry air; in addition, air also contains water vapor ($H_2O$), which makes up approximately 0.25% of the entire atmosphere. (This sounds like a small number but equals approximately $10^{16}$ kg of water, which is about four times the volume of water in all the Great Lakes!) Near the Earth's surface, the water content of air ranges from less than 1% to over 3%, depending mainly on the air temperature and the availability of liquid water in the vicinity. Water vapor in the air *decreases* the average density of the atmosphere, because the molecular mass of water is 18, which is smaller than the molecular or atomic mass of almost all the other atmospheric gases.

### 19.4 In-Class Exercise

What is the oxygen pressure in the Earth's atmosphere at sea level?

a) 0

b) 0.21 atm

c) 1 atm

d) 4.8 atm

e) 20.9 atm

## 19.3 Equipartition Theorem

We have been discussing the macroscopic properties of gases, including volume, temperature, and pressure. To explain these properties in terms of the constituents of a gas, its molecules (or atoms), requires making several assumptions about the behavior of these molecules in a container. These assumptions, along with the results derived from them, are known as the **kinetic theory of an ideal gas.**

Suppose we have a gas in a container of volume $V$, and the gas fills this container evenly. We make the following assumptions:

- The number of molecules, $N$, is large, but the molecules themselves are small, so the average distance between molecules is large compared to their size. All molecules are identical, and each has mass $m$.
- The molecules are in constant random motion on straight-line trajectories. They do not interact with one another and can be considered point particles.

- The molecules have elastic collisions with the walls of the container.
- The volume, $V$, of the container is large compared to the size of the molecules. The container walls are rigid and stationary.

The kinetic theory explains how the microscopic properties of the gas molecules give rise to the macroscopic observables of pressure, volume, and temperature as related by the Ideal Gas Law. For now, we are mainly interested in what the average kinetic energy of the gas molecules is and how it relates to the temperature of the gas. This connection is called the **equipartition theorem.** In the next section, we use this theorem to derive expressions for the specific heats of gases. We'll then see that the equipartition theorem is intimately connected to the idea of degrees of freedom of the motion of the gas molecules. Finally, in Section 19.6, we'll return to the kinetic theory, considering the distribution of kinetic energies of the molecules (instead of just the average kinetic energy), and discuss the limitations of the concept of an ideal gas.

First, we obtain the average kinetic energy of the ideal gas by simply averaging the kinetic energies of the individual gas molecules:

$$K_{\text{ave}} = \frac{1}{N}\sum_{i=1}^{N} K_i = \frac{1}{N}\sum_{i=1}^{N}\frac{1}{2}mv_i^2 = \frac{1}{2}m\left(\frac{1}{N}\sum_{i=1}^{N}v_i^2\right) = \frac{1}{2}mv_{\text{rms}}^2. \qquad (19.14)$$

Thus, the **root-mean-square speed** of the gas molecules, $v_{\text{rms}}$, is defined as

$$v_{\text{rms}} = \sqrt{\frac{1}{N}\sum_{i=1}^{N}v_i^2}. \qquad (19.15)$$

Note that the root-mean-square speed is *not the same* as the average speed of the gas molecules. But it can be considered a suitable average, because it immediately relates to the average kinetic energy, as seen in equation 19.14.

Derivation 19.2 shows how the average kinetic energy of the molecules in an ideal gas relates to the temperature of the gas. We'll see that the temperature of the gas is simply proportional to the average kinetic energy, with a proportionality constant $\frac{3}{2}$ times the Boltzmann constant:

$$K_{\text{ave}} = \tfrac{3}{2}k_{\text{B}}T. \qquad (19.16)$$

This is the equipartition theorem. Thus, when we measure the temperature of a gas, we are determining the average kinetic energy of the molecules of the gas. This relationship is one of the key insights of kinetic theory and will be very useful in the rest of this chapter and in later chapters.

Let's see how to derive the equipartition theorem starting from Newtonian mechanics and using the ideal gas law.

## DERIVATION 19.2 Equipartition Theorem

The pressure of a gas in a container is determined by the interaction of the gas molecules with the walls of the container, that is, by the change in their momentum, a force. The pressure of the gas is a consequence of elastic collisions of the gas molecules with the walls of the container. When a gas molecule hits a wall, it bounces back with the same kinetic energy it had before the collision. We assume that the wall is stationary, and thus the component of the momentum of the gas molecule perpendicular to the surface of the wall is reversed in the collision (see Section 7.5).

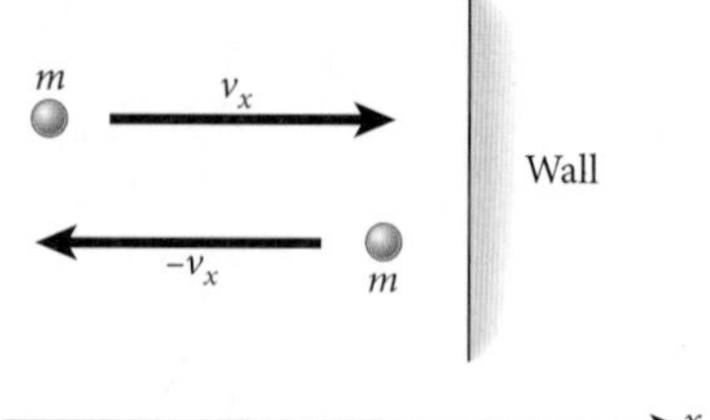


**FIGURE 19.10** Gas molecule with mass $m$ and velocity $v_x$ bouncing off the wall of a container.

Consider a gas molecule with mass $m$ and an $x$-component of velocity, $v_x$, traveling perpendicular to the wall of a container (Figure 19.10). The gas molecule bounces off the wall and moves in the opposite direction with velocity $-v_x$. (Remember, we assumed elastic collisions between molecules and walls.) The change in the $x$-component of the momentum, $\Delta p_x$, of the gas molecule during the collision with the wall is

$$\Delta p_x = p_{\text{f},x} - p_{\text{i},x} = (m)(-v_x) - (m)(v_x) = -2mv_x.$$

To calculate the time-averaged force exerted by the gas on the wall of the container, we need to know not only how much the momentum changes during the collision but also how often a collision of a gas molecule with a wall occurs. To get an expression for the time between collisions, we assume the container is a cube with side $L$ (Figure 19.11). Then a collision like that shown in Figure 19.10 occurs every time the gas molecule makes a complete trip from one side of the cube to the other and back again. This round trip covers a distance $2L$. We can express the time interval, $\Delta t$, between collisions with the wall as

$$\Delta t = \frac{2L}{v_x}.$$

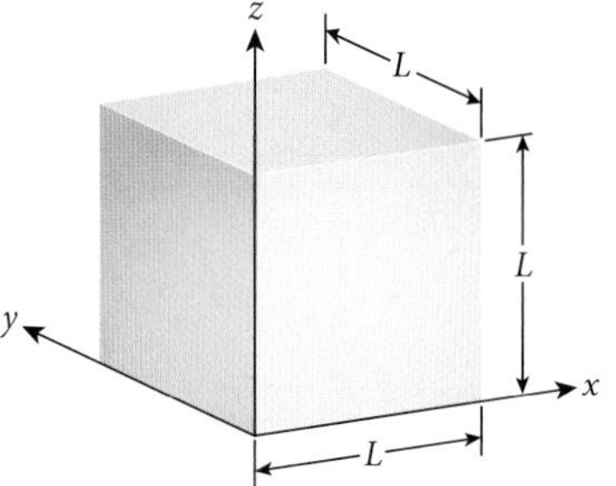


**FIGURE 19.11** A cubical container with side length $L$. Three of the faces of the cube are in the $xy$-, $xz$-, and $yz$-planes.

We know from Newton's Second Law that the $x$-component of the force exerted by the wall on the gas molecule, $F_{\text{wall},x}$, is given by

$$F_{\text{wall},x} = \frac{\Delta p_x}{\Delta t} = \frac{-2mv_x}{(2L/v_x)} = -\frac{mv_x^2}{L},$$

where we have used $\Delta p_x = -2mv_x$ and $\Delta t = 2L/v_x$.

The $x$-component of the force exerted by the gas molecule on the wall, $F_x$, has the same magnitude as $F_{\text{wall},x}$ but is in the opposite direction; that is, $F_x = -F_{\text{wall},x}$, which is a direct consequence of Newton's Third Law (see Chapter 4). If there are $N$ gas molecules, we can state the total force due to these gas molecules on the wall as

$$F_{\text{tot},x} = \sum_{i=1}^{N} \frac{mv_{x,i}^2}{L} = \frac{m}{L}\sum_{i=1}^{N} v_{x,i}^2. \qquad \text{(i)}$$

This result for the force on the wall located at $x = L$ is independent of the $y$- and $z$-components of the velocity vectors of the gas molecules. Since the molecules are in random motion, the average square of the $x$-components of the velocity is the same as the average square of the $y$-components or the $z$-components: $\sum_{i=1}^{N} v_{x,i}^2 = \sum_{i=1}^{N} v_{y,i}^2 = \sum_{i=1}^{N} v_{z,i}^2$. Since $v_i^2 = v_{x,i}^2 + v_{y,i}^2 + v_{z,i}^2$, we have $\sum_{i=1}^{N} v_{x,i}^2 = \frac{1}{3}\sum_{i=1}^{N} v_i^2$, or the average square of the $x$-components of the velocity is $\frac{1}{3}$ of the average square of the velocity. This is the reason for the name of the equipartition theorem: Each Cartesian velocity component has an equal part, $\frac{1}{3}$, of the overall kinetic energy. We can use this fact to rewrite equation (i) for the $x$-component of the force of the gas on the wall:

$$F_{\text{tot},x} = \frac{m}{3L}\sum_{i=1}^{N} v_i^2.$$

Repeating this force analysis for the other five walls, or faces of the cube, we find that each experiences a force of the same magnitude, given by $F_{\text{tot}} = \frac{m}{3L}\sum_{i=1}^{N} v_i^2$. To find the pressure exerted by the gas molecules, we divide this force by the area of the wall, $A = L^2$:

$$p = \frac{F_{\text{tot}}}{A} = \frac{\frac{m}{3L}\sum_{i=1}^{N} v_i^2}{L^2} = \frac{m\sum_{i=1}^{N} v_i^2}{3L^3} = \frac{m}{3V}\sum_{i=1}^{N} v_i^2,$$

where we have used the cube's volume, $V = L^3$, in the last step. Using equation 19.15 for the root-mean-square speed of the gas molecules, we can express the pressure as

$$p = \frac{Nmv_{\text{rms}}^2}{3V}. \qquad \text{(ii)}$$

This result holds for each face of the cube, and it can be applied to a volume of any shape. Multiplying both sides of equation (ii) by the volume leads to $pV = \frac{1}{3}Nmv_{\text{rms}}^2$, and since

*Continued—*

the Ideal Gas Law (equation 19.8) states that $pV = Nk_BT$, we have $Nk_BT = \frac{1}{3}Nmv_{rms}^2$. Canceling out the factor of $N$ on both sides leads to

$$3k_BT = mv_{rms}^2. \quad \text{(iii)}$$

Since $\frac{1}{2}mv_{rms}^2 = K_{ave}$, the average kinetic energy of a gas molecule, we have arrived at the desired result, equation 19.16:

$$K_{ave} = \tfrac{3}{2}k_BT.$$

Note that we can solve $3k_BT = mv_{rms}^2$, equation (iii) of Derivation 19.2, for the root-mean-square speed of the gas molecules:

$$v_{rms} = \sqrt{\frac{3k_BT}{m}}. \quad (19.17)$$

### EXAMPLE 19.4 Average Kinetic Energy of Air Molecules

Suppose a roomful of air is at a temperature of 22.0 °C.

**PROBLEM**

What is the average kinetic energy of the molecules (and atoms) in the air?

**SOLUTION**

The average kinetic energy of the air molecules is given by equation 19.16:

$$K_{ave} = \tfrac{3}{2}k_BT = \tfrac{3}{2}\left(1.38 \cdot 10^{-23}\text{ J/K}\right)\left(273.15\text{ K} + 22.0\text{ K}\right) = 6.11 \cdot 10^{-21}\text{ J}.$$

Often, the average kinetic energy of air molecules is given in electron-volts (eV) rather than joules:

$$K_{ave} = \left(6.11 \cdot 10^{-21}\text{ J}\right)\frac{1\text{ eV}}{1.602 \cdot 10^{-19}\text{ J}} = 0.0381\text{ eV}.$$

A useful approximation to remember is that the average kinetic energy of air molecules at sea level and room temperature is about 0.04 eV.

### 19.5 In-Class Exercise

The average kinetic energy of molecules of an ideal gas doubles if the _______ doubles.

a) temperature

b) pressure

c) mass of the gas molecules

d) volume of the container

e) none of the above

## 19.4 Specific Heat of an Ideal Gas

Chapter 18 addressed the specific heat of materials, which we now consider for gases. Gases can consist of atoms or molecules, and the internal energy of gases can be expressed in terms of their atomic and molecular properties. These relationships lead to the molar specific heats of ideal gases.

Let's begin with **monatomic gases,** which are gases in which atoms are not bound to other atoms. Monatomic gases include helium, neon, argon, krypton, and xenon (the noble gases). We assume that all the internal energy of a monatomic gas is in the form of translational kinetic energy. The average translational kinetic energy depends only on the temperature, as stated in equation 19.16.

The internal energy of a monatomic gas is the number of atoms, $N$, times the average translational kinetic energy of one atom in the gas:

$$E_{int} = NK_{ave} = N\left(\tfrac{3}{2}k_BT\right).$$

The number of atoms of the gas is the number of moles, $n$, times Avogadro's number, $N_A$: $N = nN_A$. Since $N_A k_B = R$, we can express the internal energy as

$$E_{int} = \tfrac{3}{2}nRT. \tag{19.18}$$

Equation 19.18 shows that the internal energy of a monatomic gas depends only on the temperature of the gas.

## Specific Heat at Constant Volume

Suppose an ideal monatomic gas at temperature $T$ is held at constant volume. If heat, $Q$, is added to the gas, it has been shown empirically (see Chapter 18) that the temperature of the gas changes according to

$$Q = nC_V\Delta T \quad \text{(at constant volume)},$$

where $C_V$ is the molar specific heat at constant volume. Because the volume of the gas is constant, it can do no work. Thus, we can use the First Law of Thermodynamics (see Chapter 18) to write

$$\Delta E_{int} = Q - W = nC_V\Delta T \quad \text{(at constant volume)}. \tag{19.19}$$

Remembering that the internal energy of a monatomic gas depends only on its temperature, we use equation 19.18 and obtain

$$\Delta E_{int} = \tfrac{3}{2}nR\Delta T.$$

Combining this result with $Q = nC_V\Delta T$ gives $nC_V\Delta T = \tfrac{3}{2}nR\Delta T$. Canceling the terms $n$ and $\Delta T$, which appear on both sides, we get an expression for the molar specific heat of an ideal monatomic gas at constant volume:

$$C_V = \tfrac{3}{2}R = 12.5 \text{ J/(mol K)}. \tag{19.20}$$

This value for the molar specific heat of an ideal monatomic gas agrees well with measured values for monatomic gases at standard temperature and pressure. These gases are primarily the noble gases. The molar specific heats for **diatomic gases** (gas molecules with two atoms) and **polyatomic gases** (gas molecules with more than two atoms) are higher than the molar specific heats for monatomic gases, as shown in Table 19.2. We'll discuss this difference later in this section.

For any ideal gas, we can use equation 19.18 to write

$$E_{int} = nC_V T, \tag{19.21}$$

which means that the internal energy of an ideal gas depends only on $n$, $C_V$, and $T$.

**Table 19.2 Some Typical Molar Specific Heats Obtained for Different Types of Gases**

| Gas | $C_V$[J/(mol K)] | $C_p$[J/(mol K)] | $\gamma = C_p/C_V$ |
|---|---|---|---|
| Helium (He) | 12.5 | 20.8 | 1.66 |
| Neon (Ne) | 12.5 | 20.8 | 1.66 |
| Argon (Ar) | 12.5 | 20.8 | 1.66 |
| Krypton (Kr) | 12.5 | 20.8 | 1.66 |
| Hydrogen ($H_2$) | 20.4 | 28.8 | 1.41 |
| Nitrogen ($N_2$) | 20.7 | 29.1 | 1.41 |
| Oxygen ($O_2$) | 21.0 | 29.4 | 1.41 |
| Carbon dioxide ($CO_2$) | 28.2 | 36.6 | 1.29 |
| Methane ($CH_4$) | 27.5 | 35.9 | 1.30 |

A generalization of equation 19.21, $\Delta E_{\text{int}} = nC_V\Delta T$, applies to all gases as long as the appropriate molar specific heat is used. According to this equation, the change in internal energy of an ideal gas depends only on $n$, $C_V$, and the change in temperature, $\Delta T$. The change in internal energy does not depend on any corresponding pressure or volume change.

## Specific Heat at Constant Pressure

Now let's consider the situation in which the temperature of an ideal gas is increased while the pressure of the gas is held constant. The added heat, $Q$, has been shown empirically to be related to the change in temperature as follows:

$$Q = nC_p\Delta T \quad \text{(for constant pressure)}, \tag{19.22}$$

where $C_p$ is the molar specific heat at constant pressure. The molar specific heat at constant pressure is greater than the molar specific heat at constant volume because energy must be supplied to do work as well as to increase the temperature.

To relate the molar specific heat at constant volume to the molar specific heat at constant pressure, we start with the First Law of Thermodynamics: $\Delta E_{\text{int}} = Q - W$. We substitute from equation 19.19 for $\Delta E_{\text{int}}$ and from equation 19.22 for $Q$. In Chapter 18, the work, $W$, done when the pressure remains constant was expressed as $W = p\Delta V$. With these substitutions into $\Delta E_{\text{int}} = Q - W$, we obtain

$$nC_V\Delta T = nC_p\Delta T - p\Delta V.$$

### 19.1 Self-Test Opportunity

Using the data in Table 19.2, verify the relationship given in equation 19.23 between the molar specific heat at constant pressure and the molar specific heat at constant volume for an ideal gas.

The Ideal Gas Law (equation 19.7) relates the change in volume to the change in temperature if the pressure is held constant: $p\Delta V = nR\Delta T$. Substituting this expression for the work, we finally have

$$nC_V\Delta T = nC_p\Delta T - nR\Delta T.$$

Every term in this equation contains the common factor $n\Delta T$. Dividing it out gives the very simple relation between the specific heats at constant volume and at constant pressure:

$$C_p = C_V + R. \tag{19.23}$$

## Degrees of Freedom

As you can see in Table 19.2, the relationship $C_V = \frac{3}{2}R$ holds for monatomic gases but not for diatomic and polyatomic gases. This failure can be explained in terms of the possible degrees of freedom of motion for various types of molecules.

In general, a **degree of freedom** is a direction in which something can move. For a point particle in three-dimensional space, there are three orthogonal directions, which are independent of each other. For a collection of $N$ point particles, there are $3N$ degrees of freedom for translational motion in three-dimensional space. But if the point particles are arranged in a solid object, they cannot move independently of each other, and the number of degrees of freedom is reduced to the translational degrees of freedom of the center of mass of the object (3), plus independent rotations of the object around the center of mass. In general, there are three possible independent rotations, giving a total of six degrees of freedom, three translational and three rotational.

He

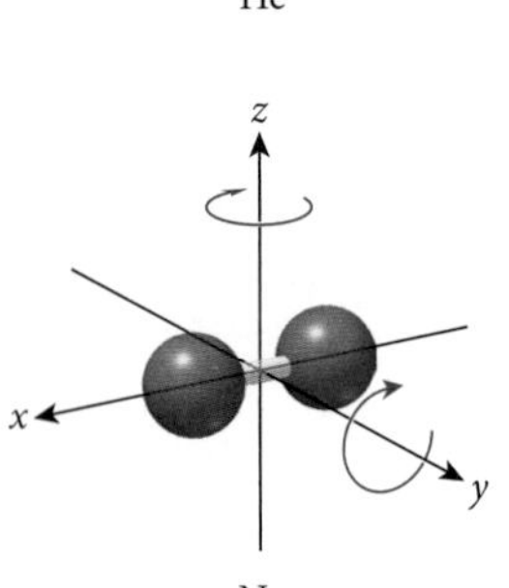


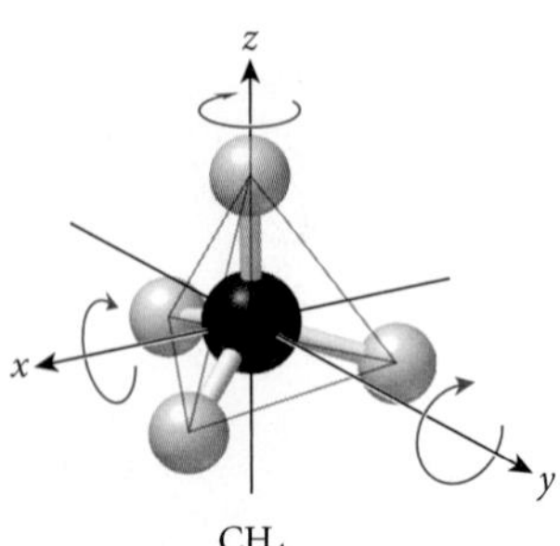


**FIGURE 19.12** Rotational degrees of freedom for a helium atom (He), a nitrogen molecule ($N_2$), and a methane molecule ($CH_4$).

Let's consider three different kinds of gases (Figure 19.12). The first is a monatomic gas, such as helium (He). The second is a diatomic gas, represented by nitrogen ($N_2$). The third is a polyatomic molecule, for example, methane ($CH_4$). The molecule of the polyatomic gas can rotate around all three coordinate axes and thus has three rotational degrees of freedom. The diatomic molecule shown in Figure 19.12 is aligned with the $x$-axis, and rotation about this axis will not yield a different configuration. Therefore, this molecule has only two rotational degrees of freedom, shown in the figure as being associated with rotations about the $y$- and $z$-axes. The monatomic gas consists of spherically symmetrical atoms, which have no rotational degrees of freedom.

The various ways in which the internal energy of an ideal gas can be allocated are specified by the principle of **equipartition of energy,** which states that gas molecules in thermal equilibrium have the same average energy associated with each independent degree of freedom of their motion. The average energy per degree of freedom is given by $\frac{1}{2}k_BT$ for each gas molecule.

The equipartition theorem (equation 19.16) for gas molecules' average kinetic energy associated with temperature, $K_{\text{ave}} = \frac{3}{2}k_BT$, is consistent with the equipartition of the molecules' translational kinetic energy among three directions (or three degrees of freedom), $x$, $y$, and $z$. The translational kinetic energy has three degrees of freedom, the average kinetic energy per translational degree of freedom is $\frac{1}{2}k_BT$, and the average kinetic energy of the gas molecules is given by $3\left(\frac{1}{2}k_BT\right)=\frac{3}{2}k_BT$.

Thus, the molar specific heat at constant volume predicted by the kinetic theory for monatomic gases by assuming three degrees of freedom for the translational motion is consistent with reality. However, the observed molar specific heats at constant volume are higher for diatomic gases than for monatomic gases. For polyatomic gases, the observed molar specific heats are even higher.

### Specific Heat at Constant Volume for a Diatomic Gas

The nitrogen molecule in Figure 19.12 is lined up along the $x$-axis. If we look along the $x$-axis at the molecule, we see a point and thus are not able to discern any rotation around the $x$-axis. However, if we look along the $y$-axis or the $z$-axis, we can see that the nitrogen molecule can have appreciable rotational motion about either of those axes. Therefore, there are two degrees of freedom for rotational kinetic energy of the nitrogen molecule and, by extension, all diatomic molecules. Therefore, the average kinetic energy per diatomic molecule is $(3+2)\left(\frac{1}{2}k_BT\right)=\frac{5}{2}k_BT$.

This result implies that calculation of the molar specific heat at constant volume for a diatomic gas should be similar to that for a monatomic gas, but with two additional degrees of freedom for the average energy. Thus, we modify equation 19.20 to obtain the molar specific heat at constant volume for a diatomic gas:

$$C_V=\frac{3+2}{2}R=\frac{5}{2}R=\frac{5}{2}\left[8.31\ \text{J}/\left(\text{mol K}\right)\right]=20.8\ \text{J}/\left(\text{mol K}\right).$$

Comparing this value to the actual measured molar specific heats at constant volume given in Table 19.2 for the diatomic gases hydrogen, nitrogen, and oxygen, we see that this prediction from the kinetic theory of ideal gases agrees reasonably well with those measurements.

### Specific Heat at Constant Volume for a Polyatomic Gas

Now let's consider the polyatomic gas, methane ($CH_4$), shown in Figure 19.12. The methane molecule is composed of four hydrogen atoms arranged in a tetrahedron, bonded to a carbon atom at the center. Looking along any of the three axes, we can discern rotation of this molecule. Thus, this particular molecule and all polyatomic molecules have three degrees of freedom related to rotational kinetic energy.

Thus, calculation of the molar specific heat at constant volume for a polyatomic gas should be similar to that for a monatomic gas, but with three additional degrees of freedom for the average energy. We modify equation 19.20 to obtain the molar specific heat at constant volume for a polyatomic gas:

$$C_V=\frac{3+3}{2}R=\frac{6}{2}R=3\left[8.31\ \text{J}/\left(\text{mol K}\right)\right]=24.9\ \text{J}/\left(\text{mol K}\right).$$

Comparing this value to the actual measured molar specific heat at constant volume given in Table 19.2 for the polyatomic gas methane, we see that the value predicted by the kinetic theory of ideal gases is close to, but somewhat lower than, the measured value.

What accounts for the differences between predicted and actual values of the molar specific heats for diatomic and polyatomic molecules is the fact that these molecules can have internal degrees of freedom, in addition to translational and rotational degrees of freedom. The atoms of these molecules can oscillate with respect to one another. For example, imagine that the two oxygen atoms and one carbon atom of a carbon dioxide molecule are connected by springs. The three atoms could oscillate back and forth with respect to each other, which corresponds to extra degrees of freedom. The molar specific heat at constant volume would then be larger than the 24.9 J/(mol K) predicted for a polyatomic gas.

A certain minimum energy is required to excite the motion of internal oscillation in molecules. This fact cannot be understood from the viewpoint of classical mechanics; Chapters 36 through 38 will provide an understanding of this fact. For now, all you need to know is that a threshold effect is observed for vibrational degrees of freedom. In fact, the rotational degrees of freedom of diatomic and polyatomic gas molecules also exhibit this threshold effect. At low temperatures, the internal energy of the gas may be too low to enable the molecules to rotate. However, at standard temperature and pressure, most diatomic and polyatomic gases have a molar specific heat at constant volume that is consistent with their molecules' translational and rotational degrees of freedom.

### 19.6 In-Class Exercise

The ratio of specific heat at constant pressure to specific heat at constant volume, $C_p/C_V$, for an ideal gas

a) is always equal to 1.

b) is always smaller than 1.

c) is always larger than 1.

d) can be smaller or larger than 1, depending on the degrees of freedom of the gas molecules.

### Ratio of Specific Heats

Finally, it is convenient to take the ratio of the specific heat at constant pressure to that at constant volume. This ratio is conventionally denoted by the Greek letter $\gamma$:

$$\gamma \equiv \frac{C_p}{C_V}. \tag{19.24}$$

Inserting the predicted values for the specific heats of ideal gases with different degrees of freedom gives $\gamma = C_p/C_V = \frac{5}{2}R/\frac{3}{2}R = \frac{5}{3}$ for monatomic gases, $\gamma = C_p/C_V = \frac{7}{2}R/\frac{5}{2}R = \frac{7}{5}$ for diatomic gases, and $\gamma = C_p/C_V = \frac{8}{2}R/\frac{6}{2}R = \frac{4}{3}$ for polyatomic gases.

Table 19.2 also lists values for $\gamma$ obtained from empirical data. Despite the fact that the empirical values of the specific heats for the gases other than the noble gases are not very close to the calculated values for an ideal gas, the values of the ratio $\gamma$ are in very good agreement with our theoretical expectations.

## 19.5 Adiabatic Processes for an Ideal Gas

As we saw in Chapter 18, an adiabatic process is one in which the state of a system changes and no exchange of heat with the surroundings takes place during the change. An adiabatic process can occur when the system change occurs quickly. Thus, $Q = 0$ for an adiabatic process. From the First Law of Thermodynamics, we then have $\Delta E_{\text{int}} = -W$, as shown in Chapter 18.

Let's explore how the change in volume is related to the change in pressure for an ideal gas undergoing such an adiabatic process, that is, a process in the absence of heat transfer. The relationship is given by

$$pV^{\gamma} = \text{constant} \quad \text{(for an adiabatic process)}, \tag{19.25}$$

where $\gamma = C_p/C_V$ is the ratio of the specific heat at constant pressure and the specific heat at constant volume for the ideal gas.

### DERIVATION 19.3 Pressure and Volume in an Adiabatic Process

In order to derive the relationship between pressure and volume for an adiabatic process, let's consider small differential changes in the internal energy during such a process: $dE_{\text{int}} = -dW$. In Chapter 18, we saw that $dW = p\,dV$, and therefore for the adiabatic process, we have

$$dE_{\text{int}} = -p dV. \tag{i}$$

However, equation 19.21 says that the internal energy is related in general to the temperature: $E_{\text{int}} = nC_VT$. Since we want to calculate $dE_{\text{int}}$, we take the derivative of this equation with respect to the temperature:

$$dE_{\text{int}} = nC_V\,dT. \tag{ii}$$

Combining equations (i) and (ii) for $dE_{\text{int}}$ then yields

$$-p dV = nC_V\,dT. \tag{iii}$$

We can take the derivative of the Ideal Gas Law (equation 19.7), $pV = nRT$ (with $n$ and $R$ constant):

$$p\,dV + V\,dp = nR\,dT. \tag{iv}$$

Solving equation (iii) for $dT$ and substituting that expression into equation (iv) gives us

$$p\,dV + V\,dp = nR\left(-\frac{p\,dV}{nC_V}\right).$$

Now we can collect the terms proportional to $dV$ on the left-hand side, giving

$$\left(1+\frac{R}{C_V}\right)p\,dV + V\,dp = 0. \tag{v}$$

Since $R = C_p - C_V$ from equation 19.23, we see that $1+R/C_V = 1+(C_p-C_V)/C_V = C_p/C_V$. Equation 19.24 represents this ratio of the two specific heats as $\gamma$. With this, equation (v) becomes $\gamma p\,dV + V\,dp = 0$. Dividing both sides by $pV$ lets us finally write

$$\frac{dp}{p} + \gamma\frac{dV}{V} = 0.$$

Integration then yields

$$\ln p + \gamma \ln V = \text{constant}. \tag{vi}$$

Using the rules of logarithms, equation (vi) can also be written as $\ln(pV^\gamma) = \text{constant}$. We can exponentiate both sides of this equation to get the desired result:

$$pV^\gamma = \text{constant}.$$

Another way of stating the relationship between pressure and volume for an adiabatic process involving an ideal gas is

$$p_f V_f^\gamma = p_i V_i^\gamma \quad \text{(for an adiabatic process)}, \tag{19.26}$$

where $p_i$ and $V_i$ are the initial pressure and volume, respectively, and $p_f$ and $V_f$ are the final pressure and volume, respectively.

Figure 19.13 plots the behavior of a gas during three different processes as a function of pressure and volume. The red curve represents an adiabatic process, in which $pV^\gamma$ is constant, for a monatomic gas with $\gamma = \frac{5}{3}$. The other two curves represent isothermal (constant $T$) processes. One isothermal process occurs at $T = 300$ K, and the other occurs at $T = 400$ K. The adiabatic process shows a steeper decrease in pressure as a function of volume than the isothermal processes.

We can write an equation relating the temperature and volume of a gas in an adiabatic process by combining equation 19.25 and the Ideal Gas Law (equation 19.7):

$$pV^\gamma = \left(\frac{nRT}{V}\right)V^\gamma = (nR)TV^{\gamma-1} = \text{constant}.$$

Assuming the gas is in a closed container, $n$ is constant, and we can rewrite this relationship as $TV^{\gamma-1} = \text{constant}$ (for an adiabatic process), or

$$T_f V_f^{\gamma-1} = T_i V_i^{\gamma-1} \quad \text{(for an adiabatic process)}, \tag{19.27}$$

where $T_i$ and $V_i$ are the initial temperature and volume, respectively, and $T_f$ and $V_f$ are the final temperature and volume, respectively.

An adiabatic expansion as described by equation 19.27 occurs when you open a container that holds a cold, carbonated beverage. The carbon dioxide from the carbonation and the water vapor make up a gas with a pressure above atmospheric pressure. When this gas expands and some comes out of the container's opening, the system must do work. Because

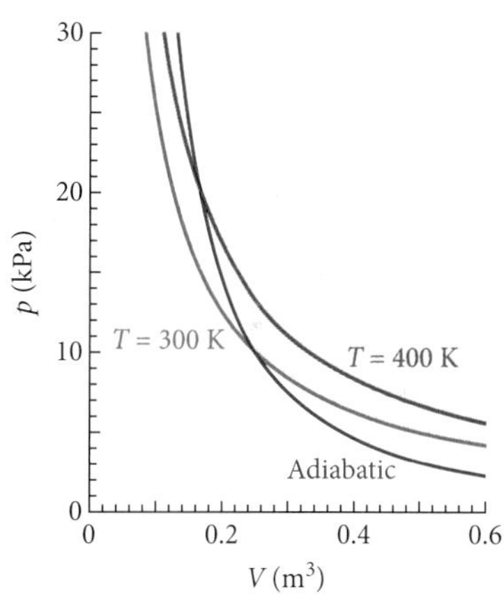


**FIGURE 19.13** Plot of the behavior of a monatomic gas during three different processes on a $pV$-diagram. One process is adiabatic (red). The second process is isothermal with $T = 300$ K (green). The third process is also isothermal with $T = 400$ K (blue).

### 19.7 In-Class Exercise

A monatomic ideal gas occupies volume $V_i$, which is then decreased to $\frac{1}{2}V_i$ via an adiabatic process. Which relationship is correct for the pressures of this gas?

a) $p_f = 2p_i$

b) $p_f = \frac{1}{2}p_i$

c) $p_f = 2^{5/3}p_i$

d) $p_f = (\frac{1}{2})^{5/3}p_i$

### 19.8 In-Class Exercise

A monatomic ideal gas occupies volume $V_i$, which is then decreased to $\frac{1}{2}V_i$ via an adiabatic process. Which relationship is correct for the temperatures of this gas?

a) $T_f = 2^{2/3}T_i$

b) $T_f = 2^{5/3}T_i$

c) $T_f = (\frac{1}{2})^{2/3}T_i$

d) $T_f = (\frac{1}{2})^{5/3}T_i$

the process happens very quickly, no heat can be transferred to the gas; thus, the process is adiabatic, and the product of the temperature and the volume raised to the power $\gamma-1$ remains constant. The volume of the gas increases, and the temperature must decrease. Thus, condensation occurs around the opening of the container.

## SOLVED PROBLEM 19.1 Bicycle Tire Pump

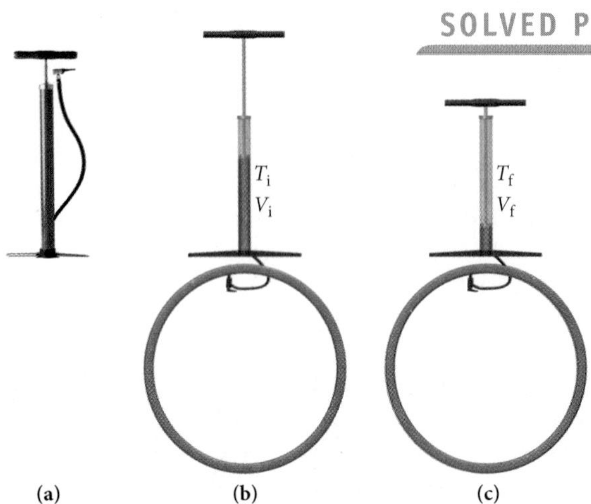


**FIGURE 19.14** (a) A hand-operated tire pump used to inflate a bicycle tire. (b) The pump and tire before the compression stroke. (c) The pump and tire after the compression stroke.

A hand-operated tire pump (Figure 19.14a) is used to inflate a bicycle tire. The tire pump consists of a cylinder with a piston. The tire pump is connected to the tire with a small hose. There is a valve between the pump and the tire that allows air to enter the tire, but not to exit. In addition, there is a valve that allows air to enter the pump after the compression stroke is complete and the piston is retracted to its original position. Figure 19.14b shows the pump and tire before the compression stroke. Figure 19.14c shows the pump and tire after the compression stroke.

The inner radius of the cylinder of the pump is $r_p = 1.00$ cm, the height of the active volume before the compression stroke is $h_i = 60.0$ cm, and the height of the active volume after the compression stroke is $h_f = 10.0$ cm. The tire can be considered to be a cylindrical ring with inner radius $r_1 = 66.0$ cm, outer radius $r_2 = 67.5$ cm, and thickness $t = 1.50$ cm. The temperature of the air in the pump and tire before the compression stroke is $T_i = 295$ K.

### PROBLEM

What is the temperature, $T_f$, of the air after one compression stroke?

### SOLUTION

#### THINK

The volume and temperature of the air before the compression stroke are $V_i$ and $T_i$, respectively, and the volume and temperature of the air after the compression stroke are $V_f$ and $T_f$, respectively. The compression stroke takes place quickly, so no heat can be transferred into or out of the system, and this process can be treated as an adiabatic process. The volume of air in the pump decreases when the pump handle is pushed down while the volume of air in the tire remains constant.

#### SKETCH

The dimensions of the tire pump and the tire are shown in Figure 19.15.

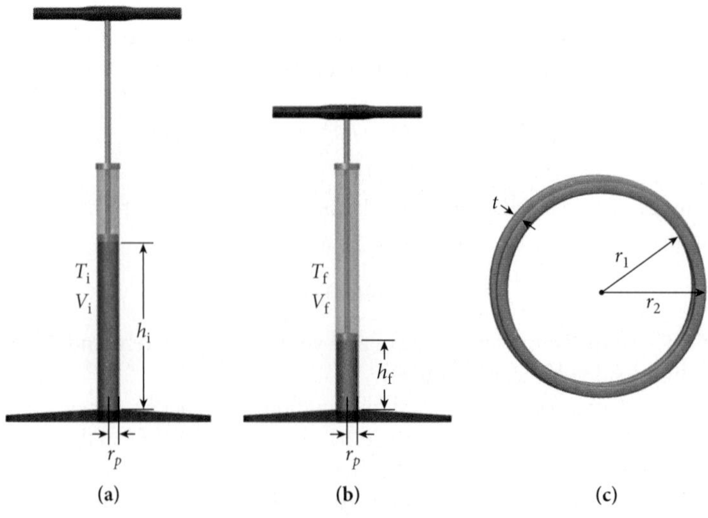


**FIGURE 19.15** (a) Dimensions of the tire pump before the compression stroke. (b) Dimensions of the tire pump after the compression stroke. (c) Dimensions of the bicycle tire.

**RESEARCH**

The volume of air in the tire is

$$V_{\text{tire}} = t\left(\pi r_2^2 - \pi r_1^2\right).$$

The volume of air in the pump before the compression stroke is

$$V_{\text{pump,i}} = \pi r_{\text{p}}^2 h_{\text{i}}.$$

The volume of air in the pump after the compression stroke is

$$V_{\text{pump,f}} = \pi r_{\text{p}}^2 h_{\text{f}}.$$

The initial volume of air in the pump-tire system is $V_{\text{i}} = V_{\text{pump,i}} + V_{\text{tire}}$, and the final volume of air in the pump-tire system is $V_{\text{f}} = V_{\text{pump,f}} + V_{\text{tire}}$. We ignore the volume of air in the hose between the pump and the tire.

This process is adiabatic, so we can use equation 19.27 to describe the system before and after the compression stroke:

$$T_{\text{f}} V_{\text{f}}^{\gamma-1} = T_{\text{i}} V_{\text{i}}^{\gamma-1}. \qquad \text{(i)}$$

**SIMPLIFY**

We can solve equation (i) for the final temperature:

$$T_{\text{f}} = T_{\text{i}} \frac{V_{\text{i}}^{\gamma-1}}{V_{\text{f}}^{\gamma-1}} = T_{\text{i}} \left(\frac{V_{\text{i}}}{V_{\text{f}}}\right)^{\gamma-1}.$$

We can then insert the expressions for the initial volume and the final volume:

$$T_{\text{f}} = T_{\text{i}} \left(\frac{\pi r_{\text{p}}^2 h_{\text{i}} + t\left(\pi r_2^2 - \pi r_1^2\right)}{\pi r_{\text{p}}^2 h_{\text{f}} + t\left(\pi r_2^2 - \pi r_1^2\right)}\right)^{\gamma-1} = T_i \left(\frac{r_{\text{p}}^2 h_{\text{i}} + t\left(r_2^2 - r_1^2\right)}{r_{\text{p}}^2 h_{\text{f}} + t\left(r_2^2 - r_1^2\right)}\right)^{\gamma-1}.$$

**CALCULATE**

Air is mostly diatomic gases (78% $N_2$ and 21% $O_2$, see Table 19.1), so we use $\gamma = \frac{7}{5}$, from Section 19.4. Putting in the numerical values then gives us

$$T_{\text{f}} = (295 \text{ K}) \left(\frac{(0.0100 \text{ m})^2 (0.600 \text{ m}) + (0.0150 \text{ m})\left((0.675 \text{ m})^2 - (0.660 \text{ m})^2\right)}{(0.0100 \text{ m})^2 (0.100 \text{ m}) + (0.0150 \text{ m})\left((0.675 \text{ m})^2 - (0.660 \text{ m})^2\right)}\right)^{(7/5)-1}$$

$$= 313.162 \text{ K}.$$

**ROUND**

We report our result to three significant figures:

$$T_{\text{f}} = 313 \text{ K}.$$

**DOUBLE-CHECK**

The ratio of the initial volume of air in the pump-tire system to the final volume is

$$\frac{V_{\text{i}}}{V_{\text{f}}} = 1.16.$$

During the compression stroke, the temperature of the air increases from 295 K to 313 K, so the ratio of final temperature to initial temperature is

$$\frac{T_{\text{f}}}{T_{\text{i}}} = \frac{313 \text{ K}}{295 \text{ K}} = 1.06,$$

which seems reasonable because the final temperature should be equal to the initial temperature times the ratio of the initial and final volume raised to the power $\gamma$–1: $(1.16)^{\gamma-1} = 1.16^{0.4} = 1.06$. If you depress the piston many times in a short period of time, each compression stroke will increase the temperature of the air in the tire by about 6%. Thus, the air in the tire and the pump gets warmer as the tire is inflated.

### Work Done by an Ideal Gas in an Adiabatic Process

We can also find the work done by a gas undergoing an adiabatic process from an initial state to a final state. In general, the work is given by

$$W = \int_{\mathrm{i}}^{\mathrm{f}} p\,dV.$$

For an adiabatic process, we can use equation 19.25 to write $p = cV^{-\gamma}$ (where $c$ is a constant), and the integral determining the work becomes

$$W = \int_{V_\mathrm{i}}^{V_\mathrm{f}} cV^{-\gamma}dV = c\int_{V_\mathrm{i}}^{V_\mathrm{f}} V^{-\gamma}dV = c\left[\frac{V^{1-\gamma}}{1-\gamma}\right]_{V_\mathrm{i}}^{V_\mathrm{f}} = \frac{c}{1-\gamma}\left(V_\mathrm{f}^{1-\gamma} - V_\mathrm{i}^{1-\gamma}\right).$$

The constant $c$ has the value $c = pV^{\gamma}$. Using the Ideal Gas Law in the form $p = nRT/V$, we have

$$c = \left(\frac{nRT}{V}\right)V^{\gamma} = nRTV^{\gamma-1}.$$

Now we insert this constant into the expression for the work and use equation 19.27. We finally find the work done by a gas undergoing an adiabatic process:

$$W = \frac{nR}{1-\gamma}\left(T_\mathrm{f} - T_\mathrm{i}\right). \tag{19.28}$$

## 19.6 Kinetic Theory of Gases

As promised earlier, this section examines some important results from the kinetic theory of gases beyond the equipartition theorem of Section 19.3.

### Maxwell Speed Distribution

Equation 19.17 gives the root-mean-square speed of gas molecules at a given temperature. This is the average speed, but what is the distribution of speeds? That is, what is the probability that a gas molecule has some given speed between $v$ and $v+dv$? The speed distribution, called the **Maxwell speed distribution,** or sometimes the *Maxwell-Boltzmann speed distribution,* is given by

$$f(v) = 4\pi\left(\frac{m}{2\pi k_\mathrm{B}T}\right)^{3/2} v^2 e^{-mv^2/2k_\mathrm{B}T}. \tag{19.29}$$

**19.2 Self-Test Opportunity**

Show explicitly that integrating the Maxwell speed distribution from $v = 0$ to $v = \infty$ yields 1, as required for a probability distribution.

The units of the Maxwell speed distribution are $(\mathrm{m/s})^{-1}$. As required for a probability distribution, integration of this distribution with respect to $dv$ yields 1: $\int_0^{\infty} f(v)dv = 1$.

A very important feature of probability distributions for observable quantities, like the one in equation 19.29 for the molecular speed, is that all kinds of physically meaningful averages can be calculated from them by simply multiplying the quantity to be averaged by the probability distribution and integrating over the entire range. For example, we can obtain the average speed for the Maxwell speed distribution by multiplying the speed, $v$, by the probability distribution, $f(v)$, and integrating this product, $vf(v)$, over all possible values of the speed from zero to infinity. The average speed is then given by

$$v_\mathrm{ave} = \int_0^{\infty} vf(v)dv = \sqrt{\frac{8k_\mathrm{B}T}{\pi m}},$$

where we used the definite integral $\int_0^{\infty} x^3 e^{-x^2/a}dx = a^2/2$ in performing the integration.

To find the root-mean-square speed for the Maxwell speed distribution, we first find the average speed squared:

$$\left(v^2\right)_\mathrm{ave} = \int_0^{\infty} v^2 f(v)dv = \frac{3k_\mathrm{B}T}{m},$$

where we employed the definite integral $\int_0^\infty x^4 e^{-x^2/a} dx = \frac{3}{8} a^{5/2} \sqrt{\pi}$. We then take the square root:

$$v_{\text{rms}} = \sqrt{\left(v^2\right)_{\text{ave}}} = \sqrt{\frac{3k_{\text{B}}T}{m}}.$$

As required, this agrees with equation 19.17 for the speed of gas molecules in an ideal gas.

The most probable speed for the Maxwell speed distribution—that is, the value at which $f(v)$ has a maximum—is calculated by taking the derivative of $f(v)$ with respect to $v$, setting that derivative equal to zero, and solving for $v$, which gives

$$v_{\text{mp}} = \sqrt{\frac{2k_{\text{B}}T}{m}}.$$

Figure 19.16 shows the Maxwell speed distribution for nitrogen ($N_2$) molecules in air at a temperature of 295 K (22 °C), with the most probable speed, the average speed, and the root-mean-square speed. You can see in Figure 19.16 that the speeds of the nitrogen molecules are distributed around the average speed, $v_{\text{ave}}$. However, the distribution is not symmetrical around $v_{\text{rms}}$. A tail in the distribution extends to high velocities. You can also see that the most probable speed corresponds to the maximum of the distribution.

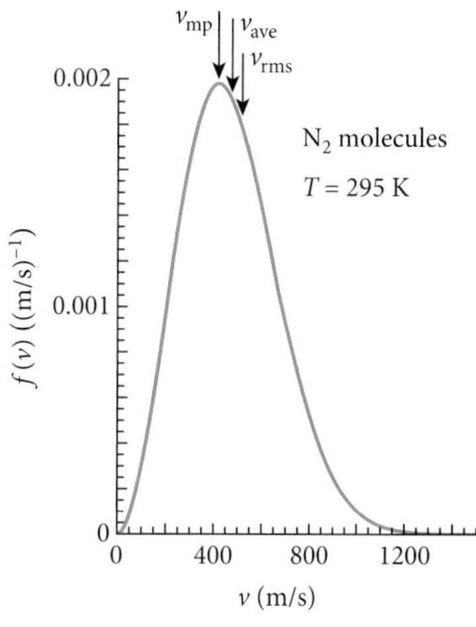


**FIGURE 19.16** Maxwell speed distribution for nitrogen molecules with a temperature of 295 K plotted on a linear scale.

Figure 19.17 shows the Maxwell speed distribution for nitrogen molecules at four different temperatures. As the temperature is increased, the Maxwell speed distribution gets wider. Consequently, the maximum value of $f(v)$ has to get lower with increasing temperature, because the total area under the curve always equals 1, as required for a probability distribution.

In addition, at a given temperature, the Maxwell speed distribution for lighter atoms or molecules is wider and has more of a high-speed tail than the distribution for heavier atoms or molecules (Figure 19.18). Figure 19.18a shows the Maxwell speed distributions for hydrogen and nitrogen molecules plotted on a logarithmic scale at a temperature of 22 °C, which is characteristic of temperatures found at the surface of the Earth. The fact that this curve has no high-speed tail is important to the composition of Earth's atmosphere. As we saw in Chapter 12, the escape speed, $v_e$, for any object near the surface of the Earth is 11.2 km/s. Thus, any air molecule that has at least this escape speed can leave Earth's atmosphere. No gas atoms or molecules are near the escape speed at this temperature. However, hydrogen gas is light compared to the other gases in the Earth's atmosphere and tends to move upward.

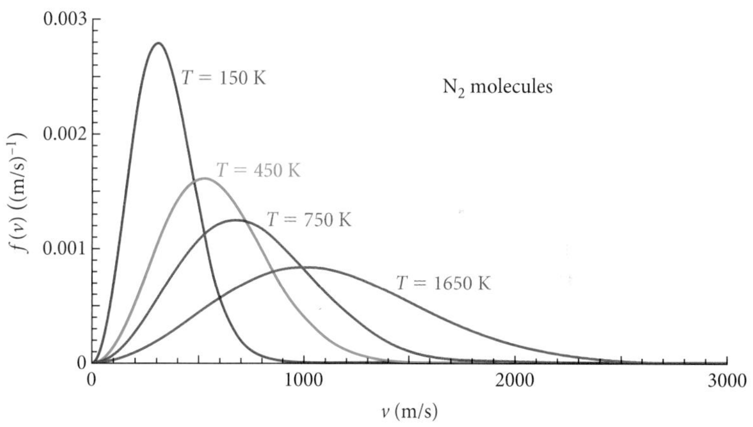


**FIGURE 19.17** Maxwell speed distribution for $N_2$ molecules at four different temperatures.

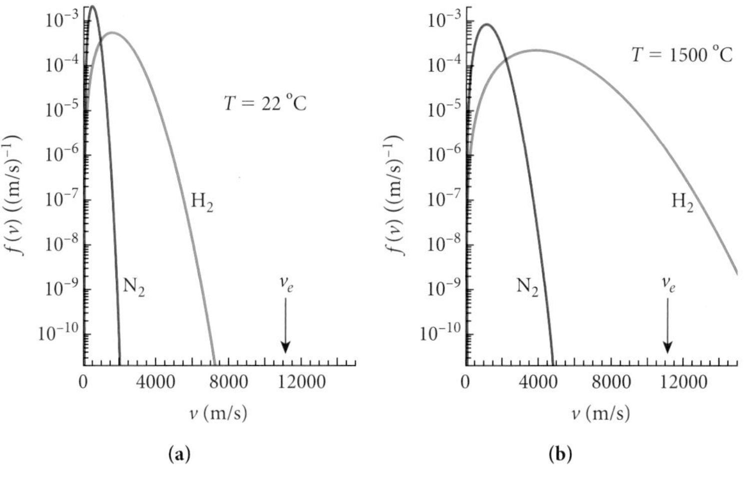


**FIGURE 19.18** Maxwell speed distributions for hydrogen molecules and nitrogen molecules in the Earth's atmosphere, plotted on a logarithmic scale: (a) at a temperature of 22 °C; (b) at a temperature of 1500 °C.

The temperature of Earth's atmosphere depends on the altitude. At an altitude of 100 km above the surface, the Earth's atmosphere starts to become very warm and can reach temperatures of 1500 °C. This layer of the atmosphere is called the *thermosphere.*

Figure 19.18b shows the Maxwell speed distributions for hydrogen and nitrogen molecules plotted on a logarithmic scale at a temperature of 1500 °C. We can calculate the fraction of hydrogen molecules that will have a speed higher than the escape speed by evaluating the integral for the Maxwell speed distribution from the escape speed to infinity:

$$\text{Fraction of } H_2 \text{ escaping} = \int_{v=11.2 \text{ km/s}}^{\infty} f(v)dv = 6\cdot 10^{-4} = 0.06\%.$$

Only a small fraction of the hydrogen molecules in the thermosphere have speeds above the escape speed. The hydrogen molecules with speeds above the escape speed will go out into space. The hydrogen molecules remaining behind have the same temperature as the rest of the thermosphere, and the speed distribution will reflect this temperature. Thus, over time on the geologic scale, most hydrogen molecules in the Earth's atmosphere rise to the thermosphere and eventually leave the Earth.

What about the rest of the atmosphere, which consists mainly of nitrogen and oxygen? Essentially no nitrogen molecules have speeds above the escape speed. The ratio of the value of the Maxwell speed distribution at the escape speed for nitrogen relative to that for hydrogen is $f_N(11.2 \text{ km/s})/f_H(11.2 \text{ km/s}) = 3\cdot 10^{-48}$. Since the entire atmosphere has a mass of $5\cdot 10^{18}$ kg and contains approximately $10^{44}$ nitrogen molecules, you can see that virtually no nitrogen molecule has a speed above the escape speed. Thus, there is very little hydrogen in Earth's atmosphere, while nitrogen and heavier gases (oxygen, argon, carbon dioxide, neon, and so on) remain in the atmosphere permanently.

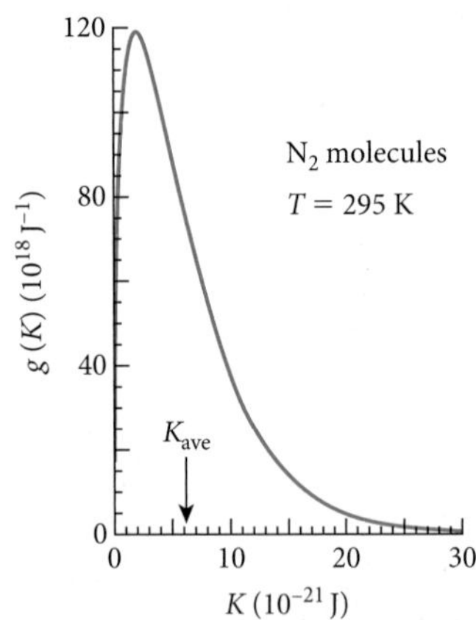


**FIGURE 19.19** Maxwell kinetic energy distribution for nitrogen molecules at a temperature of 295 K, plotted on a linear scale.

## Maxwell Kinetic Energy Distribution

Just as gas molecules have a distribution of speeds, they also have a distribution of kinetic energies. The **Maxwell kinetic energy distribution** (sometimes called the *Maxwell-Boltzmann kinetic energy distribution*) describes the energy spectra of gas molecules and is given by

$$g(K) = \frac{2}{\sqrt{\pi}}\left(\frac{1}{k_B T}\right)^{3/2}\sqrt{K}e^{-K/k_B T}. \tag{19.30}$$

The Maxwell kinetic energy distribution times $dK$ gives the probability of observing a gas molecule with a kinetic energy between $K$ and $K+dK$. Integration of this distribution with respect to $dK$ yields 1, $\int_0^{\infty} g(K)dK = 1$, as required for any probability distribution. The unit for the Maxwell kinetic energy distribution is J$^{-1}$.

Figure 19.19 shows the Maxwell kinetic energy distribution for nitrogen molecules at a temperature of 295 K. The kinetic energies of the nitrogen molecules are distributed around the average kinetic energy. Like the Maxwell speed distribution, the Maxwell kinetic energy distribution is not symmetrical around $K_{ave}$, but has a significant tail at high kinetic energies. This high-kinetic-energy tail can carry information about the temperature of the gas.

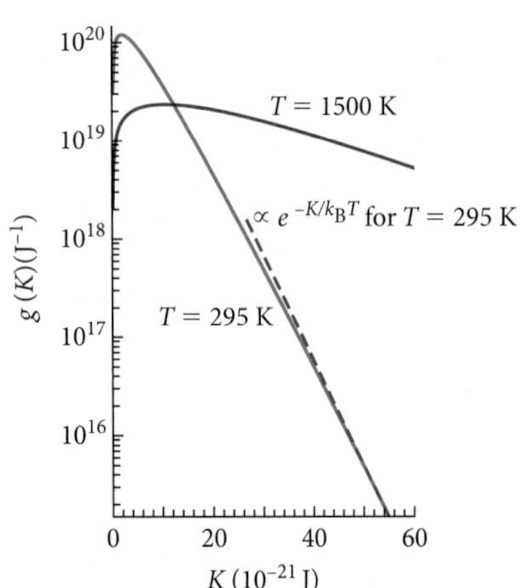


**FIGURE 19.20** Maxwell kinetic energy distribution for nitrogen molecules at $T$ = 295 K and $T$ = 1500 K, plotted on a logarithmic scale.

Figure 19.20 shows the Maxwell kinetic energy distribution for nitrogen molecules at two temperatures, $T$ = 295 K and $T$ = 1500 K. The kinetic energy distribution at $T$ = 1500 K is much flatter and extends to much higher kinetic energies than the kinetic energy distribution at $T$ = 295 K. The high-kinetic-energy tail for $T$ = 295 K in Figure 19.20 approaches a simple exponential distribution of kinetic energy with a slope of $-1/k_B T$, where $T$ = 295 K, as illustrated by the dashed line.

Note that the formulations given here for the Maxwell speed distribution and the Maxwell kinetic energy distribution do not take the effects of relativity into account. For example, the speed distribution extends up to infinite velocities, and no object can travel faster than the speed of light (to be discussed in detail in Chapter 35 on relativity). However, the relevant velocities for ideal gases are small compared with the speed of light. The highest speed shown in Figure 19.18 is $v = 1.5\cdot 10^4$ m/s (for hydrogen molecules at

$T$ = 1500 K), which is only 0.005% of the speed of light. The Maxwell speed and kinetic energy distributions are also based on the assumption that the molecules of the gas do not interact. If the gas molecules interact, then these distributions do not apply. For gas molecules moving in the Earth's atmosphere, the assumption of no interactions is justified, so many of the properties of the Earth's atmosphere can be described using the Maxwell speed distribution and the Maxwell kinetic energy distribution.

### EXAMPLE 19.5 Temperature of the Quark-Gluon Plasma

Measuring the kinetic energy distribution of the constituents of a gas can reveal the temperature of the gas. This technique is used in many applications, including the study of gases consisting of elementary particles and nuclei. For example, Figure 19.21 shows the kinetic energy distributions of antiprotons and pions produced in relativistic heavy ion collisions of gold nuclei. (Antiprotons and pions will be discussed extensively in Chapter 39.) The data were produced by the STAR collaboration and are shown as red dots. Data points for low kinetic energies are missing, because the experimental apparatus cannot detect those energies. The blue lines represent fits to the data, using equation 19.30 with the best-fit value for the temperature, $T$. Values of $k_BT = 6.71 \cdot 10^{-11}$ J = 417 MeV for the antiprotons and $k_BT = 3.83 \cdot 10^{-11}$ J = 238 MeV for the pions were extracted, meaning that the temperatures of the antiprotons and pions are $4.86 \cdot 10^{12}$ K and $2.77 \cdot 10^{12}$ K, respectively. (These temperatures are several hundred million times higher than the temperature at the surface of the Sun!) These observations and others enable researchers to infer that a quark-gluon plasma with a temperature of $2 \cdot 10^{12}$ K is created for a very short time of approximately $10^{-22}$ s in these collisions.

*Note:* The connection between the temperature of the quark-gluon plasma and the temperature values extracted from the particle energy distributions is not entirely straightforward, because the collisions of the gold nuclei are explosive events. Also, an ideal gas is not the correct model. In addition, the kinetic energies in Figure 19.21 were determined with the relativistically correct formulation, which will be introduced in Chapter 35. Nevertheless, the degree to which the observed kinetic energy distributions follow the Maxwell distribution is astounding.

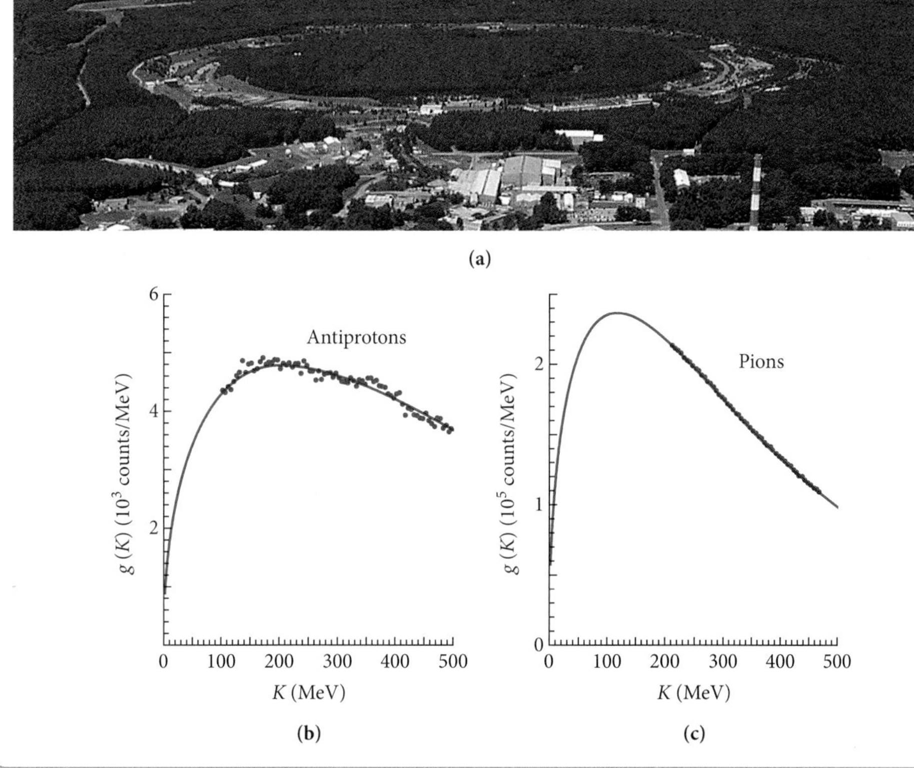


**FIGURE 19.21** (a) The 3.83-km circumference ring of the Relativistic Heavy Ion Collider at Brookhaven National Laboratory. Experimental data on the kinetic energy distribution of (b) antiprotons and (c) pions produced in collisions of gold nuclei. These collisions become by far the hottest place in the Solar System for the short times they exist and produce a state of matter, the quark-gluon plasma, which existed for a fraction of a second after the Big Bang. The blue lines represent fits to the kinetic energy distributions using equation 19.30.

To end this subsection, let's consider nuclear fusion in the center of the Sun. This nuclear fusion is the source of energy that causes all of the Sun's radiation and thus makes life on Earth possible. The temperature at the center of the Sun is approximately $1.5 \cdot 10^7$ K. According to equation 19.16, the average kinetic energy of the hydrogen ions in the Sun's center is $\frac{3}{2}k_BT = 3.1 \cdot 10^{-16}\text{ J} = 1.9\text{ keV}$. However, for two hydrogen atoms to undergo nuclear fusion, they have to come close enough to each other to overcome their mutual electrostatic repulsion. (These topics will be discussed in detail in Chapters 21 on electrostatics and 39 on nuclear physics.) The minimum energy required to overcome the electrostatic repulsion is on the order of 1 MeV, or a few hundred times higher than the average kinetic energy of the hydrogen ions in the Sun's center. Thus, only hydrogen ions on the extreme tail of the Maxwell distribution have sufficient energy to participate in fusion reactions.

## Mean Free Path

One of the assumptions about an ideal gas is that the gas molecules are point particles and do not interact with each other. But real gas molecules do have (very small, but nonzero) sizes and so have a chance of colliding with each other. Collisions of gas molecules with each other produce a scattering effect that causes the motion to be random and the distribution of their speeds to become the Maxwell distribution speed very quickly, independently of any initial speed distribution given to them.

How far do real gas molecules travel before they encounter another molecule? This distance is the **mean free path,** $\lambda$, of molecules in a gas. The mean free path is proportional to the temperature and inversely proportional to the pressure of the gas and the cross-sectional area of the molecule:

$$\lambda = \frac{k_BT}{\sqrt{2}\left(4\pi r^2\right)p}. \tag{19.31}$$

### DERIVATION 19.4 Mean Free Path

To estimate the mean free path, we start with the geometric definition of a cross section for scattering. We assume that all the molecules in the gas are spheres with radius $r$. If two molecules touch, they will have an elastic collision, in which total momentum and total kinetic energy are conserved (see Chapter 7). Let's imagine that a moving gas molecule encounters another gas molecule at rest, as shown in Figure 19.22a.

The impact parameter $b$, as shown in Figure 19.22a, is the perpendicular distance between two lines connecting the centers of the two molecules and oriented parallel to the direction of the velocity of the moving molecule. The moving molecule will collide with the stationary molecule if $b \le 2r$. Thus, the cross section for scattering is given by a circle with area

$$A = \pi\left(2r\right)^2 = 4\pi r^2,$$

as illustrated in Figure 19.22b. In a time period $t$, molecules with the average speed $\bar{v}$ will sweep out a volume

$$V = A\left(\bar{v}t\right) = \left(4\pi r^2\right)\left(\bar{v}t\right) = 4\pi r^2\bar{v}t,$$

as shown in Figure 19.22c. Thus, the mean free path, $\lambda$, is the distance the molecule travels divided by the number of molecules it will collide with in that distance. The number of molecules that the molecule will encounter is just the volume swept out times the number of molecules per unit volume, $n_V$. The mean free path is then

$$\lambda = \frac{\bar{v}t}{Vn_V} = \frac{\bar{v}t}{\left(4\pi r^2\bar{v}t\right)n_V} = \frac{1}{\left(4\pi r^2\right)n_V}.$$

This result needs to be modified because we assumed that one moving molecule was incident on a stationary molecule. In an ideal gas, all the molecules are moving, so we must use the relative velocity, $v_{\text{rel}}$, between the molecules rather than the average velocity of the molecules.

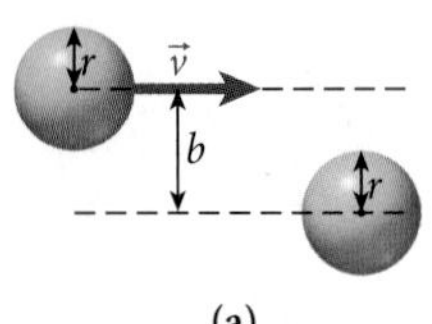


(a)

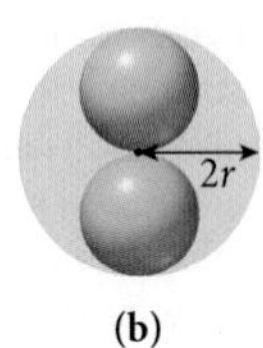


(b)

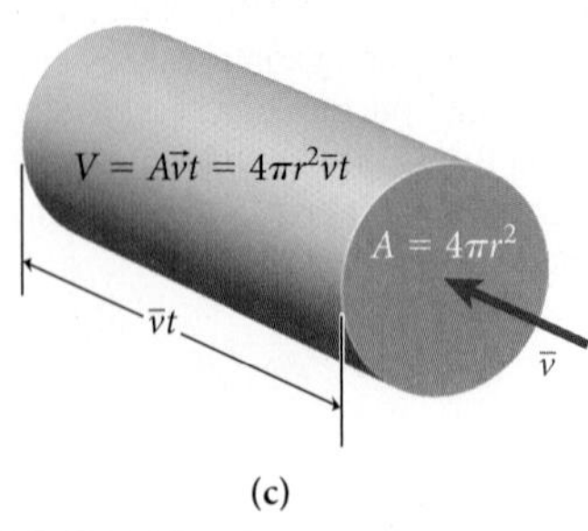


(c)

**FIGURE 19.22** (a) A moving gas molecule approaches a stationary gas molecule with impact parameter $b$. (b) The cross section for scattering is represented by a circle with radius $2r$. (c) The volume swept out by molecules inside the cross-sectional area $A$ in a time $t$ if traveling at speed $\bar{v}$.

It can be shown that $v_{rel} = \bar{v}\sqrt{2}$; so the volume swept out by the molecules changes to $V = 4\pi r^2 v_{rel} t = 4\pi r^2 \bar{v}\sqrt{2}t$. Thus, the expression for the mean free path becomes

$$\lambda = \frac{1}{\sqrt{2}\left(4\pi r^2\right)n_V}.$$

For diffuse gases, we can replace the number of molecules per unit volume with the ideal gas value, using equation 19.8:

$$n_V = \frac{N}{V} = \frac{p}{k_B T}.$$

The mean free path for a molecule in a real diffuse gas is then

$$\lambda = \frac{1}{\sqrt{2}\left(4\pi r^2\right)\left(\frac{p}{k_B T}\right)} = \frac{k_B T}{\sqrt{2}\left(4\pi r^2\right)p}.$$

### 19.9 In-Class Exercise

The mean free path of a gas molecule doubles if the ________ doubles.

a) temperature

b) pressure

c) diameter of the gas molecules

d) none of the above

Now that we have derived the formula for the mean free path, it is instructive to study a real case. Let's look at the most important gas, the Earth's atmosphere. An ideal gas has an "infinitely" long mean free path relative to the spacing between molecules in the gas. How closely a real gas approaches the ideal gas is determined by the ratio of the mean free path to the intermolecular spacing in the gas. Let's see how close Earth's atmosphere is to being an ideal gas.

## EXAMPLE 19.6 Mean Free Path of Air Molecules

### PROBLEM

What is the mean free path of a gas molecule in air at normal atmospheric pressure of 101.3 kPa and a temperature of 20.0 °C? Since air is 78.1% nitrogen, we'll assume for the sake of simplicity that air is 100% nitrogen. The radius of a nitrogen molecule is about 0.150 nm. Compare the mean free path to the radius of a nitrogen molecule and to the average spacing between the nitrogen molecules.

### SOLUTION

We can use equation 19.31 to calculate the mean free path:

$$\lambda = \frac{\left(1.381\cdot 10^{-23}\text{ J/K}\right)\left(293.15\text{ K}\right)}{\sqrt{2}\left(4\pi\right)\left(0.150\cdot 10^{-9}\text{ m}\right)^2\left(101.3\cdot 10^3\text{ Pa}\right)} = 9.99\cdot 10^{-8}\text{ m}.$$

The ratio of the mean free path to the molecular radius is then

$$\frac{\lambda}{r} = \frac{9.99\cdot 10^{-8}\text{ m}}{0.150\cdot 10^{-9}\text{ m}} = 666.$$

The volume per particle in an ideal gas can be obtained using equation 19.8:

$$\frac{V}{N} = \frac{k_B T}{p}.$$

The average spacing between molecules is then the average volume to the $\frac{1}{3}$ power:

$$\left(\frac{V}{N}\right)^{1/3} = \left(\frac{k_B T}{p}\right)^{1/3} = \left(\frac{\left(1.381\cdot 10^{-23}\text{ J/K}\right)\left(293.15\text{ K}\right)}{\left(101.3\cdot 10^3\text{ Pa}\right)}\right)^{1/3} = 3.42\cdot 10^{-9}\text{ m}.$$

*Continued—*

Thus, the ratio between the mean free path and the average spacing between molecules is

$$\frac{9.99 \cdot 10^{-8}\text{ m}}{3.42 \cdot 10^{-9}\text{ m}} = 29.2.$$

The value of this ratio demonstrates that treating air as an ideal gas is a good approximation. Figure 19.23 compares the diameter of nitrogen molecules, the average distance between molecules, and the mean free path.

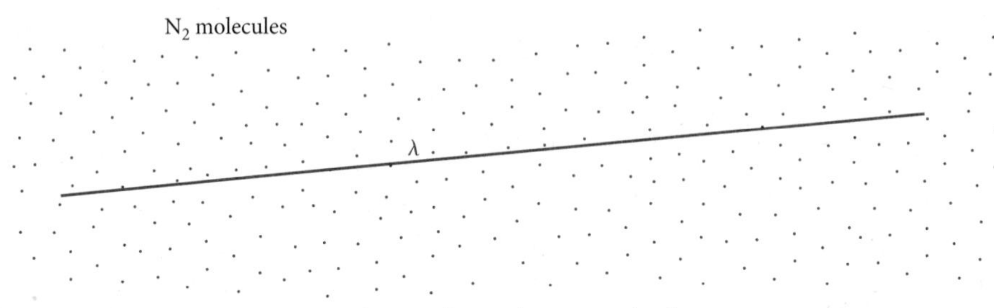


**FIGURE 19.23** Scale drawing showing the size of nitrogen molecules, the mean free path of the molecules in the gas, and the average distance between molecules.

## WHAT WE HAVE LEARNED | EXAM STUDY GUIDE

- The Ideal Gas Law relates the pressure $p$, the volume $V$, the number of moles $n$, and the temperature $T$ of an ideal gas, $pV = nRT$, where $R = 8.314$ J/(mol K) is the universal gas constant. (Boyle's, Charles's, Gay-Lussac's, and Avogadro's Laws are all special cases of the Ideal Gas Law.)
- The Ideal Gas Law can also be expressed as $pV = Nk_\text{B}T$, where $N$ is the number of molecules in the gas and $k_\text{B} = 1.381 \cdot 10^{-23}$ J/K is the Boltzmann constant.
- Dalton's Law states that the total pressure exerted by a mixture of gases is equal to the sum of the partial pressures of the gases in the mixture:
$$p_\text{total} = p_1 + p_2 + p_3 + \cdots + p_n.$$
- The work done by an ideal gas at constant temperature in changing from an initial volume $V_\text{i}$ to a final volume $V_\text{f}$ is given by $W = nRT\ln(V_\text{f}/V_\text{i})$.
- The equipartition theorem states that the average translational kinetic energy of the molecules of a gas is proportional to the temperature: $K_\text{ave} = \frac{3}{2}k_\text{B}T$.
- The root-mean-square speed of the molecules of a gas is $v_\text{rms} = \sqrt{3k_\text{B}T/m}$, where $m$ is the mass of each molecule.
- The molar specific heat of a gas at constant volume, $C_V$, is defined by $Q = nC_V\Delta T$, where $Q$ is the heat, $n$ is the number of moles of gas, and $\Delta T$ is the change in temperature of the gas.
- The molar specific heat of a gas at constant pressure, $C_p$, is defined by $Q = nC_p\Delta T$, where $Q$ is the heat, $n$ is the number of moles of gas, and $\Delta T$ is the change in temperature of the gas.
- The molar specific heat of a gas at constant pressure is related to the molar specific heat of a gas at constant volume by $C_p = C_V + R$.
- The molar specific heat at constant volume is $C_V = \frac{3}{2}R =$ 12.5 J/(mol K) for a monatomic gas and $C_V = \frac{5}{2}R =$ 20.8 J/(mol K) for a diatomic gas.
- For a gas undergoing an adiabatic process, $Q = 0$, and the following relationships hold: $pV^\gamma =$ constant and $TV^{\gamma-1} =$ constant, where $\gamma = C_p/C_V$. For monatomic gases, $\gamma = \frac{5}{3}$; for diatomic gases, $\gamma = \frac{7}{5}$.
- The Maxwell speed distribution is given by $f(v) = 4\pi\left(\frac{m}{2\pi k_\text{B}T}\right)^{3/2} v^2 e^{-mv^2/2k_\text{B}T}$, and $f(v)\,dv$ describes the probability that a molecule has a speed between $v$ and $v + dv$.
- The Maxwell kinetic energy distribution is given by $g(K) = \frac{2}{\sqrt{\pi}}\left(\frac{1}{k_\text{B}T}\right)^{3/2}\sqrt{K}e^{-K/k_\text{B}T}$, and $g(K)\,dK$ describes the probability that a molecule has a kinetic energy between $K$ and $K + dK$.
- The mean free path, $\lambda$, of a molecule in an ideal gas is given by $\lambda = \frac{k_\text{B}T}{\sqrt{2}\left(4\pi r^2\right)p}$, where $r$ is the radius of the molecule.

## KEY TERMS

gas, p. 615
mole, p. 616
Boyle's Law, p. 616
Charles's Law, p. 616
Gay-Lussac's Law, p. 617
Avogadro's Law, p. 617
Avogadro's number, p. 617
Ideal Gas Law, p. 617
universal gas constant, p. 617
Boltzmann constant, p. 619
Dalton's Law, p. 622
partial pressure, p. 622
mole fraction, p. 623
kinetic theory of an ideal gas, p. 623
equipartition theorem, p. 624
root-mean-square speed, p. 624
monatomic gas, p. 626
diatomic gas, p. 627
polyatomic gas, p. 627
degree of freedom, p. 628
equipartition of energy, p. 628
Maxwell speed distribution, p. 634
Maxwell kinetic energy distribution, p. 636
mean free path, p. 638

## NEW SYMBOLS AND EQUATIONS

$N_A = (6.02214179 \pm 0.00000030) \cdot 10^{23}$, Avogadro's number

$pV = nRT$, Ideal Gas Law, for number of moles

$pV = Nk_BT$, Ideal Gas Law, for number of molecules

$R = (8.314472 \pm 0.000015)$ J/(mol K), the universal gas constant

$k_B = (1.38106504 \pm 0.0000024) \cdot 10^{-23}$ J/K, the Boltzmann constant

$K_{ave} = \frac{3}{2}k_BT$, average translational kinetic energy of a gas molecule

$\lambda$, mean free path of a molecule in a gas

$C_V$, the molar specific heat of a gas at constant volume

$C_p$, the molar specific heat of a gas at constant pressure

$\gamma = C_p/C_V$, ratio of molar specific heats

## ANSWERS TO SELF-TEST OPPORTUNITIES

**19.1** Compare $C_p$ to $(C_V + R)$

| Gas | $C_V$ [J/(mol K)] | $C_p$ [J/(mol K)] | $(C_V + R)$ [J/(mol K)] |
|---|---|---|---|
| Helium (He) | 12.5 | 20.8 | 20.8 |
| Neon (Ne) | 12.5 | 20.8 | 20.8 |
| Argon (Ar) | 12.5 | 20.8 | 20.8 |
| Krypton (Kr) | 12.5 | 20.8 | 20.8 |
| Hydrogen ($H_2$) | 21.6 | 28.8 | 29.9 |
| Nitrogen ($N_2$) | 20.7 | 29.1 | 29.0 |
| Oxygen ($O_2$) | 20.8 | 29.4 | 29.1 |
| Carbon Dioxide ($CO_2$) | 28.5 | 36.9 | 36.8 |
| Methane ($CH_4$) | 27.5 | 35.9 | 35.8 |

**19.2** Use the definite integral $\int_0^\infty x^2 e^{-x^2/a} dx = \frac{1}{4}a^{3/2}\sqrt{\pi}$.

## PROBLEM-SOLVING PRACTICE

### Problem-Solving Guidelines

**1.** It is often useful to list all the known and unknown quantities when applying the Ideal Gas Law, to clarify what you have and what you're looking for. This step includes identifying an initial state and a final state and listing the quantities for each state.

**2.** If you're dealing with moles of a gas, you need the form of the Ideal Gas Law that involves $R$. If you're dealing with numbers of molecules, you need the form involving $k_B$.

**3.** Pay close attention to units, which may be given in different systems. For temperatures, unit conversion does not just involve multiplicative constants; the different temperature scales are also offset by additive constants. To be on the safe side, temperatures should always be converted to kelvins. Also, be sure that the value of $R$ or $k_B$ you use is consistent with the units you're working with. In addition, remember that molar masses are usually given as grams; you might need to convert the masses to kilograms.

## SOLVED PROBLEM 19.2 Density of Air at STP

### PROBLEM

What is the density of air at standard temperature and pressure (STP)?

### SOLUTION

#### THINK

We know that a mole of any gas at STP occupies a volume of 22.4 L. We know the relative fractions of the gases that comprise the atmosphere. We can combine Dalton's Law, Avogadro's Law, and the fraction of gases in air to obtain the mass of air contained in 22.4 L. The density is the mass divided by the volume.

#### SKETCH

This is one of the rare occasions where a sketch is not helpful

#### RESEARCH

We consider only the four gases in Table 19.1 that constitute a significant fraction of the atmosphere: nitrogen, oxygen, argon, and carbon dioxide. We know that 1 mole of any gas has a mass of $M$ (in grams) in a volume of 22.4 L at STP. From equation 19.12, we know that the mass of each gas in 22.4 L will be the fraction by volume times the molecular mass, so we have

$$m_{\text{air}} = r_{N_2} M_{N_2} + r_{O_2} M_{O_2} + r_{\text{Ar}} M_{\text{Ar}} + r_{CO_2} M_{CO_2}$$

and

$$\rho_{\text{air}} = \frac{m_{\text{air}}}{V_{\text{air}}} = \frac{m_{\text{air}}}{22.4 \text{ L}}.$$

#### SIMPLIFY

The density of air is then

$$\rho_{\text{air}} = \frac{(0.7808 \cdot 28 \text{ g}) + (0.2095 \cdot 32 \text{ g}) + (0.0093 \cdot 40 \text{ g}) + (0.00038 \cdot 44 \text{ g})}{22.4 \text{ L}}.$$

#### CALCULATE

Calculating the numerical result gives us

$$\rho_{\text{air}} = \frac{0.0218624 \text{ kg} + 0.006704 \text{ kg} + 0.000372 \text{ kg} + 0.00001672 \text{ kg}}{22.4 \cdot 10^{-3} \text{ m}^3} = 1.29264 \text{ kg/m}^3.$$

#### ROUND

We report our result to three significant figures:

$$\rho_{\text{air}} = 1.29 \text{ kg/m}^3.$$

#### DOUBLE-CHECK

We can double check our result by calculating the density of nitrogen gas alone at STP, since air is 78.08% nitrogen. The density of nitrogen at STP is

$$\rho_{N_2} = \frac{28.0 \text{ g}}{22.4 \text{ L}} = 1.25 \text{ g/L} = 1.25 \text{ kg/m}^3.$$

This result is close to and slightly lower than our result for the density of air; so our result is reasonable.

## SOLVED PROBLEM 19.3 Pressure of a Planetary Nebula

### PROBLEM

A planetary nebula is a cloud of mainly hydrogen gas with a density, $\rho$, of 1000.0 molecules per cubic centimeter (cm$^3$) and a temperature, $T$, of 10,000 K. An example of a planetary nebula, the Ring Nebula, is shown in Figure 19.24. What is the pressure of the gas in the nebula?

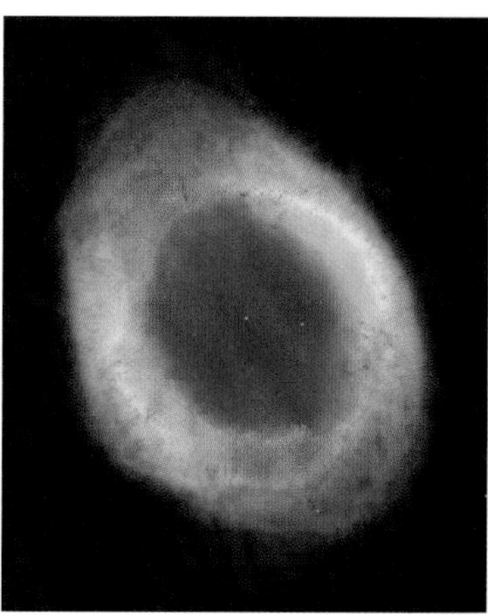

**FIGURE 19.24** The Ring Nebula, as seen by the Hubble Space Telescope.

### SOLUTION

**THINK**

Because of the very low density of the gas in a planetary nebula, the mean free path of the gas molecules (see Section 19.6) is extremely long. Therefore, it is a very good approximation to treat the nebula as an ideal gas and apply the Ideal Gas Law. We know the density, which is the number of gas molecules per unit volume, so we can replace the number of molecules in the Ideal Gas Law with the density times the volume of the nebula. The volume then cancels out, and we can solve for the pressure of the gas.

**SKETCH**

A rough sketch of a planetary nebula is shown in Figure 19.25.

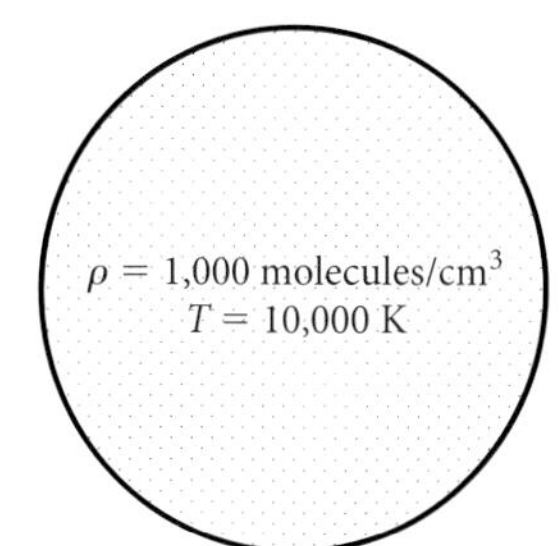


**FIGURE 19.25** An idealized planetary nebula.

**RESEARCH**

The Ideal Gas Law is given by

$$pV = Nk_BT, \qquad \text{(i)}$$

where $p$ is the pressure, $V$ is the volume, $N$ is the number of molecules of gas, $k_B$ is the Boltzmann constant, and $T$ is the temperature. The number of molecules, $N$, is

$$N = \rho V, \qquad \text{(ii)}$$

where $\rho$ is the number of molecules per unit volume in the nebula.

**SIMPLIFY**

Substituting for $N$ from equation (ii) into (i), we get

$$pV = Nk_BT = \rho V k_BT,$$

which we can simplify to get an expression for the pressure of the gas in the nebula:

$$p = \rho k_BT.$$

**CALCULATE**

Putting in the numerical values gives us

$$p = \rho k_BT = \left(\frac{1000 \text{ molecules}}{1 \text{ cm}^3}\right)\left(\frac{1 \text{ cm}^3}{10^{-6} \text{ m}^3}\right)\left(1.381\cdot 10^{-23} \text{ J/K}\right)\left(10,000 \text{ K}\right)$$

$$= 1.381\cdot 10^{-10} \text{ Pa}.$$

**ROUND**

We report our result to two significant figures:

$$p = 1.4\cdot 10^{-10} \text{ Pa}.$$

**DOUBLE-CHECK**

We can double-check our result by comparing it to the pressure of a good vacuum in a lab on Earth, which is $p_{\text{lab}} = 10^{-7}$ Pa. This pressure is about 1000 times higher than the pressure inside a planetary nebula.

## MULTIPLE-CHOICE QUESTIONS

**19.1** A system can go from state $i$ to state $f$ as shown in the figure. Which relationship is true?

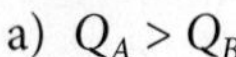

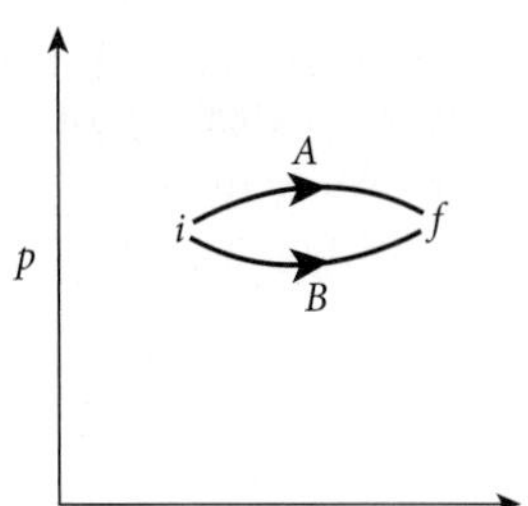


a) $Q_A > Q_B$
b) $Q_A = Q_B$
c) $Q_A < Q_B$
d) It is impossible to decide from the information given.

**19.2** A tire on a car is inflated to a gauge pressure of 32 lb/in$^2$ at a temperature of 27 °C. After the car is driven for 30 mi, the pressure has increased to 34 lb/in$^2$. What is the temperature of the air inside the tire at this point?

a) 46 °C b) 23 °C c) 32 °C d) 54 °C

**19.3** Molar specific heat at constant pressure, $C_p$, is larger than molar specific heat at constant volume, $C_v$, for

a) a monoatomic ideal gas.
b) a diatomic atomic gas.
c) all of the above.
d) none of the above.

**19.4** An ideal gas may expand from an initial pressure, $p_i$, and volume, $V_i$, to a final volume, $V_f$, isothermally, adiabatically, or isobarically. For which type of process is the heat that is added to the gas the largest? (Assume that $p_i$, $V_i$, and $V_f$ are the same for each process.)

a) isothermal process
b) adiabatic process
c) isobaric process
d) All the processes have the same heat flow.

**19.5** Which of the following gases has the highest root-mean-square speed?

a) nitrogen at 1 atm and 30 °C
b) argon at 1 atm and 30 °C
c) argon at 2 atm and 30 °C
d) oxygen at 2 atm and 30 °C
e) nitrogen at 2 atm and 15 °C

**19.6** Two identical containers hold equal masses of gas, oxygen in one and nitrogen in the other. The gases are held at the same temperature. How does the pressure of the oxygen compare to that of the nitrogen?

a) $p_O > p_N$
b) $p_O = p_N$
c) $p_O < p_N$
d) cannot be determined from the information given

**19.7** One mole of an ideal gas, at a temperature of 0 °C, is confined to a volume of 1.0 L. The pressure of this gas is

a) 1.0 atm.
b) 22.4 atm.
c) 1/22.4 atm.
d) 11.2 atm.

**19.8** One hundred milliliters of liquid nitrogen with a mass of 80.7 g is sealed inside a 2-L container. After the liquid nitrogen heats up and turns into a gas, what is the pressure inside the container?

a) 0.05 atm
b) 0.08 atm
c) 0.09 atm
d) 9.1 atm
e) 18 atm

**19.9** Consider a box filled with an ideal gas. The box undergoes a sudden free expansion from $V_1$ to $V_2$. Which of the following correctly describes this process?

a) Work done by the gas during the expansion is equal to $nRT \ln (V_2/V_1)$.
b) Heat is added to the box.
c) Final temperature equals initial temperature times $(V_2/V_1)$.
d) The internal energy of the gas remains constant.

**19.10** Compare the average kinetic energy at room temperature of a nitrogen molecule to that of a nitrogen atom. Which has the larger kinetic energy?

a) nitrogen atom
b) nitrogen molecule
c) They have the same energy.
d) It depends upon the pressure.

## QUESTIONS

**19.11** Hot air is less dense than cold air and therefore experiences a net buoyant force and rises. Since hot air rises, the higher the elevation, the warmer the air should be. Therefore, the top of Mount Everest should be very warm. Explain why Mount Everest is colder than Death Valley.

**19.12** The Maxwell speed distribution assumes that the gas is in equilibrium. Thus, if a gas, all of whose molecules were moving at the same speed, were given enough time, they would eventually come to satisfy the speed distribution. But the kinetic theory derivations in the text assumed that when a gas molecule hits the wall of a container, it bounces back with the same energy it had before the collision and that gas molecules exert no forces on each other. If gas molecules exchange energy neither with the walls of their container nor with each other, how can they ever come to equilibrium? Is it not true that if they all had the same speed initially, some would have to slow down and others speed up, according to the Maxwell speed distribution?

**19.13** When you blow hard on your hand, it feels cool, but when you breathe softly, it feels warm. Why?

**19.14** Explain why the average velocity of air molecules in a closed auditorium is zero but their root-mean-square speed or average speed is not zero.

**19.15** In a diesel engine, the fuel-air mixture is compressed rapidly. As a result, the temperature rises to the spontaneous combustion temperature for the fuel, causing the fuel to ignite. How is the temperature rise possible, given the fact that the compression occurs so rapidly that there is not enough time for a significant amount of heat to flow into or out of the fuel-air mixture?

**19.16** A cylinder with a sliding piston contains an ideal gas. The cylinder (with the gas inside) is heated by transferring the same amount of heat to the system in two different ways:

(1) The piston is blocked to prevent it from moving.

(2) The piston is allowed to slide frictionless to the end of the cylinder.

How does the final temperature of the gas under condition 1 compare to the final temperature of the gas under condition 2? Is it possible for the temperatures to be equal?

**19.17** Show that the adiabatic bulk modulus, defined as $B = -V\,(dP/dV)$, for an ideal gas is equal to $\gamma P$.

**19.18** A monatomic ideal gas expands isothermally from $\{p_1, V_1, T_1\}$ to $\{p_2, V_2, T_1\}$. Then it undergoes an isochoric process, which takes it from $\{p_2, V_2, T_1\}$ to $\{p_1, V_2, T_2\}$. Finally the gas undergoes an isobaric compression, which takes it back to $\{p_1, V_1, T_1\}$.

a) Use the First Law of Thermodynamics to find $Q$ for each of these processes.

b) Write an expression for total $Q$ in terms of $p_1$, $p_2$, $V_1$, and $V_2$.

**19.19** Two atomic gases will react to form a diatomic gas: A + B $\rightarrow$ AB. Suppose you perform this reaction with 1 mole each of A and B in a thermally isolated chamber, so no heat is exchanged with the environment. Will the temperature of the system increase or decrease?

**19.20** A relationship that gives the pressure, $p$, of a substance as a function of its density, $\rho$, and temperature, $T$, is called an *equation of state.* For a gas with molar mass $M$, write the Ideal Gas Law as an equation of state.

**19.21** The compression and rarefaction associated with a sound wave propogating in a gas are so much faster than the flow of heat in the gas that they can be treated as adiabatic processes.

a) Find the speed of sound, $v_s$, in an ideal gas of molar mass $M$.

b) In accord with Einstein's refinement of Newtonian mechanics, $v_s$ cannot exceed the speed of light in vacuum, $c$. This fact implies a maximum temperature for an ideal gas. Find this temperature.

c) Evaluate the maximum temperature of part (b) for *monatomic* hydrogen gas (H).

d) What happens to the hydrogen at this maximum temperature?

**19.22** The kinetic theory of an ideal gas takes into account not only translational motion of atoms or molecules but also, for diatomic and polyatomic gases, vibration and rotation. Will the temperature increase from a given amount of energy being supplied to a monatomic gas differ from the temperature increase due to the same amount of energy being supplied to a diatomic gas? Explain.

**19.23** In the thermosphere, the layer of Earth's atmosphere extending from an altitude 100 km up to 700 km, the temperature can reach about 1500 °C. How would air in that layer feel to one's bare skin?

**19.24** A glass of water at room temperature is left on the kitchen counter overnight. In the morning, the amount of water in the glass is smaller due to evaporation. The water in the glass is below the boiling point, so how is it possible for some of the liquid water to have turned into a gas?

## PROBLEMS

A blue problem number indicates a worked-out solution is available in the Student Solutions Manual. One • and two •• indicate increasing level of problem difficulty.

### Section 19.1

19.25 A tire has a gauge pressure of 300. kPa at 15 °C. What is the gauge pressure at 45 °C? Assume that the change in volume of the tire is negligible.

**19.26** A tank of compressed helium for inflating balloons is advertised as containing helium at a pressure of 2400 psi, which, when allowed to expand at atmospheric pressure, will occupy a volume of 244 ft$^3$. Assuming no temperature change takes place during the expansion, what is the volume of the tank in cubic feet?

•**19.27** Before embarking on a car trip from Michigan to Florida to escape the winter cold, you inflate the tires of your car to a manufacturer-suggested pressure of 33 psi while the outside temperature is 25 °F and then make sure your valve caps are airtight. You arrive in Florida two days later, and the temperature outside is a pleasant 72 °F.

a) What is the new pressure in your tires, in SI units?

b) If you let air out of the tires to bring the pressure back to the recommended 33 psi, what percentage of the original mass of air in the tires will you release?

•**19.28** A 1-L volume of a gas undergoes first an isochoric process in which its pressure doubles, followed by an isothermal process until the original pressure is reached. Determine the final volume of the gas.

•19.29 Suppose you have a pot filled with steam at 100.0 °C and at a pressure of 1.00 atm. The pot has a diameter of 15.0 cm and is 10.0 cm high. The mass of the lid is 0.500 kg. How hot do you need to heat the steam to lift the lid off the pot?

### Section 19.2

•**19.30** A quantity of liquid water comes into equilibrium with the air in a closed container, without completely evaporating, at a temperature of 25.0 °C. How many grams of water vapor does a liter of the air contain in this situation? The vapor pressure of water at 25.0 °C is 3.1690 kPa.

•**19.31** Use the equation $\rho T$ = constant and information from Chapter 16 to estimate the speed of sound in air at 40.0 °C, given that the speed at 0.00 °C is 331 m/s.

**19.32** Suppose 2 moles of an ideal gas are enclosed in a container of volume $1 \cdot 10^{-4}$ m$^3$. The container is then placed in a furnace and raised to a temperature of 400 K. What is the final pressure of the gas?

**19.33** One mole of an ideal gas is held at a constant volume of 2 L. Find the change in pressure if the temperature increases by 100 °C.

**19.34** How much heat is added to the system when 2000 J of work is performed by an ideal gas in an isothermal process? Give a reason for your answer.

•**19.35** Liquid nitrogen, which is used in many physics research labs, can present a safety hazard if a large quantity evaporates in a confined space. The resulting nitrogen gas reduces the oxygen concentration, creating the risk of asphyxiation. Suppose 1.00 L of liquid nitrogen ($\rho$ = 808 kg/m$^3$) evaporates and comes into equilibrium with the air at 21.0 °C and 101 kPa. How much volume will it occupy?

•**19.36** Liquid bromine ($Br_2$) is spilled in a laboratory accident and evaporates. Assuming the vapor behaves as an ideal gas, with a temperature 20.0 °C and a pressure of 101.0 kPa, find its density.

•**19.37** Two containers contain the same gas at different temperatures and pressures, as indicated in the figure. The small container has a volume of 1 L, and the large container has a volume of 2 L. The two containers are then connected to each other using a thin tube, and the pressure and temperature in both containers are allowed to equalize. If the final temperature is 300 K, what is the final pressure? Assume that the connecting tube has negligible volume and mass.

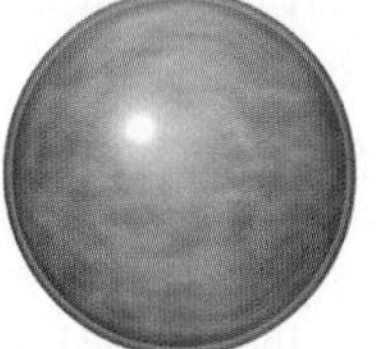


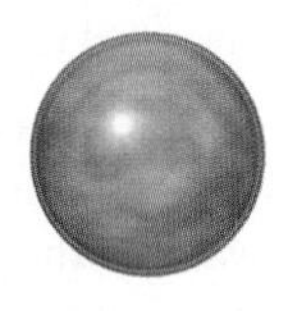


•**19.38** A sample of gas at $P$ = 1000 Pa, $V$ = 1.0 L, and $T$ = 300 K is confined in a cylinder.

a) Find the new pressure if the volume is reduced to half of the original volume at the same temperature.

b) If the temperature is raised to 400 K in the process of part (a), what is the new pressure?

c) If the gas is then heated to 600 K from the initial value and the pressure of the gas becomes 3000 Pa, what is the new volume?

•**19.39** A cylinder of cross-sectional area 12 cm$^2$ is fitted with a piston, which is connected to a spring with a spring constant of 1000 N/m, as shown in the figure. The cylinder is filled with 0.0050 mole of a gas. At room temperature (23 °C), the spring is neither compressed nor stretched. How far is the spring compressed if the temperature of the gas is raised to 150 °C?

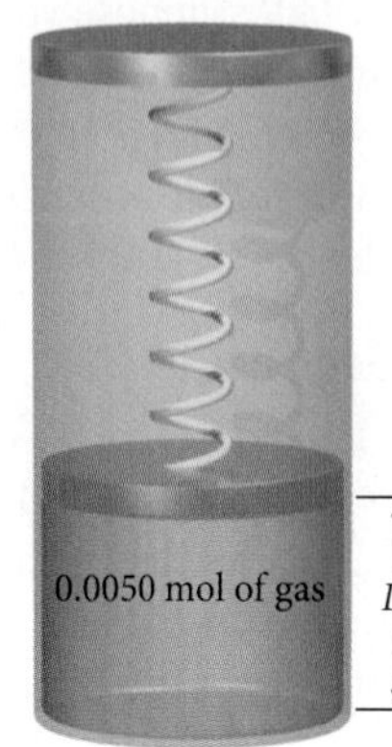


••**19.40** Air at 1 atm is inside a cylinder 20 cm in radius and 20 cm in length that sits on a table. The top of the cylinder is sealed with a movable piston. A 20-kg block is dropped onto the piston. From what height above the piston must the block be dropped to compress the piston by 1 mm? 2 mm? 1 cm?

## Section 19.3

**19.41** Interstellar space far from any stars is usually filled with atomic hydrogen (H) at a density of 1 atom/cm$^3$ and a very low temperature of 2.73 K.

a) Determine the pressure in interstellar space.

b) What is the root-mean-square speed of the atoms?

c) What would be the edge length of a cube that would contain atoms with a total of 1 J of energy?

**19.42** a) What is the root-mean-square speed for a collection of helium-4 atoms at 300 K?

b) What is the root-mean-square speed for a collection of helium-3 atoms at 300 K?

**19.43** Two isotopes of uranium, $^{235}$U and $^{238}$U, are separated by a gas diffusion process that involves combining them with flourine to make the compound $UF_6$. Determine the ratio of the root-mean-square speeds of $UF_6$ molecules for the two isotopes. The masses of $^{235}UF_6$ and $^{238}UF_6$ are 249 amu and 252 amu.

**19.44** The electrons in a metal that produce electric currents behave approximately as molecules of an ideal gas. The mass of an electron is $m_e \doteq 9.109 \cdot 10^{-31}$ kg. If the temperature of the metal is 300.0 K, what is the root-mean-square speed of the electrons?

•**19.45** In a period of 6.00 s, $9.00 \cdot 10^{23}$ nitrogen molecules strike a section of a wall with an area of 2.00 cm$^2$. If the molecules move with a speed of 400.0 m/s and strike the wall head on in elastic collisions, what is the pressure exerted on the wall? (The mass of one $N_2$ molecule is $4.68 \cdot 10^{-26}$ kg.)

•**19.46** Assuming the pressure remains constant, at what temperature is the root-mean-square speed of a helium atom equal to the root-mean-square speed of an air molecule at STP?

## Section 19.4

**19.47** Two copper cylinders, immersed in a water tank at 50.0 °C contain helium and nitrogen, respectively. The helium-filled cylinder has a volume twice as large as the nitrogen-filled cylinder.

a) Calculate the average kinetic energy of a helium molecule and the average kinetic energy of a nitrogen molecule.

b) Determine the molar specific heat at constant volume ($C_V$) and at constant pressure ($C_p$) for the two gases.

c) Find $\gamma$ for the two gases.

**19.48** At room temperature, identical gas cylinders contain 10 moles of nitrogen gas and argon gas, respectively. Determine the ratio of energies stored in the two systems. Assume ideal gas behavior.

19.49 Calculate the change in internal energy of 1.00 mole of a diatomic ideal gas that starts at room temperature (293 K) when its temperature is increased by 2.00 K.

**19.50** Treating air as an ideal gas of diatomic molecules, calculate how much heat is required to raise the temperature of the air in an 8.0 m by 10 m by 3.0 m room from 20 °C to 22 °C at 101 kPa. Neglect the change in the number of moles of air in the room.

•**19.51** What is the approximate energy required to raise the temperature of 1 L of air by 100 °C? The volume is held constant.

•**19.52** You are designing an experiment that requires a gas with $\gamma = 1.60$. However, from your physics lectures, you remember that no gas has such a $\gamma$ value. However, you also remember that mixing monatomic and diatomic gases can yield a gas with such a $\gamma$ value. Determine the fraction of diatomic molecules a mixture has to have to obtain this value.

## Section 19.5

19.53 Suppose 15.0 L of an ideal monatomic gas at a pressure of $150 \cdot 10^3$ kPa is expanded adiabatically (no heat transfer) until the volume is doubled.

a) What is the pressure of the gas at the new volume?

b) If the initial temperature of the gas was 300 K, what is its final temperature after the expansion?

**19.54** A diesel engine works at a high compression ratio to compress air until it reaches a temperature high enough to ignite the diesel fuel. Suppose the compression ratio (ratio of volumes) of a specific diesel engine is 20 to 1. If air enters a cylinder at 1 atm and is compressed adiabatically, the compressed air reaches a pressure of 66 atm. Assuming that the air enters the engine at room temperature (25 °C) and that the air can be treated as an ideal gas, find the temperature of the compressed air.

**19.55** Air in a diesel engine cylinder is quickly compressed from an initial temperature of 20 °C, an initial pressure of 1 atm, and an initial volume of 600 cm$^3$ to a final volume of 45 cm$^3$. Assuming the air to be an ideal diatomic gas, find the final temperature and pressure.

•**19.56** Six liters of a monatomic ideal gas, originally at 400 K and a pressure of 3 atm (called state 1), undergoes the following processes:

$1 \rightarrow 2$ isothermal expansion to $V_2 = 4V_1$

$2 \rightarrow 3$ isobaric compression

$3 \rightarrow 1$ adiabatic compression to its original state

Find the pressure, volume, and temperature of the gas in states 2 and 3. How many moles of the gas are there?

•19.57 How much work is done by the gas during each of the three processes in Problem 19.56 and how much heat flows into the gas in each process?

••**19.58** Two geometrically identical cylinders of inner diameter 5.0 cm, one made of copper and the other of Teflon, are immersed in a large-volume water tank at room temperature (20 °C), as shown in the figure. A frictionless Teflon piston with a rod and platter attached is placed in each cylinder. The mass of the piston-rod-platter assembly is 0.50 kg, and the cylinders are filled with helium gas so that initially both pistons are at equilibrium at 20.0 cm from the bottom of their respective cylinders.

a) A 5.0-kg lead block is *slowly* placed on each platter, and the piston is *slowly* lowered until it reaches its final equilibrium state. Calculate the final height of each piston (as measured from the bottom of its respective cylinder).

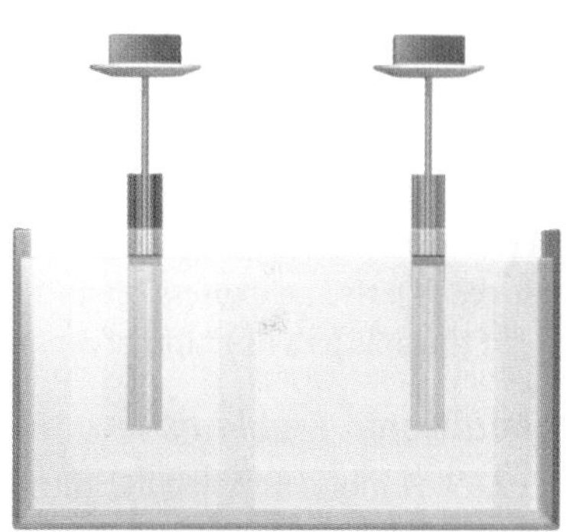

b) If the two lead blocks are *dropped suddenly* on the platters, how will the final heights of the two pistons compare?

••**19.59** Chapter 13 examined the variation of pressure with altitude in the Earth's atmosphere, assuming constant temperature—a model known as the *isothermal atmosphere.* A better approximation is to treat the pressure variations with altitude as adiabatic. Assume that air can be treated as a diatomic ideal gas with effective molar mass $M_{air} = 28.97$ g/mol.

a) Find the air pressure and temperature of the atmosphere as functions of altitude. Let the pressure at sea level be $P_0 = 101.0$ kPa and the temperature at sea level be 20.0 °C.

b) Determine the altitude at which the air pressure (and density) are half their sea-level values. What is the temperature at this altitude, in this model?

c) Compare these results with the isothermal model of Chapter 13.

## Section 19.6

**19.60** Consider nitrogen gas, $N_2$, at 20 °C. What is the root-mean-square speed of the nitrogen molecules? What is the most probable speed? What percentage of nitrogen molecules have a speed within 1 m/s of the most probable speed?

19.61 As noted in the text, the speed distribution of molecules in the Earth's atmosphere has a significant impact on its composition.

a) What is the average speed of a nitrogen molecule in the atmosphere, at a temperature of 18.0 °C and a (partial) pressure of 78.8 kPa?

b) What is the average speed of a hydrogen molecule at the same temperature and pressure?

19.62 A sealed container contains 1 mole of neon gas at STP. Estimate the number of neon atoms having speeds in the range from 200.0 m/s to 202.0 m/s. (*Hint:* Assume the probability of neon atoms having speeds between 200.0 m/s and 202.0 m/s is constant.)

•19.63 For a room at 21 °C and 1 atm, write an expression for the fraction of air molecules having speeds greater than the speed of sound, according to the Maxwell speed distribution. What is the average speed of each molecule? What is the root-mean-square speed? Assume that the air consists of uniform particles with a mass of 15 amu.

••19.64 If a space shuttle is struck by a meteor, producing a circular hole 1.0 cm in diameter in the window, how long does it take before the pressure inside the cabin is reduced to half of its original value? Assume that the cabin of the shuttle is a cube with edges measuring 5.00 m with a constant temperature of 21 °C at 1.0 atm and that the air inside consists of uniform diatomic molecules each with a mass of 15 amu. (*Hint:* What fraction of the molecules are moving in the correct direction to exit through the hole, and what is their average speed? Derive an expression for the rate at which molecules exit the hole as a function of pressure, and thus of $dP/dt$.)

## Additional Problems

19.65 A mole of molecular nitrogen gas expands in volume very quickly, so no heat is exchanged with the environment during the process. If the volume increases from 1.0 L to 1.5 L, determine the work done on the environment if the gas's temperature dropped from 22 °C to 18 °C. Assume ideal gas behavior.

19.66 Calculate the root-mean-square speed of air molecules at room temperature (22.0 °C) from the kinetic theory of an ideal gas.

19.67 At Party City, you purchase a helium-filled balloon with a diameter of 40.0 cm at 20 °C and at 1 atm.

a) How many helium atoms are inside the balloon?

b) What is the average kinetic energy of the atoms?

c) What is the root-mean-square speed of the atoms?

19.68 What is the total mass of all the oxygen molecules in a cubic meter of air at normal temperature (25 °C) and pressure ($1.01 \cdot 10^5$ Pa)? Note that air is about 21% (by volume) oxygen (molecular $O_2$), with the remainder being primarily nitrogen (molecular $N_2$).

19.69 The tires of a $3.0 \cdot 10^3$-lb car are filled to a gauge pressure (pressure added to atmospheric pressure) of 32 lb/in$^2$ while the car is on a lift at a repair shop. The car is then lowered to the ground.

a) Assuming that the internal volume of the tires does not change appreciably as a result of contact with the ground, give the absolute pressure in the tires in pascals when the car is on the ground.

b) What is the total contact area between the tires and the ground?

19.70 A gas expands at constant pressure from 3 L at 15 °C until the volume is 4 L. What is the final temperature of the gas?

19.71 An ideal gas has a density of 0.0899 g/L at 20.00 °C and 101.325 kPa. Identify the gas.

19.72 A tank of a gas consisting of 30% $O_2$ and 70% Ar has a volume of 1 m$^3$. It is moved from storage at 20 °C to the outside on a sunny day. Initially, the pressure gauge reads 1000 psi. After several hours, the pressure reading is 1500 psi. What is the temperature of the gas in the tank at this time?

19.73 Suppose 5.0 moles of an ideal monatomic gas expand at a constant temperature of 22 °C from an initial volume of 2.0 m$^3$ to 8.0 m$^3$.

a) How much work is done by the gas?

b) What is the final pressure of the gas?

19.74 Suppose 0.04 mole of an ideal monatomic gas has a pressure of 1.0 atm at 273 K and is then cooled at constant pressure until its volume is halved. What is the amount of work done by the gas?

•19.75 A 2-L bottle contains 1 mole of sodium bicarbonate and 1 mole of acetic acid. These combine to produce 1 mole of carbon dioxide gas, along with water and sodium acetate. If the bottle is tightly sealed before the reaction occurs, what is the pressure inside the bottle when the reaction has completed?

•19.76 A cylindrical chamber with a radius of 3.5 cm contains 0.12 mole of an ideal gas. It is fitted with a 450-g piston that rests on top, enclosing a space of height 7.5 cm. As the cylinder is cooled by 15 °C, how far below the initial position does the piston end up?

•19.77 Find the *most probable* kinetic energy for a molecule of a gas at temperature $T = 300.$ K. In what way does this value depend on the identity of the gas?

•19.78 A closed auditorium of volume 25,000 m$^3$ is filled with 2000 people at the beginning of a show, and the air in the space is at a temperature of 293 K and a pressure of $1.013 \cdot 10^5$ Pa. If there were no ventilation, by how much would the temperature of the air rise during the 2-h show if each person metabolizes at a rate of 70 W?

•19.79 At a temperature of 295. K, the vapor pressure of pentane ($C_5H_{12}$) is 60.7 kPa. Suppose 1.000 g of gaseous pentane is contained in a cylinder with diathermal (thermally conducting) walls and a piston to vary the volume. The initial volume is 1.000 L, and the piston is moved in slowly, keeping the temperature at 295 K. At what volume will the first drop of liquid pentane appear?

•19.80 Helium gas fills in a well-insulated cylinder fitted with a piston at 22 °C and at 1 atm. The piston is moved very slowly (adiabatic expansion) so that the volume increases to four times its original volume.

a) What is the final pressure?

b) What is the final temperature of the gas?

# 20 The Second Law of Thermodynamics

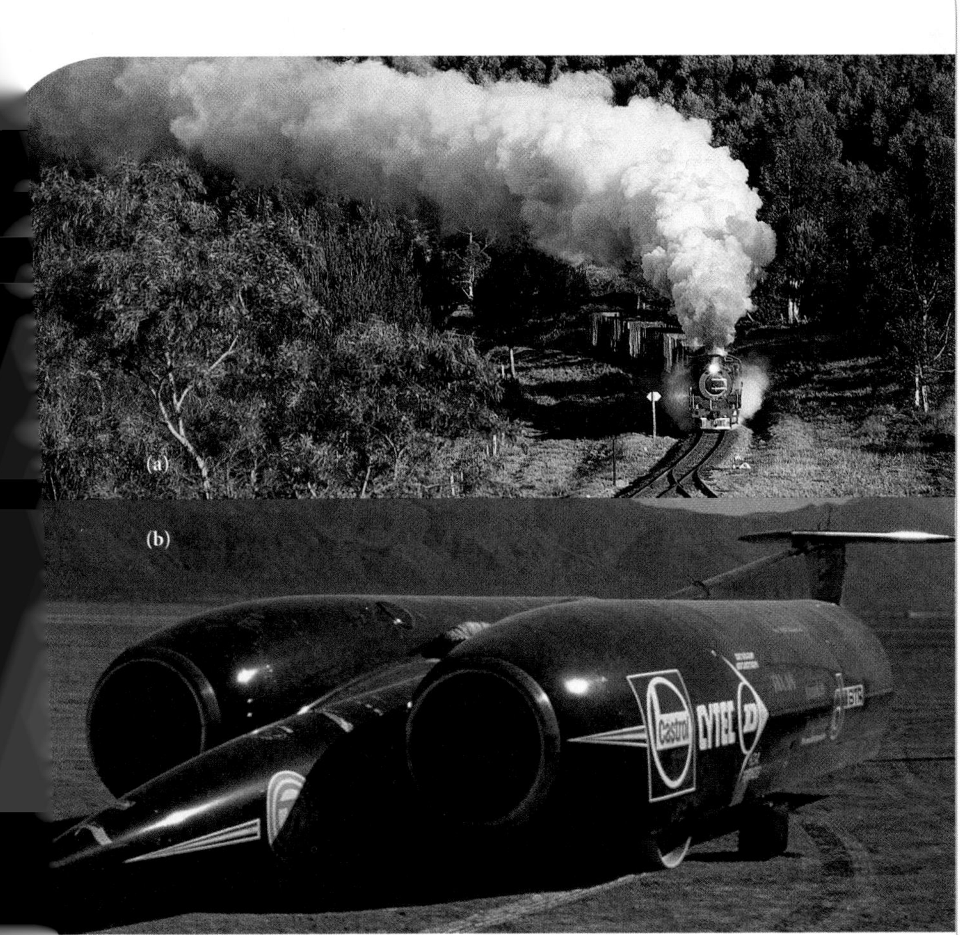


**FIGURE 20.1** (a) A steam locomotive. (b) The Thrust SSC.

## WHAT WE WILL LEARN

- Reversible thermodynamic processes are idealized processes in which a system can move from one thermodynamic state to another and back again, while remaining close to thermodynamic equilibrium.
- Many processes are irreversible; for example, the pieces of a broken coffee cup lying on the floor cannot spontaneously reassemble back into the cup. Practically all real-world thermodynamics processes are irreversible.
- Mechanical energy can be converted to thermal energy, and thermal energy can be converted to mechanical energy.
- A heat engine is a device that converts thermal energy into mechanical energy. This engine operates in a thermodynamic cycle, returning to its original state periodically.
- No heat engine can have an efficiency of 100%. The Second Law of Thermodynamics places fundamental limitations on the efficiency of a heat engine.
- Refrigerators and air conditioners are heat engines that operate in reverse.
- The entropy of a system is a measure of how far from equilibrium the system is.
- The Second Law of Thermodynamics says that the entropy of a closed system can never decrease.
- The entropy of a system can be related to the microscopic states of the constituents of the system.

The early studies of thermodynamics were motivated by the desire of scientists and engineers to discover the governing principles of engines and machines so that they could design more efficient and powerful ones. The steam engine could be used to power early locomotives and ships; however, overheated steam boilers could explode and cause damage and fatalities. Today, steam locomotives and ships are used primarily on scenic routes for tourists.

The principles that affect the efficiency of heat engines apply to all engines, from steam locomotives (Figure 20.1a) to jet engines. The Thrust SSC (supersonic car) shown in Figure 20.1b, used two powerful jet engines to set a land speed record of 763 mph in October 1997. However, for all its speed and power, it still produces more waste heat than usable energy, which is true of most engines.

In this chapter, we examine heat engines in theory and in practice. Their operation is governed by the Second Law of Thermodynamics—one of the most far-reaching and powerful statements in all of science. There are several different ways to express this law, including one involving a concept called *entropy.* The ideas discussed in this chapter have applications to practically all areas of science, including information processing, biology, and astronomy.

## 20.1 Reversible and Irreversible Processes

As noted in Chapter 18, if you pour hot water into a glass and place that glass on a table, the water will slowly cool until it reaches the temperature of its surroundings. The air in the room will also warm, although imperceptibly. You would be astonished if, instead, the water got warmer and the air in the room cooled down slightly, while conserving energy. In fact, the First Law of Thermodynamics is satisfied for *both* of these scenarios. However, it is always the case that the water cools until it reaches the temperature of its surroundings. Thus, other physical principles are required to explain why temperature changes one way and not the other.

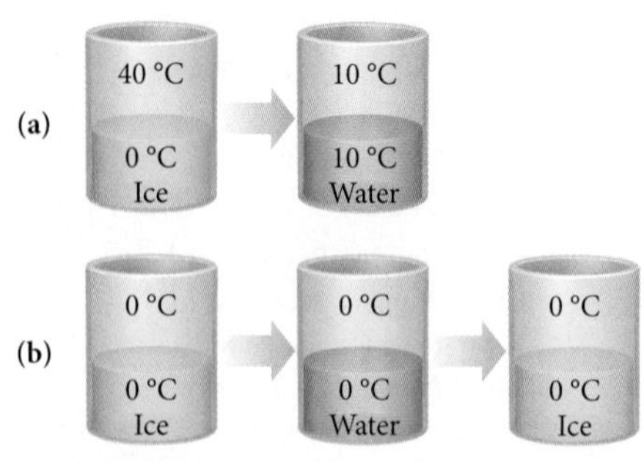


**FIGURE 20.2** Two thermodynamic processes: (a) An irreversible process in which a disk of ice at a temperature of 0 °C is placed inside a metal can at a temperature of 40 °C. (b) A reversible process in which a disk of ice at a temperature of 0 °C is placed inside a metal can at a temperature of 0 °C.

Practically all real-life thermodynamic processes are irreversible. For example, if a disk of ice at a temperature of 0 °C is placed in a metal can having a temperature of 40 °C, heat flows irreversibly from the can to the ice, as shown in Figure 20.2a. The ice will melt to water; the water will then warm and the can will cool, until the water and the can are at the same temperature (10 °C in this case). It is not possible to make small changes in any thermodynamic variable and return the system to the state corresponding to a warm can and frozen water.

However, we can imagine a class of idealized *reversible* processes. With a **reversible process,** a system is always close to being in thermodynamic equilibrium. In this case, making a small change in the state of the system can reverse any change in the thermodynamic variables of the system. For example, in Figure 20.2b, a disk of ice at a temperature of 0 °C is placed in a metal can that is also at a temperature of 0 °C. Raising the temperature of the can slightly will melt the ice to water. Then, lowering the temperature of the metal can will refreeze the water to ice, thus returning the system to its original state. (Technically speaking, the crystal structure of the ice would be different and thus the final state would be different from the initial state, but not in a way that concerns us here.)

We can think of reversible processes as equilibrium processes in which the system always stays in thermal equilibrium or close to thermal equilibrium. If a system were actually in thermal equilibrium, no heat would flow and no work would be done by the system. Thus, a reversible process is an idealization. However, for a nearly reversible process, small temperature and pressure adjustments can keep a system close to thermal equilibrium.

On the other hand, if a process involves heat flow with a finite temperature difference, free expansion of a gas, or conversion of mechanical work to thermal energy, it is an **irreversible process.** It is not possible to make small changes to the temperature or pressure of the system and cause the process to proceed in the opposite direction. In addition, while an irreversible process is taking place, the system is not in thermal equilibrium.

The irreversibility of a process is related to the randomness or disorder of the system. For example, suppose you order a deck of cards by rank and suit and then take the first five cards, as illustrated in Figure 20.3a. You will observe a well-ordered system of cards, for example, all spades, starting with the ace and going down to the ten. Next, you put those five cards back in the deck, toss the deck of cards in the air, and let the cards fall on the floor. You pick up the cards one by one without looking at their rank or suit and then take the first five cards off the top of the deck. It is highly improbable that these five cards will be ordered by rank and suit. It is much more likely that you will see a result like the one shown in Figure 20.3b, that is, with cards in random order. The process of tossing and picking up the cards is irreversible.

Another example of an irreversible process is the free expansion of a gas, which we touched on in Chapter 18 and will discuss in greater detail in Section 20.7.

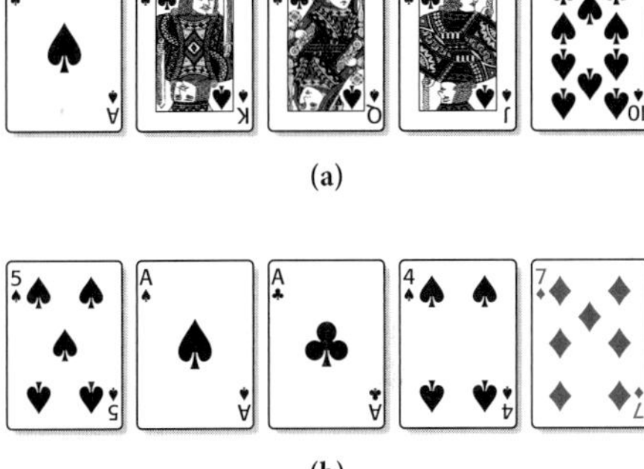

**FIGURE 20.3** (a) The first five cards of a card deck ordered by rank and suit. (b) The first five cards of a deck that has been tossed in the air and picked up randomly.

## Poincaré Recurrence Time

The common experience of déjà vu (French for "already seen") is the feeling of having experienced a certain situation sometime before. But do events really repeat? If processes leading from one event to another are truly irreversible, then an event cannot repeat exactly.

The French mathematician and physicist Henry Poincaré made an important contribution to this discussion in 1890, by famously stating his recurrence theorem. In it he postulates that certain closed dynamical systems will return to a state arbitrarily close to their initial state after a sufficiently long time. This time is known as the *Poincaré recurrence time,* or simply *Poincaré time* of the system. This time can be calculated fairly straightforwardly for many systems that have only a finite number of different states, as the following example illustrates.

### EXAMPLE 20.1 Deck of Cards

Suppose you repeat the tossing and gathering of a deck of cards performed for Figure 20.3 again and again. Eventually, there is a chance that one of the random orders in which you pick up the cards will yield the sequence ace through ten of spades for the first five cards in the deck.

**PROBLEM**

If it takes 1 min on average to toss the cards into the air and then collect them into a stack again, how long can you expect it to take, on average, before you will find the sequence of ace through ten of spades as the first five cards in the deck?

**SOLUTION**

There are 52 cards in the deck. Each of them has the same $\frac{1}{52}$ probability of being on top. So the probability that the ace of spades ends up on top is $\frac{1}{52}$. If the ace is the top card,

*Continued—*

then there are 51 cards left. The probability that the king of spades is in the top position of the remaining 51 cards is $\frac{1}{51}$. Thus, the combined probability that the ace and king of spades are the top two cards in that order is $1/(52 \cdot 51) = 1/2652$. In the same way, the probability of the ordered sequence of ace through ten of spades occurring is $1/(52 \cdot 51 \cdot 50 \cdot 49 \cdot 48) = 1/311{,}875{,}200$.

So the average number of tries needed to obtain the desired sequence is 311,875,200. Since each of them takes 1 minute, successful completion of this exercise would take approximately 593 yr.

### DISCUSSION

We can almost immediately state the average number of tries it would take to get all 52 cards in the ordered sequence ace through two, for all four suits. It is $52 \cdot 51 \cdot 50 \cdot \cdots \cdot 3 \cdot 2 \cdot 1 = 52!$ If each of these attempts took a minute, it would require an average of $1.534 \cdot 10^{62}$ yr to obtain the desired order. But even after this time, there would be no guarantee that the ordered sequence had to appear. However, since there are 52! possible different sequences of the cards in the deck, at least one of the sequences would have appeared at least twice.

## 20.2 Engines and Refrigerators

A **heat engine** is a device that turns thermal energy into useful work. For example, an internal combustion engine or a jet engine extracts mechanical work from the thermal energy generated by burning a gasoline-air mixture (Figure 20.4a). For a heat engine to do work repeatedly, it must operate in a cycle. For example, work would be done if you put a firecracker under a soup can and explode the firecracker. The thermal energy from the explosion would be turned into mechanical motion of the soup can. (However, the applications of this firecracker/soup-can engine are limited.)

An engine that operates in a cycle passes through various thermodynamic states and eventually returns to its original state. We can think of the engine as operating between two thermal reservoirs (Figure 20.4b). One thermal reservoir is at a high temperature, $T_H$, and the other thermal reservoir is at a low temperature, $T_L$. The engine takes heat, $Q_H$, from the high-temperature reservoir, turns some of that heat into mechanical work, $W$, and exhausts the remaining heat, $Q_L$ (where $Q_L > 0$), to the low-temperature reservoir. According to the First Law of Thermodynamics (equivalent to the conservation of energy), $Q_H = W + Q_L$. Thus, to make an engine operate, it must be supplied energy in the form of $Q_H$, and then it will return useful work, $W$. The **efficiency,** $\epsilon$, of an engine is defined as

$$\epsilon = \frac{W}{Q_H}. \tag{20.1}$$

A **refrigerator,** like the one shown in Figure 20.5a, is a heat engine that operates in reverse. Instead of converting heat into work, the refrigerator uses work to move heat from a low-temperature thermal reservoir to a high-temperature thermal reservoir (Figure 20.5b). In a real refrigerator, an electric motor drives a compressor, which transfers thermal energy from the interior of the refrigerator to the air in the room. An air conditioner is also a refrigerator; it transfers thermal energy from the air in a room to the air outdoors.

For this refrigerator, the First Law of Thermodynamics requires that $Q_L + W = Q_H$. It is desired that a refrigerator remove as much heat as possible from the cooler reservoir, $Q_L$,

(a)

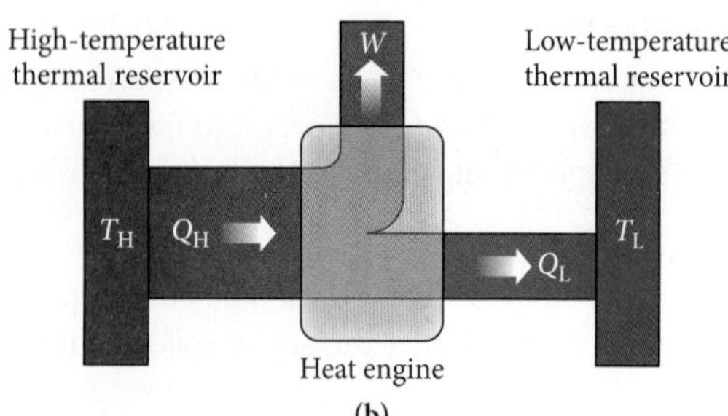


(b)

**FIGURE 20.4** (a) A jet engine turns thermal energy into mechanical work. (b) Flow diagram of a heat engine operating between two thermal reservoirs.

given the work, $W$, put into the refrigerator. The **coefficient of performance,** $K$, of a refrigerator is defined as

$$K = \frac{Q_L}{W}. \quad (20.2)$$

In the United States, refrigerators are usually rated by their annual energy usage, without quoting their actual efficiency. An efficient refrigerator is one whose energy use is comparable to the lowest energy use in its capacity class.

Air conditioners are often rated in terms of an energy efficiency rating (EER), defined as their heat removal capacity, $H$, in BTU/hour divided by the power $P$ used in watts. The relationship between $K$ and EER is

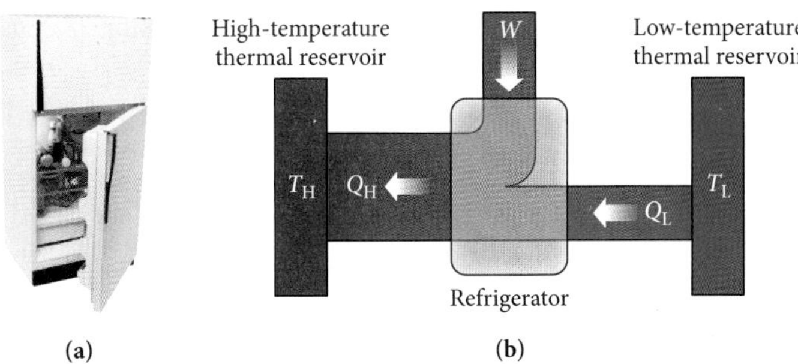


**FIGURE 20.5** (a) A household refrigerator. (b) Flow diagram of a refrigerator operating between two thermal reservoirs.

$$K = \frac{|Q_L|}{|W|} = \frac{Ht}{Pt} = \frac{H}{P} = \frac{\text{EER}}{3.41},$$

where $t$ is a time interval. The factor 1/3.41 arises from the definition of a BTU per hour:

$$\frac{1\ \text{BTU/h}}{1\ \text{watt}} = \frac{(1055\ \text{J})/(3600\ \text{s})}{1\ \text{J/s}} = \frac{1}{3.41}.$$

Typical EER values for a room air conditioner range from 8 to 11, meaning that values of $K$ range from 2.3 to 3.2. Thus, a typical room air conditioner can remove about three units of heat for every unit of energy used. Central air conditioners are often rated by a seasonally adjusted energy efficiency rating (SEER) that takes into account how long the air conditioner operates during a year.

A **heat pump** is a variation of a refrigerator that can be used to warm buildings. The heat pump warms the building by cooling the outside air. Just as for a refrigerator, $Q_L + W = Q_H$. However, for a heat pump, the quantity of interest is the heat put into the warmer reservoir, $Q_H$, rather than the heat removed from the cooler reservoir. Thus, the coefficient of performance of a heat pump is

$$K_{\text{heat pump}} = \frac{Q_H}{W}, \quad (20.3)$$

where $W$ is the work required to move the heat from the low-temperature reservoir to the high-temperature reservoir. The typical coefficient of performance of a commercial heat pump ranges from 3 to 4. Heat pumps do not work well when the outside temperature sinks below −18 °C (0 °F).

## EXAMPLE 20.2 Warming a House with a Heat Pump

A heat pump with a coefficient of performance of 3.500 is used to warm a house that loses 75,000. BTU/h of heat on a cold day. Assume that the cost of electricity is 10.00 cents per kilowatt-hour.

### PROBLEM

How much does it cost to warm the house for the day?

### SOLUTION

The coefficient of performance of a heat pump is given by equation 20.3:

$$K_{\text{heat pump}} = \frac{Q_H}{W}.$$

The work needed to warm the house is then

$$W = \frac{Q_H}{K_{\text{heat pump}}}.$$

Continued—

The house is losing 75,000 BTU/h, which we convert to SI units:

$$\frac{75{,}000 \text{ BTU}}{1 \text{ h}} \cdot \frac{1055 \text{ J}}{1 \text{ BTU}} \cdot \frac{1 \text{ h}}{3600 \text{ s}} = 21.98 \text{ kW}.$$

Therefore, the power required to warm the house is

$$P = \frac{W}{t} = \frac{Q_{\text{H}}/t}{K_{\text{heat pump}}} = \frac{21.98 \text{ kW}}{3.500} = 6.280 \text{ kW}.$$

The cost to warm the house for 24 h is then

$$\text{Cost} = (6.280 \text{ kW})(24 \text{ h})\frac{\$0.1000}{1 \text{ kWh}} = \$15.07.$$

## 20.3 Ideal Engines

An **ideal engine** is one that utilizes only reversible processes. In addition, the engine incorporates no energy-wasting effects, such as friction or viscosity. Remember, as defined in the previous section, a heat engine is a device that turns thermal energy into useful work. You might think that an *ideal* engine would be 100% efficient—would turn all the thermal energy supplied to it into useful mechanical work. However, we'll see that even the most efficient ideal engine cannot accomplish this. This fundamental inability to convert thermal energy totally into useful mechanical work lies at the very heart of the Second Law of Thermodynamics and of entropy, the two main topics of this chapter.

### Carnot Cycle

An example of an ideal engine is the **Carnot engine,** which is the most efficient engine that operates between two temperature reservoirs. The cycle of thermodynamic processes used by the Carnot engine is called the **Carnot cycle.**

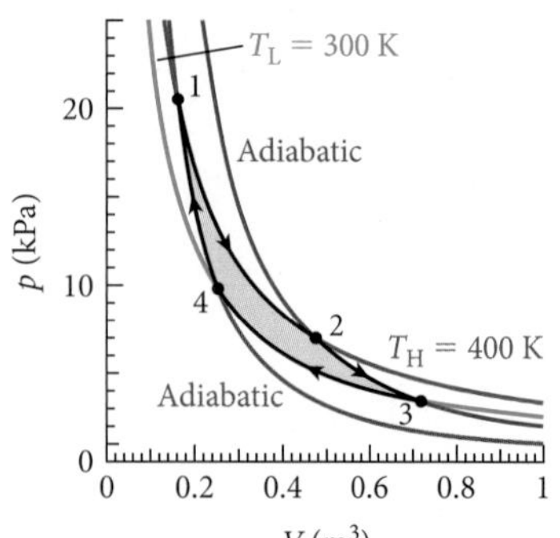


**FIGURE 20.6** The Carnot cycle, consisting of two isothermal processes and two adiabatic processes.

A Carnot cycle consists of two isothermal processes and two adiabatic processes, as shown in Figure 20.6 in a $pV$-diagram (pressure-volume diagram). We can pick an arbitrary starting point for the cycle; let's say that it begins at point 1. The system first undergoes an isothermal process, during which the system expands and absorbs heat from a thermal reservoir at fixed temperature $T_{\text{H}}$ (in Figure 20.6, $T_{\text{H}} = 400$ K). At point 2, the system begins an adiabatic expansion in which no heat is gained or lost. At point 3, the system begins another isothermal process, this time giving up heat to a second thermal reservoir at a lower temperature, $T_{\text{L}}$ (in Figure 20.6 $T_{\text{L}} = 300$ K), while the system compresses. At point 4, the system begins a second adiabatic process in which no heat is gained or lost. The Carnot cycle is complete when the system returns to point 1.

The efficiency of the Carnot engine is not 100% but instead is given by

$$\epsilon = 1 - \frac{T_{\text{L}}}{T_{\text{H}}}. \tag{20.4}$$

Remarkably, the efficiency of a Carnot engine depends only on the temperature ratio of the two thermal reservoirs. For the engine shown in Figure 20.6, for example, the efficiency is $\epsilon = 1 - (300 \text{ K})/(400 \text{ K}) = 0.25$.

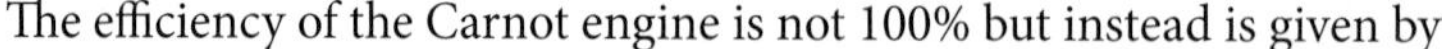

**DERIVATION 20.1 Efficiency of Carnot Engine**

We can determine the work and heat gained and lost in a Carnot cycle. To do this, we assume that the system consists of an ideal gas in a container whose volume can change. This system is placed in contact with a thermal reservoir and originally has pressure $p_1$ and volume $V_1$, corresponding to point 1 in Figure 20.6. The thermal reservoir holds the temperature of the system constant at $T_{\text{H}}$. The volume of the system is then allowed to increase until the system reaches point 2 in Figure 20.6. At point 2, the pressure is $p_2$

and the volume is $V_2$. During this isothermal process, work, $W_{12}$, is done by the system, and heat, $Q_H$, is transferred to the system. Because this process is isothermal, the internal energy of the system does not change. Thus, the amount of work done by the system and the thermal energy transferred to the system are equal, given by (see the derivation in Chapter 19):

$$W_{12} = Q_H = nRT_H \ln\left(\frac{V_2}{V_1}\right). \tag{i}$$

The system is then removed from contact with the thermal reservoir whose temperature is $T_H$. The system continues to expand adiabatically to point 3 in Figure 20.6. At point 3, the expansion is stopped, and the system is put in contact with a second thermal reservoir that has temperature $T_L$. An adiabatic expansion means that $Q = 0$. In Chapter 19, we saw that the work done by an ideal gas undergoing an adiabatic process is $W = nR(T_f - T_i)/(1 - \gamma)$, where $\gamma = C_p/C_V$ and $C_p$ is the specific heat of the gas at constant pressure and $C_V$ is the specific heat of the gas at constant volume. Thus, the work done by the system in going from point 2 to point 3 is

$$W_{23} = \frac{nR}{1-\gamma}\left(T_L - T_H\right).$$

Remaining in contact with the thermal reservoir whose temperature is $T_L$, the system is compressed isothermally from point 3 to point 4. The work done by the system and the heat absorbed by the system are

$$W_{34} = -Q_L = nRT_L \ln\left(\frac{V_4}{V_3}\right). \tag{ii}$$

(In Chapter 18, this would have been written $W_{34} = Q_L$ because $Q_L$ would be the heat added to the system and therefore $Q_L < 0$. However, in this chapter, we use the definition of $Q_L$ as the heat delivered to the low-temperature reservoir, at $T = T_L$, and hence $Q_L > 0$.)

Finally, the system is removed from contact with the second thermal reservoir and compressed adiabatically from point 4 back to the original point, point 1. For this adiabatic process, $Q = 0$, and the work done by the system is

$$W_{41} = \frac{nR}{1-\gamma}\left(T_H - T_L\right).$$

Note that $W_{41} = -W_{23}$, since the work done in the adiabatic processes depends only on the initial and final temperatures. Therefore, $W_{41} + W_{23} = 0$, and for the complete cycle, the total work done by the system is

$$W = W_{12} + W_{23} + W_{34} + W_{41} = W_{12} + W_{34}.$$

Substituting the expressions for $W_{12}$ and $W_{34}$ from equations (i) and (ii), we have

$$W = nRT_H \ln\left(\frac{V_2}{V_1}\right) + nRT_L \ln\left(\frac{V_4}{V_3}\right).$$

We can now find the efficiency of the Carnot engine. The efficiency is defined in equation 20.1 as the ratio of the work done by the system, $W$, to the amount of heat supplied to the system, $Q_H$:

$$\epsilon = \frac{W}{Q_H} = \frac{nRT_H \ln\left(\frac{V_2}{V_1}\right) + nRT_L \ln\left(\frac{V_4}{V_3}\right)}{nRT_H \ln\left(\frac{V_2}{V_1}\right)} = 1 + \frac{T_L \ln\left(\frac{V_4}{V_3}\right)}{T_H \ln\left(\frac{V_2}{V_1}\right)}. \tag{iii}$$

This expression for the efficiency of a Carnot engine contains the volumes at points 1, 2, 3, and 4. However, we can eliminate the volumes from this expression using the relation for adiabatic expansion of an ideal gas from Chapter 19, $TV^{\gamma-1} = \text{constant}$.

For the adiabatic process from point 2 to point 3, we then have

$$T_H V_2^{\gamma-1} = T_L V_3^{\gamma-1}. \tag{iv}$$

*Continued—*

Similarly, for the adiabatic process from point 4 to point 1, we have

$$T_L V_4^{\gamma-1} = T_H V_1^{\gamma-1}. \tag{v}$$

Taking the ratios of both sides of equations (iv) and (v)

$$\frac{T_H V_2^{\gamma-1}}{T_H V_1^{\gamma-1}} = \frac{T_L V_3^{\gamma-1}}{T_L V_4^{\gamma-1}} \Rightarrow \frac{V_2}{V_1} = \frac{V_3}{V_4}.$$

Using this relationship between the volumes, we can now rewrite equation (iii) for the efficiency as

$$\epsilon = 1 + \frac{-T_L \ln\left(\frac{V_3}{V_4}\right)}{T_H \ln\left(\frac{V_2}{V_1}\right)} = 1 - \frac{T_L}{T_H},$$

which matches equation 20.4.

Can the efficiency of the Carnot engine reach 100%? To obtain 100% efficiency, $T_H$ in equation 20.4 would have to be raised to infinity or $T_L$ be lowered to absolute zero. Neither option is possible. Thus, the efficiency of a Carnot engine is always less than 100%.

Note that the total work done by the system during the Carnot cycle in Derivation 20.1 is $W = W_{12} + W_{34}$, and the two contributions to the total work due to the two isothermal processes are $W_{12} = Q_H$ and $W_{34} = -Q_L$. Therefore, the total mechanical work from a Carnot cycle can also be written as

$$W = Q_H - Q_L. \tag{20.5}$$

Since the efficiency is defined in general as $\epsilon = W/Q_H$ (equation 20.1), the efficiency of a Carnot engine can thus also be expressed as

$$\epsilon = \frac{Q_H - Q_L}{Q_H}. \tag{20.6}$$

In this formulation, the efficiency of a Carnot engine is determined by the heat taken from the warmer reservoir minus the heat given back to the cooler reservoir. For this expression for the efficiency of a Carnot engine to yield an efficiency of 100%, the heat returned to the cooler reservoir would have to be zero. Conversely, if the efficiency of the Carnot engine is less than 100%, the engine cannot convert all the heat it takes in from the warmer reservoir to useful work.

French physicist Nicolas Carnot (1796–1832), who developed the Carnot cycle in the 19th century, proved the following statement, known as **Carnot's Theorem:**

No heat engine operating between two thermal reservoirs can be more efficient than a Carnot engine operating between the two thermal reservoirs.

We won't present the proof of Carnot's Theorem. However, we'll note the idea behind the proof after discussing the Second Law of Thermodynamics in Section 20.5

We can imagine running a Carnot engine in reverse, creating a "Carnot refrigerator." The maximum coefficient of performance for such a refrigerator operating between two thermal reservoirs is

$$K_{max} = \frac{T_L}{T_H - T_L}. \tag{20.7}$$

Similarly, the maximum coefficient of performance for a heat pump operating between two thermal reservoirs is given by

$$K_{\text{heat pump max}} = \frac{T_H}{T_H - T_L}. \tag{20.8}$$

### 20.1 In-Class Exercise

What is the maximum (Carnot) coefficient of performance of a refrigerator in a room with a temperature of 22.0 °C ? The temperature inside the refrigerator is kept at 2.0 °C.

a) 0.10

b) 0.44

c) 3.0

d) 5.8

e) 13.8

Note that the Carnot cycle constitutes the ideal thermodynamic process, which is the absolute upper limit on what is theoretically attainable. Real-world complications lower the efficiency drastically, as we'll see in Section 20.4. The typical coefficient of performance for a real refrigerator or heat pump is around 3.

### 20.2 In-Class Exercise

What is the maximum (Carnot) coefficient of performance of a heat pump being used to warm a house to an interior temperature of 22.0 °C? The temperature outside the house is 2.0 °C.

a) 0.15

b) 1.1

c) 3.5

d) 6.5

e) 14.8

### EXAMPLE 20.3 Work Done by a Carnot Engine

A Carnot engine takes 3000 J of heat from a thermal reservoir with a temperature $T_H = 500$ K and discards heat to a thermal reservoir with a temperature $T_L = 325$ K.

**PROBLEM**

How much work does the Carnot engine do in this process?

**SOLUTION**

We start with the definition of the efficiency, $\epsilon$, of a heat engine (equation 20.1):

$$\epsilon = \frac{W}{Q_H},$$

where $W$ is the work that the engine does and $Q_H$ is the heat taken from the warmer thermal reservoir. We can then use equation 20.4 for the efficiency of a Carnot engine:

$$\epsilon = 1 - \frac{T_L}{T_H}.$$

Combining these two expressions for the efficiency gives us

$$\frac{W}{Q_H} = 1 - \frac{T_L}{T_H}.$$

Now we can express the work done by the Carnot engine:

$$W = Q_H\left(1 - \frac{T_L}{T_H}\right).$$

Putting in the numerical values gives the amount of work done by the Carnot engine:

$$W = (3000 \text{ J})\left(1 - \frac{325 \text{ K}}{500 \text{ K}}\right) = 1050 \text{ J}.$$

### 20.1 Self-Test Opportunity

What is the efficiency of the Carnot engine in Example 20.3?

### EXAMPLE 20.4 Maximum Efficiency of an Electric Power Plant

In the United States, 85% of electricity is generated by burning fossil fuels to produce steam, which in turn drives alternators that produce electricity. Power plants can produce steam with a temperature as high as 600 °C by pressurizing the steam. The resulting waste heat must be exhausted into the environment at a temperature of 20.0 °C.

**PROBLEM**

What is the maximum efficiency of such a power plant?

**SOLUTION**

The maximum efficiency of such a power plant is the efficiency of a Carnot heat engine (equation 20.4) operating between 20.0 °C and 600 °C:

$$\epsilon = 1 - \frac{T_L}{T_H} = 1 - \frac{293 \text{ K}}{873 \text{ K}} = 66.4\%.$$

Real power plants achieve a lower efficiency of around 40%. However, many well-designed plants do not simply exhaust the waste heat into the environment. Instead, they employ *cogeneration* or *combined heat and power* (CHP). The heat that normally would

*Continued—*

**FIGURE 20.7** The coal-fired electric power plant at Michigan State University provides all the electricity used on the campus as well as providing heating on campus buildings in the winter and air-conditioning in the summer.

be lost is used to heat nearby buildings or houses. This heat can even be used to operate air conditioners that cool nearby structures, a process called *trigeneration.* By employing CHP, modern power plants can make use of up to 90% of the energy used to generate electricity. For example, the power plant at Michigan State University (Figure 20.7) burns coal to generate electricity for the campus. The waste heat is used to heat campus buildings in the winter and air-condition campus buildings in the summer.

## 20.4 Real Engines and Efficiency

A real heat engine based on the Carnot cycle is not practical. However, many practical heat engines that are in everyday use are designed to operate via cyclical thermodynamic processes. For an example of the operation of a real-world engine, let's examine the Otto cycle. Again, we assume an ideal gas as the working medium.

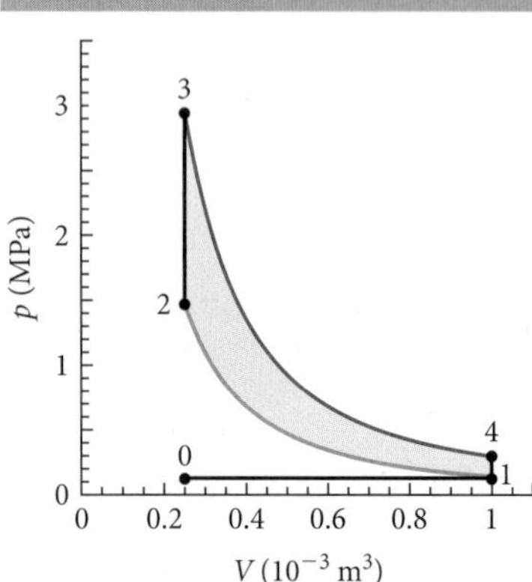


**FIGURE 20.8** The Otto cycle, consisting of two adiabatic processes and two constant-volume processes.

### Otto Cycle

The **Otto cycle** is used in the modern internal combustion engines inside automobiles. This cycle consists of two adiabatic processes and two constant-volume processes (Figure 20.8) and is the default configuration for a four-cycle internal combustion engine. The piston-cylinder arrangement of a typical internal combustion engine is shown in Figure 20.9.

The thermal energy is provided by the ignition of a fuel-air mixture. The cycle starts with the piston at the top of the cylinder and proceeds through the following steps:

- *Intake stroke.* The piston moves down with the intake valve open, drawing in the fuel-air mixture (point 0 to point 1 in Figure 20.8 and Figure 20.9a) and the intake valve closes.
- *Compression stroke.* The piston moves upward, compressing the fuel-air mixture adiabatically (point 1 to point 2 in Figure 20.8 and Figure 20.9b).
- The spark plug ignites the fuel-air mixture, increasing the pressure at constant volume (point 2 to point 3 in Figure 20.8).
- *Power stroke.* Hot gases push the piston down adiabatically (point 3 to point 4 in Figure 20.8 and Figure 20.9c).
- When the piston is all the way down (point 4 in Figure 20.8), the exhaust valve opens. This reduces the pressure at constant volume, provides the heat rejection, and moves the system back to point 1.
- *Exhaust stroke.* The piston moves up, forcing out the burned gases (point 1 to point 0 in Figure 20.8 and Figure 20.9d), and the exhaust valve closes.

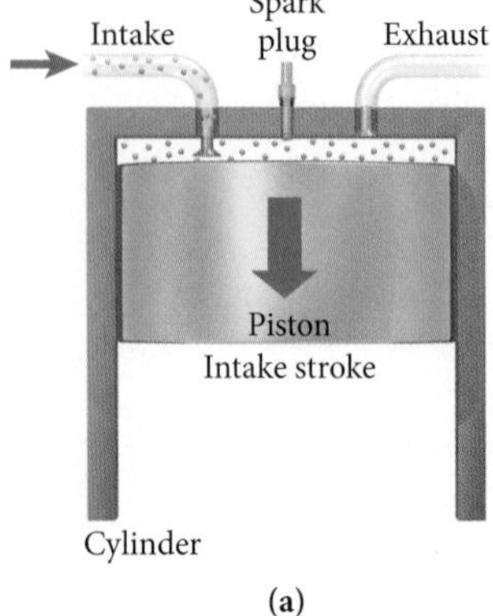

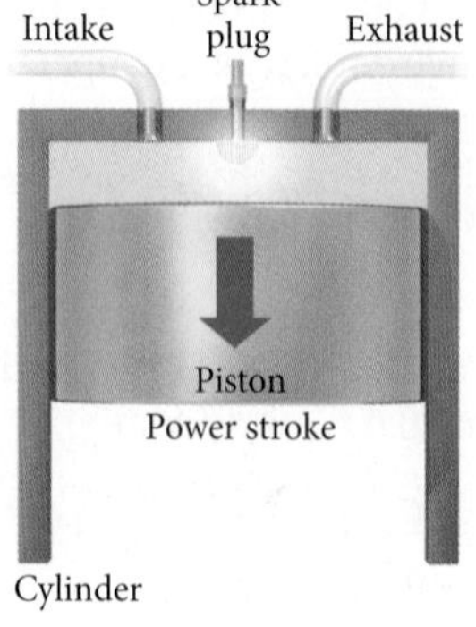


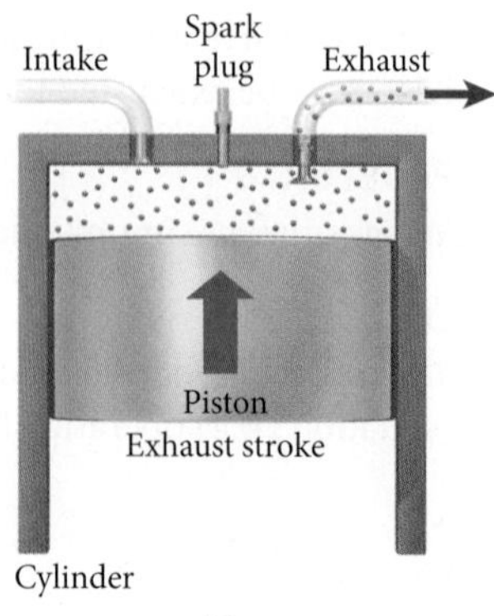


**FIGURE 20.9** The four strokes of an internal combustion engine: (a) The intake stroke, during which the fuel-air mixture is drawn into the cylinder. (b) The compression stroke, in which the fuel-air mixture is compressed. (c) The power stroke, where the fuel-air mixture is ignited by the spark plug, releasing heat. (d) The exhaust stroke, during which the burned gases are pushed out of the cylinder.

The ratio of the expanded volume, $V_1$, to the compressed volume, $V_2$, is called the *compression ratio:*

$$r = \frac{V_1}{V_2}. \tag{20.9}$$

The efficiency, $\epsilon$, of the Otto cycle can be expressed as a function of only the compression ratio:

$$\epsilon = 1 - r^{1-\gamma}. \tag{20.10}$$

(In contrast, for the Carnot cycle, equation 20.4 showed that the efficiency depends only on the ratio of two temperatures.)

## DERIVATION 20.2 Efficiency of the Otto Cycle

The Otto cycle begins at point 0 in Figure 20.8. The process from point 0 to point 1 simply draws the fuel-air mixture into the cylinder and plays no further role in the efficiency considerations. Neither does the inverse process from point 1 to point 0, the exhaust stroke, which removes the combustion products from the cylinder.

The process from point 1 to point 2 is adiabatic. The thermal energy transferred is zero, and the work done by the system (see Chapter 19) is given by

$$W_{12} = \frac{nR}{1-\gamma}(T_2 - T_1).$$

For the constant-volume process from point 2 to point 3, the work done by the system is zero, $W_{23} = 0$, and the thermal energy transferred to the system (see Chapter 18) is given by

$$Q_{23} = nC_V(T_3 - T_2).$$

The process from point 3 to point 4 is again adiabatic, and the thermal energy transferred is zero. The work done by the system is

$$W_{34} = \frac{nR}{1-\gamma}(T_4 - T_3).$$

The process from point 4 to point 1 takes place at constant volume, so the work done by the system is zero, $W_{41} = 0$, and the heat expelled from the system is

$$Q_{41} = nC_V(T_4 - T_1).$$

The total work done by the system is then simply the sum of the four contributions from the entire cycle (remember that $W_{23} = W_{41} = 0$)

$$W = W_{12} + W_{34} = \frac{nR}{1-\gamma}(T_2 - T_1) + \frac{nR}{1-\gamma}(T_4 - T_3) = \frac{nR}{1-\gamma}(T_2 - T_1 + T_4 - T_3).$$

The factor $R/(1-\gamma)$ can be expressed in terms of the specific heat capacity, $C_V$:

$$\frac{R}{1-\gamma} = \frac{C_p - C_V}{1 - \frac{C_p}{C_V}} = -C_V.$$

The efficiency of the Otto cycle is then (net work done divided by the heat gained by the system per cycle)

$$\epsilon = \frac{W}{Q_{23}} = \frac{-nC_V(T_2 - T_1 + T_4 - T_3)}{nC_V(T_3 - T_2)} = 1 - \frac{(T_4 - T_1)}{(T_3 - T_2)}. \tag{i}$$

We can use the fact that the process from point 1 to point 2 and the process from point 3 to point 4 are adiabatic to write (see Section 19.5)

$$T_1 V_1^{\gamma-1} = T_2 V_2^{\gamma-1} \Rightarrow T_2 = \left(\frac{V_1}{V_2}\right)^{\gamma-1} T_1 \tag{ii}$$

*Continued—*

and

$$T_3 V_2^{\gamma-1} = T_4 V_1^{\gamma-1} \Rightarrow T_3 = \left(\frac{V_1}{V_2}\right)^{\gamma-1} T_4, \qquad \text{(iii)}$$

where $V_1$ is the volume at points 1 and 4 and $V_2$ is the volume at points 2 and 3. Substituting the expressions for $T_2$ and $T_3$ from equations (ii) and (iii) into equation (i) for the efficiency, we obtain equation 20.10:

$$\epsilon = 1 - \frac{(T_4 - T_1)}{\left(\frac{V_1}{V_2}\right)^{\gamma-1} T_4 - \left(\frac{V_1}{V_2}\right)^{\gamma-1} T_1} = 1 - \left(\frac{V_1}{V_2}\right)^{1-\gamma} = 1 - r^{1-\gamma}.$$

### 20.2 Self-Test Opportunity

How much would the theoretical efficiency of an internal combustion engine increase if the compression ratio were raised from 4 to 15?

### 20.3 In-Class Exercise

Which of the four temperatures in the Otto cycle is the highest?

a) $T_1$

b) $T_2$

c) $T_3$

d) $T_4$

e) All four are identical.

As a numerical example, the compression ratio of the Otto cycle shown in Figure 20.8 is 4, and so, the efficiency of that Otto cycle is $\epsilon = 1 - 4^{1-7/5} = 1 - 4^{-0.4} = 0.426$. Thus, the theoretical efficiency of an engine operating on the Otto cycle with a compression ratio of 4 is 42.6%. Note, however, that this is the theoretical upper limit for the efficiency at this compression ratio. In principle, an internal combustion engine can be made more efficient by increasing the compression ratio, but practical factors prevent that approach. For example, if the compression ratio is too high, the fuel-air mixture will detonate before the compression is complete. Very high compression ratios put high stresses on the components of the engine. The actual compression ratios of gasoline-powered internal combustion engines range from 8 to 12.

## Real Otto Engines

The actual efficiency of an internal combustion engine is about 20%. Why is this so much lower than the theoretical upper limit? There are many reasons. First, the "adiabatic" parts of the cycle do not really proceed without heat exchange between the gas in the piston and the engine block. This is obvious; we all know from experience that engines heat up while running, and this temperature rise is a direct consequence of heat leaking from the gas undergoing compression and ignition. Second, the gasoline-air mixture is not quite an ideal gas and thus has energy losses due to internal excitation. Third, during the ignition and heat rejection processes, the volume does not stay exactly constant, because both processes take some time, and the piston keeps moving continuously during that time. Fourth, during the intake and exhaust strokes, the pressure in the chamber is not exactly atmospheric pressure because of gas dynamics considerations. All of these effects, together with small friction losses, combine to reduce engine efficiency. All major car companies, as well as car racing crews, government and university labs, and even a few hobbyists, are continuously seeking ways to increase engine efficiency, because a higher efficiency means higher power output and/or better gas mileage. Since the oil crisis of the mid-1970s, the engine efficiency of U.S. automobiles has steadily increased (see Section 5.7), but even better miles-per-gallon performance is necessary for the United States to reduce its greenhouse gas emissions and its dependence on foreign oil.

## SOLVED PROBLEM 20.1 Efficiency of an Automobile Engine

A car with a gasoline-powered internal combustion engine travels with a speed of 26.8 m/s (60.0 mph) on a level road and uses gas at a rate of 6.92 L/100 km (34.0 mpg). The energy content of gasoline is 34.8 MJ/L.

### PROBLEM

If the engine has an efficiency of 20.0%, how much power is delivered to keep the car moving at a constant speed?

### SOLUTION

#### THINK

We can calculate how much energy is being supplied to the engine by calculating the amount of fuel used and multiplying by the energy content of that fuel. The efficiency is the useful work divided by the energy being supplied; so, once we determine the energy supplied, we can find the useful work from the given efficiency of the engine. By dividing the work and the energy by an arbitrary time interval, we can determine the average power delivered.

## SKETCH

Figure 20.10 shows a gasoline-powered automobile engine operating as a heat engine.

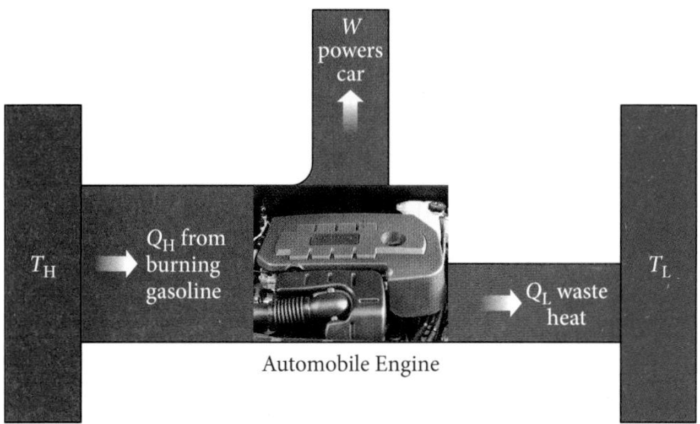


**FIGURE 20.10** A gasoline-powered automobile engine operating as a heat engine to power a car traveling at constant speed $v$ on a horizontal surface.

## RESEARCH

The car is traveling at speed $v$. The rate, $\chi$, at which the car burns gasoline, can be expressed in terms of the volume of gasoline burned per unit distance. We can calculate the volume of gasoline burned per unit time, $V_t$, by multiplying the speed of the car by the rate, $\chi$, at which the car burns gasoline:

$$V_t = \chi v. \qquad \text{(i)}$$

The energy per unit time supplied to the engine by consuming fuel is the power, $P$, given by the volume of gasoline used per unit time multiplied by the energy content of gasoline, $E_g$:

$$P = V_t E_g. \qquad \text{(ii)}$$

The efficiency of the engine, $\epsilon$, is given by equation 20.1:

$$\epsilon = \frac{W}{Q_H},$$

where $W$ is the useful work and $Q_H$ is the thermal energy supplied to the engine. If we divide both $W$ and $Q_H$ by a time interval, $t$, we get

$$\epsilon = \frac{W/t}{Q_H/t} = \frac{P_{delivered}}{P}, \qquad \text{(iii)}$$

where $P_{delivered}$ is the power delivered by the engine of the car.

## SIMPLIFY

We can combine equations (i) through (iii) to obtain

$$P_{delivered} = \epsilon P = \epsilon V_t E_g = \epsilon \chi v E_g.$$

## CALCULATE

The power delivered is

$$P_{delivered} = \epsilon \chi v E_g = (0.200)\left(\frac{6.92 \text{ L}}{100 \cdot 10^3 \text{ m}}\right)(26.8 \text{ m/s})(34.8 \cdot 10^6 \text{ J/L}) = 12{,}907.7 \text{ W}.$$

## ROUND

We report our result to three significant figures:

$$P_{delivered} = 12{,}900 \text{ W} = 12.9 \text{ kW} = 17.3 \text{ hp}.$$

*Continued—*

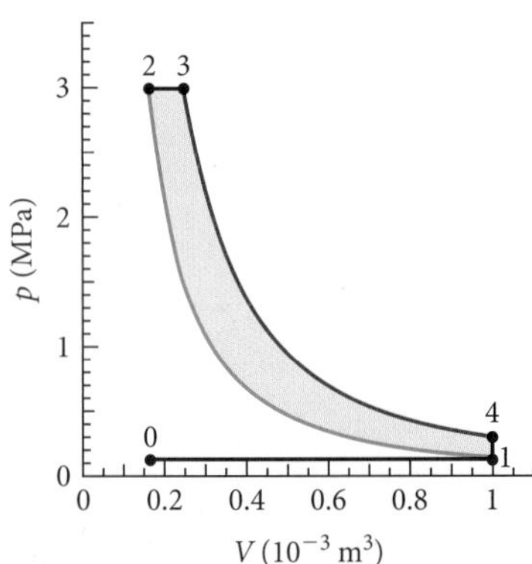

**FIGURE 20.11** A $pV$-diagram for the Diesel cycle.

### 20.4 In-Class Exercise

In which part of the Diesel cycle in Figure 20.11 is heat added?

a) on the path from point 0 to point 1

b) on the path from point 1 to point 2

c) on the path from point 2 to point 3

d) on the path from point 3 to point 4

e) on the path from point 4 to point 1

### 20.5 In-Class Exercise

During which part(s) of the Diesel cycle in Figure 20.11 is mechanical work done by the engine?

a) on the path from point 0 to point 1 and on the path from point 1 to point 0

b) on the path from point 1 to point 2

c) on the path from point 2 to point 3 and on the path from point 3 to point 4

d) on the path from point 4 to point 1

### DOUBLE-CHECK

To double-check our result, we calculate the power required to keep the car moving at 60.0 mph against air resistance. The power required, $P_{air}$, is equal to the product of the force of air resistance, $F_{drag}$, and the speed of the car:

$$P_{air} = F_{drag}v. \tag{iv}$$

The drag force created by air resistance is given by

$$F_{drag} = Kv^2. \tag{v}$$

The constant $K$ has been found empirically to be

$$K = \tfrac{1}{2}c_d A\rho, \tag{vi}$$

where $c_d$ is the drag coefficient of the car, $A$ is its front cross-sectional area, and $\rho$ is the density of air. Combining equations (iv) through (vi) gives

$$P_{air} = F_{drag}v = \left(Kv^2\right)v = \tfrac{1}{2}c_d A\rho v^3.$$

Using $c_d = 0.33$, $A = 2.2$ m$^2$, and $\rho = 1.29$ kg/m$^3$, we obtain

$$P_{air} = \tfrac{1}{2}(0.33)\left(2.2 \text{ m}^2\right)\left(1.29 \text{ kg/m}^3\right)\left(26.8 \text{ m/s}\right)^3 = 9014 \text{ W} = 9.0 \text{ kW}.$$

This result for the power required to overcome air resistance ($P_{air}$ = 9.0 kW = 12 hp) is about 70% of the calculated value for the power delivered ($P_{delivered}$ = 12.9 kW = 17.3 hp). The remaining power is used to overcome other kinds of friction, such as rolling friction. Thus, our answer seems reasonable.

## Diesel Cycle

Diesel engines and gasoline engines have somewhat different designs. Diesel engines do not compress a fuel-air mixture, but rather air only (path from point 1 to point 2 in Figure 20.11, green curve). The fuel is introduced (between point 2 and point 3) only after the air has been compressed. The thermal energy from the compression ignites the mixture (thus, no spark plug is required). This combustion process pushes the piston out at constant pressure. After combustion, the combustion products push the piston out farther in the same adiabatic manner as in the Otto cycle (path from point 3 to point 4, red curve). The process of heat rejection to the environment at constant volume (between points 4 and 1 in the diagram) also proceeds in the same way as in the Otto cycle, as do the intake stroke (path from 0 to 1) and exhaust stroke (path from 1 to 0). Diesel engines have a higher compression ratio and thus a higher efficiency than gasoline-powered four-stroke engines, although they have a slightly different thermodynamic cycle.

The efficiency of an ideal diesel engine is given by

$$\epsilon = 1 - r^{1-\gamma}\frac{\alpha^\gamma - 1}{\gamma(\alpha - 1)}, \tag{20.11}$$

where the compression ratio is again $r = V_1/V_2$ (just as for the Otto cycle), $\gamma$ is again the ratio of the specific heats at constant pressure and constant volume, $\gamma = C_p/C_V$ (introduced in Chapter 19), and $\alpha = V_3/V_2$ is called the *cut-off ratio,* the ratio between the final and initial volumes of the combustion phase. The derivation of this formula proceeds similarly to Derivations 20.1 and 20.2 for the Carnot and Otto cycles but is omitted here.

## Hybrid Cars

Hybrid cars combine a gasoline engine and an electric motor to achieve higher efficiency than a gasoline engine alone can achieve. The improvement in efficiency results from using a smaller gasoline engine than would normally be necessary and an electric motor, run off a battery charged by the gasoline engine, to supplement the gasoline engine when higher power is required. For example, the Ford Escape Hybrid (Figure 20.12) has a 99.2-kW (133-hp) gasoline engine coupled with a 70-kW (94-hp) electric motor, and the Ford Escape has a 149-kW (200-hp) gasoline engine. In addition, the gasoline engine used in the Ford Escape Hybrid, as well as in the hybrid vehicles from Toyota and Honda, applies a thermodynamic cycle different from the Otto cycle. This *Atkinson cycle* includes variable valve timing to increase the expansion phase of the process, allowing more useful work to be extracted from the energy consumed by the engine. An Atkinson-cycle engine produces less power per displacement than a comparable Otto-cycle engine but has higher efficiency.

**FIGURE 20.12** The gasoline engine and electric motor of a hybrid car.

In stop-and-go driving, hybrid cars have the advantage of using regenerative braking rather than normal brakes. Conventional brakes use friction to stop the car, which converts the kinetic energy of the car into wasted energy in the form of thermal energy (and wear on the brake pads). Regenerative brakes couple the wheels to an electric generator (which can be the electric motor driving the car), which converts the kinetic energy of the car into electrical energy. This energy is stored in the battery of the hybrid car to be used later. In addition, the gasoline engine of a hybrid car can be shut off when the car is stopped, so the car uses no energy while waiting at a stoplight.

In highway driving, hybrid cars take advantage of the fact that the gasoline engine can be run at a constant speed, corresponding to its most efficient operating speed, rather than having to run at a speed dictated by the speed of the car. Operation of the gasoline engine at a constant speed while the speed of the car changes is accomplished by a continuously variable transmission. The transmission couples both the gasoline engine and the electric motor to the drive wheels of the car, allowing the gasoline engine to run at its most efficient speed while taking power from the electric motor as needed.

## Efficiency and the Energy Crisis

The efficiencies of engines and refrigerators are not just theoretical constructs but have very important economic consequences, which are extremely relevant to solving the energy crisis. The efficiencies and performance coefficients calculated in this chapter with the aid of thermodynamic principles are theoretical upper limits. Real-world complications always reduce the actual efficiencies of engines and performance coefficients of refrigerators. But engineering research can overcome these real-world complications and provide better performance of real devices, approaching the ideal limits.

One impressive example of performance improvement has occurred with refrigerators sold in the United States, according to data compiled by Steve Chu (see Figure 20.13).

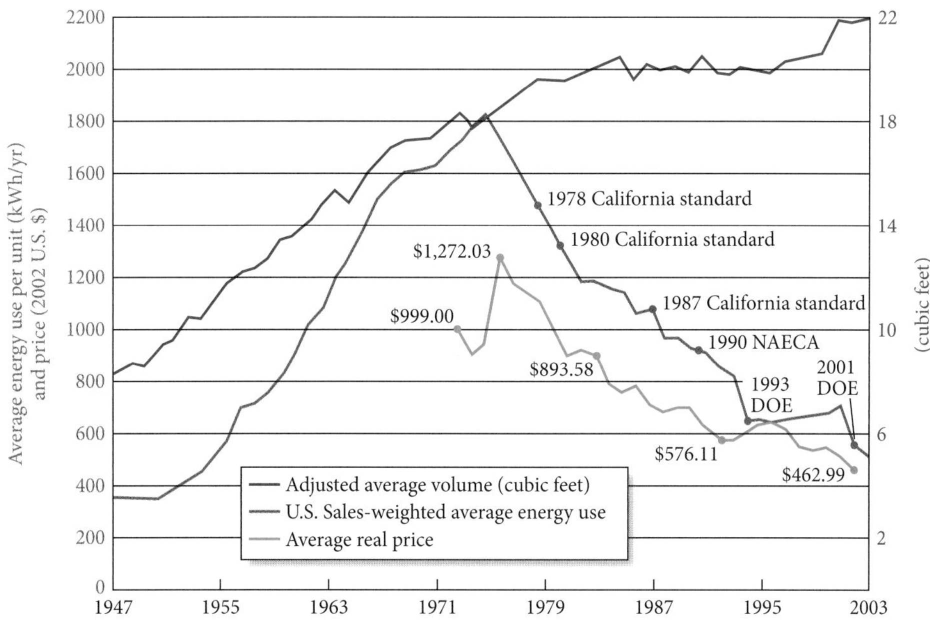


**FIGURE 20.13** Average U.S. refrigerator volume (red line), price (green line), and energy use (blue line) from 1947 to 2003.

Since 1975, the average size of a refrigerator in U.S. kitchens has increased by about 20%, but through a combination of tougher energy standards and research and development on refrigerator design and technology, the average power consumption fell by two-thirds, a total of 1200 kWh/yr, from 1800 kWh/yr in 1975 to 600 kWh/yr in 2003.

Since about 150 million new refrigerators and freezers are purchased each year in the United States, and each one saves approximately 1200 kWh/yr, a total energy savings of 180 billion kWh ($6.5 \cdot 10^{17}$ J = 0.65 EJ) is realized each year. This savings is (in 2009) approximately twice as much as the combined energy produced through the use of wind power (30 billion kWh/yr), solar energy (2 billion kWh/yr), geothermal (14 billion kWh/yr), and biomass (54 billion kWh/yr).

Energy consumption is directly proportional to energy cost, so the average U.S. household is saving a total of almost $200 each year in electricity costs from using more efficient refrigerators relative to the 1975 standard. In addition, even though refrigerators have gotten much better, their price has fallen by more than half, also indicated in Figure 20.13. So a more energy-efficient refrigerator is not more expensive. And it keeps our food and favorite beverages just as cold as the 1975 model used to!

## 20.5 The Second Law of Thermodynamics

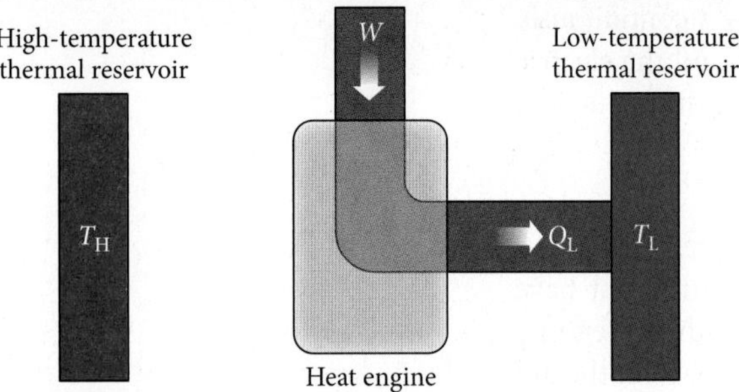


**FIGURE 20.14** A heat flow diagram showing a system that converts work completely to heat.

We saw in Section 20.1 that the First Law of Thermodynamics is satisfied whether a glass of cold water warms on standing at room temperature or becomes colder. However, the Second Law of Thermodynamics is a general principle that puts constraints on the amount and direction of thermal energy transfer between systems and on the possible efficiencies of heat engines. This principle goes beyond the energy conservation of the First Law of Thermodynamics.

Chapter 18 discussed Joule's famous experiment that demonstrated that mechanical work could be converted completely into heat, as illustrated in Figure 20.14. In contrast, experiments show that it is impossible to build a heat engine that completely converts heat into work. This concept is illustrated in Figure 20.15.

In other words, it is not possible to construct a 100% efficient heat engine. This fact forms the basis of the **Second Law of Thermodynamics:**

It is impossible for a system to undergo a process in which it absorbs heat from a thermal reservoir at a given temperature and converts that heat completely to mechanical work without rejecting heat to a thermal reservoir at a lower temperature.

This formulation is often called the *Kelvin–Planck statement of the Second Law of Thermodynamics.* As an example, consider a book sliding on a table. The book slides to a stop, and the mechanical energy of motion is turned into thermal energy. This thermal energy takes the form of random motion of the molecules of the book, the air, and the table. It is impossible to convert this random motion back into organized motion of the book. It is, however, possible to convert *some* of the random motion related to thermal energy back into mechanical energy. Heat engines do that kind of conversion.

If the Second Law were not true, various impossible scenarios could occur. For example, an electric power plant could operate by taking heat from the surrounding air, and an ocean liner could power itself by taking heat from the seawater. These scenarios do not violate the First Law of Thermodynamics because energy is conserved. The fact that they cannot occur shows that the Second Law contains additional information about how nature works, beyond the principle of conservation of energy. The Second Law limits the ways in which energy can be used.

Another way to state the Second Law of Thermodynamics relates to refrigerators. We know that heat flows spontaneously from a warmer thermal reservoir to a cooler thermal reservoir. Heat never *spontaneously* flows from a cooler thermal reservoir to a warmer ther-

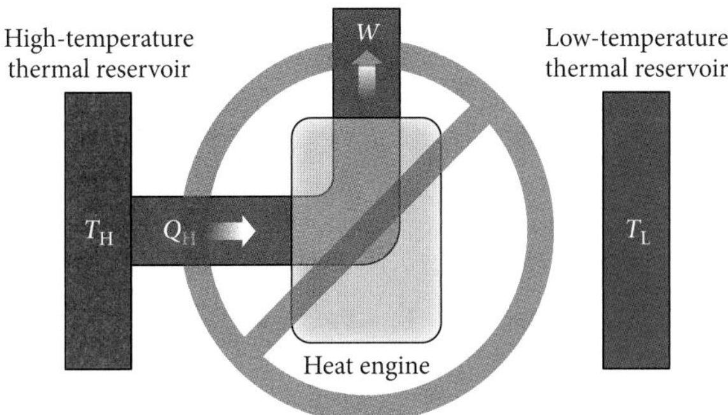


**FIGURE 20.15** A heat flow diagram illustrating the impossible process of completely converting heat into useful work without rejecting heat into a low-temperature thermal reservoir.

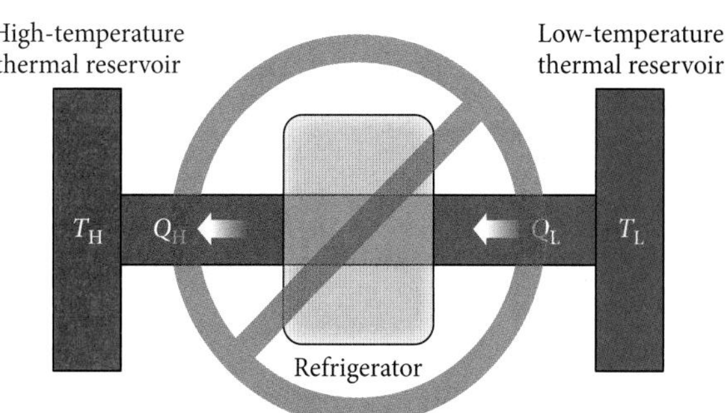


**FIGURE 20.16** A heat flow diagram demonstrating the impossible process of moving heat from a low-temperature thermal reservoir to a high-temperature thermal reservoir without using any work.

mal reservoir. A refrigerator is a heat engine that moves heat from a cooler thermal reservoir to a warmer thermal reservoir; however, energy has to be supplied to the refrigerator for this transfer to take place, as shown in Figure 20.5b. The fact that it is impossible for a refrigerator to transfer thermal energy from a cooler reservoir to a warmer reservoir without doing work is illustrated in Figure 20.16. This fact is the basis of another form of the Second Law of Thermodynamics:

> It is impossible for any process to transfer thermal energy from a cooler thermal reservoir to a warmer thermal reservoir without any work having been done to accomplish the transfer.

This equivalent formulation is often called the *Clausius statement of the Second Law of Thermodynamics.*

## DERIVATION 20.3 Carnot's Theorem

The following explains how to prove Carnot's Theorem, stated in Section 20.3. Suppose there are two thermal reservoirs, a heat engine working between these two reservoirs, and a Carnot engine operating in reverse as a refrigerator between the reservoirs, as shown in Figure 20.17.

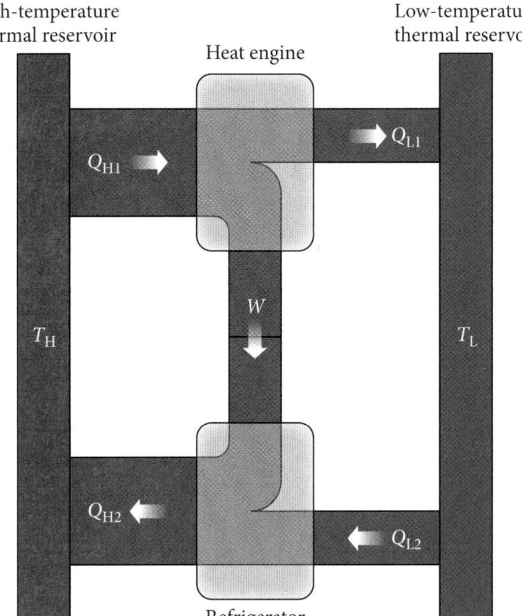


**FIGURE 20.17** A heat engine between two thermal reservoirs produces work. A Carnot engine operating in reverse as a refrigerator between two thermal reservoirs is being driven by the work produced by the heat engine.

*Continued—*

Let's assume that the Carnot refrigerator operates with the theoretical efficiency given by

$$\epsilon_2 = \frac{W}{Q_{H2}}, \quad \text{(i)}$$

where $W$ is the work required to put heat $Q_{H2}$ into the high-temperature reservoir. We also assume that this required work is provided by a heat engine operating between the same two reservoirs. The efficiency of the heat engine can be expressed as

$$\epsilon_1 = \frac{W}{Q_{H1}}, \quad \text{(ii)}$$

where $Q_{H1}$ is the heat removed from the high-temperature reservoir by the heat engine. Since equations (i) and (ii) both contain the same work, $W$, we can solve each of them for the work and then set the expressions equal to each other:

$$\epsilon_1 Q_{H1} = \epsilon_2 Q_{H2}.$$

Thus, the ratio of the two efficiencies is

$$\frac{\epsilon_1}{\epsilon_2} = \frac{Q_{H2}}{Q_{H1}}.$$

If the efficiency of the heat engine is equal to the efficiency of the Carnot refrigerator ($\epsilon_1 = \epsilon_2$), then we have $Q_{H1} = Q_{H2}$, which means that the heat removed from the high-temperature reservoir is equal to the heat added to the high-temperature reservoir. If the efficiency of the heat engine is higher than the efficiency of the Carnot refrigerator ($\epsilon_1 > \epsilon_2$), then we have $Q_{H2} > Q_{H1}$, which means that more heat is gained than lost by the high-temperature reservoir. Applying the First Law of Thermodynamics to the overall system tells us that the heat engine working together with the Carnot refrigerator is then able to transfer thermal energy from the low-temperature reservoir to the high-temperature reservoir without any work being supplied. These two devices acting together would then violate the Clausius statement of the Second Law of Thermodynamics. Thus, no heat engine can be more efficient than a Carnot engine.

The conversion of mechanical work to thermal energy (such as by friction) and heat flow from a warm thermal reservoir to a cool one are irreversible processes. The Second Law of Thermodynamics says that these processes can only be partially reversed, thereby acknowledging their inherent one-way quality.

## 20.6 Entropy

The Second Law of Thermodynamics as stated in Section 20.5 is somewhat different from other laws presented in previous chapters, such as Newton's laws, because it is phrased in terms of impossibilities. However, the Second Law can be stated in a more direct manner using the concept of **entropy.**

Throughout the last three chapters, we have discussed the notion of thermal equilibrium. If two objects at different temperatures are brought into thermal contact, both of their temperatures will asymptotically approach a common equilibrium temperature. What drives this system to thermal equilibrium is entropy, and the state of thermal equilibrium is the state of maximum entropy.

Entropy provides a quantitative measure for how close to equilibrium a system is. The direction of thermal energy transfer is not determined by energy conservation but by the change in entropy of a system. The change in entropy of a system, $\Delta S$, during a process that takes the system from an initial state to a final state is defined as

$$\Delta S = \int_{i}^{f} \frac{dQ}{T}, \quad \text{(20.12)}$$

where $Q$ is the heat and $T$ is the temperature in kelvins. The SI units for the change in entropy are joules per kelvin (J/K). It must be noted that equation 20.12 applies only to reversible processes; that is, the integration can only be carried out over a path representing a reversible process.

Note that entropy is defined in terms of its change from an initial to a final configuration. Entropy change is the physically meaningful quantity, not the absolute value of the entropy at any point. Another physical quantity for which only the change is important is the potential energy. The absolute value of the potential energy is always defined only relative to some arbitrary additive constant, but the change in potential energy between initial and final states is a precisely measurable physical quantity that gives rise to force(s). In a manner similar to the way connections between forces and potential energy changes were established in Chapter 6, this section shows how to calculate entropy changes for given temperature changes and heat and work in different systems. At thermal equilibrium, the entropy has an extremum (a maximum). At mechanical stable equilibrium, the net force is zero, and therefore the potential energy has an extremum (a minimum, in this case).

In an irreversible process in a closed system, the entropy, $S$, of the system never decreases; it always increases or stays constant. In a closed system, energy is always conserved, but entropy is not conserved. Thus, the change in entropy defines a direction for time; that is, time moves forward if the entropy of a closed system is increasing.

The definition of entropy given by equation 20.12 rests on the macroscopic properties of a system, such as heat and temperature. Another definition of entropy, based on statistical descriptions of how the atoms and molecules of a system are arranged, is presented in the next section.

Since the integral in equation 20.12 can only be evaluated for a reversible process, how can we calculate the change in entropy for an irreversible process? The answer lies in the fact that entropy is a thermodynamic state variable, just like temperature, pressure, and volume. This means that we can calculate the entropy difference between a known initial state and a known final state even for an irreversible process if there is a reversible process (for which the integral in equation 20.12 can be evaluated!) that takes the system from the same initial to the same final state. Perhaps this is the subtlest point about thermodynamics conveyed in this entire chapter.

In order to illustrate this general method of computing the change in entropy for an irreversible process, let's return to a situation described in Chapter 18, the free expansion of a gas. Figure 20.18a shows a gas confined to the left half of a box. In Figure 20.18b, the barrier between the two halves has been removed, and the gas has expanded to fill the entire volume of the box. Clearly, once the gas has expanded to fill the entire volume of the box, the system will never spontaneously return to the state where all the gas molecules are located in the left half of the box. The state variables of the system before the barrier is removed are the initial temperature, $T_i$, the initial volume, $V_i$, and the initial entropy, $S_i$. After the barrier has been removed and the gas is again in equilibrium, the state of the system can be described in terms of the final temperature, $T_f$, the final volume, $V_f$, and the final entropy, $S_f$.

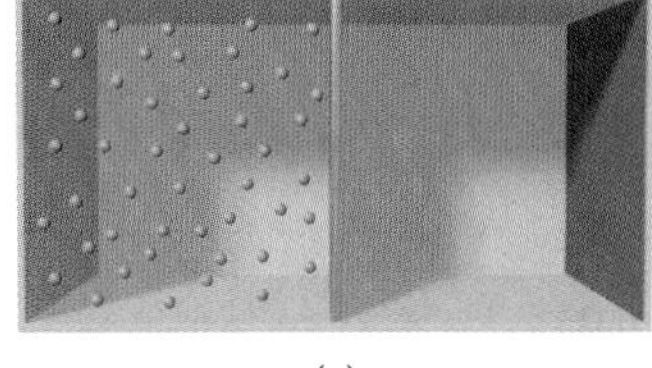
(a)

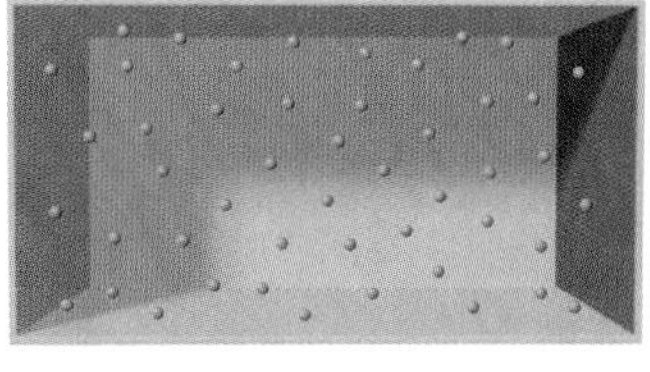
(b)

**FIGURE 20.18** (a) A gas confined to half the volume of a box. (b) The barrier separating the two halves is removed, and the gas expands to fill the entire volume.

We cannot calculate the change in entropy of this system using equation 20.12 because the gas is not in equilibrium during the expansion phase. However, the change in the properties of the system depends only on the initial and final states, not on how the system got from one to the other. Therefore, we can choose a process that the system could have undergone for which we can evaluate the integral in equation 20.12.

In the free expansion of an ideal gas, the temperature remains constant; thus, it seems reasonable to use the isothermal expansion of an ideal gas. We can then evaluate the integral in equation 20.12 to calculate the change in entropy of the system undergoing an isothermal process:

$$\Delta S = \int_i^f \frac{dQ}{T} = \frac{1}{T}\int_i^f dQ = \frac{Q}{T}. \tag{20.13}$$

(We can move the factor $1/T$ outside the integral, because we are dealing with an isothermal process, for which the temperature is constant by definition.) As we saw in Chapter 19, the work done by an ideal gas in expanding from $V_i$ to $V_f$ at a constant temperature $T$ is given by

$$W = nRT\ln\left(\frac{V_f}{V_i}\right).$$

For an isothermal process, the internal energy of the gas does not change; so $\Delta E_{\text{int}} = 0$. Thus, as shown in Chapter 18, we can use the First Law of Thermodynamics to write

$$\Delta E_{\text{int}} = W - Q = 0.$$

Consequently, for the isothermal process, the heat added to the system is

$$Q = W = nRT \ln\left(\frac{V_{\text{f}}}{V_{\text{i}}}\right).$$

The resulting entropy change for the isothermal process is then

$$\Delta S = \frac{Q}{T} = \frac{nRT \ln\left(\frac{V_{\text{f}}}{V_{\text{i}}}\right)}{T} = nR \ln\left(\frac{V_{\text{f}}}{V_{\text{i}}}\right). \tag{20.14}$$

The entropy change for the irreversible free expansion of a gas must be equal to the entropy change for the isothermal process because both processes have the same initial and final states and thus must have the same change in entropy.

For the irreversible free expansion of a gas, $V_{\text{f}} > V_{\text{i}}$, and so $\ln(V_{\text{f}}/V_{\text{i}}) > 0$. Thus, $\Delta S > 0$ because $n$ and $R$ are positive numbers. In fact, the change in entropy of any irreversible process is always positive.

Therefore, the Second Law of Thermodynamics can be stated in a third way.

The entropy of a closed system can never decrease.

## EXAMPLE 20.5 Entropy Change for the Freezing of Water

Suppose we have 1.50 kg of water at a temperature of 0 °C. We put the water in a freezer, and enough heat is removed from the water to freeze it completely to ice at a temperature of 0 °C.

**PROBLEM**

How much does the entropy of the water-ice system change during the freezing process?

**SOLUTION**

The melting of ice is an isothermal process, so we can use equation 20.13 for the change in entropy:

$$\Delta S = \frac{Q}{T},$$

where $Q$ is the heat that must be removed to change the water to ice at $T = 273.15$ K, the freezing point of water. The heat that must be removed to freeze the water is determined by the latent heat of fusion of water (ice), defined in Chapter 18. The heat that must be removed is

$$Q = mL_{\text{fusion}} = (1.50 \text{ kg})(334 \text{ kJ/kg}) = 501 \text{ kJ}.$$

Thus, the change in entropy of the water-ice system is

$$\Delta S = \frac{-501 \text{ kJ}}{273.15 \text{ K}} = -1830 \text{ J/K}.$$

Note that the entropy of the water-ice system of Example 20.5 decreased. How can the entropy of this system decrease? The Second Law of Thermodynamics states that the entropy of a closed system can never decrease. However, the water-ice system is not a closed system. The freezer used energy to remove heat from the water to freeze it and exhausted the heat into the local environment. Thus, the entropy of the environment increased more than the entropy of the water-ice system decreased. This is a very important distinction.

A similar analysis can be applied to the origins of complex life forms that have lower entropy than their surroundings. The development of life forms with low entropy is accompanied by an increase in the overall entropy of the Earth. In order for a living subsystem of Earth to reduce its own entropy at the expense of its environment, it needs a source of energy. This source of energy can be chemical bonds or other types of potential energy, which in the end arises from the energy provided to Earth by solar radiation.

Evolution toward more and more complex life forms is not in contradiction with the Second Law of Thermodynamics, because the evolving life forms do not form a closed system. A contradiction between biological evolution and the Second Law of Thermodynamics is sometimes falsely claimed by opponents of evolution in the evolution/creationism debate. From a thermodynamics standpoint, this argument has to be rejected unequivocally.

### EXAMPLE 20.6 Entropy Change for the Warming of Water

Suppose we start with 2.00 kg of water at a temperature of 20.0 °C and warm the water until it reaches a temperature of 80.0 °C.

**PROBLEM**

What is the change in entropy of the water?

**SOLUTION**

We start with equation 20.12, relating the change in entropy to the integration of the differential flow of heat, $dQ$, with respect to temperature:

$$\Delta S = \int_{i}^{f} \frac{dQ}{T}. \tag{i}$$

The heat, $Q$, required to raise the temperature of a mass, $m$, of water is given by

$$Q = cm\Delta T, \tag{ii}$$

where $c = 4.19$ kJ/(kg K) is the specific heat of water. We can rewrite equation (ii) in terms of the differential change in heat, $dQ$, and the differential change in temperature, $dT$:

$$dQ = cm\,dT.$$

Then we can rewrite equation (i) as

$$\Delta S = \int_{i}^{f} \frac{dQ}{T} = \int_{T_i}^{T_f} \frac{cm\,dT}{T} = cm \int_{T_i}^{T_f} \frac{dT}{T} = cm \ln \frac{T_f}{T_i}.$$

With $T_i = 293.15$ K and $T_f = 353.15$ K, the change in entropy is

$$\Delta S = \left[4.19 \text{ kJ/}\left(\text{kg K}\right)\right]\left(2.00 \text{ kg}\right) \ln \frac{353.15 \text{ K}}{293.15 \text{ K}} = 1.56 \cdot 10^{3} \text{ J/K}.$$

There is another important point regarding the macroscopic definition of entropy and the calculation of entropy with the aid of equation 20.12: The Second Law of Thermodynamics also implies that for all cyclical processes (such as the Carnot, Otto, and Diesel cycles)—that is, all processes for which the initial state is the same as the final state—the total entropy change over the entire cycle has to be greater than or equal to zero: $\Delta S \geq 0$.

## 20.7 Microscopic Interpretation of Entropy

In Chapter 19, we saw that the internal energy of an ideal gas could be calculated by summing up the energies of the constituent particles of the gas. We can also determine the entropy of an ideal gas by studying the constituent particles. It turns out that this microscopic definition of entropy agrees with the previous definition.

The ideas of order and disorder are intuitive. For example, a coffee cup is an ordered system. Smashing the cup by dropping it on the floor creates a system that is less ordered, or more disordered, than the original system. The disorder of a system can be described quantitatively using the concept of **microscopic states.** Another term for a microscopic state is a *degree of freedom.*

Suppose we toss $n$ coins in the air, and half of them land heads up and half of them land tails up. The statement "half the coins are heads and half the coins are tails" is a description of the *macroscopic* state of the system of $n$ coins. Each coin can have one of two *microscopic* states: heads or tails. Stating that half the coins are heads and half the coins are tails does not specify anything about the microscopic state of each coin, because there are many different possible ways to arrange the microscopic states of the coins without changing the macroscopic state. However, if all the coins are heads or all the coins are tails, the microscopic state of each coin is known. The macroscopic state consisting of half heads and half tails is a *disordered* system because very little is known about the microscopic state of each coin. The macroscopic state with all heads or the macroscopic state with all tails is an *ordered* system because the microscopic state of each coin is known.

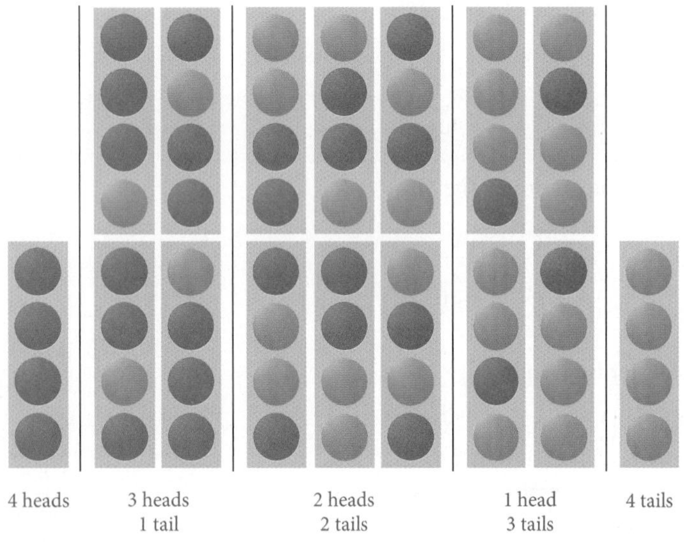


**FIGURE 20.19** The sixteen possible microscopic states for four tossed coins, leading to five possible macroscopic states.

To quantify this concept, imagine tossing four coins in the air. There is only one way to get four heads with such a toss, four ways to get three heads and one tail, six ways to get two heads and two tails, four ways to get one head and three tails, and only one way to get four tails. Thus, there are five possible macroscopic states and sixteen possible microscopic states (see Figure 20.19, where heads are represented as red circles, and tails as blue circles).

Now suppose we toss fifty coins in the air instead of four coins. There are $2^{50} = 1.13 \cdot 10^{15}$ possible microstates of this system of fifty tossed coins. The most probable macroscopic state consists of half heads and half tails. There are $1.26 \cdot 10^{14}$ possible microstates with half heads and half tails. The probability that half the coins will be heads and half will be tails is 11.2%, while the probability of having all fifty coins land heads up is 1 in $1.13 \cdot 10^{15}$.

Let's apply these concepts to a real system of gas molecules: a mole of gas, or Avogadro's number of molecules, at pressure $p$, volume $V$, and temperature $T$. These three quantities describe the macroscopic state of the gas. The microscopic description of the system needs to specify the momentum and position of each molecule of the gas. Each molecule has three components of its momentum and three components of its position. Thus, at any given time, the gas can be in a large number of microscopic states, depending on the positions and velocities of each of its $6.02 \cdot 10^{23}$ molecules. If the gas undergoes free expansion, the number of possible microscopic states increases and the system becomes more disordered. Because the entropy of a gas undergoing free expansion increases, the increase in disorder is related to the increase of entropy. This idea can be generalized as follows:

The most probable macroscopic state of a system is the state with the largest number of microscopic states, which is also the macroscopic state with the greatest disorder.

Let $w$ be the number of possible microscopic states for a given macroscopic state. It can be shown that the entropy of the macroscopic state is given by

$$S = k_B \ln w, \tag{20.15}$$

where $k_B$ is the Boltzmann constant. This equation was first written down by the Austrian physicist Ludwig Boltzmann and is his most significant accomplishment (it is chiseled into his tombstone). You can see from equation 20.15 that increasing the number of possible microscopic states increases the entropy.

The important aspect of a thermodynamic process is not the absolute entropy, but the change in entropy between an initial state and a final state. Taking equation 20.15 as the definition of entropy, the smallest number of microstates is one and the smallest entropy that can exist is then zero. According to this definition, entropy can never be negative. In practice, determining the number of possible microscopic states is difficult except for special systems. However, the change in the number of possible microscopic states can often be determined, thus allowing the change in entropy of the system to be found.

Consider a system that initially has $w_i$ microstates and then undergoes a thermodynamic process to a macroscopic state with $w_f$ microstates. The change in entropy is

$$\Delta S = S_f - S_i = k_B \ln w_f - k_B \ln w_i = k_B \ln \frac{w_f}{w_i}. \tag{20.16}$$

Thus, the change in entropy between two macroscopic states depends on the ratio of the number of possible microstates.

The definition of the entropy of a system in terms of the number of possible microstates leads to further insight into the Second Law of Thermodynamics, which states that the entropy of a closed system can never decrease. This statement of the Second Law, combined with equation 20.15, means that *a closed system can never undergo a thermodynamic process that lowers the number of possible microstates.* For example, if the process depicted in Figure 20.18 were to occur in reverse—that is, the gas underwent free contraction into a volume half its original size—the number of possible microstates for each molecule would decrease by a factor of 2. The probability of finding one gas molecule in half of the original volume then is $\frac{1}{2}$, and the probability of finding all the gas molecules in half of the original volume is $\left(\frac{1}{2}\right)^N$, where $N$ is the number of molecules. If there are 100 gas molecules in the system, then the probability that all 100 molecules end up in half the original volume is $7.9 \cdot 10^{-31}$. We would have to check the system approximately $1/(7.9 \cdot 10^{-31}) \approx 10^{30}$ times to find the molecules in half the volume just once. Checking once per second, this would take about $10^{14}$ billion years, whereas the age of the universe is only 13.7 billion years. If the system contains Avogadro's number of gas molecules, then the probability that the molecules will all be in half of the volume is even smaller. Thus, although the probability that this process will happen is not zero, it is so small that we can treat it as zero. We can thus conclude that the Second Law of Thermodynamics, even if expressed in terms of probabilities, is never violated in any practical situation.

## EXAMPLE 20.7 Entropy Increase during Free Expansion of a Gas

Let's consider the free expansion of a gas like that shown in Figure 20.18. Initially 0.500 mole of nitrogen gas is confined to a volume of 0.500 m$^3$. When the barrier is removed, the gas expands to fill the new volume of 1.00 m$^3$.

### PROBLEM

What is the change in entropy of the gas?

### SOLUTION

We can use equation 20.14 to calculate the change in entropy of the system, assuming we can treat the system as an isothermal expansion of an ideal gas:

$$\begin{aligned}\Delta S &= nR\ln\left(\frac{V_f}{V_i}\right) = nR\ln\left(\frac{1.00 \text{ m}^3}{0.500 \text{ m}^3}\right) = nR\ln 2 \\ &= (0.500 \text{ mole})[8.31 \text{ J/(mol K)}](\ln 2) \\ &= 2.88 \text{ J/K}.\end{aligned} \tag{i}$$

Another approach is to examine the number of microstates of the system before and after the expansion to calculate the change in entropy. In this system, the number of gas molecules is

$$N = nN_A,$$

*Continued—*

where $N_A$ is Avogadro's number. Before the expansion, there were $w_i$ microstates for the gas molecules in the left half of the container. After the expansion, any of the molecules could be in the left half or the right half of the container. Therefore, the number of microstates after the expansion is

$$w_f = 2^N w_i.$$

Using equation 20.16 and remembering that $nR = Nk_B$, we can express the change in entropy of the system as

$$\Delta S = k_B \ln \frac{w_f}{w_i} = k_B \ln \frac{2^N w_i}{w_i} = Nk_B \ln 2 = nR \ln 2.$$

Thus, we get the same result for the change in entropy of a freely expanding gas by looking at the microscopic properties of the system as by using the macroscopic properties of the system in equation (i).

### 20.6 In-Class Exercise

All reversible thermodynamic processes *always* proceed at

a) constant pressure.

b) constant temperature.

c) constant entropy.

d) constant volume.

e) none of the above.

## Entropy Death

The universe is the ultimate closed system. The vast majority of thermodynamic processes in the universe are irreversible, and thus the entropy of the universe as a whole is continuously increasing and asymptotically approaching its maximum. Thus, if the universe exists long enough, all energy will be evenly distributed throughout its volume. In addition, if the universe keeps expanding forever, then gravity—the only long-range force of importance—will not be able to pull objects together any more. This latter condition is the main difference from the early universe in the first moments after the Big Bang, when matter and energy were also distributed very evenly throughout the universe, as revealed by the analysis of the cosmic microwave background radiation (touched on in Chapter 17 and to be discussed in more depth in Chapter 39). But in the early universe, gravity was very strong, as a result of the concentration of matter in a very small space, and was able to develop minute fluctuations and contract matter into stars and galaxies. Thus, even though matter and energy were evenly distributed in the very early universe, the entropy was not near its maximum, and the entire universe was very far from thermal equilibrium.

In the long-term future of the universe, all stars, which currently represent sources of energy for other objects such as our planet, will eventually be extinct. Then, life will become impossible, because life needs an energy source to be able to lower entropy locally. As the universe asymptotically approaches its state of maximum entropy, every subsystem of the universe will reach thermodynamic equilibrium.

Sometimes this long-term fate of the universe is referred to as *heat death.* However, the temperature of the universe at that time will not be high, as this name may suggest, but very close to absolute zero and very nearly the same everywhere.

This is not something we have to worry about soon, because some estimates place the entropy death of the universe at about $10^{100}$ years into the future (with an uncertainty of many orders of magnitude). Obviously, the long-term future of the universe is very interesting and is currently an area of intense research. New discoveries about dark matter and dark energy may change our picture of the long-term future of the universe. But as it stands now, the universe will not go out with a bang, but with a whimper.

## WHAT WE HAVE LEARNED | EXAM STUDY GUIDE

- In a *reversible* process, a system is always close to being in thermodynamic equilibrium. Making a small change in the state of the system can reverse any change in the thermodynamic variables of the system.
- An *irreversible* process involves heat flow with a finite temperature difference, free expansion of a gas, or conversion of mechanical work to thermal energy.
- A heat engine is a device that turns thermal energy into useful work.
- The efficiency, $\epsilon$, of a heat engine is defined as $\epsilon = W/Q_H$, where $W$ is the useful work extracted from the engine and $Q_H$ is the energy provided to the engine in the form of heat.

- A refrigerator is a heat engine operating in reverse.
- The coefficient of performance, $K$, of a refrigerator is defined as $K = Q_L/W$, where $Q_L$ is the heat extracted from the cooler thermal reservoir and $W$ is the work required to extract that heat.
- An ideal heat engine is one in which all the processes involved are reversible.
- A Carnot engine uses the Carnot cycle, an ideal thermodynamic process consisting of two isothermal processes and two adiabatic processes. The Carnot engine is the most efficient engine that can operate between two thermal reservoirs.
- The efficiency of a Carnot engine is given by $\epsilon = (T_H - T_L)/T_H$, where $T_H$ is the temperature of the warmer thermal reservoir and $T_L$ is the temperature of the cooler thermal reservoir.
- The coefficient of performance of a Carnot refrigerator is $K_{max} = T_L/(T_H - T_L)$, where $T_H$ is the temperature of the warmer thermal reservoir and $T_L$ is the temperature of the cooler thermal reservoir.
- The Otto cycle describes the operation of internal combustion engines. It consists of two adiabatic processes and two constant-volume processes. The efficiency of an engine using the Otto cycle is given by $\epsilon = 1 - r^{1-\gamma}$, where $r = V_1/V_2$ is the compression ratio and $\gamma = C_p/C_V$.
- The Second Law of Thermodynamics can be stated as follows: It is impossible for a system to undergo a process in which it absorbs heat from a thermal reservoir at a given temperature and converts that heat completely to mechanical work without rejecting heat to a thermal reservoir at a lower temperature.
- The Second Law of Thermodynamics can also be stated as follows: It is impossible for any process to transfer thermal energy from a cooler thermal reservoir to a warmer thermal reservoir without any work having been done to accomplish the transfer.
- The change in entropy of a system is defined as $\Delta S = \int_i^f \frac{dQ}{T}$, where $dQ$ is the differential heat added to the system, $T$ is the temperature, and the integration is carried out from an initial thermodynamic state to a final thermodynamic state.
- A third way to state the Second Law of Thermodynamics is as follows: The entropy of a closed system can never decrease.
- The entropy of a macroscopic system can be defined in terms of the number of possible microscopic states, $w$, of the system: $S = k_B \ln w$, where $k_B$ is the Boltzmann constant.

## KEY TERMS

reversible process, p. 650
irreversible process, p. 651
heat engine, p. 652
efficiency, p. 652
refrigerator, p. 652
coefficient of performance, p. 653
heat pump, p. 653
ideal engine, p. 654
Carnot engine, p. 654
Carnot cycle, p. 654
Carnot's Theorem, p. 656
Otto cycle, p. 658
Second Law of Thermodynamics, p. 664
entropy, p. 666
microscopic states, p. 670

## NEW SYMBOLS AND EQUATIONS

$\epsilon = \frac{W}{Q_H}$, efficiency of a heat engine

$K = \frac{Q_L}{W}$, coefficient of performance of a refrigerator

$\Delta S = \int_i^f \frac{dQ}{T}$, entropy change of a system

$S = k_B \ln w$, microscopic definition of entropy

$W$, number of possible microscopic states

## ANSWERS TO SELF-TEST OPPORTUNITIES

**20.1** The efficiency is given by

$$\epsilon = 1 - \frac{T_L}{T_H} = 1 - \frac{325 \text{ K}}{500 \text{ K}} = 0.35 \quad \text{(or 35\%)}.$$

**20.2** The ratio of the two efficiencies is

$$\frac{\epsilon_{15}}{\epsilon_{10}} = \frac{1 - 15^{-0.4}}{1 - 4^{-0.4}} = \frac{66.1\%}{42.6\%} = 1.55, \text{ which is a 55\% improvement.}$$

## PROBLEM-SOLVING PRACTICE

### Problem-Solving Guidelines

**1.** With problems in thermodynamics, you need to pay close attention to the signs of work ($W$) and heat ($Q$). Work done by a system is positive, and work done on a system is negative; heat that is transferred to a system is positive, and heat that is emitted by a system is negative.

**2.** Some problems involving heat engines deal with power and rate of thermal energy transfer. Power is work per unit time ($P = W/t$) and rate of thermal energy transfer is heat per unit time ($Q/t$). You can treat these quantities much like work and heat, but remember they involve a time unit.

**3.** Entropy may seem a bit like energy, but these are very different concepts. The First Law of Thermodynamics is a conservation law—energy is conserved. The Second Law of Thermodynamics is not a conservation law—entropy is not conserved; it always either stays the same or increases and never decreases in a closed system. You may need to identify initial and final states to calculate an entropy change, but remember that entropy can change.

**4.** Remember that the entropy of a closed system always stays the same or increases (never decreases), but be sure a problem situation is really about a closed system. Entropy of a system that is not closed can decrease if offset by a larger entropy increase of the surroundings; don't assume that all entropy changes must be positive.

**5.** Do not apply the formula $\Delta S = \int_i^f \frac{dQ}{T}$ for entropy change without being sure you are dealing with a reversible process. If you need to calculate entropy change for an irreversible process, you first need to find an equivalent reversible process that connects the same initial and final states.

### SOLVED PROBLEM 20.2 Cost to Operate an Electric Power Plant

An electric power plant operates steam turbines at a temperature of 557 °C and uses cooling towers to keep the cooler thermal reservoir at a temperature of 38.3 °C. The operating cost of the plant for 1 yr is $52.0 million. The plant managers propose using the water from a nearby lake to lower the temperature of the cooler reservoir to 8.90 °C. Assume that the plant operates at maximum possible efficiency.

#### PROBLEM

How much will the operating cost of the plant be reduced in 1 yr as a result of the change in reservoir temperature? Assume that the plant generates the same amount of electricity.

#### SOLUTION

**THINK**

We can calculate the efficiency of the power plant assuming that it operates at the theoretical Carnot efficiency, which depends only on the temperatures of the warmer thermal reservoir and the cooler thermal reservoir. We can calculate the amount of thermal energy put into the warmer reservoir and assume that the cost of operating the plant is proportional to that thermal energy. This thermal energy might come from the burning of fuel (coal, oil, or gas) or perhaps from a nuclear reactor. Lowering the temperature of the cooler reservoir will increase the efficiency of the power plant, and thus lower the required amount of thermal energy and the associated cost. The cost savings are then the operating cost of the original plant minus the operating cost of the improved plant.

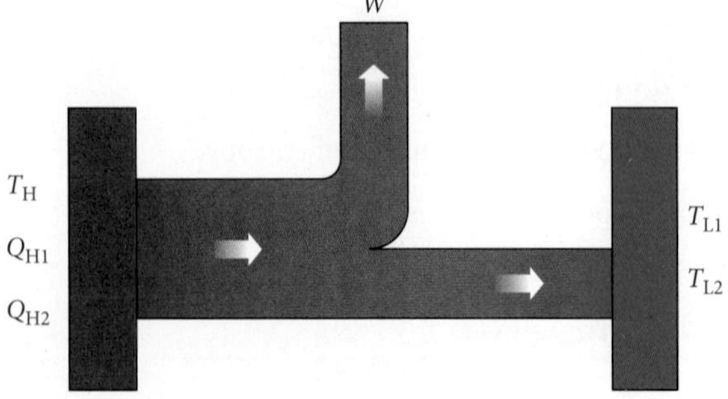


**FIGURE 20.20** The thermodynamic quantities related to the operation of an electric power plant.

**SKETCH**

Figure 20.20 is a sketch relating the thermodynamic quantities involved in the operation of an electric power plant. The temperature of the warmer reservoir is $T_H$, the thermal energy supplied to the original power plant is $Q_{H1}$, the thermal energy supplied to the improved power plant is $Q_{H2}$, the useful work done by the power plant is $W$, the original temperature of the cooler reservoir is $T_{L1}$, and the temperature of the improved cooler reservoir is $T_{L2}$.

**RESEARCH**

The maximum efficiency of the power plant when operating with the cooling towers is given by the theoretical Carnot efficiency:

$$\epsilon_1 = \frac{T_H - T_{L1}}{T_H}. \quad \text{(i)}$$

The efficiency of the power plant under this condition can also be expressed as

$$\epsilon_1 = \frac{W}{Q_{H1}}, \qquad \text{(ii)}$$

where $Q_{H1}$ is the thermal energy supplied when the cooling towers are used. We can write similar expressions for the efficiency of the power plant when the lake is used to lower the temperature of the cooler reservoir:

$$\epsilon_2 = \frac{T_H - T_{L2}}{T_H} \qquad \text{(iii)}$$

and

$$\epsilon_2 = \frac{W}{Q_{H2}}. \qquad \text{(iv)}$$

The cost to operate the power plant is proportional to the thermal energy supplied to the warmer reservoir. We can thus equate the ratio of the original cost, $c_1$, to the lowered cost, $c_2$, with the ratio of the thermal energy originally required, $Q_{H1}$, to the thermal energy subsequently required, $Q_{H2}$:

$$\frac{c_1}{c_2} = \frac{Q_{H1}}{Q_{H2}}.$$

**SIMPLIFY**

We can express the new cost as

$$c_2 = c_1 \frac{Q_{H2}}{Q_{H1}}.$$

We can get an expression for the thermal energy originally required using equations (i) and (ii):

$$Q_{H1} = \frac{W}{\epsilon_1} = \frac{W}{\left(\dfrac{T_H - T_{L1}}{T_H}\right)} = \frac{T_H W}{T_H - T_{L1}}.$$

We can get a similar expression for the thermal energy required after the improvement using equations (iii) and (iv):

$$Q_{H2} = \frac{W}{\epsilon_2} = \frac{W}{\left(\dfrac{T_H - T_{L2}}{T_H}\right)} = \frac{T_H W}{T_H - T_{L2}}.$$

We can now express the new cost as

$$c_2 = c_1 \frac{Q_{H2}}{Q_{H1}} = c_1 \frac{\left(\dfrac{T_H W}{T_H - T_{L2}}\right)}{\left(\dfrac{T_H W}{T_H - T_{L1}}\right)} = c_1 \frac{T_H - T_{L1}}{T_H - T_{L2}}.$$

Note that the work $W$ done by the power plant cancels out because the plant generates the same amount of electricity in both cases. The cost savings are then

$$c_1 - c_2 = c_1 - c_1 \frac{T_H - T_{L1}}{T_H - T_{L2}} = c_1 \left(1 - \frac{T_H - T_{L1}}{T_H - T_{L2}}\right) = c_1 \frac{T_{L1} - T_{L2}}{T_H - T_{L2}}.$$

**CALCULATE**

Putting in the numerical values, we get

$$c_1 - c_2 = c_1 \frac{T_{L1} - T_{L2}}{T_H - T_{L2}} = (\$52.0 \text{ million}) \frac{311.45 \text{ K} - 282.05 \text{ K}}{830.15 \text{ K} - 282.05 \text{ K}} = \$2.78927 \text{ million}.$$

*Continued—*

### ROUND

We report our result to three significant figures:

$$\text{Cost savings} = \$2.79 \text{ million.}$$

### DOUBLE-CHECK

To double-check our result, we calculate the efficiency of the power plant using cooling towers:

$$\epsilon_1 = \frac{T_H - T_{L1}}{T_H} = \frac{830.15 \text{ K} - 311.45 \text{ K}}{830.15 \text{ K}} = 62.5\%.$$

The efficiency of the power plant using the lake to lower the temperature of the cooler reservoir is

$$\epsilon_2 = \frac{T_H - T_{L2}}{T_H} = \frac{830.15 \text{ K} - 282.05 \text{ K}}{830.15 \text{ K}} = 66.0\%.$$

These efficiencies seem reasonable. To check further, we verify that the ratio of the two efficiencies is equal to the inverse of the ratio of the two costs, because a higher efficiency means a lower cost. The ratio of the two efficiencies is

$$\frac{\epsilon_1}{\epsilon_2} = \frac{62.5\%}{66.0\%} = 0.947.$$

The inverse ratio of the two costs is

$$\frac{c_2}{c_1} = \frac{\$52 \text{ million} - \$2.79 \text{ million}}{\$52 \text{ million}} = 0.946.$$

These ratios agree to within rounding error. Thus, our result seems reasonable.

## SOLVED PROBLEM 20.3 Freezing Water in a Refrigerator

Suppose we have 250 g of water at 0.00 °C. We want to freeze this water by putting it in a refrigerator operating in a room with a temperature of 22.0 °C. The temperature inside the refrigerator is maintained at –5.00 °C.

### PROBLEM

What is the minimum amount of electrical energy that must be supplied to the refrigerator to freeze the water?

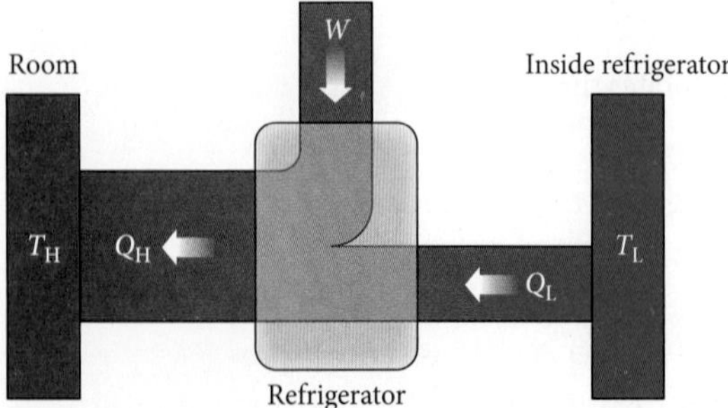


**FIGURE 20.21** A heat flow diagram for a refrigerator that takes heat from inside the refrigerator and exhausts it into the room using a source of electrical power.

### SOLUTION

### THINK

The amount of heat that must be removed depends on the latent heat of fusion and the given mass of water. The most efficient refrigerator possible is a Carnot refrigerator, so we will use the theoretical maximum coefficient of performance of such a refrigerator. Knowing the amount of heat to be removed from the low-temperature reservoir and the coefficient of performance, we can calculate the minimum energy that must be supplied.

### SKETCH

A heat flow diagram for the refrigerator is shown in Figure 20.21.

**RESEARCH**

The most efficient refrigerator possible is a Carnot refrigerator. The maximum coefficient of performance of a Carnot refrigerator is given by equations 20.2 and 20.7:

$$K_{max} = \frac{Q_L}{W} = \frac{T_L}{T_H - T_L}, \quad \text{(i)}$$

where $Q_L$ is the heat removed from inside the refrigerator, $W$ is the work (in terms of electrical energy) that must be supplied, $T_L$ is the temperature inside the refrigerator, and $T_H$ is the temperature of the room. The amount of heat that must be removed to freeze a mass $m$ of water is given by (see Chapter 18)

$$Q_L = mL_{fusion}, \quad \text{(ii)}$$

where $L_{fusion}$ = 334 kJ/kg is the latent heat of fusion of water (which can be found in Table 18.2).

**SIMPLIFY**

We can solve equation (i) for the energy that must be supplied to the refrigerator:

$$W = Q_L \frac{T_H - T_L}{T_L}.$$

Substituting the expression for the heat removed from equation (ii), we get

$$W = \left(mL_{fusion}\right)\frac{T_H - T_L}{T_L}.$$

**CALCULATE**

Putting in the numerical values gives

$$W = \frac{mL_{fusion}}{K} = \left(0.250 \text{ kg}\right)\left(334 \text{ kJ/kg}\right)\frac{295.15 \text{ K} - 268.15 \text{ K}}{268.15 \text{ K}} = 8.41231 \text{ kJ}.$$

**ROUND**

We report our result to three significant figures:

$$W = 8.41 \text{ kJ}.$$

**DOUBLE-CHECK**

To double-check our result, let's calculate the heat removed from the water:

$$Q_L = mL_{fusion} = \left(0.250 \text{ kg}\right)\left(334 \text{ kJ/kg}\right) = 83.5 \text{ kJ}.$$

Using our result for the energy required to freeze the water, we can calculate the coefficient of performance of the refrigerator:

$$K = \frac{Q_L}{W} = \frac{83.5 \text{ kJ}}{8.41 \text{ kJ}} = 9.93.$$

We can compare this result to the maximum coefficient of performance of a Carnot refrigerator:

$$K_{max} = \frac{T_L}{T_H - T_L} = \frac{268.15 \text{ K}}{295.15 \text{ K} - 268.15 \text{ K}} = 9.93.$$

Thus, our result seems reasonable.

## MULTIPLE-CHOICE QUESTIONS

**20.1** Which of the following processes always results in an increase in the energy of a system?

a) The system loses heat and does work on the surroundings.

b) The system gains heat and does work on the surroundings.

c) The system loses heat and has work done on it by the surroundings.

d) The system gains heat and has work done on it by the surroundings.

e) None of the above.

**20.2** What is the magnitude of the change in entropy when 6.00 g of steam at 100 °C is condensed to water at 100 °C?

a) 46.6 J/K
b) 52.4 J/K
c) 36.3 J/K
d) 34.2 J/K

**20.3** The change in entropy of a system can be calculated because

a) it depends only on the initial and final states.
b) any process is reversible.
c) entropy always increases.
d) none of the above.

**20.4** An ideal gas undergoes an isothermal expansion. What will happen to its entropy?

a) It will increase.
b) It will decrease.
c) It's impossible to determine.
d) It will remain unchanged.

**20.5** Which of the following processes (all constant-temperature expansions) produces the most work?

a) An ideal gas consisting of 1 mole of argon at 20 °C expands from 1 L to 2 L.

b) An ideal gas consisting of 1 mole of argon at 20 °C expands from 2 L to 4 L.

c) An ideal gas consisting of 2 moles of argon at 10 °C expands from 2 L to 4 L.

d) An ideal gas consisting of 1 mole of argon at 40 °C expands from 1 L to 2 L.

e) An ideal gas consisting of 1 mole of argon at 40 °C expands from 2 L to 4 L.

**20.6** A heat engine operates with an efficiency of 0.5. What can the temperatures of the high-temperature and low-temperature reservoirs be?

a) $T_H = 600$ K and $T_L = 100$ K
b) $T_H = 600$ K and $T_L = 200$ K
c) $T_H = 500$ K and $T_L = 200$ K
d) $T_H = 500$ K and $T_L = 300$ K
e) $T_H = 600$ K and $T_L = 300$ K

**20.7** The number of macrostates that can result from rolling a set of $N$ six-sided dice is the number of different totals that can be obtained by adding the pips on the $N$ faces that end up on top. The number of macrostates is

a) $6^N$.
b) $6N$.
c) $6N - 1$.
d) $5N + 1$.

**20.8** What capacity must a heat pump with a coefficient of performance of 3 have to heat a home that loses heat energy at a rate of 12 kW on the coldest day of the year?

a) 3 kW
b) 4 kW
c) 10 kW
d) 30 kW
e) 40 kW

**20.9** Which of the following statements about the Carnot cycle is(are) incorrect?

a) The maximum efficiency of a Carnot engine is 100% since the Carnot cycle is an ideal process.

b) The Carnot cycle consists of two isothermal processes and two adiabatic processes.

c) The Carnot cycle consists of two isothermal processes and two isentropic processes (constant entropy).

d) The efficiency of the Carnot cycle depends solely on the temperatures of the two thermal reservoirs.

**20.10** Can a heat engine with the parameter specified in the figure operate?

a) yes
b) no
c) would need to know the specific cycle used by the engine to answer
d) yes, but only with a monatomic gas
e) yes, but only with a diatomic gas

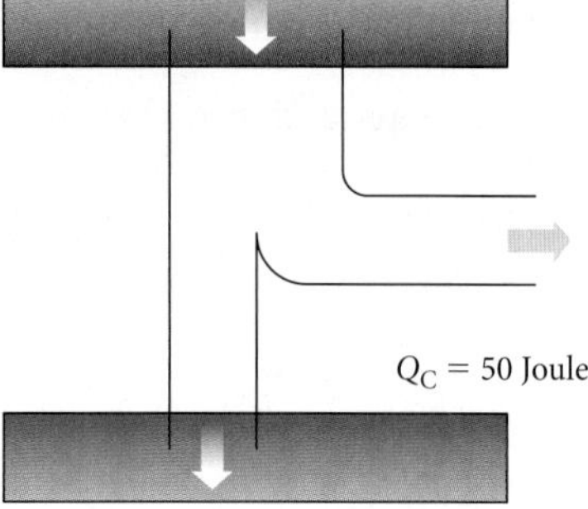


# QUESTIONS

**20.11** One of your friends begins to talk about how unfortunate the Second Law of Thermodynamics is, how sad it is that entropy must always increase, leading to the irreversible degradation of useful energy into heat and the decay of all things. Is there any counterargument you could give that would suggest that the Second Law is in fact a blessing?

**20.12** While looking at a very small system, a scientist observes that the entropy of the system spontaneously decreases. If true, is this a Nobel-winning discovery or is it not that significant?

**20.13** Why might a heat pump have an advantage over a space heater that converts electrical energy directly into thermal energy?

**20.14** Imagine dividing a box into two equal parts, part A on the left and part B on the right. Four identical gas atoms, numbered 1 through 4, are placed in the box. What are most probable and second most probable distributions (for example, 3 atoms in A, 1 atom in B) of gas atoms in the box? Calculate the entropy, $S$, for these two distributions. Note that the configuration with 3 atoms in A and 1 atom in B and that with 1 atom in A and three atoms in B count as different configurations.

**20.15** A key feature of thermodynamics is the fact that the internal energy, $E_{\text{int}}$ of a system and its entropy, $S$, are *state variables;* that is, they depend only on the thermodynamic state of the system and not on the processes by which it reached that state (unlike, for example, the heat content, $Q$). This means that the differentials $dE_{\text{int}} = T\,dS - p\,dV$ and $dS = T^{-1}dE_{\text{int}} + pT^{-1}dV$, where $T$ is temperature (in kelvins), $p$ is pressure, and $V$ is volume, are exact differentials as defined in calculus. What relationships follow from this fact?

**20.16** Other state variables useful for characterizing different classes of processes can be defined from $E_{int}$, $S$, $P$, and $V$. These include the enthalpy, $H \equiv E_{int} + pV$, the Helmholtz free energy, $A \equiv E_{int} - TS$, and the Gibbs free energy, $G \equiv E_{int} + pV - TS$.

a) Write the differential equations for $dH$, $dA$, and $dG$.

b) All of these are also exact differentials. What relationships follow from this fact?

Use the First Law to simplify.

**20.17** Prove that Boltzmann's microscopic definition of entropy, $S = k_B \ln w$, implies that entropy is an *additive* variable: Given two systems, A and B, in specified thermodynamic states, with entropies $S_A$ and $S_B$, respectively, show that the corresponding entropy of the combined system is $S_A + S_B$.

**20.18** Explain how it is possible for a heat pump like that in Example 20.2 to operate with a power of only 6.28 kW and heat a house that is losing thermal energy at a rate of 21.98 kW.

**20.19** The temperature at the cloud tops of Saturn is approximately 50. K. The atmosphere of Saturn produces tremendous winds; wind speeds of 600. km/h have been inferred from spacecraft measurements. Can the wind chill factor on Saturn produce a temperature at (or below) absolute zero? How, or why not?

**20.20** Is it a violation of the Second Law of Thermodynamics to capture all the exhaust heat from a steam engine and funnel it back into the system to do work? Why or why not?

**20.21** You are given a beaker of water. What can you do to increase its entropy ? What can you do to decrease its entropy?

## PROBLEMS

A blue problem number indicates a worked-out solution is available in the Student Solutions Manual. One • and two •• indicate increasing level of problem difficulty.

### Section 20.2

**20.22** With each cycle, a 2500-W engine extracts 2100 J from a thermal reservoir at 90 °C and expels 1500 J into a thermal reservoir at 20 °C. What is the work done for each cycle? What is the engine's efficiency? How much time does each cycle take?

•20.23 A refrigerator with a coefficient of performance of 3.8 is used to cool 2 L of mineral water from room temperature (25 °C) to 4 °C. If the refrigerator uses 480 W, how long will it take the water to reach 4 °C? Recall that the heat capacity of water is 4.19 kJ/(kg K), and the density of water is 1 g/cm$^3$. Assume the other contents of the refrigerator are already at 4 °C.

•**20.24** The burning of fuel transfers $4 \cdot 10^5$ W of power into the engine of a 2000-kg vehicle. If the engine's efficiency is 25%, determine the maximum speed the vehicle can achieve 5 s after starting from rest.

•**20.25** A heat engine consists of a heat source that causes a monatomic gas to expand, pushing against a piston, thereby doing work. The gas begins at a pressure of 300 kPa, a volume of 150 cm$^3$, and room temperature, 20 °C. On reaching a volume of 450 cm$^3$, the piston is locked in place, and the heat source is removed. At this point, the gas cools back to room temperature. Finally, the piston is unlocked and used to isothermally compress the gas back to its initial state.

a) Sketch the cycle on a $pV$-diagram.

b) Determine the work done on the gas and the heat flow out of the gas in each part of the cycle.

c) Using the results of part (b), determine the efficiency of the engine.

•**20.26** A heat engine cycle often used in refrigeration, is the *Brayton cycle,* which involves an adiabatic compression, followed by an isobaric expansion, an adiabatic expansion, and finally an isobaric compression. The system begins at temperature $T_1$ and transitions to temperatures $T_2$, $T_3$, and $T_4$ after respective parts of the cycle.

a) Sketch this cycle on a $pV$-diagram.

b) Show that the efficiency of the overall cycle is given by $\epsilon = 1 - (T_4 - T_1)/(T_3 - T_2)$.

•20.27 Suppose a Brayton engine (see Problem 20.26) is run as a refrigerator. In this case, the cycle begins at temperature $T_1$, and the gas is isobarically expanded until it reaches temperature $T_4$. Then the gas is adiabatically compressed, until its temperature is $T_3$. It is then isobarically compressed, and the temperature changes to $T_2$. Finally, it is adiabatically expanded until it returns to temperature $T_1$.

a) Sketch this cycle on a $pV$-diagram.

b) Show that the coefficient of performance of the engine is given by $K = (T_4 - T_1)/(T_3 - T_2 - T_4 + T_1)$.

### Section 20.3

**20.28** It is desired to build a heat pump that has an output temperature of 23 °C. Calculate the maximum coefficient of performance for the pump when the input source is (a) outdoor air on a cold winter day at –10.0 °C and (b) ground water at 9.0 °C.

**20.29** Consider a Carnot engine that works between thermal reservoirs with temperatures of 1000.0 K and 300.0 K. The average power of the engine is 1.00 kJ per cycle.

a) What is the efficiency of this engine?

b) How much energy is extracted from the warmer reservoir per cycle?

c) How much energy is delivered to the cooler reservoir?

**20.30** A Carnot refrigerator is operating between thermal reservoirs with temperatures of 27 °C and 0 °C.

a) How much work will need to be input to extract 10 J of heat from the colder reservoir?

b) How much work will be needed if the colder reservoir is at –20 °C?

20.31 It has been suggested that the vast amount of thermal energy in the oceans could be put to use. The process would rely on the temperature difference between the top layer of the ocean and the bottom; the temperature of the water at the bottom is fairly constant, but the water temperature at the surface changes depending on the time of day, the season, and the weather. Suppose the difference between top and bottom water temperatures at the location of the proposed thermal-energy plant was 3 °C. Assuming that the plant operated with the highest possible efficiency, is there a limit to how much of the ocean's thermal energy could be extracted? If so, what is this limit?

•20.32 A Carnot engine takes an amount of heat $Q_H = 100$ J from a high-temperature reservoir at temperature $T_H = 1000$ °C, and exhausts the remaining heat into a low-temperature reservoir at $T_L = 10$ °C. Find the amount of work that is obtained from this process.

•20.33 A Carnot engine operates between a warmer reservoir at a temperature $T_1$ and a cooler reservoir at a temperature $T_2$. It is found that increasing the temperature of the warmer reservoir by a factor of 2 while keeping the same temperature for the cooler reservoir increases the efficiency of the Carnot engine by a factor of 2 as well. Find the efficiency of the engine and the ratio of the temperatures of the two reservoirs in their original form.

•20.34 A certain refrigerator is rated as being 32% as efficient as a Carnot refrigerator. To remove 100 J of heat from the interior at 0 °C and eject it to the outside at 22 °C, how much work must the refrigerator motor do?

## Section 20.4

20.35 A refrigerator has a coefficient of performance of 5.0. If the refrigerator absorbs 40.0 cal of heat from the low-temperature reservoir in each cycle, what is the amount of heat expelled into the high-temperature reservoir?

20.36 A heat pump has a coefficient of performance of 5.0. If the heat pump absorbs 40.0 cal of heat from the cold outdoors in each cycle, what is the amount of heat expelled to the warm indoors?

20.37 An Otto engine has a maximum efficiency of 20%; find the compression ratio. Assume that the gas is diatomic.

•20.38 An outboard motor for a boat is cooled by lake water at 15 °C and has a compression ratio of 10. Assume that the air is a diatomic gas.

a) Calculate the efficiency of the engine's Otto cycle.

b) Using your answer to part (a) and the fact that the efficiency of the Carnot cycle is greater than that of the Otto cycle, estimate the maximum temperature of the engine.

•20.39 A heat engine uses 100 mg of helium gas and follows the cycle shown in the figure.

a) Determine the pressure, volume, and temperature of the gas at points 1, 2, and 3.

b) Determine the engine's efficiency.

c) What would be the maximum efficiency of the engine to be able to operate between the maximum and minimum temperatures?

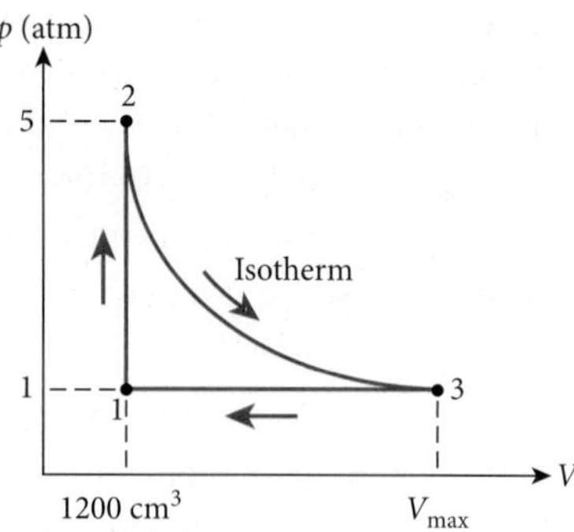


••20.40 Gas turbine engines like those of jet aircraft operate on a thermodynamic cycle known as the *Brayton cycle.* The basic Brayton cycle, as shown in the figure, consists of two adiabatic processes—the compression and the expansion of gas through the turbine—and two isobaric processes. Heat is transferred to the gas during combustion in a constant-pressure process (path from point 2 to point 3) and removed from the gas in a heat exchanger during a constant-pressure process (path from point 4 to point 1). The key parameter for this cycle is the pressure ratio, defined as $r_p = p_{max}/p_{min} \equiv p_2/p_1$, and this is the only parameter needed to calculate the efficiency of a Brayton engine.

a) Determine an expression for the efficiency.

b) Calculate the efficiency of a Brayton engine that uses a diatomic gas and has a pressure ratio of 10.

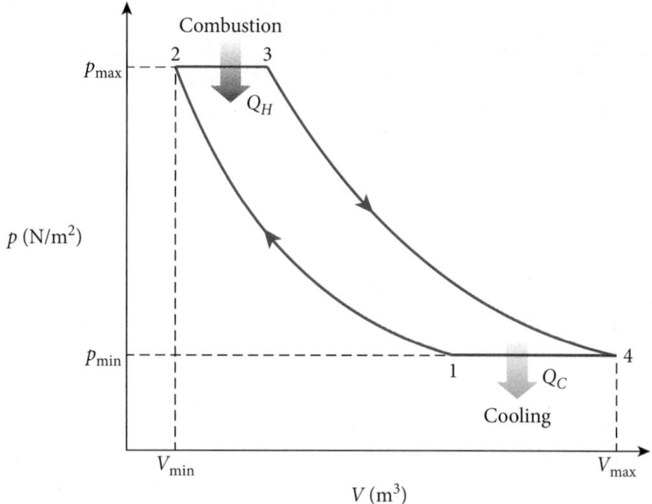


## Sections 20.6 and 20.7

20.41 One end of a metal rod is in contact with a thermal reservoir at 700 K, and the other end is in contact with a thermal reservoir at 100 K. The rod and reservoirs make up a closed system. If 8500 J are conducted from one end of the rod to the other uniformly (no change in temperature along the rod) what is the change in entropy of (a) each reservoir, (b) the rod, and (c) the system?

20.42 The entropy of a macroscopic state is given by $S = k_B \ln w$, where $k_B$ is the Boltzmann constant and $w$ is the number of possible microscopic states. Calculate the change in entropy when $n$ moles of an ideal gas undergo free expansion to fill the entire volume of a box after a barrier between the two halves of the box is removed.

20.43 A proposal is submitted for a novel engine that will operate between 400 K and 300 K.

a) What is the theoretical maximum efficiency of the engine?

b) What is the total entropy change per cycle if the engine operates at maximum efficiency?

**20.44** A 10-kg block initially slides at 10 m/s on a flat rough surface and eventually stops. If the block cools to the ambient temperature, which is 27 °C, what is the entropy change of the system?

**20.45** Suppose an atom of volume $V_A$ is inside a container of volume $V$. The atom can occupy any position within this volume. For this simple model, the number of states available to the atom is given by $V/V_A$. Now suppose the same atom is inside a container of volume $2V$. What will be the change in entropy?

•**20.46** An ideal gas is enclosed in a cylinder with a movable piston at the top. The walls of the cylinder are insulated, so no heat can enter or exit. The gas initially occupies volume $V_1$ and has pressure $p_1$ and temperature $T_1$. The piston is then moved very rapidly to a volume of $V_2 = 3V_1$. The process happens so rapidly that the enclosed gas does not do any work. Find $p_2$, $T_2$, and the change in entropy of the gas.

•20.47 Electrons have a property called *spin* that can be either up or down, analogous to the way a coin can be heads or tails. Consider five electrons. Calculate the entropy, $S_{5up}$, for the state where the spins of all five electrons are up. Calculate the entropy, $S_{3up}$, for the state where three spins are up and two are down.

•**20.48** A 0.545-kg iron bar is taken from a forge at 1000.0 °C and dropped into 10.00 kg of water at 22.0 °C. Assuming that no energy is lost as heat to the surroundings as the water and bar reach their final temperature, determine the total entropy change of the water-bar system.

••**20.49** If the Earth is treated as a spherical blackbody of radius 6371 km, absorbing heat from the Sun at a rate given by the solar constant (1370. W/m$^2$) and immersed in space with an approximate temperature of $T_{sp} = 50.0$ K, it radiates heat back into space at an equilibrium temperature of 278.9 K. (This is a slight refinement of the model in Chapter 18.) Estimate the rate at which the Earth gains entropy in this model.

••**20.50** Suppose a person metabolizes 2000. kcal/day.

a) With a core body temperature of 37.0 °C and an ambient temperature of 20.0 °C, what is the maximum (Carnot) efficiency with which the person can perform work?

b) If the person could work with that efficiency, at what rate, in watts, would they have to shed waste heat to the surroundings?

c) With a skin area of 1.50 m$^2$, a skin temperature of 27.0 °C, and an effective emissivity of $e = 0.600$, at what *net* rate does this person radiate heat to the 20.0 °C surroundings?

d) The rest of the waste heat must be removed by evaporating water, either as perspiration or from the lungs. At body temperature, the latent heat of vaporization of water is 575 cal/g. At what rate, in grams per hour, does this person lose water?

e) Estimate the rate at which the person gains entropy. Assume that all the required evaporation of water takes place in the lungs, at the core body temperature of 37.0 °C.

## Additional Problems

20.51 A nonpolluting source of energy is geothermal energy, from the heat inside the Earth. Estimate the maximum efficiency for a heat engine operating between the center of the Earth and the surface of the Earth (use Table 17.1).

**20.52** In some of the thermodynamic cycles discussed in this chapter, one isotherm intersects one adiabatic curve. For an ideal gas, by what factor is the adiabatic curve steeper than the isotherm?

**20.53** The internal combustion engines in today's cars operate on the Otto cycle. The efficiency of this cycle, $\epsilon_{Otto} = 1 - r^{1-\gamma}$, as derived in this chapter, depends on the compression ratio, $r = V_{max}/V_{min}$. Increasing the efficiency of an Otto engine can be done by increasing the compression ratio. This, in turn, requires fuel with a higher octane rating, to avoid self-ignition of the fuel-air mixture. The following table shows some octane ratings and the maximum compression ratio the engine must have before self-ignition (knocking) sets in.

| Octane Rating of Fuel | Maximum Compression Ratio Without Knocking |
|---|---|
| 91 | 8.5 |
| 93 | 9.0 |
| 95 | 9.8 |
| 97 | 10.5 |

Calculate the maximum theoretical efficiency of an internal combustion engine running on each of these four types of gasoline and the percentage increase in efficiency between using fuel with an octane rating of 91 and using fuel with an octane rating of 97.

**20.54** Consider a room air conditioner using a Carnot cycle at maximum theoretical efficiency and operating between the temperatures of 18 °C (indoors) and 35 °C (outdoors). For each 1.00 J of heat flowing out of the room into the air conditioner:

a) How much heat flows out of the air conditioner to the outdoors?

b) By approximately how much does the entropy of the room decrease?

c) By approximately how much does the entropy of the outdoor air increase?

20.55 Assume that it takes 0.070 J of energy to heat a 1.0-g sample of mercury from 10.0 °C to 10.5 °C and that the heat capacity of mercury is constant, with a negligible change in volume as a function of temperature. Find the change in entropy if this sample is heated from 10 °C to 100 °C.

**20.56** What is the minimum amount of work that must be done to extract 500.0 J of heat from a massive object at a temperature of 27.0 °C while releasing heat to a high-temperature reservoir with a temperature of 100.0 °C?

**20.57** Consider a system consisting of rolling a six-sided die. What happens to the entropy of the system if an additional die is added? Does it double? What happens to the entropy if the number of dice is three?

**20.58** An inventor claims that he has created a water-driven engine with an efficiency of 0.20 that operates between thermal reservoirs at 4 °C and 20 °C. Is this claim valid?

**20.59** If liquid nitrogen is boiled slowly—that is, reversibly—to transform it into nitrogen gas at a pressure $P = 100.0$ kPa, its entropy increases by $\Delta S = 72.1$ J/(mol K). The latent heat of vaporization of nitrogen at its boiling temperature at this pressure is $L_{vap} = 5.568$ kJ/mol. Using these data, calculate the boiling temperature of nitrogen at this pressure.

**20.60** A 1200-kg car traveling at 30.0 m/s crashes into a wall on a hot day (27 °C). What is the total change in entropy?

•**20.61** A water-cooled engine produces 1000 W of power. Water enters the engine block at 15 °C and exits at 30 °C. The rate of water flow is 100 L/h. What is the engine's efficiency?

•**20.62** Find the net change in entropy when 100 g of water at 0 °C is added to 100 g of water at 100 °C.

•**20.63** A coal-burning power plant produces 3000 MW of thermal energy, which is used to boil water and produce supersaturated steam at 300 °C. This high-pressure steam turns a turbine producing 1000 MW of electrical power. At the end of the process, the steam is cooled to 30 °C and recycled.

a) What is the maximum possible efficiency of the plant?

b) What is the actual efficiency of the plant?

c) To cool the steam, river water runs through a condenser at a rate of $40 \cdot 10^6$ gal/h. If the water enters the condenser at 20 °C, what is its exit temperature?

•**20.64** Two equal-volume compartments of a box are joined by a thin wall as shown in the figure. The left compartment is filled with 0.050 mole of helium gas at 500 K, and the right compartment contains 0.025 mole of helium gas at 250 K. The right compartment also has a piston to which a force of 20 N is applied.

a) If the wall between the compartments is removed, what will the final temperature of the system be?

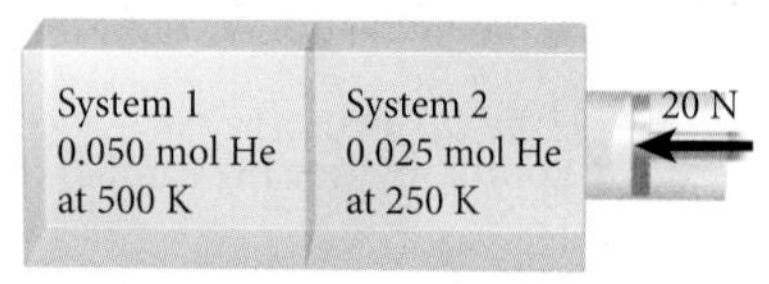


b) How much heat will be transferred from the left compartment to the right compartment?

c) What will be the displacement of the piston due to this transfer or heat?

d) What fraction of the heat will be converted into work?

•**20.65** A volume of 6 L of a monatomic ideal gas, originally at 400 K and a pressure of 3 atm (called state 1), undergoes the following processes, all done reversibly:

$1 \rightarrow 2$ isothermal expansion to $V_2 = 4V_1$

$2 \rightarrow 3$ isobaric compression

$3 \rightarrow 1$ adiabatic compression to its original state

Find the entropy change for each process.

•**20.66** The process shown in the $pV$-diagram is performed on 3 moles of a monatomic gas. Determine the amount of heat input for this process.

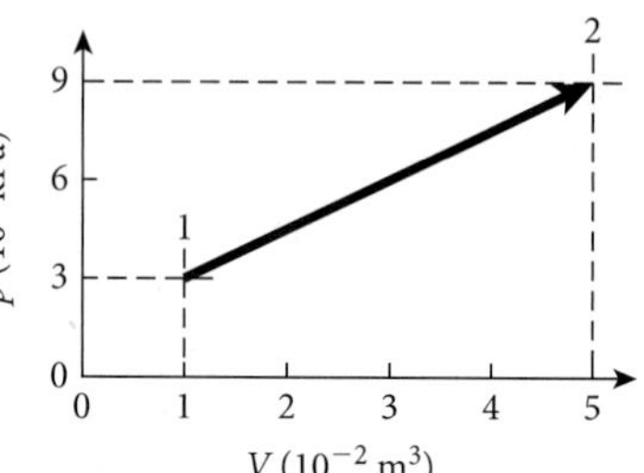


•**20.67** One mole of a monatomic ideal gas at a pressure of 4 atm and a volume of 30 L is isothermically expanded to a pressure of 1 atm and a volume of 120 L. Next, it is compressed at a constant pressure until its volume is 30 L, and then its pressure is increased at the constant volume of 30 L. What is the efficiency of this heat engine cycle?

••**20.68** Two cylinders, A and B, have equal inside diameters and pistons of negligible mass connected by a rigid rod. The pistons can move freely. The rod is a short tube with a valve. The valve is initially closed (see the figure).

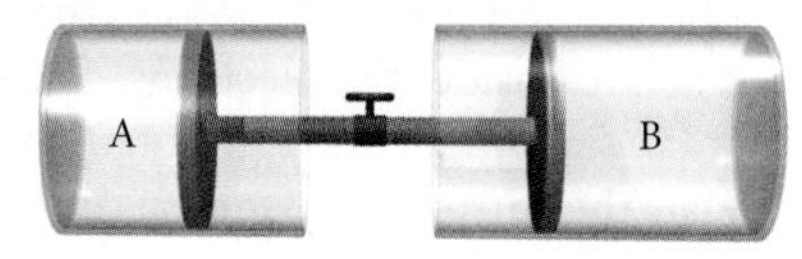


Cylinder A and its piston are thermally insulated, and cylinder B is in thermal contact with a thermostat, which has temperature $\theta = 27$ °C. Initially, the piston of cylinder A is fixed, and inside the cylinder is a mass, $m = 0.32$ kg, of argon at a pressure higher than atmospheric pressure. Inside cylinder B, there is a mass of oxygen at normal atmospheric pressure. When the piston of cylinder A is freed, it moves very slowly, and at equilibrium, the volume of the argon in cylinder A is eight times higher, and the density of the oxygen has increased by a factor of 2. The thermostat receives heat $Q' = 747.9 \cdot 10^4$ J.

a) Based on the kinetic theory of an ideal gas, show that the thermodynamic process taking place in the cylinder A satisfies $TV^{2/3}$ = constant.

b) Calculate $p$, $V$, and $T$ for the argon in the initial and final states.

c) Calculate the final pressure of the mixture of the gases if the valve in the rod is opened.

The molar mass of argon is $\mu = 40$ g/mol.

# Appendix A

## Mathematics Primer



**Notation:**

The letters $a$, $b$, $c$, $x$, and $y$ represent real numbers.

The letters $i$, $j$, $m$, and $n$ represent integer numbers.

The Greek letters $\alpha$, $\beta$, and $\gamma$ represent angles, which are measured in radians.

## 1. Algebra

### 1.1 Basics

Factors:

$$ax + bx + cx = (a + b + c)x \tag{A.1}$$

$$(a + b)^2 = a^2 + 2ab + b^2 \tag{A.2}$$

$$(a - b)^2 = a^2 - 2ab + b^2 \tag{A.3}$$

$$(a + b)(a - b) = a^2 - b^2 \tag{A.4}$$

Quadratic equation:
An equation of the form

$$ax^2 + bx + c = 0 \tag{A.5}$$

for given values of $a$ , $b$ , and $c$ has the two solutions:

$$x = \frac{-b + \sqrt{b^2 - 4ac}}{2a}$$

and (A.6)

$$x = \frac{-b - \sqrt{b^2 - 4ac}}{2a}$$

The solutions of this quadratic equation are called *roots*. The roots are real numbers if $b^2 \geq 4ac$.

## 1.2 Exponents

If $a$ is a number, $a^n$ is the product of $a$ with itself $n$ times:

$$a^n = \underbrace{a \times a \times a \times \cdots \times a}_{n \text{ factors}} \tag{A.7}$$

The number $n$ is called the *exponent*. However, an exponent does not have to be a positive number or an integer. Any real number $x$ can be used as an exponent.

$$a^{-x} = \frac{1}{a^x} \tag{A.8}$$

$$a^0 = 1 \tag{A.9}$$

$$a^1 = a \tag{A.10}$$

Roots:

$$a^{1/2} = \sqrt{a} \tag{A.11}$$

$$a^{1/n} = \sqrt[n]{a} \tag{A.12}$$

Multiplication and division:

$$a^x a^y = a^{x+y} \tag{A.13}$$

$$\frac{a^x}{a^y} = a^{x-y} \tag{A.14}$$

$$\left(a^x\right)^y = a^{xy} \tag{A.15}$$

## 1.3 Logarithms

The logarithm is the inverse function of the exponential function of the previous section:

$$y = a^x \Leftrightarrow x = \log_a y \tag{A.16}$$

The notation $\log_a y$ indicates the logarithm of $y$ with respect to the base $a$. Since the exponential and logarithm are inverse functions of each other, we can also write the identity:

$$x = \log_a(a^x) = a^{\log_a x} \qquad \text{(for any base } a\text{)} \tag{A.17}$$

The two bases most commonly used are base 10, the common logarithm base, and base $e$, the natural logarithm base. The numerical value of $e$ is

$$e = 2.718281828\ldots \tag{A.18}$$

Base 10:

$$y = 10^x \Leftrightarrow x = \log_{10} y \tag{A.19}$$

Base $e$:

$$y = e^x \Leftrightarrow x = \ln y \tag{A.20}$$

This book follows the convention of using ln to indicate the logarithm with respect to the base $e$.

The rules for calculating with logarithms follow from the rules of calculating with exponents:

$$\log(ab) = \log a + \log b \tag{A.21}$$

$$\log\left(\frac{a}{b}\right) = \log a - \log b \tag{A.22}$$

$$\log(a^x) = x \log a \tag{A.23}$$

$$\log 1 = 0 \tag{A.24}$$

Since these rules are valid for any base, the subscript indicating the base is omitted.

### 1.4 Linear Equations

The general form of a linear equation is

$$y = ax + b \tag{A.25}$$

where $a$ and $b$ are constants. The graph of $y$ versus $x$ is a straight line; $a$ is the slope of this line, and $b$ is the $y$-intercept. See Figure A.1.

The slope of the line can be calculated by inserting two different values, $x_1$ and $x_2$, into the linear equation and calculating the resulting values, $y_1$ and $y_2$:

$$a = \frac{y_2 - y_1}{x_2 - x_1} = \frac{\Delta y}{\Delta x} \tag{A.26}$$

If $a = 0$, then the line will be horizontal; if $a > 0$, then the line will rise as $x$ increases as shown in the example of Figure A.1; if $a < 0$, then the line will fall as $x$ increases.

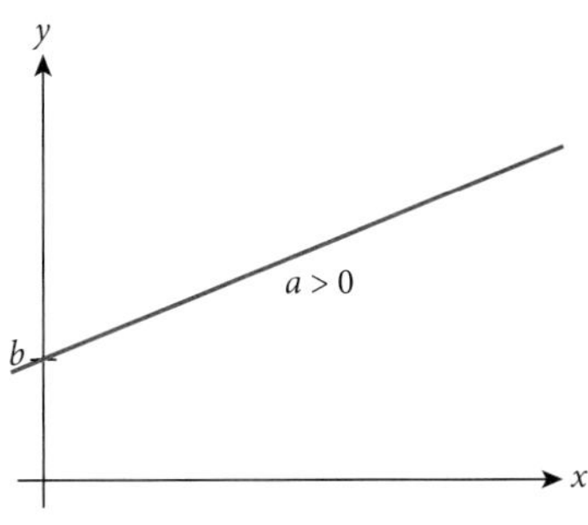


**FIGURE A.1** Graphical representation of a linear equation.

## 2. Geometry

### 2.1 Geometrical Shapes in Two Dimensions

Figure A.2 lists the area, $A$, and perimeter length or circumference, $C$, of common two-dimensional geometrical objects.

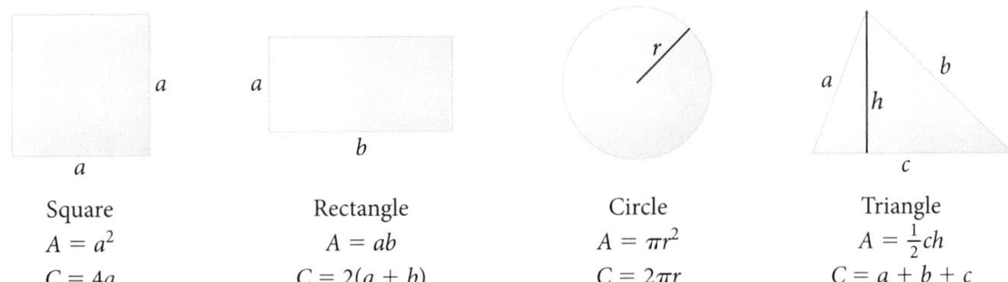


Square $A = a^2$ $C = 4a$; Rectangle $A = ab$ $C = 2(a + b)$; Circle $A = \pi r^2$ $C = 2\pi r$; Triangle $A = \frac{1}{2}ch$ $C = a + b + c$

**FIGURE A.2** Area, $A$, and perimeter length, $C$, for square, rectangle, circle, and triangle.

### 2.2 Geometrical Shapes in Three Dimensions

Figure A.3 lists the volume, $V$, and surface area, $A$, of common three-dimensional geometrical objects.

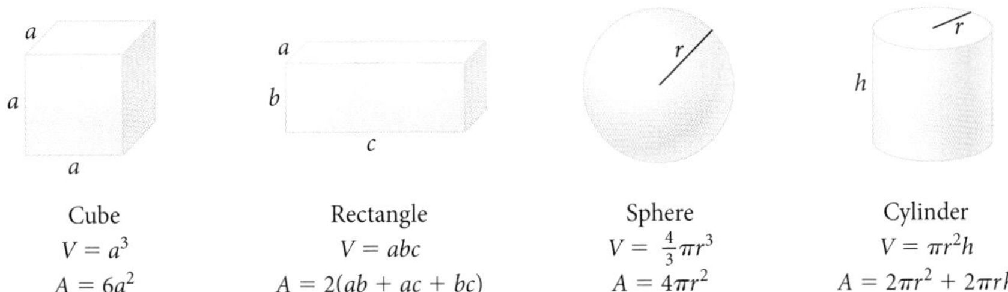


Cube $V = a^3$ $A = 6a^2$; Rectangle $V = abc$ $A = 2(ab + ac + bc)$; Sphere $V = \frac{4}{3}\pi r^3$ $A = 4\pi r^2$; Cylinder $V = \pi r^2 h$ $A = 2\pi r^2 + 2\pi rh$

**FIGURE A.3** Volume, $V$, and surface area, $A$, for cube, rectangular box, sphere, and cylinder.

## 3. Trigonometry

It is important to note that for the following all angles need to be measured in radians.

### 3.1 Right Triangles

A right triangle is a triangle for which one of the three angles is a right angle, that is, an angle of exactly 90° ($\pi/2$ rad) (indicated by the small square in Figure A.4). The hypotenuse is the side opposite the 90° angle. Conventionally, one uses the letter $c$ to mark the hypotenuse.

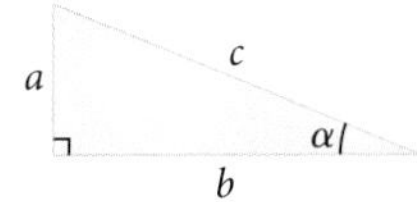


**FIGURE A.4** Definition of the side lengths $a$, $b$, $c$, and angles for the right triangle.

Pythagorean Theorem:

$$a^2 + b^2 = c^2 \tag{A.27}$$

Definition of trigonometric functions (see Figure A.5):

$$\sin\alpha = \frac{a}{c} = \frac{\text{opposite side}}{\text{hypotenuse}} \tag{A.28}$$

$$\cos\alpha = \frac{b}{c} = \frac{\text{adjacent side}}{\text{hypotenuse}} \tag{A.29}$$

$$\tan\alpha = \frac{\sin\alpha}{\cos\alpha} = \frac{a}{b} \tag{A.30}$$

$$\cot\alpha = \frac{\cos\alpha}{\sin\alpha} = \frac{1}{\tan\alpha} = \frac{b}{a} \tag{A.31}$$

$$\csc\alpha = \frac{1}{\sin\alpha} = \frac{c}{a} \tag{A.32}$$

$$\sec\alpha = \frac{1}{\cos\alpha} = \frac{c}{b} \tag{A.33}$$

Inverse trigonometric functions (the notations $\sin^{-1}$, $\cos^{-1}$, etc., are used in this book):

$$\sin^{-1}\frac{a}{c} \equiv \arcsin\frac{a}{c} = \alpha \tag{A.34}$$

$$\cos^{-1}\frac{b}{c} \equiv \arccos\frac{b}{c} = \alpha \tag{A.35}$$

**FIGURE A.5** The trigonometric functions sin, cos, tan, and cot.

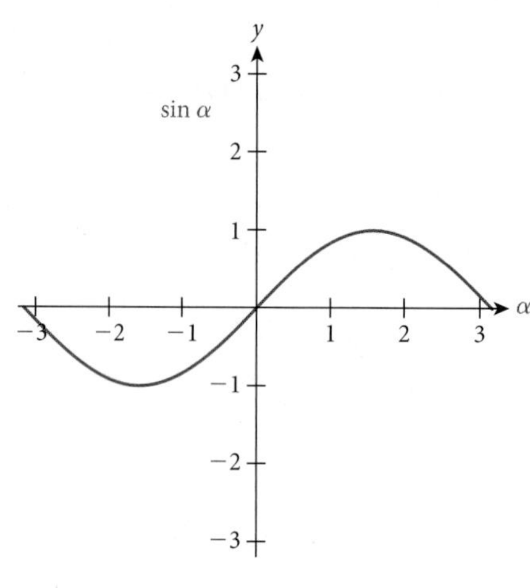


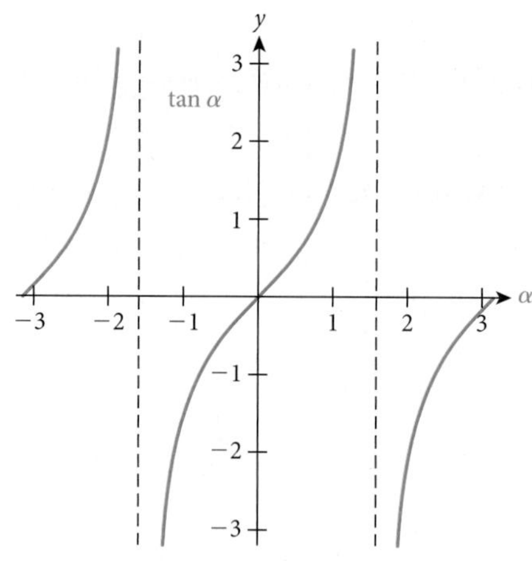


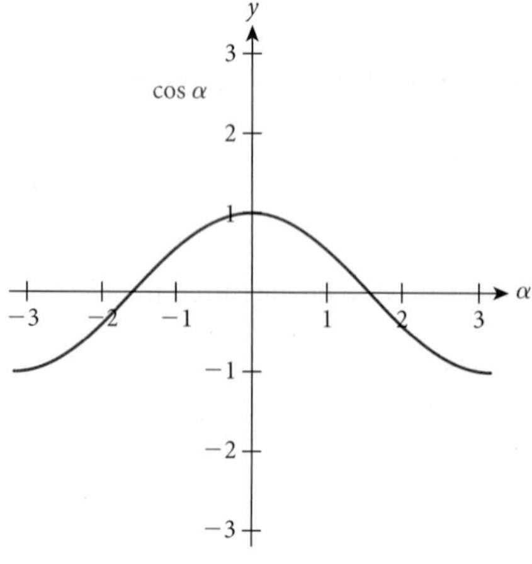


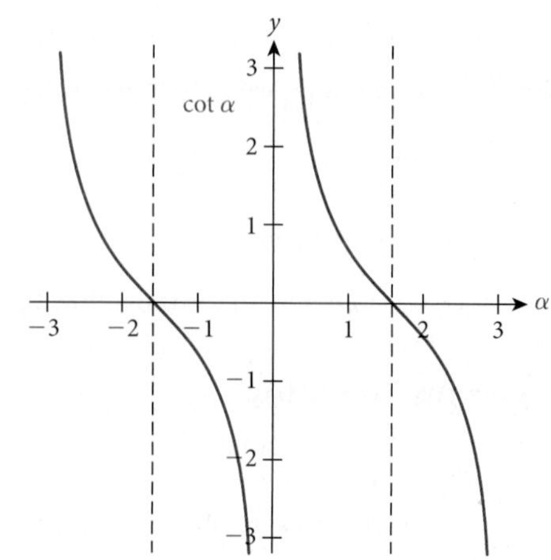

$$\tan^{-1}\frac{a}{b} \equiv \arctan\frac{a}{b} = \alpha \tag{A.36}$$

$$\cot^{-1}\frac{b}{a} \equiv \operatorname{arccot}\frac{b}{a} = \alpha \tag{A.37}$$

$$\csc^{-1}\frac{c}{a} \equiv \operatorname{arccsc}\frac{c}{a} = \alpha \tag{A.38}$$

$$\sec^{-1}\frac{c}{b} \equiv \operatorname{arcsec}\frac{c}{b} = \alpha \tag{A.39}$$

All trigonometric functions are periodic:

$$\sin(\alpha + 2\pi) = \sin\alpha \tag{A.40}$$

$$\cos(\alpha + 2\pi) = \cos\alpha \tag{A.41}$$

$$\tan(\alpha + \pi) = \tan\alpha \tag{A.42}$$

$$\cot(\alpha + \pi) = \cot\alpha \tag{A.43}$$

Other relations between the trigonometric functions:

$$\sin^2\alpha + \cos^2\alpha = 1 \tag{A.44}$$

$$\sin(-\alpha) = -\sin\alpha \tag{A.45}$$

$$\cos(-\alpha) = \cos\alpha \tag{A.46}$$

$$\sin(\alpha \pm \pi/2) = \pm\cos\alpha \tag{A.47}$$

$$\sin(\alpha \pm \pi) = -\sin\alpha \tag{A.48}$$

$$\cos(\alpha \pm \pi/2) = \mp\sin\alpha \tag{A.49}$$

$$\cos(\alpha \pm \pi) = -\cos\alpha \tag{A.50}$$

Addition formulas:

$$\sin(\alpha \pm \beta) = \sin\alpha\cos\beta \pm \cos\alpha\sin\beta \tag{A.51}$$

$$\cos(\alpha \pm \beta) = \cos\alpha\cos\beta \mp \sin\alpha\sin\beta \tag{A.52}$$

Small-angle approximations:

$$\sin\alpha \approx \alpha - \tfrac{1}{6}\alpha^3 + \cdots \quad (\text{for } |\alpha| \ll 1) \tag{A.53}$$

$$\cos\alpha \approx 1 - \tfrac{1}{2}\alpha^2 + \cdots \quad (\text{for } |\alpha| \ll 1) \tag{A.54}$$

For small angles, for $|\alpha| \ll 1$, it is often acceptable to use the small-angle approximations $\cos\alpha = 1$ and $\sin\alpha = \tan\alpha = \alpha$.

## 3.2 General Triangles

The three angles of any triangle add up to $\pi$ rads (see Figure A.6):

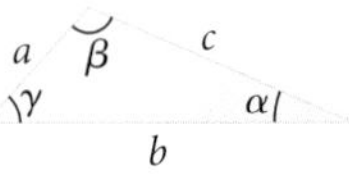

**FIGURE A.6** Definition of the sides and angles for a general triangle.

$$\alpha + \beta + \gamma = \pi \tag{A.55}$$

Law of cosines:

$$c^2 = a^2 + b^2 - 2ab\cos\gamma \tag{A.56}$$

(This is a generalization of the Pythagorean Theorem for the case where the angle $\gamma$ has a value that is not 90°, or $\pi/2$ rads.)

Law of sines:

$$\frac{\sin\alpha}{a} = \frac{\sin\beta}{b} = \frac{\sin\gamma}{c} \tag{A.57}$$

# 4. Calculus

## 4.1 Derivatives

Polynomials:

$$\frac{d}{dx}x^n = nx^{n-1} \tag{A.58}$$

Trigonometric functions:

$$\frac{d}{dx}\sin(ax) = a\cos(ax) \tag{A.59}$$

$$\frac{d}{dx}\cos(ax) = -a\sin(ax) \tag{A.60}$$

$$\frac{d}{dx}\tan(ax) = \frac{a}{\cos^2(ax)} \tag{A.61}$$

$$\frac{d}{dx}\cot(ax) = -\frac{a}{\sin^2(ax)} \tag{A.62}$$

Exponentials and logarithms:

$$\frac{d}{dx}e^{ax} = ae^{ax} \tag{A.63}$$

$$\frac{d}{dx}\ln(ax) = \frac{1}{x} \tag{A.64}$$

$$\frac{d}{dx}a^x = a^x \ln a \tag{A.65}$$

Product rule:

$$\frac{d}{dx}\left(f(x)g(x)\right) = \left(\frac{df(x)}{dx}\right)g(x) + f(x)\left(\frac{dg(x)}{dx}\right) \tag{A.66}$$

Chain rule:

$$\frac{dy}{dx} = \frac{dy}{du}\frac{du}{dx} \tag{A.67}$$

## 4.2 Integrals

All indefinite integrals have an additive integration constant, $c$.
Polynomials:

$$\int x^n dx = \frac{1}{n+1}x^{n+1} + c \qquad (\text{for } n \neq -1) \tag{A.68}$$

$$\int x^{-1} dx = \ln|x| + c \tag{A.69}$$

$$\int \frac{1}{a^2+x^2} dx = \frac{1}{a}\tan^{-1}\frac{x}{a} + c \tag{A.70}$$

$$\int \frac{1}{\sqrt{a^2+x^2}} dx = \ln\left|x + \sqrt{a^2+x^2}\right| + c \tag{A.71}$$

$$\int \frac{1}{\sqrt{a^2-x^2}} dx = \sin^{-1}\frac{x}{|a|} + c \equiv \tan^{-1}\frac{x}{\sqrt{a^2-x^2}} + c \tag{A.72}$$

$$\int \frac{1}{\left(a^2+x^2\right)^{3/2}} dx = \frac{1}{a^2} \frac{x}{\sqrt{a^2+x^2}} + c \tag{A.73}$$

$$\int \frac{x}{\left(a^2+x^2\right)^{3/2}} dx = -\frac{1}{\sqrt{a^2+x^2}} + c \tag{A.74}$$

Trigonometric functions:

$$\int \sin(ax)dx = -\frac{1}{a}\cos(ax) + c \tag{A.75}$$

$$\int \cos(ax)dx = \frac{1}{a}\sin(ax) + c \tag{A.76}$$

Exponentials:

$$\int e^{ax} dx = \frac{1}{a} e^{ax} + c \tag{A.77}$$

## 5. Complex Numbers

We are all familiar with real numbers, which can be sorted along a number line in order of increasing value, from $-\infty$ to $+\infty$. These real numbers are embedded in a much larger set of numbers, called the *complex numbers*. Complex numbers are defined in terms of their real part and their imaginary part. The space of complex numbers is a plane, for which the real numbers form one axis, labeled $\Re(z)$ in Figure A.7. The imaginary part forms the other axis, labeled $\Im(z)$ in Figure A.7. (It is conventional to use the Old German script letters R and I to represent the real and imaginary parts of complex numbers.)

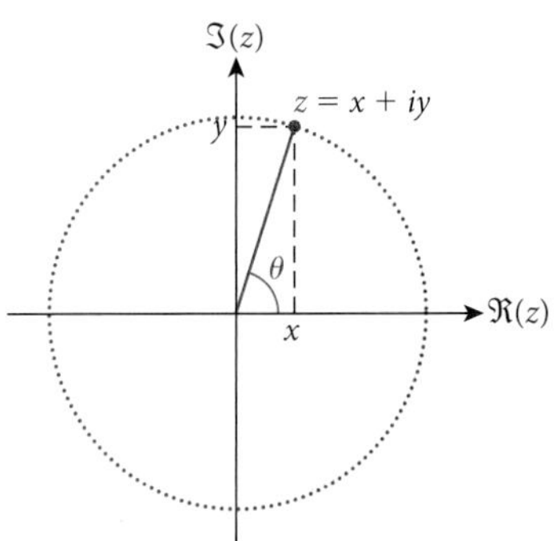

**FIGURE A.7** The complex plane. The horizontal axis is formed by the real part of complex numbers and the vertical axis by the imaginary part.

A complex number $z$ is defined in terms of its real part, $x$, its imaginary part, $y$, and Euler's constant, $i$:

$$z = x + iy \tag{A.78}$$

Euler's constant is defined as:

$$i^2 = -1 \tag{A.79}$$

Both the real part, $x = \Re(z)$, and the imaginary part, $y = \Im(z)$, of a complex number are real numbers. Addition, subtraction, multiplication, and division of complex numbers are defined in analogy to those operations for real numbers, with $i^2 = -1$:

$$(a+ib)+(c+id)=(a+c)+i(b+d) \tag{A.80}$$

$$(a+ib)-(c+id)=(a-c)+i(b-d) \tag{A.81}$$

$$(a+ib)(c+id)=(ac-bd)+i(ad+bc) \tag{A.82}$$

$$\frac{a+ib}{c+id}=\frac{(cd+bd)+i(bc-ad)}{c^2+d^2}. \tag{A.83}$$

For each complex number $z$ there exists a complex conjugate $z^\star$, which has the same real part, but an imaginary part with the opposite sign:

$$z = x+iy \Leftrightarrow z^\star = x-iy \tag{A.84}$$

We can then express the real and imaginary parts of a complex number in terms of the number and its complex conjugate:

$$\Re(z)=\tfrac{1}{2}(z+z^\star) \tag{A.85}$$

$$\Im(z)=\tfrac{1}{2}i(z^\star-z). \tag{A.86}$$

Just like a two-dimensional vector, a complex number, $z = x + iy$, has the magnitude $|z|$ as well as angle $\theta$ with respect to the axis of positive real numbers, as indicated in Figure A.7:

$$|z|^2 = zz^* \tag{A.87}$$

$$\theta = \tan^{-1}\frac{\Im(z)}{\Re(z)} = \tan^{-1}\frac{i(z^* - z)}{(z^* + z)} \tag{A.88}$$

We can thus write the complex number $z = x + iy$ in terms of the magnitude and the "phase angle":

$$z = |z|(\cos\theta + i\sin\theta) \tag{A.89}$$

An interesting and most useful identity is *Euler's formula*:

$$e^{i\theta} = \cos\theta + i\sin\theta \tag{A.90}$$

With the aid of this identity, we can write, for any complex number, $z$,

$$z = |z|e^{i\theta} \tag{A.91}$$

We can thus take a complex number $z$ to any power $n$:

$$z^n = |z|^n e^{in\theta} \tag{A.92}$$

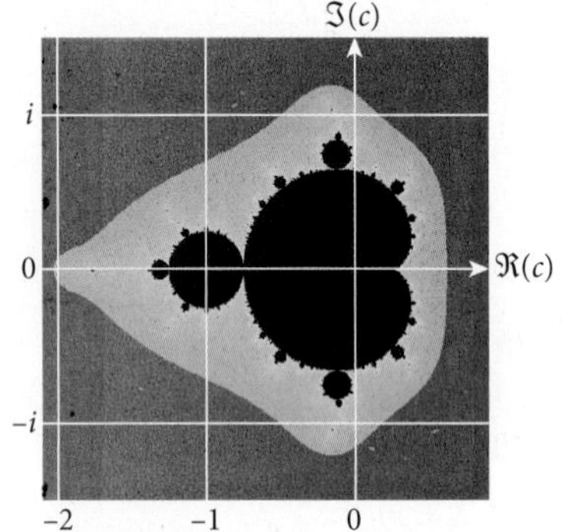


**FIGURE A.8** Mandelbrot set in the complex plane.

## EXAMPLE A.1 Mandelbrot Set

We can put our knowledge of complex numbers and their multiplication to good use by examining the *Mandelbrot set,* defined as the set of all points $c$ in the complex plane for which the series of iterations

$$z_{n+1} = z_n^2 + c, \qquad \text{with } z_0 = c$$

does not escape to infinity; that is, for which $|z_n|$ remains finite for all iterations.

This iteration prescription is seemingly simple. For example, we can see that any number for which $|c| > 2$ cannot be part of the Mandelbrot set. However, if we plot the points of the Mandelbrot set in the complex plane, a strangely beautiful object emerges. In Figure A.8, the black points are part of the Mandelbrot set, and the remaining points are color-coded by how fast $z_n$ escapes to infinity.

# Appendix B

## *Isotope Masses, Binding Energies, and Half–Lives*

Only isotopes with half–lives longer than 1 h are listed.

| *Z* | *N* | Sym | *m* (amu) | *B* (MeV) | Spin | % | $\tau$(s) |
|---|---|---|---|---|---|---|---|
| 1 | 0 | H | 1.007825032 | 0.000 | 1/2+ | 99.985 | stable |
| 1 | 1 | H | 2.014101778 | 1.112 | 1+ | 0.0115 | stable |
| 1 | 2 | H | 3.016049278 | 2.827 | 1/2+ | | 3.89E+08 |
| 2 | 1 | He | 3.016029319 | 2.573 | 1/2+ | 0.0001 | stable |
| 2 | 2 | He | 4.002603254 | 7.074 | 0+ | 100 | stable |
| 3 | 3 | Li | 6.0151223 | 5.332 | 1+ | 7.5 | stable |
| 3 | 4 | Li | 7.0160040 | 5.606 | 3/2– | 92.41 | stable |
| 4 | 3 | Be | 7.0169292 | 5.371 | 3/2– | | 4.59E+06 |
| 4 | 5 | Be | 9.0121821 | 6.463 | 3/2– | 100 | stable |
| 4 | 6 | Be | 10.0135337 | 6.498 | 0+ | | 4.76E+13 |
| 5 | 5 | B | 10.01293699 | 6.475 | 3+ | 19.9 | stable |
| 5 | 6 | B | 11.00930541 | 6.928 | 3/2– | 80.1 | stable |
| 6 | 6 | C | 12 | 7.680 | 0+ | 98.89 | stable |
| 6 | 7 | C | 13.00335484 | 7.470 | 1/2– | 1.11 | stable |
| 6 | 8 | C | 14.0032420 | 7.520 | 0+ | | 1.81E+11 |
| 7 | 7 | N | 14.003074 | 7.476 | 1+ | 99.632 | stable |
| 7 | 8 | N | 15.0001089 | 7.699 | 1/2– | 0.368 | stable |
| 8 | 8 | O | 15.99491463 | 7.976 | 0+ | 99.757 | stable |
| 8 | 9 | O | 16.999131 | 7.751 | 5/2+ | 0.038 | stable |
| 8 | 10 | O | 17.999163 | 7.767 | 0+ | 0.205 | stable |
| 9 | 9 | F | 18.0009377 | 7.632 | 1+ | | 6.59E+03 |
| 9 | 10 | F | 18.99840322 | 7.779 | 1/2+ | 100 | stable |
| 10 | 10 | Ne | 19.99244018 | 8.032 | 0+ | 90.48 | stable |
| 10 | 11 | Ne | 20.99384668 | 7.972 | 3/2+ | 0.27 | stable |
| 10 | 12 | Ne | 21.9913855 | 8.080 | 0+ | 9.25 | stable |
| 11 | 11 | Na | 21.9944368 | 7.916 | 3+ | | 8.21E+07 |
| 11 | 12 | Na | 22.9897697 | 8.111 | 3/2+ | 100 | stable |
| 11 | 13 | Na | 23.9909633 | 8.063 | 4+ | | 5.39E+04 |
| 12 | 12 | Mg | 23.9850419 | 8.261 | 0+ | 78.99 | stable |
| 12 | 13 | Mg | 24.9858370 | 8.223 | 5/2+ | 10 | stable |
| 12 | 14 | Mg | 25.9825930 | 8.334 | 0+ | 11.01 | stable |
| 12 | 16 | Mg | 27.9838767 | 8.272 | 0+ | | 7.53E+04 |
| 13 | 13 | Al | 25.98689169 | 8.150 | 5+ | | 2.33E+13 |
| 13 | 14 | Al | 26.9815384 | 8.332 | 5/2+ | 100 | stable |
| 14 | 14 | Si | 27.97692653 | 8.448 | 0+ | 92.23 | stable |
| 14 | 15 | Si | 28.9764947 | 8.449 | 1/2+ | 4.6832 | stable |
| 14 | 16 | Si | 29.9737702 | 8.521 | 0+ | 3.0872 | stable |
| 14 | 17 | Si | 30.97536323 | 8.458 | 3/2+ | | 9.44E+03 |
| 14 | 18 | Si | 31.97414808 | 8.482 | 0+ | | 5.42E+09 |
| 15 | 16 | P | 30.9737615 | 8.481 | 1/2+ | 100 | stable |
| 15 | 17 | P | 31.9739072 | 8.464 | 1+ | | 1.23E+06 |
| 15 | 18 | P | 32.9717253 | 8.514 | 1/2+ | | 2.19E+06 |
| 16 | 16 | S | 31.9720707 | 8.493 | 0+ | 94.93 | stable |
| 16 | 17 | S | 32.97145876 | 8.498 | 3/2+ | 0.76 | stable |
| 16 | 18 | S | 33.9678668 | 8.583 | 0+ | 4.29 | stable |
| 16 | 19 | S | 34.9690322 | 8.538 | 3/2+ | | 7.56E+06 |
| 16 | 20 | S | 35.96708076 | 8.575 | 0+ | 0.02 | stable |
| 16 | 22 | S | 37.9711634 | 8.449 | 0+ | | 1.02E+04 |
| 17 | 18 | Cl | 34.9688527 | 8.520 | 3/2+ | 75.78 | stable |
| 17 | 19 | Cl | 35.9683069 | 8.522 | 2+ | | 9.49E+12 |
| 17 | 20 | Cl | 36.9659026 | 8.570 | 3/2+ | 24.22 | stable |
| 18 | 18 | Ar | 35.96754511 | 8.520 | 0+ | 0.3365 | stable |
| 18 | 19 | Ar | 36.966776 | 8.527 | 3/2+ | | 3.02E+06 |
| 18 | 20 | Ar | 37.96273239 | 8.614 | 0+ | 0.0632 | stable |
| 18 | 21 | Ar | 38.9643134 | 8.563 | 7/2– | | 8.48E+09 |
| 18 | 22 | Ar | 39.9623831 | 8.595 | 0+ | 99.6 | stable |
| 18 | 23 | Ar | 40.9645008 | 8.534 | 7/2– | | 6.56E+03 |
| 18 | 24 | Ar | 41.963046 | 8.556 | 0+ | | 1.04E+09 |
| 19 | 20 | K | 38.9637069 | 8.557 | 3/2+ | 93.258 | stable |
| 19 | 21 | K | 39.9639987 | 8.538 | 4– | 0.0117 | 4.03E+16 |
| 19 | 22 | K | 40.9618254 | 8.576 | 3/2+ | 6.7302 | stable |
| 19 | 23 | K | 41.962403 | 8.551 | 2– | | 4.45E+04 |
| 19 | 24 | K | 42.960716 | 8.577 | 3/2+ | | 8.03E+04 |
| 20 | 20 | Ca | 39.96259098 | 8.551 | 0+ | 96.941 | stable |
| 20 | 21 | Ca | 40.9622783 | 8.547 | 7/2– | | 3.25E+12 |
| 20 | 22 | Ca | 41.9586183 | 8.617 | 0+ | 0.647 | stable |
| 20 | 23 | Ca | 42.95876663 | 8.601 | 7/2– | 0.135 | stable |
| 20 | 24 | Ca | 43.9554811 | 8.658 | 0+ | 2.086 | stable |
| 20 | 25 | Ca | 44.956186 | 8.631 | 7/2– | | 1.40E+07 |
| 20 | 26 | Ca | 45.9536928 | 8.669 | 0+ | 0.004 | stable |
| 20 | 27 | Ca | 46.9545465 | 8.639 | 7/2– | | 3.92E+05 |
| 20 | 28 | Ca | 47.9525335 | 8.666 | 0+ | 0.187 | 1.89E+26 |

*Continued—*

| *Z* | *N* | Sym | *m* (amu) | *B* (MeV) | Spin | % | $\tau$(s) |
|---|---|---|---|---|---|---|---|
| 21 | 22 | Sc | 42.9611507 | 8.531 | 7/2− | | 1.40E+04 |
| 21 | 23 | Sc | 43.9594030 | 8.557 | 2+ | | 1.41E+04 |
| 21 | 24 | Sc | 44.9559102 | 8.619 | 7/2− | 100 | stable |
| 21 | 25 | Sc | 45.9551703 | 8.622 | 4+ | | 7.24E+06 |
| 21 | 26 | Sc | 46.9524080 | 8.665 | 7/2− | | 2.89E+05 |
| 21 | 27 | Sc | 47.952231 | 8.656 | 6+ | | 1.57E+05 |
| 22 | 22 | Ti | 43.9596902 | 8.533 | 0+ | | 1.89E+09 |
| 22 | 23 | Ti | 44.9581243 | 8.556 | 7/2− | | 1.11E+04 |
| 22 | 24 | Ti | 45.9526295 | 8.656 | 0+ | 8.25 | stable |
| 22 | 25 | Ti | 46.9517638 | 8.661 | 5/2− | 7.44 | stable |
| 22 | 26 | Ti | 47.9479471 | 8.723 | 0+ | 73.72 | stable |
| 22 | 27 | Ti | 48.9478700 | 8.711 | 7/2− | 5.41 | stable |
| 22 | 28 | Ti | 49.9447921 | 8.756 | 0+ | 5.18 | stable |
| 23 | 25 | V | 47.9522545 | 8.623 | 4+ | | 1.38E+06 |
| 23 | 26 | V | 48.9485161 | 8.683 | 7/2− | | 2.85E+07 |
| 23 | 27 | V | 49.9471609 | 8.696 | 6+ | 0.25 | 4.42E+24 |
| 23 | 28 | V | 50.9439617 | 8.742 | 7/2− | 99.75 | stable |
| 24 | 24 | Cr | 47.95403032 | 8.572 | 0+ | | 7.76E+04 |
| 24 | 26 | Cr | 49.94604462 | 8.701 | 0+ | 4.345 | 4.10E+25 |
| 24 | 27 | Cr | 50.9447718 | 8.712 | 7/2− | | 2.39E+06 |
| 24 | 28 | Cr | 51.9405119 | 8.776 | 0+ | 83.789 | stable |
| 24 | 29 | Cr | 52.9406513 | 8.760 | 3/2− | 9.501 | stable |
| 24 | 30 | Cr | 53.9388804 | 8.778 | 0+ | 2.365 | stable |
| 25 | 27 | Mn | 51.9455655 | 8.670 | 6+ | | 4.83E+05 |
| 25 | 28 | Mn | 52.9412947 | 8.734 | 7/2− | | 1.18E+14 |
| 25 | 29 | Mn | 53.9403589 | 8.738 | 3+ | | 2.70E+07 |
| 25 | 30 | Mn | 54.9380471 | 8.765 | 5/2− | 100 | stable |
| 25 | 31 | Mn | 55.9389094 | 8.738 | 3+ | | 9.28E+03 |
| 26 | 26 | Fe | 51.948114 | 8.610 | 0+ | | 2.98E+04 |
| 26 | 28 | Fe | 53.9396127 | 8.736 | 0+ | 5.845 | stable |
| 26 | 29 | Fe | 54.9382980 | 8.747 | 3/2− | | 8.61E+07 |
| 26 | 30 | Fe | 55.93493748 | 8.790 | 0+ | 91.754 | stable |
| 26 | 31 | Fe | 56.93539397 | 8.770 | 1/2− | 2.119 | stable |
| 26 | 32 | Fe | 57.93327556 | 8.792 | 0+ | 0.282 | stable |
| 26 | 33 | Fe | 58.9348880 | 8.755 | 3/2− | | 3.85E+06 |
| 26 | 34 | Fe | 59.934072 | 8.756 | 0+ | | 4.73E+13 |
| 27 | 28 | Co | 54.942003 | 8.670 | 7/2− | | 6.31E+04 |
| 27 | 29 | Co | 55.9398439 | 8.695 | 4+ | | 6.68E+06 |
| 27 | 30 | Co | 56.936296 | 8.742 | 7/2− | | 2.35E+07 |
| 27 | 31 | Co | 57.935757 | 8.739 | 2+ | | 6.12E+06 |
| 27 | 32 | Co | 58.93319505 | 8.768 | 7/2− | 100 | stable |
| 27 | 33 | Co | 59.9338222 | 8.747 | 5+ | | 1.66E+08 |
| 27 | 34 | Co | 60.9324758 | 8.756 | 7/2− | | 5.94E+03 |
| 28 | 28 | Ni | 55.94213202 | 8.643 | 0+ | | 5.25E+05 |
| 28 | 29 | Ni | 56.939800 | 8.671 | 3/2− | | 1.28E+05 |
| 28 | 30 | Ni | 57.9353462 | 8.732 | 0+ | 68.077 | stable |
| 28 | 31 | Ni | 58.9343516 | 8.737 | 3/2− | | 2.40E+12 |
| 28 | 32 | Ni | 59.93078637 | 8.781 | 0+ | 26.223 | stable |
| 28 | 33 | Ni | 60.93105603 | 8.765 | 3/2− | 1.1399 | stable |
| 28 | 34 | Ni | 61.92834512 | 8.795 | 0+ | 3.6345 | stable |
| 28 | 35 | Ni | 62.9296729 | 8.763 | 1/2− | | 3.19E+09 |
| 28 | 36 | Ni | 63.92796596 | 8.777 | 0+ | 0.9256 | stable |
| 28 | 37 | Ni | 64.9300880 | 8.736 | 5/2− | | 9.06E+03 |
| 28 | 38 | Ni | 65.92913933 | 8.739 | 0+ | | 1.97E+05 |
| 29 | 32 | Cu | 60.9334622 | 8.715 | 3/2− | | 1.20E+04 |
| 29 | 34 | Cu | 62.92959747 | 8.752 | 3/2− | 69.17 | stable |
| 29 | 35 | Cu | 63.9297679 | 8.739 | 1+ | | 4.57E+04 |
| 29 | 36 | Cu | 64.9277929 | 8.757 | 3/2− | 30.83 | stable |
| 29 | 38 | Cu | 66.9277503 | 8.737 | 3/2− | | 2.23E+05 |
| 30 | 32 | Zn | 61.93432976 | 8.679 | 0+ | | 3.31E+04 |
| 30 | 34 | Zn | 63.9291466 | 8.736 | 0+ | 48.63 | stable |
| 30 | 35 | Zn | 64.929245 | 8.724 | 5/2− | | 2.11E+07 |
| 30 | 36 | Zn | 65.92603342 | 8.760 | 0+ | 27.9 | stable |
| 30 | 37 | Zn | 66.92712730 | 8.734 | 5/2− | 4.1 | stable |
| 30 | 38 | Zn | 67.92484949 | 8.756 | 0+ | 18.75 | stable |
| 30 | 40 | Zn | 69.9253193 | 8.730 | 0+ | 0.62 | stable |
| 30 | 42 | Zn | 71.926858 | 8.692 | 0+ | | 1.68E+05 |
| 31 | 35 | Ga | 65.93158901 | 8.669 | 0+ | | 3.42E+04 |
| 31 | 36 | Ga | 66.9282049 | 8.708 | 3/2− | | 2.82E+05 |
| 31 | 37 | Ga | 67.92798008 | 8.701 | 1+ | | 4.06E+03 |
| 31 | 38 | Ga | 68.9255736 | 8.725 | 3/2− | 60.108 | stable |
| 31 | 40 | Ga | 70.9247013 | 8.718 | 3/2− | 39.892 | stable |
| 31 | 41 | Ga | 71.9263663 | 8.687 | 3− | | 5.08E+04 |
| 31 | 42 | Ga | 72.92517468 | 8.694 | 3/2− | | 1.75E+04 |
| 32 | 34 | Ge | 65.93384345 | 8.626 | 0+ | | 8.14E+03 |
| 32 | 36 | Ge | 67.92809424 | 8.688 | 0+ | | 2.34E+07 |
| 32 | 37 | Ge | 68.927972 | 8.681 | 5/2− | | 1.41E+05 |
| 32 | 38 | Ge | 69.92424 | 8.722 | 0+ | 20.84 | stable |
| 32 | 39 | Ge | 70.9249540 | 8.703 | 1/2− | | 9.88E+05 |
| 32 | 40 | Ge | 71.92207582 | 8.732 | 0+ | 27.54 | stable |
| 32 | 41 | Ge | 72.92345895 | 8.705 | 9/2+ | 7.73 | stable |
| 32 | 42 | Ge | 73.92117777 | 8.725 | 0+ | 36.28 | stable |
| 32 | 43 | Ge | 74.92285895 | 8.696 | 1/2− | | 4.97E+03 |
| 32 | 44 | Ge | 75.92140256 | 8.705 | 0+ | 7.61 | stable |
| 32 | 45 | Ge | 76.92354859 | 8.671 | 7/2+ | | 4.07E+04 |
| 32 | 46 | Ge | 77.922853 | 8.672 | 0+ | | 5.29E+03 |
| 33 | 38 | As | 70.92711243 | 8.664 | 5/2− | | 2.35E+05 |
| 33 | 39 | As | 71.92675228 | 8.660 | 2− | | 9.33E+04 |
| 33 | 40 | As | 72.92382484 | 8.690 | 3/2− | | 6.94E+06 |
| 33 | 41 | As | 73.92392869 | 8.680 | 2− | | 1.54E+06 |
| 33 | 42 | As | 74.92159648 | 8.701 | 3/2− | 100 | stable |
| 33 | 43 | As | 75.92239402 | 8.683 | 2− | | 9.31E+04 |

| Z | N | Sym | m (amu) | B (MeV) | Spin | % | τ(s) |
|---|---|---|---|---|---|---|---|
| 33 | 44 | As | 76.92064729 | 8.696 | 3/2– | | 1.40E+05 |
| 33 | 45 | As | 77.92182728 | 8.674 | 2– | | 5.44E+03 |
| 34 | 38 | Se | 71.92711235 | 8.645 | 0+ | | 7.26E+05 |
| 34 | 39 | Se | 72.92676535 | 8.641 | 9/2+ | | 2.57E+04 |
| 34 | 40 | Se | 73.92247644 | 8.688 | 0+ | 0.89 | stable |
| 34 | 41 | Se | 74.92252337 | 8.679 | 5/2+ | | 1.03E+07 |
| 34 | 42 | Se | 75.9192141 | 8.711 | 0+ | 9.37 | stable |
| 34 | 43 | Se | 76.91991404 | 8.695 | 1/2– | 7.63 | stable |
| 34 | 44 | Se | 77.91730909 | 8.718 | 0+ | 23.77 | stable |
| 34 | 45 | Se | 78.9184998 | 8.696 | 7/2+ | | 2.05E+13 |
| 34 | 46 | Se | 79.9165213 | 8.711 | 0+ | 49.61 | stable |
| 34 | 48 | Se | 81.9166994 | 8.693 | 0+ | 8.73 | 2.62E+27 |
| 35 | 40 | Br | 74.92577621 | 8.628 | 3/2– | | 5.80E+03 |
| 35 | 41 | Br | 75.924541 | 8.636 | 1– | | 5.83E+04 |
| 35 | 42 | Br | 76.92137908 | 8.667 | 3/2– | | 2.06E+05 |
| 35 | 44 | Br | 78.91833709 | 8.688 | 3/2– | 50.69 | stable |
| 35 | 46 | Br | 80.9162906 | 8.696 | 3/2– | 49.31 | stable |
| 35 | 47 | Br | 81.9168047 | 8.682 | 5– | | 1.27E+05 |
| 35 | 48 | Br | 82.915180 | 8.693 | 3/2– | | 8.64E+03 |
| 36 | 40 | Kr | 75.9259483 | 8.609 | 0+ | | 5.33E+04 |
| 36 | 41 | Kr | 76.92467 | 8.617 | 5/2+ | | 4.46E+03 |
| 36 | 42 | Kr | 77.9203948 | 8.661 | 0+ | 0.35 | 6.31E+28 |
| 36 | 43 | Kr | 78.920083 | 8.657 | 1/2– | | 1.26E+05 |
| 36 | 44 | Kr | 79.9163790 | 8.693 | 0+ | 2.25 | stable |
| 36 | 45 | Kr | 80.9165923 | 8.683 | 7/2+ | | 7.22E+12 |
| 36 | 46 | Kr | 81.9134836 | 8.711 | 0+ | 11.58 | stable |
| 36 | 47 | Kr | 82.9141361 | 8.696 | 9/2+ | 11.49 | stable |
| 36 | 48 | Kr | 83.911507 | 8.717 | 0+ | 57 | stable |
| 36 | 49 | Kr | 84.9125270 | 8.699 | 9/2+ | | 3.40E+08 |
| 36 | 50 | Kr | 85.91061073 | 8.712 | 0+ | 17.3 | stable |
| 36 | 51 | Kr | 86.9133543 | 8.675 | 5/2+ | | 4.57E+03 |
| 36 | 52 | Kr | 87.914447 | 8.657 | 0+ | | 1.02E+04 |
| 37 | 44 | Rb | 80.918996 | 8.645 | 3/2– | | 1.65E+04 |
| 37 | 46 | Rb | 82.915110 | 8.675 | 5/2– | | 7.45E+06 |
| 37 | 47 | Rb | 83.91438482 | 8.676 | 2– | | 2.83E+06 |
| 37 | 48 | Rb | 84.9117893 | 8.697 | 5/2– | 72.17 | stable |
| 37 | 49 | Rb | 85.91116742 | 8.697 | 2– | | 1.61E+06 |
| 37 | 50 | Rb | 86.9091835 | 8.711 | 3/2– | 27.83 | 1.50E+18 |
| 38 | 42 | Sr | 79.92452101 | 8.579 | 0+ | | 6.38E+03 |
| 38 | 44 | Sr | 81.918402 | 8.636 | 0+ | | 2.21E+06 |
| 38 | 45 | Sr | 82.9175567 | 8.638 | 7/2+ | | 1.17E+05 |
| 38 | 46 | Sr | 83.91342528 | 8.677 | 0+ | 0.56 | stable |
| 38 | 47 | Sr | 84.9129328 | 8.676 | 9/2+ | | 5.60E+06 |
| 38 | 48 | Sr | 85.9092602 | 8.708 | 0+ | 9.86 | stable |
| 38 | 49 | Sr | 86.9088793 | 8.705 | 9/2+ | 7 | stable |
| 38 | 50 | Sr | 87.9056143 | 8.733 | 0+ | 82.58 | stable |
| 38 | 51 | Sr | 88.9074529 | 8.706 | 5/2+ | | 4.37E+06 |
| 38 | 52 | Sr | 89.907738 | 8.696 | 0+ | | 9.08E+08 |
| 38 | 53 | Sr | 90.9102031 | 8.664 | 5/2+ | | 3.47E+04 |
| 38 | 54 | Sr | 91.9110299 | 8.649 | 0+ | | 9.76E+03 |
| 39 | 46 | Y | 84.91643304 | 8.628 | (1/2)– | | 9.65E+03 |
| 39 | 47 | Y | 85.914886 | 8.638 | 4– | | 5.31E+04 |
| 39 | 48 | Y | 86.9108778 | 8.675 | 1/2– | | 2.88E+05 |
| 39 | 49 | Y | 87.9095034 | 8.683 | 4– | | 9.21E+06 |
| 39 | 50 | Y | 88.9058483 | 8.714 | 1/2– | 100 | stable |
| 39 | 51 | Y | 89.90715189 | 8.693 | 2– | | 2.31E+05 |
| 39 | 52 | Y | 90.907305 | 8.685 | 1/2– | | 5.06E+06 |
| 39 | 53 | Y | 91.9089468 | 8.662 | 2– | | 1.27E+04 |
| 39 | 54 | Y | 92.909583 | 8.649 | 1/2– | | 3.66E+04 |
| 40 | 46 | Zr | 85.91647359 | 8.612 | 0+ | | 5.94E+04 |
| 40 | 47 | Zr | 86.91481625 | 8.624 | (9/2)+ | | 6.05E+03 |
| 40 | 48 | Zr | 87.9102269 | 8.666 | 0+ | | 7.21E+06 |
| 40 | 49 | Zr | 88.908889 | 8.673 | 9/2+ | | 2.83E+05 |
| 40 | 50 | Zr | 89.9047037 | 8.710 | 0+ | 51.45 | stable |
| 40 | 51 | Zr | 90.90564577 | 8.693 | 5/2+ | 11.22 | stable |
| 40 | 52 | Zr | 91.9050401 | 8.693 | 0+ | 17.15 | stable |
| 40 | 53 | Zr | 92.9064756 | 8.672 | 5/2+ | | 4.83E+13 |
| 40 | 54 | Zr | 93.90631519 | 8.667 | 0+ | 17.38 | stable |
| 40 | 55 | Zr | 94.9080426 | 8.644 | 5/2+ | | 5.53E+06 |
| 40 | 56 | Zr | 95.9082757 | 8.635 | 0+ | 2.8 | 1.23E+27 |
| 40 | 57 | Zr | 96.9109507 | 8.604 | 1/2+ | | 6.08E+04 |
| 41 | 48 | Nb | | | (9/2+) | | 6.84E+03 |
| 41 | 48 | Nb | 88.9134955 | 8.617 | (1/2)– | | 4.25E+03 |
| 41 | 49 | Nb | 89.911265 | 8.633 | 8+ | | 5.26E+04 |
| 41 | 50 | Nb | 90.9069905 | 8.671 | 9/2+ | | 2.14E+10 |
| 41 | 51 | Nb | 91.9071924 | 8.662 | 7+ | | 1.09E+15 |
| 41 | 52 | Nb | 92.90637806 | 8.664 | 9/2+ | 100 | stable |
| 41 | 53 | Nb | 93.9072839 | 8.649 | 6+ | | 6.40E+11 |
| 41 | 54 | Nb | 94.9068352 | 8.647 | 9/2+ | | 3.02E+06 |
| 41 | 55 | Nb | 95.9081001 | 8.629 | 6+ | | 8.41E+04 |
| 41 | 56 | Nb | 96.9080971 | 8.623 | 9/2+ | | 4.32E+03 |
| 42 | 48 | Mo | 89.9139369 | 8.597 | 0+ | | 2.04E+04 |
| 42 | 50 | Mo | 91.9068105 | 8.658 | 0+ | 14.84 | stable |
| 42 | 51 | Mo | 92.90681261 | 8.651 | 5/2+ | | 1.26E+11 |
| 42 | 52 | Mo | 93.9050876 | 8.662 | 0+ | 9.25 | stable |
| 42 | 53 | Mo | 94.9058415 | 8.649 | 5/2+ | 15.92 | stable |
| 42 | 54 | Mo | 95.90467890 | 8.654 | 0+ | 16.68 | stable |
| 42 | 55 | Mo | 96.90602147 | 8.635 | 5/2+ | 9.55 | stable |
| 42 | 56 | Mo | 97.9054078 | 8.635 | 0+ | 24.13 | stable |
| 42 | 57 | Mo | 98.90771187 | 8.608 | 1/2+ | | 2.37E+05 |

*Continued—*

| Z | N | Sym | m (amu) | B (MeV) | Spin | % | τ(s) | Z | N | Sym | m (amu) | B (MeV) | Spin | % | τ(s) |
|---|---|---|---|---|---|---|---|---|---|---|---|---|---|---|---|
| 42 | 58 | Mo | 99.90747734 | 8.605 | 0+ | 9.63 | 3.78E+26 | 48 | 58 | Cd | 105.9064594 | 8.539 | 0+ | 1.25 | stable |
| 43 | 50 | Tc | 92.91024898 | 8.609 | 9/2+ | | 9.90E+03 | 48 | 59 | Cd | 106.9066179 | 8.533 | 5/2+ | | 2.34E+04 |
| 43 | 51 | Tc | 93.9096563 | 8.609 | 7+ | | 1.76E+04 | 48 | 60 | Cd | 107.9041837 | 8.550 | 0+ | 0.89 | stable |
| 43 | 52 | Tc | 94.90765708 | 8.623 | 9/2+ | | 7.20E+04 | 48 | 61 | Cd | 108.904982 | 8.539 | 5/2+ | | 4.00E+07 |
| 43 | 53 | Tc | 95.907871 | 8.615 | 7+ | | 3.70E+05 | 48 | 62 | Cd | 109.9030056 | 8.551 | 0+ | 12.49 | stable |
| 43 | 54 | Tc | 96.90636536 | 8.624 | 9/2+ | | 1.33E+14 | 48 | 63 | Cd | 110.9041781 | 8.537 | 1/2+ | 12.8 | stable |
| 43 | 55 | Tc | 97.90721597 | 8.610 | (6)+ | | 1.32E+14 | 48 | 64 | Cd | 111.9027578 | 8.545 | 0+ | 24.13 | stable |
| 43 | 56 | Tc | 98.90625475 | 8.614 | 9/2+ | | 6.65E+12 | 48 | 65 | Cd | 112.9044017 | 8.527 | 1/2+ | 12.22 | 2.93E+23 |
| 44 | 51 | Ru | 94.91041293 | 8.587 | 5/2+ | | 5.91E+03 | 48 | 66 | Cd | 113.9033585 | 8.532 | 0+ | 28.73 | stable |
| 44 | 52 | Ru | 95.90759784 | 8.609 | 0+ | 5.54 | stable | 48 | 67 | Cd | 114.905431 | 8.511 | 1/2+ | | 1.93E+05 |
| 44 | 53 | Ru | 96.9075547 | 8.604 | 5/2+ | | 2.51E+05 | 48 | 68 | Cd | 115.9047558 | 8.512 | 0+ | 7.49 | 9.15E+26 |
| 44 | 54 | Ru | 97.90528713 | 8.620 | 0+ | 1.87 | stable | 48 | 69 | Cd | 116.9072186 | 8.489 | 1/2+ | | 8.96E+03 |
| 44 | 55 | Ru | 98.9059393 | 8.609 | 5/2+ | 12.76 | stable | 49 | 60 | In | 108.9071505 | 8.513 | 9/2+ | | 1.51E+04 |
| 44 | 56 | Ru | 99.90421948 | 8.619 | 0+ | 12.6 | stable | 49 | 61 | In | 109.9071653 | 8.509 | 7+ | | 1.76E+04 |
| 44 | 57 | Ru | 100.9055821 | 8.601 | 5/2+ | 17.06 | stable | 49 | 62 | In | 110.90511 | 8.522 | 9/2+ | | 2.42E+05 |
| 44 | 58 | Ru | 101.9043493 | 8.607 | 0+ | 31.55 | stable | 49 | 64 | In | 112.904061 | 8.523 | 9/2+ | 4.29 | stable |
| 44 | 59 | Ru | 102.9063238 | 8.584 | 3/2+ | | 3.39E+06 | 49 | 66 | In | 114.9038785 | 8.517 | 9/2+ | 95.71 | 1.39E+22 |
| 44 | 60 | Ru | 103.9054301 | 8.587 | 0+ | 18.62 | stable | 50 | 60 | Sn | 109.9078428 | 8.496 | 0+ | | 1.48E+04 |
| 44 | 61 | Ru | 104.9077503 | 8.562 | 3/2+ | | 1.60E+04 | 50 | 62 | Sn | 111.9048208 | 8.514 | 0+ | 0.97 | stable |
| 44 | 62 | Ru | 105.9073269 | 8.561 | 0+ | | 3.23E+07 | 50 | 63 | Sn | 112.9051734 | 8.507 | 1/2+ | | 9.94E+06 |
| 45 | 54 | Rh | 98.9081321 | 8.580 | 1/2− | | 1.39E+06 | 50 | 64 | Sn | 113.9027818 | 8.523 | 0+ | 0.66 | stable |
| 45 | 55 | Rh | 99.90812155 | 8.575 | 1− | | 7.49E+04 | 50 | 65 | Sn | 114.9033424 | 8.514 | 1/2+ | 0.34 | stable |
| 45 | 56 | Rh | 100.9061636 | 8.588 | 1/2− | | 1.04E+08 | 50 | 66 | Sn | 115.9017441 | 8.523 | 0+ | 14.54 | stable |
| 45 | 57 | Rh | 101.9068432 | 8.577 | 2− | | 1.79E+07 | 50 | 67 | Sn | 116.9029517 | 8.510 | 1/2+ | 7.68 | stable |
| 45 | 58 | Rh | 102.9055043 | 8.584 | 1/2− | 100 | stable | 50 | 68 | Sn | 117.9016063 | 8.517 | 0+ | 24.22 | stable |
| 45 | 60 | Rh | 104.9056938 | 8.573 | 7/2+ | | 1.27E+05 | 50 | 69 | Sn | 118.9033076 | 8.499 | 1/2+ | 8.59 | stable |
| 46 | 54 | Pd | 99.90850589 | 8.564 | 0+ | | 3.14E+05 | 50 | 70 | Sn | 119.9021966 | 8.505 | 0+ | 32.58 | stable |
| 46 | 55 | Pd | 100.9082892 | 8.561 | (5/2+) | | 3.05E+04 | 50 | 71 | Sn | 120.9042369 | 8.485 | 3/2+ | | 9.76E+04 |
| 46 | 56 | Pd | 101.9056077 | 8.580 | 0+ | 1.02 | stable | 50 | 72 | Sn | 121.9034401 | 8.488 | 0+ | 4.63 | stable |
| 46 | 57 | Pd | 102.9060873 | 8.571 | 5/2+ | | 1.47E+06 | 50 | 73 | Sn | 122.9057208 | 8.467 | 11/2− | | 1.12E+07 |
| 46 | 58 | Pd | 103.9040358 | 8.585 | 0+ | 11.14 | stable | 50 | 74 | Sn | 123.9052739 | 8.467 | 0+ | 5.79 | stable |
| 46 | 59 | Pd | 104.9050840 | 8.571 | 5/2+ | 22.33 | stable | 50 | 75 | Sn | 124.907785 | 8.446 | 11/2− | | 8.33E+05 |
| 46 | 60 | Pd | 105.9034857 | 8.580 | 0+ | 27.33 | stable | 50 | 76 | Sn | 125.9076533 | 8.444 | 0+ | | 3.15E+12 |
| 46 | 61 | Pd | 106.9051285 | 8.561 | 5/2+ | | 2.05E+14 | 50 | 77 | Sn | 126.9103510 | 8.421 | (11/2−) | | 7.56E+03 |
| 46 | 62 | Pd | 107.9038945 | 8.567 | 0+ | 26.46 | stable | 51 | 66 | Sb | 116.9048359 | 8.488 | 5/2+ | | 1.01E+04 |
| 46 | 63 | Pd | 108.9059535 | 8.545 | 5/2+ | | 4.93E+04 | 51 | 68 | Sb | 118.9039465 | 8.488 | 5/2+ | | 1.37E+05 |
| 46 | 64 | Pd | 109.9051533 | 8.547 | 0+ | 11.72 | stable | 51 | 69 | Sb | 119.905072 | 8.476 | 8− | | 4.98E+05 |
| 46 | 66 | Pd | 111.9073141 | 8.521 | 0+ | | 7.57E+04 | 51 | 70 | Sb | 120.9038180 | 8.482 | 5/2+ | 57.21 | stable |
| 47 | 56 | Ag | 102.9089727 | 8.538 | 7/2+ | | 3.96E+03 | 51 | 71 | Sb | 121.9051754 | 8.468 | 2− | | 2.35E+05 |
| 47 | 57 | Ag | 103.9086282 | 8.536 | 5+ | | 4.14E+03 | 51 | 72 | Sb | 122.9042157 | 8.472 | 7/2+ | 42.79 | stable |
| 47 | 58 | Ag | 104.9065287 | 8.550 | 1/2− | | 3.57E+06 | 51 | 73 | Sb | 123.9059375 | 8.456 | 3− | | 5.20E+06 |
| 47 | 60 | Ag | 106.905093 | 8.554 | 1/2− | 51.839 | stable | 51 | 74 | Sb | 124.9052478 | 8.458 | 7/2+ | | 8.70E+07 |
| 47 | 62 | Ag | 108.9047555 | 8.548 | 1/2− | 48.161 | stable | 51 | 75 | Sb | 125.9072482 | 8.440 | (8−) | | 1.08E+06 |
| 47 | 64 | Ag | 110.9052947 | 8.535 | 1/2− | | 6.44E+05 | 51 | 76 | Sb | 126.9069146 | 8.440 | 7/2+ | | 3.33E+05 |
| 47 | 65 | Ag | 111.9070048 | 8.516 | 2(−) | | 1.13E+04 | 51 | 77 | Sb | 127.9091673 | 8.421 | 8− | | 3.24E+04 |
| 47 | 66 | Ag | 112.9065666 | 8.516 | 1/2− | | 1.93E+04 | 51 | 78 | Sb | 128.9091501 | 8.418 | 7/2+ | | 1.58E+04 |

| Z | N | Sym | m (amu) | B (MeV) | Spin | % | τ(s) |
|---|---|---|---|---|---|---|---|
| 52 | 64 | Te | 115.9084203 | 8.456 | 0+ | | 8.96E+03 |
| 52 | 65 | Te | 116.90864 | 8.451 | 1/2+ | | 3.72E+03 |
| 52 | 66 | Te | 117.9058276 | 8.470 | 0+ | | 5.18E+05 |
| 52 | 67 | Te | 118.9064081 | 8.462 | 1/2+ | | 5.77E+04 |
| 52 | 68 | Te | 119.9040202 | 8.477 | 0+ | 0.09 | stable |
| 52 | 69 | Te | 120.9049364 | 8.467 | 1/2+ | | 1.45E+06 |
| 52 | 70 | Te | 121.9030471 | 8.478 | 0+ | 2.55 | stable |
| 52 | 71 | Te | 122.9042730 | 8.466 | 1/2+ | 0.89 | 1.89E+22 |
| 52 | 72 | Te | 123.9028180 | 8.473 | 0+ | 4.74 | stable |
| 52 | 73 | Te | 124.9044285 | 8.458 | 1/2+ | 7.07 | stable |
| 52 | 74 | Te | 125.9033095 | 8.463 | 0+ | 18.84 | stable |
| 52 | 75 | Te | 126.905217 | 8.446 | 3/2+ | | 3.37E+04 |
| 52 | 76 | Te | 127.9044631 | 8.449 | 0+ | 31.74 | 2.43E+32 |
| 52 | 77 | Te | 128.906596 | 8.430 | 3/2+ | | 4.18E+03 |
| 52 | 78 | Te | 129.9062244 | 8.430 | 0+ | 34.08 | 8.51E+28 |
| 52 | 80 | Te | 131.9085238 | 8.408 | 0+ | | 2.77E+05 |
| 53 | 67 | I | 119.9100482 | 8.424 | 2− | | 4.86E+03 |
| 53 | 68 | I | 120.9073668 | 8.442 | 5/2+ | | 7.63E+03 |
| 53 | 70 | I | 122.9055979 | 8.449 | 5/2+ | | 4.78E+04 |
| 53 | 71 | I | 123.9062114 | 8.441 | 2− | | 3.61E+05 |
| 53 | 72 | I | 124.9046242 | 8.450 | 5/2+ | | 5.13E+06 |
| 53 | 73 | I | 125.9056242 | 8.440 | 2− | | 1.13E+06 |
| 53 | 74 | I | 126.9044727 | 8.445 | 5/2+ | 100 | stable |
| 53 | 76 | I | 128.9049877 | 8.436 | 7/2+ | | 4.95E+14 |
| 53 | 77 | I | 129.9066742 | 8.421 | 5+ | | 4.45E+04 |
| 53 | 78 | I | 130.9061246 | 8.422 | 7/2+ | | 6.93E+05 |
| 53 | 79 | I | 131.9079945 | 8.406 | 4+ | | 8.26E+03 |
| 53 | 80 | I | 132.9078065 | 8.405 | 7/2+ | | 7.49E+04 |
| 53 | 82 | I | 134.91005 | 8.385 | 7/2+ | | 2.37E+04 |
| 54 | 68 | Xe | 121.9085484 | 8.425 | 0+ | | 7.24E+04 |
| 54 | 69 | Xe | 122.908480 | 8.421 | (1/2)+ | | 7.49E+03 |
| 54 | 70 | Xe | 123.9058942 | 8.438 | 0+ | 0.09 | stable |
| 54 | 71 | Xe | 124.906398 | 8.431 | (1/2)+ | | 6.08E+04 |
| 54 | 72 | Xe | 125.9042736 | 8.444 | 0+ | 0.09 | stable |
| 54 | 73 | Xe | 126.905184 | 8.434 | 1/2+ | | 3.14E+06 |
| 54 | 74 | Xe | 127.9035313 | 8.443 | 0+ | 1.92 | stable |
| 54 | 75 | Xe | 128.9047794 | 8.431 | 1/2+ | 26.44 | stable |
| 54 | 76 | Xe | 129.903508 | 8.438 | 0+ | 4.08 | stable |
| 54 | 77 | Xe | 130.9050824 | 8.424 | 3/2+ | 21.18 | stable |
| 54 | 78 | Xe | 131.9041535 | 8.428 | 0+ | 26.89 | stable |
| 54 | 79 | Xe | 132.905906 | 8.413 | 3/2+ | | 4.53E+05 |
| 54 | 80 | Xe | 133.9053945 | 8.414 | 0+ | 10.44 | stable |
| 54 | 81 | Xe | 134.90721 | 8.398 | 3/2+ | | 3.29E+04 |
| 54 | 82 | Xe | 135.9072188 | 8.396 | 0+ | 8.87 | 2.93E+27 |
| 55 | 72 | Cs | 126.9074175 | 8.412 | 1/2+ | | 2.25E+04 |
| 55 | 74 | Cs | 128.9060634 | 8.416 | 1/2+ | | 1.15E+05 |
| 55 | 76 | Cs | 130.9054639 | 8.415 | 5/2+ | | 8.37E+05 |
| 55 | 77 | Cs | 131.906430 | 8.406 | 2+ | | 5.60E+05 |
| 55 | 78 | Cs | 132.9054469 | 8.410 | 7/2+ | 100 | stable |
| 55 | 79 | Cs | 133.9067134 | 8.399 | 4+ | | 6.51E+07 |
| 55 | 80 | Cs | 134.905972 | 8.401 | 7/2+ | | 7.25E+13 |
| 55 | 81 | Cs | 135.907307 | 8.390 | 5+ | | 1.14E+06 |
| 55 | 82 | Cs | 136.9070895 | 8.389 | 7/2+ | | 9.48E+08 |
| 56 | 70 | Ba | 125.9112502 | 8.380 | 0+ | | 6.01E+03 |
| 56 | 72 | Ba | 127.90831 | 8.396 | 0+ | | 2.10E+05 |
| 56 | 73 | Ba | 128.9086794 | 8.391 | 1/2+ | | 8.03E+03 |
| 56 | 74 | Ba | 129.9063105 | 8.406 | 0+ | 0.106 | stable |
| 56 | 75 | Ba | 130.9069308 | 8.399 | 1/2+ | | 9.94E+05 |
| 56 | 76 | Ba | 131.9050562 | 8.409 | 0+ | 0.101 | stable |
| 56 | 77 | Ba | 132.9060024 | 8.400 | 1/2+ | | 3.32E+08 |
| 56 | 78 | Ba | 133.9045033 | 8.408 | 0+ | 2.417 | stable |
| 56 | 79 | Ba | 134.9056827 | 8.397 | 3/2+ | 6.592 | stable |
| 56 | 80 | Ba | 135.9045701 | 8.403 | 0+ | 7.854 | stable |
| 56 | 81 | Ba | 136.905824 | 8.392 | 3/2+ | 11.232 | stable |
| 56 | 82 | Ba | 137.9052413 | 8.393 | 0+ | 71.698 | stable |
| 56 | 83 | Ba | 138.908836 | 8.367 | 7/2− | | 4.98E+03 |
| 56 | 84 | Ba | 139.91060 | 8.353 | 0+ | | 1.10E+06 |
| 57 | 75 | La | 131.910110 | 8.368 | 2− | | 1.73E+04 |
| 57 | 76 | La | 132.908218 | 8.379 | 5/2+ | | 1.41E+04 |
| 57 | 78 | La | 134.9069768 | 8.383 | 5/2+ | | 7.02E+04 |
| 57 | 80 | La | 136.90647 | 8.382 | 7/2+ | | 1.89E+12 |
| 57 | 81 | La | 137.9071068 | 8.375 | 5+ | 0.09 | 3.31E+18 |
| 57 | 82 | La | 138.9063482 | 8.378 | 7/2+ | 99.91 | stable |
| 57 | 83 | La | 139.9094726 | 8.355 | 3− | | 1.45E+05 |
| 57 | 84 | La | 140.910958 | 8.343 | (7/2+) | | 1.41E+04 |
| 57 | 85 | La | 141.9140791 | 8.321 | 2− | | 5.46E+03 |
| 58 | 74 | Ce | 131.9114605 | 8.352 | 0+ | | 1.26E+04 |
| 58 | 75 | Ce | 132.911515 | 8.350 | 9/2− | | 1.76E+04 |
| 58 | 75 | Ce | 132.9115515 | 8.350 | 1/2+ | | 5.83E+03 |
| 58 | 76 | Ce | 133.9089248 | 8.366 | 0+ | | 2.73E+05 |
| 58 | 77 | Ce | 134.9091514 | 8.362 | 1/2(+) | | 6.37E+04 |
| 58 | 78 | Ce | 135.907172 | 8.373 | 0+ | 0.185 | stable |
| 58 | 79 | Ce | 136.9078056 | 8.367 | 3/2+ | | 3.24E+04 |
| 58 | 80 | Ce | 137.9059913 | 8.377 | 0+ | 0.251 | stable |
| 58 | 81 | Ce | 138.9066466 | 8.370 | 3/2+ | | 1.19E+07 |
| 58 | 82 | Ce | 139.905434 | 8.376 | 0+ | 88.45 | stable |
| 58 | 83 | Ce | 140.908271 | 8.355 | 7/2− | | 2.81E+06 |
| 58 | 84 | Ce | 141.909241 | 8.347 | 0+ | 11.114 | 1.58E+24 |
| 58 | 85 | Ce | 142.9123812 | 8.325 | 3/2− | | 1.19E+05 |
| 58 | 86 | Ce | 143.913643 | 8.315 | 0+ | | 2.46E+07 |

*Continued—*

| $Z$ | $N$ | Sym | $m$ (amu) | $B$ (MeV) | Spin | % | $\tau$(s) |
|---|---|---|---|---|---|---|---|
| 59 | 78 | Pr | 136.910687 | 8.341 | 5/2+ | | 4.61E+03 |
| 59 | 80 | Pr | 138.9089384 | 8.349 | 5/2+ | | 1.59E+04 |
| 59 | 82 | Pr | 140.9076477 | 8.354 | 5/2+ | 100 | stable |
| 59 | 83 | Pr | 141.910041 | 8.336 | 2– | | 6.88E+04 |
| 59 | 84 | Pr | 142.9108122 | 8.329 | 7/2+ | | 1.17E+06 |
| 59 | 86 | Pr | 144.9145069 | 8.302 | 7/2+ | | 2.15E+04 |
| 60 | 78 | Nd | 137.91195 | 8.325 | 0+ | | 1.81E+04 |
| 60 | 80 | Nd | 139.90931 | 8.338 | 0+ | | 2.91E+05 |
| 60 | 81 | Nd | 140.9096099 | 8.336 | 3/2+ | | 8.96E+03 |
| 60 | 82 | Nd | 141.907719 | 8.346 | 0+ | 27.2 | stable |
| 60 | 83 | Nd | 142.90981 | 8.330 | 7/2– | 12.2 | stable |
| 60 | 84 | Nd | 143.910083 | 8.327 | 0+ | 23.8 | 7.22E+22 |
| 60 | 85 | Nd | 144.91257 | 8.309 | 7/2– | 8.3 | stable |
| 60 | 86 | Nd | 145.913116 | 8.304 | 0+ | 17.2 | stable |
| 60 | 87 | Nd | 146.916096 | 8.284 | 5/2– | | 9.49E+05 |
| 60 | 88 | Nd | 147.916889 | 8.277 | 0+ | 5.7 | stable |
| 60 | 89 | Nd | 148.920145 | 8.255 | 5/2– | | 6.22E+03 |
| 60 | 90 | Nd | 149.920887 | 8.250 | 0+ | 5.6 | 3.47E+26 |
| 61 | 82 | Pm | 142.9109276 | 8.318 | 5/2+ | | 2.29E+07 |
| 61 | 83 | Pm | 143.912586 | 8.305 | 5– | | 3.14E+07 |
| 61 | 84 | Pm | 144.9127439 | 8.303 | 5/2+ | | 5.58E+08 |
| 61 | 85 | Pm | 145.914696 | 8.289 | 3– | | 1.74E+08 |
| 61 | 86 | Pm | 146.9151339 | 8.284 | 7/2+ | | 8.27E+07 |
| 61 | 87 | Pm | 147.9174746 | 8.268 | 1– | | 4.64E+05 |
| 61 | 88 | Pm | 148.91833 | 8.262 | 7/2+ | | 1.91E+05 |
| 61 | 89 | Pm | 149.92098 | 8.244 | (1–) | | 9.65E+03 |
| 61 | 90 | Pm | 150.921207 | 8.241 | 5/2+ | | 1.02E+05 |
| 62 | 80 | Sm | 141.9151976 | 8.286 | 0+ | | 4.35E+03 |
| 62 | 82 | Sm | 143.911998 | 8.304 | 0+ | 3.07 | stable |
| 62 | 83 | Sm | 144.913407 | 8.293 | 7/2– | | 2.94E+07 |
| 62 | 84 | Sm | 145.913038 | 8.294 | 0+ | | 3.25E+15 |
| 62 | 85 | Sm | 146.914894 | 8.281 | 7/2– | 14.99 | 3.34E+18 |
| 62 | 86 | Sm | 147.914819 | 8.280 | 0+ | 11.24 | 2.21E+23 |
| 62 | 87 | Sm | 148.91718 | 8.263 | 7/2– | 13.82 | 6.31E+22 |
| 62 | 88 | Sm | 149.9172730 | 8.262 | 0+ | 7.38 | stable |
| 62 | 89 | Sm | 150.919929 | 8.244 | 5/2– | | 2.84E+09 |
| 62 | 90 | Sm | 151.9197282 | 8.244 | 0+ | 26.75 | stable |
| 62 | 91 | Sm | 152.922097 | 8.229 | 3/2+ | | 1.67E+05 |
| 62 | 92 | Sm | 153.9222053 | 8.227 | 0+ | 22.75 | stable |
| 62 | 94 | Sm | 155.9255279 | 8.205 | 0+ | | 3.38E+04 |
| 63 | 82 | Eu | 144.9162652 | 8.269 | 5/2+ | | 5.12E+05 |
| 63 | 83 | Eu | 145.91720 | 8.262 | 4– | | 3.97E+05 |
| 63 | 84 | Eu | 146.916742 | 8.264 | 5/2+ | | 2.08E+06 |
| 63 | 85 | Eu | 147.91815 | 8.254 | 5– | | 4.71E+06 |
| 63 | 86 | Eu | 148.917930 | 8.254 | 5/2+ | | 8.04E+06 |
| 63 | 87 | Eu | 149.9197018 | 8.241 | 5(–) | | 1.16E+09 |
| 63 | 88 | Eu | 150.919848 | 8.239 | 5/2+ | 47.81 | stable |
| 63 | 89 | Eu | 151.921744 | 8.227 | 3– | | 4.27E+08 |
| 63 | 90 | Eu | 152.921229 | 8.229 | 5/2+ | 52.19 | stable |
| 63 | 91 | Eu | 153.922976 | 8.217 | 3– | | 2.71E+08 |
| 63 | 92 | Eu | 154.92289 | 8.217 | 5/2+ | | 1.50E+08 |
| 63 | 93 | Eu | 155.9247522 | 8.205 | 0+ | | 1.31E+06 |
| 63 | 94 | Eu | 156.9254236 | 8.200 | 5/2+ | | 5.46E+04 |
| 64 | 82 | Gd | 145.9183106 | 8.250 | 0+ | | 4.17E+06 |
| 64 | 83 | Gd | 146.919090 | 8.243 | 7/2– | | 1.37E+05 |
| 64 | 84 | Gd | 147.918110 | 8.248 | 0+ | | 2.35E+09 |
| 64 | 85 | Gd | 148.919339 | 8.239 | 7/2– | | 8.02E+05 |
| 64 | 86 | Gd | 149.9186589 | 8.243 | 0+ | | 5.64E+13 |
| 64 | 87 | Gd | 150.9203485 | 8.231 | 7/2– | | 1.07E+07 |
| 64 | 88 | Gd | 151.919789 | 8.233 | 0+ | 0.2 | 3.41E+21 |
| 64 | 89 | Gd | 152.9217495 | 8.220 | 3/2– | | 2.09E+07 |
| 64 | 90 | Gd | 153.9208623 | 8.225 | 0+ | 2.18 | stable |
| 64 | 91 | Gd | 154.922619 | 8.213 | 3/2– | 14.8 | stable |
| 64 | 92 | Gd | 155.922122 | 8.215 | 0+ | 20.47 | stable |
| 64 | 93 | Gd | 156.9239567 | 8.204 | 3/2– | 15.65 | stable |
| 64 | 94 | Gd | 157.924103 | 8.202 | 0+ | 24.84 | stable |
| 64 | 95 | Gd | 158.9263861 | 8.188 | 3/2– | | 6.65E+04 |
| 64 | 96 | Gd | 159.9270541 | 8.183 | 0+ | 21.86 | stable |
| 65 | 82 | Tb | 146.9240446 | 8.207 | (1/2+) | | 6.12E+03 |
| 65 | 83 | Tb | 147.9242717 | 8.204 | 2– | | 3.60E+03 |
| 65 | 84 | Tb | 148.9232459 | 8.210 | 1/2+ | | 1.48E+04 |
| 65 | 85 | Tb | 149.9236597 | 8.206 | (2)– | | 1.25E+04 |
| 65 | 86 | Tb | 150.9230982 | 8.209 | 1/2(+) | | 6.34E+04 |
| 65 | 87 | Tb | 151.9240744 | 8.202 | 2– | | 6.30E+04 |
| 65 | 88 | Tb | 152.9234346 | 8.205 | 5/2+ | | 2.02E+05 |
| 65 | 89 | Tb | 153.9246862 | 8.197 | 0(+) | | 7.74E+04 |
| 65 | 90 | Tb | 154.9235052 | 8.203 | 3/2+ | | 4.60E+05 |
| 65 | 91 | Tb | 155.924744 | 8.195 | 3– | | 4.62E+05 |
| 65 | 92 | Tb | 156.9240212 | 8.198 | 3/2+ | | 2.24E+09 |
| 65 | 93 | Tb | 157.9254103 | 8.189 | 3– | | 5.68E+09 |
| 65 | 94 | Tb | 158.9253431 | 8.189 | 3/2+ | 100 | stable |
| 65 | 95 | Tb | 159.9271640 | 8.177 | 3– | | 6.25E+06 |
| 65 | 96 | Tb | 160.9275663 | 8.174 | 3/2+ | | 5.94E+05 |
| 66 | 86 | Dy | 151.9247140 | 8.193 | 0+ | | 8.57E+03 |
| 66 | 87 | Dy | 152.9257647 | 8.186 | 7/2(–) | | 2.30E+04 |
| 66 | 88 | Dy | 153.9244220 | 8.193 | 0+ | | 9.46E+13 |
| 66 | 89 | Dy | 154.9257538 | 8.184 | 3/2– | | 3.56E+04 |
| 66 | 90 | Dy | 155.9242783 | 8.192 | 0+ | 0.06 | stable |
| 66 | 91 | Dy | 156.9254661 | 8.185 | 3/2– | | 2.93E+04 |
| 66 | 92 | Dy | 157.924405 | 8.190 | 0+ | 0.1 | stable |
| 66 | 93 | Dy | 158.925736 | 8.182 | 3/2– | | 1.25E+07 |
| 66 | 94 | Dy | 159.925194 | 8.184 | 0+ | 2.34 | stable |

| Z | N | Sym | m (amu) | B (MeV) | Spin | % | τ(s) |
|---|---|---|---|---|---|---|---|
| 66 | 95 | Dy | 160.926930 | 8.173 | 5/2+ | 18.91 | stable |
| 66 | 96 | Dy | 161.926795 | 8.173 | 0+ | 25.51 | stable |
| 66 | 97 | Dy | 162.928728 | 8.162 | 5/2– | 24.9 | stable |
| 66 | 98 | Dy | 163.9291712 | 8.159 | 0+ | 28.18 | stable |
| 66 | 99 | Dy | 164.93170 | 8.144 | 7/2+ | | 8.40E+03 |
| 66 | 100 | Dy | 165.9328032 | 8.137 | 0+ | | 2.94E+05 |
| 67 | 94 | Ho | 160.9278548 | 8.163 | 7/2– | | 8.93E+03 |
| 67 | 96 | Ho | 162.9287303 | 8.157 | 7/2– | | 1.44E+11 |
| 67 | 98 | Ho | 164.9303221 | 8.147 | 7/2– | 100 | stable |
| 67 | 99 | Ho | 165.9322842 | 8.135 | 0– | | 9.63E+04 |
| 67 | 100 | Ho | 166.933127 | 8.130 | 7/2– | | 1.12E+04 |
| 68 | 90 | Er | 157.9298935 | 8.148 | 0+ | | 8.24E+03 |
| 68 | 92 | Er | 159.92908 | 8.152 | 0+ | | 1.03E+05 |
| 68 | 93 | Er | 160.93 | 8.146 | 3/2– | | 1.16E+04 |
| 68 | 94 | Er | 161.928775 | 8.152 | 0+ | 0.14 | stable |
| 68 | 95 | Er | 162.9300327 | 8.145 | 5/2– | | 4.50E+03 |
| 68 | 96 | Er | 163.929198 | 8.149 | 0+ | 1.61 | stable |
| 68 | 97 | Er | 164.930726 | 8.140 | 5/2– | | 3.73E+04 |
| 68 | 98 | Er | 165.9302900 | 8.142 | 0+ | 33.61 | stable |
| 68 | 99 | Er | 166.932046 | 8.132 | 7/2+ | 22.93 | stable |
| 68 | 100 | Er | 167.9323702 | 8.130 | 0+ | 26.78 | stable |
| 68 | 101 | Er | 168.9345881 | 8.117 | 1/2– | | 8.12E+05 |
| 68 | 102 | Er | 169.935461 | 8.112 | 0+ | 14.93 | stable |
| 68 | 103 | Er | 170.938026 | 8.098 | 5/2– | | 2.71E+04 |
| 68 | 104 | Er | 171.9393521 | 8.090 | 0+ | | 1.77E+05 |
| 69 | 94 | Tm | 162.9326500 | 8.125 | 1/2+ | | 6.52E+03 |
| 69 | 96 | Tm | 164.932433 | 8.126 | 1/2+ | | 1.08E+05 |
| 69 | 97 | Tm | 165.9335541 | 8.119 | 2+ | | 2.77E+04 |
| 69 | 98 | Tm | 166.9328516 | 8.123 | 1/2+ | | 7.99E+05 |
| 69 | 99 | Tm | 167.9341728 | 8.115 | 3+ | | 8.04E+06 |
| 69 | 100 | Tm | 168.934212 | 8.114 | 1/2+ | 100 | stable |
| 69 | 101 | Tm | 169.9358014 | 8.106 | 1– | | 1.11E+07 |
| 69 | 102 | Tm | 170.936426 | 8.102 | 1/2+ | | 6.05E+07 |
| 69 | 103 | Tm | 171.9384 | 8.091 | 2– | | 2.29E+05 |
| 69 | 104 | Tm | 172.9396036 | 8.084 | 1/2+ | | 2.97E+04 |
| 70 | 94 | Yb | 163.9344894 | 8.109 | 0+ | | 4.54E+03 |
| 70 | 96 | Yb | 165.9338796 | 8.112 | 0+ | | 2.04E+05 |
| 70 | 98 | Yb | 167.9338969 | 8.112 | 0+ | 0.13 | stable |
| 70 | 99 | Yb | 168.9351871 | 8.104 | 7/2+ | | 2.77E+06 |
| 70 | 100 | Yb | 169.934759 | 8.107 | 0+ | 3.04 | stable |
| 70 | 101 | Yb | 170.936323 | 8.098 | 1/2– | 14.28 | stable |
| 70 | 102 | Yb | 171.9363777 | 8.097 | 0+ | 21.83 | stable |
| 70 | 103 | Yb | 172.938208 | 8.087 | 5/2– | 16.13 | stable |
| 70 | 104 | Yb | 173.9388621 | 8.084 | 0+ | 31.83 | stable |
| 70 | 105 | Yb | 174.941273 | 8.071 | 7/2– | | 3.62E+05 |
| 70 | 106 | Yb | 175.942571 | 8.064 | 0+ | 12.76 | stable |
| 70 | 107 | Yb | 176.9452571 | 8.050 | 9/2+ | | 6.88E+03 |
| 70 | 108 | Yb | 177.9466467 | 8.043 | 0+ | | 4.43E+03 |
| 71 | 98 | Lu | 168.937649 | 8.086 | 7/2+ | | 1.23E+05 |
| 71 | 99 | Lu | 169.9384722 | 8.082 | 0+ | | 1.74E+05 |
| 71 | 100 | Lu | 170.93791 | 8.085 | 7/2+ | | 7.12E+05 |
| 71 | 101 | Lu | 171.9390822 | 8.078 | 4– | | 5.79E+05 |
| 71 | 102 | Lu | 172.938927 | 8.079 | 7/2+ | | 4.32E+07 |
| 71 | 103 | Lu | 173.940334 | 8.071 | (1)– | | 1.04E+08 |
| 71 | 104 | Lu | 174.94077 | 8.069 | 7/2+ | 97.41 | stable |
| 71 | 105 | Lu | 175.9426824 | 8.059 | 7– | 2.59 | 1.29E+18 |
| 71 | 106 | Lu | 176.9437550 | 8.053 | 7/2+ | | 5.82E+05 |
| 71 | 108 | Lu | 178.9473274 | 8.035 | 7/2(+) | | 1.65E+04 |
| 72 | 98 | Hf | 169.939609 | 8.071 | 0+ | | 5.76E+04 |
| 72 | 99 | Hf | 170.940492 | 8.066 | 7/2+ | | 4.36E+04 |
| 72 | 100 | Hf | 171.9394483 | 8.072 | 0+ | | 5.90E+07 |
| 72 | 101 | Hf | 172.940513 | 8.066 | 1/2– | | 8.50E+04 |
| 72 | 102 | Hf | 173.940044 | 8.069 | 0+ | 0.16 | 6.31E+22 |
| 72 | 103 | Hf | 174.9415024 | 8.061 | 5/2– | | 6.05E+06 |
| 72 | 104 | Hf | 175.941406 | 8.061 | 0+ | 5.26 | stable |
| 72 | 105 | Hf | 176.9432207 | 8.052 | 7/2– | 18.6 | stable |
| 72 | 106 | Hf | 177.9436988 | 8.049 | 0+ | 27.28 | stable |
| 72 | 107 | Hf | 178.9458161 | 8.039 | 9/2+ | 13.62 | stable |
| 72 | 108 | Hf | 179.94655 | 8.035 | 0+ | 35.08 | stable |
| 72 | 109 | Hf | 180.9490991 | 8.022 | 1/2– | | 3.66E+06 |
| 72 | 110 | Hf | 181.9505541 | 8.015 | 0+ | | 2.84E+14 |
| 72 | 111 | Hf | 182.9535304 | 8.000 | (3/2–) | | 3.84E+03 |
| 72 | 112 | Hf | 183.9554465 | 7.991 | 0+ | | 1.48E+04 |
| 73 | 100 | Ta | 172.94354 | 8.044 | 5/2– | | 1.13E+04 |
| 73 | 101 | Ta | 173.944256 | 8.040 | 3(+) | | 3.78E+03 |
| 73 | 102 | Ta | 174.9437 | 8.044 | 7/2+ | | 3.78E+04 |
| 73 | 103 | Ta | 175.944857 | 8.039 | 1– | | 2.91E+04 |
| 73 | 104 | Ta | 176.9444724 | 8.041 | 7/2+ | | 2.04E+05 |
| 73 | 105 | Ta | 177.9457782 | 8.034 | 7– | | 8.50E+03 |
| 73 | 106 | Ta | 178.94593 | 8.034 | 7/2+ | | 5.74E+07 |
| 73 | 107 | Ta | 179.9474648 | 8.026 | 1+ | 0.012 | 2.93E+04 |
| 73 | 108 | Ta | 180.9479958 | 8.023 | 7/2+ | 99.988 | stable |
| 73 | 109 | Ta | 181.9501518 | 8.013 | 3– | | 9.89E+06 |
| 73 | 110 | Ta | 182.9513726 | 8.007 | 7/2+ | | 4.41E+05 |
| 73 | 111 | Ta | 183.954008 | 7.994 | (5–) | | 3.13E+04 |
| 74 | 102 | W | 175.945634 | 8.030 | 0+ | | 9.00E+03 |
| 74 | 103 | W | 176.946643 | 8.025 | (1/2–) | | 8.10E+03 |
| 74 | 104 | W | 177.9458762 | 8.029 | 0+ | | 1.87E+06 |
| 74 | 106 | W | 179.9467045 | 8.025 | 0+ | 0.12 | stable |
| 74 | 107 | W | 180.9481972 | 8.018 | 9/2+ | | 1.05E+07 |

*Continued—*

| Z | N | Sym | m (amu) | B (MeV) | Spin | % | τ(s) | Z | N | Sym | m (amu) | B (MeV) | Spin | % | τ(s) |
|---|---|---|---|---|---|---|---|---|---|---|---|---|---|---|---|
| 74 | 108 | W | 181.9482042 | 8.018 | 0+ | 26.5 | stable | 78 | 109 | Pt | 186.960587 | 7.941 | 3/2– | | 8.46E+03 |
| 74 | 109 | W | 182.950223 | 8.008 | 1/2– | 14.31 | stable | 78 | 110 | Pt | 187.9593954 | 7.948 | 0+ | | 8.81E+05 |
| 74 | 110 | W | 183.9509312 | 8.005 | 0+ | 30.64 | stable | 78 | 111 | Pt | 188.9608337 | 7.941 | 3/2– | | 3.91E+04 |
| 74 | 111 | W | 184.9534193 | 7.993 | 3/2– | | 6.49E+06 | 78 | 112 | Pt | 189.9599317 | 7.947 | 0+ | 0.014 | 2.05E+19 |
| 74 | 112 | W | 185.9543641 | 7.989 | 0+ | 28.43 | stable | 78 | 113 | Pt | 190.9616767 | 7.939 | 3/2– | | 2.47E+05 |
| 74 | 113 | W | 186.9571605 | 7.975 | 3/2– | | 8.54E+04 | 78 | 114 | Pt | 191.961038 | 7.942 | 0+ | 0.782 | stable |
| 74 | 114 | W | 187.9584891 | 7.969 | 0+ | | 6.00E+06 | 78 | 115 | Pt | 192.9629874 | 7.934 | 1/2– | | 1.58E+09 |
| 75 | 106 | Re | 180.9500679 | 8.004 | 5/2+ | | 7.16E+04 | 78 | 116 | Pt | 193.9626803 | 7.936 | 0+ | 32.967 | stable |
| 75 | 107 | Re | 181.9512101 | 7.999 | 7+ | | 2.31E+05 | 78 | 117 | Pt | 194.9647911 | 7.927 | 1/2– | 33.832 | stable |
| 75 | 107 | Re | | | 2+ | | 4.57E+04 | 78 | 118 | Pt | 195.9649515 | 7.927 | 0+ | 25.242 | stable |
| 75 | 108 | Re | 182.9508198 | 8.001 | 5/2+ | | 6.05E+06 | 78 | 119 | Pt | 196.9673402 | 7.916 | 1/2– | | 7.16E+04 |
| 75 | 109 | Re | 183.9525208 | 7.993 | 3– | | 3.28E+06 | 78 | 120 | Pt | 197.9678928 | 7.914 | 0+ | 7.163 | stable |
| 75 | 110 | Re | 184.952955 | 7.991 | 5/2+ | 37.4 | stable | 78 | 122 | Pt | 199.9714407 | 7.899 | 0+ | | 4.50E+04 |
| 75 | 111 | Re | 185.9549861 | 7.981 | 1– | | 3.21E+05 | 78 | 124 | Pt | 201.97574 | 7.881 | 0+ | | 1.56E+05 |
| 75 | 112 | Re | 186.9557531 | 7.978 | 5/2+ | 62.6 | 1.37E+18 | 79 | 112 | Au | 190.9637042 | 7.925 | 3/2+ | | 1.14E+04 |
| 75 | 113 | Re | 187.9581144 | 7.967 | 1– | | 6.13E+04 | 79 | 113 | Au | 191.964813 | 7.920 | 1– | | 1.78E+04 |
| 75 | 114 | Re | 188.959229 | 7.962 | 5/2+ | | 8.75E+04 | 79 | 114 | Au | 192.9641497 | 7.924 | 3/2+ | | 6.35E+04 |
| 76 | 105 | Os | 180.953244 | 7.983 | 1/2– | | 6.30E+03 | 79 | 115 | Au | 193.9653653 | 7.919 | 1– | | 1.37E+05 |
| 76 | 106 | Os | 181.9521102 | 7.990 | 0+ | | 7.96E+04 | 79 | 116 | Au | 194.9650346 | 7.921 | 3/2+ | | 1.61E+07 |
| 76 | 107 | Os | 182.9531261 | 7.985 | 9/2+ | | 4.68E+04 | 79 | 117 | Au | 195.9665698 | 7.915 | 2– | | 5.33E+05 |
| 76 | 108 | Os | 183.9524891 | 7.989 | 0+ | 0.02 | stable | 79 | 118 | Au | 196.9665687 | 7.916 | 3/2+ | 100 | stable |
| 76 | 109 | Os | 184.9540423 | 7.981 | 1/2– | | 8.09E+06 | 79 | 119 | Au | 197.9682423 | 7.909 | 2– | | 2.33E+05 |
| 76 | 110 | Os | 185.9538382 | 7.983 | 0+ | 1.59 | 6.31E+22 | 79 | 120 | Au | 198.9687652 | 7.907 | 3/2+ | | 2.71E+05 |
| 76 | 111 | Os | 186.9557505 | 7.974 | 1/2– | 1.96 | stable | 80 | 112 | Hg | 191.9656343 | 7.912 | 0+ | | 1.75E+04 |
| 76 | 112 | Os | 187.9558382 | 7.974 | 0+ | 13.24 | stable | 80 | 113 | Hg | 192.9666654 | 7.908 | 3/2– | | 1.37E+04 |
| 76 | 113 | Os | 188.9581475 | 7.963 | 3/2– | 16.15 | stable | 80 | 114 | Hg | 193.9654394 | 7.915 | 0+ | | 1.64E+10 |
| 76 | 114 | Os | 189.958447 | 7.962 | 0+ | 26.26 | stable | 80 | 115 | Hg | 194.9667201 | 7.909 | 1/2– | | 3.79E+04 |
| 76 | 115 | Os | 190.9609297 | 7.951 | 9/2– | | 1.33E+06 | 80 | 116 | Hg | 195.9658326 | 7.914 | 0+ | 0.15 | stable |
| 76 | 116 | Os | 191.9614807 | 7.948 | 0+ | 40.78 | stable | 80 | 117 | Hg | 196.9672129 | 7.909 | 1/2– | | 2.34E+05 |
| 76 | 117 | Os | 192.9641516 | 7.936 | 3/2– | | 1.08E+05 | 80 | 118 | Hg | 197.966769 | 7.912 | 0+ | 9.97 | stable |
| 76 | 118 | Os | 193.9651821 | 7.932 | 0+ | | 1.89E+08 | 80 | 119 | Hg | 198.9682799 | 7.905 | 1/2– | 16.87 | stable |
| 77 | 107 | Ir | 183.957476 | 7.959 | 5– | | 1.11E+04 | 80 | 120 | Hg | 199.968326 | 7.906 | 0+ | 23.1 | stable |
| 77 | 108 | Ir | 184.956698 | 7.964 | 5/2– | | 5.18E+04 | 80 | 121 | Hg | 200.9703023 | 7.898 | 3/2– | 13.18 | stable |
| 77 | 109 | Ir | 185.9579461 | 7.958 | 5+ | | 5.99E+04 | 80 | 122 | Hg | 201.970643 | 7.897 | 0+ | 29.86 | stable |
| 77 | 109 | Ir | | | 2– | | 7.20E+03 | 80 | 123 | Hg | 202.9728725 | 7.887 | 5/2– | | 4.03E+06 |
| 77 | 110 | Ir | 186.9573634 | 7.962 | 3/2+ | | 3.78E+04 | 80 | 124 | Hg | 203.9734939 | 7.886 | 0+ | 6.87 | stable |
| 77 | 111 | Ir | 187.9588531 | 7.955 | 1– | | 1.49E+05 | 81 | 114 | Tl | 194.9697743 | 7.891 | 1/2+ | | 4.18E+03 |
| 77 | 112 | Ir | 188.9587189 | 7.956 | 3/2+ | | 1.14E+06 | 81 | 115 | Tl | 195.9704812 | 7.888 | 2– | | 6.62E+03 |
| 77 | 113 | Ir | 189.960546 | 7.948 | (4)+ | | 1.02E+06 | 81 | 116 | Tl | 196.9695745 | 7.893 | 1/2+ | | 1.02E+04 |
| 77 | 114 | Ir | 190.960594 | 7.948 | 3/2+ | 37.3 | stable | 81 | 117 | Tl | 197.9704835 | 7.890 | 2– | | 1.91E+04 |
| 77 | 115 | Ir | 191.962605 | 7.939 | 4(+) | | 6.38E+06 | 81 | 118 | Tl | 198.969877 | 7.894 | 1/2+ | | 2.67E+04 |
| 77 | 116 | Ir | 192.9629264 | 7.938 | 3/2+ | 62.7 | stable | 81 | 119 | Tl | 199.9709627 | 7.890 | 2– | | 9.42E+04 |
| 77 | 117 | Ir | 193.9650784 | 7.928 | 1– | | 6.89E+04 | 81 | 120 | Tl | 200.9708189 | 7.891 | 1/2+ | | 2.62E+05 |
| 77 | 118 | Ir | 194.9659796 | 7.925 | 3/2+ | | 9.00E+03 | 81 | 121 | Tl | 201.9721058 | 7.886 | 2– | | 1.06E+06 |
| 78 | 107 | Pt | 184.960619 | 7.940 | 9/2+ | | 4.25E+03 | 81 | 122 | Tl | 202.9723442 | 7.886 | 1/2+ | 29.524 | stable |
| 78 | 108 | Pt | 185.9593508 | 7.947 | 0+ | | 7.20E+03 | 81 | 123 | Tl | 203.9738635 | 7.880 | 2– | | 1.19E+08 |

| Z | N | Sym | m (amu) | B (MeV) | Spin | % | τ(s) |
|---|---|---|---|---|---|---|---|
| 81 | 124 | Tl | 204.9744275 | 7.878 | 1/2+ | 70.476 | stable |
| 82 | 116 | Pb | 197.972034 | 7.879 | 0+ | | 8.64E+03 |
| 82 | 117 | Pb | 198.9729167 | 7.876 | 3/2− | | 5.40E+03 |
| 82 | 118 | Pb | 199.9718267 | 7.882 | 0+ | | 7.74E+04 |
| 82 | 119 | Pb | 200.9728845 | 7.878 | 5/2− | | 3.36E+04 |
| 82 | 120 | Pb | 201.9721591 | 7.882 | 0+ | | 1.66E+12 |
| 82 | 121 | Pb | 202.9733905 | 7.877 | 5/2− | | 1.87E+05 |
| 82 | 122 | Pb | 203.9730436 | 7.880 | 0+ | 1.4 | 4.42E+24 |
| 82 | 123 | Pb | 204.9744818 | 7.874 | 5/2− | | 4.83E+14 |
| 82 | 124 | Pb | 205.9744653 | 7.875 | 0+ | 24.1 | stable |
| 82 | 125 | Pb | 206.9758969 | 7.870 | 1/2− | 22.1 | stable |
| 82 | 126 | Pb | 207.9766521 | 7.867 | 0+ | 52.4 | stable |
| 82 | 127 | Pb | 208.9810901 | 7.849 | 9/2+ | | 1.17E+04 |
| 82 | 128 | Pb | 209.9841885 | 7.836 | 0+ | | 7.03E+08 |
| 82 | 130 | Pb | 211.9918975 | 7.804 | 0+ | | 3.83E+04 |
| 83 | 118 | Bi | 200.977009 | 7.855 | 9/2− | | 6.48E+03 |
| 83 | 119 | Bi | 201.9777423 | 7.852 | 5+ | | 6.19E+03 |
| 83 | 120 | Bi | 202.976876 | 7.858 | 9/2− | | 4.23E+04 |
| 83 | 121 | Bi | 203.9778127 | 7.854 | 6+ | | 4.04E+04 |
| 83 | 122 | Bi | 204.9773894 | 7.857 | 9/2− | | 1.32E+06 |
| 83 | 123 | Bi | 205.9784991 | 7.853 | 6+ | | 5.39E+05 |
| 83 | 124 | Bi | 206.9784707 | 7.854 | 9/2− | | 9.95E+08 |
| 83 | 125 | Bi | 207.9797422 | 7.850 | (5)+ | | 1.16E+13 |
| 83 | 126 | Bi | 208.9803987 | 7.848 | 9/2− | 100 | stable |
| 83 | 127 | Bi | 209.9841204 | 7.833 | 1− | | 4.33E+05 |
| 83 | 129 | Bi | 211.9912857 | 7.803 | 1(−) | | 3.63E+03 |
| 84 | 120 | Po | 203.9803181 | 7.839 | 0+ | | 1.27E+04 |
| 84 | 121 | Po | 204.9812033 | 7.836 | 5/2− | | 5.98E+03 |
| 84 | 122 | Po | 205.9804811 | 7.841 | 0+ | | 7.60E+05 |
| 84 | 123 | Po | 206.9815932 | 7.837 | 5/2− | | 2.09E+04 |
| 84 | 124 | Po | 207.9812457 | 7.839 | 0+ | | 9.14E+07 |
| 84 | 125 | Po | 208.9824304 | 7.835 | 1/2− | | 3.22E+09 |
| 84 | 126 | Po | 209.9828737 | 7.834 | 0+ | | 1.20E+07 |
| 85 | 122 | At | 206.9857835 | 7.814 | 9/2− | | 6.48E+03 |
| 85 | 123 | At | 207.98659 | 7.812 | 6+ | | 5.87E+03 |
| 85 | 124 | At | 208.9861731 | 7.815 | 9/2− | | 1.95E+04 |
| 85 | 125 | At | 209.9871477 | 7.812 | 5+ | | 2.92E+04 |
| 85 | 126 | At | 210.9874963 | 7.811 | 9/2− | | 2.60E+04 |
| 86 | 124 | Rn | 209.9896962 | 7.797 | 0+ | | 8.71E+03 |
| 86 | 125 | Rn | 210.9906005 | 7.794 | 1/2− | | 5.26E+04 |
| 86 | 136 | Rn | 222.0175777 | 7.694 | 0+ | | 3.30E+05 |
| 86 | 138 | Rn | 224.02409 | 7.671 | 0+ | | 6.41E+03 |
| 88 | 135 | Ra | 223.0185022 | 7.685 | 3/2+ | | 9.88E+05 |
| 88 | 136 | Ra | 224.0202118 | 7.680 | 0+ | | 3.16E+05 |
| 88 | 137 | Ra | 225.0236116 | 7.668 | 1/2+ | | 1.29E+06 |
| 88 | 138 | Ra | 226.0254098 | 7.662 | 0+ | | 5.05E+10 |
| 88 | 140 | Ra | 228.0310703 | 7.642 | 0+ | | 1.81E+08 |
| 88 | 142 | Ra | 230.0370564 | 7.622 | 0+ | | 5.58E+03 |
| 89 | 135 | Ac | 224.0217229 | 7.670 | 0− | | 1.04E+04 |
| 89 | 136 | Ac | 225.0232296 | 7.666 | (3/2−) | | 8.64E+05 |
| 89 | 137 | Ac | 226.0260981 | 7.656 | (1−) | | 1.06E+05 |
| 89 | 138 | Ac | 227.0277521 | 7.651 | 3/2− | | 6.87E+08 |
| 89 | 139 | Ac | 228.0310211 | 7.639 | 3(+) | | 2.21E+04 |
| 89 | 140 | Ac | 229.0330152 | 7.633 | (3/2+) | | 3.78E+03 |
| 90 | 137 | Th | 227.0277041 | 7.647 | 3/2+ | | 1.62E+06 |
| 90 | 138 | Th | 228.0287411 | 7.645 | 0+ | | 6.03E+07 |
| 90 | 139 | Th | 229.0317624 | 7.635 | 5/2+ | | 2.49E+11 |
| 90 | 140 | Th | 230.0331338 | 7.631 | 0+ | | 2.38E+12 |
| 90 | 141 | Th | 231.0363043 | 7.620 | 5/2+ | | 9.18E+04 |
| 90 | 142 | Th | 232.0380553 | 7.615 | 0+ | 100 | 4.43E+17 |
| 90 | 144 | Th | 234.0436012 | 7.597 | 0+ | | 2.08E+06 |
| 91 | 137 | Pa | 228.0310514 | 7.632 | (3+) | | 7.92E+04 |
| 91 | 138 | Pa | 229.0320968 | 7.630 | (5/2+) | | 1.30E+05 |
| 91 | 139 | Pa | 230.0345408 | 7.622 | (2−) | | 1.50E+06 |
| 91 | 140 | Pa | 231.035884 | 7.618 | 3/2− | 100 | 1.03E+12 |
| 91 | 141 | Pa | 232.0385916 | 7.609 | (2−) | | 1.13E+05 |
| 91 | 142 | Pa | 233.0402473 | 7.605 | 3/2− | | 2.33E+06 |
| 91 | 143 | Pa | 234.0433081 | 7.595 | 4+ | | 2.41E+04 |
| 91 | 148 | Pa | 239.05726 | 7.550 | (3/2)(−) | | 6.37E+03 |
| 92 | 138 | U | 230.0339398 | 7.621 | 0+ | | 1.80E+06 |
| 92 | 139 | U | 231.0362937 | 7.613 | (5/2) | | 3.63E+05 |
| 92 | 140 | U | 232.0371562 | 7.612 | 0+ | | 2.17E+09 |
| 92 | 141 | U | 233.0396352 | 7.604 | 5/2+ | | 5.01E+12 |
| 92 | 142 | U | 234.0409521 | 7.601 | 0+ | 0.0055 | 7.74E+12 |
| 92 | 143 | U | 235.0439299 | 7.591 | 7/2− | 0.72 | 2.22E+16 |
| 92 | 144 | U | 236.045568 | 7.586 | 0+ | | 7.38E+14 |
| 92 | 145 | U | 237.0487302 | 7.576 | 1/2+ | | 5.83E+05 |
| 92 | 146 | U | 238.0507882 | 7.570 | 0+ | 99.275 | 1.41E+17 |
| 92 | 148 | U | 240.056592 | 7.552 | 0+ | | 5.08E+04 |
| 93 | 141 | Np | 234.042895 | 7.590 | (0+) | | 3.80E+05 |
| 93 | 142 | Np | 235.0440633 | 7.587 | 5/2+ | | 3.42E+07 |
| 93 | 143 | Np | 236.0465696 | 7.579 | (6−) | | 4.86E+12 |
| 93 | 143 | Np | | | 1(−) | | 8.10E+04 |
| 93 | 144 | Np | 237.0481734 | 7.575 | 5/2+ | | 6.75E+13 |
| 93 | 145 | Np | 238.0509464 | 7.566 | 2+ | | 1.83E+05 |
| 93 | 146 | Np | 239.052939 | 7.561 | 5/2+ | | 2.04E+05 |
| 93 | 147 | Np | 240.0561622 | 7.550 | (5+) | | 3.72E+03 |
| 94 | 140 | Pu | 234.0433171 | 7.585 | 0+ | | 3.17E+04 |
| 94 | 142 | Pu | 236.046058 | 7.578 | 0+ | | 9.01E+07 |
| 94 | 143 | Pu | 237.0484097 | 7.571 | 7/2− | | 3.91E+06 |

*Continued—*

| *Z* | *N* | Sym | *m* (amu) | *B* (MeV) | Spin | % | τ(s) |
|---|---|---|---|---|---|---|---|
| 94 | 144 | Pu | 238.0495599 | 7.568 | 0+ | | 2.77E+09 |
| 94 | 145 | Pu | 239.0521634 | 7.560 | 1/2+ | | 7.60E+11 |
| 94 | 146 | Pu | 240.0538135 | 7.556 | 0+ | | 2.07E+11 |
| 94 | 147 | Pu | 241.0568515 | 7.546 | 5/2+ | | 4.53E+08 |
| 94 | 148 | Pu | 242.0587426 | 7.541 | 0+ | | 1.18E+13 |
| 94 | 149 | Pu | 243.0620031 | 7.531 | 7/2+ | | 1.78E+04 |
| 94 | 150 | Pu | 244.0642039 | 7.525 | 0+ | | 2.55E+15 |
| 94 | 151 | Pu | 245.0677472 | 7.514 | (9/2−) | | 3.78E+04 |
| 94 | 152 | Pu | 246.0702046 | 7.507 | 0+ | | 9.37E+05 |
| 94 | 153 | Pu | 247.07407 | 7.494 | 1/2+ | | 1.96E+05 |
| 95 | 142 | Am | 237.049996 | 7.561 | 5/2(−) | | 4.39E+03 |
| 95 | 143 | Am | 238.0519843 | 7.556 | 1+ | | 5.87E+03 |
| 95 | 144 | Am | 239.0530245 | 7.554 | 5/2− | | 4.28E+04 |
| 95 | 145 | Am | 240.0553002 | 7.547 | (3−) | | 1.83E+05 |
| 95 | 146 | Am | 241.0568291 | 7.543 | 5/2− | | 1.36E+10 |
| 95 | 147 | Am | 242.0595492 | 7.535 | 1− | | 5.77E+04 |
| 95 | 148 | Am | 243.0613811 | 7.530 | 5/2− | | 2.32E+11 |
| 95 | 149 | Am | 244.0642848 | 7.521 | (6−) | | 3.64E+04 |
| 95 | 150 | Am | 245.0664521 | 7.515 | (5/2)+ | | 7.38E+03 |
| 96 | 142 | Cm | 238.0530287 | 7.548 | 0+ | | 8.64E+03 |
| 96 | 143 | Cm | 239.054957 | 7.543 | (7/2−) | | 1.04E+04 |
| 96 | 144 | Cm | 240.0555295 | 7.543 | 0+ | | 2.33E+06 |
| 96 | 145 | Cm | 241.057653 | 7.537 | 1/2+ | | 2.83E+06 |
| 96 | 146 | Cm | 242.0588358 | 7.534 | 0+ | | 1.41E+07 |
| 96 | 147 | Cm | 243.0613891 | 7.527 | 5/2+ | | 9.18E+08 |
| 96 | 148 | Cm | 244.0627526 | 7.524 | 0+ | | 5.71E+08 |
| 96 | 149 | Cm | 245.0654912 | 7.516 | 7/2+ | | 2.68E+11 |
| 96 | 150 | Cm | 246.0672237 | 7.511 | 0+ | | 1.49E+11 |
| 96 | 151 | Cm | 247.0703535 | 7.502 | 9/2− | | 4.92E+14 |
| 96 | 152 | Cm | 248.0723485 | 7.497 | 0+ | | 1.07E+13 |
| 96 | 153 | Cm | 249.0759534 | 7.486 | 1/2+ | | 3.85E+03 |
| 96 | 154 | Cm | 250.078357 | 7.479 | 0+ | | 3.06E+11 |
| 97 | 146 | Bk | 243.0630076 | 7.517 | (3/2−) | | 1.62E+04 |
| 97 | 147 | Bk | 244.0651808 | 7.511 | (1−) | | 1.57E+04 |
| 97 | 148 | Bk | 245.0663616 | 7.509 | 3/2− | | 4.27E+05 |
| 97 | 149 | Bk | 246.0686729 | 7.503 | 2(−) | | 1.56E+05 |
| 97 | 150 | Bk | 247.0703071 | 7.499 | (3/2−) | | 4.35E+10 |
| 97 | 151 | Bk | 248.073086 | 7.491 | (6+) | | 2.84E+08 |
| 97 | 151 | Bk | | | 1(−) | | 8.53E+04 |
| 97 | 152 | Bk | 249.0749867 | 7.486 | 7/2+ | | 2.76E+07 |
| 97 | 153 | Bk | 250.0783165 | 7.476 | 2− | | 1.16E+04 |
| 98 | 148 | Cf | 246.0688053 | 7.499 | 0+ | | 1.29E+05 |
| 98 | 149 | Cf | 247.0710006 | 7.493 | (7/2+) | | 1.12E+04 |
| 98 | 150 | Cf | 248.0721849 | 7.491 | 0+ | | 2.88E+07 |
| 98 | 151 | Cf | 249.0748535 | 7.483 | 9/2− | | 1.11E+10 |
| 98 | 152 | Cf | 250.0764061 | 7.480 | 0+ | | 4.12E+08 |
| 98 | 153 | Cf | 251.0795868 | 7.470 | 1/2+ | | 2.83E+10 |
| 98 | 154 | Cf | 252.0816258 | 7.465 | 0+ | | 8.34E+07 |
| 98 | 155 | Cf | 253.0851331 | 7.455 | (7/2+) | | 1.54E+06 |
| 98 | 156 | Cf | 254.0873229 | 7.449 | 0+ | | 5.23E+06 |
| 98 | 157 | Cf | 255.091046 | 7.438 | (9/2+) | | 5.04E+03 |
| 99 | 150 | Es | 249.076411 | 7.474 | 7/2(+) | | 6.12E+03 |
| 99 | 151 | Es | 250.078612 | 7.469 | (6+) | | 3.10E+04 |
| 99 | 151 | Es | | | 1(−) | | 7.99E+03 |
| 99 | 152 | Es | 251.0799921 | 7.466 | (3/2−) | | 1.19E+05 |
| 99 | 153 | Es | 252.0829785 | 7.457 | (5−) | | 4.07E+07 |
| 99 | 154 | Es | 253.0848247 | 7.453 | 7/2+ | | 1.77E+06 |
| 99 | 155 | Es | 254.088022 | 7.444 | (7+) | | 2.38E+07 |
| 99 | 156 | Es | 255.0902731 | 7.438 | (7/2+) | | 3.44E+06 |
| 99 | 157 | Es | | | (8+) | | 2.74E+04 |
| 100 | 151 | Fm | 251.081575 | 7.457 | (9/2−) | | 1.91E+04 |
| 100 | 152 | Fm | 252.0824669 | 7.456 | 0+ | | 9.14E+04 |
| 100 | 153 | Fm | 253.0851852 | 7.448 | 1/2+ | | 2.59E+05 |
| 100 | 154 | Fm | 254.0868542 | 7.445 | 0+ | | 1.17E+04 |
| 100 | 155 | Fm | 255.0899622 | 7.436 | 7/2+ | | 7.23E+04 |
| 100 | 156 | Fm | 256.0917731 | 7.432 | 0+ | | 9.46E+03 |
| 100 | 157 | Fm | 257.0951047 | 7.422 | (9/2+) | | 8.68E+06 |
| 101 | 155 | Md | 256.094059 | 7.420 | (0−,1−) | | 4.69E+03 |
| 101 | 156 | Md | 257.0955414 | 7.418 | (7/2−) | | 1.99E+04 |
| 101 | 157 | Md | 258.0984313 | 7.410 | (8−) | | 4.45E+06 |
| 101 | 157 | Md | 256.09360 | | (1−) | | 3.60E+03 |
| 101 | 158 | Md | 259.100509 | 7.405 | (7/2−) | | 5.76E+03 |
| 101 | 159 | Md | 260.103652 | 7.396 | | | 2.40E+06 |
| 103 | 159 | Lr | 262.109634 | 7.374 | | | 1.30E+04 |

# Appendix C

## *Element Properties*

| | |
|---|---|
| $Z$ | Charge number (number of protons in the nucleus = number of electrons) |
| $\rho$ | Mass density at standard temperature (20 °C = 293.15 K) and pressure (1 atmosphere) |
| $m$ | Standard atomic weight (average mass of an atom, abundance-weighted average of the isotope masses) |
| $T_{melting}$ | Temperature of melting point (transition point between solid and liquid phase) at 1 atm pressure |
| $T_{boiling}$ | Temperature of boiling point (transition point between liquid and gas phase) at 1 atm pressure |
| $L_m$ | Heat of melting/fusion |
| $L_v$ | Heat of vaporization |
| $E_1$ | Ionization energy (energy to remove least bound electron) |

| $Z$ | Sym | Name | Electron Configuration | $\rho$(g/cm$^3$) | $m$(g/mol) | $T_{melting}$ (K) | $T_{boiling}$ (K) | $L_m$ (kJ/mol) | $L_v$ (kJ/mol) | $E_1$(eV) |
|---|---|---|---|---|---|---|---|---|---|---|
| 1 | H | Hydrogen$_{gas}$ | $1s^1$ | $8.988 \cdot 10^{-5}$ | 1.00794 | 14.01 | 20.28 | 0.117 | 0.904 | 13.5984 |
| 2 | He | Helium$_{gas}$ | $1s^2$ | $1.786 \cdot 10^{-4}$ | 4.002602 | — | 4.22 | — | 0.0829 | 24.5874 |
| 3 | Li | Lithium | $[He]2s^1$ | 0.534 | 6.941 | 453.69 | 1615 | 3.00 | 147.1 | 5.3917 |
| 4 | Be | Beryllium | $[He]2s^2$ | 1.85 | 9.012182 | 1560 | 2742 | 7.895 | 297 | 9.3227 |
| 5 | B | Boron | $[He]2s^2\,2p^1$ | 2.34 | 10.811 | 2349 | 4200 | 50.2 | 480 | 8.2980 |
| 6 | C | Carbon$_{graphite}$ | $[He]2s^2\,2p^2$ | 2.267 | 12.0107 | 3800 | 4300 | 117 | 710.9 | 11.2603 |
| 7 | N | Nitrogen$_{gas}$ | $[He]2s^2\,2p^3$ | $1.251 \cdot 10^{-3}$ | 14.0067 | 63.1526 | 77.36 | 0.72 | 5.56 | 14.5341 |
| 8 | O | Oxygen$_{gas}$ | $[He]2s^2\,2p^4$ | $1.429 \cdot 10^{-3}$ | 15.9994 | 54.36 | 90.20 | 0.444 | 6.82 | 13.6181 |
| 9 | F | Fluorine$_{gas}$ | $[He]2s^2\,2p^5$ | $1.7 \cdot 10^{-3}$ | 18.998403 | 53.53 | 85.03 | 0.510 | 6.62 | 17.4228 |
| 10 | Ne | Neon$_{gas}$ | $[He]2s^2\,2p^6$ | $9.002 \cdot 10^{-4}$ | 20.1797 | 24.56 | 27.07 | 0.335 | 1.71 | 21.5645 |
| 11 | Na | Sodium | $[Ne]3s^1$ | 0.968 | 22.989770 | 370.87 | 1156 | 2.60 | 97.42 | 5.1391 |
| 12 | Mg | Magnesium | $[Ne]3s^2$ | 1.738 | 24.3050 | 923 | 1363 | 8.48 | 128 | 7.6462 |
| 13 | Al | Aluminum | $[Ne]3s^2\,3p^1$ | 2.70 | 26.981538 | 933.47 | 2792 | 10.71 | 294.0 | 5.9858 |
| 14 | Si | Silicon | $[Ne]3s^2\,3p^2$ | 2.3290 | 28.0855 | 1687 | 3538 | 50.21 | 359 | 8.1517 |
| 15 | P | Phosphorus$_{white}$ | $[Ne]3s^2\,3p^3$ | 1.823 | 30.973761 | 317.3 | 550 | 0.66 | 12.4 | 10.4867 |
| 16 | S | Sulfur | $[Ne]3s^2\,3p^4$ | 1.92–2.07 | 32.065 | 388.36 | 717.8 | 1.727 | 45 | 10.3600 |
| 17 | Cl | Chlorine | $[Ne]3s^2\,3p^5$ | $3.2 \cdot 10^{-3}$ | 35.453 | 171.6 | 239.11 | 6.406 | 20.41 | 12.9676 |
| 18 | Ar | Argon | $[Ne]3s^2\,3p^6$ | $1.784 \cdot 10^{-3}$ | 39.948 | 83.80 | 87.30 | 1.18 | 6.43 | 15.7596 |
| 19 | K | Potassium | $[Ar]4s^1$ | 0.89 | 39.0983 | 336.53 | 1032 | 2.4 | 79.1 | 4.3407 |

| $Z$ | Sym | Name | Electron Configuration | $\rho$(g/cm$^3$) | $m$(g/mol) | $T_{melting}$ (K) | $T_{boiling}$ (K) | $L_m$ (kJ/mol) | $L_v$ (kJ/mol) | $E_1$(eV) |
|---|---|---|---|---|---|---|---|---|---|---|
| 20 | Ca | Calcium | [Ar]$4s^2$ | 1.55 | 40.078 | 1115 | 1757 | 8.54 | 154.7 | 6.1132 |
| 21 | Sc | Scandium | [Ar]$3d^1\,4s^2$ | 2.985 | 44.955910 | 1814 | 3109 | 14.1 | 332.7 | 6.5615 |
| 22 | Ti | Titanium | [Ar]$3d^2\,4s^2$ | 4.506 | 47.867 | 1941 | 3560 | 14.15 | 425 | 6.8281 |
| 23 | V | Vanadium | [Ar]$3d^3\,4s^2$ | 6.0 | 50.9415 | 2183 | 3680 | 21.5 | 459 | 6.7462 |
| 24 | Cr | Chromium | [Ar]$3d^5\,4s^1$ | 7.19 | 51.9961 | 2180 | 2944 | 21.0 | 339.5 | 6.7665 |
| 25 | Mn | Manganese | [Ar]$3d^5\,4s^2$ | 7.21 | 54.938049 | 1519 | 2334 | 12.91 | 221 | 7.4340 |
| 26 | Fe | Iron | [Ar]$3d^6\,4s^2$ | 7.874 | 55.845 | 1811 | 3134 | 13.81 | 340 | 7.9024 |
| 27 | Co | Cobalt | [Ar]$3d^7\,4s^2$ | 8.90 | 58.933200 | 1768 | 3200 | 16.06 | 377 | 7.8810 |
| 28 | Ni | Nickel | [Ar]$3d^8\,4s^2$ | 8.908 | 58.6934 | 1728 | 3186 | 17.48 | 377.5 | 7.6398 |
| 29 | Cu | Copper | [Ar]$3d^{10}\,4s^1$ | 8.94 | 63.546 | 1357.77 | 2835 | 13.26 | 300.4 | 7.7264 |
| 30 | Zn | Zinc | [Ar]$3d^{10}\,4s^2$ | 7.14 | 65.409 | 692.68 | 1180 | 7.32 | 123.6 | 9.3942 |
| 31 | Ga | Gallium | [Ar]$3d^{10}\,4s^2\,4p^1$ | 5.91 | 69.723 | 302.9146 | 2477 | 5.59 | 254 | 5.9993 |
| 32 | Ge | Germanium | [Ar]$3d^{10}\,4s^2\,4p^2$ | 5.323 | 72.64 | 1211.40 | 3106 | 36.94 | 334 | 7.8994 |
| 33 | As | Arsenic | [Ar]$3d^{10}\,4s^2\,4p^3$ | 5.727 | 74.92160 | 1090 | 887 | 24.44 | 34.76 | 9.7886 |
| 34 | Se | Selenium | [Ar]$3d^{10}\,4s^2\,4p^4$ | 4.28–4.81 | 78.96 | 494 | 958 | 6.69 | 95.48 | 9.7524 |
| 35 | Br | Bromine$_{liquid}$ | [Ar]$3d^{10}\,4s^2\,4p^5$ | 3.1028 | 79.904 | 265.8 | 332.0 | 10.571 | 29.96 | 11.8138 |
| 36 | Kr | Krypton$_{gas}$ | [Ar]$3d^{10}\,4s^2\,4p^6$ | $3.749 \cdot 10^{-3}$ | 83.798 | 115.79 | 119.93 | 1.64 | 9.08 | 13.9996 |
| 37 | Rb | Rubidium | [Kr]$5s^1$ | 1.532 | 85.4678 | 312.46 | 961 | 2.19 | 75.77 | 4.1771 |
| 38 | Sr | Strontium | [Kr]$5s^2$ | 2.64 | 87.62 | 1050 | 1655 | 7.43 | 136.9 | 5.6949 |
| 39 | Y | Yttrium | [Kr]$4d^1\,5s^2$ | 4.472 | 88.90585 | 1799 | 3609 | 11.42 | 365 | 6.2173 |
| 40 | Zr | Zirconium | [Kr]$4d^2\,5s^2$ | 6.52 | 91.224 | 2128 | 4682 | 14 | 573 | 6.6339 |
| 41 | Nb | Niobium | [Kr]$4d^4\,5s^1$ | 8.57 | 92.90638 | 2750 | 5017 | 30 | 689.9 | 6.7589 |
| 42 | Mo | Molybdenum | [Kr]$4d^5\,5s^1$ | 10.28 | 95.94 | 2896 | 4912 | 37.48 | 617 | 7.0924 |
| 43 | Tc | Technetium | [Kr]$4d^5\,5s^2$ | 11 | (98) | 2430 | 4538 | 33.29 | 585.2 | 7.28 |
| 44 | Ru | Ruthenium | [Kr]$4d^7\,5s^1$ | 12.45 | 101.07 | 2607 | 4423 | 38.59 | 591.6 | 7.3605 |
| 45 | Rh | Rhodium | [Kr]$4d^8\,5s^1$ | 12.41 | 102.90550 | 2237 | 3968 | 26.59 | 494 | 7.4589 |
| 46 | Pd | Palladium | [Kr]$4d^{10}$ | 12.023 | 106.42 | 1828.05 | 3236 | 16.74 | 362 | 8.3369 |
| 47 | Ag | Silver | [Kr]$4d^{10}\,5s^1$ | 10.49 | 107.8682 | 1234.93 | 2435 | 11.28 | 250.58 | 7.5762 |
| 48 | Cd | Cadmium | [Kr]$4d^{10}\,5s^2$ | 8.65 | 112.411 | 594.22 | 1040 | 6.21 | 99.87 | 8.9938 |
| 49 | In | Indium | [Kr]$4d^{10}\,5s^2\,5p^1$ | 7.31 | 114.818 | 429.7485 | 2345 | 3.281 | 231.8 | 5.7864 |
| 50 | Sn | Tin$_{white}$ | [Kr]$4d^{10}\,5s^2\,5p^2$ | 7.365 | 118.710 | 505.08 | 2875 | 7.03 | 296.1 | 7.3439 |
| 51 | Sb | Antimony | [Kr]$4d^{10}\,5s^2\,5p^3$ | 6.697 | 121.760 | 903.78 | 1860 | 19.79 | 193.43 | 8.6084 |
| 52 | Te | Tellurium | [Kr]$4d^{10}\,5s^2\,5p^4$ | 6.24 | 127.60 | 722.66 | 1261 | 17.49 | 114.1 | 9.0096 |
| 53 | I | Iodine | [Kr]$4d^{10}\,5s^2\,5p^5$ | 4.933 | 126.90447 | 386.85 | 457.4 | 15.52 | 41.57 | 10.4513 |
| 54 | Xe | Xenon$_{gas}$ | [Kr]$4d^{10}\,5s^2\,5p^6$ | $5.894 \cdot 10^{-3}$ | 131.293 | 161.4 | 165.03 | 2.27 | 12.64 | 12.1298 |
| 55 | Cs | Cesium | [Xe]$6s^1$ | 1.93 | 132.90545 | 301.59 | 944 | 2.09 | 63.9 | 3.8939 |
| 56 | Ba | Barium | [Xe]$6s^2$ | 3.51 | 137.327 | 1000 | 2170 | 7.12 | 140.3 | 5.2117 |
| 57 | La | Lanthanum | [Xe]$5d^1\,6s^2$ | 6.162 | 138.9055 | 1193 | 3737 | 6.20 | 402.1 | 5.5769 |
| 58 | Ce | Cerium | [Xe]$4f^1\,5d^1\,6s^2$ | 6.770 | 140.116 | 1068 | 3716 | 5.46 | 398 | 5.5387 |
| 59 | Pr | Praseodymium | [Xe]$4f^3\,6s^2$ | 6.77 | 140.90765 | 1208 | 3793 | 6.89 | 331 | 5.473 |
| 60 | Nd | Neodymium | [Xe]$4f^4\,6s^2$ | 7.01 | 144.24 | 1297 | 3347 | 7.14 | 289 | 5.5250 |
| 61 | Pm | Promethium | [Xe]$4f^5\,6s^2$ | 7.26 | (145) | 1315 | 3273 | 7.13 | 289 | 5.582 |

| $Z$ | Sym | Name | Electron Configuration | $\rho$(g/cm$^3$) | $m$(g/mol) | $T_{melting}$ (K) | $T_{boiling}$ (K) | $L_m$ (kJ/mol) | $L_v$ (kJ/mol) | $E_1$(eV) |
|---|---|---|---|---|---|---|---|---|---|---|
| 62 | Sm | Samarium | [Xe]$4f^6\,6s^2$ | 7.52 | 150.36 | 1345 | 2067 | 8.62 | 165 | 5.6437 |
| 63 | Eu | Europium | [Xe]$4f^7\,6s^2$ | 5.264 | 151.964 | 1099 | 1802 | 9.21 | 176 | 5.6704 |
| 64 | Gd | Gadolinium | [Xe]$4f^7\,5d^1\,6s^2$ | 7.90 | 157.25 | 1585 | 3546 | 10.05 | 301.3 | 6.1498 |
| 65 | Tb | Terbium | [Xe]$4f^9\,6s^2$ | 8.23 | 158.92534 | 1629 | 3503 | 10.15 | 293 | 5.8638 |
| 66 | Dy | Dysprosium | [Xe]$4f^{10}\,6s^2$ | 8.540 | 162.500 | 1680 | 2840 | 11.06 | 280 | 5.9389 |
| 67 | Ho | Holmium | [Xe]$4f^{11}\,6s^2$ | 8.79 | 164.93032 | 1734 | 2993 | 17.0 | 265 | 6.0215 |
| 68 | Er | Erbium | [Xe]$4f^{12}\,6s^2$ | 9.066 | 167.259 | 1802 | 3141 | 19.90 | 280 | 6.1077 |
| 69 | Tm | Thulium | [Xe]$4f^{13}\,6s^2$ | 9.32 | 168.93421 | 1818 | 2223 | 16.84 | 247 | 6.1843 |
| 70 | Yb | Ytterbium | [Xe]$4f^{14}\,6s^2$ | 6.90 | 173.04 | 1097 | 1469 | 7.66 | 159 | 6.2542 |
| 71 | Lu | Lutetium | [Xe]$4f^{14}\,5d^1\,6s^2$ | 9.841 | 174.967 | 1925 | 3675 | 22 | 414 | 5.4259 |
| 72 | Hf | Hafnium | [Xe]$4f^{14}\,5d^2\,6s^2$ | 13.31 | 178.49 | 2506 | 4876 | 27.2 | 571 | 6.8251 |
| 73 | Ta | Tantalum | [Xe]$4f^{14}\,5d^3\,6s^2$ | 16.69 | 180.9479 | 3290 | 5731 | 36.57 | 732.8 | 7.5496 |
| 74 | W | Tungsten | [Xe]$4f^{14}\,5d^4\,6s^2$ | 19.25 | 183.84 | 3695 | 5828 | 52.31 | 806.7 | 7.8640 |
| 75 | Re | Rhenium | [Xe]$4f^{14}\,5d^5\,6s^2$ | 21.02 | 186.207 | 3459 | 5869 | 60.3 | 704 | 7.8335 |
| 76 | Os | Osmium | [Xe]$4f^{14}\,5d^6\,6s^2$ | 22.61 | 190.23 | 3306 | 5285 | 57.85 | 738 | 8.4382 |
| 77 | Ir | Iridium | [Xe]$4f^{14}\,5d^7\,6s^2$ | 22.56 | 192.217 | 2739 | 4701 | 41.12 | 563 | 8.9670 |
| 78 | Pt | Platinum | [Xe]$4f^{14}\,5d^9\,6s^1$ | 21.45 | 195.078 | 2041.4 | 4098 | 22.17 | 469 | 8.9588 |
| 79 | Au | Gold | [Xe]$4f^{14}\,5d^{10}\,6s^1$ | 19.3 | 196.96655 | 1337.33 | 3129 | 12.55 | 324 | 9.2255 |
| 80 | Hg | Mercury$_{liquid}$ | [Xe]$4f^{14}\,5d^{10}\,6s^2$ | 13.534 | 200.59 | 234.32 | 629.88 | 2.29 | 59.11 | 10.4375 |
| 81 | Tl | Thallium | [Xe]$4f^{14}\,5d^{10}\,6s^2\,6p^1$ | 11.85 | 204.3833 | 577 | 1746 | 4.14 | 165 | 6.1082 |
| 82 | Pb | Lead | [Xe]$4f^{14}\,5d^{10}\,6s^2\,6p^2$ | 11.34 | 207.2 | 600.61 | 2022 | 4.77 | 179.5 | 7.4167 |
| 83 | Bi | Bismuth | [Xe]$4f^{14}\,5d^{10}\,6s^2\,6p^3$ | 9.78 | 208.98038 | 544.7 | 1837 | 11.30 | 151 | 7.2855 |
| 84 | Po | Polonium | [Xe]$4f^{14}\,5d^{10}\,6s^2\,6p^4$ | 9.320 | (209) | 527 | 1235 | 13 | 102.91 | 8.414 |
| 85 | At | Astatine | [Xe]$4f^{14}\,5d^{10}\,6s^2\,6p^5$ | ? | (210) | ? | ? | ? | ? | ? |
| 86 | Rn | Radon | [Xe]$4f^{14}\,5d^{10}\,6s^2\,6p^6$ | $9.73 \cdot 10^{-3}$ | (222) | 202 | 211.3 | 3.247 | 18.10 | 10.7485 |
| 87 | Fr | Francium | [Rn]$7s^1$ | 1.87 | (223) | ~300 | ~950 | ~2 | ~65 | 4.0727 |
| 88 | Ra | Radium | [Rn]$7s^2$ | 5.5 | (226) | 973 | 2010 | 8.5 | 113 | 5.2784 |
| 89 | Ac | Actinium | [Rn]$6d^1\,7s^2$ | 10 | (227) | 1323 | 3471 | 14 | 400 | 5.17 |
| 90 | Th | Thorium | [Rn]$6d^2\,7s^2$ | 11.7 | 232.0381 | 2115 | 5061 | 13.81 | 514 | 6.3067 |
| 91 | Pa | Protactinum | [Rn]$5f^2\,6d^1\,7s^2$ | 15.37 | 231.03588 | 1841 | ~4300 | 12.34 | 481 | 5.89 |
| 92 | U | Uranium | [Rn]$5f^3\,6d^1\,7s^2$ | 19.1 | 238.02891 | 1405.3 | 4404 | 9.14 | 417.1 | 6.1941 |
| 93 | Np | Neptunium | [Rn]$5f^4\,6d^1\,7s^2$ | 20.45 | (237) | 910 | 4273 | 3.20 | 336 | 6.2657 |
| 94 | Pu | Plutonium | [Rn]$5f^6\,7s^2$ | 19.816 | (244) | 912.5 | 3505 | 2.82 | 333.5 | 6.0260 |
| 95 | Am | Americium | [Rn]$5f^7\,7s^2$ | 12 | (243) | 1449 | 2880 | 14.39 | 238.5 | 5.9738 |
| 96 | Cm | Curium | [Rn]$5f^7\,6d^1\,7s^2$ | 13.51 | (247) | 1613 | 3383 | ~15 | ? | 5.9914 |
| 97 | Bk | Berkelium | [Rn]$5f^9\,7s^2$ | ~14 | (247) | 1259 | ? | ? | ? | 6.1979 |
| 98 | Cf | Californium | [Rn]$5f^{10}\,7s^2$ | 15.1 | (251) | 1173 | 1743 | ? | ? | 6.2817 |
| 99 | Es | Einsteinium | [Rn]$5f^{11}\,7s^2$ | 8.84 | (252) | 1133 | ? | ? | ? | 6.42 |
| 100 | Fm | Fermium | [Rn]$5f^{12}\,7s^2$ | ? | (257) | 1800 | ? | ? | ? | 6.50 |
| 101 | Md | Mendelevium | [Rn]$5f^{13}\,7s^2$ | ? | (258) | 1100 | ? | ? | ? | 6.58 |
| 102 | No | Nobelium | [Rn]$5f^{14}\,7s^2$ | ? | (259) | ? | ? | ? | ? | 6.65 |

*Continued—*

| $Z$ | Sym | Name | Electron Configuration | $\rho$(g/cm$^3$) | $m$(g/mol) | $T_{melting}$ (K) | $T_{boiling}$ (K) | $L_m$ (kJ/mol) | $L_v$ (kJ/mol) | $E_1$(eV) |
|---|---|---|---|---|---|---|---|---|---|---|
| 103 | Lr | Lawrencium | [Rn]$5f^{14}\,7s^2\,7p^1$ | ? | (262) | ? | ? | ? | ? | 4.9 |
| 104 | Rf | Rutherfordium | [Rn]$5f^{14}\,6d^2\,7s^2$ | ? | (261) | ? | ? | ? | ? | 6 |
| 105 | Db | Dubnium | [Rn]$5f^{14}\,6d^3\,7s^2$ | ? | (262) | ? | ? | ? | ? | ? |
| 106 | Sg | Seaborgium | [Rn]$5f^{14}\,6d^4\,7s^2$ | ? | (266) | ? | ? | ? | ? | ? |
| 107 | Bh | Bohrium | [Rn]$5f^{14}\,6d^5\,7s^2$ | ? | (264) | ? | ? | ? | ? | ? |
| 108 | Hs | Hassium | [Rn]$5f^{14}\,6d^6\,7s^2$ | ? | (277) | ? | ? | ? | ? | ? |
| 109 | Mt | Meitnerium | [Rn]$5f^{14}\,6d^7\,7s^2$ | ? | (276) | ? | ? | ? | ? | ? |
| 110 | Ds | Darmstadtium | *[Rn]$5f^{14}\,6d^9\,7s^1$ | ? | (281) | ? | ? | ? | ? | ? |
| 111 | Rg | Roentgenium | *[Rn]$5f^{14}\,6d^9\,7s^2$ | ? | (280) | ? | ? | ? | ? | ? |
| 112 | | | *[Rn]$5f^{14}\,6d^{10}\,7s^2$ | ? | (285) | ? | ? | ? | ? | ? |
| 113 | | | *[Rn]$5f^{14}\,6d^{10}\,7s^2\,7p^1$ | ? | (284) | ? | ? | ? | ? | ? |
| 114 | | | *[Rn]$5f^{14}\,6d^{10}\,7s^2\,7p^2$ | ? | (289) | ? | ? | ? | ? | ? |
| 115 | | | *[Rn]$5f^{14}\,6d^{10}\,7s^2\,7p^3$ | ? | (288) | ? | ? | ? | ? | ? |
| 116 | | | *[Rn]$5f^{14}\,6d^{10}\,7s^2\,7p^4$ | ? | (293) | ? | ? | ? | ? | ? |
| 118 | | | *[Rn]$5f^{14}\,6d^{10}\,7s^2\,7p^6$ | ? | (294) | ? | ? | ? | ? | ? |
| | | | | | (longest-lived isotope) | | | | | |

*Predicted

# Answers to Selected Questions and Problems

## Chapter 1: Overview

### Multiple Choice

**1.1** c. **1.3** d. **1.5** a. **1.7** b. **1.9** c.

### Problems

**1.29** (a) Three. (b) Four. (c) One. (d) Six. (e) One. (f) Two. (g) Three. **1.31** 6.34. **1.33** $1 \cdot 10^{-7}$ cm. **1.35** $1.94822 \cdot 10^{6}$ inches. **1.37** $1 \cdot 10^{6}$ mm. **1.39** 1 μPa. **1.41** 2420 cm$^2$. **1.43** 356,000 km = 221000 miles; 407,000 km = 253000 miles. **1.45** $x_{\text{total}} = 5 \cdot 10^{-1}$ m; $x_{\text{avg}} = 9 \cdot 10^{-2}$ m. **1.47** 120 millifurlongs/microfortnight. **1.49** 76 times the surface area of Earth. **1.51** 1.56 barrels is equivalent to $1.10 \cdot 10^{4}$ cubic inches. **1.53** (a) $V_S = 1.41 \times 10^{27}$ m$^3$. (b) $V_E = 1.08 \times 10^{21}$ m$^3$. (c) $\rho_S = 1.41 \times 10^{3}$ kg/m$^3$. (d) $\rho_E = 5.52 \times 10^{3}$ kg/m$^3$. **1.55** 100 cm. **1.57** $\Delta x = 20$ m and $\Delta y = 30$ m.

**1.59** $\vec{A} = 2.5\hat{x} + 1.5\hat{y}$

$\vec{B} = 5.5\hat{x} - 1.5\hat{y}$

$\vec{C} = -6\hat{x} - 3\hat{y}$

**1.61** (2,–3). **1.63** $\vec{D} = 2\hat{x} - 3\hat{y}$.
**1.65** $\vec{A} = 65\hat{x} + 38\hat{y}$, $\vec{B} = -57\hat{x} + 20\hat{y}$, $\vec{C} = -15\hat{x} - 20\hat{y}$, $\vec{D} = 80\hat{x} - 41\hat{y}$. **1.67** 3.27 km. **1.69** $-20\hat{x} + 50\hat{y} - 3\hat{z}$, length of 60 paces. **1.71** $f = 16°$, $\alpha = 41°$ and $\theta = 140°$.

### Additional Problems

**1.73** $2 \cdot 10^{8}$. **1.75** $|\vec{A}| = 58.3$ m, $|\vec{B}| = 58.3$ m.
**1.77**

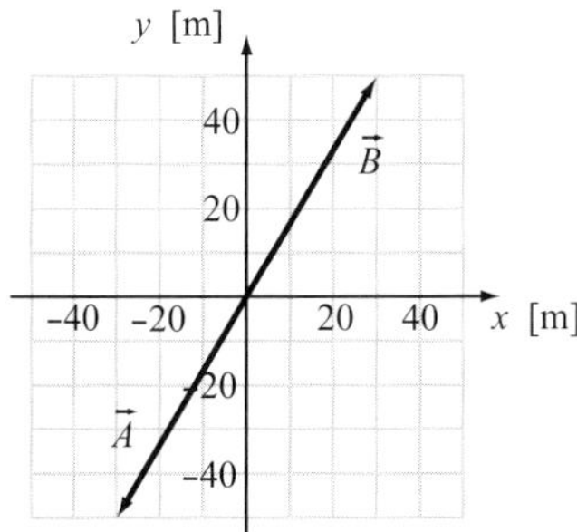


$|\vec{A}| = 58.3$ m, $|\vec{B}| = 58.3$ m.

**1.79** 63.7 m at –57.1° or 303° (equivalent angles). **1.81** $\vec{A} = 63.3$ at 68.7°; $\vec{B} = 175$ at –59.0°. **1.83** 446 at 267°. **1.85** (a) $1.70 \cdot 10^{3}$ at 295.9°. (b) $1.61 \cdot 10^{3}$ at 292°. **1.87** 1000 N. **1.89** (a) 125 miles. (b) 240° or –120° (from positive $x$-axis or E). (c) 167 miles. **1.91** 4 km at 20° W of N. **1.93** $5.62 \cdot 10^{7}$ km. **1.95** 9629 inches. **1.97** $1.4 \cdot 10^{11}$ m, 18° from the Sun.

## Chapter 2: Motion in a Straight Line

### Multiple Choice

**2.1** e. **2.3** c. **2.5** e. **2.7** d. **2.9** a.

### Problems

**2.25** distance = $7 \cdot 10^{4}$ m; displacement = $3 \cdot 10^{4}$ meters south. **2.27** 0 m/s. **2.29** (a) 4 m/s. (b) –0.2 m/s. (c) 1 m/s. (d) 2 : 1. (e) [–5,–4], [1,2], and [4,5]. **2.31** 4.0 m/s. **2.33** average speed: 40 m/s; average velocity: 30 m/s. **2.35** 2.4 m/s$^2$ in the backward direction. **2.37** 10.0 m/s$^2$. **2.39** $-1.0 \cdot 10^{2}$ m/s$^2$. **2.41** (a) 650 m/s. (b) 0.66 s and –0.98 s. (c) 8.3 m/s$^2$. (d)

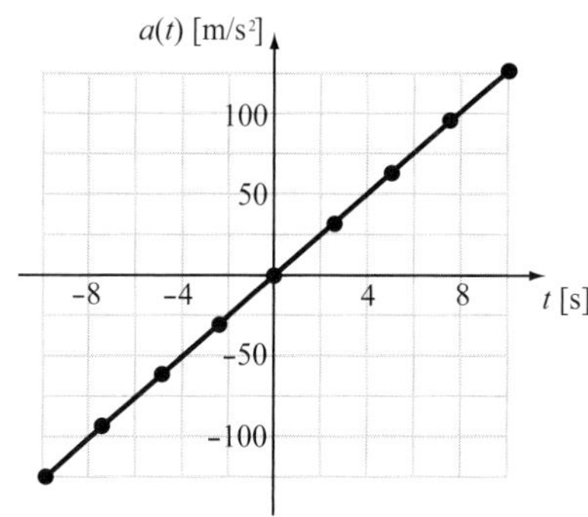


**2.43** –1200 cm. **2.45** $x = 23$ m. **2.47** $x = 18$ m. **2.49** (a) At 4.00 s, the displacement is 20. m/s. At 14.0 s, the displacement is 12 m/s. (b) 230 m. **2.51** (a) 17.7 s. (b) –1.08 m/s$^2$. **2.53** 33.3 m/s. **2.55** 20.0 m. **2.57** (a) 2.50 m/s. (b) 10.0 m. **2.59** (a) 5.1 m/s. (b) 3.8 m. **2.61** 2.33 s. **2.63** $v_{\frac{1}{2}y} = \sqrt{gy}$. **2.65** (a) 20 s. (b) 0.8 m/s$^2$. **2.67** (a) 60 m. (b) 8 s. **2.69** (a) 0.97 s. (b) $t_M = 1.6t_E$. (c) 4.6 m. **2.71** 29 m/s. **2.73** (a) 3.52 s. (b) 0.515 s.

### Additional Problems

**2.75** 395 m. **2.77** 2.85 s. **2.79** 40 m. **2.81** (a) 30 m. (b) –4 m/s. **2.83** The trains collide. **2.85** 570 m. **2.87** 290 m/s. **2.89** (a) 2.46 m/s$^2$. (b) 273 m. **2.91** (a) $v = 1.7\cos(0.46t/\text{s} - 0.31)$m/s – 0.2 m/s, $a = (-0.80\sin(0.46t/\text{s} - 0.31)$m/s$^2$. (b) 0.67 s, 7.5 s, 14 s, 21 s and 28 s. **2.93** (a) 18 hours. (b)

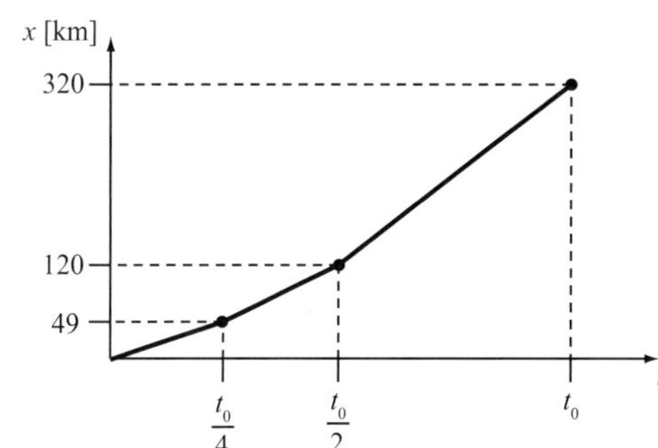

**2.95** (a) 37.9 m/s. (b) –26.8 m/s. (c) 1.13 s. **2.97** 692 m.

# Chapter 3: Motion in Two and Three Dimensions

## Multiple Choice

**3.1** c. **3.3** d. **3.5** c. **3.7** a. **3.9** a. **3.11** a. **3.13** a.

## Problems

**3.33** 2.8 m/s. **3.35** 3 km 68° north of east. **3.37** (a) 174 m. (b) 21.8 m/s, 44.6° north of west. **3.39** 30 m/s horizontally and 20 m/s vertically. **3.41** 4.69 s. **3.43** 4:1. **3.45** (a) 7.3 m. (b) 9.1 m/s. **3.47** 7 m. **3.49** (a) 60 m. (b) 75°. (c) 31 m. **3.51** initial: 20 m/s at 50°; final: 20 m/s at 30°. **3.53** 81 m/s. **3.55** 3 m/s. **3.57** 5. **3.59** (a) 60 m/s. (b) 60 m/s. **3.61** 14.3 s. **3.63** 3.94 m/s. **3.65** (a) 17.7°. (b) 7.62 s. (c) 0°. (d) 7.26 s. (e) $(0.0001\hat{x} + 5.33\hat{y})$ m/s. **3.67** 95.4 m/s.

## Additional Problems

**3.69** 26.0 m/s. **3.71** 25° below the horizontal. **3.73** helicopter: 10 m/s; box: 100 m/s. **3.75** 40 m/s at 80° above the horizontal. **3.77** 7. **3.79** 9.07 s. **3.81** 1 m/s$^2$. **3.83** 2.7 s. **3.85** (a) 19 m. (b) 2.0 s. **3.87** No. After the burglar reaches a horizontal displacement of 5.5 m, he has dropped 8.4 m from the first rooftop and cannot reach the second rooftop. **3.89** (a) Yes. (b) 49.0 m/s at 57.8° above the horizontal. **3.91** 9.2 m/s. **3.93** 9 km before the target; 0.2 s window of opportunity. **3.95** (a) 80 m/s at 50° below the horizontal. (b) 200 m. (c) 60 m/s at 40° below the horizontal.

# Chapter 4: Force

## Multiple Choice

**4.1** d. **4.3** d. **4.5** a. **4.7** c. **4.9** a. **4.11** b.

## Problems

**4.23** (a) 0.167 N. (b) 0.102 kg. **4.25** 229 lb. **4.27** 4.32 m/s$^2$. **4.29** (a) 21.8 N. (b) 14.0 N. (c) 7.84 N. **4.31** 183 N. **4.33** (a) 1.1 m/s$^2$. (b) 4.4 N. **4.35** $m_3 = 0.050$ kg; $\theta = 220°$. **4.37** (a) 400 N. (b) 500 N. **4.39** (a) 3 m/s$^2$. (b) 0.3 N. **4.41** 49.2°. **4.43** (a) 500 N. (b) 400 N. **4.45** Left: 44 N; right: 57 N. **4.47** 0.69 m/s$^2$ downward. **4.49** 280 N. **4.51** 807 N. **4.53** 80 m. **4.55** 5.84 N. **4.57** (a) 300 N. (b) 500 N. (c) Initially the force of friction is 506 N. After the refrigerator is in motion, the force of friction is the force of kinetic friction, 407 N. **4.59** 18 m/s. **4.61** 2.30 m/s$^2$. **4.63** 4.56 m/s$^2$.

## Additional Problems

**4.65** (a) 4.22 m/s$^2$. (b) 26.7 m. **4.67** (a) 59 N. (b) 77 N. **4.69** 2.45 m/s$^2$. **4.71** (a) 243 N. (b) 46.4 N. (c) 3.05 m/s. **4.73** 1.40 m/s$^2$. **4.75** 6800 N. **4.77** (a) 30 N. (b) 0.75. **4.79** 9.2°. **4.81** (a) $1.69 \cdot 10^{-5}$ kg/m. (b) 0.0274 N. **4.83** (a) 19 N. (b) $a_1 = 6.1$ m/s$^2$; $a_2 = 2.5$ m/s$^2$. **4.85** 1.72 m/s$^2$. **4.87** (a) $a_1 = 5.2$ m/s$^2$; $a_2 = 3.4$ m/s$^2$. (b) 35 N. **4.89** (a) 32.6°. (b) 243 N. **4.91** (a) 3.34 m/s$^2$. (b) 6.57 m/s$^2$.

# Chapter 5: Kinetic Energy, Work, and Power

## Multiple Choice

**5.1** c. **5.3** b. **5.5** e. **5.7** c. **5.9** b.

## Problems

**5.15** (a) $4.50 \cdot 10^3$ J. (b) $1.80 \cdot 10^2$ J. (c) $9.00 \cdot 10^2$ J. **5.17** $4 \cdot 10^6$ J. **5.19** 10 m/s. **5.21** $3.50 \cdot 10^3$ J. **5.23** –9.52 m/s. **5.25** 8 J. **5.27** $5 \cdot 10^2$ J. **5.29** 1.25 J. **5.31** 0.1. **5.33** 44 m/s. **5.35** 16 J. **5.37** (a) 2000 J. (b) 60 m/s. **5.39** $2.40 \cdot 10^4$ N/m. **5.41** 17.6 m/s. **5.43** 3.43 m/s. **5.45** $2.00 \cdot 10^6$ W; The power released by the car. **5.47** 450 W. **5.49** $3 \cdot 10^4$ J.

## Additional Problems

**5.51** 9.12 kJ. **5.53** 42 kW = 56 hp. **5.55** 63 hp. **5.57** 44 m/s. **5.59** 5.1 m/s. **5.61** 25 N. **5.63** 366 kJ. **5.65** 20 m/s. **5.67** 35.3°.

# Chapter 6: Potential Energy and Energy Conservation

## Multiple Choice

**6.1** a. **6.3** e. **6.5** d. **6.7** d. **6.9** a.

## Problems

**6.27** 29 J. **6.29** 0.0869 J. **6.31** $2 \cdot 10^6$ J. **6.33** 12 J. **6.35** (a) $F(y) = 2by - 3ay^2$. (b) $F(y) = -cU_0 \cos(cy)$. **6.37** 10 m/s. **6.39** 19 m/s. **6.41** (a) 8.7 J. (b) 18 m/s. **6.43** 30 m/s. **6.45** (a) 1 J. (b) 0 J. **6.47** (a) 4 J. (b) 3 m/s. **6.49** 5 m/s. **6.51** 16.9 kJ. **6.53** 39 kJ. **6.55** 8 J. **6.57** (a) 8.9 m/s. (b) 4.1 m/s. (c) –8.9 J. **6.59** $x = 40$ m, $y = 24$ m. **6.61** (a) 16 m/s. (b) 12 m/s. (c) $2.0 \cdot 10^{-1}$ m; 6.7 m.

## Additional Problems

**6.63** $41.0 \cdot 10^4$ J. **6.65** $2.0 \cdot 10^8$ J. **6.67** 500 J. **6.69** 1.6 m. **6.71** 3.8 m/s. **6.73** 8.85 m/s. **6.75** $1.27 \cdot 10^2$ m. **6.77** 2.21 kJ. **6.79** (a) $-1 \cdot 10^{-1}$ J (lost to friction). (b) $1 \cdot 10^2$ N/m. **6.81** (a) 12 J. (b) 9.4 J. (c) 12 J. (d) by a factor of 1/4. (e) by a factor of 1/4. **6.83** (a) 2.5 J. (b) 2.2 m/s. (c) 22 cm. **6.85** (a) $\dfrac{v}{\mu_k g}$. (b) $\dfrac{v^2}{2\mu_k g}$. (c) $mv^2/2$. (d) $-mv^2/2$. **6.87** (a) 700 J. (b) 700 J. (c) 700 J. (d) 0 J. (e) 0 J.

# Chapter 7: Momentum and Collisions

## Multiple Choice

**7.1** b. **7.3** b, d. **7.5** e. **7.7** c. **7.9** c.

## Problems

**7.21** (a) 1.5. (b) 1.0. **7.23** $p_x = 3.51$ kg m/s, $p_y = 5.61$ kg m/s. **7.25** $3 \cdot 10^4$ N, 0.87 s. **7.27** (a) 700 N s opposite to $v$. (b) 700 N s opposite to $v$. (c) 100 kg m/s opposite to $v$. **7.29** 0.01 m/s; 2 m/s; 10 m/s; 800 months. **7.31** (a) $3.2 \cdot 10^9$ m/s. (b) $5.5 \cdot 10^7$ m/s. **7.33** (a) –810. m/s. (b) 43.0 km. **7.35** 4.77 m/s. **7.37** –0.22 m/s. **7.39** 1 m/s. **7.41** 20 m/s at an angle of 40° above the horizontal.

**7.43** –34.5 km/s. **7.45** $v_B = 0.4$ m/s, $v_A = -1$ m/s.
**7.47** Position-Time:

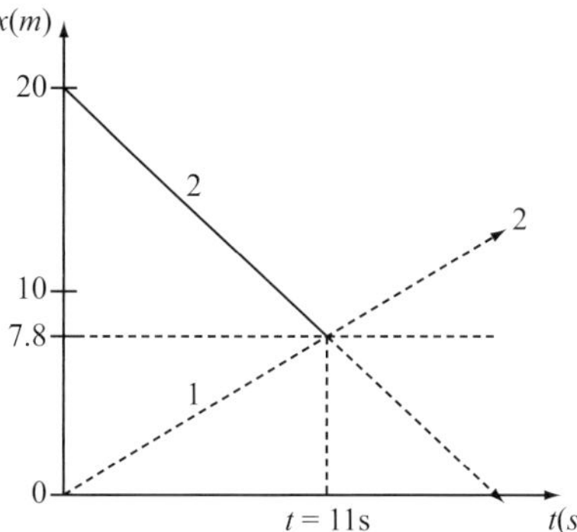


Velocity-Time:

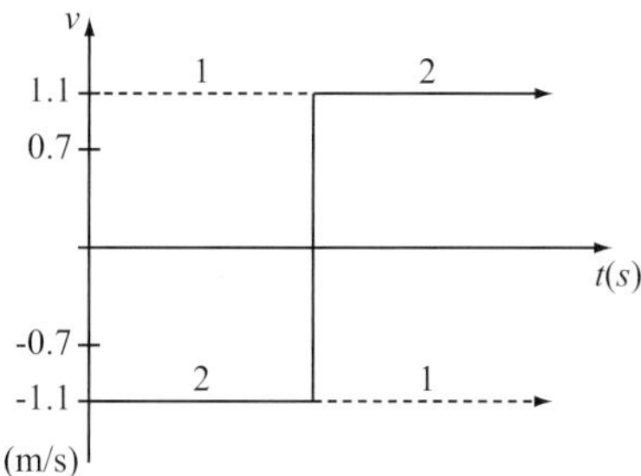


Force-Time:

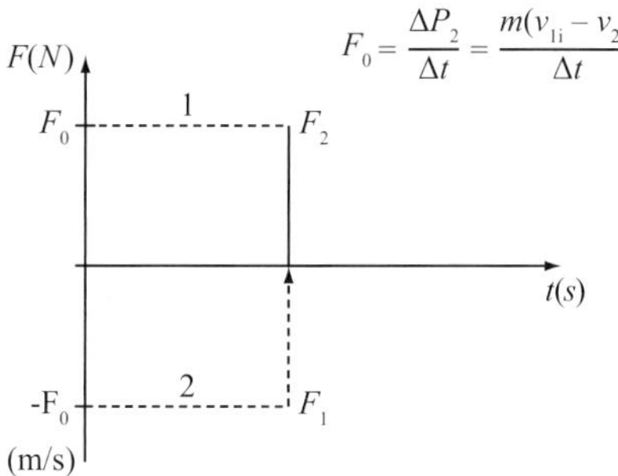


**7.49** $3.94 \cdot 10^4$ m/s. **7.51** 0.93 m/s; –24°. **7.53** $v_{1f} = 5.8 \cdot 10^2$ m/s in the positive $y$-direction, and $v_{2f} = 4.2 \cdot 10^2$ m/s at 36° below the positive $x$-axis. **7.55** Betty: 206 J; Sally: 121 J; The ratio $K_f/K_i$ is not equal to one, so the collision is inelastic. **7.57** 6.0 m/s. **7.59** smaller car: –48$g$; larger car: 16$g$. **7.61** 42 m/s. **7.63** 7 m/s.
**7.65**

| Object | $h_i$ [cm] | $h_f$ [cm] | $\epsilon$ |
|---|---|---|---|
| Range golf ball | 85.0 | 62.6 | 0.858 |
| Tennis ball | 85.0 | 43.1 | 0.712 |
| Billiard ball | 85.0 | 54.9 | 0.804 |
| Hand ball | 85.0 | 48.1 | 0.752 |
| Wooden ball | 85.0 | 30.9 | 0.603 |
| Steel ball bearing | 85.0 | 30.3 | 0.597 |
| Glass marble | 85.0 | 36.8 | 0.658 |
| Ball of rubber bands | 85.0 | 58.3 | 0.828 |
| Hollow, hard plastic balls | 85.0 | 40.2 | 0.688 |

**7.67** 40.7°. **7.69** 0.7 m. **7.71** $\epsilon = 0.69$, $K_f/K_i = 0.61$.

## Additional Problems

**7.73** (a) 0.63 m/s. (b) No. **7.75** 2 s. **7.77** $2.99 \cdot 10^5$ m/s. **7.79** –0.190 m/s. **7.81** 52400 N; 97.0$g$. **7.83** 30 kg m/s. **7.85** The momentum vector of the formation is 0.09 kg m/s $\hat{x}$ + 3 kg m/s $\hat{y}$. The 115-g bird would have a speed of 30 m/s at 2° east of north. **7.87** (a) The distance between the first ball and the pallina is 2.0 m and the distance between the second ball and the pallina is 1.7 m. (b) The distance between the first ball and the pallina is 0.98 m and the distance between the second ball and the pallina is 0.76 m. **7.89** The speed is $2v_0$ in the direction of 10° below the horizontal. **7.91** 15.9°.
**7.93**

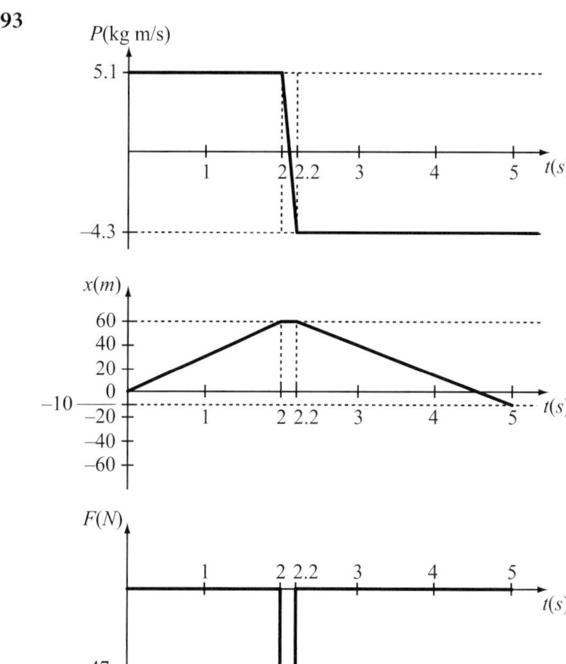


**7.95** At least four keys; key ring: 0.12 m/s; phone: 0.92 m/s. **7.97** 22 m/s at 0.22° to the right of the initial direction. **7.99** 1.37 N; 0.20 s. **7.101** (a) –(14.9 m/s) $\hat{x}$. (b) $E_{mec,f} = E_{mec,i}$ =14.1 kJ. **7.103** (a) 1.7 m/s. (b) 100°. **7.105** $1.16 \cdot 10^{-25}$ kg.

# Chapter 8: Systems of Particles and Extended Objects

## Multiple Choice

**8.1** c. **8.3** d. **8.5** e. **8.7** b. **8.9** b. **8.11** a.

## Problems

**8.25** (a) 4670 km. (b) 742,200 km. **8.27** (–0.5 m,–2 m ). **8.29** (a) 2.5$\hat{x}$ m/s. (b) Before the collision, $\vec{v}_t' = 1.5\hat{x}$ m/s and $\vec{v}_c' = -2.5\hat{x}$ m/s. After the collision, $\vec{v}_t' = -1.5\hat{x}$ m/s and $\vec{v}_c' = 2.5\hat{x}$ m/s with respect to the center of mass. **8.31** (a) –1 m/s (to the left). (b) 1 m/s (to the right). (c) –1 m/s (to the left). (d) 2 m/s (to the right). **8.33** 0.006$c$. **8.35** 100 N in the direction of the water's velocity. **8.37** (a) $J_{spec,\,toy} = 80$ s. (b) $J_{spec,\,chem} = 408$ s, $J_{spec,\,toy} = 0.2 J_{spec,\,chem}$. **8.39** 5.52 h. **8.41** (a) 11100 kg/s. (b) $1.63 \cdot 10^4$ m/s. (c) 88.4 m/s$^2$. **8.43** (20 cm,20 cm). **8.45** (6.67 cm,11.5 cm). **8.47** $(0.26\,\hat{x} + 0.12\,\hat{y})$ m. **8.49** $\frac{11x_0}{9}$.

## Additional Problems

**8.51** $6.5 \cdot 10^{-11}$ m. **8.53** 0.14 ft/s away from the direction the cannons fire. **8.55** (a) 0.87 m/s. (b) 55 J. **8.57** 9 m/s horizontally. **8.59** 4.1 km/s. **8.61** (a) 2 m/s$^2$. (b) 30 m/s$^2$. (c) 3000 m/s. **8.63** (12 cm,5 cm).
**8.65** $\left(\frac{6a}{8+\pi}, \frac{(4+3\pi)a}{8+\pi}\right)$.

**8.67** 69.3 s. **8.69** (a) $(-5.0\hat{i} + 12\hat{j})$ kg m/s. (b) $(4.0\hat{i} + 6.0\hat{j})$ kg m/s. (c) $(-9.0\hat{i} + 6.0\hat{j})$ kg m/s. **8.71** 88.2 N.

# Chapter 9: Circular Motion

## Multiple Choice

**9.1** d. **9.3** b. **9.5** c. **9.7** c. **9.9** a. **9.11** a.

## Problems

**9.27** $\frac{\pi}{2}$ rad ≈ 1.57 rad. **9.29** $3.07 \cdot 10^6$ m at 13.94° below the surface of the Earth. **9.31** (a) 0.7 rad/s$^2$. (b) 9 rad. **9.33** $\Delta d_{12} = 4.8$ m, $\Delta d_{23} = 3.6$ m. **9.35** (a) $v_A = 266.44277$ m/s, $v_B = 266.44396$ m/s, are in the direction of the Earth's rotation eastward. $\Delta v = 1.19$ mm/s. (b) $1.19 \cdot 10^{-4}$ rad/s. (c) 14.6 hours. (d) At the equator there is no difference between the velocities at $A$ and $B$, so the period is $T_R = \infty$. This means the pendulum does not rotate. **9.37** −6.5 rad/s$^2$. **9.39** −4.8 rad/s$^2$. **9.41** (a) $2.1 \cdot 10^{-1}$ rad/s. (b) $2.0 \cdot 10^{-2}$ m/s. (c) $\omega_1 = 2.0$ rad/s$^2$; $\omega_2 = 4.0 \cdot 10^{-1}$ rad/s$^2$. (d) $\alpha_2 = 2.0 \cdot 10^{-2}$ rad/s$^2$; $\alpha_3 = 1.0 \cdot 10^{-2}$ rad/s$^2$. **9.43** (a) 7.0 rad/s. (b) 2.5 rad/s. (c) $-2.5 \cdot 10^{-3}$ rad/s$^2$. **9.45** (a) 3 rad. (b) $a = 3$ m/s$^2$, $v = 9$ m/s, $x = 1$ m from the center at an angle of 3 rad, in the direction of the acceleration. **9.47** $3.43 \cdot 10^3$ N.

**9.49** $\Delta t = \sqrt{\dfrac{d}{2\mu g}}$. **9.51** 9.4 m/s. **9.53** 15°.

**9.55** (a) $v = \sqrt{rg\cos\theta\sin\theta}$. (b) $v = \sqrt{rg\cos\theta\left(\sin\theta + \mu_s\cos\theta\right)}$. (c) $v_{min} = 24.3$ m/s, $v_{max} = 57.8$ m/s.

## Additional Problems

**9.57** (a) 0.52 rad/s. (b) −0.17 rad/s$^2$. (c) 1.6 rad m/s$^2$. **9.59** (a) 54.3 revolutions. (b) 12.1 rev/s. **9.61** −0.1 rad/s$^2$, $2 \cdot 10^4$ rad. **9.63** 4.2 rotations. **9.65** $5.93 \cdot 10^{-3}$ m/s$^2$. **9.67** (a) 32.6 s$^{-1}$. (b) −2.10 s$^{-2}$. (c) 278 m. **9.69** −80 N. **9.71** (a) 60 m/s$^2$, $5 \cdot 10^3$ N. (b) 6000 N. **9.73** 51 m. **9.75** (a) 0.5 rad/s$^2$. (b) 40 m/s$^2$, 0.5 rad/s$^2$. (c) 40 m/s$^2$ at 2°. **9.77** (a) 5.1 m/s. (b) 740 N.

# Chapter 10: Rotation

## Multiple Choice

**10.1** b. **10.3** b. **10.5** c. **10.7** c. **10.9** b. **10.11** b. **10.13** e. **10.15** c.

## Problems

**10.35** $1.1 \cdot 10^3$ kg m$^2$. **10.37** (a) The solid sphere reaches the bottom first. (b) The ice cube travels faster than the solid ball at the base of the incline. (c) 5 m/s. **10.41** 5 m. **10.43** 8 m/s$^2$. **10.45** (a) 2 N m. (b) 50 rad/s. (c) 200 J. **10.47** (a) $7 \cdot 10^{-2}$ kg m$^2$. (b) 2 s. **10.49** (a) $t_{throw} = 7.7$ s. (b) 5.4 m/s. (c) $t_{throw} = 7.7$ s, 5.4 m/s. **10.51** 3 rad/s$^2$. **10.53** (a) 0.577 m. (b) 0.184 m. **10.55** (a) 6.0 rad/s$^2$. (b) 150 N m. (c) 1900 rad. (d) 280 kJ. (e) 280 kJ. **10.57** (a) $9.704 \cdot 10^{37}$ kg m$^2$. (b) $3.7 \cdot 10^{24}$ kg m$^2$. (c) $1.2 \cdot 10^{18}$ kg m$^2$. (d) $1.2 \cdot 10^{-20}$.

**10.59** $\omega = \dfrac{5J\left(R-h\right)^2}{2MR^3}$. (b) $h_0 = \left(\dfrac{8-\sqrt{44}}{10}\right)R$.

**10.61** (a) 0.2 rad/s. (b) 0.9 m/s **10.63** 0.2 rad/s.

## Additional Problems

**10.65** $E = 1.01 \cdot 10^7$ J, $\tau = 20000$ N m. **10.67** (a) $1.95 \cdot 10^{-46}$ kg m$^2$. (b) $2.06 \cdot 10^{-21}$ J. **10.69** 29 rad/s. **10.71** 10 m/s. **10.73** 0.34 kg m$^2$. **10.75** $4 \cdot 10^4$ s. **10.77** (a) 260 kg. (b) −15 N m. (c) 3.6 revolutions. **10.79** $c = 0.38$.

# Chapter 11: Static Equilibrium

## Multiple Choice

**11.1** c. **11.3** b .**11.5** a. **11.7** c.

## Problems

**11.23** $\dfrac{1000L - 800x_1}{2x_2}$ N.

**11.25** 368 N each. **11.27** 42.1 N clockwise. **11.29** (a) $6m_1g$. (b) 7/5. **11.31** 300 N; 900 N. **11.33** The forces are: 740 N on the end farthest from the bricks, 1200 N on the end nearest the bricks. Both forces are upward. **11.35** 777 N downward.

**11.37** (a) $\dfrac{mgl^2}{L\sqrt{L^2+l^2}}$. (b) $T$. (c) $\dfrac{T\sqrt{L^2+l^2}}{l}$.

**11.39** 88.5 N. **11.41** 29 N. **11.43** $T = 2000$ N; (1000 N,8000 N). **11.45** 30 kg. **11.47** 0.25. **11.49** 32.0°. **11.51** (a) The right support applies an upward force of 132 N. The left support applies an upward force of 279 N. (b) 2.14 m from the left edge of the board. **11.53** (a) 0.206 m. (b) 0.088 m. **11.55** 3 m. **11.57** unstable: $x_0. = 0$; stable: $x = \pm b$. **11.59** Person $A$ can stand at the far edge of the board without tipping it.

## Additional Problems

**11.61** 2.5 m. **11.63** 27.0°. **11.65** 3 m. **11.67** 0.7 kg. **11.69** (a) rope: 200 N, cable: 330 N. (b) $f = 690$ N. **11.71** $m_1 = 0.030$ kg, $m_2 = 0.030$ kg, $m_3 = 0.096$ kg. **11.73** 99.6 N in the right chain and 135 N in the left chain. **11.75** (a) 61 N. (b) 13° above the horizontal.

# Chapter 12: Gravitation

## Multiple Choice

**12.1** a. **12.3** c. **12.5** c. **12.7** e. **12.9** a. **12.11** c.

## Problems

**12.27** 3 mm.

**12.29** $x = \dfrac{L}{1+\sqrt{\dfrac{M_2}{M_1}}}$.

**12.31** $4 \cdot 10^{-8}$ N in the positive $y$-direction. **12.33** This weight of the object on the new planet is $\sqrt[3]{2}$ the weight of the object on the surface of the Earth. **12.35** (a) 3190 km. (b) 0.055%. **12.37** 1. **12.39** $\sqrt{2}$. **12.43** 2.5 km. **12.45** (a) $4.72 \cdot 10^{-5}$ m/s. (b) $1.07 \cdot 10^{-7}$ J. **12.47** 7140 s = 1.98 hours.

**12.49** (a) $E = \dfrac{1}{2}mv^2 - \dfrac{GMm}{r}$. (b) $E = \dfrac{L^2}{2mr^2} - \dfrac{GMm}{r}$.

(c) $r_{min} = \dfrac{-GMm + \sqrt{G^2M^2m^2 + \dfrac{2L^2E}{m}}}{2E}$;

$r_{max} = \dfrac{-GMm - \sqrt{G^2M^2m^2 + \dfrac{2L^2E}{m}}}{2E}$.

(d) $a = -\frac{GMm}{2E}$ and $e = \sqrt{1 + \frac{2L^2E}{G^2M^2m^3}}$.
**12.51** 8 km/s, 2 hours. **12.53** before: 7770 m/s, after: 8150 m/s. **12.55** (a) 5.908 km. (b) 2.954 km. (c) 8.872 mm. **12.57** (a) $3.132 \cdot 10^9$ J. (b) $3.121 \cdot 10^9$ J.

## Additional Problems

**12.59** The force on the Moon due to the Sun is $4.38 \cdot 10^{20}$ N toward the Sun. The force on the Moon due to the Earth is $1.98 \cdot 10^{20}$ N toward the Moon. The total force on the Moon is $2.40 \cdot 10^{20}$ N toward the Sun. **12.61** $3 \cdot 10^{-7}$ N. The ratio of the ball forces to the weight of the small balls is $4 \cdot 10^{-8}$ :1. **12.63** $2 \cdot 10^{10}$ J. **12.65** $2.02 \cdot 10^7$ m. **12.67** (a) 9.78$m$ N. (b) 30.1° above the horizontal. (c) $45.0° = \frac{\pi}{4}$ radians. **12.69** (a) increases by 3.287%. (b) decreases by 3.287%. **12.71** $1.9890 \cdot 10^{30}$ kg. **12.73** (a) $6 \cdot 10^4$ m/s, which is very fast. The speed about the equator of the Earth is 120 times smaller. (b) $3 \cdot 10^{12}$ m/s$^2$. (c) $3 \cdot 10^{11}$ times bigger than on Earth. (d) $3 \cdot 10^6$. (e) $2 \cdot 10^6$ m. **12.75** For the new orbit, the distance at perihelion is $6.72 \cdot 10^3$ km, the distance at aphelion is $1.09 \cdot 10^4$ km, and the orbital period is $8.23 \cdot 10^3$ s = 2.28 hours.

# Chapter 13: Solids and Fluids

## Multiple Choice

**13.1** a. **13.3** d. **13.5** $F_3 < F_1 < F_2$. **13.7** d. **13.9** e. **13.11** b.

## Problems

**13.23** $1.3 \cdot 10^{22}$. **13.25** 0.08 mm. **13.27** 0.3 m. **13.29** density: $1.06 \cdot 10^3$ kg/m$^3$; pressure: $1.10 \cdot 10^8$ Pa. **13.31** 294 mm. **13.33** (a) 42 cm. (b) 43 cm. (c) 49 cm. **13.35** 6233 m. **13.37** 1.1 m. **13.39** 10. pennies. **13.41** (a) $9.3 \cdot 10^3$ N. (b) $4.9 \cdot 10^2$ N. (c) 49 N. **13.43** $3 \cdot 10^{-5}$ m$^3$ = 30 cm$^3$. **13.45** (a) 50 N. (b) 5 kg. (c) 0.8 m/s$^2$. **13.47** (a) $1.089 \cdot 10^6$ N. (b) $9.248 \cdot 10^5$ N, which is a 17.7% increase from using hydrogen rather than helium. **13.49** $10^6$ N/m$^2$. **13.51** (a) The flow rate is 2 m$^3$/s. (b) $6 \cdot 10^2$ m/s$^2$. (c) $6 \cdot 10^3$ m/s$^2$. **13.53** The velocity of the water at the valve is 4.5 m/s and the velocity of a drop of water from rest is 4.4 m/s. **13.55** (a) $v_2 = 20.5$ m/s. (b) $p_2 = 95.0$ kPa. (c) $h = 2.14$ m. **13.57** $2 \cdot 10^{-7}$ m$^3$.

## Additional Problems

**13.59** $3.2 \cdot 10^4$ N. **13.61** (a) 10.4 N. (b) Seawater is slightly denser than freshwater (1030 kg/m$^3$ vs. 1000 kg/m$^3$), so the pressure inside the barrel will be slightly increased compared to the fresh water case. The cork would have flown out of the barrel somewhat before it became full. **13.63** 12.5 m/s. **13.65** (a) 0.683 g/cm$^3$. (b) 0.853 g/cm$^3$. **13.67** 1.32 MPa. **13.69** (a) 300 m/s. (b) $P' - P$ =10 kPa. (c) 600 kN. **13.71** 0.448 m/s. **13.73** 0.39. **13.75** 200 kPa.

# Chapter 14: Oscillations

## Multiple Choice

**14.1** c. **14.3** b. **14.5** b. **14.7** a. **14.9** $\omega_a > \omega_b > \omega_c > \omega_d = \omega_e$. **14.11** b.

## Problems

**14.21** 125 N/m. **14.23** 20. Hz. **14.25** (a) 20 N/m. (b) 1 Hz.
**14.27** $f = \frac{1}{2\pi}\sqrt{\frac{g}{h}}$.
**14.29** 2 cm. **14.31** (a) 2.0 s. (b) 1.8 s. (c) 17 s. (d) There is no period, or, $T = \infty$. **14.33** 1 s. **14.35** (a) $x = L/\sqrt{12}$. (b) $x = 0$. **14.37** (a) $I = \frac{43}{30}ML^2$. (b) $T = 2\pi\sqrt{\frac{86L}{45g}}$. (c) 52 cm. **14.39** (a) $x(0)$ = 1.0 m, $v(0)$= 2.721 m/s, $a(0)$ = –2.467 m/s$^2$. (b) $K(t) = (2.5)\pi^2\cos^2\left(\frac{\pi}{2}t + \frac{\pi}{6}\right)$. (c) 1.667 s. **14.41** (a) 4.0 J. (b) 1.9 m/s. **14.43** (a) 3 m. (b) 0.7 s.
**14.45** (a) $r_0 = \left(\frac{2A}{B}\right)^{1/6}$. (b) $\omega = \frac{3}{\sqrt{m}}\left(\frac{4B^7}{A^4}\right)^{1/6}$.
**14.47** 17 s. **14.49** 1.82 s **14.51** (a) 100 N/m. (b) 10 m/s. (c) 30 s. **14.53** (a) 0.15 m. (b) 2.0 m. (c) 0.039 m. **14.55** The amplitude of the mass's oscillation will be greatest at a frequency of 2.0 s$^{-1}$ with an amplitude of 0.43 m. At a frequency of 15 s$^{-1}$ the amplitude will be half of the maximum.

## Additional Problems

**14.57** $x = \frac{x_{\max}}{\sqrt{2}}$.
**14.59** (a) $x(t)$ = (1m)sin((1 rad/s)$t$). (b) $x(t)$ = (1.11 m)sin((1 rad/s)$t$ + 0.46 rad). **14.61** $1.0 \cdot 10^3$ N/m. **14.63** 8.8 m/s$^2$. **14.65** $n$ = 561; $t$ = 1130 s. **14.67** 2 Hz. **14.69** Q = 4100. **14.71** 5060 s.
**14.73** (a) $T = 2\int_{-A}^{A}\left[\frac{c}{2m}\left(A^4 - x^4\right)\right]^{-1/2} dx$. (b) The period is inversely proportional to $A$. (c) $T = BA^{(1-\alpha/2)}$, where $B$ is a constant. The period is proportional to $A^{(1-\alpha/2)}$.

# Chapter 15: Waves

## Multiple Choice

**15.1** a. **15.3** c. **15.5** d. **15.7** a.

## Problems

**15.17** The resolution time in air is $t_{\max}$ = 0.20 m/(343 m/s) = $5.8 \cdot 10^{-4}$ s. The resolution time in water is $t_{\max}$ = 0.20 m/(1500 m/s) = $1.3 \cdot 10^{-4}$ s. If an individual can only resolve a time difference of $5.8 \cdot 10^{-4}$ s, they will not be able to distinguish a time difference of $1.3 \cdot 10^{-4}$ s. **15.19** (a) 0.00200 m. (b) 6 waves. (c) 127 cycles. (d) 0.157 m. (e) $4.00 \cdot 10^5$ m/s. **15.21** (a) The force on the first spring is: $F_1 = k(-2x_1 + x_2)$. Similarly, the force on the last spring is $F_n = k(x_{n-1} - 2x_n)$. The force acting on the second particle to the $n$–1 particle obeys: $F_2 = k(x_{j-1} - 2x_j + x_{j+1})$.
(b) $2\omega\sin\left(\frac{K\pi}{2(n+1)}\right)$, for $K$ = 1, ..., $n$. **15.23** $y(x,t)$ = (8.9 mm)sin(10.5 m$^{-1}x$ – 100$\pi t$ + 2.67). **15.25** They require the same amount of time. **15.27** 2.3 times longer. **15.29** The sound in the air reaches Alice 0.255 seconds before the sound from the wire does.
**15.31** (b)

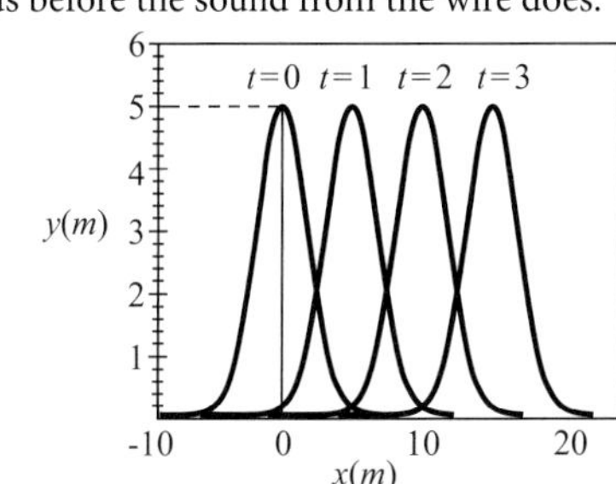

**15.33** 280 km. **15.35** 360 W. **15.37** (a) 0. (b) –4.0 rad. (c) 53 m/s. (d) 8.4 m. (e) 6.4 Hz. (f) 4.3 W. **15.39** (a) 173 m/s. (b) 86.5 Hz. **15.41** 700 N. **15.43** $f(x,t) = (1.00 \text{ cm})\sin((20.0 \text{ m}^{-1})x + (150. \text{ s}^{-1})t)$, $g(x,t) = (1.00 \text{ cm})\sin((20.0 \text{ m}^{-1})x - (150. \text{ s}^{-1})t)$; 7.50 m/s.
**15.45**

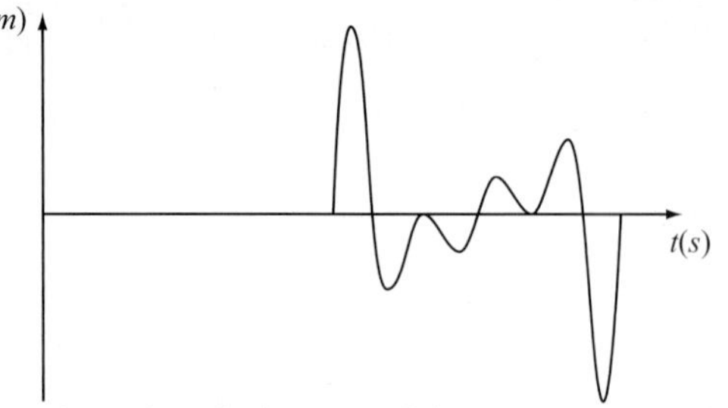


$z(t)$ does not depend on the location of the wave sources at the edges of the pool.

## Additional Problems

**15.49** (a) 69.23 Hz. (b) 54.73 Hz. **15.51** (a) 80 ms. (b) 7.9 m/s. (c) 620 m/s². **15.53** 136 m/s. **15.55** 0.0627. **15.57** $f_2 = f_1$; $v_2 = v_1/\sqrt{3}$; $\lambda_2 = \lambda_1/\sqrt{3}$. **15.59** (a) 1.27 Hz. (b) 1.60 m/s. (c) 0.26 N. **15.61** $v = 410.$ m/s; $v_{max} = 6.43$ m/s .**15.63** 310 Hz.
**15.65** (a)

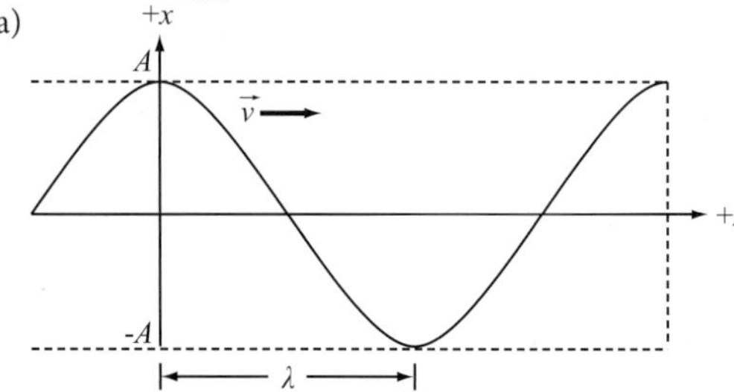


(b) 100 m/s. (c) 31.4 rad/m. (d) 300 N.
(e) $D(z,t) = (0.0300 \text{ m})\cos(10.0\pi \text{ rad/m})x - 2\pi \text{ rad})(500 \text{ s}^{-1})t)$.

# Chapter 16: Sound

## Multiple Choice

**16.1** b. **16.3** c. **16.5** c. **16.7** a. **16.9** b.

## Problems

**16.21** 200 m. **16.23** (a) 343 m/s. (b) 20 °C. **16.25** $1.0 \cdot 10^{20}$ N/m²; This value is some nine orders of magnitude larger than the actual value. Light waves are electromagnetic oscillations that do not require the motion of glass molecules, or the hypothetical ether for transmission. **16.27** 6.3 Pa. **16.29** $6.2 \cdot 10^{-8}$ W/m². **16.31** $3 \cdot 10^3$ m. **16.33** 0.700 m. **16.35** 20. m. **16.37** (a) 0.34 W/m². (b) 120 dB. (c) 0.046 m. **16.39** (a) 2.2°. (b) 10.°. **16.41** (a) 33 m/s. (b) 1300 Hz. **16.43** (a) 50°. (b) 20 km. **16.45** (a) 900 Hz. (b) 30 m/s. (c) 20 m. **16.47** 0.78 m. **16.49** 8.2 cm. **16.51** $7.1 \cdot 10^2$ Hz.

## Additional Problems

**16.53** 425 Hz. **16.55** 2 s.
**16.57**

| Note | Frequency (Hz) | Length (m) |
|---|---|---|
| G4 | 392 | 0.438 |
| A4 | 440 | 0.390 |
| B4 | 494 | 0.347 |
| F5 | 698 | 0.246 |
| C6 | 1046 | 0.164 |

**16.59** 80 dB. **16.61** 6 m/s. **16.63** 37 Hz. **16.65** 109 Hz.
**16.67** (a) In the Young's modulus case, $\delta p(x,t) = Y\frac{\partial}{\partial x}\delta x(x,t)$.
In the bulk modulus case, $\delta p(x,t) = B\frac{\partial}{\partial x}\delta x(x,t)$.
(b) $\delta p(x,t) = -B\frac{\partial}{\partial x}A\cos(\kappa x - \omega t) = B\kappa A\sin(\kappa x - \omega t)$;
The pressure amplitude for the Young's modulus case is $\delta p_{max} = -Y\kappa A$ and is $\delta p_{max} = -B\kappa A$ in the bulk modulus case. **16.69** $P_{0.00} = 2.87 \cdot 10^{-5}$ Pa, $P_{120.} = 28.7$ Pa, $A_{0.00} = 1.11 \cdot 10^{-11}$ m, $A_{120.} = 1.11 \cdot 10^{-5}$m.

# Chapter 17: Temperature

## Multiple Choice

**17.1** a. **17.3** c. **17.5** d. **17.7** a. **17.9** c.

## Problems

**17.23** –21.8 °C. **17.25** –89 °C. **17.27** (a) 194 K. (b) –110 °F. **17.29** 574.59 °F = 574.59 K. **17.31** 7780 kg/m³. **17.33** 500 °C. **17.35** 4 mm. **17.37** 100 °C. **17.39** 3.0 m. **17.41** 23 h and 59 min. **17.43** 200 °C. **17.45** 0.3 m. **17.47** $3.4 \cdot 10^{-2}$ mm. **17.49** 46 μm downward. **17.51** (a) $\beta(T) = \frac{-4.52 \cdot 10^{-5} + 11.36 \cdot 10^{-6}\,T}{1.00016 - 4.52 \cdot 10^{-5}\,T + 5.68 \cdot 10^{-6}\,T^2}$. (b) $1.82 \cdot 10^{-4}$/°C.

## Additional Problems

**17.53** 300 cm³. **17.55** 0.2 L. **17.57** 10. mm. **17.59** 1.5 mm. **17.61** 6.8 mm. **17.63** 10. °C. **17.65** 0.3%. **17.67** 9 kN. **17.69** $6 \cdot 10^{-6}$/°C. **17.71** (a) 97.2 Hz. (b) 93.7 Hz. (c) 97.1 Hz.

# Chapter 18: Heat and the First Law of Thermodynamics

## Multiple Choice

**18.1** d. **18.3** b. **18.5** a. **18.7** b. **18.9** f.

## Problems

**18.23** (a) $9.8 \cdot 10^4$ J. (b) If the body converts 100% of food energy into mechanical energy, then the number of doughnuts needed is 0.094. The body usually converts only 30% of the energy consumed. This corresponds to 0.31 of a doughnut. **18.25** 40 J.
**18.27**

| Material | Specific Heat (KJ/kgK) | Density (g/cm³) | Final Temperature °C |
|---|---|---|---|
| Lead | 0.129 | 11.34 | 22.684 |
| Copper | 0.386 | 8.94 | 22.290 |
| Steel | 0.448 | 7.85 | 22.284 |
| Aluminum | 0.900 | 2.375 | 22.468 |
| Glass | 0.840 | 2.5 | 22.476 |
| Ice | 2.22 | 0.9167 | 22.491 |
| Water | 4.19 | 1.00 | 22.239 |
| Steam | 2.01 | $5.974 \cdot 10^{-4}$ | 8350 |

**18.29** 280 K. **18.31** 25800 J. **18.33** 130 J/(kg K); The brick is made of lead. **18.35** 34 °C. **18.37** 330 g. **18.39** 20000 s; 2090 s. **18.41** (a) None of the water boils away. (b) The aluminum will completely solidify. (c) 45 °C. (d) No, it is not possible without the specific heat of aluminum in its liquid phase. **18.43** 291 g. **18.45** 400 W/(m K). **18.47** 5780 K. **18.49** 86 s. **18.51** 1.8 kW. **18.53** (a) 592 W/m². (b) 384 K.

**18.55** (a) $f = 5.88 \cdot 10^{10}\ T$ Hz/K. (b) $3.53 \cdot 10^{14}$ Hz. (c) $1.608 \cdot 10^{11}$ Hz. (d) $2 \cdot 10^{13}$ Hz.

### Additional Problems

**18.57** 11 ft² °F h/BTU. **18.59** 4.2 K. **18.61** $2.0 \cdot 10^{-2}$ W/(m K). **18.63** $6.0 \cdot 10^{2}$ W/m². **18.65** (a) 7 °C. (b) 0 °C. **18.67** $4 \cdot 10^{13}$ W. **18.69** 90:1. **18.71** (a) $4.0 \cdot 10^{6}$ J. (b) $12.

# Chapter 19: Ideal Gases

### Multiple Choice

**19.1** a. **19.3** c. **19.5** a. **19.7** b. **19.9** d.

### Problems

**19.25** 342 kPa. **19.27** (a) 38 psi. (b) 8.8%. **19.29** 374 K. **19.31** 354 m/s. **19.33** 400 kPa. **19.35** 699 L. **19.37** 200 kPa. **19.39** 2 cm. **19.41** (a) $4 \cdot 10^{-17}$ Pa. (b) 260. m/s. (c) 300 km. **19.43** The rms speed of $^{235}UF_6$ is 1.01 times that of $^{238}UF_6$. **19.45** 28.1 kPa. **19.47** (a) He: $6.69 \cdot 10^{-21}$ J; N: $1.11 \cdot 10^{-20}$ J. (b) He: constant volume: 12.471708 J/mol·K; constant pressure: 20.786180 J/mol·K; N: constant volume: 20.786180 J/mol·K; constant pressure: 29.100652 J/mol·K. (c) $\gamma_{He} = 5/3$; $\gamma_{N_2} = 7/5$. **19.49** 41.6 J. **19.51** 90 J. **19.53** (a) 47 kPa. (b) 200 K. **19.55** 40 atm; 800 K. **19.57** $Q_{12} = W_{12} = 3$ kJ, $W_{23} = -0.8$ kJ, $Q_{23} = -2$ kJ, $W_{31} = -1$ kJ, $Q_{31} = 0$.

**19.59** (a) $\rho(h) = \rho_0 \left(1 - \frac{\gamma - 1}{\gamma} \cdot \frac{Mgh}{RT_0}\right)^{\frac{1}{\gamma - 1}}$; (b) 5.39 km, 241 K; 7.27 km, 222 K. (c) 5945.4 m, 293.2 K. **19.61** (a) 469.2 m/s. (b) 1749 m/s. **19.63** rms speed: 700 m/s; average speed: 640 m/s.

### Additional Problems

**19.65** 80 J. **19.67** (a) $8.39 \cdot 10^{23}$ atoms $T(h) = T_0 - \frac{\gamma - 1}{\gamma} \cdot \frac{Mgh}{R}$. (b) $6.07 \cdot 10^{-21}$ J. (c) 1350 m/s. **19.69** (a) $3.2 \cdot 10^{5}$ Pa. (b) 410 cm². **19.71** 2.16 m/mol, likely hydrogen gas. **19.73** (a) 20 kJ. (b) 1.5 kPa. **19.75** 12.2 atm. **19.77** $2 \cdot 10^{-21}$ J; The energy depends only on temperature, not on the identity of the gas. **19.79** 0.560 L.

# Chapter 20: The Second Law of Thermodynamics

### Multiple Choice

**20.1** d. **20.3** a. **20.5** c. **20.7** d. **20.9** a.

### Problems

**20.23** 100 s.
**20.25** (a)

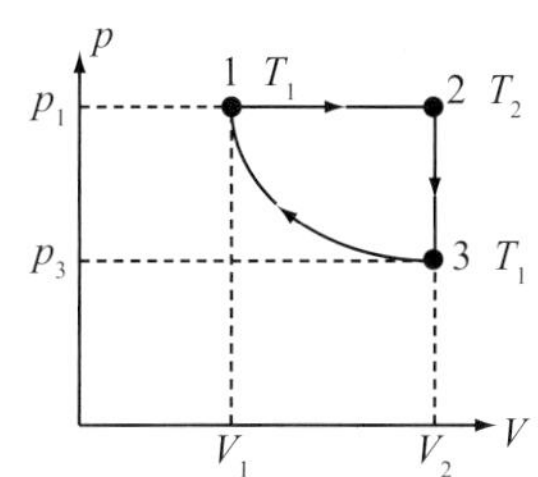


(b) $W_{12} = -90$ J, $Q_{12} = -200$ J, $W_{23} = 0$ J, $Q_{23} = 100$ J, $W_{31} = 50$ J, $Q_{31} = 50$ J. (c) 0.2.
**20.27** (a)

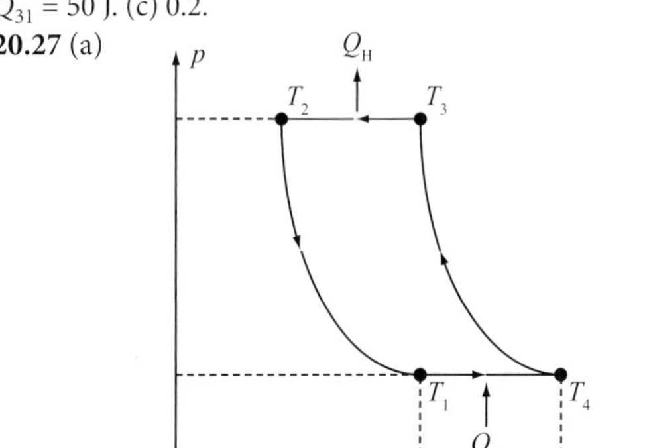


(b) $K = (T_4 - T_1)/(T_3 - T_2 - T_4 + T_1)$.
**20.29** (a) 0.7000. (b) 1430 J. (c) 430 J. **20.31** 1%. **20.33** efficiency: 0.33; $T_2 : T_1 = 2:3$. **20.35** 48 cal. **20.37** 2. **20.39** (a) point 1: 100 kPa, $1 \cdot 10^{-3}$ m³, 600 K; point 2: 500 kPa, $1 \cdot 10^{-3}$ m³, 3000 K; point 3: 100 kPa, $6 \cdot 10^{-3}$ m³, 3000 K. (b) 0.3. (c) 0.8. **20.41** (a) $\Delta S_H = -10$ J/K, $\Delta S_L = 90$ J/K. (b) 0. (c) 70 J/K. **20.43** (a) 0.3. (b) 0. **20.45** $k_B \ln 2$. **20.47** $S_{5\ up} = 0$ (exact); $S_{3\ up} = 3.18 \cdot 10^{-23}$ J/K. **20.49** $5.936 \cdot 10^{14}$ W/K.

### Additional Problems

**20.51** 0.96. **20.53** 5.96%. **20.55** 0.04 J/K. **20.57** For two dice the entropy is doubled, and for three dice the entropy is tripled. **20.59** 77.2 K. **20.61** 1. **20.63** (a) 0.5. (b) 0.3. (c) 30 °C. **20.65** $\Delta S_{isotherm} = 5$ J/K, $\Delta S_{isobar} = -5$ J/K, $\Delta S_{adiabatic} = 0$ (exact). **20.67** 0.3.

# Credits

## Photographs

### About the Authors

Photo courtesy of Okemos Studio of Photography.

### The Big Picture

**Figure 1:** © M. F. Crommie, C. P. Lutz, and D. M. Eigler, IBM Almaden Research Center Visualization Lab, http://www.almaden.ibm.com/vis/stm/images/stm15.jpg. Image reproduced by permission of IBM Research, Almaden Research Center. Unauthorized use not permitted; **2:** © M. Feig, Michigan State University; **3a-b:** STAR collaboration, Brookhaven National Laboratory; **4:** © CERN; **5:** © Vol. 54 PhotoDisc/Getty Images RF; **6:** Andrew Fruchter (STScI) et al., WFPC2, HST, NASA.

### Chapter 1

**Figure 1.1:** NASA/JPL-Caltech/L. Allen (Harvard Smithsonian CfA); **1.3:** © Digital Vision/Getty Images RF; **1.4:** © Photograph reproduced with permission of the BIPM; **1.5:** Courtesy NASA/JPL-Caltech; **1.6:** Courtesy National Institute of Standards and Technology; **1.7a:** © Gemini Observatory-GMOS Team; **1.7b:** © BananaStock/PunchStock RF; **1.7c:** Dr. Fred Murphy, 1975, Centers for Disease Control and Prevention; **1.8c:** © Hans Gelderblom/Stone/Getty Images; **1.8d:** © Wolfgang Bauer; **1.8e:** © Edmond Van Hoorick/Getty Images RF; **1.8f:** © Digital Vision/Getty Images RF; **1.8g:** © Gemini Observatory-GMOS Team; **1.8h:** NASA, ESA, and The Hubble Heritage Team (STScI/AURA); Hubble Space Telescope ACS; STScI-PRC05-20; **p. 23:** © Stockbyte/PunchStock RF; **p. 34:** NASA.

### Chapter 2

**Figure 2.1:** © Royalty-Free/Corbis; **2.2a-b:** © W. Bauer and G. D. Westfall; **2.8:** © Ted Foxx/Alamy RF; **2.9:** © Ryan McVay/Getty Images RF; **2.17:** © Larry Caruso/WireImage/Getty Images; **2.18, 2.20:** © W. Bauer and G. D. Westfall; **2.21:** © Royalty-Free/Corbis; **2.31-p. 65:** © W. Bauer and G. D. Westfall.

### Chapter 3

**Figure 3.1:** © Terry Oakley/The Picture Source; **3.3a-c:** © W. Bauer and G. D. Westfall; **3.6:** © David Madison/Photographer's Choice/Getty Images; **3.7, 3.9:** © W. Bauer and G. D. Westfall; **3.13:** © Rim Light/PhotoLink/Getty Images RF; **3.15, 3.18, 3.19:** © W. Bauer and G. D. Westfall; **3.21-3.22:** © Edmond Van Hoorick/Getty Images RF.

### Chapter 4

**Figure 4.1:** NASA; **4.2a:** © Dex Image/Corbis RF; **4.2b:** © Richard McDowell/Alamy; **4.2c:** Photo by Lynn Betts, USDA Natural Resources Conservation Service; **4.3a:** © PhotoDisc/Getty Images RF; **4.3b:** © W. Cody/Corbis; **4.3c:** © Boston Globe/David L. Ryan/Landov; **4.4:** © W. Bauer and G. D. Westfall; **4.5a:** © Radius Images/Alamy RF; **4.5c:** © The McGraw-Hill Companies, Inc./Joe DeGrandis, photographer; **4.5d:** © W. Bauer and G. D. Westfall; **4.5e:** © Brand X Pictures/PunchStock RF; **4.5f:** © W. Bauer and G. D. Westfall; **4.5g:** R. Stockli, A. Nelson, F. Hasler, NASA/GSFC/NOAA/USGS; **4.6-4.8:** © W. Bauer and G. D. Westfall; **4.9:** © Tim Graham/Getty Images; **4.11a-b:** © Ryan McVay/Getty Images RF; **4.15a, 4.16a-d, 4.20:** © W. Bauer and G. D. Westfall; **4.21a:** © Digital Vision/Getty Images RF; **4.21b:** © Brand X Pictures/JupiterImages RF; **4.22:** Used with permission by D. J. Spaanderman, precision engineer. FOM Institute for Atomic and Molecular Physics, Kruislaan 407, 1098 SJ Amsterdam, the Netherlands.

### Chapter 5

**Figure 5.1:** NASA/Goddard Space Flight Center Scientific Visualization Studio; **5.2:** © Malcolm Fife/Getty Images RF; **5.3a:** © Bettmann/Corbis; **5.3b:** © Mike Goldwater/Alamy; **5.4a:** © Royalty-Free/Corbis; **5.4b:** © Geostock/Getty Images RF; **5.5b:** © Cre8tive Studios/Alamy RF; **5.5c:** © Creatas/PunchStock RF; **5.5d:** © Royalty-Free/Corbis; **5.5e:** © General Motors Corp. Used with permission, GM Media Archives; **5.5f:** Courtesy National Nuclear Security Administration, Nevada Site Office; **5.5g:** © Royalty-Free/Corbis; **5.5h:** NASA, ESA, J. Hester and A. Loll (Arizona State University); **5.16a-c:** © W. Bauer and G. D. Westfall.

### Chapter 6

**Figure 6.1, 6.10, 6.11, 6.13a-d, 6.16:** © W. Bauer and G. D. Westfall; **6.20:** © Royalty-Free/Corbis; **6.22:** © W. Bauer and G. D. Westfall; **p. 203:** © The Texas Collection, Baylor University, Waco, Texas.

### Chapter 7

**Figure 7.1:** © Getty Images RF; **7.2a-d:** © Sandia National Laboratories/Getty Images RF; **7.5:** Reproduced by permission of the News-Gazette, Inc. Permission does not imply endorsement; **7.6:** © Photron; **7.7:** © W. Bauer and G. D. Westfall; **p. 215:** © Photron; **7.9a-c, 7.13a-b:** © W. Bauer and G. D. Westfall; **7.19:** © Delacroix/The Bridgeman Art Library/Getty Images; **p. 240 (golfball):** © Stockdisc/PunchStock RF; **p. 240 (basketball):** © PhotoLink/Getty Images RF.

### Chapter 8

**Figure 8.1:** NASA; **8.9:** © AP Photo; **8.10a-b:** © W. Bauer and G. D. Westfall; **8.11:** © Boeing; **8.14, 8.22a:** © W. Bauer and G. D. Westfall; **8.22b:** © Comstock Images/Alamy RF; **p. 274:** © W. Bauer and G. D. Westfall.

### Chapter 9

**Figure 9.1:** © EAA Photo by Craig VanderKolk; **9.2:** © Royalty-Free/Corbis; **9.7a-b:** © Baokang Bi, Michigan State University; **9.8:** © The McGraw-Hill Companies, Inc./Mark Dierker, photographer; **9.9:** © W. Bauer and G. D. Westfall; **9.13:** © Andrew Meehan; **9.14a-d, 9.15:** © W. Bauer and G. D. Westfall; **9.17:** © Royalty-Free/Corbis; **9.25:** © Patrick Reddy/Getty Images.

### Chapter 10

**Figure 10.1:** © DreamPictures/Stone/Getty Images; **10.2a:** © Stock Trek/Getty Images RF; **10.2b:** © Gemini Observatory-GMOS Team; **10.13, 10.14, 10.16a-d:** © W. Bauer and G. D. Westfall; **10.17-10.18:** © The McGraw-Hill Companies, Inc./Mark Dierker, photographer; **10.23-10.24:** © W. Bauer and G. D. Westfall; **10.26, 10.28a-b:** © The McGraw-Hill Companies, Inc./Mark Dierker, photographer; **10.29:** © Don Farrall/Getty Images RF; **10.30a-c:** © Otto Greule Jr./Allsport/Getty Images; **10.31:** © T. Bollenbach; **p. 341 (top):** © Mark Thompson/Getty Images; **10.33:** © Royalty-Free/Corbis; **10.35:** © W. Bauer and G. D. Westfall.

### Chapter 11

**Figure 11.1a:** © Digital Vision/Alamy RF; **11.1b:** © Guillaume Paumier/Wikimedia Commons; **11.1c:** © Skyscraper Source Media Inc. http://SkyscraperPage.com; **11.2-11.6a, 11.8a, 11.12a-b:** © W. Bauer and G. D. Westfall; **11.13:** © Scott Olson/Getty Images; **11.14a-b:** © W. Bauer and G. D. Westfall; **11.20:** © Segway, Inc.; **p. 376 (top):** © W. Bauer and G. D. Westfall; **p. 376 (bottom):** © Creatas Images/JupiterImages RF; **p. 378 (left)-(right):** © W. Bauer and G. D. Westfall.

## Chapter 12

**Figure 12.1:** © JAXA/NHK; **12.2:** © NASA/JPL-Caltech/S. Stolovy (SSC/Caltech); **12.3, 12.7, 12.13a-b:** NASA; **12.14:** U.S. Geological Survey; **12.20:** Courtesy of Prof. Andrea Ghez (UCLA); **12.22:** NASA; **12.27:** © Vera Rubin; **12.28a-b:** Courtesy of NASA, ESA, M. J. Jee and H. Ford (Johns Hopkins University); **12.29:** X-ray: NASA/CXC/CfA/M.Markevitch et al.; Optical: NASA/STScI, Magellan/U.Arizona/D.Clowe et al.; Lensing Map: NASA/STScI; ESO WFI, Magellan/U. Arizona/D.Clowe et al.

## Chapter 13

**Figure 13.1:** © Science Source/Photo Researchers, Inc.; **13.2b:** © Royalty-Free/Corbis; **13.2d:** © Image Source/Getty Images RF; **13.5a:** © PhotoLink/Getty Images RF; **13.5b:** © Don Farrall/Getty Images RF; **13.6a:** © C. Borland/PhotoLink/Getty Images RF; **13.6b:** © BananaStock/PunchStock RF; **13.6c:** © Comstock Images/Alamy RF; **13.14:** © W. Bauer and G. D. Westfall; **13.15, 13.19, 13.24a:** © Royalty-Free/Corbis; **13.25a-c:** © W. Bauer and G. D. Westfall; **13.26:** © PhotoLink/Getty Images RF; **13.27a-d, 13.29a:** © W. Bauer and G. D. Westfall; **13.29b:** © Don Farrall/Getty Images RF; **13.29c, 13.31, 13.35:** © W. Bauer and G. D. Westfall; **13.37a-b:** © SHOTFILE/Alamy RF; **13.38:** © The McGraw-Hill Companies, Inc./Ken Karp, photographer; **13.40:** © W. Bauer and G. D. Westfall; **13.44:** © Siede Preis/Getty Images RF; **13.45:** © Royalty-Free/Corbis; **13.46a:** © Kim Steele/Getty Images RF; **13.46b:** © W. Bauer and G. D. Westfall; **13.47:** © *The Astrophysical Journal,* Fryer & Warren 2002 (Apj, 574, L65). Reproduced by permission of the AAS.

## Chapter 14

**Figure 14.1a:** © W. Bauer and G. D. Westfall; **14.1b:** Courtesy National Institute of Standards and Technology; **14.2-14.3, 14.12, 14.14a, 14.18a-b, 14.24:** © W. Bauer and G. D. Westfall; **14.29:** © University of Washington Libraries, Special Collections, UW 21422.

## Chapter 15

**Figure 15.1:** © Matt King/Getty Images; **15.2:** © Don Farrall/Getty Images RF; **15.4:** © W. Bauer and G. D. Westfall; **15.11a:** © Ingram Publishing/Alamy RF; **15.11b:** Image courtesy of Cornell University; **15.17:** NOAA/PMEL/Center for Tsunami Research; **15.26a-d:** © W. Bauer and G. D. Westfall; **p. 512:** Courtesy Vera Sazonova and Paul McEuen; **15.28a-c, p. 513a-c (bottom):** © W. Bauer and G. D. Westfall; **15.29:** M. F. Crommie, C. P. Lutz, and D. M. Eigler, IBM Almaden Research Center Visualization Lab, http://www.almaden.ibm.com/vis/stm/images/stm15.jpg. Image reproduced by permission of IBM Research, Almaden Research Center. Unauthorized use not permitted; **15.30a-d:** Images courtesy of LIGO Laboratory, Supported by the National Science Foundation; **15.31:** Henze/NASA; **15.33a:** © W. Bauer and G. D. Westfall.

## Chapter 16

**Figure 16.1:** © Kristy McDonald/AP Photo; **16.3:** © Royalty-Free/Corbis; **16.5:** © R. Morely/Photolink/Getty Images RF; **16.7:** © Steve Allen/Getty Images RF; **16.8:** © Arnold Song and Jose Iriarte-Diaz; **16.11:** Photo by Petty Officer 3rd Class John Hyde, U.S. Navy; **16.15, 16.19a-b:** © W. Bauer and G. D. Westfall; **16.20:** Courtesy Wake Radiology; **16.22:** Courtesy U.S. Nuclear Regulatory Commission; **16.23:** © PhotoLink/Getty Images RF; **16.24:** © Stockbyte/Getty Images RF; **16.25:** © W. Bauer and G. D. Westfall; **16.27:** © Annie Reynolds/PhotoLink/Getty Images RF.

## Chapter 17

**Figure 17.1:** © Antonio Fernandez Sanchez; **17.4a;** © Royalty-Free/Corbis; **17.4b:** © W. Bauer and G. D. Westfall; **17.4c:** © Digital Vision/Getty Images RF; **17.4d:** The STAR Experiments, Brookhaven National Laboratory; **17.4e:** © MSU National Superconducting Cyclotron Laboratory; **17.4f:** NASA/WMAP Science Team; **17.4g:** NOAA; **17.4h:** © ITER; **17.6:** © Siede Preis/Getty Images RF; **17.7a-b:** © Janis Research Co., Inc. www.janis.com; **17.8:** © H. Mark Helfer/NIST; **17.9:** © ITER; **17.11:** © Michigan Department of Transportation Photography Unit; **17.13a-c, 17.14, 17.16a-d:** © W. Bauer and G. D. Westfall; **17.19, 17.23:** NASA/WMAP Science Team.

## Chapter 18

**Figure 18.1:** © Kirk Treakle/Alamy; **18.8:** © Jed Jacobsohn/Getty Images; **18.11:** © W. Bauer and G. D. Westfall; **18.12:** NASA, ESA and J. Hester (ASU); **18.14:** © Skip Brown/National Geographic/Getty Images; **18.16a:** © W. Bauer and G. D. Westfall; **18.16b:** © Owens Corning PINK Fiberglass Insulation; **18.18:** NASA; **18.19a-b:** NASA Glenn Research Center; **18.20:** NASA Goddard Space Flight Center, Visible Earth; **18.23:** Courtesy of www.EnergyEfficientSolutions.com; **18.25a-b:** NASA/IPAC/Caltech; **18.27:** © David J. Green - technology/Alamy RF.

## Chapter 19

**Figure 19.1:** © image 100/PunchStock RF; **19.8-19.9:** © W. Bauer and G. D. Westfall; **19.14a:** © PhotoDisc/Getty Images RF; **19.21a:** Brookhaven National Laboratory; **19.24:** The Hubble Heritage Team (AURA/STScI/NASA).

## Chapter 20

**Figure 20.1a:** © Frans Lemmens/Getty Images; **20.1b:** © Keith Kent/Photo Researchers, Inc.; **20.4a:** © Royalty Free/Corbis; **20.5a:** © PhotoDisc/PunchStock RF; **20.7, 20.10:** © W. Bauer and G. D. Westfall; **20.12:** © Markus Matzel/Peter Arnold, Inc.

# Line Art and Text

## Chapter 4

**Figure 4.22:** Figure used with permission by D. J. Spaanderman, precision engineer. FOM Institute for Atomic and Molecular Physics, Kruislaan 407, 1098 SJ Amsterdam, the Netherlands.

## Chapter 7

**Figure 7.16:** Based on information from DZero collaboration and Fermi National Accelerator Laboratory, Education Office.

## Chapter 17

**Figure 17.4:** RHIC figure courtesy of Brookhaven National Laboratory. Cosmic microwave background image credit NASA/WMAP Science Team. ITER machine © ITER Organization; **Figure 17.20:** From data compiled by the National Oceanic and Atmospheric Administration, Brohan et al., J. Geophys. Res., 111, D12106 (2006); **Figure 17.21:** Data from the Vostok ice core, J.R. Petit et al., Nature 399, 429–436 (1999); Problem 17.49: Figure and problem based on V.A. Henneken et al., J. Micromech. Microeng. 16 (2006) S107–S115; **Problem 17.50:** Figure and problem based on V.A. Henneken et al., J. Micromech. Microeng. 16 (2006) S107–S115.

## Chapter 18

**Figure 18.26:** (a) Concentration of carbon dioxide in the atmosphere from 1832 to 2004. The measurements from 1832 to 1978 were done using ice cores in Antarctica: ride line based on information from D.M. Etheridge, L.P. Steele, R.L. Langenfelds, R.J. Francey, J.-M. Barnola and V.I. Morgan,1998. The measurements from 1959 to 2004 were carried out in the atmosphere on Mauna Loa in Hawaii: blue line based on information from C.D. Keeling and T.P. Whorf, 2005; (b) Historical carbon dioxide records from the Law Dome DE08, DE08-2, and DSS ice cores, *Trends: A Compendium of Data on Global Change,* Carbon Dioxide Information Analysis Center, Oak Ridge National Laboratory, U.S. Department of Energy, Oak Ridge, Tenn., U.S.A. Carbon dioxide concentrations for the past 420,000 years, extracted using ice cores in Antarctica based on information from J.M. Barnola, et. al., 2003.

## Chapter 19

**Figure 19.21 (b) and (c):** Data courtesy of the STAR collaboration.

## Chapter 20

**Figure 20.13:** Image data courtesy of Steve Chu, U.S. Secretary of Energy.

## Appendix B

**Data source:** David R. Lide (ed.), Norman E. Holden in *CRC Handbook of Chemistry and Physics* 85th Edition, online version. CRC Press. Boca Raton, Florida (2005). Section 11, Table of the Isotopes.

## Appendix C

**Data sources:** http://physics.nist.gov/PhysRefData/PerTable/periodic-table.pdf

http://www.wikipedia.org/ and Generalic, Eni. "EniG. Periodic Table of the Elements." 31 Mar. 2008. KTF-Split. <http://www.periodni.com/>.

## Inside Front Cover

Fundamental Constants from National Institute of Standards and Technology, http://physics.nist.gov/constants.

Other Useful Constants from National Institute of Standards and Technology, http://physics.nist.gov/constants.

# Index

*Page references followed by f and t refer to figures and tables, respectively*

## A

## B



## C

## D



## E

# F

## G



## H

# I



# J



# K

## L



## M

## N



## O

## P



## Q



## R

## S

# T

## U



## V

## W



## Y



## Z